2001

# The
# EVERGLADES, FLORIDA BAY,
## and
# CORAL REEFS
## of the
# FLORIDA KEYS

## An Ecosystem Sourcebook

# The EVERGLADES, FLORIDA BAY, and CORAL REEFS of the FLORIDA KEYS

## An Ecosystem Sourcebook

Edited by
### JAMES W. PORTER
### KAREN G. PORTER

*Photographs by*
Clyde Butcher

## CRC PRESS

Boca Raton  London  New York  Washington, D.C.

On the cover: "Doghouse Key" by Clyde Butcher.

**Library of Congress Cataloging-in-Publication Data**

The Everglades, Florida Bay, and coral reefs of the Florida Keys: An ecosystem
sourcebook / edited by James W. Porter and Karen G. Porter; photographs by Clyde Butcher.
    p. cm.
Includes bibliographical references.
ISBN 0-8493-2026-7 (alk. paper)
    1. Ecohydrology—Florida. 2. Wetland ecology—Florida—Everglades. 3. Marine
ecology—Florida—Florida Bay. 4. Coral reef ecology—Florida—Florida Keys. I. Porter,
James W. (James Watson) 1946- II. Porter, Karen G.

QH105.F6 C67 2001
577.6'09759—dc21                                                                          2001035649

# Dedication

*This book is dedicated to the memory of Marjory Stoneman Douglas, who, by speaking out for the preservation of the Everglades, started a process of environmental awareness that now ripples well beyond her beloved River of Grass.*

# Foreword

This volume of contributed papers is the outgrowth of a special symposium, "Linkages Between Ecosystems: The South Florida Hydroscape," convened at a jointly sponsored meeting of the American Society of Limnology and Oceanography and The Ecological Society of America. These premier scientific societies involved in basic and applied ecological research in marine, terrestrial, and freshwater research in North America had recognized the emerging importance of a "whole watershed" approach to ecosystems that included the coastal marine environment. Both societies were also actively involved in influencing environmental policy and management by encouraging an integrative approach based on natural rather than political boundaries. This was reflected in their education and outreach activities. The joint meeting, titled "The Land-Water Interface: Science for a Sustainable Biosphere," was held in St. Louis, June 7–12, 1999. It focused on research at the land-water interface of both freshwater and saltwater systems with a goal of strengthening connections between research and management.

What area could be more appropriate for a special symposium at that meeting than South Florida, a series of extensive freshwater and marine ecosystems linked by hydrology and requiring a large-scale management approach? The open call for contributed papers was also an ideal way to bring together a broad range of data and viewpoints from diverse research and management programs without bias or regionalism.

The session was highly successful and participants felt that their information should be available in one place rather than scattered in various journals and publications in their fields of specialty. Numerous individuals who could not attend the meeting also wanted to contribute to a published format for their work. We proposed the book project to John Sulzycki, an editor representing St. Lucie/CRC Press at that meeting. St. Lucie Press had published the only major ecological source book available on the region, *Everglades: Its Conservation and Restoration*, edited by S.M. Davis and J.C. Ogden in 1994. It would provide a good background for our proposed volume.

The contents of this volume represent an expansion of the original symposium topic to include additional authors, an extensive policy and management section, and an example on a smaller scale, of another coastal wetland ecosystem linked by surface and subsurface water movement to nearshore reefs.

We are grateful to John Sulzycki for his persistent encouragement and support of this very large project and to the staff of CRC Press for their help in production of this volume. We also thank Ron Carroll, Director of the Institute of Ecology, for his support. Our time, and much of the mailing, copying and telephone resources were provided by the Institute of Ecology. Additional funding and encouragement were provided by Fred McManus of the U.S. Environmental Protection Agency, Region IV. Their support, and the open call for papers of the ESA/ASLO symposium venue, allowed us to be inclusive and to welcome contributions by any and all sides to an issue. We are especially grateful to Clyde Butcher for providing his evocative photographs that add a stunning aesthetic dimension to the book. Our thanks also to the participants who endured, some directly, all indirectly, the effects of Hurricane Georges on slowing the production of this volume. We believe it improved with time.

Since the conceptualization of this book, South Florida has grown in importance. The Everglades and the Florida Keys are nationally and internationally recognized as the most extensive and expensive restoration initiative to date. To quote former Secretary of Interior, Bruce Babbitt,

"If we can not get it right here, we can not get it right anywhere." We hope this volume will be a major resource for anyone trying to understand, manage, and restore the South Florida hydroscape and similar coastal ecosystems worldwide.

**Karen G. Porter**
*Big Pine Key, Florida*

# About the Editors

**James W. Porter** is Professor of Ecology and Marine Sciences at the University of Georgia. He received his B.S. from Yale in 1969 and his Ph.D. from Yale in 1973. After teaching at the University of Michigan from 1973 to 1977, he joined the faculty of the Institute of Ecology at the University of Georgia.

Dr. Porter has received numerous research awards from the National Science Foundation, the Environmental Protection Agency, the Smithsonian Institution, the National Park Service, and the National Oceanic and Atmospheric Administration for his studies on the biology and ecology of corals. His results have been published in the scientific journals *Science*, *Nature*, *Ecology*, and Marine Biology and have been written about in *Time Magazine*, *The New York Times*, and *London Times*. His award-winning photographs have appeared in *Life*, *Oceans*, and *The New York Times*. He was named editor of the professional journals *Ecology* and *Ecological Monographs* in 1977 and served in that post for four years.

Dr. Porter is an elected Fellow of the Association for the Advancement of Science, the Society of American Naturalists, and the Great Barrier Reef Committee. In 1983, he received the University of Georgia's Creative Research Award and in 1987 the University's Outstanding Teacher Award. He is one of only a few faculty members to have received awards in both teaching and research. From 1992 to 1997, he served as the Associate Director of the Institute of Ecology and as Graduate Coordinator for the Ecology Ph.D. Program. In 1997, Dr. Porter delivered the Marjory Stoneman Douglas Lecture at Wellesley College, Ms. Douglas' *alma mater*.

Dr. Porter is a marine ecologist specializing in the biology, ecology, and assessment of Floridian and Caribbean coral reefs. He is considered an international expert on the evaluation and measurement of coral reef health and, in 1991, was a Plenary Speaker at the United Nations Symposium on Global Climate Changes held in Venice, Italy. Dr. Porter's studies on coral reef decline in Florida have created intense national and international interest. He has testified before Congress three times, most recently on the effects of global climate change on coral reefs.

**Karen G. Porter** graduated from Vassar College in 1968 with an A.B. in Biology and spent that summer as a student in the Marine Ecology Course at the Marine Biological Laboratory in Woods Hole, MA. She went on to receive a Ph.D. in Biology from Yale University in 1973, studying freshwater food webs with G. E. Hutchinson. She became an Assistant Professor in the Department of Zoology of the University of Michigan, Ann Arbor, Michigan, and then an Associate and Full Professor in the Institute of Ecology at the University of Georgia, Athens, GA.

Dr. Porter's training in both marine and freshwater ecology gives her a basis to compare and contrast the two aquatic systems. Her research initially focused on selective feeding and differential digestion of algae by zooplankton and expanded to include microbial food webs in lakes, ponds, and wetlands. She has worked on numerous lakes in the Northeast, Midwest, and the Southeast where she focussed on a model system, Lake Oglethorpe, GA, and also studied the Okeefenokee Swamp. She is also familiar with tropical aquatic ecosystems such as Lake Gatun, Panama. With James, she has studied the plankton of coral reefs in the Caribbean and Indopacific and community structure and diseases of Caribbean coral reefs. Together they have taught courses in aquatic and marine sciences, general ecology, and have coordinated the Marine Ecology Course at the Marine Biological Laboratory, Woods, Hole, MA. She has served on the Editorial Board of *Limnology and Oceanography*, Board of Directors of the American Institute of Biological Sciences, and the Scientific Advisory Board of Reef Relief.

# Contributors

**Carlos Alvarez-Zarikian**
Division of Marine Geology and Geophysics
University of Miami
Miami, Florida

**Eric Annis**
Darling Marine Center
University of Maine
Walpole, Maine

**Tom Armentano**
Daniel Beard Research Center
Everglades National Park
Homestead, Florida

**Sydney T. Bacchus**
Applied Environmental Services
Athens, Georgia

**Peter J. Barile**
Division of Marine and Environmental
   Systems
Florida Institute of Technology
Melbourne, Florida

**Oron L. Bass, Jr.**
South Florida Natural Resource Center
Everglades National Park
Homestead, Florida

**Sarah Bellmund**
Biscayne National Park
Homestead, Florida

**Robert E. Bennetts**
Florida Cooperative Wildlife Research Unit
University of Florida
Gainesville, Florida

**Robin D. Bjork**
Department of Fisheries and Wildlife
Oregon State University
Corvallis, Oregon

**Courtney Black**
Negril Coral Reef Preservation Society
Negril, Jamaica

**Pat Blackwelder**
Division of Marine Geology and Geophysics
University of Miami
Miami, Florida

**Joseph N. Boyer**
Southeast Environmental Research Center
Florida International University
Miami, Florida

**Larry E. Brand**
Division of Marine Biology and Fisheries
University of Miami
Rosenstiel School of Marine and Atmospheric
   Science
Miami, Florida

**Mike Brill**
Department of Biology
University of Charleston
Charleston, South Carolina

**Chris Buzzelli**
Southeast Environmental Research Center
Florida International University
Miami, Florida

**Hunter J. Carrick**
Watershed Research and Planning Division
South Florida Water Management District
West Palm Beach, Florida

**Billy D. Causey**
Florida Keys National Marine Sanctuary
Marathon, Florida

**John H. Chick**
Illinois Natural History Survey
Great Rivers Field Station
Brighton, Illinois

**Daniel L. Childers**
Southeast Environmental Research Center
Florida International University
Miami, Florida

**E. Jane Comiskey**
Department of Ecology and Evolutionary
 Biology
University of Tennessee
Knoxville, Tennessee

**Clayton B. Cook**
Harbor Branch Oceanographic Institution
Fort Pierce, Florida

**Eurico J. D'Sa**
Office of Research and Applications
NOAA/NESDIS
Washington, D.C.

**Susan Dailey**
Southeast Environmental Research Center
Florida International University
Miami, Florida

**Donald L. DeAngelis**
U.S. Geological Survey
Biological Resources Division
University of Miami
Miami, Florida

**Robert Doren**
Southeast Environmental Research Center
Florida International University
Miami, Florida

**John Dotten**
Florida Keys National Marine Sanctuary
Marathon, Florida

**Michael J. Durako**
Center for Marine Science
University of North Carolina
 at Wilmington
Wilmington, North Carolina

**Phillip Dustan**
Department of Biology
University of Charleston
Charleston, South Carolina

**David Eaken**
South Florida Regional Laboratory
Florida Fish & Wildlife Conservation
 Commission
Marathon, Florida

**Adrienne L. Edwards**
Southeast Environmental Research Center
Florida International University
Miami, Florida

**Geoffrey Ellis**
Division of Marine Geology and Geophysics
University of Miami
Miami, Florida

**Charles Featherstone**
NOAA-AOML-OCD
Miami, Florida

**M. Drew Ferrier**
Department of Biology
Hood College
Frederick, Maryland

**Tom D. Fontaine**
South Florida Water Management District
West Palm Beach, Florida

**James W. Fourqurean**
Department of Biological Sciences
Southeast Environmental Research Center
Florida International University
Miami, Florida

**Webster Gabbidon**
Negril Coral Reef Preservation Society
Negril, Jamaica

**Evelyn E. Gaiser**
Southeast Environmental Research Program
Florida International University
Miami, Florida

**Ginger Garte**
NOAA-AOML-OCD
Miami, Florida

**Dale E. Gawlik**
South Florida Water Management District
West Palm Beach, Florida

**Linval Getten**
Negril Coral Reef Preservation Society
Negril, Jamaica

**Thomas J. Goreau**
Global Coral Reef Alliance
Chappaqua, New York

**Louis J. Gross**
Department of Ecology and Evolutionary
  Biology
University of Tennessee
Knoxville, Tennessee

**Keith Hackett**
Florida Marine Research Institute
St. Petersburg, Florida

**Margaret O. Hall**
Florida Marine Research Institute
St. Petersburg, Florida

**C. Drew Harvell**
Ecology and Evolutionary Biology
Cornell University
Ithaca, New York

**Karl E. Havens**
Watershed Research and Planning Division
South Florida Water Management District
West Palm Beach, Florida

**Lee N. Hefty**
Miami-Dade Department of Environmental
  Resources Management
Miami, Florida

**Todd D. Hickey**
U.S. Geological Survey
St. Petersburg, Florida

**Terri Hood**
Division of Marine Geology
  and Geophysics
University of Miami
Miami, Florida

**Michael A. Huston**
Oak Ridge National Laboratory
Oak Ridge, Tennessee

**Jacqueline K. Huvane**
Duke University Wetland Center
Nicholas School of the Environment
Durham, North Carolina

**Arthur Itkin**
Statistical Consultants
Islamorada, Florida

**Russell Ives**
NOAA/NOS
Silver Spring, Maryland

**Walter C. Jaap**
Florida Marine Research Institute
St. Petersburg, Florida

**Krish Jayachandaran**
Southeast Environmental Research Center
Florida International University
Miami, Florida

**Elizabeth Johns**
NOAA/AOML
Miami, Florida

**Ronald D. Jones**
Southeast Environmental Research Center
Florida International University
Miami, Florida

**Frank Jordan**
Department of Biological Sciences
Loyola University
New Orleans, Louisiana

**Karen L. Kandl**
Department of Biological Sciences
University of New Orleans
New Orleans, Louisiana

**Woo-Jun Kang**
Division of Marine and Environmental
  Systems
Florida Institute of Technology
Melbourne, Florida

**Brian D. Keller**
Florida Keys National Marine Sanctuary
Marathon, Florida

**Anne Kenne**
Southeast Environmental Research Center
Florida International University
Miami, Florida

**Kiho Kim**
Biology Department
American University
Washington, D.C.

**Wiley M. Kitchens**
U.S. Geological Survey
Florida Cooperative Fish and Wildlife Refuge
Gainesville, Florida

**Vladimir Kosmynin**
Institute of Ecology
University of Georgia
Athens, Georgia

**William L. Kruczynski**
U.S. Environmental Protection Agency
Florida Keys National Marine Sanctuary
Marathon, Florida

**Brian E. Lapointe**
Division of Marine Science
Harbor Branch Oceanographic Institution
Fort Pierce, Florida

**David Lee**
Southeast Environmental Research Center
Florida International University
Miami, Florida

**Thomas N. Lee**
Rosenstiel School of Marine and Atmospheric
  Science
University of Miami
Miami, Florida

**Matthew T. Lewin**
Southeast Environmental Research Center
Florida International University
Miami, Florida

**Janet A. Ley**
Faculty of Fisheries and Marine Environment
Australian Maritime College
Beauty Point
Tasmania, Australia

**William F. Loftus**
Biological Resources Division
U.S. Geological Survey
Homestead, Florida

**Jerome J. Lorenz**
National Audubon Society Science Center
Tavernier, Florida

**Matt Lybolt**
Florida Marine Research Institute
St. Petersburg, Florida

**Marguerite Madden**
Center for Remote Sensing and Mapping
  Science
University of Georgia
Athens, Georgia

**Douglas M. Marcinek**
Department of Fisheries and Aquatic Sciences
University of Florida
Gainesville, Florida

**Darlene Marley**
South Florida Water Management District
West Palm Beach, Florida

**William R. Matzie**
Division of Marine Science
Harbor Branch Oceanographic Institution
Fort Pierce, Florida

**Paul V. McCormick**
The Nature Conservancy
Klamath Falls, Oregon

**Thomas C. McElroy**
Department of Biological Science
Florida International University
Miami, Florida

**Carole C. McIvor**
U.S. Geological Survey
Biological Resources Division
Center for Coastal Geology
St. Petersburg, Florida

**Fred McManus**
U.S. Environmental Protection Agency
Atlanta, Georgia

Christopher McVoy
South Florida Water Management
   District
West Palm Beach, Florida

John F. Meeder
Southeast Environmental Research Center
Florida International University
Miami, Florida

Ouida W. Meier
Department of Biology
Western Kentucky University
Bowling Green, Kentucky

Manuel Merello
Florida Marine Research Institute
St. Petersburg, Florida

Simone Metz
Division of Marine and Environmental
   Systems
Florida Institute of Technology
Melbourne, Florida

ShiLi Miao
South Florida Water Management
   District
West Palm Beach, Florida

Jerry L. Miller
Naval Research Laboratory
Stennis Space Center, Mississippi

Wolf M. Mooij
Netherlands Institute of Ecology
Center for Limnology
Nieuwersluis, The Netherlands

Erich M. Mueller
Center for Tropical Research
Mote Marine Laboratory
Summerland Key, Florida

Terry A. Nelsen
NOAA-AOML-OCD
Miami, Florida

Susan Newman
South Florida Water Management
   District
West Palm Beach, Florida

M. Phillip Nott
Department of Ecology and Evolutionary
   Biology
University of Tennessee
Knoxville, Tennessee

Mike O'Neal
Department of Geology
Indiana University–Purdue University
   at Indianapolis
Indianapolis, Indiana

John C. Ogden
South Florida Water Management
   District
West Palm Beach, Florida

Joseph F. Pachut
Department of Geology
Indiana University–Purdue University
   at Indianapolis
Indianapolis, Indiana

Kathryn L. Patterson
Institute of Ecology
University of Georgia
Athens, Georgia

Matt Patterson
Biscayne National Park
Homestead, Florida

Joseph H.K. Pechmann
Southeast Environmental Research
   Center
Florida International University
Miami, Florida

Patrick A. Pitts
Harbor Branch Oceanographic
   Institution
Fort Pierce, Florida

Delene W. Porter
Institute of Ecology
University of Georgia
Athens, Georgia

James W. Porter
Institute of Ecology
University of Georgia
Athens, Georgia

**Karen G. Porter**
Institute of Ecology
University of Georgia
Athens, Georgia

**George V.N. Powell**
World Wildlife Fund–U.S.
Monteverde
Puntarenas, Costa Rica

**DeeVon Quirolo**
Reef Relief
Key West, Florida

**Craig Quirolo**
Reef Relief
Key West, Florida

**K. Ramesh Reddy**
Soil and Water Science Department
University of Florida
Gainesville, Florida

**Christopher D. Reich**
U.S. Geological Survey
St. Petersburg, Florida

**Amy Renshaw**
Southeast Environmental Research Center
Florida International University
Miami, Florida

**Jennifer Richards**
Department of Biological Sciences
Florida International University
Miami, Florida

**Laurie L. Richardson**
Department of Biological Sciences
Florida International University
Miami, Florida

**Michael S. Ross**
Southeast Environmental Research Center
Florida International University
Miami, Florida

**David Rudnick**
South Florida Water Management
  District
West Palm Beach, Florida

**Michael Rugge**
Southeast Environmental Research Center
Florida International University
Miami, Florida

**Leonard J. Scinto**
Southeast Environmental Research Center
Florida International University
Miami, Florida

**Eugene A. Shinn**
U.S. Geological Survey
St. Petersburg, Florida

**Fred Sklar**
South Florida Water Management
  District
West Palm Beach, Florida

**Catherine Slouch**
Department of Geology
Indiana University-Purdue University
  at Indianapolis
Indianapolis, Indiana

**Ned P. Smith**
Harbor Branch Oceanographic Institution
Fort Pierce, Florida

**Alan D. Steinman**
Watershed Research and Planning
  Division
South Florida Water Management
  District
West Palm Beach, Florida

**Pierre Sterling**
Southeast Environmental Research Center
Florida International University
Miami, Florida

**Peter Swart**
Division of Marine Geology
  and Geophysics
University of Miami
Miami, Florida

**Ken Tarboton**
South Florida Water Management
  District
West Palm Beach, Florida

**Leonore Tedesco**
Department of Geology
Indiana University–Purdue University
  at Indianapolis
Indianapolis, Indiana

**Katy Thacker**
Negril Coral Reef Preservation Society
Negril, Jamaica

**Ann B. Tihansky**
U.S. Geological Survey
Tampa, Florida

**Jennifer I. Tougas**
Institute of Ecology
University of Georgia
Athens, Georgia

**John H. Trefry**
Division of Marine and Environmental Systems
Florida Institute of Technology
Melbourne, Florida

**Joel C. Trexler**
Department of Biological Science
Florida International University
Miami, Florida

**Chris P. Tsokos**
Department of Mathematics
University of South Florida
Tampa, Florida

**Will Van Gelder**
Southeast Environmental Research Center
Florida International University
Miami, Florida

**Randy VanZee**
Water Supply Division
South Florida Water Management District
West Palm Beach, Florida

**Harold R. Wanless**
Department of Geological Sciences
University of Miami
Miami, Florida

**Roy Welch**
Center for Remote Sensing and Mapping
  Science
University of Georgia
Athens, Georgia

**Jennifer L. Wheaton**
Florida Marine Research Institute
St. Petersburg, Florida

**Elizabeth Williams**
Rosenstiel School of Marine and Atmospheric
  Science
University of Miami
Miami, Florida

**Doug Wilson**
NOAA/AOML
Miami, Florida

**Wilfried F. Wolff**
ICG6 Forschungszentrum Juelich
Juelich, Germany

**George Yanev**
University of South Florida
St. Petersburg, Florida

**Charles S. Yentsch**
Bigelow Laboratory of Ocean Sciences
West Boothbay Harbor, Maine

**James B. Zaitzeff**
NOAA/NESDIS_E/RA3
Office of Research and Applications
Washington, D.C.

**William Zamboni de Mello**
Departmento de Geoquimica
Universidade Federal Fluminense
Niterói, Brazil

**Paul V. Zimba**
U.S.D.A.
Agricultural Research Service
Stoneville, Mississippi

# Table of Contents

# B. Biota

## Section III — Florida Reef Tract
### A. Water Movements and Nutrients

# Introduction: The Everglades, Florida Bay, and Coral Reefs of the Florida Keys: An Ecosystem Sourcebook

"The Everglades are one thing, one vast unified harmonious whole, in which the old subtle balance, which has been destroyed, must somehow be replaced. It was never a local problem, to be settled in makeshift bits and pieces."

*The Everglades, River of Grass*
**Marjory Stoneman Douglas, 1947**

"There are very few laws in ecology. Two of them are:
    I. Everything is connected to everything else.
    II. Everything goes somewhere."

**Barry Commoner**
**First Earth Day Celebration, 1972**

The central organizing theme of this book is that the ecosystems of South Florida are connected. What evidence do we have that they are? While it seems obvious that they should be, defining and quantifying these connections is not simple. This book is about linkages between ecosystems in the South Florida hydroscape. In South Florida, water integrates both the natural and the human environment. Water is the lifeblood of healthy ecosystems and a healthy economy. In the most simplistic terms, this book is dedicated to the proposition that water flows downhill. All of humanity lives downstream. And while we know that water flow connects adjacent ecosystems, we also know that, in general, political and jurisdictional boundaries are not defined by watersheds. Recognizing the importance of water movement, and the extraordinary modifications that have occurred to water flow in South Florida, the U.S. Army Corps of Engineers delivered to Congress the "Central and Southern Florida Comprehensive Restudy Plan." This plan is illustrated in Figure 1, contrasting a triptych of historic flow (left panel), the current flow (middle panel), and the plan flow (right panel). This book offers a fourth view (Figure 2) which emphasizes connectivity throughout both the marine and freshwater hydroscapes of South Florida. This book will make it impossible in the future to undertake management or restoration activities in South Florida without *also* considering their downstream effects. By highlighting these connections, this book is intended to forever change how we manage ecosystems.

In these rich and varied chapters, the reader will find both truth and consequences. The changes that human beings have wrought upon the land and, more specifically in Florida's case, in the water, create serious challenges to both environmental and human health. None of these authors shies away from confronting these challenges. These chapters contain controversy. Not all of the authors agree with one another about the origin of the problems nor the most efficacious way to solve them. In some cases, these disagreements probably result from looking at different parts of

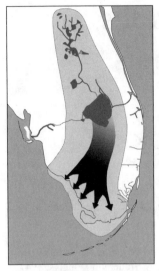
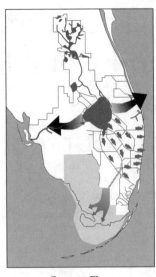
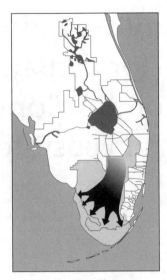

|       Historic Flow       |       Current Flow       |       The Plan Flow       |

**FIGURE 1** Recognizing the importance of water movement, and the extraordinary modifications that have occurred to water flow in South Florida, the U.S. Army Corps of Engineers delivered to Congress the "Central and Southern Florida Comprehensive Restudy Plan." This plan is illustrated by these three panels which contrast: Historic Flow (left panel), the Current Flow (middle panel), and The Plan Flow (right panel). (Redrawn from the South Florida Water Management District (1999) Plan Delivered to Congress. *Central & South Florida Restudy Update* 4:1–4.)

the same elephant. In others, the authors do not even agree whether it is or is not an elephant. Science advances by controversy. Disagreement is the best part of science, not the worst. By stating their differences clearly, these authors have set the stage for future investigations which will resolve how the system really works, and therefore how best to protect it. All of our authors have in common that they seek the truth. Each chapter in this book has been carefully reviewed by a panel of impartial external reviewers. The facts presented are not in dispute; what the facts *mean* often is.

This book is divided into five broadly defined sections. These include chapters on 1) The Florida Everglades, with subsections on (a) hydrology and nutrients, (b) the biota, and (c) ecosystem modeling; 2) Florida Bay, with subsections on (a) water movement and chemistry and (b) biota; 3) The Florida Reef Tract, with subsections on (a) water movement and nutrients and (b) biota; 4) Policy, Management, and Conservation; and finally 5) International Analogies.

## THE FLORIDA EVERGLADES: HYDROLOGY AND NUTRIENTS

In the first chapter, *The Past, Present, and Future Hydrology and Ecology of Lake Okeechobee and Its Watersheds*, Steinman et al. lay out the fundamental problem associated with water movement in the headwaters of the South Florida hydroscape: "Although most ecosystems are influenced to some degree by the presence and population growth of humans, few drainage basins in North America have had their hydrology reconfigured and replumbed as substantially as South Florida. Although the efforts to control floods have been successful, the environment has paid a high price in the process. Reducing the hydrologic connectivity throughout the Kissimmee–Okeechobee–Everglades ecosystem has resulted in a series of compartmentalized systems that interact largely through man-made structures and whose hydroperiods rarely operate as they did prior to human involvement."

In Chapter 2, *The Effects of Altered Hydrology on the Ecology of the Everglades*, Sklar et al. point out that these changes also have far-reaching influence downstream on "the ecological

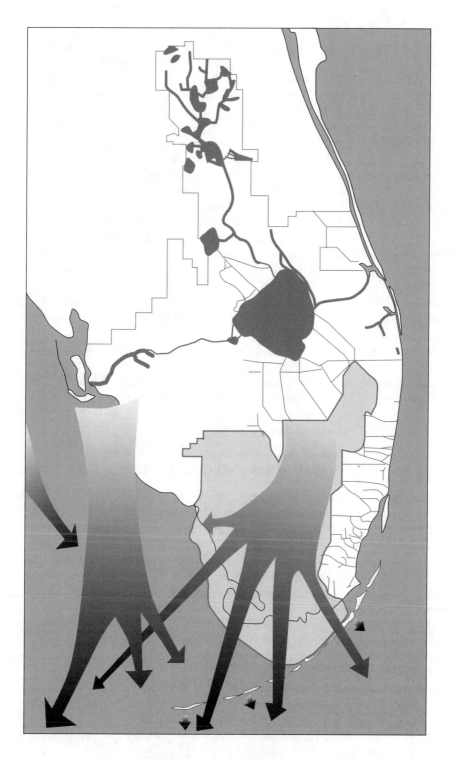

# The Reality

**FIGURE 2** This book offers a fourth view which emphasizes connectivity throughout both the marine and freshwater hydroscapes of South Florida.

characteristics of estuaries. For the coastal areas of the Everglades, decreases in historical supplies of freshwater changed the salinity regime of the areas and also changed the movement of nutrients and sediments through these areas. . . . It does appear that N inputs to the Bay could increase with increased freshwater flow. The consequences of N inputs are not certain. These areas are generally much more sensitive to P inputs. However, in western Florida Bay and in *adjacent ocean waters*, both N and P concentrations are low. Increased N inputs to these regions could stimulate productivity, including the production of algal blooms." These authors conclude with an admonition that is echoed in several chapters in this volume: "It is important that scientists and environmental managers study, and be cognizant of, the water quality linkages to hydrology, else sustainable management and restoration of the Everglades will not succeed."

The last two chapters in this section focus on phosphorus. McCormick et al. point out in Chapter 3, *Effects of Anthropogenic Phosphorus Inputs on the Everglades*, that "the Everglades ecosystem developed under extremely low rates of P supply." Data presented by Childers et al. in Chapter 4 on *Quantifying the Effects of Low-Level Phosphorus Additions on Unenriched Everglades Wetlands with In Situ Flumes and Phosphorus Dosing,* demonstrate that "the Everglades Agricultural Area (EAA) is a major source of P to these oligotrophic marshes, often by point-source canal inputs," and that "large changes in the Everglades community structure are occurring in areas subject to agricultural runoff." McCormick et al. point out that "the severity and extent of these impacts was recognized in 1988 when the federal government sued the State of Florida for allowing P-enriched discharges and associated impacts to occur in the Everglades. Settlement of this lawsuit eventually resulted in the enactment of the Everglades Forever Act by the Florida Legislature in 1994, which required the State of Florida to derive a numeric water quality criterion for P discharges into the Everglades that would *prevent ecological imbalances in natural populations of flora or fauna."* Clearly, if this criterion is to be applied to the downstream end-members of this hydroscape (that is, Florida Bay and coral reefs of the Florida Keys), this landmark phosphorous legislation might have to be amended to include reductions in nitrogen as well.

## THE FLORIDA EVERGLADES: BIOTA

In this section of the book, all of the authors demonstrate the intimate relationship between water flow and the abundance and distribution of species. Not surprisingly, these authors also sound the alarm regarding anthropogenic changes to South Florida and the survival of these same species in this increasingly highly modified hydroscape. In Chapter 5, *Ecological Scale and Its Implications for Freshwater Fishes in the Florida Everglades,* Trexler et al. state clearly that "the system of canals and levees that now directs the flow of water through the Everglades has changed the spatial arrangement of long- and short-hydroperiod marshes and associated aquatic communities." In Chapter 6, Kitchens et al., in their work on *Linkages Between the Snail Kite Population and Wetland Dynamics in a Highly Fragmented South Florida Hydroscape,* illustrate this general principle on snail kite populations. They point out that "the endangered snail kite is the only avian species that occurs throughout the South Florida freshwater hydroscape and whose population is restricted to these wetlands." Noting that "habitat quality for this species is entirely dependent on hydrology and wetland plant communities that support its principal food source, the apple snail," they conclude that the survival of this endangered species will be assured only if the restoration efforts succeed in "maintaining the spatial extent and natural hydrologic variability of the South Florida hydroscape." This means that the timing and the volume of water releases are both critical elements of a successful restoration project.

The final Chapter 7 of this section, Ross et al., in *Multi-Taxon Analysis of the "White Zone," A Common Ecotonal Feature of South Florida Coastal Wetlands,* describe a "region of low productivity characterized by low vegetation cover and canopy height," which from remotely sensed images appears as a reflective white band. They demonstrate that "over the past 50 years, the interior boundary of the white zone has encroached inland by an average of 1.5 km." Not surprisingly, "maximum shifts occurred in areas cut off by canals from upstream freshwater input." Noting that "these variables reflect the balance between ambient sea level and freshwater discharge," their conclusion is inescapable: upstream changes in the watershed are having such a profound effect on the distribution of aquatic organisms in South Florida that these changes can now be seen from outer space.

## THE FLORIDA EVERGLADES: MODELING

Good science can make accurate predictions. When a system is well understood, its response under differing perturbations can be forecast. Ecosystem modeling is as close as ecology comes to a predictive science. In Chapter 8, *Modeling Ecosystem and Population Dynamics on the South Florida Hydroscape*, DeAngelis et al. clearly state the problem: "During the past several decades, human impact has caused changes in land use and in the hydrologic pattern in South Florida. Concomitant with these changes have been precipitous declines in the abundance of some species, such as wading birds, Florida panthers, and Cape Sable seaside sparrows." Noting that "the close dependence of many biota of southern Florida on the natural hydrologic patterns, and the limited ranges of some of these species, raise concerns that such declines may continue," these modelers present a newly developed Across-Trophic-Level-Ecosystem Simulation (ATLSS) to demonstrate that "changes in the spatio-temporal patterns of water depth across the landscape make significant differences for some of the five species populations evaluated." Having shown that "it is possible to link spatially explicit temporally changing information on vegetation and hydrology at landscape scale to assess the consequences for biotic components of interest in the system," they challenge managers to use these new ecosystem forecasting tools to improve ecosystem restoration activities in South Florida.

In Chapter 9, *Maps and GIS Databases for Environmental Studies of the Everglades*, Welch et al. describe geographic information system (GIS), remote sensing, and global positioning system (GPS) techniques which they use to address environmental problems caused by "the dramatic increase in off road vehicle (ORV) use within Big Cypress National Preserve." Their remote sensing database "delineates some 37,346 km of ORV trails in Big Cypress National Preserve," giving park officials the numbers they need to assess this visitor impact.

The final chapter of this section, *Nitrous Oxide, Methane, and Carbon Dioxide Fluxes from South Florida Habitats During the Transition from Dry to Wet Seasons: Potential Impacts of Everglades Drainage and Flooding on the Atmosphere*, by Goreau and Zamboni de Mello puts South Florida management decisions in a global context. Starting with the knowledge that "South Florida hydrology has changed drastically in the past 100 years in ways that should strongly affect the strength of its habitats as sources or sinks of atmospheric trace gases," these authors predict that "the impacts of current restoration of Everglades flows on greenhouse gases is also likely to be large." Specifically, they predict that "reflooding the Everglades will basically reverse historical trends by converting old drained peat soils back into increasingly wet conditions. This will greatly increase the release of $CH_4$ from the environment and reduce the release of $CO_2$. The likely net impact would be to increase global warming." They conclude by reminding managers that "wetland ecosystems are important sources of climatically active atmospheric gases, which could be profoundly altered by land and water management practices as well as by normal seasonal cycles."

## FLORIDA BAY: WATER MOVEMENT AND CHEMISTRY

Chapters in this section lay the foundation for the realization that "although the strong circulation linkages within the South Florida coastal waters and remote upstream regions may help to maintain the strength and variety of the ecosystems, they can also provide a conduit for the input of pollutants or waters of degraded quality" (Chapter 11, Lee et al., *Transport Processes Linking South Florida*

*Coastal Ecosystems*). Lee et al.'s data are especially important in addressing the debate over water quality in Florida Bay and the Florida Keys National Marine Sanctuary: "A critical concern for water quality management in South Florida coastal waters is the input of nutrients from natural and anthropogenic sources. There is considerable effort under way to determine the nutrient budgets for the region and to predict future impacts on the carrying capacity of the ecosystem. There are many developed areas of the Keys where nutrient concentrations in inshore areas are consistently high. These are typically regions in interior canals and lagoons that have poor exchange with the coastal waters and therefore a relatively weak offshore nutrient flux. The same can be said for freshwater inputs from the Everglades. Since the freshwater volume flow is small (typically less than 10 m³/s), then the nutrient flux to the coastal waters is also small even with elevated nutrient concentrations. As an example, a relatively high concentration of dissolved nitrate in Shark River discharge of 1.0 μM in a peak discharge of 10 m³/s gives a nutrient flux of 10 μMm³/s. This is compared to the mean flow coupling the southwest Florida shelf to the Keys coastal waters through the passages in the Middle Keys, which is on the order of 1000 m³/s, if all passages are included. Even with a weak nitrate concentration of 0.1, the nitrate flux in this flow would be 100 μM m³/s, or a factor of 10 larger than the Everglades flow. The point of this comparison is that the largest continuous input of nutrients to south Florida coastal waters appears to come from the west Florida shelf and its adjacent rivers. These west Florida shelf inputs can at times also cause the transport of harmful algal blooms to South Florida waters, such as red tide organisms can result in local fish kills."

Smith and Pitts, in Chapter 12, *Regional-Scale and Long-Term Transport Patterns in the Florida Keys*, point out that "with few exceptions, tidal channels between Keys show a net flow out of Florida Bay and into the Atlantic.... The magnitude of across-channel motions in Hawk Channel is generally a small fraction of the along-channel component, but it is the flow across the channel that provides the direct link between the tidal channels between the Keys and the reef tract." We can see this phenomenon in Figure 3, which illustrates the presence of water from Florida Bay which has slipped under Hawk Channel and out onto the offshore reef. Smith and Pitts use these kinds of observations to conclude that "as a direct consequence of these transport pathways, the Florida Keys are a physically interconnected ecosystem and any region along the path can have a direct impact on other regions located downstream." This general physical oceanographic knowledge leads to the conclusion (Lee et al.) that "the sustainability of ecosystems in South Florida waters is dependent on water management policies of the entire region, as well as those of upstream regions in the eastern Gulf of Mexico."

In his tightly reasoned contribution (Chapter 13, *The Transport of Terrestrial Nutrients to South Florida Coastal Waters*) Brand presents and evaluates several alternative hypotheses to "examine if there is a causal link between the increase in the human population and associated activities in South Florida, and the ecological changes observed in the downstream coastal waters." Stressing both groundwater transport mechanisms as well as surface flow patterns, Brand hypothesizes that "the algal blooms in Florida Bay appear to develop from N-rich freshwater runoff mixing with P-rich seawater in western Florida Bay. It is hypothesized that much of the P comes from Miocene phosphorite deposits by way of Peace River erosion and subsequent coastal current transport along the southwest coast, and by way of groundwater through phosphorite-rich quartz sand deposits underneath northwestern Florida Bay. It is argued that this P source has not changed significantly over the past few decades. It appears that much of the N comes from freshwater runoff from agricultural lands through the Everglades. It is hypothesized that changes in water management practices in the past two decades have led to an increase in N inputs to eastern Florida Bay. Mixing of this water from the east with P-rich water from the west has led to the large algal blooms that have developed in north-central Florida Bay, altering the entire ecosystem." Finally, this water "is transported to the Middle and Lower Florida Keys, where it may be adversely affecting the coral reefs and other oligotrophic ecosystems there." Brand's data and ideas call into serious question

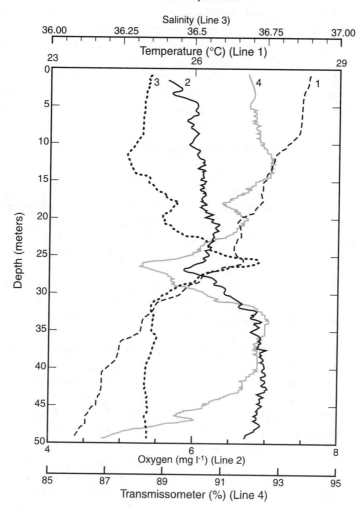

**Tennessee Reef, Channel 5**
Florida Keys 6/16/97

FIGURE 3 This hydrographic depth profile demonstrates the existence of a warm (Profile Line 1), oxygen-deplete (Profile Line 2), hypersaline (Profile Line 3), turbid (Profile Line 4) wedge of Florida Bay water on Tennessee Reef on 16 June, 1997. These data prove that coral reefs of the Florida Keys are indeed downstream of Florida Bay, and, by extension, that perturbations to water quality in Florida Bay will impact the reef.

the efficacy of the Everglades Forever Act, which regulates phosphorus pollution coming off agricultural fields in South Florida, but is silent on the issue of nitrogen.

The last two chapters in this section of the book focus on sediments and remote sensing of salinity patterns as indicators of the effect of water management practices in the region. In their fascinating Chapter 14, Nelsen et al. (*Linkages Between the South Florida Peninsula and Coastal Zone: A Sediment-Based History of Natural and Anthropogenic Influences*), demonstrate that a simple analysis of sediment cores from Florida Bay can reveal which water management plan was in use during a particular month in any given year: "After construction of the Water Conservation Areas, correlations between regional rainfall and flow in Shark River Slough varied with water management practices, ranging from essentially no correlation during the Monthly Allocation Plan to correlation approaching premanagement levels for the Rainfall Plan." Further, these cores revealed that "major storms, such as the great Labor Day hurricane of 1935 and Hurricane Donna

(1960), accounted for both immediate and long-term impacts on the ecosystem…. In central Florida Bay, decade-long trends of alternating sediment organic C buildup and reduction coincide, respectively, with periods of infrequent and more frequent hurricane activity. This supports major storm frequency as a potential mechanism for carbon storage and removal from Florida Bay." Finally, D'Sa et al., in Chapter 15, *Rapid Remote Assessments of Salinity and Ocean Color in Florida Bay*, show that "a salinity map of the central interior bay reveals a low salinity regime due to freshwater discharge from the Everglades." These real-time data, "together with monitoring of nutrients and ocean color are an essential water management tool that can be used for estimating freshwater budgets to aid the restoration efforts in Florida Bay."

## FLORIDA BAY: BIOTA

Altered water flow means altered biotic communities. In Chapter 16, Richardson and Zimba (*Spatial and Temporal Patterns of Phytoplankton in Florida Bay: Utility of Algal Accessory Pigments and Remote Sensing to Assess Bloom Dynamics*) review the literature on the extensive phytoplankton blooms in Florida Bay, which have "initiated cascading biological disturbances that have resulted in sponge and lobster mortalities and contributed to the massive dieoff of seagrasses within the bay." They describe advances in remote sensing with the result that, "in Florida Bay, algal populations can now be remotely sensed by AVRIS."

In Chapter 17 (*Modern Diatom Distributions in Florida Bay: A Preliminary Analysis*), Huvane demonstrates that "diatoms can be effectively used as proxies for salinity conditions in Florida Bay in sediment core studies," thus providing another monitoring tool to assess the effects of water management practices in South Florida.

Fourqurean et al. (Chapter 18, *Seagrass Distribution in South Florida: A Multi-Agency Coordinated Monitoring Program*) contrasts the vast area of healthy seagrasses with the smaller area involved in the seagrass dieoff: "The 14,622 km² of seagrasses in South Florida rank this area among the most expansive documented seagrass beds on earth, comparable to the back-reef environment of the Great Barrier Reef in Australia and the Miskito Bank of Nicaragua." They comment further that "while the area of seagrass dieoff was small (ca. 4000 ha) when compared to the total amount of seagrass habitat in South Florida, the ramifications of the loss were great." As with other studies in this volume, these authors are acutely aware that managing natural resources in South Florida is especially hard because "the components of the hydroscape do not respect political boundaries; many resources occur across multiple jurisdictions. Further, the environmental factors controlling the distribution of resources do not respect jurisdictional boundaries."

Durako et al. (Chapter 19, *Patterns of Change in the Seagrass-Dominated Florida Bay Hydroscape*) concur that "despite significant decreases in abundance in western Florida Bay from spring 1995 to spring 1998, *Thalassia testudinum* continues to be the dominant seagrass species in the Bay." These authors consider seagrass communities to be "at the end of the watershed pipe, and their status may reflect larger, landscape-scale problems" upstream. Noting that "much of the focus of management and restoration efforts in the South Florida hydroscape have been directed toward landscape-scale modifications to the existing flood-control system to increase the quantity of freshwater delivered to northeast Florida Bay via C-111, Taylor River Slough, and along the Everglades/Florida Bay land/sea interface," these authors emphasize that water quality will be as important as water quantity if "the metrics by which the public will judge the success of these restoration efforts are increases in seagrass abundance and water clarity."

Proper environmental restoration in the Everglades and Florida Bay will not kill corals in the Florida Keys. Healthy reefs existed in the Keys for thousands of years before humankind arrived in any numbers. Figure 4 presents a series of three schematics to compare and contrast alternative ecosystem effects of differing water management practices. The three panels, which we call the Florida Bay Hypothesis (Porter et al., 1999), depict different freshwater inputs that result in different downstream effects. In each case, two downstream water characteristics are measured: salinity and turbidity/nutrient concentrations. Salinity is not the stressor; instead, salinity simply tells you where the water mass came from. Variably saline (top panel), hypersaline (middle panel), and hyposaline water (bottom panel) always have coastal origins. Ecosystem stress arises from material carried in the water mass, not from its salinity *per se*. Turbid water with elevated nutrients, regardless of whether it is hyper- or hyposaline, threatens ecosystem function and integrity (bottom two panels). Water management practices will influence all three flow patterns, but the goal of South Florida ecosystem restoration should be to mimic the top panel as closely as possible. Restoring the South Florida landscape will not cause harm to the South Florida seascape.

Animal distributions are also good indicators of ecosystem health. In their chapter, Ley and McIvor (Chapter 20, *Linkages Between Estuarine and Reef Fish Assemblages: Enhancement by the Presence of Well-Developed Mangrove Shorelines*) demonstrate conclusively that significantly "greater densities of juvenile blue-striped grunts, gray snapper, and great barracuda occurred at sites where 1) water was deeper, clearer, and more marine in salinity regime; 2) mangrove structure was more highly developed; and 3) submerged vegetation was more abundant." This leads to the inescapable conclusion that "well-developed mangrove sites near the reef may function as staging habitats for juveniles of barracuda, grunts, and snappers, potentially enhancing both reef and estuarine fish assemblages."

In the last chapter in this section (Chapter 21, *Nesting Patterns of Roseate Spoonbills in Florida Bay 1935–1999: Implications of Landscape-Scale Anthropogenic Impacts*), Lorenz et al. point out that "the impact of the Keys' human population on the Florida Bay ecosystem is currently a contentious issue, and is the subject of a large-scale multiagency planning process (Keys Carrying Capacity Study). However, the impact of the loss of wetland habitat on animal populations has largely been overlooked." Using the historical and modern distribution of the roseate spoonbill, these authors contrast the loss of nesting sites due to "filling of wetlands for urban development in the upper Florida Keys" with the loss of feeding and breeding grounds due to misguided water management practices: "The most striking implication of these findings is that water management practices in the southern Everglades resulted in an ecological degradation of the coastal wetlands in northeastern Florida Bay to such a degree that the response by spoonbills was similar to that caused by the total destruction of wetlands on the Keys." These authors "hypothesize that plans to restore more natural surface and groundwater flows into the mainland estuaries north of Florida Bay should [therefore] positively influence spoonbill nesting patterns in the Bay. This perceived sensitivity of spoonbills, particularly to the effects of water management practices on northeastern Florida Bay, make them a highly appropriate indicator species for biological monitoring," and by extension, as an indicator of the success of ecosystem restoration efforts.

## FLORIDA REEF TRACT: WATER MOVEMENT AND NUTRIENTS

There is almost no consensus on the sources of nutrients in Florida Bay and waters of the Florida Keys National Marine Sanctuary (Figure 5). The "Natural Hypothesis" suggests that nitrogen and phosphorous come from natural processes such as resuspension of nutrient-rich sediments in Florida Bay, nitrogen fixation in the mangroves, and upwelling along the outer coast line. In contrast, the "Anthropogenic Hypothesis" suggests that the nutrients are coming from a variety of land-based sources of pollution under control of human activities in the watershed. In Chapter 22, Boyer and Jones (*A View from the Bridge: External and Internal Forces Affecting the Ambient Water Quality of the Florida Keys National Marine Sanctuary* [FKNMS]) state that "distinguishing internal from external sources of nutrients in the FKNMS is a difficult task. The finer discrimination of internal sources into natural and anthropogenic inputs is even more difficult." While noting that "the source of $NO_3$ to Florida Bay is the Taylor Slough and the C-111 basin, while the Shark River Slough impacts the west coast mangrove rivers and out onto the Shelf," they "speculate that in both cases elevated $NO_3$ concentrations are the result of $N_2$ fixation/nitrification within the mangroves." Lapointe (Chapter 23, *Biotic Phase-Shifts in Florida Bay and Fore Reef Communities of the Florida*

Florida Bay Hypothesis

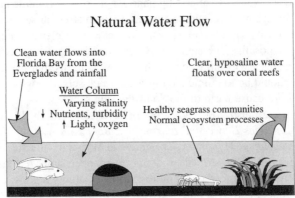

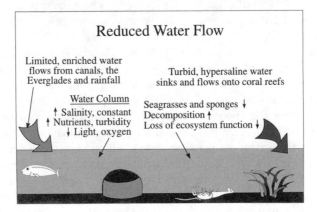

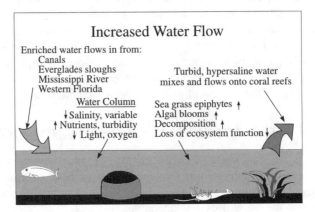

**FIGURE 4** The Florida Bay Hypothesis has been used to contrast alternative ecosystem effects of differing water management practices. The three panels depict different freshwater inputs that result in different downstream effects. In each case, two downstream water characteristics are measured: salinity and turbidity/nutrient concentrations. *Salinity is rarely, if ever, a stress.* Instead, salinity is simply the oceanographic marker that tells you where the water came from: variably saline (top panel), hypersaline (middle panel), and hyposaline water (bottom panel) always come from the land. What the water running off the land brings with it creates a stress for the ecosystem, not the slightly altered salinity *per se*. Turbid water with elevated nutrients, regardless of whether it is hyper- or hyposaline, threatens ecosystem function and integrity (bottom two panels). (Redrawn from Porter, J.W. et al., 1999. The effect of multiple stressors on the Florida Keys coral reef ecosystem: a landscape hypothesis and a physiological test. *Limnol. Oceanogr.* 44:941–949.)

*Keys: Linkages with Historical Freshwater Flows and Nitrogen Loading from Everglades Runoff*) disagrees and argues that "the Everglades have very limited capacity for removal of N compared to P and represent an external source of N to Florida Bay. The increase in the N:P ratio between the Everglades Agricultural Area and Shark River Slough is diagnostic of nitrogen contamination in carbonate systems, and is indicative of nonpoint source nitrogen enrichment associated with agricultural stormwater runoff." Boyer and Jones demonstrate that, "in the last 7 years, turbidities in Florida Bay have increased dramatically in the northeast and central regions potentially as a consequence of destabilization of the sediment from seagrass dieoff." Lapointe disagrees with the resuspension theory as the main source of reduced water clarity, and instead believes that "the series of ecological events over the past several decades in Florida Bay and the Florida Reef Tract follows, with few exceptions, the predictable responses of shallow subtropical seagrass and coral reef ecosystems to escalating nutrient loading." Boyer and Jones do note that "the least developed portion of the Upper Keys in Biscayne National Park and uninhabited Loggerhead Key (in the Dry Tortugas) exhibited lowest $NO_3^-$ and $NH_4^+$ concentrations," and acknowledge that this could be "evidence of a local anthropogenic source for both of these variables along the ocean side of the Upper, Middle, and Lower Keys. This pattern of decline implies an onshore N source which is diluted with distance from land by low nutrient Atlantic Ocean waters." They conclude, however that "it becomes clear from this analysis that the ambient water quality in the Lower Keys and Marquesas is most strongly influenced by water quality of the southwest Florida Shelf, the Middle Keys by southwest Florida Shelf and Florida Bay transport, the backcountry by internal nutrient sources, and the Upper Keys from Florida Current intrusion." While agreeing about the influence of the west coast of Florida as a source of nutrients, Lapointe argues forcefully that " 'getting the science right and getting the right science' are the two critical aspects of any environmental risk assessment, and the omission of these two steps in the development of the Everglades Restoration Plan was a key factor enabling water quality degradation of Florida Bay and Florida Keys as described in this chapter."

These alternative hypotheses result in profoundly different public policies and land use decisions. Acceptance of the "Natural Hypothesis," for instance, suggests that the Everglades Forever Act and current South Florida ecosystem restoration efforts are sufficient to protect the downstream waters of Florida Bay and the Florida Reef Tract. Acceptance of the "Anthropogenic Hypothesis," however, suggests that the Everglades Forever Act and other current attempts at South Florida ecosystem restoration are completely inadequate to protect Florida Bay and reefs of the Florida Keys. As this book goes to press, the management community seems committed to the "Natural Hypothesis," believing that its current environmental laws and conservation/restoration activities will succeed in protecting the marine waters of the Florida Keys. Such confidence should be based on a holistic "nutrient budget," one in which real numbers replace the question marks on the pie diagram in Figure 5. The "Holy Grail" for ecosystem science is a nutrient budget. This tells you the percent of the problem coming from each of the potential sources (Figure 5). The question of origin of the nutrients is also not, "Where was a particular atom of nitrogen or phosphorous *3 days* ago," but rather "Where was that atom *3 years* ago, and how did it get from there to here?" To date, no such nutrient budget exists for Florida Bay and the Florida Keys.

Chapter 24, by Keller and Itkin (*Shoreline Nutrients and Chlorophyll a in the Florida Keys, 1994-1997: A Preliminary Analysis*) also demonstrates that, even after 3 years of careful monitoring, natural variability in the data make a rigorous comparison of alternative hypotheses relating to nutrient concentrations near developed and natural shorelines difficult at best.

There is no way to read the last two chapters of this section and not feel that we are missing big pieces of the water management puzzle. Specifically, Chapter 25 by Reich et al. (*Tidal and Meteorological Influences on Shallow Marine Groundwater Flow in the Upper Florida Keys*) and Chapter 26 by Bacchus (*The "Ostrich" Component of the Multiple Stressor Model: Undermining South Florida*) ask basic questions such as: where is the groundwater coming from, where is it going, and what material that humankind has injected into the subsurface layers is it carrying as

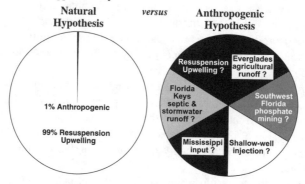

# Alternative Hypotheses
## Sources of Nutrients in Florida Bay and the Florida Keys National Marine Sanctuary

**NATURAL**
**Local**
• Resuspension of nutrient-rich Florida Bay sediments
**Regional**
• Upwelling of nutrient-rich Atlantic Ocean water

**ANTHROPOGENIC**
**Local**
• Florida Keys septic systems and stormwater runoff
**Regional**
• Everglades Agricultural Area runoff
• Southwest Florida phosphate mining and agricultural runoff
• Shallow-well injection
**Global**
• Mississippi River input

**FIGURE 5** There is almost no consensus on the sources of nutrients in Florida Bay and waters of the Florida Keys National Marine Sanctuary. The "Natural Hypothesis" suggests that a vast majority of nitrogen and phosphorous atoms have come from "natural" processes such as resuspension of nutrient-rich sediments in Florida Bay and upwelling onto the reef tract of nutrient-rich deep oceanic water. By contrast, the "Anthropogenic Hypothesis" suggests that the nutrients are coming from a variety of land-based sources of pollution under control of human activities in the watershed. These alternative hypotheses result in profoundly different public policies and land management decisions. Acceptance of the "Natural Hypothesis," for instance, suggests that the Everglades Forever Act and current South Florida ecosystem restoration efforts are sufficient to protect the downstream waters of Florida Bay and the Florida Reef Tract. Acceptance of the "Anthropogenic Hypothesis," however, suggests that the Everglades Forever Act and other current attempts at South Florida ecosystem restoration are completely inadequate to protect Florida Bay and reefs of the Florida Keys. Confidence in either hypothesis should be based on a holistic "nutrient budget," one in which real numbers replace the question marks on this pie diagram. To date, no such nutrient budget exists.

it moves? These general questions are made tangible by recent data from the Biscayne National Park (Figure 6), which show that groundwater is seeping onto offshore coral reefs. These salinity traces are especially important because they demonstrate the groundwater linkage between onshore sources of freshwater and offshore fully marine coral reef resources. These data (Figure 6) also strongly suggest that pollution pumped into freshwater coastal aquifers could ultimately travel to and impact offshore coral reefs via this groundwater connection.

From their research, Reich et al. conclude that "these data demonstrate that, whatever sewage treatment systems are installed in the Florida Keys, what must be taken into account is that both surface discharge and shallow subsurface injection will almost certainly find their way into the nearshore marine waters." Bacchus is equally definite about the policy implications of her findings: "Expensive federal initiatives such as the $213 million designated to support the Florida Keys Water Quality Improvement Act and to *increase the volume of injected effluent in the Keys*, ironically

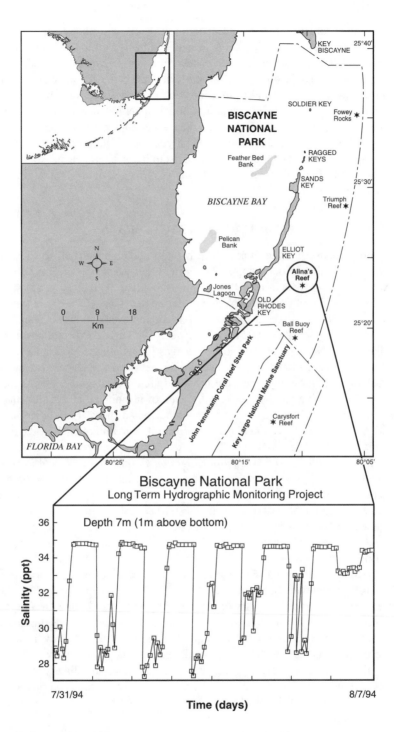

**FIGURE 6** Data from the Biscayne National Park Long-Term Hydrographic Monitoring Project shows clearly that groundwater is seeping onto offshore coral reefs on a tidal cycle. These data, provided by Mr. Richard Curry, BNP Resource Manager, are especially important because they demonstrate the groundwater linkage between onshore sources of freshwater and offshore fully marine coral reef resources. These data strongly suggest that pollution pumped into freshwater coastal aquifers will ultimately impact offshore coral reefs via this groundwater connection.

would exacerbate the problem rather than ameliorate it. Likewise, other costly federal initiatives such as the $7.8 billion Everglades Restoration Project and the Harmful Algal Blooms and Hypoxia Research and Control Act of 1998, which designated more than $52 million to address the problem of harmful algal blooms and hypoxia in the Gulf of Mexico, are being circumvented by the continued and expanded groundwater mining and aquifer injection of fluid wastes in South Florida, proposed and implemented by federal, state, regional and local government agencies." The information presented in this section suggests that Figure 4 needs to be redrawn, with subsurface groundwater flows added to each of the three diagrams. With the data presented, it is hard to disagree with Dr. Bacchus' final comment: "Although the multidisciplinary approach is gaining recognition, recent agency decisions and actions in south Florida suggest a severely limited understanding of these critical interrelationships."

## FLORIDA REEF TRACT: BIOTA

As with the biota and ecosystems of Florida Bay, habitats and organisms of the Florida Reef Tract are also in serious difficulty. In our Chapter 27 (Porter et al., *Detection of Coral Reef Change by the Florida Keys Coral Reef Monitoring Project*), we show that "1) there was a significant loss of coral species richness in the Florida Keys National Marine Sanctuary (67% of the stations surveyed lost coral species, and 80% of all coral species were found in fewer stations at the end of the survey [2000] than at the beginning [1996], and that 2) between 1996 and 2000 coral cover in the Florida Keys declined by 38%." Noting that "Hurricane Georges (which hit the Florida Keys in September 1998) continued a downward trend in coral species richness and abundance, but did not start it," we simply conclude that "these loss rates are high and are unsustainable." This is especially true because "there is a disturbing parallel between the lack of juvenile coral recruitment measured in this study and the high death rate of adult corals documented in the Florida Keys. With a high death rate and a low recruitment rate, the overall coral community in Florida is in trouble," (Chapter 29, Tougas and Porter, *Differential Coral Recruitment Patterns in the Florida Keys*).

In Chapter 28 (*The Influence of Nearshore Waters on Corals of the Florida Reef Tract*), Cook et al. demonstrate that coral growth is reduced by water of elevated turbidity coming out of Florida Bay: "Coral reefs of the Florida Reef Tract receive variable inputs of inshore waters and those of Florida Bay. Reef development is typically greatest where islands of the Florida Keys impede exchange of these waters, and is least where passes permit such exchange.... Corals at inshore and offshore sites exhibited no differences in linear growth rates, but skeletal deposition of $CaCO_3$ was 40% greater in offshore corals. We ascribe this effect to increased turbidity at the inshore site."

Chapter 30 (*Aspergillosis of Sea Fan Corals: Disease Dynamics in the Florida Keys*) by Kim and Harvell documents another source of stress to corals. In their paper on disease they note that "although disease prevalence did not vary by site, the impact of the disease (severity) was greater among sea fans near Key West." In attempting to answer the question, "Why is this area an apparent disease 'hot spot,'" they suggest that "although the findings of this study must be taken as preliminary, given the limited amount of data available for analysis, one possibility is that poor water quality at these sites, as indicated by higher N concentrations and slightly lower water clarity, exacerbated disease severity." These studies are beginning to identify the first critical links between human influences over physico/chemical properties of seawater and the survival of coral reefs in the Florida Keys.

## POLICY, MANAGEMENT, AND CONSERVATION

In Chapter 31 (*Water Quality Concerns in the Florida Keys: Sources, Effects, and Solutions*), Kruczynski and McManus review the published literature on water quality problems in the Florida Keys. These authors point out that even "properly functioning septic tank systems remove very little nutrients (4% N, 15% P) from wastewater and, depending upon their location, effluent from septic tank drain fields can rapidly migrate to surface waters." Echoing themes developed in Chapters 25 and 26 on groundwater, Kruczynski and McManus state that "Disposal of wastewater from package treatment plants or on-site disposal systems into Class V injection wells results in nutrient enrichment of the groundwater. However, it is not known whether discharges into Class V wells results in substantial nutrient loading to surface waters. This question is currently under investigation." Despite the fact that "there are no definitive studies on the geographic extent of the impact of anthropogenic nutrients," these authors conclude that "scientists agree that canal and other nearshore waters are affected by human-derived nutrients from sewage." Having worked their entire professional lives in the field of environmental protection, and knowing the costs of effective wastewater treatment, these authors nevertheless conclude "the costs of water quality improvements are a small fraction of the long-term asset value that natural resources such as reefs, hard bottoms, and seagrasses provide to the economy of the Florida Keys.... If sources of nutrient enrichment continue unabated, it is likely that the ecological balance of nearshore communities of the Keys will be changed. Since the economy of the Keys is directly linked to a healthy ecosystem, it is imperative that sources of excessive nutrients to this ecosystem be eliminated."

Successful ecosystem managing in South Florida means "getting the water right" (Chapter 32, by Causey, *The Role of the Florida Keys National Marine Sanctuary in the South Florida Ecosystem Restoration Initiative*). "Getting the water right means restoring more natural hydrologic functions, while also providing adequate water supplies and flood mitigation. To do this we need to address: the quantity of water flowing through the ecosystem; the quality of water, the timing and duration of water flows and levels; and the distribution of water through the system." Noting that "the greatest threat to the environment, the natural resources of the Keys, and the Keys' economy has been the degradation of water quality over the past two decades," Causey makes it clear that the stakes are extremely high: "The Florida Keys National Marine Sanctuary attracts 3 million tourists who spend $1.2 billion dollars annually." Because of the developing understanding that "system-wide management means a holistic, systematic approach to address issues regionally, not locally," Causey readily acknowledges that "the old paradigm of managing just within the boundaries of one's marine protected area does not and cannot succeed." The key to success will include a change in the paradigm by which local, state, and federal agencies interact to "establish a comprehensive boundary for the ecosystem based on natural and physical processes, and not political or jurisdictional boundaries."

Causey emphasizes that "sound restoration decisions must be based on sound science." The editors of this book believe that this poses a special challenge to those who control the purse-strings of scientific funding: If there is *serious disagreement* about how a system functions, then the only appropriate response from management must be to *fund both sides* of the debate, and not to suppress the least "popular" point of view. The scientific method, that is the falsification of clearly stated hypotheses, can work for management, but only if the alternative testable hypotheses are actually tested. The scientific method does not allow us to pick and choose between the truths that we reveal, no matter how uncomfortable they make the proponents of "established wisdom" feel.

Causey closes by emphasizing a point that is too often lost in scientific debates: "It is absolutely essential to bring socioeconomic information into the planning process as a foundation for informed participation at an early phase. Treat this discipline with the level of importance that you would give the natural or physical sciences."

In the final chapter of this section (Chapter 33, *The Role of a Nonprofit Organization, Reef Relief, in Protecting Coral Reefs*), Quirolo, director of Reef Relief, the premier nongovernmental organization in the Florida Keys focused exclusively on the conservation of coral reefs, places the challenge in perspective: "Although the Florida Keys coral reef is only one-tenth the size of Australia's Great Barrier Reef, it is visited by ten times more people annually." Quirolo outlines the continuing challenges facing the conservation community in this region. She summarizes by asking the toughest question, "Reef Relief's efforts have made a big difference in the Florida Keys and elsewhere. But have they succeeded? If the benchmark is the health of coral reefs, then the answer is no."

## INTERNATIONAL ANALOGY: AN INTEGRATED WATERSHED
## APPROACH TO CORAL REEF MANAGEMENT
## IN NEGRIL, JAMAICA

The last two chapters in the book are significant for several reasons. First, they constitute a cautionary tale about a tropical watershed in which land-based sources of pollution have killed the coral reef. Second, this section puts to rest an academic debate about whether nutrient enrichment or sea urchin dieoff has caused coral reef decline in this area of Jamaica: while sea urchin dieoff may be a proximate cause, nutrient addition is the ultimate cause.

In their Chapter 34, K.G. Porter et al. (*Patterns of Coral Reef Development in the Negril Marine Park: Necessity for a Whole-Watershed Management Plan*) demonstrate that "poor water quality is killing Negril's coral reefs and that action, rather than further debate, is warranted. The best and only hope for the survival of Negril's coral reefs is whole watershed management in an attempt to reduce the flow of nutrients into the coastal zone." Lapointe and Thacker (Chapter 35, *Community-Based Water Quality and Coral Reef Monitoring in the Negril Marine Park, Jamaica: Land-Based Nutrient Inputs and Their Ecological Consequences*) reach the same conclusions and state, "The significant increases in $NH_4^+$ and soluble reactive phosphorous concentrations in the South Negril River since 1997 appear to be linked largely to the increased nutrient loads associated with the effluent from the sewage treatment ponds that have become operational in the past 2 years. However, proper treatment of human sewage alone may not be adequate to restore water quality in the Negril Marine Park to the levels necessary for coral reef restoration. Our data and observations indicate that deforestation and associated topsoil loss, agriculture, fertilizers, and animal feed lots represent other major contributors to nutrient enrichment of the Negril Environmental Protected Area and downstream waters of the Negril Marine Park. Improved management and moderation of all these human activities and nutrient sources could lead to monetary savings (e.g., avoiding unnecessary or excessive use of fertilizers by farmers) and significant improvements in water quality."

Porter et al.'s chapter includes an extended analogy comparing western Jamaican and South Floridian coral reefs. This comparison should give managers serious pause for reflection. Although there are several significant differences between the two environments, "coral reefs of both Negril and the Florida Keys are hydrologically linked to terrestrially based sources of pollution and nutrient enrichment. The porous nature of karstic geology means that both surface and subsurface flows can transport dissolved and suspended materials in freshwater into the marine environment. In both places, rapid population and tourist development is concentrated along the shoreline. In both cases, behind this intense coastal zone development is intensive agriculture. In both cases, sugar cane is the dominant agricultural crop, involving massive expropriation of land from natural habitats and massive annual fertilization of the crop." With less than 1% live coral cover on many of Negril's once beautiful coral reefs, the consequence of a lack of an effective whole watershed management plan tells a cautionary tale for South Florida.

The editors of this book believe that we probably have only one human generation to reverse the disturbing trends documented in this volume. After that, ecosystem restoration will have to occur over geological time scales, not human time scales.

"Time moves for the Everglades, not in ages and in centuries, but as man knows it, in hours and in days, the small events of his own lifetime."

*The Everglades, River of Grass*
**Marjory Stoneman Douglas, 1947**

**James W. Porter**
Athens, Georgia/Big Pine Key, Florida
July 2001

# Section I

## Florida Everglades

*Ochopee*
by
Clyde Butcher

# 1 The Past, Present, and Future Hydrology and Ecology of Lake Okeechobee and Its Watersheds

*Alan D. Steinman, Karl E. Havens,*
*Hunter J. Carrick, and Randy VanZee*
South Florida Water Management District, West Palm Beach, Florida

## CONTENTS

## INTRODUCTION

The human population of South Florida is growing at a rapid rate; by the year 2050, it is estimated that the population will double to approximately 12 million people. Competing demands for adequate water supply by industry, agriculture, urban development, and environmental restoration place Lake Okeechobee at the center of a potential socio-economic maelstrom. It is critical to understand the hydrology and ecology of this lake if it is to meet these competing needs and still maintain some of its natural structure and functions.

0-8493-2026-7/02/$0.00+$1.50
© 2002 by CRC Press LLC

Lake Okeechobee sits at the center of the South Florida hydroscape (Figure 1.1). It has been referred to in the past as the liquid heart of South Florida (FEEC, 1913) because of both its central location in the region and its role as a major source of water for the Everglades and lower east and west coast estuaries of Florida. However, its physical, chemical, and biological characteristics have changed over the past century, along with most natural areas in South Florida, and some now suggest that Lake Okeechobee requires major watershed and in-lake restoration projects to prevent its ecological collapse.

Lake Okeechobee is the largest body of freshwater in the southeastern United States, with a surface area of 1730 km$^2$. Changes in land use from the 1950s through the 1970s resulted in cultural eutrophication of the lake which, in turn, resulted in blooms of cyanobacteria in the 1980s (Jones, 1987). These blooms and the associated death of macroinvertebrates, combined with taste and odor problems, provided a catalyst for increased research on the lake. Prior to the 1980s, there was a surprising absence of published scientific literature on Lake Okeechobee, given its size and regional role. There now is a growing body of information available on Lake Okeechobee, including a compilation of studies that emerged from a 5-year comprehensive study in the late 1980s (cf. Aumen and Wetzel, 1995), and several reviews (Havens et al., 1996; Steinman et al., 1999). However, none has used hydrology as the conceptual driver. In this chapter, we adopt that approach.

South Florida's subtropical climate can be subdivided into dry (November through May) and wet (June through October) seasons (Duever et al., 1994). This natural hydroperiod has been altered, such that water levels in the lake often are lowered prior to, and during, the wet season to ensure that water levels do not reach a point that threatens the integrity of the dike that surrounds Lake Okeechobee in the event of extreme rainfall and/or hurricane-associated surges; conversely, water, if available, is purposefully stored in the lake, prior to the dry season to ensure sufficient supply for a growing human population during the winter months, when water demand is greatest in the

**Inflows and Outflows of Lake Okeechobee**

**FIGURE 1.1** Map showing current condition of Lake Okeechobee with regional hydrography, including major inflows and outflows and municipalities surrounding the lake. Inset shows position of lake within the state of Florida.

region. This altered hydroperiod has had profound impacts on the natural ecosystems of South Florida (Light and Dineen, 1994), as it does elsewhere in river floodplains (Dahm et al., 1995), lakes (Webster et al., 1996), and wetlands (Bedford, 1996).

In this chapter, we provide an overview regarding how Lake Okeechobee and its watershed have changed over time, and we project how they may appear in the future using hydrologic changes and associated ecological impacts as a means to compare and contrast these changes.

## A HISTORICAL PERSPECTIVE

The underlying geology for the Okeechobee basin consists of a thick sequence of Miocene clay (Brooks, 1974). Today's lake formed about 6000 years ago, due to differential subsidence resulting from compaction of the underlying clay sequence (Brooks, 1974). This geologic configuration has changed little since then, but the morphometry and volume of the lake have varied considerably over time because of the shallow basin of the lake and the strong influence of climate on its hydrology. Historically, the margins of the lake consisted of gently sloping plains so that, at higher stages, the surrounding land was inundated and water spilled over into the Everglades to the south and likely into Lake Hicpochee to the west (Brooks, 1974) (Figure 1.2, past conditions). The name of the lake derives from the Seminole Indian language, where "oki" refers to water and "chubi" means big (Bloodworth, 1959).

Early reports of American pioneers and travelers to South Florida paint a portrait of a vast, mysterious wilderness (Vignoles, 1823; as cited in Blake, 1980). Indeed, few of these travelers ever saw Lake Okeechobee, although they heard rumors of a vast lake referred to as Lake Mayaimi, Lake Mayaco, or Lake Macaco, depending on the source, but known only to Indians (Blake, 1980; Tebeau, 1984). From almost the beginning of European settlement in the region, there were plans to modify the hydrology of the lake, either to use it as part of a cross-Florida waterway or to drain it and the surrounding Everglades for the purposes of development. Interruptions from the Civil War, economic depressions, and general haggling between the major land corporations and the railroads resulted in a number of aborted drainage attempts, but the first serious effort at draining the system began at the start of the 20th century.

Lake Okeechobee's role in draining the Everglades was critical because it was viewed as its major source of water. Within one month of his election as governor of Florida in 1904, Napolean Bonaparte Broward cruised Lake Okeechobee by steamboat and concluded that the lake level could be lowered by six feet without impeding navigation. In turn, this would result in the drainage of 6 million acres of land, both north and south of the lake (Blake, 1980). Five years later, a state legislative committee approved a plan to build or improve upon eight canals to shunt water from Lake Okeechobee east, south, or west. These canals provided the foundation for the vast network of levees, pumps, and canals that now cover South Florida. However, as a means to drain the Everglades, this plan, as the others before it, was doomed to failure because it was predicated on the assumption that excess water in the Everglades derived only from overflows from Lake Okeechobee. It was not appreciated that rainfall and groundwater inputs are major sources of water to the Everglades. In addition, much of the region is at a similar elevation to the sea and thus not prone to move by gravity alone.

## PAST CONDITIONS

### WATERSHED

The total watershed of Lake Okeechobee exceeds 10,400 km², with the Kissimmee River and floodplain accounting for two thirds of that area (Brooks, 1974). Most historical reports on the watershed deal with the Kissimmee River, Taylor Creek, and Fisheating Creek, which were the

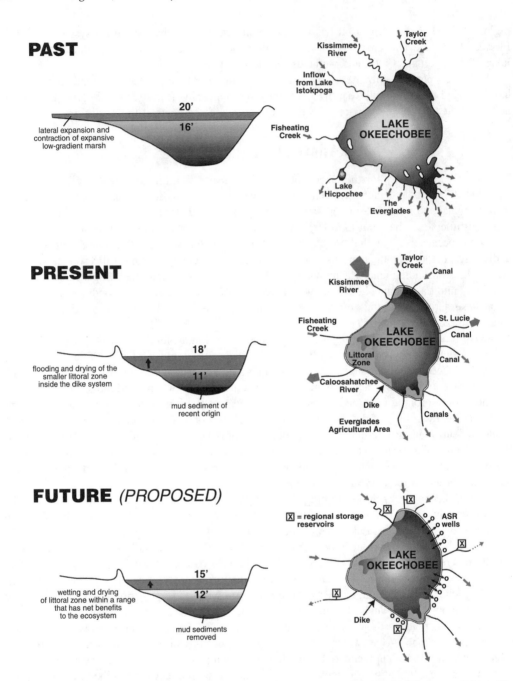

**FIGURE 1.2** Schematic representations of Lake Okeechobee during past, present, and future conditions. The left side of the panels shows generalized cross-section of the lake; the right side of the panels shows plan view of the lake. Reconstruction of past condition is a composite based on the best available information. Projection of future condition is not definitive and is subject to change.

major inflows into the lake and where navigation, commerce, and settlement were focused. These accounts often are anecdotal and give the impression of a generally flat and wet landscape (Ober, 1874a,b; Heilprin, 1887).

The earliest reports of the region are based on military accounts, associated with the second and third Seminole Wars (1835–1842 and 1855–1858, respectively). The geomorphology of the

Kissimmee River, in particular, has changed considerably due to its channelization in the 1960s, losing much of its sinuosity (see below also). Fort Basinger, built in 1837 during the second Seminole War, was located 18 miles due north of the river, but was 55 miles distant by river (Tebeau, 1984). Lieut. Rodgers explored the region in 1842, and wrote:

> The Kissimmee is a deep, rapid stream, generally running through a marshy plain, but sometimes the pine land approaches its borders, and sometimes beautiful live oak hammocks fringe its banks. ... The Kissimmee is, I think, the natural drain of the immense plains which form this part of the country; but, though deep and rapid, it is quite narrow. It is something strange that very often the surface of the river is covered by floating grass and weeds, so strongly matted together that the men stood upon the mass, and hauled the boats over it, as over shoals. The Kissimmee runs into the Oke-chobee, which filters through its spongy sides into the Everglades, whose waters finally, by many streams, empty into the ocean. (quoted in Sprague, 1848)

Rodgers also reported on Fisheating Creek from the same survey:

> The Thlo-thlo-pop-ka or Fish-Eating Creek runs through an open prairie, to which it serves as a drain. As might be expected, it gives evidence of being in the wet season a large stream, but when I examined it, the volume of water it discharged was very small. This stream is very tortuous, and sometimes, swells into a river, and dwindles to a brook. Its head is in a marshy prairie, where a number of streamlets run together about twenty miles, in a straight line due east to the Oke-cho-bee, but following the course of the creek about twice that distance. The banks of Fish-eating Creek are covered with game, and its waters filled with fish. (quoted in Sprague, 1848)

Heilprin (1887) noted that the key difference between the Kissimmee River and Taylor Creek, at least at their openings to Lake Okeechobee, was that the Kissimmee was located in "a grass country" and therefore difficult to find, "whereas the boundaries of Taylor's Creek are sharply defined by opposing walls of noble cypresses, which from their great height, 125 feet or more, present the appearance from a distance of low bluffs."

## LAKE

The morphometry of Lake Okeechobee in the 1800s (and before) has changed considerably. Some of the earliest known maps of the lake were generated by Captain E. Backus of the U.S. Army in 1837 and Frederick Oder, sponsored by the weekly magazine *Forest and Stream*, who undertook an expedition in 1874. Unlike today, where there is a dike completely encircling the lake (Figure 1.2; see "Present Conditions" section), the natural lake had a large littoral marsh surrounding the open water and the southern portion of this marsh was contiguous with the Everglades (see also Havens et al., 1996). A survey of the lake by the U.S. Army Corps of Engineers (U.S. Congress, 1913), conducted several years after draining began, indicated that the lake area was 733 square miles and the elevation at low water level was 20.6 feet above sea level (approximately 19.2 feet NGVD, National Geodetic Vertical Datum). When the lake level exceeded 22 feet above sea level (approximately 20.6 feet NGVD), water would spill over a natural bank to the south and enter as sheetflow directly into the Everglades.

Early reports from explorers provide descriptions of the lake ecosystem. Backus wrote a brief note directly on his map:

> Many small streams flow into Okee-cho-bee on the North-East and West Side, and also on the South-East side — but is not known that it has any outlet, and probably has none, except at high water, when Grassy Lake and Okee-cho-bee are united, and probably empty through some Streams into the Atlantic.

Grassy Lake presumably refers to the Everglades. Heilprin (1887) noted that the "lake proper is a clear expanse of water, apparently entirely free of mud shallows, and resting ... on a firm bed of sand" (cited from Brooks, 1974). It also was noted that "the border line of the lake is in most places not absolutely defined, owing to a continuous passage of the open waters into those of the Everglades" (Heilprin, 1887). Will (1964), based on observations from living along the lake since the early 1900s, noted that along the southwestern shore of the lake,

> ...the lake bottom was so flat, grown over with flags, bonnets and high grass, that when the lake rose after the summer rains, the shoreline could move back for half a dozen miles or so. But the south shore and halfway up the eastern side was something else again.

Presumably the presence of a natural muck sill and thick stands of custard apples presented a barrier for water movement unless lake stage exceeded approximately 22 feet (Matson and Sanford, 1913; Brooks, 1974). Some reports indicate the presence of up to eight rivers, hidden by dense vegetation and therefore not consistently identified, draining the southern portion of the lake (Brooks, 1974). However, Ober (1874a,b) claimed that he and his party could not find any obvious river or creek draining Lake Okeechobee. Rather, "the accumulated drainage of thousands of square miles of territory slowly percolates through the Everglades by thousands of channels with countless ramifications."

## CONNECTIONS TO THE ESTUARIES

As late as the 1870s, there were no direct connections from Lake Okeechobee to either the Caloosahatchee River/Estuary to the west or the St. Lucie River/Estuary to the east. Brooks (1974) noted that Lake Okeechobee likely spilled over into Lake Hicpochee at high lake stages, and that Lake Hicpochee served as the headwaters of the Caloosahatchee River. Prior to dredging and channelization of the Caloosahatchee (see below), the upper river had high banks and was covered with "a dense growth of large trees, principally pines, palmettoes, and oaks" (U.S. Congress, 1887). The early expeditions to Lake Okeechobee rarely make reference to the St. Lucie River, as most were initiated from the west up the Caloosahatchee, or south up the Everglades. Interestingly, a report from the 1770s, when Florida was occupied by the British, claimed that the St. Lucie River originated from Lake Okeechobee, an observation presumably based on the large freshwater discharge reaching the estuary (Romans, 1775; cited in Tebeau, 1984). It is most likely that discharge from the watershed, not the lake, accounted for the reduced salinity, as there is no evidence of Lake Okeechobee and the St. Lucie River being connected naturally (Storch and Taylor, 1969).

## HYDROLOGY

The hydrology of the pre-drained south Florida hydroscape was estimated using output from the Natural Systems Model (NSM) (SFWMD, 1998). This model simulates the hydrologic response of the region by replacing the current network of canals, structures, and levees with the rivers, creeks, and wetlands thought to be present prior to drainage. The NSM does not attempt to simulate the pre-drainage climate, as the input data to simulate the pre-drained hydrologic response do not exist. Rather, more recent climatic data, including rainfall, potential evapotranspiration, and tidal and inflow boundaries, are used which allow for meaningful comparisons between the current, managed system and the natural system. The NSM is closely linked to the South Florida Water Management Model (SFWMM), a regional-scale hydrologic model that simulates the hydrology of the highly managed water system in south Florida (SFWMD, 1997b). Because traditional calibration/verification methods cannot be applied to the NSM, model parameters were based on the calibrated and verified SFWMM (see below). Additional information on the SFWMM can be found at http://www.sfwmd.gov/org/pld/hsm/models/nsm.

Based on the NSM, the largest inputs of water to Lake Okeechobee were from rainfall, the Kissimmee River, Fisheating Creek, and Lake Istokpoga (Table 1.1). The NSM encompasses part of the Lake Okeechobee watershed but does not account for much of the floodplain of the Kissimmee

**TABLE 1.1**
**Inflows to Lake Okeechobee**[a]

| Inflow | Past[b] (1000 acre-feet) | 1995 Base[c] (1000 acre-feet) | 2050 Base[d] (1000 acre-feet) | Alt D13[e] (1000 acre-feet) |
|---|---|---|---|---|
| Rainfall | 1688.6 (51.6) | 1683.7 (42.1) | 1683.7 (42.6) | 1683.7 (39.9) |
| MDS positive | — | 949 (23.7) | 949 (24.0) | 949 (22.5) |
| Kissimmee River | 990.2 (30.2) | 979.6 (24.5) | 931.9 (23.6) | 931.9 (22.1) |
| Lake Istokpoga | 278.8 (8.5) | (included in MDS) | (included in MDS) | (included in MDS) |
| Fisheating Creek | 230.2 (7.0) | (included in MDS) | (included in MDS) | (included in MDS) |
| Taylor Creek/Nubbin Slough | 87.8 (2.7) | 126.2 (3.2) | 126.2 (3.2) | 23.44 (0.6) |
| Taylor Creek (reservoir) | 0 | 0 (0) | 0 (0) | 93.18 (2.2) |
| ASR | 0 | 0 (0) | 0 (0) | 135.9 (3.2) |
| St. Lucie (backflow) | 0 | 81.81 (2.0) | 99.02 (2.5) | 92.85 (2.2) |
| Caloosahatchee (backflow) | 0 | 13.89 (0.3) | 28.37 (0.7) | 5.7 (0.1) |
| EAA (backpump) | 0 | 46.13 (1.2) | 32.57 (0.8) | 0.83 (0.0) |
| C43 (reservoir) | 0 | 0 (0) | 0 (0) | 152.77 (3.6) |
| North storage | 0 | 0 (0) | 0 (0) | 50.97 (1.2) |
| L8 (backflow) | 0 | 53.91 (1.3) | 13.78 (0.3) | 4.45 (0.1) |
| S309 (backflow) | 0 | 0 (0) | 48.66 (1.2) | 65.08 (1.5) |
| Others | — | 63.24 (1.6) | 35.96 (0.9) | 34.69 (0.8) |
| Total | 3275.6 | 3997.5 | 3949.2 | 4224.5 |

[a] Numbers in parentheses refer to percentages within each column.
[b] Simulated past condition.
[c] Present condition.
[d] Future condition without changes.
[e] Future conditions with Restudy modifications.

River. Flow vectors show a major source from Lake Istokpoga, northeast of Lake Okeechobee (Figure 1.3). Inflows that occur outside the model boundary, such as in much of the Fisheating Creek, Kissimmee River, Taylor Creek, and Nubbin slough basins, are applied directly to the lake (Table 1.1) but are not reflected in the vector diagram.

Outputs of water from Lake Okeechobee were lumped together at a coarse geographic scale. Evapotranspiration was the largest loss of water from the lake, followed by outflows toward the southeast and then toward the southwest (Table 1.2). Flow vectors show virtually no water flowing due east and relatively little water flowing west into Lake Hicpochee, with most of the water being routed to the south and particularly southeast into the Everglades. This is consistent with anecdotal reports of a short river flowing from the lake at its southeast bay (Figure 1.2).

## PRESENT CONDITIONS

### WATERSHED

Even in pre-drainage times, agriculture was present in the region north of Lake Okeechobee. Since the 1800s, land use in the watershed north of the lake has been dominated by cattle ranching (Flaig and Havens, 1995). Higher density ranching, along with an increased number of dairy operations, developed in the 1950s as farmers started using improved breeds, and improved drainage resulted in a switch from native range cattle grazing to high-production improved pastures (Gatewood and Cornwell, 1976). Beef cattle populations increased from 20,000 individuals in 1925 to 160,000 in 1950; milking cows rose in number from 6000 in 1950 to 48,000 in 1969 (Flaig and Havens, 1995). Perhaps not surprisingly, increased ranching activity was accompanied by increased drainage

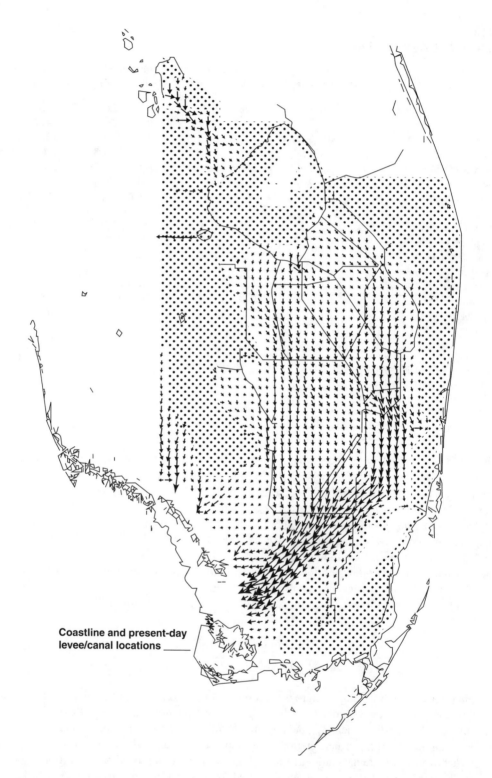

Coastline and present-day
levee/canal locations _____

**FIGURE 1.3** Map showing projected flow vectors according to output from the Natural Systems Model (NSM).

**TABLE 1.2**
**Outflows from Lake Okeechobee[a]**

| Outflow | Past[b] (1000 acre-feet) | 1995 Base[c] (1000 acre-feet) | 2050 Base[d] (1000 acre-feet) | Alt D13[e] (1000 acre-feet) |
|---|---|---|---|---|
| Evapotranspiration | 2381.2 (73.3) | 2360.7 (59.5) | 2329.7 (59.4) | 2382.4 (56.9) |
| MDS negative | — | 280.5 (7.2) | 280.5 (7.2) | 280.5 (6.7) |
| South-East LO | 605.5[f] (18.6) | — | — | — |
| South-West LO | 261.8[f] (8.1) | — | — | — |
| St. Lucie (regulatory) | — | 126 (3.2) | 87.78 (2.2) | 10.76 (0.3) |
| St. Lucie (agricultural demands) | — | 23.13 (0.6) | 18.93 (0.5) | 24.89 (0.6) |
| St. Lucie (estuary) | — | 0 (0) | 0 (0) | 19.09 (0.5) |
| Caloosahatchee (regulatory) | — | 289.78 (7.3) | 206.09 (5.3) | 13.00 (0.3) |
| Caloosahatchee agricultural demands | — | 70.74 (0) | 89.99 (2.3) | 36.73 (0.9) |
| Caloosahatchee (estuary) | — | 0 (0) | 0 (0) | 21.67 (0.5) |
| EAA storage | — | 0 (0) | 0 (0) | 278.75 (6.7) |
| WCAs | — | 61.57 (1.6) | 110.52 (2.8) | 101.33 (2.4) |
| St. Lucie reservoir | — | 0 (0) | 0 (0) | 0 (0) |
| Caloosahatchee reservoir | — | 0 (0) | 0 (0) | 19.13 (0.5) |
| EAA water supply | — | 377.09 (9.5) | 329.33 (8.4) | 174.44 (4.2) |
| LEC water supply | — | 39.75 (1.0) | 117.52 (3.0) | 68.00 (1.6) |
| Glades water supply | — | 165 (4.2) | 161.56 (4.1) | 148.82 (3.6) |
| STAs water supply | — | 0 (0) | 3.53 (0.1) | 3.32 (0.1) |
| L8 agricultural demands | — | 20.22 (0.5) | 0.64 (0) | 0.73 (0.0) |
| North storage | — | 0 (0) | 0 (0) | 126.71 (3.0) |
| ASRs | — | 0 (0) | 0 (0) | 263.13 (6.3) |
| Others | — | 155.47 (3.9) | 184.61 (4.7) | 210.61 (5.0) |
| Total | 3248.5 | 3969.95 | 3920.71 | 4183.99 |

[a] Numbers in parentheses refer to percentages within each column.
[b] Simulated past condition.
[c] Present condition.
[d] Future condition without changes.
[e] Future conditions with Restudy modifications.
[f] Lumped categories specific to NSM.

capacity. The total linear feet of canals in the most intensively used drainage basins in the watershed north of Lake Okeechobee increased by 14% between the mid-1970s and 1995 (Steinman and Rosen, 2000). Current land use is partitioned among agriculture (50%, of which 30% is pasture), wetlands (18%), upland forest (10%), open water (10%), rangeland (7%), urban (3%), and other (2%) (Dames and Moore, 1997).

One of the most conspicuous changes in the Lake Okeechobee watershed is the channelization of the Kissimmee River. Historically, this river meandered for approximately 165 km, within a 1.5- to 3-km-wide floodplain, before emptying into Lake Okeechobee (Koebel, 1995). Channelization of the river took place between 1962 and 1971, largely for flood control purposes, transforming the Kissimmee River into a canal 90 km long, 10 m deep, and 100 m wide, and resulting in the loss of 12,000 to 14,000 ha of wetland habitat (Koebel, 1995). Currently, a massive restoration project is underway which involves restoring 70 km of river channel and 11,000 ha of wetland habitat. Associated with drainage changes in the watershed was the change in outflows from Lake Istokpoga. In addition, a variable amount of water from Lake Istokpoga is now routed east through the Istokpoga Canal and into the Kissimmee River, instead of reaching Lake Okeechobee directly, as it was presumed to do in the past.

## LAKE

Prior to 1930, the borders of Lake Okeechobee expanded and contracted depending on water surface elevation. At present, the lake is encircled by the Herbert Hoover Dike (Figure 1.1). A series of hurricanes in the late 1920s resulted in massive flooding along the south shore of Lake Okeechobee and the loss of approximately 2000 lives in that area. Although earthen levees had been constructed prior to these hurricanes, this last tragedy provided the impetus for Congressional approval of funding to construct two 34-foot-high levees around the south and north shores of the lake. Construction of this phase of the "Hoover Dike" began in 1930 and was completed in 1938. It was not until the 1960s that the dike completely encircled Lake Okeechobee. Today, all the inflows and outflows, except one (Fisheating Creek), are regulated by water-control structures (Figure 1.1).

Construction of the Hoover Dike completed the alteration of the hydrology of the lake that was begun in the early 1900s; it also changed the ecology of the lake. The large littoral, marsh zone that extended north, west, and south from the lake under high water conditions was constrained by the levee (Figure 1.2). A new, much smaller littoral zone formed along the western shoreline of the lake, within the constraints of the dike (Havens et al., 1996). Today, the lake consists of two main parts that are adjacent but largely uncoupled under normal hydrologic conditions: a large, eutrophic pelagic zone (1330 km$^2$) and a smaller, oligotrophic marsh zone (400 km$^2$) (Steinman et al., 1999).

Lake stage now is influenced by a prescribed regulation schedule. This schedule determines the timing and quantity of water that is to be released from the lake when the stage exceeds a defined level, which varies with season (SFWMD, 1997a). According to the current schedule (WS/E, or water supply/environmental), a stage level above 13.5 ft NGVD on May 31 (and start of the wet season) would require water to be released until levels dropped below that trigger level. The volume of water released would depend on stage height, with larger releases (water control gates opened further) at higher lake levels. The trigger level for release slowly increases in the schedule from 13.5 ft NGVD on May 31 to 15.5 ft NGVD on September 30. Releases are made to the south through the Miami, New River, and Hillsboro Canals; to the east through the St. Lucie and West Palm Beach canals; and to the west through the Caloosahatchee River (Figure 1.1). Excessive freshwater releases to the east and west coasts create environmental problems in the estuaries (see below) and also represent a loss of valuable freshwater to tide. Excessive releases to the south also may create problems because of flooding to either natural areas (e.g., nesting and wildlife habitat) in the Everglades or agricultural areas. One of the innovative features of the WS/E regulation schedule is that it takes advantage of long-range and short-range climate forecasting; embedded within the schedule is the opportunity to release water slowly in anticipation of *El Niño* events and to hold water in anticipation of *La Niña* events. Additional information on WS/E can be found at http://www.sfwmd.gov/org/pld/hsm/reg_app/lok_reg/wse-faq.html.

The environmental resources of Lake Okeechobee are being threatened by three main elements: nutrient enrichment, altered hydroperiod, and expansion of nuisance and exotic species (Table 1.3). Since the early 1970s, total phosphorus concentrations have doubled in the lake, algal bloom frequency has increased, more phosphorus has accumulated in lake sediments, and the food web of the lake has been altered (Havens et al., 1996; Brezonik and Engstrom, 1998; Steinman et al., 1999). A mud bottom has replaced the natural sandy bottom over approximately 45% of the lake bed (Mehta, 1993), and its spatial extent appears to be increasing (Havens and James, 1999). These changes coincide with increased agricultural activity in the lake's watershed.

Although mean lake stage was higher in the pre-drainage period, water had the ability to flood surrounding lands, preventing lake levels from reaching extreme heights. Today, excessive water in the lake either is released as massive discharges to the estuaries or piles up, resulting in higher stage, reduced light transmittance through the water column, and a reduction in submerged vegetation (Steinman et al., 1997). At normal water levels, circulation patterns appear to isolate the central region from the marsh zone (cf. Carrick et al., 1994). However, at high water levels,

**TABLE 1.3**
**The Main Human-Induced Threats to the Ecological Health of Lake Okeechobee and Proposed Solutions to Those Threats**

| Threats | Problems | Proposed Solutions |
|---|---|---|
| Nutrient enrichment | Increased phosphorus concentrations in the lake | Enhanced source control |
| | | Enhanced best management practices |
| | Increased algal blooms | Reservoir-assisted stormwater treatment areas |
| | | Restoration of isolated wetlands |
| Altered hydroperiod | Excessive high waters (loss of submerged aquatic vegetation/ fisheries) | New lake regulation schedule (WS/E) |
| | | Discretionary operational protocols |
| | Excessive low waters (spread of exotic species; water supply risks) | Increased water storage in the watershed (above-ground reservoirs, ASR wells) |
| Exotic species | Spread of torpedograss | Prescribed burns |
| | Spread of *Melaleuca* | Herbicide applications |
| | | Additional research on control |

resuspended muds from the central pelagic region are transported into nearshore zones that traditionally have displayed clearer water (Havens, 1997; Havens and James, 1999). In addition, water levels above 15 ft NGVD almost totally inundate the marsh zone (Figure 1.4). Under these conditions, nutrient-rich water in the pelagic region may be transported into the low nutrient marsh region, resulting in changes in periphyton and macrophyte community structure (cf. McCormick and O'Dell, 1996; Havens et al., 1999). There is a general concern that the pelagic region of the lake has reached a new ecological steady state, one with nutrient-rich and turbid water characteristics, resistant to return to its historical clear water state, because of internal loading from resuspended sediments (Havens et al., 1996).

Finally, the littoral zone of the lake has been invaded by the exotic species *Melaleuca quinquenervia* (Australian broadleaf paper-bark tree) and *Panicum repens* (torpedograss). Torpedograss is a perennial amphibious grass with extensive rhizomes and has already displaced 14,000 acres of native marsh (Smith et al., 1998). Observational data suggest that this species can grow regardless of hydrologic regime but spreads most rapidly under dry conditions. This could be particularly problematic at lake stages of 11 ft and below, when almost the entire marsh region is exposed (Figure 1.4). Unlike the native vegetation, torpedograss grows in dense mats, with little open water between shoots, resulting in poor habitat for sport fish (Smith et al., 1993, 1999).

## CONNECTIONS TO THE ESTUARIES

Lake Okeechobee now is connected to the Caloosahatchee River to the west and to the St. Lucie River to the east. Connecting the lake to the Caloosahatchee was started in earnest by Hamilton Disston in the early 1880s, when Disston and his associates were awarded a contract to drain all lands overflowed by Lake Okeechobee, the Kissimmee River, and contiguous lakes (Blake, 1980). The excess water was first to be diverted into Lake Okeechobee, followed by a lowering of the lake by digging canals and draining the water through the canals. In 1882, one of Disston's company vessels made the first steamship trip from Fort Myers, at the mouth of the Caloosahatchee River, to the lake and up to Kissimmee City, some 90 miles north of Lake Okeechobee (Blake, 1980). Although economic depressions and corporate overexpansion eventually doomed Disston's venture, the connection of the Caloosahatchee to Lake Okeechobee persisted. Today, flow is controlled by three combination locks and dams on the Caloosahatchee River, with one located directly at the connection to the lake. At the highest discharge levels, flow through these structures can exceed 10,000 cfs.

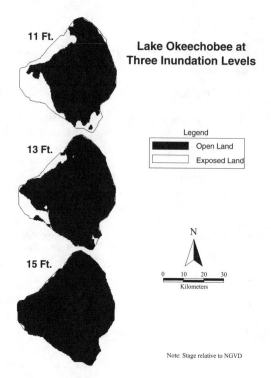

**FIGURE 1.4**  Area of littoral zone in Lake Okeechobee exposed when lake level is at 11 ft, 13 ft, and 15 ft NGVD.

The connection of Lake Okeechobee eastwards to the St. Lucie River was recommended by the Florida Everglades Engineering Commission in a 1913 report, an idea that had been considered 8 years earlier by Governor Broward and rejected. This connection not only would reduce the lake stage and prevent flooding of low-lying farmland, but also would provide (1) a cross-Florida navigation pathway linking the St. Lucie River, Lake Okeechobee, and the Caloosahatchee River; and (2) flow for a hydroelectric plant at the eastern end of the St. Lucie Canal (FEEC, 1913). Funding problems delayed the construction of the St. Lucie Canal connection, but the South Fork of the St. Lucie was connected to the lake in 1924 via the St. Lucie Canal, although at a smaller design than originally specified (Blake, 1980). Today, water control structures are located at the connection between the lake and the canal, as well as at the interface of the canal with the South Fork of the St. Lucie River.

As a result of the current canal system, both the Caloosahatchee and St. Lucie Estuaries experience deviations in the magnitude and timing of freshwater discharge, relative to historic conditions. The connections to the lake, as well as the expanding network of canals that ramify through the watersheds to drain agricultural lands, both contribute to the problem of excessive freshwater discharge during the wet season. Conversely, basin demands for irrigation and urban water during prolonged dry periods, when regulatory discharges from Lake Okeechobee do not occur, may result in insufficient inflows. In the Caloosahatchee, discharges greater than 6000 cfs result in an almost entirely freshwater estuary (Chamberlain and Doering, 1998). Analysis of historical data resulted in a preliminary estimate of an optimum range for freshwater discharge of 300 to 2800 cfs to promote estuary health (Chamberlain and Doering, 1998). The ecology of the St. Lucie Estuary has changed dramatically; where seagrass and oysters were once abundant, the system is now phytoplankton dominated, with high chlorophyll-*a* concentrations, high nutrient levels, and hypoxic and anoxic events in the bottom waters (Chamberlain and Hayward, 1996).

## HYDROLOGY

The present-day hydrology was derived using output from the South Florida Water Management Model (SFWMD, 1997b), using the 1995 base run (i.e., based on 1995 water demands from urban and agricultural areas). The model simulates the major components of the hydrologic cycle in South Florida, including rainfall, evapotranspiration, infiltration, overland and groundwater flow, canal flow, canal–groundwater seepage, levee seepage, and groundwater pumping. It simulates hydrology at a daily time step based on climatic data for the 1965–1995 period (SFWMD, 1997b). Whereas most of the model domain in the SFWMM is a distributed system of 2-mile by 2-mile grid cells, Lake Okeechobee is simulated as a lumped hydrologic system. The lake's hydrology is simulated by a mass balance approach using the modified delta storage (MDS) methodology (Trimble, 1986). The MDS term represents the total contribution of water budget components that are assumed not to change from what has happened historically. Unaccounted-for inflow or outflow is accounted for in the MDS term (Tables 1.1, 1.2). Additional information on the SFWMM can be found at http://www.sfwmd.gov/org/pld/hsm/models/sfwmm.

Water levels in Lake Okeechobee are managed through both regulatory (flood control) and nonregulatory releases. The regulatory releases are made according to a regulation schedule (see above), as set forth by the U.S. Army Corps of Engineers. Nonregulatory releases are made to meet water supply requirements of urban and regional users, agricultural and irrigation demands in the Lake Okeechobee area, and environmental needs of the St. Lucie and Caloosahatchee Estuaries, the Water Conservation Areas within the Everglades, and Everglades National Park. The SFWMM has the capability to simulate lake releases based on meeting minimum flows and levels (SFWMD, 1997b).

Based on the SFWMM, the largest inflows to Lake Okeechobee come from rainfall and the Kissimmee River (Table 1.1). The MDS component also represents a major source of inflow, indicating that a large fraction of flow was not directly accounted for. It is assumed that at least a portion of this component is similar to the volume that occurred under pre-development or natural system conditions, such as the sources from Lake Istokpoga and Fisheating Creek. However, even if those historical sources are summed together (509,000), this still leaves a difference of 440,000 acre-feet from the MDS under the 1995 base conditions (Table 1.1). Thus, other sources also are contributing to this MDS term under 1995 base conditions, such as overestimation of rainfall (no gauges are located directly over the lake), underestimation of evapotranspiration, seepage, flow through structures that is not accounted for, or groundwater infiltration. Overall inflows were approximately 18% lower in past conditions compared to present ones, but because 8% of the 1995 base inflows are attributable to backflows from surrounding areas (see Table 1.1), the effective net change was only approximately 10%.

According to the SFWMM, the largest loss of water out of Lake Okeechobee is to evapotranspiration (Table 1.2). Other substantial losses include flows south to the Everglades Agricultural Area (EAA) and for water supply to local municipalities, west to the Caloosahatchee for regulatory purposes, and east to the St. Lucie, also for regulatory purposes (all <10%; Table 1.2). Unaccounted-for losses (i.e., modified delta storage) are small relative to the unaccounted-for inflows. Direct comparisons of outflows between the past and present conditions are confounded by the different categories used by the NSM and SFWMM. Although evapotranspiration is the major loss in both models, the greater loss under past conditions most likely reflects the ability of the water to spread into adjacent floodplains and the larger littoral marsh (Figure 1.2). With inundated floodplains, the surface area-to-volume ratio of the lake increases, thereby enhancing evaporation. In addition, a larger marsh area increases the coverage of emergent vegetation, thereby enhancing transpiration.

## FUTURE CONDITIONS

The future conditions are based on projects contained within the Comprehensive Everglades Restoration Plan (CERP), which emerged from recommendations in the Central and Southern Florida Project Comprehensive Review Study, commonly called the Restudy. The Restudy was authorized by the U.S. Congress, who directed the U.S. Army Corps of Engineers to review the Central and Southern Florida Project, which was the original water management project for the region, dating from 1948. The goal of CERP is to implement changes to the original project to restore and preserve South Florida's natural ecosystems, while improving water supplies and maintaining flood protection. The future conditions outlined below may be modified depending on constraints associated with the implementation of these projects.

The projects identified in CERP that impact Lake Okeechobee and its watersheds include the following: (1) construction of water storage reservoirs to hold water that otherwise would be sent to tide; (2) construction of water preserve areas to treat urban runoff, store water, reduce seepage, and improve wetlands; (3) management of Lake Okeechobee as a natural resource by changing the regulation schedule to minimize extreme high and low water levels and by incorporating climate forecasting into the schedule (see above); (4) improved water delivery to the estuaries to maintain the salinity envelopes within a range considered healthy for the biota; (5) construction of underground water storage wells, known as aquifer storage and recovery, to store water but prevent water loss to evaporation that would otherwise occur in surface reservoirs; (6) construction of artificial wetlands to treat urban and agricultural runoff water before it is discharged to natural areas; and (7) delivery of water to the Everglades with a more natural hydroperiod. More information can be obtained about CERP at the following website: http://www.evergladesplan.org/.

### WATERSHED

CERP contains several specific projects in the watershed (Figure 1.2):

1. Construction of an above-ground reservoir, capable of storing approximately 200,000 acre-feet (20,000 acres at a 10-foot depth) of water in the Kissimmee River region. This reservoir would detain water during wet periods for later use during dry periods. This reservoir also would help reduce the duration and frequency of extreme high and low water levels in Lake Okeechobee.
2. Construction of an above-ground reservoir, capable of storing approximately 50,000 acre-feet of water, and an artificial wetland with a capacity of approximately 20,000 acre-feet in the Taylor Creek/Nubbin Slough Basin, north of the lake. Phosphorus loads from this basin are some of the highest in the watershed (SFWMD, 1997a). This storage and treatment area would attenuate flows to Lake Okeechobee, and reduce nutrient loads to the lake by increasing assimilation.
3. Construction of two combined reservoir-constructed wetland treatment areas and plugging of select local drainage ditches. The two reservoir-constructed wetland treatment areas are projected to be approximately 2000-acre facilities each, and the plugged drainage ditches would restore wetland conditions to approximately 3500 acres throughout the Lake Okeechobee watershed. These features would attenuate peak flows and retain phosphorus before entering the lake.

### LAKE

The two main features in CERP that deal specifically with Lake Okeechobee are a revised lake schedule (WS/E; discussed above) and construction of a series of aquifer storage and recovery (ASR) wells ringing the lake (Figure 1.2). The initial ASR proposal calls for 200 wells, each with the capacity of 5 million gallons per day, although this number is far from certain. These wells would provide additional storage while both reducing evaporative losses and maintaining current land use (e.g., agriculture), increase the ability of the lake to meet the competing demands for water supply, reduce harmful regulatory discharges to the Caloosahatchee and St. Lucie Estuaries, and maintain and enhance the existing level of flood control. Treated lake water would be injected into these wells when the climate-based inflow model forecasts that the lake water level could rise to levels deemed harmful to the marsh zone of the lake.

Output from the SFWMM indicates that under the recommended plan (D13R) of the Restudy, there will be a reduction in both the duration and frequency of high and low water levels in the lake, relative to "without project" (i.e., unchanged) conditions (Figure 1.5). These changes are expected to benefit the biological communities of the lake by minimizing periods of both extreme high water levels that cause declines in submerged plant communities and extreme low water levels that promote expansion of exotic plants, but still maintaining a variable hydropattern considered beneficial for the ecology of the lake (Smith et al., 1995).

## CONNECTIONS TO THE ESTUARIES

The estuary-related features proposed in CERP are designed to reduce the amount of freshwater discharge that is currently lost to the sea and to regulate its distribution to the estuary to mimic a more natural hydroperiod. The goal is to reduce the magnitude and frequency of high freshwater discharge events, during which the estuaries become oligohaline all the way to their mouths, as well as to maintain some semblance of freshwater flow during the dry seasons, when high salinity conditions can reach far up into both estuaries. Ideally, creating a more natural hydroperiod will provide the conditions necessary for restoration of seagrass and oyster beds in the St. Lucie Estuary and their enhancement in the Caloosahatchee Estuary.

For the St. Lucie watershed, storage reservoirs are proposed for the C-44 (St. Lucie Canal) basin and the basin just north of the canal (C-23/C-24/C-25). The storage capacity of the reservoir in the C-44 would be approximately 40,000 acre-feet, whereas the reservoirs in the north and south forks of the St. Lucie Estuary in the C-23/C-24/C-25 basin would be approximately 39,000 and 10,000 acre-feet, respectively. These reservoirs are designed to capture local runoff to attenuate peak flows to the estuary and provide environmental water supply benefits.

For the Caloosahatchee watershed, an above-ground storage reservoir with a proposed capacity of 160,000 acre-feet is proposed, as well as ASR wells with capacity of 220 million gallons per day. These features will capture excess runoff from the basin and releases from Lake Okeechobee,

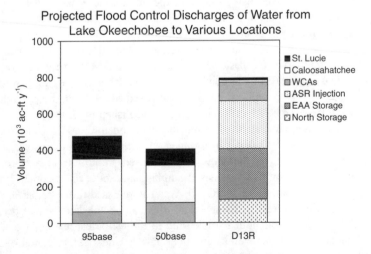

**FIGURE 1.5** Mean annual flood control releases from Lake Okeechobee for a 31-year (1965–1995) simulation for the 1995 base, 2050 base, and Alt D13 scenarios. See text for explanation of conditions for each scenario.

thereby providing environmental water supply and water quality benefits to the estuary. In addition to the reservoir and ASR wells, a constructed wetland capable of storing 20,000 acre-feet of water is proposed in the watershed. This would receive water once estuary and agricultural/urban water demands have been met and when water levels in the storage reservoir exceed 6.5 ft above grade. At this point, a series of pump stations would backpump excess water from the reservoir and the basin to Lake Okeechobee, after the water was treated through the constructed wetland.

## HYDROLOGY

Future hydrological conditions were modeled using the SFWMM. Two separate scenarios are provided, one based on future demands in the year 2050 but *without* inclusion of the CERP features (2050 Base) and one based on future demands in the year 2050 but *with* the CERP features incorporated (Alt D13). In both scenarios, rainfall is the major source of water, accounting for approximately 40% of total inflow, followed by MDS and the Kissimmee River (Table 1.1). Total inflow in Alt D13 was approximately 6.5% greater than 2050 Base, largely because of increased sources from the constructed reservoirs and ASR wells. Inflows from Taylor Creek/Nubbin Slough and EAA backpumping are reduced in Alt D13 compared to 2050 Base, which would help improve the water quality of Lake Okeechobee because those areas are major sources of phosphorus loads (SFWMD, 1997a).

Evapotranspiration is the major loss in both scenarios, accounting for approximately 60% (Table 1.2). All other outflows account for less than 10%, but they are partitioned quite differently between Alt D13 and 2050 Base. Whereas outflows to storage and ASRs are relatively important in Alt D13, outflows for water supply, regulatory demands, and agricultural demands are greater in 2050 Base (Table 1.2). A 31-year simulation of mean annual flood control releases from Lake Okeechobee clearly shows that the amount of excess water released to the estuaries is dramatically reduced under Alt D13 compared to either the 1995 or 2050 Base conditions (Figure 1.5).

## SUMMARY

The hydrology and ecology of Lake Okeechobee, its watershed, and the connecting estuaries have changed dramatically over the past 150 years. Although most ecosystems are influenced to some degree by the presence and population growth of humans, few drainage basins in North America have had their hydrology reconfigured and replumbed as substantially as South Florida. Although the efforts to control floods have been successful, the environment has paid a high price in the process. Reducing the hydrologic connectivity throughout the Kissimmee–Okeechobee–Everglades ecosystem has resulted in a series of compartmentalized systems that interact largely through man-made structures and whose hydroperiods rarely operate as they did prior to human involvement.

In Table 1.4, we summarize the past, present, and projected future impacts of extreme high rainfall (flood) and low rainfall (drought) conditions for Lake Okeechobee and its watershed. Under the high rainfall scenario, past conditions involved increased runoff from the watershed, especially from Lake Istokpoga, Kissimmee River, Taylor Creek, and Fisheating Creek, and rising lake levels that eventually spilled over into the surrounding marshes and low gradient lands. At lake levels above approximately 20 ft NGVD, water overtopped an organic berm at the south end of the lake and entered into the Everglades as sheetflow. Under present conditions, runoff reaches the lake even faster than in the past because of increased drainage and reduced detention. Lake stage rises and floods the littoral marsh above 15 ft NGVD. Water is released to the estuaries, depending on conditions associated with the regulation schedule. Environmental damage can result because of increased nutrient transport throughout the lake and to the estuaries, less light reaching the submerged aquatic plants, and reduced salinity in the estuaries. Under future conditions, more inflow will be detained with the restoration of wetlands in the watershed. Water will not rise to levels that

**TABLE 1.4**
**Summary of Hydrologic and Environmental Impacts to Lake Okeechobee Under Extreme Dry and Wet Conditions for Past, Present, and Future Conditions**

| High Rainfall | | |
|---|---|---|
| **Past** | **Present** | **Future** |
| Increased surface inflows to lake | Increased surface inflows to lake | Inflows detained in watershed |
| Increased lake levels; marsh extends over low gradient landscape | Increased lake levels; marsh flooded once stage exceeds 15 ft NGVD | Inflows detained in reservoirs or removed from lake into ASR wells |
| At very high lake levels, water exits to south and west as broad sheetflow into Everglades and Lake Hicpochee | Water is released to east and west coast estuaries according to lake regulation schedule | High lake stages do not occur; no regulatory releases to estuaries |
| | Damage occurs within lake due to loss of submergent and emergent vegetation, increased nutrient transport into marsh, and increased movement of mud | Ecological damage within lake minimized |

| Low Rainfall | | |
|---|---|---|
| **Past** | **Present** | **Future** |
| Low surface water inputs and high evapotranspiration | Low surface water inputs, high evapotranspiration, *and* increased demands on lake for urban/agricultural water supply | As lake stage declines, water inputs augmented by storage in reservoirs and ASR wells |
| Lake levels recede and marsh "migrates" downslope | Lake levels recede and at 11 ft NGVD; marsh area is 100% exposed | Extreme low lake stages do not occur because of extra water supplies |
| Salinity increases upstream in estuaries but is buffered by slow release of natural discharge in basins | Salinity increases in estuaries up to control structures; oligohaline species threatened | Salinity increases upstream in estuaries but is buffered by slow release of discharge from Restudy features in basins |
| No substantial loss of habitat | Net loss of habitat due to drought, but some increased habitat with submerged plants at south end of lake; exotic plant coverage expands | Ecological damage within lake minimized |

imperil the ecology of the system because water can be stored in reservoirs and ASR wells and regulatory releases to the estuaries will be minimized.

Under the low rainfall scenario, past conditions involved reduced runoff from the watershed, resulting in lake levels falling. As the upper marsh started to dry out and recede, the lower marsh migrated lakeward, following the lower lake level. Salinities started to creep up the estuaries. Unless drought conditions became severe, it was unlikely that there were substantial losses of habitat. Under present conditions, not only is there reduced runoff, but increased demands for water supply also exacerbate the low rainfall situation. At 11 ft NGVD, the marsh is 100% exposed, although submerged aquatic plants may expand their range lakeward, with more light reaching the benthos. Exotic species that favor dry conditions, such as *Melaleuca* and torpedograss, may also expand their range. Salinity quickly increases within the estuaries, threatening some species. Under future conditions, water inputs to the lake will be augmented by storage reservoirs and ASR wells. This will prevent or reduce extreme low lake stages. Releases from storage facilities in the east and west coast basins will reduce the rate of salinity increase.

The Okeechobee ecosystem inhabited by the Calusa and Seminole Indians, and visited by the likes of Taylor, Rodgers, Ober, and Heilprin, is long gone. As long as humans populate this region, the hydrology and ecology of the system will, by necessity, be altered. In this chapter, we have projected how the system functioned before intervention, and denoted what aspects of the system currently function well and which ones do not. This information allows us to evaluate how a future system might work better. Hydrologic remedies include detaining water for longer times in the watershed, avoiding extreme high water and low water levels in the lake, and providing for a more natural hydroperiod to the estuaries and the Everglades. In addition, other projects, conducted in concert with hydrologic restoration, also will help improve water quality in Lake Okeechobee and its tributaries. These projects include enhanced phosphorus source control via the implementation of best management practices or alternative technologies on beef cattle and dairy ranches, restoration of wetlands in the watershed of the lake, and possibly the complete or partial removal of mud sediments from the lake.

## ACKNOWLEDGMENTS

The authors gratefully acknowledge the advice of Paul Trimble, Luis Cadavid, and Christopher McVoy with respect to the hydrologic modeling output. We thank Mark Brady for his assistance with the graphics. The comments of Tom James, Paul Trimble, Todd Tisdale, Jim Porter, and one anonymous referee helped improve the manuscript.

## REFERENCES

Aumen, N.G. and R.G. Wetzel. 1995. Ecological studies on the littoral and pelagic systems of Lake Okeechobee, Florida (USA). *Arch. Hydrobiol. Beih. Ergebn. Limnol.* 45:1–356.

Bedford, B.L. 1996. The need to define hydrologic equivalence at the landscape scale for freshwater wetland mitigation. *Ecol. Applic.* 6:57–68.

Blake, N.M. 1980. *Land into Water — Water into Land*. University Presses of Florida,

Bloodworth, B.E. 1959. Florida Place Names, Ph.D. thesis, University of Florida, Gainesville.

Brooks, H.K. 1974. Lake Okeechobee. *Mem. Miami Geol. Soc.* 2:256–285.

Brezonik, P.L. and D.R. Engstrom. 1998. Modern and historic accumulation rates of phosphorus in Lake Okeechobee, Florida. *J. Paleolimnol.* 20:31–46.

Carrick, H.J., D. Worth, and M.L. Marshall. 1994. The influence of water circulation on chlorophyll–turbidity relationships in Lake Okeechobee as determined by remote sensing. *J. Plankton Res.* 16:1117–1135.

Chamberlain, R.H. and P.H. Doering. 1998. Preliminary estimate of optimum freshwater inflow to the Caloosahatchee Estuary: a resource-based approach. In S.F. Treat, Ed., *Proceedings of the Charlotte Harbor Public Conference and Technical Symposium*, Charlotte Harbor National Estuary Program Technical Report 98-02, pp. 21–130.

Chamberlain, R. and D. Hayward. 1996. Evaluation of water quality and monitoring in the St. Lucie Estuary, Florida. *Water Res. Bull.* 32:681–695.

Dahm, C.N., K.W. Cummins, M. Valett, and R.L. Coleman. 1995. An ecosystem view of the restoration of the Kissimmee River. *Restoration Ecol.* 3:225–238.

Dames and Moore. 1997. Land use/cover database development project: interpretive definitions and reference document. Deliverable to the South Florida Water Management District.

Duever, M.J., J.F. Meeder, L.C. Meeder, and J.M. McCollum. 1994. The climate of south Florida and its role in shaping the Everglades ecosystem. In S.M. Davis and J.C. Ogden, Eds., *Everglades. The Ecosystem and Its Restoration*. St. Lucie Press, Boca Raton, FL.

FEEC (Florida Everglades Engineering Commission). 1913. Report of the Florida Everglades Engineering Commission to the Board of Commissioners of the Everglades Drainage District and the Trustees of the Internal Improvement Fund, State of Florida.

Flaig, E.G. and K.E. Havens. 1995. Fate of phosphorus in the Lake Okeechobee watershed, Florida, USA: overview and recommendations. *Ecol. Eng.* 5:127–142.

Gatewood, S.E. and E.W. Cornwell. 1976. *An Analysis of Cattle Ranching in the Kissimmee River Basin*, Eco-Impact, Inc., Gainesville, FL.

Havens, K.E. 1997. Water levels and total phosphorus in Lake Okeechobee. *Lake Reserv. Manage.* 13:16–25.

Havens, K.E., N.G. Aumen, R.T. James, and V.H. Smith. 1996. Rapid ecological changes in a large subtropical lake undergoing cultural eutrophication. *Ambio* 25:150–155.

Havens, K.E., T.L. East, S.-J. Hwang, A.J. Rodusky, B. Sharfstein, and A.D. Steinman. 1999. Algal responses to experimental nutrient addition in the littoral community of a subtropical lake. *Freshwater Biol.* 42:329–344.

Havens, K.E. and R.T. James. 1999. Localized changes in transparency linked to mud sediment expansion in Lake Okeechobee, Florida: ecological and management implications. *Lake Reserv. Manage.* 15:54–69.

Heilprin, A. 1887. *Explorations of the West Coast of Florida and in the Okeechobee Wilderness*, Wagner Free Institute of Science of Philadelphia.

Jones, B. 1987. Lake Okeechobee eutrophication research and management. *Aquatics* 9:21–26.

Koebel, J.W. 1995. An historical perspective on the Kissimmee River restoration project. *Restoration Ecol.* 3:149–159.

Light, S.S. and J.W. Dineen. 1994. Water control in the Everglades: a historical perspective. *In* S.M. Davis and J.C. Ogden, Eds., Everglades. *The Ecosystem and Its Restoration*, St. Lucie Press, Boca Raton, FL, pp. 47–84.

Matson, G.C. and S. Sanford. 1913. *Geology and Groundwaters of Florida*, U.S. Geol. Survey Water Supply Paper 319.

McCormick, P.V. and M.B. O'Dell. 1996. Quantifying periphyton responses to phosphorus in the Florida Everglades: a synoptic-experimental approach. *J. N. Am. Benthol. Soc.* 15:450–468.

Mehta, A.J. 1993. *Lake Okeechobee Phosphorus Dynamics Study. Vol. IX.* Sediment characterization-resuspension and deposition rates. Deliverable to the South Florida Water Management District.

Ober, F.A. 1874a. Lake Okeechobee I. *Appletons J.* Oct. 31: 559–564.

Ober, F.A. 1874b. Lake Okeechobee II. *Appletons J.* Nov. 7: 591–594.

SFWMD. 1997a. Surface water improvement and management (SWIM) plan — update for Lake Okeechobee. Planning Document, South Florida Water Management District, West Palm Beach, FL.

SFWMD. 1997b. South Florida Water Management Model, Version XX, Documentation. Hydrologic Systems Modeling Division, South Florida Water Management District, West Palm Beach, FL.

SFWMD. 1998. Natural System Model. Version 4.5 Documentation. Hydrologic Systems Modeling Division, South Florida Water Management District, West Palm Beach, FL.

Smith, B.E., D.G. Shilling, W.T. Haller, and G.E. MacDonald. 1993. Factors influencing the efficacy of glyphosate on torpedograss (*Panicum repens* L.). *Weed Res.* 29:441–448.

Smith, B.E., K.A. Langeland, and C. Hanlon. 1998. Comparison of various glyphosate application schedules to control torpedograss. *Aquatics* 20:4–11.

Smith, B.E., K.A. Langeland, and C.G. Hanlon. 1999. Influence of foliar exposure, adjuvants, and rain-free period on the efficacy of glyphosate for torpedograss control. *J. Aquat. Plant Manage.* 37:13–16.

Smith, J.P., J.R. Richardson, and M.W. Collopy. 1995. Foraging habitat selection among wading birds (Ciconiiformes) at Lake Okeechobee, Florida, in relation to hydrology and vegetative cover. *Arch. Hydrobiol. Beih. Ergebn. Limnol.* 45:247–285.

Sprague, J.T. 1848. *The Origin, Progress, and Conclusion of the Florida War*, University of Florida Press, Gainesville.

Steinman, A.D. and B.H. Rosen. 2000. Lotic-lentic linkages associated with Lake Okeechobee, Florida. *J. N. Am. Benthol. Soc.* 19:733–741.

Steinman, A.D., R.H. Meeker, A.J. Rodusky, W.P. Davis, and S-J. Hwang. 1997. Ecological properties of charophytes in a large, subtropical lake. *J. N. Am. Benthol. Soc.* 16:781–793.

Steinman, A.D., K.E. Havens, N.G. Aumen, R.T. James, K.-R. Jin, J. Zhang, and B. Rosen. 1999. Phosphorus in Lake Okeechobee: sources, sinks, and strategies. *In* K.R. Reddy, G.A. O'Connor, and C.L. Schelske, Eds., *Phosphorus Biogeochemistry of Subtropical Ecosystems: Florida as a Case Example*, Lewis Publishers, Boca Raton, FL, pp. 527–544.

Storch, W.V. and R.L. Taylor. 1969. Some environmental effects of drainage in Florida. *J. Irrigation Drainage Div., Proc. Am. Soc. Civil Eng.* IR1:139–151.

Tebeau, C.W. 1984. Exploration and early descriptions of the Everglades, Lake Okeechobee and the Kissimmee River. *In* Gleason, P.J., Ed., *Environments of South Florida Present and Past*, Miami Geological Society.

Trimble, P. 1986. *South Florida Regional Routing Model*, South Florida Water Management District Technical Publication 86-3, West Palm Beach, FL.

U.S. Congress. 1887. House of Representatives. Report of the Chief of Engineers in Report of the Secretary of War, Vol. 2, Part 2. Improvement of Caloosahatchee [sic] River, Florida. Appendix O – Report of Capt. Black. 50th Congress, House document 87.

U.S. Congress. 1913. House of Representatives. Intracoastal waterway: across Florida section. Letter of transmittal from the Secretary of War with a letter from the Chief of Engineers, report on survey of the across-Florida section of the proposed continuous inland waterway from Boston, Mass., to the Rio Grande. 63rd Congress. House document 233.

Webster, K.E., T.K. Kratz, C.J. Bowser, J.J. Magnuson, and W.J. Rose. 1996. The influence of landscape position on lake chemical response to drought in northern Wisconsin, USA. *Limnol. Oceanogr.* 41:977–984.

Will, L.E. 1964. *A Cracker History of Okeechobee*, Great Outdoors Press, St. Petersburg, FL.

# 2 The Effects of Altered Hydrology on the Ecology of the Everglades

*Fred Sklar, Christopher McVoy, Randy VanZee,*
*Dale E. Gawlik, Ken Tarboton, David Rudnick,*
*and ShiLi Miao*
South Florida Water Management District

*Tom Armentano*
Everglades National Park, Daniel Beard Research Center

## CONTENTS

0-8493-2026-7/02/$0.00+$1.50
© 2002 by CRC Press LLC

**39**

# INTRODUCTION

Drainage of the Everglades changed South Florida from a subtropical wetland to a human- dominated landscape with a strong retirement, tourism, and agricultural economy. As a result, the Everglades is half its original size, water tables have dropped, hydroperiods have been altered, flows have been diverted, wetlands have been impounded, wildlife has been reduced, water quality has been degraded, and habitats have been invaded by nonindigenous plants. All of these impacts are caused directly or indirectly by an altered hydrology. Previous reviews of the ecological impacts of altered hydrology in the Everglades (Davis, 1943; Loveless, 1959; Craighead, 1971; McPherson et al., 1976; Gleason, 1984; Tropical BioIndustries, 1990; Gunderson and Loftus, 1993; Davis and Ogden, 1994; Sklar and Browder, 1998) have done much to increase public and scientific awareness of problems associated with altered hydrologic regimes and drainage. We will update this natural history by taking an ecologically comprehensive approach and highlighting current scientific studies.

It is not always easy to show direct cause-and-effect relationships between altered drainage and ecosystem disturbance. It is difficult because, one, a long period of record is required in order to filter out changes due to climatic variability, and, two, many factors are associated with an altered hydrologic regime. Factors such as nutrient loading can be as important as hydrology in shaping community structure. The goals for this paper are to review scientific understanding of historical hydropatterns (i.e., hydrologic durations and depths of inundation) in the Everglades and summarize research documenting the impacts of altered hydrology on the ecological structure and function of the Everglades.

It is recognized by wetland ecologists around the world that source, timing, duration, and depth of water will influence biogeochemical processes in soils and water, physiological processes of plant growth and decomposition, and reproduction and migration of fauna (Sharitz and Gibbons, 1989; Patten, 1990; Mitsch and Gosselink, 1993; Mitsch, 1994, to name a few). Soils, plants, and animals in turn, affect the hydrology. These ecological feedbacks allow for self-organization and succession (Odum, 1983). Usually, environmental restoration programs are attempts to redirect an altered rate or direction of succession. However, succession is affected by the available gene pool, climate, and antecedent conditions. It is clear that the decreased extent of the Everglades and surrounding uplands, changes in the soil and topography, exotic species, and the current system of canals and levees all constitute constraints on restoration to pre-drainage (pre-1880) conditions. The challenge facing science and society is to determine which key ecological driving forces will be restored to guide future succession in the remaining Everglades. The direction taken in the past has been deemed inappropriate for society and legally indefensible by the Federal Everglades Settlement Agreement of 1991 and the Everglades Forever Act of 1994. Now, to redirect succession, we must develop a better ecological understanding of differences between the current system and the pre-drainage Everglades.

This paper should be viewed as a brief anthology of many works and not a synthesis for setting hydrologic targets for restoration. It is a result of the collaboration to summarize the data and findings regarding the "ecological and hydrological needs of the Everglades" as required by Florida Law (Section 373.4592(4)(d)5,F.S., of the Everglades Forever Act). As a result, only "major" findings are presented and only organisms directly impacted by hydrology are discussed. This paper is divided into three sections: (1) past and present hydrologic change, (2) the effects of altered hydrology on Everglades ecology, and (3) information gaps and future research needs. The hydrology section will examine the sequence of events that have led up to the current water management system and will present historical accounts of the ecological affects associated with this sequence. The second section, on the impacts of altered hydrology, will examine current accounts of ecological impacts of altered hydrology in the Everglades gleaned from experiments and ongoing research programs. The third is a recommendation for adaptive management and new experiments as a way to assess the influence of water management on soils, plants, and animals in the Everglades.

## HISTORIC HYDROLOGIC CHANGE

### Pre-Central and South Florida Project

The first major efforts to drain the Everglades began in 1880, yet the earliest comprehensive and systematic depictions of the system were not produced until the 1940s. By that time, however, the system was already substantially altered by preceding drainage activities. Between 1880 and 1940, water tables declined as much as 9 feet, large areas of organic soils decomposed and subsided, and topographic changes of 1 to 5 feet due to subsidence actually reversed the direction of surface water flow (Davis, 1943b). The extent of these changes in the Everglades is impressive given that they all preceded the construction of the extensive Central and South Florida (C&SF) Project for Flood Control and Other Purposes, that began in 1947.

The first drainage projects were initiated by developer Hamilton Disston in 1880 with the construction of canals connecting the Kissimmee chain of lakes, construction of the 13-Mile Canal into the Sawgrass Plains of the northern Everglades (Light and Dineen, 1994), and most significantly for the Everglades, completion of continuous canals from Lake Okeechobee to the Caloosahatchee River, with drainage to the west coast of Florida (Harney, 1884). In the Kissimmee Valley, Disston most certainly altered the timing, but probably not the total volume, of flow into Lake Okeechobee and the Everglades. In contrast, comparison of early elevations suggests that the Caloosahatchee canals, completed in 1883, achieved their intended purpose — the lowering of Lake Okeechobee to reduce flooding in the Caloosahatchee Valley (Meigs, 1879). The lowering appears to have been about 2 to 3 feet, from perhaps 24 feet above a sea-level datum at Punta Rassa (Meigs, 1879; Slattery, 1913) to 22 or 21 feet above the same datum after drainage (Matson and Sanford, 1913). Lowering of the lake levels by 2 to 3 feet would have been significant relative to the custard apple rim along the southern shore, and would have reduced overflows from the lake into the Everglades. Anthropogenic reduction of an important inflow into the Everglades may therefore have started within a few years of 1883.

The second wave of drainage activity, from 1906 to the 1920s (Table 2.1, also see Color Figure 2.1*), affected the Everglades directly. Three of the four major canals that cut through the Everglades — the North New River, Hillsboro, and Miami — were begun in the 1910s and opened by 1915. The fourth, the West Palm Beach Canal, was likely finished within the following five years. By the late 1920s, the new St. Lucie Canal, with drainage to the east coast, and an expanded Caloosahatchee Canal further lowered water levels in Lake Okeechobee. Although subsidence of the soil surface south of the Lake theoretically would have increased the frequency of overflows, drainage efforts successfully avoided this by lowering the Lake level more quickly than the soil

---

* Color figures follow page 648.

**TABLE 2.1**

**Approximate Dates of Initial Construction and Open for Service of Major Hydraulic Works Affecting the Everglades**

| Canal | Initiated | Opened | Citation |
|---|---|---|---|
| Small, local truck farming canals related to railroad | 1896 | 1897 | Jones et al. (1948) |
| North New River Canal | 1906 | 1912 | Interbureau Committee (1930) Clayton (1936) |
| South New River Canal | 1906–1909 | 1913 | Marston et al. (1927) Florida Everglades Engineering Commission (1914) |
| Miami Canal | 1910 | 1913[a] | Marston et al. (1927) Florida Everglades Engineering Commission (1914) |
| Hillsboro Canal | 1910 | 1915 | Marston et al. (1927) Jones et al., (1948) |
| West Palm Beach Canal | 1913–1917 | 1920 | Marston et al. (1927) |
| St. Lucie Canal | 1916 | 1926 | Marston et al. (1927) |
| Caloosahatchee Canal | 1915 | 1925[b] | Marston et al. (1927) Herr (1943) |
| Lake Okeechobee South Shore Levee[c] | 1921 | 1926 | Marston et al. (1927) |
| Lake Okeechobee Levee[d] | 1932 | 1938 | Parker et al. (1955) |
| Tamiami Trail and Canal | 1916 | 1928 | Tamiami Trail Commission (1928) |

*Note:* Opening dates are less well-defined, as improvements, redredging, widening, etc. of canals often continued for numerous years.

[a] DuPuis (1954), a doctor who settled in the Lemon City area north of Miami in 1898, notes an earlier date for the first opening of the Miami Canal locks: "However, when the Miami Canal locks were opened in the early part of 1911, garden vegetables in the edge of the Everglades died and some of the driven wells, 25 feet deep, went dry as far east as the Florida East Coast Railway, a distance of four miles."

[b] According to Herr (1943), Chief Engineer of the Okeechobee Flood Control District, 1929–1944: "The Caloosahatchee Canal had been in existence prior to that time (1925), but its capacity was small and it had little effect on the lake elevations." A second project, increasing its capacities, was completed in 1938.

[c] This was a low muck levee. It was seriously breached during the hurricanes of 1926 and 1928.

[d] This is a much more solid levee constructed by the U.S. Army Corps of Engineers with a top elevation of 32 to 34 ft above mean sea level (Herr, 1943).

subsided. The last unimpeded surface overflow into the Everglades appears to have occurred in late 1915. At about the same time as this separation of Lake Okeechobee from the northern Everglades, Tamiami Trail, a road designed to connect the east and west coasts, was being constructed — in effect, separating the southern Everglades from the northern Everglades.

The hydrologic isolation of surface water in the Everglades from Lake Okeechobee, completion of the four major canals, and construction of the Tamiami Trail all strongly affected Everglades hydrology. Water tables were lowered throughout the Everglades basin. Figure 2.1 depicts water levels adjacent to the New River Canal, measured 25 years apart. Differences primarily reveal the effect of drainage, as both sets of measurements were made in years of similar antecedent rainfall, and at a similar point during the dry season (February–March). Water levels had declined dramatically, from just 1 foot below ground surface in 1915, to some 5 feet below ground by 1939.

These drops in water level, though drastic, are less surprising when one considers the changes in the annual water budget of the Everglades caused by the four major canals. Modeling (SFWMD,

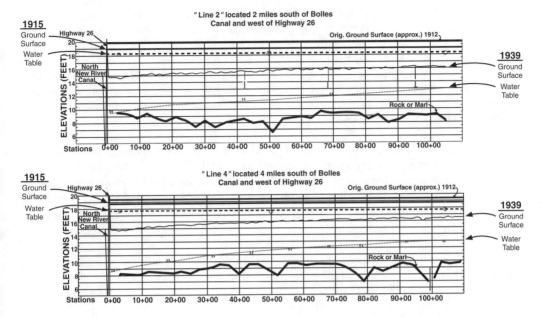

**FIGURE 2.1**   Cross-sectional views of two 2-mile long (10,000-ft) transects adjacent to the northern portion of the North New River Canal, showing elevations of bedrock, water table, and ground surface. February–March 1915 water table data measured by Baldwin and Hawker (1915). March 1939 data measured by Clayton et al. (1942). Historic weather data indicated that both periods of time had similar quantities of below-average precipitation.

1998) and discharge measurements (Parker et al., 1955) suggest that these four canals collectively discharged on the order of 1.5 million acre-feet per year, and drained approximately 1.5 million acres. The average annual removal of water was therefore 1 foot (12 in.). The magnitude of this removal can be put in perspective by comparing it with net annual precipitation (total precipitation minus evapotranspiration) in the Everglades. Using an annual average of 50 in. of total precipitation and 45 in. of evapotranspiration (SFWMD, 1998), one is left with an average of 5 in. of net annual precipitation. This means that more than twice (12 in. divided by 5 in., or 240%) of the net annual precipitation formerly available to the Everglades was now being removed by these canals. It is not surprising that water tables dropped precipitously.

The hydrological and ecological effects of the canals became apparent shortly after canal completion. John King worked in the early 1900s as a civil engineer and surveyor for Miami developer Capt. Jaudon, accumulating field experience in the lower Everglades west and southwest of Miami (King, 1917a–e; Larned, 1917). In early 1917, just a few years after completion of the first major canals, King already noticed definite changes in the Everglades:

> ...the drying up of the 'Glades, due to the various canals, is playing havoc with the birds here. The finer ones are fast disappearing. They lack feeding grounds. There are, occasionally, in the southern portion, a few green leg white herons as well as small blue and Louisiana blues, but five years has made a marked change. Of the food birds, the limpkin are found only occasionally. A guide told me his record was two in a season.... . (Larned, 1917)

As the northern and central portions of the Everglades were being drained by the four major canals, the canal and levee associated with the Tamiami Trail were draining the southern Everglades. Captain Jaudon and other promoters of the Tamiami Trail had extensive interests in several townships of Everglades land south of the proposed Tamiami Trail and intended to drain the lands for farming (King, 1917a–e). Blockage of southward flow of Everglades water and diversion to the sea appears to have been an explicit goal of the Tamiami Trail promoters:

> The idea actuating the Dade County Commissioners was that the drainage of the Everglades would be promoted by the construction of the proposed road, because it was the plan to dig a canal and use the rock excavated from the canal for the road bed. The canal would constitute a waterway of value in draining the adjacent lands and the drainage thus affected would enhance their value to the State. (Tamiami Trail Commissioners, 1928)

There was public opposition to this Tamiami Trail because of a concern about flooding upstream from the proposed road. These concerns were well founded. Only a few years after the initiation of dredging and levee construction in 1916, flooding due to blockage of southward flow across the Tamiami Trail was noted. In a written response to a 1923 complaint by the Pennsylvania Sugar Company (Pennsuco), F.C. Elliot, Chief Engineer of the Everglades Drainage District, agreed that the Tamiami Trail,

> ...act[ed] as a continuous dam across the Everglades preventing the natural flow of water and jeopardizing [by flooding] the lands East and Northwest along the Tamiami Trail and Miami Canal. (June 23, 1923, letter from F.C. Elliot; in Graham, 1951)

As early as 1915, vegetation changes due to drainage were already apparent, in this case in the part of the Everglades from Lake Okeechobee south to the latitude of Ft. Lauderdale:

> The drainage of the Everglades has proceeded sufficiently to induce noticeable changes in the character of the vegetation in certain places. In the interior of the glades, along edges of the sloughs which once supported a luxuriant growth of water lilies the lowering of the water table is accompanied by the invasion of saw grass and Sagittaria on the lower ground. Two or three miles south of the lake willows are gradually encroaching upon ground which under former conditions of poorer drainage supported a heavy growth of saw grass. In the 'lower glades' saw grass is giving way to myrtle, maiden cane and fennel. (Baldwin and Hawker, 1915)

As water levels dropped under the influence of drainage, organic soils (peats) of the Everglades were exposed to air for progressively longer periods each year. Prior to drainage, these soils had been protected from aerobic decomposition by a year-round or nearly year-round inundation. With drainage, the surface elevation of the soils of the Everglades began to subside, partly due to physical compaction and actual burning, but mostly due to oxidation (Clayton, 1936). Few understood the implications. Forty years after drainage began, a comment by Stephens and Johnson (1951) gives a sense of the enormity of the peat loss due to the lack of early understanding of peat subsidence:

> In making plans for the original drainage of the Everglades, apparently the main causes of subsidence were misunderstood. The original shrinkage of the peat due to drainage was considered, but the continuing losses by slow oxidation were not taken into account. Had the true nature and causes of subsidence losses been fully understood in the earlier days, the original plans might have been modified so as to have saved a large portion of the waste which has occurred since that time. (Stephens and Johnson, 1951)

Soil subsidence has been well documented along several subsidence lines set up in 1916 by the U.S. Department of Agriculture in the northern Everglades (Clayton, 1936; Clayton et al., 1942; Stephens and Johnson, 1951; Shih et al., 1979a,b,c). These subsidence lines consisted of regularly resurveyed transects (Figure 2.2). Exponential declines in soil surface were typical. The rate constant was found by Clayton (1936) and by Stephens and Johnson (1951) to be a linear function of the water table depth. Inflection points such as the one seen in Figure 2.2 reflect a change in the

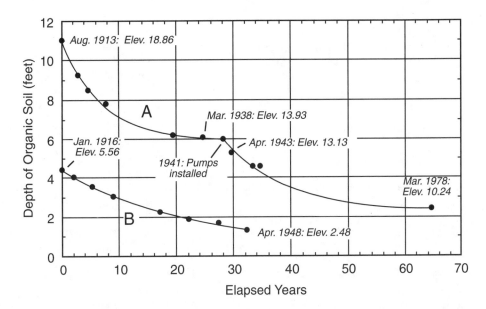

**FIGURE 2.2**  Subsidence of organic soils along lines established 1913–1916 by the U.S. Dept. of Agriculture, in the northern (A) and eastern (B) Everglades. (Data from Stephens and Johnson, 1951; and Shih et al., 1979.

drained water table, often due to a switch from gravity-driven to pumped drainage. Between 1912 and 1940, as much as 6 to 7 feet of soil were lost in the Lake Okeechobee area (Stephens and Johnson, 1951). Across the northern Everglades, the average subsidence rate during this period was approximately one inch per year. As a result of regional drainage in the northern Everglades, by the 1940s, the originally flat surface of these wetlands had transformed into a concave, sunken basin (Stephens and Johnson, 1951). A third to nearly half the depth of the original 10 to 12 feet of peat soil in the area directly south of Lake Okeechobee was lost. So much so, that by 1940 the original slope of the land was reversed, descending northward and toward, rather than away from, Lake Okeechobee (Stephens, 1942).

Changes in peat thickness in the central and southern Everglades were measured less systematically than in the northern Everglades. A U.S. Department of Agriculture subsidence line for the portion of the Everglades near the New River (Ft. Lauderdale) was established at Davie in 1913. Pre-construction surface elevations were measured along the main canal routes (Fla. Everglades Eng. Comm., 1914). Various measurements of peat thickness from original engineering drawings and land surveys recorded shortly after the onset of canal drainage are also available. All of these measurements were compared with 1940 estimates of peat thickness (Jones et al., 1948). Table 2.2 shows peat thickness measured close to the onset of canal drainage, again in the 1940s and, where available, at a post-1940 date. With a few exceptions, Table 2.2 indicates that peat thickness in the central and southern Everglades was less in the 1940s than in the period just after drainage (1911 to 1918). The most severe peat losses occurred along the eastern edge of the Everglades. The severity was partly because pre-drainage peat thicknesses were less there than in the northern Everglades and partly because the underlying sand allowed water tables to be drawn down below the peat horizon. Some locations lost all of the original organic soil, leaving only the underlying sand.

## POST-CENTRAL AND SOUTH FLORIDA PROJECT

The 1940s marked an important second phase in the evolution of Everglades water management. Recognition of over-drainage and the accompanying loss of soils by oxidation and fire was by then widespread in certain quarters. In 1942 and again in 1943 the Soil Science Society of Florida passed

**TABLE 2.2**
**Approximate Peat Thickness at Different Points in Time[a]**

| Location | Approx. Peat Thickness (ft) | | |
|---|---|---|---|
| | 1918 | 1940s | 1986 |
| Tamiami Canal (R37–R40) | 4.6 | 3 | 3 |
| | 5.3 | 4.3 | 2.9 |
| | 3.6 | 3.0 | 1.9 |
| | 3.7 | 1.0 | |
| Snapper Creek, 4 mi. south of Tamiami Canal (R37–R39) | 4.4 | 2.2 | — |
| | 3.1 | 2.5 | — |
| | 0.9 | 0.9 | — |
| | 1911 | 1940s | 1954 |
| Miami Canal, southern part | 6 | 3 | 2 |
| | 1916 | 1940s | |
| Eastern Everglades at Davie (T50 R41 Sec. 34) | 4.4 | 0.4 | |
| | 1912 | 1940s | |
| Eastern Everglades (T 51 R 41) | 4.0 | 0.5 | |
| Miami Canal Area (T 53 R 40) | 2.8 | 0.7 | |
| Eastern Tamiami Canal (T 54 R 39) | 1.9 | 1.4 | |

[a] R = range; T = township.

resolutions addressed to the Governor and his Cabinet indicating "extremely serious conditions pertaining to soil and water conservation of the Everglades Area" and calling for a "comprehensive plan of conservation and development" (Allison, 1943). An experienced drainage engineer for the U.S. Sugar Corp. wrote:

> The zealousness of the past has finally accomplished, in a way, the drainage of the 'Glades. In fact, the insidious loss of water through the years has virtually overdrained much of the area to a disconcerting extent. The coastal settlements, which are dependent on the back country seepage for their water supplies, are already complaining about lack of water, and large areas of idle peat and muck land have been seriously burned and the soils have generally subsided. These effects are as serious as the soil erosion problems of other states… ." (Bestor, 1942)

A report by an Advisory Committee to the Everglades Drainage District (Advisory Committee, 1944) laid out an influential general plan of improvement. After noting the obvious physical problems caused by the financial collapse of the Everglades Drainage District in 1931 (e.g., canal obstruction by hyacinths and locks out of service), the Committee distinguished between developed lands (primarily within the present Everglades Agricultural Area) and undeveloped lands, pointing out that, under wet conditions, the canal system, even if improved, did not have the capacity to drain both developed and undeveloped lands. Additionally, runoff from the undeveloped lands occurred first, filling the canals, and preventing timely drainage of the developed agricultural land. Under dry conditions different problems arose: As an uncontrolled system, too much water drained off the undeveloped areas, exposing the soil to great losses by oxidation and fire.

The solution proposed by the Advisory Committee later became the basis for the C&SF Project constructed by the U.S. Army Corps of Engineers (USACE). The recommendation was simply to impound the undeveloped lands by building levees along most and between several of the arterial canals. Such impoundments would allow water tables to be raised to protect soils and create wildlife habitat in the undeveloped lands while simultaneously "increasing the efficiency of the present [canal] system in service of [the developed] lands in agricultural use" (Advisory Committee, 1944).

Although not recognized at the time, it is important to note that this solution was to bring both environmental loss as well as gain to the "undeveloped lands." Without doubt, the impoundments made it possible to raise water tables above ground, reversing what the chairman of the Soils Department, University of Florida, called a "man-made desert," specifically, "a sky filled with acrid smoke from the burning of the soil itself; [and] vast areas of baking, cracking organic soils in the undeveloped sections of the Glades that are literally screaming for water and ready to burst into flame at the drop of a match" (Allison, 1943). At the same time, the impoundments transformed what was, prior to any drainage, a flowing system into a nonflowing, ponded one. Even where flow continued, it was primarily through canals, from or to point sources, instead of being spread across the landscape in the network of "broad sloughs that once carried the main flow of surface water" (Advisory Committee, 1944). If water flow was the pre-drainage driving force needed to maintain a ridge and slough landscape in the Everglades — a hypothesis to be tested — then elimination of flow in the series of impoundments was a significant step.

The C&SF Project (Color Figure 2.1), authorized by Congress in 1948 after the great hurricane of 1947, gave the USACE funds to construct modifications along the lines described above, and gave the Central and Southern Florida Flood Control District (now known as the South Florida Water Management District, SFWMD) the authority to manage it. The first major earthworks of the C&SF Project reinforced the existing north–south line, and extended it along the edge of the flatwoods in Palm Beach County (Color Figure 2.1). These levees were completed during 1952 to 1954 and became the eastern boundaries of what would become the Water Conservation Areas (WCAs). During the period 1954 to 1959, levees 5, 6, and 7 were constructed, forming the northern and western borders of the WCAs. Construction of additional levees (1 through 4 and 28) completed the partitioning off of 283,000 ha (700,000 acres) of deep muck lands that became known as the Everglades Agricultural Area (EAA). Flood protection for the EAA was enhanced by the construction of large-capacity pump stations (S-5A). Other flood protection activities during the 1954 to 1959 period included the deepening of the Hillsboro, North New River, and Miami canals in the EAA, and construction of water control structures (S-11A, S-11B, and S-11C) that moved water from WCA-2 to WCA-3. Pump stations (S-9 and private pump stations) were also constructed to move water west from urban areas into the WCAs.

Between 1960 and 1963, additional levees and structures were constructed, completing the impoundment of the WCAs. Diagonal levees (L-35, L-67A and C) were added to WCA-2 and WCA-3, allowing the newly isolated southeastern corners (2B and 3B) to be held at lower elevations, thus reducing seepage into the Biscayne Aquifer. Levee 29 and four large water control structures (S-12A, 12B, 12C, and 12D), located west of the L-67 levee, were constructed along the northern edge of Everglades National Park (ENP). Additional structures built along the Miami Canal improved discharge capacity to coastal areas and improved flood control for western portions of the EAA (S-7 and S-8). Finally, the L-28 was completed in two sections, to form the western boundary for WCA-3A. The middle section was left open to allow unimpeded flow between Big Cypress and WCA-3A.

The creation of WCA-3B deserves special mention due to its central location within the pre-drainage ridge and slough landscape. Although created primarily to reduce seepage, the ecological effects of WCA-3B have been far-reaching. Bedrock topography (Parker et al., 1955), depths of organic soil (Jones et al., 1948), and vegetation (Davis, 1943a,b,c) all indicate that the centerline of the deep and hydrologically critical Shark River Slough ran NE–SW south of Tamiami Trail, between the L-67 Extension and L-31 (Color Figure 2.1). Tree islands visible on satellite imagery clearly indicate that the pre-drainage flowlines into Shark River Slough originally ran directly through WCA-3B. Thus, the L-29 levee at the bottom of WCA-3B eliminated natural overland flow into Shark River Slough. On the north side of WCA-3B, the L67-A and C levees were constructed almost perpendicular to the pre-drainage flowlines, blocking water that originally flowed southeast and south. The blocked flow was forced southwestward. The L-67A canal actually became a shunt, bypassing the network of sloughs and running water rapidly toward the ENP. Water that had previously flowed across a 20-mile section of Tamiami Trail was now funneled westward into a section only half as wide, and into an area not originally a part of Shark River Slough (Leach et al., 1971). In addition, the L-29 levee and the S-12 structures replaced natural overland flow through sloughs with a series of four large point discharges.

During 1965 to 1973, parts of the C&SF Project were reworked or added in order to satisfy water requirements of ENP. These water requirements were codified by Congress in 1970 and were designed to meet minimum monthly deliveries that would have been expected in the 1940s and 1950s. Minimum water deliveries, defined for Shark River Slough, Taylor River Slough, and the eastern panhandle of the ENP, were achieved by enlarging the L-67A canal as well as creating a 10-mile long extension from Tamiami Trail southward into Shark River Slough. The L-67 Ext. levee and canal extend N–S almost completely across the full width of Shark River Slough, exacerbating the blockage caused by WCA-3B. Also during this period, the ENP - South Dade Conveyance System (most canals south of L-31) was authorized to provide water not only to the ENP but also for the expanding agricultural and urban needs of Miami–Dade County. As part of this system, S-332 furnished water to the headwaters of Taylor River Slough, while S-18C delivered water via the C-111 canal to the ENP panhandle area.

**Water Conservation Areas**

The C&SF Project established six primary hydrologic units: Big Cypress, Lake Okeechobee, WCA-1, WCA-2, WCA-3, and the ENP. Before drainage began in the 1880s, all three water conservation areas and Shark River Slough, the deeper water portion of ENP, were part of a ridge and slough landscape (SFWMD, 2000). The WCAs are currently managed by a set of regulation schedules that determine when flood control releases are to be made and when water levels are to be maintained. One way to estimate differences between natural (pre-drainage) and managed water level fluctuations in the WCAs is by comparing the output from two models of regional hydrology, the Natural Systems Model (NSM) and the South Florida Water Management Model (SFWMM), as shown in Figure 2.3. Both models simulate surface and groundwater movement, water depths, and evapotranspiration (Bales et al., 1997; Sklar et al., 2001). The two models are typically run for a 31-year period, using the same set of measured 1965–1995 weather data (rainfall and potential evapotranspiration) and tidal boundary conditions, to permit direct comparison of the model outputs. Other than fixed seasonal variation in evapotranspiration-related coefficients, model coefficients remain constant throughout the simulation period; topography and vegetation are assumed constant.

The SFWMM v3.5 (MacVicar et al., 1984; SFWMD, 1997) simulates both overland flow and managed flow through canals and structures. Spatial inputs include present land use or vegetation and topography. In the simulations, 1995 hydrologic management decisions are assumed to have been applied uniformly throughout the 31-year period. The NSM v4.5 (SFWMD, 1998), based on the SFWMM, represents pre-drainage conditions by removing all simulation of managed canal and structure flow. Inputs include spatial estimates of pre-drainage topography and vegetation and estimates of roughness and evapotranspiration coefficients.

Water Conservation Area 1, an area of 572 km$^2$ (221 mi$^2$), is part of the Arthur R. Marshall Loxahatchee National Wildlife Refuge (Refuge), managed by the U.S. Fish and Wildlife Service (USFWS). Pump station S-5A at the north end of WCA-1 discharges agricultural drainage water

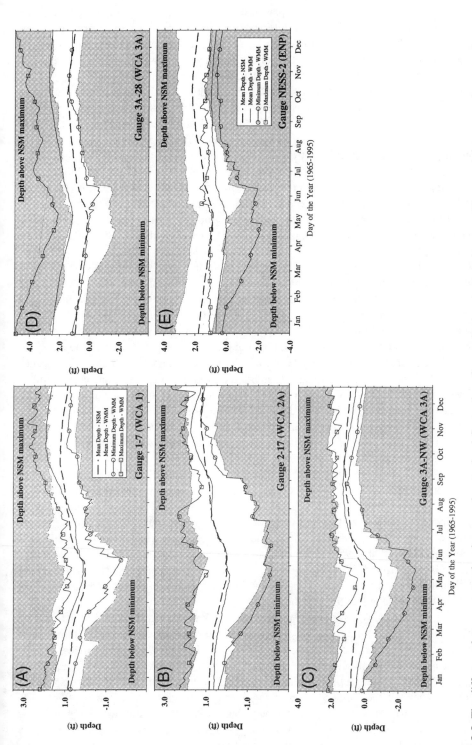

**FIGURE 2.3** The difference between natural and managed water level fluctuations for a calendar year are shown as 30-year minimum, maximum, and average water levels for the Natural Systems Model v4.5 (NSM) and the South Florida Water Management Model v3.5 (SFWMM), respectively, at selected gauge locations (SFWMD, 1998). The SFWMM uses the regulation schedules and water control structure rules of 1995 to simulate 1965–1995

from the West Palm Beach Canal into a peripheral canal that encircles WCA-1. The Hillsboro Canal, via pump station S-6, discharges water into the southwestern portion. Historical data suggest that WCA-1 was originally wetter when it was part of the Hillsboro Lakes region of the Everglades (Davis, 1943a; Parker et al., 1955). Soils data (Gleason and Spackman, 1974), long-term vegetation studies (Alexander and Crook, 1974) and hydrologic models (Fenemma et al., 1994) also suggest that WCA-1 was wetter prior to the construction of drainage canals. Based on this information, in 1992, the USFWS proposed a change in the WCA-1 regulation schedule to provide deeper water with longer hydroperiods. Under the former regulation schedule (1965–1982), the northern and central portions of the Refuge dried out almost every year, allowing terrestrial and exotic vegetation to invade. A comparison of NSM simulations with SFWMM simulations made using the more recent regulation schedule (Figure 2.3A) suggests that the new schedule is generally deeper, or the same as natural water levels, particularly from October to February. During May to September, mean water levels for both simulations are within 0.5 ft, with somewhat less variability in the SFWMM. The impacts of this new regulation schedule are not yet clear. Cattail (*Typha* sp.) along the edges of WCA-1 and sloughs and tree islands within the interior seem unaffected at this time. An important difference between managed and pre-drainage conditions is that the impoundment of WCA-1 and flow through the peripheral canal together have greatly reduced overland flow across the present landscape.

Water Conservation Area 2, at 543 km$^2$ (210 mi$^2$), is the smallest of the three WCAs. In 1961, the L-35B levee divided the area into two smaller units, WCA-2A (448 km$^2$) and WCA-2B (95 km$^2$), in an effort to reduce seepage and improve the water storage capabilities of WCA-2A. In contrast to WCA-1 and WCA-3, which receive most of their water from direct rainfall, WCA-2A receives the majority of its water (59%) from surface water inflows which include drainage from the EAA and outflows from WCA-1 (SFWMD 1992). Prior to drainage, WCA-2 was part of the extensive ridge and slough landscape. Except in very dry times, the sloughs supported aquatic species such as white water lily (*Nymphaea odorata*) and spatterdock (*Nuphar luteum*). After termination of overflows from Lake Okeechobee into the Everglades and completion of the major canals, much of the ridge and slough landscape was over-drained, exposing the soil to microbial oxidation and extensive, long-burning peat fires. Peat on many tree islands was burned, significantly lowering island ground surfaces. Vegetation of the landscape also changed, with emergent wet prairie species filling in the formerly aquatic sloughs (Andrews, 1957; Loveless, 1959).

During the 1960s, after construction of the peripheral levees and canals, water levels were raised again, creating "high water" conditions relative to what had been present in the 1950s. With higher water, the wet prairies reverted to sloughs but there was a destructive overtopping or "drowning" (Dineen, 1972; Worth, 1988) of those tree islands whose elevations had been lowered by the previous peat fires. There have since been a number of downward adjustments to the WCA-2A regulation schedule so that 1995 water management was more similar to that predicted by the NSM (Figure 2.3B). In northern WCA-2A, the sawgrass ridges have expanded sufficiently to engulf the formerly aquatic sloughs, creating a plain of sawgrass and suggesting that water depths in the slough areas had been too shallow to keep them open.

The largest of the WCAs, WCA-3 covers an area of 2369 km$^2$ (915 mi$^2$), all but a small portion of which was originally part of the ridge and slough landscape. Parts of the area still retain aspects of that landscape: a strongly directional, linear network of alternating sawgrass ridges, aquatic sloughs, and tree islands, all aligned in the pre-drainage direction of water flow (Parker et al., 1955). In some parts of WCA-3, water depths that are likely shallower than prior to drainage appear to have transformed some of the formerly open sloughs into more closed wet prairies filled with emergent species. The western edge may have always been somewhat shallower, with wet prairie species present, as suggested by an 1883 expedition: "We find ourselves in a species of grassy waters, bounded on each side by a thick wall of saw grass. …In other words, a water course a hundred yards wide, with a thin species of marsh grass covering it … we have no trouble in propelling our boats…" (Wintringham, 1964). A cypress forest borders WCA 3 from the L-28 gap south to Tamiami Trail. In 1962, WCA-3 was divided into WCA-3A and WCA-3B (2037 and 332 km², respectively) by the construction of two interior levees (L-67A and L-67C) so that water losses due to levee seepage and groundwater flows could be reduced. WCA-3A is also partially divided by the Miami Canal and by canals bordering both sides of Interstate Highway I-75. Major inflows include the Miami Canal, which drains the EAA; the S-9 pump station, which drains urban areas east of the Everglades; and the S-11 structures, which drain agricultural areas and WCA-2A.

Levees and drainage of the EAA eliminated overland flow from the former sawgrass plains to the north, causing over-drainage of northern WCA-3A. Soil subsidence and peat fires may have eliminated tree islands (Zaffke, 1983; Schortemeyer, 1980); wet prairies appear to have increased at the expense of sloughs (Loveless, 1959). Recent aerial photographs of the area north of Highway I-75 compared with ones from 1940 (USDA-SCS, 1940) indicate severe loss of the original linear ridge and slough pattern. The relative importance of water depths, fire, and the lack of overland water flow in obliterating the original landscape pattern has not yet been determined. Water level differences between the NSM and SFWMM in northern WCA-3A are considerable (Figure 2.3C). Average SFWMM water levels at the 3A-NW gauge are nearly always below that of the NSM. Average and minimum NSM values are 1.0 to 1.5 ft higher than SFWMM values during the dry season.

The southern section of WCA-3 has experienced a quite different hydrologic regime. Funneling of water by the L-67s into a narrower cross-section and impoundment by the L-29 levee and the S-12 structures have reduced flow, altered hydroperiod timing, and created deeper water (relative to WCA-3B and ENP) in southern WCA-3A. The water depths have promoted aquatic slough vegetation similar to pre-drainage descriptions (Zaffke, 1983; King, 1917c). At the same time, impoundment, reduced water flow, and altered flow directions may also have contributed to a blurring of the originally sharp distinction between ridges and sloughs. Average SFWMM water levels for southern WCA-3A are 0.75 to 1.75 ft greater than NSM average levels (Figure 2.3D). Nevertheless, observations of pre-drainage water depths suggest that managed water depths may not be excessive, on average, compared to actual pre-drainage conditions. The degradation of landscape pattern may instead reflect the influence of impoundment and cessation of water flow.

## Everglades National Park

Everglades National Park encompasses some 5698 km² (2200 mi²) of former ridge and slough landscape (freshwater sloughs, sawgrass ridges, and tree islands), marl-forming prairies on adjacent higher ground, mangrove forests, and saline tidal flats. Although low (large areas below 1 m NGVD) and flat in many parts, both local and regional variations in topography play a determining role in hydrology, vegetation, and soil-formation. ENP is the second largest national park in the United States. It is also one of the nation's 10 most endangered parks. The decline in the biological resources of the ENP has been primarily linked to changes in the volume, timing, and distribution of water inflows. Even as early as 1938 it was noted that areas of the future park were already damaged and in need of substantial restoration (Beard, 1938). Flows through the southern outflow of the pre-drainage ridge and slough landscape (that is, through Shark River Slough) probably first began decreasing when Lake Okeechobee levels decreased sufficiently to reduce overflows into the Everglades. Flows through Shark River Slough certainly decreased when the North New River, Hillsboro, West Palm Beach, and Miami canals began lowering water depths in the Everglades. Flows were further reduced by the completion, in 1928, of the first east–west highway and accompanying borrow canal across the Everglades (Tamiami Trail).

Completion of the L-67 and L-29 levees of the C&SF Project physically blocked flow into Shark River Slough, pushing flow west of the L-67 Extension to discharge instead onto the higher elevation, western flank of the Shark River Slough, the Ochopee marl-forming prairie (Leach et al., 1971). From 1970 to 1983, when flows to ENP were governed by a minimum delivery schedule, annual surface flows averaged 430,000 ac-ft, with all of the surface water passing onto the Ochopee marl-forming prairie via the S-12 structures (National Park Service, 1995), and essentially none into Northeast Shark River Slough (Figure 2.4). This is in contrast to natural conditions when more than half of the flows passed through Northeast Shark River Slough. Note also the higher flows into Northeast Shark River Slough prior to completion of the L-29 in 1960. The shift of water westward shortened hydroperiods, decreased water depths, and reduced if not eliminated water flow in Northeast Shark River Slough, while at the same time lengthening hydroperiods on the Ochopee marl-forming prairie. In Northeast Shark Slough, the area dominated by sawgrass stands has increased while aquatic slough communities have declined (Alexander and Crook, 1975; Davis and Ogden, 1994; Olmsted and Armentano, 1997). Populations of fish and crayfish, essential prey for wading birds, were reduced (Loftus et al., 1990). The effects of reduced inflows in the current system most strongly affects areas near L-31. Average SFWMM water levels in Northeast Shark River Slough (NESS-2 gauge) are 0.7 to 1.25 ft less than the average NSM water levels for most of the calendar year (Figure 2.3E).

Construction of the South Dade Conveyance System, along the southeastern boundary of the ENP, after completion of the C&SF Project, added further hydrologic changes. It fostered agricultural and urban development in the east Everglades, decreasing the extent of marl-forming prairies. It compartmentalized the remaining marl prairie/Rocky Glades wetlands through a network of levees and canals, and it interfered with freshwater flows through the second most important flowway of the ENP, Taylor River Slough. Table 2.3 shows the effects of the C&SF Project and the South Dade Conveyance System on the water levels in the headwaters of the Taylor River Slough (Van Lent and Johnson, 1993). Weekly average stages from 1957, the first year of monitoring, through 1989 are shown for Gauge-789. This site is located on the L-31N canal and is representative of water levels in the northern Taylor River Slough basin.

### Florida Bay

The development of south Florida and the associated alteration of regional hydrology have not only impacted the freshwater wetlands of the Everglades, but also the coastal wetlands and estuaries of the region. Some of these coastal areas, such as the Caloosahatchee and St. Lucie estuaries, have been subjected to large and unnatural increases in freshwater inputs from flood control structures.

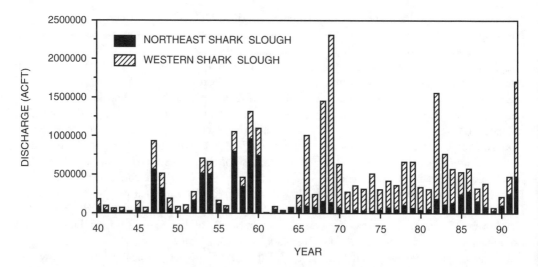

**FIGURE 2.4** Annual surface water inflows to Northeast Shark River Slough and the western Shark River Slough sections of Everglades National Park (1940–1993), based upon a hydrologic year of June through the following May (South Florida Natural Resources Center, 1994).

**TABLE 2.3**
**Summary of the Water Level and Hydroperiod Changes at Gauge 789 (elevation: 5.0 ft) in the Rocky Glades, the Headwater Region of the Taylor River Slough**

| Period | Average October Water Levels (ft) | Average April Water Levels (ft) |
|---|---|---|
| 1957–1962 | 5.82 | 3.22 |
| 1963–1989 | 4.35 | 2.03 |

*Source:* Data from Van Lent and Johnson (1993) and USACE (1994).

For the coastal areas of the Everglades, however, freshwater input has diminished. Much of the freshwater that naturally flowed through the Taylor River Slough and Shark River Slough to the coast has been diverted by canals to other coastal areas (Light and Dineen, 1994).

An important goal of Everglades restoration is to restore the hydrologic conditions and ecological characteristics of the coastal zone. Florida Bay restoration is of particular concern, because it has been subject to drastic ecological changes during the past 10 to 100 years (Boesch et al., 1993; Fourqurean and Robblee, 1999). These changes include increased seagrass mortality and algal blooms and decreased water clarity. It is generally thought that decreased freshwater inflow from the Everglades and resultant increases in salinity have contributed to these ecological changes.

Restoration of the southern Everglades and Florida Bay is currently under way. Structural and operational changes that have been made during the past five years include: (1) SFWMD's acquisition of the Frog Pond agricultural area adjacent to the eastern boundary of the ENP that allows for increased water deliveries through Taylor River Slough to Florida Bay, and (2) removal of the levee on the south side of the lower C-111 canal, thus increasing the flow of water through the southeast Everglades (ENP panhandle) and to the northeast corner of Florida Bay. With these changes in water management in the southern Everglades and high rainfall in the mid-1990s, more freshwater has been flowing toward Florida Bay. Figure 2.5 shows that water discharged into Taylor River Slough and the ENP panhandle has, relative to rainfall, increased since 1993. From 1993 through 1996, average annual discharge was more than double the average annual discharge from 1979 through 1992. Average annual rainfall only increased 18% between these two time periods. Furthermore, with increased freshwater flow, salinity in northeastern Florida Bay has decreased (McIvor et al., 1994; Boyer and Jones, 1999).

## Groundwater

Previous geologic investigations provided only limited interpretations of groundwater/surface water interactions in the WCAs (Parker, 1942, 1955; Schroeder, 1958; Leach et al., 1971; Florida Geological Survey, 1991). Interactions between canals and aquifers in WCAs and groundwater flow beneath levees were addressed by the USACE in several design memorandums (1951 and 1968). Few previous investigations addressed the impacts of the WCA compartmentalization on regional interactions between groundwater and surface water (regional recharge or discharge by vertical fluxes through wetlands). Little attention has been given to groundwater geochemistry and the possible relation to ecological processes in the WCAs. The hydrogeological framework defined by Florida's Groundwater Quality Monitoring Program (Florida Geological Survey, 1992) shows large data deficiencies in the Everglades.

Although monitoring wells in the northern Everglades are limited in number and aerial coverage, these data suggest that compartmentalization of the remnant Everglades has a significant effect on the groundwater hydrology. For example, WCA-1 is maintained at approximately 16 ft NGVD,

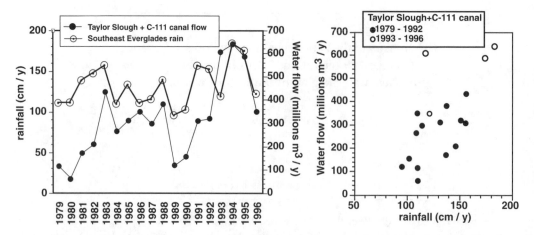

**FIGURE 2.5**    Annual regional rainfall and the annual discharge of water into Everglades National Park through its eastern boundary into Taylor River Slough and the wetlands south of the C-111 canal (left). While water discharge generally increases with increasing rainfall, relatively more water has flowed into the park since 1993 (right).

while the adjacent WCA-2A is maintained at approximately 12 ft NGVD. This head difference of 4 ft across the levee causes subsurface leakage from WCA-1 beneath the levee and discharges into WCA- 2A as evidenced by ionic signatures of the different water masses (Krupa et al., 1998). These horizontal and vertical flows enhanced by compartmentalization are not representative of the natural Everglades system that existed prior to levee and canal construction (Krupa et al., 1998). Subsurface flows take on even greater importance in the southern Everglades because the hydraulic conductivity of the upper unconfined aquifer can be approximately 2 orders of magnitude greater than the northern portion of WCA-1 (SFWMD, 1997).

## Water Budget Comparisons

The average annual water movement in and out of the ENP and WCAs from 1965 through 1995 that would have occurred under natural conditions (simulated by NSM v4.5) and under the 1995-base managed conditions (simulated by SFWMM v3.5), are summarized in Color Figures 2.2 and 2.3, respectively. Differences between natural and managed surface and groundwater flows are striking. Beginning with Lake Okeechobee at the top, the first most obvious hydrologic alteration is the complete elimination of overland flows to the EAA and Caloosahatchee Basin due to construction of the Lake Okeechobee levee and water-control structures. In place of some 868,000 ac-ft of overland outflow that would have flowed south and west had the system not been altered, 989,000 ac-ft of channelized outflow to the south, west, and east occurs for urban and agricultural supply and flood control. In addition, 216,000 ac-ft is discharged back into Lake Okeechobee to prevent regional flooding adjacent to the Lake. The second most obvious hydrologic alteration to Lake Okeechobee is discharge of 149,000 ac-ft to the St. Lucie Basins.

In the EAA south of Lake Okeechobee, which was once dense sawgrass with a narrow rim of pond-apple forest, the NSM water budget (Color Figure 2.2) indicates that 155,000 ac-ft of net precipitation plus surface water from Lake Okeechobee and adjacent basins would have combined to form a southerly overland flow of 1,024,000 ac-ft. Under the 1995-base managed condition (Color Figure 2.3) and the same climate, flows to the WCAs are channelized from agricultural drainage (917,000 ac-ft), flood control discharge (62,000 ac-ft), and water supply to meet urban, agriculture, and environmental needs (205,000 ac-ft). Some of this water passes through the WCAs and is supplemented by Everglades rainfall to meet urban and agricultural needs in the Lower East Coast (172,000 ac-ft), and some WCA water seeps back into the EAA as groundwater (36,000 ac-ft).

The vast majority of this water enters the WCAs as point-source inputs and not as overland flow as in the NSM water budget.

Downstream of the EAA and the lake, the three WCAs act, in part, as large reservoirs for Lower East Coast flood control and the storage of EAA drainage water (Color Figure 2.3). This storage helps create a large groundwater flow (677,000 ac-ft) to the Lower East Coast. This flow contributes to the huge movement of surface water (2,843,000 ac-ft) to the Atlantic Ocean. In the 1995-base case, water is pumped into the WCAs from the Lower East Coast (238,000 ac-ft) and from the EAA (917,000 ac-ft) to prevent agricultural and urban flooding. Water also moves through control structures to meet the environmental needs of the ENP (358,000 ac-ft) and to control WCA-3A flooding (421,000 ac-ft). The only remaining exchange of surface water in the WCAs is the inflow from Big Cypress Preserve along the western boundary of WCA-3A. This managed system is in sharp contrast to the NSM's predicted exchange of surface water, the NSM's lack of groundwater movement, and the NSM's low amount of freshwater flow to the Atlantic Ocean (Color Figure 2.3).

Finally, in the NSM water budget, the vast majority of all inflows and outflows in the ENP are overland flows (Color Figure 2.2). This is not true for the 1995-base SFWMM water budget (Color Figure 2.3). According to the 1995-base managed conditions, inflows from the Lower East Coast into the ENP are only 18% of what was predicted to occur by NSM. As a result, Shark Slough outflow from ENP to the southwest is 48% of what was predicted to occur by the NSM. A big difference between the NSM and the 1995-base SFWMM is the amount of annual ground-water seepage to the Lower East Coast. In the NSM it is only 28,000 ac-ft, but in the SFWMM it is 306,000 ac-ft. Another difference is the amount of freshwater flow toward Florida Bay via the ENP and the Lower East Coast. The NSM total average discharge is 156,000 ac-ft; however, the SFWMM average discharge is 249,000 ac-ft. This increase over NSM reflects an increase in the water supply to Florida Bay that was initiated in 1993 and incorporated into the 1995-base managed condition. Note that the 1995-base is a fixed Everglades infrastructure with stationary operational rules (circa 1995) and thus is not indicative of the amount of actual freshwater delivered to Florida Bay by the C&SF Project from 1965 to 1995.

## ECOLOGICAL EFFECTS OF ALTERED HYDROLOGY

### SOIL RESPONSES TO HYDROLOGY

Over the majority of the Everglades, the pre-drainage surface topography was created by accumu-lation of organic (peat) soils. However, anthropogenic drainage has altered peat accumulation and has exposed the soils to oxidation. This oxidative subsidence appears to have significantly influenced Everglades wetlands at two different scales: (1) regional (100s of km$^2$), by affecting slope and flow of surface water; and (2) local, by reducing elevation differences between elements of the ridge and slough landscape. The latter form of differential subsidence may have "flattened" the landscape. As the rate of oxidation is proportional to the distance between soil surface and the water table (Stephens and Johnson, 1951), the higher landscape elements — peat-based tree islands and sawgrass ridges — would be subjected to more oxidation and subsidence than the lower elevation sloughs.

Rates of regional soil subsidence in the Everglades have been approximately 2.5 cm/yr in drained agricultural areas (Snyder et al., 1978) and somewhat higher in undeveloped, "natural" areas (Stephens and Johnson, 1951). Radiocarbon dating of peats near Belle Glade (McDowell et al., 1969) suggest a peat accretion rate during the last 1000 years of about 0.16 cm/yr, less than one tenth the subsidence rate. Recent measurements in areas with extended hydroperiod and/or phosphorus (P) enrichment, such as northern WCA-2A, suggest a maximum accretion rate of 1.1 cm/yr (Craft and Richardson, 1993; Reddy et al., 1993). Accretion rates are lowest (0.04 to 0.28 cm/yr) in areas of reduced hydroperiod, such as northern WCA-3A (Craft and Richardson, 1993; Reddy et al., 1993; Robbins et al., 1996).

Although Jones et al. (1948) offered the first soil map of the entire Everglades, an earlier soil mapping effort by Baldwin and Hawker (1915) described a 6-mile-wide transect through the middle of the Everglades. They mapped, in detail, 3 miles on either side of the full length of the North New River canal, thus passing from the Atlantic Coastal Ridge to the Pond Apple Zone at the shore of Lake Okeechobee. Figure 2.6 illustrates the principal soil profiles seen in 1915 and in 1940. In 1915, all organic soil present was uniformly a brown, fibrous, slightly decomposed peat. By 1940, between mileposts 47 and 54, the brown fibrous peat was completely gone, leaving only a trace of organic matter in the top portion of the remaining sand. Between mileposts 3 and 47, the upper layers of the original brown fibrous peat had decomposed into a black, nonfibrous material. In addition to these changes, the soil surface had subsided and the vegetation had changed. In 1915, the North New River canal traversed 20 miles of ridge and slough landscape before reaching the Atlantic Coastal Ridge. By 1940, no ridge and slough landscape remained, having been completely replaced by the expansion of the sawgrass plains and a short stretch of grassland toward the coast.

The widespread appearance of a new surface layer of black, finely fibrous peat is highly significant. All evidence points to oxidation as the cause of this layer: the loss of structure, the increase in density and mineral content, and the black coloration. Soil scientists and drainage engineers at the Everglades Agricultural Experiment Station in Belle Glade, who studied these soils during the 1930s, in fact clearly state that the oxidized surface layer was a result of lowered water levels (Clayton et al., 1942). The black surface layer visible in all the areas mapped by Jones et al. (1948) as Everglades Peats, reflects the widespread oxidation caused by lowered water tables throughout the Everglades.

The limited literature examining the effects of flooding and drying on natural Everglades soils suggests that during dry periods, some nutrients are made more available as a result of increased decomposition rates. These nutrients are subsequently released into the overlying water upon reflooding of the area (Worth, 1981, 1988). Preliminary laboratory studies show that naturally drained soil cores from WCA-2A had P fluxes into the overlying water column of 2.45 to 10.12 mg P $m^{-2} d^{-1}$ in soils from an enriched site and 0.005 to 0.016 mg P $m^{-2} d^{-1}$ in soil cores from an unenriched site (Newman and Reddy, unpublished data).

## CHANGE IN THE NATURE OF FIRE

There is little doubt that fires occurred in the Everglades with some frequency even prior to drainage. Modern scientific studies confirm that the periodic occurrence of moderate surface fires would not have destroyed the sawgrass landscape. Loveless (1959) and Forthman (1973) found that sawgrass regenerates rapidly after fire, provided the soil is sufficiently wet or inundated to prevent sawgrass culms from being killed. Hofstetter (1984) suggests that periodic fires may even be beneficial, reducing the build-up of flammable leaf litter. If water levels decline far enough below ground to dry out the surface of the organic soils, a different type of fire, the so-called peat or muck fires, can occur. Extensive peat fires, covering tens to hundreds of square miles (Bender, 1943), spread during dry periods beginning in the 1920s (Robertson, 1953) and continued into the 1950s (FGFWFC, 1956; Wallace et al., 1960). Such fires could burn for months and even through the wet seasons of multiple years (Bender, 1943). Water tables in the peat that are lower than the normal annual dry season minima of 4 to 6 in. below ground surface permit peat fires to occur (Cornwell and Hutchinson, 1974). An increase in wildfire intensity and frequency is attributed by many investigators to overdrainage of ENP wetlands (Robertson, 1955; Craighead, 1971; Wade et al., 1980). Davis (1943) described the effect of lowered water levels and burning on the Rockland Marl Marsh area:

> Excessive artificial drainage has recently created drier conditions promoting these fires and also causing the shallow organic soils to become oxidized and subside until now some areas once soil covered are returning to rockland conditions.

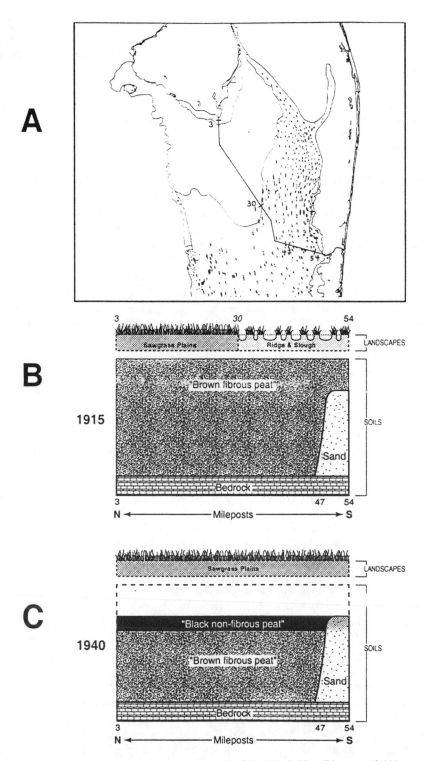

**FIGURE 2.6** Generalized soil profiles along the length of the North New River canal (A), as mapped in 1915 (B) and again in 1940 (C). Note southward expansion of the Sawgrass Plains landscape, elimination of the ridge and slough landscape, a general loss of elevation due to soil oxidation, and conversion of the surface layer from brown, fibrous peat to black, nonfibrous peat.

These fires affected not only the sawgrass, but the higher lying tree islands as well. Bay heads, one of the higher elevation vegetation types, dried sufficiently in the ENP to suffer repeated burns (Robertson, 1953). Farther north in WCA-3, Loveless (1959) noted similarly drastic effects of fire on the higher lying communities:

> These [early summer of 1956] fires completely destroyed many tree island communities by burning the peat substrate out from under the tree growth. Some of these areas are now open water ponds devoid of any type of emergent vegetation while others support sparse strands of sawgrass, water-lily, floating heart and other aquatic species.

Fire is an important ecological process shaping the Everglades vegetation patterns (Robertson, 1954; White, 1970; Cohen, 1974; Duever et al., 1976; Wade et al., 1980; Wu et al., 1996). It can determine inland expansion of mangroves, tree island growth (Davis, 1940; Egler, 1952), and plays an important role in preventing cypress trees from extending into marshes. Without moderate fires, cypress domes in the Big Cypress Preserve can be replaced, perhaps permanently, by mixed hardwoods (Wade et al., 1980). Moderate fire is a natural process of leaf-burn with many vital functions (Wright and Heinselman, 1973). It influences nutrient cycling, stimulates net primary production, and may even be the evolutionary process that selects for sawgrass and slough environments (Wade et al., 1980). The mean moderate fire interval for a sawgrass marsh is about 9 years (Gunderson and Snyder, 1994; Wu et al., 1996). A hydrologic regime that prevents this 9-year fire cycle (i.e., a regime extremely wet or extremely dry) can increase the amount of willow or other woody vegetation which, in turn, can create an environment that is more fire tolerant (Wu et al., 1996).

Altered hydroperiods, when combined with altered nutrient inflows, can also affect fire (Wu et al., 1996). Where ponding and nutrients have led to the encroachment of cattail (Newman et al., 1998), the Everglades Landscape Fire Model (ELFM) found that moderate fire-spreading characteristics (i.e., fires that do not burn peat) were reduced due to the expansion of cattail (Wu et al., 1996). A 30% increase in cattails reduced the average annual fire frequency by 21% and mean annual area burned by 23%. The ELFM predicted an average of 3.4 moderate fires a year in sawgrass marshes and only 2.7 moderate fires a year in marshes with a significant biomass of cattails.

## VEGETATION RESPONSES TO HYDROLOGY

The pre-drainage Everglades was composed of a directionally patterned mosaic of tree islands, shrubs, sawgrass ridges, and aquatic sloughs (King 1917c,d; Parker et al., 1955). To varying degrees, the same elements are still present (Kushlan, 1990). In this low-nutrient environment, water depth (and the associated variable hydroperiod) is considered to be a key determinant of vegetation (Kolipinski and Higer, 1969; Gunderson and Snyder, 1994). Slight changes in the depth (±10 cm) and period of inundation (±90 days), over long periods of time (5 to 10 years), influence the presence of certain species and can shift the spatial distribution of plant communities. Note that water depth is simply water elevation minus ground elevation, so in the peat-based portions of the Everglades water depths can be altered either by changes in water elevation or by any changes in elevation of the peat surface.

Drainage and water management of the Everglades has created hydrologic subunits (the WCAs) with differing histories of water depths and flows. Various scenarios in comparison with pre-drainage conditions exist: areas that became drier, others that became wetter, and still others that first became drier and later became wetter. Similarly, water flow amounts have variously increased, decreased, or been forced to change direction. In the ENP, shortened hydroperiods and reduced water depths have allowed woody species to encroach into marshes in the East Everglades, Taylor River Slough (Olmsted et al., 1980), and Shark River Slough (Kolipinski and Higer, 1969). Encroaching species in the sloughs are mostly native, but in the East Everglades (which includes margins of Northeast Shark River Slough), much of the expansion is by invasive exotic trees (*Schinus terebinthifolius*,

*Melaleuca quinquenervia*, and *Causuarina equisetifolia*) that in some areas now dominate (Loope and Urban, 1980; DeVries, 1995).

In WCA-2A, initial over-drainage probably caused a conversion of pre-drainage sloughs to shallower wet prairies (Loveless, 1959; Goodrick, 1974), as well as increasing the rate of peat and elevation loss on tree islands. Subsequently, average water elevations may have increased some 2.5 ft (0.75 m) under managed conditions between 1955 and 1969. During this period wet prairies converted back to more closely resemble pre-drainage aquatic sloughs (Loveless et al., 1970). At the same time, many of the tree islands in WCA-2A were lost (Dineen, 1974), most likely because the increasing water elevations covered the previously subsided tree island peat.

Similarly, increased average water elevations in WCA-3A in the early 1980s increased obligate wetland species such as *Sagittaria lancifolia* and slough species such as *Nymphaea odorata* and *Utricularia* spp. (David, 1997), while at the same time reducing the extent and diversity of wet prairies. Zaffke (1983) reported replacement of wet prairies in southern WCA-3A by aquatic sloughs due to extended hydroperiods and increased water depth. Wood and Tanner (1990) did not find at least 13 species that Loveless (1959) encountered in wet prairies 30 years earlier in the same area. These observations suggest that many wet prairie species only germinate in areas with an annual spring dry period (Goodrick, 1974).

Vegetative growth is the dominant mechanism of expansion for most Everglades sedges and grasses. However, long-term genetic diversity is preserved through sexual reproduction. Hydrologic factors, including continuous water cover, limit population distributions by affecting the survival of seeds, seedlings, and saplings (van der Valk and Davis, 1978; Moore and Keddy, 1988; Brown and Bedford, 1997). Ponzio et al. (1995) found sawgrass seed germination was highest in water-saturated soil, intermediate in soil with 5 cm water above the surface, and lowest in soil with 10 cm above the surface. The more aquatic slough species such as *Nymphaea*, *Nuphar*, or *Utricularia* either do not require dry soil to germinate or only do so during exceptionally dry periods. Colonization of the different Everglades microtopographies is therefore influenced by hydrologic regime, and contributes to the maintenance of characteristic mosaics and genetic diversity.

In the northeastern portion of WCA-2A, an expanding front of cattails (*Typha* sp.), invading and displacing sawgrass, is well documented (Swift and Nicholas, 1987; Davis and Ogden, 1994; Rutchey and Vilchek, 1994; Jensen et al., 1995). A 6-year (1986 to 1991) study (Urban et al., 1993) in WCA-2A supports the idea that two factors led to the spread of cattail: nutrient enrichment and prolonged hydroperiods. Studies of cattail cover, soil nutrient concentrations, topography, and fire history in the Holeyland and Rotenberger Wildlife Management areas located in the northern Everglades suggest that causal factors for cattail expansion are also site specific. Cattail proliferation in the Holeyland management area has been found to be largely controlled by hydrology and elevated nutrients, whereas cattail distribution in Rotenberger is primarily determined by historic muck fires (creating new unvegetated areas) and elevated nutrients (Newman et al., 1998). It seems that shortened hydroperiods in Rotenberger have resulted in excessive soil oxidation and an increased frequency of wildfires, processes that mobilize P stored in the soils (Wade et al., 1980). This increase in bioavailable P can spur the colonization of cattails (Urban et al., 1993; Davis, 1991).

Cattail and sawgrass have different hydrologic tolerance, because they have very different leaf morphology, anatomy, and physiology. Sawgrass has coarse, gray-green, sawtoothed leaves with several xeromorphic characteristics such as a thick waxy cuticle, numerous bands of lignified fibers, and marginal spines on the leaf surface (Miao and Sklar, 1998). On the other hand, cattail has wide, smooth, light-green, and spongy leaves. Cattail plants can produce more adventitious roots (i.e., oxygen-gathering structures), create more air spaces in the stems, and transport more oxygen to the soil rhizosphere than sawgrass grown under the same conditions (Kludze and DeLaune, 1996). Despite the stress of low-oxygen conditions, cattail net photosynthesis and biomass tend to be greater than sawgrass (Pezeshki et al., 1996; Figure 2.7). These differences are due to contrasting internal gas transportation systems. Cattail has a pressurized ventilation system that can actively transport oxygen to the roots when growing in deep-water habitats, whereas sawgrass has only molecular diffusion for oxygen transport (Brix et al., 1992; Chanton et al., 1993). The dominance of cattail with increased flooding levels has been demonstrated experimentally. When sawgrass, cattail, and *Eleocharis* were grown in a low-nutrient, low-water mixture, Newman et al. (1996) found no dominance by any species. However, cattail biomass was significantly greater relative to the other two species when grown in a low-nutrient, high-water mixture (Figure 2.8). It should be noted that the observed cattail dominance did not result from a significantly increased growth of cattail plants but from significant decreases in the growth of sawgrass in deep water.

## Periphyton

Periphyton is an important primary producer in the Everglades marshes, particularly in the sloughs (Browder et al., 1994). Typically, the native periphyton exhibits a thick, calcareous (white, creamy)

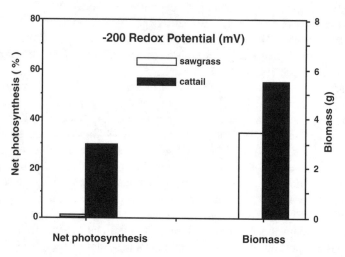

**FIGURE 2.7** Changes in net photosynthesis and total biomass in sawgrass (open bars) and cattail (closed bars) in response to a soil redox condition of –200 mV (i.e., low-oxygen). (Modified from Pezeshki et al., 1996.

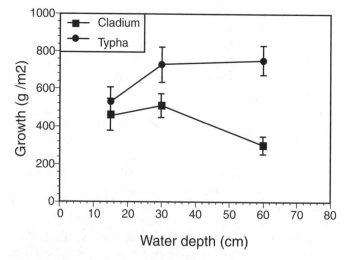

**FIGURE 2.8** Growth of sawgrass (*Cladium*) and cattail (*Typha*) mixtures grown in outdoor tanks for 2 years (final–initial biomass). (Adapted from Newman et al., 1996).

appearance with a layered structure (Gleason and Spackman, 1974; Browder et al., 1994). The calcareous periphyton community has a high calcite content and is usually dominated by filamentous blue-green algae of the genera *Scytonema* and *Schirothrix*. Periphyton communities require an aquatic environment to survive, grow, and reproduce. They have adapted to cope with seasonal drydowns and periodic droughts by either surviving under low metabolism or entering a state of dormancy. Individual species differ in their resistance to desiccation and ability to recolonize. Thus, periphyton community composition may respond to hydroperiod variations. Frequent and prolonged drying may promote dominance of calcareous periphyton, while year-round flooding may alter water quality and thus favor noncalcareous periphyton (Van Meter-Kasanof, 1973; Browder et al., 1982, 1994).

The effects of hydrology on periphyton are not clear. There is some evidence that periphyton can be negatively affected by high water depths. Swift and Nicholas (1987) found a statistically significant negative correlation between water depth and cell volumes for *Scytonema* and *Schizothrix*. Gleason and Spackman (1974) observed calcareous periphyton developed better in the upper 0.67 m of the water column. Browder et al. (1994) suggest that calcareous periphyton mats will not develop in deep, open water environments. For example, in Taylor River Slough, bottom sediments of deep ponds have more organic material and less calcite mud than sediments in surrounding wet prairies (Browder et al., 1994). Studies conducted in the ENP reported an average depth of 0.75 ft for about 7 to 10 months as the optimum depth and duration of flooding for marl-producing periphyton communities (Tropical BioIndustries, 1990).

## Tree Islands

Within the matrix of wetland types that make up most of the greater Everglades are small topographical highs or tree islands. These islands have historically provided habitat for a wide variety of terrestrial plants and animals. Gawlik and Rocque (1998) found that tree islands support about twice as many species of birds than do marshes in the central Everglades. Because the maximum elevations of the highest tree islands are only slightly above mean annual maximum water levels, tree islands, with their less flood-tolerant vegetation, are very sensitive to changes in hydrology. In the 1960s, the number of tree islands in portions of WCA-2A and the northern portions of WCA-3A declined significantly (Schortemeyer, 1980). In WCA-2A (Color Figure 2.4), tree islands were lost because relatively high water levels were sustained for many years after prolonged drainage had caused subsidence of the tree island peat. The tree islands in WCA-2A lost their trees and shrubs between 1960 and 1970, a period of relatively prolonged high water levels (Dineen, 1974; Worth, 1988).

Prolonged dry conditions can also result in a loss of tree islands. Prolonged low water levels in the northern section of WCA-3A resulted in tree island destruction because peat fires removed as much as 25 cm of their elevation (Zaffke, 1983). When surrounding water levels returned to normal conditions, subsequent flooding, with lower tree island elevations, probably resulted in water depths too great for shrub and tree recolonization.

The conservation of healthy tree islands should be a goal of restoration and a criterion for determining appropriate water flows and levels in the Everglades. However, a review of the available literature has provided little insight into what constitutes a healthy tree island. Studies on tree islands in the WCAs have been restricted almost exclusively to descriptions of their vegetation (Davis, 1943b,c; Loveless, 1959; McPherson, 1973; Alexander and Crook, 1974; Zaffke, 1983). Almost no studies have been done to test the various hypotheses of tree island sustainability and development (Loveless, 1959). An unanswered question remains. If all tree islands have lost significant elevation, how can water managers restore the pre-drainage hydrology and preserve tree islands at the same time? Answers may be forthcoming in van der Valk and Sklar (in press).

## Muhly Grass

Freshwater prairies of muhly grass (*Muhlenbergia filipes*) support the federally endangered Cape Sable Seaside Sparrow as well as a number of other bird species in the southern Everglades (Richter and Myers, 1993). Increases in hydroperiod have caused shifts away from the muhly-dominated prairie, preferred by the sparrow, and toward a sawgrass-dominated community (Pimm et al., 1995). With a typical hydroperiod of only 1 to 5 months, muhly prairies have shorter inundation periods and shallower flooding depths than typical sawgrass marshes (Olmsted and Loope, 1984). Hydrologic plans to reduce hydroperiods may re-establish areas once dominated by muhly grass. However, other than documenting the rates of biomass recovery (2 to 3 years) after annual prescribed fuel-reduction burns (Herdon and Taylor, 1986), little is known about the effects of hydrology on muhly grass ecology.

## Invasive Exotics

Hydrology can affect the ability of exotic or normally restricted plant species to expand into a variety of Everglades habitats. Austin (1976) reported that wet prairie sites, disturbed by fire, are particularly susceptible to invasion by *Melaleuca quinquenervia*. It was also observed that Brazilian pepper (*Schinus terebinthifolius*) is often a major component of postburn vegetation (Wade et al., 1980). In South Florida, these and other exotic plants tend to establish in disturbed areas — abandoned farm land, along roadways, canals, and drainage ditches, as well as in wetlands that have been cleared or have been stressed due to hydroperiod changes.

Melaleuca (*Melaleuca quinquenervia*) is a pioneering Australian species that has been spreading since its introduction to South Florida in the early 1900s. It was thought to have a very high transpiration rate and an ability to drain the Everglades. It had neither. Transpiration of melaleuca is no more than that of other forest types of the same density (Woodall, 1981, 1984), and theoretical models suggest that total evapotranspiration from melaleuca stands is likely to be only slightly greater than that of short-canopy, native vegetation replaced by melaleuca invasions (Chin, 1998). In its native range, it grows in low-lying flooded areas and is especially well adapted to ecosystems that are periodically swept by fire. These are common conditions in South Florida, making it an ideal habitat for colonization. Melaleuca grows equally well in the deep peat soil of WCA-1 and the inorganic, calcareous soil of the ENP. In general, wetland areas such as sawgrass prairies are more susceptible than drier, upland areas. However, increasing hydroperiod length in some wetlands can slow the spread of invasive plant species such as melaleuca by limiting suitable germination sites.

Melaleuca is very responsive to fire. It forms dense stands with higher fuel loadings than a typical sawgrass marsh — about 20 kg/m$^2$ (Conde et al,. 1980) compared to 2.8 kg/m$^2$, respectively (Hofstetter, 1976). Heat value for the fuel is as high as 11,200 btu/lb compared to less than 8000 btu/lb for herbaceous fuel (Huffman, 1980). As a result, fires in melaleuca stands burn hotly and intensively (Flowers, 1991). After a fire, melaleuca has vigorous regeneration of sprouts and prolific seed release (Burkhead, 1991). Hydrologic management was not able to deter the spread of melaleuca. Before state and federal control operations with herbicides were initiated in 1990, melaleuca was distributed throughout South Florida. Today, large untreated monocultures of melaleuca are limited to WCA-2B and the wetlands just east of the southern Everglades.

Unlike melaleuca, Brazilian pepper (*Schinus terebinthifolius*) is not a good fuel for fire and can function as a firebreak. It thrives on disturbed soils and is especially invasive in areas affected by drainage. It cannot establish in deep wetlands and can rarely grow on sites inundated longer than 3 to 6 months. In the Everglades, Brazilian pepper is mainly restricted to levee berms and other disturbed areas. However, a large area of the ENP along the Gulf of Mexico has Brazilian pepper and it has also established itself on tree islands in northern WCA-3A. Some form of chemical and hydrologic management may reverse this trend.

The newest of the invasive exotics is Old World climbing fern (*Lygodium microphyllum*). It is an exotic twining fern that was first found in southern Florida in the late 1950s. It is now spreading rapidly throughout the region. Old World climbing fern overtops and smothers Everglades tree islands, pinelands, and cypress swamps and spreads across open wetland marshes. It also forms dense mats of rachis plant material. These thick, spongy mats are slow to decompose, exclude native understory plants, and can act as a site for additional fern colonization. Significant infestations have been found recently in WCA-1A, Lake Okeechobee, and Big Cypress National Preserve. Increased hydroperiod does not seem to have an effect on this species as it has expanded greatly in areas that have experienced higher than normal water levels over the last few years. This plant also alters fire ecology. Burning mats of the lightweight fern break free during fires and are kited away by heat plumes, leading to distant fire-spotting. Additionally, the plant acts as a flame ladder — carrying fire high into native tree canopies.

Finally, there is the problem of torpedograss (*Panicum repens*). The origin of torpedograss is uncertain. Its tolerance to desiccation (Wilcut et al., 1988), its ability to store high levels of carbohydrate in its rhizomes (Manipura and Somaratne, 1974), and its survival in 2 to 4 feet of water (Tarver, 1979) are the characteristics that made torpedograss a desirable cattle foraging plant for early Florida farmers. It now occurs in 70% of Florida's public waters and has displaced 14,000 acres of native marsh plants in Lake Okeechobee (Schardt, 1994). Unfortunately, most herbicides used for torpedograss control are indiscriminate for grasses and can damage non-target woody plants (Langeland et al., in review). The control of torpedograss requires the destruction of the below-ground rhizome system (Chandrasena, 1990).

## WILDLIFE RESPONSES TO HYDROLOGY

### Avian Dynamics — Wading Birds

A decline in populations of some wading bird species was one of the first, and most noticeable, pieces of evidence that the Everglades ecosystem was being degraded. The recovery of these birds has now been identified as a key component of a successful Everglades restoration (Walters et al., 1992). Many aspects of wading bird reproductive and foraging ecology are influenced by water depth and water recession rate. Shallow receding water levels are associated with good reproductive and foraging success, whereas a reversal in water recession, especially late in the nesting cycle, and an associated increase in water depth have a negative effect. Although there is an association between water recession and foraging and nesting conditions, the mechanisms by which receding water leads to good foraging and nesting conditions have not been identified. Hypothesized mechanisms have focused either on a reduction in water depth whereby prey are easier to capture (Frederick and Collopy, 1989; Bancroft et al., 1990), or an increase in prey abundance due to their release from predation following a drydown (Kushlan, 1976a, 1987). The distinction is important because water management strategies differ between the two mechanisms. For example, long periods without drydowns may increase overall fish abundance (Loftus and Ekland, 1994) but the fish may not become vulnerable to capture by birds until water levels recede.

To determine if wading birds respond more strongly to high prey vulnerability (i.e., shallow water depths) than to high fish abundance, a series of foraging experiments was conducted from 1996 to 1998 in a set of SFWMD experimental ponds located just west of the Refuge. These experiments examined use of feeding sites by seven species of wading birds in response to water depth (10, 19, and 28 cm) and fish density (3 or 10 fish per m$^2$) treatments (Gawlik, 1996). Wood Storks (*Mycteria americana*) and White Ibis (*Eudocimus albus*) were unable to exploit all the ponds due to the wide range of depths and left the ponds early. In contrast, Great Egrets (*Casmerodius albus*) persisted because of their ability to forage at greater depths. There were noticeable differences among treatments in the rate of fish consumption and giving-up densities (fish densities at which bird species left the ponds), suggesting that the cost of foraging deeper was highest for Wood

Storks, White Ibis, and Snowy Egrets (*Egretta thula*). The results of this experiment, as they pertain to hydrology of the Everglades, suggest that water depths, in the ranges examined, influenced the selection of foraging sites for seven of the eight species examined. Only the Great Blue Heron (*Ardea herodias*) was unaffected by a water depth of 28 cm. Although leg length appeared to constrain the use of deep water by most species, leg length did not equate to maximum foraging depth as previously reported. For example, the Little Blue Heron (*Egretta caerulea*) was equally abundant in all water depths, whereas the longer legged Wood Stork was restricted to the shallow and medium depth treatments. One mechanism used by wading birds to exploit deep water was behavioral plasticity manifested through the use of aerial foraging. This foraging behavior was particularly common for Snowy Egrets, but also has been observed in other species. Although some birds have the ability to use aerial foraging, this behavior is energetically more costly than wading and thus requires higher prey intake rates to become energetically profitable. It also requires the presence of at least one perch within approximately 10 m of a foraging site. The temporal dynamics and foraging behaviors exhibited by birds in this experiment indicate that species most likely to be impacted by unusually high water conditions in the Everglades include the Wood Stork, White Ibis, and Snowy Egret, which were clearly constrained by both water depth and fish density.

The effects of hydrology on the distribution and abundance of wading birds were also apparent from the Systematic Reconnaissance Flight program (Porter and Smith, 1984; Bancroft et al., 1992, 1994; Hoffman et al., 1994) that was developed during the 1980s. Transects were established in a grid such that all of the northern and central Everglades were divided into 4-km² cells. Data reported here are from WCA-1 and WCA-2A. The abundance in each cell of Great Blue Herons, Great Egrets, Wood Storks, and White Ibises in response to water depth and vegetation was contrasted between years with different water levels. The year 1988 was considered a wet year and 1989 was considered a drought year (Bancroft et al., 1994). The results indicated that water depth and the vegetation community in a 4-km² cell influenced the abundance of wading birds in that area, but that the relationship among those variables differed between a wet and drought year. In the wet year, there was a water depth threshold above which bird abundance was predicted to decline. The depth threshold varied among species and ranged up to 76 cm for the Great Blue Heron. In the drought year, the relationship between bird abundance and water depth was positive and linear.

## Avian Dynamics — Snail Kite

The Snail Kite (*Rostrhamus sociabilis*) is a raptor of the tropics and subtropics that, in North America, occurs only in central and south Florida. The Snail Kite is classified as federally endangered and consumes primarily apple snails. Populations in Florida have fluctuated widely since the early 1900s and it has been hypothesized that many of these changes were in response to human-caused changes to the hydrology of South Florida.

From 1992 to 1994, 282 radio transmitters were attached to 271 individual Kites to estimate survival and movement among wetlands (Bennetts and Kitchens, 1997). Adult survival was fairly high and ranged from 0.90 to 0.92. As expected, juvenile survival was lower, ranging from 0.50 to 0.71. Survival of juveniles, but not adults, varied among years, suggesting that juveniles are more sensitive to environmental fluctuations than adults (Bennetts et al., 1999). Snail Kites moved frequently among wetlands within Florida, but movement did not appear to be directly related to water levels. It is possible, however, that hydrology may influence movement only during droughts, which did not occur during this study (Bennetts and Kitchens, 1997). If so, then the tendency of individual Kites to range widely among hydrologic basins may mediate a loss of food due to dry conditions in any single basin. In essence, Kites are using the ecosystem at a much larger spatial scale than individual hydrologic basins, such as WCAs. The same is likely true for wading birds (Strong et al., 1997). In both cases, large-scale movements are one mechanism to deal with living in a variable ecosystem.

## Avian Dynamics — Cape Sable Seaside Sparrow

The Cape Sable Seaside Sparrow is a small, secretive bird that inhabits short-hydroperiod, freshwater prairies in the southern Everglades, particularly those dominated by the bunchgrass muhly (*Muhlenbergia filipes*). Sparrow populations occur both east and west of the Shark River Slough, south of Tamiami Trail. The bird is classified as a federally endangered species and populations have declined dramatically between 1992 and 1997 (Lockwood et al., 1997).

Two factors that appear to limit breeding potential are suitable vegetation and water levels (Nott et al., 1998). Observational data suggest that water levels must be below 10 cm in order for the birds to breed (Pimm et al., 1995; Lockwood et al., 1997). If water levels are not less than 10 cm by April the birds will not initiate breeding. Likewise, a water level rise to depths over 10 cm usually marks the end of the breeding season, typically in July. In addition to the obvious role of hydrology via water depths during the breeding season, hydrology affects the plant community and therefore the presence of suitable breeding habitat. In drained areas northeast of the Shark River Slough, shrubs have increased and severe fires have burned away peat, thus rendering the habitat unsuitable (Pimm et al., 1995). It should be noted that the role of fire in maintaining suitable habitat is not fully understood, and it is likely that some amount of burning is essential. It has been suggested that increases in hydroperiod, particularly west of the Shark River Slough, have caused shifts in the vegetation away from the muhly-dominated prairie, preferred by the sparrow, toward a sawgrass-dominated community (Pimm et al., 1995). Hydrologic management recommendations for the sparrow issued in a U.S. Fish and Wildlife Service biological opinion (Pimm et al., 1995) were to increase water flows to Northeast Shark River Slough and reduce flows west of the Shark River Slough.

## Fish

There are two contradictory ideas of how hydrology affects the extant Everglades freshwater fish community. The first hypothesis suggests that fish density in the southern Everglades is highest when the marsh is managed for frequent drydowns (Kushlan, 1976b). Data collected from 1965 to 1972 using pull-traps (described in Higer and Kolipinski, 1967) indicated that under a regime of frequent drydowns small 2-cm omnivorous fishes (Kushlan, 1976b) dominated the fish community. As the length of time without a drydown increased, the community shifted and became dominated by larger carnivorous sunfishes and catfish. The results suggested a shift in the structure of the community as a result of hydroperiod.

The second hypothesis suggests that both small and large fish densities increase during periods without drydowns and there is no shift in community dominance toward large fishes as a function of increased hydroperiod (Loftus and Eklund, 1994). This hypothesis is based on data collected with throw-traps (described in Kushlan, 1974) from 1977 to 1985, an 8-year period without a drydown. Throw-traps do not disturb the sampling site as much as pull-traps with repeated sampling; therefore, throw-traps do not produce biased data inherent with pull-traps. Loftus and Eklund (1994) argued persuasively that the first hypothesis is without merit, because it is based on biased data.

Historically, freshwater flowed to Florida Bay via the Taylor River Slough. The northern mangrove fringe of Florida Bay experienced reduced freshwater flow and increased salinity levels because of the construction of a canal network upstream in the 1960s. Based on fish samples taken from 1991 to 1996 in this area, fish density was negatively correlated with increasing salinity (related to freshwater flow) and, to a lesser degree, positively correlated with water depth (Lorenz, 1997).

## Apple Snails

The Florida apple snail (*Pomacea paludosa*) is an important component of the Everglades food web. It is the primary food source for Snail Kites and Limpkins (*Aramus guarauna*), and it is consumed by alligators and turtles (Darby et al., 1997). Data on snail movements based on 58 individuals with radio transmitters (Darby et al., 1997) revealed that apple snails do not seek out deep-water refuge during a drydown. Their movements become restricted when water depths recede to 10 cm or less, and if they become stranded their response is to conserve moisture by closing their operculum tightly. Snails survived dry conditions for up to 12 weeks in the field, but these animals tended to be juveniles (Darby et al., 1997). Adults had high mortality following reproduction, regardless of hydrologic conditions. Their typical life span was 12 to 18 months.

Although the timing of a drydown may have little effect on adult survival, it can affect population size by reducing reproductive output. If a drydown occurs before egg laying, which typically peaks in April, then recruitment will be reduced (Darby et al., 1997). However, drydowns after that time will have less impact on the population because young of the year are more adept at surviving dry conditions than adults, and adults are likely to die within a few weeks, regardless of the drydown (Darby et al. 1997). Thus, a slight shift in the timing of drydowns has the potential to have a large impact on snail recruitment, which presumably will resonate throughout the food web.

### Alligators

Alligators (*Alligator mississippiensis*) are dependent on marsh hydrology for many aspects of their life history. Nesting success is probably more closely linked to water levels than any other parameter. Nests will often fail if, during the 60- to 65-day incubation period, water levels get so low as to allow raccoons and other predators access to the nest or so high as to flood the nest (Mazzotti and Brandt, 1994). Alligators adapt to fluctuating water levels by adjusting the height of their nest cavity at egg-laying time to spring water levels, which historically were good predictors of maximum water levels. However, water management practices such as late-season regulatory water releases have reduced the predictability of water level rise and thereby increased nesting losses (Kushlan and Jacobsen, 1990). The once large nesting alligator population of Grossman's Slough, at the southern end of Northeast Shark River Slough in the ENP (T. Armentano, National Park Service, pers. comm.), may have declined due to this unpredictability of water management.

### Florida Bay Response to Changing Hydrology

The hydrological conditions of watersheds strongly influence the ecological characteristics of estuaries (Sklar and Browder, 1998). Changing freshwater flow affects estuaries directly by altering the magnitude and variability of salinity over time and space and by altering the exchange of energy and materials across the terrestrial–estuarine boundary. During this century, the diversion of freshwater from the Everglades toward the Atlantic Ocean and Gulf of Mexico (Color Figures 2.2 and 2.3) has resulted in decreased freshwater flow toward the Everglades' southern boundary via the Shark River Slough and Taylor River Slough (McIvor et al., 1994). For Florida Bay, which lies along this boundary, this has resulted in increased salinity and coincident changes in ecological structure and function (Fourqurean and Robblée, 1999; McIvor et al., 1994).

### Salinity History

The history of salinity levels in Florida Bay during the past century has been reconstructed from chemical and biological indicators preserved in coral skeletons and sediments. Such reconstructions are possible using corals because when corals grow they deposit skeleton layers in annual rings that can be counted and dated, much like with tree rings (Swart et al., 1996). Salinity for a given year can be estimated from the analysis of the oxygen isotope ratio of carbonate within a ring. This isotope ratio varies as a function of water evaporation and thus generally reflects salinity conditions at the time of carbonate deposition. Using this approach, Swart et al. (1996, 1999) have determined that the salinity of southern Florida Bay waters was more variable, with more frequent periods of low salinity during the 1800s than during the 1900s (Figure 2.9). A striking finding of

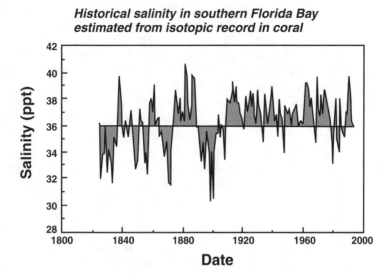

**Historical salinity in southern Florida Bay estimated from isotopic record in coral**

**FIGURE 2.9** Historical salinity of southern Florida Bay for the last 150 years, as estimated by Swart et al. (1999) from the isotopic record of a coral skeleton. This record shows that southern Florida Bay was strongly influenced by the construction of the Flagler Railway around 1910. Salinity in this century has been higher and less variable than during the last century.

these studies is that an abrupt change in the salinity regime of the southern Bay coincided with the construction of a railway through the Florida Keys by Henry Flagler around 1910, prior to major canal construction in South Florida. With railway construction, passes between the Keys were filled, changing the circulation patterns of the bay and apparently also changing the salinity of the bay.

Historical reconstructions of Florida Bay salinity have also been made from the analysis of Foraminifera and mollusc species assemblages in sediment cores from the northeastern and central regions of the bay (Brewster-Wingard and Ishman, 1999). These analyses indicate marked changes in both the regional salinity regime and local benthic vegetation coverage, with distinct shifts occurring around 1900 to 1920 and around 1940. These shifts coincide with construction of the Flagler railway and water management structures. As in Swart et al. (1999), it is inferred from the sedimentary record that salinity has generally been higher in the 1900s than the 1800s. However, unlike the findings of Swart et al. (1999), the variance of salinity in this century, and particularly since 1940, appears to be greater than in the last century. This difference may be the consequence of spatially complex water circulation patterns in Florida Bay. The sediment coring sites were close to the Everglades and probably reflect the history of changing water management and decreased hydrological buffering by the diminished Everglades wetlands.

Additional evidence of historical salinity change has been found in southern Everglades near the boundary of Florida Bay. Meeder et al. (1996) have estimated the rate at which these former freshwater wetlands have become saline marshes. Florida Bay waters have intruded into the marshes because of decreased freshwater inflow from the north and rising sea level. During the past 50 years, the interface of freshwater and saline zones in northeast Florida Bay has moved northward between 3 and 4 km, while this interface has moved only about 0.5 km into the Taylor River Slough.

## Salinity Effects

The physiological status of all estuarine organisms is affected by changing salinity. Thus, alterations of the historical salinity regime of Florida Bay have undoubtedly caused alterations of ecological structure and function in the bay. For the dominant biota of Florida Bay, the specific effects of historical salinity change are not certain. However, the major ecological changes that have been observed in the bay during the past decade, including turtle grass (*Thalassia testudinum*) mass mortality, phytoplankton blooms, and the decline of pink shrimp and other upper trophic level species, have all been at least partially attributed to salinity changes (Fourqurean and Robblee, 1999). In northeastern Florida Bay and saline ponds along the northern coast of the bay, increases in widgeon grass (*Ruppia maritima*) and the alga *Chara* have been associated with increasing freshwater flow into these areas during the mid-1990s. In the nearshore zone where these species and *Halodule wrightii* are dominant, submersed vegetation biomass has been found to be negatively correlated with salinity variance (Montague and Ley, 1993). High salinity has been experimentally shown to negatively impact *Thalassia* primary productivity at salinities above 40 ppt, with greatly decreased productivity at salinities near 60 ppt (M. Durako, pers. comm.). However, *Thalassia* appears to be surprisingly tolerant of salinity in the range of 10 to 20 ppt (Jones et al., 1998; M. Durako, pers. comm.). These results indicate that increased freshwater flow to the bay will not result in the sudden mortality of turtle grass beds, which are thought to have become more dominant in the bay in recent decades because of decreased freshwater flow (Zieman et al., 1989).

Upper trophic level species in Florida Bay have been affected by salinity changes both directly via physiological mechanisms and indirectly via changes in habitat (e.g., seagrass bed coverage). For example, a decline in magnitude of the pink shrimp harvest occurred during the 1980s and this decline has been attributed to habitat loss and environmental factors, including salinity (Browder et al., 1999; Ehrhardt and Legault, 1999). Pink shrimp have been found to have higher mortality rates under hypersaline conditions than under marine conditions (Browder et al., 1999). Furthermore, the magnitude of the shrimp fishery harvest has been found to be positively correlated with freshwater depths, and presumably freshwater discharge, from the Everglades (Browder, 1985).

### Nutrient Loading Effects

Changing the input of freshwater to Florida Bay from the Everglades has concurrently changed the transport of various materials in these waters. Both dissolved materials (mineral nutrients, organic matter, metals, pesticides) and particulate materials (detritus, soils) are transported through canals, wetlands, creeks, rivers, and groundwater. The historical effects of changing freshwater flow on nutrient inputs to Florida Bay are unknown. Measurements of freshwater flow into the bay did not begin until the mid-1990s. However, based on this relatively short period of record, it is clear that there is a strong relationship between both phosphorus and nitrogen loads to Florida Bay and the quantity of water that is discharged into the southern Everglades (Rudnick et al., 1999). Because nutrient loading depends upon rates of freshwater flow, water management in this century has altered the quantity and spatial distribution of these loads. With historical decreases in water flow, nutrient loads to Florida Bay may have also decreased; however, increases in fertilizer use and other anthropogenic nutrient sources in Florida may have more than compensated for decreased water discharge. Did regional nutrient enrichment produce a long-term increase in nutrient loads to Florida Bay in this century? Based on flow-weighted mean concentrations of nutrients, such a trend of increasing inputs to the southern Everglades has not been evident since 1990 for phosphorus and a trend of decreasing nitrogen inputs has occurred since the mid-1980s (Rudnick et al., 1999).The ecological significance to Florida Bay of changing nutrient inputs from the Everglades can be considered by comparing the magnitude of this source of nutrients relative to other external sources, such as the Gulf of Mexico, the atmosphere, and the effluents from the Florida Keys. The nutrient budget, estimated by Rudnick et al. (1999), concluded that phosphorus contributions to Florida Bay from the Everglades were less than 3% of all external P contributions to the bay. The dominant P source appears to be the Gulf of Mexico. Given minute P inputs from the Everglades, the extent to which these P inputs have changed historically probably has had little impact on Florida Bay. In contrast to P, the Everglades could be an important source of nitrogen for Florida Bay, contributing up to 13% of external nitrogen (N) contributions to the bay as a whole and nearly

50% of N contributions to the eastern region of the bay (Rudnick et al., 1999). Because N inputs from the Everglades, which are largely in the form of dissolved organic nitrogen, do appear to be sensitive to changing freshwater flow, it is possible that changes in these inputs have affected trophic structure and function in the bay. It has been suggested that these inputs may be responsible for stimulating algal blooms in Florida Bay (Lapointe and Tomasko, 1995). Bioassays of the nutrient limitation of phytoplankton do indicate that inorganic N is a limiting nutrient in the western bay and, to a lesser extent, in central Florida Bay (Tomas et al., 1999), while P is clearly the limiting nutrient in the eastern bay (Fourqurean et al., 1992, 1993; Tomas et al., 1999). The lability of dissolved organic N from the Everglades and the extent to which this N may become available to phytoplankton is unknown.

Based on preliminary results, it appears unlikely that increasing freshwater flow to Florida Bay as part of the effort to restore the Everglades will significantly affect the input of P to Florida Bay; however, it does appear that N inputs to the bay could increase with increased freshwater flow (Rudnick et al., 1999). The magnitude and consequences of any increased N inputs are uncertain. Anthropogenic nitrogen increases are likely to be minimized by wetland stormwater treatment areas, which remove and retain both P and N. The efficiency of N retention in Everglades wetlands is high for inorganic N (Rudnick et. al 1999) and, based on results from a 5-year pilot study, total N retention within treatment areas is likely to be moderate (about 50%; J. Newman, pers. comm.). At this time, it is premature to conclude that efforts to restore Florida Bay by increasing freshwater flow will cause any harm to any part of the bay or its adjacent waters.

## SUMMARY OF THE ECOLOGICAL EFFECTS OF ALTERED HYDROLOGY

Historical alterations and large-scale development have isolated large segments of the original Everglades from the natural system. Davis and Ogden (1994) estimate that more than half of the original Everglades system has been lost to drainage and development. Today, these developed areas support a variety of land uses, ranging from intensively managed agriculture in the EAA to rapidly spreading urban areas adjacent to WCAs. Together, their impacts on Everglades hydrology have been dramatic and include:

1. *Loss of water storage, overland flow, and spatial extent:* A loss of over 50% of the original ecosystem has reduced the water storage capacity of the ecosystem considerably. A loss of the ability to store large volumes of water that slowly move as overland flow across the Everglades landscape increases the susceptibility of the system to the effects of flood and drought. A reduction of the original ecosystem by half as a result of land development activities has placed a fundamental limitation on its capacity to support populations of wading birds, alligators, and panthers that once used the area in much greater numbers (Davis and Ogden, 1994; USACE, 1994; Science Subgroup, 1996).
2. *Fragmentation of the Everglades:* The historical system has been subdivided by the construction of canals, levees, and roads, resulting in the loss of connections between the central Everglades and adjacent transitional wetlands. Everglades wildlife communities and the long-term sustainability of the ecosystem may be impaired by this separation and isolation. Compartmentalization has significantly altered the amount of overland flow. The construction of canals and levees and the impoundment of the WCAs have caused overdrainage of some areas and excessive flooding in other areas (USACE, 1994).
3. *Changes in timing, distribution, and quantity of water:* Altered discharges into, within, and between freshwater wetlands and estuaries have led to the destruction, loss, or degradation of native Everglades plant communities and associated wildlife habitat (District, 1992; Davis et al., 1994; Davis and Ogden, 1994) and the timing of Wood Stork nesting (Ogden, 1994; Ogden et al., 1997).

4. *Altered fire regimes*: Fire is a natural process (Wright and Heinselman, 1973) that can influence the spread of exotics (Austin, 1976), prevent encroachment of emergent vegetation into sloughs (Forthman, 1973; Hofstetter, 1984), restrict the spread of herbaceous groundcover (Wade et al., 1980), and influence the expansion of hardwood species into marsh habitats (Duever et al., 1976; Koch, 1996). Fires can be rejuvenating or destructive.

5. *Invasion of native plant communities by exotic species:* Species such as melaleuca, Australian pine, and Brazilian pepper have displaced native plants (Bodle et al., 1994) and degraded or destroyed wildlife habitat.

6. *Freshwater discharge to estuaries:* The quantity, timing, and distribution of water discharged from the freshwater areas of the ENP and the WCAs downstream to Florida Bay and Biscayne Bay may have contributed to the altered biological integrity of these estuarine ecosystems (Boesch et al., 1993; Fenemma et al., 1994; Sklar and Browder, 1998).

## INFORMATION GAPS AND FUTURE RESEARCH NEEDS

Florida has now entered an important third phase in the evolution of the Everglades. Although the media may call this the "restoration" phase, it is really more of a "sustainability" phase with restoration of particular components where it is feasible. This paper has summarized the results of various scientific programs designed to support decision-making, improve sustainability of the current system, and stop the further degradation of the Everglades. Better predictive understanding and increased knowledge of hydrologic assessments to support Everglades sustainability will be obtained if the following topics are pursued.

### UNDERSTANDING OVERLAND FLOW

Historic connectivity can never be restored completely, as only half of the original Everglades remains and society has spent hundreds of millions of dollars building and maintaining canals and flow-control structures for flood control and water supply. The hydrologic flows of the past created an environment of slow overland biogeochemical processes. Most significantly, there were no point-sources of water, drying periods were short lived, and peat and marl soils accumulated slowly everywhere. The resulting landscape was partitioned into a mosaic of regions that functioned in ways not yet fully understood. The pond apple (*Annona glabra*) forest that once fringed Lake Okeechobee may have functioned as an upstream sediment and nutrient trap for overland flow south of the lake. A more comprehensive landscape approach to managing water requires further study, and, although the exact mechanisms are not well understood, it is likely that the overland flow once present throughout most of the Everglades was a significant ecological driving force.

### UNDERSTANDING THE IMPACTS OF GROUNDWATER

The vertical connectivity of the Everglades is potentially as important as the horizontal connectivity. Diffusion, soil porosity, seepage, sink holes, and aquifer dynamics are but a few of the complex groundwater factors that influence water supply and subsidence for the entire South Florida region. Urbanization and compartmentalization of the Everglades has changed surface and groundwater exchanges in ways never before possible. Deep canals, ones that cut through carbonate rock, may serve to bring mineral water up to the surface and distribute it to habitats adapted to soft, low pH, organic water. Little is being done to document these changes and their impacts on deep subsidence, water chemistry, and community structure in the Everglades. Doing so will require a new system for monitoring water quantity and water quality. It would be of great utility to obtain groundwater gradients, flow directions, and velocities, as well as potentiometric surface maps of transitional areas to determine the extent of groundwater/surface water exchanges.

## Quantifying Flows to Florida Bay

The hydrological restoration of Florida Bay requires information on the amount of freshwater that is now flowing to the bay, and the relationship between salinity levels in the bay and freshwater flow. For the past three years, the U.S. Geological Survey has measured the amount of freshwater flowing into the bay through the main creeks along the north coast of the bay. However, there are other sources of freshwater that enter the bay from the Everglades, including small creeks, overland flow, and subsurface seepage. Estimating this total freshwater input requires hydrological modeling of the southern Everglades and mangrove forest. Furthermore, estimating how any given rate of freshwater input affects the salinity of the bay also requires hydrodynamic modeling. Both wetland hydrological and bay hydrodynamic modeling projects are under way as part of the cooperation interagency science program.

Finally, the hydrological restoration of Florida Bay requires information on the ecological response of the bay to changing salinity. This not only entails continuing to monitor water quality and biological (e.g., seagrass) changes, but also completing research on the cause-and-effect relationship between salinity change and the ecological response.

## Controlling the Spread of Invasive Species

Scientists know more about the biology of some of the invasive plant species of the Everglades than they do about most of the endemic ones. Nevertheless, more must be done to understand how hydrology may enhance or reduce the many plant species that are encroaching upon the Everglades.

## Restoring Natural Fire Regimes

It is unlikely that the natural ecosystems of the Everglades can be restored without also restoring some semblance of the natural fire regimes. The Everglades is composed of a heterogeneous mosaic of tree islands, shrubs, sawgrass, sloughs, and open water. The mixed patches of vegetation, sloughs, and open water in the Everglades function as natural fire breaks. Preserving this natural patchiness is critical to maintaining the landscape biodiversity as well as the natural fire regimes. Fire behavior, and its role in shaping the structure of each habitat, must be investigated. Little is known about the importance of crown fires, muck fires, and wet-season fires. Water managers need to know the long-term impacts of prescribed burns on the vegetation patterns, soil dynamics, and nutrient cycling in the Everglades. It is particularly important that soil dynamics be restored to compensate for soils lost or altered by managing the hydrology. Experiments on interactions among fire, hydrologic soil, and vegetation need to be designed.

## Understanding Hydrologic Impacts on Flora and Fauna

Recovery of plant and animal communities will require a better understanding of macrophyte and periphyton biology, fish population dynamics, plant–animal interactions, and spatial processes in relation to hydrology. Because of their prominence in setting hydrologic restoration targets, it is critical to identify clearly the effects of hydrology on wading bird feeding and nesting patterns. Large-scale monitoring will provide general trends related to hydrology, but short-term experiments are needed to provide more precise parameter estimates and to support complex simulation models.

## Sustaining Biodiversity

Tree islands and tropical hammocks are hotspots of biodiversity in the Everglades. They are part of a very heterogeneous landscape. This heterogeneous landscape is an environment where fish, birds, plant seeds, and nutrients move from one habitat to another (not all on the same time scale). This movement and the movement of water create natural gradients. However, little is know about these natural gradients and their use by different species. It is known that hydrology can alter the biodiversity of the Everglades. Almost all the tree islands in WCA-2A have disappeared. As a result, local and landscape-scale biodiversity has declined. What are the implications to the greater Everglades and natural gradients? Sustaining biodiversity will require small-scale experiments to evaluate food webs and habitat health in relation to hydrology. Sustaining biodiversity will also require large-scale computer models to evaluate animal movements and landscape heterogeneity in relation to hydrology.

## SUMMARY

Drainage of the Everglades began in 1880 and in some locations reduced water tables up to 9 feet, reversed the direction of surface water flows, altered vegetation, created abnormal fire patterns, and induced high rates of subsidence. Most of these changes were caused by four major canals (Miami, North New River, Hillsboro, and West Palm Beach), constructed between 1910 and 1920. Other hydrologic alterations of major significance include the levee around Lake Okeechobee and construction of the Tamiami Trail.

The initiation of the C&SF Project for Flood Control in 1947 created a system of levees and borrow canals, essentially complete by 1963, that continued to alter water tables and surface flows by creating a highly compartmentalized landscape. Compartmentalization induced ponding in the southern regions of each WCA, increased the frequency and intensity of peat/muck fires in northern regions of the WCAs, and disrupted "overland flow" by creating a hydrologic environment dominated by flows along levee edges, in borrow canals and through water control structures.

Low water tables, drainage, and droughts have altered the balance between peat accretion and peat oxidation — a balance that maintains wetland elevation. An increase in microbial oxidation and frequency of peat fires as a result of drainage has contributed to peat subsidence and lowered ground level elevations. Decreased wetland elevation has made the WCAs more vulnerable to excessive ponding and the ENP more vulnerable to saltwater intrusion from rising sea levels.

Hydrology can influence the success of invasive and exotic vegetation. Cattails are competitive in deep-water environments because they can actively transport oxygen to their roots. Brazilian pepper has become established in areas of disturbed soil horizons where water tables have been lowered and inundation is now less than 4 to 6 months per year.

Animals are also affected by altered hydrology. The vulnerability of fish to predation by wading birds is directly affected by water depth. Even when fish densities are high, water depths greater than 28 cm (during February, March, and April) can decrease foraging and the feeding of fledglings by Tricolored Herons, White Ibis, Snowy Egrets, and Wood Storks. Altered hydrology can also have a direct impact on fish. However, there are two contradictory ideas of how hydrology affects fish populations. One hypothesis is that fish densities are highest when the marsh is managed for frequent drydowns, while the other stipulates that fish densities are lowest when the marsh is managed for frequent drydowns.

Foraging by endangered Snail Kites for apple snails and nesting by endangered Cape Sable Seaside sparrows in muhly grass have conflicting hydrologic requirements. Drydowns to less than 10 cm during March and April impede the movement of the apple snail, thus interfering with peak snail reproduction behavior and the production of fertile egg clusters. In contrast, if water levels are not less than 10 cm by April, the seed- and insect-eating Cape Sable Seaside Sparrow will not initiate breeding. The effects of altered hydrology on other animal populations, particularly amphibians and reptiles, are less well understood.

This anthology of historical information and hydrologic studies conducted over the last 100 years and covering millions of hectares has found that the hydrology of the Everglades has been fundamentally altered. Landscape fragmentation, reduction in overland flow, and habitat loss have resulted. These ecosystem changes have, in turn, affected plant and animal population structure and dynamics. The goals of an Everglades restoration program must address these findings.

Although this chapter disassociates water quality from hydrology in an attempt to address water management needs, it is important that scientists and environmental managers be cognizant of the water quality linkages to hydrology, else sustainable management and restoration of the Everglades will not succeed.

## ACKNOWLEDGMENTS

This research would not have been possible without the support of Thomas Fontaine, Director of the Everglades Systems Research Division of the South Florida Water Management District. A special thanks is extended to the Hydrologic Modeling Division for our use of the NSM and SFWMM and to W. ENP for extensive research on historical changes. Significant contributions were also made by D. Swift, C. Fitz, Y. Wu, A. Ferriter, S. Krupa, K. Rutchey, Q. Dong, S. Newman, and C. Hanlon. GIS and graphic support was provided by L. Gulick and K. Jacobs. Reviews by T. Fontaine, G. Goforth, G. Redfield, B. Bedford, and C. Stevenson were greatly appreciated. We are especially grateful to K. Jacobs for her dedication to the 1999 Everglades Interim Report and to Jan Johansen for her editorial assistance.

## REFERENCES

Advisory Committee. 1944. Report by Advisory Committee on the present drainage system in relation to water control requirements of Everglades Drainage District. May 1, 1944. Everglades Drainage District, Miami, FL, 41 pp.

Alexander, T.R. and A.G. Crook. 1973. Recent and Long-Term Vegetation Changes and Patterns in South Florida, Final Report, Part 1, EVER-N-51, U.S. National Park Service, Coral Gables, FL, 215 pp.

Alexander, T.R. and A.G. Crook. 1974. Recent vegetational changes in southern Florida, in *Environments of South Florida: Present and Past*, P.J. Gleason, Ed., Miami Geological Society, pp. 61–72.

Alexander, T.R. and A.G. Crook. 1975. Recent and Long-Term Vegetation Changes and Patterns in South Florida, Final Report, Part 2, EVER-N-51, U.S. National Park Service, Coral Gables, FL, 856 pp.

Allison, R.V. 1943. The need of the Everglades for a specific plan of development based on the physical and chemical characteristics of its soils and a rational handling of its natural water supply, *Soil Sci. Soc. Fla. Proc.*, 5-A:126–131.

Andrews, R. 1957. Vegetative cover-types of Loxahatchee [Wildlife Refuge] and their principal components, in *Central and Southern Florida Project for Flood Control and Other Purposes*. Part I. *Agricultural and Conservation Areas*, Suppl. 25, General Design Memorandum, Plan of Regulation for Conservation Area No. 1, U.S. Army Corps of Engineers, Office of the District Engineer, Jacksonville, FL, pp. B25–B33.

Austin, D.F. 1976. Vegetation of southeastern Florida – I. Pine Jog. *Fl. Sci.*, 39(4):230–235.

Bailey, M. and S.D. Jewell. 1997. A.R.M. Loxahatchee National Wildlife Refuge, in *South Florida Wading Bird Report*, Vol. 3, D.E. Gawlik, Ed., South Florida Water Management District, West Palm Beach, FL.

Baldwin, M. and H.W. Hawker. 1915. Soil survey of the Fort Lauderdale area, Florida, in *Field Operations of the Bureau of Soils, 1915*, U.S. Department of Agriculture, pp. 751–798.

Bales, J.D., J.M. Fulfors, and E. Swain. 1997. *Review of Selected Features of the Natural System Model, and Suggestions for Applications in South Florida*, Water Resources Investigations Report 97-4039, U.S. Geological Survey, Denver, CO.

Bancroft, G.T, J.C. Ogden, and B.W. Patty. 1988. Wading bird colony formation and turnover relative to rainfall in the Corkscrew Swamp area of Florida during 1982 through 1985, *Wilson Bull.*, 100:50–59.

Bancroft, G.T., S.D. Jewell and A.M. Strong. 1990. Foraging and Nesting Ecology of Herons in the Lower Everglades Relative to Water Conditions. Final report to the South Florida Water Management District, West Palm Beach, FL.

Bancroft, G. T., W. Hoffman, R.J. Sawicki, and J.C. Ogden. 1992. The importance of the Water Conservation Areas in the Everglades to the Endangered Wood Stork (*Mycteria americana*), *Conservation Biol.*, 6:392–398.

Bancroft, G. T., A.M. Strong, R.J. Sawicki, W. Hoffman, and S.D. Jewell. 1994. Relationships among wading bird foraging patterns, colony locations, and hydrology in the Everglades, in *Everglades: The Ecosystem and Its Restoration*, S.M. Davis and J.C. Ogden, Eds., St. Lucie Press, Delray Beach, FL, pp. 615–657.

Barko, J.W. and R.M. Smart. 1986. Sediment-related mechanisms of growth limitation in submersed macrophytes, *Ecology*, 67:1328–1340.

Barry, J.M. 1997. *Rising Tide: The Great Mississippi Flood of 1927 and How it Changed America*, Simon and Schuster, New York.

Bass, S. and L. Oberhofer. 1997. Everglades National Park, *South Florida Wading Bird Report*, Vol. 3, D.E. Gawlik, Ed., South Florida Water Management District, West Palm Beach, FL.

Beard, D.B. 1938. *Wildlife Reconnaissance*, Everglades National Park Project, U.S. Dept. of Interior, National Park Service, Washington, D.C., 106 pp.

Bender, G.J. 1943. The Everglades Fire Control District, *Soil Sci. Soc. Fla. Proc.*, 5-A:149–152.

Bennetts, R.E. and W.M. Kitchens. 1997. The Demography and Movements of Snail Kites in Florida, U.S.G.S. Biological Resources Division, Florida Cooperative Fish & Wildlife Research Unit, Tech. Report No. 56.

Bennetts, R.E., V.J. Dreitz, W.M. Kitchens, J.E. Hines, and J.D. Nichols. 1999. Annual survival of snail kites in Florida: radio telemetry versus capture-resighting data, *Auk*, 116:435–447.

Bestor, H.A. 1942. The principle elements of a long time soil and water conservation plan for the Everglades, *Soil Sci. Soc. Fla. Proc.*, 4-A:90–99.

Bildstein, K.L., W. Post, J. Johnston, and P. Frederick. 1990. Freshwater wetlands, rainfall and the breeding ecology of White Ibises (*Eudocimus albus*) in coastal South Carolina, *Wilson Bull.*, 102:84–98.

Bodle, J.M. 1994. Does the scourge of the south threaten the Everglades?, in *An Assessment of Invasive Non-Indigenous Species in Florida's Public Lands*, D.C. Schmitz and T.C. Brown, Eds., Tech. Report TSS-94-100, Florida Dept. of Environmental Protection, Tallahassee, 303 pp.

Boesch, D.F., N.E. Armstrong, C.F. D'Elia, N.G. Maynard, H.W. Paerl, and S.W. Williams. (1993). Deterioration of the Florida Bay Ecosystem: An Evaluation of the Scientific Evidence. South Florida Water Management District, West Palm Beach, FL, 29 pp.

Boyer, J.N. and R.D. Jones. 1999. Effects of freshwater inputs and loading of phosphorus and nitrogen on the water quality of eastern Florida Bay, in *Phosphorus Biogeochemistry of Subtropical Ecosystems: Florida as a Case Example*, K.R. Reddy, G.A. O'Connor, and C.L. Schelske, Eds., Lewis Publishers, Boca Raton, FL, pp. 545–564.

Brandt, L.A. 1997. Spatial and Temporal Changes in Tree Islands of the Arthur R. Marshall Loxahatchee National Wildlife Refuge, Ph.D. dissertation, University of Florida, Gainesville.

Brewster-Wingard, G.L. and S.E. Ishman. 1999. Historical trends in salinity and substrate in central Florida Bay: a paleoecological reconstruction using modern analogue data, *Estuaries*, 22:369–383.

Brix, H., B.K. Sorrell, and P.T. Orr. 1992. Internal pressurization and convective gas flow in some emergent freshwater macrophytes, *Limnol. Oceanogr.*, 37:1420–1433.

Browder, J.A. 1985. Relationship between pink shrimp production on the Tortugas grounds and water flow patterns in the Florida Everglades, *Bull. Mar. Sci.*, 37:839–856.

Browder, J.A., D. Cottrell, M. Brown, M. Newman, R. Edwards, J. Yuska, M. Browder, and J. Krakoski. 1982. *Biomass and Primary Production of Microphytes and Macrophytes in Periphyton Habitats of the Southern Everglades*, Report T-662, South Florida Research Center, Homestead, FL.

Browder, J.A., P.J. Gleason, and D.R. Swift. 1994. Periphyton in the Everglades: spatial variation, environmental correlates and ecological implications, in *Everglades: The Ecosystem and Its Restoration*, S.M. Davis and J.C. Ogden, Eds., St. Lucie Press, Delray Beach, FL.

Browder, J.A., O. Bass, J. Gebelein, and L. Oberhoffer. 1997. Florida Bay, in *South Florida Wading Bird Report*, Vol. 3, D.E. Gawlik, Ed., South Florida Water Management District, West Palm Beach, FL, pg. 5.

Browder, J.A., V.R. Restrepo, J.K. Rice, M.B. Robblee, and Z. Zein-Eldin. 1999. Environmental influences on potential recruitment of pink shrimp, *Penaeus duorarum*, from Florida Bay nursery grounds, *Estuaries*, 22:484–499.

Brown, S.C. and B L. Bedford. 1997. Restoration of wetland vegetation with transplanted wetland soil: an experimental study, *Wetlands*, 17:424–437.

Burkhead, R.R. 1991. Melaleuca control in Big Cypress National Preserve, in *Proceedings of the Symposium on Exotic Pest Plants*, USDI. Nat. Park Serv.

Chandrasena, J.N.N.R. 1990. Torpedograss (*Panicum repens* L.) control with lower rates of glyphosate, *Trop. Pest. Manage.*, 36:336–342.

Chanton, J. P., G.J. Whiting, J.D. Happell, and G. Gerard. 1993. Contrasting rates and diurnal patterns of methane emission from emergent aquatic macrophytes, *Aquatic Botany*, 46:111–128.

Chin, D.A. 1998. Evapotranspiration of melaleuca forest in south Florida, *J. Hydrologic Eng.*, 3(2):131–139.

Clayton, B.S. 1936. *Subsidence of Peat Soils in Florida*, Report No. 1070, U.S. Dept. of Agriculture, Bureau of Agricultural Engineering, Belle Glade, FL, 15 pp.

Clayton, B.S., J.R. Neller, and R.V. Allison. 1942. *Water Control in the Peat and Muck Soils of the Florida Everglades*, Bull. 378, University of Florida Agricultural Experiment Station and U.S. Dept. of Agriculture Soil Conservation Service, Gainesville, FL, 74 pp.

Cohen, A.D. 1974. Evidence of fires in the ancient Everglades and coastal swamps of south Florida, in *Environments of South Florida: Present and Past*, P.J. Gleason, Ed., Miami Geological Society, Miami, FL, pp. 213–218.

Comiskey, E.J., L.J. Gross, D.M. Fleming, M.A. Huston, O.L. Bass, H.-K. Luh, and Y. Wu. 1997. A spatially explicit individual-based simulation model for Florida panther and white-tailed deer in the Everglades and Big Cypress landscapes, in *Proceedings of the Florida Panther Conference*, D. Jorden, Ed., Fort Myers, FL, Nov. 1–3, U.S. Fish and Wildlife Service, pp. 494–503.

Conde, L.F., D.L. Rockwood, and R.F. Fisher. 1980. Growth studies on melaleuca, *Proceedings of Melaleuca Symposium*, U.S.D.A. Forest Service, Sept. 23–24, p. 23–28.

Conner, W.H., J.R. Toliver, and F.H. Sklar. 1986. Natural regeneration of cypress in a Louisiana swamp, *Forest Ecol. Manage.*, 14:305–317.

Cornwell, G. and E.C. Hutchinson. 1974. An Ecological Analysis of an Everglades Township in Southwestern Palm Beach County, Florida, a report to Gulf & Western Industries and Gulf & Western Food Products Company. Ecoimpact, Inc., Gainesville, FL, 30 pp.

Craft, C.B. and C.J. Richardson. 1993. Peat accretion and N, P and organic C accumulation in nutrient-enriched and unenriched Everglades peatlands, *Ecol. Appl.*, 3:446–458.

Craighead, F.C., Sr. 1971. *The Trees of South Florida*, Univ. of Miami Press, Coral Gables, FL, 212 pp.

Darby, P.C., J.D. Croop, H.F. Percival, and W.M. Kitchens. 1997. Ecological Studies of Apple Snails (*Pomacea paludosa*). 1995 Report to South Florida Water Management District and St. Johns River Water Management District, Florida Cooperative Fish and Wildlife Research Unit, Gainesville, FL.

David, P.G. 1997. Changes in plant communities relative to hydrological conditions in the Florida Everglades, *Wetlands*, 16:15–23.

Davis, J.H. 1940. *The Ecology and Geologic Role of Mangroves in Florida*, Carnegie Institute, Washington, D.C., 412 pp.

Davis, J.H. 1943a. The natural history of South Florida, *Fla. Geol. Surv. Bull.*, 517(25).

Davis, J.H. 1943b. *The Natural Features of Southern Florida*, Geol. Bull. 25, Florida Geological Survey, Tallahassee, FL, 311 pp.

Davis, J.H. 1943c. Vegetation Map of Southern Florida (Fig. 4-71 of Bulletin 25), 1:400,000, Florida Geological Survey, Tallahassee, FL.

Davis, S.M. 1982. Patterns of Radio Phosphorus Accumulation in the Everglades after Its Introduction into Surface Water, Report 82-2, South Florida Water Management District, West Palm Beach, FL.

Davis, S.M. 1991. Growth, decomposition and nutrient retention of *Cladium jamaicense* Crantz and *Typha domingensis* Pers. in the Florida Everglades, *Aquatic Botany*, 40:203–224.

Davis, S.M. and J.C. Ogden, Eds. 1994. *Everglades: The Ecosystem and its Restoration*, St. Lucie Press, Delay Beach, FL.

DeAngelis, D.L., L.J. Gross, M.A. Huston, W.F. Wolff, D.M. Fleming, E.J. Comiskey, and S.M. Sylvester. 1998. Landscape modeling for Everglades ecosystem restoration, *Ecosystems*, 1:64–75.

DeLaune, R.D., R.J. Buresh, and W.H.J. Patrick. 1979. Relationship of soil properties to standing crop biomass of *Spartina alterniflora* in a Louisiana marsh, *Estuar. Coast. Mar. Sci.*, 8:477–487.

DeVries, D. 1995. East Everglades Exotic Plant Control Project, Annual Report to Dade County, Everglades National Park.

Dineen, J.W. 1972. Life in the tenacious Everglades: Central and South Florida Flood Control District, *In Depth Report* 1(5), 112 pp.

Dineen, J.W. 1974. Examination of water management alternatives in Conservation Area 2A, *In Depth Report* 2(3):1–11.

DiToro, D., P. Paquin, K. Subbaramu, and D. Gruber, 1990. Sediment oxygen deman model: methans and ammonia oxidation, *J. Environ. Eng.*, 116:945–986.

Duever, M.J., J.E. Carlson, L.A. Riopelle et al. 1976. Ecosystem analysis of Corkscrew Swamp, in *Cypress Wetlands for Water Management: Recycling and Conservation*, H.T. Odum et al., Eds., Third Annual Report to National Science Foundation and the Rockefeller Foundation Center for Wetlands, University of Florida, Gainesville, pp. 707–737.

DuPuis, J.G. 1954. *History of Early Medicine in Dade County, Florida*, publ. privately by the author, Miami, 140 pp.

Egler, F.E. 1952. Southeast saline Everglades vegetation: Florida and its management, *Veg. Acta Geobotanica*, 3:213–265.

Ehrhardt, N.M. and C.M. Legault. 1999. Pink shrimp, *Farfantepenaeus duorarum*, recruitment variability as an indicator of Florida Bay dynamics, *Estuaries*, 22: 471–483.

Fennema, R.J., C.J. Neidrauer, R.A. Johnson, T.K. MacVicar, and W.A. Perkins. 1994. A computer model to simulate natural Everglades hydrology, in *Everglades: The Ecosystem and Its Restoration*, S.M. Davis and J.C. Ogden, Eds., St. Lucie Press, Delray Beach, FL, pp. 249–289.

Feunteun, E. and L. Marion. 1994. Assessment of Grey Heron predation on fish communities: the case of the largest European colony, *Hydrobiologia*, 279/280:327–344.

FGFWFC. 1956. Recommended program for Conservation Area 2 by Florida Game and Fresh Water Fish Commission, letter of transmittal signed A.D. Aldrich, Oct. 31, 1956, Florida Game and Fresh Water Fish Commission, Tallahassee, FL, 120 pp. + 2 maps.

Fitz, H.C. and Sklar, F.H. 1999. Ecosystem analysis of phosphorus impacts and altered hydrology in the Everglades: a landscape modeling approach, in *Phosphorus Biogeochemistry in Subtropical Ecosystems*, Reddy, K.R. et al., Eds., Lewis Publishers, Boca Raton, FL, pp. 585–620.

Fitz, H.C., R. Costanza, and E. Reyes. 1993. The Everglades Landscape Model (ELM): Summary Report of Task 2, Model Development, report to South Florida Water Management District, West Palm Beach, FL, 109 pp.

Fitz, H.C., E.B. DeBellevue, R. Costanza, R. Boumans, T. Maxwell, L. Wainger, and F.H. Sklar. 1996. Development of a general ecosystem model for a range of scales and ecosystems, *Ecological Modeling*, 88:263–295.

Fitzpatrick, J., D. DiToro, M. Meyers, and M.Z. Moustafa. 1997. A Framework for Evaluating the Impact of Nutrient Inputs on Emergent Vegetation in South Florida, presented at the American Geophysical Union, May 27–30, 1997, Baltimore, MD.

Fleming, D.M., W.F. Wolff, and D.L. DeAngelis. 1994. Importance of landscape heterogeneity to wood storks in Florida Everglades, *Environ. Manage.*, 18(5):743–757.

Florida Everglades Engineering Commission. 1914. Florida Everglades, Report of the Florida Everglades Engineering Commission to the Board of Commissioners of the Everglades Drainage District and the Trustees of the Internal Improvement Fund. State of Florida, Senate Doc. No. 379, 63rd Congress, 2nd Session, U.S. Government Printing Office, Washington, D.C., 148 pp.

Florida Geological Survey. 1991. *Florida's Ground Water Quality Monitoring Program — Hydrogeological Framework,* Special Publ. 32, Tallahassee, FL, 97 pp.

Florida Geological Survey. 1992. *Florida's Ground Water Quality Monitoring Program — Background Geochemistry,* Special Publ. 34, Tallahassee, FL, 363 pp.

Flowers, J.D. 1991. Subtropical fire suppression in *Melaleuca quinquenervia, Proceedings of the Symposium on Exotic Pest Plants,* U.S. Dept. of the Interior, National Park Service, pp. 151–158.

Forthman, C.A. 1973. The Effects of Prescribed Burning on Sawgrass, *Cladium jamaicense* Crantz, in South Florida, M.S. thesis, University of Miami, Coral Gables, FL, 83 pp.

Fourqurean, J.W. and M.B. Robblee. 1999. Florida Bay: a history of recent ecological changes, *Estuaries,* 22:345–357.

Fourqurean, J.W., J.C. Zieman, and G.V.N. Powell. 1992. Phosphorus limitation of primary production in Florida Bay: evidence from the C:N:P ratios of the dominant seagrass *Thalassia testudinum, Limnol. Oceanogr.,* 37:162–171.

Fourqurean, J.W., R.D. Jones, and J.C. Zieman. 1993. Processes influencing water column nutrient characteristics and phosphorus limitation of phytoplankton biomass in Florida Bay, FL, USA: inferences from spatial distributions, *Estuarine Coastal Shelf Sci.,* 36:295–314.

Frederick, P. and D. Battaglia. 1997. Water Conservation Areas 2 and 3, *South Florida Wading Bird Report,* Vol. 3, in D.E. Gawlik, Ed., South Florida Water Management District, West Palm Beach, FL, pp. 2–3.

Frederick, P.C. and G.V.N. Powell. 1994. Nutrient transport by wading birds in the Everglades, in *Everglades: The Ecosystem and Its Restoration,* S.M. Davis and J.C. Ogden, Eds., St. Lucie Press, Delray Beach, FL.

Frederick, P.C. and M.G. Spalding. 1994. Factors affecting reproductive success of wading birds (Ciconiiformes) in the Everglades ecosystem, in *Everglades: The Ecosystem and Its Restoration,* S.M. Davis and J.C. Ogden, Eds., St. Lucie Press, Delray Beach, FL, pp. 659–692.

Frederick, P.C. and M.W. Collopy. 1989. Nesting success of five Ciconiiform species in relation to water conditions in the Florida Everglades, *Auk,* 106:625–634.

Gawlik, D.E. 1996. Influence of water depth on the selection of foraging sites by wading birds. Proceedings of the Workshop Ecological Assessment of the 1994–95 High Water Levels in the Everglades, Miami, FL.

Gawlik, D.E. 1997. Summary, in *South Florida Wading Bird Report,* Vol. 3, D.E. Gawlik, Ed., South Florida Water Management District, West Palm Beach, FL, p. 1.

Gawlik, D.E. and D.A. Rocque. 1998. Avian communities in bayheads, willowheads and sawgrass marshes of the central Everglades, *Wilson Bull.,* 110:45–55.

Gawlik, D.E. and J.C. Ogden, Eds. 1996. *1996 Late-Season Wading Bird Nesting Report for South Florida,* South Florida Water Management District, West Palm Beach, FL.

Gleason, P.J., Ed. 1984. *Environments of South Florida: Past and Present II,* Miami Geological Society, Coral Gables, FL.

Gleason, P.J. and Spackman, W., Jr. 1974. Calcareous periphyton and water chemistry in the Everglades, in *Environments of South Florida: Present and Past,* P.J. Gleason, Ed., Miami Geological Society, Coral Gables, FL, pp. 146–181.

Goodrick, R L. 1974. The wet prairies of the northern Everglades, in *Environments of South Florida: Present and Past,* P. Gleason, Ed., Miami Geological Society, Coral Gables, FL, pp. 47–52.

Graham, W.A. 1951. The Pennsuco sugar experiment, *Tequesta, Fl.,* 11:27–49.

Gunderson, L.H. and W.F. Loftus. 1993. The Everglades, in *Biodiversity of the Southeastern United States,* W.H. Martin, S.G. Boyce, and A.C. Echternacht, Eds., John Wiley & Sons, New York, pp. 199–255.

Gunderson, L.H. and J.R. Snyder. 1994. Fire patterns in the southern Everglades, in *Everglades: The Ecosystem and Its Restoration,* S.M. Davis and J.C. Ogden, Eds., St. Lucie Press, Delray Beach, FL, pp. 291–306.

Hamrick, J.M. and M.Z. Moustafa. 1998a. Development of the Everglades Wetland Hydrodynamic Model, Part I: Model formulation, *Water Resources Res.* (In review.)

Hamrick, J. M., and M.Z. Moustafa. 1998b. Development of the Everglades Wetland Hydrodynamic Model, Part II: Model implementation, *Water Resources Res.* (In review.)

Harney, W.W. 1884. The drainage of the Everglades, *Harper's,* 68(406):598–605.

Harvey, J.W., S.L. Krupa, R.H. Mooney, and P. Succeeder. 1998a. Are ground-water and surface water connected by vertical hydrologic fluxes through peat? *EOS Trans. Am. Geophys. Union,* 79(17):S87.

Harvey, J.W., S.L. Krupa, S.L. Mooney, and C. Gefvert, 1998b. Interactions Between Ground-water and Surface Water in the Everglades Nutrient Removal Area that Affect Mercury Transport and Transformation, Annual Meeting of South Florida Mercury Science Program Investigators, West Palm Beach, FL, May 18–20.

Harvey, J.W., R.H. Mooney, and S.L. Krupa, 1998c. Use of Seepage Meters To Determine the Vertical Flux of Ground Water and Surface Water through Wetland Peat in the Florida Everglades, U.S. Geological Survey Water Resources Investigations Report 98.

Heilprin, A. 1887. Explorations on the west coast of Florida and in the Okeechobee wilderness, *Trans. Wagner Free Inst. Sci. Philadelphia,* 1:365–506 + 21 plates.

Herdon, A. and D. Taylor. 1986. Response of a Muhlenbergia Prairie to Repeated Burning: Changes in Above-Ground Biomass, South Florida Research Center Report, SFRC-86/05, Everglades National Park.

Herr, B. 1943. A report to the Board of Commissioners of Okeechobee Flood Control District on the Activities of the District and on Lake Okeechobee, 1929–1943, Okeechobee Flood Control District, West Palm Beach, FL, 99 pp.

Heyes, A., C.C. Gilmour, and J.M. Benoit, Controls on the distribution of methylmercury in the Florida Everglades, *EOS Trans. Am. Geophys. Union*, 79(17):S67.

Higer, A.L. and M.C. Kolipinski. 1967. Pull-up trap: a quantitative device for sampling shallow water animals, *Ecology*, 48:1008–1009.

Hoffman, W., G.T. Bancroft, and R.J. Sawicki. 1994. Foraging habitat of wading birds in the Water Conservation Areas of the Everglades, in *Everglades: The Ecosystem and Its Restoration*, S.M. Davis and J.C. Ogden, Eds., St. Lucie Press, Delray Beach, FL, pp. 585–614.

Hofstetter, R.H. 1976. The ecological role of fire in south Florida, *Fl. Nat. Arpr.*, pp. 2–9.

Hofstetter, R.H. 1984. The effect of fire on the pineland and sawgrass communities of southern Florida, in *Environments of South Florida: Past and Present II*, P. Gleason, Ed., Miami Geological Society, Coral Gables, FL, 551 pp.

HSM. 1997. Documentation Update for the South Florida Water Management Model, Hydrologic Systems Modeling Division, South Florida Water Management District, West Palm Beach, FL, 239 pp.

Huffman, J.B. 1980. Melaleuca wood and bark utilization research — a progress report, in *Proceedings of Melaleuca Symposium*, USDA Forest Service, Sept. 23–24, pp. 37–116.

HydroQual. 1989. Development and Calibration of a Sediment Flux Model of Chesapeake Bay, final report prepared for U.S. Army Engineers, Waterways Experiment Station, Vicksburg, MS, by HydroQual, Inc., Mahwah, NJ.

Interbureau Committee. 1930. Memorandum regarding agricultural conditions in the Everglades of Florida and the effect on those conditions of proposed navigation and flood-protection improvements. House Doc. No. 71-47, 71st Congress, 2nd Session, Washington, D.C.

Jensen, J.R., K. Rutchey, M.S. Koch, and S. Narumaiani, 1995. Inland wetland change detection in the Everglades Water Conservation Area 2A: using a time series of normalized sensed data, *Photogrammetric Eng. Remote Sensing*, 61(2):199–209.

Jones, L.A. et al. 1948. *Soils, Geology and Water Control in the Everglades Region*, Agricultural Experiment Station Bull. 442. University of Florida Agricultural Experiment Station, Gainesville.

Jones, L.M., C.L. Montague, and E. Chipouras. 1998. Modified transplantations of four submersed macrophyte species throughout two salinity gradient areas in northeastern Florida Bay (abstract), in *Proceedings of the 1998 Florida Bay Science Conference*, University of Miami, FL.

King, J.W. 1917a. Report of Progress upon Preliminary Survey for Reclaiming Lands within the Southern Drainage District, unpublished manuscript dated Sept. 17, 1917; "Copy No. 3", Jaudon Collection, Box 16, Historical Museum of South Florida, Miami.

King, J.W. 1917b. Contemplated Location and Dimension of the Miami Drainage Canals in the Southern Drainage District, unpublished blueprint to accompany report dated Sept. 17, 1917, no scale; approx. 1 1/16 in. = 1 mile (1:60,000), Jaudon Collection, Box 16, Historical Museum of South Florida, Miami.

King, J.W. 1917c. Report of Exploration: Examination and Reconnaissance of the Lands of the Tamiami Trail in Dade County, Florida, unpublished manuscript dated Mar. 23, 1917, Jaudon Collection, Box 16, Historical Museum of South Florida, Miami.

King, J.W. 1917d. Map Showing Results of Examination of the Tamiami Trail Lands in Dade County Florida, unpublished blueprint to accompany report dated Mar. 23, 1917, Report item #18), 4 in. = 1 mile (1:15,840), Jaudon Collection, Box 16, Historical Museum of South Florida, Miami.

King, J.W. 1917e. Profiles Showing Soil Conditions on Tamiami Trail Lands, unpublished blueprint to accompany report dated Mar. 23, 1917, report item #17, horiz. scale approx. 1:9,700, vert. scale approx. 1:48, Jaudon Collection, Box 16, Historical Museum of South Florida, Miami.

Kludze, H.K. and R.D. DeLaune 1996. Soil redox intensity effects on oxygen exchange and growth of cattail and sawgrass, *Soil. Sci. Am. J.*, 60:616–621.

Koch, M.S. 1996. Resource Availability and Abiotic Stress Effects on *Rhizophora mangle* L. (Red Mangrove) Development in South Florida, Ph.D. dissertation, University of Miami, Coral Gables, FL.

Kolipinski, M.C. and A.L. Higer. 1969. Some Aspects of the Effects of the Quantity and Quality of Water on Biological Communities in Everglades National Park, Open File Rep. FL-69007. U.S. Geological Survey, Tallahassee, FL, 97 pp.

Kreamer, J.M. 1892. Map of Hic-po-chee and Okeechobee Sugar Lands, Lee and De Soto counties, Florida; embracing 175,000 acres of land available for sugar cultivation; 60 chains = 1 inch; 36 × 52 inches.

Krupa. S.L., J.W. Harvey, C. Gefvert, and J. Giddings, 1998. Geologic and anthropogenic influences on ground-water–surface water interactions in the Northern Everglades, *EOS Trans. Am. Geophys. Union*, 79(17):S176.

Kushlan, J.A. 1974. Quantitative sampling of fish populations in shallow, freshwater environments. Transactions of the American Fisheries Society, 103:348–352.

Kushlan, J.A. 1976a. Wading bird predation in a seasonally fluctuating pond, *Auk*, 93:464–476.

Kushlan, J.A. 1976b. Environmental stability and fish community diversity, *Ecology*, 57:821–825.

Kushlan, J.A. 1986. Responses of wading birds to seasonally fluctuating water levels: strategies and their limits, *Colonial Waterbirds*, 9:155–162.

Kushlan, J.A. 1987. External threats and internal management: the hydrologic regulation of the Everglades, Florida, U.S.A., *Environ. Manage.*, 1:109–119.

Kushlan, J.A. 1990. Freshwater marshes, in *Ecosystems of Florida*, R.L. Myers and J.J. Ewel, Eds., University of Central Florida Press, Orlando, pp. 324–363.

Kushlan, J.A. and T. Jacobsen. 1990. Environmental variability and reproductive success of Everglades alligators, *J. Herpetol.*, 24:176–184.

Kushlan, J.A. and D.A. White. 1977. Nesting wading bird populations in southern Florida, *Fla. Sci.*, 40:65–72.

Langeland, K., B. Smith, and C. Hanlon. (In review.) Torpedograss — forage gone wild, *Wetland Weeds*.

Lapointe, B.E. and D. Tomasko. 1995. Evidence of long-term, large-scale eutrophication in Florida Bay and the Florida Reef tract (abstract), Estuarine Research Federation 13th Biennial International Conference.

Larned, W.L. 1917–1918. Lost in the wastes of the Everglades, *Forest Stream Rod Gun*, 87(12):585.

Leach, S.D., H. Klein, and E.R. Hampton. 1971. Hydrologic Effects of Water Control and Management of Southeastern Florida, Open File Report 71005, U.S. Geological Survey, Tallahassee, FL, 193 pp.

Light, L.L., L.H. Gunderson, and C.S. Holling. 1995. The Everglades: evolution of management in a turbulent ecosystem, in *Barriers and Bridges to the Renewal of Ecosystems and Institutions*, L.H. Gunderson, C.S. Holling, and S.S. Light, Eds., Columbia University Press, New York, pp. 103–168.

Light, S.S. and J.W. Dineen. 1994. Water control in the Everglades: a historical perspective, in *Everglades the Ecosystem and Its Restoration*, S.M. Davis and J.C. Ogden, Eds., St. Lucie Press, Delray Beach, FL, pp. 47–84.

Light, S.S. and W.J. Dineen. 1994. Water control in the Everglades: a historical perspective, in *Everglades: The Ecosystem and Its Restoration*, S.M. Davis and J.C. Ogden, Eds., St. Lucie Press, Delray Beach, FL, 826 pp.

Lockwood, J.L., K.H. Fenn, J.L. Curnutt, D. Rosenthal, K.L. Balent, and A.L. Mayer. 1997. Life history of the endangered Cape Sable Seaside Sparrow, *Wilson Bull.*, 109:720–731.

Lodge, T.E. 1998. *The Everglades Handbook*, St. Lucie Press, Boca Raton, FL.

Loftus, W.F. and A. Eklund. 1994. Long-term dynamics of an Everglades small-fish assemblage, in *Everglades: The Ecosystem and Its Restoration*, S.M. Davis and J.C. Ogden, Eds., St. Lucie Press, Delray Beach, FL, pp. 461–484.

Loftus, W.F, J.D. Chapman, and R. Conrow 1990. Hydroperiod effects on Everglades marsh food webs with relation to marsh restoration efforts, in *Fisheries and Coastal Wetlands Research*, Vol. 6, Larson, M. and M. Soukup, Eds., Proc. 1986 Conf. on Science in National Parks, Ft. Collins, CO, pp. 1–22.

Loope, L.L. and N. Urban. 1980. Fire Effects in Tree Islands of the East Everglades Region of Everglades National Park, South Florida Research Center Tech. Report, Everglades National Park, Homestead FL.

Lorenz, J.J. 1997. The Effects of Hydrology on Resident Fishes of the Everglades Mangrove Zone, final report to Everglades National Park, National Audubon Society.

Loveless, C.M. 1959. A study of vegetation of the Florida Everglades, *Ecology*, 40(1):1–9.

Loveless, C.M., G.W. Cornwell, R.L. Downing, A.R. Marshall, and J.N. Layne. 1970. Report of the Special Study Team on the Florida Everglades, South Florida Water Management District, West Palm Beach, FL, 42 pp.

MacVicar, T.K., T. VanLent, and A. Castro. 1984. South Florida Water Management Model: Documentation Report, South Florida Water Management District, West Palm Beach, FL, 123 pp.

Manapura, W.V. and A. Somaratne. 1974. Some effects of manual and chemical defoliation on the growth and carbohydrate reserves of (*Panicum repens* L. Beauv.), *Weed Res.*, 14:167–172.

Marston, A., S.H. McCrory, and G.B. Hills. 1927. Report of Everglades Engineering Board of Review to Board of Commissioners of Everglades Drainage District, May, 1927, Everglades Drainage District, State of Florida, Tallahassee.

Matson, G.C. and S. Sanford. 1913. Geology and Ground Waters of Florida, Water Supply Paper 319, U.S. Geological Survey, Washington, D.C., 445 pp.

Mazzotti, F.J. and L.A. Brandt. 1994. Ecology of the American alligator in a seasonally fluctuating environment, in *Everglades: The Ecosystem and Its Restoration*, S.M. Davis and J.C. Ogden, Eds., St. Lucie Press, Delray Beach, FL, pp. 485–505.

McDowell, L.L., J.C. Stephens, and E.H. Stewart. 1969. Radiocarbon chronology of the Florida Everglades peat, *Soil Sci. Soc. Am. Proc.*, 33:743–745.

McIvor, C.C., J.A. Ley, and R.D. Bjork. 1994. Changes in freshwater inflow from the Everglades to Florida Bay including effects on biota and biotic processes: a review, in *Everglades: The Ecosystem and Its Restoration*, S.M. Davis and J.C. Ogden, Eds., St. Lucie Press, Delray Beach, FL, pp. 117–146.

McPherson, B.F. 1973. Vegetation in Relation to Water Depth in Conservation Area 3, Florida, U.S. Geological Survey Open File Report #73025.

McPherson, B.F., G.Y. Hendrix, H. Klein, and H.M. Tyus. 1976. The environment of south Florida — a summary report, U.S. Geological Survey Professional Paper 1011, 82 pp.

Meeder, J.F., M.R. Ross, G. Telesnicki, P.L. Ruiz, and J.P. Sah. 1996. Vegetation analysis in the C-111/ Taylor Slough Basin, final report to the District (Contract C-4244).

Meigs, J.L. 1879. Examination of Caloosahatchee River, Florida, Annual Report of the Chief of Engineers, U.S. Army Corps of Engineers, pp. 863–870.

Miao, S.L. and F.H. Sklar. 1998. Biomass and nutrient allocation of sawgrass and cattail along a nutrient gradient in Florida Everglades, *Wetland Ecosystem Manage.*, 5:245–264.

Miller, W.L. 1988. *Description and Evaluation of the Effects of Urban and Agricultural Development on the Surficial Aquifer System, Palm Beach County, Florida*, Water Resources Investigation Report 88-4056, U.S. Geological Survey, Tallahassee, FL, 58 pp.

Mitsch, W.J. 1994. *Global Wetlands: Old World and New*, Elsevier, Amsterdam.

Mitsch, W.J. and J.G. Gosselink. 1993. Wetlands, 2nd ed., Van Nostrand-Reinhold, New York.

Montague C.L. and J.A. Ley. 1993. A possible effect of salinity fluctuation on abundance of benthic vegetation and associated fauna in northeastern Florida Bay, *Estuaries*, 16:703–717.

Moore, D.R.J. and P.A. Keddy. 1988. Effects of a water-depth gradient on the germination of lakeshore plants, *Can. J. Botany*, 66:548–552.

Moustafa, M.Z. and J. Hamrick. 1998. Calibration of the Wetland Hydrodynamic Model to the Everglades Nutrient Removal Project, abstract presented at the Society of Wetlands Scientist Annual Meeting in Anchorage, AK, June 1998.

Myers, R.L. 1975. The relationship of site conditions to the invading capability of *Melaleuca quinquenervia* in southwest Florida, Master's thesis, University of Florida, Gainesville, 151 pp.

National Park Service. 1995. *A Hydrological Evaluation of the Everglades National Park Experimental Water Delivery Program Test 6 Iteration*, Everglades National Park, South Florida Natural Resources Center, 90 pp.

Nealon, D. 1984. *Ground-Water Quality Study of the Water Conservation Areas*, South Florida Water Management District, West Palm Beach, FL, p. 27.

Newman, S., J.B. Grace, and J.W. Koebel. 1996. The effects of nutrients and hydroperiod on mixtures of *Typha domingensis*, *Cladium jamaicense* and *Eleocharis interstincta*: implications for Everglades restoration, *Ecol. Appl.*, 6:774–783.

Newman, S., J. Schuette, J.B. Grace, K. Rutchey, T. Fontaine, K.R. Reddy, and M. Pietrucha.1998. Factors influencing cattail abundance in the northern Everglades, *Aquat. Botany*, 60:265–280.

Noble, C.V., R.W. Drew, and J.D. Slabaugh. 1996. Soil Survey of Dade County Area, U.S. Dept. of Agriculture Natural Resources Conservation Service, 116 pp.

Nott, M.P., O.L. Bass, D.M. Fleming, S.E. Killeffer, N. Fraley, L. Manne, J.L. Cornutt, T.M. Brooks, R. Powell, and S.L. Pimm. 1998. Water levels, rapid vegetation changes, and the endangered Cape Sable Seaside Sparrow, *Animal Conservation*, 1:23–32.

Odum, H.T. 1983. *Systems Ecology: An Introduction*, John Wiley & Sons, New York.

Ogden, J.C. 1991. Wading bird colony dynamics in the central and southern Everglades. An annual report. South Florida Research Center, Everglades National Park, Homestead, FL.

Ogden, J.C. 1994. A comparison of wading bird nesting colony dynamics (1931–1946 and 1974–1989) as an indication of ecosystem conditions in the southern Everglades, in *Everglades: The Ecosystem and Its Restoration*, S.M. Davis and J.C. Ogden, Eds., St. Lucie Press, Delray Beach, FL, pp. 533–570.

Ogden, J.C. 1996. Wood Stork nesting patterns in *Everglades and Big Cypress colonies, in Proceedings of the Conference: Ecological Assessment of the 1994–1995 High Water Conditions in the Southern Everglades*, T.V. Armentano, Ed., Held under the auspices of the Science Sub-Group of the South Florida Management and Coordination Working Group, 219 pp.

Ogden, J.C., G.T. Bancroft, and P.C. Frederick. 1997. Ecological success indicators: reestablishment of healthy wading bird populations, in *Ecologic and Precursor Success Criteria for South Florida Ecosystem Restoration*, a Science Sub-group report to the Working Group of the South Florida Ecosystem Restoration Task Force, U.S. Army Corps of Engineers, Jacksonville, FL.

Olmsted I.C. and T.V. Armentano. 1997. *Vegetation of Shark Slough, Everglades National Park*, South Florida Natural Resources Center, Everglades National Park, 41 pp.

Olmsted, I.C. and L.L. Loope. 1984. Plant communities of Everglades National Park, in *Environments of South Florida, Past and Present II*, P.J. Gleason, Ed., Miami Geological Society, Coral Gables, FL, pp. 167–184.

Olmsted, I.C., L.L. Loope, and R.E. Rintz. 1980. *A Survey and Baseline Analysis of Aspects of the Vegetation of Taylor Slough, Everglades National Park*, SFNC Report T-586, Everglades National Park, Homestead, FL.

Orem, W.H., A.L. Bates, H.E. Lerch, and J.W. Harvey, 1998. Sulfur Geochemistry of the Everglades: Sources, Sinks and Biogeochemical Cycling. Annual Meeting of South Florida Mercury Science.

Parker, G.G. 1942. Notes on geology and groundwater of the everglades in Southern Florida, *Proc. Florida Soil Sci. Soc.*, 4.

Parker, G.G. 1955. Water Resources of Southeastern Florida, Water Supply Paper 1255, U.S. Geological Survey, Tallahassee, FL.

Parker, G.G., G.E. Ferguson, S.K. Love et al. 1955. Water Resources of Southeastern Florida, with Special Reference to the Geology and Ground Water of the Miami Area, Water Supply Paper #1255, U.S. Geological Survey, Tallahassee, FL, 965 pp.

Patten, B.C. 1990. *Wetlands and Shallow Continental Water Bodies*, Vol. 1, SPB Academic Publishing, The Hague, The Netherlands.

Pezeshki, S.R., R.D. DeLaune, H.K. Kludze and H.S. Choi. 1996. Photosynthetic and growth responses of cattail (*Typha domingensis*) and sawgrass (*Cladium jamaicense*) to soil redox conditions, *Aquatic Botany*, 54:25–35.

Pimm, S.L., T. Brooks, J.L. Curnutt, J. Lockwood, L. Manne, A. Mayer, M.P. Nott, and G. Russell. 1995. Population Ecology of the Cape Sable Sparrow (Ammodramus maritima mirabilis). Annual Report 1995, NBS/NPS, Everglades National Park, Homestead, FL.

Ponzio, K.J., S.J. Miller, and M.A. Lee, 1995. Germination of sawgrass, *Cladium jamaicense* Crantz, under varying hydrologic conditions, *Aquatic Botany*, 51:115–12.

Porter, K.M. and A.R.C. Smith. 1984. Evaluation of Sampling Methodology — Systematic Flight/Pilot Wading Bird Survey, Technical Report, Everglades National Park, Homestead, FL.

Powell, G.V.N. 1987. Habitat use by wading birds in a subtropical estuary: implications of hydrography, *AUK*, 104:740–749.

Program Investigators, West Palm Beach, FL, May 18–20, 1998. Abstract published with meeting program.

Reddy, K.R., R.D. DeLaune, W.F. Debusk, and M.S. Koch. 1993. Long-term nutrient accumulation rates in the Everglades, *Soil Sci. Soc. Am. J.*, 1 57:1147–1155.

Richardson, D.R. 1977. Vegetation of Atlantic Coastal Ridge of Palm Beach County, Florida, *Florida Sci.*, 40(4): 281–330.

Richardson, J.R., W.L. Bryant, W.M. Kitchens, J.E. Mattson, and K.R. Pope. 1990. An Evaluation of Refuge Habitats and Relationships to Water Quality, Quantity and Hydroperiod, a report to the Arthur R. Marshall Loxahatchee National Wildlife Refuge, Boynton Beach, FL.

Richter, W. and E. Myers. 1993. The birds of a short-hydroperiod, Muhlenbergia-dominated wetland prairie in southern Florida, *Florida Field Naturalist*, 21(1):1–10.

Robbins, J.A., X. Wang, and R.W. Rood. 1996. Sediment Core Dating (Everglades WCAs), report to South Florida Water Management District, West Palm Beach, FL.

Robertson, W.B., Jr. 1955. An Analysis of the Breeding Bird Populations of Tropical Florida in Relation to the Vegetation, Ph.D. thesis, University of Illinois, 599 pp.

Robertson, W.B. 1953. A Survey of the Effects of fire in Everglades National Park, submitted Feb. 15, 1953, U.S. Dept. of Interior National Park Service, 169 pp.

Robertson, W.B., Jr. 1954. Everglades fires — past, present, and future, *Everglades Nat. Hist.*, 2(1): 9–16.

Rudnick, D.T., Z. Chen, D.L. Childers, J.N. Boyer, and T.D. Fontaine III. 1999. Phosphorus and nitrogen inputs to Florida Bay: the importance of the Everglades watershed, *Estuaries*, 22:398–416.

Rutchey, K. and L. Vilchek. 1994. Development of an Everglades vegetation map using a SPOT image and the global positioning system, *Photogr. Eng. Remote Sensing*, 60:767–775.

Rutchey, K. and L. Vilchek. 1998. Air photo-interpretation and satellite imagery analysis techniques for mapping cattail coverage in a Northern Everglades Impoundment, *Photogr. Eng. Remote Sensing*, 64:1–7.

Schardt, J.D. 1994. Florida Aquatic Plant Survey 1992, Tech. Report 942-CGA, Florida Dept. of Environmental Protection, Tallahassee, FL, 83 pp.

Schortemeyer, J.L. 1980. *An Evaluation of Water Management Practices for Optimum Wildlife Benefits in Conservation Area 3A*, Florida Game and Fresh Water Fish Commission, Tallahassee, FL.

Schroeder, M., H. Klein, and N. Hoy, 1958. *Biscayne Aquifer of Dade and Broward Counties, Florida*, Report of Investigation No. 17, Florida Geological Survey, Tallahassee, Florida. 56 pp.

Science Subgroup. 1996. *South Florida Ecosystem Restoration: Scientific Information Needs*, report to the Working Group of the South Florida Ecosystem Restoration Task Force, U.S. Army Corps of Engineers, Jacksonville, FL.

Science Subgroup. 1997. *Ecologic and Precursor Success Criteria for South Florida Ecosystem Restoration*, report to the Working Group of the South Florida Ecosystem Restoration Task Force, U.S. Army Corps of Engineers, Jacksonville, FL.

SFWMD. 1992. *Surface Water Improvement and Management Plan for the Everglades*, supporting information document, South Florida Water Management District, West Palm Beach, FL.

SFWMD. 1996. *Florida Everglades Program: Everglades Construction Project*, Final Programmatic Environmental Impact Statement, Vol. III, Appendices, South Florida Water Management District, West Palm Beach, FL.

SFWMD. 1997. *Documentation for the South Florida Water Management Model*, Hydrologic Systems Modeling Division, Planning Department, South Florida Water Management District, West Palm Beach, FL.

SFWMD. 1998. *Natural Systems Model Version 4.5 Documentation*, Planning Department, South Florida Water Management District, West Palm Beach, FL.

SFWMD, 2000. *The Everglades Interim Report 2000*, Water Resources Evaluation Department, South Florida Water Management District, West Palm Beach, FL.

Sharitz, R.R. and J.W. Gibbons, 1989. *Freshwater Wetlands and Wildlife*, U.S. Dept. of Energy, Office of Health and Environmental Research, National Technical Information Service (DE90005384), Springfield, VA.

Shih, S.F., E.H. Stewart, L.H. Allen, Jr., and J.W. Hilliard. 1979a. An Investigation of Subsidence Lines in 1978, Belle Glade AREC Research Report EV-1979-1, Belle Glade Agricultural Research and Education Center, Belle Glade, FL, 28 pp.

Shih, S.F., E.H. Stewart, L.H. Allen, Jr., and J.W. Hilliard. 1979b. Variability of depth to bedrock in Everglades organic soil, *Soil Crop Sci. Soc. Fla.*, 38:6671.

Shih, S.F., J.W. Mishoe, J.W. Jones, and D.L. Myhre. 1979c. Subsidence related to land use in Everglades agricultural area, *Trans. Am. Soc. Agric. Eng.*, 22(3):560-563, 568.

Sklar, F.H. and J.A. Browder. 1998. Coastal environmental impacts brought about by alterations to freshwater flow in the Gulf of Mexico, *Environ. Manage.*, 22(4):547–562.

Sklar, F.H., R. Costanza, and J.W. Day, Jr. 1990. Model conceptualization, in *Wetlands and Shallow Continental Water Bodies*, Vol. 1, B.C. Patten et al., Eds., SPB Academic Publishing, The Hague, The Netherlands.

Sklar, F.H., H.C. Fitz, Y. Wu, R. Van Zee, and C. McVoy. 2001. The design of ecological landscape models for Everglades restoration, *Ecological Economics*, 37(3): 379–402.

Slattery, C.J.R. 1913. Drainage Map, Kissimmee and Caloosahatchee Rivers and Lake Okeechobee, Florida. Part of House Doc. No. 137, 63rd Congress, 1st Session.

Snyder, G.H. and J.M. Davidson, 1994. Everglades agriculture — past, present and future, in *Everglades: The Ecosystem and Its Restoration*, S.M. Davis and J.C. Ogden, Eds., St. Lucie Press, Delray Beach, FL, pp. 85–115.

Snyder, G.H., H.W. Burdine, J.R. Crockett, G.J. Gascho, D.S. Harrison, G. Kidder, J.W. Mishoe, D.L. Myhre, F.M. Pate, and S.F. Shih. 1978. Water table management for organic soil conservation and crop production in the Florida Everglades, University of Florida Agricultural Experiment Station Bull. 801.

South Florida Natural Resources Center. 1994. *Restoration of Northeast Shark Slough and the Rocky Glades*, Everglades National Park, Homestead, FL.

Stephens, J.C. 1942. The principal characteristics of the Kissimmee–Lake Okeechobee Watershed. B. The Everglades, *Soil Sci. Soc. Fla. Proc.*, 4-A:24–28.

Stephens, J.C. and L. Johnson. 1951. Subsidence of organic soils in the Upper Everglades region of Florida, *Soil Sci. Soc. Fla. Proc.*, 11:191–237.

Strong, A.M., G.T. Bancroft, and S.D. Jewell. 1997. Hydrological constraints on tricolored heron and snowy egret resource use, *Condor*, 99:894-905.

Swart, P.K., G.F. Healy, R.E. Dodge, P. Kramer, J.H. Hudson, R.B. Halley, and M.B. Robblee. 1996. The stable oxygen and carbon isotopic record from a coral growing in Florida Bay: a 160 year record of climatic and anthropogenic influence, *Palaeogeogr. Palaeoclimatol. Palaeocol.*, 123:219–237.

Swart, P.K., G. Healy, L. Greer, M. Lutz, A. Saied, D. Anderegg, R.E. Dodge, and D. Rudnick. 1999. The use of proxy chemical records in coral skeletons to ascertain past environmental conditions in Florida Bay, *Estuaries*, 22:384–397.

Swift, D.R. and R.B. Nicholas. 1987. *Periphyton and Water Quality Relationships in the Everglades Water Conservation Areas, 1978–1982*, Technical Publ. 87-2, South Florida Water Management District, West Palm Beach, FL.

Tamiami Trail Commissioners. 1928. History of the Tamiami Trail and a Brief Review of the Road Construction Movement in Florida, The Tamiami Trail Commissioners and the County Commissioners of Dade County, Miami, FL, 27 pp.

Tarver, D.P. 1979. Torpedo grass (*Panicum repens* L.), *Aquatics*, 1:5–6.

Tomas, C.R., B. Bendis, and K. Johns. 1999. Role of nutrients in regulating plankton blooms in Florida Bay, in *The Gulf of Mexico Large Marine Ecosystem*, H. Kumpf, K. Steidinger, and K. Sherman, Eds., Blackwell Scientific, Oxford, pp. 323–337.

Trexler, J.C., R.C. Tempe, and J. Travis. 1994. Size-selective predation of sailfin mollies by two species of heron, *Oikos*, 69:250–258.

Tropical BioIndustries. 1990. Hydroperiod Conditions of Key Environmental Indicators of Everglades National Park and Adjacent East Everglades Area as Guide to Selection of an Optimum Water Plan for Everglades National Park, Florida, Tropical BioIndustries, Inc., Miami, FL.

USACE. 1951. Central and Southern Florida Project. Part I. Agricultural and Conservation Areas. *Geology and Soils*, Supp. I, No. 21, U.S. Army, Washington, D.C.

USACE, Jacksonville District. 1957. Central and Southern Florida Project for Flood Control and Other Purposes. Part I. Agricultural and Conservation Areas, Suppl. 25, General Design Memorandum, Plan of Regulation for Conservation Area No. 1, Nov. 29, Corps of Engineers, U.S. Army, Office of the District Engineer, Jacksonville, FL.

USACE. 1968. Water Resources for Central and Southeastern Florida, Main Report No. 56, U.S. Dept. of Army, Washington, D.C.

USACE. 1980. *Regulation Schedule Review: Water Conservation Area No. 3A*, U.S. Army Corps of Engineers, Washington, D.C.

USACE. 1994. *Final Integrated General Reevaluation Report and Environmental Impact Statement, Canal 111 (C-111), South Dade County, Florida*, U.S. Army Corps of Engineers, Washington, D.C.

U.S. Dept. of Agriculture. Soil Conservation Service (USDA–SCS). 1940. Aerial Photography, Everglades Area Florida. "Photographed 1940 by Aero Service Corp., Philadelphia. Index compiled 6-5-40. Project AIS 20674." Aerial negative scale 1:40,000. U.S. Dept. Agric., Soil Conserv. Service, Washington, D.C.

Urban, N.H., S.M. Davis, and N.G. Aumen. 1993. Fluctuations in sawgrass and cattail in Everglades Water Conservation Area 2A under varying nutrient, hydrologic and fire regimes, *Aquatic Botany*, 46:203–223.

Van der Valk, A.G. and C.B. Davis. 1978. The role of seed banks in the vegetation dynamics of prairie glacial marshes, *Ecology*, 59:322–335.

Van der Valk, A.G. and F.H. Sklar, Eds. In press. *Tree Islands of the Everglades*, Kluwer Academic, Norwell, MA.

Van Lent, T. and R. Johnson 1993. *Towards the Restoration of Taylor Slough*, South Florida Natural Resources Center, Everglades National Park, Homestead, FL.

Van Meter-Kasanof, N. 1973. Ecology of the microalgae of the Florida Everglades. Part I. Environment and some aspects of freshwater periphyton, 1959-1963, *Nova Hedwigia*, 24:619–664.

Wade, D.D., J.J. Ewel, and R. Hofstetter. 1980. *Fire in South Florida Ecosystems*, U.S. Dept. of Agriculture Forest Service General Technical Report SE-17, Southeast Forest Experimental Station, Asheville, NC.

Walker, W.W., Jr. and R.H. Kadlec. 1996. A Model for Simulating Phosphorus Concentrations in Water and Soils Downstream of Everglades Stormwater Treatment Areas, report to the National Park Service and Everglades National Park, Homestead, FL.

Wallace, H.E., W.H. Herke, N.F. Schlaack, F.J. Ligas, and C.M. Loveless. 1960. Recommended Program for Conservation Area 3 by Florida Game and Fresh Water Fish Commission, Florida Game and Fresh Water Fish Commission, Vero Beach, FL.

Walters, C., L. Gunderson, and C.S. Holling. 1992. Experimental policies for water management in the Everglades, *Ecol. Appl.*, 2:189–202.

Weller, M.W. 1995. Use of two waterbird guilds as evaluation tools for the Kissimmee River restoration, *Restoration Ecol.*, 3:211–224.

White, W.A. 1970. *The Geomorphology of the Florida Peninsula*, Florida Bureau of Geology Bull. 51, Tallahassee, FL, 164 pp.

Wilcut, J.W., R.R. Truelove, and D.E. Davis. 1988. Factors limiting the distribution of cogongrass, *Imperata cylindrica* and torpedograss, *Panicum repens*, *Weed Sci.*, 36:577–582.

Wintringham, M.D., Ed. 1964. North to south through the Everglades in 1883. Part II. Tequesta 24:59–93.

Wood, J.M. and G.W. Tanner. 1990. Graminod community composition and structure within four Everglades management areas, *Wetlands*, 10:127–149.

Woodall, S.L. 1981. Evapotranspiration and melaleuca, in *Proceedings of Melaleuca Symposium*, R.K. Geiger, Ed., Florida Department of Agricultural and Consumer Services, Division of Forestry, September 23–24, 1980.

Woodall, S.L. 1984. Rainfall Interception Losses from Melaleuca Forest in Florida, Research Note SE-323, U.S. Dept. of Agriculture Forest Service, Southeastern Forest Experiment Station, Lehigh Acres, FL.

Worth, D. 1981. *Preliminary Environmental Responses to Marsh De-Watering and Reduction in Water Regulation Schedule in Water Conservation Area 2A*, South Florida Water Management District, West Palm Beach, FL.

Worth, D.F. 1988. *Environmental Response of WCA-2A to Reduction in Regulation Schedule and Marsh Draw-Down*, South Florida Water Management District, West Palm Beach, FL.

Wright, H.E. and M.L. Heinselman. 1973. The ecological role of fire in natural conifer forests of western and northern North America, *Quaternary Res.*, 3(3):319–328.

Wu, Y., F.H. Sklar, and K. Rutchey. 1996. Fire simulations in the Everglades landscape using parallel programming, *Ecol. Modeling*, 93(1996):113–124.

Wu, Y., F.H. Sklar, and K. Rutchey. 1997. Analysis and simulations of fragmentation patterns in the Everglades, *Ecol. Appl.*, 7(1):268–276.

Zaffke, M. 1983. *Plant Communities of Water Conservation Area 3A: Base-Line Documentation Prior to the Operation of S-339 and S-340*, Technical Memorandum DRE-164, South Florida Water Management District, West Palm Beach, FL.

Zieman, J.C., J.W. Fourqurean, and R.L. Iverson. 1989. Distribution, abundance and productivity of seagrasses and macroalgae in Florida Bay, *Bull. Mar. Sci.*, 44:292–311.

# 3 Effects of Anthropogenic Phosphorus Inputs on the Everglades

*Paul V. McCormick, Susan Newman, ShiLi Miao, Dale E. Gawlik, and Darlene Marley*
Everglades Department, South Florida Water Management District

*K. Ramesh Reddy*
Soil and Water Science Department, University of Florida

*Tom D. Fontaine*
Environmental Monitoring and Assessment Division, South Florida Water Management District

## CONTENTS

0-8493-2026-7/02/$0.00+$1.50
© 2002 by CRC Press LLC

## INTRODUCTION

Phosphorus (P) is a key element controlling aquatic productivity, and the widespread use of this nutrient to increase soil fertility is responsible for the eutrophication of freshwater ecosystems worldwide (Tiessen, 1995). The ecological effects of increased P loading on lakes and rivers have been well-documented and include excessive productivity, reduced dissolved oxygen, changes in species composition, and reduced biodiversity (National Academy of Sciences, 1969; Likens, 1972; Harper, 1992; Havens and Steinman, 1995). The relationship between increased P loading and wetland change has not been widely investigated as most attention has focused on the ability of wetlands to transform and remove pollutants rather than on the ecological impacts associated with this removal process (e.g., Howard-Williams, 1985; Moshiri, 1993; Olson, 1993). The need for national water quality standards for wetlands has been recognized (USEPA, 1990) in an effort to afford these ecosystems the same level of protection as other freshwaters.

The Everglades ecosystem (Figure 3.1) developed under extremely low rates of P supply. Increased rates of P loading from agricultural sources over the past several decades have been associated with changes in ecological conditions that are indicative of cultural eutrophication. The severity and extent of these impacts was recognized in 1988 when the federal government sued the State of Florida for allowing P-enriched discharges and associated impacts to occur in the Everglades. Settlement of this lawsuit eventually resulted in the enactment of the Everglades Forever Act by the Florida Legislature in 1994, which required the State of Florida to derive a numeric water-quality criterion for P discharges into the Everglades that would "prevent ecological imbalances in natural populations of flora or fauna." These legal and legislative events spawned numerous research and monitoring efforts to document the process of eutrophication in the Everglades and to identify control measures necessary to reverse ecological impacts caused by this process.

The objective of this chapter is to synthesize current information on the effects of anthropogenic P loading on the Everglades in order to:

- Define reference (i.e., historical) conditions for P in the Everglades.
- Characterize the ecological impacts caused by increased P loading.
- Identify marsh P concentrations that produce these impacts.

## REFERENCE CONDITIONS OF THE ECOSYSTEM

Defining historical or reference conditions of the Everglades is a critical first step in determining the extent of P enrichment and associated ecological impacts. Reconstructions of reference conditions typically are based on (1) available historical records; (2) sampling in minimally impacted locations (i.e., reference sites) that are believed to best portray historical conditions in different regions and habitats; and (3) paleoecological assessments, which entail the collection and dating of soil cores and the analysis of nutrient content and preserved materials (e.g., pollen, algal cell walls, charcoal) that are diagnostic of past conditions and events. Unfortunately, with the exception

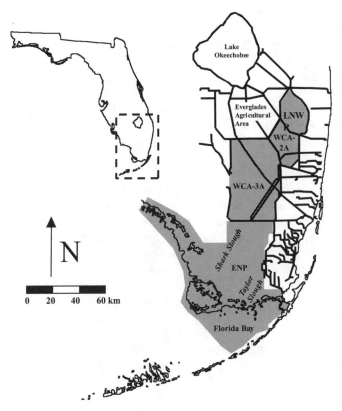

**FIGURE 3.1** Major hydrologic units of the remnant Everglades–Florida Bay ecosystem (shaded areas). The canal and levee system is represented as solid black lines. ENP = Everglades National Park, LNW = Loxahatchee National Wildlife Refuge.

of large-scale vegetation patterns (e.g., Davis, 1943; Loveless, 1959), historical descriptions of the system are scarce, and most are qualitative and lack critical detail. Similarly, relatively little paleoecological evidence is currently available for reconstructing predrainage conditions (but see Cooper and Goman, 2001; Willard et al., 2001); therefore, much of what is known about the ecological characteristics of the predrainage Everglades is based on relatively recent data from minimally impacted locations in the marsh interior.

## Water Chemistry

### Phosphorus Loading

Nutrient inputs to the Everglades are derived primarily from atmospheric deposition (rainfall and dry fallout), which is typically low in P. Estimates of annual atmospheric P inputs in South Florida and reconstructions of P accumulation in Everglades soils (see below) indicate that historical loading rates were extremely low, probably averaging less than 0.1 g P $m^{-2}$ $y^{-1}$ (SFWMD, 1992). Atmospheric inputs of P were augmented by inflows from Lake Okeechobee, which was connected by surface-water flows to the northern Everglades during periods of high water (Parker et al., 1955). These inflows were probably enriched in P compared with the marsh as paleoecological evidence indicates that Lake Okeechobee has been eutrophic for several thousand years (Gleason and Stone, 1975). However, the effects of this enrichment appear to have been limited to a zone of pond apple (*Annona glabra* L.) and sawgrass (*Cladium jamaicense* Crantz.) along the lake's southern fringe (Snyder and Davidson, 1994). This assumption of rapid P removal is consistent with current estimates of the P removal capacity of the Everglades marsh (Walker, 1995). Therefore, it is unlikely that inflows from Lake Okeechobee exerted a strong influence on the P dynamics across much of the predrainage Everglades.

The accuracy of atmospheric P loading estimates based on contemporary sampling is affected by several confounding factors. Current sampling of rainfall and dry deposition reflects regional agricultural and urban sources as well as background inputs and, consequently, overestimates historical inputs (SFWMD, 1997). Sampling typically is conducted using ground-based collection systems, which are susceptible to contamination from local P sources (e.g., guano, insects) that do not represent new inputs of P to the ecosystem. As shown for rainfall samples collected across the Everglades (Figure 3.2), P concentrations in rainfall samples vary greatly, even among samples collected at the same station, and the data are highly skewed due to a small number of extremely high values that are likely caused by contamination. Most samples contain total phosphorus (TP) in concentrations near or below the limits of analytical detection (4 μg L$^{-1}$) for state government laboratories; thus, the low end of the distribution is truncated and cannot be measured precisely. Average rainfall TP concentrations used to calculate loading rates have been estimated from these data as the mean concentration (e.g., Waller and Earle, 1975; Davis, 1994), which consistently overestimates the central tendency of the data, sometimes by a wide margin. The median P concentration, a more appropriate estimator of the average value, ranges between 4 and 7 μg L$^{-1}$ across all South Florida Water Management District (SFWMD) collection stations, whereas mean concentrations are consistently higher and considerably more variable (Figure 3.2). Alternative data analyses techniques (e.g., Ahn, 1999) also show that average P concentrations in rainfall are consistently below 10 μg L$^{-1}$ across the Everglades. Dryfall probably accounts for 50 to 75% of atmospheric P inputs to the Everglades. However, accurate measurement of rates of dry deposition has proven even more troublesome than that of rainfall and no accurate estimates are currently available (SFWMD, 1997).

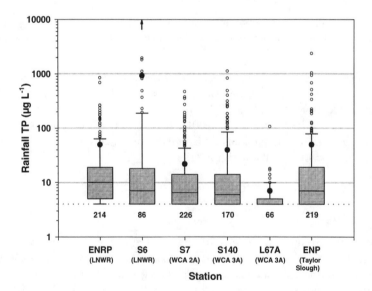

**FIGURE 3.2**    Range of TP concentrations in rainfall samples collected from SFWMD atmospheric deposition collection stations throughout the Everglades between 1987 and 1997. The top, mid-line, and bottom of each box represent the 75th, 50th (median), and 25th percentiles of data, respectively. The vertical lines represent the 10th and 90th percentiles, respectively. Large closed circle is the arithmetic mean; small open circles (O) are observations outside the 90th percentiles. Numbers below each box are the number of samples collected at that site. The dashed line represents the analytical detection limit for TP (4 μg L$^{-1}$); concentrations at or below the detection limit were analyzed as 4 μg L$^{-1}$.

## Water-Column P Concentrations

Interior areas of the Everglades generally retain the oligotrophic characteristics of the predrainage ecosystem and, thus, provide the best contemporary information on historical P concentrations. Available water-column TP data are summarized for several interior sampling stations that are distant to canal inflows and, therefore, are believed to best reflect the reference condition for nutrients in different areas of the marsh (Figure 3.3). These data may represent an upper estimate of historical TP concentrations in the Everglades, as several stations are located in areas that either have been (1) excessively drained (e.g., northern WCA-3A), a condition that promotes soil oxidation and P release; or (2) so heavily exposed to canal inflows (e.g., WCA-2A) that some P inputs have likely intruded even into interior areas.

Median water-column TP concentrations at reference stations throughout the Everglades range between 4 and 10 μg L$^{-1}$ and are lowest in southern areas of the marsh (e.g., Taylor River Slough and the C-111 basin), which have been least affected by anthropogenic nutrient loads. Mean concentrations are more variable among stations and exceed 10 μg L$^{-1}$ in some areas, largely as a result of periodic concentration excursions during periods of low water and/or marsh drying. High P concentrations at reference stations can be attributed to P released as a result of oxidation of exposed soils (Swift and Nicholas, 1987), increased fire frequency during droughts (Forthman, 1973), and difficulties in collecting water samples that are not contaminated by flocculent marsh sediments when water depths are low. Localized enrichment in the pristine Everglades also is associated with bird rookeries, alligator holes, and other areas of natural disturbance (Davis, 1994). However, all these forms of enrichment (e.g., soil oxidation, fecal inputs) represent recycling of existing P and, therefore, do not affect the total amount of P stored in the marsh.

## Other Key Water Chemistry Features

Interior areas of the Everglades differ with respect to pH and concentrations of major ions such as bicarbonate, calcium, and sodium. Surface waters across much of the Everglades, including

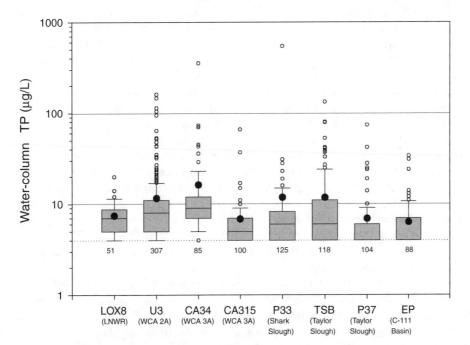

**FIGURE 3.3** Range of TP concentrations in surface water samples collected from interior marsh sampling stations (see Figures 3.5, 3.7–3.10 for exact locations).

WCA-2A, WCA-3A, and Everglades National Park (ENP), are slightly basic and highly mineralized (i.e., high ionic content). In contrast, the interior of the Arthur R. Marshall Loxahatchee National Wildlife Refuge (LNWR) is slightly acidic and contains extremely low concentrations of major ions, a condition that reflects the rainfall-driven hydrology of this area. These background differences among Everglades marshes influence species composition (e.g., Swift and Nicholas, 1987) and the water-quality and biological impacts caused by canal inflows (e.g., Gleason et al., 1975).

Water-column concentrations of total nitrogen (TN) (1000 to 2500 µg L$^{-1}$) and other macronutrients are relatively high in interior areas compared with those for P (generally <10 µg L$^{-1}$). Ratios of TN:TP in reference areas generally exceed 100:1 and are indicative of strong P limitation (Swift and Nicholas, 1987; McCormick et al., 1996). Concentrations of micronutrients (e.g., copper, iron, silica, zinc) vary but are generally not considered limiting compared with the extremely low P concentrations in the marsh interior (e.g., Steward and Ornes, 1983).

Interior Everglades habitats exhibit characteristic diel fluctuations in water-column dissolved oxygen (DO), although aerobic conditions are generally maintained throughout much or all of the diel cycle (Belanger et al., 1989; McCormick et al., 1997). High daytime concentrations in open-water habitats (i.e., sloughs, wet prairies) are a product of photosynthesis by periphyton and other submerged vegetation. These habitats may serve as oxygen sources for adjacent sawgrass stands, where submerged productivity is low (Belanger et al., 1989). Oxygen concentrations decline rapidly during the night due to periphyton and sediment microbial respiration and generally fall below the 5 mg L$^{-1}$ standard for Class III Florida waters (Criterion 17-302.560(21), F.A.C.). However, these diurnal excursions are characteristic of reference areas throughout the Everglades (McCormick et al., 1997) and are not considered a violation of the Class III standard (Nearhoof, 1992).

## SOIL AND POREWATER

In the early 1900s the Everglades organic soils were categorized into three groups based on the overlying vegetation (Baldwin and Hawker, 1915). Soils close to Lake Okeechobee were classified as custard apple muck, based on the dominance of custard apple, also known as pond apple (*Annona glabra* L.). Farther away from the lake, land was dominated by willow (*Salix* spp.) and elderberry (*Sambucus canadensis* L.), and the soils were classified as willow and elder soils. However, the predominant organic soil in the Everglades was sawgrass peat, with sawgrass (*Cladium jamaicense* Crantz.) as the dominant vegetation. Of these various soil types, custard apple soils were considered to have a greater native fertility than sawgrass soils, which formed under lower nutrient conditions (Snyder and Davidson, 1994). By the late 1940s, extensive soil surveys were completed (Davis, 1946; Jones, 1948), and the peat soils of the Everglades were classified into two main groups: Everglades peat (derived mostly from sawgrass) and Loxahatchee peat (derived primarily from slough species such as water lily).

Although the majority of the soil data collected in the early 1900s were descriptive, there were a few accounts of soil samples collected and analyzed for nutrients. An examination of these records showed that at sites adjacent to Lake Okeechobee, P concentrations in the surface 30 cm of uncultivated organic soil averaged 0.61% $P_2O_5$ (2700 mg P kg$^{-1}$)(Hammar, 1929). In contrast, at sites near the Tamiami Trail, representative of the Everglades interior, P concentrations averaged 0.1% $P_2O_5$ (~400 mg P kg$^{-1}$) in muck, marl, and hammock areas (King, 1917). While analytical techniques of the early 1900s may be less accurate and precise than modern techniques, these values cover the range found today.

Present-day soil nutrient data may also be used to estimate reference conditions by evaluating nutrient concentrations in areas that are outside the zone of nutrient impact. Total P concentrations in the surface 0 to 10 cm of soil in interior areas of the Holeyland, WCAs, and Rotenberger Tract range between 200 and 500 mg kg$^{-1}$ (DeBusk et al., 1994; Reddy et al., 1994a; Newman et al., 1997; Richardson et al., 1997a; Newman et al., 1998). In ENP, soil TP concentrations downstream of canal inflows ranged from 1420 mg kg$^{-1}$ near the inflows to as low as 320 mg kg$^{-1}$ farther downstream (Doren et al., 1997), suggesting that background TP concentrations for ENP were $< 400$ mg kg$^{-1}$.

In addition to variations in nutrient concentration per soil mass (mg kg$^{-1}$), nutrient content of the soils also changes per unit volume, as a function of changing soil bulk density. The typical bulk density of flooded Everglades peat soils is approximately 0.08 g cm$^{-1}$, whereas soils subjected to extended dry out and oxidation can have bulk densities greater than 0.2 g cm$^{-3}$ (Newman et al., 1998). Studies in other wetlands have shown that plant growth increases linearly with increases in bulk density from 0.1 to 0.4 g cm$^{-3}$ (DeLaune et al., 1979; Barko and Smart, 1986). Thus, the expression of soil nutrient data on a volumetric (bulk density corrected) basis may provide greater information relative to plant growth. Following correction for the varying bulk densities in the peat soils of the Everglades, a historical TP concentration of $<40$ μg cm$^{-3}$ may be applicable for most regions (DeBusk et al., 1994; Reddy et al., 1994a, 1998; Newman et al. 1997, 1998). In the LNWR, most of the interior area has soil TP $< 20$ μg TP cm$^{-3}$ (Newman et al., 1997).

The most biologically available form of soil P is that found in the porewater. Surface 0 to 10 cm soil porewater collected from interior areas of the Everglades typically has soluble reactive P (SRP) concentrations of $<50$ μg L$^{-1}$, and frequently these values are at or below analytical limits of detection (4 μg L$^{-1}$) (DeBusk et al., 1994; Koch-Rose et al., 1994; Reddy et al. 1994a, 1998; Newman et al., 1997).

## SOIL MICROBES AND BIOGEOCHEMICAL CYCLES

Wetlands host complex microbial communities, including bacteria, fungi, protozoa, and viruses. The size and diversity of these communities are directly related to the quality and quantity of available resources. Microbes typically respond rapidly to external nutrient loading, and this response is often measured as an increase in the rate of various microbially mediated processes. These processes regulate biogeochemical cycles and nutrient availability in wetlands and, therefore, play an important role in determining water quality and ecosystem productivity.

Relatively little information is available on microbial populations and processes in the Everglades. However, it has been established that many of the microbial activities and associated biogeochemical reactions are P limited in interior areas of the marsh (Reddy et al., 1999). Microorganisms in these areas are dependent on P derived from decomposition of organic substrates, as external loading rates (described earlier) are extremely low. Hydrolysis of organic compounds to release bioavailable P is catalyzed by phosphatase enzymes. High levels of phosphatase activity (PA) in the Everglades interior (Wright and Reddy, 2001a) are indicative of severe P limitation. Activities of other enzymes (B-D-glucosidase and protease) involved in carbon and nitrogen cycling are low in interior areas, suggesting that microbial enzyme production is nutrient limited (Wright and Reddy, 2001a).

Heterotrophic microbial activities in the soil and detrital layers have been shown to be nutrient limited in different parts of the interior Everglades (Amador and Jones, 1993, 1995; Drake et al., 1996; DeBusk and Reddy, 1998; Wright and Reddy, 2001b). Aerobic heterotrophic microbial activity in the interior sites dominated by sawgrass was 64 mg $CO_2$–C kg$^{-1}$ h$^{-1}$ in the detrital layer and decreased to 49 mg $CO_2$–C kg$^{-1}$ h$^{-1}$ in the 0 to 10-cm soil layer. Under anaerobic conditions, rates decreased to 48 and 25 mg $CO_2$–C kg$^{-1}$ h$^{-1}$ in detritus and 0 to 10-cm soil, respectively (Wright and Reddy, 2001b). Rates of all microbially mediated reactions involved in nitrogen cycling are low in sawgrass stands in the interior of WCA-2A (Reddy et al., 1999). Low heterotrophic microbial activity and slow rates of associated biogeochemical processes result in slow rates of organic matter turnover and nutrient release, thus maintaining oligotrophic conditions in the marsh interior. Under resource-limited conditions, nutrients are rapidly assimilated as they are released during decomposition. Thus, nutrients are held in tight cycles and utilized within the system, suggesting that microbial immobilization of nutrients is the dominant process under background conditions.

## PERIPHYTON

Aquatic vegetation and other submerged surfaces in the Everglades interior are covered with periphyton, a community of algae, bacteria, and other microorganisms. As in other wetlands, Everglades periphyton exhibits three growth forms: (1) epipelon (overlying the soil surface), (2) epiphyton (attached to the stems of rooted vegetation), and (3) metaphyton (growing in the water-column or at the water surface, sometimes in association with other floating vegetation such as *Utricularia purpurea*). All three forms of periphyton are abundant in oligotrophic areas of the Everglades and account for a significant portion of marsh primary productivity. Periphyton represent an important habitat for invertebrate populations and, along with macrophyte detritus, forms the base of the Everglades food web (Browder et al., 1994; Rader, 1994). These mats store large amounts of P (approaching 1 kg TP m$^{-2}$) and, thus, may play a critical role in maintaining low P concentrations in reference areas of the marsh (McCormick et al., 1998; McCormick and Scinto, 1999).

Periphyton abundance and productivity exhibit predictable spatial and seasonal patterns in the oligotrophic Everglades. Periphyton typically accounts for much of the vegetative biomass and primary productivity in sloughs and wet prairies (Wood and Maynard, 1974; Browder et al., 1982; McCormick et al., 1998), habitats characterized by sparse emergent macrophyte cover and high light penetration to the water surface. Periphyton productivity is low in sawgrass stands in these same oligotrophic areas due to reduced light availability (Grimshaw et al., 1997; McCormick et al., 1998). Consequently, sawgrass stands are characterized by low DO and a predominance of heterotrophic activity in the water column compared with periphyton-dominated sloughs (Belanger et al., 1989; Rader, 1994).

Periphyton biomass and productivity peak towards the end of the wet season (August through October) and reach a minimum during the colder months of the dry season (January through March). Periphyton biomass in open-water habitats can exceed 1 kg m$^{-2}$ during the wet season (Wood and Maynard, 1974; Browder et al., 1982; McCormick et al., 1998) when floating mats can become so dense as to cover the entire water surface (Color Figure 3.1*). Periphyton growth rates in open-water habitats are as much as 20-fold higher during the wet season compared with the dry season (Swift and Nicholas, 1987; McCormick et al., 1996). Seasonal fluctuations in periphyton gross primary productivity are similar but less dramatic, ranging between 3 and 8.5 g carbon (C) m$^{-2}$ d$^{-1}$ during the wet season compared with 1.6 to 6.7 g C m$^{-2}$ d$^{-1}$ during the dry season (Browder et al., 1982; Belanger et al., 1989; McCormick et al., 1997, 1998).

The chemical composition of periphyton in the oligotrophic Everglades is indicative of severe P limitation. Periphyton samples from reference areas of major hydrologic units within the Everglades are characterized by an extremely low P content (generally <0.05%) and extremely high N:P ratios (generally >60:1 w:w). This observational evidence for P limitation is supported by experimental fertilization studies that have shown that (1) periphyton responds more strongly to P enrichment than to enrichment with other commonly limiting nutrients such as nitrogen (Scheidt et al., 1989; Vymazal et al., 1994); (2) periphyton changes in response to experimental P enrichment mimic those that occur along nutrient gradients in the marsh (McCormick and O'Dell, 1996). Thus, it is well established that periphyton is strongly P limited in reference areas of the Everglades.

Oligotrophic areas of the Everglades contain a characteristic periphyton flora adapted to low P availability and the ionic content of the surface water in a particular area (Swift and Nicholas, 1987). Mineral-rich waters, such as those found WCA-2A and Taylor River Slough (ENP), support a periphyton assemblage dominated by a few species of calcium-precipitating cyanobacteria and diatoms. This assemblage appears to be favored by waters that are both low in P and at or near saturation with respect to calcium carbonate (CaCO$_3$) (Gleason and Spackman, 1974), the latter condition reflecting the influence of the limestone geology of the region. In contrast, the interior

---

* Color figures follow page 648.

of the LNWR contain an assemblage of desmid green algae and diatoms adapted to the extremely low mineral content of waters in this marsh. Waters across much of the southern Everglades (WCA-3A, Shark River Slough) tend to be intermediate with respect to mineral content and contain some of the taxa from both of the above assemblages.

## VEGETATION

Located in the transition zone between temperate and tropical areas, the Everglades flora has many representatives from these two areas (39% and 61%, respectively) (Ewel, 1986; Gunderson, 1994), as well as taxa that are endemic to the region (Long, 1974). The vegetation communities characteristic of the pristine Everglades are dominated by species adapted to low P, seasonal patterns of wetting and drying, and periodic natural disturbances such as fire, drought, and occasional freezes (Davis, 1943; Parker, 1974; Steward and Ornes, 1983; Duever et al., 1994). Major wetland habitats that have been altered by P enrichment include sawgrass marshes, wet prairies, and sloughs (Loveless, 1959; Gunderson, 1994). The spatial arrangement of these habitats is constantly changing as a result of temporal and spatial variation in environmental factors such as fire, water depth, nutrient availability, and local topography (Loveless, 1959).

Sawgrass is the dominant macrophyte in the Everglades, and stands of this species compromise approximately 65 to 70% of the total vegetation cover of the Everglades (Loveless, 1959). Two types of sawgrass marshes have been identified in the Everglades interior: (1) dense stands of tall plants, and (2) sparse stands of short plants (Loveless, 1959; Wood and Tanner, 1990; Gunderson, 1994; Miao and Sklar, 1998). Tall, dense stands are generally monotypic, whereas sparse stands can be mixed with a variety of other sedges, grasses, herbs, and attached emergent or floating aquatic plants. The distribution of these two types of sawgrass marshes may be determined by the combined effects of soil type, nutrient availability, and fire frequency.

Wet prairies include a collection of low-stature, graminoid marshes occurring on both peat and marl soils (Gunderson, 1994). Dominant macrophyte taxa in these habitats include *Rhynchospora*, *Panicum*, and *Eleocharis* (Loveless, 1959; Craighead, 1971). Wet prairies over marl occur in the southern Everglades on the east and west margins of the Shark River Slough and Taylor River Slough, where bedrock elevations are slightly higher and hydroperiods shorter (Gunderson, 1994).

Sloughs are deeper water habitats that remain wet most or all of the year and are characterized by floating macrophytes such as fragrant white water lily (*Nymphaea odorata*), floating hearts (*Nymphoides aquaticum*), and spatterdock (*Nuphar advena*) (Loveless, 1959; Gunderson, 1994). Submerged aquatic plants, primarily bladderworts (*Utricularia foliosa* and *U. purpurea*, in particular), also can be abundant in these habitats and, in the case of *U. purpurea*, provide a substrate for the formation of dense periphyton mats.

Historically, cattail (*Typha* spp.) was one of several minor macrophyte species native to the Everglades marsh (Davis, 1943; Loveless, 1959). In particular, cattail is believed to have been associated largely with areas of disturbance such as alligator holes and recent burns (Davis, 1994). Analyses of Everglades peat deposits reveal no evidence of cattail peat, although the presence of cattail pollen indicates its presence historically in some areas (Gleason and Stone, 1994; Davis et al., 1994; Bartow et al., 1996). Findings such as these confirm the historical presence of cattail in the Everglades but provide no evidence for the existence of dense cattail stands covering large areas (Wood and Tanner, 1990) as now occurs in the northern Everglades. In contrast, sawgrass and water lily peats have been a major freshwater component of Everglades histosols for approximately 4000 years (McDowell et al., 1969).

## FAUNA

Everglades fauna comprises a diversity of animals that depend on marsh primary production and ranges from microscopic invertebrates to top predators such as wading birds and alligators. While these organisms do not respond directly to increased P loading, they are affected by P-related changes in periphyton and vegetation and associated habitat modifications.

Aquatic invertebrates (e.g., insects, snails, crayfish) represent a key intermediate position in energy flow through the Everglades food web as these taxa are the principal consumers of marsh primary production and, in turn, are consumed by vertebrate predators. The macroinvertebrate fauna of the Everglades is fairly diverse (approximately 200 taxa identified) and is dominated by Diptera (49 taxa), Coleoptera (48 taxa), Gastropoda (17 taxa), Odonata (14 taxa), and Oligochaeta (11 taxa) (Rader, 1999). Most studies have focused on a few conspicuous species (e.g., crayfish and apple snails) considered to be of special importance to vertebrate predators, and relatively little is known about the distribution and environmental tolerances of most taxa. An assemblage of benthic micro-invertebrates (meiofauna) dominated by Copepoda and Cladocera is also present in the Everglades (Loftus et al., 1986), but even less is known about the distribution and ecology of these organisms.

Invertebrates occupy several functional niches within the Everglades food web; however, most taxa are direct consumers of periphyton and/or plant detritus. For example, almost 80% of the invertebrates collected from interior sloughs in WCA-2A were classified as consumers of either one or both of these food resources (Rader and Richardson, 1994). In contrast, relatively few taxa consume living macrophyte tissue. Rader (1994) sampled both periphyton and macrophyte habitats in this same area and, based on the proportional abundance of different functional groups, suggested that grazer (periphyton) and detrital (plant) pathways contributed equally to energy flow in the pristine Everglades food web.

Invertebrates are not distributed evenly among Everglades habitats but instead tend to be concentrated in periphyton-rich habitats such as sloughs. In an early study, Reark (1961) noted that invertebrate densities in ENP were higher in periphyton habitats compared with sawgrass stands. Rader (1994) reported similar findings in the northern Everglades and found mean annual inverte-brate densities to be more than sixfold higher in sloughs than in sawgrass stands. Invertebrate assemblages in sloughs were more species rich and contained considerably higher densities of most dominant invertebrate groups (Figure 3.4). Functionally, slough invertebrate assemblages contained similar densities of periphyton grazers and detritivores, compared with a detritivore-dominated assemblage in sawgrass stands. Higher invertebrate densities in sloughs were attributed primarily to abundant growths of periphyton and submerged vegetation, which provide oxygen and a source of high-quality food.

The Everglades fish community in spikerush and sawgrass habitats contains about 30 species dominated by killifishes, livebearers, and juvenile sunfishes (Loftus and Eklund, 1994). Species in deeper open-water alligator holes include Florida gar, yellow bullhead, and adult sunfishes. A comparison of fish biomass estimates between the Everglades and those of other freshwater marshes revealed that the Everglades has some of the lowest values (Turner et al., 1999). Averaged across three to five seasons and nine sites, average fish biomass was 0.61 g m$^{-2}$ in the Everglades. Published mean values for other wetlands ranged from 1.4 to 513 g m$^{-2}$ (Turner et al., 1999).

The breeding bird community in the central Everglades has fewer species than those in more northern wetlands (Brown and Dinsmore, 1986) or in Texas coastal marshes (Weller, 1994). The average number of species in the Everglades was 2.3 per site, and the number of individuals averaged 4.3 per site (Gawlik and Rocque, 1998). In contrast to the depauperate breeding bird community, the Everglades does support a large number of winter residents and may provide critical habitat for many species of trans-gulf migrants that winter in the tropics.

Wading birds are one group of birds that were historically very abundant in the Everglades as compared to other regions. Populations of some species are reported to have declined 90% since the 1930s (Ogden, 1994). Loss of habitat and changes in hydrology are two of the most often cited reasons for the declines. It is thought that even though the Everglades is an oligotrophic system it was able to support large numbers of top predators because of seasonal drydowns in water levels, which concentrated fish and other aquatic prey items from large areas into small pools of receding water (Kushlan, 1986).

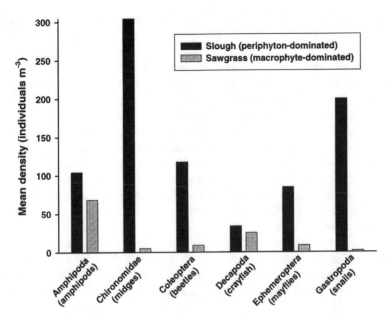

**FIGURE 3.4**    Densities of dominant groups of macroinvertebrate in periphyton-dominated (sloughs) and macrophyte-dominated (sawgrass stands) habitats in interior, oligotrophic areas of WCA-2A. (Adapted from Rader, 1994.)

## PATTERNS OF P ENRICHMENT IN THE MARSH

While atmospheric deposition remains the primary source of P for the Everglades, canal inputs have contributed an increasing percentage of annual P inputs to all areas of the Everglades in recent decades (Fitz and Sklar, 1999). These discharges have been associated with distinct zones of enrichment in several areas of the marsh.

### MARSH WATER-COLUMN TP CONCENTRATIONS

Canal discharges into the Everglades are elevated in P and other elements (see McCormick et al., 1996) compared with interior areas of the marsh. Changes in marsh P concentrations have been documented downstream of these discharges in several parts of the Everglades. As described below, the degree and spatial extent of P enrichment varies among areas depending on the source and location of inflows, topography, and presence of interior canals.

### WCA-2A

The largest database of marsh P exists for WCA 2A, where spatially intensive sampling has been conducted during the past two decades. Canal waters originating from Lake Okeechobee and the Everglades Agricultural Area (EAA) enter this marsh through the S10 structures located along the northern levee (Figure 3.5A) and flow southward to create a P gradient that currently extends as far as 7 km into the marsh (McCormick et al., 1996; Smith and McCormick, 2001). Smaller volumes of EAA drainage water are discharged into the southwestern portion of this marsh through the S7 pump station and are also associated with P enrichment in that area of the marsh.

Background TP concentrations in WCA-2A are illustrated by data collected from five sampling stations in the marsh interior (Figure 3.5B). Mean TP concentrations near 10 μg L$^{-1}$, and median concentrations between 7 and 8 μg L$^{-1}$ are maintained throughout this area. The most complete data set for this marsh exists for station U3 and includes several outliers, samples with extremely high TP concentrations. Many of these high TP values likely resulted from sample contamination at low water depths and, therefore, may not reflect typical water-column concentrations in this area of the marsh.

Water chemistry changes along the nutrient gradient south of the S10s currently are being monitored with a network of 15 fixed sampling stations located 0 to 14 km downstream of canal discharges (Figure 3.5A). McCormick et al. (1996) summarized patterns of P enrichment among these stations during 1994 and 1995 (Figure 3.5C). The mean TP concentration during this period was 104 μg L$^{-1}$ immediately downstream of canal discharges compared with mean concentrations of ≤11 μg L$^{-1}$ at interior stations >8 km downstream. Declines in soluble reactive P, an indicator of bioavailable P, were of a similar magnitude and averaged 48 μg L$^{-1}$ just downstream of the canal compared with ≤4 μg L$^{-1}$ (detection limit) at interior marsh stations. These declines in water-column P concentrations were substantially higher than could be explained by dilution alone and reflected biological and chemical removal of this limiting nutrient by the soils and marsh biota as discussed elsewhere in this chapter.

Long-term changes in water-column TP in this marsh between 1978 and 1997 were assessed using data from 49 marsh stations (including those described above) that had been sampled for varying periods of time during the past two decades (Smith and McCormick, 2001). Total P concentrations, both in canal discharges and the marsh, generally increased through the mid-1980s

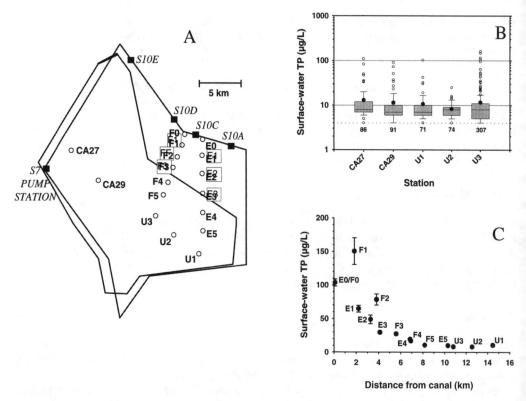

**FIGURE 3.5** (A) Permanent marsh and canal (E0 and F0) stations currently being sampled to document water chemistry in the marsh interior and along a nutrient-enrichment gradient in WCA-2A. Major inflow structures (■) are italicized. (B) Average water-column TP concentrations and ranges for sampling stations in the marsh interior (see Figure 3.2 for interpretation of box plots). (C) Mean (±1 SE) water-column TP concentrations at stations downstream of the S10s during a 20-month period between 1994 and 1995. (From McCormick, P.V. et al., *J. N. Am. Bentholog. Soc.*, 15, 433, 1996. With permission.)

and then decreased into the early 1990s. These patterns are consistent with trends detected for canal discharges across the Everglades (Walker, 1999), suggesting a linkage to changes at the watershed scale (e.g., implementation of agricultural best management practices, weather patterns). In the marsh, this trend was inversely correlated with marsh stage and rainfall, both of which were low during drought years in the mid- and late 1980s, and relatively high in the late 1970s and 1990s. Thus, changes in the P gradient may be influenced by interannual changes in marsh hydrology. Results of this analysis also indicated that temporal changes in the marsh varied as a function of distance from the canal. Specifically, movement of a water-column P front into the marsh during the 1980s (generally low water years) and a recession during the early 1990s (higher water years) were evident at sampling stations closer to the canal, whereas stations further into the marsh showed little if any decline during the 1990s and did not approach the 1970s concentrations (Figure 3.6).

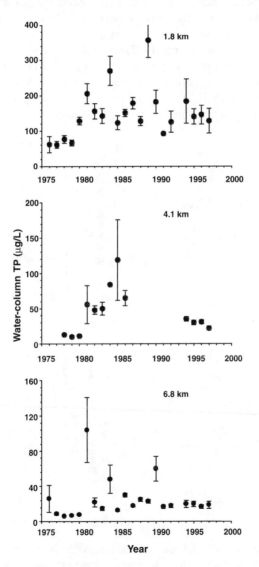

**FIGURE 3.6**   Interannual variation in water-column TP concentrations (mean ±1 SE) at three long-term sampling stations located 1.8, 4.1, and 6.8 km downstream of the S10s. (Based on Smith and McCormick, 2001.)

## Loxahatchee National Wildlife Refuge

The LNWR is exposed to the same EAA drainage that has caused extensive P enrichment in WCA-2A. However, whereas these discharges enter WCA-2A as sheet flow, intrusions of P-enriched waters into the LNWR generally are restricted to the marsh perimeter. Data collected from a network of 14 SFWMD sampling stations in the marsh illustrate this pattern (Figure 3.7A). These stations, all of which are located >1 km into the marsh, generally maintain water-column TP concentrations <10 µg L$^{-1}$ (Figure 3.7B).

Changes within perimeter areas of the LNWR were documented monthly by the SFWMD beginning in April 1996 at nine marsh stations and two canal stations in proximity to the S6 pump station (Figure 3.7A). Water chemistry changes in this area reflect the combined effects of S5A and S6 discharges. As documented in WCA-2A, water-column P concentrations decrease exponentially with increasing distance from the canal (Figure 3.7C). However, the gradient in the LNWR is rather steep, indicating that canal waters seldom intrude far into the marsh. Mean TP concentrations between April 1996 and October 1997 were 44 and 51 µg L$^{-1}$ at the two canal stations compared with concentrations around 10 µg L$^{-1}$ at marsh sites >2 km from the canal. Higher mean TP at the most interior site (Z4) during 1997 was the result of a single extreme measurement (130 µg L$^{-1}$) recorded on March 25, 1997. Soluble reactive P averaged between 15 and 19 µg L$^{-1}$ in canal waters, and decreased to ≤4 µg L$^{-1}$ in the marsh interior.

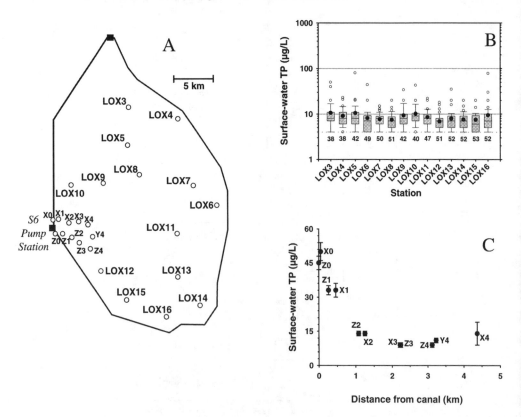

**FIGURE 3.7**    (A) Permanent marsh and canal (X0 and Z0) stations currently being sampled to document water chemistry changes in the marsh interior and along a nutrient-enrichment gradient in the LNWR. Major inflow structures (■) are italicized. (B) Average water-column TP concentrations and ranges for sampling stations in the marsh interior sampled since 1993 (see Figure 3.2 for interpretation of box plots). (C) Mean (±1 SE) water-column TP concentrations at stations in proximity to the S6 pump station during the period between April 1996 and October 1997.

## WCA-3A

This area is bisected by the Miami Canal, which serves as a conduit for canal waters discharged through the S8 pump station (Figure 3.8A). The presence of this canal has caused northern portions of this marsh to become severely overdrained but has reduced the flow of P-enriched canal waters into these same areas. Enrichment in northern WCA-3A is limited to areas immediately adjacent to the Miami Canal, particularly near the S339 and S340 structures, where water is detained and spills over into the marsh. Similarly, urban and agricultural drainage into southeastern WCA-3A through S9 tends to be diverted away from the marsh by interior canals. Other sources of agricultural drainage water include the S11 and S150 structures in the north, which discharge from WCA-2A and the EAA, respectively, and the L28 interceptor canal, which drains farmland to the west.

Relatively few data are available to characterize P gradients in WCA-3A. Surface waters in the marsh interior are generally low in P as illustrated for SFWMD sampling stations that are distant from inflows (Figure 3.8B). Sampling downstream of selected water-control structures (S9 and S339) in 1997 indicated modest TP gradients extending as far as 3 km into the marsh (Figure 3.8C). However, the full extent of P enrichment within WCA-3A has not been characterized as thoroughly as for WCA-2A and the LNWR.

## EVERGLADES NATIONAL PARK

Water enters ENP in three principal locations (Figure 3.9A): (1) Shark River Slough through the S12 and S333 structures located along Tamiami Trail; (2) Taylor River Slough through S332; and

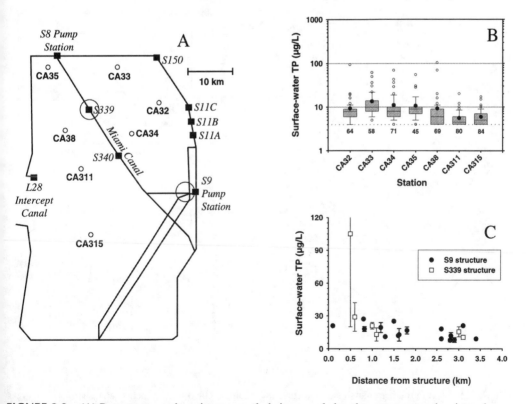

**FIGURE 3.8** (A) Permanent marsh stations currently being sampled to document water chemistry changes in the marsh interior in WCA-3A. Major inflow structures (■) are italicized. B. Average water-column TP concentrations and ranges for sampling stations in the marsh interior sampled since 1994 (see Figure 3.2 for interpretation of box plots). (C) Mean (±1 SE) water-column TP concentrations at stations in proximity to the S9 and S339 structures (circled areas) on two sampling events during the fall of 1997.

(3) the C111 canal in the east. The source of these inflows include WCA-3A (S12s) and a network of canals draining agricultural and urban lands in the east (S332, C111). Phosphorus concentrations in these inflows are considerably lower than those entering the northern Everglades but still tend to be elevated compared with sampling stations in the interior of ENP, which exhibit some of the lowest water-column TP concentrations in the Everglades (Figure 3.9B). Phosphorus loads and flow-weighted TP concentrations in waters entering Shark River Slough are fivefold and nearly twofold higher, respectively, than those released into Taylor River Slough and the C111 basin (Rudnick et al., 1999). As for the northern Everglades, TP concentrations generally peaked during the drought years of the mid-1980s and have declined during the 1990s (Walker, 1999).

The network of marsh sampling stations in ENP is less extensive than in northern areas of the Everglades and, consequently, patterns of enrichment are more difficult to define. Data collected downstream of the S12 structures during 1997 showed no discernible gradient in water-column TP (Figure 3.9C). However, loads have increased in past years as indicated by elevated soil TP concentrations downstream of these inflows (Raschke, 1993). Temporal and spatial changes in water-column P concentrations in different parts of ENP are discussed in greater detail by Walker (1997) and Rudnick et al. (1999).

## Marsh Soil and Porewater P Concentrations

Since the early 1990s, an extensive soil coring effort has produced soil nutrient maps for various regions in the Everglades. These maps show that TP concentrations in soils near canals or other

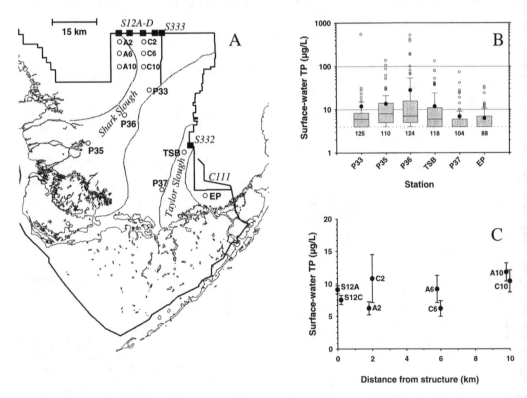

**FIGURE 3.9**  (A) Permanent marsh stations currently being sampled to document water chemistry changes in the marsh interior of freshwater areas of ENP. Major inflow structures (■) are italicized. (B) Average water-column TP concentrations and ranges for sampling stations in the marsh interior (see Figure 3.2 for interpretation of box plots). (C) Mean (±1 SE) water-column TP concentrations at stations downstream of the S12 structures during 1997.

management structures are more than twofold higher than concentrations in interior areas (Figure 3.10) (Koch and Reddy, 1992; DeBusk et al., 1994; Reddy et al., 1994a,b; Newman et al., 1997). In the LNWR and WCA-2A, elevated TP concentrations are associated primarily with increased P loading from canal waters. By contrast, increased soil TP in Holeyland, Rotenberger, and northern WCA-3A largely appear to be a function of overdrainage of these soils with resultant soil compaction and nutrient concentration as illustrated by increased soil bulk densities and elevated nutrients (Newman et al., 1998).

Analysis of spatial soils data from the LNWR, WCAs, and Holeyland suggests that the influence of P loading on Everglades soils is generally restricted to a distance of ~5 km from inflow structures or canals (Reddy et al., 1998). In most cases, these spatial soils data represent a single point in time; however, an intensive study of WCA-2A has resulted in the establishment of temporal as well as spatial responses to external P loads. Total P concentrations in surficial soils in WCA-2A have increased over threefold since the 1970s. Soil cores encompassing the surface 0- to 10-cm soil depths were collected at sites 1.6, 3.2, and 6.4 km south of the Hillsboro Canal in 1975 and 1976 (Davis, 1989). Total P concentrations in soils within 1.6 km ranged from 420 to 440 mg

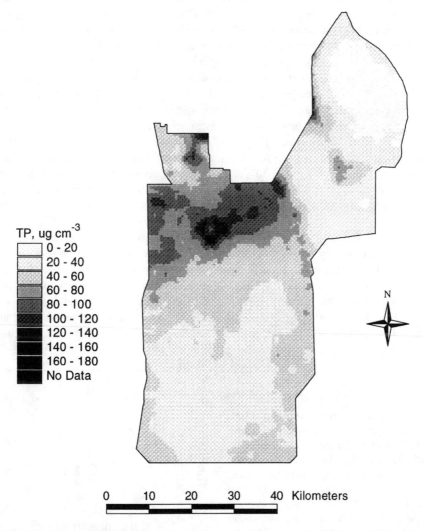

TP, $\mu$g cm$^{-3}$
- 0 - 20
- 20 - 40
- 40 - 60
- 60 - 80
- 80 - 100
- 100 - 120
- 120 - 140
- 140 - 160
- 160 - 180
- No Data

N

0    10    20    30    40  Kilometers

**FIGURE 3.10**  Distribution of TP content in the surface soils throughout the WCAs and Holeyland.

TP kg⁻¹, while cores at distances 3.2 and 6.4 km from inflow ranged between 310 and 340 mg TP kg⁻¹ (Davis, 1989). Soils collected at the same depth from similar locations in 1990 showed that TP concentrations were >1500 mg kg⁻¹ at a distance of 1.4 km, 1100 mg kg⁻¹ at 3.5 km, and only decreased to values consistently <400 mg kg⁻¹ at sites more than 9 km from the canal (Reddy et al., 1991; Koch and Reddy, 1992). These soils samples were not collected in identical locations nor analyzed using identical methods; therefore, differences in accuracy or precision certainly may exist. However, based on the magnitude of the change, it is still apparent that both soil TP concentrations in WCA-2A and the area influenced by external P loads have increased in recent decades. Recent soil coring indicates continuing enrichment between 1990 and 1996 in soils 2 to 7 km from the canal compared with no increase in soils >7 km from these inflows (Figure 3.11) (Reddy et al., 1998). The spatial extent of P enrichment also increased in the last decade. Geostatistical analyses and a comparison of 1990 spatial soils data with that collected by in 1998 showed that in 1998 over 73% of WCA-2A had soil TP concentrations >500 mg kg⁻¹, compared to 48% observed in 1990 (DeBusk et al., 2001).

In general, the effect of P loading is restricted to the surface 30 cm of soil depth (Reddy et al., 1998). Vertical gradients of TP within the soils revealed that TP was highest in the surface soils and decreased with soil depth, with the steepest vertical gradient at sites closest to canal inflows (Koch and Reddy, 1992; Reddy et al., 1998). External P loading also has increased the rate of soil accumulation in areas so enriched. The vertical accretion of peat has been estimated by measuring the location of the ¹³⁷Cs peak within a soil depth profile. The ¹³⁷Cs peak corresponds to the soil surface in 1964; the average post-1964 accumulation is then calculated as the depth of soils to the peak divided by the difference between the soil collection date and 1964. Peat accretion rates in WCA-2A reached a maximum of 1.1 cm y⁻¹ at a distance of 0.3 km from inflow and decreased logarithmically with distance to less than 0.25 cm y⁻¹ in interior areas of the marsh (Craft and Richardson, 1993a,b; Reddy et al., 1993). These peat accretion rates produced corresponding P accumulation rates of 0.46 to 1.1 g m⁻² y⁻¹ in cattail-dominated (i.e., enriched) soils and 0.06 to 0.25 g m⁻² y⁻¹ in unenriched soils. In the LNWR, soil accumulation rates range from 0.07 to 0.42 cm y⁻¹ in unenriched and enriched soils, while much lower accumulation rates were observed in the northern end of WCA-3A (0.04 to 0.28 cm y⁻¹) (Craft and Richardson, 1993a, 1998; Robbins et al., 1996).

The influence of external P loads on Everglades soil chemistry is a slow process. The component first and most impacted by elevated P loads is the flocculent layer of material resting on the soil surface, which is comprised of unconsolidated plant detritus and/or benthic periphyton. A field P loading experiment conducted in WCA-2A showed elevated P concentrations in the benthic periphyton within one month of the start of P addition (S. Newman, unpub. data). In contrast, increased P concentrations in the surficial (0 to 3 cm) soil layer were observed after one year, and only at the highest loading rate of 12.8 g P m⁻² y⁻¹. Similar results were obtained by Richardson and

**FIGURE 3.11**   Total phosphorus content of soils (0- to 10-cm depth) downstream of the S10s in WCA-2A. Samples were collected July 1990, February and August 1996, and March 1997. Dashed lines show the distance range where significant ($0.050 < p < 0.001$) increases in soil TP were detected. (From Reddy, K.R., Wang, Y., DeBusk, W.F., Fisher, M.M., and Newman, S., *Soil Science Society of America Journal*, 62, 1134, 1998.)

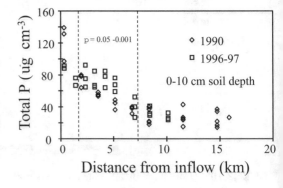

Vaithiyanathan (1995), who found that TP concentrations in the surficial sediment–periphyton layer increased twofold in P-enriched flume channels compared with unenriched controls after two years of dosing, whereas no increase in soil TP was observed. Following the disappearance of the benthic periphyton in response to P loading, it is anticipated that external P loads will penetrate further into the soil profile.

The ability of Everglades soils to act as a sink or source for P is dependent on the forms of P that accumulate. The most conventional approach to the identification of P in soils is through extraction with different chemicals (Newman and Robinson, 1999). Using fractionation procedures, it has been shown that P is primarily stored as organic P, with approximately one third of TP stored as inorganic P (primarily Ca- and Mg-bound P) (Qualls and Richardson, 1995a; Reddy et al., 1998). As discussed previously, peat accreted faster under P-enriched areas; therefore, different P forms also accumulated faster in enriched areas. A comparison of the accumulation of different P forms along the nutrient gradient in WCA-2A revealed that organic P compounds accumulated at rates 7 to 8.2 times higher in enriched than unenriched areas, Ca-bound P accumulated at a rate 6.7 times higher under enriched conditions, while accretion of microbial biomass and Fe/Al-bound inorganic P was two- to threefold higher (Qualls and Richardson, 1995a).

The fractionation procedures also identify forms as labile (i.e., easily transformed or exchanged) or resistant. The labile forms will have a significant influence on the P concentrations in porewater in equilibrium with the soils. As a result, porewater P concentrations follow the same trends as those observed in soils, with an exponential decrease in concentrations with increasing distance from inflow. Unlike soils, however, porewater P concentrations tend to show high seasonal variability at enriched sites but no seasonal variability at unenriched sites (Koch-Rose et al., 1994). Enriched areas of the LNWR and WCA-2A have been shown to have SRP concentrations $>1000$ μg L$^{-1}$ in the surface 0- to 10-cm fraction (Koch and Reddy, 1992; DeBusk et al., 1994; Koch-Rose et al., 1994; Newman et al., 1997).

## ECOLOGICAL RESPONSES TO P ENRICHMENT

### Soil Microbes and Biogeochemical Processes

Increased nutrient loading to the Everglades has resulted in a gradient in the quality and quantity of organic matter, rates of nutrient accumulation, microbial biomass and community composition, and biogeochemical cycling downstream of canal discharges. This gradient has been documented in WCA-2A, WCA-1, Holeyland, and WCA-3A (Reddy et al., 1998). In reference areas of the Everglades, P is the primary factor limiting the microbial processes that control decomposition and nutrient cycling rates. Compared with reference areas of the marsh, nutrient-enriched areas are characterized by the rapid turnover of organic matter and by open elemental cycling, where nutrient inputs often exceed demand (Reddy et al., 1999). These changes have important environmental and ecological consequences including: (1) a conversion from a P-limited to an N-limited system because of high P availability and increased biological demand for N, and (2) an accumulation of low N:P ratio detritus and accelerated rates of decomposition and nutrient cycling. Many biogeochemical processes that affect plant productivity and water chemistry are accelerated by P enrichment, resulting in the release of other plant nutrients such as N. The accumulation of P and other nutrients in the soils and biota coupled with accelerated cycling rates may maintain eutrophic conditions in already enriched areas for some time following load reductions.

The most extensive measurements of biogeochemical processes have been conducted along the nutrient gradient in WCA-2A. All forms of P in the litter and surface (0- to 10-cm) soil layers increased with increasing P enrichment in WCA-2A (Reddy et al., 1999). Both the C:P ratio and potentially mineralizable organic P (PMP) of litter and soils were elevated at enriched sites compared with reference areas in the marsh. The effects of P enrichment on the size of the soil microbial pool has also been investigated. For example, in the litter layer of WCA-2A soils, microbial biomass C was approximately sixfold higher in enriched areas compared with reference locations (Reddy et al., 1999). Both, microbial C and N were found to be higher in soils with elevated TP concentrations along the 10-km transect (DeBusk and Reddy, 1998; White and Reddy, 2000). This hypothesis was further tested using the results of a year-long P-dosing mesocosm experiment conducted in an open slough in unenriched soil (McCormick and O'Dell, 1996). Surface soils were found to have significantly higher microbial biomass C and N with increased P-dosing rates, lending additional support that the microbial pool was P limited (White and Reddy, 2000). The microbial biomass P pool was also affected by the allochthonous P enrichment. Significantly higher microbial biomass P was seen at P-enriched stations in WCA-2A compared with the marsh interior (Reddy et al., 1999). An increase in the size of the microbial pool in response to nutrient enrichment can potentially affect nutrient biogeochemical cycling and ecosystem productivity.

The microbial community controls key ecosystem processes including nutrient cycling. Therefore, many microbially mediated processes that control C, N, and P dynamics and ultimately affect water quality, sediment chemistry, and plant productivity are accelerated. Addition of P to unenriched Everglades soil (230 mg P kg$^{-1}$) stimulated microbial respiration, measured as organic C mineralization, and resulted in a shift towards anaerobic respiration (Bachoon and Jones, 1992; Amador and Jones, 1993). Correspondingly, enriched soils contained between $10^3$- and $10^4$-fold higher numbers of anaerobes, including methanogens, sulfate reducers, and acetate producers, than soils in reference areas (Drake et al., 1996). Addition of electron acceptors such as $O_2$, $NO_3^-$, and $SO_4^{2-}$ also accelerated microbial respiration in WCA-2A soils, indicating that microbial activity is limited by the availability of suitable electron acceptors as well as P (Wright and Reddy, 1998). Consumption of these electron acceptors during microbial respiration was higher in soils collected from P-enriched area of WCA-2A than in reference areas (Fisher, 1997), providing further evidence of the stimulation of various microbial pathways by surface water P inputs.

Microbes produce a wide range of extracellular enzymes that catalyze the decomposition of organic matter (Sinsabaugh, 1994). Among the enzymes measured in detrital and soil layers in WCA-2A, alkaline phosphatase activity (APA), an indicator of P mineralization, decreased with P enrichment, while B-D-glucosidase, a measure of C mineralization, increased (Wright and Reddy, 1996). Arylsulfatase and phenol oxidase activities were unrelated to P loading. Enzyme activity was highest in the detrital layer and decreased with increasing soil depth. The close relationship between phosphatase activity and canal inflows indicated the controlling influence of P loading on microbial processes (Wright and Reddy, 1996). The use of APA as an indicator of P enrichment is discussed further under periphyton responses (see below).

Increased P loading to the northern Everglades is associated with dramatic changes in the cycling of other biologically important elements such as N. Organic N mineralization occurs through (1) hydrolytic deamination of amino acids and peptides, (2) degradation of nucleotides, and (3) metabolism of methylamines by methanogenic bacteria (King et al., 1983). Potentially mineralizable N (PMN) was estimated to be 3.5- and 1.7-fold higher for litter and 0- to 10-cm soil, respectively, collected from the P-enriched sites of WCA-2A (White and Reddy, 2000). Unenriched soil amended with P also demonstrated significantly higher rates of inorganic N release from these peat soils (White and Reddy 2000). Higher rates of N mineralization were suggested in enriched areas and were attributed to low detritus C:N ratios and P-non-limiting conditions (Koch-Rose et al., 1994).

Nitrification is a key process in the N budget of wetland systems, since the $NO_3$ formed is available for macrophyte and periphtyon growth. More importantly, nitrification provides the substrate ($NO_3^-$) for denitrification, whereby N is lost as gaseous end products (e.g., $N_2O$ and $N_2$) to the atmosphere. Nitrification rates appear to be limited by P in reference areas and are elevated both in detrital and surface soil layers in P-enriched locations (Reddy et al., 1999). Denitrification rates, as indicated by the activity of denitrifying enzymes, are also higher in enriched areas near canal inflows (White and Reddy 1999). However, this response is more likely related to high $NO_3^-$ loads at these inflow points than to P enrichment.

Phosphorus loading to wetlands increases the P concentration of periphyton mats and can result in a shift towards N limitation (see periphyton responses below). Such a shift was suggested along a nutrient gradient in WCA-2A, as rates of biological N fixation in periphyton mats were higher in enriched areas than in the marsh interior (Inglett, 1999).

## PERIPHYTON

Periphyton is sensitive to changes in water-column P concentrations and rapidly accumulates P from the water as it becomes available (Davis, 1982; McCormick et al., 1998). Thus, a strong relationship between P concentrations in the water and in the periphyton is maintained along marsh P gradients downstream of canal inflows (Figure 3.12A) (Grimshaw et al., 1993; McCormick et al., 1996). Field dosing experiments also have shown that periphyton accumulates P in proportion to the loading rate, and that this accumulation can be detected within weeks at loading rates similar to those in highly enriched areas of the Everglades. Periphyton P increased rapidly and by as much as 10-fold within field enclosures receiving weekly pulses of dissolved inorganic P (loading rate = 0.25 g m$^{-2}$ wk$^{-1}$) (Figure 3.12B), while no changes in soil or macrophyte P content were detected even after several months of dosing (McCormick and Scinto, 1999). These findings are consistent with earlier work by Davis (1982), who found that periphyton mats concentrated radiolabeled SRP at a faster rate than either macrophytes or soils on a biomass-specific basis.

Physiological changes in the periphyton mat occur rapidly as internal P concentrations increase. Two physiological responses to P enrichment that have been well documented in the Everglades are a decrease in phosphatase activity and an increase in cell metabolism (McCormick and Scinto, 1999; Newman et al., in press). Phosphatases are enzymes that allow microbes to scavenge P from the surrounding environment. Algal and bacterial production of these enzymes decreases as internal stores of P increase in response to P enrichment. Such decreases have been documented along P gradients in the Everglades (Bush and Richardson, 1995) and in field P-dosing experiments (Newman et al., in press), indicating that the limiting influence of P on periphyton metabolism and growth is reduced near canal inflows.

Relaxation of P limitation stimulates algal photosynthesis and microbial respiration, the processes that allow for the fixation and utilization of energy for growth. The primary productivity of periphyton mats in the oligotrophic interior of WCA-2A increased by as much as threefold within three weeks in response to weekly P additions to field mesocosms (Figure 3.13A) (McCormick and Scinto, 1999). These same changes have been documented along P gradients in this marsh (Figure 3.13B). Similarly, periphyton growth rates on artificial substrata are correlated strongly with increases in water-column TP and can be more than 10-fold higher in highly enriched areas of the marsh compared with the oligotrophic interior (Swift and Nicholas, 1987; McCormick et al., 1996).

Elevated P loads and concentrations result in the loss of species adapted to survival under P-limited conditions and their replacement by species capable of higher growth rates under P-enriched conditions. One of the most pronounced changes involves the loss of the calcareous assemblage of cyanobacteria and diatoms which is seasonally abundant in oligotrophic, mineral-rich waters that cover large areas of the Everglades. This oligotrophic assemblage is replaced by a eutrophic assemblage of filamentous cyanobacteria, filamentous green algae, and diatoms in enriched areas of the marsh (Table 3.1) (Swift and Nicholas, 1987; McCormick and O'Dell, 1996). Surveys of periphyton and water quality throughout the Everglades (Swift and Nicholas, 1987; Raschke, 1993; McCormick et al., 1996; Pan et al., 1997) have found consistently strong correlations between water-column P concentrations and the abundance of several diatom species. Many of these species are recognized as reliable indicators of eutrophication in other freshwater ecosystems

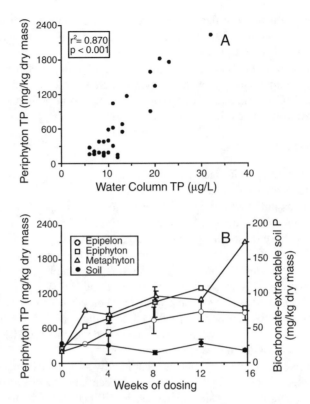

**FIGURE 3.12**   (A) Relationship (Pearson's product–moment correlation coefficient) between water-column TP concentrations and the P content of floating periphyton mats collected downstream of the S10s in WCA-2A. (B) Accumulation of P in periphyton components and soils in response to weekly P additions (0.25 g P m$^{-2}$ wk$^{-1}$) to experimental plots in an oligotrophic slough in the interior of WCA-2A. (From McCormick, P.V. and Scinto, L.J., in *Phosphorus Biogeochemistry of Subtropical Ecosystems*, Reddy, K.R. et al., Lewis Publishers, Boca Raton, FL, p. 301. With permission.)

(e.g., Palmer, 1969; Lowe, 1974; Lange-Bertalot, 1979) and, similarly, can be used to identify areas of the marsh affected by P-enrichment.

Controlled dosing studies have provided experimental evidence that species changes documented downstream of canal inflows result primarily from P enrichment. In an early study in ENP, Flora et al. (1986) found that the assemblage of calcareous cyanobacteria and diatoms indicative of oligotrophic conditions was lost from experimental dosing channels within several weeks in response to P concentrations of ≤20 µg L$^{-1}$ SRP. McCormick and O'Dell (1996) compared taxonomic changes downstream of canal inflows into WCA-2A to those produced by P additions to experimental enclosures in the oligotrophic marsh interior. In the marsh, the calcareous assemblage that existed at low water-column P concentrations (TP = 5 to 7 µg L$^{-1}$) was replaced by a filamentous green algal assemblage at moderately elevated concentrations (TP = 10 to 28 µg L$^{-1}$) and by eutrophic cyanobacteria and diatoms species at even higher concentrations (TP = 42 to 134 µg L$^{-1}$). Taxonomic changes in response to experimental P enrichment were similar to those documented along the marsh gradient (Figure 3.14), thereby providing causal evidence that changes in the periphyton assemblage were largely a product of P enrichment. Similar conclusions were reached independently by Pan et al. (1997), who studied diatom changes along marsh and experimental P gradients in this same marsh.

Whereas increased P loading affects periphyton directly by increasing biomass-specific productivity and favoring species with higher growth rates, other ecological changes caused by P enrichment act to reduce periphyton abundance and its contribution to marsh primary productivity

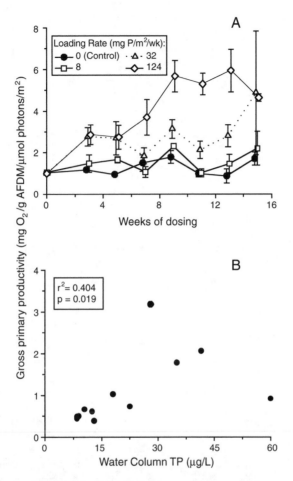

**FIGURE 3.13**   (A) Changes in the productivity of floating periphyton mats in experimental slough plots in WCA-2A exposed to different P loading rates. (From McCormick, P.V. and Scinto, L.J., *Phosphorus Bio-geochemistry of Subtropical Ecosystems*, Reddy, K.R., O'Connor, G.A., and Schelske, C.L., Eds., CRC/Lewis Publishers, Boca Raton, FL, 1999, 301.) (B) The relationship between floating mat productivity and water-column TP downstream of the S10s in WCA-2A (McCormick, unpubl. data).

---

**TABLE 3.1**
**Periphyton Species Indicative of Low and High Phosphorus Availability in the Everglades**

| Group | Low P Availability | High P Availability |
|---|---|---|
| Cyanobacteria (blue-green algae) | *Oscillatoria limnetica, Schizothrix calcicola, Scytonema hofmannii* | *Oscillatoria princeps* |
| Bacillariophyta (diatoms) | *Amphora lineolata, Anomoeoneis serians, Anomoeoneis vitrea, Cymbella lunata, Cymbella turgida, Synedra synegrotesca* | *Amphora veneta, Epithemia adnata, Gomphonema parvulum, Navicula confervacea, Navicula minima, Nitzschia amphibia, Nitzschia fonticola, Nitzschia palea, Rhopalodia gibba* |
| Chlorophyta (green algae) | — | *Spirogyra* spp. |

*Source:* From McCormick, P.V. and Stevenson, R.J., *J. Phycol.*, 34, 726, 1998.

---

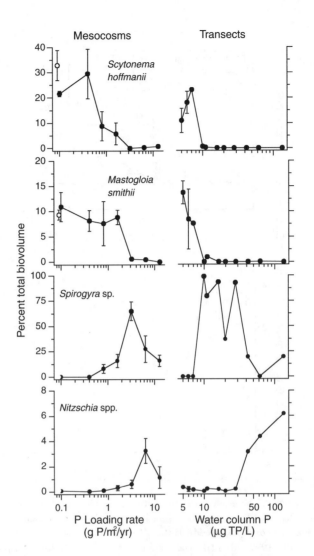

**FIGURE 3.14**    Changes in the relative abundance (percent biovolume) of dominant algal taxa in experimental mesocosm enclosures dosed weekly with different P loads (left panels) and along an enrichment gradient (right panels) in the same marsh (WCA-2A). (From McCormick, P.V. and O'Dell, M.B., *J. N. Am. Bentholog. Soc.*, 15, 450, 1996. With permission.)

in enriched areas. One of the most dramatic effects of P enrichment in the Everglades is to increase the growth and coverage of emergent macrophytes, particularly cattail. Dense cattail stands dominate enriched areas of the marsh and reduce light penetration to levels that inhibit periphyton growth (Grimshaw et al., 1997). Consequently, areal periphyton productivity is sharply lower in enriched areas of the marsh compared with oligotrophic areas (McCormick et al., 1998). These investigators found that productivity in enriched open-water habitats in WCA-2A equaled or exceeded that in oligotrophic open waters (sloughs and wet prairies). However, open water accounted for less than 4% of areal coverage in the enriched marsh compared with 30% in oligotrophic areas of WCA-2A. Periphyton productivity was negligible in the cattail stands that covered more than 90% of the enriched marsh. Consequently, areal periphyton productivity in enriched areas of the marsh averaged six- and 30-fold lower than in oligotrophic areas during the wet and dry seasons, respectively (Figure 3.15). Independent measurements of aquatic community

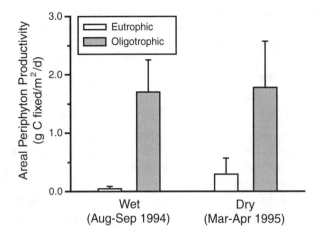

**FIGURE 3.15**    Areal periphyton productivity in oligotrophic and highly enriched areas of WCA-2A during the wet and dry seasons of 1994–1995. Values are means (±1 SE) of two sampling locations in each area and are habitat weighted to account for shifts in the areal coverage of different vegetative habitats caused by enrichment. (See McCormick et al., 1998, for more details.)

metabolism also have found extremely low submerged (including periphyton) productivity in enriched areas (Belanger et al., 1989; McCormick et al., 1997).

Periphyton responses to P enrichment appear to be greatest at relatively low water-column P concentrations, and available evidence suggests a shift away from P limitation in highly enriched areas of the marsh. The relationship between periphyton species composition and P in WCA-2A was strongest in areas of the marsh where water-column TP was <30 µg $L^{-1}$ and mass N:P ratios exceeded 50:1 (McCormick et al., 1996). In this same study, enrichment bioassays indicated that limitation by nutrients other than P, particularly N, occurred periodically at sites with higher TP concentrations. Similarly, Vymazal et al. (1994) documented higher periphyton biomass in marsh plots fertilized with both N and P than in those fertilized with P alone. This shift from P to N limitation may explain why experimental P-enrichment studies have simulated periphyton species changes observed in the marsh at lower (e.g., 10 to 30 µg $L^{-1}$) water-column TP concentrations but fail to reproduce changes observed at much higher concentrations (e.g., McCormick and O'Dell, 1996).

## COMMUNITY METABOLISM AND DISSOLVED OXYGEN CONCENTRATIONS

Phosphorus enrichment causes a shift in the balance between autotrophy and heterotrophy as a result of contrasting effects on periphyton productivity and microbial respiration. Rates of aquatic primary productivity (P) and respiration (R) are approximately balanced (P:R ratio = 1) across the diel cycle in minimally impacted sloughs throughout the Everglades (Belanger et al., 1989; McCormick et al., 1997). In contrast, respiration rates exceed productivity by a considerable margin (P:R ratio << 1) at enriched locations. This change is related primarily to a reduction in areal periphyton productivity coupled with increased detrital inputs that stimulate microbial respiration (such as sediment oxygen demand) (e.g., Belanger et al., 1989).

The shift from autotrophy to heterotrophy with P enrichment, in turn, affects dissolved oxygen (DO) concentrations in enriched areas of the marsh. For example, DO concentrations at an enriched site in WCA-2A rarely exceeded 2 mg $L^{-1}$ compared with concentrations as high as 12 mg $L^{-1}$ at reference locations (Figure 3.16A) (McCormick et al., 1997). Depressed water-column DO concentrations have subsequently been documented at several enriched marsh locations in WCA-2A and the LNWR and confirmed in experimental P-enrichment studies (McCormick and Laing, accepted). Declines in DO along the marsh P gradient were steepest within a range of water-column TP concentrations roughly between 10 and 30 µg $L^{-1}$ (Figure 3.16B). Lower DO in enriched areas of the marsh are associated with other changes, including an increase in anaerobic microbial processes and a shift in invertebrate species composition toward species tolerant of low DO, as described later in this chapter.

## Marsh Vegetation

There is considerable evidence to show that vegetation patterns in the Everglades have been affected by P enrichment. As for periphyton, enrichment initially stimulates the growth of existing vegetation as evidenced by increased plant P content, photosynthesis, and biomass production. Persistent enrichment and/or enrichment above certain concentrations or loads eventually produces a shift in vegetation composition toward species better adapted to rapid growth and expansion under conditions of high P availability. Current understanding of the progression of vegetation

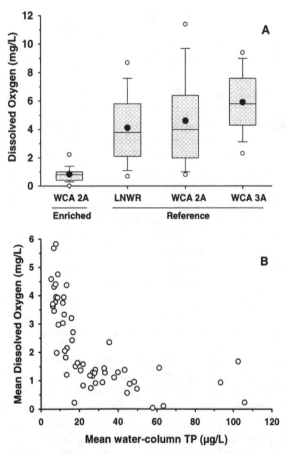

**FIGURE 3.16** (A) Average water-column DO concentrations and ranges at three reference marsh stations in the WCAs and an enriched marsh station in WCA-2A based on periodic diel sampling between 1979 and 1985. Measurements were collected hourly for 1 to 8 consecutive diel cycles using Hydrolab Datasondes® suspended at mid-depth in the water-column. See Figure 3.2 for interpretation of box plots. (B) Changes in mean daily water-column DO along a canal P gradient in WCA-2A. Each point is the mean DO value at 13 sampling stations along the gradient during five sampling periods between 1995 and 1998. Measurements were collected at 15 to 30-minute intervals for three to four consecutive diel cycles using Hydrolab Datasondes® suspended at mid-depth in the water-column. See McCormick and Laing (accepted) for more details. (Part A from McCormick, P.V. et al., *Archive für Hydrobiologie*, 140, 117, 1997. With permission.)

changes associated with P enrichment is based on: (1) elucidation of the life-history strategies of dominant species; (2) patterns of physiological, population, and community change along marsh P gradients; and (3) experimental studies that have documented macrophyte responses to controlled-P enrichment. Unfortunately, experimental data are less conclusive than for periphyton because macrophyte responses take longer to occur and few experiments have been conducted long enough to document noticeable shifts in macrophyte species composition (but see Craft et al., 1995). Current models suggest that time lags between the onset of P enrichment and vegetation responses may be as long as several decades (e.g., Walker and Kadlec, 1996). Thus, much of the evidence for P-related changes in Everglades vegetation is based on correlative and observational evidence from field studies, which have been corroborated by small-scale greenhouse and field experiments that provide mechanistic explanations (e.g., differential changes in germination and growth rates) for these changes.

## P-Limited Nature of Everglades Macrophytes

It is generally accepted that macrophyte communities in the Everglades are P limited. Sawgrass is adapted to the low-P conditions indicative of the pristine Everglades (Steward and Ornes, 1975b, 1983). During field and greenhouse manipulations, sawgrass responded to P enrichment either by increasing the rate of growth or P uptake (Steward and Ornes, 1975a, 1983; Craft et al., 1995; Miao et al., 1997, Daoust and Childers, 1999). Furthermore, additions of N alone had no effect on sawgrass or cattail growth under low-P conditions (Steward and Ornes, 1983; Craft et al., 1995). Recent experimental evidence in the Everglades National Park (Daoust and Childers, 1999) has shown that other native vegetation associations such as wet prairie communities are limited by P as well.

## Changes in Sawgrass Habitats

Sawgrass populations in the Everglades have life-history characteristics indicative of plants adapted to low-nutrient environments (Davis, 1989, 1994; Miao and Sklar, 1998). Compared with cattail, sawgrass plants display slow growth, extended life cycles, low reproductive yield, and an inability to alter biomass allocation (e.g., storage vs. photosynthetic tissues) in response to changes in the resource environment (Table 3.2) (Davis, 1989, 1994; Miao and Sklar, 1998; Goslee and Richardson, 1997). As expected in a P-limited ecosystem, the P concentration of sawgrass tissue increases with increases in soil and water-column P concentrations (Figure 3.17) (Koch and Reddy, 1992; Craft and Richardson, 1997; Miao and Sklar, 1998, Richardson et al., 1997b). This P accumulation corresponded with an increase in plant biomass, P storage, and annual leaf production and turnover rates along a P gradient in WCA-2A (Davis, 1989; Craft and Richardson, 1997; Miao and Sklar, 1998). Population dynamics also are affected by P accumulation as indicated by changes in plant density and size, with a higher density of smaller plants in reference areas and lower densities of larger plants in enriched locations (Miao and Sklar, 1998). In addition, P enrichment enhanced sawgrass seed production by increasing both the yield and the number of seeds produced (Goslee and Richardson, 1997; Miao and Sklar, 1998).

## Changes in Slough/Wet Prairie Vegetation

Sloughs and wet prairies harbor much of the biodiversity and secondary production of the Everglades and provide critical foraging habitats for top predators, such as wading birds (Belanger et al., 1989; Hoffman et al., 1994). Available evidence indicates that these habitats are particularly sensitive to P enrichment and are replaced by cattail stands in enriched areas of the marsh (e.g., Doren et al., 1997; Daoust, 1998). This transition represents a fundamental shift in both community structure and function. Changes occur in two stages: (1) alteration of existing vegetation, and (2) invasion by cattails.

**TABLE 3.2**
**Summary of Differences in Life History Characteristics Between Sawgrass and Cattail**

|  | Sawgrass | Cattail |
|---|---|---|
| *Growth* | | |
| Leaf growth rate | Low | High |
| Leaf turnover rate | Low | High |
| Net production | Low | High |
| Biomass accumulation | Slow | Fast |
| *Physiology* | | |
| Photosynthetic rate | Low | High |
| Stomatal conductance | Low | High |
| Leaf vs. root biomass allocation | Not flexible | Flexible |
| *Reproduction* | | |
| Seed yield (g) | Low | High |
| Seed number | Few | Many |
| Seed size | Large | Small |
| Flowering | March–May | January–February |
| Fruiting | June–August | April–June |
| Timing of dispersal | July–August | May–July |
| *Germination* | | |
| Min. days required for germination | 14–22 days | 2–7 days |
| Germination % | Low | High |
| Germination duration | Long | Short |
| *Morphology and anatomy* | | |
| Air space in leaves | Small | Large |
| Gas transport | Diffusion | Bulk flow ventilation |

*Source:* Data from Chanton et al. (1993); Davis (1989, 1991, 1994); Goslee and Richardson (1997); Miao and DeBusk (1999); Miao and Sklar (1998); Stewart et al. (1997).

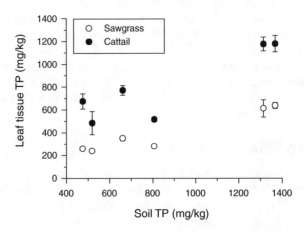

**FIGURE 3.17** Changes in sawgrass and cattail P content (leaf tissue TP) along a canal P gradient in WCA-2A.

Slough habitats in the northern Everglades are characterized by *Nymphaea odorata* and *Utricularia purpurea*. These habitats intermix with wet prairies, which are typically dominated by *Eleocharis* spp. Changes in species composition and biomass and the gradual disappearance of these communities have been documented along a nutrient gradient downstream of the S10 structures in WCA-2A. Satellite imagery has indicated a decline in open-water habitats and a corresponding increase in cattail coverage in enriched areas of the marsh (Rutchey and Vilchek, 1994).

The process of slough enrichment and replacement by cattail indicated by satellite imaging is supported by ground-based sampling methods (McCormick et al., 1999) that have documented changes in slough vegetation and encroachment on these habitats by cattail as far as 7 km downstream of canal inflows into WCA-2A where soil TP concentrations averaged between 400 and 600 mg kg$^{-1}$ (Figure 3.18), and water-column TP averaged as low as 20 µg L$^{-1}$. Whereas *Eleocharis* declined in response to increased soil P, *Nymphaea* was stimulated by enrichment and was dominant in slightly enriched sloughs. Increased occurrence of cattail in sloughs was associated with a decline in *Nymphaea*, probably as a result of increased shading of the water surface. These findings are consistent with those of Vaithiyanathan et al. (1995) who documented a decline in slough habitats along this same nutrient gradient and the loss of sensitive taxa such as *Eleocharis* at locations where soil TP exceeded 700 mg kg$^{-1}$. Patterns of response to P enrichment in slough–wet prairie communities in the southern Everglades are somewhat different from those in the northern Everglades. In a field dosing experiment in Shark River Slough, Scheidt et al. (1989) documented a shift from an *Eleocharis–Utricularia* marsh to one dominated by *Sagittaria* sp. and *Panicum* sp. at an average water-column SRP concentration of 33 µg L$^{-1}$ (5.5 times background concentrations of 6 µg L$^{-1}$).

Vegetation changes documented along marsh nutrient gradients are supported by experimental P-enrichment studies. *Nymphaea* P accumulation and leaf growth rates in field enclosures in WCA-2A increased in response to loads of 6.4 to12.8 g P m$^{-2}$ y$^{-1}$ within 2 years and to loads of 3.2 g P m$^{-2}$ y$^{-1}$ within 3 years (S. Miao, unpub. data). Similarly, experimental P enrichment of an *Eleocharis* wet prairie in ENP resulted in an increase in net above-ground primary productivity and an accelerated rate of biomass turnover (Daoust, 1998). Persistent additions of P resulted in a decline in those species deemed to be sensitive to P in the gradient studies discussed earlier. One year of P additions to slough plots at a loading rate of 4.8 g P m$^{-2}$ y$^{-1}$ resulted in a significant decline in the existing *Utricularia purpurea*–periphyton community (Craft et al., 1995). Similar outcomes have been documented in slough P-enrichment studies in WCA-2A and WCA-3B (Steward and Ornes, 1975a; McCormick and O'Dell, 1996; Newman, unpubl. data). Two years of P additions to experimental channels in WCA-2A resulted in the decline or loss of *U. purpurea*, *Eleocharis cellulosa*, and *Eleocharis vivipara* (Qualls and Richardson, 1995b).

Cattail invasion rarely has been documented in slough enrichment experiments. As discussed below, this lack of experimental confirmation of gradient trends may relate to the relatively short duration of these enrichment studies. For example, cattail became established in P-enriched dosing channels in ENP only after several years following the cessation of dosing (R. Jones, Florida International University, pers. comm.). Enrichment of slough plots in WCA-2B with P also resulted in cattail invasion after existing vegetation had been cleared (Richardson et al., 1995). The preferential pattern of cattail encroachment into open-water habitats is consistent with the life-history characteristics of this species and its response to P enrichment, as determined by experimental manipulations discussed below.

## Changes in Cattail Distribution and Its Relationship to P Enrichment

The southern cattail, *Typha domingensis* (Pers.), a species common to wetlands in warmer climates, is believed to be a natural component of the Everglades ecosystem, occurring largely in scattered diffuse stands (Davis, 1994). However, a rapid increase in the spatial distribution of cattail has been

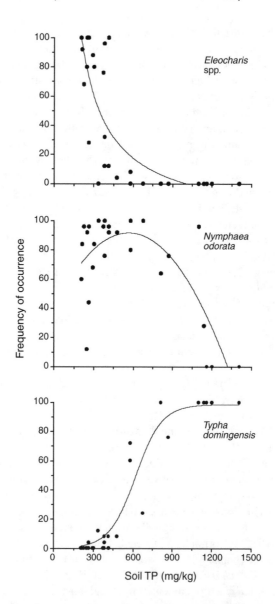

**FIGURE 3.18**    Changes in abundance (frequency of occurrence) of dominant macrophyte species in sloughs along a canal P gradient in WCA-2A in 1997. Frequency of occurrence was measured based on presence or absence in 25 1-m² quadrants along permanent 50-m transects at 27 sampling stations along the gradient. Soil TP values are means of three soil cores taken to a depth of 10 cm at each station.

documented across the Everglades in recent decades (Rutchey and Vilchek, 1994, 1999; Jensen et al., 1995; Newman et al., 1998). The most dramatic expansion has occurred in WCA-2A, where the total area of the landscape impacted by cattail increased from 422 ha in 1991 to 1646 ha in 1995 (Figure 3.19) (Rutchey and Vilchek, 1999).

Several studies have shown that cattail expansion in the Everglades is associated with both fertile and disturbed environments (Davis, 1991; Urban et al., 1993; Craft and Richardson, 1997; Richardson et al., 1997b; Miao and Sklar, 1998; Miao and DeBusk, 1999), and that a combination of elevated nutrients and increased flooding will allow cattail to outcompete sawgrass and slough vegetation (Newman et al., 1996). Vegetation and soils analyses along nutrient gradients show that cattail populations are extensive in areas with elevated soil P concentrations (DeBusk et al., 1994;

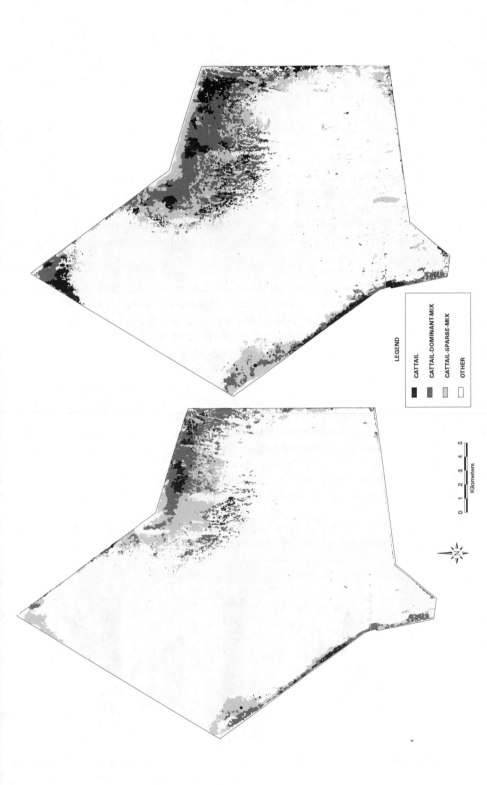

**FIGURE 3.19** Changes in the coverage of cattail within WCA-2A between 1991 and 1995. Coverage estimates are based on ground-truthed aerial photography as described by Rutchey and Vilchek (1999).

Craft and Richardson, 1997; Doren et al., 1997; Newman et al., 1997; Miao and DeBusk, 1999). Using a Markov transition probability model to quantify the dynamics of the rapid cattail expansion in WCA-2A, Wu et al. (1997) suggested that cattail expansion will be accelerated at soil TP concentrations >650 mg kg[-1]. Furthermore, due to high P concentrations already present in soils downstream of the S10 and S7 structures, this model predicts that cattail will continue to expand and occupy approximately 30% of WCA-2A in the next 20 years even in the absence of continued P inputs. These predictions are consistent with modeling efforts by Walker and Kadlec (1996) that indicate a lag between soil P enrichment and cattail invasion.

Field experiments support a close relationship between cattail growth and expansion and P enrichment. Cattail plants transplanted to enriched and unenriched sites and allowed to grow for six months exhibited significantly different growth responses. Plants grown at the enriched site had approximately 170% greater relative growth rate and produced over 10-fold more biomass than those grown at the unenriched site (Miao and DeBusk, 1999). Plants at the enriched site produced an average of approximately seven shoots each, while almost no shoot production occurred at the unenriched site (Figure 3.20). After 2.5 years, the transplanted cattail plants expanded and filled in all open areas (approximately 560 m²) in the enriched site, while no expansion occurred at the unenriched site.

Factors other than P may influence cattail expansion in some areas. For example, cattail expansion was found to be correlated with hydrologic changes in the Holeyland and with severe muck fires in Rotenberger Tract (Newman et al., 1998). However, soils in these areas also were shown to be high in P, on a volumetric basis, prior to cattail expansion. When soil TP was corrected for bulk density and soil depth, both Holeyland and Rotenberger were found to have elevated soil TP compared to other areas of the Everglades. These findings suggest that both areas have sufficient P to support rapid cattail expansion, which may have contributed to a greater initial growth rate for cattail in Holeyland relative to other regions of the northern Everglades (Figure 3.21). Thus, rapid cattail expansion appears contingent upon high P availability. However, while developing this hypothesis, it is recognized that the form of P in the soils will influence its bioavailability. Holeyland soils have inorganic P values twofold higher than any other northern Everglades soils. Also, inorganic P is correlated positively with cattail cover. Inorganic P is taken up directly by higher plants and is readily available to support growth (Marschner, 1986). Organic P, the primary form of P stored in Everglades soils, must be mineralized to inorganic P before it can be utilized for growth. Elevated levels of inorganic P are also associated with the enriched, cattail-dominated zone in WCA-2A (DeBusk et al., 1994).

Cattail is characterized by a high growth rate, a short life cycle, high reproductive output, and other traits that confer a competitive advantage under enriched conditions (Table 3.2) (Davis, 1989,

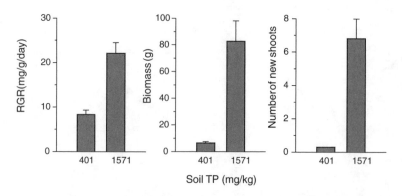

**FIGURE 3.20**   Relative growth rate, biomass production, and the number of new shoots produced by cattail plants transplanted to enriched and unenriched sites in WCA-2A. Bars are means (±1 SE) of measurements taken from 12 plants at each site.

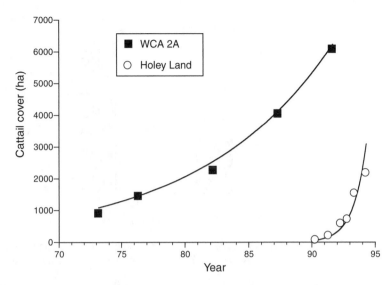

**FIGURE 3.21**    Increases in cattail coverage over time in Holeyland and WCA-2A. (From Newman, S. et al., *Aquatic Botany*, 60, 265, 1998. With permission.)

1994; Goslee and Richardson, 1997; Miao and Sklar, 1998). For example, higher photosynthetic rates, an indicator of potential growth rate, allow cattail to outcompete sawgrass under enriched conditions. Although the two species showed similar photosynthesis rates in unenriched areas, where soil TP averaged near 450 mg kg$^{-1}$, rates for cattail were approximately 47% greater than sawgrass in areas where soil TP concentrations exceeded 500 mg kg$^{-1}$ (Miao and DeBusk, 1999). Higher photosynthetic rates are associated with greater leaf production by cattail plants compared with sawgrass in enriched areas (Davis, 1989).

Experimental studies also indicate that cattail is a competitively superior species under enriched conditions. For example, the rate of regrowth following leaf removal (as might be caused by fire) was similar for cattail and sawgrass grown under unenriched conditions, but was approximately 75% faster for cattail when the two species were grown in P-enriched soils (Figure 3.22) (Miao and DeBusk, 1999). Findings such as these indicate the ability of cattail to recover and expand more quickly than sawgrass following certain types of disturbance (e.g., surface fires, herbivory) in P-enriched areas.

The relationship between P enrichment and cattail expansion has been clouded by the results of field enrichment studies (e.g., Craft et al., 1995; Qualls and Richardson, 1995b; Daoust, 1998), which have not found P enrichment to lead to cattail establishment in sloughs. These studies are relatively short term (<5 years) compared to the history of enrichment in the Everglades (>30 years) and may not span a sufficient timeframe. This raises the question of whether there is a lag time between enrichment and establishment and the factors (e.g., seed dispersal, marsh drying) that might contribute to this lag.

Like most clonal plants, cattail can spread by two methods: (1) seed dispersal and germination, and (2) vegetative growth via rhizomes. Whereas expansion via vegetative growth is slow and requires an existing vegetation stand, seed dispersal and germination allow for new stands to become established at distant locations. High seed production and wind dispersal are characteristics of cattail that should enhance this species' ability to invade new locations (McNaughton, 1966; Grace and Wetzel, 1981; Wilcox et al., 1985; Grace, 1987; Stewart et al., 1997). Initial invasion of a new location is dependent upon seed availability followed by successful germination and seedling survival. Densities of viable cattail seeds in the surface soils of the northern and central Everglades are generally quite low (Van der Valk and Rosburg, 1997), and cattail seed banks are restricted largely to areas where cattail stands are the dominant vegetation (Van der Valk and Rosburg, 1997).

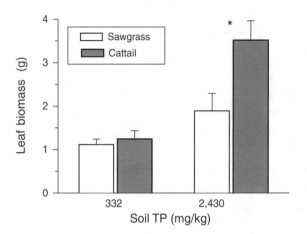

**FIGURE 3.22** Leaf regrowth (biomass produced) following leaf removal for cattail and sawgrass plants at enriched and unenriched site in WCA-2A. Bars are means (±1 SE) of measurements taken from 25 plants at each site.

Even in an area with a viable seed bank, cattail establishment appears to be quite slow under low P conditions, due to low frequencies of seed germination and seedling survival. Greenhouse studies have shown that while cattail seeds germinate rather quickly (within two to five days), initial seedling growth is slower (four to five months) than for sawgrass seedlings, particularly under low-nutrient conditions (Goslee and Richardson, 1997). The survival and growth of cattail seedlings were most successful when grown under high soil nutrient concentrations and saturated as opposed to flooded soils (Miao et al., 2000). Thus, while established cattail stands are extremely tolerant of a wide range of environmental conditions, the successful colonization of new locations via seed dispersal appears dependent upon specific hydrologic and nutrient conditions.

## LOWER TROPHIC LEVELS

Invertebrate responses to P enrichment appear to be driven primarily by P-induced changes in: (1) the quantity and quality of different food resources, (2) water-column DO, and (3) the availability of suitable substrata as habitat. Much of the existing evidence for invertebrate changes in response to P enrichment have come from transect studies along the nutrient gradient in WCA-2A. Rader and Richardson (1994) sampled invertebrate assemblages in open-water habitats along this gradient produced by canal inflows through the S10 structures. Sampling was conducted on six dates between 1990 and 1991 using sweep nets (mesh size 2.0 to 2.5 mm) and sediment cores. Invertebrate species richness was highest at the most enriched sites, while Shannon's diversity was highest at sites exposed to intermediate levels of enrichment. Although species shifts were observed along the gradient, most dominant taxonomic groups reached their highest densities at enriched sites. No shift in the functional composition of the assemblage was noted along the gradient. Sampling was limited to a single habitat (open-water), the coverage of which is dramatically less at enriched compared with unenriched sites (see McCormick et al., 1998). Therefore, conclusions concerning areal changes in invertebrate densities and production in response to enrichment are limited. Available evidence suggests that invertebrate densities and species richness may be extremely low in cattail stands (Davis, 1994), which account for more than 90% of areal coverage in enriched areas of WCA-2A.

Sampling along the same nutrient gradient in WCA-2A was repeated in 1994 and 1995 using sweep nets of smaller mesh size (0.35 mm) to increase the efficiency of invertebrate sampling (McCormick et al., 2001). Rather than concentrating solely on open-water habitats, sampling was conducted on a habitat-weighted basis to account for changes in vegetation along the gradient.

Macroinvertebrate densities increased nearly twofold with enrichment, but no change in species richness or diversity was detected. However, significant changes in taxonomic composition were detected as taxa tolerant of low DO became increasingly dominant with enrichment. This response was consistent with substantially lower DO concentrations at these sites compared with reference areas as discussed earlier. Differences in conclusions between McCormick et al. (2001) and Rader and Richardson (1994) likely are related in part to the habitat-weighted sampling employed in the former study.

Few experiments have investigated the relationship between invertebrate abundance and composition and P enrichment in the Everglades. The macroinvertebrate assemblage was sampled on artificial substrata (Hester–Dendy samplers) in flume dosing channels that had been exposed to different P loads for several years (Zahina and Richardson, 1997). Invertebrate colonization of these substrata was extremely variable between channels of the same treatment, apparently due to pronounced differences in vegetation between replicate channels. Consequently, few discernible trends in invertebrate density and taxonomic or functional composition were detected. However, there was weak evidence that a few taxonomic groups (oligochaetes and ostracods) responded positively to enrichment in a manner similar to that documented along P-nutrient gradients in the marsh (Rader and Richardson, 1994). Additional experimental studies are required to further establish invertebrate changes induced by P enrichment and the mechanisms (e.g., oxygen depletion, changes in food base) underlying such responses.

## POTENTIAL IMPACTS ON HIGHER TROPHIC LEVELS

The few studies that have been conducted on the effects of nutrient additions on fish communities consistently show that fish densities increase with enrichment. In WCA-2A, Rader and Richardson (1994) found similar fish species composition between enriched and unenriched sites, but fish densities were two- to threefold higher in enriched areas. Likewise, Turner et al. (1999) found that fish biomass estimates in enriched sites in WCA-2A and ENP averaged over 1 g m$^{-2}$ and were consistently higher than unenriched sites. A similar pattern was observed in the Okefenokee Swamp, where local nutrient enrichment from a wading bird colony was associated with increased fish biomass (Oliver and Schoenberg, 1989).

The effects of nutrient enrichment on bird communities can occur through two different pathways. The first is by altering the vegetation structure and the second is by changing the foodweb and affecting food density. In the Everglades, one of the dominant changes in vegetation structure as a result of nutrient enrichment is a shift from sawgrass ridges and open sloughs toward tall, dense stands of cattail with a higher biomass per unit area (Miao and Sklar, 1998; Miao and DeBusk, 1999). Avian community surveys in WCA-2A show that there are more individuals per site (all species pooled) but not more species per site at enriched marsh sites compared with unenriched sites (Crozier and Gawlik, in review). Three of the four most numerous species, red-winged blackbird (*Agelaius phoeniceus*), boat-tailed grackle (*Quiscalus major*), and common yellowthroat (*Geothlypis trichas*), rely on the vegetation for both nesting and foraging. The first two species are more abundant in enriched areas, perhaps because the dense vegetation in these areas offers more foraging and nesting locations per unit area compared with unenriched sites. However, the common yellowthroat (*Geothlypis trichas*) was less abundant in enriched areas even though it uses the vegetation in similar ways as the boat-tailed grackle and red-winged blackbird. Thus, it is likely that the common yellowthroat is responding to nutrients via a different mechanism than these other species. The other abundant species, the common moorhen (*Gallinula chloropus*), uses the vegetation primarily as a food source and is likely being affected by changes in food density (see below).

In some cases, bird abundance and species richness are not linearly related to vegetation density but rather reach a peak at moderate densities and then decline thereafter. For example, in freshwater marshes, maximum species richness was attained when the ratio of open water to vegetated wetlands reached 50:50 (Weller and Spatcher, 1965). Likewise, a statistical analysis of wading bird abundance in relation to vegetation in the northern Everglades indicated that wading bird abundance increased with cattail cover up to a point and then declined (Bancroft et al., in review).

The effects of nutrient enrichment on birds via the food web also tend to result in a higher density of birds. In Tunisia, nutrient enrichment caused increased plant biomass, which in turn resulted in increased herbivorous bird use (Tamisier and Boudouresque, 1994). Similarly, nutrient enrichment of ponds in New York led to higher insect biomass which in turn resulted in a higher density of tree swallows (*Tachycineta bicolor*), which rely on insects for food (McCarty, 1997). As mentioned above, this pathway (increased food availability with enrichment) is the likely mechanism by which the common moorhen, a herbivorous bird in the Everglades, achieves highest densities in enriched areas.

## CURRENT UNDERSTANDING, CRITICAL KNOWLEDGE GAPS, AND FUTURE RESEARCH NEEDS

The Everglades ecosystem developed under conditions of extreme P limitation, and it is clear that anthropogenic P loads have altered this unique resource. As described here, the results of experimental studies and measurements conducted along P gradients have increased our understanding of the progression of ecological responses to P enrichment in the Everglades. This information will serve as the basis for determining the concentrations and loads that cause these various responses. Available evidence indicates the following:

1. The Everglades marsh is extremely sensitive to P enrichment, although the ecological changes caused by enrichment manifest themselves over different time scales ranging from days and weeks (e.g., microbial and periphyton changes) to several years or decades (e.g., vegetation changes such as cattail expansion).
2. Changes in Everglades populations and ecosystem processes have been documented in areas of the marsh where water-column TP concentrations exceed background levels of approximately 10 $\mu$g L$^{-1}$.
3. Although ecological responses to P enrichment occur at different rates, most of these changes ultimately occur within a relatively narrow range of water-column TP concentrations between approximately 10 and 30 $\mu$g L$^{-1}$.

This range of P concentrations within which many ecological responses occur in the Everglades is consistent with eutrophication patterns previously established for other freshwater ecosystems (e.g., Sawyer, 1947; Likens, 1972; Vollenweider, 1976; Harper, 1992). However, given the extremely P-limited condition of the pristine Everglades, it is possible that even smaller changes in P availability may cause substantial shifts in native populations (see Chapter 4).

Uncertainty remains over the extent of spatial and temporal variation in ecosystem responses to P enrichment and the rate of recovery following reductions in P inputs. Ongoing work by several research groups should help to address the issues described below.

### SPATIAL COVERAGE OF SAMPLING AND EXPERIMENTATION

The Everglades is a heterogeneous ecosystem at several spatial scales and previous research has not been distributed equitably among geographic regions or habitat types. Most studies have been conducted in WCA-2A due to accessibility and the presence of a pronounced nutrient gradient. Considerably less research has been conducted in the LNWR and ENP, and no major field studies have been conducted in WCA-3A. Field dosing studies have focused largely on slough and wet prairie habitats (Scheidt et al., 1989; Richardson et al., 1995, 1997; McCormick and O'Dell, 1996; McCormick and Scinto, 1999; Chapter 4, this volume). Information on the response of sawgrass stands to P enrichment has come largely from observational studies (but see experimental findings from Craft et al., 1995; Miao and DeBusk, 1999), and little is known concerning the response of marl prairies, which cover large areas of the southern Everglades and represent some of the most oligotrophic habitats in the ecosystem.

A largely untested hypothesis is that different habitat types may vary in their sensitivity to P enrichment. Evidence from the Everglades and from other ecosystems indicates that the persistence of natural populations of flora and fauna are linked closely to the heterogeneous nature of the Everglades landscape (see Davis and Ogden, 1994). Maintenance of this habitat diversity requires that the ecosystem be protected from human disturbances such as P enrichment that drive it towards spatial homogeneity (e.g., cattail expansion).

## TEMPORAL PATTERNS OF P ACCUMULATION AND ECOLOGICAL RESPONSES

Intensive research over the last few years provides a reasonably good basis for predicting the progression of marsh responses to eutrophication; however, differences in the rate and extent of response to different P loading rates are still poorly understood. Given the short-term nature of most experimentation, such studies provide limited evidence for long-term changes (e.g., decades) that are of interest to water managers and regulators. Monitoring along marsh nutrient gradients provides the clearest evidence of the long-term impacts caused by different P loads, even though conditions are less controlled and loads and concentrations can vary over time. Correlative relationships between marsh P concentrations and ecological change, when combined with controlled experimentation to support cause-and-effect relationships, provide the best available evidence for identifying P concentrations and loads that produce long-term ecological changes in the Everglades.

## INTERACTIONS BETWEEN P ENRICHMENT AND OTHER NATURAL AND ANTHROPOGENIC FACTORS

Both natural (e.g., droughts, fires) and human-induced (e.g., altered hydroperiods and water depths) disturbances affect ecological patterns in the Everglades, sometimes in a manner that may be similar to the effects of anthropogenic P loading (e.g., Newman et al., 1998). Interactions between these disturbances and increased P loading may intensify or, in certain cases, counteract the effects of P enrichment alone. Similarly, inputs of other nutrients such as N may intensify the effects of P enrichment in areas of the marsh that are highly enriched. However, while these factors may alter the rate of ecosystem response to P enrichment, currently there is little evidence to suggest that human-induced changes in water depth, hydroperiod, or fire frequency determine the ecological changes that eventually occur in areas of the marsh receiving external P loads.

## RATES OF MARSH RECOVERY FOLLOWING REDUCTIONS IN P LOADS

Most research emphasis has been on documenting the ecological impacts caused by P enrichment and the P concentrations and loads that cause various changes in the Everglades. An understanding of the rate of ecosystem recovery following load reductions is required to develop realistic time frames and expectations for restoration. Key questions related to recovery include the following:

1. How quickly will water-column and soil P concentrations decline in enriched areas of the marsh following P load reductions?
2. Will P and associated ecological changes (e.g., *Typha* expansion into sloughs) continue to spread across the marsh as a result of internal P loading even after external loads are reduced?
3. How quickly will P-impacted areas of the marsh recover and what level of recovery can be expected?

While these questions have yet to be answered for the Everglades, evidence from other ecosystems indicate that the rate of ecosystem recovery from eutrophication may be considerably slower than the initial enrichment process (Perry and Vanderklein, 1996).

## REFERENCES

Ahn, H., Statistical modeling of total phosphorus concentrations measured from south Florida rainfall, *Ecological Modeling*, 116, 33, 1999.

Amador, J.A. and Jones, R.D., Nutrient limitations on microbial respiration in peat soils with different total phosphorus content, *Soil Biology and Biochemistry*, 25, 793, 1993.

Amador, J.A. and Jones, R.D., Carbon mineralization in pristine and phosphorus-enriched peat soils of the Florida Everglades, *Soil Science*, 159, 129, 1995.

Bachoon, D. and Jones, R.D., Potential rates of methanogenesis in sawgrass marshes with peat and marl soils in the Everglades, *Soil Biology and Biochemistry*, 24, 21, 1992.

Baldwin, M. and Hawker, H.W., *Cumulose Soils*, USDA Bureau of Soil Soil Survey of the Ft. Lauderdale Area, FL, 1915.

Bancroft, G.T., Gawlik, D.E., and Rutchey, K., Abundance of wading birds relative to vegetation and water depths in the northern Everglades, *Waterbirds*, in review.

Barko, J.W. and Smart, R.M., Sediment-related mechanisms of growth limitation in submersed macrophytes, *Ecology*, 67, 1328, 1986.

Bartow, S.M., Craft, C.B., and Richardson, C.J., Reconstructing historical changes in Everglades plant community composition using pollen distributions in peat, *Journal of Lake and Reservoir Management*, 12, 313, 1996.

Belanger, T.V., Scheidt, D.J., and Platko, J.R. II., Effects of nutrient enrichment on the Florida Everglades, *Lake and Reservoir Management*, 5, 101, 1989.

Browder, J.A., Cottrell, D., Brown, M., Newman, M., Edwards, R., Yuska, J., Browder, M., and Krakoski, J., *Biomass and Primary Production of Microphytes and Macrophytes in Periphyton Habitats of the Southern Everglades*, Report T-662, South Florida Research Center, Homestead, FL, 1982.

Browder, J.A., Gleason, P.J., and Swift, D.R., Periphyton in the Everglades: spatial variation, environmental correlates, and ecological implications, in *Everglades: The Ecosystem and Its Restoration*, Davis, S.M. and Ogden, J.C., Eds., St. Lucie Press, Delray Beach, FL, 1994, 379.

Brown, M. and Dinsmore, J.J., Implications of marsh size and isolation for marsh bird management, *Journal of Wildlife Management*, 50, 392, 1986.

Bush, M. and Richardson, C.J., Phosphatase as a biochemical indicator of phosphorus availability, in *Effects of Phosphorus and Hydroperiod Alterations on Ecosystem Structure and Function in the Everglades*, Richardson, C.J., Craft, C.B., Qualls, R.G., Stevenson, R.J., Vaithiyanathan, P., Bush, M., and Zahina, J., Eds., Duke Wetland Center Publ. No. 95-05, report submitted to Everglades Agricultural Area Environmental Protection District, 1995, 224.

Chanton, J.P., Whiting, G.J., Happell, J.D., and Gerard, G., Contrasting rates and diurnal patterns of methane emission from emergent aquatic macrophytes, *Aquatic Botany*, 46, 111, 1993.

Cooper, S.R. and Goman, M., Historical changes in water quality and vegetation in WCA-2A as determined by paleoecological analyses, in *An Integrated Approach to Wetland Ecosystem Science: The Everglades Experiments*, Richardson, C.J., Ed., Springer-Verlag, New York, 2001.

Craft, C.B. and Richardson, C.J., Peat accretion and N, P, and organic C accumulation in nutrient-enriched and unenriched Everglades peatlands, *Ecological Applications*, 3, 446, 1993a.

Craft, C.B. and Richardson, C.J., Peat accretion and phosphorus accumulation along a eutrophication gradient in the northern Everglades, *Biogeochemistry*, 22, 133, 1993b.

Craft, C.B. and Richardson, C.J., Relationships between soil nutrients and plant species composition in Everglades peatlands, *Journal of Environmental Quality*, 26, 224, 1997.

Craft, C.B. and Richardson, C.J., Recent and long-term organic soil accretion and nutrient accumulation in the Everglades, *Soil Science Society of America Journal*, 62, 834, 1998.

Craft, C.B., Vymazal, J., and Richardson, C.J., Response of Everglades plant communities to nitrogen and phosphorus additions, *Wetlands*, 15, 258, 1995.

Craighead, F.C., *The Trees of South Florida*, Vol. I. *The Natural Environments and Their Succession*, University of Miami Press, Coral Gables, FL, 1971.

Crozier, G.E. and Gawlik, D.E., Avian response to nutrient loading in the northern Everglades, *Auk*, in review.

Daoust, R.J., Investigating How Phosphorus Controls Structure and Function in Two Everglades Wetland Plant Communities, M.S. thesis, Florida International University, Miami, 1998.

Daoust, R.J. and Childers, D.L., Controls on emergent macrophyte composition, abundance, and productivity in freshwater Everglades wetland communities, *Wetlands*, 19, 262, 1999.

Davis, J.H., Jr., *The Natural Features of South Florida, Especially the Vegetation, and the Everglades*, Bull. No. 25, Florida Geological Survey, Tallahassee, FL, 1943.

Davis, J.H., Jr., *The Peat Deposits of Florida*, Bull. No. 30, Florida Geological Survey, Tallahassee, FL, 1946.

Davis, S.M., *Patterns of Radiophosphorus Accumulation in the Everglades after Its Introduction into Surface Water*, Tech. Publ. 82-2, South Florida Water Management District, West Palm Beach, FL, 1982.

Davis, S.M., Sawgrass and cattail production in relation to nutrient supply in the Everglades, in *Freshwater Wetlands and Wildlife*, Sharitz, R.R., and Gibbons, J.W., Eds., U.S. Dept. of Energy, Office of Scientific and Technical Information, Oak Ridge, TN, 1989, 325.

Davis, S.M., Growth, decomposition and nutrient retention of *Cladium jamaicense* Crantz. and *Typha domingensis* Pers. in the Florida Everglades, *Aquatic Botany*, 40, 203, 1991.

Davis, S.M., Phosphorus inputs and vegetation sensitivity in the Everglades, in *Everglades: The Ecosystem and Its Restoration*, Davis, S.M., and Ogden, J.C., Eds., St. Lucie Press, Delray Beach, FL, 1994, 357.

Davis, S.M. and Ogden, J.C., Eds., *Everglades: The Ecosystem and Its Restoration*, St. Lucie Press, Delray Beach, FL, 1994.

Davis, S.M., Gunderson, L.H., Park, W.A., Richardson, J., and Mattson, J., Landscape dimension, composition, and function in a changing Everglades ecosystem, in *Everglades: The Ecosystem and Its Restoration*, Davis, S.M., and Ogden, J.C., Eds., St. Lucie Press, Delray Beach, FL, 1994, 419.

DeBusk, W.F., Reddy, K.R., Koch, M.S., and Wang, Y., Spatial distribution of soil nutrients in a northern Everglades marsh — Water Conservation Area 2A, *Soil Science Society of America Journal*, 58, 543, 1994.

DeBusk, W.F. and Reddy, K.R., Turnover of detrital organic carbon in a nutrient-impacted Everglades marsh, *Soil Science Society of America Journal*, 62, 1460, 1998.

DeBusk W.F., Newman, S., and Reddy, K.R., Spatio-temporal patterns of soil phosphorus enrichment in Everglades WCA-2A, *Journal of Environmental Quality*, 30, 2001.

DeLaune, R.D., Buresh, R.J., and Patrick, W.H., Relationship of soil properties to standing crop biomass of *Spartina alterniflora* in a Louisiana marsh, *Estuarine and Coastal Marine Science*, 8, 477, 1979.

Doren, R.F., Armentano, T.V., Whiteaker, L.D., and Jones, R.D., Marsh vegetation patterns and soil phosphorus gradients in the Everglades ecosystem, *Aquatic Botany*, 56, 145, 1997.

Drake, H.L., Aumen, N.G., Kuhner, C., Wagner, C., Grießhammer, A., and Schmittroth, M., Anaerobic microflora of Everglades sediments: effects of nutrients on population profiles and activities, *Applied Environmental Microbiology*, 62, 486, 1996.

Duever, M.J., Meeder, J.F., Meeder, L.C., and McCollom, J.M., The climate of south Florida and its role in shaping the Everglades ecosystem, in *Everglades: The Ecosystem and Its Restoration*, Davis, S.M. and Ogden, J.C., Eds., St. Lucie Press, Delray Beach, FL, 1994, 225.

Ewel, J.J., Invasibility: lessons from south Florida, in *Ecology of Biological Invasions of North America and Hawaii*, Mooney, H. A., and Drake, J. A., Eds., Springer-Verlag, New York, 1986, 214.

Fisher, M.M., Estimating Landscape Flux of Phosphorus Using Geographic Information Systems (GIS), M.S. thesis, University of Florida, Gainesville, 1997.

Fitz, H.C. and Sklar, F.H., Ecosystem analysis of phosphorus impacts and altered hydrology in the Everglades: a landscape modeling approach, in *Phosphorus Biogeochemistry of Subtropical Ecosystems*, Reddy, K.R., O'Connor, G.A., and Schelske, C.L., Eds., Lewis Publishers, Boca Raton, FL, 1999, 585.

Flora, M.D., Walker, D.R. Burgess, D.R., Schiedt, D.J., and Rice, R.G., *The Response of Experimental Channels in Everglades National Park to Increased Nitrogen and Phosphorus Loading*, Water Resources Report No. 86-6, National Park Service, Water Resources Division, Colorado State University, Fort Collins, 1986.

Forthman, C.A., The Effects of Prescribed Burning on Sawgrass, *Cladium jamaicense*, Crantz, in South Florida. M.S. Thesis, University of Miami, Coral Gables, FL, 1973.

Gawlik, D.E. and Rocque, D.A., Avian communities in Bayheads, willowheads, and sawgrass marshes of the central Everglades, *Wilson Bulletin*, 110, 45, 1998.

Gleason, P.J. and Spackman, W., Calcareous periphyton and water chemistry in the Everglades, in *Environments of South Florida: Past and Present*, Gleason, P.J., Ed., Memoir No. 2. Miami Geological Society, Coral Gables, FL, 1974, 225.

Gleason, P.J. and Stone, P.A., *Prehistoric Trophic Level Status and Possible Cultural Influences on the Enrichment of Lake Okeechobee*, Central and Southern Flood Control District, West Palm Beach, FL, 1975.

Gleason, P.J. and Stone, P.A., Age, origin, and evolution of the Everglades peatland, in *Everglades: The Ecosystem and Its Restoration*, Davis, S.M. and Ogden, J.C., Eds., St. Lucie Press, Delray Beach, FL, 1994, 149.

Gleason, P.J., Stone, P.A., Hallett, D., and Rosen, M., Preliminary Report on the Effect of Agricultural Runoff on the Periphytic Algae of Conservation Area 1. Unpublished report, Central and Southern Flood Control District, West Palm Beach, FL, 1975.

Goslee, S.C. and Richardson, C.J., Establishment and seedling growth of sawgrass and cattail from the Everglades, in *Effects of Phosphorus and Hydroperiod Alterations on Ecosystem Structure and Function in the Everglades*, Richardson, C.J., Craft, C.B., Qualls, R.G., Stevenson, R.J., Vaithiyanathan, P., Bush, M., and Zahina, J., Eds., Duke Wetland Center Publ. No. 97-05, Report to the Everglades Agricultural Area Environmental Protection District, 1997, chap. 13.

Grace, J.B., The impact of preemption on the zonation of two *Typha* species along lakeshores, *Ecological Monographs*, 57, 283, 1987.

Grace, J.B. and Wetzel, R.G., Habitat partitioning and competitive displacement in cattails (*Typha*): experimental field studies, *The American Naturalist*, 118, 463, 1981.

Grimshaw, H.J., Rosen, M., Swift, D.R., Rodberg, K., and Noel, J.M., Marsh phosphorus concentrations, phosphorus content and species composition of Everglades periphyton communities, *Archive für Hydrobiologie*, 128, 257, 1993.

Grimshaw, H.J., Wetzel, R.G., Brandenburg, M., Segerblom, M., Wenkert, L.J., Marsh, G.A., Charnetzky, W., Haky, J.E., and Carraher, C., Shading of periphyton communities by wetland emergent macrophytes: decoupling of algal photosynthesis from microbial nutrient retention, *Archive für Hydrobiologie*, 139, 17, 1997.

Gunderson, L.H., Vegetation of the Everglades: determinants of community composition, in *Everglades: The Ecosystem and Its Restoration*, Davis, S.M., and Ogden, J.C., Eds., St. Lucie Press, Delray Beach, FL, 1994, 323.

Hammar, H.E., The chemical composition of Florida Everglades peat soils, with special reference to their inorganic constituents, *Soil Science*, 28, 1, 1929.

Harper, D., *Eutrophication of Freshwater: Principles, Problems, and Restoration*, Chapman & Hall, London, 1992.

Havens, K.E. and Steinman, A.D., Aquatic systems, in *Soil Amendments: Impacts on Biotic Systems*, Rechcigl, J.E., Ed., Lewis Publishers, Boca Raton, FL, 1995, 121.

Hoffman, W., Bancroft, G.T., and Sawicki, R.J., Foraging habitat of wading birds in the Water Conservation Areas of the Everglades, in *Everglades: The Ecosystem and Its Restoration*, Davis, S.M. and Ogden, J.C., Eds., St. Lucie Press, Delray Beach, FL, 1994, 585.

Howard-Williams, C., Cycling and retention of nitrogen and phosphorus in wetlands: a theoretical and applied perspective, *Freshwater Biology*, 15, 391, 1985.

Inglett, P.W., Spatial and Temporal Patterns of Periphyton $N_2$ Fixation in a Nutrient-Impacted Everglades Ecosystem, M.S. thesis, University of Florida, Gainesville, 1999.

Jensen, J.R., Rutchey, K., Koch, M.S., and Narumalani, S., Inland wetland change detection in the Everglades Water Conservation Area 2A using time series of normalized remotely sensed data, *Photogrammetric Engineering Remote Sensing*, 61, 199, 1995.

Jones, L.A., *Soils, Geology, and Water Control in the Everglades Region*, Bull. 442, University of Florida Agricultural Experiment Station, Belle Glade, FL, 1948

King, G.M., Klug, M.J., and Lovely, D.R., Metabolism of acetate, methanol, and methylated amines in intertidal sediments of Lowes Cove, Maine, *Applied Environmental Microbiology*, 45, 1848, 1983.

King, J.W., *Exploration: Examination and Reconnaissance of the Lands of the Tamiami Trail in Dade County, Florida*, The Dade County Commissioners Commission of Dade County, 1917.

Koch, M.S. and Reddy, K.R., Distribution of soil and plant nutrients along a trophic gradient in the Florida Everglades, *Soil Science Society of America Journal*, 56, 1492, 1992.

Koch-Rose, M.S., Reddy, K.R., and Chanton, J.P., Factors controlling seasonal nutrient profiles in a subtropical peatland of the Florida Everglades, *Journal of Environmental Quality*, 23, 526, 1994.

Kushlan, J., Responses of wading birds to seasonally fluctuating water levels: strategies and their limits, *Colonial Waterbirds*, 9, 155, 1986.

Lange-Bertalot, H., Pollution tolerance of diatoms as a criterion for water quality estimation, *Nova Hedwigia*, 64, 285, 1979.

Likens, G.E., Ed., *Nutrients and Eutrophication*, Vol. 1, Special symposium, American Society of Limnology and Oceanography, Allen Press, Lawrence, KS, 1972.

Loftus, W.F. and Eklund, A.M., Long-term dynamics of an Everglades small-fish assemblage, in *Everglades: The Ecosystem and Its Restoration*, Davis, S.M. and Ogden, J.C., Eds., St. Lucie Press, Delray Beach, FL, 1994, 461.

Loftus, W.F., Chapman, J.D., and Conrow, R., Hydroperiod effects on Everglades marsh food webs, with relation to marsh restoration efforts, in *Fisheries and Coastal Wetlands Research*, Larson, G., and Soukup, M., Eds., Vol. 6 of the Proceedings of the 1986 Conference on Science in National Parks, U.S. National Parks Service and The George Wright Society, Ft. Collins, CO, 1986, 1.

Long, R.W., The vegetation of southern Florida, *Florida Science*, 37, 33, 1974.

Loveless, C.M., A study of the vegetation of the Florida Everglades, *Ecology*, 40, 1, 1959.

Lowe, R.L., *Environmental Requirements and Pollution Tolerance of Freshwater Diatoms*, EPA-670/4-74-005, Office of Research and Development, U.S. Environmental Protection Agency, Cincinnati, OH, 1974.

Marschner, H., *Mineral Nutrition of Higher Plants*, Academic Press, New York, 1986.

McCarty, J.P., Aquatic community characteristics influence the foraging patterns of tree swallows, *Condor*, 99, 213, 1997.

McCormick, P.V. and Laing, J., Effects of increased phosphorus loading on dissolved oxygen in a subtropical wetland, the Florida Everglades, *Wetlands Ecology and Management*, accepted.

McCormick, P.V. and O'Dell, M.B., Quantifying periphyton responses to phosphorus enrichment in the Florida Everglades: a synoptic-experimental approach, *Journal of the North American Benthological Society*, 15, 450, 1996.

McCormick, P.V. and Scinto, L.J., Influence of phosphorus loading on wetlands periphyton assemblages: a case study from the Everglades, in *Phosphorus Biogeochemistry of Subtropical Ecosystems*, Reddy, K.R., O'Connor, G.A., and Schelske, C.L., Eds., CRC/Lewis Publishers, Boca Raton, FL, 1999, 301.

McCormick, P.V. and Stevenson, R.J., Periphyton as a tool for ecological assessment and management in the Florida Everglades, *Journal of Phycology*, 34, 726, 1998.

McCormick, P.V., Chimney, M.J., and Swift, D.R., Diel oxygen profiles and water column community metabolism in the Florida Everglades, U.S.A., *Archive für Hydrobiologie* 140, 117, 1997.

McCormick, P.V., Newman, S., Miao, S.L., Reddy, K.R., Gawlik, D., Fitz, C., Fontaine, T.D., and Marley, D., Ecological needs of the Everglades, in *Everglades Interim Report*, South Florida Water Management District, West Palm Beach, FL, 1999, chap. 3.

McCormick, P.V., Rawlik, P.S., Lurding, K., Smith, E.P., and Sklar, F.H., Periphyton–water quality relationships along a nutrient gradient in the northern Everglades, *Journal of the North American Benthological Society*, 15, 433, 1996.

McCormick, P.V., Shuford, R.B.E., III, Backus, J.B., and Kennedy, W.C., Spatial and seasonal patterns of periphyton biomass and productivity in the northern Everglades, Florida, USA, *Hydrobiologia*, 362, 185, 1998.

McCormick, P.V., Shuford, R.B.E., III, and Rawlik, P.S., *Macroinvertebrate Responses to Phosphorus Enrichment in the Northern (WCA 2A) Everglades*, Tech. publ. EMA-392, South Florida Water Management District, West Palm Beach, FL, 2001.

McDowell, L.L., Stephens, J.C., and Stewart, E.H., Radiocarbon chronology of the Florida Everglades peat, *Soil Science Society of America Proceedings*, 33, 743, 1969.

McNaughton, S.J., Ecotype function in the *Typha*-community type, *Ecological Monographs*, 36, 297, 1966.

Miao, S.L. and DeBusk, W.F., Effects of phosphorus enrichment on structure and function of sawgrass and cattail communities in Florida wetlands, in *Phosphorus Biogeochemistry of Subtropical Ecosystems*, Reddy, K. R., O'Connor, G.A., and Schelske, C.L., Eds., Lewis Publishers, Boca Raton, FL, 1999, 275.

Miao, S.L. and Sklar, F.H., Biomass and nutrient allocation of sawgrass and cattail along a nutrient gradient in the Florida Everglades, *Wetlands Ecology and Management*, 5, 245, 1998.

Miao, S.L., Borer, R.E., and Sklar, F.H., Sawgrass seedling responses to transplanting and nutrient additions, *Restoration Ecology*, 5, 162, 1997.

Miao, S.L., McCormick, P.V., Newman, S., and Rajagopalan, S., Interactive effects of seed availability, hydrology, and phosphorus enrichment on cattail colonization in the Florida Everglades, *Wetlands Ecology and Management*, 9, 1, 2001.

Moshiri, G.A., Ed., *Constructed Wetlands for Water Quality Improvement*, Lewis Publishers, Boca Raton, FL, 1993.

National Academy of Sciences, *Eutrophication: Causes, Consequences, Correctives*, National Academy of Sciences Press, Washington, D.C., 1969.

Nearhoof, F., *Nutrient-Induced Impacts and Water Quality Violations in the Florida Everglades*, Water Quality Technical Series 3(24), Bureau of Water Facilities Planning and Regulation, Department of Environmental Regulation, Tallahassee, FL, 1992.

Newman, S., and Robinson, J.S., Forms of organic phosphorus in water, soils, and sediments, in *Phosphorus Biogeochemistry of Subtropical Ecosystems*, Reddy, K.R., O'Connor, G.A., and Schelske, C.L., Eds., Lewis Publishers, Boca Raton, FL, 1999, 207.

Newman, S., Grace, J.B., and Koebel, J.W., Effects of nutrients and hydroperiod on *Typha*, *Cladium*, and *Eleocharis*: implications for Everglades restoration, *Ecological Applications*, 6, 774, 1996.

Newman, S., Reddy, K.R., DeBusk, W.F., Wang, Y., Shih, G., and Fisher, M.M., Spatial distribution of soil nutrients in a northern Everglades marsh: Water Conservation Area 1, *Soil Science Society of America Journal*, 61, 1275, 1997.

Newman, S., Schuette, J., Grace, J.B., Rutchey, K., Fontaine, T., Reddy, K.R., and Pietrucha, M., Factors influencing cattail abundance in the northern Everglades, *Aquatic Botany*, 60, 265, 1998.

Newman, S., McCormick, P.V., and Backus, J.G., Phosphatase activity as an early warning indicator of wetland eutrophication: problems and prospects, in *Phosphatases in the Environment*, Whitton, B.A., Ed., Kluwer Academic, Norwell, MA, in press.

Ogden, J.C., 1994. A comparison of wading bird nesting colony dynamics (1931–1946 and 1974–1989) as an indication of ecosystem conditions in the southern Everglades, in *Everglades: The Ecosystem and Its Restoration*, Davis, S.M., and Ogden, J.C., Eds., St. Lucie Press, Delray Beach, FL, 1994, 533.

Oliver, J.D. and Schoenberg, S.A., Residual influence of macronutrient enrichment on the aquatic food web of Okefenokee Swamp abandoned bird rookery, *Oikos*, 55, 175, 1989.

Olson, R.K., Ed., *Created and Natural Wetlands for Controlling Nonpoint Source Pollution*, CRC Press, Boca Raton, FL, 1993.

Palmer, C.M., A composite rating of algae tolerating organic pollution, *Journal of Phycology*, 5, 78, 1969.

Pan, Y., Stevenson, R.J., Vaithiyanathan, P., Slate, J., and Richardson, C.J., Using experimental and observational approaches to determine causes of algal changes in the Everglades, in *Effects of Phosphorus and Hydroperiod Alterations on Ecosystem Structure and Function in the Everglades*, Richardson, C.J., Craft, C.B., Qualls, R.G., Stevenson, R.J., Vaithiyanathan, P., Bush, M., and Zahina, J., Eds., Duke Wetland Center Publ. No. 97-05, Report to the Everglades Agricultural Area Environmental Protection District, 1997, chap. 2.

Parker, G.G., Hydrology of the pre-drainage system of the Everglades in southern Florida, in *Environments of South Florida: Past and Present*, Gleason, P.J., Ed., Memoir No. 2, Miami Geological Society, Coral Gables, FL, 1974, 18.

Parker, G.G., Ferguson, G.E., and Love, S.K., *Water Resources of Southeastern Florida with Special Reference to the Geology and Groundwater of the Miami Area*, Water Supply Paper 1255, U.S. Geological Survey, U.S. Government Printing Office, Washington, D.C., 1955.

Perry, J. and Vanderklein, E., *Water Quality: Management of a Natural Resource*, Blackwell Science, Cambridge, MA, 1996.

Qualls, R.G. and Richardson, C.J., Forms of soil phosphorus along a nutrient enrichment gradient in the northern Everglades, *Soil Science*, 160, 183, 1995a.

Qualls, R.G. and Richardson, C.J., Long-term response of Everglades macrophyte communities to varying concentrations of PO4 in experimental mesocosms, in *Effects of Phosphorus and Hydroperiod Alterations on Ecosystem Structure and Function in the Everglades*, Richardson, C.J., Craft, C.B., Qualls, R.G., Stevenson, R.J., Vaithiyanathan, P., Bush, M., and Zahina, J., Eds., Duke Wetland Center Publ. No. 95-05, report submitted to Everglades Agricultural Area Environmental Protection District, 1995b, 185.

Rader, R.B., Macroinvertebrates of the northern Everglades: species composition and trophic structure, *Florida Scientist*, 57, 22, 1994.

Rader, R.B., The Florida Everglades: natural variability, invertebrate diversity, and foodweb stability, in *Freshwater Wetlands of North America: Ecology and Management*, Batzer, D.P., Rader, R.B., and Wissinger, S.A., Eds., John Wiley & Sons, New York, 1999, 25.

Rader, R.B. and Richardson, C.J., Response of macroinvertebrates and small fish to nutrient enrichment in the northern Everglades, *Wetlands*, 14, 134, 1994.

Raschke, R.L., Diatom (Bacillariophyta) community responses to phosphorus in the Everglades National Park, USA, *Phycologia*, 32, 48, 1993.

Reark, J.B., *Ecological Investigations in the Everglades*, 2nd annual report to the Superintendent of Everglades National Park, University of Miami, Coral Gables, FL, 1961.

Reddy, K.R., DeBusk, W.F., Wang, Y., DeLaune, R.D., and Koch, M., *Physico-Chemical Properties of Soils in the Water Conservation Area 2 of the Everglades*, report to South Florida Water Management District, University of Florida, Gainesville, 1991.

Reddy, K.R., DeLaune, R.D., Debusk, W.F., and Koch, M.S., Long-term nutrient accumulation rates in the Everglades, *Soil Science Society of America Journal*, 57, 1147, 1993.

Reddy, K.R., Wang, Y., DeBusk, W.F., and Newman, S., *Physico-Chemical Properties of Soils in the Water Conservation Area 3 (WCA-3) of the Everglades*, report to South Florida Water Management District, University of Florida, Gainesville, 1994a.

Reddy, K.R., Wang, Y., Olila, O.G., Fisher, M.M., and Newman, S., *Influence of Flooding on Physico-Chemical Properties and Phosphorus Retention of the Soils in the Holeyland Wildlife Management Area*, report to South Florida Water Management District, University of Florida, Gainesville, 1994b.

Reddy, K.R., Wang, Y., DeBusk, W.F., Fisher, M.M., and Newman, S., Forms of soils phosphorus in selected hydrologic units of Florida Everglades ecosystems, *Soil Science Society of America Journal*, 62, 1134, 1998.

Reddy, K.R., White, J.R., Wright, A.L., and Chua, T., Influence of phosphorus loading on microbial processes in soil and water column of wetlands, in *Phosphorus Biogeochemistry of Subtropical Ecosystems*, Reddy, K.R., O'Connor, G.A., and Schelske, C.L., Eds., Lewis Publishers, Boca Raton, FL, 1999, 249.

Richardson, C.J. and Vaithiyanathan, P., Nutrient profiles in the Everglades: examination along the eutrophication gradient, in *Effects of Phosphorus and Hydroperiod Alterations on Ecosystem Structure and Function in the Everglades*, Richardson, C.J., Craft, C.B., Qualls, R.G., Stevenson, J., Vaithiyanathan, P., Bush, M., and Zahina, J., Eds., Duke Wetland Center Publ. No. 95-05, report submitted to Everglades Agricultural Area Environmental Protection District, 1995, chap. 6.

Richardson, C.J., Craft, C.B., Qualls, R.G., Stevenson, J., Vaithiyanathan, P., Bush, M., and Zahina, J., *Effects of Phosphorus and Hydroperiod Alterations on Ecosystem Structure and Function in the Everglades*, Duke Wetland Center Publ. No. 95-05, report to the Everglades Agricultural Area Environmental Protection District, 1995.

Richardson, C.J., Craft, C.B., Qualls, R.G., Stevenson, J., Vaithiyanathan, P., Bush, M., and Zahina, J., *Effects of Phosphorus and Hydroperiod Alterations on Ecosystem Structure and Function in the Everglades*, Duke Wetland Center Publ. No. 97-05, report to the Everglades Agricultural Area Environmental Protection District, 1997a.

Richardson, C.J., Vaithiyanathan, P., Romanowicz, E.A., and Craft, C.B., Macrophyte community responses in the Everglades with an emphasis on cattail (*Typha domingensis*) and sawgrass (*Cladium jamaicense*) interactions along a gradient of long-term nutrient additions, altered hydroperiod and fire, in *Effects of Phosphorus and Hydroperiod Alterations on Ecosystem Structure and Function in the Everglades*, Richardson, C.J., Craft, C.B., Qualls, R.G., Stevenson, J., Vaithiyanathan, P., Bush, M., and Zahina, J., Duke Wetland Center Publ. No. 97-05, report to the Everglades Agricultural Area Environmental Protection District, 1997b, chap. 14.

Robbins, J.A., Wang, X., and Rood, R.W., *Sediment Core Dating (Everglades WCAs)*, report to the South Florida Water Management District, West Palm Beach, FL, 1996.

Rudnick, D.T., Chen, Z., Childers, D.L., Boyer, J.N., and Fontaine, T.D., III., Phosphorus and nitrogen inputs into Florida Bay: the importance of the Everglades watershed, *Estuaries*, 22, 398, 1999.

Rutchey, K. and Vilchek, L., Development of an Everglades vegetation map using a SPOT image and the global positioning system, *Photogrammetric Engineering and Remote Sensing*, 60, 767, 1994.

Rutchey, K. and Vilchek, L., Air photo-interpretation and satellite imagery analysis techniques for mapping cattail coverage in a northern Everglades impoundment, *Photogrammetric Engineering and Remote Sensing*, 65, 185, 1999.

Sawyer, C.N., Fertilization of lakes by agricultural and urban drainage, *Journal of the New England Water Works Association*, 61, 109, 1947.

Scheidt, D.J., Flora, M.D., and Walker, D.R., Water quality management for Everglades National Park, *American Water Resources Association*, September, 377, 1989.

SFWMD, Draft Surface Water Improvement and Management Plan for the Everglades, Supporting Information Document, South Florida Water Management District, West Palm Beach, FL, 1992.

SFWMD, Atmospheric Deposition in South Florida, Advisory panel final report to the South Florida Water Management District, West Palm Beach, FL, 1997.

Sinsabaugh, R.L., Enzymatic analysis of microbial pattern and processes, *Biology and Fertility of Soils*, 17, 69, 1994.

Smith, E.P. and McCormick, P.V., Long-term relationship between phosphorus inputs and wetland phosphorus concentrations in a northern Everglades marsh, *Environmental Monitoring and Assessment*, 68, 153, 2001.

Snyder, G.H. and Davidson, J.M., Everglades agriculture — past, present, and future, in *Everglades: The Ecosystem and Its Restoration*, Davis, S.M., and Ogden, J.C., Eds., St. Lucie Press, Delray Beach, FL, 1994, 85.

Steward, K.K. and Ornes. W.H., Assessing a marsh environment for wastewater renovation, *Journal of the Water Pollution Control Federation*, 47, 1880, 1975a.

Steward, K.K. and Ornes, W.H., The autecology of sawgrass in the Florida Everglades, *Ecology*, 56, 162, 1975b.

Steward, K.K. and Ornes, W.H., Mineral nutrition of sawgrass (*Cladium jamaicense* Crantz.) in relation to nutrient supply, *Aquatic Botany*, 16, 349, 1983.

Stewart, H., Miao, S.L., Colbert, M., and Carraher, C.E., Jr., Seed germination of two cattail (*Typha*) species as a function of Everglades nutrient levels, *Wetlands*, 17, 116, 1997.

Swift, D.R. and Nicholas, R.B., *Periphyton and Water Quality Relationships in the Everglades Water Conservation Areas, 1978–1982*, Tech. Publ. 87-2, South Florida Water Management District, West Palm Beach, FL, 1987.

Tamisier, A. and Boudouresque, C., Aquatic bird populations as possible indicators of seasonal nutrient flow at Ichkeul Lake, Tunisia, *Hydrobiologia*, 279/280, 149, 1994.

Tiessen, H., Ed., *Phosphorus in the Global Environment: Transfers, Cycles, and Management*, SCOPE Vol. 54, John Wiley & Sons, New York, 1995.

Turner, A.M., Trexler, J.C., Jordan, C.F., Slack, S.J., Geddes, P., Chick, J.H., and Loftus, W., Targeting ecosystem features for conservation: standing crops in the Everglades, *Conservation Biology*, 13, 898, 1999.

Urban, N.H., Davis, S.M., and Aumen, N.G., Fluctuations in sawgrass and cattail densities in Everglades Water Conservation Area 2A under varying nutrient, hydrologic and fire regimes, *Aquatic Botany*, 46, 203, 1993.

USEPA, Water Quality Standards for Wetlands: National Guidance, EPA-440/S-90-011, Office of Water Regulations and Standards, U.S. Environmental Protection Agency, Washington, D.C., 1990.

Vaithiyanathan, P., Zahina, J., and Richardson, C. J., Macrophyte species changes along the phosphorus gradient, in *Effects of Phosphorus and Hydroperiod Alterations on Ecosystem Structure and Function in the Everglades*, Richardson, C.J., Craft, C.B., Qualls, R.G., Stevenson, J., Vaithiyanathan, P., Bush, M., and Zahina, J., Eds., Duke Wetland Center Publ. No. 95-05, report submitted to Everglades Agricultural Area Environmental Protection District, 1995, 273.

Van der Valk, A.G. and Rosburg, T.R., Seed bank composition along the phosphorus gradients in the northern Florida Everglades, *Wetlands*, 17, 228, 1997.

Vollenweider, R.A., Advances in defining critical loading levels for phosphorus in lake eutrophication, *Mem. Inst. Ital. Idrobiol.*, 33, 53, 1976.

Vymazal, J., Craft, C.B., and Richardson, C.J., Periphyton response to nitrogen and phosphorus additions in the Florida Everglades, *Algological Studies*, 73, 75, 1994.

Walker, W.W., Jr., Design basis for Everglades Stormwater Treatment Areas, *Water Research Bulletin*, 31, 671, 1995.

Walker, W.W., Jr., *Analysis of Water Quality and Hydrologic Data from the C-111 Basin*, draft report to the U.S. Department of the Interior, Everglades National Park, 1997.

Walker, W.W., Jr., Long-term water quality trends in the Everglades, in *Phosphorus Biogeochemistry of Subtropical Ecosystems*, Reddy, K.R., O'Connor, G.A., and Schelske, C.L., Eds., Lewis Publishers, Boca Raton, FL, 1999, 447.

Walker, W.W., Jr., and Kadlec, R.H., *A Model for Simulating Phosphorus Concentrations in Waters and Soils Downstream of Everglades Stormwater Treatment Areas*, draft document prepared for the U.S. Department of Interior, 1996.

Waller, B.G. and Earle, J.E., *Chemical and Biological Quality of Water in Part of the Everglades, Southeastern Florida*, Water Resources Investigations 56-75, U.S. Geological Survey, Tallahassee, FL, 1975.

Weller, M.W., Bird-habitat relationships in a Texas estuarine marsh during summer, *Wetlands*, 14, 293, 1994.

Weller, M.W. and Spatcher, C.S., *Role of Habitat in the Distribution and Abundance of Marsh Birds*, Special Report No. 43, Agriculture and Home Economics Experiment Station, Iowa State University, 1965.

White, J.R. and Reddy, K.R., The influence of nitrate and phosphorus loading on denitrification enzymes activity in Everglades wetland soils, *Soil Science Society of America Journal*, 63, 1945, 1999.

White, J.R. and Reddy, K.R., The effects of phosphorus loading on organic nitrogen mineralization of soils and detritus along a nutrient gradient in the northern Everglades, Florida, *Soil Science Society of America Journal*, 64, 1525, 2000.

Wilcox, D.A., Apfelbaum, S.I., and Hiebert, R.D., Cattail invasion of sedge meadows following hydrologic disturbance in the Cowles Bog Wetland Complex, Indiana Dunes National Lakeshore, *Wetlands*, 4, 115, 1985.

Willard, D.A., Weimer, L.M., and Riegel, W.L., Pollen assemblages as paleoenvironmental proxies in the Florida Everglades, *Review of Paleobotany and Palynology*, 113, 213, 2001.

Wood, E.J.F. and Maynard, N.G., Ecology of the micro-algae of the Florida Everglades, in *Environments of South Florida: Past and Present*, Gleason, P.J., Ed., Memoir No. 2, Miami Geological Society, Coral Gables, FL, 1974, 123.

Wood, J.M. and Tanner, G.W., Graminod community composition and structure within four Everglades management areas, *Wetlands*, 10, 127, 1990.

Wright, A.L. and Reddy, K.R., Influence of nutrient loading on extracellular enzyme activity in soils of the Florida Everglades, *Soil Science Society of America Journal*, 65:588–595, 2001.

Wright, A.L. and Reddy, K.R., Heterotrophic microbial activities in northern Everglades wetland soils, *Soil Science Society of America Journal*, in press.

Wu, Y., Sklar, F.H., and Rutchey, K.R., Analysis and simulations of fragmentation patterns in the Everglades, *Ecological Applications*, 7, 268, 1997.

Zahina, J. and Richardson, C.J., Preliminary assessment of changes in macroinvertebrate community structure resulting from phosphorus dosing in an Everglades slough, in *Effects of Phosphorus and Hydroperiod Alterations on Ecosystem Structure and Function in the Everglades*, Richardson, C.J., Craft, C.B., Qualls, R.G., Stevenson, J., Vaithiyanathan, P., Bush, M., and Zahina, J., Eds., Duke Wetland Center Publ. No. 97-05, report to the Everglades Agricultural Area Environmental Protection District, 1997, chap. 16.

# 4 Quantifying the Effects of Low-Level Phosphorus Additions on Unenriched Everglades Wetlands with *In Situ* Flumes and Phosphorus Dosing

*Daniel L. Childers, Ronald D. Jones, Joel C. Trexler,*
*Chris Buzzelli, Susan Dailey, Adrienne L. Edwards,*
*Evelyn E. Gaiser, Krish Jayachandaran, Anna Kenne,*
*David Lee, John F. Meeder, Joseph H.K. Pechmann,*
*Amy Renshaw, Jennifer Richards, Michael Rugge,*
*Leonard J. Scinto, Pierre Sterling, and Will Van Gelder*
Florida International University

## CONTENTS

1-8493-2026-7/02/$0.00+$1.50
© 2002 by CRC Press LLC

## INTRODUCTION

The Everglades, an International Biosphere Reserve, is one of the largest freshwater wetland landscapes in North America. These wetlands are highly oligotrophic, and phosphorus (P) is the macronutrient that limits ecosystem productivity and biomass accumulation (Swift, 1981; Amador and Jones, 1993). Total P concentrations typically range from 0.15 to 0.3 $\mu$M (5–10 ppb; Walker, 1991). Phosphorus follows a sedimentary cycle in ecological systems; thus, hydrologic inputs presumably dictate the supply and transport of this limiting nutrient in Everglades wetlands. Furthermore, there are no biotically mediated mechanisms for removal of P from ecosystems in the way that, for example, wetlands may remove excess nitrogen via denitrification. This lack of internal control of P cycling suggests that the oligotrophic wetlands of the Everglades are particularly susceptible to changes in P loading (Lutz, 1977; Swift, 1981; Swift and Nicholas, 1987; Koch and Reddy, 1992; Caraco, 1993; Grimshaw et al., 1993; Davis, 1994; Amador and Jones, 1993; McCormick et al., 1996; Daoust, 1998).

The Everglades Agricultural Area (EAA) is a major source of P to the oligotrophic marshes, often by point-source canal inputs (Coale et al., 1994a,b; Davis, 1994; Doren et al., 1997; Wu et al., 1997). Large changes in Everglades community structure are occurring in areas subject to agricultural runoff (e.g., Doren et al., 1997; Wu et al., 1997). These major inputs of allochthonous nutrients are complicated by the current hydrologic regime of the Everglades system, which is characterized by dramatic alterations to flow, inundation, and the timing of both. Lateral and temporal hydroperiod gradients have been greatly reduced from historical patterns (Light and Dineen, 1994). Because the Water Conservation Areas (WCAs) are hydrologically isolated from each other, only vestiges of the original sheetflow remain. In areas where the hydrologic regime is less impacted (e.g., Everglades National Park, ENP), the source of the water is no longer upstream wetlands but canals. For example, Walker (1991) reported that total phosphorus (TP) concentrations of water entering ENP through the S-12 control structures (the four structures that control all water flow into Shark River Slough from the Tamiami Canal; see Light and Dineen, 1994, for details) increased 4 to 21% per year in the mid- to late 1980s. Phosphorus levels are always high in marsh soils near canal inputs (Belanger et al., 1989; Craft and Richardson, 1993a; Davis, 1994; Doren et al., 1997), and ENP marshes near the S-12 water inflows show familiar patterns of impact (Scheidt et al., 1989; Doren et al., 1997). A major problem facing researchers and managers is the question of how much of this P being loaded to Everglades wetlands is excessive. Thus, it is critical that water quality decisions for Everglades wetlands be based on empirical knowledge of how these systems respond to nutrient additions, not on extrapolation of knowledge from other wetland ecosystems. In this chapter, we present a unique experimental study designed to generate just this type of empirical information.

In 1991, Florida and the U.S. Department of Justice entered into a Settlement Agreement to resolve a federal lawsuit brought against the state. The purpose of this litigation was to end discharges of nutrient-enriched canal water into the Loxahatchee National Wildlife Refuge (LOX) and ENP. As part of the Settlement Agreement and the 1994 Florida Everglades Forever Act, all parties agreed to conduct research to determine the Narrative Class III nutrient water quality criterion for the Everglades. This criterion stated that nutrients may not "be altered so as to cause an imbalance in natural populations of aquatic flora or fauna." A major component of this research is field dosing studies in unimpacted, pristine Everglades wetlands. In this chapter, we describe the P dosing study that we are currently conducting as part of this nutrient threshold research. The results of this study will provide valuable information to the Florida Environmental Regulatory Commission by the deadline set forth in the Everglades Forever Act and will help assign a water quality standard for Everglades wetlands.

Much past research in the Everglades has focused on the WCAs near the EAA, or in regions near canal sources of nutrient inputs. A great deal is known about WCA-2A, where marshes have been receiving large P loads for nearly 30 years (SFWMD, 1992; Davis, 1994). Nearly half of

WCA-2A is now impacted by these long-term P additions (Koch and Reddy, 1992). Soil P levels in WCA-2A decline with distance from canal sources but are high throughout relative to less impacted Everglades marshes. WCA-2A soils contain 500 to over 1600 µg P g$^{-1}$ dry soil (Reddy et al., 1991; Craft and Richardson, 1993b) compared to ENP marshes, where soil P levels are typically 200 to 300 µg P g$^{-1}$ dry soil (Amador and Jones, 1993; Doren et al., 1997). Areas receiving direct canal inputs show dramatic shifts in plant species composition from diverse assemblages to rapidly growing monospecific stands of *Typha* sp. (Belanger et al., 1989; Davis, 1991; Doren et al., 1997). Rates of methane release are several times higher compared to unenriched marshes (Bachoon and Jones, 1992; Chanton et al., 1993). Relative densities of dipteran and coleopteran insects and ostracods are higher while relative densities of fishes, decapod, cladoceran, and amphipod species are lower in enriched aquatic faunal communities (Rader and Richardson, 1992, 1994; Turner et al., 1999). Our field dosing experiment in ENP and LOX is unique and important to management and restoration of the Everglades landscape because of the need for these unenriched areas of the Everglades to be represented when water quality standards are determined.

Our approach utilizes flow-through flumes, three in Shark River Slough (ENP) and one in south-central LOX. Each flume has four 100-m channels — one control channel and three treatment channels (Figures 4.1 and Color Figure 4.1*). We add P to increase TP concentrations in the experimental channels by 5, 15, and 30 ppb (0.17, 0.5, and 1.0 µM) above ambient. We sampled all major ecosystem components in these flumes for 1 year before experimental P additions began. Sampling will continue for 3 years of P additions and for a minimum of 1 year after P additions end. Our high-P treatment is roughly equivalent to the experimental P concentration used in a previous (1983–84) ENP flume study (Flora et al., 1986; Scheidt et al., 1988; Walker et al., 1989), but we expect our experiment to be considerably more informative than this prior flume study for a number of reasons:

1. The 1983–84 study had channels at only one location in ENP (and only one channel per treatment), whereas ours has three flumes located along a transect across Shark River Slough, giving us spatial replication of all treatments.
2. The 1983–84 flume had no control channel, while each of our flumes has a control channel.
3. Water flux through the 1983–84 channels was quantified every 2 weeks, and nutrient addition rates for the following 2 weeks were based on this single measurement while we are using electronics to continuously measure water flux through the flumes and instantaneously calculate the amount of P necessary to maintain desired concentrations.
4. The 1983–84 study used fixed-wall flumes that may have caused extensive shading and edge effects, whereas our flumes have walls specially designed to minimize shading.
5. Phosphorus additions to the 1983–84 flumes were based on a constant P load, thus experimental P concentrations varied with water depth; our technique maintains constant levels of P addition by concentration.

Our central objective is to quantify the water quality standard for Everglades wetlands. We will thus identify the P concentration above which "an imbalance in flora or fauna occurs." These imbalances should occur as predictable and statistically identifiable changes through time in various biotic and abiotic components of the Everglades wetland ecosystem (Figure 4.2). Notably, only some of these changes will be visible, but the ability to visually observe a floral/faunal imbalance of an ecosystem component is not a measure of the importance of that component. Our central hypotheses are that:

---

* Color figures follow page 648.

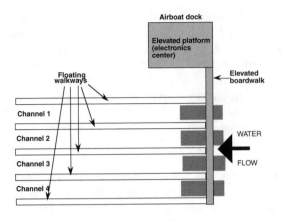

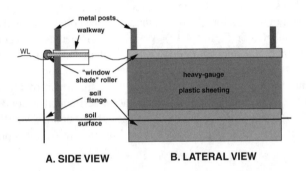

**A. SIDE VIEW**          **B. LATERAL VIEW**

**FIGURE 4.1**   (A) Simplified blueprint of the flume setup, including four 100-m experimental channels each 3 m wide and separated by floating walkways, the connecting boardwalk, and the instrumentation platform and airboat dock. The 10-m mixing areas located at the head of each channel are shown as gray. (B) Diagram of flume wall construction using heavy-gauge black plastic sheeting joined to the aluminum flange inserted into the soil. Walls hang on "window shade" rollers attached to the floating walkways. The entire floating walkway–channel wall structure is held in place with metal posts set firmly into the underlying bedrock.

1. Changes in P cycling, initiated by low-level additions of P to the water column, induce an ecosystem state change in Everglades wetlands that leads to the types of disturbed ecosystems discussed above.
2. This ecosystem state change occurs as a trophic cascade of response first measurable in the microbial components (Figure 4.2).
3. Ecosystem state change occurs more rapidly when the concentration of experimentally added P is greater, but the endpoint of this ecosystem state change will not be affected by the concentration of experimentally added P (in other words, there is no P threshold. Time, not P concentration, is the independent variable of singular importance).
4. During ecosystem state change, the system may become N limited and may actually be a source of P to downstream wetlands. This hypothesis is based on recent evidence that aquatic systems, both freshwater and marine, tend to be P limited when oligotrophic but N limited when mesotrophic to eutrophic and that the switch from P to N limitation indicates a critical state change in trophic status (Vitousek and Reiners, 1975; Vitousek and Howarth, 1991; Vitousek et al., 1993).

Our hypotheses are unique to this type of research being driven by the need to define a water quality standard for Everglades wetlands. Given this, and the fact that our expected results are relevant to all Everglades wetlands, we feel that our flume study is of interest to the full range of Everglades scientists and managers alike.

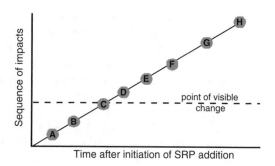

FIGURE 4.2    Conceptual diagram of how P addition affects different ecosystem components of an Everglades wetland through floral and faunal changes. A = increased TP concentrations in water column; B = changes in microbial community structure and function; C = changes in periphyton community structure and function; D = increased P content in soils; E = changes in microinvertebrate community; F = increased production by emergent macrophytes; G = changes in macroinvertebrate and fish communities; H = changes in macrophyte community. Note that the first three changes are not observable by the naked eye.

## FIELD EXPERIMENTAL DESIGN

Flume techniques have been used in a variety of coastal environments (see Childers, 1994, and Childers et al., 1998, for reviews) including saltmarshes (Chalmers et al., 1985; Wolaver et al., 1985; Childers and Day, 1988; Spurrier and Kjerfve, 1988; Childers et al., 1993); brackish marshes (Wolaver and Zieman, 1983; Childers and Day, 1990a,b); mangroves (Twilley, 1985; Rivera-Monroy et al., 1995); coral reefs (Rogers, 1979; Atkinson and Bilger, 1992); intertidal seagrass banks (Childers, unpubl. data); and oyster and mussel reefs (Dame et al., 1984; Asmus et al., 1992). Flumes have also been used to quantify faunal utilization of wetland surfaces in saltmarshes (Peterson and Turner, 1994) and tidal freshwater marshes (McIvor and Odum, 1986). The basic design involves walls built parallel to natural water flow to assure that "water in = water out." Edge effects of the channel walls are possible problems with *in situ* flumes. These effects have two major components: hydrologic turbulence and shading. Flow rates through our flumes are extremely slow and therefore we assume that edge turbulence is not a factor. Most other flume studies have dealt with potential wall shading by simply removing the walls between samplings. We minimize shading effects by attaching the channel walls to rollers on the edges of floating walkways that separate each channel (Figure 4.1b). As a result, the walls never extend more than about 10 cm above the water surface. Virtually all of our sampling is carried out in the central 1 m of the 3-m-wide channels to further reduce any possible wall effects. The walls attach at the bottom to aluminum flanges inserted 30 to 40 cm into the soil (Figure 4.1b). These flanges prevent cross-channel movement of constituents by root or rhizome translocation or interstitial water movement.

Wetlands of the Shark River Slough and LOX are considered to be relatively unaffected by nutrient additions compared to other regions of the Everglades (Scheidt et al., 1989). By locating our study in these wetlands we assume that our experiments are not beginning at some intermediate stage of response to P additions. In Shark River Slough, we spatially replicated our design at three locations. Each flume has four experimental channels 3 m wide and 100 m long built parallel to the direction of natural water flow (Color Figure 4.1). These three sites are all in long hydroperiod slough marshes with peat soils and similar vegetation, including *Eleocharis*-dominated marsh, extensive periphyton mats, and relatively high plant species diversity. Flume walls allow us to assume that a water parcel is "treated" by processes occurring in the 300 m² portion of marsh enclosed by a channel during passage through the channel. Within each flume, we randomized the P treatments by channel.

Phosphorus additions to our flume channels are electronically controlled and based on instantaneous water volume flux to maintain consistent experimental concentrations. The first 10 m of each flume channel is the nutrient mixing area and the zone where we measure water flux (Figure 4.1 and Color Figure 4.1). This mixing area is devoid of vegetation and has a solid fiberglass "floor" fixed to the soil surface. Highly sensitive acoustic Doppler flow sensors, which are capable of measuring flow rates down to 1–2 mm sec⁻¹ and integrate flow across the width and depth, are set up in these mixing areas. They transmit channel-specific flow rates in real time to an on-site computer which also receives real-time water level data from a pressure transducer. The computer calculates channel-specific flux and determines the rate at which P must be pumped into that channel, based on its programmed treatment level (Color Figure 4.1). Phosphorus is added from a reservoir filled with $NaH_2PO_4 + Na_2HPO_4$ (to make a pH of 7.0). We are thus adding P as soluble reactive P (SRP), which is quickly taken up into particulate fractions (measured as TP) in the mixing areas. Each flume is also instrumented with a full meteorological station that includes wet and dry deposition collectors for sampling atmospheric nutrient inputs. The on-site computers are all interfaced to cellular communications, allowing us to access water flux data, met station data, pump status, and system diagnostics at any time (Color Figure 4.1). All on-site electronics are powered by solar panels and a bank of deep-cycle batteries.

In our flume study, we are following ecosystem responses in space (along flume channels) and time (over 3 years of P additions) simultaneously without modifying either the hydrodynamics or geomorphology (Bothwell, 1985; Atkinson and Bilger, 1992; Peterson et al., 1993a,b; Peterson and Turner, 1994). We expect changes in various ecosystem components to be time-lagged in different channels, depending on the experimental P concentrations. Furthermore, as the experiment progresses we expect these transitions to progress down-channel, with the rate of progression a function of P treatment. This interaction of space and time at a range of scales is both an asset and a complicating liability (Thorhallsdóttir, 1990; Hoekstra et al., 1991; Holling, 1992). Controlled, small-scale nutrient enrichment experiments (e.g., laboratory microcosms, field mesocosms) are able to detect temporal response but give little insight into spatial aspects of those responses. On the other hand, spatial analysis of areas already impacted by nutrient additions (e.g., gradient analysis, transect analysis) are temporally static unless carried out over long time frames. While capturing change in space (at scales ranging from centimeters to kilometers) and time (at scales ranging from minutes to years), we are also observing change at a range of trophic and energetic scales (Farmer and Adams, 1989; Martinez, 1993; Pahl-Wostl, 1993; Peterson et al., 1993b). This combination of spatiotemporal integration and trophic/energetic articulation is a unique feature of our study.

## STATISTICAL EXPERIMENTAL DESIGN

Each flume is considered to be one experimental block with four treatments: three levels of P addition and a control. We are using four separate flumes (three in ENP and one in LOX) to increase the robustness of our results based on our knowledge of natural variability in Everglades ecosystems. A preliminary analysis of variability in selected dependent variables in Shark River Slough provided insight into the spatial scales at which community composition varies in these habitats (Table 4.1). These analyses indicated marked patchiness at spatial scales above 10 m. Using the coefficient of variation (CV) of samples taken across several spatial scales, we found that variability was roughly constant or increased for most dependent variables when comparing areas separated by 1 to 10 m (~0.01 ha), 10 to 100 m (~1 ha), and 1 to 5 km (~1000 ha; Table 4.1). These three levels of spatial variability are roughly equivalent to variation within a channel, between channels in a given flume, and between flumes, respectively. These patterns suggest that communities are, to some extent, responding to local conditions. Thus, the outcome of experiments conducted in any single location may be subject to unique and uncontrolled local conditions. For this reason, we replicated our *in situ* experiment across Shark River Slough and included a flume in LOX.

Flume data are analyzed using a generalized linear model (GLM) as a blocked repeated measures analysis of variance (ANOVA; McCullagh and Nelder, 1989). We chose a GLM analysis

**TABLE 4.1**
**Coefficient of Variation (%) at Three Levels of Spatial Scale for Selected Parameters Measured in Shark River Slough**

| Parameter | Spatial Scale | | |
| --- | --- | --- | --- |
| | 0.01 ha | 1 ha | 1000 ha |
| Alkaline phosphatase activity | — | 8.1 | 12.9 |
| Floating periphyton mat volume | 25.8 | 51.2 | 69.5 |
| *Eleocharis* stem density | 60.6 | 30.9 | 42.1 |
| Freshwater prawn density | 44.3 | 38.5 | 53.9 |
| Fish density | 34.3 | 43.9 | 21.3 |

*Note:* Variance was estimated from samples representing the indicated spatial scale (e.g., the 0.01-ha samples were taken from the same 10 m × 10 m area, the 1-ha samples were taken from plots separated by 20 m along a 100-m transect, and 1000-ha samples were taken from transects separated by >1 km). Analyses are of unpublished data gathered in the area proposed for this study.

that employs a model fit to data using a "linking function" transformation. The best known "linking function" assumes that data are normally distributed, which may require some simple transformation (i.e., log, or log($x + 1$) transformations). However, the GLM is equally appropriate for dependent variables that are best described by many standard distributions such as the binomial (i.e., multiway contingency tables and logistic regression) and cumulative normal (i.e. probit analysis). We analyze individual dependent variables using appropriate linking functions; however, the basic design of independent variables remains constant. We also use GLM to develop statistical models of the response surfaces generated by our experimental results, using maximum likelihood estimation to obtain parameter values and estimate values of dependent variables at different levels of the independent variables.

Measurements made at sampling locations within a given channel may not be statistically independent. The degree of non-independence, or spatial autocorrelation, can vary among our dependent variables. For example, on a scale of plots separated by 20 m, *Eleocharis* stem density is considerably less spatially autocorrelated than is the floating periphyton mat volume (Trexler, unpubl. data). Microbial parameters display even less spatial autocorrelation with this spatial scale. Therefore, we continually test for independence among sampling locations within the same channel and adjust our statistical design whenever appropriate. If this test of independence fails, we use profile analysis (a form of repeated measures ANOVA) to assess the pattern of variation within channels and permit tests of pattern difference among channels (Von Ende, 1993). This approach is analyzed as a univariate ANOVA if assumptions of homoscedasticity and constancy of correlations among the repeated measures are met (Tabachnick and Fidell, 1983). If not, standard multivariate techniques are available for this analysis. We prefer univariate ANOVA, however, because of slightly improved power.

As a check on our statistical design, we calculated the number of degrees of freedom and F-values for testing differences in normally distributed variables (Table 4.2). Power analysis of a simplified design confirmed that realistically obtainable treatment effects will lead to rejection of our null hypothesis of no effect (Table 4.3). The ability to detect significant treatment effects at conventionally accepted probability levels is not likely to be identical for all dependent variables. We calculated *a priori* estimates of sampling variability for a few representative variables using data from past Shark River Slough research efforts (Table 4.3). We estimated that P additions must create an increase in the among-channel variance of dependent variables by at least 4.7 times the within-channel variability in order to reject the null hypothesis (Table 4.2). For alkaline phosphatase activity, the variance among sites with and without P impacts is over 6 times greater than variance

**TABLE 4.2**
**ANOVA Table for Experimental Design**

| Source | df | MS | F |
|---|---|---|---|
| P (phosphorus) | 3 | | $MS_P/MS_{P \times B}$ |
| B (block) | 2 | | |
| P × B | 6 | | |
| D (distance) | d | | $MS_D/MS_{B \times D}$ |
| B × D | 2d | | |
| T (time) | t | | $MS_T/MS_{B \times T}$ |
| B × T | 2t | | |
| P × D | 3d | | $MS_{P \times D}/MS_{P \times B \times D}$ |
| B × P × D | 6d | | |
| P × T | 3t | | $MS_{P \times T}/MS_{P \times B \times T}$ |
| B × P × T | 6t | | |
| D × T | dt | | $MS_{D \times T}/MS_{B \times D \times T}$ |
| B × D × T | 2dt | | |
| P × D × T | 3dt | | $MS_{P \times D \times T}/MS_{P \times B \times D \times T}$ |
| B × P × D × T | 6dt | | |

*Note:* MS = mean squares; subscript abbreviations: P = levels of P addition (ambient concentration +0, +0.17 μ$M$, +0.5 μ$M$, and +1.0 μ$M$); B = block, or flume/site; D,d = distance within a given flume channel; T,t = time. Lowercase letters indicate the number of levels of each factor minus one.

**TABLE 4.3**
**Minimum P Treatment Effects Detectable by Experimental Design and Natural Variability Anticipated Among Flume Sites**

| Dependent Variable | "Pristine" Site (% change) | Natural Site Variability Site Comparison (% difference) |
|---|---|---|
| Alkaline phosphatase activity | 12.6 | 5–8 (78) |
| Floating periphyton mat (mL/m²) | 47.5 | 28–78 |
| *Eleocharis* stem density (#/m²) | 102.8 | 6–71 |
| Freshwater prawn density (#/m²) | 75.1 | 10–64 |
| Fish density (#/m²) | 58.2 | 5–42 |

*Note:* Percent change at "pristine" site indicates a conservative difference between treatment and control channels that must be exceeded to be detectable at $\alpha = 0.05$ level with $(1–ß) = 0.80$. These estimates are conservative because they are based on two sample $t$-tests with comparable replication to two of our channels. Our design, with more than two levels of P treatment, actually require slightly less change for rejection of the null hypothesis. These estimates are for main effects only; power tests of interactions will differ. Natural site variability is among "pristine" sites in Shark River Slough separated by more than 1 km. Variation among P-impacted sites near S12 structures and "pristine" sites in Shark River Slough (shown in parentheses) is indicative of differences anticipated among treatment and control channels for alkaline phosphatase activity. No comparable data are available for other dependent variables.

within these sites (R. Jones, unpubl. data; Tables 4.1 and 4.3). Note that for other variables, natural variability among sites is of the same order of magnitude as minimal detectable percent change within sites (Table 4.3). Thus, our design allows detection of P addition effects at the standard $\alpha = 0.05$ level.

Following our omnibus tests of P treatment effects, we avoid using contrasts to ascertain the P level with a minimum detectable effect. Failure to reject the null hypothesis of no effect for a given P level does not prove that there is no effect at that level; rather, this may be due to insufficient statistical power. We then emphasize statistical response surface modeling of the magnitude of response over time at different levels of P addition. Use of these dose–response curves focuses attention on the issue of what degree of change constitutes an imbalance in flora and fauna rather than on statistical hypothesis tests.

If hypothesis testing for a threshold P level is desired (assuming that this threshold exists), we will employ a relatively new approach called bioequivalence testing (Dixon and Garrett, 1993). This approach shifts the burden of proof from testing the null hypothesis that the difference in a response variable, between control and treatment, is zero to testing the null hypothesis that this difference exceeds some predetermined amount. Thus, the null hypothesis must be rejected to conclude that a given P concentration had no significant effect. In order to conduct a bioequivalence test, one must first decide what degree of difference will be considered equivalent to no effect. This difficult philosophical and biological question must be addressed before a threshold level of P can be set, even if bioequivalence statistics are not used.

## METHODS

The organization of this study closely follows a simplified conceptual model of how P cycles in Everglades wetland ecosystems (Figure 4.3). Notably, this model is also the blueprint for our modeling effort and project integration/synthesis. Key Element Groups quantify important structural parameters and process rates associated with their component of the ecosystem (including the microbiota, soils, periphyton, macrophytes, and fauna; Table 4.4). We also have groups responsible for: (1) sampling and quantifying waterborne nutrient concentrations, fluxes, and budgets; (2) tracking visible change in the flumes using photography, image analysis, and GIS, and; (3) project integration, synthesis, modeling, and budgeting. The personnel responsible for field operations and maintenance, and for data management and quality assurance/quality control (QA/QC) are independent of these Key Element Groups.

The sheer volume and diversity of datasets being generated by this project necessitate a well-organized data management program headed by a project Data Manager. The QA/QC process is

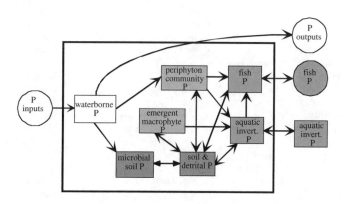

**FIGURE 4.3** A simple conceptual model of phosphorus cycling in Everglades marsh ecosystems.

performed independently of the data management either by analytical lab personnel or by individual investigators. All data are compiled, archived, and stored in redundant external formats. Project principal investigators (PIs) may request any portion of the project dataset for analysis. Site-specific electronic data are regularly downloaded. These data, including meteorological and hydrologic information, are available to all PIs upon request.

Each Key Element Group follows their own sampling and analytical protocols (see Table 4.4 for a summary). Repeatable and nondestructive sampling methods are being used by all groups, and destructive sampling is minimized. We carry out all sampling from either the floating walkways or from portable bridges over the channels (Color Figure 4.1). The field and lab techniques used by each group are detailed below, with particular emphasis on the structural parameters and process rates being quantified.

**TABLE 4.4**
**Detailed Listing of the Spatial and Temporal Sampling Regimes of All Major Parameters by Key Element Group**

| | | | Sampling Frequency | |
| Parameter | Principal Investigator | Locations | Temporal, Initial | Temporal, Long Term |
|---|---|---|---|---|
| A. Physical and environmental parameters | | | | |
| 1. Water level, water velocity | | One per channel | Continuous | Continuous |
| 2. Air and water temperature | | One per flume | Continuous | Continuous |
| 3. Wind speed and direction | | One per flume | Continuous | Continuous |
| 4. Rainfall (amount, intensity, frequency, content) | | One per flume | Continuous | Continuous |
| 5. Solar irradiance (incident) | | One per flume | Continuous | Continuous |
| 6. Solar irradiance (water surface, sediment surface) | | One per channel | Quarterly | Quarterly |
| 7. Water level gauge surveys | | One per flume | Annually | Annually |
| 8. Flume channel microtopographic surveys | | One per channel | Annually | Annually |
| B. Biogeochemical components | Scinto, Jones | | | |
| 1. Flume flux (regular) | | 0 and 85 m | Monthly | Monthly |
| 2. Flume flux (around regions of state change) | | Variable | Variable | Variable |
| 3. Ammonium, nitrate, nitrite, SRP, TN, DOC, DON | | 0 and 85 m | Daily | Daily |
| 4. Total phosphorus | | 0 and 85 m | Daily | Daily |
| 5. Silicate | | 0 and 85 m | Weekly | Weekly |
| 6. Conductivity | | 0 and 85 m | Weekly | Weekly |
| 7. Turbidity, pH | | 0 and 85 m | Weekly | Weekly |
| C. Microbial components | Scinto, Jones | | | |
| 1. Trace gas production, soil | | 1, 3, 8, 33, 58, 83 m | Monthly | Quarterly |
| 2. Biomass production, soil and water | | 1, 3, 8, 33, 58, 83 m | Monthly | Quarterly |
| 3. Nitrogen fixation, soil | | 1, 3, 8, 33, 58, 83 m | Quarterly | Seasonally |
| 4. Exoenzymes, soil | | 1, 3, 8, 33, 58, 83 m | Monthly | Quarterly |
| 5. Alkaline phosphatase activity | | 1, 3, 8, 33, 58, 83 m | Monthly | Quarterly |
| D. Soil Components | Jayachandaran, Meeder, Scinto | | | |
| 1. Whole soil analysis | | 1, 3, 8, 33, 58, 83 m | Monthly | Quarterly |
| 2. TN, TP, TC, % OM, bulk density | | 1, 3, 8, 33, 58, 83 m | Monthly | Quarterly |
| 3. Redox (Eh), pH | | Variable (Hydrolabs) | Monthly | Quarterly |
| 4. Porewater chemistry | | 0.5, 1, 2, 5, 10, 25, 50, 75 m | Monthly | Quarterly |
| 5. Marker horizon analysis | | 6, 37, 67 m | Quarterly | Quarterly |

**TABLE 4.4 (CONTINUED)**
**Detailed Listing of the Spatial and Temporal Sampling Regimes of All Major Parameters by Key Element Group**

| Parameter | Principal Investigator | Locations | Temporal, Initial | Temporal, Long Term |
|---|---|---|---|---|
| | | Sampling Frequency | | |
| 6. Sediment elevation | | 6, 37, 67 m | Quarterly | Quarterly |
| 7. Long-term depositional rates (Cs analysis) | | 6, 37, 67 m | Once | Once |
| 8. Paleochronologic and paleoecological analysis | | 6, 37, 67 m | Once | Once |
| 9. Below-ground production — minirhizotrons | | 6, 37, 67 m | Monthly | Quarterly |
| 10. Below-ground production — screen-intercept | | 6, 37, 67 m | Quarterly | Quarterly |
| E. Periphyton components | Gaiser, Richards | | | |
| 1. Production, community composition — periphytometers | | 3, 42, 79 m | Every 6–8 weeks | Every 6–8 weeks |
| 2. Production — dowel growth | | Every 8.5 m | Every 6–8 weeks | Every 6–8 weeks |
| 3. Production — oxygen diurnals | | 3, 42, 79 m | Quarterly | Quarterly |
| 4. Community composition — HPLC pigment analysis | | 3, 42, 79 m | Every 6–8 weeks | Every 6–8 weeks |
| 5. Biovolume and biomass measures | | 3, 42, 79 m | Every 6–8 weeks | Every 6–8 weeks |
| 6. Periphyton tissue nutrient content | | 3, 42, 79 m | Every 6–8 weeks | Quarterly |
| 7. Utricularia growth rates | | | Monthly | Quarterly |
| 8. Utricularia tissue nutrient content | | | Every 6-8 weeks | Quarterly |
| F. Macrophyte components | Edwards, Lee | | | |
| 1. Above-ground biomass | | 5, 18.5, 36.5, 66.5 m | Monthly | Monthly |
| 2. Above-ground productivity | | 5, 18.5, 36.5, 66.5 m | Monthly | Monthly |
| 3. Species composition | | 5, 18.5, 21.5, 36.5, 49, 66.5 m | Monthly | Monthly |
| 4. Aboveground physiology | | 5, 18.5, 21.5, 36.5, 49, 66.5 m | Quarterly | Quarterly |
| 5. C, N, P content of aboveground biomass | | 5, 18.5, 21.5, 36.5, 49, 66.5 m | Monthly | Quarterly |
| 6. Genetic variation | | 5, 18.5, 21.5, 36.5, 49, 66.5 m | Twice | Twice |
| G. Faunal components | Pechman, Trexler | | | |
| 1. Microinvertebrates — inverted funnel traps | | 10-13, 44–47, 73–77 m | Twice a year | Twice a year |
| 2. Macroinvertebrates — sweep nets | | 10-13, 44–47, 73–77 m | Twice a year | Twice a year |
| 3. Nekton — throw traps | | 10-13, 44–47, 73–77 m | Twice a year | Twice a year |
| 4. Food web relationships (caging studies) | | 10-13, 44–47, 73–77 m | None | Twice a year |
| H. Spatial analysis | Childers, Rugge | | | |
| 1. 1-m² photopoints | | 0, 1, 9, 19, 37, 71 m | Every 2 months | Every 2 months |
| 2. Oblique whole-channel photography | | One per channel | Every month | Every month |
| 3. Aerial (helicopter) photography | | One per flume | Every 2 months | Every 2 months |

*Note:* Actual sampling area of each flume channel is 85 m long, excluding the 10-m mixing area at the channel head and the terminal 5 m of channel. Temporal Initial Frequency refers to the sampling scheme that took place in the first 3 months after P dosing began.

## BIOGEOCHEMICAL COMPONENTS

Biogeochemical analysis of the flumes involves quantifying key process rates of the P, C, and N cycles, diurnal and seasonal patterns in these cycles, and how both rates and patterns are affected by P addition. Four times per year (twice during the wet season and twice during the dry season) we collect water samples hourly for a 24-hour period at the upstream and downstream ends of each flume. These samples are analyzed for soluble inorganic nutrients (SRP, $NH_4$, $NO_3^-$ $NO_2$) by standard colorimetric methods (U.S.E.P.A. 1983, methods 365.1, 353.2, 353.2, respectively) and for soluble organic nutrients (DOP, DON, DOC), and total nutrients (TP, TOC, TN). Total P is analyzed by the method of Solorzano and Sharp (1980), and DOP is determined by the difference between TP and SRP. Total organic C and DOC are analyzed by EPA standard method 415.1(U.S.E.P.A. 1983). Total N and DON are determined using an Total N Analyzer (ANTEK Instruments Model 7000; Houston, TX). We use real-time channel water flux rates to calculate whole channel flux rates for these constituents in all flume channels simultaneously. During these same diurnal sampling events we also measure several environmental parameters, including dissolved oxygen, pH, electrical conductivity, and temperature using Hydrolab Multiprobes (Hydrolab Corp., Austin TX). In addition to these intensive diurnal flux measurements, we collect water samples at several locations (1, 3, 8, 33, 58, and 83 m downstream) along each channel four times per year. Quantifying nutrient spiraling lengths is one application of these spatially intensive flux data (Newbold et al., 1981, 1983). Shifts in the relative proportions of soluble inorganic to soluble organic to total organic forms with distance (i.e., time) are being used to calculate nutrient spiraling lengths and cycling rates.

The Biogeochemical Group overlaps significantly with several other Key Element Groups, including the Soils Group and the Microbial Group. As a result, samples necessary for the determination of chemical cycles, soil, soil porewater, and flocculent material (floc) chemical characteristics are often collected in conjunction with the sampling of other Key Element Groups.

## MICROBIAL COMPONENTS

Microbial populations and activities in the water column and soils are the ecosystem components that change most rapidly in response to P addition, according to our central hypothesis. The microbiological properties we are quantifying include: (1) enzyme activities and bacterial densities in the soil, the flocculent layer (floc), and the water; (2) production of trace gases ($CO_2$ and $CH_4$) in the soil and the floc; and (3) nitrogen fixation and denitrification in the soil and the floc. We intensively sampled a number of these indices immediately after P dosing began (on day 7, day 14, and after 1, 2, and 3 months). After the third month, we began conducting these analyses quarterly. Phosphatase, glucosidase, glucosaminidase, and sulfatase enzyme activities are determined using the model fluorescent substrate 4-methylumbelliferone (Sinsabaugh et al., 1997). We determine bacterial densities using a fluorometric assay based on the double-stranded DNA stain Pico-Green (Molecular Probes, Inc.; Eugene OR)(Tranvik, 1997). Nitrogen fixation, denitrification, and the production of $CO_2$ and $CH_4$ are quantified on homogenized soil and floc slurries. Denitrification is measured as the increase in $N_2O$ content of headspace gas by the acetylene blockage technique using a gas chromatograph (GC; Hewlett-Packard 5890) fitted with an electron-capture detector (ECD; Gordon et al., 1986). Nitrogen fixation, or more correctly nitrogenase activity, is measured via acetylene reduction to ethylene on a GC equipped with a flame-ionization detector (FID; Stal, 1988). The production of $CO_2$ and $CH_4$ are measured as the content of these gases appearing in $CO_2$-free air-purged headspace after 72-h incubation at 25°C. Carbon dioxide and $CH_4$ are measured by GC equipped with both FID (for $CH_4$) and ECD (for $CO_2$) according to the protocols of Amador and Jones (1993). We are also measuring the oxidation/reduction (redox) status of the soils, with probes interfaced to the Hydrolab Multiprobes units, as an integrative index of microbial functioning. The microbial processes that we are measuring are primarily responsible for the cycling of macronutrients in the system; thus, the information from the Microbial Group greatly aids the work of the Biogeochemical Group.

## Soil Components

Changes in the chemical characteristics of the soil and floc material are being studied at several locations (1, 3, 8, 33, 58, and 83 m downstream) within each flume channel. Our schedule of initial (immediately after P dosing began) and routine sampling are as described above in the Microbial Group methods. Volumetric sampling of floc and the surface 10 cm of soil is done by removing small diameter (<3 cm) intact cores. The floc and soils are carefully separated in the field and returned to the lab for analysis. Soil and floc are homogenized and analyzed for pH (soil pH on 1:1 soil:H2O slurry, floc by direct insertion of electrode), dry weight (ASTM, 1995, D2216-80), bulk density (ASTM, 1995, D4531-86), ash free dry weight (ASTM, 1995, D2974-87), total P (Solorzano and Sharp, 1980), total C, and total N. Total C and N content of the soil and floc is measured with a N/C analyzer (model NA 1500, Carlo Erba Instruments, S.p.a.; Milan, Italy; Nelson and Sommers, 1996). Additionally the soil is analyzed for total extractable P. This is the P released following $CHCl_3$ fumigation and $NaHCO_3$ extraction. Soil porewater is collected at 0.5, 1, 2, 5, 10, 25, 50, and 75 m downstream from a 10-cm soil depth using porewater microsippers, and samples are analyzed for soluble nutrients as above.

We are tracking nutrient storage in flume channel soils by determining the relative abundance of soil components vertically through the soil column. These soil components are based on the three dominant soil forming processes: (1) calcium carbonate precipitation produced by periphyton activity, (2) fibrous organics or root material produced by vascular plants, and (3) amorphous organics produced from decomposition of periphyton and vascular plants. We are quantifying rates of deposition of organic floc using clay marker horizons (Bloom, 1964; Boumans and Day, 1993). The clay horizons were applied as a clay slurry to square-meter$^2$ areas of marsh enclosed in baffle boxes. A survey to the top of the marker clay allows the separation of above marker accretion from below marker soil production.

The distribution of root production and the rates of growth and senescence are being determined using minirhizotrons (Hendrick and Pregitzer, 1992) and soil screen ingrowth methods. Minirhizotrons were installed at 5, 36.5, and 66 m and are photographed semiannually. Soil screens were also inserted and are sampled semiannually for root growth by depth. We retain root tissue from the soil screens for biomass and for C, N, and P analysis. Floc and soil composition and bulk density are measured at intervals to quantify any change in soil nutrient concentrations or rates of deposition. The relative percent of algal vs. vascular vegetation making up the organic floc is determined by staining and microscopic point counting methods.

## Periphyton Components

Periphyton is an integrated community of submersed macrophytes, attached microalgae, bacteria, fungi, and microconsumers. In Shark River Slough, periphyton forms a thick, calcareous, floating mat of cyanobacteria and other microbes and is typically associated with purple bladderwort (*Utricularia purpurea*). Many studies have shown substantial alterations in periphyton community composition and structure in response to even low levels of P addition, including complete disintegration of the floating mat (Swift and Nicholas, 1987; Raschke, 1993; Vymazal et al., 1994; McCormick et al., 1996; McCormick and O'Dell, 1996). We employ several different methodological tools to quantify periphyton community structure and function, and how both are affected by P additions (Table 4.5).

We use periphytometers (artificial substrates; plastic flow-through chambers that contain 20 horizontal glass microslides) to assess algal composition and growth rates. Although periphytometers may produce compositionally biased communities, they offer the advantage of a substrate

**TABLE 4.5**
**Sampling Protocol for Periphyton-Related Parameters at the Dosing Facilities**

| Substrate | Interval | Location (m) | Measurements |
|---|---|---|---|
| *Utricularia* | 8 wk | 5, 19, 39, 65 | Growth rates, CNP, stable isotopes |
| | | All | Densities (*U. foliosa*) |
| Floating mat | 8 wk | 5, 19, 39, 65 | Cover, cell counts, ash-free dry mass, chlorophyll *a*, CNP, HPLC, dry mass, % *Utricularia*, animals, N-fixation, productivity |
| Benthos | 3 mo | 5, 19, 39, 65 | Cell counts, ash-free dry mass, chlorophyll *a*, HPLC, dry mass, CNP |
| Periphytometers | 2 wk | 3 | Cell counts |
| | 6 wk | 3, 47, 78 | Cell counts, ash-free dry mass, chlorophyll *a*, HPLC, dry wt. |

that is consistent in composition and area and does not vary among treatments. We deploy three floating periphytometers in each channel at 3, 47, and 78 m. To assess short-term successional trends in species composition, we initially harvested periphytometer slides after 2, 4, 6, and 8 weeks of incubation. Our pre-dose biomass data, described below, suggested that 6 weeks is the optimal incubation interval. Thus, every 6 weeks we remove 14 slides and analyze them for cell composition, density, biovolume, chlorophyll *a* content, accessory pigment content, dry weight, ash-free dry weight, and nutrient content.

The floating periphyton mat is sampled bimonthly to assess algal cell composition, density, biovolume, chlorophyll *a* and accessory pigment content, dry weight, ash-free dry weight, nutrient content (total C, N, and P) and *Utricularia* biomass. We extract small (4.2 cm$^2$) cores from the floating periphyton so as not to significantly disturb the integrity of the mat. Fifteen cores are taken at 5, 19, 39, and 65 m and composited for a 1-m$^2$ area at each site. We used pre-dose sampling of periphyton and *Utricularia* dry weight biomass to determine that 15 small cores provides an adequate representation of the periphyton mat in a 1-m$^2$ area. We expand these biomass estimates to the entire flume channel using periphyton cover estimates calculated with image analysis of digital photographs taken of each quadrat bimonthly. We are sampling benthic periphyton directly from floc samples collected by the Microbial and Soils Key Element Groups. These samples are analyzed for chlorophyll *a*, accessory pigments, living cell densities and biovolumes, biomass (as dry weight and ash-free dry mass), and nutrient content.

Productivity is measured quarterly in the lab as $O_2$ and $CO_2$ exchange using light and dark bottle incubations of periphyton cores. We also measure whole-system $O_2$ production in the field using Hydrolabs that generate continuous and simultaneous time-series datasets for all channels. We also quantify N fixation by periphyton mat cores quarterly using similar laboratory incubations that are interfaced to a $N_2$ gas chromatograph.

Periphyton often forms round mats around the dead macrophyte stems — parochially termed "sweaters." During the pre-dose period, these periphyton sweaters were a minor component of the community in all flume channels. When sweaters appear, we sample three stems (clipping the entire stem) of each of the dominant sweater-containing macrophytes and analyze the periphyton as above. This destructive sampling is restricted to the 1-m$^2$ quadrants described above.

Three species of *Utricularia* are found in the flumes: *U. purpurea, U. foliosa, and U. gibba.* *Utricularia purpurea* provides an important substrate for the periphyton mat but is also found free-floating. *U. foliosa* grows under or through the mat and in open water. *U. gibba* (a filamentous *Utricularia*) is found at low densities in the mat or in open water. We measure growth rates of free-floating *U. purpurea* and *U. foliosa* bimonthly by marking internodes above the most recently matured leaf with strings attached to styrofoam floats, then counting the number of new nodes produced after 2 to 6 weeks. We make these measurements on 48 (*U. purpurea*) or 36 (*U. foliosa*) plants, and the sampling interval varies between species and among seasons. Tissue samples are collected for C, N, and P content and stable isotope analysis bimonthly. For *U. purpurea*, 10 free-floating plants adjacent to the 5-, 19-, 39-, and 65-m quadrants are collected. We composite tissue from the most recently matured node and internode of these plants for nutrient analysis and composite shoot tip tissue for C and N stable isotope analysis. For *U. foliosa*, we sample a half-expanded leaf for stable isotope analysis. The most recently matured leaf on a single individual is collected from plants adjacent to the 5-, 19-, 39-, and 65-m quadrants for nutrient analysis.

Biomass (as chlorophyll *a* content, dry mass, and ash-free dry mass) and nutrient content are determined using standard procedures for periphyton (APHA, 1989). Accessory pigments are analyzed by high-performance liquid chromatography (HPLC). For cell counts, a proportion of each sample is oxidized for the identification of diatoms. Another aliquot is removed for the enumeration and identification of living diatoms and soft algae. Taxa are identified to the lowest possible taxonomic unit using standard literature keys. Voucher specimens and digital images of each taxon are housed in a working research collection.

## Macrophyte Components

To characterize how P additions affect the community dynamics of emergent macrophytes, we monitored above-ground biomass and net primary productivity. At each site, we randomly located six 1-m² quadrants, with the locations replicated across all flume channels. Each month, the standing biomass of individual species is estimated nondestructively by counting individuals in each quadrant, measuring the heights and diameters of a subsample of those species, and applying these measurements to allometric regressions of biomass (Daoust and Childers, 1998). Net primary productivity is assessed by tracking a subsample of marked individuals through time. At the ENP flumes, the common species that we count and measure include *Eleocharis cellulosa*, *E. elongata*, *Panicum hemitomom*, *Sagittaria lancifolia*, and *Paspalidium geminatum*. Less abundant species, including *Nymphaea odorata*, *Nymphoides aquatica*, *Pontederia cordata*, *Rhynchospora tracyii*, and *Leersia hexandra*, are also counted during monthly surveys within plots and are counted throughout each flume channel every 3 months. The emergent community at the LOX flume is dominated by *E. elongata*, *N. odorata*, and *N. aquatica*, with *R. tracyii*, *S. lancifolia*, *P. cordata*, and *Peltandra virginiana* occurring less frequently.

As a measure of the physiological responses to increased P, we measure photosynthesis (Pmax), respiration, and transpiration rates of both individual plants and entire plots to provide estimates of C fixation rates. These measurements, performed in the field with a CID gas analyzer, are coordinated with tissue nutrient sampling in order to quantify nutrient use efficiencies. The P and N content of the leaves of each species is determined 4 times per year using tissue from plants immediately adjacent to our randomly located plots. We chose to sample only quarterly because there is little intermonthly variability in tissue C:N:P ratios for these dominant species (Daoust and Childers, 1999). We also assess photosynthetically active radiation (PAR), red:far-red light quality, and ultraviolet (UV)-B irradiance at different canopy heights both above and below the water surface. Since stand density affects radiation in the water column, and could indirectly affect other processes, we are measuring the spectral distribution of radiation (300 to 900 nm) with an Ocean Optics spectrometer. We are thus calculating PAR, red:far-red ratios (R:fR), and UV-B fluxes at different canopy heights both above and below the water surface and along gradients of P influence found at any point in time in our flume channels.

## Faunal Components

Our sampling protocols for fish, invertebrates, and amphibians are based on: (1) our comparison of sampling techniques (Turner and Trexler, 1997); (2) background data provided by our studies in areas near the flumes and elsewhere in the Everglades (Trexler et al., 1998; Turner et al., 1999); and (3) experience gained in pre-dosing sampling in the flumes. We employ three techniques, which together comprehensively sample the most prominent taxa: inverted funnel traps, sweep nets, and 1-m$^2$ throw traps (Turner and Trexler, 1997). Throw traps best sample fish, crayfish, odonate nymphs, and grass shrimp; sweep nets best sample amphipods, mites, and dipteran larvae; inverted funnel traps best sample cladocerans, dipteran larvae, ostracods, and copepods (Turner and Trexler, 1997).

We routinely sample twice a year — once in the dry season (March) and once in the wet season (late September–early October). Samples are collected at three locations along each flume channel: 10 to 13 m, 44 to 47 m, and 73 to 76 m. In addition, we collect samples at these same locations along a reference transect parallel to and about 10 m from each flume. For each technique, we collect duplicate samples at each location within the experimental channels (one from each side), and one sample at each location along the reference transects. Baseline (pre-dosing) samples were collected in August–September 1997 in the ENP flumes. We also conduct studies in which densities of certain faunal groups are manipulated within cages placed in the flume channels. These caging studies began in Fall 1999 and are providing data for interpretation of our routine sampling results.

## SPATIAL ANALYSIS

The spatial analysis group uses photography at three different scales (1-m$^2$ quadrant, channel, and site) to document visual changes occurring under the different levels of P addition. We use a Canon EOS-1 with either a fixed 28-mm or a 28-mm–105-mm zoom lens for all of our photographs. At the smallest scale, we are taking photos of 1-m$^2$ photopoint quadrants at six set locations in the center of each channel: 0–1 m, 1–2 m, 9–10 m, 19–20 m, 37–38 m, and 71–72 m. All photopoints are photographed three to four times each year. We are using panchromatic slide film for all six channel photopoints and are also taking color infrared (IR) slides of those photopoints that correspond to channel regions where visible responses are occurring. We convert the slides to digital format with a high-resolution slide scanner, then classify the digital images and estimate percent cover for periphyton, open water, and vegetation using image analysis-GIS software (including IDRISI, ARC-INFO, and ERDAS-IMAGINE). We create GIS overlays of each photopoint in which sequential photographs are data layers. This allows us to easily calculate rates of change in our classified categories and to animate change through time at a given location.

Monthly, we take oblique panchromatic photos of each flume channel, looking downstream from the upstream end, from a height of 5 to 6 m above water level. These whole-channel images are designed to document change at the scale of an entire flume channel. Every two months, we augment these whole-channel oblique photos with whole-flume aerial photos shot from a helicopter at set hover points located using GPS, and at 100-m altitude. These aerial photos cover an entire flume site and some of the surrounding landscape, including the area just downstream of the flumes.

## INTEGRATION AND MODELING

In a multifaceted, multidimensional project with many investigators, integration and synthesis are extremely important. Our approach to project integration involves both modeling and whole-system field experiments. We are actively developing and using simulation models to continually synthesize data and to "adaptively manage" our research effort (including identifying parameters or processes that are not being adequately quantified, verifying temporal and spatial scales of sampling, and eliminating redundancies of effort). This adaptive management is an important objective of our modeling component. We are also using whole-system field experiments to generate data at a holistic level that are important to whole-project data integration but are not being addressed by other research groups.

We are developing and calibrating simulation models in parallel with field empirical research. A typical model spatial domain is a 3 m × 85 m portion of slough habitat in Everglades National Park (i.e., one flume channel), and a typical integration interval (dt) is 0.125 day for simulations spanning 1 to 30 years. Ultimately, simulation modules will simulate the dynamics of macrophytes, biogeochemistry, consumers, periphyton, and inorganic chemistry. Our first efforts have focused on developing the forcing function and periphyton modules. The forcing function module, which drives the others, includes water temperature, photosynthetically active radiation, water depth, current flow, and rainfall. Our periphyton module assumes a floating periphyton mat that is 2 cm thick and covers 50% of the marsh area. The model simulates C and P biomass, net productivity, and P uptake. Biotic variables include periphyton C and P biomass, and water column TP. Periphyton productivity, thus C and P uptake, is controlled by TP concentration and the periphyton tissue C:P ratio (Buzzelli et al., 2000).

Our whole-system field manipulation is designed to explore important linkages among three key components: periphyton, emergent macrophytes, and soil. To do so, we are conducting an experiment in which we have three different systems manipulations: (1) periphyton removal, (2) macrophyte removal, and (3) all components present. To study the effects of P enrichment on linkages among ecosystem components, we are adding P at a monthly rate of 33.3 mg m$^{-2}$ (400 mg$^{-2}$ m$^{-2}$ yr$^{-1}$) to nutrient treatment plots. The experiment is being conducted in 1-m$^2$ experimental mesocosms that are replicated at the three ENP flume sites. At each site, we have six total mesocosms — two with periphyton removed, two with macrophytes removed, and two undisturbed, with nutrients added to one of each. We are using intensive nondestructive techniques to quantify a range of responses (Table 4.6; see Dailey, 2000, for details).

**TABLE 4.6**
**Details of the Whole-System Field Experiment Being Used To Test Linkages Among Ecosystem Components and To Synthesize Project Data**

| Parameter | Sampling Frequency | Analysis |
|---|---|---|
| Periphyton C, N, P | Monthly | ANOVA repeated measures, element budget analysis |
| Macrophyte C, N, P | Monthly | ANOVA repeated measures, Element budget analysis |
| Soil and flocculent layer C, N, P; Eh, pH | Bi-yearly | ANOVA repeated measures, element budget analysis |
| DO mg/L instantaneous | Monthly | ANOVA repeated measures |
| DO mg/L, %DO saturation, pH Hydrolab data | Quarter annually | Empirical models, curve integration |
| EOS/digital photos | Monthly | GIS software analysis for % cover, ANOVA repeated measures |
| Macrophyte turnover rates | Monthly | ANOVA repeated measures |
| Macrophyte total counts and phenotypic measurements | Monthly | ANOVA repeated measures |
| Porewater nutrients NH$_4^+$, SRP, NO$_3^-$, NO$_2^-$, TOC | Monthly | ANOVA repeated measures |
| New biomass in manipulation plots | Monthly | Empirical models — carbon budget |

## PRE-DOSING RESULTS

As we noted above, we began P additions in October 1998. In the year preceding the initiation of the experiment, we sampled all major ecosystem components to generate a pre-dosing picture of our system. These data were collected for comparison to data from our experimental P dosing, but they also shed new light on the structure and function of Everglades wetlands in general, because our focus is ENP and LOX wetlands and not those in the WCAs. We briefly provide a sample of the pre-experimental data generated by some Key Element Groups below.

## SOIL COMPONENTS

The pre-dosing vertical soil constituent analyses revealed several trends that are consistent between cores. In general, the organic content was quite high throughout the cores, except near the limestone bedrock where the carbonate content increased dramatically (Figure 4.4). We also noted a lack of recognizable above-ground biomass (leaves and stems) below 8 cm, suggesting either that aboveground biomass is tightly recycled and relatively rapidly decomposed, or that soil accretion rates are slow. During the pre-dose period, floc accumulated at a mean rate of 6.04 cm yr$^{-1}$ ($\pm 1.52$) for all flumes combined (Table 4.7). There are no statistical differences or trends between sites or between individual channels. Our pre-dose sampling of the root ingrowth screens elucidated some sampling problems; we replaced the screens using a different mesh and larger diameter coring device.

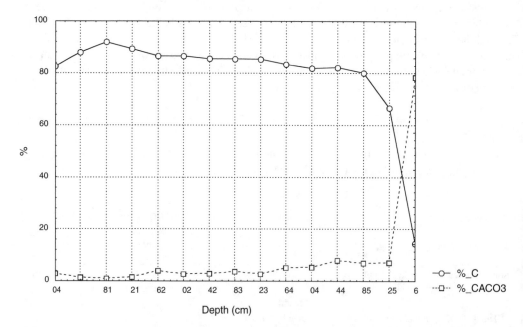

**FIGURE 4.4**   Comparison between organic carbon and calcium carbonate (CaCO$_3$) in a vertical profile from core B1-65.

**TABLE 4.7**
**Summary of Floc Accretion Rate Data**

|  | Site | | | | | |
|---|---|---|---|---|---|---|
|  | Channel 1 | Channel 2 | Channel 3 | Channel 4 | Mean (cm/year) | Standard Deviation |
| Site A | 5.30 | 5.68 | 8.35 | 9.23 | 7.14 | 1.944350961 |
| Site B | 6.64 | 4.53 | 4.44 | 6.17 | 5.44 | 1.127830268 |
| Site C | 4.36 | 6.41 | 6.41 | 5.03 | 5.55 | 1.027897832 |
| Channel mean (cm/yr) | 5.43 | 5.54 | 6.40 | 6.81 | 6.04 | 0.668914447 |
| Standard deviation | 1.15 | 0.95 | 1.96 | 2.17 | 0.95 | 1.524187236 |

## Periphyton Components

Periphyton communities on periphytometer slides sampled in the ENP flumes were dominated numerically by diatoms and taxonomically by desmids (Table 4.8). Common periphyton mat taxa, including *Schizothrix hoffmani* and *Scytonema calcicola,* were present in low abundance on the slides. The diatom taxa *Mastogloia smithii, Fragilaria synegrotesca, Encyonema* sp. 1, and *Brachysira vitrea* were dominant. Biomass measured using periphytometers was consistently higher in the Loxahatchee flume channels compared to the ENP flumes (Table 4.9). We found no consistent differences in periphytometer biomass among channels or quadrants within flumes, but benthic periphytometers contained significantly lower biomass than floating slides (< 80 µg dry mass cm$^{-2}$). We observed a marked seasonality in productivity and taxonomic composition from periphytometers (Figure 4.5a). Biomass was high in late summer and decreased into early spring. Accumulation rates on periphytometer slides were highest (~5 µg dw cm$^{-2}$ day$^{-1}$) in August and October and decreased to a minimum (< 1 µg dw cm$^{-2}$ day$^{-1}$) in April. Chlorophyll *a* followed a similar pattern. There were also clear differences in algal community composition among 8-wk incubation episodes, denoting significant seasonal preferences for many algal taxa.

In winter and spring 1998, we followed algal accumulation on periphytometer slides throughout the 8-wk incubation period (Figure 4.5b). Although biomass usually increased throughout the 8-wk period, we occasionally found that slides collected after 8 wk contained significantly fewer cells than slides incubated for only 6 wk (e.g., August and December 1997; Figure 4.5b). We attributed this decline to episodes of sloughing when accumulation rates were highest. In these cases, an 8-wk sampling interval underestimated biomass and explained the low dry mass and chlorophyll *a*

---

### TABLE 4.8
**Taxonomic Composition of Algae Accumulating on Periphytometer Slides Incubated for 8 wk at Sites A, B, and C[a]**

|  | Blue-Greens | Greens | Desmids | Diatoms | Other |
|---|---|---|---|---|---|
| Abundance (mean percent) | 21 | 3 | 25 | 50 | 1 |
| Richness (number of taxa) | 21 | 33 | 83 | 54 | 11 |

[a] Everglades National Park, June 1997 to June 1998.

---

### TABLE 4.9
**Cell Density, Dry Mass, Percent Ash-Free Dry Mass (% AFDM) and Chlorophyll *a* Estimates of Periphyton Biomass on Glass Slides Incubated for 8 wk at Sites A, B, C, and L**

| Flume | Cell Density (cm$^{-2}$ x 100) | Dry Mass (µg/cm$^2$) | % AFDM | Chlorophyll a (µg/cm$^2$) |
|---|---|---|---|---|
| A | 1204 | 147 | 78 | 0.50 |
| B | 1657 | 147 | 72 | 0.75 |
| C | 1584 | 133 | 67 | 0.51 |
| L | — | 200 | 89 | 0.81 |
| Significant differences (*p* <0.05) | A < B, C | A, B, C < L | C < A, B  A, B, C < L | A, C < B, D |

[a] Everglades National Park, June 1997 to June 1998.

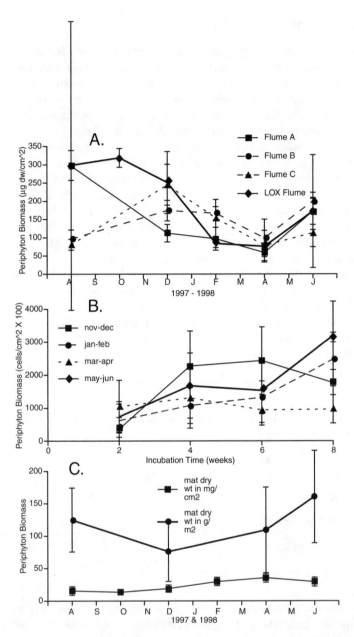

**FIGURE 4.5** Periphytometer data showing periphyton dry weight accumulated on glass slides. (A) Incubated for 8 weeks at all four flumes (L = LOX); (B) as means (±SD) for all flumes during four sampling intervals, plotted at time-incremented accumulation; (C) seasonal changes in dry weight biomass of the periphyton mat as means of all flumes and all channels (±SD), in mg cm$^{-2}$ and extrapolated to g m$^{-2}$ using estimates of mat cover.

estimates for several ENP flumes during the August 1997 sampling period (Table 4.9). Although an 8-wk sampling interval may be best for the lower productivity dry season, the risk of underestimating production during the wet season is considerable. Ordinations of diatom assemblages from 2-, 4-, 6-, and 8-wk incubations showed little compositional difference between 6- and 8-wk incubations. We also anticipate that sloughing in this 6- to 8-wk time frame may be an even greater problem in experimental channels receiving P additions. For these reasons, we changed our standard periphytometer sampling interval to 6 wk.

The floating periphyton mat contains a matrix of *Utricularia purpurea*, cyanobacterial filaments (mostly *Schizothrix hofmanni* and *Scytonema calcicola*), other associated algae, bacteria, fungi, and invertebrate animals. The organic portion of the mat made up an average of 41% of the dry biomass (24 mg cm$^{-2}$) and consisted of mostly senescent material. ENP Site C had the highest periphyton cover and biomass, but a higher percentage of this material was inorganic (Table 4.10). Notably, these are floating mat periphyton data from which *Utricularia* fragments and animals were manually removed. Periphyton mat coverage was highest in August 1997 (81% channel coverage). Cover subsequently decreased through the dry season to a minimum of 30% in June 1998. These changes appeared to be synchronized with seasonal changes in the growth rate of the dominant substrate, *Utricularia purpurea* (Figure 4.6b). It is important to note that our LOX flume has little or no floating periphyton mat because of the soft water found in this peat-rich, ombrotrophic system.

**TABLE 4.10**
**Summary of Biomass Estimates of the Periphyton Mat at Sites A, B, and C[a]**

| | A | | B | | C | |
|---|---|---|---|---|---|---|
| | Mean | Std. Dev. | Mean | Std. Dev. | Mean | Std. Dev. |
| Percent cover | 48 | 27 | 54 | 31 | 53 | 25 |
| Dry wt (mg/cm$^2$) | 19.9 | 7.8 | 21.6 | 9.3 | 30 | 10.7 |
| Dry wt (g/m$^2$)[b] | 86.8 | 44.7 | 105.4 | 60.4 | 160.4 | 66.3 |
| Ash-free dry mass/dry wt | 0.47 | — | 0.43 | — | 0.33 | — |
| Chlorophyll *a* (μg/cm$^2$) | 7.3 | 5.9 | 7.5 | 6.0 | 7.5 | 6.3 |
| Chlorophyll *a* (mg/m$^2$)[b] | 128.1 | 137.1 | 214.9 | 204.9 | 185.2 | 211.8 |
| Chlorophyll *a* (μg/mg dry wt) | 0.4 | 0.3 | 0.4 | 0.3 | 0.4 | 0.3 |

[a] Everglades National Park, August 1997 to June 1998.
[b] These values were expanded to a square-meter basis using percent cover estimates.

No substantial benthic periphyton mat has been observed in any of our flume channels. Benthic periphyton biomass ranged from 12.1 to 24 mg dw cm$^{-2}$. Interestingly, the benthic floc contained a larger average proportion of chlorophyll *a* than the floating periphyton mat when we normalized values to dry weight (mean 1.1 μg chlorophyll *a* mgdw$^{-1}$). However, these samples were taken in April 1998, when floating periphyton cover was lowest (Figure 4.5c), permitting greater light penetration through the water column.

Growth of both *Utricularia purpurea* and *U. foliosa* was greatest in the summer and reduced by approximately half in the winter. *U. foliosa* tissue had twice the P content per g dw of *U. purpurea* (Figure 4.6a) and *U. foliosa* added new leaves three to four times as fast as *U. purpurea*, regardless of season (Figure 4.6b). *U. purpurea* comprised 9% ± 4% of the periphyton mat on a dry weight basis, but was a relatively P-rich component of the mat, as it contained 1.4 times more P than did the periphyton component.

## MACROPHYTE COMPONENTS

Preliminary analyses of the pre-dosing macrophyte data showed significant differences in total biomass between flume sites. These differences reflected differences in species abundances among sites, even among the flumes within ENP. The combined biomass of *Eleocharis cellulosa*, *E. elongata*, *Panicum hemitomom*, and *Paspalidium germanicum* ranged from 24.5 to 67.8 gdw m$^{-2}$ per site (Site A = 24.5 ± 3.32 gdw m$^{-2}$, Site B = 67.8 ± 5.74 gdw m$^{-2}$, Site C = 33.8 ± 4.48 gdw m$^{-2}$). Although *Sagittaria lancifolia* was not abundant by stem count, estimates of total biomass increased dramatically when it was included in biomass calculations (i.e. when it represented more than 10%

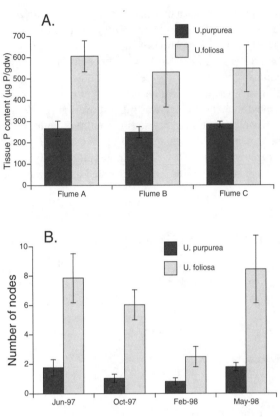

**FIGURE 4.6** (A) Total phosphorus content of *Utricularia purpurea* and *Utricularia foliosa* in February 1998. (B) Number of new nodes per week for *U. purpurea* and *U. foliosa* from June 1997 to May 1998.

of the total number of stems; Site A = 134 ± 20.8 gdw m$^{-2}$, Site B = 424.4 ± 68.0 gdw m$^{-2}$, Site C = 33.8 ± 3.86 gdw m$^{-2}$).

The P content of macrophyte tissue in our ENP flumes showed some interesting patterns. We found no significant difference in P content of *E. cellulosa*, *E. elongata*, *P. hemitomom*, and *P. geminatum* — they ranged from approximately 500 to 800 µg P gdw$^{-1}$. *S. lancifolia*, however, contained significantly more P than these other dominant species, with tissue P content values as high as 1200 to 1400 µg P gdw$^{-1}$. This difference is interesting because Daoust and Childers (1999) found no differences in the N:P ratios of *E. cellulosa*, *E. elongata*, *P. hemitomom*, or *S. lancifolia* growing in eastern Shark River Slough. Furthermore, when *S. lancifolia* was present in our biomass plots it was often the dominant species by biomass. Thus, in some situations the vast majority of P contained in macrophyte tissues is in *S. lancifolia*. These preliminary data on tissue P content and biomass led us to target *S. lancifolia* as a key to helping understand how the marsh macrophyte community responds to added P.

## FAUNAL COMPONENTS

Pre-dosing sampling has provided information on spatial variability of fish, amphibian, and invertebrate densities in our ENP flumes and on possible effects of flume wall enclosures on these patterns. Densities of several common fish and macroinvertebrates were significantly higher at Flume A than at Flumes B or C. Grass shrimp (*Palaemonetes paludosus*) densities illustrate both patterns well ($F_{2,63}$ = 13.13, P = 0.0001; Figure 4.7). Also, there were significant differences among up-channel, mid-channel, and down-channel densities for several dominant taxa (e.g., grass shrimp,

$F_{2,63}$ = 5.40, P = 0.007, with no significant interaction between flume and location within channel). A common pattern was for densities to be higher at the ends than in the center of the channels, suggesting that the flume walls may be restricting the movement of animals into channels (Figure 4.7). These faunal gradients may, in turn, affect other ecosystem components and processes. Our pre-dosing data, along with data from control channels, enable us to separate spatial patterns related to P additions from those unrelated to our experimental treatments. Notably, we found no significant differences between densities in flume channels and densities in the nearby reference sites (for dominant fish and macroinvertebrates collected by throw trap) with the exception of creeping water bugs (*Pelocoris femoratus*) and mosquitofish (*Gambusia holbrooki*) densities, which were significantly higher in the reference areas ($F_{1,75}$ = 17.82, P = 0.0001 and $F_{1,75}$ = 12.76, P = 0.0006, respectively).

## SUMMARY AND CONCLUSIONS

In this chapter, we have described a study that is providing critical baseline information needed to establish Class III water quality standards for the Everglades Protection Area. We also present pattern and process data on a host of ecosystem components from the little-studied Everglades marsh systems. Our methodology entails four 4-channel flumes in the *Eleocharis* "wet prairie" marshes of Everglades National Park and Loxahatchee National Wildlife Refuge. Our central objective is to quantify the effects of low-level P addition on unimpacted Everglades wetlands. Each flume has a control channel and three experimental channels to which we continuously add P to enhance ambient concentrations by 5, 15, and 30 ppb (0.17, 0.5, and 1 μM P). Our working model of ecosystem response to this P addition hypothesizes that the endpoint of Everglades ecosystem change is ultimately not dependent on the concentration of P added. That is, the concentration increase in our experimental channels determines the *rate* at which P accumulates in the system and the *time* from P addition until imbalances in flora or fauna are observed, but it does not determine the final ecosystem state. We expect that all of our experimental flume channels will change to a similar ecosystem state (endpoint), but the time course of that transition will be negatively related to the concentration of P.

We are quantifying ecosystem responses at several hierarchical levels. We quantify all major processes associated with macronutrient cycling and calculate full biogeochemical budgets for each. We follow the structural and functional dynamics of the microbial, algal, macrophyte, and faunal communities. Food web structure and energy flow are included in our analyses, including how both change in response to P additions. A host of soil variables are also monitored throughout the study. We work at this wide range of biological response because we anticipate that floral and faunal imbalances caused by P additions are manifest first in the microbiota of the soils and periphyton,

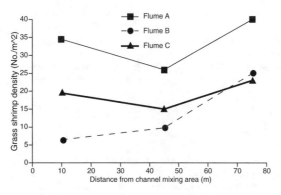

**FIGURE 4.7**    Density of grass shrimp along the length of the channels at the three Shark River Slough flumes.

and thus precede the changes in macrophytes and macrofauna that are observable with the naked eye. The interaction of ecosystem-level changes in space and time is critical to understanding and predicting the effects of P additions on Everglades wetlands. This study thus represents a comprehensive, integrative, system-based analysis of effects of water quality changes on the Everglades wetland landscape.

## REFERENCES

Amador, J.A. and R.D. Jones. 1993. Nutrient limitations on microbial respiration in peat soils with different total phosphorus content, *Soil Biol. Biochem.*, 25:793–801.

APHA. American Water Works Association, and the Water Pollution Control Federation. 1989. *Standard Methods for the Examination of Water and Wastewater*, 117th ed., American Public Health Association, Washington, D.C.

Asmus, H., R.M. Asmus, T.C. Prins, N. Dankers, G. Frances, B. Maaß, and K. Reise. 1992. Benthic–pelagic flux rates on mussel beds: tunnel and tidal flume methodology compared, *Helgo. Meeres.*, 46:341–361.

ASTM. 1995. *Book of ASTM Standards*, Vol. 04.08: *Soil and Rock (I)*, N.C. Furcola, Ed., American Society for Testing and Materials, Philadelphia, PA.

Atkinson, M.J. and R.W. Bilger. 1992. Effects of water velocity on phosphate uptake in coral reef-flat communities. *Limnol. Oceanogr.* 37(2):273–279.

Bachoon, D. and R.D. Jones. 1992. Potential rates of methanogenesis in sawgrass marshes with peat and marl soils in the Everglades, *Soil Biol. Biochem.*, 24(1):21–27.

Belanger, T.V., D.J. Scheidt, and J.R. Platko. 1989. Effects of nutrient enrichment on the Florida Everglades, *J. Lake Reservoir Manage.*, 5(1):101–112.

Bloom, A.L. 1964. Peat accumulation and compaction in a Connecticut coastal marsh, *J. Sedimentary Petrol.*, 34:599–603.

Bothwell, M.L. 1985. Phosphorus limitation of lotic periphyton growth rates: an intersite comparison using continuous-flow troughs, *Limnol. Oceanogr.*, 30:527–542.

Boumans, R.M.J. and J.W. Day, Jr. 1993. High precision measurements of sediment elevation in shallow coastal areas using a sediment-erosion table, *Estuaries*, 16:375–380.

Buzzelli, C.P., D.L. Childers, Q. Dong, and R. Jones, 2000. Simulation of periphyton phosphorus dynamics in Everglades National Park, *Ecol. Mod.*, 134(1):103–115.

Caraco, N.F. 1993. Disturbance of the phosphorus cycle: a case of indirect effects of human activity, *Trends Ecol. Evolution*, 8(2):51–54.

Carpenter, S.R. and J.F. Kitchell. 1993. *The Trophic Cascade in Lakes*, Cambridge Univ. Press, Cambridge, U.K., 386.

Chalmers, A.G., R.G. Wiegert, and P.L. Wolf. 1985. Carbon balance in a salt marsh: interactions of diffusive export, tidal deposition, and rainfall-caused erosion, *Estuarine Coastal Shelf Sci.*, 21:757–771.

Chanton, J.P., G.J. Whiting, J.D. Happell, and G. Gerard. 1993. Contrasting rates and diurnal patterns of methane emission from emergent aquatic macrophytes, *Aquatic Botany*, 46:111–128.

Childers, D.L. 1994. Fifteen years of marsh flumes: a review of marsh–water column interactions in Southeastern USA estuaries, in *Wetlands of the World*, Mitsch, Ed., Elsevier Press, Amsterdam.

Childers, D.L. and J.W. Day, Jr. 1988. A flow-through flume technique for quantifying nutrient and materials fluxes in microtidal estuaries, *Estuarine Coastal Shelf Sci.*, 27(5):483–494.

Childers, D.L. and J.W. Day, Jr. 1990a. Marsh:water column interactions in two Louisiana estuaries. I. Sediment dynamics, *Estuaries*, 13(4):393–403.

Childers, D.L. and J.W. Day, Jr. 1990b. Marsh:water column interactions in two Louisiana estuaries. II. Nutrient dynamics, *Estuaries*, 13(4):404–417.

Childers, D.L., S. Cofer-Shabica, and L. Nakashima. 1993. Spatial and temporal variability in marsh–water column interactions in a Georgia salt marsh, *Mar. Ecol. Prog. Ser.*, 95(1,2):25–38.

Childers, D.L., S. Davis, V. Rivera-Monroy, and R.R. Twilley. 1998. Wetland–water column interactions and the biogeochemistry of estuary-watershed coupling around the Gulf of Mexico, in *Biogeochemistry of Gulf of Mexico Estuaries*, Bianchi, T.S. and Twilley, R., Eds., John Wiley & Sons, New York, pp. 211–235.

Coale, F.J., F.T. Inuzo, and A.B. Bottcher. 1994a. Phosphorus in drainage water from sugarcane in the Everglades Agricultural Area as affected by drainage rate, *J. Environ. Qual.*, 23(1):121–126.

Coale, F.J., F.T. Inuzo, and A.B. Bottcher. 1994b. Sugarcane production impact on N and P in drainage water from an Everglades histosol, *J. Environ. Qual.*, 23(1):116–120.

Craft, C.B. and C.J. Richardson. 1993a. Peat accretion and P accumulation along a eutrophication gradient in the northern Everglades, *Biogeochemistry*, 22:133–156.

Craft, C.B. and C.J. Richardson. 1993b. Peat accretion and N, P, and organic C accumulation in nutrient-enriched and unenriched Everglades peatlands, *Ecol. Appl.*, 3(3):446–458.

Dailey, S.K. 2000. Phosphorus Enrichment Effects on Interactions Among the Ecosystem Components in a Long Hydro-period Oligotrophic Marsh in Everglades National Park, Ph.D. dissertation, Florida International University, Miami.

Dame, R.F., R.G. Zingmark, and E. Haskin. 1984. Oyster reefs as processors of estuarine materials, *J. Exp. Mar. Biol. Ecol.*, 83:239–247.

Daoust, R. 1998. Determination of the Effects of Phosphorus Addition on Several Everglades Marsh Communities, M.S. thesis, Florida International University, Miami.

Daoust, R. and D.L. Childers. 1998. Quantifying aboveground biomass and estimating productivity in nine Everglades wetland macrophytes using a non-destructive allometric approach, *Aquatic Botany*, 62:115–133.

Daoust, R. and D.L. Childers. 1999. The importance of considering intra-annual variability in vascular tissue nutrient content when assessing ecosystem nutritional status using C:N:P ratios, *Wetlands*, 19(1):262–275.

Davis, S.M. 1994. P inputs and vegetation sensitivity in the Everglades, in *Everglades: The Ecosystem and Its Restoration*, Davis, S.M. and Ogden, J.C., Eds., St. Lucie Press, FL, pp. 357–378.

Dixon, P.M. and K.A. Garrett. 1993. Statistical issues for field experimenters, in *Wildlife Toxicology and Population Modelling*, Kendall, R.J. and Lacher, T.E., Eds., Lewis Publishers, Boca Raton, FL.

Doren, R.F., T.V. Armentano, L.D. Whiteaker, and R.D. Jones. 1997. Marsh vegetation patterns and soil P gradients in the Everglades ecosystem, *Aquatic Botany*, 56:145–163.

Farmer, A.M. and M.S. Adams. 1989. A consideration of the problems of scale in the study of the ecology of aquatic macrophytes, *Aquatic Botany*, 33:177–189.

Flora, M.D., D.R. Walker, K.A. Burgess, D.J. Sheidt, and R.G. Rice. 1986. *The Response of Experimental Channels in Everglades National Park to Increased N and P Loading. Data Report: Chemistry and Primary Productivity*, Water Resources Rep. 86-6, Water Res. Div., NPS, Ft. Collins, CO.

Gordon, A.S., W.J. Cooper, and D.J. Scheidt. 1986. Denitrification in marl and peat sediments in the Florida Everglades, *Appl. Environ. Microbiol.*, 52(5):987–991.

Grimshaw, H.J., M. Rosen, D. Swift, K. Rodberg, and J.M. Noel. 1993. Marsh phosphorus concentrations, phosphorus content, and species composition of Everglades periphyton communities, *Arch. Hydrobiol.*, 128(3):257–276.

Hendrick, R.L. and K.S. Pregitzer. 1992. The demography of fine roots in a northern hardwood forest, *Ecology*, 73: 1094–1104.

Hoekstra, T.W., T.F.H. Allen, and C.H. Flather. 1991. Implicit scaling in ecological research, *Bioscience*, 41(3):148–154.

Holling, C.S. 1992. Cross-scale morphology, geometry, and dynamics of ecosystems, *Ecol. Monogr.*, 62(4):447–502.

Koch, M.S. and K.R. Reddy. 1992. Distribution of soil and plant nutrients along a trophic gradient in the Florida Everglades, *Soil Sci. Soc. Am. Jour.*, 56:1492–1499.

Light, S.S. and J.W. Dineen. 1994. Water control in the Everglades: a historical perspective, in *Everglades: The Ecosystem and Its Restoration*, Davis, S.M. and Ogden, J.C., Eds., St. Lucie Press, FL.

Lutz, J.R. 1977. *Water Quality and Nutrient Loadings of the Major Inflows from the Everglades Agricultural Area to the Conservation Areas, SE Fl*, Tech. Pub. 77-6, South Florida Water Management District, West Palm Beach, FL.

Martinez, N.D. 1993. Effect of scale on food web structure, *Science*, 260:242–243.

McCormick, P.V., R.S. Rawlik, K. Lurding, E.P. Smith, and F.H. Sklar. 1996. Periphyton-water quality relationships along a nutrient gradient in the northern Everglades, *J. N. Am. Benthol. Soc.*, 15:433–449.

McCormick, P.V. and M.B. O'Dell. 1996. Quantifying periphyton responses to phosphorus enrichment in the Florida Everglades: a synoptic-experimental approach, *J. N. Am. Benthol. Soc.*, 15:450–468.

McCullagh, P. and J.A. Nelder. 1989. *Generalized Linear Models*, 2nd ed., Chapman & Hall, New York.

McIvor, C.C. and W.E. Odum. 1986. The flume net: a quantitative method for sampling fishes and macrocrustaceans on tidal marsh surfaces, *Estuaries*, 9(3):219–224.

Nelson, D.W. and L.E. Sommers. 1996. Total carbon, organic carbon, and organic matter, in *Methods of Soil Analysis*, Part 3, Sparks, D.L., Ed., SSSA Book Series 5. Soil Science Society of America, Madison, WI, pp. 961–1010.

Newbold, J.D., J.W. Elwood, R.V. O'Neill, and W. Van Winkle. 1981. Measuring nutrient spiralling in streams, *Can. J. Fish. Aquat. Sci.*, 38:860–863.

Newbold, J.D., J.W. Elwood, R.V. O'Neill, and A.L. Sheldon. 1983. Phosphorus dynamics in a woodland stream ecosystem: a study of nutrient spiralling, *Ecology*, 64:1249–1265.

Pahl-Wostl, C. 1993. Food webs and ecological networks across temporal and spatial scales, *Oikos*, 66:415–432.

Peterson, B.J., B. Fry, L. Deegan, and A. Hersey. 1993a. The trophic significance of epilithic algal production in a fertilized tundra river ecosystem, *Limnol. Oceanogr.*, 38(4):872–878.

Peterson, B.J., L. Deegan, J. Helfrich, J. Hobbie, M. Hullar, B. Moller, T. Ford, A. Hersey, A. Hiotner, G. Kipphut, M. Lock, D. Fiebig, V. McKinley, M. Miller, J. Vestal, R. Ventullo, and G. Volk. 1993b. Biological responses of a tundra river to fertilization, *Ecology*, 74(3):653–672.

Peterson, G.W. and R.E. Turner. 1994. The value of salt marsh edge vs. interior as a habitat for fish and decapod crustaceans in a Louisiana tidal marsh, *Estuaries*, 17(1B):235–262.

Rader, R.B. and C.J. Richardson. 1992. The effects of nutrient enrichment on algae and macroinvertebrates in the Everglades: a review, *Wetlands*, 12(2):121–135.

Rader, R.B. and C.J. Richardson. 1994. Response of macroinvertebrates and small fish to nutrient enrichment in the northern Everglades, *Wetlands*, 14(2):134–146.

Raschke, R.L. 1993. Diatom (Bacillariophyta) community responses to phosphorus in the Everglades National Park, USA, *Phycologia*, 32:48–58.

Reddy, K.R., W.F. DeBusk, Y. Wang, R. DeLaune, and M. Koch. 1991. *Physico-Chemical Properties of Soils in Water Conservation Area 2 of the Everglades*, final report to the South Florida Water Management District, West Palm Beach, FL, 118 pp.

Rivera-Monroy, V.H., J.W. Day, R.R. Twilley, F. Vera-Herrera, and C. Coronado-Molina. 1995. Flux of nitrogen and sediment in a fringe mangrove forest in Terminos Lagoon, Mexico, *Estuarine Coastal Shelf Sci.*, 40:139–160.

Rogers, C.S. 1979. The productivity of San Cristobal reef, Puerto Rico, *Limnol. Oceanogr.*, 24(2):342–349.

Scheidt, D.J., M.D. Flora, D.R. Walker, R.G. Rice, and D.H. Landers. 1988. *The Response of the Everglades Marsh to Increased N and P Loading*, Part I, South Florida Research Center, Everglades National Park, Homestead, FL.

Scheidt, D.J., M.D. Flora, and D.R. Walker. 1989. Water quality management for Everglades National Park, in *Wetlands: Concerns and Successes*, American Water Resources Association, Bethesda, MD, pp. 377–390.

SFWMD. 1992. *Draft Surface Water Improvement and Management Plan for the Everglades*, Supporting Information Doc., South Florida Water Management District, West Palm Beach, FL, 472 pp.

Sinsabaugh, R.L., S. Findlay, P. Franchini, and D. Fischer. 1997. Enzymatic analysis of riverine bacterioplankton production, *Limnol. Oceanogr.*, 42(1):29–38.

Solorzano, L. and J.H. Sharp. 1980. Determination of total dissolved P and particulate P in natural waters, *Limnol. Oceanogr.*, 25:754–758

Spurrier, J.D. and B. Kjerfve. 1988. Estimating the net flux of nutrients between a salt marsh and a tidal creek, *Estuaries*, 11(1):10–14.

Stal, L.J. 1988. Nitrogen fixation in cyanobacterial mats, in *Methods in Enzymology: Cyanobacteria*, Packer, L. and Glazer, A.N., Eds., Academic Press, San Diego, CA, pp. 474–484.

Swift, D.R. 1981. *Preliminary Investigations of Periphyton and Water Quality Relationships in the Everglades WCAs*, Tech. Rep. 81-5, South Florida Water Management District, West Palm Beach, FL.

Swift, D.R. and R.B. Nicholas. 1987. *Periphyton and Water Quality Relationships in the Everglades WCAs, 1978–1982*, Tech. Pub. 87-2, South Florida Water Management District, West Palm Beach, FL.

Tabachnick, B.G. and L.S. Fidell. 1983. *Using Multivariate Statistics*, Harper & Row, New York.

Thorhallsdóttir, T.E. 1990. The dynamics of a grassland community: a simultaneous investigation of spatial and temporal heterogeneity at various scales, *J. Ecol.*, 78:884–908.

Tranvik, L.J. 1997. Rapid fluorometric assay of bacterial density in lake water and seawater, *Limnol. Oceanogr.*, 42(7):1629–1634.

Trexler, J.C., J.H.K. Pechmann, and J. H. Chick. 1998. *Fish and Aquatic Macroinvertebrate Monitoring Studies in Taylor Slough*, Second Annual Report, Cooperative Agreement CA-5280-6-9011 between Everglades National Park, U.S. Department of the Interior, and Florida International University, 84 pp.

Turner, A.M. and J.C. Trexler. 1997. Sampling aquatic invertebrates from marshes: evaluating the options, *J. N. Am. Benthol. Soc.*, 16:694–709.

Turner, A.M., J.C. Trexler, F. Jordan, S.J. Slack, P. Geddes, J. Chick, and W. Loftus. 1999. Targeting ecosystem features for conservation: standing crops of the Florida Everglades, *Conserv. Biol.*, 13(4):898–911.

Twilley, R. 1985. The exchange of organic carbon in basin mangrove forests in a southwest Florida estuary, *Estuarine Coastal Shelf Sci.*, 20:543–557.

U.S.E.P.A. 1983. *Methods for the Chemical Analysis of Water and Wastes*, U.S. Environment Protection Agency, Cincinnati, OH.

Vitousek, P.M. and W.A. Reiners. 1975. Ecosystem succession and nutrient retention: a hypothesis, *Bioscience*, 25(6):376–381.

Vitousek, P.M. and R.W. Howarth. 1991. Nitrogen limitation on land and in the sea: how can it occur? *Biogeochemistry*, 13:87–115.

Vitousek, P.M., L.R. Walker, L.D. Whiteaker, and P.A. Matson. 1993. Nutrient limitations to plant growth during primary succession in Hawaii Volcanoes National Park. Biogeochem. 23:197–215.

Von Ende, C.N. 1993. Repeated measures analysis: growth and other time-dependent measures, in *Design and Analysis of Ecological Experiments*, Scheiner, S.M. and Gurevitch, J., Eds., Chapman & Hall, New York, pp. 113–137.

Vymazal, J., C.B. Craft, and C.J. Richardson. 1994. Periphyton response to nitrogen and phosphorus additions in the Florida Everglades, *Algol. Stud.*, 73:75–97.

Walker, D.R., M.D. Flora, R.G. Rice, and D.J. Scheidt. 1989. *The Response of the Everglades Marsh to Increased N and P Loading*, Part II, South Florida Research Center, Everglades National Park, Homestead, FL.

Walker, W.W. 1991. Water quality trends at inflows to Everglades National Park, *Water Res. Bull.*, 27(1):59–72.

Wolaver, T.G. and J.C. Zieman. 1983. Effect of water column, sediment, and time over the tidal cycle on the chemical composition of tidal water in a mesohaline marsh, *Mar. Ecol. Prog. Ser.*, 12:123–130.

Wolaver, T., G. Whiting, B. Kjerfve, J. Spurrier, H. McKellar, R. Dame, T. Chrzanowski, R. Zingmark, and T. Williams. 1985. The flume design — a methodology for evaluating material fluxes between a vegetated salt marsh and the adjacent tidal creek, *J. Exp. Mar. Biol. Ecol.*, 91:281–291.

Wu, Y., F.H. Sklar, and K. Rutchey. 1997. Analysis and simulations of fragmentation patterns in the Everglades, *Ecol. Appl.*, 7(1):268–276.

# 5 Ecological Scale and Its Implications for Freshwater Fishes in the Florida Everglades

*Joel C. Trexler*
Florida International University

*William F. Loftus*
U.S. Geological Survey

*Frank Jordan*
Loyola University

*John H. Chick*
Illinois Natural History Survey

*Karen L. Kandl*
University of New Orleans

*Thomas C. McElroy*
Florida International University

*Oron L. Bass, Jr.*
South Florida Natural Resources Center

## CONTENTS

0-8493-2026-7/02/$0.00+$1.50
© 2002 by CRC Press LLC

## INTRODUCTION

In her influential 1947 book, Marjory Stoneman Douglas coined the phrase "river of grass" to describe the Florida Everglades. Historically, marshes and sloughs of the Everglades extended uninterrupted in the southerly river-like flow of fresh water from Lake Okeechobee to Florida Bay. However, since the late 1800s that flow has been increasingly disrupted and the Everglades ecosystem is now broken into a number of compartments separated by levees and canals. Has this relatively new spatial structure affected the aquatic fauna of the Everglades? Has it disrupted historical linkages between northern and southern portions of the Everglades drainage basin? We examine patterns of spatial variation in fish communities to ask if their density, relative abundance, or population genetic structure has been marked by the relatively new spatial divisions imposed by human alterations of this wetland. Almost no information exists about historical aquatic communities of the Everglades, yet the planned restoration effort must benefit its aquatic communities if wading birds and other predators are to benefit (Frederick and Spalding, 1994; Ogden, 1994) and ecosystem function is to be restored. Our analysis of present-day communities generates hypotheses about the historical ecological functions sought by restoration and provides benchmarks to evaluate future system alterations.

There remains some disagreement about patterns in the structure of fish communities in the Everglades and ecological factors creating them. Published studies on the ecology of Everglades fishes emphasize the interaction of seasonal hydrological variation, vegetation dynamics, and the dynamics of fish communities in its freshwater marshes (Kushlan 1976a, b, 1981; Loftus and Eklund, 1994; Jordan, 1996). Much of this work has emphasized the role of alligator ponds as dry-season refuges determining fish demography and dynamics each year (Kushlan, 1974, 1976b; Karr and Freemark, 1985). However, alligator ponds vary in frequency across the Everglades and are sparsely distributed in many regions. Recent simulation models call into question the capacity of alligator pond refuges to explain the dynamics of Everglades fish communities (DeAngelis et al., 1997). It has been suggested that the density of small fishes has declined in the Everglades because water management extended the period between system-wide drydown, permitting an increase in the abundance of large piscivorous fishes (Kushlan, 1987). However, the pattern of fishes at sites experiencing different hydroperiods is not consistent with this hypothesis (Loftus and Eklund, 1994). With the increasing emphasis on a scientific basis for Everglades management and restoration, recent studies have provided new information that can be brought to bear on these issues.

The spatial scale of community dynamics, local topographic relief, and spatial variation in nutrient biogeochemistry may play a larger role in determining the structure and function of Everglades fish communities than previously appreciated (Frederick and Spalding, 1994; Chick and McIvor, 1994; DeAngelis, 1994; Jordan, 1996; Jordan et al., 1998). Here, we review the results of several ongoing studies of the spatial scale of fish community structure in the Everglades. Our analysis groups the community into two classes based on body size — large (>8 cm standard length) and small (<8 cm). This break is necessary because of size-based bias in sampling techniques but is also biologically defensible because body size influences the way fish respond to their environment. Compared to small fishes, larger fishes require larger and deeper dry-season refuges and can move longer distances in searching for appropriate habitat, and many are potential predators of smaller fish. Thus, the expedient size-based division of our analyses likely corresponds to size-related behavioral responses of fishes to their environment. The break roughly separates the cyprinodontiform segment of the fish community, and juveniles of all species, from adult piscivorous species, including centrarchids, bullhead catfish, gar, and bowfin (Loftus and Kushlan, 1987). Adult lake chubsuckers, with a diverse diet that ranges from detritus to larval fish, are also included in the large-fish group. The adults of most introduced fishes fall into the large-fish category, though these species are not, at present, particularly common in marshes distant from either canals or the mangrove zone (Trexler et al., 2000).

We address three questions related to ecological scaling: What is the spatial scale of variation in fish abundance? Are spatial patterns consistent over time? What is the spatial scale of genetic variation in selected fishes from the Everglades? Our analyses are at three spatial scales arranged in a nested design: among water-management units, within units among sites separated by 10 to 15 km, and among plots separated by approximately 1 km within those sites. The majority of data reported here were gathered between 1996 and 1997, but we also report aspects of a long-term sampling effort conducted at a subset of the study sites in the Everglades National Park. We examine the spatial scale of variability in fish density and relative abundance at these three levels. Then, with a slightly different sampling design, we will examine the pattern of genetic diversity in two species of fish from the same study regions. From these data, we develop hypotheses about the role of environmental fluctuation in setting the scale of fish-community structure in the Everglades.

## MATERIALS AND METHODS

### NATURAL HISTORY: THE EVERGLADES FISH FAUNA

The native fish community of the Everglades is relatively species poor and lacks endemic species. There are about 30 freshwater-fish species inhabiting the ecosystem (Loftus and Kushlan, 1987), a relatively small number for the spatial area of the drainage basin (Swift et al., 1986; Trexler, 1995). A historical explanation for the paucity of freshwater fish species is compelling. The freshwater Everglades ecosystem is relatively young, 5000 years (Gleason and Stone, 1994), and is found at the tip of a peninsula. However, the low topography and karstic limestone geology limit aquatic habitat diversity and exclude members of the ichthyofauna found in northern Florida (Swift et al., 1986; Gilbert, 1987). In most of North America, cyprinids play a prominent role in fish community composition. However, species composition of the Everglades fauna is dominated by cyprinodontiforms, especially poeciliids and fundulids derived from the U.S. southeastern coastal plain (Loftus and Kushlan, 1987). Cyprinodontiforms are also numerically dominant (Loftus and Kushlan, 1987; Jordan, 1996; Jordan et al., 1998). The Everglades also has low standing stocks (g/m$^2$) of fishes when compared to other wetland, floodplain, or littoral zone ecosystems (Turner et al., 1999). Several species of introduced fishes are naturalized members of the fauna, including at least five cichlids, and a poeciliid (Fuller et al., 1999; Loftus, 2000). At present, these species comprise less than 10% of the fishes (numerically or by biomass) in most habitats, with the exception of canals, very short-hydroperiod wetlands peripheral to the sloughs, and the mangrove zone downstream of the sloughs (Trexler et al., 2000). Their ecological impact as predators and competitors of native species in the Everglades is, as yet, unclear, especially in slough habitats.

## THE STUDY AREA

Our study is limited to the central and southern Everglades Protection Area encompassing Water Conservation Areas (WCAs) 3A and 3B, and Shark River Slough (SRS) and Taylor Slough (TS) in Everglades National Park (ENP) (Figure 5.1). WCA-3A is subdivided by canals and levees into four separate compartments that we treat individually (Figure 5.1: WCA-3A-nw, -ne, -se, and -sw). Water moves through these compartments roughly from north to south. WCA-3B has been completely isolated from the north–south flow way by barrier levees since 1963 (Light and Dineen, 1994). SRS receives water from rainfall and flow from the WCAs to the north through a series of flood-control gates (S-12 structures). TS, also in ENP, had historical connections with the SRS

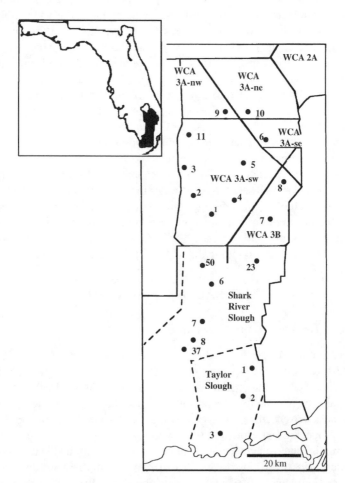

**FIGURE 5.1**   Map of the study area showing sampling sites. Solid lines indicate canal or levee boundaries and dashed lines indicate the boundaries of Shark River and Taylor River sloughs in Everglades National Park. Sites 1 and 2 in TS are each bordered to the east and west by short-hydroperiod plots not visible on the scale of this figure. We treat these short-hydroperiod plots as separate sites to yield a total of five sites in TS.

(Tabb, 1987) but currently functions as a separate drainage basin. The historical headwaters of TS are now drained and developed; since 1968, all water other than rainfall entering TS has come from the South Dade Water Conveyance canal system via pumps (Light and Dineen, 1994).

The phrase "river of grass" is an apt description of the Everglades. The large drainage basin that includes the WCAs and SRS and the smaller drainage basin of TS have shallow but hydrologically significant contours parallel to the roughly north–south direction of flow. Thus, as the water table recedes in the dry season, habitats to the east and west sides of the flow-way dry first while those near the center remain inundated the longest. Marshes of the central flow-way remain inundated year-round in most years, sustaining a peat-forming environment (Gleason and Stone, 1994). We identified study sites as short-hydroperiod marsh if they experience an estimated average inundation of <300 days per year, and a long-hydroperiod marsh if they experience an average inundation of >300 days per year. These designations were based on simulations run by the South Florida Water Management Model (Fennema et al., 1994) and from personal experience at the study sites. There is marked interannual variation in local hydroperiod that results from local and regional rainfall patterns, and water-management decisions.

All sites were located in wet prairie habitats that are dominated by spikerush, mostly *Eleocharis cellulosa* (Gunderson, 1994; Jordan et al., 1997b). Sampling sites were restricted to this habitat for both practical and biological reasons. Wet prairies are typically 10 to 20 cm lower than adjacent dense sawgrass stands (Jordan et al., 1997b). Thus, these habitats retain water and fishes longer in the dry season than adjacent sawgrass habitats. Sampling in sawgrass, the other dominant habitat, is difficult because of its density and abrasive character. Also, stratifying across habitat type was not logistically feasible because of the number of samples required. Throw-trap samples collected in dense sawgrass stands adjacent to some of our spikerush-dominated wet prairie study sites (six sites, eight samples per habitat per site) yielded fewer fish in the sawgrass habitat than in the spikerush (mean number of fish m$^{-2}$: sawgrass = 20.6, spikerush = 39.1; paired t$_5$ = –3.13, P = 0.03). In a more extensive study, Jordan (1996) also found fewer fishes in sawgrass than in spikerush habitats in the Loxahatchee National Wildlife Refuge in the northern Everglades.

## SAMPLING DESIGN

We measured fish density in a nested sampling design to permit analysis at three spatial scales: water-management units, sites, and plots. The water-management units corresponded to areas bounded by canals and levees (e.g., WCAs) or natural drainage features (e.g., TS). We chose sites haphazardly to cover the spatial area of water-management units, and the number of sites per management unit varied roughly with the size of that unit.

Quantitative study of the aquatic-animal fauna of the Everglades is inherently difficult because of the dense aquatic vegetation, extensive unbounded habitats, and logistics of access to the habitats, particularly in the dry season. Sampling fishes in the Everglades is complicated by a problem of scale: The small-bodied cyprinodontiforms are relatively abundant at the 1-m scale, while the larger bodied species (e.g., centrarchids) usually occur at such low densities that few specimens of any age/size class appear in samples collected at the 1-m scale. Thus, we applied two approaches for sampling fishes in the Everglades that targeted different components of the community: throw trapping for small fishes (<8 cm standard length) and boat electrofishing for larger fishes (>8 cm).

## Throw Trapping

We obtained density estimates of small fishes with a 1-m$^2$ throw trap. Selection of this technique is justified by several comparative studies that examined a variety of alternative techniques and trap sizes (Kushlan, 1981; Chick et al., 1992; Loftus and Eklund, 1994). For sampling over a large spatial area as in this study, we must determine if sampling bias varies across the range of habitat variables encountered at different sites. Jordan et al. (1997a) did that by comparing clearing efficiency and sampling accuracy of throw traps in wet prairies with a range of stem densities.

They conducted throw trapping in block nets with a known number of marked fish added and estimated the "true" density by rotenone sampling within the block nets. They concluded that, on average, 63% of the fishes present were collected. More important, they found a high correlation ($r = 0.82$) between the throw-trap estimated density and block-net estimated density. No effect of water depth, emergent plant stem density, canopy height, or periphyton volume was detected on throw-trap collection accuracy or clearing efficiency.

We selected three plots within each site, where we marked off 100 m × 100 m grids divided by 10-m increments. Throw-trap samples were taken at randomly selected coordinates in the grids. For logistical reasons, seven samples were taken per visit in the ENP and five samples per visit in the WCAs. Plots were separated by 0.5 to 2.0 km, while nearest-neighbor sites within water-management units were separated by 5 to 10 km in SRS and TS, and by 8 to 17.5 km in WCA-3A.

Fish sampling has been conducted for different lengths of time at the study sites indicated on Figure 5.1. Sites 6, 23, and 50 in the ENP have been sampled monthly or approximately bimonthly by throw trap since 1977, a period encompassing a range of hydrological conditions from very dry to very wet years. For this chapter, we report analysis of a 10-year interval from 1985 to 1995, which we will call the "long-term study." The more spatially extensive effort to sample 20 sites was begun in mid-1995 and continues to the present. This study period has included only wet years with relatively long hydroperiods at each site. Data collected from July 1996 to April 1997 were analyzed, and those analyses will be referred to as the "short-term study." Routine electrofishing at a subset of the study sites was begun in 1997. We visited the study sites five times per year during February, April, July, October, and December. These months coincide with important transitional phases of Everglades wet–dry season hydrology (Fennema et al., 1994). Not all sites can be sampled by one or both of these techniques in any particular month because water recession depends on local topography, and water depth may exceed that amenable to sampling by throw trap (>1.1 m). In this paper, analysis of subsets of the data that are most useful for illustrating the issues of spatial scale in fish-community composition and dynamics are reported.

## Electrofishing

We used airboat electrofishing to examine patterns in the abundance and composition of large fishes (standard length > 8 cm) across space and time. Electrofishing catch per unit effort (CPUE) is a reliable index of fish density (measured by block nets) in freshwater marshes of the Everglades (Chick et al., 1999). Electrofishing samples were collected quarterly (10/97, 2/98, 4/98, 7/98, 10/98) from wet prairies in the vicinity of a subset of the throw-trap sites. We sampled four sites within WCA-3A (1, 3, 4, 5), four sites within SRS (6, 7, 8, 37), and three sites within the smaller TS (1, 2, 3) (Figure 5.1). At each site, we conducted three 5-min electrofishing transects in the vicinity of each of the three 1-ha plots permanently established at each site. To improve our sampling consistency, we standardized electrofishing power (wattage = voltage × amperage) at 1500 W over different temperature and conductivity conditions using the methods described by Burkhardt and Gutreuter (1995).

For each transect, we identified the major vegetation types present, visually estimated the percentage cover of periphyton, measured water depth (cm), temperature (°C), and specific conductance (μS/cm). Each transect was separated by a 50-m buffer and covered approximately 150 to 250 m of marsh with the airboat running at idle speed. We also conducted electrofishing within alligator ponds once during the dry (4/98) and wet (10/98) seasons. We located 10 to 12 ponds in each water management unit and electrofished for 5 minutes within each pond.

## POPULATION GENETIC ANALYSIS

Spatial patterns of genetic variation are sensitive to migration and gene flow, population size and sex ratio, and mating patterns of individuals (Slatkin, 1985). Thus, analysis of population genetic structure provides a useful surrogate measure for fish migration because direct analysis of movement is not practical for the abundant small fishes of the Everglades. Analysis of population structure has other benefits: It is reflective of average movement and mating patterns over multiple generations and is less subject to the idiosyncratic behavior of individuals than direct measures of movement. We examined allozymic diversity in eastern mosquitofish, *Gambusia holbrooki,* and spotted sunfish, *Lepomis punctatus*, from population samples taken across the study area (Trexler et al., unpubl. data). In 1996, 50 individuals from each of 52 populations of mosquitofish and 12 to 24 individuals from each of 20 populations of spotted sunfish were genotyped at 10 to 12 loci, respectively. This permitted us to examine patterns of heterozygosity and partition genetic variation at three spatial scales using Wright's *F*-statistics comparable to partitioning variance by analysis of variance (Hartl and Clark, 1997): among water-management units ($F_{pt}$), among local sites within units ($F_{sp}$), and within individuals in local sites ($F_{is}$). Most of the locations where eastern mosquitofish were collected were re-sampled in 1999, a year when much of the ecosystem dried and aquatic animals were forced into local alligator ponds and other aquatic refuges.

## A Note on Statistical Techniques

The analyses reported generally used standard statistical methods not requiring detailed discussion here. However, two decisions regarding analysis of the community-structure data warrant mention. We used nested analysis of variance (ANOVA) and multiple analysis of variance (MANOVA) to partition variance into different spatial and temporal sources. We have opted not to employ repeated-measures designs to test for temporal effects. Use of a repeated-measures design would not alter our partitioning of variance to spatial and general temporal effects, but would affect partitioning of variance among interactions of space and time. These temporal effects are not the focus of discussion in this work. Also, there is no general agreement on consideration of "plot" as a repeated factor because random sampling is applied within plots and the same fish are not sampled each visit. Detailed analysis of temporal autocorrelation in fish communities from the long-term study plots have found that it is limited to adjacent bimonthly samples from long-hydroperiod sites (Trexler and Loftus, 2000), probably because of the marked inter-year variance in hydropattern at each study site. Little or no temporal autocorrelation is seen in throw-trap samples collected bimonthly from short-hydroperiod marshes.

We used nonmetric multidimensional scaling (NMDS) to ordinate the Bray–Curtis dissimilarity indices (Kuskal and Wish, 1978) and to identify latent patterns in our species composition data. This technique configures samples along axes based on the proximity of their dissimilarity scores. The location of samples on each axis is independent from that on other axes, and can thus be used as dependent variables in analyses with environmental data. A stress statistic is produced as a measure of goodness-of-fit of the newly created axes and the ordering of the dissimilarity matrix (Kuskal and Wish, 1978). The preferred ordination has the minimum number of axes necessary to describe the data; this is indicated by a Monte Carlo analysis of ordinations starting with one axis and incrementally adding more until the stress no longer differs from the randomized data. NMDS was chosen because it is robust to nonlinearities in the patterns of density among species (Faith et al., 1987; Minchin, 1987; Legendre and Legendre, 1998: 413).

## RESULTS

### What Is the Spatial Scale of Small-Fish Abundance Patterns? (Short-Term Study)

We addressed this question by analysis of absolute abundance and relative abundance of fishes collected by throw-trap between July 1996 and April 1997 from our 20 study sites scattered across WCA-3A, WCA-3B, SRS, and TS. We examined 37,718 specimens of 33 different species (Table 5.1).

**TABLE 5.1**
**Fishes Collected by Throw-Trap Between July 1996 and April 1997
from Everglades Marsh Study Plots**

| Species | Common Name | Total Collected | Maximum No. per m$^2$ |
|---|---|---|---|
| *Lepisosteus platyrhincus* | Florida gar | 1 | 1 |
| *Esox niger* | Chain pickerel | 5 | 1 |
| *Notropis maculatus* | Taillight shiner | 1 | 4 |
| *Notropis petersoni* | Coastal shiner | 60[b] | 9 |
| *Erimyzon sucetta* | Lake chubsucker | 145[b] | 5 |
| *Ameiurus natalis* | Yellow bullhead | 29 | 3 |
| *Noturus gyrinus* | Tadpole madtom | 8 | 1 |
| *Clarias batrachus* (I)[a] | Walking catfish | 4 | 3 |
| *Cyprinodon variegatus* | Sheepshead minnow | 41[b] | 11 |
| *Fundulus chrysotus* | Golden topminnow | 1,844[b] | 10 |
| *Fundulus confluentus* | Marsh killifish | 87[b] | 4 |
| *Fundulus seminolis* | Seminole killifish | 1 | 1 |
| *Jordanella floridae* | Flagfish | 1,783[b] | 34 |
| *Lucania goodei* | Bluefin killifish | 8,391[b] | 33 |
| *Lucania parva* | Rainwater killifish | 1 | 1 |
| *Belonesox belizanus* (I)[a] | Pike killifish | 3 | 1 |
| *Gambusia holbrooki* | Eastern mosquitofish | 9,825[b] | 98 |
| *Heterandria formosa* | Least killifish | 12,713[b] | 65 |
| *Poecilia latipinna* | Sailfin molly | 1,699[b] | 48 |
| *Labidesthes sicculus* | Brook silverside | 5 | 2 |
| *Elassoma evergladei* | Everglades pygmy sunfish | 487[b] | 9 |
| *Enneacanthus gloriosus* | Bluespotted sunfish | 238[b] | 6 |
| *Chaenobryttus gulosus* | Warmouth | 18 | 3 |
| *Lepomis macrochirus* | Bluegill | 6 | 1 |
| *Lepomis marginatus* | Dollar sunfish | 14 | 1 |
| *Lepomis microlophus* | Redear sunfish | 55[b] | 4 |
| *Lepomis punctatus* | Spotted sunfish | 197[b] | 11 |
| *Lepomis* sp. | Unidentified sunfish | 16 | 2 |
| *Micropterus salmoides* | Largemouth bass | 4 | 3 |
| *Etheostoma fusiforme* | Swamp darter | 2 | 1 |
| *Cichlasoma bimaculatum* (I)[a] | Black acara | 7 | 1 |
| *Cichlasoma urophthalmus* (I)[a] | Mayan cichlid | 21 | 1 |
| *Tilapia mariae* (I) | Spotted tilapia | 4 | 1 |
| Total | | 37,718 | |

[a] (I) = introduced species.
[b] species used in NMDS.

## Small-Fish Density

Most of the variation in fish density was among our study sites within water-management units (Table 5.2). Inter-site variation explained over 23% of the total variation in fish density (all species pooled), and no additional variation could be attributed to sources among water management units. Variation among plots within sites explained less than 3% of the total variation, and temporal variation explained a little over 16% of the total variation. Low variance among plots indicates that they tend to be similar; this pattern is illustrated in the high spatial autocorrelation seen at the 3- and 5-km scale (Figure 5.2). Spatial autocorrelation becomes negative with comparisons at the 18-km to 43-km scale, and then positive at longer scales. Positive correlations at the extreme comparisons may be because the nutrient-enriched study sites were at the extremes of the study area, particularly sites 11 and 37 (see environmental factors discussion, below).

**TABLE 5.2**
**Partitioning Variation in Transformed Fish Density $(\sqrt{\#fish/m^2})$ from Throw-Trap Samples Collected July 1996 through April 1997**

| Source | SS | df | MS | Den | F | p | CD |
|---|---|---|---|---|---|---|---|
| Spatial variation | | | | | | | |
| Units | 74.561 | 3 | 24.854 | 2 | 0.267 | 0.80 | — |
| Site(units) | 1118.263 | 12 | 93.189 | 3 | 24.408 | <0.001 | 23.9 |
| Plot(site) | 114.545 | 30 | 3.818 | 8 | 3.097 | <0.001 | 2.5 |
| Temporal variation | | | | | | | |
| Time | 0.608 | 3 | 0.203 | 6 | 0.030 | >0.80 | — |
| Time × unit | 163.196 | 17 | 9.600 | 6 | 1.433 | 0.10 | 3.5 |
| Time × site (unit) | 281.306 | 42 | 6.698 | 7 | 2.310 | <0.001 | 6.0 |
| Time × plot (site) | 327.746 | 113 | 2.900 | 8 | 2.352 | <0.001 | 7.0 |
| Error | 1787.220 | 1449 | 1.233 | — | — | — | — |

*Note:* Sources listed in parentheses denote nested factors. "Den" indicates the MS used as the denominator for each F statistic. Data were collected from *Eleocharis*-dominated sloughs and wet prairies. Using type III sums of squares, the coefficients of determination for each separate factor do not sum to the total variance explained for the full model; for the full model, $R^2 = 0.68$; CD = coefficient of determination.

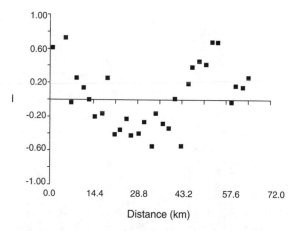

**FIGURE 5.2** Spatial autocorrelation measured by Moran's *I* relative to the distance separating pairs of sampling sites. Estimated annual average density of all fish species was the dependent variable. The value of *I* for groups of sites separated by a given distance is plotted. Only values derived from at least 20 pairs of mean density are plotted.

## Species Patterns

We identified three independent dimensions that cumulatively explained 95% of the variation in small-fish community composition at our 22 study sites (NMDS: three-axes mean stress = 8.330, $p = 0.020$; axis 1, $R^2 = 0.27$; axis 2, $R^2 = 0.49$; axis 3, $R^2 = 0.19$). Of the 14 species included in this analysis, only the coastal shiner and redear sunfish were not significantly correlated with at least one axis in this analysis (Figure 5.3). The composition of small-fish communities differed among water-management units (MANOVA: Region Wilk's lambda = 0.199, P < 0.001), primarily because differences in the species correlated to axes 2 and 3 (univariate analyses: axis 1, $F_{2,19} = 1.798$, $p = 0.193$; axis 2, $F_{2,19} = 15.124$, $p < 0.001$; axis 3, $F_{2,19} = 1.219$, P = 0.009). In particular, TS had relatively more golden topminnows and bluefin killifish, and fewer mosquitofish, least killifish and sailfin mollies than SRS or the WCAs; sites from WCA and SRS tended to ordinate in a mixed cluster. The source of regional variation can be seen in more detail in the patterns of individual species (Figure 5.4). Three species were collected in only one or two of the water-management units. Marsh killifish and coastal shiners were only collected in SRS and TS (Figure 5.5), and sheepshead minnows were only collected in SRS. Further, sheepshead minnows were only collected at sites 6, 23, and 50, the northern SRS sites. Marsh killifish and coastal shiners were collected more widely throughout the water-management units, though marsh killifish were restricted to three sites in SRS. Flagfish were routinely collected in SRS and TS and were found at only three of the five sites in WCA-3A-sw; they were not collected in the other northern study water-management units. Because these species were never caught in large numbers, they are likely to occur in the areas where we failed to collect them, but at densities below our detection limits.

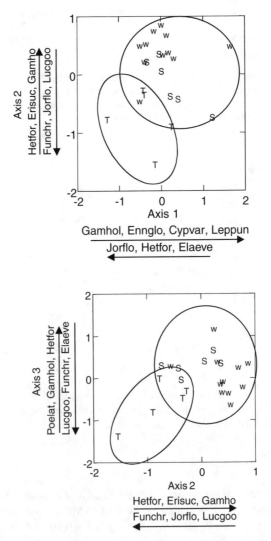

**FIGURE 5.3**    NMDS ordination of fish species composition at study sites. Fish species correlated with each axis ($r > 0.2$) are shown, and arrows indicate positive or negative correlations. Species names are represented by the first three letters of each genus and first three letters of species names, and the order of listing correspond to the relative strength of the correlation with each axis. Letters indicating the water-management region where each site is located are plotted: S = SRS, T = TS, and W = WCA-3.

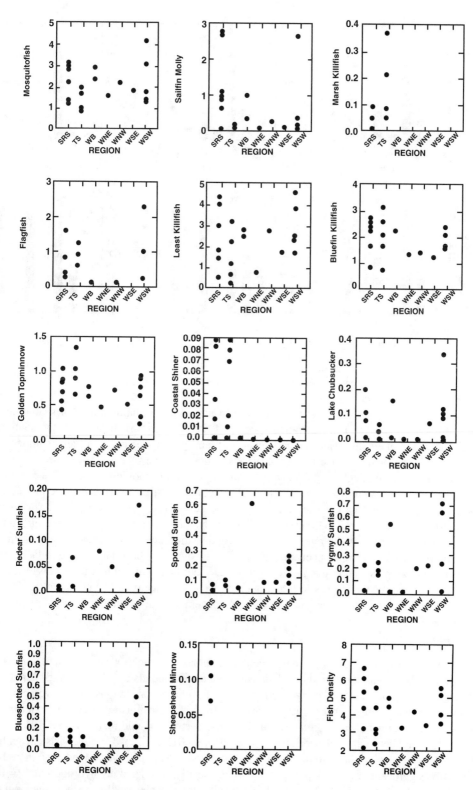

**IGURE 5.4** Mean density (number/m²) of each species and all species summed collected in each water-anagement unit. Annual means are plotted for each site.

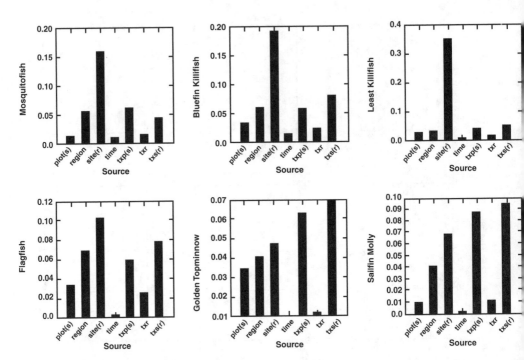

**FIGURE 5.5**   Partitioning of variance in the six most abundant species for 1996–1997 collections. Heigh of the bars indicates the percentage of total variance attributable to each source.

Variation among sites within water-management units was the major source of spatial hetero geneity for four of the six most common species of fish (Figure 5.5). In general, the sources o spatial variance ranked site, water-management unit, and plot in decreasing order of importance In no case did seasonal variance explain a marked amount of the total variation (time as a main effect, Figure 5.5). For flagfish, golden topminnows, and sailfin mollies, there was a substantia amount of variance attributable to interactions of temporal and spatial variance at the plot and site scales. The time by plot-within-site variance may be best considered as sampling error because each plot-time combination is based on only five to seven throw-trap samples. These interaction. explain the most variance in the rarest species whose densities are near our lower limit of sampling efficiency. If densities of these species fluctuate slightly, they may become too rare to be capturec in five to seven samples, yielding many zero estimates and inflating the temporal variance estimates These results reinforce the impression that much of the heterogeneity in fish communities is at the inter-site scale (tens of kilometers) within water-management units, rather than among units.

### Environmental Factors

We tested for correlations among environmental variables (hydroperiod and nutrient status) an measures of fish community structure (fish density and relative abundance). We analyzed fish densit averaged across plots and sampling events to test for these correlations. This averaging wa appropriate based on the partitioning of variance from the previous section. Hydroperiod an nutrient status are related in Everglades marshes. The percent organic matter in our soil sample was highly correlated with hydroperiod ($r = 0.84$, $P < 0.001$, $n = 21$). This is consistent with th use of soil type (peat vs. marl) to distinguish hydroperiod. Also, there was a tendency, though no statistically significant, for total phosphorus to be positively related to hydroperiod ($r = 0.3$. $P = 0.15$, $n = 21$). Sites 11 and 37 had the highest soil total phosphorus of our study sites, but the had moderate hydroperiods. If those sites are omitted, the correlation of TP and hydroperio increases ($r = 0.41$, P = 0.08, $n = 19$). Those sites were uniquely affected by their location: site 11 received nutrient-enriched runoff from a nearby canal and site 37 is at the estuarine–freshwater interface and abuts natural creeks (Rudnick et al., 1999).

Both hydroperiod and nutrient status influenced the density of fish at our study sites, though the correlation of these factors makes separating their effects difficult. In general, increasing hydroperiod or nutrient status (over the range sampled) led to greater densities of throw-trap collected fishes (Figure 5.6). When all 20 study sites were considered, soil-nutrient status explained more variation than hydroperiod and was the only significant factor in the model (TP, $F_{1,17} = 7.31$, P = 0.015; hydroperiod $F_{1,17} = 2.85$, P = 0.161; TP × hydroperiod $F_{1,17} = 1.24$, P = 0.28; $R^2 = 0.45$). However, site 11 is indicated as an outlier with a Studentized residual = 3.77. Because both sites 11 and 37 are influenced by unique circumstances, we repeated the analysis, dropping those sites. Hydroperiod then became the only significant factor (TP, $F_{1,15} = 0.48$, P = 0.50; hydroperiod $F_{1,15} = 4.58$, P = 0.049; TP × hydroperiod $F_{1,15} = 2.01$, P = 0.18; $R^2 = 0.55$). This suggests that in spikerush marshes across the Everglades, hydrology is the primary factor explaining variation in fish density unless variations in nutrient levels are exaggerated by natural or anthropogenic sources.

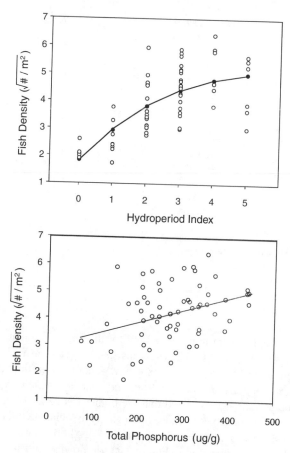

**FIGURE 5.6** Annual average fish density (square-root transformed) estimated from throw-trap samples relative to hydroperiod index and nutrient status in 1997. (A) The hydroperiod index is based on estimates of the average number of days the marsh surface in an area was inundated between 1970 and 1990. 0 = <120 days, 1 = 120–180 days, 2 = 180–240 days, 3 = 240–300 days, 4 = 300–340 days, and 5 = 340–365 days. (B) Nutrient status is estimated by the average total phosphorus in soil samples taken at each study site. A composite sample from each study plot was analyzed.

## WHAT IS THE SPATIAL SCALE OF LARGE-FISH ABUNDANCE PATTERNS? (SHORT-TERM STUDY)

We addressed this question by analysis of airboat-electrofishing CPUE from October 1997 through October 1998, collected from 11 study sites scattered across WCA-3A, WCA-3B, SRS, and TS. We collected 583 specimens from marshes and ponds in our electrofishing study representing 16 species (Table 5.3).

---

**TABLE 5.3**
**Fishes Collected by Electrofishing Between October 1997 and October 1998 from Everglades Marsh Study Sites**

| Species | Common Name | Total Captured | Maximum CPUE (per plot) |
|---|---|---|---|
| Lepisosteus platyrhincus | Florida gar | 121 | 2.00 |
| Amia calva | Bowfin | 42 | 0.67 |
| Esox americanus | Grass pickerel | 9 | 1.00 |
| Esox niger | Chain pickerel | 8 | 0.67 |
| Erimyzon sucetta | Lake chubsucker | 187 | 4.00 |
| Ameiurus natalis | Yellow bullhead | 23 | 0.67 |
| Clarias batrachus (I)[a] | Walking catfish | 7 | 0.33 |
| Centropomus undecimalis | Snook | 2 | 0.33 |
| Micropterus salmoides | Largemouth bass | 45 | 1.33 |
| Chaenobryttus gulosus | Warmouth | 40 | 1.33 |
| Lepomis punctatus | Spotted sunfish | 40 | 1.00 |
| Lepomis macrochirus | Bluegill | 12 | 0.67 |
| Lepomis microlophus | Redear sunfish | 6 | 0.67 |
| Cichlasoma urophthalmus (I)[a] | Mayan cichlid | 25 | 1.00 |
| Oreochromis aureus (I)[a] | Blue tilapia | 15 | 1.00 |
| Tilapia mariae (I)[a] | Spotted tilapia | 1 | 0.33 |
| Total | | 583 | 9.33 |

*Note:* Maximum CPUE is the average of three 5-minute transects in one plot area.

[a] (I) = introduced species.

---

### Large-Fish Abundance

Variation among water-management units accounted for the largest amount of variance in electrofishing CPUE (Table 5.4). Large-fish CPUE was consistently greater in WCA-3A and SRS than in TS, although the magnitude of these differences was variable through time (Figure 5.7). This was consistent with our expectations of short-hydroperiod marshes which experience more severe drying. Over long time periods, we expected that the higher frequency of severe drydowns would limit the population size of large fishes because of mortality associated with crowding in deepwater refuges (Loftus and Kushlan 1987, Loftus and Eklund 1994). Although water depths were relatively similar in SRS and TS during our study (Figure 5.7), drydowns tend to be more severe in TS (Fennema et al. 1994). Water depths in WCA-3A were greater than both SRS and TS, and it is interesting to note that CPUE for alligator ponds in the dry season was substantially lower for this water management unit (9.1 vs. 25.4 and 25.6 fish/5 min).

Differences in CPUE among sites within water-management units were also significant, both as a main effect and over time (Table 5.4; Figure 5.8). Large fishes are known to congregate in canals and the marsh areas adjacent to these canals (Loftus and Kushlan, 1987; Turner et al., 1999).

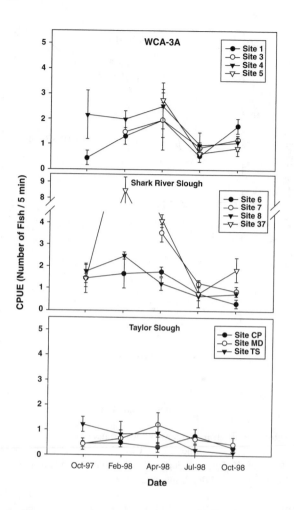

**FIGURE 5.7** Variation in electrofishing CPUE among sites by date. Data were collected from three water-management units within the southern Everglades: WCA-3A, SRS, and TS. Error bars are ±1 standard error.

For example, in October 1997, we collected a series of 27 electrofishing transects in marshes adjacent to canals and compared these to 73 transects from central marsh localities. CPUE was significantly greater ($F_{1,98} = 25.16$, P = 0.0001, $R^2 = 0.20$) for canal-adjacent marshes than for central marshes (3.44 vs. 1.59 fish per 3 min).

## Species Patterns

In addition to these differences in abundance, we also observed conspicuous differences in the composition of large fishes among regions. The lake chubsucker, *Erimyzon sucetta*, was the dominant species in SRS. This species also was abundant in WCA-3A, but was far less abundant in TS (Figure 5.9). As with overall CPUE, the lower abundance in TS marshes may be related to hydroperiod. Two species of pickerel, *Esox americanus* and *Esox niger*, reach the southern limits of their range in or just south of WCA-3A. We never captured these species in either SRS or TS. *Esox niger* has occasionally been observed in ENP (see Loftus and Kushlan 1987), but *Esox americanus* has not. The relative abundance of cichlids was far greater in SRS and TS than in WCA-3A (Figure 5.9). Within SRS and TS, we captured these species most frequently in marshes adjacent to deep-water refuges such as natural streams or man-made canals.

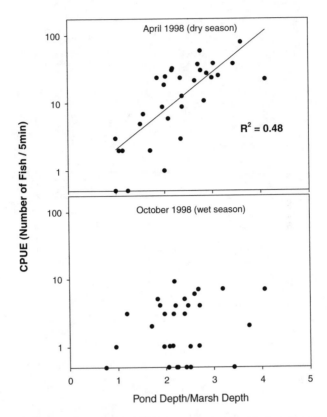

**FIGURE 5.8**    Electrofishing CPUE from alligator ponds compared with the ratio of pond depth to marsh depth. Data were collected from 10 to 12 ponds located within three water-management units of the southern Everglades: WCA-3A, SRS, and TS. The line depicted in the dry season was fitted through linear regression: $\log_{10}(\text{CPUE} + 1) = 0.187 + 0.403$ (pond depth/marsh depth).

## Environmental Factors

Two variables that appeared to influence the patterns of CPUE variation among sites were soil nutrients and proximity to natural streams or man-made canals. No short-hydroperiod marshes were sampled because of the difficulty of electrofishing in such shallow, rocky conditions. The greatest CPUE levels found in our study occurred in site 37 within SRS, the southernmost site. The marshes at this site are adjacent to natural streams and have the greatest concentration of soil nutrients of any of our sites. By comparison, the southernmost site within TS, site 3, was characterized by low CPUE and the lowest concentration of soil nutrients among our sites.

Temporal variability of our CPUE data appeared to largely reflect seasonal variation in water depth (Table 5.4). For example, within WCA-3A and SRS, CPUE varied substantially among sites for all sampling dates except July 1998 (Figure 5.8). The July sample was taken at the end of a mild, but protracted dry season, apparently before most large fishes had returned to the marshes from deep-water refuges. Even though this dry season was not severe, the abundance of large fish within alligator ponds was still substantially greater than during the following wet season. Comparing our alligator pond data from the end of April 1998 (dry season) and October 1998 (wet season), CPUE of large fishes was almost 10-fold greater in April for ponds within TS (25.6 vs 2.6 fish/5 min), almost 6-fold greater in April for ponds within SRS (25.4 vs. 5.91 fish/5 min), and 6.5-fold greater in April for ponds in WCA-3A (9.1 vs. 1.4 fish/5 min). During this dry season CPUE was positively related to the ratio of pond depth to marsh depth ($F_{1,34} = 4.05$, $P = 0.0001$ $R^2 = 0.48$). This indicated that large fishes were more abundant in ponds that provided greater refuge (Figure 5.10). This relationship was not significant during the wet season.

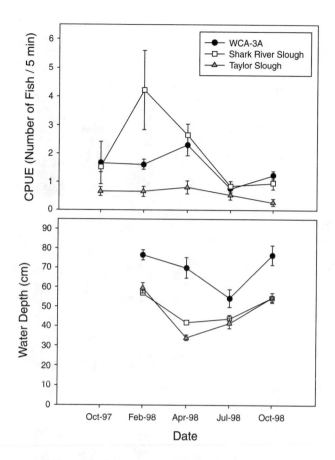

**FIGURE 5.9** Variation in electrofishing CPUE and water depth among water-management units by date. Data are averaged across three or four sites sampled within each of three water-management units of the southern Everglades: WCA-3A, SRS, and TS. Error bars are ±1 standard error.

**TABLE 5.4**
**Results of an Analysis of Variance Examining Patterns in Electrofishing CPUE (Number of Fish per 5-min Electrofishing) Through Time, Across Water-Management Units (WCA-3A, SRS, TS), and Sites Within Water-Management Units**

| Source | SS | DF | MS | Den | F | p | CD |
|---|---|---|---|---|---|---|---|
| Spatial variation | | | | | | | |
| Water management unit | 6.99 | 2 | 3.495 | 2 | 10.10 | 0.006 | 0.23 |
| Site(unit) | 2.77 | 8 | 0.346 | 6 | 3.76 | 0.001 | 0.09 |
| Temporal variation | | | | | | | |
| Time | 5.75 | 4 | 1.435 | 5 | 7.93 | <0.001 | 0.19 |
| Time × unit | 2.22 | 8 | 0.278 | 5 | 1.54 | 0.188 | |
| Time × site (unit) | 5.07 | 28 | 0.181 | 6 | 1.97 | 0.025 | 0.17 |
| Error | 8.55 | 93 | 0.0919 | | | | |

*Note:* Data were collected from *Eleocharis*-dominated sloughs and wet prairies from October 1997 to October 1998. "Den" indicates the MS used as the denominator for each *F* statistic. CD = coefficient of determination; Model $R^2 = 0.72$.

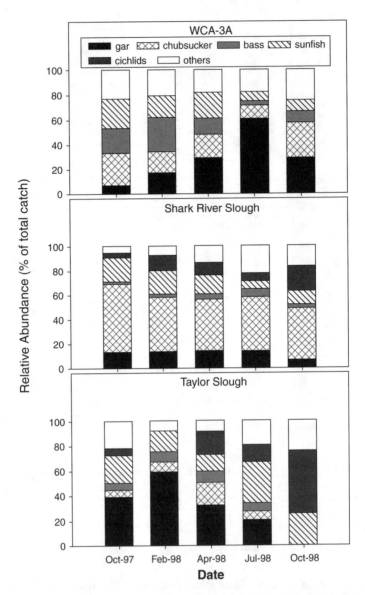

**FIGURE 5.10**    Variation in the relative abundance of fishes captured by electrofishing among water-management units by date. Data were collected from three water-management units within the southern Everglades: WCA-3A, SRS, and TS.

## Are Local Patterns in Small-Fish Abundance and Composition Consistent Over Time? (Long-Term Study)

We examined throw-trap data collected at sites 6, 23, and 50 (Figure 5.1) between 1985 and 1995 to determine how community structure varied among these sites during this time period. The three study sites were chosen to represent three levels of hydroperiod from very short (site 50 experienced prolonged dry periods in 7 of 10 years between 1985 and 1995) to long (site 6 dried briefly in 3 years during that period). Hydroperiod was intermediate at site 23. There was a 2-year drought from 1989 to 1990 when all of the Everglades were dry for some period of time. At site 6, the marsh surface was exposed for a short time only in the two drought years of this 10-year interval, while sites 23 and 50 went dry in 6 and 7 of the 10 years, respectively.

## Small-Fish Abundance

Results from the short-term study indicated the importance of hydroperiod to spatial variation in fish abundance and species composition over the course of a year. A similar result was noted for a comparison of sites 6, 23, and 50 in SRS between 1985 and 1995. The 10-year average density of fishes at these three sites by month is consistently ranked from short hydroperiod with the fewest fishes to long with the most fishes (Figure 5.11; Loftus and Eklund, 1994).

## Small-Fish Species Composition

The nonmetric multidimensional scaling of angular-transformed relative abundance indicated that these sites differed consistently in the composition of their fish communities. Two dimensions were observed that explained over 84% of the total variance in relative abundance (NMDS: two-axes mean stress = 24.653, P = 0.020; axis 1, $R^2$ = 0.40; axis 2, $R^2$ = 0.44). When the scores from the six factors were analyzed in a MANOVA, significant differences were noted among sites and over time within sites (site: Wilks' lambda = 0.944, P < 0.001; year: Wilks' lambda = 0.419, P < 0.001; site × year: Wilks' lambda = 0.533, P = 0.011). Site 50 was distinguished from site 6 in having relatively more marsh killifish and sheepshead minnows and fewer bluefin killifish, spotted sunfish, and mosquitofish, while site 23 was intermediate (Figure 5.12). Perhaps more interesting was the effect of the drought years 1989 and 1990 on the patterns of relative abundance. Fish species composition was affected by the drought event at site 6 (Figure 5.12: pre-drought 1985–88 vs. drought years 1989–90 vs. post-drought 1991–95: Wilks' lambda = 0.028, P < 0.001) and site 23 (Wilks' lambda = 0.151, P = 0.007), but not site 50 (Wilks' lambda = 0.596, P = 0.424). The pre- and post-drought years differed in species composition at site 6, but not at site 23. Thus, the severity of effect of the drought on small-fish species composition appeared to be related to the long-term hydroperiod at the site. Much of the shift at site 6 in 1989 was related to a greater relative abundance of marsh killifish and sheepshead minnows and fewer bluefin killifish, least killifish, and mosqui-tofish. This pattern was also observed after droughts at site 6 in previous years (Loftus and Eklund, 1994). The fish community during the post-drought years at site 6 could be characterized as having relatively greater density of fishes typical of longer hydroperiod sites in the Everglades (bluefin killifish, spotted sunfish, least killifish). This is probably linked to a series of very wet years in the 1990s (Trexler and Loftus, 2001).

Recovery of species composition after a drought is influenced by patterns of recovery for individual species (Table 5.5). The average time for mosquitofish and flagfish to reappear in samples

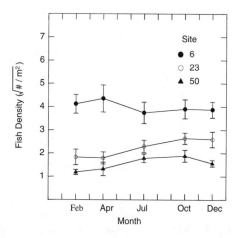

**FIGURE 5.11**  Average density of all fishes between 1985 and 1995 plotted by month. Sites 6, 23, and 50 are plotted.

following a drydown was more than one census period (≈2.5 months). On the other hand, bluefin killifish took, on average, over 9 months to reappear in samples from the long-hydroperiod site and much longer at the short-hydroperiod site. It appears that the relative abundance patterns are site characteristics that are regained following hydrological perturbation over a time period set by

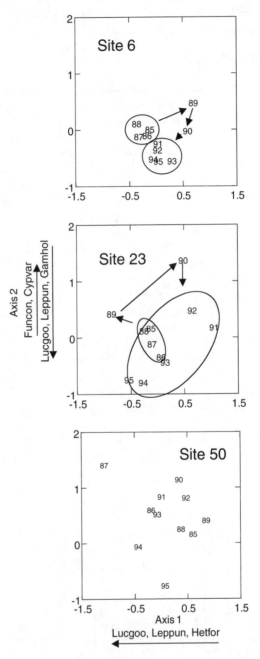

**FIGURE 5.12**    NMDS ordination of inter-annual variation in fish species composition plotted by study sites. Fish species correlated with each axis (*r* > 0.2) are shown, and arrows indicate positive or negative correlations. Species names are represented by the first three letters of each genus and first three letters of species names, and the order of listing corresponds to the relative strength of the correlation with each axis. Numbers indicate the year when each collection of samples was made. Circles enclose years in pre-drought and post-drought periods except for site 50, where there were no differences among the three time periods.

the recolonization abilities of the species living there. Slow times could be the result of slow dispersal rates or low reproductive, and therefore population growth, capacities. It is also possible that mosquitofish, an early colonizer, could resist the re-invasion of some species by predation on immigrating juveniles or by resource competition (Hurlbert et al., 1972; Meffe, 1985).

**TABLE 5.5**
**The Average Time Following a Drydown Before Each Species Reappears in Samples and Their Rank Density at That Site**

| Species | Long Hydroperiod (Site 6) | | Short Hydroperiod (Site 50) | |
| --- | --- | --- | --- | --- |
| | Rank Density | No. of Surveys | Rank Density | No. of Surveys |
| Mosquitofish | 1 | 1.3 | 1 | 1.4 |
| Bluefin killifish | 2 | 3.4 | 4 | 17.6 |
| Least killifish | 3 | 3.0 | 7 | 4.5 |
| Flagfish | 4 | 1.3 | 2 | 1.4 |
| Golden topminnow | 5 | 2 | 6 | 7 |
| Sailfin molly | 6 | 5.7 | 8 | 9.5 |
| Marsh killifish | 8 | 3.4 | 3 | 1.3 |
| Sheepshead minnow | — | — | 5 | 4 |

*Note:* Ranks were based on the average density of each species between 1985 and 1995. Approximately 2.5 months passed between surveys, on average.

## WHAT IS THE SPATIAL SCALE OF GENETIC VARIATION?

In both species studied, the level of genetic differentiation was greatest among local sites within regions, although, in general, the total amount of differentiation was low (Table 5.6A). This is an unusual pattern of differentiation; we did not observe the typical isolation by distance (Slatkin, 1985) indicated when allele frequency variation is correlated with geographic distance separating pairs of sampling sites. Our result may indicate that, at the time of our sampling, local fish populations were not at genetic equilibrium following a major drought event in 1989–1990, a period of about 14 mosquitofish generations and three to seven spotted sunfish generations prior to our collections. We hypothesize that major population events, fish concentration, and mortality resulting from the drought, followed by marsh recolonization and rapid population growth, may have taken place at spatial scales shorter than the distance among sampling sites (approximately 2 to 10 km), because that was the spatial scale of greatest genetic variation. In addition, we hypothesize that the overall low levels of genetic differentiation were the result of routine movements of individuals in the post-drought years, which homogenized the local pattern generated by the drought years prior to sampling in 1996. This is supported by low heterozygosity of mosquitofish from WCA-3B relative to the rest of the study area (Table 5.6B); WCA-3B is completely isolated from the rest of the Everglades by levees and lacked significant access to large aquatic refuges such as major canals in droughts such as the one in 1989–1990.

In 1999, the Everglades experienced a drydown event not as severe as in 1989–1990 but one that affected patterns of heterozygosity in mosquitofish consistent with those hypothesized for the drought. In 1999, we observed overall deviation from Hardy–Weinberg expectations and a significant deficiency of heterozygosity as might be expected from a rapid diminution of population size. For example, observed heterozygosity at highly polymorphic loci were consistent with Hardy–Weinberg expectations, but loci with low observed polymorphism displayed significant deviations from Hardy–Weinberg expectations. Deviation from Hardy–Weinberg expectations at loci with low polymorphism resulted primarily from the lack of rare-allele heterozygotes. There was no pattern of deviation related to hydrology of the sites sampled. The level of genetic differentiation was again

## TABLE 5.6
## Patterns of Genetic Diversity in Mosquitofish and Spotted Sunfish Collected from the Southern Everglades

### A. Population Structure Estimates[a]

|  | Mosquitofish (1996) | | Mosquitofish (1999) | | Spotted Sunfish (1996) | |
|---|---|---|---|---|---|---|
|  | $F_{SP}$ | $F_{PT}$ | $F_{SP}$ | $F_{PT}$ | $F_{SP}$ | $F_{PT}$ |
| Estimate | 0.0106 | 0.0033 | 0.0072 | 0.0004 | 0.0203 | −0.0076 |
| Upper bound | 0.0199 | 0.0069 | 0.0167 | 0.0024 | 0.0579 | 0.0006 |
| Lower bound | 0.0034 | 0.0004 | 0.0033 | −0.0010 | −0.0044 | −0.0180 |

### B. Average Heterozygosity ± Standard Error[b]

|  | Mosquitofish (1996) | Mosquitofish (1999) | Spotted Sunfish (1996) |
|---|---|---|---|
| WCA-3A | 0.158 + 0.004 (25) | 0.091 + 0.005 (11) | 0.102 + 0.001 (3) |
| WCA-3B | 0.117 + 0.001 (3) | 0.101 + 0.006 (4) | 0.140 (1) |
| SRS | 0.149 + 0.004 (16) | 0.087 + 0.004 (15) | 0.101 + 0.029 (3) |
| TS | 0.137 + 0.004 (8) | 0.106 + 0.007 (13) | N/A |
| Canal | N/A | N/A | 0.125 + 0.006 (8) |

*Note:* For mosquitofish, the same locations were sampled in 1996 and 1999 whenever possible.

[a] Spatial partitioning of genetic variation for fishes. $F_{SP}$ indicates the genetic variance in samples relative to among samples, and $F_{PT}$ indicates the genetic variance among samples relative to the total study. Upper and lower bound estimates are 95% confidence interval estimates from bootstrap sampling.
[b] Heterozygosity of fishes collected from the Everglades. Number of populations sampled for both *F*-statistics and heterozygosity estimates are indicated in parentheses.

greatest among local sites within regions and the total amount of differentiation was low (Table 5.6A). Also, differences in allele frequencies did not generally increase with greater geographic distances separating pairs of sampling sites. Mosquitofish sampled in 1999 displayed less heterozygosity than mosquitofish sampled in 1996 (Table 5.6B). Also, unlike the 1996 samples, the levels of heterozygosity did not significantly differ among regions in the 1999 mosquitofish samples. The observed reduction in genetic variation is consistent with predicted effects of a drought yielding reduction in mosquitofish population sizes.

Long-hydroperiod marshes and canals may serve as source locations for colonists to nearby short-hydroperiod marshes following drydown events. We tested this hypothesis following the method of McCauley et al. (1995) who noted that source populations should be less genetically differentiated than nearby populations experiencing frequent extinction and recolonization. Some support was found for this hypothesis in mosquitofish, and possibly sunfish, though with less statistical power. The genetic variation among samples of mosquitofish collected in 1996 from short-hydroperiod sites was greater than that among nearby long-hydroperiod sites, which in turn was greater than among canal sites (combining results from WCA-3A, SRS, and TS, $z_p = 1.498$, P = 0.067). Also, samples of spotted sunfish from small ponds in the interior Everglades tended to be more heterogeneous than samples from canals and natural stream sites, though the trend was not statistically significant.

Preliminary analyses using a new maximum-likelihood method that estimates population parameters by integrating over all possible genealogies and migration events between subpopulations (Beerli and Felsenstein, 1999) indicated that two uncorrelated factors affect gene flow in TS. These are distance between the sites and the presence of a road or barrier. The estimate of gene flow between pairs of sites located along a road was significantly greater than gene flow between other sites in TS long-hydroperiod marshes, or between the slough and the road. These data suggest that barriers to fish movement imposed by human alterations do shape fish migration across the land-scape and could affect population structure. The new analytical protocol is promising because it does not make ecologically unrealistic assumptions of equal population sizes and symmetric migration inherent in traditional $F_{ST}$-based estimates of gene flow. However, it is also computer intensive and cannot be used for simultaneous analysis of our entire data set until a parallel computing version is completed (Beerli, pers. comm.). We are proceeding with re-analyses of our data one water management region at a time; however, analysis of the TS data alone required 5 weeks of computer time!

# DISCUSSION

We addressed three questions regarding the scale of variation in different aspects of the Everglades fish community: scale of density and species composition, temporal stability of that scale, and scale of genetic variation in selected species. We found that the major scale of variation in both community structure (measured as density and relative abundance) and genetic variation of small fishes is in the approximately 10-km range. Patterns of small-fish community structure appear to be related to local environmental parameters, especially hydroperiod, and they may vary over time as local hydroperiod changes. This raises an important caveat: Our short-term study of spatial variation in small and large fishes was conducted during a period of wetter-than-average years when water levels were high across the ecosystem. Our long-term study, conducted over a smaller spatial area and only on small fishes, underscores that spatial relationships of community structure at our short-term study sites are likely to change during periods of drier years. Given this limitation, what generalities and hypotheses can we draw from our results?

## THE SCALE OF COMMUNITY DYNAMICS IN THE EVERGLADES

The spatial scale of fish density and community composition differs for small and large fishes. We have compared density and species-composition data at three spatial scales for small fishes and at two scales for large fishes: 1-km (among plots, small fishes only), 10-km (among sites within water-management units), and among water-management units. Two classes of explanations are possible for this difference: sampling artifacts and meaningful biological effects. We have strived to document and minimize the sampling biases of these two techniques, a measure of sampling accuracy (Jordan et al., 1997a; Chick et al., 1999). It is more difficult to equalize the variance about our estimates (a measure of precision) from such different techniques. Thus, the statistical power to detect patterns probably differs between our two datasets. We have repeated our small-fish analyses on a limited data set that included only those sites at which large fishes were also sampled, and obtained the same results as analysis of the full dataset. Potential differences in precision cannot explain the rank differences in partitioning of variation that we noted in our two datasets, so we conclude that the differences are biological in origin.

While regional distinctions could be identified in the small-fish data, most of the variation was among sites within water-management units in fish density and species composition. These results held up when two high-nutrient sites were removed. On the other hand, our large-fish data indicated consistent differences among the water-management units that greatly exceeded the inter-site variation. For both large and small fish, the WCAs and SRS were more similar to each other than either was to TS. The WCAs and SRS were historically part of a single drainage basin while TS was historically a separate drainage. TS has been severely drained in recent years such that recent management may also explain its distinctiveness. However, TS had relatively more golden topmin-nows and bluefin killifish than SRS and WCAs, a pattern not consistent with it having a shorter hydroperiod; these species tend to increase in abundance in wetter periods in SRS.

The different spatial scales of variability in small and large fish could be explained by the way body size influences the dispersal ecology and generation time of fishes. Our "large fishes" group includes species with longer generation times and generally slower population growth capacities than most of the species in our "small fishes" group. Also, dispersal distance and swimming performance scale positively to body size in fishes (Ware, 1978), indicating that our large fishes can probably cover a larger area within their lifespan. Thus, small fishes may recover more quickly following a drought by way of *in situ* reproduction, while larger fishes may locate aquatic refuges more effectively than smaller ones. Either or both of these factors could contribute to the difference in major scale of variation we observed.

Much of the variation we noted in fish density and species relative abundance of both small and large fishes is well explained by hydrological variation. For example, TS has lower small- and large-fish density, probably because it lacks the extensive long-hydroperiod marshes present in SRS and WCAs. Overall, we noted marked differences in density and species composition between short- and long-hydroperiod sites in the small-fish data. These distinctions indicate that a better understanding of fish response to hydrological variation is needed. In the following section, we propose two scale-based conceptual models of fish response to drought.

Our genetic studies of mosquitofish and spotted sunfish are consistent with these results. They are consistent with the interpretation that drydown events leave lasting effects on the population structure of these species, probably reflecting a history of population collapse and expansion from local refuges. Also, the spatial scale of this dynamic for mosquitofish appears to be at a similar one to the patterns of small-fish density. Spotted sunfish are not abundant enough in marshes to sample at the same scale as mosquitofish. However, our results support the hypothesis that alligator ponds are local refuges for spotted sunfish and that they move widely when inhabiting canals. Less well supported by the genetic data, but intriguing, are hints that long-hydroperiod marshes and canals serve as important refuges for population recovery in short-hydroperiod marshes by mosquitofish. Also, it appears that barriers to mosquitofish movement across the landscape do affect patterns of gene flow and reshape their dispersal ecology. This result is well supported by a recent sampling study in the same study area (Loftus, unpubl. data).

## ALTERNATIVE MODELS FOR MARSH DYNAMICS AND RECOLONIZATION FOLLOWING DROUGHT

Seasonal displacement of fish across the Everglades by fluctuating water level is thought to have profound effects on the dynamics of its fish communities (Kushlan, 1974, 1976a; Karr and Freemark, 1985; Loftus and Kushlan, 1987). For example, this may provide the concentrations of fish that feed wading birds (Bancroft et al., 1994; Frederick and Spalding, 1994; Ogden, 1994). To what extent do density fluctuations reflect changes in local population size vs. different levels of population concentration? Alternatively, we may ask over what spatial scales do fishes interact and how are those scales affected by seasonal cycles in water level? Are long-hydroperiod marshes stable sites that serve as sources of individuals (*sensu* Pulliam, 1988) for short-hydroperiod marshes, where they die in the next season?

Though the topographic relief of the Everglades is slight, small variations in elevation can have marked effects on the vegetation of an area and its hydroperiod. Marshes in the Everglades are a complex mixture of vegetatively defined habitats that appear to reflect local elevation (Jordan et al., 1997b). Wet prairies are often dominated by spikerush (mostly *Eleocharis cellulosa*), though they may be dominated by maidencane, *Panicum hemitomon*, or beakrush, *Rynchospora tracyi*, in local low sites where water is last to dry (Busch et al., 1998; McPherson, 1973). Wet prairies form braided habitats subdivided by patches of sawgrass-dominated marsh. Sawgrass (*Cladium jamaicense*) often dominates areas with locally high topography and can grow in very dense stands that may hamper or preclude passage of fishes. Sawgrass marshes typically form strands, with a long axis parallel to the direction of water flow (Figure 5.13), that form a maze-like barrier for animals attempting to move perpendicular to water flow.

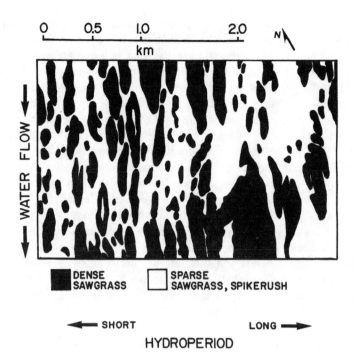

**FIGURE 5.13**   Map of the distribution of dense sawgrass strands in the SRS. (Redrawn from Olmsted et al., 1982.)

The matrix of wet prairies and sawgrass marshes has the potential to limit the spatial scale of fish concentration during dry-season water recession, as well as expansion upon reflooding. Two extreme patterns of fish movement are possible. Fish may concentrate at a regional scale with fish from short-hydroperiod marshes on the eastern and western fringes of the ecosystem moving toward the central flow-way and its long-hydroperiod marshes as water recedes (Regional Concentration Model, Figure 5.14). An alternative model is that fish become stranded in local wet prairies as water recedes, precluded from long-distance movement at the boundary by sawgrass marshes (Local Concentration Model, Figure 5.14). These extremes predict two different magnitudes of fish concentration as drydown proceeds, with potentially important implications for food availability for wading birds. Individual species probably differ in the propensity to disperse across the landscape, and so we may not expect a single scale of dynamics to describe the entire community. This is, in fact, what we see when comparing our large- and small-fish datasets.

Unlike some fishes in seasonal tropical wetlands and floodplains (Lowe-McConnell, 1987), we have not detected directed long-distance migrations of fishes as water recedes in the Everglades.

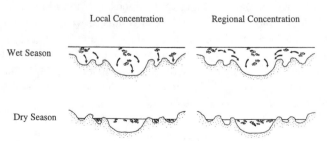

**FIGURE 5.14**   Conceptual model of alternative hypotheses for fish concentration during Everglades dry seasons.

However, as water rises, the surviving fishes of all size classes are at times seen "streaming" in mass-directed movements into recently reflooded habitats. It is not clear if these fishes are leaving dry-season refuges because water conditions have degraded, or if they are actively invading recently accessible habitats in search of food. We hypothesize that water recession in the Everglades is generally too slow, and topographic gradients are too shallow, for small fishes to respond in mass, long-distance migrations to deep-water refuges. In contrast, the expansion of aquatic habitats and reflooding of marshes in the early wet season are often rapid (one day or less) as a result of dramatic rainfall events; both small and large fishes appear to respond actively to these dramatic changes in habitat availability. This pattern supports the local concentration model for small-fish population dynamics and the hypothesis that long-hydroperiod marshes serve as source habitats in drought years while short-hydroperiod regions are generally sinks in those years. Local concentration of small fish is indicated by the lack of an increase of fish density in long-hydroperiod marshes during the early dry season, as nearby short-hydroperiod marshes are drying. The concentration of fishes in ponds, while at times dramatic, may be explained by local movements from adjacent marshes (Turner et al., 1999). Large fishes, on the other hand, may be successful at moving across the landscape to locate alligator ponds and other deep-water refuges as water recedes. This was demonstrated by the increase of large fishes in alligator ponds even in years when the marsh surface did not dry (see electrofishing study by Nelson and Loftus, 1996). Thus, large fishes may be better described by the regional concentration model.

### EVERGLADES RESTORATION AND MANAGEMENT: WHAT DOES A SPATIAL VIEW ADD?

The largest, most costly environmental restoration project in history has been authorized by the United States and the state of Florida, which will dramatically redesign the hydrology of the Everglades ecosystem (U.S. Army Corps of Engineers, 1998). Although uncertainties exist about the extent of restoration possible given the constraints of a smaller ecosystem, competing demands for water, and maintenance of water quality, a major goal of the Everglades restoration project is to restore the ecological integrity of the remaining Everglades. Our fish-community data serve three major roles in the restoration process. These data will be used for ecological backcasting to estimate fish community structure at locations within the "historic" Everglades by use of hydrological and ecological simulation models (DeAngelis et al., 1997) and to set restoration targets and success criteria. Also, the statistical relationships between environmental and fish community parameters we generate lead to hypotheses about the origins of community structure that provide a motivation for both experimental work and improvement of simulation models used in hydrological management. Finally, the network of sampling sites provides baseline data, and their continuation is essential to monitor conditions that result from future hydrological restoration.

Our results at the local and regional scales suggest that hydrological restoration of over-drained areas of the Everglades will lead to changes in the abundance and composition of fish communities (see also Loftus and Eklund, 1994). Patterns of drying have clear implications for management, and choices made have great potential to influence the food base for wading birds and other predators (Kushlan, 1987; Frederick and Spalding, 1994). Our results provide a clear indication that hydrological pattern is a major factor in shaping fish communities and presumably in shaping historical Everglades communities. We also found that increases in nutrient status can lead to marked changes in the local fish community. Ongoing and future research will address these changes in more detail, but a clear lesson is that hydrological management must not ignore nutrient effects from natural and anthropogenic sources (Turner et al., 1999).

## CONCLUSIONS

The system of canals and levees that now directs the flow of water through the Everglades has changed the spatial arrangement of long- and short-hydroperiod marshes and their associated aquatic communities. For example, the main route of water into the Everglades National Park now lies substantially westward of its historical path. Our data indicate that this water-management system has imposed a structure to the fish communities there, but possibly in more subtle ways than we might have guessed. Our studies indicate that communities of small fishes in the Everglades are most variable at the 10- to 15-km scale in density, relative abundance, and genetic variation (though more species need to be examined). Large-fish density and relative abundance appear to vary more across larger spatial scales. There were notable differences among the water-management units in small-fish species composition, mostly in the rare taxa, and density, mostly in TS. The complete isolation of WCA-3B may explain the modest, but significant, decrease in genetic variation of mosquitofish there. Our large-fish data displayed more marked differences among management units. Local environmental factors, especially hydroperiod, are important in determining these patterns in both small and large fishes. The effects of hydrological fluctuation may vary with fish size because larger fish can move greater distances in response to gradual drying than smaller ones. We propose alternative conceptual models of community dynamics based on local and regional scales of fish concentration in the dry season to explain the differences between our small- and large-fish data. Future research will test these models with direct study of fish movement. Ultimately, this information will help in assessing the plans and future activities of restoration of the Florida Everglades.

## ACKNOWLEDGMENTS

We thank Jon Moulding, Sue Perry, Dale Gawlik, G. Ronnie Best, Tom Fontaine, and Tom Armentano for providing assistance and financial support for this effort. Also, we thank the many field technicians, graduate students, and postdoctoral research associates who have helped to gather and process the samples and data. This work was funded by cooperative agreements 5280-6-9011 and 5280-8-9003 (a CESI project) issued to Trexler by the Everglades National Park, contract C-E6636 issued to Trexler and Jordan by the South Florida Water Management District, and cooperative agreement 1445-CA09-95-0112 issued to Trexler by the U.S. Geological Survey. The long-term study was supported by the U.S. National Park Service through funding assistance provided to Loftus and Bass.

## REFERENCES

Bancroft, G.T., A.M. Strong, R.J. Sawicki, W. Hoffman, and S.D. Jewell. 1994. Relationships among wading bird foraging patterns, colony locations, and hydrology in the Everglades. p. 615–657, in *Everglades: The Ecosystem and Its Restoration*, Davis, S.M. and Ogden, J.C., Eds., St. Lucie Press, Delray Beach, FL.

Beerli, P. and J. Felsenstein. 1999. Maximum-likelihood estimation of migration rates and effective population numbers in two populations using a coalescent approach, *Genetics*, 152:763–773.

Burkhardt, R.W. and S. Gutreuter. 1995. Improving electrofishing catch consistency by standardizing power., *N. Amer. J. Fish. Manage.*, 15:375–381.

Busch, D.E., W.F. Loftus, and O.L. Bass, Jr. 1998. Long-term hydrologic effects on marsh plant community structure in the southern Everglades, *Wetlands*, 18:230–241.

Chick, J.H. and C.C. McIvor. 1994. Patterns in the abundance and composition of fishes among different macrophytes: viewing a littoral zone as a landscape, *Can. J. Fish. Aquatic Sci.*, 51:2873–2882.

Chick, J.H., F. Jordan, J.P. Smith, and C.C. McIvor. 1992. A comparison of four enclosure traps and methods used to sample fishes in aquatic macrophytes, *J. Freshwater Ecol.*, 7:353–361.

Chick, J.H., S. Coyne, and J.C. Trexler. 1999. Effectiveness of airboat electrofishing for sampling large fishes in shallow vegetated habitats, *N. Amer. J. Fish. Manage.*, 19:957–967.

DeAngelis, D.L. 1994. Synthesis: spatial and temporal characteristics of the environment, pp. 307–320, in *Everglades: The Ecosystem and Its Restoration*, Davis, S.M. and Ogden, J.C., Eds., St. Lucie Press, Delray Beach, FL.

DeAngelis, D.L., W.F. Loftus, J.C. Trexler, and R.E. Ulanowicz. 1997. Modeling fish dynamics in a hydrologically pulsed ecosystem, *J. Aquatic Ecosyst. Stress Recovery*, 6:1–13.

Faith, D.P., P.R. Minchin, and L. Belbin. 1987. Compositional dissimilarity as a robust measure of ecological distance, *Vegetatio*, 69:57–68

Fennema, R.J., C.J. Neidrauer, R.A. Johnson, T.K. MacVicar, and W.A. Perkins. 1994. A computer model to simulate natural Everglades hydrology, pp. 249–289, in *Everglades: The Ecosystem and Its Restoration*, Davis, S.M. and Ogden, J.C., Eds., St. Lucie Press, Delray Beach, FL.

Frederick, P.C., and M.G. Spalding. 1994. Factors affecting reproductive success of wading birds (Ciconiiformes) in the Everglades ecosystem, pp. 149–197, in *Everglades: The Ecosystem and Its Restoration*, Davis, S.M. and Ogden, J.C., Eds., St. Lucie Press, Delray Beach, FL.

Fuller P.L., L.G. Nico, and J.D. Williams. 1999. *Nonindigenous Fishes: Introduction to Inland Waters of the United States*, Special Publ. No. 27, American Fisheries Society, Bethesda, MD, 613 pp.

Gilbert, C.R. 1987. Zoogeography of the freshwater fish fauna of southern Georgia and peninsular Florida, *Brimleyana*, 13:25–54.

Gleason, P.J. and P. Stone. 1994. Age, origin, and landscape evolution of the Everglades peatland, pp. 149–197, in *Everglades: The Ecosystem and Its Restoration*, Davis, S.M. and Ogden, J.C., Eds., St. Lucie Press, Delray Beach, FL.

Gunderson, L. 1994. Vegetation of the Everglades: Determinants of community composition, pp. 323–340, in *Everglades: The Ecosystem and Its Restoration*, Davis, S.M. and Ogden, J.C., Eds., St. Lucie Press, Delray Beach, FL.

Hartl, D.L. and A.G. Clark. 1997. *Principles of Population Genetics*, 3rd ed., Sinauer, Sunderland, MA.

Hurlbert, S.H., J. Zedler, and D. Fairbanks. 1972. Ecosystem alteration by mosquitofish (*Gambusia affinis*) predation, *Science*, 175:639–541.

Jordan, F. 1996. Spatial Ecology of Decapods and Fishes in a Northern Everglades Wetland Mosaic, Ph.D. dissertation, University of Florida, Gainesville.

Jordan, C.F., S. Coyne, and J.C. Trexler. 1997a. Sampling fishes in heavily vegetated habitats: the effects of habitat structure on sampling characteristics of the 1-m$^2$ throw trap, *Trans. Am. Fish. Soc.*, 126:1012–1020.

Jordan, F., H.L. Jelks, and W.M. Kitchens. 1997b. Habitat structure and plant species composition in a northern Everglades wetland landscape, *Wetlands*, 17:275–283.

Jordan, F., K.J. Babbitt, and C.C. McIvor. 1998. Seasonal variation in habitat use by marsh fishes, *Ecol. Freshwater Fish*, 7:159–166.

Karr, J.R. and K.E. Freemark. 1985. Disturbance and vertebrates: an integrative perspective, pp. 153–168, in *The Ecology of Natural Disturbance and Patch Dynamics*, Pickett, S.T.A. and White, P.S., Eds., Academic Press, New York.

Kruskal, J.B. and M. Wish. 1978. *Multidimensional Scaling*, Sage University Paper Series on Quantitative Applications in the Social Sciences, 07-0111, Sage Publications, Beverly Hills, CA.

Kushlan, J.A. 1974. Effects of a natural fish kill on the water quality, plankton, and fish population of a pond in the Big Cypress Swamp, Florida, *Trans. Am. Fish. Soc.*, 103:235–243.

Kushlan, J.A. 1976a. Environmental stability of fish community diversity, *Ecology*, 57:821–825.

Kushlan, J.A. 1976b. Wading bird predation in a seasonally fluctuating pond, *Auk*, 93:464–476.

Kushlan, J.A. 1981. Sampling characteristics of enclosure fish traps, *Trans Am. Fish. Soc.*, 110:557–562.

Kushlan, J.A. 1987. External threats and internal management: the hydrologic regulation of the Everglades, Florida, *Environ. Manage.*, 11:109–119.

Legendre, P. and L. Legendre. 1998. *Numerical Ecology*, 2nd Edition. Elsevier, New York.

Light, S.S. and J.W. Dineen. 1994. Water control in the Everglades: a historical perspective, pp. 47–84, in *Everglades: The Ecosystem and Its Restoration*, Davis, S.M. and Ogden, J.C., Eds., St. Lucie Press, Delray Beach, FL.

Loftus, W.F. 2000. Inventory of fishes of Everglades National Park, *Florida Scientist*, 63:27–47.

Loftus, W.F. and A. Eklund. 1994. Long-term dynamics of an Everglades small-fish assemblage, pp. 461–484, in *Everglades: The Ecosystem and Its Restoration*, Davis, S.M. and Ogden, J.C., Eds., St. Lucie Press, Delray Beach, FL.

Loftus, W.F. and J.A. Kushlan. 1987. Freshwater fishes of southern Florida, *Bull. Florida State Mus. Biol. Sci.*, 31:147–344.

Lowe-McConnell, R.H. 1987. *Ecological Studies in Tropical Fish Communities*, Cambridge University Press, Cambridge, U.K.

McCauley, D.E., J. Raveill, and J. Antonovics. 1995. Local founding events as determinants of genetic structure in a plant metapopulation, *Heredity*, 75:630–636.

McPherson, B.F. 1973. *Vegetation in Relation to Water Depth in Conservation Area 3, Florida*, U.S. Geological Survey report 73025.

Meffe, G.K. 1985. Predation and species replacement in American southwestern fishes: a case study, *Southwest. Nat.*, 30:173–187.

Minchin, P.R. 1987. An evaluation of the relative robustness of techniques for ecological ordination, *Vegetatio*, 69: 89–107

Nelson, C.M. and W.F. Loftus. 1996. Fish communities of alligator ponds in the high water periods of 1983–1985 and 1994–1996, pp. 89–101, in *Proceedings of the Conference: Ecological Assessment of the 1994–1995 High Water Conditions in the Southern Everglades*, Armentano, T.V., Ed., South Florida Management and Coordination Working Group.

Ogden, J.C. 1994. A comparison of wading bird colony nesting dynamics (1931–1946 and 1974–1989) as an indication of ecosystem conditions in the southern Everglades, pp. 533–570, in *Everglades: The Ecosystem and Its Restoration*, Davis, S.M. and Ogden, J.C., Eds., St. Lucie Press, Delray Beach, FL.

Olmsted, I.C., V.L. Dunevitz, and J.M. Johnson. 1982. Vegetation map of Shark River Slough, Everglades National Park. South Florida Research Center, U.S. National Park Service, Everglades National Park.

Pulliam, H.R. 1988. Sources, sinks, and population regulation, *Am. Nat.*, 132:652–661.

Rudnick, D.T., Z. Chen, D.L. Childers, J.N. Boyer, and T.D. Fontaine, III. 1999. Phosphorus and nitrogen inputs to Florida Bay: the importance of the Everglades watershed, *Estuaries*, 22:398–416.

Slatkin, M. 1985. Gene flow in natural populations, *Ann. Rev. Ecol. Syst.*, 16:393–430.

Swift, C.C., C.R. Gilbert, S.A. Bartone, G.H. Burgess, and R.W. Yerger. 1986. Zoogeography of the freshwater fishes of the southeastern United States: Savannah River to Lake Pontchartrain, pp. 213–265, in *Zoogeography of North American Freshwater Fishes*, Hocutt, C.H. and Wiley, E.O., Eds., John Wiley & Sons, NY.

Tabb, D.C. 1987. *Key Environmental Indicators of Ecological Well-Being, Everglades National Park and the East Everglades, Dade County, Florida*, Report to the U.S. Army Corps of Engineers, Jacksonville, FL (Tropical Bioindustries, Miami, FL, Contract DACW 17-84-C-0031).

Trexler, J.C. 1995. Restoration of the Kissimmee River: a conceptual model of past and present fish communities and its consequences for evaluating restoration success. Rest. Ecol. 3:195–210

Trexler, J.C. and W.F. Loftus. 2001. *Analysis of Relationships of Everglades Fish with Hydrology Using Long-Term Databases from the Everglades National Park*, Report to Everglades National Park, Cooperative Agreement CA, 5280-8-9003.

Trexler, J.C., W.F. Loftus, F. Jordan, J.J. Lorenz, J.H. Chick, and R.M. Kobza. 2000. Empirical assessment of fish introductions in southern Florida, U.S.A.: an evaluation of contrasting views, *Biol. Invasions*, 2:265–277.

Turner, A.M., J.C. Trexler, F. Jordan, S.J. Slack, P. Geddes, J.H. Chick, and W.F. Loftus. 1999. Targeting ecosystems features for conservation: standing crops in the Everglades, *Conservation Biol.*, 13:898–911.

U.S. Army Corps of Engineers. 1998. Draft Integrated Feasibility Report and Programmatic Environmental Impact Statement, Central and Southern Florida Project — Comprehensive Review Study, Jacksonville District (2 CD-ROMs).

Ware, D.M. 1978. Bioenergetics of pelagic fish: theoretical changes in swimming speed and ration with body size, *J. Fish. Res. Board Can.*, 35:220–228.

# 6 Linkages Between the Snail Kite Population and Wetland Dynamics in a Highly Fragmented South Florida Hydroscape

*Wiley M. Kitchens, Robert E. Bennetts, and Donald L. DeAngelis*
U.S. Geological Survey, Biological Resources Division

## CONTENTS

## INTRODUCTION

The endangered snail kite (*Rostrhamus sociabilis*) is the only avian species that occurs throughout the South Florida freshwater hydroscape and whose population is restricted to these wetlands. Habitat quality for this species is dependent on hydrology and wetland plant communities that support its principal food source, the apple snail (*Pomacea paludosa*). Consequently, there is direct linkage between changes in hydrology and changes in habitat quality for the kite.

The pre-drainage or "natural" South Florida freshwater hydroscape, dominated by the Everglades, was characterized by: (1) its spatial extent, in excess of 3.6 million ha; (2) its spatially continuous sheet flow, emanating variously south of Lake Okeechobee and flowing almost imperceptibly (36 m/hr; Holling et al., 1994) downslope on the flat gradient of about 2 cm/km (Leach et al., 1972); and (3) the resultant heterogeneous landscape mosaic of water depths and habitat types (Davis and Ogden, 1994). The current or "managed" system has been reduced to 1.8 million ha, approximately half its original size. The system is now severely fragmented by compartmentalization into a series of interconnected impoundments designated Water Conservation Areas (WCAs), consisting of 100+ water control structures and approximately 2000 km of dike works and canals. Entire wetland landscapes have been virtually eliminated (Davis et al., 1994), while others have been severely altered as a result of water management practices (Science Subgroup, 1993).

Under the existing management regime for kites, droughts have been portrayed as demographic catastrophes in which both survival and reproduction plummet (Beissinger and Takekawa, 1983; Beissinger, 1986, 1995). Management recommendations have consequently focused primarily on increasing the interval between drydowns and maintaining permanent water levels (e.g., Stieglitz and Thompson, 1967; Martin and Doebel, 1973; Beissinger, 1983, 1995). This effectively results in "freezing" or stabilizing specific hydrologic conditions in large-scale impounded wetland units (principally the WCAs) to provide a critical habitat for the maintenance of this species. However, while this approach may confer some short-term benefits, it is counter to the natural dynamics under which these wetland systems evolved (Leach et al., 1972). It essentially promotes the continued maintenance of a fragmented system of fixed components rather than the interconnected dynamic mosaic targeted by recent restoration efforts (Science Subgroup, 1993).

In this paper we frame the issues of water management needs for snail kites in the context of an adaptive management paradigm (e.g., Holling, 1978; Walters, 1986; Foin et al., 1998). We begin with the management assumptions stated above and follow a progression of kite and habitat responses at various spatial scales, ranging from local, specific sites to the entire region, detailing the subsequent consequences. We conclude with recommendations for a new management paradigm, one that more closely reflects the historic "natural" system within which the kite proliferated (Sykes, 1984).

## LINKAGES BETWEEN BIRDS AND HABITAT AT MULTIPLE SCALES

The snail kite is a wetland-dependent species whose population dynamics are a function of changes in carrying capacity of the wetland hydroscape comprising its range. This carrying capacity is subject to acute declines during regional droughts. As stated previously, it is widely held that droughts drastically impact both survival and reproduction of the kite. Our concern is that simplistic single-species management schemes that seek to counter or solve the drought "dilemma" by creating drought refugia within selected prime wetland areas may well compromise the habitat values critical for population recovery and long-term sustainability. This concern is particularly amplified when aspects of temporal and spatial scale are ignored (Bennetts and Kitchens, 1997a).

Most severe droughts are localized to part of the snail kite's range, rather than extending across its entirety. While droughts resulting in drydowns can cause acute reductions in carrying capacity of local wetland sites, the ability of the snail kite population to sustain itself through time is a function of the total habitat base across the region. The continued persistence of the snail kite population in the face of the periodic droughts that characterize South Florida will largely depend on the maintenance of a critical network of *viable* wetlands dispersed across the entire regional hydroscape (Bennetts and Kitchens, 1997a,b). The term *viable* is emphasized to bring attention to the fact that wetlands, by definition, are areas subjected to periodic wet and dry conditions. They are occupied by plant and animal populations specifically adapted to these conditions. These wetland communities are highly sensitive to alterations in hydroperiods, particularly those extending or stabilizing water depths. Recovery time from the deleterious effects of such alterations can be extended.

The general problem concerning current management can be considered a failure to discriminate between survival of individual kites and the requirements for long-term sustainability of the population. Key to this oversight is inadequate consideration of either the bird or the habitat at multiple spatial and temporal scales. It is only in the context of cross-scale linkages that the effects of habitat management on the kite populations become apparent (Bennetts and Kitchens, 1997b). The following narrative is a descriptive account linking the biology of snail kites and their habitat at different scales, followed by a consideration of the response of the habitat, and the birds, at these different scales to current management.

## MACROSCALE: RANGE OR REGIONAL SCALE

The entire population of snail kites in the U.S. is currently confined to that area of Florida lying south of an east–west line crossing the state at the approximate latitude of Orlando. The area is characterized climatically as subtropical with pronounced wet and dry seasons. Precipitation is approximately 100 cm/year, with the bulk occurring during June to August. While the precipitation averaged over this region is seasonally predictable (Duever et al., 1994), it is spatially variable (MacVicar and Lin, 1984), with some parts of the region experiencing localized droughts while other areas are receiving normal or even above normal precipitation. The landscape is largely a flattened limestone basin overlain to the north by shallow sands and to the south by peat. The regional topographical gradient is extremely flat, declining to the south approximately 2 cm/km with localized topographic discrepancies approaching 1.5 m (Gunderson, 1994). Given its abundant, spatially variable precipitation and slight drainage gradients, much of the landscape is dominated by a mosaic of water and wetlands, or a "hydroscape."

The hydroscape used by kites is encompassed within several watersheds, including: (1) its two principal watersheds, the Kissimmee/Okeechobee/Everglades (KOE) and Upper St. Johns River (USJR) basins; (2) portions of the Big Cypress Basin; (3) the Loxahatchee River Basin, principally Loxahatchee Slough and Pal Mar; and (4) portions of the Caloosahatchee River Basin. The Kissimmee chain of lakes (Tohopekaliga, East Tohopekaliga, Kissimmee, Marion, Tiger, Pierce, Jackson, and Hatchineha), Lake Okeechobee, the Water Conservation Areas (3A, 3B, 2A, 2B, and 1 or A.R.M. Loxahatchee National Wildlife Refuge), the Holeyland Wildlife Management Area, and the Everglades National Park are among those areas used by kites in the KOE. The Blue Cypress Marsh Water Conservation Area, Blue Cypress Water Management Area of the USJR, and the West Palm Beach Water Catchment Area of the Loxahatchee River watershed are other important areas used by kites. Numerous other smaller wetlands and water retention units or "peripheral" areas scattered throughout the region are critically important areas in the event of region-wide droughts. Although termed "peripheral," they are regularly utilized. Use is not restricted to drought use only, as been implied by Beissinger and Takekawa (1983) and Takekawa and Beissinger (1989).

The wetlands and associated water bodies composing this hydroscape vary in size from a few hectares to several thousand hectares, totaling more than 1.2 million ha. These wetlands include lacustrine, palustrine, and riverine types and occur within several major physiographic provinces and features. The palustrine wetlands include mosaics of extensive gramminoid marsh strands dominated by sawgrass (*Cladium jamaicense*), with tree islands on isolated topographic highs. Tree islands are dominated variously by woody species, including red bay (*Persea palustris*), dahoon holly (*Ilex cassine*), and willow (*Salix carolinana*). Wet prairies, dominated by spike and beak rushes (*Eleocharis* and *Rhynchospora* spp.), and aquatic sloughs, dominated by fragrant water lily (*Nymphaea odorata*) and submerged aquatics including bladderwort (*Utricularia* spp.), occupy the topographic "lows" or troughs interspersed within the strands of sawgrass and tree islands. The lacustrine types occur as littoral areas dominated by extensive coenoclines of spatterdock (*Nuphar luteum*), and maidencane (*Panicum* spp.) and are bordered by bullrushes (*Scirpus* spp.) on the lakeward edge and willow or cattail (*Typha* spp.) toward the shore.

The snail kite population inhabiting the region is currently estimated at around 3000 birds (Dreitz et al., in press), having steadily increased over time from apparent population lows in the 1950s and 1960s following major droughts. Annual counts since 1969 have varied between 65 and 996, but generally support an increasing population trend over this period, even when adjusted for observer and other biases (Bennetts et al., 1999). Observations of birds outfitted with radio trans- mitters (Bennetts and Kitchens, 1997a,b) suggest that kites use the regional hydroscape as an extensive network of habitats. At this scale, the snail kites are quite nomadic and move frequently throughout the region (Figure 6.1). Our data indicate that, on average, birds move from one distinct wetland area to another several times per year. Thus, rather than a meta-population consisting of distinct subpopulations, snail kites in Florida are more of a panmictic population utilizing the entire spatial extent of their range. These extensive movements of individual kites among the various wetland areas and types within the region serve to effectively "connect" these rather spatially and hydrologically fragmented systems in a networked fashion. In addition to the nomadic movement of individuals, the birds congregate variously throughout this network with dramatic temporal shifts

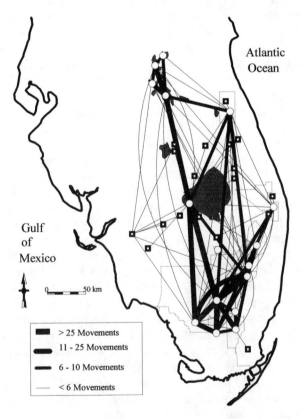

**FIGURE 6.1**   South Florida showing a summary of inter-wetland movements (solid lines) of adult, radio- tagged snail kites during 1992 and 1993. These movements illustrate a basic habitat network used by snail kites (circles represent major wetland units, squares represent peripheral areas).

in numbers and locations. Depending upon the season, the resultant pattern is a continuum of individual and communal foraging and nesting activities, highly variable in time and space.

## Microscale: Individual Birds (and Snails) at the Local or Site-Specific Scale

At the level of individual birds, we focus primarily on foraging and nesting ecology. Snail kites feed almost exclusively on apple snails. Snail kites are visual hunters and must locate their prey within the water column. They capture their prey by flying low over the marsh and, after sighting a snail, drop to pluck their prey from the top few centimeters of water. Foraging is widely dispersed in areas with sparse vegetation and concentrated along boundaries distinguishing various emergent macrophytes, where patches of coenoclines are interspersed. The latter areas are typically extensive and usually delineate plant communities of different hydroperiod preferences (e.g., spike rushes and water lilies in the Everglades marshes, or maidencane and rushes in the littoral reaches of lakes). In general, foraging areas are shallow and provide an emergent stem density sufficiently sparse to enable kites to see their prey, yet dense enough to provide an emergent substrate for snails concentrating at or near the surface in numbers that attract foraging birds. The snails climb plant stems to the water surface in order to breathe, particularly during warmer periods when water-column oxygen concentrations are reduced. There may be a threshold depth beyond which it is energetically inefficient for snails to make the excursion between the surface and bottom waters for breathing, thus limiting the snails to shallow marsh areas. Likewise, because water depth and temperature vary inversely during the warmer months, snails in deeper waters emerge to breathe less frequently and thus are less often exposed to foraging activities (Darby et al., 1997).

## Mesoscale: Habitat or Landscape Scale

When viewed at an expanded spatial scale, foraging habitats tend to occur in three general categories (Bennetts et al., 1994). For the Everglades marshes, foraging is generally confined to two general vegetation types classified by Loveless (1959) as wet prairies and aquatic sloughs. The wet prairies are further divided into *Eleocharis*- and *Rhynchospora*-dominated prairies, although foraging kites tend to use the former more frequently. Functionally, these are three different coenoclinal types, spatially arrayed or interspersed along a gradient of hydroperiods ranging from 70 to 99% inundated. Precise hydroperiod discriminations tend to be highly site specific but follow a general trend of 50 to 82% for *Rhynchospora* prairies (peaking at 75%), and 50 to 94% for *Eleocharis* prairies (peaking at 85%) in the Water Conservation Areas (Zaffke, 1983). Aquatic sloughs are dominated principally by fragrant water lily, floating heart, *Nymphaea aquatica*, and submerged aquatics, including bladderwort. These sloughs tend to be more or less permanently inundated except during exceptionally dry years as defined by Bennetts and Kitchens (1997a) and Bennetts et al. (1999). Foraging in the lake environments outside the Everglades is generally confined to the reaches of the littoral system dominated by extensive stands of maidencane and spike rushes (Bennetts et al., 1994) with hydroperiods ranging from 50 to 90% (Sincock and Powell, 1957). Foraging is focused in the deeper areas of this zone. Deeper areas are dominated by spatterdock that is either a poor substrate for visual hunting or unsuitable for snails. Shallower areas are dominated by cattail or flag (*Sagittaria*) in stands too densely vegetated for visual hunting, and snail densities are reduced.

Nesting habitat has somewhat different constraints. Snail kites tend to nest in woody substrates that are located over water. Nests may occur singularly in isolated woody shrubs such as pond apple (*Annona glabra*), willow, and wax myrtle (*Myrica cerifera*) on microtopographic mounds in the marshes. Communal nests are more often found in the shrub/scrub strands or "tree islands" that are either completely surrounded by water or occur in the trailing edge or "tails" of tree islands where hydroperiods are extended. In the absence of woody vegetation, kites will nest in herbaceous vegetation such as cattail, although nest failure under these circumstances tends to be higher due to nest collapse, especially after wind events (Chandler and Anderson, 1984; Snyder et al., 1989).

This occurs more often in the lacustrine habitats (e.g., Lake Okeechobee), typically at times when shorelines have migrated downslope due to excessively low water conditions, or in lake systems with littoral systems confined by levees and subject to elevated water stages. Kites tend to avoid areas without standing water during the nesting season, presumably due to vulnerability to increased terrestrial predation pressures.

Habitat requirements of the kite can be summarized graphically in relation to a hydrological gradient (Figure 6.2), where foraging habitats tend to occur in coenoclines occupying the wetter areas, and nesting habitat tends to occupy patches in relatively drier habitats. When viewed from a landscape perspective, expanding in spatial extent to another scale, the interspersion and proximal locations of these habitat types become apparent (Figure 6.3). Obviously, foraging habitat is always required, and nesting habitat is required during periods of nesting. One without the other in proximal association will not be suitable habitat, at least for some life-history stages. Given the existing hydrology management paradigms promoting stable, specific water regimes in selected areas of the system, this interspersion of foraging and nesting habitat is at risk of decline. The following sections elaborate this concern.

## HYDROSCAPE, HABITAT, AND KITES: MANAGEMENT AT MULTIPLE SCALES

### THE HYDROLOGICALLY ALTERED SYSTEM

As stated earlier, the South Florida hydroscape has been dramatically changed in the past 50 years, with wetland areas being reduced from 3.6 to 2.4 million ha (Davis and Ogden, 1994). Water levels and flows that once fluctuated seasonally in response to rainfall and runoff have been dramatically altered in an attempt to reduce effects of flooding and droughts on human activities in South Florida (McPherson and Halley, 1996). The inter- and intra-annual variability of water stages in the remaining principal wetland systems supporting kites has been severely altered. Both peak and minimal stages have been lowered, and the range between peak and minimal stages has been considerably decreased in several areas (Figure 6.4). Due to dams, dikes, and levees, for the first time in the geological history of the Everglades, stages and flows are no longer coupled to rainfall (Figure 6.5). The KOE no longer exists as a single hydrologic unit. Instead, it is subdivided into the Upper Kissimmee Chain of Lakes, Lake Okeechobee (with dredged outlets to both coasts), the remaining Everglades partitioned into three major impoundments (WCAs 1, 2, and 3, serving principally as flood protection and water supply reservoirs), and the Everglades National Park. The massive sheet-flow system that characterized the pre-drainage Everglades (Science Subgroup, 1993; Davis and Ogden, 1994) has been eliminated and replaced by this series of impoundments that effectively step the water level stages from north to south in a fashion that exaggerates the vertical component of the hydroperiod (Figure 6.6). This alteration has resulted in excessive ponding in the southern end of impoundments while over-draining the northern ends (Dineen, 1972; Light and Dineen, 1994).

### ALTERED HYDROLOGY FOR KITES: CONSEQUENCES OF THE EXISTING PARADIGM

Descriptive accounts of snail kite populations during the late 1800s and early 1900s indicated that snail kites were relatively abundant in Florida, at least at some locations (Scott, 1881; Bailey, 1884; Wayne, 1895; Howell, 1932). By the 1920s and continuing through the 1950s, virtually all reports described snail kites in Florida as declining or rare (Howell, 1932; Sprunt, 1945; Wachenfeld, 1956). Declines were attributed primarily to widespread drainage that occurred throughout Florida (Howell, 1932; Sprunt, 1945; Sykes, 1983a; Bennetts et al., 1994). Since that time snail kite numbers have rebounded from what was considered to be a low of about 50 to 100 birds (Sprunt, 1945) to a current estimate of approximately 3000 birds (V.J. Dreitz, in press). This rebound has been largely

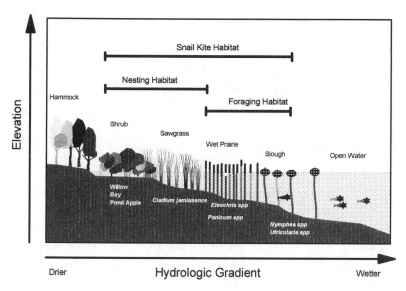

**FIGURE 6.2**    Hypothetical cross-section of South Florida hydroscape depicting kite foraging and nesting habitat.

**FIGURE 6.3**    Oblique aerial photograph of portion of South Florida hydroscape depicting spatial interspersion of wet prairies, sloughs, and tree islands (WCA-3).

attributed to the creation of the impounded Water Conservation Areas that provided long-hydrope-riod marshes in areas substantially affected by drainage (e.g., Sykes, 1983b).

This concept has been extended further into a management paradigm that critical habitat for kites requires maintaining continuous inundation of specific wetland units as refugia from drought (e.g., Stieglitz, 1965; Stieglitz and Thompson, 1967; Beissinger, 1988). This approach emerged largely from assumptions that reduced counts during a systematic annual survey implied population reduction in response to water levels in the surveyed areas (e.g., Sykes, 1979, 1983a; Beissinger, 1988, 1995) and the conclusion of Snyder et al. (1989) and Beissinger (1986) that nest success was

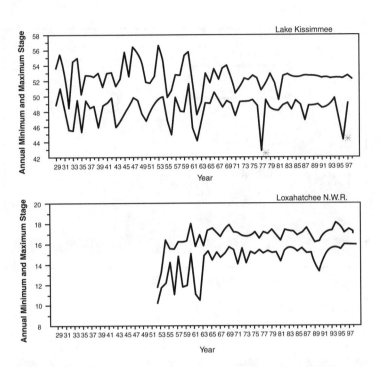

**FIGURE 6.4**   Annual minimum and maximum stages for Lake Kissimmee and Arthur R. Marshall Loxa-hatchee National Wildlife (WCA-1). Note the dampening of the range of stages, as well as the decreased maxima and elevated minima after 1960 (coincident with completion of the Central and South Florida Project) for both areas. The minimal stages for 1977 and 1996 in Lake Kissimmee were management drawdowns for consolidation of bottom substrates. (Data courtesy of South Florida Water Management District.)

lower during years of low water. It was more recently suggested by a population viability analysis (Beissinger 1995) that the Florida population of snail kites would not be viable unless the interval between successive "droughts" exceeded 4.3 years. However, this model used the annual survey as a primary basis for verification and as a source of data for its most sensitive parameter, survival during drought years. The annual survey has been critically analyzed and its value in estimating annual changes in population size has been discredited (Bennetts et al., 1999). Regardless, the general perception has emerged that the occurrence of drought or drying implies a demographic catastrophe for kites and their forage base. The underlying foundation for this view is another perception that apple snails undergo massive mortality in response to drying (e.g., Beissinger, 1988). Recent work by Darby et al. (1997) has also discounted this generalization. This work indicates that the effect of drying on apple snails depends largely on its timing and duration and in many cases may not even affect survival of apple snails.

As a result, protection has centered on critical habitat/wetland units almost entirely within the Everglades and Lake Okeechobee watersheds (WCA-3A and the littoral system of Lake Okeechobee) as drought refugia (*Federal Register* 42 [155]: 40685-40688; 50 CAR Ch. 1; 10-1-94 ed.). Bennetts and Kitchens (1997b) presented the following argument critiquing this approach: (1) this spatial configuration of protected habitat ignores the vulnerability to simultaneous low water conditions in these proximal areas during droughts, and (2) continuous inundation in these units, while beneficial over the short term, will erode habitat quality over the long term. We agree that regionally widespread drydowns may affect the snail kite population and that the frequency should not be anthropogenically increased. However, the problem is that inferences regarding the return frequency of widespread regional droughts have frequently been interpreted to mean that the occurrence of local drying events should also be decreased, by whatever artificial means available. Local and widespread regional drying events, however, are not the same. Their ecological implications are vastly different. The critical issue that has been largely ignored by these

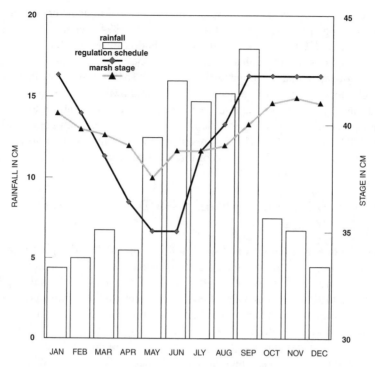

**FIGURE 6.5** Average monthly rainfall (1970–1984) and surface water elevations (gauges 1–9) for Arthur R. Marshall Loxahatchee National Wildlife Refuge (WCA-1) (1977–1985). Stage regulation schedule also indicated. (Data courtesy of South Florida Water Management District.)

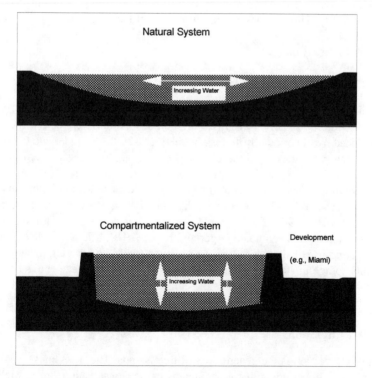

**FIGURE 6.6** A conceptual illustration of how compartmentalization of wetlands with levees increases depth by decreasing the potential for lateral movement of water.

recommendations (however, see Sykes, 1983a) is that the lack of periodic drying can detrimentally affect the kites' nesting and foraging habitat (Bennetts et al., 1994; Bennetts and Kitchens, 1997a). Bennetts and Kitchens (1997b) use the shifts in distribution of nesting birds in WCA-3A over a 30-year period to demonstrate this point. Birds originally nested in the longest hydroperiod portion of the marsh following initial impoundment in the 1960s. However, nest distributions steadily shifted up the elevational gradient through time toward shallower and consequently shorter hydroperiod areas. In this system, while the traditional nesting area is still subject to some inter-annual variability, the area is subject to increased hydroperiods and ponding depths as a result of impoundment (Zaffke, 1983; David, 1996). This has resulted in major shifts in both foraging and nesting habitat composition and quality. The nature of these habitat conversions, and a concern regarding recovery times, is detailed below.

## HABITAT RESPONSES TO INCREASED HYDROPERIODS OR DECREASED RETURN INTERVALS BETWEEN DRYDOWNS

A discussion of the tolerances of the plant species that comprise the principal foraging and nesting habitats of the kite must span several temporal and spatial scales. Lists of wetland species found in these habitats are presented in Loveless (1959) for the Everglades and Pesnell and Brown (1977) and Ager and Kerce (1970) for lakes. Our discussion is limited to the dominant or indicator species. Tables 6.1 and 6.2 summarize numerous studies examining relationships between species occurrence and hydrology in the South Florida hydroscape. Four conclusions are immediately obvious: (1) there is considerable variability among studies as to the preferences or tolerances regarding hydroperiod for any given species, which probably reflects differences in soils, site histories, and disturbance regimes; (2) the relative hydroperiod distinction for these community types tends to remain relatively constant within sites; (3) virtually none of the species can tolerate continuous inundation; and (4) practically all studies document detrimental habitat conversions over time resulting from the excessive depths and hydroperiods following impoundment intended to reduce hydrological variability. This last observation warrants further discussion.

### Foraging Habitat Alterations

The predominant conversion noted in these studies is the transformation of wet prairie communities to slough communities in areas subjected to excessive depths or hydroperiods (e.g., WCA 1, 2 and 3) (U.S.D.I., 1972; McPherson, 1973; Dineen, 1974; Worth, 1983; Zaffke, 1983; Wood and Tanner, 1990; David, 1996), or to sawgrass communities in the Everglades Park and other sites subjected to reduced hydroperiods (Kolipinski and Higer, 1969; Alexander and Crook, 1974; Davis et al., 1994), or flow velocities (hypothesized by Davis et al., 1994) as a result of impoundment. Extended hydroperiod affects were discernible at some sites in transition from one type to another. For example, Wood and Tanner (1990) attributed a significant reduction in species in wet prairie communities in sites in WCA-3A to extended hydroperiods when comparing the results of their studies to that of Zaffke (1983) for the same sites. Similar conversions in the littoral system of Lake Okeechobee have also been noted (Ager and Kerce, 1970; Milleson, 1987). The net result of conversion of wet prairie to sloughs is the *loss* of prime snail kite foraging habitat. Although snail kites will readily forage in slough communities (Sykes, 1983a; Bennetts et al., 1994; Sykes et al., 1995), data from more than 5000 locations of radio-tagged birds (Bennetts and Kitchens, unpubl. data) indicate that wet prairie communities are used to a far greater extent than sloughs for foraging.

### Nesting Habitat Alterations

Another notable conversion is the general loss of tree islands. These declines are generally attributable to extended hydroperiods or increased depths in the Conservation Areas (McPherson, 1973; Dineen, 1974; Alexander and Crook, 1974; Worth, 1983). Milleson (1987) noted a reduction in

**TABLE 6.1**
**Summary of Vegetation/Hydroperiod Studies in South Florida**

| Investigators; Location | Sloughs | | Habitat Type<br>Wet Prairies | | | Sawgrass | Tree Islands |
| | Nymphaea/Nuphar | Utricularia | Eleocharis | Rhynchospora | Panicum | Cladium | Salix |
| --- | --- | --- | --- | --- | --- | --- | --- |
| David, 1996;<br>WCA-3 | 96% | 95% | 75%[b] | 91%[b] | 61% | 31–84%[a] | 68%[a] |
| Zaffke, 1983;<br>WCA-3 | 95–99% | | 50–94% | 50–82% | N/A[c] | N/A[c] | N/A[c] |
| Lowe, 1986;<br>Upper St. Johns Marsh | N/A[c] | | 75–90% | 94% | 80–87% | 46–86% | 72–80% |
| Goodrick, 1984;<br>WCA-2 and WCA-3 | 88%[a] | | 74%[a] | | | N/A[c] | |
| McPherson, 1973;<br>WCA-3 | 87–98%[a] | | 78–85%[a] | | | 76–93%[a] | 69–75%[a] |

*Note:* Hydroperiods are expressed as percentages of time inundated. Data are reported either for habitats or for indicator species for habitat.

[a] Range of mean values for different stations.
[b] Mean value for range.
[c] N/A = data not reported.

**TABLE 6.2**
**Studies Documenting Habitat Conversions in South Florida**

| Investigator(s); Location | Conversion | Cause |
| --- | --- | --- |
| McPherson, 1973; WCA-3 | Tree island recession | Flooding due to impoundment |
| Alexander and Crook, 1984; South Florida | Tree island losses Beak rush lost in WCA-3 | Fire and flooding |
| Dineen, 1974; WCA-2 | Losses of wet prairie species Loss of tree islands | Impoundment Flooding to permanent pool |
| Davis, et al., 1994; South Florida | Wet prairie losses to sawgrass | |
| Wood and Tanner, 1990; WCA-3, ENP, NESS | Wet prairies in converting to slough communities | Impoundment, extended hydroperiods |
| Kolipinski and Higer, 1969; ENP | Decrease in wet prairie Increase in sawgrass | Decreased hydroperiods due to impounding upstream |
| Dineen, 1972; WCA-2 | Wet prairie species reduced Tree islands water damaged | Area continuously flooded for 10 yr |
| Schortemeyer, 1980 | Tree islands damaged Wet prairies reduced | Extended hydroperiods and depths due to impoundment |
| Ager and Kerce, 1970; Lake Okeechobee | Reduced wet prairie species | Elevated water stages in lake |
| Milleson, 1987; Lake Okeechobee | Reduced extent of Salix | Elevated water stages |

willow heads in Lake Okeechobee's littoral zone following prolonged elevated stages, between 1973 and 1981. This trend continued through 1989, according to Richardson and Harris (1995). Given the extended time frames required for recovery of these tree species and the vulnerability of seedlings to fire or flooding (Gunderson et al., 1988; Gunderson, 1994), these losses are significant. The losses have occurred fairly rapidly, in 6 to 7 years (Ager and Kerce, 1970; U.S.D.I., 1972), but recovery to comparable age structure and stature may require decades. Thus, losses appear to occur within time periods of flooding that correspond roughly to the lower limit of drydowns suggested by Beissinger. The conflicting nature of these time frames is not trivial. Gunderson (1994) suggests this mimics the temporal domain of decades for natural phenomena (*El Niño*) that occur on similar time scales that are capable of causing considerable detrimental impacts to these tree communities. The prospect for rapid recovery is thus further compounded by the impacts of natural flood/drought frequencies operating in the South Florida hydroscape. Losses of historical colonial roosts and nesting sites resulting from the impoundment and stabilization of hydrologic regimes no doubt have occurred in the past and can only be expected to be exacerbated by continuing these same regimes.

## Up Against the Wall

While kites continue to nest in numbers in WCA-3, the trend has been for nesting to shift toward higher elevations and shorter hydroperiods (Bennetts et al., 1988). In fact, many of the adults that nest in the southern portion of the Water Conservation Area actively forage the marshes of nearby Everglades National Park, where the hydroperiods are considerably shorter and the hydrology considerably more variable (R.E. Bennetts, pers. observ.). As noted earlier, Zaffke (1983), Wood and Tanner (1990), and David (1996) have documented the conversion of wet prairies to aquatic sloughs in that area, with concurrent losses of interspersed herbaceous and woody species. The immense size of the area (237,000 ha) and the inability to completely stabilize its hydrologic regime are mitigating factors.; however, in time the area could continue to convert to less desirable habitat. Given that nesting is already occurring in the highest elevations within WCA-3A (south of Alligator Alley), there may be nowhere else for shifts to occur within this unit if the current trend continues. That is, kites are up against the "hydrologic wall."

The tendency for loss under extended flooding, of emergent species and entire communities or zones comprising coenoclines has also been observed in other wetland ecosystems (Van der Valk and Welling, 1988; Van der Valk, 1991). Van der Valk and Welling (1988) documented drastic reductions in percent cover and distribution and abundances of species within the coenocline of experimental units with increased water depths in as little as three years. This phenomenon is particularly acute in impounded systems where depth and hydroperiod become controlling variables as the water mass is constrained to vertical rather than lateral expansions/retractions in response to volumetric changes (Figure 6.6). In the natural state, microtopographic relief played at least as influential a role as landscape gradient in maintaining the dynamic, interspersed mosaic of the Everglades hydroscape (Gunderson, 1994). In an impounded system, water depths exceed the tolerances, and the variability attributable to local topography can no longer provide conditions favorable for emergent species comprising kite habitat (Figure 6.7). From all evidence (Zaffke, 1983; Wood and Tanner, 1990; David, 1996), the shifting mosaic associated with inter- and intra-annual hydroperiod and depth variability in WCA-3 is virtually "drowned out," with succession tending progressively toward a more aquatic phase. Thus, vegetation change has become unidirectional, rather than a dynamic switch between slough and wet prairie phases in response to environmental variability. This unidirectional trend is increasingly degrading this important habitat resource.

## HYDROLOGY AND HABITAT QUALITY

Both kites and wetlands respond to changes in hydrological regimes. *In situ* plant communities respond through successional processes at time scales spanning months to several decades. Kites are capable of two types of responses: (1) behavioral responses typically on the order of daily time scales (e.g., simply dispersing to a more suitable location in response to unfavorable hydrologic conditions at a particular locale), or (2) population responses over a period of months or years (e.g., changes in survival or reproduction) (Bennetts and Kitchens, 1997b). The following is a conceptual model (Bennetts and Kitchens, 1998) that integrates critical aspects of wetland and kite responses to hydrology.

## Depth

The empirical relationship between snail kites' use of a given habitat and water depth has been well recognized and has been illustrated by the distribution of nests or foraging birds with respect to water depth (e.g., Stieglitz and Thompson, 1967; Sykes, 1987; Bennetts et al., 1988). We have observed that kites typically abandon foraging areas after water depths reach a critical depth of < 10 cm.

## Drydown Intervals

Drying events result in periodic reductions in the availability, although not necessarily abundance of snails as kite food. Consequently, the response of kites is simple. They either move or die. Bennetts and Kitchens (1997a) observed that the numbers of kites using a particular area were reduced by about 50% in the year following a drydown. Incremental recovery to pre-drydown numbers occurred within three years. Recovery of snails and the return of kites probably depend considerably on the timing and severity (i.e., magnitude and duration) of a given drying event (Darby et al., 1997).

NATURAL SYSTEM

IMPOUNDED SYSTEM

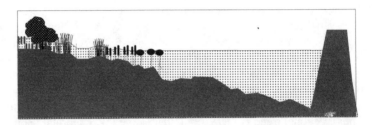

**FIGURE 6.7**   A conceptual illustration of how impounding of wetlands with levees increases depth by buildup of water behind the levee. In the "natural system" example, water flows downslope and depth is regulated by gradient of water surface elevations.

## Long-Term Hydroperiods

Although the occurrence of drying may temporarily decrease local prey availability, the absence of drying results in changes in plant communities as detailed above. Recovery of herbaceous species subjected to "over-flooding" conditions can occur in as little as one year (Worth, 1983). However, recovery of suitable stature and structure of woody species for optimal nesting habitat has been documented to require more than 9 years in the littoral system of Lake Okeechobee (Milleson, 1987).

We hypothesize that overall habitat quality is a function of the combined suitability of each of these hydrologic variables or scales. Each regulates a different aspect of the environment important to kites. In combination, these factors regulate: (1) the behavior of apple snails, and consequently their availability to kites; (2) apple snail population dynamics; and (3) plant community change. Habitat quality therefore requires alignment of suitable conditions for each of these factors. This alignment (Figure 6.8) can be viewed as a dynamic "window" of hydrologic conditions at a given point in time and space in which snail kites occur (Bennetts and Kitchens, 1997a).

## PARADIGMS: EXISTING AND PROPOSED

Past perception regarding management of snail kites is counter to the wetland dynamics under which this system and snail kites have evolved and persisted. The "dynamic landscape hypothesis" (Bennetts and Kitchens, 1998), better reflects the natural dynamics of the South Florida hydroscape. Table 6.3 outlines this proposed paradigm and contrasts it with misperceptions of the existing paradigm. The existing paradigm is a hypothesis based on the notion that droughts are catastrophic to kite populations and must be mitigated through the creation and maintenance of selected permanently inundated, drought-free refugia. This hypothesis/paradigm is flawed in that it neglects important aspects of both temporal and spatial scale on the persistence of kites in South Florida.

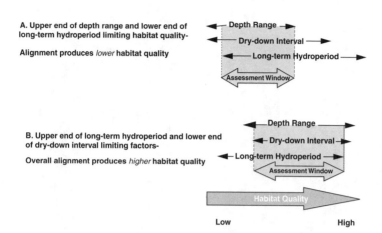

**FIGURE 6.8**　A conceptual illustration of dynamic "window" of hydrologic variables regulating habitat quality for kites. Ideal habitat conditions exist when favorable values for all three factors are coincident, or aligned, as in Example B

Evidence is now clear that this strategy eventually erodes the quality of both the primary and refugia habitats rending these areas ineffective in the long term. We propose a new hypothesis/paradigm based on the fact that temporal and spatial variability is, and always has been, an intrinsic feature of the environment of the South Florida hydroscape, under which snail kites have persisted.

There are at least two critical management implications of this dynamic landscape view for the snail kite (Bennetts and Kitchens, 1997a). The first is that attempts to create artificially stable habitat by reducing hydrologic variability will erode the quality of habitat over longer time scales. The second is that spatial extent and distribution of habitat across the region are critical. Providing adequate refugia from drought should be accomplished by maintaining high-quality habitat across a large enough area (e.g., in several watersheds) to encompass regional climatic variability, rather than by prolonging local inundation. A broadly distributed spatial extent enables areas to incur periodic drying (necessary for plant communities) on a stochastic basis through climatic variability, rather than trying to "counteract" natural rainfall patterns with artificial hydrology (Bennetts and Kitchens, 1997b). Any reduction in the spatial extent and region-wide distribution of snail kite habitat effectively reduces the range of the kites and increases the probability of droughts affecting the entire range (Bennetts and Kitchens, 1997a).

## ADAPTIVE MANAGEMENT RECOMMENDATIONS

The management of the South Florida hydroscape is an excellent opportunity for the protection, management, and conservation of snail kites as an adaptive management exercise; however, adaptive management must be more than just iterative trials of different management scenarios. A key feature of a sound adaptive management program is that learning is incorporated as an explicit goal of the adaptive management process (Lee, 1993). Iterative trial of management strategies without an explicit learning goal is haphazard, not adaptive. According to Lee (1993), adaptive management embodies a simple concept: "Policies are experiments; learn from them." Further, Lee (1993) contends that an effective adaptive management program must: (1) be explicit about what to expect from management policies that are implemented (models help in this regard), (2) have monitoring schemes in place that are capable of evaluating system responses to management actions, and (3) allow participants to transform the resulting comparisons into learning, make corrections in their actions, and change actions and plans accordingly.

**TABLE 6.3**
**Key Features of a Proposed New Management Paradigm Presented with Contrasting Elements of the Existing Paradigm**

| Proposed Paradigm | Existing Paradigm | Key Management Considerations |
|---|---|---|
| Variability in snail kite distribution at a local level is to be expected. Snail kites are nomadic and shift frequently throughout a habitat network. | Although the nomadic tendencies of snail kites has been previously recognized, periodic reductions in local bird numbers is frequently interpreted as a local reduction in habitat quality. | When local reductions occur during nesting, it may even be considered as a "take" as defined under the endangered species act when actions can be taken to prevent the response from occurring (e.g., holding water in impoundments). While we agree that an anthropogenic management action that results in reductions of kites locally should constitute a "take", natural drying that can be "prevented" from occurring should not. |
| Responses of snail kites to "local" drying events is largely behavioral (i.e., they move), whereas a "numerical" response occurs primarily in cases of "droughts" with a spatial extent covering all or most of the kite's range in Florida. | Numerical responses of kites occur in response to "droughts" regardless of their spatial extent. | Under the old paradigm the spatial extent of "droughts" has never been specified. Consequently, there has been no distinction between periodic "local" drying, which is necessary to maintain the habitat and widespread regional droughts, which occur rarely (e.g., once every 10–15 years) and whose demographic effects can be minimized by a broad spatial extent of quality habitat. Under the new paradigm, spatial extent of habitat is deemed critical to viability of snail kites in Florida. |
| Local drying events, at natural frequencies, are considered as an essential component of a functioning ecosystem. | Drying events are considered as demographic catastrophes in which survival and reproduction of kites plummet. | The old paradigm views drying events as requiring stabilization. We agree with previous authors (e.g., Sykes, 1979; Beissinger, 1988; Bennetts et al., 1988, Beissinger, 1995; Sykes et al., 1995) that suitable snail kite habitat is inundated for relatively long periods (e.g., 1–5 yr average return interval of drying events, with considerable variability). However, excessive stabilization results in a slow but steady conversion of wetlands to a more homogeneous aquatic state, degrading their habitat value for snail kites over long time scales. |
| Hydrologic variability, both temporal and spatial, is and always has been an intrinsic feature of the environment of snail kites in southern and central Florida. | Variability in hydrology is potentially detrimental to snail kites and water levels should be kept stable and high. | It is the hydrologic variability that likely enables the coexistence of species (e.g., snail kites and wood storks) with seemingly different hydrologic requirements. The habitats for each can be maintained, although they do not necessarily occur at the same sites at the same time. |
| The fitness of individual birds is enhanced by an asynchronous and variable environment. Birds may be less sensitive to localized disturbance events because of their ability to escape such events (Wiens, 1989). Asynchrony of disturbance would help to ensure that some refugia are available during most disturbance events. | The fitness individual birds is threatened by a variable environment. | The behavioral responses of snail kites to drying events appear well adapted to cope with natural climatic variability. Snail kites exhibit behavior quite adapted to hydrologic variability. When environmental conditions of the wetland site they currently occupy deteriorate, the birds respond by moving to more favorable sites. |
| Periodic disturbance events such as fire, hurricanes, and local and regional droughts are integral parts of southern Florida's landscape patterns (Davis et al., 1994). Attempting to increase stability in a dynamic ecosystem is not only difficult but undesirable ecologically. | Periodic disturbance, especially droughts, should be minimized so as to create a "stable" environment. | In virtually every ecosystem where disturbance processes have been markedly reduced, there has been a subsequent realization of their ecological importance (Pickett and White, 1985). |

We have begun the process for the first component required for an effective adaptive management program by explicitly outlining the key contrasting features of the existing and proposed paradigms. This outline can, and should, be viewed as a series of alternative hypotheses to be tested in an adaptive management program. General predictions can easily be derived from the outline we have provided. For example, we would predict that the response of snail kites to "local" drying events is primarily behavioral, consisting of a redistribution of kites to alternative habitats. Only in cases of "widespread regional droughts" occurring throughout all or most of the watersheds occupied by kites would we predict a numerical response at the population level. The existing paradigm does not identify spatial extent as an important component of "drought" response by snail kites. References are often to drying of individual wetlands (e.g., WCA-3A) (Takekawa and Beissinger, 1989). Thus, under this hypothesis, a numerical response would be predicted, regardless of the spatial extent of a given drying event. Predictive models can take this a step further and generate specific predictions about the response of the system to various management actions. However, these models must incorporate the best ecological knowledge concerning the ecosystem in order to make reliable predictions about how the ecosystem and its components will respond to a proposed management action. A predictive model for the snail kites, using an individual-based, spatially explicit approach, has been developed as part of the Across Trophic Level Simulation System (DeAngelis, 1997) being developed for evaluating hydrologic alternatives within the South Florida Ecosystem Restoration Initiative (Science Subgroup, 1993). The current version of the model is being tested against historic survey data on the snail kite (Mooij et al., in press). On the basis of such models and the management goals that have been set, a particular set of management options is chosen.

By explicitly stating our hypotheses and predictions, we also can begin the second primary component, monitoring. All too often, monitoring is viewed as something required by legal or political mandates, rather than as an opportunity to understand the ecological responses of the system being managed. Several key components are required for an effective monitoring program — it must be (1) capable of distinguishing among the predictions of explicitly stated hypotheses; (2) based on the best available science and established sufficiently prior to the management implementation so that changes caused by particular management options are properly measured and sorted out from the noise of natural environmental variability; and (3) be designed flexibly enough to incorporate appropriate temporal and spatial scales required to accomplish the above.

For example, the response of snail kites to drying events has been much debated. We suggest (predict) that a behavioral response is far more common than a numerical response. Thus, any monitoring program intended to evaluate this hypothesis must be able to distinguish between behavioral and numerical responses. Local and/or short-term evaluations that monitor only a portion of the population at one location have a high probability of producing spurious conclusions. Behavioral (i.e., movement) and demographic (i.e., numerical) responses in these evaluations are easily confounded. The capture–mark–resighting technique presently in place for the snail kite resolves this distinction when carried out for at least 2 to 3 years beyond the event of interest. This approach explicitly estimates decreased resighting of birds temporarily inhabiting drought refugia, from decreased survival. This program also has now been in place for several years prior to the initiation of upcoming restoration policies.

We have suggested that habitat quality erodes under a stabilized high-water regime. We hypothesize herein, based on strong evidence to date, that prolonged flooding of a site, without occasional drydowns, leads to vegetation changes reducing suitability as a nesting area. Thus, we would predict that the distribution of snail kites within a given wetland under prolonged or continuous inundation will shift toward sites with higher elevations and shorter hydroperiods. Under the existing paradigm, no upper limit of drying frequencies is suggested, and it is often implied the longer the inundation, the better (Beissinger, 1995). We would predict under this hypothesis that habitat quality and distribution of the birds should remain relatively stable. This hypothesis can be tested if the nesting distribution at individual sites is monitored regularly, provided that spatial information is recorded (e.g., using a GPS). Simply monitoring total population size for the whole region will not give specific information relevant to site quality. The total population may be doing relatively well, even while nesting habitats are stressed by long hydroperiods to the point that they will inevitably decline in the future. Spatial and temporal heterogeneity are key attributes of the system that must be explicitly included in the monitoring, modeling, and management decision-making. However, for this hypothesis we emphasize from the outset that the time scales over which a response is expected are quite long. Monitoring must be of sufficient duration and/or repeated for several years at periodic intervals of longer time scales to enable evaluation of shifts at the time scale of decades. Year to year variation is to be expected, and the appropriate measure or gauge is general trends over the long term rather than a direct annual association between water levels and distribution.

Beissinger (1995) recently predicted that kite populations would not become stable until the frequency between successive droughts (no spatial extent identified) exceeds 4.3 years. We suggest that this hypothesis, as stated, has to some extent already been tested. Most wetland habitats used by snail kites in Florida have had drying return frequencies less than 4.3 years over the past three decades (Bennetts et al., 1988; Bennetts and Kitchens, 1997a). During this time, the population has been steadily increasing (Sykes, 1979, 1983b; Bennetts et al., 1994, 1999), rather than decreasing as predicted by the Beissinger hypothesis. However, we fully recognize that proponents of the existing paradigm will not have had an opportunity to express agreement with the predictions as we have expressed them. Perhaps Beissinger (1995) intended his hypothesis to imply widespread regional droughts, rather than local drying, although no spatial extent was defined. In this respect, we would encourage all proponents of alternative viewpoints to state their hypotheses explicitly along with their corresponding predictions, for only if such hypotheses and predictions are made explicit (e.g., the spatial extent of droughts, if that is deemed important) can we reliably evaluate their predictions and thereby protect this endangered species.

## CONCLUSION

The endangered snail kite is the only species occurring throughout the South Florida freshwater hydroscape and whose entire U.S. population is restricted to the these wetlands. Habitat quality for this species is dependent on hydrology and wetland plant communities that support their principal food source, the apple snail. Consequently, there is direct linkage between changes in hydrology to changes in habitat quality for the kite. Previous management recommendations have focused on "freezing" or stabilizing specific hydrologic conditions in large-scale impounded wetland units to provide critical habitat for the maintenance of this species. This is counter to the natural dynamics under which these wetland systems evolved, resulting in a fragmented system of fixed components rather than a spatially continuous mosaic. Intra- and inter-wetland heterogeneity is thus reduced by constraining water to vertical rather than lateral expansions/retractions, in response to increases or decreases in water volume. Both kites and wetlands respond to changes in resulting hydrological regimes. Kites are capable of behavioral responses in time scales on the order of days or population responses on the order of months to years. On the other hand, the *in situ* plant communities respond through successional processes at time scales spanning months to several decades. Consequently, the reversal of these successional changes in vegetation, unfavorable to the kite, that have resulted from continuous flooding may require time frames that exceed the life span of individual kites. Thus, a focus is needed that addresses both snail kite population dynamics and plant community dynamics at spatial and temporal scales that are pertinent to both. In this chapter, we described habitat quality for snail kites in the context of fulfilling their foraging and reproductive requirements. We further discussed the linkages between the population dynamics of this species and the ecosystem processes that regulate habitat quality. We suggested that the adaptive behavioral responses of kites, while essential to their persistence, are less effective without a parallel adaptive management strategy for these wetlands as a whole. This strategy is translated into a proposed paradigm based on maintaining the spatial extent and natural hydrologic variability of the South Florida hydroscape.

## ACKNOWLEDGMENTS

Financial support was provided by the National Park Service, U.S. Fish and Wildlife Service, U.S. Army Corps. of Engineers, U.S. Geological Survey/Biological Resources Division, South Florida Water Management District, and St. Johns River Water Management District through the Florida Cooperative Fish and Wildlife Research Unit cooperative agreement No. 14-16-0007-1544, RWO90. We appreciate the helpful comments of Victoria Dreitz and Cynthia Loftin and the editorial assistance of Janell Brush. This is contribution No. R-07953 of the Florida Agricultural Experiment Station Journal Series, Institute of Food and Agricultural Sciences, University of Florida.

## REFERENCES

Ager, H.A. and K.E. Kerce. 1970. Vegetation changes associated with water level stabilization in Lake Okeechobee, Florida, 24th Ann. Conf. of S.E. Assoc. Game and Fish Comm., pp. 338–351.

Alexander, T.R. and A.G. Crook. 1974. Recent vegetation changes in southern Florida, pp. 199–210, in *Environments of South Florida: Present and Past*, Memoir 2. Gleason, P.J., Ed., Miami Geological Society.

Bailey, H.B. 1884. Breeding habits of the Everglade Kite, *Auk*, 1:95.

Beissinger, S.R. 1983. *Nest Failure and Demography of the Snail Kite: Effects of Everglades Water Management*, Annual Report to the U.S. Fish and Wildlife Service.

Beissinger, S.R. 1986. Demography, environmental uncertainty, and the evolution of mate desertion in the Snail Kite, *Ecology*, 67:1445–1459.

Beissinger, S.R. 1995. Modeling extinction in periodic environments: Everglades water levels and Snail Kite population viability, *Ecol. Appl.*, 5:618–631.

Beissinger, S.R. 1988. The snail kite, pp. 148–165, in *Handbook of North America Birds*, Vol. IV, Palmer, R.S., Ed., Yale University Press, New Haven, CT.

Beissinger, S.R. and J.E. Takekawa. 1983. Habitat use and dispersal by snail kites in Florida during drought conditions, *Florida Field Nat.*, 11:89–106.

Bennetts, R.E. and W.M. Kitchens. 1997a. *The Demography and Movements of Snail Kites in Florida*, Fl. Coop. Fish Wildl. Res. Unit. Tech. Rep. No. 56, U.S. Geological Survey/Biological Resources Division, Gainesville, FL.

Bennetts, R.E. and W.M. Kitchens. 1997b. Population dynamics and conservation of Snail Kites in Florida: the importance of spatial and temporal scale, *Colonial Waterbirds*, 20:324–329.

Bennetts, R.E. and W.M. Kitchens. 1998. Recovery of the snail kite in Florida: beyond a reductionist paradigm, pp. 486–501, in *Transactions of the 63rd North American Wildlife and Natural Resources Conference*, Wadsworth, K.G., Wildlife Management Institute.

Bennetts, R.E., M.W. Collopy, and S.R. Beissinger. 1988. *Nesting Ecology of Snail Kites in Water Conservation Area 3A*, Fl. Coop. Fish Wildl. Res. Unit. Tech. Rep. No. 31, University of Florida, Gainesville.

Bennetts, R.E., M.W. Collopy, and J.A. Rodgers, Jr. 1994. The Snail Kite in the Florida Everglades: a food specialist in a changing environment, pp. 507–532, in *Everglades: The Ecosystem and Its Restoration*, Davis, S.M. and Ogden, J.C., Eds., St. Lucie Press, Delray Beach, FL.

Bennetts, R.E., W.A. Link, J.R. Sauer, and P.W. Sykes, Jr. 1999. Sources of variability and estimation of the trajectory from a 26-year annual survey of snail kites in Florida, *Auk*, 116:312–323.

Chandler, R. and J.M. Anderson. 1984. Notes on Everglades kite, *Am. Birds*, 28:856–858.

Darby, P.C., P.L. Valentine-Darby, R.E. Bennetts, J.D. Croop, H.F. Percival, and W.M. Kitchens. 1997. Ecological Studies of Apple Snails (*Pomacea paludosa*, Say), final report prepared for South Florida Water Management District and St. Johns River Water Management District, Contract # E-6609, Fl. Coop. Fish Wildl. Res. Unit, Gainesville, FL.

David, P.G. 1996. Changes in plant communities relative to hydrologic conditions in the Florida Everglades, *Wetlands*, 16:15–23.

Davis, S.M. and J.C. Ogden, Eds. 1994. *Everglades: The Ecosystem and Its Restoration*, St. Lucie Press, Delray Beach, FL.

Davis, S.M., L.H. Gunderson, W.A. Park, J.R. Richardson, and J.E. Mattson. 1994. Landscape dimension, composition, and function in a changing Everglades ecosystem, pp. 419–444, in *Everglades: The Ecosystem and Its Restoration*, Davis, S.M. and Ogden, J.C., Eds., St. Lucie Press, Delray Beach, FL.

DeAngelis, D.L. 1997. ATLSS: Across-Trophic-Level System Simulation: An Approach to Analysis of South Florida Ecosystems, progress report, U.S. Geological Survey/BRD, Miami FL.

Dineen, J.W. 1972. Life in the tenacious Everglades, *In Depth Report. Central and Southern Florida Flood Control District*, 1:1–12.

Dineen, J.W. 1974. Examination of water management alternatives in Water Conservation Area 2A, *In Depth Report. Central and Southern Florida Flood Control District*, 2:1–10.

Dreitz, V.J., J.D. Nichols, J.E. Hines, R.E. Bennetts, W.M. Kitchens, and D.L. DeAngelis. (In press.) Use of resighting data to estimate the rate of population growth of the snail kite in Florida, *J. Appl. Stat.*

Duever, M.J., J.F. Meeder, L.C. Meeder, and J.M. McCollom. 1994. The climate of South Florida and its role in shaping the Everglades ecosystem, pp. 255–248, in *Everglades: The Ecosystem and Its Restoration*, Davis, S.M. and Ogden, J.C., Eds., St. Lucie Press, Delray Beach, FL.

Foin, T.C., S.P.D. Riley, A.L. Pawley, D.R. Ayres, T.M. Carlsen, P.J. Hodum, and P.V. Switzer. 1998. Improving recovery planning for threatened and endangered species, *Bioscience*, 48:177–184.

Goodrick, R.L. 1984. The wet prairies of the northern Everglades, pp. 185–190, in *Environments of South Florida: Past and Present II*, Gleason, P.J., Ed., Miami Geological Society.

Gunderson, L.H. 1994. Vegetation of the Everglades: determinants of community, pp. 323–340, in *Everglades: The Ecosystem and Its Restoration*, Davis, S.M. and Ogden, J.C., Eds., St. Lucie Press, Delray Beach, FL.

Gunderson, L.H., J.R. Stenberg, and A.K. Herndon. 1988. Tolerance of five hardwood species to flooding regimes in South Florida, pp. 1099–1111, in *Freshwater Wetlands and Wildlife*, Sharitz, R.R. and Gibbons, J.W., Eds., Ninth Ann. Symp., Savannah River Ecology Laboratory. U.S. Dept. of Energy.

Holling, C.S. 1978. *Adaptive Environmental Assessment and Management*, John Wiley & Sons, London.

Holling, C.S., L.H. Gunderson, and C.J. Walters. 1994. The structure and dynamics of the Everglades system: guidelines for ecosystem restoration, pp. 741–756, in *Everglades: The Ecosystem and Its Restoration*, Davis, S.M. and Ogden, J.C., Eds., St. Lucie Press, Delray Beach, FL.

Howell, A.H. 1932. *Florida Bird Life*, Coward-McMann, Inc., New York.

Kolipinski, M.C. and A.L. Higer. 1969. Vegetative changes in Shark River Slough, Everglades National Park, Florida, 1940–64, pp. 15–34, in *Some Aspects of the Effects of the Quantity and Quality of Water on Biological Communities in Everglades National Park*, Open File Report, U.S. Geological Survey, Tallahassee, FL.

Leach, S.D., H. Klein, and E.R. Hampton. 1972. Hydrologic effects of water control and management of southeastern Florida, *Florida Bureau of Geology Report of Investigations*, 60:115.

Lee, K.N. 1993. *Compass and Gyroscope: Integrating Science and Politics for the Environment*, Island Press, Washington, D.C.

Light, S.S. and J.W. Dineen. 1994. Water control in the Everglades: a historical perspective, pp. 47–84, in *Everglades: The Ecosystem and Its Restoration*, Davis, S.M. and Ogden, J.C., Eds., St. Lucie Press, Delray Beach, FL.

Loveless, C.M. 1959. A study of the vegetation in the Florida Everglades, *Ecology*, 40:1–9.

Lowe, E.F. 1986. The relationship between hydrology and vegetational pattern within the floodplain marsh of a subtropical, Florida lake, *Florida Scientist*, 49:212–233.

MacVicar, T.K. and S.S.T. Lin. 1984. Historical rainfall activity in central and southern Florida: average, return period estimates and selected extremes, pp. 477–509, in *Environments of South Florida: Past and Present II*, Gleason, P.J., Ed., Miami Geological Society.

Martin, T.W. and J.H. Doebel. 1973. Management techniques for the Everglade Kite, preliminary report, *Proc. Ann. Conf. Southeastern Assoc. Game and Fish Comm.*, 27:225–236.

McPherson, B.F. 1973. *Vegetation in Relation to Water Depth in Conservation Area 3, Florida*, Open File Report, U.S. Geological Survey, Tallahassee, FL.

McPherson, B.F. and R. Halley. 1996. *South Florida Environment — A Region Under Stress*, U.S. Geological Survey Circular 1134, Tallahassee, FL.

Milleson, J.T. 1987. Vegetation changes in the Lake Okeechobee littoral zone 1972–1982. Tech. Publ. No. 87-3, South Florida Water Management District, West Palm Beach, FL.

Mooij, W.M., R.E. Bennetts, D.L. DeAngelis, and W.M. Kitchens. (In press.) Exploring the effects of drought extent on the Florida Snail Kite: interplay between spatial and temporal scales. *Ecological Modelling*.

Pesnell, G.L. and P.T. Brown III. 1977. The major plant communities of Lake Okeechobee, Florida, and their associated inundation characteristics as determined by gradient analysis. Tech. Publ. 77-1, South Florida Water Management District, West Palm Beach, FL.

Pickett, S.T.A. and P.S. White. (Eds.). 1985. *The Ecology of Natural Disturbance and Patch Dynamics*. Academic Press, New York.

Richardson, J.R. and T.T. Harris. 1995. Vegetation mapping and change detection in the Lake Okeechobee marsh ecosystem, *Arch. Hydrobiol. Spec. Issues Advanc. Limnol.*, 45:17–39.

Schortemeyer, J.L. 1980. An Evaluation of Water Management Practices for Optimum Wildlife Benefits in Conservation Area 3A. Florida Game and Freshwater Fish Commission Report.

Science Subgroup. 1993. Federal Objectives for the South Florida Ecosystem Restoration. Science Subgroup of the South Florida Management and Working Group, Federal Task Force.

Scott, W.E.D. 1881. On birds observed in Sumpter, Levy, and Hillsborough counties, *Florida Bull. Nuttall Ornithol. Club*, 6:14–21.

Sincock, J.L. and J.A. Powell. 1957. An ecological study of waterfowl areas in central Florida, Trans. 22nd North Am. Wildl. Conf. Wildlife Management Inst. pp. 220–236.

Snyder, N.F.R., S.R. Beissinger, and R. Chandler. 1989. Reproduction and demography of the Florida Everglade (Snail) Kite, *Condor*, 91:300–316.

Sprunt, A., Jr. 1945. The phantom of the marshes, *Florida Naturalist*, 4:57–63.

Stieglitz, W.O. 1965. The Everglades Kite (*Rostrhamus sociabilis plumbeus*), report for the Bureau of Sport Fisheries and Wildlife, Division of Refuges, Washington, D.C.

Stieglitz, W.O. and R.L. Thompson. 1967. Status and Life History of the Everglade Kite in the United States, special science report, Wildl, No. 109., U.S. Department of the Interior, Bureau of Sport Fisheries and Wildlife, Washington, D.C.

Sykes, P.W., Jr. 1979. Status of the Everglade Kite in Florida, 1968–1978. *Wilson Bull.*, 91:495–511.

Sykes, P.W., Jr. 1983a. Snail Kite use of the freshwater marshes of South Florida, *Florida Field Naturalist*, 11:73–88.

Sykes, P.W., Jr. 1983b. Recent population trends of the Snail Kite in Florida and its relationship to water levels, *J. Field Ornithol.*, 54:237–246.

Sykes P.W., Jr. 1984. The range of the snail kite and its history in Florida, *Bull. Fl. St. Mus.*, 29:211–264.

Sykes, P.W., Jr. 1987. Snail Kite nesting ecology in Florida, *Florida Field Naturalist*, 15:57–70.

Sykes, P.W., Jr., J.A. Rodgers, Jr., and R.E. Bennetts. 1995. Snail Kite (*Rostrhamus sociabilis*), pp. 1–32, in *Birds of North America*, Poole, A. and Gill, F., Eds., Academy of Natural Science and American Ornithological Union, Philadelphia.

Takekawa, J.E. and S.R. Beissinger. 1989. Cyclic drought, dispersal, and conservation of the Snail Kite in Florida: lessons in critical habitat, *Cons. Biol.*, 3:302–311.

U.S.D.I. 1972. A Preliminary Investigation of the Effects of Water Levels on Vegetative Communities of Loxahatchee National Wildlife Refuge, Florida. U.S. Department of Interior Bureau of Sport Fisheries and Wildlife.

Van der Valk, A.G. 1991. Response of wetland vegetation to a change in water level, pp. 7–16, in *Wetlands Management and Restoration*, Finlayson, C.M. and Larsson, T., Eds., Workshop Proc. of the Swedish Environmental Protection Agency.

Van der Valk, A.G. and C.H. Welling. 1988. The Development of zonation in freshwater wetlands: an experimental approach, pp.145–158, in *Diversity and Pattern in Plant Communities*, During, H.J., Werger, M.J., and Willems, J.H., Eds., SPB Academic Publishing, The Hague.

Wachenfeld, A.W. 1956. Present status of the Everglade Kite, *Linnaean Newsletter*, 10:1.

Walters, C.A. 1986. *Adaptive Management of Natural Resources*, Macmillan, New York.

Wayne, A.T. 1895. Notes on the birds of the Wacissa and Aucilla River regions of Florida, *Auk*, 12:362–367.

Wiens, J.A. 1989. *The Ecology of Bird Communities*. Cambridge University Press, New York.

Wood, J.M. and G.W. Tanner. 1990. Graminoid community composition and structure within four Everglades management areas, *Wetlands*, 10:127–149.

Worth, D. 1983. Preliminary responses to marsh dewatering and reduction in water regulation schedule in Water Conservation Area 2A. Tech. Publ. 83-6, South Florida Water Management District, West Palm Beach, FL.

Zaffke, M. 1983. Plant communities of Water Conservation Area 3A: base-line documentation prior to the operation of S-339 and S-340. Technical Memorandum, South Florida Water Management District, West Palm Beach, FL.

# 7 Multi-Taxon Analysis of the "White Zone," a Common Ecotonal Feature of South Florida Coastal Wetlands

*Michael S. Ross, Evelyn E. Gaiser,*
*John F. Meeder, and Matthew T. Lewin*
Southeast Environmental Research Center, Florida International University

## CONTENTS

## INTRODUCTION

Coastal ecosystems of South Florida are arranged in predictable zonal patterns that have been described by numerous authors (e.g., Harshberger, 1914; Davis, 1940; Stephenson and Stephenson, 1950; Egler, 1952; Craighead, 1971; Ross et al., 1992, 2000). These studies illustrate that local variations in biotic zonation are influenced by a number of interacting physical factors, including bedrock and surface topography, sediment type and depth, coastal exposure to wave energy, salinity and tidal amplitude of the adjacent marine system, and magnitude of upstream sources of fresh water. Coastal gradients in South Florida generally begin in fringing mangrove swamps or rock barrens and terminate in upland forests or freshwater marshes. Gradient length is highly variable, ranging from tens of meters to more than ten kilometers.

A common feature of many South Florida coastal areas is a zone of low plant cover, clearly recognizable as a white band on black-and-white or color infrared photographs, sandwiched between more densely vegetated fringing mangrove and interior wetland ecosystems. Egler (1952) described such a "white zone" midway within the Southeast Saline Everglades (SESE), a broad expanse of marine, brackish, and freshwater wetlands extending south and east from the Atlantic Coastal Ridge. Stephenson and Stephenson (1950) also distinguished a white zone along rocky shorelines in the Florida Keys, where the zone is the most interior and least frequently inundated of five coastal bands recognizable on the basis of substrate color. With more limited exposure to waves, mangroves colonize such coastlines, but the white zone frequently remains as a reflective band beyond the reach of most non-storm tides, surrounding the broadleaved hardwood hammocks of the interior (Ross et al. 1992). The presence of a band of low apparent plant productivity midway along a coastal gradient is not restricted to Florida coastlines. Carter (1988) presented general marsh profiles for Britain and eastern North America in which the zone of sparsest vegetation occupied a region near the intersection of salt- and freshwater bodies.

What is known of the nature and dynamics of the white zone? In his monograph, Egler (1952) did not speculate on the causes of the sparse plant cover of the white zone, focusing instead on the transitional character of its vegetation, which included both graminoid elements from the interior marshes and dwarf mangrove forms from the coastal swamps. He argued that the mixture of interior and coastal taxa in the white zone reflected a balance between several types of disturbances. For instance, the periodic occurrence of fire and freezing temperatures was seen to limit the invasion of the fire- and cold-sensitive mangrove species into the interior. In contrast, major tidal events associated with hurricanes and tropical storms dispersed mangrove propagules well into the interior and raised soil salinities above the tolerance of some freshwater species. Ross and others followed on Egler's pioneering work, sampling vegetation, soils, and site characteristics within a diffuse network of sites throughout the SESE, then examining aerial photographs of 1940 and 1994 for evidence of changes in the position of the white zone that might be associated with sea-level rise and/or water management. They found that the inner boundary of the white zone had shifted toward the interior by an average of 1.5 km during the period, with maximum change in areas cut off from upstream water sources by canals or roads. Surprisingly, the plant species composition of the expanded white zone had also changed in the half century since the earlier study; that is, there was a general increase in the relative abundance of mangroves, at the expense of graminoids more commonly associated with brackish or freshwater conditions (Ross et al., 2000).

These results suggest that, while the position of the sparsely vegetated white zone appears to provide an effective indication of the extent of marine influence, the gradient in plant species composition responds independently to the same coastal influences. To clarify the significance of the white zone within coastal wetlands and to elucidate its relationship with the local biota, we undertook a detailed examination of a single transect in the easternmost portion of the SESE, bordering on southwestern Biscayne Bay (Figure 7.1). Since the white zone demonstrates a potential for indicating environmental change, it is important to determine if its physical appearance is linked to biotic pattern. The relatively fine detail of the data collected along this transect allowed us to offer some more refined inferences about the relationships of several important environmental variables with the distributions of species (and assemblages) within three taxonomic groupings and with their apparent productivity. These groups include the two main primary producers, plants and algae (primarily diatoms), and one group of consumers, the benthic invertebrates (predominantly mollusks). We hoped that understanding the ecological sensitivities of these assemblages might aid in developing indices of changing environmental conditions. Our analyses of these data focus on the following questions:

1. What structural, compositional, or physical features define the white zone within SESE coastal wetlands?
2. How are individual species and species assemblages arranged along important environmental gradients? Do the three taxonomic groups differ in their apparent sensitivity to the measured environmental variables?
3. Is it possible to define distinct compositional ecotones, or boundaries, within the coastal gradient for any of the three groups of organisms? If so, do the positions of these ecotones correspond among groups? Do they correspond to the structurally defined boundaries of the white zone?

**FIGURE 7.1** Remotely sensed image of the Turkey Point study area, showing position of transect sampling locations, and our photointerpretation of the white zone.

# METHODS

## STUDY AREA

The study area selected was south of Turkey Point Nuclear Power Plant in southeastern Miami–Dade County, Florida (Figure 7.1). These swamps contain the most extensive remaining example of the coastal wetlands of southern Biscayne Bay; although they have been impacted by drainage and other hydrologic modifications, they are less disturbed than sites farther north within the SESE. Previous investigations of Egler (1952) and Ross et al. (2000) suggested that the white zone is well defined in the vicinity.

## FIELD SAMPLING

During a two-day period in April 1997, a transect was established roughly perpendicular to the Biscayne Bay shoreline. Circular plots of radius 5 m were established at 100-m intervals, beginning 50 m from the coast and extending to the L-31E levee, ~3900 m distant. Shoot cover of all macrophyte species was estimated in octave categories within each plot, and the height and species of the tallest plant were recorded. At the center of every second plot, a 10-cm diameter PVC tube was pushed into the sediment to wall off surface water from entering, and a soil core was extracted within the walls of the tube with a 4-cm diameter × 30-cm deep bucket auger. The core hole was then pumped free of water and allowed to refill from the sides. A sample of approximately 30 ml of porewater was drawn, and its specific conductivity was determined in the field with a Hanna 8733 conductivity meter.

A topographic survey was completed in May 1997. Elevations of the ground surface at the vegetation plots and intermediate points were determined by theodolite, beginning from a U.S. Geological Survey (USGS) benchmark of known elevation adjacent to the L-31E levee. Soil profiles were examined intermittently along the transect, and the depth of sediment deposited by Hurricane Andrew was determined at 100-m intervals by trenching and by probing with a 1-cm diameter aluminum rod.

In April 1998 the transect was visited for a third time to determine the coastal distributions of diatoms and benthic invertebrates. Plots were established at 100-m intervals, and porewater conductivity was determined by methods identical to those used in 1997. Two additional reference plots were established on the west side of the canal, approximately 4020 and 4120 m from the coast. Benthic invertebrates and diatoms were sampled from the surface sediments. At each plot, the presence and appearance of a periphyton mat were noted. For diatoms, five 1-cm deep cores were taken through the mat or surface sediments with a 4.2-cm$^2$ coring device and combined into a single sample. For invertebrates, the upper 1 cm of soil was collected from a 750-cm$^2$ area, immediately washed through a 2-mm mesh sieve, and frozen upon return to the lab. Previous work determined that sediment accretion rates at Turkey Point averaged approximately 2 mm per year (Meeder et al., 1996); on this basis, we estimated that our collection represented about 5 years of accumulation.

## LABORATORY PROCEDURES

In the laboratory, surface soil samples were thawed for invertebrate analysis, and live and dead individuals were identified to species by using regional literature (Pilsbry, 1946; Tabb and Manning, 1961; Moore, 1964; Abbott, 1972; Thompson, 1984). For diatoms, surface sediment cores were thawed, homogenized with a blender, and quantitatively subsampled. A subsample was dried to constant weight in a 60°C oven and again in a muffle furnace at 500°C to determine dry mass and ash-free dry mass, respectively. A second subsample was cleaned for diatom analysis using a combination of strong acids and chemical oxidizers to remove calcite and organic material. Permanent slides were prepared and 500 diatom valves per slide were counted and identified on random transects of measured area. Identifications were made using standard and regional literature (Schmidt, 1874–1959; Hustedt, 1927–1966; Hustedt, 1930; Patrick and Reimer, 1966, 1975; Navarro, 1982; Foged, 1984; Podzorski, 1985; Round et al., 1990; Krammer and Lange-Bertalot, 1986–1997; Lange-Bertalot, 1993). Digital photographs were taken of each taxon and stored in an Image Pro® database linked to corresponding ecological information.

## Analytical Procedures

Global positioning satellite (GPS) coordinates of the coastal and interior ends of the transect were determined and superimposed on a USGS digital orthophoto quarter-quadrangle (1:12,000) of January 1994. The photo was also the basis for identifying and digitizing the interior and coastal boundaries of the white zone in the vicinity of the transect.

We examined plant, benthic invertebrate, and diatom compositional patterns with respect to the measured physical variables (e.g., distance to shore, porewater conductivity, and elevation). Species abundances for the three groups (percent cover for macrophytes; total density for living and dead invertebrates and diatoms) were analyzed individually and at the community level. For the community analysis, the abundance of species that occurred in two plots or more were relativized and then fourth-root transformed to reduce the influence of the most abundant taxa. We examined trends in compositional similarity among samples using nonmetric multidimensional scaling (NMDS) ordination in PC-ORD version 3.11 (McCune and Mefford, 1995) incorporating the Bray–Curtis dissimilarity metric (Clarke and Warwick, 1994).

The strength of association between the physical variables and community composition was assessed in several ways. Using the BIO-ENV procedure in the Primer program (Clarke and Warwick, 1994), we calculated a weighted Spearman rank correlation between the site-by-site dissimilarity matrices (Bray–Curtis index) based on the transformed species abundances and the individual dissimilarity matrices for distance to coast, porewater conductivity, and elevation.

We applied a weighted-averaging (WA) approach to estimate each species' optimum (abundance-weighted mean) and tolerance (weighted standard deviation of optimum) for each environmental variable based on its abundance along the gradient, using WACALIB 3.3 (Line et al., 1994). In this procedure, a species-based estimate of each environmental variable at each site is calculated as the average of the optima of the species present, weighted by their abundances and possibly tolerances (ter Braak, 1987). Environmental values calculated from species abundances are then compared to the observed values using regression. If the relationship is strong, the calibration model can then be applied to modern monitoring or fossil biotic data to infer environmental quality with a measured degree of confidence. Because taxa had unequal occurrences, we followed the recommendations of Birks et al. (1990) and used the number of occurrences to adjust the tolerance assigned to each taxon. By incorporating tolerances into the model, we weighted species with narrow distributions along the gradient more heavily than species with broadly dispersed or erratic distributions. We used classical regression (Birks et al., 1990) to eliminate shrinkage in the range of inferred values, because it resulted in the most evenly distributed residuals. We used bootstrapping to calculate the root mean squared error of prediction (RMSE; Birks et al., 1990). The resulting bootstrapped estimates of the environmental variable were then plotted against the observed values and a regression coefficient calculated. Residuals from the regression of measured vs. species-inferred distance values were examined for trends, then plotted against measured conductivity and elevation values to determine if distance estimates were biased.

We used two methods to examine the continuity of the biotic assemblages along the transect and to identify any breaks or ecotones within the gradient. For all three groups, we visually identified peaks in the sequence of dissimilarity scores between adjacent sampling stations, using a five-plot (500-m) moving window to smooth the curves (modified from Whittaker, 1960).

We also assessed the clustering of species boundaries (lower boundary, most coastward occurrence; upper boundary, most interior occurrence) by comparing the observed distribution with a null expected distribution derived through a Monte Carlo simulation and obtaining a test statistic (Auerbach and Shmida, 1993). This analysis was only applied to diatoms, because the number of plant or mollusk species boundaries was too small to provide adequate statistical power.

# RESULTS

## DELINEATION OF THE WHITE ZONE

Aerial photographs from 1992 indicate that the white zone was a broad and conspicuous band throughout the coastal swamps south of the Turkey Point power plant (Figure 7.1). The outer boundary was very distinct, intersecting our transect at about 170 m from shore. The inner boundary, however, was less clearly defined. Beginning at approximately 2760 m, a speckled transitional zone of approximately 350-m width separated the white zone from distinctly darker signatured vegetation. This transitional band, which may signal an incipient expansion of the white zone was variable throughout the Turkey Point region, and was entirely absent in some areas.

## PHYSICAL VARIABLES

Soil profiles along the transect were affected by marine sediments deposited by Hurricane Andrew (August, 1992). Deposits thinned from 12 to 16 cm adjacent to the coast to 2 cm or less at 2500 m. Beneath the storm deposits, sediments were characterized as peaty marls within 500 m of the coast, but farther inland they were predominantly marls (calcitic silts and muds) precipitated under freshwater conditions. West of 3800 m the marls are organic rich, approaching peats in areas where sawgrass was dominant. Elevation generally increased from coast to interior, ranging from 15 cm above sea level (ASL) at the coast to 40 cm ASL at the base of the L-31E levee (Figure 7.2). Local depressions (ephemeral drainages) were present between 1000 and 2000 m, but the overall correlation of elevation with distance was very strong ($r = 0.95$). In contrast, porewater conductivity — the third measured physical variable — was only weakly correlated with distance to the coast and elevation in 1998 ($r = -0.18$ and $r = -0.29$, respectively). Through 2000 m, conductivity was similar to nearshore surface water (35–45 mS/cm/sec; Figure 7.2). Beyond 2000 m, conductivity gradually rose, reaching a plateau of 50–65 mS/cm/sec at 2900 m, then dropping sharply to near-freshwater conditions (<5 mS/cm/sec) within a few hundred meters of the levee. Similar spatial patterns in 1997 (not shown) suggest that the conductivity profile in Figure 7.2 is representative of conditions during the late South Florida spring, when osmotic moisture stress is apt to be highest during the year.

## VEGETATION PATTERN

Total macrophyte species cover decreased from about 50% adjacent to Biscayne Bay to <10% between 900 and 2000 m (Figure 7.3A). Over the same sequence, the height of the tallest plant (R. mangle in all plots) decreased from 125 cm to about 50 cm. Cumulative species cover increased sharply beyond 2000 m, exceeding 100% at several points near the L-31E levee. Maximum plant height in the interior half of the transect was variable and difficult to interpret because the tallest species differed among plots. The white zone at Turkey Point is therefore characterized by vegetation cover of <50% and maximum vegetation height of <1 m.

Spatial patterns in plant community composition are summarized in the NMDS ordination (Figure 7.4A), with sites labeled according to their distance from the coast. The general decrease in axis 1 scores with distance from Biscayne Bay indicates strong coastal zonation in species composition. Stations 2100 to 3300 were also distinguished from stations 3400 to 4100 by their positions on axis 1. The relative isolation of the latter group in the upper left portion of the graph suggests that the freshwater community was somewhat distinct from the rest of the gradient.

The distribution of individual species along the transect provides further detail about the zonal nature of the vegetation (Figure 7.5; Appendix). Two species (R. mangle and A. germinans) were

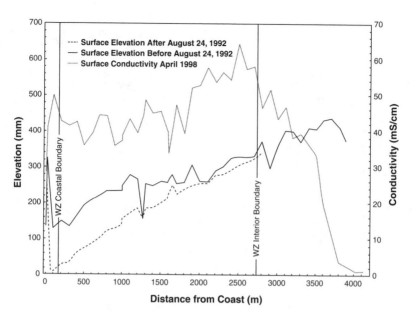

**FIGURE 7.2** Profile of surface elevation before and after Hurricane Andrew (August 24, 1992) and surface conductivity (measured in April 1998) along the Turkey Point transect. The location of the white zone (WZ) boundaries shown in Figure 7.1 are indicated.

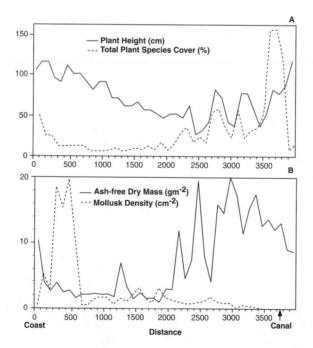

**FIGURE 7.3** Structure of plant, diatom, and mollusk assemblages along the Turkey Point transect. (A) Maximum plant height and total plant species cover. (B) Periphyton ash-free dry mass and mollusk density.

found throughout all portions of the transect east of the levee, but the other nine species had ranges of 2500 m or less (Figure 7.6). Within the first kilometer or so of Biscayne Bay, *R. mangle* and *A. germinans* were the sole vascular plant species present, with the former much higher in abundance. Moving inland to about 1300 m, *L. racemosa* and *D. spicata* became present at low

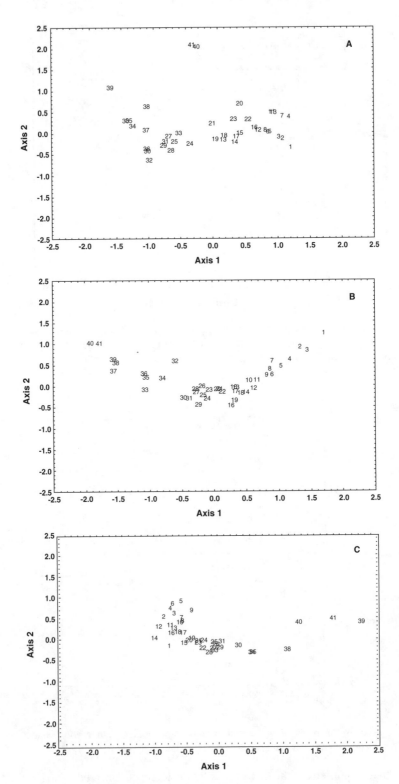

**FIGURE 7.4**   Nonmetric multidimensional scaling (NMDS) ordinations of plots along the Turkey Point transect for (A) plant, (B) diatom, and (C) mollusk assemblages. Samples are numbered sequentially by distance from the coast (i.e., 1 = 0–99 m, 2 = 100–199 m, ...).

abundances, and the cover of *A. germinans* increased. *Juncus romoerianus* was first sampled at about 2000 m from shore but became much more abundant after 2400 m, coincident with a decline in the importance of *R. mangle*. Little subsequent change in the species mix was evident until about 500 m from the L-31E levee, where *Conocarpus erecta* and *Cladium jamaicense* became common in the marsh, and *D. spicata* cover increased sharply. In general, plant species richness increased from coast to interior (Figure 7.7), with new species entering the community throughout and only *Ruppia maritima* dropping out entirely east of the L-31E levee. Over the same gradient, shrubs were replaced by graminoids as the dominant growth form. The two disjunct sites in the marsh west of the L-31E levee were similar in species composition to much of the freshwater Everglades, with *C. jamaicense* being the dominant species and *Eleocharis cellulosa* in a subordinate canopy position (Figure 7.5).

The strong zonation in species composition illustrated in Figures 7.4A and 7.5 is further supported by results of the BIO-ENV and weighted averaging regression procedure. Plant species composition showed a stronger correlation with distance than with elevation or soil conductivity in Spearman rank correlations ($r = 0.75$ vs. 0.59 and 0.17, respectively). This was confirmed by the weighted-averaging procedure, which also detected a strong fit between observed distance and the distance inferred from species composition ($R^2 = 0.87$; RMSE = 526; Figure 7.8A). Residuals from this model exhibited little pattern with distance beyond the first 1 km of the transect, where

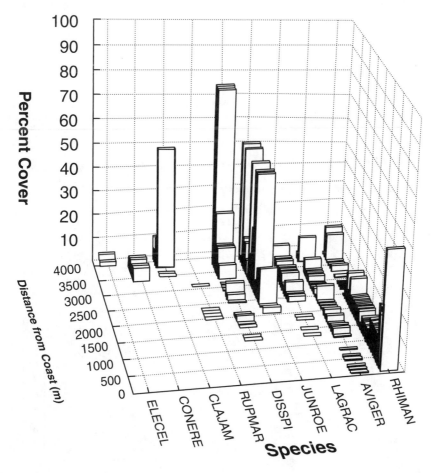

**FIGURE 7.5** Cover profiles of plant species occurring in more than one plot along the Turkey Point transect. Species codes are identified in the Appendix.

vegetation composition was very homogeneous (Figure 7.8D). More importantly, the residuals were uncorrelated with either porewater conductivity or elevation.

The sequence of dissimilarities in plant species composition among adjacent points indicates a pair of compositional breaks at 1200 and 1700 m from shore (Figure 7.9A). These represent a relatively broad ecotonal zone associated with the shift from a virtually pure *R. mangle* community

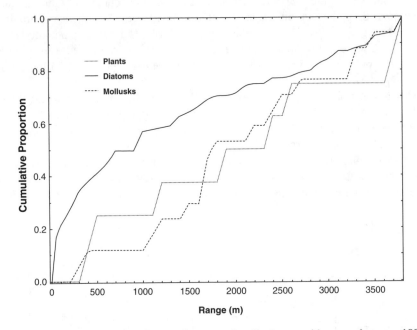

**FIGURE 7.6**    Cumulative proportion of plant, diatom, and mollusk taxa with ranges between 100 and 4100 meters along the Turkey Point transect. A taxon's range is defined as the distance between its most coastal and most interior appearance.

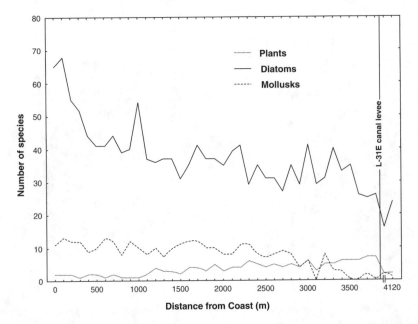

**FIGURE 7.7**    Species richness profiles for plants, diatoms, and mollusks along Turkey Point transect.

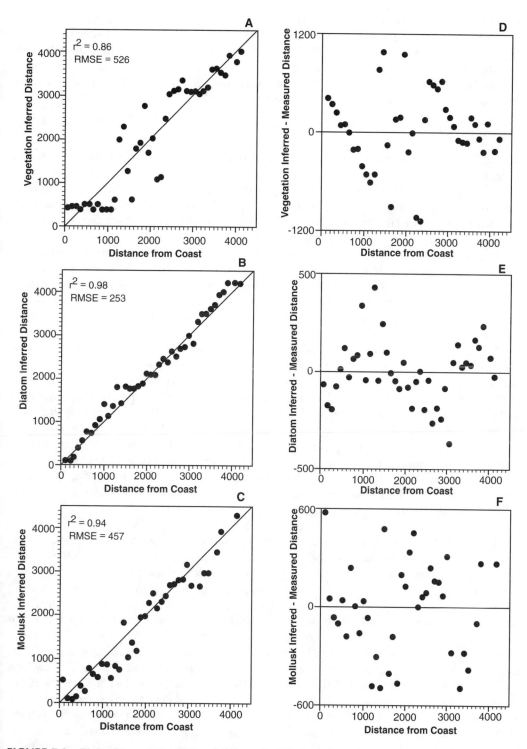

**FIGURE 7.8** Plots of measured vs. inferred distance for (A) plant, (B) diatom, and (C) mollusk assemblages. Distance was inferred from assemblages using a weighted-average (WA) regression model with tolerance downweighting for plants and diatoms and without downweighting for mollusks. Residuals from each WA calibration model are plotted against distance for (D) plant, (E) diatom, and (F) mollusk assemblages.

to one in which dominance was shared more evenly with several other halophytic species (e.g., *L. racemosa, D. spicata, A. germinans,* and *Juncus romoerianus*). A second ecotone is indicated between 3200 and 3500 m. High dissimilarities in this zone mark the transition from the halophytic assemblage described above to a *C. jamaicense*-dominated marsh community usually associated with freshwater conditions. This inner ecotone is also associated with a shift from *R. mangle/A. germinans* to *L. racemosa/C. erecta* dominance among woody plants.

## DIATOM PATTERN

Diatoms were incorporated in a coherent benthic periphyton mat in some sections of the Turkey Point transect, but these were for the most part interior to the white zone. No mat was present until 1800 m from shore, where patches of green filamentous algae were first observed at the base of some of the larger mangrove shrubs. Between 2200 and 3000 m, the sediment intermittently included a surface (<1 cm) crust that was reddish-yellow in color. This surface algal mat became continuous after 3000 m, thickening to 2 cm or more by 3600 m. The thickest algal mats were found at the stations west of the L-31E levee. Ash-free dry mass (AFDM) estimates of periphyton biomass reflected these visible changes in mat thickness (Figure 7.3B). AFDM ranged from 10 to 20 g/m$^2$ in interior sites (2200 to 4120 m) and dropped to a low of 2 to 4 g/m$^2$ in the eastern portions of the white zone (200 to 2200 m). An increased organic component in coastal sites (0 and 100 m) may reflect an increase in mangrove-derived fibrous peat in the surface sediments rather than an increase in periphyton biomass.

The NMDS diatom ordination demonstrated a zonation with distance from the Biscayne Bay coast that was even stronger than that exhibited in the vegetation (Figure 7.4B). Sites were essentially ordered by distance along axis 1. Sites representing the inner two thirds of the white zone (1200 to 3000 m) were relatively similar in species composition, resulting in a clumped distribution in the center of axis 1. Interior (3100 to 4120 m) and coastal (0 to 1100 m) sites were distinct from the rest of the transect, yet a high degree of variation in their assemblages was denoted in axis 2.

A total of 154 diatom taxa, comprising 40 genera, were identified from the 41 sampling locations at Turkey Point. Because the flora of tropical coasts has been poorly examined taxonomically, we had difficulty assigning definitive names to many taxa and resorted to a numbering system for species and varieties of questionable identity. Pictures of some of these taxa, assembled according to ecological affinities along the coastal gradient, are shown in Figure 7.10.

Distance optima and tolerances estimated for each taxon by weighted averaging (Appendix) can be used to discriminate cosmopolitan taxa from those with more definite range restrictions. In comparison to plants, relatively few diatom species were cosmopolitan in range, and progressively smaller ranges were represented by increasing numbers of species (Figure 7.6). Most cosmopolitan taxa were also relatively common in individual samples. For instance, *Navicula rhynchocephala* var. 01, *Encyonema evergladianum, Navicula* sp. 02, and *Nitzschia amphibia* were common at most sites along the coastal gradient. However, most taxa had narrower ranges, including *Brachysira neoexilis, Nitzschia palea* var. *debilis, N. serpentiraphe, Mastogloia smithii* var. *lacustris, Encyonema vulgare,* and *E. silesiacum* which were restricted to freshwater sites adjacent to the L-31E canal. Moving coastward, several taxa including *Caponea caribbea, Caloneis* sp. 01, *Amphora* sp. 01, *Rhopalodia acuminata,* and *Mastogloia erythreae* attained highest abundance in the center of the gradient, thereby characterizing the white zone. The coastal flora was most distinctive — 55 taxa were restricted to the intertidal environment. Of these, several occurred in high abundance, including *Tryblionella granulata, Psammodictyon mediterraneum, Tryblionella debilis* var. 01, *Caloneis permanga, Caloneis* sp. 02, *Campylodiscus* sp. 01, *Surirella fastuosa, Mastogloia decipiens, Cocconeis scutellum,* and *Achnanthes submarina.*

In contrast to the pattern among plants, diatom species richness was highest near the coast, with a maximum of 68 species at 100 m. Diversity declined rapidly over the first 500 m of the

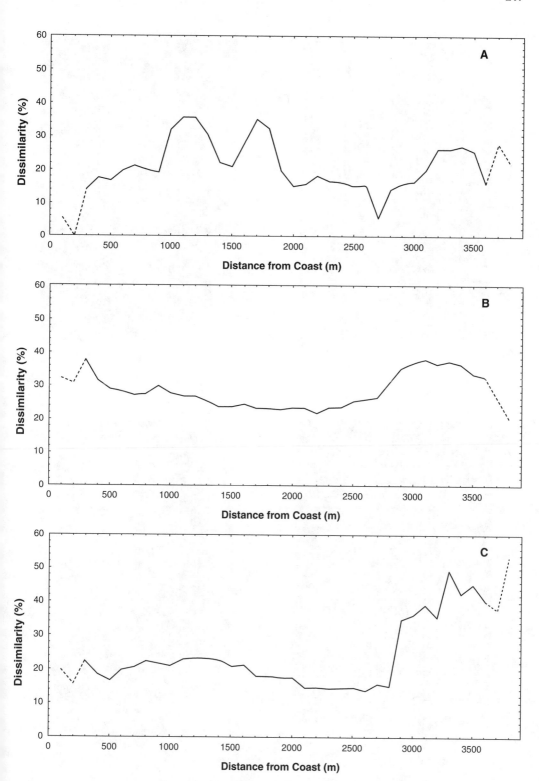

**FIGURE 7.9**  Bray–Curtis dissimilarities between adjacent sampling locations along Turkey Point transect. Points west of the L-31E levee are not included. Data points connected by solid lines are smoothed using a five-plot (500-m) moving window, while end points connected by broken lines are unsmoothed.

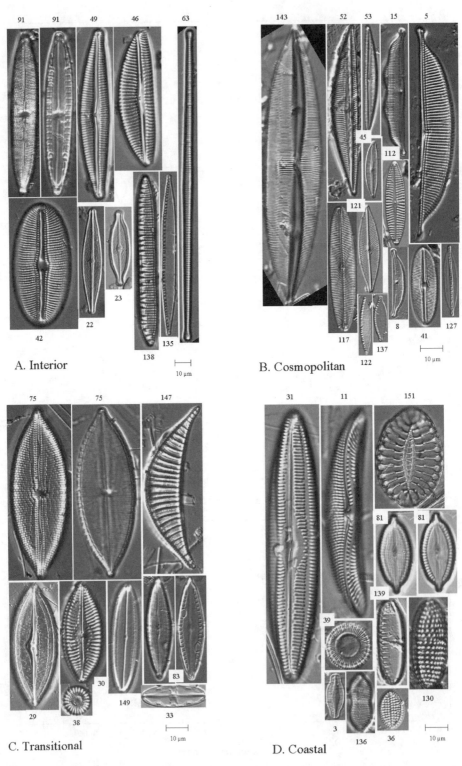

**FIGURE 7.10** Digital photographs of selected diatom taxa from (A) interior, (B) cosmopolitan, (C) white zone, and (D) coastal sites. All photographs were taken at 1250× magnification; 10-μm scale bar at the bottom of the page applies to all figures unless otherwise noted in the individual photo. Numbers are cross-referenced to taxon names in the Appendix.

transect, then more gradually. Species richness was lowest in the freshwater marsh west of the levee, where only 15 to 25 species were found in each plot (Figure 7.7).

The BIO-ENV procedure detected a strong correlation of diatom compositional dissimilarity with distance and elevation ($r = 0.83$ and 0.74, respectively), but little relationship to porewater conductivity ($r = 0.23$). The WA model also produced an excellent fit of diatom-inferred distance to observed distance ($r^2 = 0.98$, RMSE = 253; Figure 7.8B). Nevertheless, residuals from the WA model predictions exhibited conspicuous patterning (Figure 7.8E). Distances were generally over-estimated between 400 and 1700 m and beyond 3000 m, and underestimated adjacent to the coast and between 1800 and 2900 m. Further analyses of these residuals revealed no correlation with elevation, but a significant correlation with porewater conductivity ($r = -0.39$; $p < 0.02$); that is, underestimates of distance were associated with zones of high conductivity, and overestimates were associated with fresher water.

Two pronounced discontinuities in species composition were evident in the dissimilarity sequence for diatoms (Figure 7.9B). The first was a relatively sharp peak at 300 m and the second was a broader area of change between 3000 and 3300 m. These two ecotones reflected the presence of sizable groups of exclusively coastal and freshwater species, respectively, that were unable to thrive in the interior of the coastal gradient. The diatom dissimilarity sequence thus resembled the macrophyte sequence, except that the coastal ecotone in the diatom gradient was positioned about 1 km closer to Biscayne Bay.

The analysis of diatom species boundaries confirmed the presence of two ecotonal zones, the first within a few hundred meters of Biscayne Bay and a second about 3 km inland (Figure 7.11). Table 7.1 shows the site locations having significantly more or fewer than expected first or final appearances (boundaries) of species than would be expected in a random model.

## BENTHIC INVERTEBRATES

Mollusks dominated the benthic invertebrate assemblage. Only one species of foraminiferan, *Penoroplis planatus*, comprised more than 1% of any sample. The other non-molluscan invertebrates sampled were crabs (*Uca*, only six specimens), annelids (two specimens), and a second estuarine foraminiferan, *Ammonia beccarii,* that was usually below our minimum size limit. Density of invertebrates decreased precipitously with distance from the coast (Figure 7.12). Maximum density of live invertebrates of all taxa was 4600 m$^{-2}$ at 500 m, and invertebrates were entirely absent at several sites beyond 3400 m. More than 99% of the invertebrates sampled were dead specimens, which also exhibited an identical trend of declining density with distance from shore. Live invertebrates were predominantly immature forms. The most frequently found live specimens, in decreasing order, were *Batillaria minima, Cyrenoida floridana*, and *Certhidea beattyi* near the coast, and *Littoridinops monroensis* closer to the interior (Figure 7.12).

The NMDS invertebrate ordination resembled the diatom ordination in general pattern (Figure 7.4C). The samples generally arc from sites closest to the coast to sites farthest from it, with a diffuse coastal arm (sites that include marine transported gastropods) and a more compressed interior arm (sites associated with freshwater pulmonates). The tightly grouped core of sites near the middle of the transect were characterized by *Battilaria minima, Cyrenoida floridana*, and *Cerithidea beattyi*. However, the placement of sites with respect to distance was not as orderly as in the diatom analysis, partly because the coastal invertebrate assemblage is not well developed at this location and also because of influx and mixing of offshore marine invertebrates with the resident species.

Fourteen of the 17 invertebrate taxa were present within 100 m of the coast, but only three taxa extended beyond 3500 m (Figure 7.12). Among the former group, the foraminiferan *Peneroplis planatus* and the mollusks *Acteocina canalicolata, Triphora* cf. *nigrocincta*, and *Rissoina catesbyana* typically inhabit seagrass beds in nearshore marine areas. The result is a rather constant decline in species richness with distance (Figure 7.7). This pattern resembles the diatom gradient,

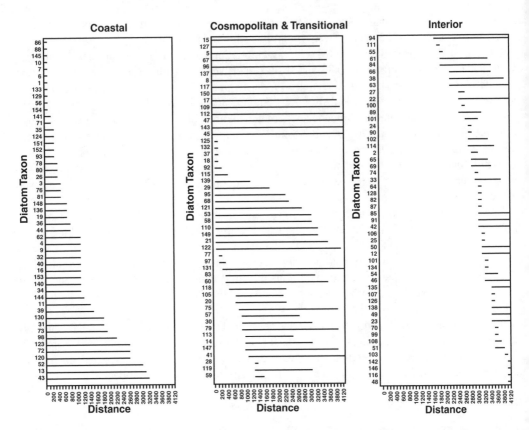

**FIGURE 7.11**   Ranges for 154 diatom taxa along the Turkey Point transect, arranged according to most coastal appearance. Species codes are identified in the Appendix.

---

**TABLE 7.1**
**Sites with Significantly More or Fewer Upper (Interior)**
**and Lower (Coastal) Diatom Range Boundaries**
**Than Expected**

|  | Distance from Coast (m) |
| --- | --- |
| More upper boundaries | 100, 200, 3000 |
| More lower boundaries | 100, 3100, 3200, 3300, 3500, 3600 |
| Fewer upper boundaries | No sites |
| Fewer lower boundaries | No sites |

---

but invertebrates differ in the absence of species distributed exclusively in the freshwater portions of the transect. In general, invertebrate species are more broadly distributed along the transect than diatoms or plants, with only a single taxon (*Triphora* cf. *nigrocincta*) whose range was less than 500 m (Figure 7.6).

The correlation between distance and invertebrate species composition calculated by BIO-ENV was somewhat lower than for plants and diatoms ($r = 0.68$); little correlation was found between invertebrate assemblages and conductivity or elevation ($r = 0.27$ and $0.53$, respectively). However, based on the weighted-averaging regression, the invertebrates were intermediate between plants

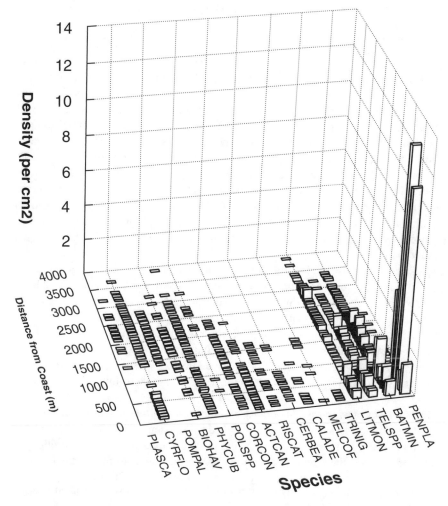

**FIGURE 7.12** Density profiles (total live + dead) for invertebrate species occurring in more than one plot at Turkey Point. Species codes are identified in the Appendix.

and diatoms in the strength of their relationship to distance. Unlike the other two groups, the fit of the unweighted WA model is superior to that of the tolerance-weighted model ($R^2 = 0.94$, RMSE = 457 and $R^2 = 0.90$, RMSE = 546, respectively; Figure 7.8C). Between 1800 and 3000 m, there is some evidence of patterning in the residuals from the unweighted model (Figure 7.8F), but this pattern is uncorrelated with conductivity or elevation.

Dissimilarities between neighboring invertebrate assemblages are low and relatively constant within the coastal two thirds of the transect, though there is a plateau of slightly more rapid change between 1200 and 1500 m (Figure 7.9C). Dissimilarity increases sharply between 2800 and 3200 m, and remains high thereafter. This is the region where the freshwater pulmonates and the hydrobiid, *Littordinops monroensis,* become abundant. West of the levee, pulmonate density and diversity increases to typical Everglades marsh levels (Meeder et al., 1996). The increase in point-to-point variability in the inner one-third of the transect is not associated with the addition of new species that are absent from coastal locations, but rather with minor species dropping out entirely from an already speciose-poor mollusk community and dominant taxa becoming much less abundant than in coastal sites.

# DISCUSSION

## THE NATURE OF THE WHITE ZONE

At Turkey Point, the white zone begins abruptly at about 200 m from the coast, but its interior boundary at 2.5 to 3 km distance is less easily defined. This pattern parallels that observed by Egler (1952: 256), who described the interior boundary as "... distinct, but not knife-edge, definite enough so that the changeover usually occurs within half a kilometer...." Our analyses suggest that its white appearance on aerial photographs is attributable to sparse cover of low-growing plants, in conjunction with the reflective quality of the exposed marl soils or the fresh storm deposits that cover them. The absence of a continuous periphyton mat in the white zone at Turkey Point may also be a contributing factor, because the chlorophyll present in algae is expected to produce a red color on color-infrared photos. However, the spectral signature of periphyton is complicated by the depth and quality of surface water, if present, and perhaps by the structure, composition, and calcite content of the mat itself (e.g., Rutchey and Vilcheck 1999). Moreover, we have observed a well-developed periphyton mat within the white zone in other South Florida locations, though that is not generally the rule.

The sparse cover and dwarfed morphological forms that characterize the vegetation of the white zone suggest that these are very unproductive wetland ecosystems. Ross et al. (2001) recently estimated productivity in a nearby dwarf mangrove basin in Biscayne National Park (BNP), where maximum plant height and total plant cover were considerably higher than midway in the Turkey Point white zone. Mangrove aboveground production was 8600 kg ha$^{-1}$ yr$^{-1}$, which was 29% that of an adjacent coastal mangrove forest. At Turkey Point, productivity was undoubtedly much less. In the current study, we were not able to directly address the causes of low productivity in the white zone, and can only suggest a few factors that may combine to limit biological activity:

1. *Wide seasonal fluctuations in salinity and moisture content:* As in the relatively steep, compressed Florida Keys gradients described by Stephenson and Stephenson (1950), the white zone of the more gradual mainland coast occupies a supratidal position; that is, saltwater reaches the interior only during spring tides and storms, and even its coastal portions are not wetted regularly by the semidiurnal tides. Once tidal waters do enter the white zone, they drain slowly and unevenly, tending to pool in local depressions, where they evaporate over days or weeks. By April of a dry year, the process may result in both surface desiccation and saline or hypersaline porewater. These environmental conditions contrast sharply with those encountered during the fall months, when the combination of high tides and heavy rains often causes persistent flooding with fresh or brackish water. While certain plants may grow reasonably well in one or the other of these conditions, adaptations that allow them to persist in both may require physiological trade-offs which result in reduced productivity (Ball, 1988). Environmental variability may be moderated at the interior boundary of the white zone by freshwater influx and at the coastal end by recurrent tidal flushing.

2. *Absence of freshwater input from upstream sources:* In our recent examination of historical changes in SESE coastal wetlands over the past five decades (Ross et al., 2000), we found minimal change in the position of the white zone in Taylor Slough and north of Long Sound, but much more dramatic landward shifts north of Joe Bay, in the triangular area between Card Sound Road and U.S. 1, and south of Turkey Point. We attribute this pattern to differences in access to upstream freshwater sources: The latter three basins were cut off or deprived of water as a result of water management activities, while water supply to the first two basins remained high during the period. The mechanism for this effect is not entirely clear, but two possibilities surface: (a) by supplementing the nutrient supply to coastal marshes, upstream water sources may enhance

productivity, thereby narrowing the width of the white zone; and (b) maintenance of connectivity between freshwater and marine water sources may increase production by buffering ecotonal areas from the seasonal fluctuations discussed above.

3. *Heavy marl soils that may contribute to phosphorus limitation:* White zone soils in much of the SESE are marls of the Perrine series, which are silt loam in texture (USDA, 1996). As such, these soils present both physical and chemical impediments to plant growth. Because of their heavy texture, they drain slowly and therefore may undergo long periods of anaerobiosis. Moreover, owing to their calcareous nature and high pH (typically 7.4 to 7.8), phosphorus availability may be limited. In the Biscayne National Park dwarf mangrove ecosystem described above, where soils were similar to those of the Turkey Point white zone, Jacobson et al. (unpubl. manuscript) found porewater with very high alkaline phosphatase exoenzyme activity, indicating phosphorus limitation. Inasmuch as red mangrove leaf phosphorus content was also reduced in these locations in comparison to the coastal fringe, there is strong support for the notion that phosphorus availability limits white zone productivity.

4. *Sporadic occurrence of natural disturbances:* Egler (1952) believed that fire was an important factor in maintaining a mixture of graminoids and mangroves in the white zone. Today the large graminoids capable of carrying a fire (e.g., *C. jamaicense* and *J. romoerianus*) are absent or of low abundance in most of the white zone, and it is difficult to imagine these areas burning. Periodic freezes, however, may be more important in controlling the expansion of sensitive mangrove species such as *R. mangle* and *L. racemosa*. While recovery of the pre-disturbance canopy is probably achieved in less than 5 years in relatively productive mangrove communities, freeze events may occur frequently enough to limit mangrove cover in the white zone, where recovery is likely to be slower. South Florida coastal areas have been affected by killing freezes at least five times in the last two decades (Olmsted et al., 1993; Ross pers. obs.).

## Species and Community Responses to Coastal Gradients

Species of each taxonomic group sort predictably into bands that parallel the SESE coastline. Using abundance-weighted optima and tolerances for distance from the coast, spatially explicit niches can be defined for most species. Detection of a strong assemblage response to the coastal distance gradient is to be expected, since sites are by definition spatially autocorrelated. However, the strong linear relationship observed for all three groups is striking, and suggests that distance effectively incorporates trends in physiologically significant environmental variables. Our sampling strategy was to measure several environmental factors (i.e., salinity and elevation, which is related to water depth) during the dry season, when we expected osmotic stress to be maximal; however, we found little or no correlation of these variables with the assemblage data. This suggests that species distributions within these tidal areas reflect the entire annual range of variation in such variables, rather than a level measured during any single time of year. By monitoring a suite of parameters along a coastal gradient regularly, one could characterize sites by parameter ranges instead of averages and perhaps define ecological forcing factors more directly.

Strong and analogous responses to distance are apparent among the three groups of species, indicating an overriding effect of this composite variable on assemblage composition. Nonetheless, the groups vary in the spatial scale at which they respond to variation in the coastal environment. By exploring these differences, we may identify ecological constraints that are imposed on these groups by their disparate natural histories.

### Vascular Plants

Although several vegetation analyses show that plant community composition changes in a regular manner with distance from the shore at Turkey Point (Figures 7.4A, 7.5, 7.8A), this clinal sequence is also marked by prominent discontinuities at 1200 to 1700 m and at about 3200 m from shore (Figure 7.9A). The more interior of these coincides well with the inland boundary of the white zone, but the coastal discontinuity is located well within the white zone borders. These ecotonal zones do not reflect wholesale shifts in species composition, but instead signify changes in the relative abundance of taxa, as well as the addition of new species to the assemblage with increasing distance from the coast. The observed increase in plant diversity along this coastal transect represents the lower end of a long gradient that ends in upland forests, especially pine rocklands, where species richness reaches its South Florida maximum (Ross et al., 1992; USFWS, 1999).

Not surprisingly, the increase in plant species diversity along the transect is accompanied by an increase in structural complexity, as graminoids of both rhizomatous and bunch-forming morphologies occupy the openings among the mangroves. Because the changeover to a mixed assemblage of graminoids and shrubs occurs midway within the white zone, it may be divided into inner and outer subzones on the basis of vegetation. While these subzones appear to share a low production potential, the contrast in their structural organization and physical setting suggests differences in disturbance regime (fires and freezes more frequent in inner zone, storm-associated sedimentation more common in outer zone), as well as in other ecosystem properties, including nutrient cycling characteristics, and faunal and algal associations. Thus, the diatom and mollusk assemblages we studied may be expected to respond to vegetation structure insofar as it affects light availability, protection from desiccation, and the types of substrate accessible for colonization and growth.

All of the wetland plants sampled at Turkey Point are rooted, emergent, woody, or herbaceous perennials. Most or all are characterized by seeds that are readily dispersed by water. Once they have germinated, however, they must survive through several to many annual cycles at a fixed location, though many are able to spread laterally through vegetative means. As a result of their life span and size, plants and plant assemblages may display a relatively coarse-scale response to environmental variation (i.e., integrating over relatively long time periods and wide spaces). This may explain, in part, the lack of association between plant species composition and dry-season salinity observed in this study.

### Diatoms

According to several measures, including weighted-averaging regression/calibration (Figure 7.8B), BIO-ENV correlation, and range analysis (Figure 7.6), diatoms are more sensitive to the coastal distance gradient than either plants or mollusks. This may be attributed to their considerable species richness, which is an order of magnitude larger than the other two groups. The quality of WA regression models has been shown to be positively correlated with the number of species entering the model (ter Braak et al., 1993). Besides their diversity, however, diatoms are also several orders of magnitude smaller than the other organisms examined here; if their perceived microenvironments are scaled correspondingly, the spatial range at which they integrate environmental variation is also expected to be much smaller. In addition, because diatoms are single-celled organisms with short generation times (~1 to 2 divisions per day), they respond quickly to environmental changes that directly affect them physiologically. Over the course of the millions of years of competition, diversification, and specialization, these qualities have facilitated the evolution of diatoms into what we now consider to be an ideal indicator organism.

Compared to other regions of North America, the diatom flora of the southeastern U.S. has been poorly explored. In South Florida, the flora of the freshwater Everglades has recently received attention related to the management and restoration of this fragile ecosystem (Swift and Nicholas, 1987; Slate, 1998; McCormick and O'Dell, 1996), yet very little of this research has had a taxonomic focus, and virtually none has emphasized the coastal saline flora. As a result, the flora presented here includes many species that are new to the North American literature and several taxa for which we have been unable to find descriptions.

Like other biota in South Florida, the coastal diatom flora contains a mixture of temperate and tropical taxa (Long and Lakela, 1971; Witkowski et al., 2000). Many of the taxa have not been previously reported in standard North American or European literature developed for predominantly temperate floras but can be found in tropical floras, particularly of the Caribbean basin (DeFelice, 1978; Montgomery, 1978; Sullivan, 1981; Navarro, 1982; Reimer, 1996). The Turkey Point flora is genus rich, typical of tropical coastlines that experience large fluctuations in salinity (Østrup, 1913; Mann, 1925; Hagelstein, 1938; Sullivan, 1981; Foged, 1984). It shares the same dominant genera — *Mastogloia, Amphora, Nitzschia,* and *Caloneis* — with the studies cited above. There is substantial overlap between the Turkey Point flora and the diatom floras described for Cuba (55 taxa; Foged, 1984), coral reefs in the Florida Keys (41 taxa; Montgomery, 1978), and mangroves in the Bahamas (28 taxa, Sullivan 1981; and 15 taxa, Reimer, 1996), Southwest Florida (30 taxa; Navarro, 1982), and Jamaica (15 taxa; Podzorski, 1985).

Based on their distance response, diatoms of the SESE can be sorted into four categories: (1) coastal, (2) transitional, (3) interior, or (4) cosmopolitan. The sharpness of the boundaries between the zones varies. The coastal–transitional boundary is graded, with species terminating in clumps spanning several hundred meters, while the transitional–interior boundary is more abrupt with more range terminations than expected over a short distance (Figure 7.11). The exterior limit to the coastal zone is fixed at 0 m; the interior boundary appears to fall between 1100 and 1600 m, which includes the upper and lower boundaries of 12 and 13 taxa, respectively. Another boundary within the coastal zone occurs between 100 and 300 m (see Table 7.1), where the range of many (18) of the marine taxa ends. At the other end of the transect, the interior zone has a diffuse lower limit that falls between 2600 and 3300 m, where 15 and 29 taxa appear and disappear, respectively. Over 70% of these taxa have short ranges (less than 300 m), suggesting that this is a zone of transition along the gradient (also shown in the dissimilarity diagram, Figure 7.9B). The rest of the taxa of the interior zone continue into the freshwater marsh west of the canal. The transitional zone can be defined by default as the upper and lower boundaries of the coastal and interior zones, respectively, and includes the visually defined white zone. The fourth group contains taxa that are cosmopolitan, which we arbitrarily defined as having ranges greater than 60% of the total transect length (2500 m). A final group may be considered that contains remaining taxa with short ranges that extend into more than one zone (only six taxa).

The coastal zone is well defined in terms of diatom composition, containing many (55) species that are not found elsewhere along the gradient. Diatoms inhabiting the coastal zone must be able to withstand tidal fluctuations in water depth and salinity and associated changes in dissolved gasses and nutrient availability. This zone is also characterized by thick and unconsolidated sediments that were deposited during Hurricane Andrew. These sediments and associated diatom communities are prone to redistribution and scouring during storms and high tides. Because of this, diatom communities of intertidal zones usually comprise a varied assemblage of marine planktonic taxa, loosely attached epiphytes that have been transported into the shallow sediments, and benthic taxa that reside within the sediments themselves (Vos and de Wolf, 1993). The dominant genera in the coastal portion of the Turkey Point transect (i.e., *Mastogloia, Amphora, Nitzschia, Tryblionella*) are common constituents of soft-intertidal sediments in the tropics (Foged, 1984; Navarro, 1982). Many of the taxa were also recorded from coral reef communities in the Florida Keys (Montgomery, 1978). Several notably marine taxa were probably transported from the ocean to coastal sediments during storm tides (i.e., *Cyclotella striata, Catacombas gaillonii, Biddulphia* spp., *Terpsinoë musica*).

Whereas several species extend into the transitional zone from adjacent zones, only a few (19) relatively uncommon taxa are restricted to it. The decline in species richness in the transitional zone parallels an apparent decrease in algal productivity in the white zone. Few taxa thrive in this portion of the gradient. The diatom assemblage of the transitional zone (~1100 to 3300 m) is dominated by *Navicula rhynchocephala* var. 01, *Nitzschia amphibia, Amphora holsatica, Navicula scopuloroides,* and *Synedra filiformis* var. *exilis*. These taxa are wide-ranging and are found in abundance throughout the transect. Narrowly distributed taxa that are diagnostic of the transitional zone include *Caponea caribbea*, *Mastogloia fallax* var. 01, *Amphora* sp. 01, *Rhopalodia acuminata*, *Mastogloia erythreae*, and *Fragilaria exigua* var. 01. These taxa must be capable of surviving large fluctuations in salinity and frequent desiccation. Of these, complementary ecological data could only be found for *Caponea caribbea*, which was recently described from collections by Podzorski (1984) from algal mats in hypersaline pools along the Broad River in western Jamaica.

The interior zone is characterized by 44 taxa, most of which are common elsewhere in the freshwater Everglades system. This flora has been described in Swift and Nicholas (1987), Slate (1998), and McCormick and O'Dell (1996). Typical dominant species are *Mastogloia smithii* var. *lacustris*, *Nitzschia serpentiraphe*, *Fragilaria synegrotesca*, and *Brachysira neoexilis*. These diatoms are most frequently found in association with the thick cyanobacterial mats that typify Everglades marshes (Browder et al., 1994). Their decrease in abundance toward the coast reflects the disorganization of this complex mat community, as well as the coastward increase in osmotic stress. More unexpected in the interior zone were taxa such as *Plagiotropis lepidoptera* var. *proboscidea*, several *Mastogloia* spp., and *Navicula scopuloroides*, which are classically considered halophilic (Patrick and Reimer, 1975; Hustedt, 1927–1966). In our transect, they increased in abundance toward the coast, but their presence in interior sites suggests that (1) salinity in interior regions is frequently elevated, and/or (2) there is substantial physical mixing during high tides that transports diatoms between zones.

The cosmopolitan group contains 26 taxa that are found throughout the gradient, including *Encyonema evergladianum*, *Navicula rhynchocephala* var. 01, *Navicula trivialis*, *Navicula digitoradiata*, *Amphora veneta*, *Nitzschia amphibia*, and *Mastogloia* cf. *quinquecostata*. These were also among the most abundant diatom species in each sample, suggesting both superior dispersal capability and an ability to thrive in a variety of environmental conditions. Most of these taxa have been frequently encountered in other areas of the Everglades (i.e., *E. evergladianum*, *A. veneta*; see Slate, 1998) and in the Caribbean basin (*N. rhynchocephala*, *N. trivialis*, *N. digitoradiata*; see Foged, 1984; Podzorski, 1984), indicating that they are ecological generalists.

The assortment of most diatoms into one of three zones along the coastal gradient suggests taxon-specific differences in physiological tolerances to several variables with linear or complex relationships to the distance gradient (i.e., salinity, nutrient availability, pH, and water depth or hydroperiod). Of the variables that we were able to measure, conductivity significantly biased the results of the weighted-averaging distance model, indicating a diatom response to pockets of high or low salinity that are unrelated to the distance gradient. The WA distance model predicted more interior locations for assemblages between 1000 to 2000 m, where relatively low salinity conditions prevailed; conversely, the model predicted more coastal locations for assemblages between 2500 and 3100 m, a band where several of our highest conductivity readings were obtained (Figure 7.8E). More generally, it can be said that coastal portions of the transect contained mostly halophilic taxa that have been described from benthic collections elsewhere in the tropics, whereas interior regions contained mostly taxa with freshwater affinities. Within the white zone, taxa ranged from halophilic to halophobic, and there were many generalists with an apparent ability to withstand a broad range of osmotic stress. Thus, the boundaries of the white zone are best defined by the taxa comprising well-defined zones adjacent to it, rather than by the presence of diagnostic forms within the zone itself.

### Invertebrates

Like other Everglades consumer assemblages, mollusks numerically dominate the large-bodied invertebrate fauna of the SESE (Rader, 1994). The abundance of periphyton available for grazing taxa and particulate organics for filter feeding species contributes to the production and pattern of composition of primary consumers in the Everglades and other coastal environments. In addition, mollusks are highly sensitive to salinity and typically sort predictably along salinity gradients (Moore, 1964). While our study found a strong correlation of distance and mollusk assemblage composition (Figure 7.8b), the effect of salinity was negligible. However, only a short portion of the gradient beyond 3300 m was characterized by very low salinity (Figure 7.2). Freshwater invertebrates thrived in this zone while salt-tolerant species occurred throughout the rest of the transect. Occasional appearance of larvae or dead adults of marine taxa in freshwater zones probably results from intermittent high or storm tides that transport these individuals landward. In general, mollusk species are distributed in one of three ways along the coastal gradient: (1) abundant only in coastal regions of the transect, (2) present in low abundance in the center of the transect (i.e., the white zone), or (3) present in low abundance throughout the transect. The plotted change in compositional dissimilarity between adjacent samples (Figure 7.9C) shows a major transition between 3000- and 3300 m that marks the interior extent of coastal and white zone taxa. Samples from this part of the transect also appear as outgroups in the NMDS ordination (Figure 7.4C). A less obvious transition that occurs between 1100 and 1700 m (Figure 7.12) reflects the exterior extent of transitional zone taxa and interior extent of several coastal species.

Invertebrate diversity increases toward the coast to a maximum of 13 taxa. Like the diatom assemblages, the coastal mollusk fauna includes several species that live within the marsh and others that have been transported into the shallow intertidal zone by storms. The latter group is characterized by species that are normally found as epifauna on seagrass blades (Abbott, 1972) including *Acteocina caralicolata, Triphora* cf. *nigrocincta*, and *Rissoina catesbyana*, as well as the foraminiferan species *Peneroplis planatus*. These marine taxa tolerate short episodes of low salinity, but not desiccation. They were likely transported into the coastal wetlands as adults by storm tides (Wilbur and Yonge, 1966). The inland extent of *P. planatus* probably represents the maximum distance of marine transport, because the species' small size, light weight, and discoid shape are more conducive to transport than heavier mollusk shells. Other larger taxa are found as far inland as 1600 to 2000 m, which represents the eastern extent of the supratidal mollusk zone. In addition to allochthonous marine species, several taxa were found living in the surface sediments of the intertidal zone. These include *Cyrenoida floridana*, the only filter feeder observed, and the algal browsing gastropods *Cerithidea beattyi* and *Batillaria minima*. These three species have broad salinity tolerances, thriving in brackish to saline conditions, and are able to withstand periods of desiccation. During drying episodes, these species aestivate by closing their shells tightly and secreting a mucous-like substance that prevents water loss. *B. minima* is the most tolerant of these species to salinity and desiccation (Abbott, 1972). In coastal samples (0 to 1100 m) it was the most abundant taxon in the adult size range (15 mm). Several common coastal invertebrates, including *Littorina, Melampus*, oysters, and the fiddler crab *Uca*, were notably rare or absent at Turkey Point. These taxa are more characteristic of more productive coastlines with larger trees and peat soils.

The transitional or white zone is characterized by a composite of marine and more euryhaline "freshwater" species. Only two species, *Planorbella scalaris* and *Pomaceae palundosa*, are restricted to the white zone. They are normally considered to be freshwater, salt-intolerant gastropods and may colonize the white zone during the rainy season. All but four of the coastal taxa were also found in the white zone, but in low abundance. Based on the observed lack of adults and living specimens, most of these are marine taxa that are probably transported into the white zone during high and storm tides. The exception is *B. minima*, which had the highest abundance of live individuals of the white zone taxa.

Only three taxa were collected in the freshwater marsh between 3200 and 4120 m, including two sites west of the L-31E canal. These included the pulmonate gastropod *Physella cubensis*, the hydrobiid *Littordinops monroensis*, and the clam *Cyrenoida floridana*. *P. cubensis* is common in the freshwater Everglades, but *L. monroensis* and *C. floridana* are brackish-water species more typical of the lower saline Everglades (Thompson, 1984). They are capable of tolerating low salinity for short periods of time and may disperse from coastal populations into the interior marsh as larvae. The coastward boundary of this zone (~2800 to 3300 m) also demarcates the western extent of many of the transitional and intertidal taxa. It is an area of sharp change in salinity (Figure 7.2), which may account for the transition in biota.

## PLANNING AND EVALUATING COASTAL WETLAND RESTORATION

The coastal wetland gradient we studied at Turkey Point was vastly different from its condition of only a century ago, due to the combined effects of sea level rise, drainage, and interference with freshwater sheet flow. A visible sign of change is evident from a comparison of aerial photos of 1940 and 1994, which indicate a landward movement in the interior boundary of the white zone of about 1800 m. Further evidence is contained in dated sediment cores along the transect, which show a consistent upcore increase in the salt-tolerance of the fossil invertebrate assemblage (Meeder et al., 1996). These signals of ecological change are not unique to our study area; they represent a widespread problem of saltwater intrusion and mangrove encroachment throughout South Florida. Restoration of coastal wetlands is an essential component of the effort to restore the greater Everglades ecosystem, especially as the condition of these areas is strongly linked to the health of adjacent water bodies, as well as the viability of populations of higher trophic organisms, such as fish and wading birds (Ogden, 1994; Lorenz, 1999). The relationships described above may aid in setting goals for coastal wetland restoration, as well as in assessing the success of restoration efforts once they are undertaken.

Despite their altered state in comparison to the pre-development condition, Turkey Point wetlands offer well-organized coastal gradients in plant, diatom, and invertebrate assemblages. These gradients are characterized for the most part by clinal change in species composition with distance from the shore but may be further differentiated into coastal, transitional, and interior assemblages. Knowledge of these distributions may, for a given coastal basin, provide four possible indicators of wetland status that all reflect the hydrologic balance between rising sea level and changes in the magnitude of terrestrial water sources (i.e., freshwater sheet flow and groundwater discharge) and only indirectly reflect on wetland health *per se*: (1) the position of the white zone, especially its leading edge; and (2–4) the position or range of coverage of the coastal, transitional, and interior compositional units for plants, diatoms, and invertebrates. All four geographic variables may be used to develop rational restoration objectives based on retrospective (i.e., paleoecologic) analyses, and in tracing the trajectory of coastal change following restoration through direct sampling. The potentials of each are discussed briefly below.

### The Position of the White Zone

Given the availability of appropriate aerial imagery, comparisons in position of the white zone over different time periods is an expeditious means of characterizing changes in the coastal gradient over large areas. Interpretation and comparison of photographs using modern GIS techniques necessarily involves some subjectivity and may be complicated by short-term phenomena such as standing water or a recent history of freeze, fire, or storm that may obscure the boundaries of the white zone. The retrospective use of aerial photographs is limited by the inability to field-truth the photointerpretation, and because the first comprehensive photos of the South Florida coastline date only to the 1930s, while Everglades drainage began several decades earlier. Nevertheless, the value of the spatial perspective predicted by aerial photographs cannot be overemphasized.

### Zonation of Vascular Plants

Delineation of the position of vegetation units along selected coastal transects is a relatively straightforward field and analytic task and could form the heart of a network directed at tracking coastal change. Utilization of aerial photographs to monitor plant species composition throughout the South Florida coastal zone at the necessary level of detail is currently problematic. However, this may soon become feasible as the availability of appropriate remotely sensed data improves, along with our ability to interpret them. Changes in rooted vegetation may track saltwater encroachment in a coarse-scale manner (perhaps associated with disturbances) because, once established, most plant species can tolerate more severe conditions than during colonization. Thus, changes in plant communities may reflect more long-term and permanent changes in wetland status than mollusks or diatoms, whose distributions may shift in response to finer scale variation. In considering the use of plant zonation as an indicator of coastal change, one should recognize that these patterns differ regionally. For instance, several species that were well represented in the supratidal zone at Turkey Point (e.g., *Avicennia germinans*, *Distichlis spicata*, and *Juncus romoerianus*) are relatively uncommon in SESE wetlands adjoining northeastern Florida Bay (Meeder et al., 1996).

## Diatom Zonation

Of the three taxonomic groups we studied, diatom assemblages were most closely aligned with coastal distance, yet also exhibited anomalies that appeared to be associated with salt tolerance (Figure 7.8E). We therefore expect their community composition to exhibit a relatively fine-scale sensitivity to saltwater intrusion (i.e., responding on an annual or even seasonal basis). They should indicate on a short-term basis the ecological effect of management efforts to increase freshwater flows to the coast. Diatoms have frequently been used for reconstructing coastal paleoenvironments and sea-level change (Gell and Gasse, 1990; Juggins, 1992; Denys and de Wolf, 1999); their sensitivity to salinity variation has been documented in a variety of settings (Ehrlich, 1975; McIntire and Moore, 1977; Fritz, 1990). Although the quality of diatoms preserved in coastal sediments is often compromised by severe physicochemical stresses (Sherrod et al., 1989), more thorough paleoecological reconnaissance surveys in the region are needed before ruling out the application of diatoms in reconstructing saltwater encroachment in the SESE. If vertical sequences of diatoms can be extracted from sediments, the WA distance-transfer function provided here could be applied to fossil assemblages to document past rates of coastal transgression. Sequences that pre-date modern landscape modifications (primarily canalization) would be helpful in more completely assessing anthropogenic influences on the rate of saltwater encroachment into the SESE. In addition, because there is a high degree of overlap between the SESE and Florida Bay diatom flora (DeFelice and Lynts, 1978; Chapter 17, this volume), species affinities defined in this study could be applied to paleosalinity reconstructions elsewhere in the South Florida marine environment.

## Invertebrate Zonation

Invertebrates were intermediate between diatoms and plants in the strength of their association with coastal distance, suggesting that they can be good indicators of coastal wetland status under future management scenarios. They can also be extremely useful in paleoecologic reconstructions, because they remain well-preserved in both carbonate and peat sediments. In conjunction with [210]Pb-based estimates of sediment accretion rates, Meeder et al. (1996) applied published mollusk habitat affinities to species profiles in soil cores, thereby documenting an increase in the rate of saltwater intrusion in SESE within the twentieth century. In several basins, including Turkey Point, the increase coincided with management actions that reduced freshwater supply.

The four elements discussed above complement one another as monitoring/planning tools for restoration of coastal wetlands. As a group, they provide broad ranges in (1) expected speed of response to saltwater intrusion, (2) time periods for which precise information about the historical coastal gradient may be provided, and (3) cost of data acquisition. As discussed earlier, they are descriptors of the position of the gradient, not of the condition of ecosystems within it. Perhaps maintaining or restoring the spatial arrangement of coastal plant, algal, and invertebrate communities, based on knowledge of current species–environment relationships and historical species distributions, is a first step toward ensuring adequate ecosystem function.

## SUMMARY

The white zone is an important ecological phenomenon of South Florida coastal ecosystems, yet its potential as an environmental monitoring tool remains untapped. By investigating factors that influence the position of its boundaries and its relationship with several multispecies groups, we have taken a first step in exploring the capacity of the white zone as an environmental indicator. Our results are summarized as follows:

1. The white zone is a region of low productivity characterized by low vegetation cover and canopy height (<50% and <1 m, respectively). From remotely sensed images, it appears as a reflective white band, resulting mainly from marl, fresh storm deposits, and, in some areas, periphyton. The majority of the white zone corresponds to the supratidal region of the coast, which is irregularly flooded by tidal waters. Some areas, however, are more regularly inundated by semidiurnal tides.

2. Over the past 50 years, the interior boundary of the white zone has encroached inland by an average of 1.5 km. Maximum shifts occurred in areas cut off by canals from upstream fresh water input (1.8 km at Turkey Point).

3. Plant, diatom, and mollusk species assemblages correlate strongly with the coastal gradient and may be separated into coastal, transitional and interior units. For all groups, the transitional–interior interface corresponds strongly with the inner boundary of the white zone, but no correlation exists between the coastal–transitional interface and the coastal boundary of the white zone.

4. The coastal gradient is characterized by a host of environmental variables (salinity, elevation, wave energy, wind speed, etc.). Of the environmental variables we measured (salinity and elevation), diatoms are weakly correlated with elevation and show significant correlation with salinity after accounting for spatial autocorrelation. Neither plants nor mollusks show any correlation with salinity or elevation.

5. The position of the white zone and coverage of the coastal, transitional, and interior compositional units of plants, diatoms, and mollusks may provide an indirect assessment of wetland status. These variables reflect the balance between ambient sea level and freshwater discharge, though each is likely to respond in a unique fashion.

6. Future investigations should address other influential environmental variables (e.g., soil nutrients, hydrologic variation, disturbance) to further the understanding of the factors controlling species composition. Also, studies should emphasize monitoring change in the position of the white zone and species distributions over varied time scales. This could include long-term monitoring or retrospective studies involving paleoecological methods.

## ACKNOWLEDGMENTS

We thank the South Florida Water Management District, especially Rick Alleman and Janet Ley, for their support of our coastal research over the last decade. Dave Reed, Pablo Ruiz, Jay Sah, and Guy Telesnicki provided field and lab assistance. This is SERC Contribution #153.

## REFERENCES

Abbott, R.T. 1972. *American Seashells*, Van Nostrand, Princeton, NJ.

Auerbach, M. and Shmida, A. 1993. Vegetation changes along an altitudinal gradient on Mt. Hermon, Israel — no evidence for discrete communities, *J. Ecology*, 81:25–33.

Ball, M. 1988. Ecophysiology of mangroves, *Trees*, 2:129–142.

Birks, H.J.B., Line, J.M., Juggins, S., Stevenson, A.C., and ter Braak, C.J.F. 1990. Diatoms and pH reconstruction, *Phil. Trans. R. Soc. Lond. B*, 327:263–278.

Browder, J.A., Gleason, P.J., and Swift, D.R. 1994. Periphyton in the Everglades: spatial variation, environmental correlates, and ecological implications, in *Everglades: The Ecosystem and Its Restoration*, Davis, S.M. and Ogden, J.C., Eds., St. Lucie Press, Delray Beach, FL.

Carter, R.W.G. 1988. *Coastal Environments*, Academic Press, London, U.K., 617 pp.

Clarke, K.R. and Warwick, R.M. 1994. *Change in Marine Communities: An Approach to Statistical Analysis and Interpretation*, Natural Environment Research Council, U.K.

Craighead, F.C., Sr. 1971. *The Trees of South Florida*, Vol. 1: *The Natural Environments and Their Succession*, University of Miami.

Davis, J.H., Jr. 1940. The ecology and geologic role of mangroves in Florida, *Pap Tortugas Lab.*, 32:304–412 (Carnegie Institute, Wash. Publ. No. 517).

DeFelice, D.R. 1975. Model Studies of Epiphytic and Epipelic Diatoms of Upper Florida Bay and Associated Sounds, M.S. thesis, Duke University, Durham, NC.

DeFelice, D.R. and Lynts, G.W. 1978. Benthic marine diatom associations: Upper Florida Bay (Florida) and associated sounds, *J. Phycol.*, 14:25–33.

Denys, L. and de Wolf, H. 1999. Diatoms as indicators of coastal paleo-environments and relative sea-level change, in *The Diatoms*, Stoermer, E.F. and Smol, J.P., Eds., Applications for the Environmental and Earth Sciences, Cambridge University Press, Cambridge, U.K.

Egler, F.E. 1952. Southeast saline Everglades vegetation, Florida, and its management, *Veg. Acta Geobot.*, 3:213–265.

Ehrlich, A. 1975. The diatoms from the surface sediments of the Bardawil Lagoon (Northern Sinai) — Paleoecological significance. *Beihefte zu Nova Hedwigia*, 53:253–277.

Foged, N. 1984. Freshwater and littoral diatoms from Cuba, *Bibliotheca Diatomologica*, 5:243 pp.

Fritz, S.C. 1990. Twentieth-century salinity and water-level fluctuations in Devil's Lake, North Dakota: test of a diatom-based transfer function, *Limnol. Oceanogr.*, 35:1771–1781.

Gell, P. A. and Gasse, F. 1990. Relationships between salinity and diatom flora from some Australian saline lakes. Proceedings of the 11th International Diatom Symposium, pp. 631–647.

Hagelstein, R. 1938. The Diatomaceae of Porto Rico and the Virgin Islands, *N.Y. Acad. Sci. Scientific Survey of Porto Rico and the Virgin Islands*, 8:313–450.

Harshberger, J.W. 1914. The vegetation of south Florida Trans., *Wagner Free Inst. Sci. Philos.*, 3:51–189.

Hustedt, F. 1927–1966. Die Kieselalgen Deutschlands, Österrichs und der Schweiz, in *Kryptogamen-Flora von Deutschland, Österriech und der Schweiz*, Rabenhorst, L., Ed., 7. Akademische Verlagsgesellschaft.

Hustedt, F. 1930. Bacillariophyta (Diatomaceae), in *Die Süßwasser-flora Mitteleuropas*, Pascher, A., Ed., Gustav Fischer Verlag.

Janzen, D.H. 1985. Mangroves: where's the understory? *J. Trop. Ecol.*, 1:89–92.

Juggins, S. 1992. Diatoms in the Thames estuary, England: ecology, paleoecology and salinity transfer function, *Bibliotheca Diatomologica*, 25:1–26.

Krammer, K. and Lange-Bertalot, H. 1986-1997. Bacillariophyceae, in *Süßwasserflora von Mitteleuropa*, Ettl, H., Gerloff, J., Heynig, H., and Molenhauer, D., Eds., Gustav Fischer Verlag, 2:1–4.

Lange-Bertalot, H. 1993. 85 Neue Taxa und über 100 weitere neu definierte Taxa ergänzend zur Süßwasserflora von Mitteleuropa, *J. Cramer*, 2:1–4.

Line, J.M., ter Braak, C.J.F., and Birks, H.J.B. 1994. WACALIB version 3.3 — a computer program to reconstruct environmental variables from fossil assemblages by weighted averaging and to derive sample-specific errors of prediction, *J. Paleolimnol.*, 10:147–152.

Long, R.W. and Lakela, O. 1971. *A Flora of Tropical Florida*, University of Miami, 962 pp.

Lorenz, J.J. 1999. The response of fishes to physical–chemical changes in the mangroves of northeast Florida Bay, *Estuaries*, 22:500–517.

Mann, A. 1925. Marine diatoms of the Philippine Islands, *U.S. Smithson. Inst. Bull.*, 100:1–182.

McCormick, P.V. and O'Dell, M.B. 1996. Quantifying periphyton responses to phosphorus in the Florida Everglades: a synoptic-experimental approach, *J. No. Am. Benthol. Soc.*, 15:450–468.

McCune, B. and Mefford, M.J. 1995. *PC-ORD: Multivariate Analysis of Ecological Data, Version 2.0*, MJM Software Design, Gleneden Beach, OR, 126 pp.

McIntire, C.D. and Moore, W.W. 1977. Marine littoral diatoms: ecological considerations, in *The Biology of Diatoms*, Werner, D., Ed., University of California Press, Berkeley.

Meeder, J.F., Ross, M.S., Telesnicki, G., Ruiz, P.L., and Sah, J.P. 1996. *Vegetation Analysis in the C-111/Taylor Slough Basin*, final report to the South Florida Water Management District, Contract C-4244, West Palm Beach, FL.

Montgomery, R.T. 1978. Environmental and Ecological Studies of the Diatom Communities Associated with the Coral Reefs of the Florida Keys. Ph. D. dissertation, Florida State University, Tallahassee, FL.

Moore, D.R. 1964. Mollusca of the Mississippi coastal waters, *Gulf Coast Bull.*, 1, Biloxi, MS.

Navarro, N. 1982. Marine diatoms associated with mangrove prop roots in the Indian River, Florida, USA, *Bibliotheca Phycologica*, 61:1–154.

Ogden, J.C. 1994. A comparison of wading bird nesting colony dynamics (1931–1946 and 1974–1989) as an indication of ecosystem conditions in the southern Everglades, in *Everglades: the Ecosystem and Its Restoration*, Davis, S.M. and Ogden, J.C., Eds., St. Lucie Press, Delray Beach, FL.

Olmsted, I., Dunevitz, H., and Platt, W.J. 1993. Effects of freezes on tropical trees in Everglades National Park, Florida, USA, *Trop. Ecol.*, 34:17–34.

Østrup, E. 1913. Diatomaceae ex Insulis Danicis Indiae Occidentalis imprimis a F. Børgesen lectae, *Dansk Bot. Arkiv*, 1:1–29.

Patrick, R. and Reimer, C.W. 1966. The diatoms of the United States, I, *Acad. Nat. Sci. Philadelphia, Monogr.*, 13:1–688.

Patrick, R. and Reimer, C.W. 1975. The diatoms of the United States, II, part 1, *Acad. Nat. Sci. Philadelphia, Monogr.*, 13:1–213.

Pilsbry, H.A. 1946. Land mollusca of North America (north of Mexico), *Acad. Nat. Sci. Phila. Monogr.*, 3:237–302.

Podzorski, A.C. 1984. *Caponea caribbea* Podzorski, a structurally unique new diatom from Jamaica, *Nova Hedwigia*, 40:1–8.

Podzorski, A.C. 1985. An illustrated and annotated check-list of diatoms from the Black River Waterways, St. Elizabeth, Jamaica, *Bibliotheca Diatomologica*, 7:1–177.

Rader, R.B. 1994. Macroinvertebrates of the northern Everglades: species composition and trophic structure, *Florida Scientist*, 57:22–33.

Reimer, C.W. 1996. Diatoms from some surface waters on Great Abaco Island in the Bahamas (Little Bahama Bank), *Beiheft zu Nova Hedwigia*, 112:343–354.

Ross, M.S., O'Brien, J.J., and Flynn, L.J. 1992. Ecological site classification of Florida Keys terrestrial habitats, *Biotropica*, 24:488–502.

Ross, M.S., Meeder, J.F., Sah, J.P., Ruiz, P.L., and Telesnicki, G.J. 2000. The Southeast Saline Everglades revisited: a half-century of coastal vegetation change, *J. Vegetation Sci.*, 11:101–112.

Ross, M.S., Ruiz, P.L., Telesnicki, G.J., and Meeder, J.F. 2001. Aboveground biomass and production in mangrove communities of Biscayne National Park, Florida (USA), following Hurricane Andrew, *Wetlands Ecol. Manage.*, 9:27–37.

Round, F.E., Crawford, R.M., and Mann, D.G. 1990. *The Diatoms: Biology and Morphology of the Genera*, Cambridge University Press, Cambridge, U.K.

Rutchey, K. and Vilchek, L. 1999. Air photo-interpretation and satellite imagery analysis techniques for mapping cattail coverage in a northern Everglades impoundment, *J. Photogrammetric Eng. Remote Sensing*, 65:185–191.

Schmidt, A. et al. 1874–1959. *Atlas der Diatomaceenkunde*, R. Reisland, Ascherleben.

Sherrod, B.L., Rolins, H.B., and Kennedy. 1989. Subrecent intertidal diatoms from St. Catherines Island, Georgia: Taphonomic complications, *J. Coastal Res.*, 5:665–677.

Slate, J. 1998. Inference of Present and Historical Environmental Conditions in the Everglades with Diatoms and other Siliceous Microfossils, Ph.D. dissertation, University of Louisville, KY.

Stephenson, T.A. and Stephenson, A. 1950. Life between tide-marks in North America: I. The Florida Keys, *J. Ecol.*, 38:354–402.

Sullivan, M.J. 1981. Community structure of diatoms epiphytic on mangroves and *Thalassia* in Bimini Harbour, Bahamas, in *Proceedings of the 6th International Diatom Symposium, Budapest*, Ross, R., Ed., O. Koeltz Publ., Roenigstein, Germany.

Swift, D.R. and Nicholas, R.B. 1987. *Periphyton and Water Quality Relationships in the Everglades Water Conservation Areas*, Tech. Publ. 87-2, South Florida Water Management District, West Palm Beach, FL.

Tabb, D.C. and Manning, R.B. 1961. A checklist of the flora and fauna of northern Florida Bay and adjacent brackish waters of the Florida mainland collected during the period July 1957 through September 1960, *Bull. Mar. Sci.*, 1:550–647.

ter Braak, C.J.F. 1987. Calibration, pp. 78–90, in *Data Analysis in Community and Landscape Ecology*, Jongman, R.H.G., ter Braak, C.J.F., and van Tongeren, O.F.R., Ed., Wageningen, Pudoc.

ter Braak, C.J.F., Juggins, S., Birks, H.J.B., and van der Voet, H. 1993. Weighted averaging partial least squares (WA-PLS): definition and comparison with other methods for species–environmental calibration, in Patil, G.P. and Rao, C.R., Eds., *Multivariate Environmental Statistics*, North-Holland, Amsterdam.

Thompson, F.G. 1984. *The Freshwater Snails of Florida: A Manual for Identification*, University of Florida, Gainesville.

U.S.D.A. 1996. *Soil Survey of Dade County Area, Florida*, U.S. Department of Agriculture, Washington, D.C.

U.S.F.W.S. 1999. *South Florida Multi-Species Recovery Plan*, U.S. Fish and Wildlike Service, Atlanta, GA.

Vos, P.C. and de Wolf, H. 1993. Diatoms as a tool for reconstructing sedimentary environments in coastal wetlands; methodological aspects, *Hydrobiologia*, 269/270:285–296.

Whittaker, R.H. 1960. Vegetation of the Siskiyou Mountains, Oregon and California, *Ecol. Monogr.*, 30:279–338.

Wilbur, K.M. and Yonge, C.M. 1966. *Physiology of Molluska*, Vol. 2. Academic Press, New York.

Witkowski, A., Lange-Bertalot, H., and Metzeltin, D. 2000. Diatom flora of marine coasts, I. *Iconographica Diatomologica*, 7:1–924.

# APPENDIX

**Plant, Diatom, and Mollusk Taxa Collected from the Coastal Transect at Turkey Point, South Florida**

| No./I.D. | Taxa | Distance Optimum | Distance Tolerance | Number of Sites | Mean Abundance |
|---|---|---|---|---|---|
| | **Plants** | | | | |
| RHIMAN | *Rhizophora mangle* L. | 1162 | 1175 | 37 | 8.1 |
| RUPMAR | *Ruppia maritima* L. | 2333 | 435 | 6 | 0.5 |
| AVIGER | *Avicennia germinans* (L.) | 2625 | 995 | 30 | 1.9 |
| LAGRAC | *Laguncularia racemosa* Gaertn. f. | 3150 | 527 | 21 | 3.5 |
| JUNROE | *Juncus roemerianus* Scheele | 3214 | 526 | 19 | 20.0 |
| DISSPI | *Distichilis spicata* (L.) | 3634 | 454 | 20 | 11.0 |
| CONERE | *Conocarpus erecta* L. | 3684 | 200 | 6 | 4.8 |
| CLAJAM | *Cladium jamaicense* Crantz | 3894 | 149 | 5 | 13.5 |
| MUHCAP | *Muhlenbergia capillaris* (Lam.) Trin. | 3900 | —a | 1 | 50.0 |
| TYPDOM | *Typha domingensis* Pers. | 3900 | —a | 1 | 0.5 |
| ELECEL | *Eleocharis cellulosa* Torr. | 4087 | 47 | 2 | 3 |
| | **Diatoms** | | | | |
| 1 | *Achnanthidium delicatulum* Kütz. | 50 | —a | 1 | 0.3 |
| 6 | *Amphora* sp. 02 | 50 | —a | 1 | 0.6 |
| 7 | *Amphora fontinalis* Hust. | 50 | —a | 1 | 6.8 |
| 10 | *Amphora pediculus* (Kütz.) Grun. | 50 | —a | 1 | 10.0 |
| 85 | *Mastogloia lanceolata* Thwaites ex. W. Sm. | 50 | —a | 1 | 6.4 |
| 87 | *Mastogloia pusilla* Grun. | 50 | —a | 1 | 0.4 |
| 128 | *Nitzschia fluminensis* Grun. | 50 | —a | 1 | 0.2 |
| 132 | *Nitzschia capitellata* Hust. | 50 | —a | 1 | 0.3 |
| 144 | *Pleurosigma* sp. 01 | 50 | —a | 1 | 1.2 |
| 56 | *Biddulphia* sp. 01 | 50 | —a | 1 | 0.3 |
| 154 | *Terpsinoë musica* Ehr | 50 | —a | 1 | 0.3 |
| 140 | *Nitzschia wuellerstorfi* Lange-Bert. | 83 | 71 | 2 | 0.8 |
| 70 | *Gomphonema parvulum* Kütz. | 100 | 71 | 2 | 0.3 |
| 55 | *Eunotia flexuosa* (Bréb.) Kütz. | 132 | 144 | 4 | 0.4 |
| 34 | *Cocconeis* sp. 01 | 146 | 88 | 3 | 2.7 |
| 3 | *Achnanthes submarina* Hust. | 149 | 183 | 5 | 1.6 |
| 18 | *Anomeoneis sphaerophora f. costata* (Kütz.) Schmidt | 150 | —a | 1 | 0.3 |

## Plant, Diatom, and Mollusk Taxa Collected from the Coastal Transect at Turkey Point, South Florida

| No./I.D. | Taxa | Distance Optimum | Distance Tolerance | Number of Sites | Mean Abundance |
|---|---|---|---|---|---|
| | **Diatoms** | | | | |
| 36 | Cocconeis scutellum Ehr. | 150 | —a | 1 | 2.6 |
| 93 | Mastogloia splendida (Greg.) Cl. var. 01 | 150 | 82 | 3 | 0.3 |
| 124 | Nitzschia amphibia Grun. var. 02 | 150 | —a | 1 | 1.3 |
| 131 | Nitzschia linearis (Ag. ex. W. Sm.) W. Sm. | 150 | —a | 1 | 2.5 |
| 150 | Navicula scopuloroides Hust. | 150 | 100 | 3 | 2.6 |
| 79 | Mastogloia braunii Grun. | 154 | 103 | 4 | 1.8 |
| 123 | Nitzschia amphibia Grun. var. 01 | 160 | 96 | 3 | 0.9 |
| 77 | Mastogloia barbadensis (Grev.) Cl. | 170 | 137 | 4 | 0.3 |
| 151 | Surirella fastuosa (Ehr) Kütz. | 174 | 79 | 3 | 0.3 |
| 135 | Nitzschia palea var. debilis (Kütz.) Grun. | 196 | 219 | 6 | 5.4 |
| 91 | Mastogloia smithii var. lacustris Thwaites | 200 | 71 | 2 | 8.7 |
| 75 | Mastogloia cf. quinquecostata Grun. | 213 | 147 | 5 | 3.9 |
| 80 | Mastogloia crucicula (Grun.) Cl. | 245 | 143 | 5 | 1.2 |
| 76 | Mastogloia asperuloides Hust. | 250 | —a | 1 | 0.5 |
| 114 | Navicula seminulum Grun. | 250 | 155 | 3 | 0.4 |
| 26 | Caloneis liber (W. Sm.) Cl. | 268 | 144 | 4 | 0.6 |
| 35 | Cocconeis placentula Ehr. var. 01 | 273 | 286 | 8 | 4.5 |
| 96 | Mastogloia sp. 03 | 300 | 70 | 2 | 1.1 |
| 4 | Entomoneis cf. paludosa (W. Sm.) Reimer | 301 | 358 | 7 | 1.0 |
| 19 | Amphora obtusa Greg. | 310 | 241 | 5 | 0.3 |
| 147 | Rhopalodia acuminata Krammer | 335 | 273 | 7 | 1.4 |
| 33 | Caponea caribbea Podz. | 347 | 280 | 10 | 0.8 |
| 9 | Amphora libyca Ehr. | 370 | 292 | 8 | 1.0 |
| 43 | Diploneis puella (Schumann) Cl. var. 01 | 388 | 249 | 7 | 2.8 |
| 152 | Auricula cf. complexa (Greg.) Cl. | 389 | 304 | 10 | 1.5 |
| 39 | Cyclotella striata (Kütz.) Grun. | 450 | 367 | 6 | 0.5 |
| 31 | Oestrupia sp. 02 | 472 | 357 | 8 | 0.5 |
| 143 | Plagiotropis lepidoptera var. proboscidea (Cl.) Reim. | 483 | 227 | 10 | 1.6 |
| 139 | Tryblionella debilis Arnott var. 01 | 546 | 254 | 10 | 0.8 |
| 61 | Catacombas gaillonii (Bory) Williams & Round | 621 | 707 | 2 | 0.4 |

| | Species | | | | |
|---|---|---|---|---|---|
| 16 | *Amphora turgida* Greg. | 627 | 343 | 9 | 0.6 |
| 38 | *Cyclotella meneghiniana* Kütz. | 628 | 519 | 9 | 0.9 |
| 30 | *Oestrupia* sp. 03 | 664 | 453 | 14 | 0.6 |
| 11 | *Amphora proteus* Greg. | 675 | 450 | 10 | 0.6 |
| 138 | *Nitzschia serpentiraphe* Lange-Bert. | 752 | 270 | 8 | 1.4 |
| 119 | *Proschkinia bulheimii* (Grun.) Karayeva | 862 | 658 | 12 | 1.0 |
| 51 | *Encyonema delicatula* Kütz. | 863 | 560 | 28 | 0.3 |
| 129 | *Trybionella graeffii* (Grun. ex. Cl.) Mann | 885 | 321 | 14 | 0.3 |
| 29 | *Caloneis permanga* (Bailey) Cl. | 912 | 451 | 15 | 0.6 |
| 72 | *Gyrosigma nodiferum* (Grun.) Reimer | 948 | 566 | 16 | 0.5 |
| 8 | *Amphora holsatica* Hust. | 1233 | 1034 | 37 | 5.2 |
| 28 | *Caloneis molaris* (Grun.) Krammer | 1250 | —[a] | 1 | 0.4 |
| 97 | *Mastogloia submarginata* Cl. & Grun. | 1320 | 454 | 20 | 0.3 |
| 13 | *Amphora ovalis* (Kütz.) | 1323 | 807 | 24 | 0.9 |
| 71 | *Gomphonema* sp. 01 | 1339 | 669 | 19 | 0.3 |
| 58 | *Synedra filiformis* var. *exilis* Cl.-Eul. | 1380 | 112 | 2 | 10.8 |
| 109 | *Navicula pseudocrassirostris* Hust. | 1432 | 1239 | 13 | 1.7 |
| 117 | *Navicula johannrossii* Giffen | 1498 | 407 | 15 | 8.2 |
| 67 | *Scoliopleura* sp. 01 | 1512 | 1153 | 4 | 0.9 |
| 120 | *Navicula subhamulata* Grun. var. 01 | 1526 | 660 | 27 | 0.4 |
| 104 | *Navicula digitoradiata* (Greg.) Ralfs. | 1527 | 459 | 7 | 0.8 |
| 15 | *Amphora tetragibba* Cl. var. 01 | 1539 | 1043 | 23 | 1.1 |
| 142 | *Pinnularia* sp. 01 | 1541 | 820 | 32 | 1.8 |
| 20 | *Brachysira* sp. 01 | 1561 | 463 | 15 | 1.3 |
| 111 | *Navicula ramossissima* (Ag.) Cl. var. 01 | 1580 | 1050 | 41 | 0.4 |
| 95 | *Mastogloia gibbosa* Brun. | 1606 | 903 | 33 | 0.5 |
| 66 | *Frustulia rhomboides* Hust. var. 01 | 1607 | 918 | 29 | 0.4 |
| 94 | *Mastogloia* sp. 01 | 1630 | 729 | 10 | 0.4 |
| 122 | *Nitzschia amphibia* Grun. | 1639 | 714 | 11 | 2.4 |
| 126 | *Tryblionella punctata* W. Sm. | 1687 | 963 | 34 | 0.3 |
| 136 | *Psammodictyon mediterraneum* (Hust.) Mann | 1710 | 933 | 30 | 1.4 |
| 149 | *Rossia linearis* Voigt. | 1727 | 839 | 34 | 0.5 |
| 42 | *Diploneis parma* Cl. var. 01 | 1736 | 615 | 30 | 0.9 |
| 112 | *Navicula rhynchocephala* Kütz. var. 01 | 1775 | 373 | 8 | 13.4 |
| 153 | *Diatom* sp. 01 | 1777 | 624 | 12 | 1.7 |
| 110 | *Fallacia pygmaea* (Kütz.) Stickle & Mann | 1850 | —[a] | 1 | 0.5 |

# Plant, Diatom, and Mollusk Taxa Collected from the Coastal Transect at Turkey Point, South Florida

| No./I.D. | Taxa | Distance Optimum | Distance Tolerance | Number of Sites | Mean Abundance |
|---|---|---|---|---|---|
| | **Diatoms** | | | | |
| 54 | Encyonema sp. 02 | 1950 | 0 | 1 | 0.4 |
| 57 | Fragilaria exigua (Grun.) var. 01 | 1970 | 595 | 27 | 2.9 |
| 82 | Mastogloia elliptica (Ag.) Cl. | 2010 | 840 | 24 | 4.1 |
| 148 | Rhopalodia acuminata Krammer var. 01 | 2026 | 1028 | 12 | 1.4 |
| 17 | Amphora veneta Kütz. | 2037 | 1273 | 38 | 2.4 |
| 21 | Brachysira aponina Kütz. | 2320 | 947 | 28 | 1.7 |
| 78 | Mastogloia binotata (Grun.) Cl. | 2329 | 797 | 25 | 0.4 |
| 14 | Amphora sp. 01 | 2349 | 539 | 21 | 2.6 |
| 121 | Navicula trivialis Lange-Bert. | 2433 | 673 | 30 | 4.2 |
| 52 | Seminavis sp. 01 | 2472 | 535 | 19 | 9.3 |
| 5 | Amphora caroliniana Giffen | 2489 | 1037 | 26 | 1.1 |
| 108 | Navicula molestiformis Hust. | 2506 | 788 | 30 | 0.3 |
| 60 | Fragilaria sp. 01 | 2526 | 585 | 4 | 2.7 |
| 27 | Caloneis linearis (Grun.) Boyer | 2550 | 141 | 2 | 0.4 |
| 99 | Navicula angusta Grun. | 2650 | —[a] | 1 | 0.7 |
| 46 | Encyonema vulgare Krammer var. 01 | 2658 | 1450 | 11 | 1.6 |
| 59 | Synedra filiformis Grun. var. 01 | 2673 | 708 | 25 | 4.9 |
| 74 | Hantzschia sp. 01 | 2682 | 838 | 33 | 0.3 |
| 118 | Navicula sp. 03 | 2711 | 562 | 9 | 1.5 |
| 40 | Diploneis interrupta (Kütz.) Cl. | 2762 | 800 | 27 | 0.4 |
| 83 | Mastogloia erythraea v. grunowi Foged | 2841 | 532 | 8 | 2.2 |
| 103 | Navicula pseudoarvensis Hust. | 2842 | 118 | 3 | 0.4 |
| 24 | Caloneis clevei (Lager.) Cl. | 2850 | —[a] | 1 | 0.4 |
| 65 | Fragilaria lanceolata (Kütz.) Reich. | 2850 | 848 | 2 | 1.6 |
| 89 | Mastogloia pusilla Grun. var. 01 | 2850 | —[a] | 1 | 0.4 |
| 88 | Mastogloia punctifera Brun. | 2973 | 273 | 3 | 0.3 |
| 2 | Achnanthes jamaicense Podz. var. 01 | 2983 | 70 | 2 | 0.5 |
| 146 | Rhopalodia gibba (Ehr.) O. Müll. | 3025 | 739 | 17 | 0.3 |
| 37 | Hyalodiscus sp. 01 | 3043 | 567 | 13 | 0.3 |
| 73 | Gyrosigma sp. 01 | 3050 | —[a] | 1 | 1.0 |
| 116 | Navicula sp. 01 | 3068 | 625 | 33 | 0.3 |

| 93 | *Mastogloia undulata* Grun. var. 01 | 3098 | 716 | 8 | 0.6 |
|----|----|----|----|----|----|
| 101 | *Navicula cryptocephala* Hust. | 3131 | 189 | 5 | 0.4 |
| 63 | *Fragilaria synegrotesca* Lange-Bert. | 3150 | —[a] | 1 | 2.8 |
| 81 | *Mastogloia decipiens* Hust. | 3150 | —[a] | 1 | 0.7 |
| 86 | *Mastogloia paradoxa* Grun. | 3150 | —[a] | 1 | 0.6 |
| 113 | *Navicula rhynchocephala* Kütz. var. 02 | 3150 | 494 | 2 | 0.8 |
| 127 | *Nitzschia dissipata* (Kütz.) Grun. | 3150 | —[a] | 1 | 3.9 |
| 68 | *Gomphonema clavatum* Ehr. | 3179 | 249 | 4 | 0.4 |
| 44 | *Diploneis suborbicularis* (Greg.) Cl. | 3196 | 1074 | 38 | 0.4 |
| 64 | *Tabularia* sp. 01 | 3201 | 246 | 3 | 0.2 |
| 25 | *Caloneis westii* (W. Sm.) Hendey | 3250 | —[a] | 1 | 0.5 |
| 105 | *Navicula durrenbergiana* Hust. var. 01 | 3250 | —[a] | 1 | 0.4 |
| 62 | *Neosynedra provincialis* (Grun.) Williams & Round | 3300 | 490 | 18 | 0.4 |
| 100 | *Navicula complanatoides* Hust. | 3350 | —[a] | 1 | 0.4 |
| 133 | *Nitzschia lorenziana* Grun. | 3350 | —[a] | 1 | 0.3 |
| 130 | *Tryblionella granulata* (Grun.) Mann | 3460 | 876 | 25 | 1.3 |
| 53 | *Seminavis* sp. 02 | 3468 | 146 | 3 | 3.1 |
| 32 | *Campylodiscus* sp. 01 | 3480 | 235 | 7 | 0.5 |
| 84 | *Mastogloia fallax* Cl. var. 01 | 3538 | 220 | 10 | 0.6 |
| 106 | *Navicula gallica* var. *perpusilla* (Grun.) Lange-Bert. | 3550 | —[a] | 1 | 0.5 |
| 125 | *Nitzschia capitellata* Hust. | 3550 | —[a] | 1 | 0.3 |
| 49 | *Encyonema silesiacum* (Bleisch) Mann | 3574 | 355 | 6 | 0.6 |
| 69 | *Gomphonema insigne* Greg. | 3650 | —[a] | 1 | 0.4 |
| 98 | *Melosira* sp. 01 | 3650 | —[a] | 1 | 1.2 |
| 107 | *Navicula jaagii* Meist. | 3700 | 70 | 2 | 0.3 |
| 50 | *Encyonema* sp. 01 | 3750 | 100 | 3 | 0.8 |
| 48 | *Encyonema paucistriata* (Cl.-Eul.) Mann | 3753 | 170 | 5 | 0.3 |
| 12 | *Amphora* sp. 03 | 3774 | 233 | 9 | 3.4 |
| 41 | *Diploneis oblongella* (Naegeli) Cl.-Eul. | 3820 | 279 | 9 | 1.7 |
| 22 | *Brachysira neoexilis* Lange-Bert. | 3847 | 360 | 10 | 1.9 |
| 134 | *Nitzschia obtusa* W. Sm. | 3876 | 214 | 7 | 0.4 |
| 90 | *Mastogloia rigida* Hust. | 3927 | 215 | 9 | 1.4 |
| 91 | *Mastogloia smithii* var *lacustris* Grun. | 3927 | 215 | 9 | 8.8 |
| 23 | *Brachysira neoexilis* Lange-Bert. var. 01 | 3933 | 157 | 5 | 2.0 |
| 45 | *Encyonema evergladianum* Kr. | 3953 | 236 | 7 | 5.9 |
| 137 | *Nitzschia rosenstockii* Lange-Bert. | 3954 | 186 | 5 | 1.3 |

**Plant, Diatom, and Mollusk Taxa Collected from the Coastal Transect at Turkey Point, South Florida**

| No./I.D. | Taxa | Distance Optimum | Distance Tolerance | Number of Sites | Mean Abundance |
|---|---|---|---|---|---|
| **Diatoms** | | | | | |
| 102 | Navicula diffluens A. Schm. | 4020 | —[a] | 1 | 1.3 |
| 47 | Encyonema microcephala Gr. | 4120 | —[a] | 1 | 0.5 |
| 115 | Navicula schroeterii Meist. | 4120 | —[a] | 1 | 0.4 |
| 141 | Opephora sp. 01 | 4120 | —[a] | 1 | 0.5 |
| 145 | Pleurosigma affine Grun. | 4120 | —[a] | 1 | 0.6 |
| **Invertebrates** | | | | | |
| TRINIG | Triphora cf. nigrocincta (C. B. Adams) | 168 | 127 | 4 | 7.5 |
| CALADE | Calliostoma cf. adelae Schwengel | 515 | 453 | 7 | 5.9 |
| PENPLA | Peneroplis planatus (Fichter & Moll) | 768 | 525 | 16 | 13.8 |
| CERLIT | Cerithium litteratum (Born) | 794 | 440 | 10 | 5.6 |
| ACTCAN | Acteocina canaliculata (Say) | 933 | 621 | 18 | 7.5 |
| RISCAT | Rissoina catesbyana Orbigny | 982 | 557 | 15 | 5.9 |
| MELCOF | Melampus coffeus Linne | 1192 | 572 | 6 | 5.0 |
| CORCON | Corbula contracta Say | 1276 | 716 | 23 | 7.5 |
| TELSPP | Tellina sp. 01 | 1516 | 850 | 31 | 13.6 |
| CERBEA | Cerithidea beattyi Bequaert | 1874 | 988 | 34 | 19.0 |
| BIOHAV | Biomphalaria havanensis (Pfeiffer) | 1898 | 876 | 21 | 6.1 |
| POMPAL | Pomaceae palundosa (Say) | 2155 | 479 | 16 | 7.3 |
| POLSPP | Polygrya sp. 01 | 2312 | 895 | 8 | 5.5 |
| CYRFLO | Cyrenoida floridana (Dall) | 2360 | 1093 | 30 | 13.7 |
| LITMON | Littoridinops monroensis (Fravenfeld) | 2365 | 1124 | 36 | 18.5 |
| PLASCA | Planorbella scalaris (Jay) | 2389 | 533 | 11 | 6.0 |
| PHYCUB | Physella cubensis (Pfeiffer) | 2604 | 1362 | 33 | 12.5 |

*Note:* Taxa are ordered by distance optima (in meters from coast) calculated by weighted averaging. Also given for each taxon is its distance tolerance, number of occurrences in the 41 sites along the transect, and mean abundance when present.

[a] Tolerance not calculated for taxa present at only one location.

# 8 Modeling Ecosystem and Population Dynamics on the South Florida Hydroscape

*Donald L. DeAngelis*
U.S. Geological Survey, Biological Resources Division

*Sarah Bellmund*
Biscayne National Park

*Wolf M. Mooij*
Netherlands Institute of Ecology, Centre for Limnology

*M. Phillip Nott, E. Jane Comiskey, and Louis J. Gross*
Department of Ecology and Evolutionary Biology, The University of Tennessee

*Michael A. Huston*
Oak Ridge National Laboratory

*Wilfried F. Wolff*
ICG6 Forschungszentrum Juelich

## CONTENTS

0-8493-2026-7/02/$0.00+$1.50
© 2002 by CRC Press LLC

# INTRODUCTION

South Florida ecosystems have been losing ground to development and drainage for the past hundred years. Large-scale development in South Florida has always been premised on draining the swamp to "reclaim" land for purposes more "useful" to humans. After a century of such human interference in the natural processes of the Everglades, a large-scale restoration effort for the ecosystems of South Florida is now underway. This restoration of the natural systems of South Florida can only be accomplished with a thorough knowledge of how changes in management will alter environmental factors such as hydrology and how these factors affect the biotic components of the ecosystem.

It is no small task to predict (or, more properly, "project;" see below) how modifications in the structure and operations of dikes, levees, canals, pumps, and gates will alter water levels through the greater Everglades system. Nonetheless, hydrologists have developed large and complex landscape hydrology models, with which they can evaluate the effects of changes with impressive accuracy. What about the effects on the biotic community, however, which is the real object of the restoration? Ecologists are traditionally not as well prepared to make specific, detailed predictions about how a system will respond to given external influences. Of course, ecologists have a wealth of knowledge about the physiology and behaviors of animals and plants, and Everglades ecologists have a general understanding of how changes in both the extent of natural habitat and the abiotic factors across the landscape affect their abilities to survive and reproduce. However, it is not easy from such traditional qualitative information to make quantitative projections necessary to compare plans whose effects on the landscape may differ only in subtle ways.

We describe here a spatially explicit modeling approach to projecting the effects of different hydrologic conditions on the biotic community. We use the term "projection" here, because "prediction" might be taken to imply that we can forecast climatic and other external factors and that our ecological models contain precise relationships between these environmental factors and effects on populations. Unfortunately, neither is true. Therefore, rather than "predict" what will happen, we can make projections of how an ecological system may react to a given scenario of environmental conditions. However, the spatially explicit approach, by including as much realism regarding the landscape as is possible and relevant, has the goal of discriminating the effects of subtle changes in the same way that landscape hydrologic models can discriminate the differences in water depth caused by differences in water regulation.

A detailed landscape model of the entire Everglades and all of its components is clearly impossible. Neither the empirical data nor the computer capability exist for such a model. However, by appropriate focus on key aspects of the ecosystem, working models have been developed and applied to restoration questions. Our goal here is to outline the general ideas and some specifics of what has been done.

## THE SETTING

The historical Everglades covered an area of approximately 10,500 km². This area extended south from Lake Okeechobee over 200 km and from east to west about 80 km at the widest point (Gunderson and Loftus, 1993). This area has extremely low topographic relief, with a slope of approximately 3 cm/km.

Despite the flatness of the terrain, the Everglades shares many of the complexities of other landscapes. In fact, the region of southern Florida encompassed by Everglades National Park and Big Cypress National Preserve, which we will call the "greater Everglades" or just Everglades here, is a mosaic of many types of wetland habitats (e.g., cypress swamp, marl prairie, deep slough, mangrove estuary), as well as some upland habitats (e.g., pine flatwoods). The very flatness of the landscape means that minute differences in elevation become critical to species and communities. More than 80 vegetation communities are defined on geographic information systems (GIS) vegetation maps of the region, indicating the great vegetative heterogeneity of the Everglades.

Temporally, as well, the Everglades landscape is complex. In addition to its spatial heterogeneity, the landscape is marked by a distinct seasonality, with much of the landscape alternately flooding and drying during a year. The annual hydrologic cycle is a dominant driving force for the biotic community.

## THE PROBLEM

During the past several decades, human impact has caused changes in land use and in the hydrologic patterns in South Florida. Concomitant with these changes have been precipitous declines in the abundances of some species, such as wading birds, Florida panthers, and Cape Sable seaside sparrows. The close dependence of many biota of southern Florida on the natural hydrologic patterns and the limited ranges of some of these species raise concerns that such declines may continue.

Much of the change in hydrologic pattern has occurred through the creation of a network of levees and canals during the middle part of this century. While these water regulation structures are used for flood control and for providing water for agriculture and urban areas, their use has resulted in a net loss of water to the southern Everglades. In addition, water flows have been diverted away from the populated eastern parts of the Everglades, causing a drying of the valuable high-elevation, short-hydroperiod wetlands in that area. Water tends to be artificially ponded in some areas, such as the Water Conservation Areas (WCAs) to the north of Everglades National Park, later to be released in pulses to the park when levels become too high in these areas. These pulsed releases can cause large "reversals" of water depths in the areas of release, often harming animals whose seasonal cycles are adapted to more gradual changes in water levels. Figure 8.1 shows the main causal factors of human development, protection of water supply, and flood control and the actions these have entailed in southern Florida.

Thus, the amounts, spatial patterns, and timing of water flows have been altered from their natural ranges. The effects of the changed patterns of water on wildlife have been enormous. Many

# Human Impacts to the Landscape

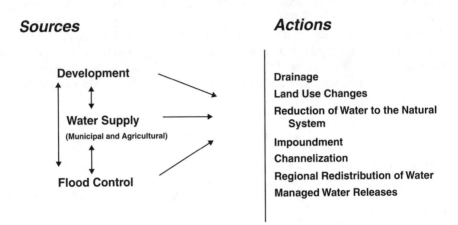

**FIGURE 8.1**    The basic causes, or sources, of human alteration of the landscape of southern Florida and the major actions that have been undertaken.

species depend on seasonal changes in water that were predictable under natural conditions. For wading birds, the gradual drop in water levels during the dry season concentrates fish prey at a rate appropriate for foraging. For the Cape Sable seaside sparrow, this drop in water level dries some areas to make nesting possible in marl prairie habitat. Alligators build their nests when water levels are falling, anticipating that there will be neither so fast a decline in water levels that nest desiccation occurs nor reversals of this decline that could flood nests and disperse prey.

Water regulation has changed this picture. Some areas of the Everglades now dry out more frequently than under natural conditions. Other areas are ponded for periods much longer than natural, so that vegetation types are changed. In addition, water is often released from upstream at inopportune times, causing water levels to rise when they should be falling, thus flooding nests of some species or making prey more difficult to catch by wading birds and alligators. Figure 8.2 shows how the anthropogenic actions have affected the landscape of southern Florida and the effects on some of the biota.

## WHAT CONSTITUTES RESTORATION?

Because of public protests over the perceived decline of the Everglades, in 1992 Congress authorized a Central and South Florida Project Comprehensive Review Study (C&SF Restudy) to study the feasibility of modifying the current water regulation structures (levees, dikes, and canals) and operations to provide benefits for the native biotic community of this region. In 1996, Vice President Gore announced a $1.5 billion initiative for rehabilitation of the Everglades and its surrounding areas, the "greater Everglades ecosystem," which includes Florida Bay, Biscayne Bay, Big Cypress National Preserve, and other areas.

Much of South Florida is occupied by urban and agricultural areas, so the landscape cannot be returned to its natural state. Restoration planners must try to determine first what changes are feasible. Feasible changes include restoring patterns of water flow and hydroperiod somewhat closer to natural, adding back some land or buffer land, and imposing quality standards on water flowing from agricultural lands to the north. Planners must then evaluate and compare among the feasible changes to determine which will do the most to restore the Everglades to its former biotic richness.

It is popular to say that the aim of restoration must be to restore the function of an ecosystem, though it is often less easy to say exactly what "function" means. One aspect of ecosystem

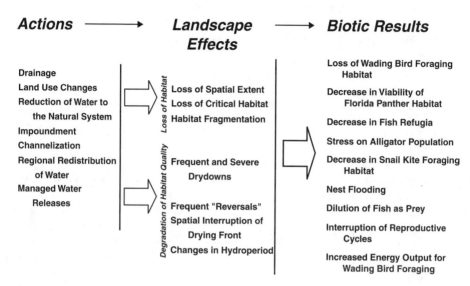

**FIGURE 8.2** Human actions have landscape effects and these, in turn, have effects on the biota; some of the most important effects on the landscape and biota are listed.

"function" is certainly the maintenance of an appropriate level of biological diversity. In particular, one measurement of the restoration of an ecosystem is the degree to which the system can support the same species in the same numbers that the natural system supported. Not enough scientific records exist to estimate reliably the numbers of most of the populations of the Everglades, and this is likely to be the case for any ecosystem restoration project. Identification of the community of species that were at least present in the natural system is somewhat easier. In particular, for the Everglades region there are a number of species that are either unique to or otherwise characteristic of the natural system. We have defined a subset of species that are representative of the variety of higher trophic level components of the natural biota and are critical in other ways, either by being on the list of threatened or endangered species or by being links in whole ecosystem function. These include the following:

- *Suite of wading birds (wood storks, white ibises, great blue herons, great egrets):* These birds integrate over a large part of the landscape and are dependent on a high production of fish and aquatic macroinvertebrates.
- *Cape sable seaside sparrow:* These birds are landscape specialists, requiring sufficient marl prairie habitat maintained by periodic fires and proper hydrologic conditions.
- *Snail kite:* These are nomadic prey specialists, depending on a spatially broad set of habitat sites but on one prey resource, the apple snail.
- *American alligator:* This is a key species in the Everglades, which, by maintaining ponds, provides refugia for many aquatic species during dry periods.
- *Florida panther:* The Florida panther is endangered, with currently fewer than 50 adults. It is a landscape generalist and depends heavily on white-tailed deer for food.

These diverse species are top consumers or have specialized diet or habitat demands. If, under a given restoration plan, all these species can be predicted to be maintained at viable levels within the Everglades ecosystem, this would be a strong indication that the restoration plan is effective in restoring ecosystem function.

This leads naturally to the question of whether accurate projections can be made, and how. An approach to the computer simulation of the Everglades ecosystem called Across-Trophic-Level System Simulation (ATLSS), has been developed with the task of assisting in the comparison of restoration plans. Below, we describe the general concepts of this modeling approach and then the details of specific models within the approach.

## METHODS

### CONCEPTUAL BASIS OF A MODELING APPROACH FOR THE EVERGLADES LANDSCAPE

The feasibility of using modeling to make useful projections about a system depends on a number of things. It depends, first, on knowing the key factors and relationships operating in a system. Second, these factors and relationships must be quantifiable to some degree. Third, one must use a modeling approach that is able to incorporate these key factors and relationships. When these conditions are met, a model can be constructed that can be used to make projections of how the system may respond under a specified set of conditions. The ATLSS program has as its goal the modeling of a set of key species of the greater Everglades region, including those mentioned in the preceding section, as well as the relevant aspects of their environment. The ATLSS models are designed to relate the details of a changing environment to possible changes in population numbers and spatial distributions. In this section, we outline the premises in this modeling approach.

Projecting the population demographics of each of the species populations mentioned above depends on accurately representing several key factors and relationships, as listed below:

1. The models must have an accurate description of the complex, dynamic changing landscape that these populations inhabit. Water levels, prey abundances, and other relevant factors vary at many spatial and temporal levels of resolution.
2. The models must incorporate physiological, behavioral, and life-history adaptations of the species. These are the attributes that allow individuals of these species to extract energy and nutrients from this landscape and to reproduce at a sufficient rate to balance mortality. These attributes may or may not be sufficient to cope with changes to the environment.
3. The models must be able to take into account variability within populations. Often, only a small part of a population survives and successfully reproduces, so the characteristics of these successful individuals should be represented.
4. Stochasticity, in both its environmental and demographic forms, plays an important role in small populations in limited areas, such as we deal with in the Everglades landscape.

Any modeling approach, to successfully answer the question of how a species will react to changes in the landscape, must include details of the behaviors and energetic constraints of individuals and how these relate to a given landscape that is dynamically changing. For example, because of the heterogeneity of habitat types and water depths on the Everglades landscape, primary food sources for wading birds, such as fish, become available to these birds in patchy spatial patterns, which can change rapidly through time. The relevant question then is, can wading birds foraging from a breeding colony successfully locate and bring back sufficient food at a continuous rate to raise their offspring to fledging? To answer this, knowing only the average production of fish across the landscape is not sufficient. One must also know the changing spatial configuration of availability of these fish, as well as the physiological and behavioral capabilities of the birds for exploiting this resource.

The causal linkages outlined above cannot be delineated by simple, aggregated models, such as the standard predator–prey models of mathematical ecology. The abstract models of mathematical ecology may at best offer only general guidelines, whereas what are needed are highly specific projections of what will happen under various, possibly subtle changes in the landscape. Useful answers can come only from detailed, spatially explicit simulation models that can incorporate the hydrologic model projections and take into account landscape heterogeneity and configuration at a resolution at which the animals and plants respond to it. In particular, one must simulate as accurately as possible how the individuals of the population use their time and energy budgets and repertoires of behaviors to feed themselves and bring back food for their offspring.

Because stochasticity is inherent and often extremely important in the small populations that characterize the Everglades, stochasticity must be built into the major activities (e.g., movement, choice of foraging sites, death) of individuals modeled. This is done technically through a pseudo-random number generator. This means that any particular simulation will give as a result only one possible outcome of a model. One can certainly not assume that this will be a precise prediction of the dynamics of the actual system. Although, as discussed earlier, particular outcomes cannot be predicted, what can be projected is a probability distribution of outcomes for a population. To obtain a probability distribution of possible outcomes, one can run the simulation model many times in a Monte Carlo manner, using the same model parameter values, the same starting conditions on the population, and the same scenario of environmental driving forces (pattern of landscape hydrology through time in the case of the ATLSS Everglades models), but with different pseudo-random number generator seeds. A probability distribution emerges for variables such as population size through time, representing the effects of demographic stochasticity.

## THE MODELING TOOLS

We have developed a combination of different simulation techniques to describe the components of the Everglades system (see DeAngelis et al., 1998 for an overview). These are described in detail at the website atlss.org. We use spatially explicit, individual-based (SEIB) models to describe the population dynamics of the set of species listed above. By *individual-based* we mean that each individual is followed through the period of time of the simulation. A simulated time period can run for a few months (e.g., a species' breeding season) or for decades. In the latter case, the model might follow many individuals through their entire lifetimes. All individuals present in the population are simulated, so they can interact directly (e.g., mating or aggression) and indirectly (e.g., exploitation of prey, using up of territories). By *spatially explicit*, we mean that individuals move on a landscape. In the ATLSS models, the details of the landscape are provided by "layers" — for example, a GIS vegetation-type map that has resolution down to 28.5 m, a hydrologic model that describes water depth on a day-by-day basis at points on a $500 \times 500$ m grid (or finer resolution, where necessary), and maps of land-use type, roads, etc. across the Everglades region.

The ATLSS models are currently designed for use in assessment of the effects on key biota of alternative water plans for regulation of water flow across the Everglades landscape. In making specific projections for such issues there is no alternative to detailed site-specific ecological modeling — in particular, the use of SEIB models for predicting how changing landscape patterns of hydrology will affect populations.

The SEIB models are expressly designed to perform detailed simulations for populations of individuals. These models assign to each individual a set of variables that may include sex, age, weight, condition, spatial location, social status (immature, breeding adult, nonbreeding adult, etc.), current activity, and whatever other variables are important to a particular problem. Some variables, such as age, change deterministically through time; however, most processes, such as day-to-day movement patterns, foraging success, reproduction, and survival are probabilistic, so the associated variables are random variables. Some processes are described by equations — for example, the weight and energetics of an individual. Others, such as whether the individual is breeding or not, whether or not and where it is foraging, etc., are outcomes of sets of behavioral "rules" that describe either deterministically or probabilistically what an animal will do in a given situation.

Each SEIB model for a particular Everglades population is linked, through the responses of individuals to habitat types and water depths, to the dynamic Everglades landscape. In addition to their dependence on these landscape characteristics, however, the key species also depend on a changing resource base on the landscape, including vegetative biomass, lower trophic level invertebrates, and decomposers. Models are thus included, where appropriate, for main forage components. Wading birds, for example, are coupled to a spatially explicit model of a size-structured fish population (see DeAngelis et al., 1997). White-tailed deer are coupled to a model of seasonally varying vegetation dynamics.

## RESULTS

The models of individual species are only briefly described below. The object here is to describe the context in which each was developed, the particular aspects of the life history encompassed by the model, the level of detail of description of the individuals and their environment, some of the assumptions of the model that have proven to be crucial, and finally some model results. Each of these models is separate but utilizes the same landscape layers of topography, hydrology, and vegetation described above. Thus, all of the models are able to respond to identical environmental scenarios.

### WADING BIRDS

#### Context of Model

The numbers and breeding success of wading birds have declined about 90% in the southern Everglades during the past few decades (Ogden, 1994). Species such as the wood stork (*Mycteria americana*) and the white ibis (*Eudocimus albus*) depend on high densities of fish and aquatic macroinvertebrates during their breeding season. In the natural Everglades system, such concentrations were produced during the dry season by the gradually receding waters. There have been at least two primary changes to the post-drainage landscape that can severely affect wading birds. First, there has been a loss, to agriculture and urban development, of much of the higher elevation land of the eastern Everglades that dries out early in the dry season, providing concentrated prey for early nesting. Second, regulatory releases of water from upstream during the dry season often cause "reversals," or upward fluctuations in water depth during the normal period of recession, diluting fish prey during the wading birds' breeding season and possibly causing loss of their offspring.

#### Scope of the Model

The objective of the model is to compare various hydrologic scenarios for the Everglades, reflecting different structures and operations for water regulation, and their effect on the success of wading bird breeding colonies. The birds modeled include the wood stork, the white ibis, the great egret (*Casmerodius albus*), and the great blue heron (*Ardea herodias*). The model, described in detail by Wolff (1994) and Fleming et al. (1994), follows an entire breeding colony, from the inception of nesting through the raising of offspring to fledging. The main components of the model are the foraging behavior of the adults, the feeding of their offspring, and the energetics of adults and nestlings. The model simulates all of the individual adults in the colony, assumed to have mates, and all of their offspring. The basic time interval is 15 minutes, allowing high temporal resolution in following the search of individual birds for foraging sites with adequate prey density, capturing a certain amount of prey biomass, and bringing it back to the nest. Many hundreds or thousands of birds can be simulated simultaneously, and interactions among individuals such as exploitative competition and flocking behavior are simulated. This level of detail is essential if one is to determine how subtle changes in the landscape affect the individual's ability to get food and hence to raise its nestlings to the fledgling stage.

The model includes a landscape and food base, assumed here to be fish (a crayfish component will be added later), the dynamics of which are modeled on 5-day time steps. The landscape is divided into $500 \times 500$ m cells, each with its own elevation, hydroperiod, and prey dynamics. When a cell is flooded, fish populations build up. Fish can move between cells to some degree, and some are able to retreat into refugia, such as alligator ponds, when cells dry out. Wading birds can utilize the fish as prey when water levels drop to less than a depth of about 35 cm during the dry season, the depth range at which most wading birds can start to feed. If the fish are available in high enough densities, the wading birds can quickly capture large numbers to take back to the nest. Effects of exploitation by wading birds of the fish in spatial cells are taken into account in the model. Figure 8.3a shows some of the main activities on the scale of a breeding season, while Figure 8.3b illustrates the basic logic of foraging within a day, as simulated in the model.

## Crucial Model Assumptions

The behaviors and physiological constraints on birds in this model require many rules, including what times of day they can forage, maximum flight speed, maximum foraging distance, amount of food that can be carried back to the nest, tendency to join other birds already feeding at a site, and depth to which they can forage. There are more than 200 rules, making this a relatively complex model. Wolff's (1994) study shows that the rule governing when to start nesting behavior is a highly important one. Wolff assumes that the readiness to initiate reproductive behavior is signaled by seasonal cues, but that the birds must feed to satiation for at least a week before they will actually

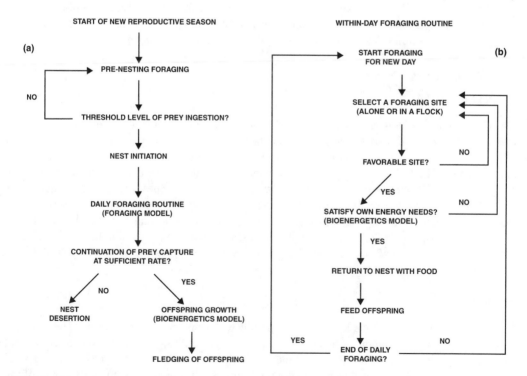

**FIGURE 8.3**   Schematics of some of the main components of the wading bird model. (a) Events on the scale of a breeding season; nesting is assumed to be triggered early in the dry season when wading birds ingest amounts of prey greater than a threshold over a period of a week. (b) The main components of model wading bird foraging within a day.

start nesting. This rule is important because it takes approximately 16 weeks for a new egg to reach the fledgling stage. A bird that delays initiation of nesting runs the risk of not having produced fledglings before the start of the next wet season, at which point prey are diluted by the higher water and become more difficult to find. Another rule that can have some significant consequences under food shortages is the order in which pairs feed their offspring. Feeding the largest (usually the oldest) nestling preferentially increases the chance that, under limited food conditions, a given nest produces at least one fledgling.

## Some Results

The model has been used to explicitly test the hypothesis that the loss of certain types of habitat would affect the success of a breeding colony (described in Fleming et al., 1994). The southern Everglades has lost substantial amounts of higher elevation marsh that was flooded for fewer than nine months of the year. This short-hydroperiod area typically provides wading birds with high concentrations of fish early in the dry season that can trigger the start of nesting. To test this hypothesis, Wolff (1994) modeled a colony of 250 individuals, located in the center of a landscape with a gentle slope, so that a spectrum of elevations was included. Wolff removed from the model various fractions of this higher elevation area (0%, 5%, 10%, 15%). For each of these four cases, 50 replicate runs were performed. A loss of 5% areal extent of short-hydroperiod wetland did not cause much change in the overall pattern, although it increased the variability in timing of the colony formation. Success rate of fledglings was still above 90% (724 out of a potential 750). Loss of 10% areal extent led to highly variable results, although with an average delay of 63 days in colony formation and a mean number of fledglings of only 162. A loss of 15% led to very small fledgling success in all replicate simulations. Thus, losses of relatively small amounts of this specific habitat appear capable of causing significant delays in the initiation of breeding and greatly reduced fledging success, a trend that has been noted in the empirical data (Ogden, 1994).

## CAPE SABLE SEASIDE SPARROW

### Context of Model

The Cape Sable seaside sparrow (*Ammodramus maritimus mirabilis*) is a race of seaside sparrow that is confined to the marl prairies of the southern Everglades. The population consisted, until 1993, of two major subpopulations of about 3000 individuals each and a few much smaller subpopulations. In 1993, one of the large subpopulations, situated to the west of Shark Slough, suffered a major decline, from which it has not recovered. This appears to be a consequence of several years of flooding of the sparrow's breeding habitat during the reproductive season. This flooding interrupted or totally prevented the breeding of the ground-nesting sparrows in this western subpopulation. The high water was ultimately due to greater than average rainfall, but during some periods it was exacerbated by regulatory releases of water through structures (S12s) north of the breeding grounds.

### Scope of the Model

The ATLSS model for the Cape Sable seaside sparrow was developed by Nott (1998). The objective of the model is to help predict whether proposed changes in the water regulation structures and operations upstream of the Cape Sable seaside sparrow population will tend to increase or decrease the vulnerability of the sparrow to flooding in the future. Because nesting is the critical part of the life cycle affected by water, a SEIB model was developed that focuses on nesting behavior. Studies of the sparrow in recent years have built up a detailed knowledge of the sparrow's life cycle, the types of vegetation used in breeding, breeding behavior and phenology, and the effects of water level on nesting. The other essential element is accurate knowledge of landscape vegetation, topography, and water levels through time. Fine-scale vegetation and topography data have been collected in this sparrow's range, and output from the South Florida Water Management Model is adapted to provide water depths on a daily basis on 500 × 500 m cells.

Each individual sparrow is modeled in detail during the reproductive period. Attributes of age, sex, and breeding status are kept track of day by day. When the breeding season starts, males choose territories of the appropriate size and habitat type in locations where water levels have receded sufficiently. Females look for unattached males. This is followed by egg laying and raising young. The adults and nestlings are exposed to mortality, which is modeled as daily probabilities, as well as to loss of the whole nest if water levels rise during nesting. Up to three broods can be raised if the nesting site remains dry for more than 80 days. During the non-nesting season, an age-specific probability of mortality is imposed. The primary activities of the breeding season are shown in Figure 8.4.

## Crucial Model Assumptions

The projections of the model are sensitive to the response of the sparrow to water levels. It is assumed that sparrows will not start nesting until water level has gone down to less than 5 cm above the ground. Once nesting has started, a rise of water levels to 15 cm during the breeding

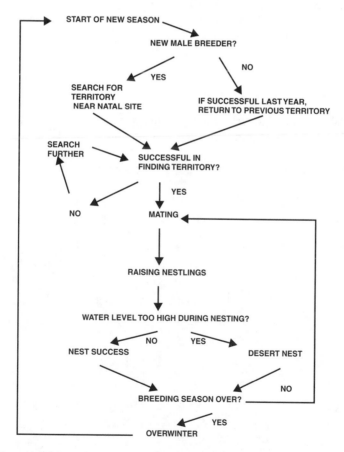

**FIGURE 8.4**  Schematic of the main components of a male Cape Sable sparrow during the breeding season. If the male has not nested before, it chooses a site near its natal site if water level is low enough. If it has nested successfully the preceding year, it chooses a site at or near the site it last used. Mating is assumed to occur within 10 days after the male claims a territory. A nest will be abandoned if flooded during the nesting period.

season is assumed to flood the nest. Clearly, then, the model is sensitive to the topography of the landscape. Another important rule is that concerning male choice of nesting sites. The model assumes that: (1) males nesting for the first time choose an available suitable site as close to their natal site as possible, and (2) males that have nested before return to the same site the next year if they were successful or move to a nearby suitable site if they were not successful the previous year. These "limited dispersal" rules seem to be justified by current information, although more data are necessary to characterize dispersal accurately. Different rules for dispersal can be shown to produce significantly different nesting success results.

## Some Results

Because of the sensitivity of the Cape Sable seaside sparrow model to topography, it has only been applied thus far to one of the two major subpopulations, the western subpopulation, for which sufficiently precise estimates of topography are available. The model has been calibrated to fit the only empirical estimate of population size prior to the 1990s, that of 1981. Using this calibration, the model projects a decline in the population during the early 1980s, but then a rise during the late 1980s and early 1990s, to fit quite well the next empirical data point in 1992 (see Figure 8.5). The model predicts very well the rapid decline in this subpopulation during the middle 1990s. Using this validated model, we have performed viability analyses on the Cape Sable seaside sparrow western subpopulation for the various proposed alternative restoration plans for the Everglades. These results played an important role in the selection of a plan.

## SNAIL KITE

### Context of the Model

The population of snail kites (*Rostrhamus sociabilis*) in Florida is isolated and probably below 2000. The snail kite is an extreme specialist, feeding almost exclusively on apple snails, so it can

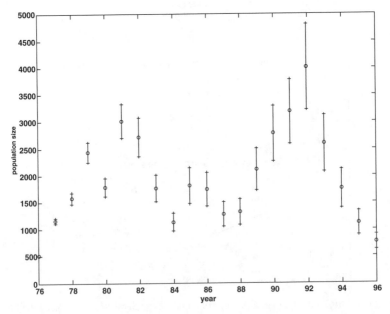

**FIGURE 8.5**   The Cape Sable seaside sparrow model is run using hydrologic data between 1976 and 1996. The starting number of sparrows is fixed so that its prediction for 1981 is calibrated against the 1981 empirical data point for total population size in the western population. The projections for 1992 to 1996 roughly match those of the empirical surveys for those years.

be affected negatively by droughts that reduce the availability of apple snails. Changes in hydrology that affect this prey can adversely affect snail kites. However, the snail kite is also highly nomadic and can move hundreds of kilometers to another habitat site if necessary. The range of the snail kite covers much of central and southern Florida. The long-range survival of the snail kite appears to depend on maintaining habitat over that full range so that droughts affecting the entire range simultaneously are rare.

## Scope of the Model

The primary foraging and reproductive habitat of the snail kite consists of several isolated wetlands, such as the littoral zone of Lake Okeechobee, lakes in the Upper Kissimmee and St. Johns River watersheds, the Water Conservation Areas, and parts of Shark Slough. There is also what can be called "peripheral wetlands," where individuals can normally forage well enough to survive but not to nest and raise offspring. Apple snails die or estivate, becoming unavailable, when a site becomes dry. After a drydown, a particular site may not be good habitat for a few years, until the apple snail population recovers. However, snail kites are highly mobile and can move to other sites. A spatially explicit, individual-based model has been developed that can follow individual birds through their lives, with particular emphasis on environmental conditions of particular sites and the rules regarding movement of the kites (Mooij, in press). Individual-based models can take into account the increased knowledge base of snail kites as they age and move around their environment. The flow chart for activities of a snail kite is shown in Figure 8.6.

The model is not based on 500 × 500 m spatial cells but instead on several (14, currently) discrete spatial sites of different areal extent. Carrying capacity for the snail kites has been estimated for each site. Time-series data on water levels are available for more than 20 years, from which the occurrences of droughts can be inferred. Effects of drought on carrying capacity and the recovery of a given site from drought are modeled. An additional factor of importance is gradual environmental change. Prolonged flooding leads to changes in vegetation type and loss of woody vegetation for nesting. The model uses an estimate of the rate of deterioration to follow these changes over the long term.

## Crucial Model Assumptions

Key model rules are those governing where a snail kite will move when a given site dries out and is no longer good habitat. It is critical to know how quickly a snail kite will be able to locate more favorable habitat. Currently, radio-tagging information indicates that most movements are to nearby sites (Bennetts and Kitchens, 1997). Carrying capacities of sites for nesting snail kites have been estimated but are not precise and need improvement. Finally, estimates of the rate of deterioration of a given site under prolonged flooding must be refined using vegetative succession modeling.

## Some Results

The model has been parameterized with the best available data on snail kites and the relevant changing conditions of their environment. To explore the properties of the model, the 14 modeled primary snail kite habitats were subjected to hypothetical climate regimes. This included various frequencies of local and global droughts. A hypothetical, though typical scenario lasting 25 years is shown in Figure 8.7. Three systemwide droughts and more than 15 local droughts occur over this simulated period. In this particular simulation, which is not unlike the situations between 1967 and 1992 in the snail kite's range in Florida, the snail kite population manages to grow relatively steadily. It is not greatly affected by droughts that are localized. This model will be used to explore a number of possible futures.

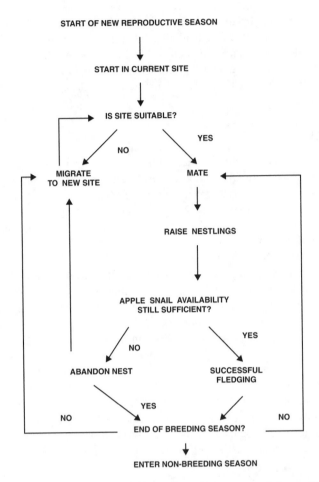

**FIGURE 8.6**   Schematic of the main components of the breeding season of the snail kite. Site suitability is mainly determined by vegetation type and water level. Drought is the main cause of site unsuitability and may cause movement to a distant habitat site; however, snail kites may also make movements when conditions in the current site are good.

## WHITE-TAILED DEER, FLORIDA PANTHER

### Context of Model

As the largest terrestrial carnivore in South Florida and an animal that depends upon large home range sizes, the Florida panther (*Felis concolor coryi*) serves as a key species for assessing the health of these ecosystems. Individual panther success (e.g., survival and reproduction) in South Florida is closely linked to a panther's ability to obtain large prey items, notably white-tailed deer (*Odocoileus virginianus seminolus*) and feral hogs. The white-tailed deer population is critically affected by the spatio–temporal hydrologic pattern. Thus, in order to predict panther success, it is essential to obtain reasonable methods to determine how hydrologic changes would affect this key prey resource and link this to the panthers.

### Scope of the Model

The objective of the model is to determine the potential long-term impacts (e.g., over 30 years or more) of changes in habitat extent and other habitat modifications (particularly hydrologic) on the

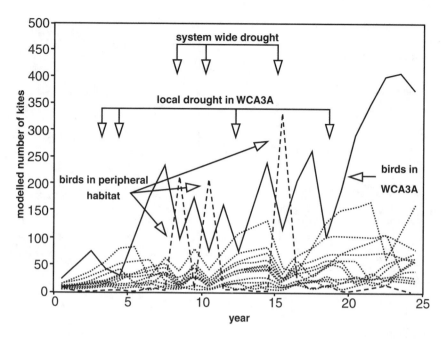

**FIGURE 8.7**   Typical simulation result showing the model snail kite population over a 25-year population. In this particular simulation, there are three global droughts and four droughts that are local to WCA-3A. The changes through time of 14 subpopulations in the various local sites are shown.

panther population. Because of the dynamic nature of the landscape and the small population size (about 50 adults in 1998) and long-distance movements of the Florida panther, a spatially explicit, individual-based modeling approach was chosen both for the panther and its primary prey, the white-tailed deer. With this modeling approach the extensive available empirical data on panthers can be used to produce a model that appropriately tracks the effects of alternative hydrologic scenarios. The individual-based model, SIMPDEL, tracks movement, growth, reproduction, and mortality of individual deer and panthers (Comiskey et al., 1997). The model operates on a daily time step. Deer and panther movements are simulated, taking account of local water conditions, forage, and prey availability. Spatially, the model makes use of vegetation data to calculate forage availability on a 100-m scale but tracks deer and panther locations on the daily time step at the 500-m scale.

Two flow charts illustrate the main features of the white-tailed deer that are modeled. Figure 8.8a illustrates the life cycle of an individual to adulthood. Figure 8.8b illustrates how energetics and mortality are figured into this life cycle.

## Crucial Model Assumptions

The primary driving variables for the white-tailed deer are water level and forage biomass. Forage biomass is a function of vegetation type in a particular 100 × 100 m cell, the season of the year, and hydrologic conditions. Three classes of forage quality — high, medium, and low — are modeled as dynamic landscape variables. Reproduction by the deer depends critically on the amount of high-quality forage. Successful reproduction also depends on there being available dry land for fawn production. Thus, the successful projection of deer dynamics depends on the accurate modeling of forage availability and water levels. The model relies on a landscape consisting of a vegetation map, daily water depth data at the 100-m scale of resolution, a land-use map, a map of feral hog density, and a road map.

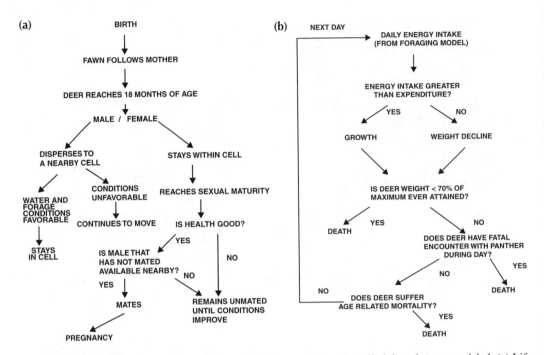

**FIGURE 8.8**    Two flow charts illustrate the main features of the white-tailed deer that are modeled. (a) Life cycle of an individual to adulthood. (b) How energetics and mortality are figured into this life cycle.

## Some Results

The model produces detailed spatial information on deer and panther distribution pattern changes over time that appear reasonable based upon consultation with experts. The model also allows computation of aggregated variables, such as age-dependent mortality, age-structure, body weight distribution, and birth rates, with available data and comparison of movement patterns of model individuals with radio-collar tracking data. A computer program allowing visualization of panthers on landscapes at various levels of resolution has been written to allow easy access to the radio-collar information available and provide a means to readily compare observed movement patterns to model output.

## DISCUSSION

The modeling approach described here became an official part of the Central and South Florida Review Study, in which different restoration alternatives were reviewed during October 1997 to June 1998. Because the deadlines on completion of the review were advanced in time, the SEIB demographic models of snail kites, wading birds, Florida panthers, and white-tailed deer were not complete. Nevertheless, it was possible to construct simpler spatially explicit species index (SESI) models for several of the species of interest. SESI models have a temporal component and are based on a common "landscape structure" that can be used to model the responses of any species in the system. These models were designed to generate color-coded landscapes representing the suitability index (a value between 0 and 1) for each landscape cell. SESI models have been developed to compute foraging condition indices (FCI) for two types of wading birds (long-legged and short-legged) and for snail kites, while breeding potential indices (BPI) have been developed for white-tailed deer and the Cape Sable seaside sparrow populations. These indices were computed for each of 31 years. This 31-year period was the interval over which rainfall data are available, so the South Florida Water Management Model for hydrology could be run with different water management structures and operations in place. The different alternatives were evaluated and compared with the situation under which no changes are made in the system (Curnutt et al., 2000).

The wading bird FCI used hydrologic model output to determine if water depths were appropriate for successful wading bird foraging (short-legged and long-legged birds considered separately) in particular spatial cells on 5-day time steps on all 500-m cells in the landscape. Fish densities in each particular cell were assumed sufficient for foraging if the cell had been flooded for a sufficiently long time period. The model computed the foraging conditions, integrated over particular subareas of the Everglades during the breeding season. Occurrences such as reversals in the drying down of water, which could interrupt breeding, contributed negatively to the index.

The white-tailed deer BPI used knowledge of how hydrologic factors affect the production and availability of food resources and the availability of dry bedding sites during the breeding season. These were calculated at a resolution of 500 m.

The Cape Sable seaside sparrow demographic model was reasonably complete for the restudy evaluations; therefore, in addition to a BPI for the sparrow, it was possible also to project population trends under each of the five alternatives. This model, in particular, provided useful information for the agencies involved in the restudy to evaluate and compare the scenarios.

All of the SEIB demographic models described here are complete and are capable of being simulated for the same landscape hydrology scenarios. The work described here shows how modeling is beginning to play a role in planning and management of the Everglades ecosystem. The models applied have been relatively complex. This reflects our viewpoint that for progress to be made in conservation biology and other applied areas of ecology, the traditional abstract models of theoretical ecology are unlikely to be a successful strategy. The objective of models applied to practical problems should be to bring to bear as much pertinent information on a problem as necessary. This will often include the use of detailed models, when they are supported by data. This is nothing new in the environmental sciences. Environmental scientists and engineers routinely use models with thousands of equations (in hydrologic and meteorological models, for example). The above examples show that the judicious use of SEIBs may be possible when approaching ecological conservation problems. These models counter some negative preconceptions about these models, such as the following.

## "THERE IS A LACK OF DATA TO SUPPORT SEIB MODELS"

Comments about SEIB models include the assertions that the amount of detail in such models cannot be supported in terms of what we can measure and parameterize and that we may need to seek less "data-hungry" modeling approaches (Wennergren et al., 1995). Models such as that for the Cape Sable seaside sparrow and the snail kite, each with relatively few parameters, show that quite interesting SEIB models often have sparse data requirements. The essential thing about SEIB models is not the amount of data they require, but their adaptability to using the types of data commonly collected at the individual level. Models can be tailored to address fairly narrow problems. Natural history and behavioral information, which usually cannot be incorporated into state-variable models, can be formulated as "rules" in SEIB models and brought to bear on a problem.

Rather than being a liability, individual behavior models increase the relevance of behavioral ecology to population ecology. These models are a means for utilizing large amounts of data already collected, often at great cost, at the individual level. The combining of behavioral and physiological information into individual behavior models also helps to reveal gaps in existing data that could stimulate more focused and useful field studies. In addition to the impressive empirical work at the individual organism level by behavioral ecologists, there is also highly developed relevant theory at the individual level, such as foraging theory (Stephens and Krebs, 1986). Judiciously used to help predict energy and time constraints on foraging, this individual-level theory can supplement empirical information where necessary and can carry one very far in predicting how well a population of individuals will do in specific situations.

## "SEIB Models Are Too Complex"

Critics often confuse mathematical complexity with ecological complexity. Individual-based models are indeed complex in the sense that they cannot be analyzed mathematically. However, they are "transparent" in that every parameter refers to something measurable and with a clear meaning. This is not true of the parameters, say, of aggregated models such as the coupled differential equations of traditional food web and community dynamics. Parameters such as "carrying capacity" and "interaction strengths" of populations are conceptually extremely complex; they have no clear meaning and cannot be directly measured. Casti (1997) has written insightfully on SEIB models. In describing a SEIB model of the vehicular traffic of Albuquerque (400,000 travelers on 30,000 road segments), he called the model "medium-simple," because of the simplicity of the set of rules, and praised its clarity. Simplicity is not simply a matter of whether a model admits of an analytic solution. Some critics of SEIB models (Wennergren et al., 1995) refer to even a modest set of equations for age and spatial structure as "unwieldy," although models of far greater size are hardly termed unwieldy by modelers in other sciences. Whereas Wennergren et al. (in regard to spatially explicit individual behavior models) state that "… the 'realism' of these models is no guarantee of their usefulness," we believe that a high degree of realism is at the very least a prerequisite in any model for it to be useful in conservation ecology.

## "There Will Be Uncertainty Multiplication in SEIB Models"

Because there can be many parameters in some SEIB models, critics often allege that the uncertainty inherent in each parameter will multiply in these models. If we examine in detail a model with many rules and parameter values, such as the model of wood storks, we can easily see most parameters add information that reduces overall the uncertainty of projections, despite the fact that there may be uncertainty in the parameter values. The rules in this model include many limits that prevent overpredictions: upper limits on flight speeds, upper limits on foraging, limits on the amount of food wood storks can carry, and limits on the number of eggs laid. To prevent underprediction, the rules include the birds' knowledge of the landscape, ability to follow other birds, and strategies for when to move to other sites. Presumably, we cannot know these perfectly, and each rule or parameter introduces some uncertainty; however, the net effect a given rule is generally to give a better answer to how effectively a wading bird can exploit resources. Perhaps more importantly, SEIB models contain an inner parallelism of many individuals acting independently. Uncertainty in parameter values does not multiply across individuals and is, in fact, buffered by the large variation in the states of the individuals. The chestnut that inclusion of more parameters necessarily adds uncertainty is incorrect.

For the above reasons SEIB models are of great importance in applied areas such as conservation and restoration ecology on landscapes such as the Everglades. In the realm of more general theory, the individual-based approach also provides an avenue for important progress in ecology. E. O. Wilson (1975) forecast that behavioral ecology and population ecology would be tightly interfaced by the end of the 20th century. Much of this interfacing, if it is to occur, will be accomplished through the extension of population models to incorporate the behavior and energetics of individual organisms in a realistic way. This will pave the way towards theory reduction, or interpreting the "higher level phenomena" of population dynamics in terms of "lower level processes" or mechanisms at the individual level (Shrader-Frechette and McCoy, 1993). Because theory reduction is one of the ultimate goals of science, and because theory reduction is a form of simplification in science, the basing of population modeling on individual behavior is a step towards the consolidation and simplification of ecological theory. The wood stork model of Wolff (1994) and the white-tailed deer/Florida panther model of Comiskey et al. (1997) are examples of how this can be done; therefore, these models are important not only from an applied viewpoint, but also from a theoretical one.

## CONCLUSIONS

Hydrology is the dominant driving force in the freshwater wetlands of southern Florida. Changes in hydrology during the last several decades are thought to have caused the observed population decline in many of the species adapted to the natural annual water cycle. To understand the possible effects of these historical changes and to predict the consequences of proposed plans for restoration of the Everglades, a landscape modeling system, Across-Trophic-Level System Simulation (ATLSS), has been developed. ATLSS is in a multimodeling approach combining ecosystem, landscape, and population modeling approaches. It uses GIS vegetation data and existing hydrology models for South Florida to provide the basic dynamic landscape. Because the models within ATLSS require finer spatial resolution than are produced by available hydrologic models of southern Florida, we have developed within ATLSS an approach for obtaining finer scale hydrology based on vegetation data and constraints on water conditions for each vegetation type. Output from a this fine-scale hydrology drives models of changes through time of vegetation and lower trophic levels, such as functional groups of fish and herpetofauna. These are used as input for higher trophic levels. These latter include several species of wading birds, the snail kite, Cape Sable seaside sparrow, Florida panther, white-tailed deer, and snail kite. One specific objective of ATLSS is to predict the responses of this suite of higher trophic level species to several proposed alterations in Everglades hydrology. Spatially explicit, individual-based computer models simulate these species. These models use detailed observational data on behavior and physiology and dynamically link these with spatially explicit abiotic information to scale up from individuals to population and community levels. The ATLSS models are currently being used to evaluate the effects on biota of changes in water regulation proposed for Everglades restoration. Here we compared a few different scenarios and showed how changes in the spatio–temporal pattern of water depth across the landscape pattern make significant differences for some of the species populations evaluated. The results show how it is possible to link spatially explicit, temporally changing information on vegetation and hydrology at a landscape scale to assess the consequences for biotic components of interest in the system.

## ACKNOWLEDGMENTS

Support for this work was provided by the U.S. Geological Survey, the Environmental Protection Agency, and the U.S. Army Corps of Engineers and Department of Interior's Critical Ecosystems Studies Initiation, a special funding initiative administered by the National Park Service (Ecological Processes Interagency Agreement #9999). M. P. Nott, E. J. Comiskey, and L. J. Gross were supported through Cooperative Agreement Number 1445-CA09-95-0094 between the U.S. Geological Survey and University of Tennessee; W. M. Mooij and W. F. Wolff were supported through Cooperative Agreement Number 1445-CA09-0111 between the U.S. Geological Survey and the University of Miami. D. L. DeAngelis was supported in part by the Florida Caribbean Science Center of the U.S. Geological Survey, Biological Resources Division. This is publication number 2758 of the Netherlands Institute of Ecology, Centre for Limnology.

## REFERENCES

Bennetts, R.E. and W.M. Kitchens. 1997. The Demography and Movements of Snail Kites in Florida, U. S. Geological Survey/Biological Resources Division, Florida Cooperative Fish & Wildlife Research Unit, Technical Report Number 56, 169 pp.

Casti, J.L. 1997. *Would-Be Worlds*, John Wiley & Sons, New York, 242 pp.

Comiskey, E.J., L.J. Gross, D.M. Fleming, M.A. Huston, O.L. Bass, H.-K. Luh, and Y. Wu. 1997. Spatially-explicit individual-based simulation model for Florida panther and white-tailed deer in the Everglades and Big Cypress, in *Proceedings of the Florida Panther Conference*, Jordan, D., Ed., Ft. Myers, FL, U.S. Fish and Wildlife Service, pp. 494–503.

Curnutt, J.L., E.J. Comiskey, M.P. Nott, and L.J. Gross. 2000. Landscape-based spatially explicit species index models for Everglades restoration, *Ecological Applications*, 10:1849–1860.

DeAngelis, D.L., L.J. Gross, M.A. Huston, W.F. Wolff, D.M. Fleming, E.J. Comiskey, and S.M. Sylvester. 1998. Landscape modeling for Everglades ecosystem restoration, *Ecosystems*, 1:64–75.

DeAngelis, D.L., W.F. Loftus, J.C. Trexler, and R.E. Ulanowicz. 1997. Modeling fish dynamics and effects of stress in a hydrologically pulsed ecosystem, *Journal of Aquatic Ecosystem Stress and Recovery*, 6:1–13.

Fleming, D.M., W.F. Wolff, and D.L. DeAngelis. 1994. Importance of landscape heterogeneity to wood storks in Florida Everglades, *Environmental Management*, 18:743–757.

Gunderson, L.H. and W.F. Loftus. 1993. The Everglades, in *Biodiversity of the Southeastern United States: Terrestrial Communities*, Martin, W.H., Boyce, S.G., and Echternacht, A.C., Eds., John Wiley & Sons, New York, chapt. 6.

Mooij, W.M., R.E. Bennetts, W.M. Kitchens, and D.L. DeAngelis. In press. Exploring the effect of drought extent and interval on the Florida snail kite: interplay between spatial and temporal scales, *Ecological Modelling*.

Nott, M.P. 1998. Effects of Abiotic Factors on Population Dynamics of the Cape Sable Seaside Sparrow and Continental Patterns of Herpetological Species Richness: An Appropriately Scaled Landscape Approach, Ph.D. dissertation, The University of Tennessee, Knoxville, 216 pp.

Ogden, J.C. 1994. A comparison of wading bird nesting colony dynamics (1931–1946 and 1974–1989) as an indication of ecosystem conditions in the southern Everglades, pp. 553–570, in *Everglades: The Ecosystem and Its Restoration*, Davis, S.M. and Ogden, J.C., Eds., St. Lucie Press, Delray Beach, FL.

Shrader-Frechette, K.S. and E.D. McCoy. 1993. *Method in Ecology*, Cambridge University Press, Cambridge, U.K.

Stephens, D.W. and J.R. Krebs. 1986. *Foraging Theory*, Princeton University Press, Princeton, NJ, 247 pp.

Wennergren, U., M. Ruckelshaus, and P. Karieva. 1995. The promise and limitations of spatial models in conservation biology, *Oikos*, 74:349–356.

Wilson, E.O. 1975. *Sociobiology: The New Synthesis*, The Belknap Press, Cambridge, MA.

Wolff, W.F. 1994. An individual-oriented model of a wading bird nesting colony, *Ecological Modelling*, 72:75–114.

# 9 Maps and GIS Databases for Environmental Studies of the Everglades

*Roy Welch and Marguerite Madden*
Center for Remote Sensing and Mapping Science, The University of Georgia

*Robert Doren*
Southeast Environmental Research Center, Florida International University

## CONTENTS

## INTRODUCTION

A clear understanding of the linkages between ecosystems in a region must be based on data that are comprehensive in scope, adequate in detail, and completely cover the area/time frame of interest. Geographic information system (GIS), remote sensing, and Global Positioning System (GPS) techniques are tools that can be used to generate maps and digital databases that encompass broad areas and geographically link the results of individual studies conducted on a local level — in order to study the region as a whole.

The Everglades is an example of a large area under investigation by diverse groups of scientists with the common goal of understanding hydrologic cycles, nutrient levels, energy exchange, dynamics of biotic communities, and predictions of the outcome of forthcoming restoration efforts (Figure 9.1). Many of these studies are focused on particular portions of the Everglades using samples to represent the system as a whole. In order to integrate the various studies and provide a continuum of data across the Everglades that can be used to project how ecosystems will respond to restoration of natural hydrologic flows, both geographic databases and detailed large-scale maps are required.

In the early 1990s, resource agencies in South Florida recognized the need for a broad-scale GIS database approach for landscape studies of the Everglades (Rose and Draughn, 1991). Funds for such an endeavor, however, did not become available until Hurricane Andrew crossed the Florida peninsula in August of 1992. The devastation of the hurricane, and in particular the extensive damage to ecologically important mangrove communities on the west coast of Florida, highlighted the need to evaluate hurricane damage to vegetation communities and made evident the fact that no current detailed maps or data on vegetation distributions existed for the Everglades (Color Figure 9.1*).

In 1994, the Center for Remote Sensing and Mapping Science at The University of Georgia began working with the National Park Service to develop the first comprehensive and detailed

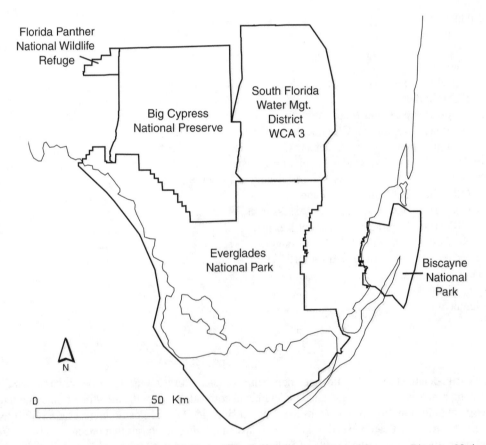

**FIGURE 9.1** Location map of Everglades National Park, Big Cypress National Preserve, Biscayne National Park, and Florida Panther National Wildlife Refuge.

vegetation database of the national parks/preserves in South Florida. Covering an area of over 10,000 km², the database was created in GIS format and used to plot 1:15,000-scale vegetation maps keyed to 80 U.S. Geological Survey (USGS) 7.5-minute topographic quadrangles within Everglades National Park, Big Cypress National Preserve, Biscayne National Park, and the Florida Panther National Wildlife Refuge (see Figure 9.1). These wetland areas will be collectively referred to as the "Parks" in this chapter.

Early work on the project was described by Welch et al. (1995) and Welch and Remillard (1996). This chapter is based on an article by Welch et al. (1999) previously published in an issue of *Photogrammetric Engineering and Remote Sensing* (*PE&RS*) devoted entirely to Everglades mapping projects.

As discussed by Doren et al. (1999), a detailed vegetation map database was required to document the status of vegetation in the Parks and provide a basis for evaluating changes in the Park lands caused by hurricanes, spread of exotic plants such as Brazilian pepper (*Schinus terebinthifolius*) and melaleuca (*Melaleuca quinquenervia*), use of off-road vehicles (ORVs), and patterns of water flow and pollution linked to population expansion and agricultural production in bordering areas. These are problems of concern to the National Park Service and to various federal, state, and local agencies such as the USGS, U.S. Army Corps of Engineers, U.S. Department of Agriculture, U.S. Environmental Protection Agency, Florida Department of Environmental Protection, South Florida Water Management District, and Dade County, to name some of the more involved groups concerned with maintenance, use, and protection of valuable land and water resources.

At the outset of the project in 1994, three major constraints were apparent:

1. The detail requirements for this vast, poorly mapped wetland area could not be met using Landsat or SPOT satellite images. Consequently, up-to-date aerial photographs were needed for mapping, as was adequate ground control. Because of cost factors, the absence of road networks, and lack of dry, firm terrain, it was not possible to pre-mark control or survey a ground control network after acquisition of the aerial photographs.
2. A vegetation classification system suitable for use with aerial photographs and sufficiently detailed to meet Parks requirements did not exist.
3. Implementation of mapping and database construction techniques would require a combination of conventional procedures and the development of new procedures, particularly involving the use of satellite images and GPS navigation techniques keyed to the mapping problems.

Consequently, this chapter is devoted to documenting the methods employed to overcome these constraints and produce detailed maps and GIS databases suitable for supporting environmental studies of the Everglades. It is anticipated that the availability of a seamless vegetation database will prove to be an important asset to scientists addressing problems of concern in South Florida.

## DATA SOURCES

The datasets needed to construct a vegetation database of the national parks/preserves in South Florida included existing maps, satellite images, aerial photographs, and information retrieved from field surveys and sampling procedures. Each of these data sources is described in the following paragraphs.

### MAPS

In most projects involving the mapping or inventory of natural resources, the ground control points (GCPs) necessary to fit detail extracted from satellite images or aerial photographs to a map coordinate system are obtained from existing map coverage of large to medium scale. However, at the initiation of the project, the best available map coverage of the Parks included approximately 80 USGS 7.5-minute quadrangles of 1:24,000 scale. Of these quadrangles, conventional line maps meeting National Map Accuracy Standards (NMAS) (i.e., ± 7.2 m $RMSE_{xy}$) were available for the land areas along the north and east margins of the Parks. The interior regions were covered by quadrangles in orthophotomap format produced from aerial photographs recorded in the 1960s and 1970s that were subjected to cartographic treatment involving the overprinting of color patterns, lettering, and symbols that tended to mask the vegetation patterns. Most importantly, however, the possibility for deriving GCPs of sufficient accuracy and density from these orthophotomaps was limited. It was virtually impossible to find stable points of detail visible on both the orthophotomaps and the 1994/1995 color infrared (CIR) aerial photographs employed for this project. For these reasons, the orthophotomaps were deemed unsatisfactory. Unfortunately, the excellent USGS Digital Orthophoto Quarter Quads (DOQQs) only became available as the vegetation mapping project neared completion. Other maps available for the project included general locational maps distributed to visitors by the Parks and earlier vegetation maps covering portions of the Parks or of limited resolution as described by Doren et al. (1999).

## SATELLITE IMAGES

Although there have been numerous applications of satellite images for thematic classification of vegetation, the spatial resolution in Landsat or SPOT multispectral images is not sufficient for the identification of Everglades wetland vegetation species and communities — or for the construction of a vegetation database having a minimum mapping unit of approximately 1 hectare (ha) (Rutchey and Vilchek, 1994; Welch et al., 1995). However, these satellite images have excellent geometric integrity and can be easily rectified to a standard map coordinate system to within ±0.5 to ±1 pixel, provided six to eight well-distributed GCPs can be identified in a given scene. Given the large area coverage per SPOT scene (~ 60 × 60 km) and the availability from the South Florida Water Management District of eight contiguous digital SPOT panchromatic images of 10-m resolution providing complete coverage of the study area in 1993, a decision was made to rectify these scenes and use them as a source of GCPs for the aerial photographs. The ground control required to rectify the SPOT images was established by conducting a differential GPS survey that extended from Flamingo at the southern land margins of Everglades National Park along existing roads northward through Everglades National Park and Big Cypress National Preserve, and then eastward to the boundary with the South Florida Water Management District's Water Conservation Area (WCA) 3A. Universal Transverse Mercator (UTM) grid coordinates referenced to the North American Datum of 1983 (NAD 83) were established for 23 road intersections and bridges identifiable on the SPOT images. By using these GCPs in combination with road intersections taken from the line maps of the more populated areas at the margins of the Parks (and with coordinates converted from NAD 27 to NAD 83), each of the SPOT scenes was rectified to an accuracy of approximately ± 0.5 to ± 0.9 pixel (± 5 to ± 9 m). These scenes were then joined to create a continuous digital mosaic of the project area. The rectified SPOT scenes and mosaic subsequently proved to be excellent sources of ground control for the CIR aerial photographs employed to build the vegetation database.

## AERIAL PHOTOGRAPHS

Color infrared aerial photographs of 1:40,000 scale obtained in January, March, and December 1994 and January and October 1995 as part of the USGS National Aerial Photography Program (NAPP) were employed as the primary source material in building the vegetation database. These photographs were recorded on Eastman Kodak SO-134 film and are registered to the 7.5-minute topographic quadrangles (Color Figure 9.2a).

In order to expedite the interpretation of the photographs, facilitate the construction of vege-tation coverages in digital format, and minimize any errors in the identification and digitizing of

GCPs, the 1:40,000-scale CIR film transparencies were enlarged and reproduced as nominal 1:10,000-scale CIR paper prints. The ground control necessary to rectify features delineated on the aerial photographs was obtained by locating points of detail (bushes, stream junctions, small ponds) common to both the SPOT images and the aerial photographs, measuring the UTM coordinates of these points of detail on the rectified digital SPOT images, and assigning the coordinates for these "control points" to their respective features on the CIR enlargements. Typically, rectification coefficients developed using a second degree polynomial for this vast area of flat terrain yielded RMSE values at independent checkpoints of approximately ± 4 to ± 5 m per photo. These results are compatible with the NMAS for 1:24,000-scale maps and verified the validity of using the rectified SPOT images as a source of control for the CIR aerial photographs.

## FIELD STUDIES

Fieldwork involved the GPS surveys mentioned above, GPS-assisted reconnaissance of areas to be mapped, and GPS-assisted accuracy checks of areas previously mapped. These GPS-assisted activities required extensive use of helicopters as well as the more traditional automobile and foot travel for the collection of field data. In particular, the GPS-assisted helicopter data collection techniques developed for this project proved invaluable in establishing the Everglades Vegetation Classification System, providing the ground-truth necessary to interpret the CIR enlargements and confirming the accuracy of the resulting database and map products.

The Bell Jet Ranger 206 helicopters employed by the National Park Service are equipped with GPS receivers that enable the pilots to pre-define their flight track, conduct real-time navigation guided by the GPS unit, and record the coordinates of landing points or features of interest. In order to maximize the advantage of this positioning technology, the SPOT image mosaic was loaded into a Dell laptop computer along with the Desktop Mapping System (DMS; *R-WEL*, Inc.) and FieldNotes (PenMetrics, Inc.) software packages. A Trimble GPS receiver with an external antenna mounted on the forward hull of the helicopter was connected to the serial port of the computer. This setup enabled a person in the rear seat of the helicopter to hold the computer on his or her lap, display the satellite image mosaic, and track in real time the flight path of the helicopter. It also provided a means of collecting ground-truth information linked to coordinates provided by the GPS receiver. Upon reaching an area of interest, the helicopter circled or landed to allow identification of plants. Species-attribute information and additional notes pertaining to hurricane damage, fire history, or exotic control measures that may have influenced the area were entered into the computer and linked with the GPS coordinates.

Back in the laboratory, the flight path of the helicopter (defined by the GPS coordinates) was plotted at 1:10,000 scale and registered to the appropriate CIR air photographs. Attributes describing the plant species collected during the flight and tagged with GPS coordinates were also registered to the photographs, providing the interpreters with a comprehensive set of notes that enabled the correlation between photo signatures and vegetation classes. Correspondingly, once a map was printed as a draft copy, it was possible to define the UTM coordinates of points to be visited as part of a map revision or accuracy check procedure and to input these to the flight navigation system of the helicopter. Overall, these procedures, when fully developed, provided a rapid and cost-effective means of obtaining the ground-truth information necessary to complete the project. They also permitted the total time spent in helicopters to be reduced by an estimated 30 to 40%. With helicopter rental rates varying from $450 to $650/hour, any reduction in flying time translated to a significant savings in cost.

## VEGETATION DATABASE AND MAP PRODUCTS

Construction of the vegetation database and associated 1:15,000-scale map products from the various source materials involved the following steps: (1) interpretation of the CIR aerial photographs and delineation of the vegetation classes on clear plastic overlays registered to the photographs; (2) scanning of the overlays; (3) data conversion to vectors in digital format; (4) editing and attributing the vector files to produce map coverages keyed to the USGS 7.5-minute topographic quadrangle boundaries; (5) quality control and accuracy checks; and (6) generation of final map plots and digital data files.

Interpretation of the aerial photographs required the consideration of two key issues. The first was whether to use scanned aerial photographs and undertake on-screen, heads-up digitizing or to employ more traditional analog photointerpretation techniques. The disadvantages of using digital technology were immediately evident. For example, in 1994, the cost associated with scanning a large number of CIR photos was prohibitive. Furthermore, in order to capture the detail in the contact-scale aerial photographs, a scanning resolution of approximately 25 to 30 μm (1000 to 800 dpi) was required. At this resolution, it was necessary to store a file between 250 and 150 megabytes per photograph, or over 40 gigabytes of image data for the study area. These large file sizes slow computer operations and were not easily accommodated in 1994. In addition to these issues, there was the problem associated with training the interpreters in image-processing techniques and of establishing a relatively expensive workstation for each interpreter. Most important, however, was the reality of overcoming the problems associated with computer screen resolution that largely eliminated the synoptic view and required the interpreter to work on very small patches of terrain — making it difficult to map the vegetation patterns within the context of surrounding features. Tests conducted to evaluate the efficiency of heads-up interpretation and digitizing confirmed that this approach offered more problems than solutions.

The alternative to using the image data in digital format was to employ aerial photographic enlargements and to transfer the data to overlays as described below. Costs were low, technology was simple, and the speed with which a interpreter could identify and delineate features on a per-photo basis was greatly enhanced.

## EVERGLADES VEGETATION CLASSIFICATION SYSTEM

The second issue was the establishment of a vegetation classification system that would provide the detail required by Parks and South Florida Water Management District personnel, yet prove suitable for use with aerial photographs. Madden et al. (1999) describe the evolution of the Everglades Vegetation Classification System derived by representatives from the Parks, Center for Remote Sensing and Mapping Science, and South Florida Water Management District. Basically, it was apparent at the onset of the database/mapping project that existing vegetation classification systems such as the *USGS Land-Use and Land-Cover Classification System for Use with Remote Sensor Data* (Anderson et al., 1976), the *U.S. Fish and Wildlife Service Cowardin System for Wetlands and Deepwater Habitats of the United States* (Cowardin et al., 1979), and the Florida Land Use and Cover Classification System (FLUCCS) employed by the Florida Department of Transportation (FLUCCS, 1985), would not be adequate for compiling a vegetation database and associated maps of the plant communities. These systems are national or statewide in scope and do not include the desired level of detail for South Florida vegetation.

Late in the first year of the Everglades vegetation mapping project, an additional classification system was released by The Nature Conservancy (TNC), Arlington, VA, and ESRI, Redlands, CA, for use in the USGS Biological Resources Division (BRD)/National Park Service Vegetation Mapping Program (The Nature Conservancy, 1994). The objective of this program is to develop a uniform hierarchical vegetation classification standard to generate vegetation maps for most of the park units under National Park Service management. Although this system was considered for use in the Everglades, several factors led to the decision to develop a new Everglades Vegetation Classification System for this mapping project: (1) the interpretation of the NAPP aerial photographs was well underway when the TNC final draft was made available; (2) the degree of community level information in the National Vegetation Classification System was not complete and required further refinement; and (3) the unique floristic composition of the South Florida Everglades warranted special attention to plant species and communities that do not occur elsewhere in the conterminous U.S. Because the TNC and Everglades Vegetation Classification Systems are similarly structured, the two classification systems can be integrated as required. Both are hierarchical and combine physiognomy at the highest level (i.e., the coarsest level is based on the height, spacing, and life form of the dominant species) and floristics at the lowest level (i.e., the finest level groups species as associations or plant communities). Plant community classes in the Everglades Vegetation Classification System are therefore compatible with the "community element" level of the TNC standardized National Vegetation Classification System.

Development of the new Everglades vegetation classification system was based on vegetation classification systems previously used by researchers mapping portions of Everglades National Park and Big Cypress National Preserve (e.g., Davis, 1943; McPherson, 1973; Gunderson and Loope, 1982; Olmsted et al., 1983; Rose and Draughn, 1991). Used in combination with detailed descriptions of Everglades vegetation such as those provided by Egler (1952), Craighead (1971), Duever et al. (1986a), and Davis and Ogden (1994), a list of possible vegetation classes was compiled. The 1:40,000-scale CIR aerial photographs were then carefully examined to determine if these classes could indeed be identified. Classes that could not be distinguished on the photographs were eliminated from the system and the remaining classes were organized hierarchically under seven major types: forest, scrub, savanna, prairies and marshes, shrublands, exotics, and additional class headings (Welch and Madden, 1999).

Each of the major classes is further divided into classes corresponding to plant communities. In cases where individual species can be discerned on the aerial photographs (e.g., red, black, and white mangrove), a third level of detail was included in the classification system. Table 9.1 illustrates the hierarchical arrangement of forest classes (e.g., mangrove, buttonwood, and subtropical forests) and subclasses, with additional detail provided within the attached footnotes. All class names are abbreviated for labeling database and map products. For example, red mangrove forest is designated FMr. Samples of the Everglades vegetation maps labeled with these classes are provided in Rutchey and Vilchek (1999) and Welch et al. (1999).

In order to accommodate the complex vegetation patterns that are found in the Everglades and generally maintain a minimum mapping unit of one hectare, a three-tiered scheme was developed for attributing vegetation polygons (Welch et al., 1995; Obeysekera and Rutchey, 1997). Using this scheme, photointerpreters can annotate each polygon with a dominant vegetation class accounting for more than 50% of the vegetation in the polygon. Secondary and tertiary vegetation classes are then added as required to describe mixed-plant communities within the polygon. In addition, one or more of 13 numerical modifiers can be attached to *each* dominant, secondary, and tertiary vegetation label to indicate factors such as human influence, hurricane damage, altered drainage, and extensive ORV use that might influence vegetation growth and distribution (Table 9.2). Other modifiers provide information about the vegetation distribution (e.g., scattered individuals) and important environmental characteristics (e.g., periphyton, numerous ponds, or exposed pinnacle rock).

Extensive fieldwork was conducted as part of this project to verify vegetation identification on the aerial photographs and, in doing so, document plant communities in the Everglades Vegetation Classification System. Between November of 1994 and February of 1997, personnel from the Center for Remote Sensing and Mapping Science, Everglades National Park, and Big Cypress National Preserve spent a total of 42 days in the field conducting ground and helicopter surveys using the computer-GPS data collection configuration described earlier. It is estimated that over 2000 field points were collected and entered into the Everglades fieldchecking database over a 3-year period. Additional helicopter and airboat surveys were conducted by South Florida Water Management District personnel in support of their mapping efforts with over 1000 sites documented in the Water Conservation Areas.

**TABLE 9.1**
**Hierarchy of Forest Vegetation Classes**
**within the Everglades Vegetation Classification System**

|  |  | Abbr. |
|---|---|---|
| I. Forest[a] |  | F |
| A. | Mangrove forest | FM |
|  | 1.  Red *(Rhizophora mangle)* mangrove | FMr |
|  | 2.  Black *(Avicennia germinans)* mangrove | FMa |
|  | 3.  White *(Laguncularia racemosa)* mangrove | FMl |
|  | 4.  Mixed mangrove[b] | FMx |
| B. | Buttonwood *(Conocarpus erectus)* forest[c] | FB |
| C. | Subtropical hardwood forest[d] | FT |
| D. | Oak–sabal forest[e] | FO |
| E. | Paurotis palm *(Acoelorrhaphe wrightii)* forest | FP |
| F. | Cabbage palm *(Sabal palmetto)* forest | FC |
| G. | Swamp forest | FS |
|  | 1.  Mixed hardwood swamp forest[f] | FSh |
|  | 2.  Cypress strands[g] | FSc |
|  | a.  Cypress domes/heads[h] | FSd |
|  | 3.  Cypress–mixed hardwoods[i] | FSx |
|  | 4.  Mixed hardwoods, cypress, and pine[j] | FSa |
|  | 5.  Cypress-pines[k] | FSCpi |
|  | 6.  Bayhead[l] | FSb |

[a] High-density stands of trees with heights over 5 m.
[b] Specific mixtures of mangrove species, when identified, will be distinguished as subgroups.
[c] *Conocarpus erectus* with variable mixtures of subtropical hardwoods.
[d] *Lysiloma latisiliquum, Quercus virginiana, Bursera simaruba, Mastichodendron foetidissimum, Swietenia mahagoni,* among others.
[e] *Quercus laurifolia, Q. virginiana, Sabal palmetto.*
[f] *Quercus virginiana, Q. laurifolia, Acer rubrum, Sabal palmetto, Fraxinus caroliniana.*
[g] *Taxodium ascendens, T. distichum*; cypress strands may contain an understory of species such as *Annona glabra, Chrysobalanus icaco,* and *Fraxinus caroliniana.*
[h] *Taxodium ascendens, T. distichum*; cypress growing in a depression such that trees in the center are tallest and give the characteristic dome shape. Domes may contain a fringe of short cypress less than 5 m.
[i] *Taxodium ascendens* and *T. distichum* with variable mixtures of subtropical and temperate hardwoods.
[j] Mixture of subtropical hardwoods with *Taxodium distichum* and occasional *Pinus elliottii* var. *densa.*
[k] *Taxodium distichum* with *Pinus elliottii* var. *densa* and a mixed hardwood scrub understory.
[l] *Magnolia virginiana, Annona glabra, Chrysobalanus icaco, Persea borbonia, Ilex cassine, Metopium toxiferum,* among others.

*Source:* From Madden, M., D. Jones, and L. Vilchek. 1999. Photointerpretation key for the Everglades Vegetation Classification System, *Photogrammetric Engineering and Remote Sensing,* 65(2):171–177. With permission.

## PHOTOINTERPRETATION AND DATABASE COMPILATION

The interpretation of aerial photographs was conducted by a team of skilled photointerpreters using the CIR photographic enlargements in paper print format. These enlargements (~ 1 × 1 m) were placed on a large light table which enabled the interpreter to employ either front or back lighting as the situation demanded. A clear polyester overlay was then registered to the photograph using the fiducials in the margins of the paper print. The locations of all GCPs were annotated, and the

**TABLE 9.2**
**Special Numeric Modifiers Added to Vegetation Labels in the Everglades Vegetation Classification System**

|  |  | Modifier Number |
|---|---|---|
| VIII. | Special Modifiers | |
| | A.  Hurricane damage classes | |
| |     1.  Low to medium (0 to 50% damage) | −1 |
| |     2.  High (>51 to 75% damage) | −2 |
| |     3.  Extreme (>75% damage) | −3 |
| | B.  Low density (scattered individuals) | −4 |
| | C.  Human influence | −5 |
| |     1.  Abandoned agriculture | −6 |
| |     2.  Altered drainage | −7 |
| |     3.  High-density ORV trails | −8 |
| | D.  Periphyton | −9 |
| | E.  Treatment damage (e.g., herbicide treatment) | −10 |
| | F.  Other damage (e.g., freeze damage) | −11 |
| | G.  Ponds | −12 |
| | H.  Exposed rock (i.e., pinnacle rock) | −13 |

*Source:* From Madden, M., D. Jones, and L. Vilchek. 1999. Photointerpretation key for the Everglades Vegetation Classification System, *Photogrammetric Engineering and Remote Sensing*, 65(2):171–177. With permission.

vegetation polygons and land/water features were viewed under magnification and delineated on the overlays. When additional detail or stereo-viewing was required, the interpreters were able to refer to the original film transparencies and examine these under high magnification using a Bausch & Lomb Zoom 95 Stereoscope.

Once all the control points, vegetation polygons, and appropriate land and water features were delineated on the large transparent photo overlays, these overlays were scanned at a resolution of 65 μm (400 dpi). The scanned overlay files in Tagged Image File Format (TIFF) were then transferred to a personal computer on which the R2V (Able Software Company) and DMS software packages were resident. The R2V software was employed to convert the raster overlay files to digital vector files. These vector files were subsequently rectified (geocoded) to an accuracy of about ± 4 to ± 5 m (RMSE$_{xy}$) using the DMS software and the previously identified GCPs in the UTM coordinate system. They were then output in ESRI Arc/Info Generate format to a Sun Unix Workstation for editing, attributing, edge-matching (of six or more overlays), and construction of vegetation map coverages corresponding to the USGS 7.5-minute topographic quadrangles. The Arc/Info workstation software, supplemented by modules developed at the Center for Remote Sensing and Mapping Science for attributing, visualization, and hardcopy output, was used for these tasks. The time required to prepare a complete dataset for a single map area corresponding to a USGS topographic quadrangle varied with the complexity of the plant communities — ranging from about 125 hours for relatively non-complex areas to over 250 hours for the complex vegetation patterns found in Big Cypress National Preserve (see Color Figure 9.3).

In the early stages of the project, draft map products were generated and checked in the field to verify classification accuracy, with revisions undertaken as necessary to create the final digital datasets and 1:15,000-scale paper maps prior to their release to the South Florida Natural Resources Center. As the interpreters gained skill and confidence, however, the extent of field verification and helicopter use was reduced to GPS-assisted reconnaissance missions. These observations were employed by the interpreters to ensure continued accurate interpretation of plant communities.

Independent checks of the vegetation classification accuracy for dominant, secondary, and tertiary vegetation at 88 random sample points on six maps scattered across Everglades National Park yielded values from 77 to 97% correct for individual quadrangles, with an average value of 90% correct. Although funds for more extensive accuracy evaluations were curtailed by the Parks, it is estimated that the average overall classification accuracy averages better than 85% for the entire study area.

## APPLICATIONS OF THE EVERGLADES VEGETATION DATABASE

The Everglades vegetation database resides at the South Florida Natural Resources Center at Everglades National Park and at Big Cypress National Preserve in digital Arc/Info format and as printed maps of 1:15,000 scale. Natural resource managers of the Parks refer to this database for baseline information on vegetation community patterns. It also serves as a geographic framework within which studies conducted at a local level (i.e., fine scale) can be spatially imbedded.

As reported in Doren et al. (1999), it was acknowledged at the onset of the Everglades vegetation mapping project that in some cases an even finer scale of vegetation detail (such as plant species distributions, plant densities, or frequency information) would be required for particular resource management applications. As a result, the production of a number of high-resolution data subsets nested within the Everglades vegetation database was considered.

Development of high-resolution datasets required special acquisition of large-scale CIR aerial photographs at a scale of 1:7000 to allow added differentiation of vegetation detail and a smaller minimum mapping unit of 0.02 hectare (or $14 \times 14$ m). Altogether, aerial photographs were taken for a total of 31 "high-resolution" sites, each approximately $1 \times 7$ km in size. Due to a change in the scope of work, however, digital databases and maps were developed for only three of the sites (Welch and Madden, 1999). McCormick (1999) discusses one of these sites used to produce 1:5000-scale maps depicting plant species distributions and information on height and density classes of the exotic tree melaleuca. This high-resolution dataset was employed to assess the effectiveness of exotic vegetation management practices and demonstrated the value of incorporating nested subsets of detailed vegetation within the broader Everglades GIS database. Other investigations originally targeted for some of the other high-resolution study sites included relating macrophyte changes to agricultural runoff and phosphorus enrichment and mapping reference plant communities for change analysis following hydrologic restoration (Doren et al., 1997a,b).

The Everglades vegetation database also was used in the development of a multimedia approach for linking vegetation information to descriptive text, images, and sound for enhanced use of the database by scientists interested in Everglades flora and fauna (Hu, 1999). This multimedia database contains hyperlinks from vegetation polygons in two quadrangles of Everglades National Park to text, scanned ground photographs, digital video clips, and audio segments highlighting the characteristics of Everglades plant communities, individual species, and invasive exotics, as well as plant–animal interactions, hurricane damage and post-fire vegetation succession. It also includes SPOT panchromatic imagery and scanned NAPP CIR aerial photographs.

In an application involving the long-term monitoring of plant communities as indicators of biogeochemical change, the Everglades vegetation database was used in support of the U.S. Environmental Protection Agency (EPA) South Florida Ecosystem Assessment Project (Welch and Madden, 2000). This project assessed vegetation patterns along a north–south corridor extending from Lake Okeechobee south to Florida Bay, an area of approximately 5600 km². The EPA randomly generated the coordinate locations for 250 environmental monitoring sites (divided into Cycle 4 and Cycle 5 of a multiphase study) distributed throughout the South Florida Water Management District Water Conservation Area 1 (WCA-1), WCA-2, and WCA-3, along with the Rotenberger/Holeyland Everglades Agricultural Area (EAA) and portions of Everglades National Park (Figure 9.2). Vegetation communities within $1 \times 1$ km (1 km²) plots centered on the EPA monitoring sites were extracted from the existing Everglades vegetation database, and a similar vegetation database created by the South Florida Water Management District in WCA-3. Vegetation patterns in areas outside of the existing databases were interpreted from USGS DOQQs produced from the same 1994/1995 NAPP CIR aerial photographs used to create the Everglades vegetation database. The classification system followed the Everglades Vegetation Classification System and included vegetation identified to the plant community, association, and species levels. Areal statistics for vegetation types within each of the 250 monitoring sites (1 km²) were provided to the EPA for further analysis and correlation with environmental data collected at the monitoring sites. The cumulative distribution of four major plant communities (i.e., cattail, sawgrass, wet prairie, and other) provided status and trend information on the range of vegetation types within regions and latitudinal zones distributed north to south throughout the Everglades system (Figure 9.3).

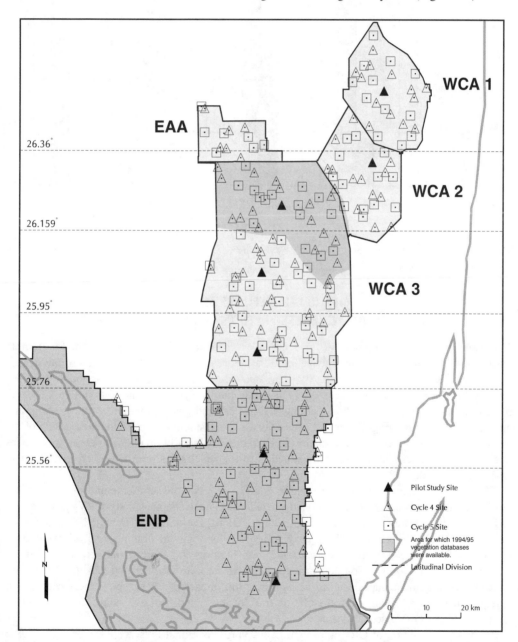

**FIGURE 9.2**   Locations of EPA South Florida Ecosystem Assessment Project monitoring sites in the Everglades.

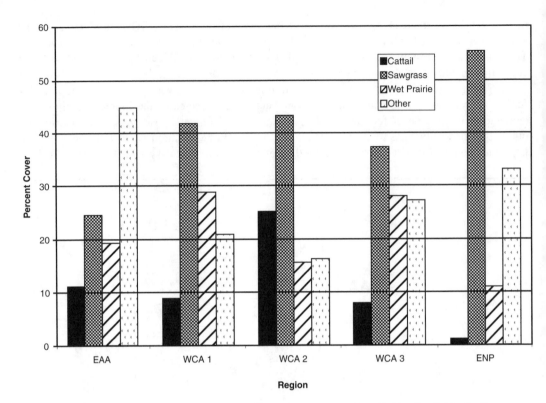

**FIGURE 9.3**   Graph of major vegetation cover by region as compiled in 250 EPA South Florida Ecosystem Assessment Project monitoring sites.

## OFF-ROAD VEHICLE TRAILS
## IN BIG CYPRESS NATIONAL PRESERVE

Concurrent with the development of the Everglades vegetation database, the Center for Remote Sensing and Mapping Science and the National Park Service undertook the mapping of ORV trails in Big Cypress National Preserve, a 2950-km$^2$ area just north of Everglades National Park (see Figure 9.1). The Preserve was created in 1974 within the framework of the National Park System to manage human activities such as hunting, fishing, oil and gas exploration, cattle grazing, and ORV use while maintaining the natural, recreational, and aesthetic values of the area. Much of the area is inaccessible by roads, and users of the Preserve rely heavily on ORVs to access private in-holdings and to partake in the recreational activities for which the Preserve was created (Tebeau, 1966; Duever et al., 1981).

The dramatic increase in ORV use within Big Cypress National Preserve and the consequent impact on the environment has generated considerable controversy among ORV user groups, environmentalists, state resource agencies, and the National Park Service (Daniel, 1999; Christian, 2000; Daerr, 2000; Wilkinson, 2001). Originally used for logging, farming, and oil exploration activities, ORVs are now primarily associated with hunting and recreational activities (Tebeau, 1966; Duever et al., 1986a). Consequently, the National Park Service has undertaken the development of a management plan designed to minimize the impact of ORV use on the environment. A requirement for the plan is information on trail locations and summary statistics documenting the magnitude of ORV use.

In their investigation of ORV impacts in Big Cypress National Preserve, Duever et al. (1981; 1986a,b) mapped ORV trails from aerial photographs acquired in 1940, 1953, and 1973. Although these maps are valuable records of historical ORV use in Big Cypress, they do not document the increased use of ORVs that has occurred within the Preserve over the past decade. Consequently, the CIR aerial photographs recorded in 1994/1995 were employed to develop a current and comprehensive ORV trail database for the entire Big Cypress area, along with summary statistics on the total length of ORV trails and an assessment of the impact of ORV trails on vegetation (Welch and Madden, 1998).

## ORV TRAIL DATABASE DEVELOPMENT

Using procedures previously developed by the Center for Remote Sensing and Mapping Science for mapping Everglades vegetation, ORV trails were interpreted from the 1:10,000-scale enlarged prints of the 1994/1995 CIR NAPP aerial photographs and delineated on the overlays (Color Figure 9.2). Ground control points were then transferred to the overlays and delineated trails were scanned, vectorized, and registered to the UTM ground coordinate system referenced to NAD 83. In order to be compatible with the Everglades vegetation database, ORV digital vector data were converted to Arc/Info format, and coverages corresponding to the 32 USGS 1:24,000-scale topographic quadrangles within Big Cypress were compiled. On a quadrangle basis, ORV trail coverages were then edited and attributed to distinguish trail classes, major roads, and canals.

During the development of the ORV trail database, trails were observed on the ground and on helicopter missions conducted in support of the ongoing Everglades vegetation mapping project. However, due to budgetary constraints in the late 1990s, funds for separate helicopter flights to verify ORV trail interpretations were not provided by the Preserve. It was determined, at that time, that any detailed ground checks deemed necessary to verify trails were to be conducted by Big Cypress National Preserve personnel.

## ORV TRAIL DATABASE ACCURACY ASSESSMENT

Continued controversy over the use of ORVs in Big Cypress National Preserve and the need to assess the reliability of the ORV trail maps following completion of the ORV trail database led to a subsequent accuracy assessment in May 2000. Areas corresponding to six USGS 7.5-minute topographic quadrangles (i.e., Burns Lake, California Slough, Deep Lake, Lostmans Trail, Monroe Station, and Thompson Pine Island) were selected as representative of the various terrain and conditions for ORV use within Big Cypress National Preserve (Figure 9.4).

Four transects, approximately 15 to 20 km in length, were defined across each quadrangle — two in a north–south direction and two in an east–west direction (Welch et al., 2001) (Figure 9.5). Each transect contains multiple 3- to 4-km segments that zig-zag and cross as many trails as possible. Surveys were conducted along each transect with the aid of a Bell Ranger 206 helicopter occupied by two Center for Remote Sensing and Mapping Science personnel, a Big Cypress National Preserve botanist who acted as an observer, and the helicopter pilot. The pilot flew along the transects defined by latitude/longitude coordinates logged into the helicopter GPS navigation unit prior to each flight. In addition, a laptop computer was linked to a Trimble GPS unit with a Coast Guard beacon antenna mounted on an exterior bracket attached to the helicopter. The computer–GPS combination provided differentially corrected coordinates accurate to within a few meters and allowed the track of the helicopter to be continuously displayed on a digital map of ORV trails.

As the helicopter passed over the intersections of the ORV trails with the transect, (i.e., the observed points), the Center for Remote Sensing and Mapping Science and Big Cypress personnel jointly determined if the trail was obviously created by ORVs, the ORVs were using an existing road, the trail appeared to be some other linear feature such as a game trail or survey line, or there was no evidence of a trail. If the trail was deemed an ORV trail, it was classified as primary, secondary, or tertiary and a note was made if the trail appeared to be abandoned.

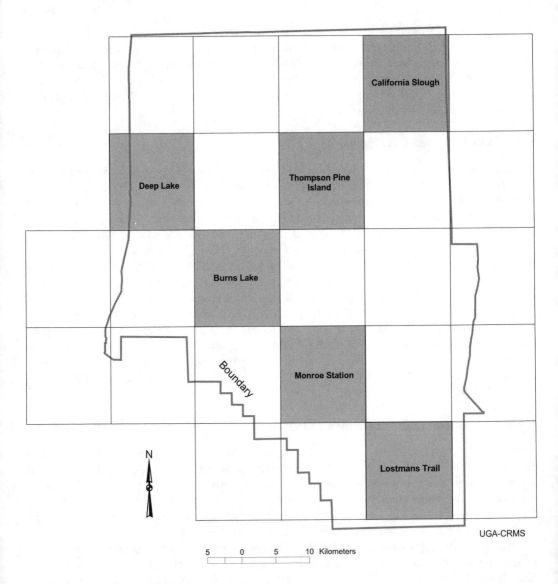

**FIGURE 9.4**    Index map of the six USGS 7.5-minute quadrangles selected for ORV trail database accuracy assessment in Big Cypress National Preserve.

All observations were stored in a digital database. During the flights and at regular intervals along the transects, photographs were taken with a 35-mm film camera and digital images with a Kodak Digital Science Field Imaging System (FIS) 265. The hand-held Kodak digital camera was connected to a Garmin III Plus GPS that "stamped" the location, time, and date on each image (Figure 9.6). These images, along with those scanned from the 35-mm film, were input to ArcView to provide a pictorial record of approximately 330 field observations that document the validation process, depict current conditions of trails in Big Cypress National Preserve, and illustrate the ORV class designations.

The majority of the trails in Big Cypress are narrow (2 to 5 m wide) tertiary trails with a double track of exposed soil and vegetation growing between the tracks (see Figure 9.6). Although these trails were usually visible beneath the helicopter, old abandoned trails that have vegetation growing in the tracks were sometimes difficult to detect. Along routes heavily used by ORVs, the trails

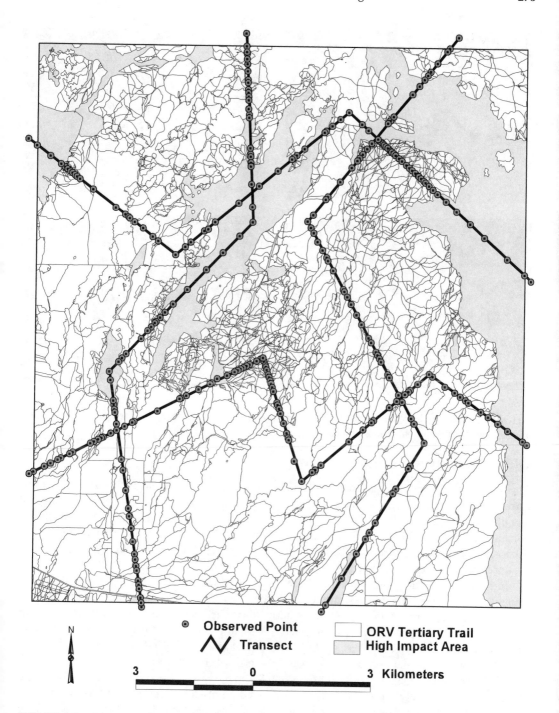

**FIGURE 9.5**  Sample of four helicopter transects created across the Burns Lake quadrangle area in Big Cypress National Preserve.

become wider (5 to 10 m), more rutted, and devoid of vegetation. In a few cases, trails were >10 m in width and were designated as primary trails. Areas of concentrated ORV use mapped as high-impact polygons in the ORV database were identified by multiple tertiary and secondary trails converging in an area or appearing as "braided" trails.

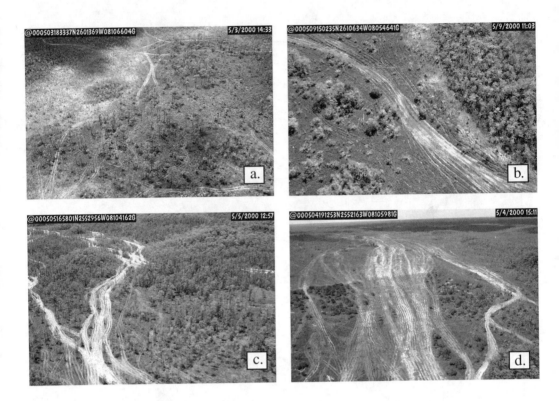

**FIGURE 9.6** Examples of ORV trail classifications: (a) tertiary trails, (b) secondary trail, (c) primary trails, and (d) braided trails in a high-impact area.

## ORV TRAIL ACCURACY ASSESSMENT RESULTS

The number of observed points along transects within each quadrangle ranged from 232 for Burns Lake to 260 for Monroe Station, for a total of 1487 observed points. Observed points were classed as ORV trails, roads used by ORVs, other trails, or no trails visible. Since only four of the six quadrangles involved in the accuracy assessment currently permit ORV use, the results of the accuracy assessment are discussed below in two sections: (1) quadrangles with ORV access, and (2) quadrangles closed to ORV access.

In areas where ORV use is permitted (Burns Lake, California Slough, Lostmans Trail, and Thompson Pine Island quadrangles) the percent of observed points that were mapped as ORV trails and noted as active or abandoned ORV trails during the helicopter surveys ranged from 93.1% for Burns Lake to 73.9% for Thompson Pine Island. A small percentage of mapped ORV trails were noted as roads used as ORV trails (e.g., 2.6% in Burns Lake and 2.8% in California Slough). Also, a small percentage of the previously mapped ORV trails were categorized as other trails, which included game trails and survey lines (11.6% and 9.8% in the case of Lostmans Trail and Thompson Pine Island, respectively). The percentage of observed points for which no ORV, roads or other trails were visible within quadrangles open to ORV use ranged from 16.3% for Thompson Pine Island (due to pine and cypress tree cover obscuring the ground) to 3.0% for Burns Lake.

Off-road vehicle use has been historically restricted in the Deep Lake and Monroe Station quadrangles and only 48.8 and 41.1% of the mapped ORV trails could be identified as ORV trails from the helicopter. Other trails — mainly game trails in Deep Lake and a mix of game trails and old survey cuts or possibly logging tram lines in Monroe Station — accounted for approximately 15% of the observed points in these quadrangles. A considerable proportion of the mapped ORV trails in Deep Lake and Monroe Station was not visible. Although further investigation is required to determine the origin of linear features identified on the aerial photographs and mapped as ORV trails within the Deep Lake and Monroe Station (inside the Loop Road) restricted areas, it is possible that very old ORV trails, survey cuts, and tram lines are visible on the 1994/1995 aerial photographs but not from the helicopter due to vegetation regrowth.

## REVISED ESTIMATE OF ORV TRAIL LENGTH

Based on the results of the ORV accuracy assessment, the total trail length of ORV trails within the six-quadrangle study area was adjusted by applying the percentage of observed points noted as ORV trails in each quadrangle (or ORV accuracy) to the total length of ORV trails depicted on the maps prepared from the 1994/1995 aerial photographs. The total length of ORV trails in the six quadrangles is therefore reduced from 16,217 km to an estimated 10,977 km.

The estimate of ORV trail length for the entire Big Cypress National Preserve was revised using two accuracy values, one for areas with ORV access and one for areas closed to ORV use. An accuracy value of 82.4% was derived for areas open to ORV access by dividing the total number of observed points noted as ORV trails (803) by the total number of observed points (975) in the four quadrangles where ORV access is permitted (Burns Lake, California Slough, Lostmans Trail, and Thompson Pine Island). For areas closed to ORVs, the number of observed points noted as ORV trails (230) divided by the total number of observed points in the Deep Lake and Monroe Station quadrangles (512), provided an accuracy value of 44.9%. These two accuracy values were then applied to ORV trail length as tallied by ORV Management Units, established by the National Park Service to manage ORV use in Big Cypress.

The ORV Management Units in which access by ORVs is permitted include the West Addition Area, Bear Island, Northeast Addition Area, Turner River, Corn Dance, and Stairsteps (Figure 9.7). Total trail length for these units as depicted on the original ORV trail maps is 42,169 km. This figure is reduced to an estimated 34,747 km when multiplied by 82.4%. Similarly, the mapped trails in the ORV Management Units closed to ORVs totaled 5789 km. Application of the 44.9% accuracy figure as noted above reduces the length of trails in the Deep Lake and Loop Units to an estimated 2599 km. A summation of these estimates (34,747 + 2599 km) provides an updated ORV trail length of 37,346 km for Big Cypress, as compared to the 47,958 km determined from the original mapping effort. While this estimate of ORV trail length may be subject to further revision based on accuracy assessments of ORV trails in the remaining 26 quadrangles, any major changes to the current estimate are thought to be unlikely.

## VEGETATION TYPES IMPACTED BY ORV TRAILS

Since both the ORV trail and vegetation datasets are cast on the same ground coordinate system, overlay commands were easily implemented to determine the spatial coincidence between the two datasets. Four representative quadrangles were selected for the analysis of ORV impact on vegetation. Summary statistics were compiled to determine the total length of primary, secondary, tertiary, and high-impact trails that passed through each vegetation community classified according to the Everglades vegetation classification system. These data were collapsed to summarize vegetation into nine vegetation types: exotics, hardwood forest, mangrove forest, cypress forest, cypress savanna, pine savanna, hardwood scrub/shrub, mangrove scrub, and prairie. Additional classes include human influence (e.g., camps) and small water bodies. An analysis of spatial coincidence between ORV trails and vegetation communities in the four-quadrangle area revealed that trails cross many different types of vegetation from hardwood forests and mangrove scrub to open prairies (Figure 9.8). Prairie communities were most often crossed, followed by cypress forest and cypress savanna.

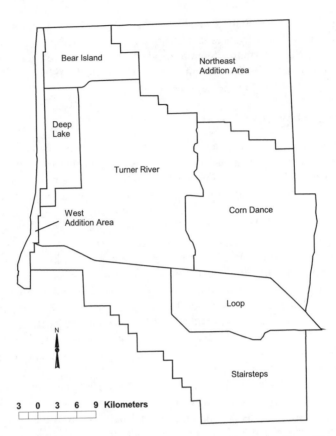

**FIGURE 9.7**    ORV Management Units in Big Cypress National Preserve. (From Welch, R., Madden, M., and R. Doren. 1999. Mapping the Everglades, *Photogrammetric Engineering and Remote Sensing*, 65(2):163–170. With permission.)

## CONCLUSION

Construction of a seamless digital vegetation map database for the 10,000-km² area occupied by Everglades National Park, Big Cypress National Preserve, Biscayne National Park, and the Florida Panther National Wildlife Refuge required the integration of GPS, remote sensing, and GIS technologies. In addition, these techniques were used to create a database delineating some 37,346 km of ORV trails in Big Cypress National Preserve and to assess its accuracy. In both instances, 1:15,000-scale map products were produced from the databases. These broad-scale databases demonstrate the possibilities for: (1) utilizing satellite images and aerial photographs to identify, map, and analyze detailed vegetation patterns over large areas; (2) developing a hierarchical vegetation classification system for use with remotely sensed data; and (3) using linked, real-time GPS positioning and image display techniques on portable computers to aid in the navigation of a helicopter and the collection of ground-truth data necessary for database development/accuracy assessment. It also clearly proved the efficiency for combining conventional analog photo interpretation techniques with computer-based scanning and data transformation procedures to create vegetation and ORV trail datasets of excellent geometric and thematic accuracy.

The databases and maps for the Everglades have established a baseline of information on the distribution patterns for vegetation within the Parks and the current status of ORV trails within Big Cypress National Preserve. The vegetation database can be used to monitor changes in vegetation patterns due to restoration in hydrology, as well as hurricanes, fire, flooding, pollution, and the invasion of exotic plant species. It will allow modeling of the impact of human activities on the

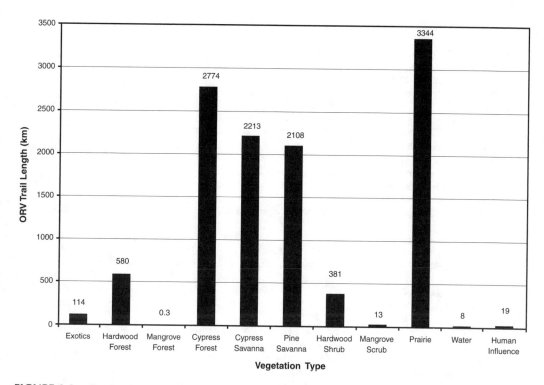

**FIGURE 9.8** Graph of the spatial coincidence between ORV trails and Everglades vegetation in four quadrangles of Big Cypress National Preserve. (From Welch, R., Madden, M., and R. Doren. 1999. Mapping the Everglades, *Photogrammetric Engineering and Remote Sensing*, 65(2):163–170. With permission.)

distribution of vegetation communities within the Parks. The ORV trail database documents the spatial patterns of ORV utilization in Big Cypress National Preserve and should prove valuable for monitoring the effectiveness of future ORV use and permitting procedures.

With increased attention being given to modeling ecosystem processes in South Florida, it is anticipated that the Everglades vegetation database will provide a geographic framework suitable for integrating the results of independent local-level studies being conducted throughout the Everglades. The GIS database can act as a data structure for spatial data storage, display, and retrieval. It can also facilitate spatial analyses required to assess the effects of pending restoration efforts. The adoption of common tools for broad-scale investigations, namely GIS, image processing, and GPS, along with the establishment of methodologies for database development and accuracy assessment will aid in the identification of ecological linkages as considered in a regional context.

## ACKNOWLEDGMENTS

This study was sponsored by the U.S. Department of Interior, National Park Service, Everglades National Park (Cooperative Agreement No. 5280-4-9006) and Big Cypress National Preserve (Cooperative Agreement Nos. 5280-7-9002 and 5120-00-001). The authors wish to express their appreciation for the devoted efforts of the staff at the Center for Remote Sensing and Mapping Science, The University of Georgia; South Florida Natural Resources Center, Everglades National Park; and Big Cypress National Preserve. Cooperation of Field Guides, Inc., Homestead, FL; the South Florida Water Management District, West Palm Beach, FL; SPOT Image Corporation, Reston, VA; and the U.S. Environmental Protection Agency Ecosystems Research Division, Athens, GA, is also gratefully acknowledged.

## REFERENCES

Anderson, J.R., E.E. Hardy, J.T. Roach, and R.E. Witmer. 1976. *A Land Use and Land Cover Classification System for Use with Remote Sensor Data*, U.S. Geological Survey, Professional Paper 964, U.S. Government Printing Office, Washington, D.C., 28 pp.

Christian, C. 2000. Hunters no longer welcome at Big Cypress, *Guns & Gear*, November: 10–11.

Cowardin, L.M., V. Carter, F.C. Golet, and E.T. LaRoe. 1979. *Classification of Wetlands and Deepwater Habitats of the United States*, Fish and Wildlife Service, U.S. Department of Interior, Washington, D.C., 103 pp.

Craighead, F.C., Sr. 1971. *The Trees of South Florida*, Vol. I, *The Natural Environments and Their Succession*, University of Miami Press, Coral Gables, FL, 212 pp.

Daerr, E.G. 2000. New ORV plan at Big Cypress, *National Parks*, 74(9–10):11.

Daniel, A. 1999. ORVs overrun Big Cypress, *National Parks*, 73(5-6):A–3.

Davis, J.H., Jr. 1943. The natural features of southern Florida, especially the vegetation of the Everglades, *Florida Geological Survey Bulletin*, 25:1–311.

Davis, S.M. and J.C. Ogden, Eds. 1994. *Everglades: The Ecosystem and Its Restoration*, St. Lucie Press, Delray Beach, FL, 826 pp.

Doren, R.F., T.V. Armentano, L.D. Whiteaker, and R.D. Jones. 1997a. Marsh vegetation patterns and soil phosphorus gradients in the Everglades ecosystem, *Aquatic Botany*, 56(199):145–163.

Doren, R.F., D.L. Childers, T.V. Armentano, W.J. Platt, C. Horvitz, M. Norland, and B. Sparkman. 1997b. *Restoring Wetlands on Abandoned Agricultural Lands in Everglades National Park: A Strategic Plan for Guiding Research, Monitoring and Management of the Hole-in-the-Donut Restoration Program*, South Florida Natural Resources Center Tech. Rep. 97-002, South Florida Natural Resources Center, Everglades National Park, National Park Service, U.S. Department of the Interior, Homestead, Fl, 35 pp.

Doren, R.F., K. Rutchey, and R. Welch. 1999. The Everglades: A perspective on the requirements and applications for vegetation map and database products, *Photogrammetric Engineering and Remote Sensing*, 65(2):155–161.

Duever, M.J., J.E. Carlson, and L.A Riopelle. 1981. *Off-Road Vehicles and Their Impacts in the Big Cypress National Preserve*, U.S. Department of Interior, National Park Service, South Florida Research Center Report T-614, 214 pp.

Duever, M.J., J.E. Carlson, J.F. Meeder, L.C. Duever, L.H. Gunderson, L.A. Riopelle, T.R. Alexander, R.L. Meyers, and D.P. Spangler. 1986a. *The Big Cypress National Preserve*, National Audubon Society, New York, 444 pp.

Duever, M.J., L.A. Riopelle, and J.M. McCollom. 1986b. *Long Term Recovery of Experimental Off-Road Vehicle Impacts and Abandoned Old Trails in the Big Cypress National Preserve*, U.S. Department of Interior, National Park Service, South Florida Research Center Report SFRC-86/09, 47 pp.

Egler, F.E. 1952. Southeast saline Everglades vegetation, Florida, and its management. *Vegetatio Acta Geobotanica*, 3:213–265.

FLUCCS. 1985. *Florida Land Use, Cover and Forms Classification System*, Department of Transportation, State Topographic Bureau, Thematic Mapping Section, Procedure No. 550-010-00101, 81 pp.

Gunderson, L.H. and L.L. Loope. 1982. *An Inventory of the Plant Communities within the Deep Lake Strand Area, Big Cypress National Preserve*, Tech. Rep. T-666, South Florida Research Center, National Park Service, U.S. Department of Interior, Homestead, FL, 39 pp.

Hu, S. 1999. Integrated multimedia approach to the utilization of an Everglades vegetation database, *Photogrammetric Engineering and Remote Sensing*, 65(2):193–198.

Madden, M., D. Jones, and L. Vilchek. 1999. Photointerpretation key for Everglades vegetation, *Photogrammetric Engineering and Remote Sensing*, 65(2):171–177.

McCormick, C. 1999. Mapping exotic vegetation in the Everglades from large-scale aerial photographs, *Photogrammetric Engineering and Remote Sensing*, 65(2):179–184.

McPherson, B.F. 1973. *Vegetation Map of Southern Parts of Subareas A and C, Big Cypress Swamp, Florida*, Hydrologic Investigations Atlas HA-492, Florida Department of Natural Resources and U.S. Geological Survey, Washington, D.C.

Nature Conservancy. 1994. *Standardized National Vegetation Classification System*, Report for the U.S. Department of Interior, U.S. Geological Survey and National Park Service Vegetation Mapping Program, Arlington, VA, 221 pp.

Obeysekera, J. and K. Rutchey. 1997. Selection of scale for Everglades landscape models, *Landscape Ecology*, 12(1):7–18.

Olmsted, I.C., W.B. Robertson, Jr., J. Johnson, and O.L. Bass, Jr. 1983. *Vegetation of Long Pine Key, Everglades National Park*, Tech. Rep. SFRC-83/05, South Florida Research Center, National Park Service, U.S. Department of Interior, Homestead, FL, 64 pp.

Rose, M. and F. Draughn. 1991. GIS applications at Everglades National Park, *GIS World*, 4(3):49–51.

Rutchey, K. and L. Vilchek. 1994. Development of an Everglades vegetation map using a SPOT image and the Global Positioning System, *Photogrammetric Engineering and Remote Sensing*, 60(6):767–775.

Rutchey, K. and L. Vilchek. 1999. Air photointerpretation and satellite imagery analysis techniques for mapping cattail coverage in a northern Everglades impoundment, *Photogrammetric Engineering and Remote Sensing*, 65(2):185–191.

Tebeau, C.W. 1966. *Florida's Last Frontier: The History of Collier County*, University of Miami Press, Coral Gables, FL, 278 pp.

Welch, R. and M. Madden. 1998. *Off-Road Vehicle Trail Database for Big Cypress National Preserve*, final report to the U.S. Department of Interior, National Park Service, Cooperative Agreement Number 5280-7-9002, Center for Re-mote Sensing and Mapping Science, University of Georgia, Athens, 30 pp.

Welch, R. and M. Madden. 1999. *Vegetation Map and Digital Database of South Florida National Park Service Lands To Assess Long-Term Effects of Hurricane Andrew*, final report to the U.S. Department of Interior, National Park Service, Cooperative Agreement Number 5280-4-9006, Center for Re-mote Sensing and Mapping Science, University of Georgia, Athens, 43 pp.

Welch R. and M. Madden. 2000. *Aerial Photo Vegetation Assessment in the Everglades Ecosystem*, final report to the U.S. Department of Interior, National Park Service, Cooperative Agreement Number 5280-4-9006, Center for Remote Sensing and Mapping Science, University of Georgia, Athens, 36 pp.

Welch, R. and M. Remillard. 1996. GPS, photogrammetry and GIS for resource mapping applications, in *Digital Photogrammetry: An Addendum to the Manual of Photogrammetry*, Greve, C.W., Ed., American Society for Photogrammetry and Remote Sensing, Bethesda, MD, pp. 183–194.

Welch, R., M. Remillard, and R. Doren. 1995. GIS database development for South Florida's National Parks and Preserves, *Photogrammetric Engineering and Remote Sensing*, 61(11):1371–1381.

Welch, R., Madden, M., and R. Doren. 1999. Mapping the Everglades, *Photogrammetric Engineering and Remote Sensing*, 65(2):163–170.

Welch, R., M. Madden, and T. Litts. 2001. *Off-Road Vehicle Trail Accuracy Assessment: Big Cypress National Preserve (BICY)*, Cooperative Agreement Number 5120-00-001, Center for Remote Sensing and Mapping Science, University of Georgia, Athens, 35 pp.

Wilkinson, T. 2001. On the beaten path, *National Parks*, 75(3–4):34–38.

# 10 Nitrous Oxide, Methane, and Carbon Dioxide Fluxes from South Florida Habitats During the Transition from Dry to Wet Seasons: Potential Impacts of Everglades Drainage and Flooding on the Atmosphere

*Thomas J. Goreau*
Global Coral Reef Alliance

*William Zamboni de Mello*
Departmento de Geoquímica, Universidade Federal-Fluminense

## CONTENTS

## INTRODUCTION

Global atmospheric concentrations of nitrous oxide ($N_2O$), methane ($CH_4$), and carbon dioxide ($CO_2$) are currently increasing (IPCC, 1992, 1994). The major sources of these gases are biological, but their poorly known sources and sinks, especially from tropical and subtropical areas, limit confidence in the quantitative details of future temperature, ozone, climate, and radiation scenarios generated by atmospheric models (IPCC, 1994).

Gaseous carbon and nitrogen releases from soils to the atmosphere are primarily a function of oxygen concentration, temperature, moisture, and the C/N/P ratios of decomposing organic matter (Dickinson and Pugh, 1974; Swift et al., 1979). Carbon dioxide is derived from respiration of plant roots, soil animals, fungi, and bacteria. Methane-producing bacteria are strictly anaerobic, while most $CH_4$ consuming bacteria are obligate aerobes. Microbial $N_2O$ production in soils can, in principle, result from either anaerobic denitrification or aerobic nitrification, with the balance of these sources unknown.

In May 1984, we made simultaneous sets of $CO_2$, $N_2O$, and $CH_4$ flux measurements along elevation, moisture, organic matter, and fertilization gradients in a wide variety of subtropical South Florida soils and waters during the end of the dry season and the beginning of the rainy season. Studies of Everglades hydrology from 1979 to 1988 (Walker, 1991) show that this was a typical year in terms of rainfall and nutrient levels. Because these sites include a wide range of drainage and flooding conditions measured under both dry and wet conditions, they span much of the range of responses of South Florida soils to both drainage and flooding and allow predictions of the impacts of water management practices on sources and sinks of greenhouse gases.

## SITES

South Florida has a strongly seasonal subtropical climate. 80% of the rainfall occurs in the May–October rainy season. Our measurements span the last weeks of a long dry season and the start of the rainy season. The area is very low lying, and small elevation differences with respect to the water table result in large vegetation, soil, and water availability differences. Study sites were chosen along an elevation gradient from Everglades swamp soils (2 m above sea level) flooded most of the year to soils atop the central coastal limestone ridge (4 to 4.5 m above sea level) which are never inundated (Figure 10.1). Sites were visually representative of several major South Florida soil and vegetation types based on initial surveys and were free of apparent disturbance except as noted.

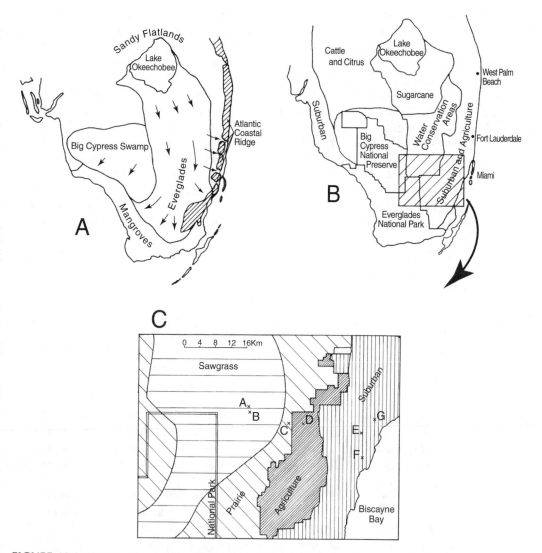

**FIGURE 10.1** (A) Map showing the distribution of major physiographic provinces of South Florida, with arrows indicating path of water flow. (After Klein et al., 1975.) (B) Map showing major land-use patterns in South Florida; the rectangle indicates the area of the expanded scale map. (C) Detail map showing location of sample sites. Habitat types after 1979 Everglades National Park Map: A = Sawgrass sites; B = Tamiami Canal and Levee sites; C = Prairies sites; D = Fallow and burned field sites, combustion gases; E = Forest, unfertilized lawn, mulch sites; F = USDA Subtropical Horticultural Station; and G = Arboretum, University of Miami, Coral Gables.

## Aquatic Sites

- *Group A:* A pond in the Gifford Arboretum of the University of Miami, whose surface was half covered with water lilies (*Nymphaea* sp.) and water hyacinth (*Eichornia crassipes*). Water was stagnant, 1 m deep, and $CH_4$ bubbles periodically broke the surface.
- *Group B:* Lake Osceola, connected by the Coral Gables Canal to the Tamiami Canal to the north and to the ocean to the south. This lake is fresh during the rainy season but brackish with a fresh surface layer and is only weakly tidally mixed in the dry season, when monitored in this study. Depth at sample location was 1 m.

- *Group C:* Coral Gables Canal, one of a network of canals that lower the water table to prevent summer flooding in South Florida. This minor connecting canal flows through the University of Miami campus and suburban Coral Gables and is not passable to boat traffic. Water levels were low and flow sluggish, conditions that promote high nutrient concentrations from septic tank leaching.
- *Group D:* The Tamiami Canal, the major east–west canal draining excess water from the Everglades to the sea via the Miami River. Fluxes were measured south of South Florida Water Management District Water Conservation Area (WCA) 3B. This location is representative of waters flowing from the Conservation Areas into the Everglades National Park and towards urban areas. At one site, water depth was 1 m, and clumps of cattail (*Typha* sp.) and pickerelweed (*Pontederia lanceolata*) grew in 20 cm of submerged Australian "pine" (*Casuarina equisetifolia*) litter overlying peat. At the other, 10 m away, water depth was 0.3 m, and no rooted aquatic plants were present. The underlying peat was covered with a dense mat of fine pink *Casuarina* roots actively growing into canal water from trees on the adjacent forested levee (these turn brown after they acquire a waterproof outer layer, which limits their water- and nutrient-absorbing capacity).

## EVERGLADES SITES

- *Group E:* These sites, near the central axis of the Everglades sawgrass ecosystem were overwhelmingly dominated by 2- to 3-m high sedge sawgrass (*Cladium jamaicense*), with minor buttonbush (*Cephalanthus occidentalis*), pickerelweed (*Pontederia lanceolata*), *Sagittaria* sp., and arrowhead sedge (*Cyperaceae*) species. Little surface litter was overlying dark-brown sawgrass peat between clumped culms. Peat at this site is about 1 m thick, overlying Pleistocene limestone. Sites are in south central WCA-3B. Water retained by levees and sluice gates extends the natural flooding period and recharges the underlying Biscayne Aquifer, source of South Florida's drinking water. At the start of the study period, the soil surface was exposed, as is typical for this time of year. The water table gradually rose to the surface during generally dry local conditions, due to seepage from rain falling earlier in the Okeechobee area to the north and scattered showers. The water surface fell below the soil surface again briefly as dry conditions continued, rising to summer flood conditions after summer thunderstorms began near the end of the study period. Only fluxes from control sites undisturbed for the entire period following initial submergence are considered in this category. Fluxes from these sites prior to soil inundation are averaged separately under group H, below.
- *Group F:* Same as group E above but fertilized with $NaNO_3$ uniformly broadcast over the 1 m$^2$ surrounding the measurement site. Fertilizer was added one day before the water table first rose to the soil surface. Fertilization amounted to 10 g N m$^{-2}$, equivalent to 100 kg N ha$^{-1}$. This fertilization rate is a low one by South Florida agriculture and lawn standards, but roughly equivalent, in total addition to the environment, to the contribution of an incontinent itinerant alligator, Florida panther, fisherman, or naturalist.
- *Group G:* Same as group F above, but with $NH_4Cl$ substituted for sodium nitrate to provide the same total nitrogen loading rate.
- *Group H:* Unfertilized control sites before submergence or fertilization. Undisturbed sites after inundation are averaged separately in group E, above.
- *Group I:* Located on the south bank of the Tamiami Canal, on a levee made of dredged sawgrass peat packed atop limestone blocks blasted and dredged during canal construction before 1928. Levee vegetation is a nearly impenetrable monospecific forest of Australian pine (*Casuarina equisetefolia*), an aggressive imported species widely planted as windbreaks along South Florida roads, canals, and beaches. Australian pine grows on very poor soil due to root symbiosis with the N-fixing Actinomycete, *Frankia* sp.

Some 5 cm of *Casuarina* needle litter graded downward into peat which was extensively penetrated by its roots. When first studied, the litter and most of the peat lay above the water table, but rising canal levels had inundated all but the top centimeter of soil at the time final measurements were made.

- *Group J:* Undisturbed sites in the Everglades prairie habitat. This sedge and grass community predominates at slightly higher elevations and shorter inundation periods than the sawgrass community. The water table at this location has been lowered by drainage canals serving the agricultural areas to the south and east, causing expansion of prairie into what had formerly been a sawgrass community. The water table rose from more than 80 cm below the surface to 25 cm below it during most of the period of study. The soil surface had a 1- to 2-mm thick marly algal crust on the surface, which had grown the previous rainy season. Periphyton growth and calcareous marl accumulation was higher at this site than in the sawgrass community despite shorter inundation periods because reduced plant biomass allows higher light penetration to the water (or soil) surface. Soil was dark-brown peat down to 12 cm, becoming lighter as it graded downward into marl and sand. Vegetation was mainly dominated by 1- to 1.5-m tall sedges, primarily *Scirpus* sp. Isolated clumps of sawgrass had greatly reduced size and abundance at this site compared to the wetter sites. These sites were in the initial stage of replacement by isolated young cajeput (*Melaleuca quinquenervia*) seedlings. Cajeput is an aggressive Australian myrtle tree that is tolerant of inundation, highly fertile, and a prolific stump and root shooter controllable only by complete (and expensive) uprooting. A monospecific forest of cajeput will soon completely dominate this site, and much of the peripheral Everglades, within decades (Alexander and Crook, 1974).
- *Group K:* Similar to group J above, fertilized at 100 kg N ha$^{-1}$ of NaNO$_3$.
- *Group L:* Similar to group J above, fertilized at 100 kg N ha$^{-1}$ of NH$_4$Cl.

## FALLOW AND BURNED SITES

- *Group M:* Sites in the drained Everglades Agricultural Area (EAA). Originally sawgrass, this area has been drained, plowed, and farmed primarily to winter vegetables such as tomatoes, beans, and peppers. Soil color is lighter than the prairie site, containing more marl and sand. Fires caused by lightning or careless people are frequent in the late dry season, and a roughly 1-km$^2$ brushfire was observed on May 16. The fire moved at a slow walking pace along a broad front across a fallow field, killing numerous lizards, insects, and tortoises. Grasses and herbs were brown and dry, burning readily with a 0.5- to 1-m high flame, bursting to 3 m over larger shrubs. The soil was not greatly heated by the passage of the fire, remaining normal to the touch beneath the hot ashes immediately after the passage of the fire front. CO$_2$, N$_2$O, and CH$_4$ contents of combustion gases in smoke were determined in syringe samples taken as little as 10 to 20 cm above the flames. Three days after the fire, fluxes were measured at soil sites in the burned field and nearby unburned control sites. Intervening rain showers had moistened the soil through. About 75 to 90% of standing vegetation had been destroyed by the fire, and remaining stumps were scorched and seemingly lifeless. Patches of a gray–white ash specked with black charcoal covered most of the surface. Fluxes were measured at sites with and without ash.
- *Group N:* Unburned part of the same field as group M above. These sites were located within 12 m of the burned sites, on the other side of the sharp line marking where the fire had burned itself out. The field had been plowed and left fallow, developing a patchy 0.5-m high stratum of mixed sedge, grass, and weedy herbs, with occasional weeds reaching 1 m. Little surface litter or ash was present, despite proximity to the edge of the fire.

## SECONDARY FOREST SITES

- *Group O:* A privately owned secondary forest on an undeveloped lot atop the limestone ridge. This site is never inundated, lying 4.5 m above sea level. Originally a Florida pine (*Pinus elliottii* var. densa) forest with saw palmetto (*Serrenoa repens*) understory cleared about 30 years before, it is being replaced by secondary forest composed of orchid tree (*Bauhinia variegata*) and woman's tongue (*Albizia lebbeck*), now up to 25 m tall. Two of the most widely planted ornamental trees in South Florida, these are prolific, rapidly growing, nitrogen-fixing leguminous trees which frequently dominate empty lots after a few years. The 0.25-ha secondary forest has a 3-cm thick litter layer that grades into 5 cm of dark-brown quartz sand soil overlying limestone bedrock.
- *Group P:* Similar to group O above, fertilized by surface broadcasting over 1 m$^2$ at a loading rate of 100 kg N ha$^{-1}$ of $NaNO_3$.
- *Group Q:* Similar to group P above, fertilized at 100 kg N ha$^{-1}$ of $NH_4Cl$.

## GRASS SITES

- *Group R:* Grass lawn adjacent to secondary forest above, a pure stand of St. Augustine grass (*Stenotaphrum secundatum*) maintained by mowing and hand weeding without fertilizers or other chemicals. This grass now makes up over 90% of South Florida lawns. Growing from fleshy aboveground runners, it overgrows other local species. The soil at lawn sites differed from adjacent forest sites in that it lacked a litter layer. It was much lighter in color and much sandier because of reduced organic matter compared to nearby forest sites. The well-drained sandy soil is only about 4 cm thick and overlies a hard but porous limestone bedrock.
- *Group S:* Similar to group R above except that the lawn had been hand weeded to produce a pure stand of *Zoysia* sp. Growing from a network of underground runners, this species grows much more slowly than St. Augustine grass. It is found in most South Florida lawns not planted to St. Augustine grass. Sites were located within 5 m of site R.
- *Group T:* Bermuda grass (*Cynodon dactylon*) lawn growing between coconut trees at the U.S. Department of Agriculture (USDA) Subtropical Horticultural Station. This grass is the third most common lawn grass in South Florida and is the slowest growing under unfertilized conditions. It is similar in form and growth habit to St. Augustine grass, but has much thinner stalks and blades. Like the other two grasses, it has a water-efficient C-4 biosynthetic pathway (Gould and Shaw, 1983). These sites were atop the coastal ridge with original soil and vegetation similar to unfertilized lawn sites, but they differed in being heavily fertilized and far more productive. Fertilizer containing 6% $NH_4^+$-N and 3% $NO_3^-$-N, plus $PO_4^{3-}$ and trace metals, was uniformly broadcast every three months. It was last applied the week before sampling began. The site was heavily mulched by bark chips piled around coconut trees and blown or spilled into the grass.

## AGRICULTURAL SITES

- *Group U:* Sugarcane (*Saccharum officinarum*) field at the National Noble Cane germ plasm collection of USDA. This site was mulched with straw and had not been fertilized or watered for over three months. Sugarcane plants appeared dormant.
- *Group V:* Cocoa (*Theobroma cacao*) at the USDA site was in a shade house maintained at high humidity by an automatic sprinkler system. It had been fertilized one month before with $NH_4NO_3$ fertilizer.

- *Group W:* Coconut (*Cocos nucifera*) field at the USDA site. Prolifically bearing Malaysian–Panamanian dwarf coconut palms of Jamaican stock with fronds reaching to 10 m. Fluxes were measured in heavily mulched soil 1 m from the base of the coconut trunk and 5 m away from the grass sites T above. The site had been heavily fertilized with ammonium nitrate and other elements one week previously.

## MULCH SITES

- *Group X:* These sites were 10 m from the unfertilized grass sites R and S above, but soil thickness was less than 1 cm. A $10 \times 20$ m, 0.5-m thick mulch pile was placed early in the dry season, 5 months prior to the start of flux measurements. Mulch material was derived from branches, boughs, and leaves freshly cut from living non-nitrogen fixing trees and shredded into smaller than centimeter sizes by a commercial tree chipping machine. About 75% of mulch consisted of *Ficus benjamina* (the most common South Florida strangler fig). The remainder was predominantly *Bischofia javanica*.
- *Group Y:* Similar to group X above, fertilized over 1 $m^2$ at 100 kg N $ha^{-1}$ of $NaNO_3$.
- *Group Z:* Similar to group X above, fertilized over 1 $m^2$ at 100 kg N $ha^{-1}$ of $NH_4Cl$.

## NITROUS OXIDE PRODUCTION AS A FUNCTION OF OXYGEN

To evaluate whether $N_2O$ is produced or consumed under aerobic and anaerobic conditions, $N_2O$ concentrations and production were measured along a vertical profile spanning aerobic to anaerobic conditions in a vertical stratified marine fjord, Saanich Inlet, Vancouver Island, British Columbia. Water samples from various depths were incubated with and without added nitrous oxide and with and without added acetylene, which is reported to be an inhibitor of denitrification.

## METHODS

Fluxes to the atmosphere from soil or water surfaces were determined in a 600-$cm^2$ area within 10 cm high by 0.04-mm thick cylindrical steel chambers. Four were deployed simultaneously, generally in pairs 1 m apart separated by 10 m. Chambers were pushed gently 1 cm into the soil, or floated on the water surface using a floating collar. A vent valve was opened to the air during emplacement and during headspace sampling so that no pressure gradient developed. The soil surface was undisturbed except that, on rare occasions rigid, recently fallen twigs which could disturb the soil surface and were lying across the edge of the chamber were removed, if this could be done without disturbance to other leaves and twigs. The chamber was not forced in at sites where roots or rocks interfered with the thin edge of the chamber, which would cause a bad seal and erratic variations in gas concentrations. This occurred only in bean fields measured at a commercial farm, where the soil contained 1 to 3 cm calcitic marl lumps; the data were rejected.

Chambers were painted white to reflect light and did not affect the temperature indicated by a thermometer lying on the soil surface inside the chamber during flux measurements. Temperatures were measured by mercury thermometers placed 2 cm into the soil between chambers during sampling. The exact sites were reoccupied each time, using the narrow "footprint" of the chamber as a guide for placement.

Fluxes were determined from the initial slope of the time series of headspace gas concentrations taken 1, 6, 11, and 21 minutes after chamber placement. 40 $cm^3$ samples were taken in doubly ground glass syringes with mated noninterchangeable barrels fitted with three-way stopcocks. Short emplacements were used to minimize flux perturbations due to alteration of gas concentration gradients in the upper soil by back-diffusion from the headspace. The same flux value was found from a site measured three times in succession with twice the sampling interval recovery time between measurements (Goreau, 1982). Gas evolution curves were fitted by straight line or by the initial slope of the concave upwards or downwards curve. Analyses were performed on a Shimadzu GC-6 gas chromatograph equipped for analysis of a single air sample by thermal conductivity, flame ionization, and electron capture detectors. Air samples stored in syringes in a room with high $CO_2$ content had increased concentrations and variance, but errors were small if syringes were dry and analysis took place within a day. Air samples taken from above the soil surface stored in syringes with a distilled water "seal" for several days generated much larger and more variable changes, presumably from growth of airborne bacteria in the sample on wet glass surfaces. Samples reported in this study were taken in mid-afternoon, stored in dry syringes, and analyzed within 24 hours of sampling, most within 12 hours. Replicate air samples had a standard deviation similar to replicate standard analyses. Minimum detectable fluxes based on the sensitivity of the detector to background levels for volumes of this size were about 0.4 mg $C$–$CO_2$ m$^{-2}$ h$^{-1}$, 1.0 µg $N$–$N_2O$ m$^{-2}$ h$^{-1}$, and 0.3 µg $C$–$CH_4$ m$^{-2}$ h$^{-1}$.

## RESULTS

The mean and standard deviation of $CO_2$, $N_2O$, and $CH_4$ fluxes to the atmosphere and the number of flux determinations at each group of sites are shown in Table 10.1 and Figures 10.2 to 10.4. These show the average fluxes computed for the entire sampling interval. Temporal trends during the study interval at specific sites resulted from changes from dry to wet conditions and addition of fertilizers and are shown in Figures 10.5 to 10.11 and discussed below. The variance between four simultaneously determined fluxes at sites within 10 m of each other was often two orders of magnitude greater than the precision with which each flux could be determined, so within-habitat flux variability primarily reflects small-scale soil heterogeneity. There is considerable overlap between many sites, so more data are needed from a number of locations to permit statistically meaningful flux comparisons between many sites. Results are discussed below by habitat type.

Soil temperatures did not change during the period of incubation, but large differences were found between sites at which the soil surface was exposed to the sun and those at which vegetation shaded the ground. Aquatic sites, forest soils, sawgrass soils, and cocoa and sugarcane sites had soil or water temperatures of 27 ± 1°C, while prairie, fallow field, burned, lawn, mulch, and coconut sites had temperatures of 35 ± 2°C with intermediate values on one overcast day. Soil temperatures were similar in burned sites covered with gray–white ashes to those in unburned fallow fields, declining from 36 ± 1°C at 2-cm depth to 26 ± 1°C at 10-cm depth in each.

### AQUATIC SITES

Submerged sites were the largest sources of $CH_4$ (Figure 10.4). The highest flux was from the stagnant lily pond and the second from Lake Osceola. Carbon dioxide production was very low from still lake and pond waters, but high from flowing canal waters. Lake Osceola was a modest source of $N_2O$ to the atmosphere, but the lily pond and the Tamiami canal slowly consumed $N_2O$ from the atmosphere. As the primary focus of this study was on soils, only the Tamiami Canal sites were measured more than once (Figure 10.5). Initially, flow was sluggish, but on the second visit water levels had risen 10 cm and flow was vigorous. On the first visit, $CO_2$ release from the cattail site was nearly twice that at the open site underlain by tree roots. On the second visit, $CO_2$ releases had dropped sharply at the first site, but had risen to very high levels at the second. Both sites released $CH_4$ on the first visit, but on the second visit only trivial amounts of $CH_4$ were produced at the first site while the other consumed it from the atmosphere. Nitrous oxide was consumed by the water at both sites on the first visit, but released on the second. Adjacent land sites on the banks of the levee, measured at the same times, showed relatively consistent patterns of $CO_2$ and $CH_4$ release but behaved in temporal patterns opposite of canal water with respect to $N_2O$: The soils

## TABLE 10.1
## Fluxes of Carbon Dioxide, Nitrous Oxide, and Methane from South Florida Soils, May 1984

| Site | | | $CO_2$[a] | | $N_2O$[b] | | $CH_4$[c] | |
|---|---|---|---|---|---|---|---|---|
| | | Water | | | | | | |
| A | 1 | Lily pond | 53.7 | (1) | −1.2 | (1) | $3.1 \times 10^5$ | (1) |
| B | 2 | Lake Osceola | 46.7 ± 19.8 | (2) | 12.6 ± 6.5 | (2) | $4.1 \times 10^3 ± 4.0 \times 10^3$ | (2) |
| C | 3 | Coral Gables Canal | 278.5 ± 64.0 | (7) | nd | | nd | |
| D | 4 | Tamiami Canal | 222.3 ± 228.8 | (4) | −2.0 ± 1.3 | (4) | 86.4 ± 94.0 | (4) |
| | | Soils | | | | | | |
| E | 5 | Flooded sawgrass controls | 75.5 ± 29.0 | (6) | 3.0 ± 1.7 | (6) | 257.6 ± 309.3 | (6) |
| F | 6 | $NO_3^-$ fertilized sawgrass | 75.9 ± 38.6 | (3) | 40.2 ± 42.5 | (3) | 51.1 ± 19.9 | (3) |
| G | 7 | $NH_4^+$ fertilized sawgrass | 84.4 ± 11.5 | (3) | 67.4 ± 75.5 | (3) | 571.4 ± 363.4 | (3) |
| H | 8 | Non–flooded sawgrass control | 153.6 ± 37.6 | (8) | 43.7 ± 25.4 | (8) | 94.3 ± 46.8 | (8) |
| I | 9 | Casuarina forested levee | 52.4 ± 35.0 | (4) | −1.2 ± 2.3 | (4) | 36.5 ± 36.7 | (4) |
| J | 10 | Prairie controls | 110.1 ± 36.7 | (14) | 19.2 ± 8.2 | (14) | 7.9 ± 9.5 | (14) |
| K | 11 | $NO_3^-$ fertilized prairie | 55.4 ± 23.0 | (3) | 18.9 ± 9.7 | (3) | 8.5 ± 3.5 | (3) |
| L | 12 | $NH_4^+$ fertilized prairie | 86.1 ± 20.9 | (3) | 204.3 ± 77.3 | (3) | 30.7 ± 12.7 | (3) |
| M | 13 | Burned field | 76.1 ± 27.5 | (4) | 6.9 ± 3.8 | (4) | 0.0 ± 0.0 | (4) |
| N | 14 | Unburned fallow field | 164.4 ± 59.4 | (4) | 20.6 ± 7.0 | (4) | 0.0 ± 1.8 | (4) |
| O | 15 | Secondary forest controls | 164.9 ± 81.3 | (22) | 15.9 ± 24.8 | (22) | −25.7 ± 34.4 | (22) |
| P | 16 | $NO_3^-$ fertilized forest | 188.5 ± 52.2 | (3) | 70.3 ± 77.8 | (3) | −12.8 ± 13.1 | (3) |
| Q | 17 | $NH_4^+$ fertilized forest | 132.2 ± 76.2 | (3) | 115.1 ± 93.1 | (3) | −1.8 ± 15.9 | (3) |
| R | 18 | St. Augustine grass | 91.3 ± 110.5 | (14) | 2.0 ± 5.0 | (14) | −9.3 ± 26.8 | (14) |
| S | 19 | Zoysia grass | 142.4 ± 168.4 | (14) | 2.3 ± 2.7 | (14) | −12.3 ± 19.7 | (14) |
| T | 20 | Fertilized Bermuda grass | 305.1 ± 127.0 | (4) | 12.0 ± 2.3 | (4) | 48.6 ± 127.2 | (4) |
| U | 21 | Sugarcane | 44.8 ± 49.4 | (8) | 1.3 ± 2.2 | (8) | −13.4 ± 11.0 | (8) |
| V | 22 | Cocoa | 106.2 ± 43.5 | (8) | 10.7 ± 22.1 | (8) | −24.3 ± 14.1 | (8) |
| W | 23 | Coconut | 254.9 ± 99.0 | (4) | 26.3 ± 29.0 | (4) | −3.2 ± 361.9 | (4) |
| X | 24 | Mulch controls | 666.3 ± 897.4 | (18) | 14.2 ± 20.1 | (18) | −6.5 ± 30.1 | (18) |
| Y | 25 | $NO_3^-$ fertilized mulch | 415.8 ± 377.3 | (3) | 22.9 ± 19.7 | (3) | −3.4 ± 2.2 | (3) |
| Z | 26 | $NH_4^+$ fertilized mulch | 396.6 ± 309.3 | (3) | 29.3 ± 26.8 | (3) | 0.0 ± 2.1 | (3) |

*Note:* Means ± standard deviations are shown; number of replicates in parentheses.

[a] Carbon dioxide (mg $C–CO_2$ m$^{-2}$ h$^{-1}$).
[b] Nitrous oxide (µg $N–N_2O$ m$^{-2}$ h$^{-1}$).
[c] Methane (µg $C–CH_4$ m$^{-2}$ h$^{-1}$).

released it when the soil was largely above the water table, but consumed it when all but the top centimeter was submerged.

## EVERGLADES SITES

The sawgrass community released twice as much $CO_2$ prior to submergence as afterwards, the prairie community having an intermediate value (Figure 10.2). Sawgrass peat soils were the largest soil $CH_4$ sources measured, especially after being waterlogged up to the surface. Prairie peat soils were consistent but modest $CH_4$ sources, presumably from deeper waterlogged horizons, as the organic-rich upper soil layers were generally dry. Both Everglades habitats were large sources of $N_2O$ to the atmosphere, but, while sawgrass soils released more than twice as much $N_2O$ as prairie soils before they were submerged, they released about one tenth as much afterwards. Nitrogen fertilization did not appear to affect $CO_2$ release, but $NH_4^+$-treated sawgrass sites released

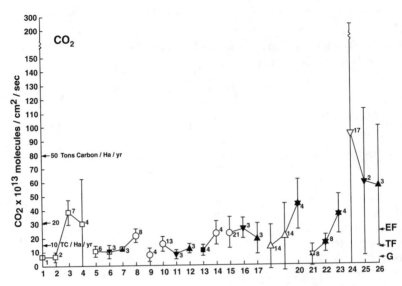

**FIGURE 10.2** Carbon dioxide fluxes (molecules/cm²/sec and C ton/ha/yr), plotted vs. site number. Because increasing sample number corresponds to drier conditions, carbon dioxide release is seen to be lowest in the wettest sites. At right, G is the global average flux, TF is the flux of temperate forests, and EF is the flux of equatorial forests.

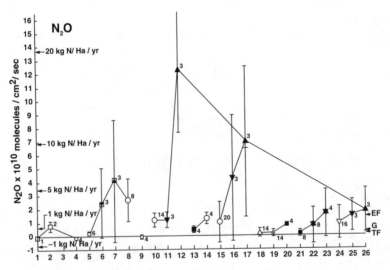

**FIGURE 10.3** Nitrous oxide fluxes (molecules/cm²/sec and N kg/ha/yr), plotted vs. site number. The highest fluxes are seen at intermediate water table levels and in response to fertilization. At right, G is the global average flux, TF is the temperate forest flux, and EF is the equatorial forest flux.

more methane than controls. The large variation in $CH_4$ release from controls at the sawgrass site suggests that local factors are responsible. Nitrogen fertilization was shortly followed by rain showers and had a dramatic effect on $N_2O$ production at both sites (Figures 10.6 and 10.7). Prairie soil $N_2O$ production shot up following $NH_4^+$ fertilization but was not affected by $NO_3^-$ fertilization. Soil $N_2O$ production was very low in unfertilized sawgrass controls. A large peak of $N_2O$ was released from the $NO_3^-$-treated site during the first submergence of the soils, but dropped following drying and was followed by a higher peak from the $NH_4^+$-treated soil when its surface was re-exposed

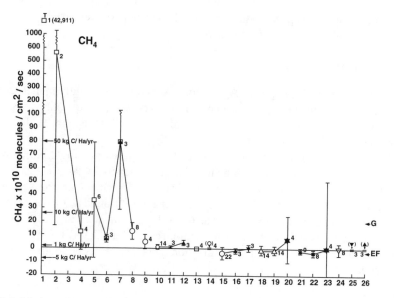

**FIGURE 10.4** Methane fluxes (molecules/cm²/sec and C kg/ha/yr), plotted vs. site number. Methane release is highest in the wettest sites, and drier sites absorb methane from the atmosphere. At right, G is the global average flux, and EF is the equatorial and temperate forest flux.

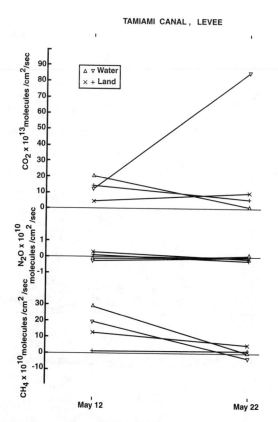

**FIGURE 10.5** Gas fluxes from flooding Tamiami Canal during low and high water stages. Fluxes were measured from canal water surfaces and adjacent land levee soils.

to the air by evaporation and seepage. Nitrous oxide release from both fertilized sites then dropped to low control levels following a second submergence.

## Fallow and Burned Sites

The fallow field in the drained Everglades Agricultural Area released slightly higher levels of $CO_2$ and $N_2O$ than prairie sites, but neither produced nor consumed significant quantities of $CH_4$. Burned sites produced around half as much $CO_2$ and one third as much $N_2O$ as unburned sites. No $CH_4$ production or consumption was detected.

Combustion gases contained elevated levels of all three gases, but rapid entrainment of air by rising flames diluted concentrations to near atmospheric levels in samples taken more than 50 cm above the flames, even in thick smoke. The $CH_4$ to $CO_2$ molecular production ratio was scattered between $2.45 \times 10^{-3}$ and $11.5 \times 10^{-3}$, with a mean of around $4.6 \times 10^{-3}$ (Figure 10.8). The $N_2O$ to $CO_2$ molecular production ratio was $6.2 \times 10^{-5}$, equivalent to $1.24 \times 10^{-4}$ atoms of nitrogen released as nitrous oxide per atom of carbon released as $CO_2$ (Figure 10.8). Numerous hydrocarbons were also measured by flame ionization detector, but these could not be identified due to lack of standards.

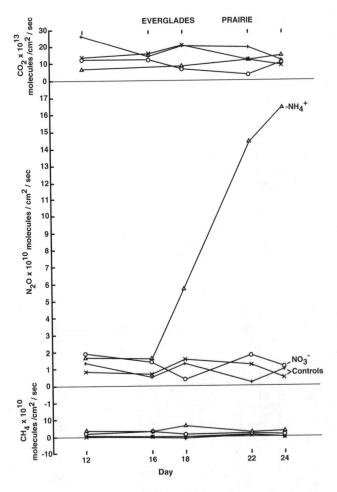

**FIGURE 10.6** Gas fluxes from Prairie sites during the late dry season and early wet season. $NH_4^+$ indicates plots fertilized with ammonium; $NO_3^-$ plots fertilized with nitrate. Control plots received no fertilizer.

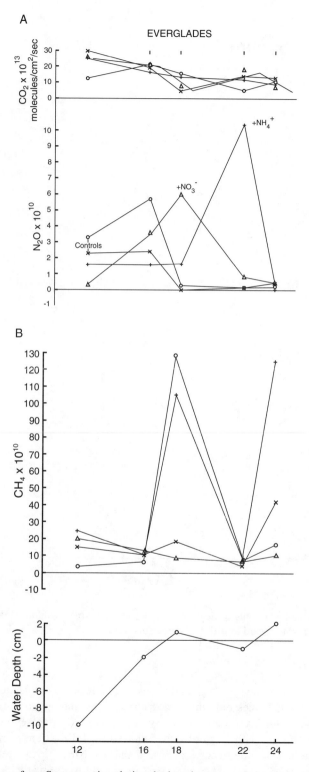

**FIGURE 10.7** Gas fluxes from Sawgrass sites during the late dry season and early wet season. Ammonium-fertilized, nitrate-fertilized, and control plots are shown. The water table level with regard to the soil surface is shown at bottom.

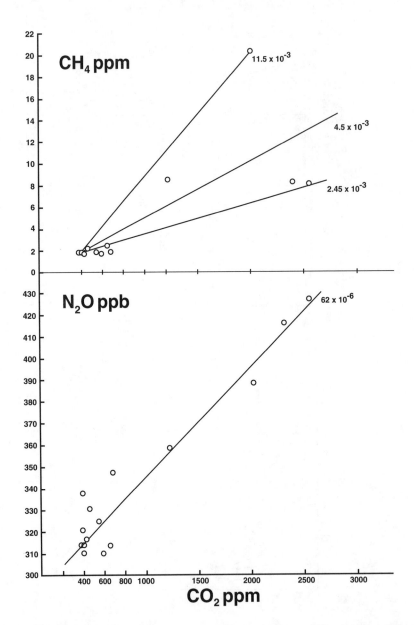

**FIGURE 10.8** Methane and nitrous oxide concentrations in smoke from wildfire in dry prairie at the end of the dry season, shown with regard to carbon dioxide concentrations.

## SECONDARY FOREST

Release of $CO_2$ from forest soil was high and appreciably affected by moisture (Figure 10.9). Light rain events during the first week only moistened surface litter and uppermost soil, after over three rainless months. Soils quickly became very dry again, and $CO_2$ releases fell. The rainy season began abruptly on May 22 with 5 cm of rain, followed by 1.3 to 9 cm of rain per day over the following week. Carbon dioxide releases increased sharply. Nitrous oxide releases also generally rose during the first moist period, dropped as soils dried, and were greatly elevated once the rainy season was established. Methane consumption from the atmosphere generally declined as the soil was wetted.

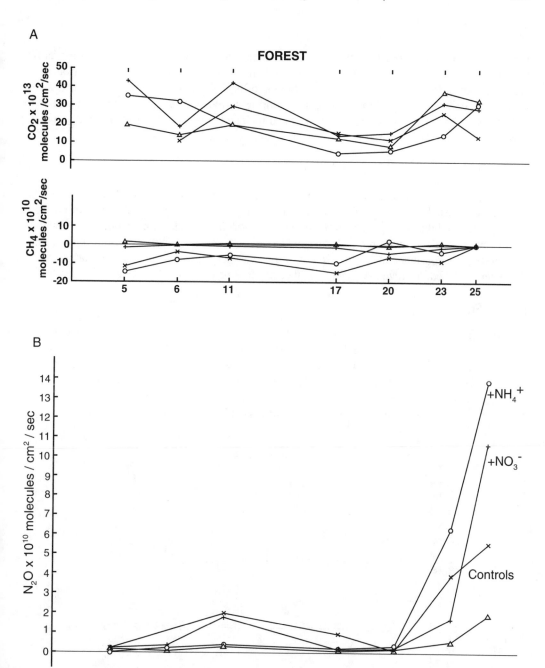

**FIGURE 10.9** Gas fluxes from secondary forest during the late dry season and early wet season. Ammonium-fertilized, nitrate-fertilized, and control plots are shown.

Fertilizer was spread uniformly over the surface on May 17, after the flux had been measured. White fertilizer powder remained visibly undisturbed on the soil surface following a rainless interval, during which the gas fluxes declined. After the first rain shower, all surface fertilizer dissolved, and gas fluxes rose. Nitrogen fertilization had little effect on either $CO_2$ or $CH_4$ fluxes, but dramatically elevated $N_2O$ release. Ammonium had a larger stimulatory effect than nitrate.

## GRASS SITES

Grass sites produced more $CO_2$ when soil was moist, accompanied by bursts of grass growth (Figure 10.10). $N_2O$ releases were very low from lawns throughout most of the period, rising somewhat towards the very end. Slow consumption of $N_2O$ from the atmosphere was occasionally noted. $CH_4$ consumption was most appreciable during periods of damp soil, but small production of $CH_4$ was also noted on occasion. Fertilized and mulched grass sites showed elevated and highly variable production of all three gases compared to unfertilized grass sites.

## AGRICULTURAL SITES

Agricultural sites released more $CO_2$ the more recently they had been fertilized (Figure 10.2). Sugarcane released the lowest values, cacao intermediate, and coconut the highest. Nitrous oxide releases followed the same pattern: Release was very low in the sugar cane, moderately low in the cacao (except for one measurement), and moderate in the coconut, with one high flux. Methane was moderately consumed at all sites except for one very large positive and one very large negative flux from mulched, fertilized coconuts. Neither of these had elevated $N_2O$ release.

## MULCH SITES

Mulch sites were the largest and most variable $CO_2$ producers (Figure 10.11). Nitrous oxide production was high but less variable, being comparable to that of secondary forest. Most sites showed no measurable $CH_4$ flux, but two moderately negative and two small positive fluxes were measured. Neither $NH_4^+$ nor $NO_3^-$ fertilized sites were appreciably elevated over controls in release of any gas. This is because the high C- to N-ratio of the mulch resulted in available N being sequestered by decomposers, without any excess available for dissimilatory microbial metabolism.

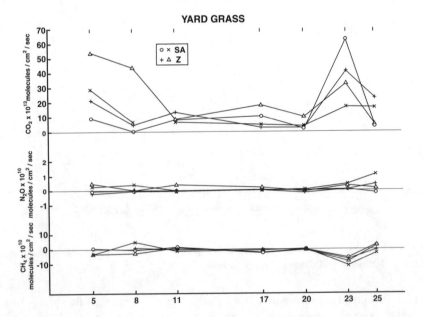

**FIGURE 10.10** Gas fluxes from unfertilized grass sites during the late dry season and early wet season. SA = St. Augustine grass sites, Z = zoysia grass sites.

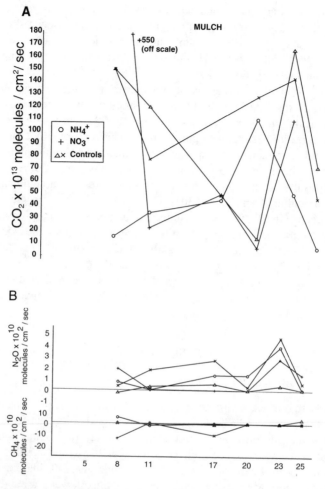

**FIGURE 10.11** Gas fluxes from mulch during the late dry season and early wet season. Ammonium-fertilized, nitrate-fertilized, and control plots are shown.

## Comparison of Fluxes Between Sites

Student's t tests have been used to compare the fluxes of the three gases between habitats and treatments:

1. Comparison between aquatic habitats (A to D) and terrestrial habitats (E to Z) showed that there were no significant differences in $CO_2$ fluxes or in $N_2O$ fluxes, but aquatic habitats had very significantly greater $CH_4$ releases ($p = 3.63 \times 10^{-8}$).
2. Flooding vs. dry conditions in the sawgrass everglades habitats showed that dry conditions had very significantly higher $CO_2$ fluxes than flooded conditions ($p = .000585$) and very significantly higher $N_2O$ fluxes ($p = .00109$) than flooded soils, but were not significantly different in $CH_4$ fluxes.
3. The effects of nitrate and ammonium fertilization on fluxes varied between habitat. In sawgrass Everglades habitat neither nitrate nor ammonium fertilization affected $CO_2$ or $CH_4$ fluxes, while $N_2O$ fluxes were significantly higher for both nitrate ($p = .0270$) and ammonium ($p = .0293$) fertilization.

4.  In dry prairie habitat, nitrate and ammonium fertilization had very different results. Nitrate addition did not affect $N_2O$ or $CH_4$ releases, but was associated with significantly lowered $CO_2$ fluxes ($p = .0137$). Ammonium fertilization in contrast did not affect $CO_2$ fluxes, but significantly raised $CH_4$ fluxes ($p = .00134$) and very strongly significantly raised $N_2O$ fluxes ($p = 2.668 \times 10^{-8}$).

5.  In forest habitat, neither nitrate nor ammonium fertilization had significant effects on $CO_2$ or $CH_4$ fluxes, and both very strongly significantly increased $N_2O$ fluxes, with ammonium having the greater effect ($p = 9.299 \times 10^{-5}$) than nitrate ($p = .00668$).

6.  In grass, fertilization resulted in significantly greater releases of all three gases.

7.  In mulch, neither ammonium nor nitrate fertilization had a significant impact on any gas.

8.  In contrasting alternative land uses on higher elevation soils, conversion of forests to agriculture resulted in no change in $CH_4$ or $N_2O$ fluxes and a significantly lower $CO_2$ flux from agricultural land ($p = .0301$). Conversion of forests to grass resulted in no significant change in $CO_2$ fluxes, a significant reduction in $CH_4$ consumption ($p = .0378$), and a very significant decrease in $N_2O$ release from grass ($p = .00285$). Conversion of agriculture to grass resulted in no significant change in $CO_2$ or $CH_4$ fluxes, but a significantly lower $N_2O$ release from grass ($p = .04612$).

9.  Effects of burning in prairie habitat showed that the burned areas had significantly lower $CO_2$ fluxes ($p = .0178$), very strongly significantly lower $N_2O$ fluxes ($p = .00689$), but no significant difference in $CH_4$ fluxes.

## DISCUSSION

### SOUTH FLORIDA AS A SOURCE OF $CO_2$

Most South Florida soils release $CO_2$ to the atmosphere at rates well above the global average (Figure 10.2). Carbon dioxide releases are larger from upland soils than from marsh soils. They are elevated by mulching, P fertilization, and drainage. Unfertilized South Florida forests, lawns, and agricultural sites released $CO_2$ at high rates, comparable to mean values from undisturbed Central Amazonian rainforest soils of 16.2 T C ha$^{-1}$ yr$^{-1}$, measured during the dry season (Goreau and de Mello, 1985, 1988). Marsh sites and burned fields had lower $CO_2$ and $N_2O$ release rates, comparable to those from undisturbed temperate hardwood forest soils at Hubbard Brook, New Hampshire, 7 T C ha$^{-1}$ yr$^{-1}$ (Goreau, 1982). Mulched sites had exceptionally high rates of $CO_2$ loss, comparable to the highest reported in the literature, 61 T C ha$^{-1}$ yr$^{-1}$ from a Costa Rica rainforest with no dry season on highly fertile volcanic soils (Schulze, 1967).

### SOUTH FLORIDA AS A SOURCE OF $CH_4$

Methane releases from aquatic areas (Figure 10.4) were comparable to mean fluxes from northern Minnesota swamps and lakes, around 1500 kg C ha$^{-1}$ yr$^{-1}$ (Harriss et al., 1985), and those found from permanently drowned Amazonian forest root mat layers, 1500 kg C ha$^{-1}$ yr$^{-1}$ (Goreau and de Mello, 1985), and from floating grass mats in the Amazonian floodplain, 980 to 1120 kg C ha$^{-1}$ yr$^{-1}$ (Bartlett et al., 1988, 1990). Everglades soils were much smaller sources per unit area but still significant compared to global average production, especially during peat submergence. Their fluxes were comparable to those from Florida Cypress bogs, 40 to 350 kg C ha$^{-1}$ yr$^{-1}$ (Harriss et al., 1985) and from freshwater wetlands in Belgium, 60 kg C ha$^{-1}$ yr$^{-1}$ (Boeckx and Van Cleemput, 1997), but generally low compared to fertilized rice paddies in California, 400 to 700 kg C ha$^{-1}$ yr$^{-1}$ (Cicerone and Shetter, 1981); in Spain, 35,000 kg C ha$^{-1}$ yr$^{-1}$ (Seiler et al., 1984), or Northern Minnesota bogs (Harriss et al., 1985). Upland soils were virtually all $CH_4$ consumers comparable to Amazon forest soils (-2.2 kg C ha$^{-1}$ yr$^{-1}$; Goreau and de Mello, 1985), the chief exception being heavily mulched and fertilized sites, which had sporadic "hot spots" of $CH_4$ production, presumably in anoxic microhabitats. Methane concentrations in peat increase with depth below the water table, but are low above it (Benstead and Lloyd, 1994). Methane release is likely to be greatest from freshwater wetlands, as brackish and marine wetlands are likely to have $CH_4$ production suppressed by competition with $SO_4^{-2}$ reducing bacteria. Reduced sulfur gas emissions in South Florida take place largely in mangrove areas, and sawgrass, which covers 42% of the area, produces only 24% of the region's reduced sulfur gases (Hines et al., 1993). In contrast, sawgrass accounts for a disproportionately large fraction of $CH_4$ releases from the area, and the high sulfur-emitting habitats are likely to be very low methane emitters. Because the $CH_4$ flux through vascular spaces in sawgrass and other aquatic vegetation can exceed fluxes from soil and water (Happell et al., 1993), the fluxes should be even higher than reported here.

## South Florida as a Source of N$_2$O

South Florida $N_2O$ releases are generally large per unit area compared to the global average, especially in unfertilized dry Everglades sites, agricultural sites, forest sites, and mulch sites (Figure 10.3). These habitats released $N_2O$ to the atmosphere at a rate comparable to Amazonian rain forest soils, one of the Earth's largest sources: 1.1 to 1.7 kg N ha$^{-1}$ yr$^{-1}$ (Goreau and de Mello, 1985; Livingston et al., 1988; Matson et al., 1990). Nitrous oxide release was very low in flooded unfertilized Everglades sites, most aquatic sites, unfertilized grass lawns, and sugarcane (also a C-4 grass) sites. These habitats are comparable in $N_2O$ fluxes to Amazonian grasslands, 0.2 kg N ha$^{-1}$ yr$^{-1}$ (Goreau and de Mello, 1985), or temperate hardwood forests at Hubbard Brook, New Hampshire, 0.2 kg N ha$^{-1}$ yr$^{-1}$ (Goreau, 1982). The comparable values for unfertilized Bermuda grass are 0.1 kg N ha$^{-1}$ yr$^{-1}$ in Spain (Slemr et al., 1984), and 1.8 to 3.6 kg N ha$^{-1}$ yr$^{-1}$ in sheep pasture, unfertilized except for manure (Christensen, 1983). Fertilized grass and pasture sites at the two grass sites were as high as 1.4 and 183 kg N ha$^{-1}$ yr$^{-1}$, respectively. Fertilized sites in the Everglades and in secondary forest released very large fluxes of $N_2O$, equivalent to values from more heavily fertilized agricultural fields on mineral soils: 3.9 kg N ha$^{-1}$ yr$^{-1}$ in Ontario corn fields vs. 0.4 for unfertilized controls (McKenney et al., 1978); 1 to 14 kg N ha$^{-1}$ yr$^{-1}$ from fertilized soybeans in Spain (Slemr et al., 1984); and up to a maximum of 90 kg N ha$^{-1}$ yr$^{-1}$ in heavily fertilized fallow German fields (Conrad et al., 1983). Nitrous oxide response to fertilization decreased in elevated soils with heavy mulch compared to less heavily mulched upland soils (dashed line on Figure 10.3). Nitrous oxide-producing nitrogen transformations may have been outcompeted for nitrogen by litter decomposers such as bacteria and fungi that sprouted abundantly after heavy rain. South Florida agricultural and urban habitats are therefore very intense hot spots of $N_2O$ release unless they are very heavily mulched, which causes nitrogen limitation.

## Sources of N$_2$O: Nitrification vs. Denitrification

These data shed much light on the controversy over the relative roles of nitrification and denitrification as sources of atmospheric $N_2O$ (Goreau et al., 1980). Denitrification takes place only in the absence of molecular oxygen, but oxygen is essential for nitrification. Denitrification has often been inferred to be the source of $N_2O$ releases from soils because denitrifying bacteria can be made to release it under laboratory conditions, but denitrification will be a source of the gas to the atmosphere only if the gas can escape before being consumed. Nitrous oxide is the energetically favored electron acceptor in denitrification, and so is rapidly consumed once oxygen disappears. Anoxic denitrifying environments removed from contact with the atmosphere are often very low or completely free of $N_2O$ and are capable of consuming large amounts of added $N_2O$ (Figure 10.12), implying that denitrification tends to act as a sink of the gas.

Nitrifying bacteria produce $N_2O$ during growth at all positive concentrations of oxygen, but do so at a sharply elevated rate at the lowest oxygen levels (Goreau et al., 1980). Much $N_2O$ evolved from soils that was formerly interpreted as being due to denitrification may actually have been

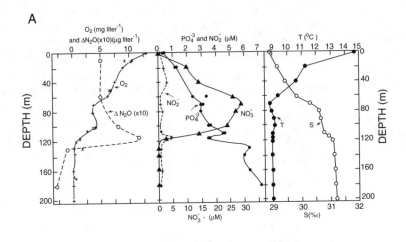

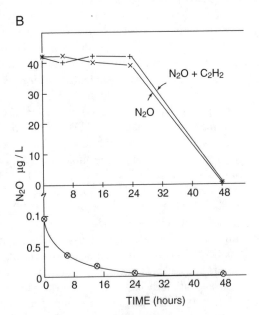

**FIGURE 10.12**    (A) Vertical profile of nitrous oxide (as a fraction of atmospheric equilibrium concentration), oxygen, nitrate, and other water properties at Saanich Inlet Fjord, southeast Vancouver Island, British Columbia. Waters are anoxic below the 90-m sill depth. Nitrous oxide is shown as the difference between the measured value and the concentration in equilibrium with the atmosphere at that temperature and pressure. Positive values indicate supersaturation of nitrous oxide, indicating a local source, while negative values show undersaturation as the result of local consumption. Only tiny traces of nitrous oxide could be measured in deep oxygen-deficient waters. (B) Nitrous oxide concentrations in incubations of near-bottom (180-m depth) anoxic waters from Saanich Inlet. Samples were incubated with added sterile nitrous oxide-saturated seawater, with and without sterile seawater saturated with acetylene, regarded to be an inhibitor of denitrification. The bottom sample is an unamended control, which contained traces of nitrous oxide that were quickly consumed. Added nitrous oxide, at levels 400 times greater than ambient, was completely consumed in anoxic waters. Nitrous oxide consumption was not blocked by acetylene or by ambient hydrogen sulfide, compounds considered potential inhibitors of nitrous oxide reduction by denitrifying bacteria. The time delay in consumption could represent the lag phase in growth of a nitrous oxide-using bacteria or the induction period for synthesis of the nitrous oxide reductase enzyme.

produced by nitrification, which is likely to be the major process producing the gas in the atmosphere (Bremner and Blackmer, 1981; Goreau, 1982). Everglades agricultural area soils have been found to release much larger amounts of nitrogen when decomposing under drained conditions than when waterlogged, and most of the nitrogen is released as nitrate under aerobic decomposition, while only traces are produced under waterlogged conditions (Hanlon et al., 1997). Nitrate release is found to increase when the water table is lowered, while ammonium and total nitrogen release is not affected (Martin et al., 1997).

Our data are consistent with the view that nitrification was the predominant source of $N_2O$ from South Florida soils. Ammonium fertilization stimulated $N_2O$ release most markedly at sites similar to South Florida agricultural habitats. Nitrate stimulated $N_2O$ release only when peat was waterlogged and in soggy forest litter in which anoxic microsites are indicated by sporadic $CH_4$ release. Nitrous oxide produced by denitrification in anoxic soil and litter microsites has a much better probability of escape to the atmosphere without being consumed than that produced in submerged peat. Production of $N_2O$ by denitrification was proposed to explain greater releases following nitrate fertilization than ammonium fertilization in Amazonian clay soils during the wet season (Keller et al., 1988), a pattern opposite to that found here.

Nitrous oxide releases from exposed soils in a temperate-zone (Cape Cod), *Spartina alterniflora* salt marsh are also suggestive of a predominant role for nitrification. Undisturbed control sites in areas of tall and short growth forms of *Spartina* were contrasted with sites that had been experimentally fertilized with sewage. Fertilized saltmarsh soils had much more lush plant growth, were more oxidized due to higher aeration by plant roots, and had thousand-fold higher nitrate concentrations and nitrifying bacteria counts than unfertilized controls (B. Bowden and J. Waterbury, pers. comm.). Nitrous oxide production rates from fertilized *Spartina* marsh soils were greatly elevated over unfertilized rates. These results strongly suggest that nitrification in aerated surface sediments and waters can be the predominant source of $N_2O$ to the atmosphere even from wet organic soils, from which denitrification has been traditionally regarded as the source.

Varying nitrogen and phosphorus ratios can also greatly affect emissions of $N_2O$. Everglades peat soils have high N contents, and ammonium produced by microbial oxidation of soils is very rapidly nitrified and leached as nitrate from drained soils (Reddy, 1982). Everglades canal and surface water nitrate concentrations are highest in the dry season and lowest in the rainy season (Klein et al., 1975). Because nitrification in seasonally flooded soils is most intense shortly after drainage (Ponnamperuma, 1984), maximal $N_2O$ production from these soils from nitrification would be expected early in the dry season rather than at its end. Although added P is almost entirely removed, the amount of N retained depends on whether P is also added (Craft and Richardson, 1993), a result of the strong P limitation found in Everglades habitats (Vaithiyanathan and Richardson, 1998). During the rainy season, total N and P levels are higher than during the dry season, but N becomes more elevated than P (Kochrose et al., 1994). Ammonium is found to stimulate $CO_2$ release from soils when P levels are high, but this release is reduced in low-P soils (Amador and Jones, 1993). Increased P fertilization has been found to increase the release of $CO_2$ and $CH_4$ from Everglades soils to which acetate, glucose, cellulose, and sawgrass residues have been added (Amador and Jones, 1995). Everglades waters have steadily rising P concentrations, while N levels have fallen (Walker, 1991, 1998). However they still remain in a P-deficient status in which N is in excess and is poorly retained and in which small amounts of P additions trigger large growth responses of eutrophic plants and algae. Phosphorus-limited soils, such as those in the Everglades, release much larger amounts of N oxides than N-limited soils (Hall and Matson, 1999). This is presumably because plants under N-limited conditions compete very effectively for available N, but under P limitation more N is available than can be used by primary producers, leaving the excess for dissimilatory N-gas-producing pathways.

## Biomass Combustion as a Source of Atmospheric Gases

The molecular production ratio of $N_2O$ to $CO_2$ averaged $6.2 \times 10^{-5}$, 34 times lower than reported values of $2.2 \times 10^{-3}$ (Crutzen et al., 1979; subsequently revised downwards one order of magnitude), or $1.5^{-3} \times 10^{-4}$ (Crutzen et al., 1984). The ratio of $CH_4$ to $CO_2$ production showed much larger variance, and was about $4.6 \times 10^{-3}$, a number smaller than the 1.6% reported by Crutzen et al. (1979) or the 1 to 1.6% reported by Crutzen et al. (1984). Forest fires in Colorado in a pinon and juniper forest with grass, juniper, and sagebrush undergrowth and in a mature spruce and fir forest were sources of the 1979 measurements, while the later figures are for burning of felled Amazonian rain forest. Our measurements were made in a secondary grass–herb community on fallow agricultural land with much smaller biomass but a larger fraction of combustible material than the forest sites so it is likely to have burned at a much lower temperature. Total N volatilization by fires under these conditions is low, rising at the increased temperatures found in woody biomass fires (Chandler et al., 1983). Further measurements of biomass combustion of vegetation with differing moisture, C/N ratios, and combustion temperatures are needed to improve current estimates of global biomass combustion as a source of $CO_2$, $CH_4$, and $N_2O$ to the atmosphere.

## Gas Fluxes, Land, and Water Management in South Florida

The highest $N_2O$ fluxes reported from soils were measured in drained agricultural areas just south of Lake Okeechobee, with averages of 48, 97, and 165 kg N ha$^{-1}$ yr$^{-1}$ for sugar cane, St. Augustine grass, and a fallow field, respectively (Terry et al., 1981; Duxbury et al., 1982). They were interpreted as being largely due to denitrification because the largest fluxes occurred during the rainy season. Microbial oxidation of drained soils has caused the peat to subside up to 7 m in 30 years and is expected to decompose most remaining soil in the agricultural area within decades (Lucas, 1982). Large amounts of nitrogen and phosphorus released into Lake Okeechobee and the South Florida Water Management District (SFWMD) Water Conservation Areas from the agricultural area cause eutrophication of the lake and canal waters (Klein et al., 1975). Rising urban water demand now forces the SFWMD to backpump agricultural waste waters into the WCAs. Backpumped waters are more enriched in N than in P (SFWMD, 1982). Added N was rapidly transformed in the system, but its fate depended on water height (Figures 10.7 and 10.8). Carbon releases were barely affected by N fertilization. Experimental work in the northern Everglades National Park shows strong P limitation of primary production (M. Flora and D. Walker, pers. comm.; Vaithiyanathan and Richardson, 1998). Our fertilization sites were not in a basin receiving backpumped waters, and we did no P additions. However, elevated soil sites receiving P and trace metals together with N had higher releases of gases than similar sites receiving only N, indicating that P fertilization practices are an important determinant of South Florida gas fluxes to the atmosphere. Fluxes from marshes fertilized by backpumping with P-rich agricultural drainage waters should be quite different than those observed with N addition alone.

The major land use in unpaved suburban areas of South Florida is lawn grass. Many lawns are unfertilized, but the sandy organic-poor soils have stimulated extensive use of St. Augustine grass growing on Everglades peat slabs which are machine cut and harvested by sod farmers south of Lake Okeechobee. Duxbury et al. (1982) reported much larger $N_2O$ fluxes from sugar cane fields and St. Augustine grass sod farms in the Okeechobee agricultural area than we found from South Florida mineral soils, so such lawns should be even more productive of $CO_2$ and $N_2O$ than the fertilized lawn studied. Because our data are confined to a short interval of a year within a limited range of sites and treatments, further data from a wider range of habitats over a complete seasonal cycle are needed to evaluate annual losses of C and N to the atmosphere and the net impact of land and water management practices upon them in South Florida.

## PAST AND FUTURE CHANGES IN TRACE GAS FLUXES IN SOUTH FLORIDA

South Florida hydrology has changed drastically in the past 100 years in ways that should strongly affect the strength of its habitats as sources or sinks of atmospheric trace gases. The impacts of current restoration of Everglades flows on greenhouse gases is also likely to be large. Reflooding the Everglades will basically reverse historical trends by converting old drained peat soils back into increasingly wet conditions. This will greatly increase the release of $CH_4$ from the environment and reduce the release of $CO_2$. The likely net impact would be to increase global warming impacts.

The situation with $N_2O$ is much less clear, because its greatest output is found in intermediate or "low" $O_2$ waters and soils. If areas which had been seasonally dry become permanently flooded, large $N_2O$ sources would be replaced by areas consuming the gas; however, new areas that had been formerly too dry to produce much $N_2O$ may now become sufficiently wet to do so. Given the hypsographic curve of the area, though, more areas will become too wet to be large sources than will be converted into it, so the likely impact would be to diminish production of this gas. However, this effect is counteracted by the high use of fertilizers for agriculture and lawns in the area, along with the strong additions from backpumping of agricultural wastewater, nutrient additions from agricultural fertilizer, runoff from pasture manure, and decomposing peat, all of which reflect a rapidly growing population and its strong effect on increasing $N_2O$ production. If fertilizer increases continue, so will $N_2O$ emissions. Because the ecosystem is P limited, any added N will be lost, as it is already present in excess, with a resulting build-up of drainage water nitrate and $N_2O$. Because their growth is restricted by a lack of P, plants become poor competitors for excess N, allowing more of it to enter microbial pathways and resulting in release of N gases to the atmosphere. Changes in land use (including restoration of Everglades flows), increased drainage of soils in agricultural and urban areas, and intense use of fertilizers will make South Florida a region of much higher than global average greenhouse gas emissions from natural and anthropogenic causes, the future trajectory of which is difficult to predict and therefore should be monitored directly. Reduction of the sugar subsidies could produce a dramatic change in rates of drainage and fertilization of the Everglades and hence in greenhouse gas emissions.

Despite lack of full year-round measurements, the measurements reported here during both the dry season and the wet season allow us to qualitatively estimate the impacts of past and future South Florida land and water management on the atmosphere. Drainage patterns that began in the early part of the 20th century lowered the water table when Everglades water was pumped out to sea, which must have progressively increased $CO_2$ emissions, greatly reduced $CH_4$ emissions, and greatly elevated $N_2O$ emissions. Reflooding would be thought to have the opposite effect, but this will only be true if the ecosystem regains its original flooded area, water levels, and nutrient status. Return to pre-anthropogenic conditions will not take place for at least four reasons:

1. Increasing amounts of water will continue to be removed for agriculture and urban uses and will not contribute to recharge, so the system will never become as wet as it was unless urban and agricultural land uses are eliminated.
2. The large amount of new nutrient inputs that have accumulated from fertilizers used in sugar cane, vegetables, and pasture; cow manure; urban runoff, and peat drainage will maintain the system in a N-rich condition in which excess N flows through bacterial pathways, leading to elevated N gas emissions to the atmosphere for a long time.
3. Increase in P relative to N (Walker, 1991, 1998) keeps the system in a state that is increasingly unable to retain N. So much previous wetland has been permanently drained and utilized that the flow will be concentrated into smaller but more intense source areas.

As a result, $N_2O$ fluxes may not decline to lower levels with increased flooding and could remain at permanently high levels in the future.

Year-round field work like that done in this brief study would shed much light on the net impacts of land and water management on the atmosphere and its potential for being restored to a better balance, as well as the long-term impacts of past and present water and land use policies. Direct monitoring of the effects of drainage changes will be needed for some years before the net impacts of Everglades reflooding on atmospheric budgets of the major climatically active gases can be estimated with confidence.

## CONCLUSIONS

Wetland ecosystems are important sources of climatically active atmospheric gases, which could be profoundly altered by land and water management practices as well as by normal seasonal cycles. Over 150 measurements of carbon dioxide ($CO_2$), nitrous oxide ($N_2O$), and methane ($CH_4$) simultaneously to the atmosphere from soils and waters were made from a wide variety of subtropical South Florida habitats during the late dry season and early wet season of 1984. These showed that South Florida habitats were major sources of atmospheric greenhouse gases, but that each gas had different patterns of environmental and anthropogenic influence. Unfertilized South Florida habitats produced $CO_2$ and $N_2O$ at rates per unit area comparable to Amazonian rain forests, and fertilized habitats yielded higher emission rates. Nitrous oxide and $CO_2$ releases were higher from drained than from submerged Everglades soils. Nitrous oxide release was sharply elevated by ammonium ($NH_4^+$) fertilization at all sites except mulch. Nitrate ($NO_3^-$) fertilization did not affect $N_2O$ release from drained Everglades soils, and elevated it to a lesser extent than equivalent $NH_4^+$ additions at waterlogged sites and from wet forest litter (presumably from anoxic microsites). The data are therefore consistent with nitrification as the major local biological source of $N_2O$. Methane was generally consumed by most soils but was produced at moderate rates from Everglades soils and at very high rates from aquatic sites. Nitrogen fertilization alone did not appreciably alter $CO_2$ and $CH_4$ fluxes because of phosphorus limitation. Current land and water management practices, such as alteration of water tables and backpumping of nutrient-enriched agricultural wastewaters into the South Florida WCAs, strongly change the rates at which the three gases are released to the atmosphere: Release of nutrient-rich waters into the Everglades increases $N_2O$ fluxes, drainage increases $CO_2$ fluxes, and flooding increases $CH_4$ fluxes. Historical drainage patterns during the 20th century have acted to increase $CO_2$ emissions and greatly reduce $CH_4$ emissions, while stimulating increased $N_2O$ release. Reflooding of the Everglades to restore a more natural flow will decrease $CO_2$ emissions and greatly increase $CH_4$ releases, but probably not to the original levels. In contrast, $N_2O$ release is likely to remain elevated due to high inputs of P and N from agricultural fertilizer and drainage. Long-term monitoring is needed to determine the impacts of water and land use practices on the area's budget of the major climatically active gases.

## ACKNOWLEDGMENTS

We thank Mark Flora and Dave Walker (Everglades Research Center, National Park Service), Dr. Robert Knight (U.S. Department of Agriculture Subtropical Horticultural Station), Dr. Breck Bowden (Yale University Forestry School), Dr. John Waterbury (Woods Hole Oceanographic Institution), and Dr. Ron Hofstetter (University of Miami) for discussions; James McAteer, for assistance with initial surveys; Prof. Steven C. Wofsy (Harvard University), for gas standards, valves, assistance modifying the gas chromatograph inlet system, and help with references; Steve Orzack (Fresh Pond Research Institute), for assistance with bibliography and software; Ron Friedland, for statistical programming; and Barbara Glaccum, for drafting. We particularly wish to thank Dr. Tom Lovejoy and the World Wildlife Fund for support of measurements in Amazonian rain forests which made this work in South Florida possible, as well as a field trial of the equipment. The late Cesare Emiliani provided laboratory space at the University of Miami in which the analyses were done.

## REFERENCES

Alexander, T.R. and A.G. Crook. 1974. Recent vegetational changes in southern Florida, in *Environments of South Florida: Present and Past*, Memoir 2, Gleason, P.J., Ed., Miami Geological Society.

Amador, J.A. and R.D. Jones. 1993. Nutrient limitations on microbial respiration in peat soils with different total phosphorus content, *Soil Biology and Biochemistry*, 25:793–801.

Amador, J.A. and R.D. Jones. 1995. Carbon mineralization in pasture and phosphorus enriched peat soils of the Florida Everglades, *Soil Science*, 159:129–141.

Bartlett, K.B., P.M. Crill. D.I. Sebacher, R.C. Harriss, J.O. Wilson, and J.M. Melack. 1988. Methane flux from the Central Amazonian floodplain, Journal of Geophysical Research, 93:1571–1582.

Bartlett, K.B., P.M. Crill, J.A. Bonassi, J.E. Richey, and R.C. Harriss. 1990. Methane flux from the Amazon river floodplain: emissions during rising water, *Journal of Geophysical Research*, 95:16773–16788.

Benstead, J. and D. Lloyd. 1994. Direct mass-spectrometer measurement of gases in peat cores, *FEMS Microbiology Ecology*, 13:233–240.

Blackmer, A.M., S.G. Robbins, and J. M. Bremner. 1982. Diurnal variability in rate of emission of nitrous oxide from soils, *Soil Science Society of America Journal*, 46:937–942.

Boeckx, P. and O. Van Cleemput. 1997. Methane emissions from a freshwater wetland in Belgium, *Soil Science Society of America Journal*, 61:1250–1256.

Bremner, J.M. and A.M. Blackmer. 1981. Terrestrial nitrification as a source of atmospheric nitrous oxide, pp. 151–170 in *Denitrification, Nitrification, and Atmospheric Nitrous Oxide*, Delwiche, C.C., Ed., Wiley, New York.

Chandler, C., P. Cheney, P. Thomas, L. Trabaud, and D. Williams. 1983. *Fire in Forestry*, Vol. 1: *Forest Fire Behaviour and Effects*, Wiley, New York.

Christensen, S. 1983. Nitrous oxide emission from a soil under permanent grass: seasonal and diurnal fluctuations as influenced by manuring and fertilization, *Soil Biology and Biochemistry*, 15:531–536.

Cicerone, R.J. and J.D. Shetter. 1981. Sources of atmospheric methane: measurements in rice paddies and a discussion, *Journal of Geophysical Research*, 86:7203–7209.

Conrad, R., W. Seiler, and G. Bunse. 1983. Factors affecting the loss of fertilizer nitrogen into the atmosphere as $N_2O$, *Journal of Geophysical Research*, 88:6709–6718.

Craft, C.B. and C.J. Richardson. 1993. Peat accretion and N, P, and organic C accumulation in nutrient-enriched and unenriched Everglades peatlands, *Ecological Applications*, 3:446–458.

Crutzen, P.J., L.E. Heidt, P.J. Krasnec, W.H. Pollock, and W. Seiler. 1979. Biomass burning as a source of atmospheric gases CO, $H_2$, $N_2O$, NO, $CH_3Cl$, and COS, *Nature*, 282:253–256.

Crutzen, P.J., A.C. Delany, J. Greenberg, P. Haagenson, L. Heidt, R. Lueb, W. Pollock, W. Seiler, A. Wartburg, and P. Zimmerman. 1984. Tropospheric chemical composition measurements in Brazil during the dry season, *Journal of Atmospheric Chemistry*, 2: 154–167.

Dickinson, C.H. and G.J.F. Pugh, Eds. 1974. *Biology of Plant Litter Decomposition*, Academic Press, London.

Duxbury, J.M., D.R. Bouldin, R.E. Terry, and R.L. Tate. 1982. Emission of nitrous oxide from soils, *Nature*, 298:462–464.

Goreau, T.J. 1982. The Biogeochemistry of Nitrous Oxide, Ph.D. thesis, Harvard University, Cambridge, MA.

Goreau, T.J., W.A. Kaplan, S.C. Wofsy, M.B. McElroy, F.W. Valois, and S.W. Watson. 1980. Production of nitrite and nitrous oxide by nitrifying bacteria at reduced concentrations of oxygen, *Applied and Environmental Microbiology*, 40:526–532.

Goreau, T.J. and W. Z. de Mello. 1985. Effects of deforestation on sources and sinks of atmospheric carbon dioxide, nitrous oxide, and methane from Central Amazonian soils and biota during the dry season: a preliminary study, in *Proc. Workshop on Biogeochemistry of Tropical Rain Forests: Problems for Research*, D. Athie, T.E. Lovejoy, and P. de M. Oyens, Eds., Centro de Energia Nuclear na Agricultura and World Wildlife Fund, Piricicaba, Sao Paulo, Brazil, pp. 51–66.

Goreau, T.J. and W.Z. de Mello. 1988. Tropical deforestation: some effects on atmospheric chemistry, *Ambio*, 17:275–281.

Gould, F.W. and R.B. Shaw. 1983. *Grass Systematics*, Texas A&M University Press, College Station, TX.

Hall, S.J. and P.A. Matson. 1999. Nitrogen oxide emissions after nitrogen additions in tropical forests, *Nature*, 400:152–155.

Hanlon, E.A., D.L. Anderson, and O.A. Diaz. 1997. Nitrogen mineralization in histosols of the Everglades Agricultural Area, *Communications in Soil Science and Plant Analysis*, 28:73–87.

Happell, J.D., J.P. Chanton, G.J. Whiting, and W.J. Showers. 1993. Stable isotopes as tracers of methane dynamics in Everglades marshes with and without active populations of methane oxidizing bacteria, *Journal of Geophysical Research*, 98:14771–14782.

Harriss, R.C., E. Gorham, D.I. Sebacher, K.B. Bartlett, and P.A. Flebbe. 1985. Methane flux from northern peatland, *Nature*, 315:652–654.

Hines, M.G., R.E. Pelletier, and P.M. Crill. 1993. Emissions of sulfur gas from marine and freshwater wetlands of the Florida Everglades: rates and extrapolation using remote sensing, *Journal of Geophysical Research*, 98:8991–8999.

IPCC. 1992. *Climate Change 1992*, supplementary report to the Intergovernmental Panel on Climate Change Scientific Assessment, Cambridge University Press, Cambridge, U.K.

IPCC. 1994. *Climate Change 1994: Radiative Forcing of Climate Change and an Evaluation of the IPCC IS92 Emission Scenarios*, Cambridge University Press, Cambridge, U.K.

Keller, M., W.A. Kaplan, S.C. Wofsy, and J.M. Da Costa. 1988. Emissions of $N_2O$ from tropical forest soils: response to fertilization with $NH_4^+$, $NO_3^-$, and $PO_4^{3-}$, *Journal of Geophysical Research*, 93:1600–1604.

Klein, H., J.T. Armbruster, B.F. McPherson, and H.J. Freiberger. 1975. *Water and the South Florida Environment*, U.S. Geological Survey, Water Resources Investigation 24-75, Dept. of the Interior, Washington, D.C.

Kochrose, M.S., K.R. Reddy, and J.P. Chanton. 1994. Factors controlling seasonal nutrient profiles in the Florida Everglades, *Journal of Environmental Quality*, 23:526–533.

Livingston, G.P., P.M. Vitousek, and P.A. Matson. 1988. Nitrous oxide flux and nitrogen transformations across a landscape gradient in Amazonia, *Journal of Geophysical Research*, 93:1593–1599.

Lucas, R.E. 1982. *Organic Soils (Histosols): Formation, Distribution, Physical and Chemical Properties and Management for Crop Production*, Research Report 435 Farm Science, Michigan State University Agricultural Experiment Station and Cooperative Extension Service, East Lansing, MI.

Martin, H.W., D.B. Ivanoff, D.A. Greetz, and K.R. Reddy. 1997. Water table effects on histosol drainage water carbon, nitrogen, and phosphorus, *Journal of Environmental Quality*, 26:1062–1071.

Matson, P.A., P.M. Vitousek, G.P. Livingston, and N.A. Swanberg. 1990. Sources of variation in nitrous oxide fluxes from Amazonian ecosystems, *Journal of Geophysical Research*, 95:16789–16798.

McKenney, D.J., D.L. Wade, and W.I. Findlay. 1978. Rates of nitrous oxide emission from N-fertilized soil, *Geophysical Research Letters*, 5:777–780.

Ponnamperuma, F.N. 1984. Effects of flooding on soils, in T.T. Kozlowski, Ed., *Flooding and Plant Growth*, Academic Press, Orlando, FL.

Reddy, K.R. 1982. Mineralization of nitrogen in organic soils, *Soil Science Society of America Journal*, 46:561–566.

Schulze, E.D. 1967. Soil respiration of tropical vegetation types, *Ecology*, 48:652–653.

Seiler, W. and P.J. Crutzen. 1980. Estimate of gross and net fluxes of carbon between the biosphere and the atmosphere from biomass burning, *Climatic Change*, 2:207–247.

Seiler, W., A. Holzapfel-Pschorn, R. Conrad, and D. Scharffe. 1984. Methane emission from rice paddies, *Journal of Atmospheric Chemistry*, 1:241–268.

SFWMD. 1982. *An Analysis of Water Supply Backpumping from the Lower East Coast Planning Area*, South Florida Water Management District, West Palm Beach, FL.

Slemr, F., R. Conrad, and W. Seiler. 1984. Nitrous oxide emissions from fertilized and unfertilized soils in a subtropical region (Andalusia, Spain), *Journal of Atmospheric Chemistry*, 1:159–169.

Swift, M.J., O.W. Heal, and J.M. Anderson. 1979. *Decomposition in Terrestrial Ecosystems*, University of California Press, Berkeley.

Terry, R.E., R.L. Tate, and J.M. Duxbury. 1981. Nitrous oxide emissions from drained cultivated organic soils of South Florida, *Journal of the Air Pollution Control Federation*, 31:1173–1176.

Vaithiyanathan, P. and C.J. Richardson. 1998. Biogeochemical characteristics of the Everglades sloughs, *Journal of Environmental Quality*, 27:1439–1450.

Walker, W.W. 1991. Water Quality trends at inflows to the Everglades National Park, *Water Resources Bulletin*, 27:59–72.

Walker, W.W. 1998. Long term water quality trends in the Everglades, Symp. on Phosphorus Biogeochemistry in Florida Ecosystems, U.S. Dept. of the Interior, http://www.shore.net/~wwwalker/clearwtr.

Winfrey, M.R. 1984. Microbial production of methane, in *Petroleum Microbiology*, Atlas, R.M., Ed., Macmillan, New York.

# Section II

## Florida Bay

*Gaskin Bay*
by
Clyde Butcher

# 11 Transport Processes Linking South Florida Coastal Ecosystems

*Thomas N. Lee and Elizabeth Williams*
University of Miami

*Elizabeth Johns and Doug Wilson*
NOAA/AOML

*Ned P. Smith*
Harbor Branch Oceanographic Lab

## CONTENTS

## INTRODUCTION

South Florida coastal ecosystems consist of a collection of distinct marine environments that are strongly connected by their circulation and exchange processes on a regional scale and by oceanic boundary currents to remote upstream regions of the Gulf of Mexico and Caribbean. South Florida coastal waters are comprised of a set of separate subregions defined by their different physical characteristics, flow properties, and species compositions (Figure 11.1). The degree of linkage

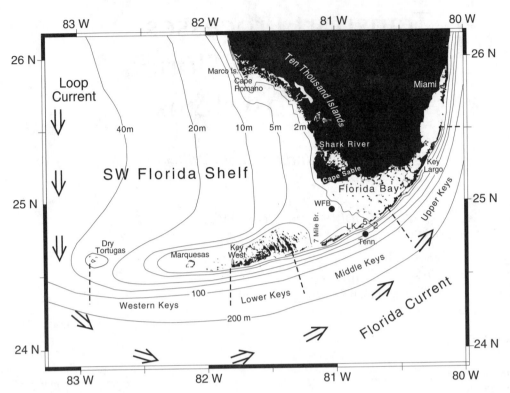

**FIGURE 11.1**  South Florida coastal system of interconnected subregions: Keys coastal zone consisting of Upper, Middle, Lower, and Western Keys subregions; Florida Bay; and the Southwest Florida Shelf. Tidal passages at Seven-Mile Bridge, Long Key Channel (LK), Channel 5 (5), and Channel 2 (2) are also shown. Depth contours are in meters.

between subregions depends on the strength of their transports and the volume of water exchanged between subregions, as well as over the whole domain.

Florida Bay is made up of a complex maze of shallow basins separated by mud banks and mangrove islands. The Bay is openly connected to the southwest Florida shelf along its wide western boundary, but exchange with the Atlantic coastal zone of the Keys is restricted to a few narrow tidal channels between the Keys island chain. The northern boundary is mangrove fringed with freshwater input in the northeastern region through Taylor Slough and Trout River. The rapid fall-off of tidal range with distance from the western boundary[1] and the dramatic increases observed in interior salinities[2] indicate poor exchange of the northeast and central portions of the Bay with adjacent subregions.

Florida Bay interacts with two very different continental shelves: the southwest Florida shelf to the west and the Keys coastal zone to the east and south. These two shelf regions interact with each other through the tidal channels between the Keys and with strong offshore boundary currents at their outer edges: the Loop Current in the Gulf and Florida Current on the Atlantic side of the Keys. The southwest Florida shelf is the southern extension of the wide, shallow west Florida shelf with its smoothly varying topography aligned in a northwest–southeast direction. The west Florida shelf undergoes seasonal stratification of the water column from changes in wind mixing, air–sea exchange, and river runoff along its eastern border[3] and has a strong, coherent response to along-shore wind forcing.[4] Freshwater runoff also occurs in the Ten Thousand Islands area through a series of small rivers: Shark River, Broad River, and Lostmans River. Loop Current eddies propagate southward along the outer edge of the shelf and develop into persistent eddy structures off the Dry Tortugas.[5–7]

The Keys coastal zone consists of a narrow, curving shelf with complex topography associated with its shallow reef tract. The curving shoreline causes regional differences in response to prevailing easterly winds and Florida Current influences. The lower Keys are normally aligned with the wind and form a downwelling coast, with onshore flow in the upper layer and offshore in the lower layer, and a westward coastal current that intensifies toward the west due to greater persistence of Florida Current eddies off the Tortugas.[8] The prevailing wind is onshore in the upper Keys, with little influence on alongshore flow, whereas the Florida Current converges nearer the outer reefs, causing strong downstream flows. The outer shelf region of the Keys is dominated by meanders of the Florida Current and the downstream propagation of eddies.[6-9]

The circulation and exchange of South Florida coastal waters depend on the interplay of local responses to different types of forcing in the different subregions. The cumulative effect of these responses also determines the fate of freshwater discharges through the Everglades, Florida Bay, and west Florida shelf. In this chapter, we present new observations to describe the circulation and exchange processes characteristic to each subregion and their effect on transport trajectories that provide the regional linkages between ecosystems for the spread of waterborne materials, as well as recruitment pathways from offshore spawning areas to nearshore nurseries.

## METHODS

The data and analyses are primarily from an ongoing study of the circulation and exchange of Florida Bay waters with surrounding regions conducted by University of Miami and National Oceanographic and Atmospheric Administration (NOAA), Atlantic Oceanographic and Meteorological Laboratory (AOML) as part of the South Florida Ecosystem Restoration Prediction and Modeling (SFERPM) Program supported by NOAA/Coastal Ocean Program (COP). Observational methods consist of a combination of bi-monthly interdisciplinary shipboard surveys, *in situ* moorings, shipboard Acoustic Doppler Current Profiler (ADCP) transport transects in the major Keys flow passages, and Lagrangian surface drifters to describe and quantify the circulation and exchange pathways as related to local forcing and coupling with the waters of the Atlantic and Gulf of Mexico.

A moored instrument array was first deployed in September 1997 to be maintained for 3 years (Figure 11.2). The array consisted of four bottom-mounted 1200-Khz RDI Workhorse ADCPs equipped with nearsurface (3-m) conductivity and temperature (C/T) sensors on the southwest Florida shelf offshore of Shark River Slough (moorings A and C are on the 6-m isobath, B and D are at 13-m depths). The ADCPs profile currents in 30-cm depth bins from about 2 m below the surface to 2 m above the bottom at the 6-m stations, and 50-cm-depth bins at the 13-m sites. Also included is a Shark River Slough plume array of nine C/T sensors to monitor changes in the Shark River Slough discharge nearfield and a single mid-depth current/C/T mooring near the western boundary of Florida Bay (WFB) in a water depth of 4 m. Three current/C/T moorings were positioned along the Florida Keys reef tract to measure interaction and exchange between the southwest shelf, Everglades discharge, Florida Bay, Keys coastal waters, and the Florida Current. These Keys moorings were positioned seaward of Tennessee (T) and Looe (L) reefs at a depth of 24 m. Currents and temperature were measured nearsurface (7 m) and nearbottom (21 m), and nearsurface conductivity measurements (4 m) were made as well. A single mooring was deployed in Hawk Channel (HC) offshore of Marathon at the 11-m isobath that was equipped with current and temperature sensors at 7 m and conductivity at 4 m. In addition, a bottom pressure array measures cross- and alongshelf pressure gradients on the southwest shelf, as well as the cross-Keys pressure gradient.

Multidisciplinary surveys of water properties of the Keys coastal waters, western Florida Bay, and southwest Florida shelf out to the Dry Tortugas were conducted bi-monthly. Satellite-tracked surface drifters of the type used in the Coastal Ocean Dynamics Experiment (CODE) were deployed within the Shark River Slough discharge plume on each of the bimonthly surveys. Nearly simultaneous surveys of similar water properties were conducted through the shallow interior parts of

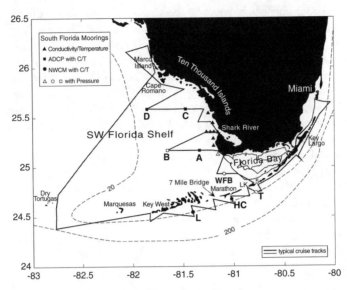

**FIGURE 11.2** Ongoing measurements of the Florida Bay Circulation and Exchange Study, part of the NOAA South Florida Ecosystem Restoration, Prediction and Modeling Program. Moored array was deployed Sept. 21, 1997, for 3-year period. Bi-monthly cruise tracks of the *RV Calanus* multidisciplinary survey vessel over the total study region and shallow catamaran, *Virginia K*, within Florida Bay are shown.

Florida Bay using a shallow-draft catamaran. Volume transports were directly measured over semidiurnal tidal cycles on six occasions at Long Key Channel, Channel 5, and Channel 2 and on a single day at Seven-Mile Bridge. A RDI 600-kHz, direct-reading, broadband ADCP with a 20-degree beam angle was mounted forward of the bow between the hulls of a shallow catamaran to make continuous vertical profiles of horizontal currents while crossing the channels. The instrument location was forward of any hull distortions of the current field.

Wind data are obtained from local Coastal Marine Automated Network (CMAN) stations in the Keys (Figure 11.2). Satellite-derived sea-surface temperature images were obtained from the Rosenstiel School of Marine and Atmospheric Science (RSMAS) satellite facility and U.S. Geological Survey South Florida website. All current and wind time series records were filtered with a 40-hour low-pass filter to remove the tidal fluctuations so that the underlying subtidal variability is clearly observed.

## RESULTS

### SALINITY PATTERNS

The typical seasonal pattern of precipitation in South Florida consists of a dry season during winter and spring followed by a wet season during summer and fall. This precipitation pattern combined with poor exchange properties results in hypersaline conditions within the interior lagoons of Florida Bay in the dry season and low salinity in the wet season.[2,10] However, the precipitation pattern of 1998 during the first year of our study was unusual in that the strong El Niño of 1997–1998 caused enhanced precipitation during the usual winter/spring dry season associated with the passage of moisture laden low-pressure systems from the Gulf of Mexico, and it decreased precipitation during the usual summer/fall wet season as the El Niño ended and La Niña developed. The salinity patterns of South Florida coastal waters show a clear response to this evaporation–precipitation cycle and to the delivery of freshwater through river discharge. A band of low-salinity water was observed along the coast between Cape Sable and Cape Romano in the fall of 1997 that continued to increase in size and fresh water content through the winter/spring wet period that ended in early summer (Figure 11.3a). The size and freshwater content of this band then decreased through the summer and fall dry period of 1998 (Figure 11.4a). A similar pattern of enlarging and shrinking freshwater distribution was also apparent within Florida Bay from discharge variations of Taylor River Slough.

Salinity isopleths on the southwest Florida shelf are generally aligned with the local isobaths in a southerly direction, suggesting a net southerly movement and preferred mixing in the direction of flow. The local minimum of salinity near the Shark River Slough mouth indicates a local source of low salinity from the river discharge. This low-salinity plume, with salinities less than 32, is typically confined to the region between Cape Sable and Cape Romano and within western Florida Bay. The plume is generally well mixed vertically except for a small region in the immediate vicinity of the river mouth where one to three changes in salinity can occur over the 3-m water depth. A much larger region of decreased salinities ranging from 32 to 36 extends from near Cape Romano to the Tortugas. Freshwater discharge through the Everglades is generally less than 10 m³/s, which is not sufficient to create this large volume of decreased salinities on the shelf. The orientation of the isopleths toward the north along the west Florida shelf indicates that discharges from larger rivers to the north on the west Florida shelf or possibly the Mississippi River are the likely sources.

The salinity of the surface waters of the Tortugas was less than oceanic salinities (36.0) during all surveys and indicates that significant dilution by river discharge into the eastern Gulf of Mexico and transport to the Tortugas region is a common occurrence. Seasonal variability of salinity in the Tortugas did not show a similar pattern as the inshore low-salinity band along the Ten Thousand Islands. Rather, in the Tortugas the surface salinity was highest (35 to 36) in the winter/spring wet season and lowest (32 to 34) in the summer dry season, again suggesting low-salinity transport from the north. A band of lower salinity waters (33 to 34) extended across the shelf in the upper

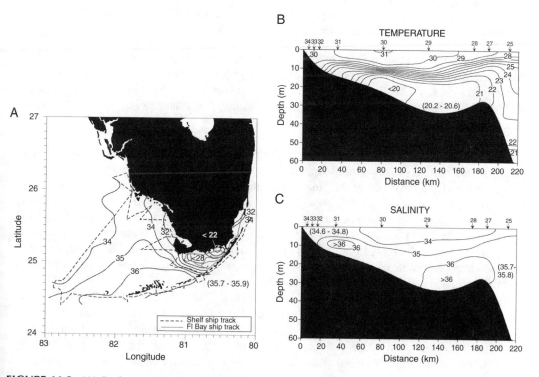

**FIGURE 11.3** (A) Surface salinity June 8–16, 1998, at the end of the El Niño-enhanced rainy season, which took place during winter instead of the more typical dry winter and wet summer seasons. (B) Marco Island to Tortugas temperature section. (C) Salinity section, June 1998.

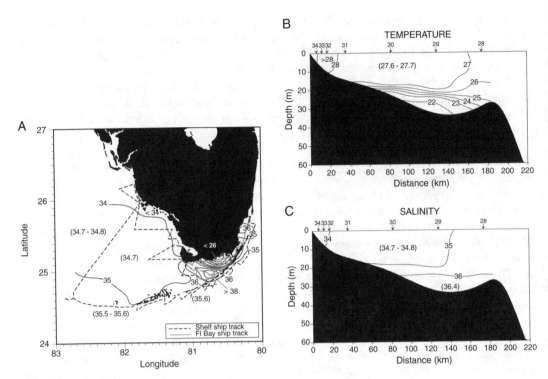

**FIGURE 11.4** (A) Surface salinity Oct. 12–22, 1998, at the end of the La Niña-enhanced dry season, which took place during summer instead of the more typical wet summer and dry winters seasons. (B) Marco Island to Tortugas temperature section. (C) Salinity section, October 1998.

10 m from near the Tortugas to the outer part of the inner shelf off Marco Island (Figures 11.3a,b). However, this band did not reach the coast, which suggests southward transport along the middle of the west Florida shelf. Higher salinities often develop north of the region between Big Pine Key and Marquesas Keys (Figures 11.3a and 11.4a). Salinities greater than 36 can occur in this region, which suggests contributions from local evaporation.

Seasonal patterns of stratification on the southwest Florida shelf are shown in Figures 11.3b and c. Stratification begins to develop in the spring with a decrease in wind mixing and increase in runoff and becomes highly stratified in summer with weak wind mixing and transport of low-salinity waters from the north. At this time, the shelf is essentially a two-layered system consisting of a shallow surface mixed layer of about 5 to 8 m, a sharp thermocline/halocline over about the 10-m strata where salinity increases from 34 to 36 and temperature decreases from 30 to 20°C, and a well-mixed cold, salty bottom layer with salinities greater than 36 and temperatures below 20°C. Increased wind mixing in fall begins to break down the stratification (Figures 11.4b and c), and by winter the middle portions of the shelf are vertically well mixed; however, near the inshore end of the section off Marco, stratification remains for most of the year due to river runoff. South of Cape Romano in water depths less than 15 m the shelf is vertically well mixed throughout the year except near river mouths in the Ten Thousand Islands region. The coastal zone of the Florida Keys is also observed to be well mixed vertically in both temperature and salinity throughout the year for depths less than 12 m. At deeper depths of 20 to 30 m in the outer shelf, transient stratification occurs due to seasonal changes, as well as nearbottom intrusions of cold layers from internal waves, bores and frontal eddies.[11,12]

The largest variations of salinity were observed within the interior of Florida Bay on seasonal time scales (Figures 11.3a and 11.4a). Following the El Niño-enhanced wet period of winter/spring 1998, the east and central portions of the Bay were diluted by freshwater runoff to less than

26 salinity and the western region was characterized by strong horizontal gradients increasing from 26 in the interior to about 34 at the western edge. The strong gradient is due to the convergence of southwest Florida shelf waters with the eastward extension of the Shark River Slough plume and interior Bay waters. The curvature of this gradient pattern toward Long Key and Channel 5 tidal passages between the Keys indicates a strong along-gradient advection toward these passages combined with weak cross-gradient mixing. During the La Niña dry period, Florida Bay waters with salinities less than 26 shrank to a small area off the mouth of Taylor River Slough, and hypersaline conditions developed throughout the central region of the Bay with salinities greater than 38, whereas in the western region salinities were typical of the southwest shelf (34 to 35) and Keys coastal zone (35 to 36). These observations clearly show that there is little exchange between the eastern and central regions of Florida Bay and between the central region with either the western region of Florida Bay or the Keys coastal waters. The large salinity changes within Florida Bay appear to be a response to local evaporation, precipitation, and freshwater discharge, whereas salinities in the western bay are buffered by significant interaction with waters of the southwest shelf, Shark River Slough plume, and Keys coastal zone. Florida Bay interior lagoons were observed to be vertically well mixed in temperature and salinity during all seasons. This indicates that wind mixing and air–sea exchanges are sufficient to mix the shallow waters (depths less than 3 m), as tides are almost nonexistent in the interior lagoons. However, wind forcing is still insufficient to cause significant exchange of interior waters with adjacent regions, due to isolation by the shallow mud banks. The low-salinity Shark River Slough plume was also found to be well-mixed vertically except for the region directly adjacent to the river mouth, where a change of 3 in salinity can occur over the 3-m water depth.

## SUBTIDAL CURRENTS

South Florida marine ecosystems are linked by currents that foster the exchange of waterborne materials between the different subregions. Current variability occurs on time scales ranging from minutes to seasons to years. Tidal currents often account for a large part of the total variability in coastal waters and are important local mixing and dispersion mechanisms. However, due to their symmetry, tidal currents are not efficient transport mechanisms on spatial scales larger than their tidal excursion lengths, which are typically only a few kilometers. Therefore, the emphasis here will be on low-frequency currents with periods greater than tidal, normally referred to as subtidal currents. Subtidal currents are the primary mechanism linking adjacent as well as remote regions to South Florida ecosystems. In South Florida coastal waters, subtidal currents are driven primarily by local winds and offshore boundary currents and are strongly influenced by bathymetry and coastline orientations.

### Southwest Florida Shelf

Location of the moored instrument array is shown in Figure 11.2. The ADCP moorings were located within the inner part of the shelf at the 6- and 13-m isobaths for a 3-year period from Sept. 1997 to Oct. 2000. These locations are within the southernmost portion of the west Florida inner shelf, bordered on the south by the Florida Keys island chain that extends from Key Largo to the Dry Tortugas. The primary subtidal current response to wind forcing on wide, shallow continental shelves, such as the west Florida shelf, is a strong alongshore current directly forced by coherent, synoptic-scale, along-shore winds and opposing cross-shelf currents in upper and lower Ekman layers.[4,13–15]

Subtidal vector time series from moorings A, B, and D at nearsurface and nearbottom depths for the first year deployment are shown in Figure 11.5, together with wind vectors from the Sombrero Light CMAN station (location shown in Figure 11.2). Subtidal current fluctuations are primarily in the alongshore (north–south) direction as a direct response to alongshore wind forcing events. Lee and Williams[8] have shown that wind variability over the Keys as measured at CMAN sites is highly coherent and useful for comparison to the currents on the southwest shelf. Coherence between wind and current fluctuations in the period band of 2 days to 2 weeks is very high at the 95% significance level, with the currents lagging the Sombrero winds by about 12 hours. Presumably this lag time would be shorter if wind records were available closer to the current measurement sites. Amplitudes of the wind-driven currents ranged from about 6 to 18 cm/s and were only slightly stronger near the surface with little change of direction with depth, indicating a barotropic or uniform response to wind forcing, as found previously for the west Florida shelf north of our study area.[4]

During the fall (September, October, and November) winds from the northeast prevail. These winds have a larger offshore than southward component on the west Florida shelf and therefore cause weak southward alongshore flows of 3 to 7 cm/s that can persist for periods ranging from 5 to 10 days (Figure 11.5a). Southward advection during these events could transport passive particles distances of 13 to 60 km toward the Keys. Several of these events together could result in significant movement of west Florida shelf water masses to the southwest shelf and western Keys. In western Florida Bay, low-frequency flow events were toward the southeast or northwest parallel to the local isobaths during this period. In winter (December, January, and February) the amplitude of low-frequency currents increases due to an increase in wind strength with the passage of cold fronts that occurred with a clockwise rotation of wind direction in the period band of 4 to 10 days. The strongest flows tend to align with the orientation of the local isobaths near each mooring site. The western Florida Bay site shows increased flow events toward the southeast (in the direction toward Long Key Channel) that are well correlated to winds from the northwest and north following the passage of cold fronts. During spring the strength of the wind forcing and the current response decreases, although the wind can be moderately strong and persistent from the east for several days. In summer, winds are weak from the southeast, and nearsurface currents at the inshore stations (WFB, A, and C) show a marked decrease in speed and a shift toward the north. The strong wind and current event in late September was produced by the passage of Hurricane Georges on September 24 and 25, 1998 (Figure 11.5b). Maximum winds at the Sombrero CMAN station reached 40 m/s from the east and caused a 1.5-m set-down of sea level in western Florida Bay and northwestward currents of 80 cm/s on the southwest shelf and western Florida Bay as observed in the unfiltered records.[16]

Interestingly, the amplitudes of southward flow events in the subtidal records during winter, spring, and early summer were significantly larger than for northward flow events, indicating the presence of a mean flow to the south (Figures 11.5a,b). This southward offset does not appear to be wind induced, for the amplitudes of southward and northward wind events were approximately equal. The net southward flow was stronger at offshore stations B and D and was observable for the period October to April in the upper layer and from October to July in the lower layer. Satellite sea-surface temperature imagery for April 12 shows a cool southward current at mid-shelf that appears to extend along the entire west Florida shelf (Figure 11.5c). The southward flow is indicated by the northward orientation of folded wave patterns along the shoreward temperature front.

Seasonal flow patterns from the total 3-years of moored array measurements are shown in Figures 11.6a-d. There occurred a southward mean flow in the upper and lower layers at the offshore stations B and D for the fall, winter, and spring seasons. During summer this flow was weak or toward the north. The inshore stations show a weaker northward flow in the upper layer during the fall, winter, and spring seasons that suggests a mean recirculation of part of the stronger southward flow occurring offshore. During summer the mean northward flow at the inshore stations was stronger than at the offshore sites. There was also an onshore component of the seasonal mean flow in the lower layer that was strongest in fall and winter and results in an "estuarine-like" circulation over the inner shelf with offshore flow in the upper layer and onshore flow in the lower layer during these seasons. Seasonal mean flows at the western Florida Bay site (WFB) were strong northward in the fall and northeastward for the other seasons.

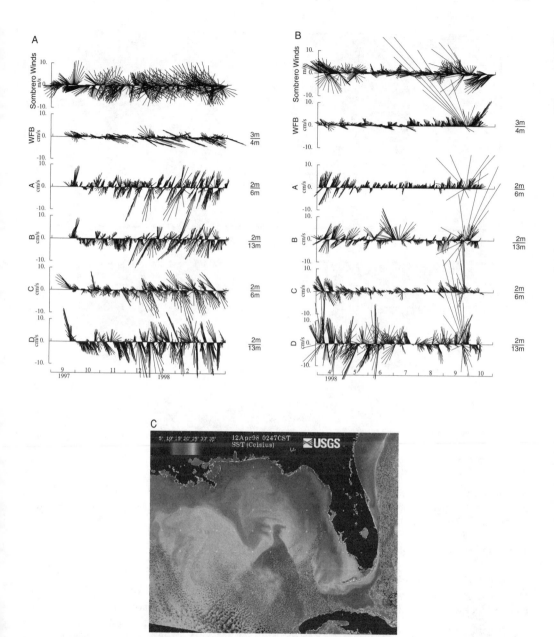

**FIGURE 11.5** (A) Vector time series of subtidal, nearsurface currents from SW shelf moored array and winds from Sombrero CMAN station for September, 1997 to March 1998. Instrument and water depths are given by the ratio at the right, and mooring identification is given on the left. (B) Vector time series of subtidal, nearsurface currents from SW shelf moored array and winds from Sombrero CMAN station for April to October 1998. Instrument and water depths are given by the ratio at the right, and mooring identification is given on the left. (C) Surface thermal patterns derived from satellite AVHRR for April 12, 1998, following a ring separation that leaves the Loop Current in a young stage with strong interaction with the west Florida shelf and possible forcing of a cold southward coastal current along the west Florida inner shelf and into southwest Florida coastal region. Also shown is offshore entrainment of cold shelf water near the Mississippi River delta along the eastern front of the recently separated ring. (Image copied from USGS website.)

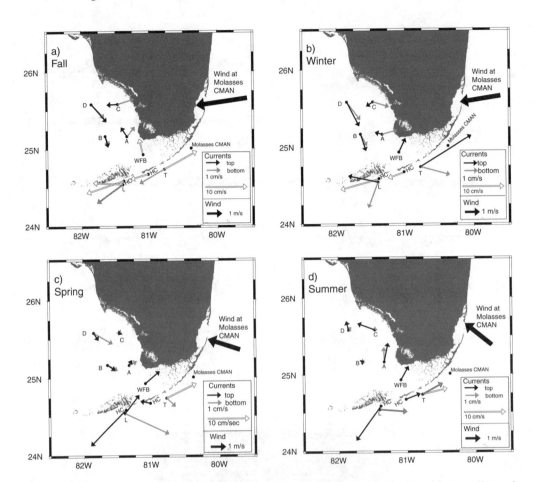

**FIGURE 11.6** Mean flow vectors from nearsurface (top) and nearbottom (bottom) positions at all mooring sites and mean wind from Molasses CMAN station for three-year period September 1997 to October 2000. Mooring I.D.; Looe Reef (L); Hawk Channel sites (HC); Tennessee Reef (T); Western Florida Bay (WFB); ADCP's (A, B, C, D). (a) Fall (September, October, November); (b) Winter (December, January, February); (c) Spring (March, April, May); (d) Summer (June, July, August).

The 3-year annual mean flows from all sites are shown in Figure 11.10. Annual mean flows were southward at the offshore stations (B, D) at about 1 cm/s. Upper layer mean flows suggest an anti-clockwise recirculation with northward return flow occurring nearshore. There also occurred a significant mean onshore flow in the lower layer at all sites. The physical mechanism controlling the strength and direction of the southward mean flow is not understood at present. However, it is quite possible that the southward flowing Loop Current along the outer part of the west Florida shelf can force a southward flow in the middle and inner shelf regions as well. The annual mean flow in western Florida Bay was toward the northeast.

## Florida Keys Coastal Zone

Subtidal current patterns of the Keys coastal zone are distinct from those on the southwest Florida shelf and are also different among subregions of the Keys. These differences are primarily associated with the curving coastline of the Keys, the narrow shelf, and the degree of interaction with the nearby Florida Current. The Keys coastal zone can be separated into four subregions. In the upper Keys, from Elliott Key to Plantation Key where the coastline is aligned more north–south, onshore winds prevail and the Florida Current is close to the shelf edge and has a controlling influence on outer shelf currents.[8] The middle Keys, from about Plantation Key to Big Pine Key where most of the coastline curvature occurs, is a transitional region in terms of the extent of wind and Florida Current forcing, and significant interaction takes place with western Florida Bay through tidal passages. The lower Keys, from Big Pine Key to Key West where the coastline is oriented east–west, is aligned with the prevailing winds to form a downwelling coast, but with increasing influence from Florida Current eddies arriving from the Tortugas. Finally, the western Keys, from Key West to the Dry Tortugas, are characterized by strong influence from persistent Tortugas eddies offshore, downwelling winds, and strong interaction with the southwest Florida shelf.

Subsurface, taut-wire current meter and temperature moorings have been maintained at sites along the outer shelf between the Dry Tortugas and Key Largo for different time periods since April 1989.[8] Mooring locations are shown in Figure 11.2. The Looe Reef mooring has been maintained continuously, and we now have over 12 years of data at this site. The Tennessee Reef and Hawk Channel sites off Marathon and near Looe Reef are presently being maintained (since September 1997) along with the Looe Reef mooring as part of the SFERPM Program.

As an example of the above-mentioned subtidal flow behavior, the vector time series from the upper layer at Tennessee Reef, Hawk Channel, and Looe Reef sites are shown in Figure 11.7 together with Sombrero CMAN winds for the same one-year time period as the measurements made on the southwest Florida shelf. All vectors have been rotated to align with isobaths in the alongshore direction. Current amplitudes are strongest at the seaward edge of the reef tract near the Looe and Tennessee Reefs, where alongshore current amplitudes range from ±50 to 60 cm/s in response to both the Florida Current and alongshore wind forcing. Alongshore current amplitudes at the Hawk Channel site ranged from ±20 to 40 cm/s and were significantly coherent with alongshore wind. Coherence of currents at the different sites was not particularly high, though many of the strong-amplitude current events are well correlated visually. Both propagating and stationary flow events occurred in the current records (Figure 11.7). Downstream flow events of late September 1997, early March 1998, and mid-April 1998 propagate downstream through the array and are most likely due to the downstream movement of an onshore meander of the Florida Current. The downstream propagation of the strong current reversals of mid-March and early April 1998 are believed to be due to the downstream movement of frontal eddies. Large stationary flow events that occur at nearly the same time at each site appear to be directly forced by local alongshore winds that are coherent over the array. Examples of current response to southwestward wind events occurred in mid-September and early November 1997 and late March and late October 1998. Examples of northeastward current responses to northward alongshore winds occurred in mid-September, late October, and early December 1997 and in late September 1998. In addition, the passage of Hurricane Georges on September 25 with winds of 40 m/s from the east resulted in maximum alongshore currents to the southwest of 100 cm/s at Hawk Channel and offshore sites as observed in unfiltered records.[16]

The annual variability of outer shelf currents in the Keys as shown by monthly mean currents (Figures 11.8a,b) appears to be directly influenced by annual cycles of alongshore wind (Figure 11.9) and Florida Current volume transport. The outer shelf site at Tennessee Reef in the middle Keys shows strong downstream flow for most of the year due to the close proximity to the Florida Current, although minimum downstream flow occurs during the winter and fall seasons of maximum easterly winds. The longer records from the Looe Reef site show seasonal maximum upstream flow with Ekman-like onshore (offshore) transports in the upper (lower) layers occurring during the fall and winter seasons of maximum upstream winds (westward in lower Keys and southwestward in upper Keys) (Figures 11.8a,b). However, during the summer when wind forcing is weak and during the spring semi-annual maximum of upstream winds, the alongshore currents are downstream and coincident with the annual maximum downstream volume transports of the Florida Current. The annual cycle of alongshore currents at the Hawk Channel site off Marathon tend to more closely follow the seasonal changes in alongshore wind with westward flow in the winter, spring, and fall periods of stronger easterly winds, and eastward flow during summer when the wind shifts to southeasterly and the downstream flow of the Florida Current is strongest (Figure 11.8b).

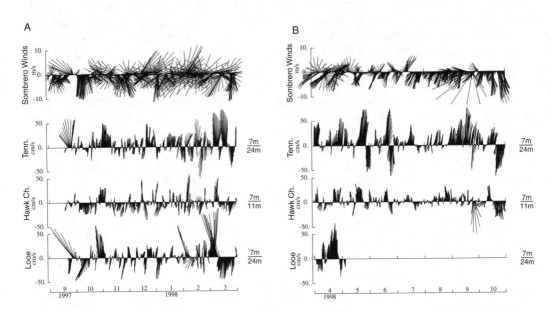

**FIGURE 11.7** (A) Vector time series of subtidal, nearsurface currents from Keys coastal zone moored array and winds from Sombrero CMAN station for September 1997 to March 1998. Instrument and water depths are given by the ratio at the right, and mooring identification is given on the left. (B) Vector time series of subtidal, nearsurface currents from Keys coastal zone moored array and winds from Sombrero CMAN station for April to October 1998. Instrument and water depths are given by the ratio at the right, and mooring identification is given on the left.

The annual temperature cycle in the nearsurface waters at the outer shelf off Looe Reef shows a maximum monthly mean of about 30°C in August and a minimum of about 24°C from January to March. The annual temperature cycle appears to follow the seasonal cycle of local air temperature. Stratification was strongest during spring and summer seasons and weakest during fall.

Seasonally averaged nearsurface currents in the Keys coastal zone during the three-year period from September 1997 to October 2000 are also shown in Figures 11.6a–d. The Tennessee Reef site shows strong downstream mean flows in the upper layer during all seasons due to the shoreward convergence of the Florida Current in the upper Keys, which brings the Florida Current closer to the reef tract in the upper Keys compared to the lower Keys. Mean lower layer flows at this site vary with seasons, southwestward in the fall, northeastward in summer and offshore in winter and spring. Thus the influence of strong downstream flow in the Florida Current has a stronger effect on the outer reef currents in the upper layer of the northern Keys. This effect has been observed to be even stronger off Key Largo (Lee and Williams[8]). Seasonal mean flows at two Hawk Channel sites off Marathon and shoreward of Looe Reef tend to follow the seasonal pattern of local wind forcing and are toward the southwest in the fall, winter, and spring when the mean winds are from the east or northeast. Strongest southwestward mean flows in Hawk Channel were in the fall at about 4 to 9 cm/s when the mean winds from the northeast are strongest. During the summer there was a divergence of mean flows in Hawk Channel with eastward flow off Marathon and southwestward flow inshore of Looe Reef. The cause of this divergence is not known, but is thought to be related to the curvature of the Keys coastline (aligned more east-west off Looe Reef) relative to the seasonal development of summer southeasterly winds and stronger Florida Current flow. This would result in stronger downstream flows at more northerly locations in the Keys and westward wind driven flows in Hawk Channel at lower Keys sites. At Looe Reef the mean flow was toward the west for fall and winter seasons from significant westward wind

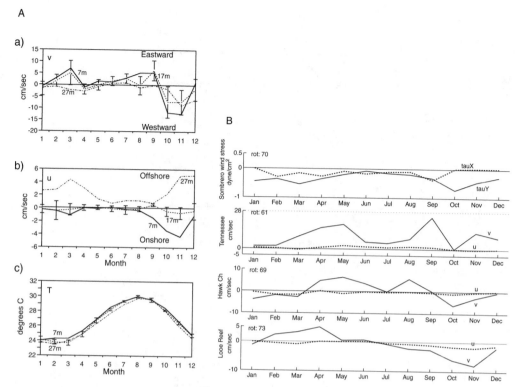

**FIGURE 11.8** (A) Monthly mean of (a) alongshore currents, (b) cross-shore currents, and (c) temperature from the Looe Reef mooring site for the 5-year period April 1989 to April 1994. Error bars display the standard error of the mean. (From Lee, T.N. and Williams, E., *Bull. Mar. Sci.*, 64: 35–56, 1999. With permission.) (B) Monthly mean of alongshelf and cross-shelf current components from upper layer (<7 m) at Keys coastal zone mooring sites and Sombrero wind-stress components for the period September 1997 to October 1998, except for Looe Reef site, where data from the 9.5-year period April 1989 to October 1998 are included. Rotation angles are listed at the left (rot:).

forcing and a more offshore position of the Florida Current associated with frequent eddy passages. During spring and summer seasons the strength of the westward winds decreased and the Florida Current had a larger effect causing eastward mean flows for those seasons at this shelf edge site.

Annual mean currents and winds in the Keys coastal zone are shown in Figure 11.10 for the three-year period. Because of the curving coastline of the Keys island chain, in the presence of a prevailing westward zonal wind, the region from Dry Tortugas to the middle Keys constitutes a downwelling coast on the mean. This results in a mean westward coastal countercurrent, with an onshore component in the upper layer and offshore flow in the lower layer. Mean speeds of the countercurrent were –2 cm/s at Looe Reef and reached –6 cm/s in Hawk Channel inshore of Looe Reef. The upper Keys, oriented in a more north/south direction, are characterized by onshore winds on the mean that result in a weak northward mean flow without significant cross-shelf Ekman transports. This response was observed in Hawk Channel off Key Largo (Lee[17], Pitts[18]). However, in the outer shelf off Key Largo, a strong northward mean baroclinic flow of 20 cm/s was reported previously (Lee[8]) near the surface with a significant onshore component and offshore component in the lower layer. This strong downstream flow is due to the close proximity to the Florida Current front. The Florida Current converges shoreward in the upper Keys following its northward turn in the middle Keys resulting in stronger downstream flows in the outer shelf of the upper Keys, which is displayed by the stronger northeastward mean flow at Tennessee Reef of about 7 cm/s (Figure 11.10).

A

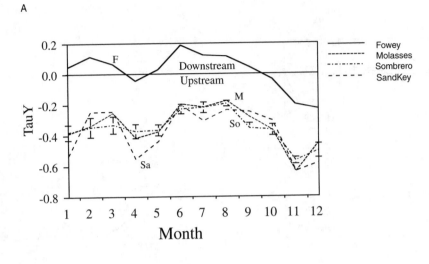

B

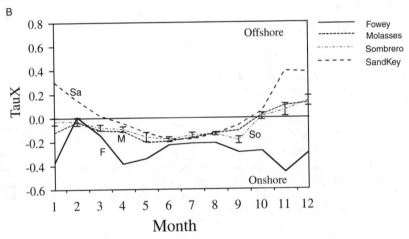

**FIGURE 11.9**    Alongshore (A) and cross-shore (B) monthly mean wind-stress components from CMAN wind stations for the period April 1989 to April 1994. Error bars display standard error of the mean. (From Lee, T.N. and Williams, E., *Bull. Mar. Sci.*, 64: 35–56, 1999. With permission.)

Large cyclonic gyres with horizontal dimensions reaching 200 km have been observed over the slope off the Tortugas.[9,12] These features form where the southward flowing Loop Current makes its eastward turn into the Straits of Florida and can persist for several months until forced to the east and replaced by an approaching Loop Current frontal eddy.[7] Recirculation in the gyres provides a retention mechanism for locally spawned larvae, and upwelling within the gyre interior estimated up to 2 m/day can support new production to aid larval survival.[9] The constant replenishment of gyres by Loop Current frontal eddies makes them the dominant physical process in the offshore waters of the western Keys. The time interval between gyres is short, usually less than one month. Their persistence appears to cause the westward intensification of the coastal countercurrent.[8] Interaction with Loop Current frontal eddies can force the gyres to the east where they undergo a decrease in size and duration as they migrate to their demise in the middle or upper Keys.[9,12] The gyres can influence flows in the lower Keys over time scales of 2 to 4 weeks, but are rarely observed in the middle and upper Keys where the primary mode of Florida Current influence on coastal current variability occurs from weekly period meanders and frontal eddies. Thus, the coastal waters of the middle Keys appear to be a transition region between a wind- and gyre-dominated domain

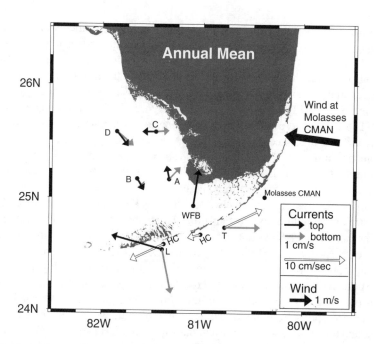

**FIGURE 11.10**  Annual mean flow vectors from nearsurface (top) and nearbottom (bottom) positions at all mooring sites and mean wind from Molasses CMAN station for a three-year period September 1997 to October 2000. Mooring I.D.: Looe Reef (L); Hawk Channel sites (HC); Tennessee Reef (T); Western Florida Bay (WFB); ADCP's (A, B, C, D).

in the lower to western Keys and a Florida Current/frontal eddy-dominated domain in the upper Keys. Due to this transition, the outer shelf of the middle Keys becomes a zone of divergence of alongshore flow, westward in the lower Keys and northward in the upper Keys. Compensation for the flow divergence may occur from strong onshore mean flows near Tennessee Reef in the middle Keys and also from a net southeastward flow from the Gulf of Mexico and Florida Bay observed through the Keys passages by Smith.[19]

## VOLUME TRANSPORTS: EXCHANGE BETWEEN GULF AND ATLANTIC

It is well known that the Gulf of Mexico is openly connected to the Atlantic through the Straits of Florida where a mean downstream slope of sea level helps to maintain the strong western boundary current system, the Loop Current/Florida Current that joins the two water bodies. However, the Gulf and Atlantic are also joined at the open passages between the Keys islands. Any slope in sea level by tidal or subtidal processes will drive a flow between the southwest Florida shelf and the Keys coastal zone. The largest passages are found in the middle Keys (Channel 5, Long Key Channel, and Seven-Mile Bridge) and western Keys (Key West and Rebecca Channel). Long Key Channel and Channel 5 connect western Florida Bay directly to the curving portion of the Keys coastal zone, whereas Seven-Mile Bridge channel connects the southwest Florida shelf to the Keys coastal zone near the transition between the middle and lower Keys subregions. Rebecca Channel connects the western Keys coastal waters to the southwest Florida shelf. Smith[19] has used current meters and sea-level recorders placed within the Keys passages to estimate volume transports. He shows that there is a persistent subtidal flow to the southeast toward the Keys coastal zone that sometimes reverses toward Florida Bay and the southwest shelf. Smith estimated the mean southeastward flows through these channels as 620 m³/s for Bahia Honda Channel during a two-month fall period and 262 m³/s for a one-year period through Long Key Channel with strongest transport toward the Keys in winter and spring months.

As part of the SFERPM Program we have made direct measurements of volume transport in the major passages of the middle Keys over semidiurnal tidal periods using a 600-Khz shipboard ADCP mounted forward of the bow between the hulls of a shallow catamaran. These measurements have been used to calibrate the flow in the passages to improve the time-series estimates of transport (Smith and Lee, unpubl. analysis). Typical examples of shipboard-derived volume transports through Channels 2, 5, and Long Key are shown in Figure 11.11 (solid triangles) on February 22 and 23, 1996. These data were fit with a cubic spline to interpolate transports every 12 minutes. Also shown are the tidal predicted transports and the residual transports determined as the difference of spline fit transport minus tidal predicted values. The average of the residual transport time series over the semidiurnal tidal cycle is the net residual transport. Positive values are toward Florida Bay and negative values are toward the reef tract. Figure 11.11 clearly shows the relative differences in magnitude between the three flow channels. Measured peak transports in Long Key Channel range from 5000 to 8000 m³/s and are two to four times larger than peak flows in Channel 5, which are two times larger than the peak flows in Channel 2. Measured transports in all three channels indicate a significant net transport out of Florida Bay toward the reef tract. This is seen in both the longer durations of the ebb cycle and greater magnitudes of the ebb flows (Figure 11.11). The tidal predicted transports indicate that tide-induced residual flows were small at this time. To correct the measured transports for any tide-induced residual flow (from the diurnal inequality) we subtract the tide-predicted transports from the measured (spline fit) values to form a residual transport time series. The residual transports show a persistent outflow from Florida Bay, resulting in a net (average over the tidal cycle) discharge of –909, –322, and –198 m³/s through Long Key, Channels 5, and 2, respectively.

The net transports through each channel for all six experiments are given in Table 11.1 for both tide-uncorrected and tide-corrected transports, the difference of which is the net tidal transport over the semidiurnal periods. The total net transports for all three channels ranged from –1420 m³/s in February 1996 to +271 m³/s in February 1997 and was directed toward the reef tract in five out of the six experiments. Net tidal-induced flows ranged from +29 to –1212 m³/s. Residual flows induced by tidal inequalities can be large when the measurements extend over only half of the diurnal cycle, requiring the removal of tide-predicted transports from the observed values before computing net residual flows.

In addition to the shipboard volume transport measurements we also made time series measurements of sea-level slope between western Florida Bay and the Keys coastal waters by recording bottom pressure variations at moorings WFB and Tennessee Reef (see Figure 11.2 for locations). The subtidal slope variations are determined by first demeaning and detrending the bottom pressure measurements and then taking the difference between the Tennessee Reef and western Florida Bay sites. These subtidal time series are plotted in Figure 11.12 together with the east–west component of the wind stress computed from Sombrero Light CMAN winds adjusted for 10-m height. The negative value of sea-level difference (SLD) is used to highlight the visual comparison with wind stress. Also shown is the volume transport time series from Long Key Channel derived from current-meter and sea-level data using the method of Smith[19] and calibrated for the channel with shipboard ADCP transects. The sea-level difference and volume transport time series have been normalized by their standard deviations for better comparison. Sea-level difference fluctuations are highly coherent with the volume transport time series and with local wind forcing. Coherence values of 0.9 and higher at the 95% significance level occurred between sea-level difference and volume transport time series in the 3- to 20-day period band. A near-linear relationship exists between sea-level difference and volume transport, as shown in Figure 11.13. Changes in sea-level difference account for about 73% of the volume transport variance. A sea-level difference of 10 cm can drive a volume transport of approximately 480 m³/s. In addition, there is a mean bias of –280 m³/s in transport that represents the mean background transport from western Florida Bay to the reef tract during this period. Applying the linear relationship of Figure 11.13 to the mean flow requires a

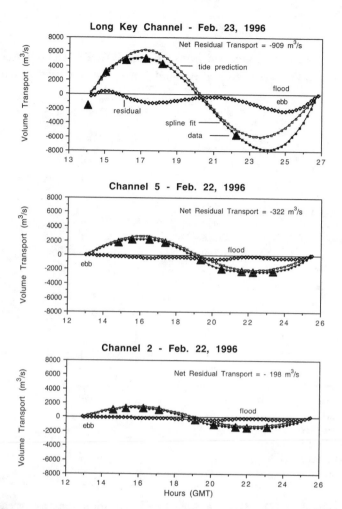

**FIGURE 11.11**   Shipboard ADCP-derived volume transports for Channels 5 and 2 on February 22, 1996 (middle and lower panels), and for Long Key Channel on February 23, 1996 (upper panel). Included are the measured transports (solid triangles), spline fit to measured transports (small solid circles), tide-predicted transports (small open squares), and residual transports (open squares).

mean sea-level slope of –5.95 cm. This sea-level slope was then added to the sea-level difference fluctuation time series in Figure 11.12. Importantly, the mean Gulf to Atlantic flow computed here for the 6-month period between September 1997 and April 1998 is in very close agreement with the one-year mean flow of –262 m³/s determined by Smith[19] for Long Key Channel over the period July 1992 to July 1993, indicating a relatively stable long-term mean.

The sea-level difference across the Keys depends on the relative variations of sea level in the Keys coastal waters and western Florida Bay. These two regions respond differently to the same wind forcing. Sea-level variations in the Keys Atlantic coastal waters have the greatest response to alongshore winds (winds toward either 245° or 65° near Long Key) due to the geostrophic require-ment for coastal sea level to balance the alongshore, wind-driven coastal current, whereas subtidal sea-level variations in western Florida Bay are more responsive to east–west wind forcing that stems from a simple set-down and set-up mechanism due to the shallow depths and semi-enclosed shape of the isobaths that open to the west (Figure 11.2). However, sea-level variations in western Florida Bay will also respond to large-scale coastal sea-level changes on the west Florida shelf from alongshore, wind-driven coastal currents (toward the NNW or SSE), as well as Loop Current

**TABLE 11.1**
**Volume Transports, $Q$ (m³/s), Derived from Shipboard ADCP Sections Across Tidal Channels in the Middle Keys with and without Correction for Tidal Diurnal Inequality**

| Date | Channel | Net $Q$ (No Tide Correction) | Net $Q$ (Tide Correction) | Difference (Net Tidal Transport) |
|------|---------|------------------------------|---------------------------|----------------------------------|
| 2/23/96 | Long Key | −987 | −909 | −78 |
| 2/22/96 | Channel 5 | −230 | −322 | 92 |
| 2/22/96 | Channel 2 | −174 | −189 | 15 |
| | Total | −1391 | −1420 | 29 |
| 6/16/96 | Long Key | −1379 | −495 | −884 |
| 6/15/96 | Channel 5 | −110 | 81 | −191 |
| 6/15/96 | Channel 2 | −177 | −40 | −137 |
| | Total | −1666 | −454 | −1212 |
| 10/24/96 | Long Key | −50 | −390 | 440 |
| 10/23/96 | Channel 5 | −52 | −148 | 96 |
| 10/23/96 | Channel 2 | −83 | −133 | 50 |
| | Total | −85 | −671 | 586 |
| 2/27/97 | Long Key | −346 | −140 | −206 |
| 2/26/97 | Channel 5 | 286 | 266 | 20 |
| 2/26/97 | Channel 2 | 144 | 145 | −1 |
| | Total | 84 | 271 | −187 |
| 5/20/97 | Long Key | −993 | −670 | −323 |
| 5/21/97 | Channel 5 | −356 | −200 | −156 |
| 5/21/97 | Channel 2 | −280 | −170 | −110 |
| | Total | −1629 | −1040 | −589 |
| 7/15/98 | Long Key | −541 | −615 | 74 |
| 7/14/98 | Channel 5 | −344 | −343 | −1 |
| 7/14/98 | Channel 2 | −306 | −282 | −24 |
| | Total | −1191 | −1240 | 49 |

variations. The combined result of these sea-level processes for a northeast or east wind (that can occur in the fall, winter, or spring) will cause sea level to set-up in the coastal waters of the middle Keys and set-down in western Florida Bay, resulting in a positive sea-level slope across the Keys that will drive a flow into western Florida Bay reaching magnitudes of 500 to 1000 m³/s and persisting for several days (Figure 11.12). The prevailing winds in the Keys are from the southeast, which is an onshore wind in the middle Keys with little influence on coastal sea level due to the narrow shelf and lack of a wind driven alongshore current. However, these same winds are aligned with the bathymetry of the west Florida shelf and will cause a northwest alongshore shelf current and a corresponding set-up of coastal sea level that will extend into western Florida Bay, thereby resulting in a negative sea-level slope across the Keys and a volume transport toward the reef tract. Over the entire record these southeastward transport events range from about −500 to −2500 m³/s and can last from 3 to 10 days. The stronger Gulf to Atlantic flow events tend to occur in the winter and spring seasons as cold front passages cause increased winds from the west and northwest that will set-up sea level in western Florida Bay relative to the Atlantic. Southeastern flows of −1500 to −2500 m³/s can occur in these seasons as sea level stands 20 to 35 cm higher in the Gulf than the Atlantic during strong wind events.

### NEARSURFACE CURRENT TRAJECTORIES

The linkages connecting South Florida coastal waters and their variability are shown by trajectories of nearsurface drifters deployed in the Shark River Slough low-salinity plume and tracked by

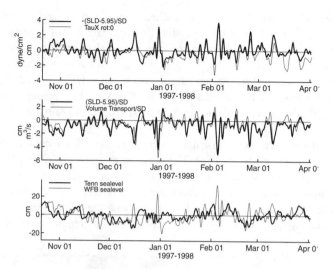

**FIGURE 11.12** Subtidal times series of demeaned and detrended sea level height at Tenn. and WFB sites (lower panel); sea-level difference (SLD: Tenn.–WFB) plus the mean SLD (–5.95 cm) divided by the standard deviation (SD) and volume transport/SD through Long Key Channel (+ is toward Florida Bay) (middle panel); and east–west wind stress (TauX) plotted with the negative of the SLD (upper panel) for October 1997 to April 1998.

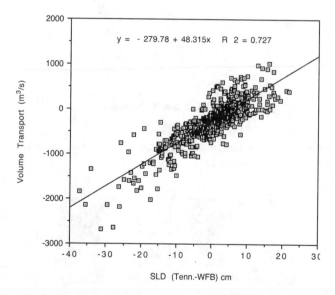

**FIGURE 11.13** Liner fit of volume transport through Long Key Channel vs. SLD (Tenn.–WFB) for the data shown in Figure 11.12, October 1997 to April 1998.

satellite. As part of the SFERPM Program, CODE-type, nearsurface drifters have been deployed off Shark River Slough and western Florida Bay on hydrographic cruises beginning in September 1994. CODE drifters are cross-shaped dacron panels about 1 m² supported at the corners by small floats tethered by thin nylon line to reduce wave and wind slippage. Except for the floats, only a small whip antenna is above the water surface. Drifter positions are determined by Argos approximately four to six times per day. A plot of the combined drifter tracks is shown in Figure 11.14. A total of 18 separate drifter deployments covered all months except July and November. Eleven of the drifters had a net movement toward the southeast and the tidal passages of the middle Keys (Long Key Channel, Channel 2, and Seven-Mile Bridge). Three of the drifters went toward the southwest and entered the western Keys coastal waters near the Dry Tortugas or Marquesas Keys. Only two of the drifters went north. One changed direction to the southwest and was entrained by the Florida Current west of the Tortugas, and the other reached as far north as 28.5° before becoming entrained by the Loop Current and transported back toward the Tortugas. All of the drifters eventually entered the Keys coastal waters where most tended to recirculate in the coastal countercurrent and eddies.

Drifter deployments were evenly distributed over seasons with six deployments in winter, four in spring, four in summer, and four in fall. The seasonal patterns of drifter trajectories are shown in Figure 11.15 and indicate a strong dependence on seasonal winds. During the winter and spring, when Gulf to Atlantic flows through the Keys are strongest, there was a preferred drifter pathway to the Keys coastal waters through western Florida Bay and the tidal passages in the middle Keys. During summer, when weaker southeast winds prevail, there was a preferred motion to the northwest

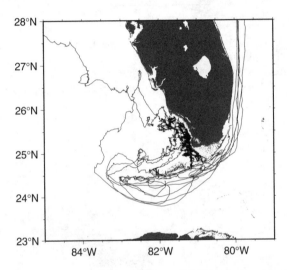

**FIGURE 11.14**  Combined plot of surface trajectories from all drifters deployed in the Shark River discharge plume from September 1994 to November 1999.

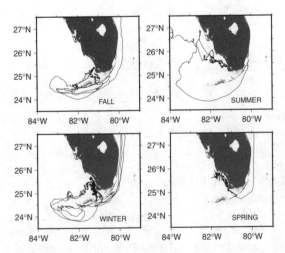

**FIGURE 11.15**  Seasonal patterns of drifter trajectories for (a) fall, (b) winter, (c) spring, and (d) summer.

along the west Florida coast. In the fall, there was a preference for southwestern movement in response to fall wind events from the northeast. The single drifter that moved toward the southwest during winter also occurred with northeast winds. Also, the southwestward part of the trajectory shown for the summer case actually occurred following summer during fall northeast winds.

Drifter trajectories show considerable subtidal variability that appears to be associated with local wind forcing. To quantify the wind influence we applied a multiple linear regression model using a least-squares approach to determine the percent of the subtidal variance of drifter-derived current components due to north–south and east–west wind forcing. The results indicate that approximately 70 to 80% of the subtidal variance of drifter movements on the southwest Florida shelf and western Florida Bay is due to local wind forcing. As an example, we show results of this analysis for a drifter deployed in the Shark River Slough plume on February 28, 1996 (Figure 11.16). This drifter first moved toward the northwest then reversed toward the southeast moving through western Florida Bay, then entered the Keys coastal waters through Long Key Channel on March 28. The north–south (v) and east–west (u) currents derived from the drifter movements and filtered to smooth tidal variations are in excellent agreement with the modeled current components derived from the regression with both u and v wind components (Figure 11.17). The modeled currents accounted for 78% of the v component variance and 41% of the u component. Presumably the correlation with the u component is reduced by the high-frequency tidal noise that was not completely removed from the observed current by the filtering technique. The remaining 20 to 30% of the drifter current variance not directly attributed to local wind forcing could be due to subtidal slope-generated currents from the relative differences in sea level on the Gulf and Atlantic sides of the Keys, or from long-term variations in the position of the Loop Current.

The drifter trajectories show that two typical exchange pathways couple the southwest Florida shelf, western Florida Bay, and the Keys coastal zone. The most persistent route is southeastward through western Florida Bay and the passages between the Keys, then westward along the reef tract to the Tortugas (Figure 11.18a). The route through western Florida Bay is driven primarily by local wind forcing (as shown above and also shown by Wang[20]), and by a mean sea-level slope between the Gulf of Mexico and Atlantic. The westward trajectory in Hawk Channel and the reef tract is sustained by local alongshore westward wind forcing in the middle and lower Keys that is enhanced toward the west by recirculating gyres and eddies north of the Florida Current.[8] This coastal countercurrent system of the Keys is further indicated in satellite thermal imagery by warm filaments from the Florida Current front that become entrained toward the west along the outer reefs at the time of the westward movement of this drifter through the lower Keys coastal region.[21] It takes one to two months for drifters released in the Shark River Slough plume to reach the Florida Keys, and then less than two weeks to reach the Tortugas region due to the increased flow in the coastal countercurrent. However, if winds have a significant southerly component, then drifters

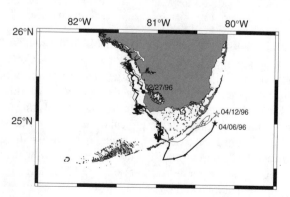

**FIGURE 11.16**  Surface trajectories of two drifters deployed simultaneously in the Shark River Slough discharge plume on February 27, 1996.

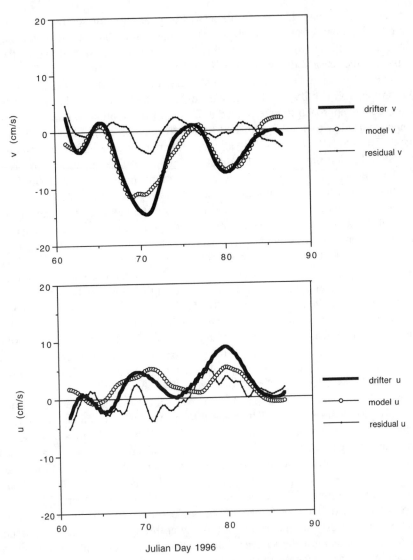

**FIGURE 11.17** Comparison of current components derived from drifter trajectories (drifter v) with modeled currents from wind components (model v) for March 1–28, 1996. Upper panel compares north–south (v) components and lower panel compares east–west (u) components. Residuals are also shown.

entering the Keys coastal waters through the tidal channels in the middle Keys will turn toward the north and may become entrained in the strong northward Florida Current (Figure 11.16).

A more direct route between Everglades discharge and the Tortugas occurs during east and northeast wind forcing when nearsurface flows on the southwest Florida shelf are toward the southwest (Figure 11.18b). The transport time scale to reach the Tortugas is approximately one month in this case. This southwest route is more typical of the fall season, when east and northeast winds prevail, but can occur during any season if east and northeast winds persist. A case in point is the anomalous northeast winds during the El Niño winter of 1998 that caused a surface drifter released in the Shark River Slough plume on February 5, 1998, to reach the Tortugas via the southwestern route in approximately two months (Figure 11.19a), whereupon it was entrained into a Tortugas gyre and recirculated south of the Tortugas and western Keys.[21]

The western Keys coastal waters are also highly connected to southwest Florida shelf waters by the combined influence of the counterclockwise circulation in the Tortugas gyre, wind-forced

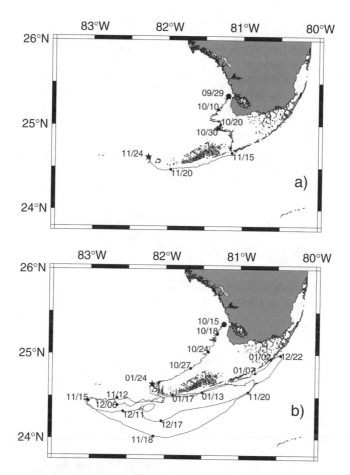

**FIGURE 11.18** (a) Surface drifter trajectories for September 29 to Nov. 24, 1997; (b) October 15, 1998 to January 21, 1999.

shelf circulation, and Loop Current influences. Surface drifter patterns reveal northward movement of gyre return flows onto the southwest Florida shelf in the vicinity of the Tortugas to be a common occurrence (Figures 11.19a,b). The continuation of the northward flow on the shelf has been observed to extend along the inner shelf to at least 26° before turning offshore and is primarily a response to the prevailing southeasterly winds.[9] The shelf offshore movement is again a result of wind forcing that shifts seasonally to northeast winds in the fall. Offshore movement continues until the drifter is entrained in the southward flow of the Loop Current and returns to the Tortugas region, whereupon it can again be entrained by the gyre circulation and repeat once again the exchange route with the southwest Florida shelf. Drifters have been observed to repeat this circuit several times over 5- to 8-month periods before becoming entrained into the Florida Current.[9]

## DISCUSSION

### SOUTH FLORIDA RECIRCULATION SYSTEM

The combination of the above-mentioned physical processes tends to form a recirculating current system that connects the entire South Florida coastal system. This recirculation system is shown schematically in Figure 11.20 and is driven by local wind-forced currents on the southwest Florida shelf, western Florida Bay, and Keys coastal zone, combined with a mean slope-driven current

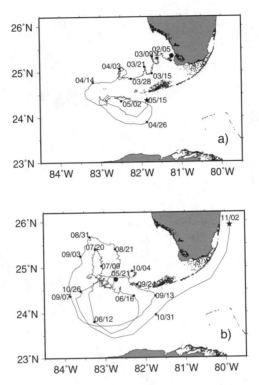

**FIGURE 11.19**    (a) Surface drifter trajectories for February 5 to May 15, 1998; (b) May 21 to November 2, 1994.

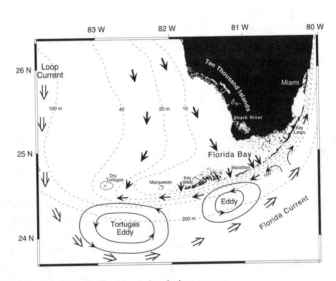

**FIGURE 11.20**    Schematic of South Florida recirculation system.

toward the southeast coupling the Gulf and Atlantic coastal waters, as well as offshore boundary current eddies that intensify countercurrents along the Keys.

Results of our first year observations show significantly different circulation patterns in the two shelf regions that interact with Florida Bay. Subtidal currents on the southwest Florida shelf and western Florida Bay are dominated by local wind-forced currents (Figures 11.5 and 11.17). A significant southward mean flow appears to be caused by the higher water levels in the Gulf compared to the Atlantic resulting in a negative sea-level slope across the Keys on the mean

(Figure 11.13). In contrast, the Keys coastal waters display a strong response to alongshore wind forcing and Florida Current interactions (Figure 11.7).[9,12] The curving shoreline of the Keys causes regional differences in response to prevailing easterly winds and Florida Current influences. The lower and western Keys are normally aligned with the wind and form a downwelling coast with onshore flow in the upper layer, offshore flow in the lower layer, and a westward coastal current that intensifies toward the west due to increased contributions from Florida Current eddies (Figures 11.6, 11.8, and 11.9). The prevailing wind is onshore in the upper Keys with little influence on alongshore flow, whereas the Florida Current is located closer to the outer reefs, causing strong downstream flows (Figure 11.6). The middle Keys represent a transition region where the coastal currents can be either toward the southwest or northeast depending on whether the local alongshore wind has a northerly or southerly component.

The different orientations of the Gulf and Atlantic shelf regimes separated by the curving island chain of the Florida Keys causes a different circulation and sea-level response to the same wind forcing, which results in a sea-level slope across the Keys and drives a subtidal flow that varies with the wind forcing (Figures 11.12 and 11.13). Southeast winds typical of the Keys region cause a set-up of sea level in Florida Bay and set-down in the middle and northern Keys coastal waters and drives a southeast flow toward the reef tract through the Keys passages and against the prevailing wind. East and northeast winds cause sea level to set-up in the Keys coastal waters and set-down in Florida Bay, driving subtidal flows toward Florida Bay through the passages.

Our vessel-mounted ADCP transects show the magnitude of the combined flows through Channels 5, 2, and Long Key Channel to range from about $-1400$ m³/s toward the southeast to 270 m³/s toward Florida Bay. The Long Key subtidal volume transport time series (Figure 11.12) gives transport variations ranging from $+1000$ to $-2500$ m³/s, and adding Channels 5 and 2 would increase this range to about $+1400$ to $-3500$ m³/s. The amplitude of these subtidal flows is equivalent in magnitude to the mean river discharge onto the southeast U.S. shelf by all the rivers between Florida and Cape Hatteras, NC. Approximately 500 to 1000 m³/s flows through Long Key Channel alone when the flow is toward the southeast, which is about 100 to 200 times greater than the peak freshwater discharge out of Shark River. Also, we found that a significant number of the surface drifters deployed near Shark River Slough were observed to move through western Florida Bay and toward the reef track through Long Key Channel (Figures 11.14, 11.15, and 11.16). Therefore, it appears that under typical conditions of southeast winds low-salinity discharge from Shark River Slough is advected and dispersed in the downstream direction through western Florida Bay and toward the Keys and provides a low-salinity input for dilution of the western part of the Bay that helps to regulate increasing salinities in the Bay interior. The advective/dispersal time scale for materials in the Shark River Slough plume to reach the reef tract is estimated from drifter trajectories at one to two months.

The trajectories of satellite-tracked surface drifters deployed in the Shark River Slough discharge plume clearly show strong linkages between western Florida Bay and the Florida Keys out to the Tortugas (Figures 11.14 and 11.15). It takes one to two months for drifters released in the Shark River Slough plume to reach the Florida Keys, and then less than two weeks to reach the Tortugas region due to the increased flow in the coastal countercurrent. However, if winds have a significant southerly component, then drifters entering the Keys coastal waters in the middle Keys will turn toward the north and may become entrained in the strong northward flow of the Florida Current.

Movement of the drifters from the southwest Florida shelf and western Florida Bay to the reef tract depends on local wind forcing plus a slope-driven current that has a subtidal variation due to the relative set-up or set-down of sea level on either side of the Keys from wind forcing, and a longer term mean contribution due to sea level standing higher in the Gulf of Mexico than the Atlantic. The comparison of volume transport time series in Long Key Channel with sea-level difference variations between our western Florida Bay mooring site (WFB, Figure 11.2) and Tennessee Reef site indicate a mean sea-level difference of $-5.95$ cm sloping down toward the

Keys that drives a mean southeast flow through Long Key Channel of $-280$ m$^3$/s. If we include Channels 5 and 2, this mean flow is increased to about $-400$ m$^3$/s, and if we also include the estimated flow through Seven-Mile Bridge at Vaca Cut the total mean Gulf to Atlantic flow approaches $-800$ m$^3$/s. TOPEX/POSEIDON altimeter measurements indicate sea-level height differences between the southwest Florida shelf and the Keys Atlantic coastal zone of about 10 to 30 cm.[22] The mean sea-level difference estimated from a general circulation model simulation of the Gulf of Mexico[23] shows the Gulf to be about 12 cm higher at the 300-m isobath on the west Florida shelf than at the same depth in the middle Keys. Therefore, our mean sea-level difference of $-6$ cm across the Keys does not seem unreasonable.

Recent comparisons of numerical model results of Gulf of Mexico circulation with TOPEX/POSEIDON altimetry and surface drifters provides new evidence that variations of Loop Current penetration into the northeast Gulf can cause significant change in the sea-surface slope between the eastern Gulf and the Keys, as well as the magnitude and shoreward extent of Loop Current-induced southward flow on the southwest Florida shelf.[22] These model/data comparisons indicate that a mature Loop Current that is fully extended into the northeast Gulf can cause trapping of the cross-shelf pressure gradient and a southward jet over the steep shelf escarpment and little penetration of southward flow onto the shelf. Whereas, a "young" Loop Current (i.e., following a recent eddy shedding) located near the southwest portion of the shelf will cause a larger along-isobath sea-level slope and the onshore penetration of southward flow into the inner shelf. The mean period of eddy shedding is estimated at 12.4 months, although with considerable variability.[24] During our first year of measurements, a large Loop Current eddy was shed in early March 1998.[24] Following this event, the Loop Current shifted to a young stage and was located off the southwest Florida shelf near the Dry Tortugas. Surprisingly, this young stage continued for approximately 9 months before regrowth of the Loop finally occurred in November 1998. The increase in southward flows was observed on the southwest shelf in the winter of 1997 and continued through the summer of 1998 in mid- to lower layers at ADCP sites B and D (Figures 11.5 and 11.6); it occurred nearly simultaneously with the prolonged young stage of the Loop Current, and provides further evidence of significant Loop Current influence on coastal currents on the southwest Florida shelf.

## CONNECTION OF SOUTH FLORIDA WATERS TO REMOTE REGIONS

Connection of South Florida coastal ecosystems with remote areas of the eastern Gulf of Mexico was recently verified by discovery of large volumes of Mississippi River (MR) water in the Florida Keys.[25,26] Multidisciplinary surveys during the period of September 10–13, 1993, revealed a band of anomalously low-salinity water embedded in the Florida Current and adjacent coastal waters extending from Key West to Miami, a distance of 260 km (Figure 11.21). Surface mixed layer salinity values as low as 31 were found, where typically surface salinity is greater than 36. The average offshore distance of the band was approximately 40 km. A Conductivity, Temperature, Depth (CTD) section off Looe Reef showed the band extended to approximately 20 m in depth.[26] The estimated volume of the band is approximately $33.3 \times 10^{10}$ m$^3$ for the Key West to Miami region, thereby requiring about $1.2 \times 10^{10}$ m$^3$ of freshwater to mix with oceanic waters to produce the low-salinity band. Biological and chemical indicators within the band, together with its large volume, suggest the 1993 MR flood as being the only conceivable source.[25] The mechanism for transport of MR water into the Straits of Florida was determined from satellite-tracked drifters and sea-surface temperature maps to be entrainment by the Loop Current.[25] It took approximately three weeks for the MR discharge waters to reach the Florida Keys and another 12 days to reach Cape Lookout, NC, where simultaneous shipboard measurements revealed anomalously low-salinity water along the Gulf Stream front and outer shelf.[27] Salinity time series from CMAN stations in the Keys indicate that the low-salinity water remained in the Keys coastal waters for approximately 3 months.[26] The transport of MR water to the Florida Keys appears to be the result of several independent processes occurring simultaneously. The unusually high discharge of MR water

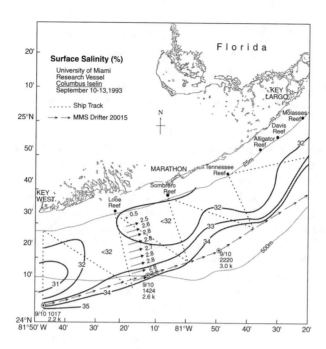

**FIGURE 11.21** Surface salinity from continuous thermosalinograph on the *R/V Columbus Iselin* for the period September 10–13, 1993. Nearsurface current vectors from shipboard ADCP are shown on the section south of Looe Reef, and the trajectory of MMS satellite-tracked surface drifter #20015 is superimposed. Date, time, and current speed are given at drifter locations.

occurred at a time when the Loop Current was in an extreme northerly position. The combination of summer heating and the river discharge vertically stratified the shelf waters near the delta, inhibiting vertical mixing of the MR plume with ambient shelf waters. Persistent, upwelling, favorable winds transported the MR plume offshore to the southeast, where it was entrained by the Loop Current and transported to the Straits of Florida along its shoreward front.

South Florida coastal waters are also linked to waters of the north coast of Cuba by the combination of wind-induced nearsurface transports, downstream Florida Current transport, and recirculation in coastal eddies and wind-driven countercurrents. Persistent easterly winds in the southern Straits of Florida will transport nearsurface waters toward the west and north from the north coast of Cuba, whereupon entrainment by the Florida Current will result in further eastward and northward transport to the vicinity of the Keys coastal waters and recirculation back to the west in eddy- and wind-supported countercurrents. Modeling experiments using a realistic mean Florida Current and easterly winds predict that it takes from 4 to 6 days for surface materials off the north coast of Cuba to reach the middle to upper Florida Keys.[28]

In addition, South Florida coastal waters are connected to remote upstream regions in the Caribbean through transport by the Caribbean Current, Loop Current, and Florida Current. The Florida Current is made up of about equal parts of waters originating in the South Atlantic and North Atlantic subtropical gyres[29,30] and is therefore an important link in both the North Atlantic Sverdrup circulation[31] and global thermohaline circulation.[32] The upper layer waters of the Florida Current with temperatures greater than 24°C are derived primarily from the South Atlantic[29] and are transported across the equator and through the Caribbean by the combined influence of the North Brazil Current and the North Atlantic wind-driven subtropical gyre. Therefore, the waters of the Florida Keys reef tract are comprised primarily of waters from the South Atlantic as part of the global thermohaline circulation, with additional contribution from eastern Gulf of Mexico shelf waters and southwest Florida river discharges.

## IMPLICATIONS FOR LARVAL RECRUITMENT

The combination of the above-mentioned physical processes tends to form a recirculating retention zone and recruitment pathway for pelagic larvae spawned in the Florida Keys coastal waters or foreign larvae transported from remote sources. This recruitment conveyor system is shown schematically in Figure 11.22. The four primary physical processes that drive the system are

1. *The Florida Current/Loop Current:* The Florida Current/Loop Current forms the offshore leg of the conveyor. Rapid downstream transport occurs in currents that can reach 100 to 200 cm/s. Larvae can be transported great distances from remote upstream sources in the eastern Gulf of Mexico and Caribbean Sea. This can be particularly significant for species with long pelagic larval stages, such as spiny lobster larvae which can remain in the plankton for 6 to 12 months.[33,34] The shoreward front of the Florida Current/Loop Current is an area of nearsurface current convergence. Therefore, both larvae and their planktonic food source will tend to be concentrated together in the frontal zone.[35] Onshore meanders of the front can transport larvae closer to the coastal zone and settlement habitat. Also, the mean shoreward displacement of the front in the middle and upper Keys will carry larvae closer to settlement habitat in this region. Larval detrainment from the front to the coastal zone can occur through small-scale cross-frontal mixing, eddy circulations, surface Ekman transports, and swimming in late-stage larvae. Increased abundances of lobster and conch larvae have been observed near the outer reefs when the Florida Current front is located in a nearshore position.[35,36] Eddy circulations in small-scale frontal eddies and Tortugas gyres have been shown to aid exchange of Florida Current and coastal species of fish larvae and pink shrimp larvae.[37,38]

2. *The Tortugas gyre:* The cyclonic circulation of the Tortugas gyre and its evolution into smaller eddies in the lower Keys provides a mechanism to entrain newly spawned larvae that can be retained in the gyre circulation for up to several months or escape the gyre on one of its shoreward circuits.[9] Pink shrimp larvae have been shown to take advantage of this pathway.[38] The Tortugas gyre also enhances food availability through increased primary production from upwelling of about 2 m/day and concentration of micro-zooplankton.[9,12] Early-stage spiny lobster larvae have been shown to be concentrated within the gyre and late stages near the Florida Current front and in the gyre interior,[35] indicating that the gyre circulation functions to retain both locally spawned lobster larvae and foreign recruits.

3. *Shoreward Ekman transports:* Onshore surface Ekman transports prevail throughout the region from the Dry Tortugas to the middle Keys due to persistent westward winds and east–west orientation of the coastal zone in this region. Spiny lobster larvae tend to be distributed in the upper mixed layer above the thermocline,[39,40] where onshore Ekman transports in this upper layer can result in concentration in the Florida Current front and detrainment into the coastal zone. Shoreward Ekman transports are further enhanced near the Florida Current axis, where surface currents and winds are opposed, significantly increasing the surface stress and onshore Ekman transport[41] and resulting in concentration of larvae along the front or in the interior of a coastal gyre.

4. *Coastal countercurrent:* The westward-flowing coastal countercurrent described above (Figures 11.6–11.8) as resulting from the combined influences of downwelling winds and coastal gyres provides the primary return leg of the recruitment conveyor. This feature can extend from the middle Keys to the Dry Tortugas. Its northern extent is limited by the curving coastline that causes the prevailing westward winds to change from an alongshore orientation in the lower Keys to onshore in the upper Keys. The maximum northward penetration of the countercurrent occurs in the fall when south-westward winds prevail. These fall winds are oriented alongshore in the middle Keys and have a southward component in the upper Keys which could result in a countercurrent

extending the entire length of the Keys from the Dry Tortugas to Key Largo. Cross-shore flows in the countercurrent are onshore in the upper layer and offshore in the lower layer in the lower Keys due to an Ekman response to the westward (downwelling) winds (Figure 11.8). However, the cross-shelf flow can shift to offshore in the Dry Tortugas region due to influence from the cyclonic circulation in the gyre. This gyre circulation also appears to cause a westward intensification in the countercurrent. Larvae that become detrained from the Florida Current front will be transported westward and shoreward by the coastal countercurrent, providing ample opportunity for recruitment to the reefs and nearshore zones.

Because of the variable nature and mix of processes that form the recruitment conveyor system, larvae are provided with many opportunities for recruitment into the Tortugas and Florida Keys on time scales ranging from hours to months. Recruitment pathways providing even longer retention times, such as required by spiny lobster larvae, are also available, especially in the Tortugas region and are shown schematically in Figure 11.22. This pathway requires movement onto the southwest Florida shelf and return to the Tortugas or Keys via either the Loop Current/Florida Current or by way of the mean south–southeastward flow on the west Florida shelf and through western Florida Bay, eventually entering the coastal countercurrent in the middle or lower Keys. The west Florida shelf/Loop Current pathway has been observed with satellite-tracked drifters and can increase larval retention to eight months.[9] The western Florida Bay/coastal countercurrent route is also quite plausible, as shown by trajectories of surface drifters deployed on the southwest Florida shelf and western Florida Bay (Figures 11.14–11.16, 11.18a), and observations of net southeastward flow through the tidal channels between the Keys (Figures 11.12 and 11.13).[19]

Seasonal changes in the conveyor system may also influence seasonal patterns in recruitment. The seasonal maximum in westward and southwestward alongshore winds in the fall can cause a seasonal maximum in the strength and northward extent of the coastal countercurrent. Together with a seasonal maximum, onshore surface Ekman transport and a seasonal minimum Florida Current downstream flow, these consequences should result in greater larval retention and enhanced opportunity for recruitment into the Tortugas and Keys coastal waters in the fall. During summer months, winds are weak and have a northward component and the Florida Current flow is

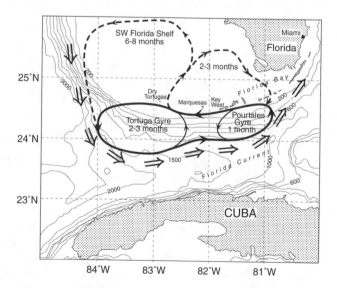

**FIGURE 11.22** Schematic of possible recruitment pathways for fish and lobster larvae spawned locally in the Dry Tortugas and Florida Keys. (From Lee, T.N. and Williams, E., *Bull. Mar. Sci.*, 64: 35–56, 1999. With permission.)

maximum, causing a reduction in the spatial extent of the coastal countercurrent. The result could be a decrease in recruitment to the upper Keys as larvae are carried out of the coastal retention zone by the Florida Current, but recruitment in the lower Keys and Tortugas should continue with little interference.

## Implications for South Florida Coastal Ecosystems

It is clear from the above discussions that South Florida coastal waters are highly connected throughout the region by local circulation patterns, as well as to remote regions of the Gulf of Mexico and Caribbean by large-scale western boundary currents of the Gulf Stream system. The recirculating current systems provide ample opportunity for recruitment of local- and foreign-spawned larvae on time scales commensurate with larval duration periods and may help explain the large abundances and high diversities of the coastal ecosystems. However, the current systems that help maintain the strength and variety of the ecosystems can also make them vulnerable to intrusions of pollutants or waters of poor quality from nearby or remote sources. Hypersaline and high-turbidity conditions occasionally develop in western Florida Bay and can be transported to the middle Keys reef tract in a matter of days to weeks and then spread westward towards the Tortugas in the wind and eddy driven coastal countercurrents. Porter et al.[42] observed a near-bottom stratified layer of high-salinity water during the summer of 1992 at Looe, Sombrero, and Tennessee Reefs. Northward transport of Florida Bay water toward upper Keys reefs is less likely due to the lower and middle Keys alignment with prevailing easterly winds.

Freshwater discharges through the Everglades in the Ten Thousand Islands area and Taylor River Slough are necessary to help moderate the development of hypersalinity in the poorly flushed regions of central Florida Bay. Recent planning for the restoration of South Florida coastal waters considers increasing the freshwater flow through the Everglades to Shark River Slough and Taylor River Sloughs. Due to the poor exchange of eastern and central Florida Bay with the surrounding waters these additional freshwater inputs at the northern and western boundaries are essential to help reduce hypersaline conditions from developing in the central bay during the dry season. Moderating seasonal salinity fluctuations may help to stabilize seagrass beds and improve water quality in western Florida Bay and thus also in the Keys reef tract, as these waters strongly interact.

A critical concern for water quality management in South Florida coastal waters is the input of nutrients from natural and anthropogenic sources. Considerable effort is underway to determine the nutrient budgets for the region and to predict future impacts on the carrying capacity of the ecosystems.[43] Many developed areas of the Keys have nutrient concentrations in inshore areas that are consistently high. These are typically regions in interior canals and lagoons that have poor exchange with the coastal waters and therefore a relatively weak offshore nutrient flux. The same can be said for freshwater inputs from the Everglades. Because the freshwater volume flow is small (typically less than 10 $m^3$/s), the nutrient flux to the coastal waters is also small even with elevated nutrient concentrations. As an example, a relatively high concentration of dissolved nitrate in Shark River Slough discharge of 1.0 $\mu M$ in a peak discharge of 10 $m^3$/s gives a nutrient flux of 10 $\mu M$ $m^3$/s. This is compared to the mean flow coupling the southwest Florida shelf to the Keys coastal waters through the passages in the middle Keys, which is on the order of 1000 $m^3$/s if all passages are included. Even with a weak nitrate concentration of 0.1, the nitrate flux in this flow would be 100 $\mu M$ $m^3$/s or a factor of 10 larger than the Everglades flow. The point of this comparison is that the largest continuous input of nutrients to South Florida coastal waters appears to come from the west Florida shelf and its adjacent rivers. These west Florida shelf inputs can at times also cause the transport of harmful algal blooms to South Florida waters, such as red tide organisms that can result in local fish kills. In addition, intrusions of large volumes of Mississippi River waters sometimes occur that invade the Keys coastal waters and extend into parts of west Florida Bay.[25] These events can decrease salinity, increase turbidity, and transport herbicides from farming areas of the Mississippi valley to the coastal waters of the Keys. The sustainability of coastal ecosystems in South Florida is dependent on water management policies. However, due to the strong connectivity of these waters over the entire region and with remote regions, it is important that water management and fisheries policies take a wide view including the eastern Gulf of Mexico and upstream regions of the Caribbean.

## CONCLUSIONS

The South Florida coastal ecosystem is made up of an interconnected set of distinct marine environments. In this chapter we have analyzed new results from a combination of recent shipboard hydrographic surveys, Eulerian current measurements, and Lagrangian drifter trajectories obtained as part of the South Florida Ecosystem Restoration Prediction and Modeling Program to show a high degree of connectivity between these subregions and with remote upstream areas of the Gulf of Mexico. Linkages between subregions are provided by circulation and exchange processes responding to both local and remote forcing.

The southwest Florida shelf and the Keys Atlantic coastal zone are connected by the passages between the islands in the middle, lower, and western Keys; however, these regions respond differently to wind forcing that is coherent over both areas due to different topographic constraints. The narrow, curving Keys coastal strip forms a downwelling coastal zone and westward currents in the lower and western Keys from prevailing easterly winds, whereas in the upper Keys region these winds are onshore. The middle Keys are a transition region. The seasonal cycle of the winds results in a seasonal change in coastal currents that is more pronounced in the upper and middle Keys, where northward flows occur in the summer during southeasterly winds and southward flows occur in fall, winter, and spring from northeast and east winds. Mean flows in the lower and western Keys are toward the west throughout the year due to the westward-oriented coastline. Water mass properties and currents in the Keys coastal zone are also highly influenced by interaction with the Florida Current and evolution of eddies that travel downstream along its frontal boundary. The larger mesoscale eddies (100 to 200 km) are initiated along the Loop Current front and can remain stationary in the Dry Tortugas region for several months before they move into the lower Keys and follow the curve of the Florida Current toward the upper Keys with decreasing size and increasing forward speed. These features intensify the westward countercurrents in the Keys and provide enhanced exchange with coastal waters that helps to maintain oceanic oligotrophic conditions required for coral reef survival. Although the eddies do not display a seasonal preference, the Florida Current annual cycle has a significant influence on mean coastal currents with maximum downstream flows in summer and minimum in fall. Also, annual mean flows in the Keys are directly related to the offshore distance of the Florida Current. Stronger downstream mean flows occur in the upper Keys due to the close proximity of the Florida Current to the outer shelf, and greater upstream (countercurrent) mean flows occur in the lower and western Keys due to the Florida Current location further offshore on the mean, together with more persistent eddies and downwelling winds.

The southwest Florida shelf represents the southern extreme of the west Florida shelf as it merges with the Florida Keys. Thus, this region is highly influenced by the processes occurring on the west Florida shelf, such as strong synoptic wind forcing, seasonal changes in wind forcing, river discharge and stratification, and Loop Current excursions into the northeast Gulf. As a result, the variability of local circulation patterns is highly dependent on synoptic-scale winds. Strongest subtidal currents are in the alongshore (north–south) direction and are a direct barotropic response to alongshore winds. Seasonal changes in wind forcing also produce seasonal differences in the strength and variability of the currents, with greater current amplitudes in winter following cold front passages and weaker currents in summer. A seasonal pattern is also present in the upper layer currents, which are more southward in the winter, spring, and fall, changing to northward in the summer with a shift of summer winds to southeasterly, whereas the lower layer currents are more persistent toward the south throughout the year. Surprisingly, the strength of the mean southward flow appears to vary with the position of the Loop Current as suggested by Hetland et al.[22] A fully extended, mature Loop Current is related to weaker southward mean flows, and a young Loop Current that remains near the Tortugas for about 9 months from March to November 1998 occurs with increased southward mean flow. Cross-shelf flows indicate an "estuarine-like" circulation occurring for most of the year with offshore flow in the upper layer and onshore in the lower layer.

The connection between the two shelf regimes is provided by the transports through the Keys passages. Our ADCP shipboard measurements across the passages of the middle Keys show subtidal transport variations ranging from about +300 to −1500 m³/s with negative values representing flow toward the reef tract and positive toward Florida Bay. Time-series estimates of volume transport and sea-level difference across the Keys show that these subtidal transport variations are due to local wind forcing that produces a different sea-level response in each shelf region, thus causing a sea-level slope across the Keys and transports ranging from +1400 to −3500 m³/s. In addition, a mean southeastward transport of about −1000 m³/s is estimated that opposes the prevailing winds and appears to be related to the mean southward transport derived from the moored current measurements on the southwest Florida shelf. This indicates that the Loop Current in the Gulf of Mexico must play an important role in maintaining the mean southeastward flows through the Keys passages that link the inner regions of the west Florida shelf to the Keys coastal zone.

The southeastward mean flow connecting the two shelf regions provides the source water for western Florida Bay and entrains the freshwater outflows from the Everglades through the Ten Thousand Islands. The magnitude of this mean southeast flow is about 100 to 200 times larger than the freshwater outflow from the Everglades, which results in a low-salinity band that is trapped along the coast of the Ten Thousand Islands and extends to the southeast into western Florida Bay. Transport of this low-salinity water to western Florida Bay provides an additional source of freshwater that, together with discharges from Taylor River Slough and Trout River in northeast Florida Bay, help to reduce hypersaline conditions in the central region of the Bay.

Trajectories of nearsurface drifters deployed in the Shark River Slough discharge plume show that three common pathways connect the entire South Florida coastal system. The primary pathways are either to the southeast and through the passages of the middle Keys, which is most common during winter and spring, or southwest to the Tortugas during the fall. Advective time scales to reach the Keys coastal zone are one to two months for these routes. The third pathway is to the northwest in the summer and eventual entrainment by the Loop Current, followed by transport to the Tortugas. This exchange route takes place over a 3- to 6-month time period. After drifters reach the Keys coastal zone, they tend to either recirculate in coastal eddies and wind-driven countercurrents for periods of 1 to 3 months or become entrained in the Florida Current and are removed from the coastal system.

Low-salinity intrusions into the South Florida coastal regions from southward transport down the west Florida shelf and entrainment along the Florida Current front show the region to be significantly linked to remote regions of the eastern Gulf of Mexico. Although the physical mechanisms providing the linkages are not well understood, the most likely causes are the Loop Current and its influence on shelf circulation.

The combination of recirculating current systems that take part in linking the different subdomains of the South Florida coastal region tend to form an effective retention zone for locally spawned larvae. The varied time scales to circuit the different-size eddies or coastal countercurrents provide the larval pathways and opportunities for recruitment from both local and foreign sources.

Although the strong circulation linkages within the south Florida coastal waters and with remote upstream regions may help to maintain the strength and variety of the ecosystems, they can also provide a conduit for the input of pollutants or waters of degraded quality. Thus, the sustainability of ecosystems in South Florida waters is dependent on water management policies of the entire region, as well as those of upstream regions in the eastern Gulf of Mexico.

## ACKNOWLEDGMENTS

We are greatly appreciative of the dedicated work of Mark Graham and Robert Jones of our Ocean Technology group in maintaining the current moorings. We thank Ryan Smith and Bob Roddy of AOML for their help in instrument exchanges and surveys. The CMAN data were collected as part of a cooperative agreement between FIO and NOAA/NDBC through the SEAKEYS Program. We appreciate the assistance of the RSMAS Marine Department and the crew of the *R/V Calanus* for their help in mooring operations and shipboard surveys. Support for this work was provided by NOAA/CIMAS through the South Florida Ecosystem Restoration Prediction and Modeling Program, Contract NA67RJ0149.

## REFERENCES

1. Wang, J., Van de Kreeke, J., Krishnan, N., and Smith, D., Wind and tide response in Florida Bay, *Bull. Mar. Sci.*, 54: 579–601, 1994.
2. Fourqurean, J.W., Jones, R.D., and Zieman, J.C., Processes influencing water column nutrient characteristics and phosphorus limitation of phytoplankton biomass in Florida Bay, USA; inferences from spatial distributions, *Estuaries, Coastal Shelf Sci.*, 36: 295–314, 1993.
3. Weisberg, R.H., Black, B.D., and Yang J., Seasonal modulation of the West Florida continental shelf circulation, *J. Geophys. Res.*, 23:2247–2250, 1996.
4. Mitchum, G.T. and Sturges W., Wind driven currents on the West Florida shelf, *J. Phys. Oceanogr.*, 12: 1310–1317, 1982.
5. Paluszkiewicz, T., Atkinson, L.P., Posmentier, E.S., and McClain, C.R., Observations of a loop current frontal Eddy intrusion onto the West Florida shelf, *J. Geophys. Res.*, 88: 9639–9652, 1983.
6. Lee, T.N., Leaman, K., Williams, E., Berger T., and Atkinson, L., Florida current meanders and gyre formation in the southern straits of Florida, *J. Geophys. Res.*, 100(C5): 8607–8620, 1995.
7. Fratantoni, P.S., Lee, T.N., Podesta, G.P., and Muller-Karger, F., The influence of loop current perturbations on the formation and evolution of tortugas eddies in the southern straits of Florida, *J. Geophys. Res.*, 103(C11): 24759–24799, 1998.
8. Lee, T.N. and Williams, E., Mean distribution and seasonal variability of coastal currents and temperature in the Florida Keys with implications for larval recruitment, *Bull. Mar. Sci.*, 64: 35–56, 1999.
9. Lee, T.N., Clarke, M.E., Williams, E., Szmant, A.F., and Berger T., Evolution of the tortugas gyre and its influence on recruitment in the Florida Keys, *Bull. Mar. Sci.*, 54(3): 621–646, 1994.
10. Robblee, M.B., Tilmant, J.T., and Emerson, J., Quantitative observations on salinity, *Bull. Mar. Sci.*, 44: 523, 1989.
11. Leichter, J.J. and Miller, S.L., Predicting high-frequency upwelling: spatial and temporal patterns of temperature anomalies on a Florida coral reef, *Cont. Shelf Res.*, 19: 911–928, 1999.
12. Lee, T.N., Rooth, C., Williams, E., McGowan, M., Szmant, A.F., and Clarke, M.E., Influence of Florida Current, gyres and wind-driven circulation on transport of larvae and recruitment in the Florida Keys coral reefs, *Cont. Shelf. Res.*, 12: 971–1002, 1992.
13. Csanady, G.T., The arrested topographic wave, *J. Phys. Oceanogr.*, 8: 47–62, 1978.
14. Beardsley, R.C. and Butman, B., Circulation on the New England continental shelf: response to strong winter storms, *Geophys. Res. Lett.*, 1: 181–184, 1974.
15. Lee, T.N., Williams, E., Wang, J., and Evans, R., Response of South Carolina continental shelf waters to wind and gulf stream forcing during winter of 1986, *J. Geophys. Res.*, 94: 10715–10754, 1989.
16. Lee, T.N., Williams, E., Johns, E., and Wilson, D., First year results from enhanced observations of circulation and exchange processes in western Florida Bay and connecting coastal waters, including effects of El Niño and hurricane Georges, in *Proceedings of Florida Bay and Adjacent Marine Systems Science Conf.*, pp. 145–147, 1999.
17. Lee, T.N., Coastal circulation in the Key Largo Coral Reef Marine Sanctuary: physics of shallow estuaries and bays, in *Lecture Notes in Coastal and Estuaries Studies*, Vol. 16, Springer-Verlag, Berlin, pp. 178–198, 1986.
18. Pitts, P.A., An investigation of near-bottom flow patterns along and across Hawk Channel, Florida Keys, *Bull. Mar. Sci.*, 54: 610–620, 1994.
19. Smith, N.P., Long-term gulf-to-Atlantic transport through tidal channels in the Florida Keys, *Bull. Mar. Sci.*, 54: 602–609, 1994.
20. Wang, J.D., Subtidal flow patterns in western Florida Bay, *Estuaries, Coastal Shelf Sci.*, 46: 901–915, 1998.
21. Lee, T.N., Johns, E., Wilson, D., and Williams, E., Site characterization for the tortugas region: physical oceanography and recruitment, *Tortugas 2000 Report by Florida Keys National Marine Sanctuary*, 1999.

22. Hetland, R.D., Hsueh, Y., Leben, R., and Niiler, P.P., A loop current-induced jet along the edge of the west Florida shelf, *Geophys. Res. Lett.*, 26: 2239–2242, 1999.

23. Hsueh, Y., Yuan, D., and Clarke, A.J., A numerical study of the DeSoto Canyon intrusion in the Northeastern Gulf of Mexico, *J. Phys. Oceanogr.*, submitted.

24. Sturges, W. and Leben, R., Frequency of ring separations from the loop current in the Gulf of Mexico: a revised estimate, *J. Phys. Oceanogr.*, 30: 1814–1819, 2000.

25. Ortner, P., Lee, T.N., Milne, P., Zika, R., Clarke, M.E., Podesta, G.P., Swart, P., Tester, P.A., Atkinson, L.P., and Johnson, W., Mississippi river flood waters that reached the gulf stream, *J. Geophys. Res.*, 100(C7): 13595–13601, 1995.

26. Gilbert, P.S., Lee, T.N., and Podesta, G.P., Transport of anomalous low-salinity waters from the Mississippi river flood of 1993 to the straits of Florida, *Cont. Shelf Res.*, 16(8): 1065–1085, 1996.

27. Tester, P.A. and Atkinson, L.P., Hydrometeorological setting, in *Coastal Oceanographic Effects of Summer 1993 Mississippi River Flooding: A Special NOAA Report*, Dowgiallo, M.J., Ed., U.S. Department of Commerce, Washington, D.C., pp. 71–75, 1994.

28. Mooers, C.N.K., South Florida oil spill research center, *Spill Sci. Technol. Bull.*, 4: 35–44, 1997.

29. Schmitz, W. and Richardson, P. L., On the sources of the Florida current, *Deep-Sea Res.*, 38(suppl.1): 5379–5409, 1991.

30. Wilson, W.D. and Johns, W.E., Velocity structure and transport in the windward island passages, *Deep-Sea Res.*, 44(3): 487–520, 1997.

31. Leetmaa, A., Niiler, P.P., and Stommel, H., Does the Sverdrup relation account for the mid-Atlantic circulation? *J. Mar. Res.*, 35: 1–10, 1977.

32. Gordon, A.L., Interocean exchange of thermocline water, *J. Geophys. Res.*, 91: 5037–5046, 1986.

33. Lewis, J.B., The phyllosoma larvae of the spiny lobster, *Panulirus argus. Bull. Mar. Sci. Gulf Carib.*, 1: 89–103, 1951.

34. Sims, H.W., Jr. and Ingle, R.M., Caribbean recruitment of Florida's spiny lobster population, *Q. J. Florida Acad. Sci.*, 29: 207–243, 1966.

35. Yeung, C., Larval Transport and Retention of Lobster Phyllosoma Larvae in the Florida Keys, Ph.D. dissertation, University of Miami, Coral Gables, FL, p. 217, 1996.

36. Stoner, A.W., Mehta, N., and Lee, T.N., Recruitment of *strombus* veligers to the Florida reef tract: relation to hydrographic events, *J. Shellfish Res.*, 16: 1–6, 1997.

37. Limouzy-Paris, C.B., Graber, H.C., Jones, D.L., Ropke, A., and Richards, W.J., Translocation of larval coral reef fishes via sub-mesoscale spin-off eddies from the Florida current, *Bull. Mar. Sci.*, 60: 966–983, 1996.

38. Criales, M.M. and Lee, T.N., Larval distribution and transport of penaeoid shrimps during the presence of the Tortugas gyre in May–June, 1991, *U.S. Fisheries Bull.*, 93: 471–482, 1995.

39. Yeung, C. and McGowan, M.F., Differences in inshore-offshore and vertical distribution of phyllosoma larvae of *Panulirus*, *Scyllarus*, and *Scyllarides* in the Florida Keys in May–June 1989, *Bull. Mar. Sci.*, 49(3): 699–714, 1991.

40. Yeung, C., Couillard, J.T., and McGowan, M.F., The relationship between vertical distribution of spiny lobster phyllosoma larvae (Crustacea: Palinuridae) and isolume depths generated by a computer model, *Rev. Biol. Trop.*, 41(1): 63–67, 1993.

41. Rooth, C. and Xie, L., Air-sea boundary dynamics in the presence of mesoscale surface currents, *J. Geophys. Res.*, 97: 14431–14438, 1992.

42. Porter, J.W., Lewis, S.K., and Porter, K.G., The effect of multiple stressors on the Florida Keys coral reef ecosystem: a landscape hypothesis and a physiological test, *Limnol. Oceanogr.*, 44: 941–949, 1999.

43. Rudnick, D., Madden, C., Sklar, F., Kelly, S., Donovan, C., Picard, K., McCauliffe, J., and Korvela, M., Nutrient cycling and transport in the Florida Bay — Everglades ecotone, in *Proc. of Florida Bay and Adjacent Marine Systems Science Conf.*, pp. 80–81, 1999.

# 12  Regional-Scale and Long-Term Transport Patterns in the Florida Keys

*Ned P. Smith and Patrick A. Pitts*
Harbor Branch Oceanographic Institution

## CONTENTS

## INTRODUCTION

Major circulation studies in the Florida Keys and Florida Bay region began in the early 1980s. Individual studies since then have generally been restricted to shelf waters on the Atlantic Ocean side of the Keys, Florida Bay on the Gulf of Mexico side of the Keys, and tidal channels that exchange Atlantic and Gulf waters. This chapter integrates results from published studies and unpublished data into a description of regional-scale transport patterns. The objective is to document the physical linkage of Gulf and Atlantic waters, starting with the southward flow of Gulf water along the southwest coast of the Florida Peninsula and ending with exchanges across the reef tract on the Atlantic side of the Keys. Two pathways are described. The first is the counterclockwise flow of water along the eastern fringe of the Loop Current in the Gulf of Mexico, then along the northern and western side of the Florida Current in the Straits of Florida. The second pathway, which is emphasized here, involves the more direct eastward or southeastward movement of water through Florida Bay, through the tidal channels between Keys, and then seaward across the shelf to the reef tract on the Atlantic side of the Keys. This linking of Gulf and Atlantic sides of the

0-8493-2026-7/02/$0.00+$1.50
© 2002 by CRC Press LLC

Keys is a response to both wind and tide forcing, which may change in relative importance with location and with the seasons.

## REGIONAL FLOW PATTERNS OUTSIDE OF FLORIDA BAY

Large-scale flow patterns in the immediate vicinity of the Florida Keys are bounded on the west by the Loop Current in the Gulf of Mexico, and on the south and east by the Florida Current in the Straits of Florida. While direct effects of the Loop Current and the Florida Current are probably minimal, the southward flow on the eastern fringe of the Loop Current and the eastward and northward flow along the western fringe of the Florida Current are important components of the regional flow field. Diurnal and semidiurnal tidal motions and low-frequency wind-driven currents, which can dominate the current at a given time and location, are superimposed onto the quasi-steady counterclockwise flow of water around the Keys.

### WEST OF FLORIDA BAY

Some of the early nautical charts, apparently based on regional wind fields, offered misleading information regarding flow patterns in shelf waters of the eastern Gulf of Mexico. Hydrographic Office Chart 10690, for example, shows eastward and northeastward flow in shelf waters directly west of Florida Bay. While this would be a logical consequence of wind forcing, it is not supported by observations. Similarly, Ichiye et al. (1973) calculated wind-driven, Ekman surface currents and depth-integrated Ekman transports for a $2° \times 2°$ square, centered approximately on Florida Bay. Calculations suggested westward or northwestward surface currents with speeds ranging from about 5 to 18 cm s$^{-1}$ from September through May, then northwestward during June–August with speeds less than 2 cm s$^{-1}$. Depth-integrated Ekman transport is northwestward from September through May, then northward but insignificant in magnitude during June–August. However, as they began to accumulate, direct measurements confirmed a southward flow along the west side of the Florida Peninsula and relegated wind-drift to secondary importance at best (Wennekens, 1959; Vukovich et al., 1979).

The southward flow of water in the eastern Gulf of Mexico as the Loop Current approaches the Straits of Florida appears to be indirectly responsible for the southward flow of water along the southwestern part of the Florida Peninsula. Wennekens (1959) noted a southward flow along and inside of the 10-m isobath off the southwest coast of the Florida Peninsula, as well as a northeastward flow into Florida Bay through the middle of the open western boundary. Vukovich et al. (1979) used satellite data to suggest that eastward meanders of the Loop Current get no closer than 75 km to the western boundary of Florida Bay at latitude 25°N. Residual flow patterns in shelf waters off the southwest coast of Florida indicate southward flow in both near-surface and near-bottom layers.

Current meter data collected at a study site located 48 km west of Whitewater Bay from December 13, 1983, to November 1, 1984 (Environmental Science and Engineering, 1987) can be used to describe flow in the near-bottom levels over the inner shelf off the southwestern tip of the Florida Peninsula (Figure 12.1, Station a). Currents were recorded 3 m above the bottom in 13 m of water (Figure 12.2). Hourly current vectors are plotted head-to-tail and presented as a progressive vector diagram (PVD). The PVD is well suited for characterizing general flow patterns at a study site over time scales of days and longer. The pattern shows a quasi-steady southeastward flow past the study site. The resultant speed over the 326-day study period was 3.7 cm s$^{-1}$ and the resultant direction was 126°. High-frequency east–west motions reflect tidal oscillations which dominated the instantaneous current at the study site.

Using salinity as a tracer, Lee et al. (1996) tracked a plume of low-salinity water from the mouth of Shark River southward toward Florida Bay. The width of the plume varied from 2 to 10 km, and vertical stratification was as strong as 1 psu m$^{-1}$ in the upper 3 m near the mouth of

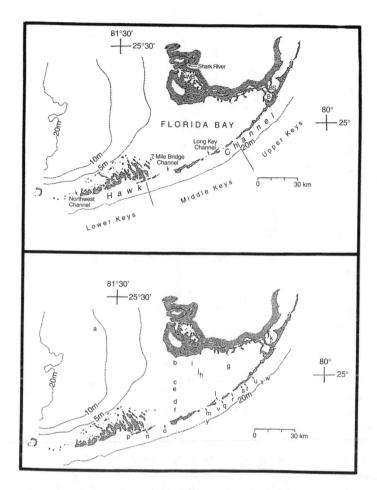

**FIGURE 12.1** Maps showing Florida Keys study area including geographic names referred to in the text (top) and locations of study sites (bottom). Whitewater Bay, Barnes Sound, and Blackwater Sound are labeled WB, BS, and B, respectively, in the top plot. Station identifiers (a–x) in the bottom plot are the same as those appearing in the text. Hatched lines show the 5-, 10-, and 20-m isobaths; the 20-m isobath on the Atlantic side of the Keys also represents the location of the reef tract.

the river. The study identified Shark River Slough as a significant source of freshwater for Florida Bay, as well as the path taken along the inner shelf off the southwest coast of Florida.

## STRAITS OF FLORIDA

On the southern and eastern sides of the Florida Keys, the relative position and behavior of the Florida Current has a profound effect on the movement of water at and immediately seaward of the reef tract. When the axis of the Florida Current is relatively close to shore, flow at the reef tract is eastward or northeastward toward Miami; when the Florida Current meanders seaward, cold, cyclonic gyres often form which drive westward flow at the reef tract off the Lower Keys (Lee et al., 1992, 1994, 1995). The heading of the Loop Current as it enters the Straits of Florida determines whether or not gyres form in the southern Straits of Florida off the Dry Tortugas. Lee et al. (1994, 1995) combined current meter data, satellite-tracked drifters, shipboard hydrographic data, and satellite-derived surface thermal imagery to describe the formation and movement of large cyclonic gyres in the southern Straits of Florida. Gyres can form when a well-developed Loop Current penetrates deeper into the northeastern Gulf of Mexico. As a result of the more southward

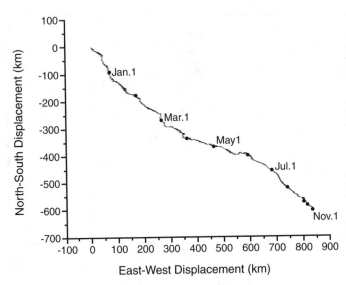

**FIGURE 12.2**    Progressive vector diagram constructed from currents recorded 3 m above the bottom in 13 m of water over the southwest Florida shelf (Figure 12.1, Station a), December 11, 1983, to November 1, 1994. Positive values on the *x* and *y* axes represent eastward and northward flow, respectively. Dots are spaced at the beginning day of each month.

heading of the Loop Current as it leaves the Gulf, the flow overshoots the entry to the Straits of Florida and the axis of the Florida Current shifts southward. This allows gyres to form just south of the Dry Tortugas. Gyres with diameters as large as 200 km can persist for time periods on the order of 100 days. During this time, they can move eastward at about 5 km day$^{-1}$ along the Atlantic side of the Lower Keys. Earlier work by Lee et al. (1992) reported a gyre, reduced in size, over the Pourtales Terrace off the Middle Keys, which has since been identified as a later stage in the downstream decomposition of a gyre formed off the Dry Tortugas. Lee et al. (1994) noted that when the Loop Current does not penetrate deeply into the Gulf, gyres do not form and the axis of the Florida Current lies much closer to the Lower Keys. Vukovich et al. (1979) reported that the northern fringe of the Florida Current lies within the 200-m isobath 75% of the time, suggesting that the flow at the reef tract should alternate between relatively strong eastward or northeastward flow when gyres are absent, and westward or southwestward when gyres are present.

Further downstream in the Straits of Florida, off the Middle Keys and Upper Keys, gyres play an increasingly minor role in flow patterns recorded at the reef tract. Navigation charts indicate that the axis of the current is only 25 to 30 km seaward of the reef tract off the Upper Keys, leaving little room to the west for gyres. The Florida Current can meander laterally across the Straits of Florida, however, and frontal eddies can form on the western fringe (Lee, 1975; Lee and Mayer, 1977). The presence or absence of frontal eddies along the western edge of the Florida Current is an important factor in determining the nature of the flow at and just seaward of the reef tract.

## FLOW THROUGH FLORIDA BAY

The alternate route linking shelf waters off the southwest coast of the Florida Peninsula with the reef tract on the Atlantic side of the Keys is the shorter pathway through Florida Bay. Topographic constraints imposed by shallow banks and mangrove islands, and by the narrow channels that connect sub-basins within the bay, greatly reduce the eastward movement of water through the northern and central parts of the bay. Also, flow through the shallow interior of the bay can be made more circuitous as a result of local wind forcing. Nevertheless, the studies and data described below indicate a generally west-to-east movement of water through the interior of the bay that eventually exits through the tidal channels between Keys on the southeastern and southern sides of the bay.

Drifter studies describe a northwest-to-southeast movement of water across the more open waters of western and southern Florida Bay. Lee et al. (1996) tracked surface drifters released in the freshwater plume leaving Shark River and followed them southward, then southeastward as they moved through south-central Florida Bay before exiting through Long Key Channel. Similarly, Wang (1998) followed surface drifters moving southeast through the bay. The time required to move through the bay was on the order of 5 to 10 days, depending on wind conditions at the time.

## FLOW ACROSS THE WESTERN BOUNDARY

The exchange of water across the open western boundary of Florida Bay, taken here to be the 81°05′W meridian, was investigated in two field studies, and both show nearly the same results (Smith, 2000). The northern and central sections of the western boundary appear to be regions of net flow into the bay, while the section south of approximately 24°52′N appears to be a region of net outflow. The earlier of the two studies, conducted in 1994–95, included current meters spaced at 14-km intervals along the 81°05′W meridian (Figure 12.1, Stations b–d). At each study site, current meters were moored at mid-depth in about 3 m of water. Current measurements were extrapolated vertically and combined with water level data to quantify depth-integrated transport. Dividing the vertically integrated transport by the mean water depth, the depth-averaged inflow at the northern and central stations averaged 3.0 and 1.4 cm $s^{-1}$, respectively, and the depth-averaged outflow at the southern station averaged 0.5 cm $s^{-1}$. Incorporating the cross-sectional area of the western boundary represented by each data point, these three values indicate a mean inflow of 1377 $m^3$ $s^{-1}$ across the northern and central sections of the western boundary and an outflow of −381 $m^3$ $s^{-1}$ across the southern section, for a net mean inflow of 996 $m^3$ $s^{-1}$.

The second field study, conducted in 1996–97 by the Army Corps of Engineers Waterways Experiment Station (WES), used acoustic doppler profilers (ADPs) at three locations along the western boundary (Figure 12.1, Stations b, e, f). Bottom-mounted ADPs quantified local transport by recording both water depth and current speed in from 2 to 12 layers, according to mean water depth, and the tidal and low-frequency rise and fall in water level. Again, the depth-averaged current speed indicated an inflow through the northern and central parts of the western boundary and an outflow through the southern part. Depth-averaged current speeds calculated from the ADP data at the northern and central study sites were 2.3 and 1.0 cm $s^{-1}$, respectively, and the depth-averaged current speed calculated for the southern site was 0.9 cm $s^{-1}$. These three data points indicate a mean inflow of 1264 $m^3$ $s^{-1}$ across the northern and central sections of the western boundary, a mean outflow of −277 $m^3$ $s^{-1}$ through the southern section, and thus a net mean inflow of 987 $m^3$ $s^{-1}$.

Harmonic constants for the principal tidal constituents from the current meter and water level data obtained from both studies can be combined to quantify the tide-induced transport of water into the bay. Results of a one-year simulation using ten diurnal, semidiurnal and shallow-water tidal constituents indicate that the tide-induced residual transport of water across the 81°05′W meridian is an inflow that averages 1470 $m^3$ $s^{-1}$. The surface area of Florida Bay east of 81°05′W is approximately 2220 $km^2$, and transport at this rate would raise the water level of the bay 5.7 cm $day^{-1}$. A quasi-steady state is presumably reached when the surface of the bay is elevated only a few centimeters at most. Wang et al. (1994) have modeled Florida Bay and used simulations to suggest that the tide-induced set-up is on the order of 1.2 cm. Results from these observational and modeling studies suggest that a tide-induced transport of water into Florida Bay elevates the water level of the bay and forces some fraction of the observed outflow through the tidal channels between Keys along the southeast side of the bay (described below) and through the southern part of the western boundary where the tide-induced inflow is weakest.

A comparison of flow across the western boundary with local winds indicates high coherence between east–west flow and winds out of the eastern sector — the prevailing wind direction in this region — over time scales longer than about 7 days (Smith, 2000). In both field studies, northwestward wind stress was most coherent with east–west transport through the northern part of the boundary, while southwestward wind stress was most effective in forcing water across the southern part. Westward winds were most closely coupled with flow across the central part of the boundary. Coherence values were 0.53 to 0.88 within the 7- to 21-day period range. Resultant wind stress was favorable for forcing water offshore across the western boundary during both study years, yet an inflow was observed across the central and northern parts. Smith (2000) reported that westward wind stress can produce a set-down of water level in the bay, thus the observed inflow at mid-depth may be in response to an eastward-directed barotropic pressure gradient.

## FLOW THROUGH THE INTERIOR

Current meter records from numerous channels connecting sub-basins in the interior of the bay show considerable variation in speed but a surprising degree of directional persistence, especially over time scales longer than a few days (Pitts and Smith, 1995). Figure 12.3 is an example of such a record. The plots show current meter data recorded in an east–west channel just north of the Calusa Keys in the east–central part of the bay (Figure 12.1, Station g). Data were recorded hourly during a 48-day study from June 10 to July 27, 1998. In the upper plot, along-channel current components have been converted to displacements by multiplying current speeds by the time interval they represent and accumulating the values. For example, a speed of 28 cm s$^{-1}$ (1 km h$^{-1}$) produces a displacement of 1 km over a one-hour time period. Positive currents represent westward flow, thus the increasingly negative cumulative displacements indicate a net eastward movement of water. The resultant eastward flow past the current meter was approximately 15 cm s$^{-1}$.

Hourly along-channel current speeds shown in the lower plot are useful for depicting the unsteady nature of the current and the relative importance of tidal and low-frequency fluctuations.

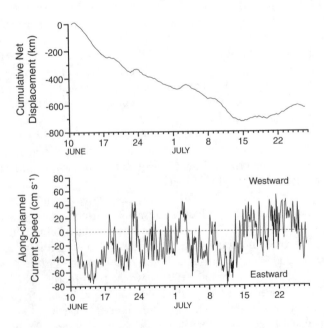

**FIGURE 12.3**   Cumulative net displacement diagram (top) and hourly along-channel currents (bottom) recorded 1 m above the bottom in 3 m of water in Calusa Keys Channel (Figure 12.1, Station g), June 10 to July 27, 1998. Positive currents and displacements indicate westward flow.

Again, positive current speeds indicate westward-flowing water. Tidal currents contribute a signif-icant fraction of the high-frequency variability. The $M_2$ and $K_1$ constituent amplitudes are 9 and 6 cm s$^{-1}$, respectively. Other diurnal and semidiurnal tidal constituents have amplitudes of 1 to 2 cm s$^{-1}$. While the ebb and flood of the tide is relatively weak through this and other channels connecting sub-basins in the northern and northeastern parts of the bay, the exchanges are contin-uous and dependable, and thus represent an important flushing mechanism that serves to maintain or re-establish water quality.

Low-frequency current speeds in Calusa Keys Channel generally range between 30 and 70 cm s$^{-1}$ over time scales of several days to about a week during the 48-day study. A comparison of winds and currents indicates a very close coupling between low-frequency fluctuations of flow through the channel and local wind stress. Wind forcing was responsible for the 11-day reversal in displacement at the end of the study period and other more transient reversals that appear earlier in the record. It is important to note, however, that the resultant wind direction during the study period was toward the northwest — nearly the opposite direction of the net eastward flow. So, while wind forcing can be shown to explain a large percentage of the low-frequency variation about the resultant flow, the forcing that produces the resultant flow itself is more complex.

## TIDAL FORCING IN THE INTERIOR OF FLORIDA BAY

The quasi-steady nature of the flow through the channel north of the Calusa Keys and other channels in the interior suggests that tidal forcing may play a role in driving water eastward and linking Gulf and Atlantic sides of Florida Bay. One possibility involves water moving into the bay across the tops of the shallow banks during the flood tide. Both semidiurnal and diurnal tidal waves entering from the Gulf of Mexico could contribute to this "overtopping" process. Near tidal channels connecting western Florida Bay with interior sub-basins high tide coincides very nearly with maximum flood, and low tide coincides very nearly with maximum ebb (Pitts and Smith, 1995). Thus, while water floods into the bay relatively easily, the shallow water over the banks at low tide inhibits the ebb tide from returning across the top of the bank due to increased frictional resistance. Connecting channels offer a path of least resistance for the outflow. Pitts and Smith (1995) calculated a 12.2 cm s$^{-1}$ mean westward flow out of the interior through Iron Pipe Channel (Figure 12.1, Station h) during a 400-day study from August 1994 to September 1995. Outflow through the channel was remarkably steady throughout the study period. Similarly, quasi-steady outflow was recorded in Conchie Channel from July 1994 to April 1995, and in Man of War Channel from late August 1994 to mid-September 1995 (Figure 12.1, Stations i and j, respectively). While a part of this outflow may be explained in terms of fresh water leaving the interior of the bay, especially during the wet season (May–November), Fourqurean et al. (1993) showed that the interior of the bay has had hypersaline conditions over extended periods of time. Also, a quasi-steady outflow was observed through these channels throughout the dry season. Thus, a local recirculation, involv-ing a relatively strong flooding of water across the banks and a correspondingly strong ebb tide outflow through the channels, is also possible.

The paradigm of a tide-induced transport of water into, through, and out of Florida Bay has been difficult to support through observations, however. If tidal forcing is important, one might expect to see outflow through these channels vary according to spring/neap or tropic/equatorial conditions. The cycling in and out of phase of the $M_2$ and $S_2$ semidiurnal constituents, and the $K_1$ and $O_1$ diurnal constituents produce "beat frequencies" with maximum amplitudes recurring approximately every two weeks. The fortnightly Mf constituent accounts for the interaction of the $K_1$ and $O_1$ constituents, and the MSf constituent accounts for the interaction of the $M_2$ and $S_2$ constituent. Several long current meter records from connecting channels in the interior of the bay and long records from tidal channels that connect Florida Bay with the Atlantic have been used to quantify fortnightly constituents (Smith, 1994, 1998). Results do not provide strong supporting evidence that tidal forcing is responsible for the net outflow from the bay observed through these channels, even though both the spring–neap and the tropic–equatorial tide sequences are important along the western side of Florida Bay. For example, when the $K_1$ and $O_1$ constituents are isolated and used to calculate the transport of water into and out of the bay, the difference between flood and ebb transport at times of tropic tide is approximately 60% greater than the difference at times of equatorial tide. Similarly, when the $M_2$ and $S_2$ constituents are isolated, the difference between flood and ebb at times of spring tide is just under 60% greater than the difference at times of neap tide.

Given the large fortnightly variability in the tide-induced residual transport of water into the bay, one would expect a detectable fortnightly variation of flow through connecting channels within the bay, and through the tidal channels that connect Florida Bay with Atlantic shelf waters. Both the Mf and MSf constituents have amplitudes on the order of only 1 to 2 cm s$^{-1}$, however. More significantly, the phase angles of the fortnightly constituents indicate that strongest flow through the channels occurs before the times of spring or tropic tide along the western boundary of the bay. In Long Key Channel, for example, the phase angle of the Mf constituent indicates a 32° phase lead (29-hour time lead) over the phase angle calculated for transport across the western boundary of the bay. In some channels, however, the harmonic constants of the fortnightly constituents may involve an interaction of both Gulf and Atlantic tides. Thus, tidal forcing cannot be ruled out as a mechanism for forcing west-to-east flow across Florida Bay on the basis of comparisons involving Gulf tides alone.

## OTHER FORCING MECHANISMS

Another mechanism for forcing water west-to-east through Florida Bay is the difference in mean sea level between the eastern Gulf of Mexico and the adjacent waters of the tropical North Atlantic Ocean. Chew et al. (1982) reported an 8 to 9-cm drop in mean sea level between Key West and Miami. Using altimeter measurements, Hetland et al. (1999) observed sea-level height differences of 10 to 30 cm between the southwest Florida shelf and shelf waters on the Atlantic side of the Keys. The same barotropic pressure gradient that explains the eastward flow of the Florida Current as it leaves the Gulf of Mexico between the Florida Keys and Cuba can be used to explain a west-to-east downslope movement of water through Florida Bay and into the Atlantic via the tidal channels along the southeastern and southern sides of the bay.

While the relative importance of sea-level differences and tidal forcing in producing a quasi-steady movement of water through Florida Bay has not yet been established, low-frequency fluctuations clearly arise in response to wind forcing. Pitts and Smith (1995) compared low-frequency variations in the flow through Iron Pipe and Man of War Channels with a series of wind-stress components to describe the wind forcing to which the nontidal circulation is most responsive. Results differed for each channel, but high coherence levels were found in each case over time scales on the order of 2 to 6 days. Highest coherence values were on the order of 0.6, suggesting that approximately 60% of the low-frequency variance in the along-channel current can be explained by wind stress. While this is a substantial fraction of the total variance, the transfer function (gain) provided by spectral analysis indicates that only 5 to 15% of the outflow observed through these channels can be accounted for by wind forcing. Thus, wind forcing serves primarily to increase or decrease the rate at which water exits the bay through these channels.

Working in the northeast corner of Florida Bay, Pitts (1998) combined 14-month current meter and water level records to investigate the long-term volume transport through Jewfish Creek (Figure 12.1, Station k) and document the linkage of Florida Bay with Biscayne Bay through Blackwater Sound and Barnes Sound. Results indicate a clearly defined seasonal fluctuation in exchanges through Jewfish Creek in response to seasonally changing wind stress. Water exits Florida Bay through Blackwater Sound during late spring and summer, then enters from Barnes Sound during fall, winter, and early spring months. Over time scales of several days to about two weeks, water level differences between Barnes Sound and Blackwater Sound force a transport of

±30–50 m³ s⁻¹ through Jewfish Creek. These low-frequency motions account for 80% of the total variance in along-channel volume transport. For comparison, the amplitude of the $M_2$ tidal constituent is 27 m³ s⁻¹. Tidal exchanges in the northeast corner of Florida Bay are forced by Atlantic tides entering Barnes Sound through Biscayne Bay, as Blackwater Sound is virtually tideless.

## FLOW THROUGH TIDAL CHANNELS

Virtually all of the major tidal channels in the Florida Keys that connect Florida Bay with Atlantic shelf waters have been investigated to determine tidal and long-term exchanges between the Gulf and Atlantic sides of the Keys (Smith, 1994, 1998, in review; Pratt and Smith, 1998). With few exceptions, tidal channels between Keys show a net flow out of Florida Bay and into the Atlantic. Smith (1994) reported a net outflow through Bahia Honda Channel, Moser Channel, and Newfound Harbor Channel in the Lower Keys (Figure 12.1, Stations n–p, respectively) that averaged 3 to 10 cm s⁻¹ during 6 to 7 month studies in 1987–88. A similar long-term outflow rate was recorded through Channel Five and Long Key Channel in the Middle Keys (Figure 12.1, Stations l and m, respectively). Low-frequency variability was more pronounced in these two channels and included temporary reversals lasting from several days to a week.

Based on drogue-drift trajectories (Lee et al., 1996, 1998) and several long current meter records (Smith 1994, 1998), Long Key Channel appears to be one of the primary conduits for linking Florida Bay with the Atlantic Ocean side of the Keys. Figure 12.4 is an example of the quasi-steady long-term outflow from Florida Bay that occurs through this channel. The plot shows a cumulative net displacement diagram constructed from currents recorded in mid-channel at mid-depth from August 29, 1994, to July 24, 1995. The descending curve results when current speeds are stronger than flood speeds, or when the ebb part of the tidal cycle lasts longer than the flood part. During this 329-day study period, outflow from the bay averaged 4.8 cm s⁻¹. Periods of net inflow, represented by an ascending curve, occur frequently but are short-lived — on the order of several days. The increased outflow rate during winter months has been noted in other time series from Long Key Channel (Smith, 1994).

Smith's (1998) work has also provided information on flow through channels in the Upper Keys, including Channel Two, Indian Key Channel, Whale Harbor Channel, Snake Creek, and

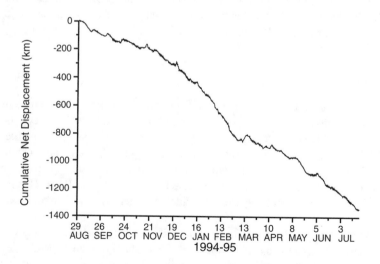

**FIGURE 12.4**  Cumulative net displacement diagram constructed from currents recorded in Long Key Channel (Figure 12.1, Station m), August 29, 1994, to July 24, 1995. Positive displacement indicates northward flow into Florida Bay.

Tavernier Creek (Figure 12.1, Stations q–u, respectively). Results indicate that, while current speeds are comparable to those found in channels of the Middle and Lower Keys, volumes of water exchanged between Gulf and Atlantic sides of the Upper Keys are substantially less due to the smaller cross-sectional areas of the channels. In terms of long-term net transport, Snake Creek and Whale Harbor Channel have shown distinct differences from other tidal channels. One study produced data that suggest a net inflow to Florida Bay from the Atlantic (Smith, 1998). Unpublished data from a second, 4-month study of Whale Harbor Channel, however, showed a net outflow, primarily as a result of strong outflow during a 2-week period within the study. These two channels could be anomalous due to nearby bottom topography. Cross Bank, just west of the channels, may serve to isolate these channels from the general west-to-east movement of water through Florida Bay. Alternately, they may be more responsive to local wind forcing, or transport patterns may have been influenced by anomalous wind conditions during the time of the study.

While Tavernier Creek is a primary link connecting northeast Florida Bay with Atlantic waters through the Upper Keys, it is a relatively small channel and does not appear to be a major pathway linking the bay and Atlantic sides of the Florida Keys. However, it may play a significant role locally, especially for renewing water quality in this isolated portion of Florida Bay. During a 3-month study in midsummer 1994, periods of quasi-steady inflow and outflow resulted in virtually no net transport (Smith, 1998). In a longer study from late October 1995 through early December 1996, however, a net outflow was clearly indicated, though periods of net inflow occurred in spring and fall months for several weeks at a time. Flow through Tavernier Creek may be substantially more sensitive to local wind forcing than are channels of the Middle and Lower Keys.

Combining current meter data and water level records with the cross-sectional area of a channel, volume transport can be estimated for long-term net outflow. Calculations from a series of studies lasting from several months to one year suggest that the long-term net outflow through Long Key Channel averaged 250 to 300 $m^3$ $s^{-1}$ (Smith, 1998; unpublished data). For comparison, the net outflow through Channel Five was 190 $m^3$ $s^{-1}$ during a 38-day study, and outflows through Channel Two and Indian Key Channel were 54 and 30 $m^3$ $s^{-1}$, respectively, during equally short study periods (Smith, 1998). More recent work (Smith, in review) combined 9-month current meter and water level time series with shorter current meter records from additional channel sites used for channel calibration to improve volume transport estimates through Long Key Channel and Seven-Mile Bridge Channel. Results indicate mean total outflows from Florida Bay through Long Key Channel and Seven-Mile Bridge Channel of 392 and 454 $m^3$ $s^{-1}$, respectively. Maximum outflow can exceed 9000 $m^3$ $s^{-1}$ through Long Key Channel and 13,000 $m^3$ $s^{-1}$ through Seven-Mile Bridge Channel from a combination of tidal and nontidal forcing.

## TIDAL EXCHANGES

Just as current meter and water level data can be combined to estimate long-term volume transport, they can be used to estimate volume transport over half tidal cycles. WES investigations quantified volumes of water exchanged in 19 channels between Jewfish Creek and Seven-Mile Bridge (Pratt and Smith, 1998). Data were collected over single tidal cycles, but for many channels this work provides the only indication of the magnitude of tidal exchanges. Results show a general decrease in the magnitude of tidal exchanges from the Middle Keys to the Upper Keys. This is due primarily to the decreasing cross-sectional areas of the tidal channels rather than to differences in the magnitude of tidal forcing.

Smith (1998) summarized results from eight study sites from Bahia Honda Channel to Tavernier Creek. Calculations indicate that volumes of Florida Bay water carried into the Atlantic with the ebb and volumes of Atlantic water entering Florida Bay on the flood far exceed the long-term average transport rates. For Long Key Channel, in particular, $M_2$ flood and ebb volume transports are approximately $52 \times 10^6$ $m^3$. $M_2$ inflow and outflow through Channel Five and Channel Two are approximately 23 and $13 \times 10^6$ $m^3$, respectively. For all other channels of the Middle and

Upper Keys, $M_2$ transport is less than $5 \times 10^6$ m³ over any half tidal cycle. These results make the important point that vigorous tidal exchanges in the Middle Keys provide an alternate mechanism for transporting dissolved and suspended material. Regardless of the magnitude or direction of the advective transport between the Gulf and Atlantic sides of the Keys, tide-induced mixing can be a significant mechanism for transporting materials from regions of higher concentration to regions of lower concentration.

Just as the interaction of the rise and fall of the tide with the ebb and flood can force water into Florida Bay through the open western boundary, the same mechanism can produce a tide-induced residual flow into the bay through the tidal channels between Keys. The amount of water entering the southeastern and southern parts of the bay is significant, and it helps set up water levels within the bay even as the outflow produced by higher bay water levels serves to drain the bay through these same channels. Tide-induced transport calculated for Channel Five, Long Key Channel, and Channel Two are 100, 69, and 57 m³ s⁻¹, respectively (Smith, 1998). The 101 m³ s⁻¹ value for Bahia Honda Channel is also significant, but Bahia Honda Channel probably does not drain the interior of Florida Bay, as do the channels in the Middle Keys. The tide-induced residual transport of water into Florida Bay through Indian Key Channel, Snake Creek, and Tavernier Creek are all less than 6 m³ s⁻¹, and thus play an insignificant role. Whale Harbor Channel tides suggest an anomalous tide-induced outflow, but again the volumes are too small to be of practical importance. Combining the tide-induced transport through all of the major tidal channels in the Upper and Middle Keys, the total inflow is about 20% of the value calculated for the inflow to the bay across western boundary.

## WIND-DRIVEN EXCHANGES

Recent work suggests a close coupling between local wind stress and low-frequency flow through channels between Keys. Smith (in review) quantified the low-frequency volume transport through Long Key Channel and Seven-Mile Bridge Channel in response to wind stress. A southeast-to-northwest wind stress component was found to be most effective for driving water through both channels. In Long Key Channel, the coherence of northwestward wind stress with along-channel flow was 0.80 to 0.87 over all periodicities longer than about three days. The gain spectrum indicated that a ±1 dyne cm⁻² low-frequency variation in the wind stress will force ±1300 to 1800 m³ s⁻¹ fluctuations in transport through Long Key Channel. Similarly, in Seven-Mile Bridge Channel coherence values for northwestward wind stress were greater than 0.8 at periodicities between 2 and 7 days, and ±1 dyne cm⁻² variations in wind stress produced fluctuations of approximately ±1000 m³ s⁻¹.

Considering the sensitivity of the two channels to northwestward wind stress, the prevailing northwestward winds that occur during summer months in this region should be favorable for forcing an inflow to Florida Bay. In fact, quasi-steady outflow occurs throughout the year, so wind forcing cannot be the primary mechanism to explain the observed outflow. However, the data do indicate that outflow is reduced during summer months, probably in response to the seasonal wind forcing. Also, Smith's (in review) study showed that highest coherence occurred when the along-channel axis lies 55° to the right of the wind stress vector, suggesting that flow through the channels may be influenced by an Ekman transport of water in Atlantic shelf waters.

## HAWK CHANNEL CIRCULATION

Water leaving Florida Bay through tidal channels between Keys enters Hawk Channel, a shallow, elongated basin that serves as the continental shelf on the Atlantic Ocean side of the Keys. Hawk Channel is approximately 10 km wide, and the reef tract lies along the seaward side. It extends 230 km from Fowey Rocks, off Biscayne Bay, to Sand Key, off Key West. Maximum water depths along the center of the channel are 7 to 8 m off the Upper Keys and 12 to 15 m off the Middle and Lower Keys. Transport processes that move water across Hawk Channel provide the final link connecting Florida Bay with the reef tract.

Studies of circulation patterns in Hawk Channel include work by Lee (1986) in Key Largo National Marine Sanctuary (Upper Keys), Lapointe et al. (1992) near Looe Key National Marine Sanctuary (Lower Keys), and Pitts (1994, 1997) off Key Largo, Bahia Honda Key, and Key West. Studies indicate that flow along Hawk Channel dominates across-channel motions, and the along-channel flow follows seasonal changes in regional wind patterns. During winter months the along-channel flow is to the southwest or west, toward Key West, when prevailing winds are predominately out of the northeast quadrant. Lee (1986) and Pitts (1994), working off Key Largo, reported a southward flow of 2 to 5 cm s$^{-1}$ during fall and winter months, while westward flow in Hawk Channel seaward of Bahia Honda in the Lower Keys averaged 6 cm s$^{-1}$ during the same season (Pitts, 1994).

Figure 12.5 is an example of Hawk Channel flow in response to fall/winter wind forcing. The progressive vector diagram represents currents recorded 2 m above the bottom in 7 m of water at a study site that was midway across Hawk Channel off Long Key (Figure 12.1, Station v). The study period was from November 20, 1992, to February 18, 1993. The plot illustrates the south-westward flow that is characteristic for this time of year. In this example, the resultant flow is 3.5 cm s$^{-1}$ toward 225° during the 90-day study period. During the same time period, the resultant wind-stress vector calculated from observations made at Molasses Reef was 0.46 dynes cm$^{-2}$ along a heading of 260°. Low-frequency fluctuations in wind forcing coincide with temporary reversals in along-channel flow.

Flow along Hawk Channel becomes more variable during summer months, because the south-easterly winds that are characteristic of this season are more perpendicular to the coastline. A purely southeasterly wind will intersect the coast at right angles near Long Key in the Middle Keys. Lee (1986) and Pitts (1994) observed relatively persistent northward flow in Hawk Channel in the Upper Keys that averaged 1 to 4 cm s$^{-1}$ during summer studies. Flow off the Middle and Lower Keys is weaker in magnitude and more variable in direction (Lapointe et al., 1992; Pitts, 1994, 1997).

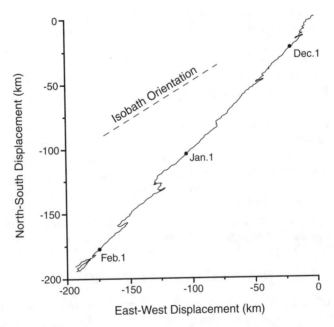

**FIGURE 12.5**  Progressive vector diagram constructed from near-bottom current measurements at Hawk Channel off Long Key (Figure 12.1, Station v), November 20, 1992, to February 18, 1993. Positive values on the *x* and *y* axes represent eastward and northward flow, respectively.

Spectral analysis indicates a close coupling between along-channel wind stress and along-channel flow in Hawk Channel at periodicities of 2 to 4 days (Pitts, 1994, 1997). Phase spectra suggest that currents lag wind stress by 0.5 to 3 days at the periodicities for which coherence was significant. Results indicate that for all headings to the left of approximately 300°, wind stress will act to drive water south and west along the length of Hawk Channel.

Pitts (1994) provided evidence suggesting that variations in flow along the axis of Hawk Channel may create a divergence that requires a compensating across-channel transport. Divergence in two forms was apparent during a 124-day study that involved current meters in the middle of Hawk Channel off Key Largo and Bahia Honda Key, 110 km apart. During late summer and early fall, the flow off the Upper Keys was northeastward, while the current off the Lower Keys was westward. During the second half of the study, resultant flow was southwestward toward Key West at both sites. The resultant speed at the downstream site was over six times the resultant speed at the upstream site, however, and divergence probably occurred as a result of an accelerating along-channel flow. The net outflow from Florida Bay through tidal channels between Keys (Smith, 1994, 1998, in review) may account for some fraction of the divergence within Hawk Channel, but Lee (1986) has noted that divergent along-channel flow off the Upper Keys can be explained by a mean onshore flow past his outer shelf study sites.

The magnitude of across-channel motions in Hawk Channel is generally a small fraction of the along-channel component, but it is the flow across the channel that provides the direct link between the tidal channels between Keys and the reef tract. It is significant that most current meter records indicate seaward near-bottom flow (Lapointe et al., 1992; Pitts, 1994, 1997). The progressive vectors shown in Figure 12.5, for example, show a resultant seaward flow of just over 1 cm s$^{-1}$ during the 90-day study period. In some instances, a seaward flow may be forced locally by a net outflow through a nearby tidal channel (Smith, 1994, 1998). Also, during periods of relatively cool or dry weather, near-bottom pulses of water can creep across Hawk Channel as density currents when hyperpycnal water leaves Florida Bay with the ebb tide (Smith and Pitts, 1998; Porter et al., 1999). A study by Pitts (1997) documented an inverse relationship between across-channel winds and across-channel currents, suggesting an upwind return flow of near-bottom currents. However, an earlier study indicated no correlation between across-channel winds and across-channel currents, although across-shelf currents were coherent with along-channel wind stress at periodicities of 2 to 3 days (Pitts, 1994). It is probable that some fraction of the near-bottom seaward movement of water is an indirect response to a landward Stokes drift forced by onshore winds or a response to a surface Ekman transport when the Keys lie to the right of the wind vector. Measurements are not available to document flow patterns in near-surface layers, but given the generally onshore winds it is likely that a landward drift in some form is present in the upper part of the water column much of the time.

Except in the immediate vicinity of tidal channels between Keys, tidal currents are of distinctly secondary importance in transporting water across Hawk Channel. $M_2$ amplitudes range from 2 to 4 cm s$^{-1}$ for along-channel current components, and across-channel component amplitudes are generally smaller (Pitts, 1994, 1997). The tidal excursion — the horizontal distance that a parcel of water is transported across Hawk Channel — may be calculated from $AT/\pi$, where $A$ is the amplitude in km h$^{-1}$ and $T$ is the period of the tidal constituent in hours. $M_2$ constituent tidal excursions corresponding to amplitudes of 1 to 2 cm s$^{-1}$, for example, are less than 0.3 km and thus a small fraction of the width of Hawk Channel. While tidal oscillations may explain a significant fraction of the variance in a current meter record (Lee, 1986; Pitts, 1994, 1997), this is due to the quasi-steady nature of the nontidal current, which contributes little to the total variance.

Data from several field studies suggest that the relative importance of tides in moving material across Hawk Channel may increase to the south and west. Pitts (1994, 1997) found that tides contribute 5% of the total across-shelf variance off the Upper Keys, 25% of the total off the Middle Keys, and 48 to 70% of the across-shelf variance off the Lower Keys. The prominent across-shelf tidal currents reported from the Lower Keys study probably reflects the influence of ebbs and floods through nearby Northwest Channel, just west of Key West.

A study designed to investigate the relative importance of turbulent mixing suggested that across-shelf transport by this mechanism is less important than across-shelf transport by advection (Smith and Pitts, 1998). Diffusion coefficients were calculated using hourly current measurements and paired with mean across-shelf nutrient concentration gradients. Results suggested that transport by diffusion was of distinctly secondary importance. Diffusion by eddies too small to be resolved by hourly measurements, however, may increase the importance of turbulent transport. Also, under special circumstances involving unusually strong across-shelf gradients of dissolved or suspended matter, across-shelf transport may be locally and temporally significant. Based on available information, however, it appears that the linkage of tidal channels with the reef tract is more by advection than by diffusion.

## EXCHANGES AT THE REEF TRACT

The coral reef tract on the Atlantic side of the Florida Keys extends for approximately 300 km from Key Largo to the Dry Tortugas. It defines the seaward side of Hawk Channel at the shelf break. Ogden et al. (1994) have described the reef tract as a critical component of the Florida Keys seascape and summarized environmental concerns stemming from natural and anthropogenic changes. Results from a series of studies provide insight to the flow of water along the reef tract, and to the exchange of water between Hawk Channel and the Straits of Florida (Lee, 1986; Lee et al., 1992, 1994; Lee and Williams, 1999). In general, the correlation of along-isobath flow at the reef tract and along-channel flow in Hawk Channel is low. Flow in Hawk Channel is more responsive to local wind forcing; flow at the reef tract responds to wind forcing, but it is also significantly influenced directly and indirectly by the Florida Current.

Lee (1986) reported results from three study sites spaced along a 70-km section of the reef tract at the 30-m isobath off the Upper Keys. Along-isobath currents were northward with speeds averaging 15 to 30 cm s$^{-1}$. Low-frequency fluctuations in the along-isobath flow occurred over time scales of approximately two days to two weeks. Significant standard deviations were attributed to a combination of lateral meandering of the axis of the Florida Current and the shedding of eddies that formed along the western fringe of the Florida Current. The passage of an eddy brings with it a temporary reduction of the northward flow at the reef tract, a local and temporary upwelling that can lower water temperature, and an exchange of water between Hawk Channel and the Straits of Florida.

Figure 12.6 is an example of currents that are typical near the reef tract off the Upper Keys. Data were recorded at mid-depth in 19 m of water near Molasses Reef off Key Largo (Figure 12.1, Station w) from August 10 to October 21, 1998. Current vectors were converted to along- and across-isobath components, and these components were passed through a numerical filter to remove high-frequency variability before plotting. The top plot shows that low-frequency currents at the study site were dominated by along-isobath flow to the northeast. The time series indicates northeastward flow as strong as 30 to 40 cm s$^{-1}$ when the Florida Current shifts westward in the Straits of Florida, and current speeds of less than 10 cm s$^{-1}$ when the Florida Current meanders eastward. The bottom plot shows that across-isobath exchanges at the reef tract are comparatively weak, generally ranging between ±2 cm s$^{-1}$, and values are relatively evenly distributed between seaward and shoreward flow.

In the absence of frontal eddies, and depending on the nature of local wind forcing and the proximity of the Florida Current, either of two mechanisms can force exchanges of water between Hawk Channel and the Straits of Florida. When the western fringe of the Florida Current is close to the reef tract, it is likely that periods of upwelling occur as a result of an ageostrophic landward flow in the benthic boundary layer (Hsueh and O'Brien, 1971). Alternately, when wind forcing is downwelling favorable, a mean landward flow at mid-depth and seaward near-bottom flow suggest

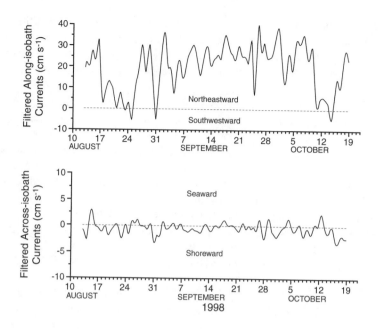

**FIGURE 12.6** Low-pass filtered along-isobath (top) and across-isobath (bottom) current speeds recorded near Molasses Reef (Figure 12.1, Station w). Positive values indicate northwestward and seaward flow in the top and bottom plots, respectively. Note the difference in the *y*-axis scale on the two plots.

that wind stress might be forcing water from the Straits of Florida landward in near-surface layers and producing a compensating return flow of Hawk Channel water in near-bottom layers (Lee, 1986).

As noted above, gyres forming south of the Dry Tortugas move eastward past the Lower Keys (Lee et al., 1992, 1994; Lee and Williams, 1999). Although gyres diminish in size and rarely affect the circulation along the reef tract off the Middle and Upper Keys, their impact on flow off the Lower Keys can be significant. In the absence of gyres, and when the axis of the Florida Current lies closer to the Keys, flow at the reef tract off the Lower Keys can be strong and eastward. In the presence of a gyre, the flow at the reef tract is westward, though it may diminish in speed.

Lee and Williams (1999) combined long current meter records from six locations between Carysfort Reef and the Dry Tortugas to describe mean and seasonally fluctuating flow patterns along the reef tract, as well as regions of divergence associated with wind forcing. Long-term resultant flow off the Upper and Middle Keys is northeastward, while the resultant flow off the Marquesas Keys and Dry Tortugas is westward. This divergence, combined with the much smaller resultant flow found at the reef tract off the Lower Keys, suggests a compensating inflow from the Straits of Florida. As noted above, the net inflow through tidal channels on the landward side of Hawk Channel may replace some fraction of the water removed by divergence at the reef tract. As a result of the curvature of the reef tract, flow both along and across the reef tract differs substantially with seasonal variations in the magnitude and direction of wind forcing. At Looe Key, for example, flow is eastward for much of the year, then westward during late fall and early winter months. Across-isobath flow varies seasonally as well, but patterns differ significantly between near-surface, mid-depth, and near-bottom levels.

Another recent reef tract study (Smith and Pitts, 1998) involved simultaneous near-surface and near-bottom current measurements near Conch Reef off the Upper Keys (Figure 12.1, Station x) and near Tennessee Reef off the Middle Keys (Figure 12.1, Station y); study sites were separated by a distance of 47 km. The analysis focused on local across-isobath flow in response to wind forcing. At both study sites the resultant near-surface and near-bottom across-isobath currents were in the same direction. Directional shear was only about 8°, with resultant near-bottom currents directed to the left of the resultant near-surface current. Across-isobath flow at the two study sites was poorly correlated. The linear correlation coefficients calculated from near-surface and near-bottom currents at the Conch Reef and Tennessee Reef study sites were +0.075 and +0.356, respectively. Along-isobath flow was highly correlated through the water column at both study sites, however.

A careful comparison of winds and across-isobath current components suggested that a purely Ekman transport does not occur, probably because of the nearby coastal boundary and because steady-state conditions are rarely met. For example, seaward across-shelf flow at the upper and lower levels at Tennessee Reef, where the local isobath orientation is 065–245°, were most highly correlated with wind stress headings of 013° and 024°, respectively. Thus, seaward flow is 142° and 131° to the right of the wind. It is noteworthy that northeastward wind stress is a rare occurrence in the Keys. More significantly, landward transport at the reef tract would be expected during winter months, when wind forcing is into the southwestern quadrant. Similarly, at Conch Reef, where local isobaths are oriented 052–232°, landward flow at near-surface and near-bottom levels at the reef tract is most highly correlated with wind stress headings of 228° and 195°. Thus, landward transport is 94° and 127° to the right of wind forcing at near-surface and near-bottom levels, respectively.

Tidal exchanges across the reef tract provide a small but dependable linkage between the Straits of Florida and Hawk Channel. Smith and Pitts (1998) reported results of harmonic analyses of across-isobath components of currents recorded at study sites near Conch Reef and near Tennessee Reef. At both locations, only the $M_2$ constituent is physically significant with amplitudes at Conch Reef and Tennessee Reef of 2.6 and 3.0 cm s$^{-1}$, respectively. The corresponding across-isobath tidal excursion is approximately 0.4 km. Although the amplitudes of the other constituents were within the precision of the current meter, the tidal excursion may increase and decrease somewhat during times of spring and neap tides.

Leichter et al. (1996) have suggested that tidal bores generated by breaking internal waves provide another mechanism for importing water from the Straits of Florida to the reef tract. This mechanism appears to be restricted to summer months, from May to November with a maximum from July to September, but semidiurnal tidal bores can decrease near-bottom water temperature as much as 5.4°C, increase salinity slightly and produce landward, upslope flow of as much as 10 to 30 cm s$^{-1}$.

## SUMMARY AND CONCLUSION

A synthesis of studies conducted on both sides of the Florida Keys and in the connecting channels over the past 15 years reveals a dynamic system, with temporal variability over a wide range of time scales arising primarily in response to tidal and wind forcing. Averaging over tidal periods and the longer time scales associated with meteorological forcing, however, reveals transport pathways that represent a clear coupling between Gulf and Atlantic sides of the Keys. Gulf-to-Atlantic transport can be either around the Keys, via the Loop Current and Florida Current, or it can involve a more complex route through Florida Bay and the tidal channels (Figure 12.7). Understandably, the linkage of Gulf and Atlantic waters through Florida Bay has taken on greater significance, as a result of questions related to water quality in Florida Bay. As a direct consequence of these transport pathways, the Florida Keys are a physically interconnected ecosystem, and any region along the path can have a direct impact on other regions located downstream.

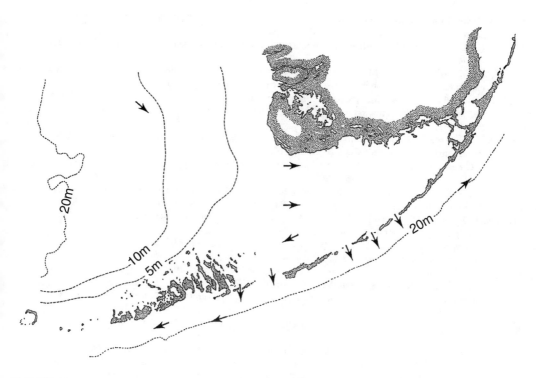

**FIGURE 12.7** Map showing general circulation through Florida Bay and the Florida Keys. Arrows were placed at sites where data indicate a consistent flow direction.

# REFERENCES

Chew, F., E. Balazs, and C. Thurlow. 1982. The slope of the mean sea level along the Straits of Florida and its dynamical implications, *Oceanologica Acta*, 5:21–30.

Environmental Science and Engineering, Inc. 1987. *Southwest Florida Shelf Ecosystems Study*, Vol. II, Data Synthesis Report. Minerals Management Service Report 87-0023, U.S. Dept. of Interior, 348 pp.

Fourqurean, J.W., R.D. Jones, and J.C. Zieman. 1993. Processes influencing water column nutrient characteristics and phosphorus limitation of phytoplankton biomass in Florida Bay, USA: Inferences from spatial distributions, *Estuarine Coastal Shelf Sci.*, 36:295–314.

Hetland, R.D., Y. Hsueh, R. Leben, and P.P. Niiler. 1999. A Loop Current-induced jet along the edge of the west Florida shelf, *Geophys. Res. Lett.*, 26:2239–2242.

Hsueh, Y. and J. O'Brien. 1971. Steady coastal upwelling induced by an along-shore current, *J. Phys. Oceanogr.*, 1:180–186.

Ichiye, T., H.-H. Kuo, and M.R. Carnes. 1973. Wind-driven currents in the eastern Gulf of Mexico, in *Assessment of Currents and Hydrography of the Eastern Gulf of Mexico*, Contract No. 601, Dept. of Oceanography, Texas A & M University, College Station, TX, 2.1–2.29.

Lapointe, B.E., N.P. Smith, P.A. Pitts, and M. Clark. 1992. *Baseline Characterizations of Chemical and Hydrographic Processes in the Water Column of Looe Key National Marine Sanctuary*, final report to U.S. Dept. of Commerce, NOAA, Office of Ocean and Coastal Resource Mgmt., Contract No. NA86AA-H-CZ071, Washington, D.C., 59 pp.

Lee, T.N. 1975. Florida Current spin-off eddies, *Deep-Sea Res.*, 22:753–765.

Lee, T.N. 1986. Coastal circulation in the Key Largo Coral Reef Marine Sanctuary, pp. 178–198, in *Physics of Shallow Estuaries and Bays*, van de Kreeke, J., Ed., Springer, New York.

Lee, T.N. and D.A. Mayer. 1977. Low-frequency current variability and spin-off eddies along the shelf off Southeast Florida, *J. Mar. Res.*, 35:193–220.

Lee, T.N. and E. Williams. 1999. Mean distribution and seasonal variability of coastal currents and temperature in the Florida Keys with implications for larval recruitment, *Bull. Mar. Sci.*, 64:35–56.

Lee, T.N., C. Rooth, E. Williams, M. McGowan, A.F. Szmant, and M.E. Clark, 1992. Influence of Florida Current, gyres and wind-driven circulation on transport of larvae and recruitment in the Florida Keys coral reefs, *Cont. Shelf Res.*, 12:971–1002.

Lee, T.N., M.E. Clarke, E. Williams, A.M. Szmant, and T. Berger. 1994. Evolution of the Tortugas gyre and its influence on recruitment in the Florida Keys, *Bull. Mar. Sci.*, 54:621–646.

Lee, T.N., K. Leaman, and E. Williams. 1995. Florida Current meanders and gyre formation in the southern Straits of Florida, *J. Geophys. Res.*, 100:8607–8620.

Lee, T.N., E. Williams, E. Johns, and N.P. Smith. 1996. Flow within Florida Bay and interaction with surrounding waters, p. 48, in *Program and Abstracts*, 1996 Florida Bay Science Conference, Florida Sea Grant, Key Largo, FL.

Lee, T.N., E. Johns, D. Wilson, and E. Williams. 1998. Florida Bay circulation and exchange study, in *Proc. 1998 Florida Bay Science Conf.*, Florida Sea Grant, Miami, FL.

Leichter, J., S. Wing, S. Miller, and M. Denny. 1996. Pulsed delivery of subthermocline water to Conch Reef (Florida Keys) by internal tidal bores, *Limnol. Oceanogr.*, 4:1490–1501.

Ogden, J.C., J.W. Porter, N.P. Smith, A.M. Szmant, W.C. Jaap, and D. Forcucci. 1994. A long-term interdisciplinary study of the Florida Keys seascape, *Bull. Mar. Sci.*, 54:1059–1071.

Pitts, P.A. 1994. An investigation of near-bottom flow patterns along and across Hawk Channel, Florida Keys, *Bull. Mar. Sci.*, 54:610–620.

Pitts, P.A. 1997. An investigation of tidal and nontidal current patterns in Western Hawk Channel, Florida Keys, *Cont. Shelf Res.*, 17:1679–1687.

Pitts, P.A. 1998. Tidal and long-term volume transport through Jewfish Creek, Florida Keys, *Bull. Mar. Sci.*, 63:559–570.

Pitts, P.A. and N.P. Smith. 1995. *Long-Term Net Transport Through Three Tidal Channels in the interior of Florida Bay*, final report to the National Park Service for Cooperative Agreement CA 5280-4-9022, 51 pp.

Porter, J.W., S.K. Lewis, and K.G. Porter. 1999. The effect of multiple stressors on the Florida Keys coral reef ecosystem: a landscape hypothesis and physiological test, *Limnol. Oceanogr.*, 44:941–949.

Pratt, T.C. and N.P. Smith. 1998. *Florida Bay Field Data Report*, draft final report, U.S. Waterways Experiment Station, Coastal and Hydraulics Laboratory, Vicksburg, MS.

Smith, N.P. 1994. Long-term Gulf-to-Atlantic transport through tidal channels in the Florida Keys, *Bull. Mar. Sci.*, 54:602–609.

Smith, N.P. 1998. Tidal and long-term exchanges through tidal channels in the Middle and Upper Florida Keys, *Bull. Mar. Sci.*, 62:199–211.

Smith, N.P. 2000. Observations of shallow-water transport and shear in western Florida Bay, *J. Phys. Oceanogr.*, 30:1802–1808.

Smith, N.P. Tidal, low-frequency and long-term mean transport through two tidal channels in the Florida Keys, *Cont. Shelf Res.* (in review).

Smith, N.P. and P.A. Pitts. 1998. *Hawk Channel Transport Study: Pathways and Processes*, final report to the U.S. Environmental Protection Agency and South Florida Water Management District under Contract No. C-6627-A1, 95 pp.

Vukovich, F.M., B.W. Crissman, M. Bushness, and W.J. King. 1979. Some aspects of the oceanography of the Gulf of Mexico using satellite and *in situ* data, *J. Geophys. Res.*, 84:7749–7757.

Wang, J.D. 1998. Subtidal flow patterns in western Florida Bay, *Estuarine Coastal Shelf Sci.*, 46:901–915.

Wang, J.D., J. van de Kreeke, N. Krishnan, and D. Smith. 1994. Wind and tide response in Florida Bay, *Bull. Mar. Sci.*, 54:579–601.

Wennekens, M.P. 1959. Water mass properties of the Straits of Florida and related waters, *Bull. Mar. Sci.*, 9:1–52.

# 13 The Transport of Terrestrial Nutrients to South Florida Coastal Waters

*Larry E. Brand*

Rosenstiel School of Marine and Atmospheric Science, University of Miami

## CONTENTS

## INTRODUCTION

With the large increase in the human population and its agricultural activities in the twentieth century, there has also been a large increase in the flow of nutrients from land into coastal waters (Nixon, 1995; Richardson and Jorgensen, 1996; Moffat, 1998). The increased mobilization of nutrients has led to the eutrophication, first, of small water bodies such as ponds, lakes, and rivers (Vollenweider, 1992a), then estuaries such as Chesapeake Bay (Cooper and Brush, 1991; Harding and Perry, 1997) and more recently large seas such as the Black Sea (Bodeanu, 1992; Mee, 1992; Cociasu et al., 1996), Baltic Sea (Larsson et al., 1985; Nehring, 1992), and Adriatic Sea

(Vollenweider, 1992b; Justic et al., 1995), as well as continental shelf areas such as the Mississippi River delta (Turner and Rabalais, 1991, 1994; Justic et al., 1995).

Before the twentieth century, relatively few humans lived in South Florida, but since then the human population and accompanying agricultural activities have increased greatly, especially after World War II. Not only are they sources of nutrients, but they also have greatly altered the hydrology of South Florida (DeGrove, 1984; Light and Dineen, 1994). The environmental effects of reduced freshwater flow into the southern Everglades and Florida Bay have been recognized and this is now being reversed to some extent (McIver et al., 1994).

Florida Bay and the Florida Keys are at the downstream end of the Kissimmee River-Lake Okeechobee-Everglades watershed (Figure 13.1). Their ecological health depends on what happens upstream. Within the past 20 years, a number of ecological changes have occurred in South Florida coastal waters. In Florida Bay, large algal blooms have developed and persisted, large areas of seagrasses and sponges have died off, and major changes have occurred in fish populations (Zieman et al., 1989; Robblee et al., 1991; Boesch et al., 1993; Durako, 1994; Thayer et al., 1994; Butler et al., 1995; McPherson and Halley,1997; Fourqurean and Robblee, 1999). In the Florida Keys, macroalgae have overgrown many coral reefs, coral diseases appear to be spreading, and many corals have died (Dustan and Halas, 1987; Porter and Meier, 1992; Ogden et al.,1994; Kuta and Richardson, 1996; Richardson, et al., 1996; Richardson, 1997; Porter et al., 2002). Many of these changes are classical indicators of nutrient enrichment (eutrophication). This chapter will examine whether there is a causal link between the increase in the human population and associated activities in South Florida, and the ecological changes observed in the downstream coastal waters.

## METHODS

Temperature and salinity measurements were made with a YSI meter. For chlorophyll measurements, three 100-ml replicate water samples were filtered (after adding 1 mg of $MgCO_3$) through GF/F glass fiber filters, and the filters were frozen until extracted (within a few days). These filters were then extracted for 30 minutes with 10 ml of dimethyl sulfoxide and then with an added 15 ml of 90% acetone at 5°C overnight and measured fluorometrically before and after acidification for the measurement of chlorophyll and phaeopigment concentrations (Burnison, 1980; Parsons et al., 1984). Fluorescence measurements were made with a Turner Designs 10-000R or 10-AU fluorometer equipped with an infrared-sensitive photomultiplier and calibrated using pure chlorophyll $a$.

Because of concerns about the intercomparison and accuracy of different methods for measuring suspended particulate chlorophyll as an indicator of phytoplankton biomass, an internal intercalibration study was conducted. Water was collected at five stations in Florida Bay. The water was filtered and extracted overnight in 100% acetone. This does not provide as good an extraction as DMSO/90% acetone, but we wanted to compare our high-performance liquid chromatography (HPLC) method with the other methods which precluded the use of DMSO. The following instruments were compared:

1. Three Turner 10-000R fluorometers equipped with infrared-sensitive photomultipliers, D-daylight bulbs, 3-66 reference filters, 5-60 excitation filters, and 2-64 emission filters.
2. Two Turner 10-AU fluorometers equipped with infrared-sensitive photomultipliers, D-daylight bulbs, 10-032 reference filters, 10-AU-045 excitation filters, and 10-AU-050 emission filters.
3. Turner 10-AU fluorometer equipped with an infrared-sensitive photomultiplier, 10-089 bulb, 10-032 reference filter, 10-113 excitation filter, and 10-115 emission filter (Welschmeyer technique).
4. Hitachi F-4500 fluorescence spectrophotometer with 2.5-nm bandwidths set at 432-nm excitation and 667-nm emission.

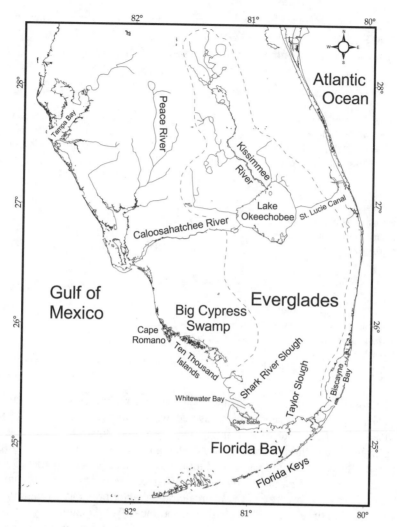

**FIGURE 13.1** Map of South Florida. Dashed line shows boundaries of the Kissimmee–Lake Okeechobee–Everglades watershed.

5. Shimadzu UV 2100U UV-VIS recording spectrophotometer.
6. Gilson HPLC system with two model 302 pumps, 811B mixer, 802B manometric module, 621 data module, and an Rheodine injector (100-µl sampling loop) with a Shimadzu SPD-6AV UV-VIS detector set at 452 nm and with an 8-µl flow-through cell and a Shimadzu CTO-6A column oven modified for cooling as well as heating. The Viadac 201TP C18 reverse-phase column was run at 34°C using a binary solvent system (80:20 methanol:0.5 $M$ ammonium acetate and 80:20 methanol:acetone).

Because of time limitations, HPLC measurements were made only on samples 4 and 5. Because of extract volume limitation, spectrophotometer measurements were also made only on samples 4 and 5. The results are shown in Table 13.1. Although there is more variability than we would like, we see no evidence of any systematic difference between the methods. The high spectrophotometric measurement on sample 5 is not surprising, as we know that method is inaccurate at low concentrations of chlorophyll.

**TABLE 13.1**
**Chlorophyll Intercomparison Conducted on February 6, 1997**

| Instrument | Sample No.[a] | | | | |
|---|---|---|---|---|---|
| | 1[b] | 2[c] | 3[d] | 4[e] | 5[f] |
| Turner 10-000R fluorometer A | 0.66 | 3.98 | 4.06 | 6.59 | 0.57 |
| Turner 10-000R fluorometer B | 0.78 | 3.83 | 3.99 | 7.07 | 0.61 |
| Turner 10-000R fluorometer D | 0.70 | 3.81 | 3.87 | 6.65 | 0.56 |
| Turner 10-AU fluorometer C | 0.73 | 3.64 | OS | 6.61 | 0.59 |
| Turner 10-AU fluorometer E | 0.74 | 3.86 | 3.88 | 6.53 | 0.56 |
| Turner 10-AU fluorometer F (Welschmeyer) | 0.65 | 3.47 | 3.60 | 5.83 | 0.57 |
| Hitachi spectrofluorometer | 0.74 | 3.18 | 3.42 | 6.31 | 0.52 |
| Shimadzu spectrophotometer | ND | ND | ND | 7.06 | 1.17 |
| Gilson HPLC | ND | ND | ND | 5.52 | 0.44 |

*Note:* OS = off scale; ND = no data.

[a] Results given in µg/l chlorophyll *a*.
[b] Rabbit Key Pass, 24°58.474′ N 80°49.594′ W.
[c] Rabbit Key Basin, 24°59.459′ N 80°52.522′ W.
[d] Oxfoot Bank, 25°00.944′ N 81°00.284′ W.
[e] South of Cape Sable, 25°05.678′ N 80°59.694′ W.
[f] Between Rankin Key and Tin Can Channel, 25°07.046′ N 80°48.442′ W.

In an Environmental Protection Agency-sponsored intercalibration study with seven other laboratories, our chlorophyll measurements agreed well with six of the laboratories. The one laboratory that reported chlorophyll concentrations 2 to 5 times lower than the other laboratories participating in the intercalibration was Florida International University (FIU); however, we use FIU chlorophyll data as recorded in the South Florida Water Management District (SFWMD) water quality database in our examination of long-term trends because it is the longest running database available. The differences observed in the intercalibration study probably explain the quantitative differences observed between our chlorophyll data and the SFWMD chlorophyll data.

For the measurement of phycocyanin as an indicator of estuarine cyanobacteria, filters were extracted with phosphate buffer ($10^{-1}$ $M$ potassium phosphate and $10^{-3}$ $M$ 2-mercaptoethanol), and the fluorescence was measured with a SPEX Fluorolog-3 spectrofluorometer calibrated with pure phycocyanin. Excitation was 580 nm with a 12-nm bandpass, and the emission peak at 639 nm was scanned and the area integrated for quantification.

Nutrient bioassays were conducted by adding no nutrients (as a control), 250 µ$M$ NO$_3$, 25 µ$M$ PO$_4$, or 250 µ$M$ NO$_3$ and 25 µ$M$ PO$_4$ (as a control) to 30-ml water samples and monitoring algal biomass daily fluorometrically for 1 to 2 months. The highest *in vivo* chlorophyll fluorescence measurements achieved in each enrichment were compared to determine if the initial water samples were originally N or P limited. An index of nutrient limitation was calculated using the equation:

$$\frac{FLN - FLP}{FLN + FLP}$$

where FLN is the highest chlorophyll fluorescence observed in the incubation with 250 µ$M$ NO$_3$ added, and FLP is the highest chlorophyll fluorescence observed in the incubation with 25 µ$M$ PO$_4$ added.

In mixing experiments, no nutrients were added. The highest chlorophyll fluorescences of two different incubated water samples were compared to a sample derived from simply mixing the two different water samples together.

The water flow, water level, and nutrient data were obtained from the hydrometeorological and water quality databases of the South Florida Water Management District.

## RESULTS AND DISCUSSION

### FLORIDA BAY

### The Problem

Florida Bay is at the downstream end of the Kissimmee River–Lake Okeechobee–Everglades watershed and can therefore be affected by any changes that occur in that hydrological system. Around the beginning of the twentieth century, construction of canals began to drain South Florida land for agriculture and human habitation (Schomer and Drew, 1982; DeGrove, 1984; SFWMD, 1992). Drainage of the Everglades was greatly accelerated after 1947 with the development of the South and Central Florida Flood Control Project, which greatly lowered the water table and reduced the flow of freshwater to Florida Bay, increasing its overall salinity and shifting Florida Bay from a brackish to often hypersaline ecosystem (Smith et al., 1989; McIver et al., 1994).

For over a decade now, large algal blooms have occurred in northcentral and northwest Florida Bay (Boesch et al., 1993; Fourqurean et al., 1993; Phlips et al., 1995, 1999; Phlips and Badylak, 1996; Fourqurean and Robblee, 1999). Average chlorophyll concentrations (Figure 13.2) show an intense bloom in northcentral Florida Bay and high concentrations in the northwest. Cyanobacteria dominate the northcentral bloom, and diatoms (primarily *Rhizosolenia*) dominate the bloom in the west. Recent data on phycocyanin concentrations (Figure 13.3) show that cyanobacteria are abundant not only in northcentral Florida Bay but also at the mouth of Shark River and along parts of the southwest coastline. A number of hypotheses have been proposed for the source of the nutrients that generate these microalgal blooms. Most prominent is the hypothesis that the large seagrass dieoff that occurred in 1987 generated large amounts of nutrients from both biomass degradation and the destabilization and resuspension of the sediments (Robblee et al., 1991; Durako, 1994; Carlson et al., 1994; Zieman et al., 1994). Undoubtedly this did occur, but the hypothesis does not account for either the spatial or temporal distribution of the microalgal blooms.

### Seagrass Dieoff as a Source of Nutrients

Although there are no quantitative data before 1989, many fishers and other boaters who are frequently in Florida Bay have been quoted by DeMaria (1996) as observing algal blooms and water "discoloration" beginning in 1981 and increasing thereafter, well before the 1987 seagrass dieoff:

> In 1981, the water got dirtier, the blooms grew, and the seagrass started dying. …From 1981 to 1986, the decline [of the bay] was gradual. In 1987, the bay started to collapse quickly. …The commercial fishermen saw the water [in western Florida Bay] changing — first by becoming dirty, then algae blooms started, and then seagrass died off.

Unfortunately, we have no quantitative data on the early development of the algal bloom, so we have to rely on the visual observations of persons who were on the bay for long periods of time. Nevertheless, these observations tend to support the hypothesis of Lapointe and Clark (1992), Tomasko and Lapointe (1994), Lapointe et al. (2002), and Lapointe and Barile (2001) that eutrophication may have led to the seagrass dieoff, not the reverse. Eutrophication has been documented

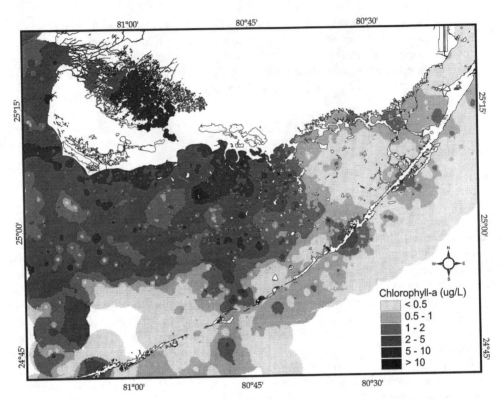

**FIGURE 13.2**   Chlorophyll concentrations measured between February 1996 and November 2000. The data (3454 measurements with three replicates each) are plotted using fixed-radius surface interpolation.

to lead to seagrass declines in many other parts of the world (Kemp et al., 1983; Orth and Moore, 1983; Cambridge and McComb, 1986; Silberstein et al., 1986; Giesen et al., 1990).

Seagrass dieoff and the resulting biomass decomposition and sediment resuspension also cannot explain the spatial distribution of the algal blooms or nutrients in Florida Bay. It is well established that the general flow of water is from the northwest (around Cape Sable) to the southeast through the passes between the Florida Keys (Smith, 1994; Wang et al, 1994; Lee et al., 2002). A comparison of the distribution of average chlorophyll concentrations from 1996 to 2000 (Figure 13.2) with the distribution of seagrass dieoff in 1987 (Figure 13.4) indicates that much of the algal bloom area is upstream, not downstream of the seagrass dieoff areas. Furthermore, a comparison of the inorganic nitrogen (N) and total phosphorus (P) spatial distributions (Figures 13.5 and 13.6, respectively) with seagrass dieoff areas reveals that the highest concentrations of nutrients are neither downstream nor at the same location as the seagrass dieoff. The data of Fourqurean et al. (1993); Frankovich and Fourqurean (1997), and Boyer et al. (1997) show the same pattern. While there is some overlap of high N and high P areas in northcentral Florida Bay where some of the seagrass died, seagrass decomposition could not generate a nutrient pattern with P upstream to the northwest and N to the northeast of the seagrass dieoff area, with currents flowing to the southeast. While seagrass dieoff most likely did result in the release of nutrients, it cannot explain the spatial pattern of nutrients or algal blooms in Florida Bay today.

Furthermore, it has been over 13 years since the seagrass dieoff, yet the bloom remains and has even gotten larger. While some of the nutrients would have been stored in the sediments for a while and released later, it seems rather unlikely that the nutrients and the resultant algal blooms would not have been flushed out by now. This is particularly true in the deep channel south of Cape Sable where tidal currents flow in and out daily and there was no significant seagrass dieoff; yet, P and the microalgal biomass are persistently high. Furthermore, much of the water in Florida

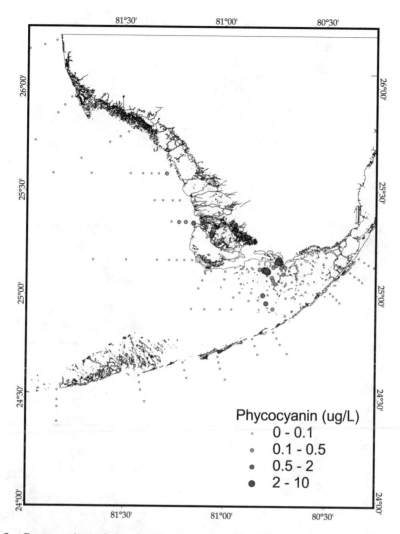

**FIGURE 13.3**  Concentrations of phycocyanin measured in October and November 2000.

Bay was flushed out during Hurricane Georges (September 25, 1998), yet the algal blooms afterward remained in the same areas (Figure 13.7) as at other times (Figure 13.2). The appearance of algal blooms in the same place as before indicates a persistent source of nutrients, not leftover nutrients from the seagrass dieoff 11 years earlier. It seems likely that other sources are quantitatively more important sources of nutrients than seagrass dieoff. Seagrass dieoff may be primarily a symptom, not a primary cause, of eutrophication in Florida Bay.

## Sewage as a Source of Nutrients

Another potential source of nutrients is sewage and other nutrient-rich freshwater runoff from land in nearby areas of high human activity. In the case of Florida Bay, this would be primarily the heavily populated Florida Keys. Considerable amounts of N and P are estimated to be generated by human activities in the Florida Keys (EPA, 1992), and some fraction of this is clearly entering the local waters (Lapointe et al., 1990, 1994; Shinn et al., 1994; Paul et al., 1995a,b). The problem with this hypothesis is that the source of the nutrients is downstream of the algal blooms, not upstream. Sewage may be causing local eutrophication and leading to algal overgrowth of the coral reefs downstream, but it cannot be the source of the nutrients generating the algal blooms

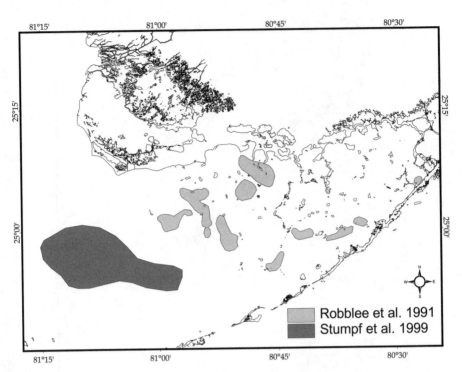

**FIGURE 13.4** Location of the major seagrass dieoff in 1987. (Redrawn from Robblee et al., 1991, and Stumpf et al., 1999).

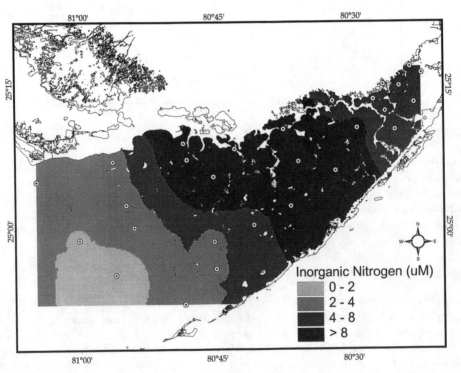

**FIGURE 13.5** Average concentrations of dissolved inorganic N measured in monthly sampling of Florida Bay from 1991 to 1998. Data are plotted using nearest neighbor surface interpolation. (Calculated from SFWMD data files.)

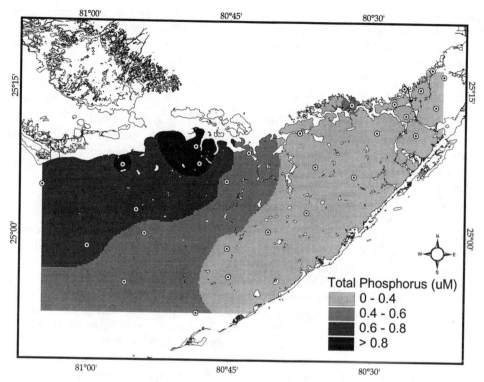

**FIGURE 13.6** Average concentrations of total P measured in monthly sampling of Florida Bay from 1991 to 1998. Data are plotted using nearest neighbor surface interpolation. (Calculated from SFWMD data files.)

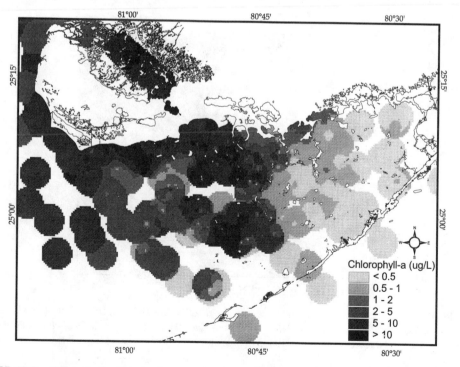

**FIGURE 13.7** Chlorophyll concentrations after Hurricane Georges measured between October 1998 and December 1998. The data (373 measurements) are plotted using fixed-radius surface interpolation.

upstream to the northwest. Concentrations of N (Figure 13.5), P (Figure 13.6), and chlorophyll (Figure 13.2) all decrease, not increase, along transects from northern Florida Bay to the Florida Keys. The algal blooms in Florida Bay are, in fact, the farthest away from the Florida Keys and the human population.

## The Gulf of Mexico as a Source of Nutrients

It has been hypothesized that the Gulf of Mexico waters on the southwest shelf of Florida may be the source of nutrients generating the blooms in Florida Bay (Fourqurean et al., 1993; Rudnick et al., 1999). Undoubtedly this is true to some extent, but it is not clear how much Gulf of Mexico water actually makes it into Florida Bay. Most water moving south along the west coast of Florida goes through the lower Florida Keys (Figure 13.8; Lee et al., 2002). Furthermore, it is not clear how relatively low P concentrations in the Gulf of Mexico could generate the high P concentrations in the area south of Cape Sable (Figure 13.9).

Seagrass beds and other shallow water ecosystem components could be trapping the nutrients as the Gulf of Mexico water flows over them, ultimately generating higher P concentrations. This does not, however, explain the absence of high P concentrations in shallow waters in the southwest part of Florida Bay or in the middle and lower Florida Keys, which also receive Gulf of Mexico waters. Indeed it appears that these areas receive a much larger volume of Gulf of Mexico water than the restricted embayments of Florida Bay, based upon the drifter data of Lee et al. (2002). Furthermore, if nutrient trapping was the primary cause of the elevated P concentrations south of Cape Sable, one would also expect elevated N concentrations, which are not observed (Figure 13.10). A Gulf of Mexico source of nutrients also cannot explain the distribution of chlorophyll in Whitewater Bay, with highest chlorophyll concentrations farthest from the entrance to the Gulf of Mexico (Figure 13.11). The Gulf of Mexico is probably a source of some nutrients, particularly P, to Florida Bay, but it is probably not the dominant source because it cannot explain the spatial patterns of N and P.

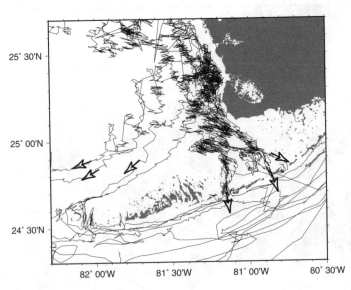

**FIGURE 13.8**    Trajectories of drifters released near the mouth of Shark River. (From Lee, T. et al., 2002, in *The Everglades, Florida Bay, and Coral Reefs of the Florida Keys: An Ecosystem Sourcebook*, Porter, J.W. and Porter, K.G., Eds., CRC Press, Boca Raton, FL. With permission.)

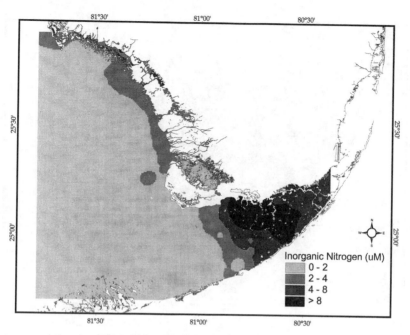

**FIGURE 13.9** Average inorganic N concentrations in South Florida measured in monthly sampling from 1997 to 1998. Data are plotted using nearest neighbor surface interpolation. (Calculated from SFWMD data files.)

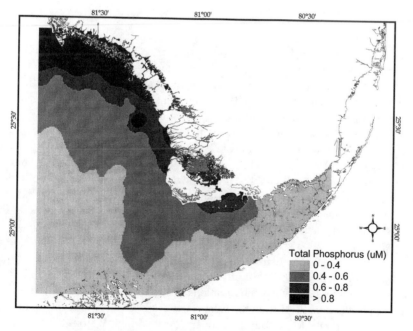

**FIGURE 13.10** Average total P concentrations in South Florida measured in monthly sampling from 1997 to 1998. Data are plotted using nearest neighbor surface interpolation. (Calculated from SFWMD data files.)

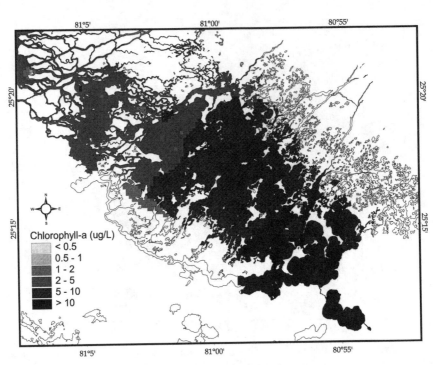

**FIGURE 13.11**   Chlorophyll concentrations measured in Whitewater Bay between April 1996 and November 1999. The data are plotted using fixed-radius surface interpolation.

## N:P Ratios

The fact that N and P have such different spatial patterns (Figures 13.5 and 13.6) suggests that each has a different source. The spatial differences generate a very large gradient in N:P ratios. Ratios of total N to total P and inorganic N to inorganic P are well above the Redfield ratio of 16 throughout Florida Bay (Fourqurean et al., 1993; Boyer et al., 1997). This has led many researchers to assume that P is the primary limiting nutrient and that inputs of N to Florida Bay are not a cause of the algal blooms. It is well known however that many organic N molecules are not readily available to phytoplankton while many organic P molecules are, due to the activity of phosphatase enzymes (Vitousek and Howarth, 1991). This reflects the fact that organic N is bound by direct carbon bonds while organic P is bound by ester bonds. Examination of the ratio of inorganic N to total P (Figure 13.12) indicates ratios greater than the Redfield ratio in eastern Florida Bay and ratios less than the Redfield ratio in western Florida Bay. This suggests the potential for P limitation in the east and N limitation in the west. The results of around 1000 nutrient bioassays conducted from March 1998 to July 2000 show mostly N limitation in the west and P limitation in the east (Figure 13.13), with a spatial distribution similar to the ratios of inorganic N to total P (Figure 13.12). Tomas et al. (1999) have also observed N limitation in western Florida Bay and P limitation in eastern Florida Bay. The observation of N limitation in western Florida Bay suggests that indeed much of the organic N from the west is not available to the phytoplankton. The largest algal blooms are in central Florida Bay (Figure 13.2) where high P from the west (Figure 13.6) meets high N from the east (Figure 13.5), and the inorganic N to total P ratio is close to the Redfield ratio (Figure 13.12). The N:P ratio changes by a factor of 10 (4 to 40) over a distance of 25 km through the algal bloom, so the location of the algal bloom would not change substantially if some of the organic N was available and/or some of the organic P was not available.

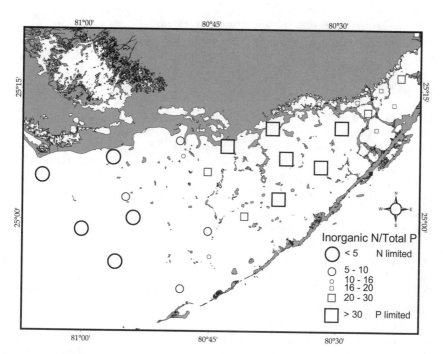

**FIGURE 13.12** Average molar ratios of dissolved inorganic N to total P measured in Florida Bay from 1991 to 1998. (Calculated from SFWMD data files.)

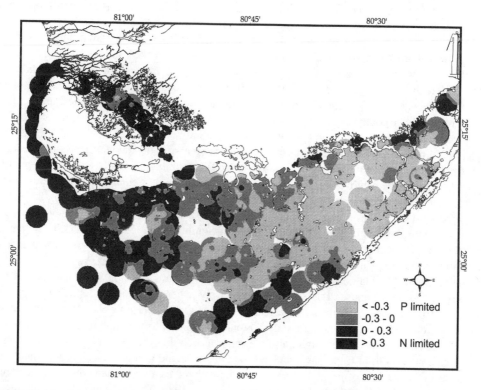

**FIGURE 13.13** Results of 1119 nutrient bioassay experiments conducted between March 1998 and July 2000. Data are plotted using fixed-radius surface interpolation.

The experimental mixing of N-rich water from the east with P-rich water from the west indeed does lead to a large increase in algal biomass (Figure 13.14). To understand the source of the algal bloom, we need to know the source of the P in the west and the source of the N in the east.

## The Source of P

South Florida coastal waters are generally P limited (Lapointe, 1987, 1989; Powell et al., 1989; Brand et al., 1991; Fourqurean et al., 1992a,b, 1993; Frankovich and Fourqurean, 1997), just as in other shallow carbonate environments such as Jamaica (D'Elia et al., 1981), Bermuda (Lapointe and O'Connell, 1989; Jensen et al., 1998), Belize (Lapointe et al., 1987), and the Bahamas (Littler et al., 1988; Short, 1987; Short et al., 1990; Lapointe et al., 1992). This is because calcium carbonate chemically scavenges phosphate from seawater (DeKanel and Morse, 1978; Kitano et al., 1978). This occurs in both groundwater flowing through limestone as well as shallow surface water in which calcareous sediments are resuspended.

It is hypothesized that phosphorite deposits created in the Miocene in northeast Florida and subsequently transported south into South Florida by Appalachian erosion may be the ultimate source of the high P observed in areas of South Florida (Brand, 1996; Brand et al., 2001). Apparently, 16 to 23 million years ago, geomorphological configurations, along with meteorological and oceanographic conditions, led to massive upwelling in the area of northeast Florida, which over time led to large deposits of phosphorite (Riggs, 1979, 1984; Mallinson et al., 1994; Compton, 1997). Subsequently, the Appalachian Mountains were uplifted, greatly increasing erosion to the south and flooding the Florida Peninsula with siliciclastic materials (Ginsburg and Shinn, 1994; Lane, 1994; Scott, 1997) mixed with phosphorite granules derived from the old upwelling site (Riggs, 1979; Warzeski et al., 1996; Cunningham et al., 1998).

Compared to the east coast of South Florida and eastern Florida Bay, P concentrations tend to be high all along the southwest coast, especially in Tampa Bay and Charlotte Harbor (Odum, 1953). P concentrations are also particularly high in the Ten Thousand Islands area and in northwest Florida Bay south of Cape Sable (Figure 13.10). The Peace River drains the area where the phosphorite deposits are close to the surface in central Florida and are currently being mined for fertilizer (Hobbie, 1974; Fraser and Wilcox, 1981; Compton, 1997). This explains the high P concentrations and low N:P ratios in the Peace River and downstream in Charlotte Harbor, and the coastal waters to the south (Figure 13.15) as a result of the coastal current flowing toward the south (Froelich et al., 1985; Doering and Chamberlain, 1998; Squires et al., 1998). This also explains the strong N limitation observed in nutrient bioassay experiments conducted along the southwest coast from the Peace River to Florida Bay (Figure 13.15). While this would help explain the relatively low N:P ratios off the southwest coast, it would not explain the particularly high concentrations of P in the Ten Thousand Islands area and in northwest Florida Bay.

Warzeski et al. (1996) and Cunningham et al. (1998) have mapped a deposit of quartz sand mixed with phosphorite granules along the length of South Florida (Figure 13.17). They hypothesized that this deposit is an old riverbed resulting from the Miocene uplift of the Appalachian Mountains and subsequent erosion to the south. These phosphorite-rich quartz sand deposits appear to underlie South Florida coastal waters in two places — south of Cape Sable and east of Cape Romano in the Ten Thousand Islands area, precisely where the highest concentrations of P (Figure 13.10) are observed. In the area south of Cape Sable, the phosphorite-rich quartz sand deposits are over 150 m thick and are only 10 to 15 m below Florida Bay. It is hypothesized that groundwater flowing upward through these phosphorite-rich deposits generates the high P concentrations observed south of Cape Sable and east of Cape Romano in the Ten Thousand Islands area.

Increasingly, the role of groundwater is being considered as a significant source of nutrients to coastal waters (D'Elia et al., 1981; Capone and Bautista, 1985; Valiela et al., 1990; Bokuniewicz and Pavlik, 1990; Oberdorfer et al., 1990; Lapointe et al., 1990; Giblin and Gaines, 1990; Simmons, 1992; Millham and Howes, 1994; Shinn et al., 1994; Moore, 1996, 1999; Cable et al., 1996;

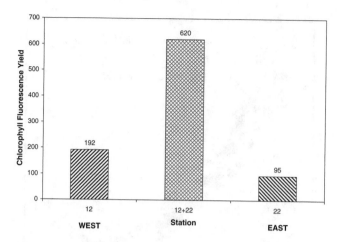

**FIGURE 13.14** Maximum chlorophyll fluorescence yield in water collected in November 2000 from eastern Florida Bay, western Florida Bay, and an experimental mixture of the two.

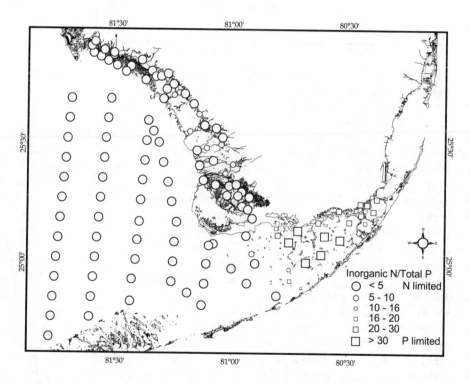

**FIGURE 13.15** Average molar ratios of dissolved inorganic N to total P measured from 1997 to 1998. (Calculated from SFWMD data files.)

Bacchus, 2002). In most cases, N has been examined, both because N is the primary limiting nutrient in most marine ecosystems (Ryther and Dunstan, 1971; Howarth, 1988) and because N flows more freely than P through various geological deposits without being chemically scavenged (Vitousek and Howarth, 1991). In South Florida, however, P is the limiting nutrient, so the plausibility of P being transported by groundwater must be examined.

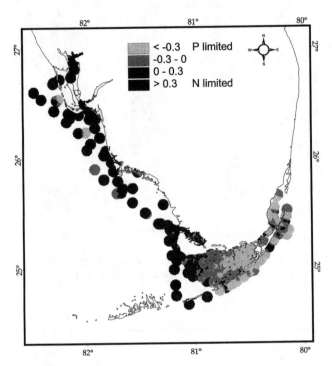

**FIGURE 13.16**   Results of 1301 nutrient bioassay experiments conducted along the southwest coast of Florida and Hawk Channel and combined with the Florida Bay bioassay data shown in Figure 13.13. Data are plotted using fixed-radius surface interpolation.

South Florida is characterized by relatively high rainfall, leading to considerable surface water runoff and groundwater flow into the adjoining coastal waters. The structure of the Floridan Aquifer System is well known for its large and numerous springs and large volume of groundwater flow. The extremely porous weathered limestone of South Florida is honeycombed with tunnels, fissures, solution holes, and caverns, allowing free flow of groundwater (Schomer and Drew, 1982; Stamm, 1994; Bacchus, 2002). The Miami Oolite formations are particularly porous (Missimer, 1984; Parker, 1984). Kreitman and Wedderburn (1984) consider the Biscayne Aquifer to be perhaps "the most permeable water table aquifer in the world" and the Floridan Aquifer System to be "one of the most extensive and prolific artesian aquifers in the United States." Before the water table was lowered, freshwater springs flowed freely in Biscayne Bay (Schomer and Drew, 1982; Parker, 1984). Bush and Johnston (1988) found that the hydraulic head of the Floridan Aquifer under Florida Bay is high enough to be capable of driving groundwater up into Florida Bay.

To test this hypothesis, tracer gases based upon the $^{238}U$–$^{206}Pb$ decay series and the decay of bomb-produced tritium have been used. Eight atoms of $^4He$ are generated as a $^{238}U$ atom radioactively decays to $^{206}Pb$. During this process, $^{222}Rn$ gas is also produced. $^3He$ is generated by the decay of $^3H$, most of which was produced by the detonation of numerous nuclear weapons in the 1950s and 1960s. As groundwater is not in contact with the atmosphere, the production of $^4He$, $^3He$, and $^{222}Rn$ underground leads to excess concentrations of these gases in groundwater, out of equilibrium with the atmosphere. Corbett et al. (1999, 2000) and Top et al. (2001) have used $^4He$, $^3He$, and $^{222}Rn$ as trace gas indicators of groundwater input into South Florida coastal waters and confirmed the input of significant amounts of groundwater into Florida Bay.

It is hypothesized that two processes could be pushing Floridan Aquifer water up through the phosphorite-rich sand deposits and up into Florida Bay. Because of the elevation of the recharge area of the Floridan Aquifer (northern Florida and the southern Appalachians), groundwater from the Floridan Aquifer has a considerable pressure head in South Florida (Bush and Johnston, 1988);

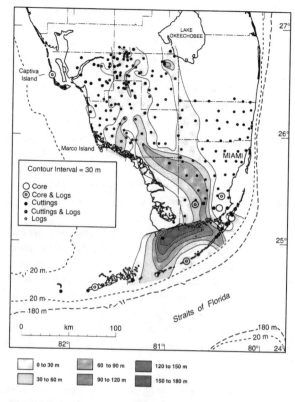

**Net thickness of coarse-grained (>1 mm) Miocene-to-Pliocene sands**

**FIGURE 13.17** Location and thickness of phosphorite-rich quartz sand deposits. (Redrawn from Cunningham, K.J. et al., *Geol. Soc. Am. Bull.*, 110, 231–258, 1998).

therefore, there is potential for general upward movement of groundwater in South Florida. It is hypothesized that this upward hydrologic pressure may push groundwater up through the phosphorite-rich sand channel deposits, generating the high P concentrations observed in the northwest and northcentral areas of Florida Bay and the Ten Thousand Islands area. The thick layers of coarse sand and gravel may make the upward flow easier than where the intermediate confining layer overlying the Florida Aquifer is composed of impervious limestone. Furthermore, quartz sand will not chemically scavenge P like calcium carbonate. This may be why P concentrations (Figure 13.10) show a good spatial correlation with the thickness of the phosphorite-rich quartz sand deposits (Figure 13.17). The much higher P concentrations in sediments in west Florida Bay than east Florida Bay (Yarbro and Carlson, 1998) are also consistent with this hypothesis. The Floridan Aquifer becomes increasingly saline toward the south along the Florida peninsula, indicating mixing with the adjacent seawater (Kreitman and Wedderburg, 1984; Katz, 1992; Miller, 1997). The increase in salinity will cause desorption of the phosphate from the limestone (Moore, 1996).

Another mechanism that could move groundwater upward through the P deposits and into the overlying coastal waters is geothermal heating, as hypothesized by Kohaut (1960, 1965, 1967), Kohaut et al. (1977), Smith and Griffin (1997), and Whitaker and Smart (1990, 1993, 1997). Thermal convection is induced in porous carbonate platforms by geothermally warmed groundwater rising and being replaced by seawater entrained into the carbonate platform from the sides. This appears to occur to a depth of around 1000 m in Florida (Whitaker and Smart, 1993). This can explain a number of warm springs in Florida and the spatial temperature patterns in groundwater in Florida (Sproul, 1977). In a few cases, Floridan Aquifer water emerges at the surface undiluted. Hot Springs in Charlotte County is full-strength seawater and emerges at 35.6°C (Nordlie, 1990). Warm Mineral Springs in Sarasota County emerges at 30°C and 9.6% seawater. Other springs from the Floridan Aquifer emerge under coastal waters. Fanning et al. (1981) argued that the prime area of geothermal heating is in the center of the Florida Platform, which today is along the west coast. Their data on elevated $^{226}$Ra and $^{222}$Rn were used as evidence that groundwater from the Floridan Aquifer was flowing up through the Hawthorn Formation rich in P and U and entering the coastal waters (Fanning et al., 1982). Their data show that this groundwater is rich in P, yielding N:P ratios well below the Redfield ratio. The data of Meyers et al. (1993) support this hypothesis. Their salt and isotope data indicate that Floridan Aquifer water does flow upward, infiltrating the surficial Biscayne Aquifer groundwater. Maddox et al. (1992) also observed the upconing of saline water into the overlying freshwater aquifer in various parts of South Florida. If this upward movement is occurring, it would be important to reconsider the practice of injecting wastewater into aquifers by deep injection wells. At the present time, approximately 800 million liters of wastewater are being pumped into the Floridan Aquifer in South Florida each day (Miller, 1997).

Whitaker and Smart (1990, 1997) have argued for geothermal circulation on the Bahamas Banks as well. This would explain the net flow of seawater out of the blue holes and lagoons such as Normans' Cay Pond (pers. observ.).

The withdrawal of significant amounts of surficial groundwater in South Florida over the past century and the subsequent drying up of freshwater springs in Biscayne Bay and Florida Bay may have enhanced the upward flow of saline groundwater (Upchurch and Randazzo, 1997). On a longer time scale, sea level rise over the past 18,000 years would have flushed seawater upward through the phosphorite deposits and increasing salinity would have caused desorption of phosphate (Moore, 1996).

Of potential significance is that the P source has existed for millions of years. Because of the large capacity for P sequestration by calcium carbonate, P generated within the last century in many cases has not made it to the coastline yet. On a long time scale and/or under more eutrophic conditions, calcium carbonate can become saturated with P (McGlathery et al., 1994; Corredor et al., 1999), allowing a greater flow of P from inland sources to coastal waters. While much work remains to be done to determine if P from the underlying Miocene phosphorite deposits is being transported into the overlying coastal waters, the data presented above are at least consistent with the hypothesis.

If this phosphorite-groundwater hypothesis is correct, it may explain a number of coastal geomorphological features in South Florida. During the last deglaciation, the South Florida coastline has retreated as sea level has risen. Approximately 3000 years ago, sea level rise continued, but at a slower rate. Since then, the rate of sea level rise has been slow enough that various biogeological processes such as the biological production of calcareous sediments and mangrove trapping of sediments have been able to stabilize the coastline and stop the retreat (Wanless et al., 1994; Davis, 1997). In two places along the coastline, however, biogeological processes have been able to actually advance the shoreline back out to sea — Cape Sable and the Ten Thousand Islands area east of Cape Romano (Wanless et al., 1994). These two areas are precisely where the phosphorite-rich quartz sand deposits underlie the coastal waters and where we today find high concentrations of P. It is hypothesized that elevated P concentrations have existed in these areas ever since the continental shelves were flooded by sea level rise. As P is the primary limiting nutrient in calcareous coastal ecosystems, it is further hypothesized that P-rich groundwater input from these phosphorite deposits has historically enhanced local biological productivity, which in turn has enhanced various biogeo-logical processes that have advanced the coastline back out to sea.

In the case of the Ten Thousand Islands area, oyster and vermetid gastropod reefs have been able to grow upward fast enough over the past 3000 years to keep up with sea level rise (Shier, 1969; Hoffmeister, 1974; Parkinson, 1989; Wanless et al., 1994). These reefs grow to the sea surface, allowing mangroves to take root on top of them. Subsequent deposition of organic matter and sediment trapping leads to the formation of islands. This complex labyrinth of such islands, known as the Ten Thousand Islands, reflects the pattern of oyster reef development over the past 3000 years. It is hypothesized that P-rich groundwater has maintained higher concentrations of phytoplankton in this area over the past 3000 years, providing more food for the oysters and promoting faster growth.

The largest calcareous mudbanks in Florida Bay are in the northwest and northcentral areas south of Cape Sable. Wanless and Tagett (1989) have shown that these mudbanks are in the area where active production and accumulation of calcareous mud occurs. These mudbanks have kept up with sea level rise over at least the past 2000 years. To the east and west of the large mudbanks, the calcareous muds are being transported and eroded away. These calcareous mudbanks are composed of materials primarily biologically produced, most on location, but some carried in from the Gulf of Mexico and deposited in northwest Florida Bay (Stockman et al., 1967; Nelson and Ginsberg, 1986; Wanless and Tagett, 1989; Bosence, 1989). Given the remarkable spatial correlation between the phosphorite-rich quartz sand deposits (Figure 13.17), high water-column P concentrations (Figure 13.10), and the largest biologically produced calcareous mud banks (Figure 13.18 in Wanless and Tagett, 1989), it is hypothesized that P transported by groundwater from the phosphorite-rich quartz sand deposits into the otherwise P-limited South Florida coastal waters has led to the particularly large biological production of calcareous mudbanks south of Cape Sable.

P-enhanced growth of benthic macroalgae and seagrasses in the area would also enhance the trapping and deposition of suspended particles (Scoffin, 1970) carried into the area from the Gulf of Mexico. Of significance is that Zieman et al. (1989), Fourqurean et al. (1992), and Frankovich and Fourqurean (1997) have found the lushest growth of seagrasses and epiphytes in Florida Bay to be in the northwest area where the phosphorite-rich quartz sand deposits and largest calcareous mudbanks occur.

Also of interest is that Cape Sable is an area where biogeological processes have been causing the coastline to advance back out to sea over the past 3000 years. The area is composed of old calcareous mud banks similar to the ones found south of Cape Sable in Florida Bay (Roberts et al., 1977; Enos and Perkins, 1977; Wanless et al., 1994). This tends to suggest that the phosphorite deposits in the Cape Sable area have been promoting the biological production of calcareous muds in the area. The northern area has become land, which is progressively moving to the south. Roberts et al. (1977) have aged the deposits on Cape Sable, demonstrating the progression to the south.

It is hypothesized that the Ten Thousand Islands, Cape Sable, and the Florida Bay calcareous mud banks are largely the result of biological production stimulated by P-enriched groundwater from underlying phosphorite deposits. The spatial correlations between the geomorphological features and the underlying phosphorite deposits are quite striking.

Although the data apparently do not exist, it is reasonable to hypothesize that phosphorite deposits were carried farther out on the western Florida continental shelf than has been mapped. Therefore, it is possible the elevated P concentrations and low N:P ratios (Figure 13.15) observed along the SW coast of Florida could be the result of groundwater coming up through phosphorite deposits throughout the southwest continental shelf and not just from freshwater runoff from rivers flowing through surface phosphorite deposits in central Florida. The $^{226}$Ra and $^{222}$Rn data of Fanning et al. (1981, 1982) are in agreement with this hypothesis. Harmful algal blooms often develop offshore on the west Florida shelf (Kusek et al., 1999), but it is not clear yet if offshore phosphorite deposits are involved in their development.

It is hypothesized that the phosphorite deposits are a persistent source of P that has not changed significantly over the past few thousand years. Actually, phosphate mining in central Florida over the past century may have increased the input of P into southwest Florida coastal waters, but there is no evidence that it has increased substantially in the past two decades. The phosphorite deposits appear to have shifted a shallow carbonate ecosystem that would normally be P limited into one that is N limited. This leads to the possibility that an increase in N inputs rather than P into Florida Bay could have led to an increase in algal blooms. Indeed, the highest chlorophyll concentrations

(Figure 13.2) are in the central bay where P in the west meets N from the east. The question becomes: Where is the N coming from and has it increased in the past two decades?

## The Source of N

It is hypothesized that most of the N is coming out of the Everglades-agricultural system. The spatial distribution of N (Figure 13.5) and its correlation with low salinity (Figure 13.18) suggests that freshwater runoff from the Everglades through Taylor Slough and the South Dade Conveyance System (SDCS) canal system (Figure 13.19) is the major source of N to Florida Bay. There are no significant algal blooms at the mouth of Taylor Slough because of strong P limitation.

Figure 13.20 shows that N and P are scavenged from the water as it moves south from Lake Okeechobee and the Everglades Agricultural Area to Florida Bay (Figure 13.21), but much N remains, as it is not scavenged by the limestone and vegetation as efficiently as P. While less than 0.1 μ*M* P enters Florida Bay, around 60 μ*M* N is entering. It is this strong depletion of P but only partial depletion of N that leaves eastern Florida Bay strongly P limited.

The natural Everglades ecosystem was an oligotrophic wetland, in which the natural source of nutrients was primarily in rainfall, and the vegetation sequestered most of the nutrients before they reached the coastal waters (Lodge, 1994). During the 20th century, a large portion of the Everglades was drained and converted to agricultural land (DeGrove, 1984; Light and Dineen, 1994). Everglades land area converted to sugar cane farms in the Everglades Agricultural Area

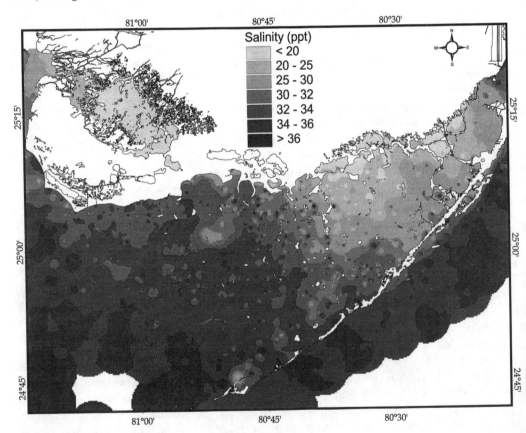

**FIGURE 13.18**   Average salinity measured with a YSI meter between February 1996 and November 2000. The data (3030 measurements) are plotted using fixed-radius surface interpolation.

(EAA) south of Lake Okeechobee (Figure 13.21) increased dramatically after the Cuban Revolution of 1959 (Figure 13.22). Initially, most of the nutrient-rich runoff from the EAA was pumped back up into Lake Okeechobee through pump stations S2 and S3 (Figure 13.21; Light and Dineen, 1994). This eventually helped lead to the eutrophication of Lake Okeechobee (Havens et al., 1996). As a result, it was decided to instead pump the nutrient-rich EAA runoff south through the extensive canal system (Figure 13.21) of the Central and South Florida Project for Flood Control and Other

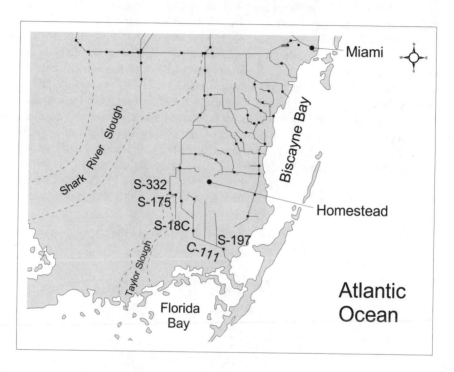

**FIGURE 13.19**  Map of the South Dade Conveyance System.

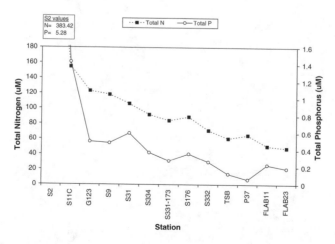

**FIGURE 13.20**  Average concentrations of total N and total P in 1998 along the canal system from Lake Okeechobee and the Everglades Agricultural Area to Florida Bay (Calculated from SFWMD data files.). Nutrient concentrations at station S2 are off scale; N is 383.42 $\mu M$ and P is 5.28 $\mu M$.

Purposes (C & SF Project), operated by the South Florida Water Management District, starting around 1979 (Lodge, 1994). As a result, a large increase in water flow to the south is observed in locations such as S2 south of Lake Okeechobee (Figure 13.23) and S151 in the central Everglades (Figure 13.24). It is that diversion of nutrient-rich water to the south into the Everglades canal system that is thought to have led to the ecological changes observed in the Everglades in the past two decades (Davis, 1994; Lodge, 1994). From the 1960s to the 1990s, N input into Water Conservation Area (WCA) 2A south of the EAA increased by a factor of 12.4 (Davis, 1994). Davis (1994) estimates that P input into Everglades National Park increased threefold as a result of the diversion of agricultural runoff.

At about the same time as agricultural runoff from the EAA began being diverted south in the early 1980s, the South Dade Conveyance System (Figure 13.19), which was designed to enhance the flow of water from the Everglades canal system into Florida Bay, was completed (Light and Dineen, 1994; SFWMD, 1992). The system was designed to transport more freshwater into Everglades National Park and Florida Bay and to lower groundwater levels to the east, allowing farmers to better drain their agricultural fields in southwest Dade county and shift from seasonal to year round farming (Ley, 1995) and allowing developers to greatly expand suburbs to the west of Miami, closer to the Everglades (Light and Dineen, 1994). This engineering was successful and more water was injected into Florida Bay, beginning in the early 1980s. Figure 13.25 shows the increase in water flowing through the South Dade Conveyance System into Florida Bay. It is hypothesized that

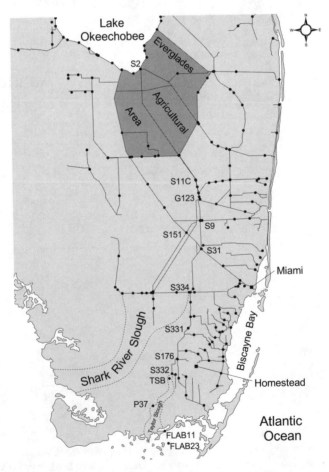

**FIGURE 13.21**    Map of the South Florida Water Management District canal system and Everglades Agricultural Area.

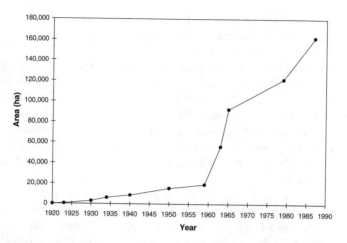

**FIGURE 13.22** Changes in sugar cane farm area in the Everglades Agricultural Area over time. (Data from Snyder and Davidson, 1994.)

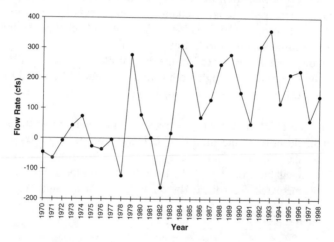

**FIGURE 13.23** Annual water flow through the S2 pump station. Negative values indicate backpumping north from the Everglades Agricultural Area into Lake Okeechobee; positive values indicate pumping south from the Everglades Agricultural Area into the Everglades. (Calculated from SFWMD data files.)

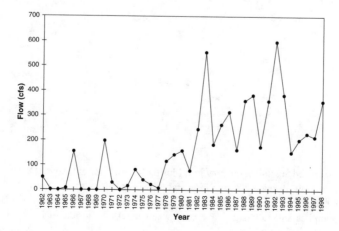

**FIGURE 13.24** Annual water flow through the S151 pump station. (Calculated from SFWMD data files.)

this led to the injection of not only more freshwater into Florida Bay, but also more N. The experimental mixing of this N-rich canal water with P-rich water from western Florida Bay generates a large increase in algal biomass (Figure 13.26).

The C111 Interim Plan implemented in the early 1990s increased flow into northeast Florida Bay even more (Ley, 1995). This was associated with a 42% increase in nitrate and 229% increase in ammonia in Florida Bay from 1989–1990 to 1991–1994, and a 42% increase in chlorophyll (Ley, 1995). The C111 canal was altered so that more water would flow farther to the west and less water would flow to the east. More water was pumped through the S332 and S175 structures (Figure 13.19) into Taylor Slough (Figure 13.27). This had the effect of injecting N-rich water closer to the area of high P in western Florida Bay and appears to have increased the size and intensity of the algal blooms.

As noted earlier, fishers and other boaters began to observe a systematic decline in water quality around 1981 (DeMaria, 1996), right after nutrient-rich agricultural runoff from the EAA was diverted south into the Everglades canal system and the South Dade Conveyance System was enlarged to enhance the flow of Everglades canal water into Florida Bay. While we do not have quantitative data on N concentrations in Florida Bay before 1989, the temporal correlation between the diversion of nutrient-rich agricultural runoff into the Shark River, Taylor Slough, and Florida Bay and the observed ecological changes is rather remarkable. It is difficult to imagine that the pre-agriculture oligotrophic Everglades wetlands would generate the high concentrations of N that are observed in northeast Florida Bay today (Figure 13.5). The larger, more dramatic increase in algal blooms that was observed in the early 1990s correlates with the large increase in water flow from the Everglades through Taylor Slough (Figure 13.27) and the diversion of the runoff farther to the west, which resulted in injecting N-rich water closer to the hypothesized natural P source.

Figure 13.28 shows the shift in water flow from the C111 canal into northeast Florida Bay over toward Taylor Slough, closer to northcentral Florida Bay. Figure 13.29 shows the resulting shift in water level in the marshes north of Florida Bay (Figure 13.30) from east to west. Although no water quality data are available from the 1980s to compare with the 1990s, the shift from east to west did continue to some extent from the early to late 1990s. As a result, one can see a shift from

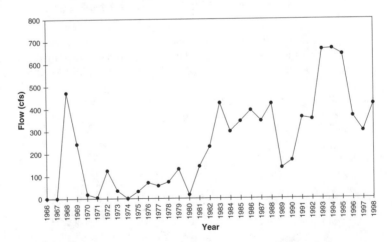

**FIGURE 13.25**   Estimated annual water flow into Florida Bay through the South Dade Conveyance System. Data are calculated as flow through S332, S175, and S18C structures minus S197. S332, and S175 pump water into the Taylor Slough system. S18C pumps water into northeast Florida Bay and Barnes Sound. S197 flow into Barnes Sound is subtracted to leave only water entering Florida Bay directly. (Calculated from SFWMD data files.)

east to west in inorganic N (Figure 13.31) and chlorophyll (Figure 13.32) from the early to late 1990s along a transect across north Florida Bay (Figure 13.30).

A comparison of the spatial extent of the bloom during the dry season when there is little input of N-rich water from Taylor Slough (Figure 13.33) and during the rainy season when there is a large input of N-rich water from Taylor Slough (Figure 13.34) also demonstrates that input of the N-rich water from the Everglades Agricultural Areas through Taylor Slough is probably the largest source of N that generates the algal blooms.

A comparison of chlorophyll concentrations over time at four stations in northcentral Florida Bay in the center of the bloom (Figure 13.35) with the flow of the N-rich water into Florida Bay through the Taylor Slough system shows a general correlation (Figure 13.36), with highest chlorophyll concentrations in 1994 and 1995, when water flow was highest. The same correlation appears when the data are examined on a monthly basis (Figure 13.37). The major source of N generating the algal bloom may be the flow of water from Taylor Slough through McCormick

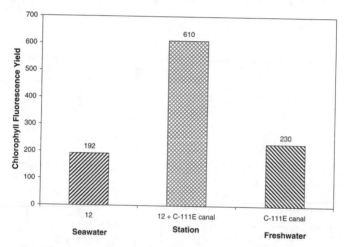

**FIGURE 13.26**   Maximum chlorophyll fluorescence yield in water collected in November 2000 from a South Dade Conveyance System canal, western Florida Bay, and an experimental mixture of the two. Station 12 is in northwestern Florida Bay.

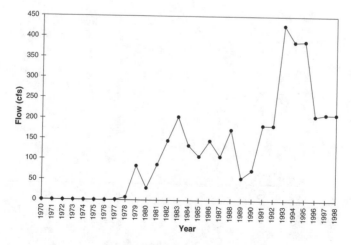

**FIGURE 13.27**   Annual water flow through the Taylor Slough system as measured at structures S332 and S175.

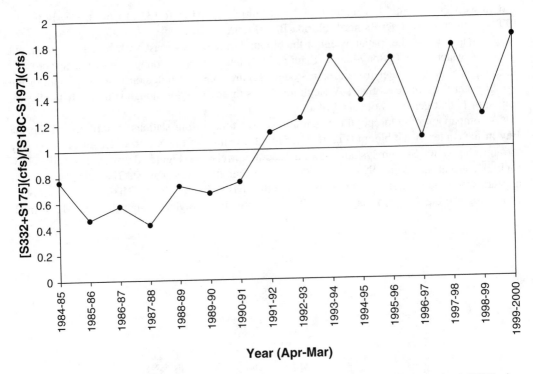

**FIGURE 13.28**    Ratio of water flow through S332 and S175 into Taylor Slough to flow through S18C minus S197 out of the C111 canal into northeast Florida Bay.

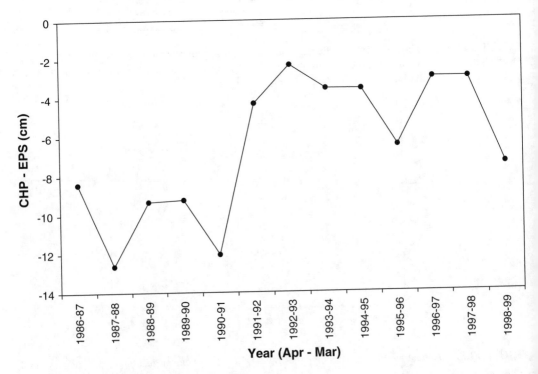

**FIGURE 13.29**    Changes in height difference in water level between stations NP-CHP and NP-EPS over time.

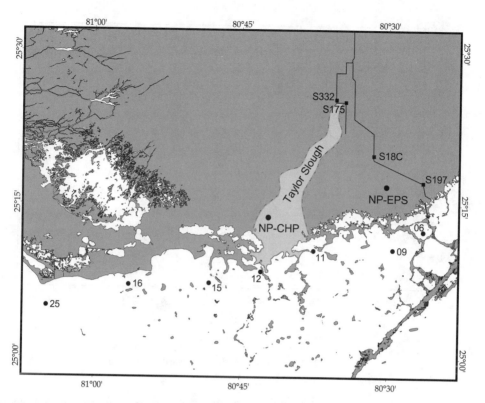

**FIGURE 13.30** Location of water control structures in the South Dade Conveyance System, water level stations in marshes north of Florida Bay, and water quality stations in Florida Bay.

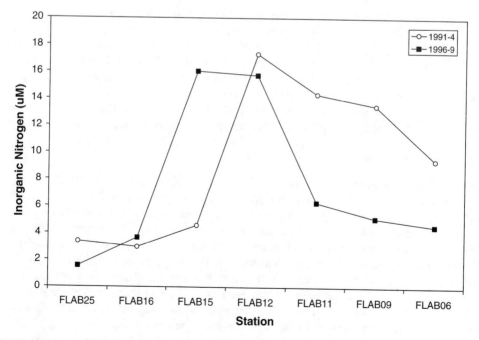

**FIGURE 13.31** Comparison of inorganic N concentrations along an east–west transect (Figure 13.30) of northern Florida Bay between April 1991 through March 1994 and April 1996 through March 1999. (Calculated from SFWMD data files.)

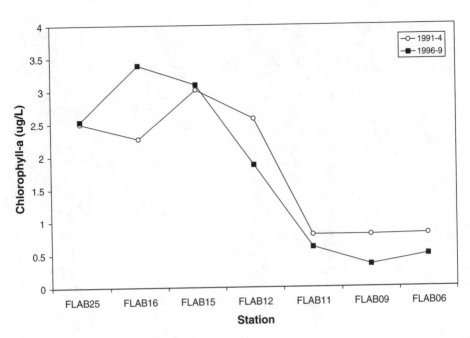

**FIGURE 13.32**    Comparison of chlorophyll concentrations along an east–west transect (Figure 13.30) of northern Florida Bay between April 1991 through March 1994 and April 1996 through March 1999. (Calculated from SFWMD data files.)

Creek and Alligator Creek (Figure 13.38) into Florida Bay, as these creeks flow directly into the high P area. The concentration of chlorophyll correlates well with the flow through McCormick Creek alone (Figure 13.39).

Boyer and Jones (1999) have used the same dataset to argue that freshwater runoff from the Everglades-agricultural system is not causing the algal blooms in Florida Bay. They, however, chose to examine chlorophyll concentrations only at the stations in northeast Florida Bay where there is little P, extreme P limitation, and no algal blooms. They ignore their chlorophyll data in northcentral Florida Bay where the major algal bloom occurs. Our examination of chlorophyll concentration variation over time in northcentral Florida Bay where the major bloom occurs and where the N from the east meets the P from the west correlates well with freshwater runoff (Figures 13.36 and 13.37), suggesting that N from the Everglades agricultural system is indeed generating the algal blooms.

## Hypothesized Scenario

The algal blooms in Florida Bay appear to develop from N-rich freshwater runoff mixing with P-rich seawater in western Florida Bay. It is hypothesized that much of the P comes from Miocene phosphorite deposits by way of Peace River erosion and subsequent coastal current transport along the southwest coast and by way of groundwater through phosphorite-rich quartz sand deposits underneath northwestern Florida Bay. It is argued that this P source has not changed significantly over the past few decades. It appears that much of the N comes from freshwater runoff from agricultural lands through the Everglades. It is hypothesized that changes in water management practices in the past two decades have led to an increase in N inputs to eastern Florida Bay. Mixing of this water from the east with the P-rich water from the west has led to the large algal blooms that have developed in northcentral Florida Bay, altering the entire ecosystem.

The following scenario is envisioned as explaining the increase in N load to Florida Bay and increasing eutrophication. Large areas of the northern Everglades were converted to agricultural

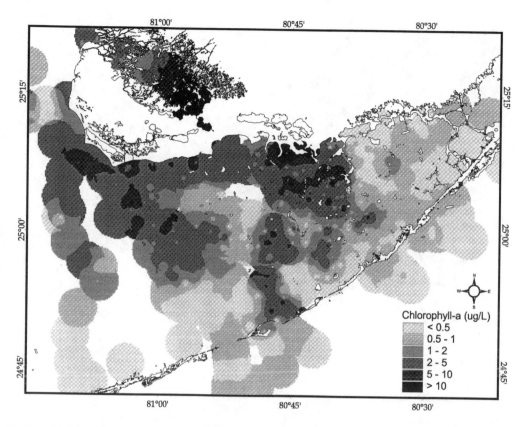

**FIGURE 13.33**   Average chlorophyll concentrations are plotted using fixed-radius surface interpolation during the dry season (February to April) from 1996 to 1999.

fields, particularly after the Cuban Revolution in 1959. Initially, much of the nutrient rich farm runoff was pumped back up into Lake Okeechobee. This eventually led to the eutrophication of Lake Okeechobee, so the farm runoff began being pumped south in the early 1980s. This led to nutrient enrichment and ecological changes in the Everglades to the south. At the same time, fishers and boaters started seeing algal blooms in western Florida Bay in 1981, probably as a result of N-rich water from Shark River mixing with the P-rich water of the western shelf and being advected into the bay. Chlorophyll concentrations in 1989–1990 were elevated in northwest Florida Bay but not yet in northcentral Florida Bay (Figure 13.40). Increasing nutrients from Taylor Slough and the C111 canal were probably initially taken up primarily by macroalgae and epiphytes on seagrasses in shallow Florida Bay. Indeed, increasing epiphytes on the seagrasses and increasing macroalgal biomass were observed in Florida Bay in the 1980s (Lapointe et al., 2002; Lapointe and Barile, 2001), probably the result of N-rich water from Taylor Slough now entering Florida Bay in increasing quantities. Fishers in the bay at the time observed that, "The first seagrass die-off occurred in the western Bay in 1985, when macroalgae laid so thick on the bottom that it denuded the bottom" (SFWMD, 1995). This eutrophication may have then led to the massive seagrass dieoff, as hypothesized by Lapointe et al. (2002) and Lapointe and Barile (2001). In the early 1990s, water flow was increased even more and the N-rich freshwater runoff was shifted from the C111 canal in the east to Taylor Slough in the west. This had the effect of moving the N loading closer to the natural P source. This would explain the large increase in the algal blooms in northcentral Florida Bay starting in 1991.

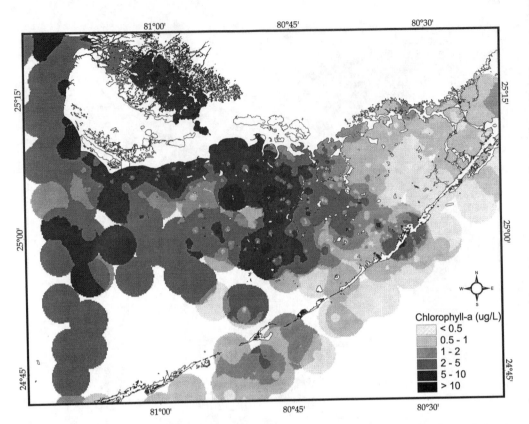

**FIGURE 13.34** Average chlorophyll concentrations are plotted using fixed-radius surface interpolation during the rainy season (September to November) from 1996 to 1999.

## THE TEN THOUSAND ISLANDS AREA

The southwest coast and Ten Thousand Islands area are characterized by relatively high P concentrations (Figure 13.10), relatively low N concentrations (Figure 13.9), and inorganic N:total P ratios less than the Redfield ratio (Figure 13.15). This results in an ecosystem that is primarily N limited, as observed in nutrient bioassay experiments (Figure 13.41). It is hypothesized (as discussed earlier) that this is the result of P enrichment from ancient phosphorite deposits. Some is eroding from surface deposits in the Peace river watershed and being transported downriver into the Charlotte Harbor system and down the coast by alongshore currents. In the Ten Thousand Islands area, it is hypothesized that groundwater may be transporting P up from phosphorite deposits below (discussed earlier). This results in the southwest coast being susceptible to eutrophication from N enrichment, as observed in western Florida Bay. Unlike Florida Bay, with the Everglades and Everglades Agricultural Areas as its watershed, the Ten Thousand Islands and southwest coast have the Big Cypress Swamp as their main watershed, which is relatively unaltered compared to the highly altered Everglades agricultural system, so major eutrophication has not yet developed. The urban areas (Naples) and agricultural areas in the watershed upstream are rapidly expanding, however, and one can expect N runoff to increase. Indeed, this probably explains the much higher chlorophyll concentrations in Henderson Creek, which drains the urban and agricultural areas, compared to Blackwater River and the Faka-Union Canal, which drain the Big Cypress Swamp (Figure 13.42).

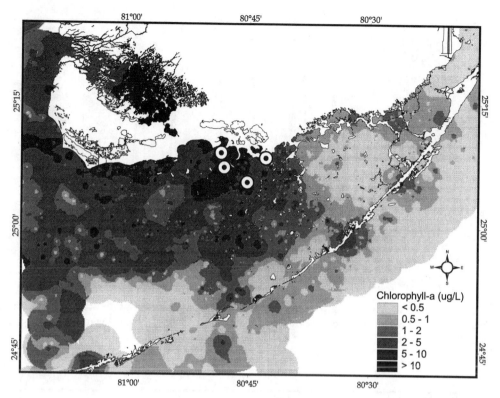

**FIGURE 13.35**   Location of four SFWMD stations superimposed on a map of chlorophyll measured as in Figure 13.2.

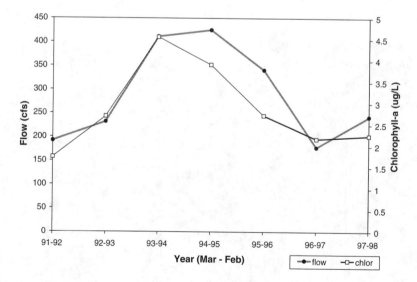

**FIGURE 13.36**   Annual water flow through the Taylor Slough system as measured at structures S332 and S175 and annual chlorophyll concentrations averaged at the four stations shown in Figure 13.35. Each year is calculated from March to the following February to avoid splitting water flow during the high-runoff period. (Calculated from SFWMD data files.)

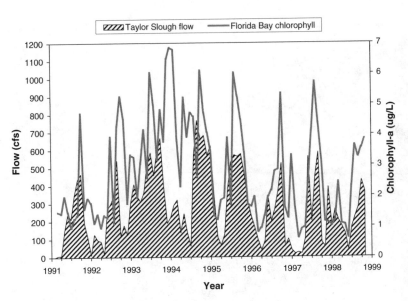

**FIGURE 13.37**   Monthly water flow through the Taylor Slough system as measured at structures S332 and S175 and monthly chlorophyll concentrations averaged at the four stations shown in Figure 13.35. (Calculated from SFWMD data files.)

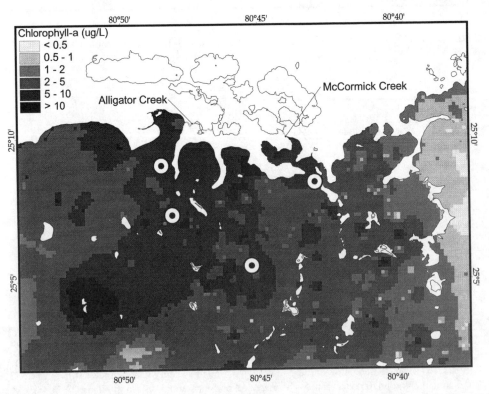

**FIGURE 13.38**   Map showing McCormick Creek and Alligator Creek, location of four stations at which chlorophyll concentrations were averaged, and overall distribution of chlorophyll as measured in Figure 13.2.

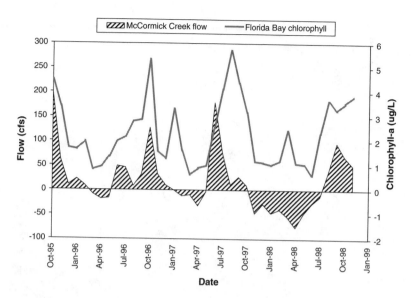

**FIGURE 13.39**   Monthly flow through McCormick Creek (from data of Clinton Hittle and Eduardo Patino on USGS website at ftp://ftp.envirobase.usgs.gov/sfep/epatino/Mcc-dv-9599.csv) and monthly chlorophyll concentrations (from SFWMD data files) averaged at the four stations shown in Figure 13.38. Negative flow indicates net tidal flow from Florida Bay up into upland lagoons.

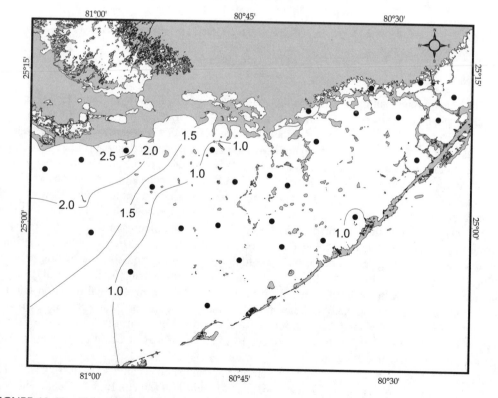

**FIGURE 13.40**   Chlorophyll concentrations measured between June 1989 and August 1990. (Redrawn from Fourqurean, J.W. et al., *Estuar. Coast. Shelf Sci.*, 36, 295–314, 1993).

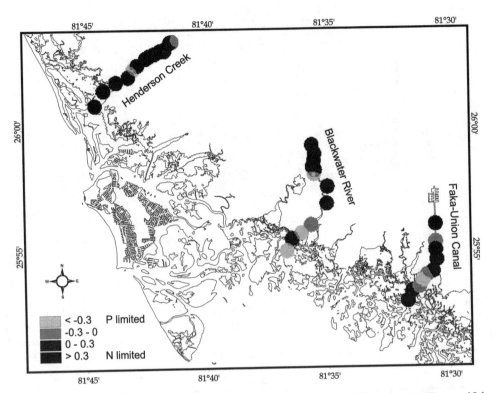

**FIGURE 13.41**   Results of nutrient bioassay experiments conducted in June 1999 in the Ten Thousand Islands area. Data are plotted using fixed-radius surface interpolation.

## Biscayne Bay

South Biscayne Bay is similar to eastern Florida Bay in that it is P limited (Figure 13.43) and has low chlorophyll concentrations (Figure 13.44) (Brand, 1988; Brand et al., 1991). This is the result of its being a carbonate environment and not having any phosphorite deposits nearby. Clearly, large amounts of N are entering south Biscayne Bay as a result of freshwater runoff (Figure 13.45), but this does not result in large algal blooms as in Florida Bay because of the absence of large sources of P such as phosphorite deposits. The only elevated chlorophyll concentrations are observed near the canal mouths (Figure 13.44). The largest algal blooms occur near the canal mouths at the beginning of the rainy season as a result of increased freshwater flow. Brand (1988) and Brand et al. (1991) have hypothesized that this is the result of P accumulating in soil and groundwater during the dry season and then being flushed out into the canals at the beginning of the rainy season. Also supporting this hypothesis is the finding of a strong negative correlation between chlorophyll and salinity (Brand, 1988; Brand et al., 1991), indicating that most nutrients in Biscayne Bay are associated with freshwater flow from land. As a result of direct canal injection into the bay and no marshes filtering the freshwater input, significant amounts of P can enter Biscayne Bay with increased freshwater runoff and stimulate algal blooms. This contrasts with eastern Florida Bay, where freshwater runoff is filtered through 10 to 20 km of marshes before entering the bay, resulting in extremely low P inputs.

North Biscayne Bay has approximately ten times as much chlorophyll (Figure 13.44), which correlates with the lower salinity (Figure 13.46) and P (Figure 13.47) there. Brand (1988) and Brand et al. (1991) attributed the much higher chlorophyll concentrations in the north to freshwater runoff entering much smaller basins and being diluted much less than in south Biscayne Bay. At the northern end of north Biscayne Bay, the ecosystem shifts toward N limitation, particularly near the mouths of the Miami and Little Rivers (Figure 13.43). This may be the result of calcium carbonate

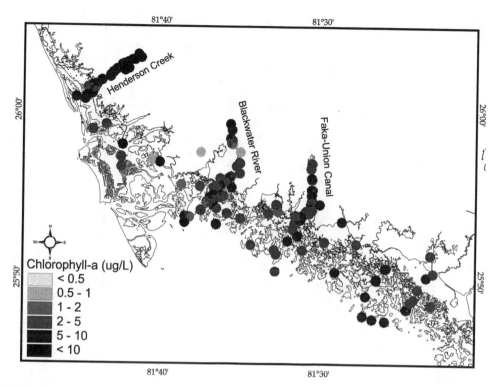

**FIGURE 13.42** Chlorophyll concentrations measured in the Ten Thousand Islands area between January 1997 and June 1999. The data (283 measurements) are plotted using fixed-radius surface interpolation.

saturation by P in the heavily eutrophied area, as hypothesized by McGlathery et al. (1994) and Corredor et al. (1999). The lack of chemical scavenging of P by saturated calcium carbonate can result in excess P and a shift to N limitation.

South Biscayne Bay is an ecosystem dominated by benthic autotrophs, primarily seagrasses and macroalgae. These plants have rather heavy loads of epiphytes on them, similar to what was observed in Florida Bay shortly before the seagrass dieoff there in the 1980s (Lapointe et al., 2002; Lapointe and Barile, 2001). This suggests that a significant increase in P loading to south Biscayne Bay could cause a major shift in the ecosystem, as happened in Florida Bay. That shift has already occurred in north Biscayne Bay, where few seagrasses are left and phytoplankton are the dominant autotrophs.

## FLORIDA KEYS COASTAL WATERS

It is generally agreed that ecological changes are occurring in the Florida Keys coastal ecosystem (EPA, 1992; Ogden et al., 1994). One of the suspected causes of some of these changes — increased nutrients — is well known to alter ecosystems. For example, increased nutrients can cause macroalgae to overgrow corals (Smith et al., 1981; Maragos et al., 1985; Bell, 1992) and epiphytes to overgrow seagrasses (Cambridge and McComb, 1986; Silberstein et al., 1986; Tomasko and Lapointe, 1991; Lapointe et al., 1994). This has been observed in many areas throughout the world.

It has been estimated that there are 25,000 cesspools and septic tanks and 182 marinas with 2707 wet slips, and 1410 live-aboard boats in the Florida Keys (EPA, 1992). These sources, along with several sewage outfalls and other human activities, are thought to be injecting nutrients into the Florida Keys coastal waters.

Many of these sources are located along canals. One observes elevated concentrations of nutrients (Lapointe et al., 1990, 1992; Lapointe and Matzie, 1997) and phytoplankton (Figure 13.48)

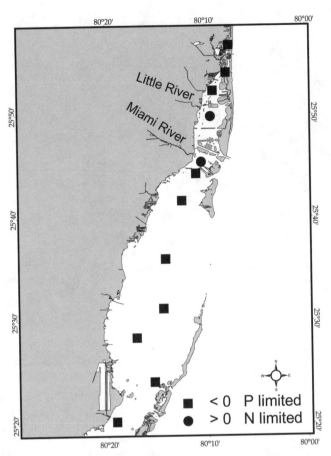

**FIGURE 13.43**  Results of nutrient bioassay experiments conducted six times between April 1986 and February 1987 in Biscayne Bay. (Data from Brand, 1988.)

in many of the canals in the Florida Keys. It appears quite likely that the source of much of the nutrients generating this algal biomass is septic tanks at homes along these canals (Lapointe et al., 1990; EPA, 1992). Shinn et al. (1994) and Paul et al. (1995a,b) have indeed documented that sewage is entering the coastal waters around the Florida Keys. It is thought that much of these nutrients from septic tanks are driven into the canals and coastal waters by rain-driven groundwater. Lapointe et al. (1990) observed a buildup of nutrients in groundwater of the Florida Keys during the dry season and an increase in nutrients in local marine waters during the wet season, suggesting that rainfall drives nutrient-rich groundwater into local marine waters. This is similar to what Brand (1988) and Brand et al. (1991) observed in Biscayne Bay. Chlorophyll concentrations usually drop dramatically not too far outside the canal mouths (Figure 13.48).

   In addition to leakage of sewage into the groundwater, there is direct deliberate pumping of sewage into the coastal waters. In a transect from the Gulf of Mexico over the Key West sewage outfall and out through Hawk Channel to Sand Key, higher chlorophyll concentrations have been observed around the Key West sewage outfall (Figure 13.49).

   In a transect along Hawk Channel between the Florida Keys and the offshore coral reefs, increasing chlorophyll has been observed from the upper Keys to the middle Keys (Figure 13.50). It is fairly clear that water from Florida Bay causes at least some of the chlorophyll increase in the middle and lower Keys, but sewage input from the Keys may also be a factor. The net flow of water in Hawk Channel is from the northeast to the southwest (Pitts, 1994, 1997), and chlorophyll concentrations appear to increase as it moves along the Keys. An increase in chlorophyll along the

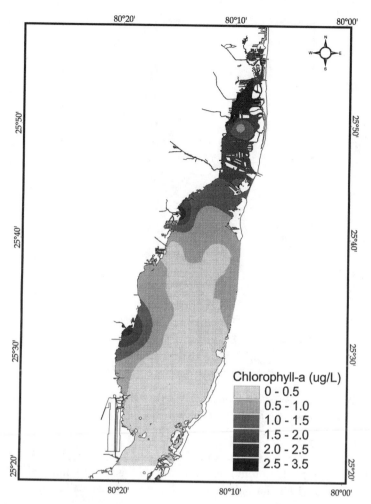

**FIGURE 13.44** Chlorophyll concentrations measured in Biscayne Bay between March 1986 and February 1987 by Brand (1988). Data are plotted using nearest neighbor surface interpolation.

transect from Elliot Key to Plantation Key (and Pacific Reef to Crocker Reef) was found where virtually no Florida Bay water would be expected to be getting into Hawk Channel.

In addition to local sewage, nutrient-rich, high-chlorophyll water from Florida Bay also enters Florida Keys coastal water. A comparison of Figures 13.33 and 13.34 shows elevated average chlorophyll concentrations on the reef side of Long Key during the rainy season. Plumes of turbid, low-salinity, nutrient-rich, high-chlorophyll water have been observed and documented being transported from Florida Bay all the way out to the coral reefs. Quite often, one observes plumes out into Hawk Channel but they do not make it out to the reefs and the water over the reefs appears quite clear. Sometimes, however, one can observe plumes of turbid water all the way out to the reefs and beyond. Figure 13.51 shows chlorophyll concentrations along a transect in Hawk Channel (Figure 13.52) along the Florida Keys from Key Biscayne to Marathon. Especially high chlorophyll concentrations are observed at stations 14 and 15 around Long Key where the largest passes between Florida Bay and Hawks Channel occur. Chlorophyll continues at somewhat elevated concentrations to the south as the coastal countercurrent transports the plume south. The data of Smith (1994) and Wang et al. (1994) confirm that this is the general flow pattern.

The data of Lee et al. (2002) show that drifters released near the outflow of Shark River usually end up flowing southeast through the passes around Long Key or Vaca Key out over the reefs or

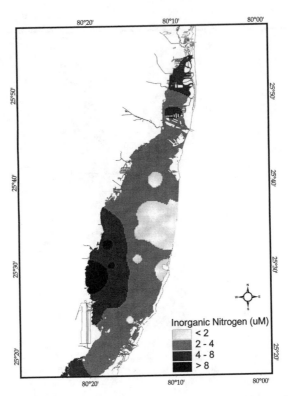

**FIGURE 13.45** Average concentrations of dissolved inorganic N measured in monthly sampling of Biscayne Bay from 1993 to 1998. Data were plotted using nearest neighbor surface interpolation. (Calculated from SFWMD data files.)

southwest along the north coast of the lower Keys and then south into the Gulf Stream (Figure 13.8). These data indicate that in addition to Florida Bay water, outflow from the Shark River makes it to the reefs. Dawes et al. (1999) have documented a decline in macroalgal species diversity and an increase in species that proliferate under eutrophic conditions in the Content Keys. This is of interest because the Content Keys are a long distance from any human habitation or sewage source but are downstream of the Shark River outflow.

On March 23, 1996, a plume of very turbid, low-salinity, high-chlorophyll water was observed out over the reefs at Looe Key, which could be seen in an uncorrected satellite image of reflectance processed by Dr. Richard Stumph (http://coastal.er.usgs.gov/flbay/html/199603/19960323_14ref.tif). Along a transect through this plume (Figure 13.53), increasing turbidity and chlorophyll were observed from Big Pine Key out to the reefs and beyond (Figure 13.54). Declining turbidity and chlorophyll were not observed until well beyond the reefs. This water probably originated near the Shark River outfall or in western Florida Bay. It appears that this water was pushed by the northerly winds into the Gulf Stream, which picked it up and carried it east out over the reefs. This plume appears to be squeezed between the Gulf Stream and the coastal countercurrent as a narrow front. Figure 13.55 shows a strong positive correlation between annual N flux through the Shark River and average annual chlorophyll concentrations measured by Brian Lapointe (Lapointe et al., 2002) in the area around Looe Key in the lower Florida Keys.

Taken together, these data suggest that nutrients from Everglades-agricultural runoff are being transported not just to Florida Bay, but also to the Florida Keys and coral reefs and contributing to their eutrophication. In the Upper Keys, which are heavily populated, there are only a few small channels where Florida Bay water could get into Hawk Channel. Furthermore, water in northeast Florida Bay is low in chlorophyll and phosphate, so what Florida Bay water might get into Hawk Channel probably would not have much effect in the Upper Keys. The Middle Keys, however, have

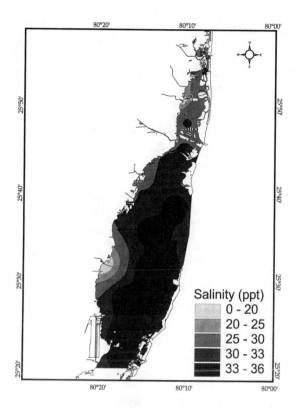

**FIGURE 13.46**   Average salinity measured in monthly sampling of Biscayne Bay from March 1986 to February 1987 by Brand (1988). Data are plotted using nearest neighbor surface interpolation.

a much smaller human population but large passes that are well documented to allow large amounts of water from Florida Bay to reach Hawk Channel (Smith, 1994). Furthermore, the Middle Keys area is downstream of the largest blooms in central Florida Bay. The Lower Keys are downstream of the Shark River outfall. It is hypothesized that sewage is the cause for the increase in chlorophyll in the Upper Keys, and Florida Bay influx is the dominant cause for the increase in the Middle Keys. In the Lower Keys, the Key West sewage outfall and water from the Shark River plume appear to have a large impact.

## CONCLUSIONS

South Florida coastal waters are mostly shallow carbonate ecosystems. Because of the chemical scavenging of P by calcium carbonate, one would expect P to be the primary limiting nutrient. This is what one observes along the east side of South Florida in Biscayne Bay and eastern Florida Bay. Anthropogenic P from land tends to be scavenged by limestone before it reaches the coastal waters, as observed along the Everglades–Taylor Slough flow pathway. Direct injection of runoff through canals, however, allows for more P to reach coastal waters, as observed in Biscayne Bay during the rainy season. The western side of South Florida has large deposits of phosphorite that were transported from the north during the Miocene. It is hypothesized that these deposits are the source of the higher P concentrations observed on the western side of South Florida. The P concentrations are high enough that the coastal ecosystems are shifted to N limitation. Because inorganic N compounds (unlike P) flow more readily through soil and limestone without chemical scavenging, the western side of South Florida is more susceptible to anthropogenic N inputs. This is clearly observed downstream of the N-rich Everglades Agricultural Area where freshwater runoff meets P-rich waters from western Florida Bay.

The flux of N to Florida Bay increased dramatically in the 1980s as runoff from the Everglades Agricultural Area began being pumped south into the Everglades instead of north into Lake Okeechobee, and the South Dade Conveyance System was expanded to allow more water to be pumped into Florida Bay. This was followed by algal blooms, seagrass dieoff, increased turbidity, and sponge dieoff. The algal blooms increased further as N-rich water from the Everglades agricultural system was pumped farther west, closer to the P source.

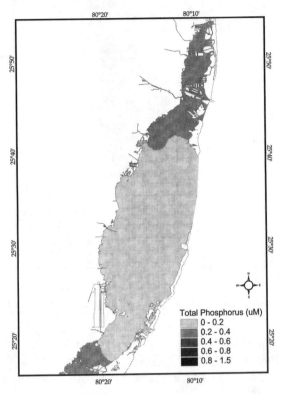

**FIGURE 13.47**    Average concentrations of total P measured in monthly sampling of Biscayne Bay from 1993 to 1998. Data were plotted using nearest neighbor surface interpolation. (Calculated from SFWMD data files.)

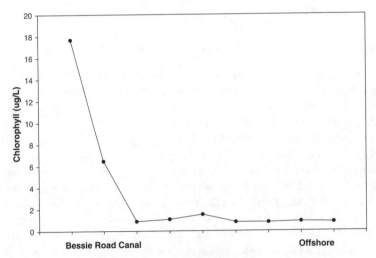

**FIGURE 13.48**    Chlorophyll concentrations along a 4-km transect from a Tavernier canal out into Hawk Channel.

Some of this nutrient-enriched water from Florida Bay and the Shark River is transported to the middle and lower Florida Keys, where it may be adversely affecting the coral reefs and other oligotrophic ecosystems there.

South Biscayne Bay is relatively oligotrophic, but increased direct runoff through canals could add more P to this P-limited ecosystem, causing more eutrophication and a possible shift from a seagrass-dominated ecosystem to a phytoplankton-dominated ecosystem. North Biscayne Bay is already eutrophic, having approximately ten times higher algal concentrations and relatively few seagrasses.

The Ten Thousand Islands area appears to be naturally somewhat eutrophic because of the high P concentrations, hypothesized to come from phosphorite deposits. As this results in N limitation, it is predicted that eutrophication will increase in the Ten Thousand Islands area as urban and agricultural areas upstream expand and result in larger N loads downstream.

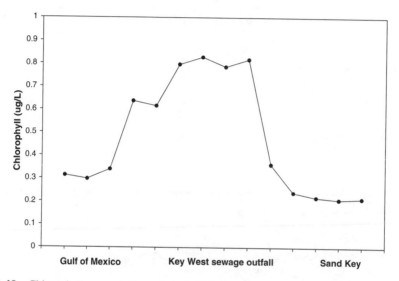

**FIGURE 13.49**   Chlorophyll concentrations along a 33-km transect over the Key West sewage outfall.

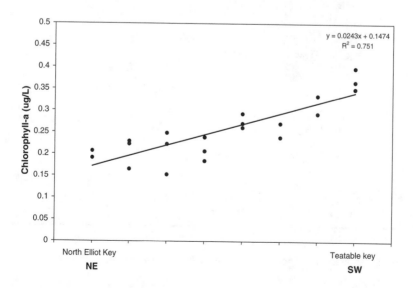

**FIGURE 13.50**   Chlorophyll concentrations along a transect in Hawk Channel between the Florida Keys and the coral reefs from the northeast to the southwest.

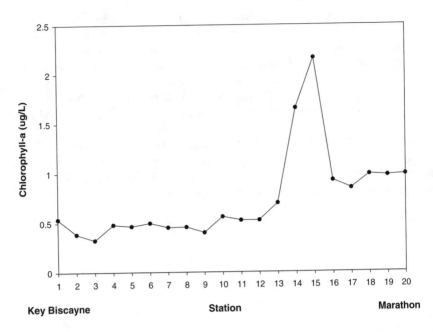

**FIGURE 13.51**   Chlorophyll concentrations at stations shown in Figure 13.52 on November 4, 1997.

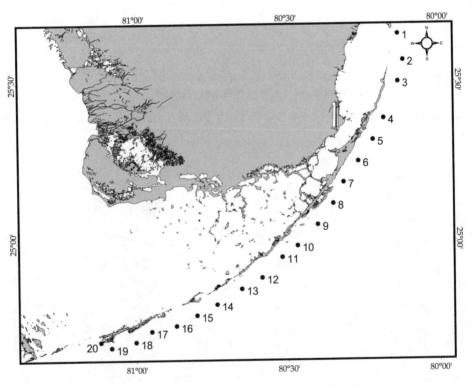

**FIGURE 13.52**   Station locations where samples were taken November 4, 1997.

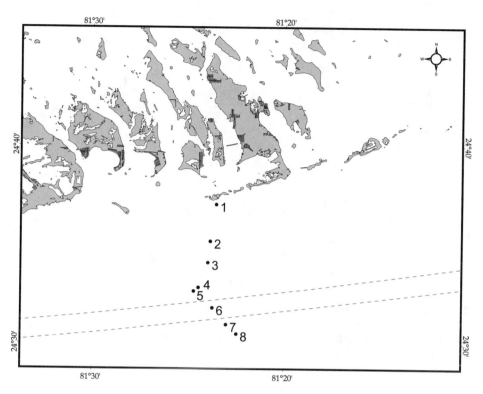

**FIGURE 13.53** Location of a transect from Big Pine Key out to the Gulf Stream. Dashed line indicates the boundary of a plume with reflectances greater than 5% (based upon satellite imagery of Dr. Richard Stumpf at http://coastal.er.usgs.gov/flbay/html/199603/19960323_14ref.tif).

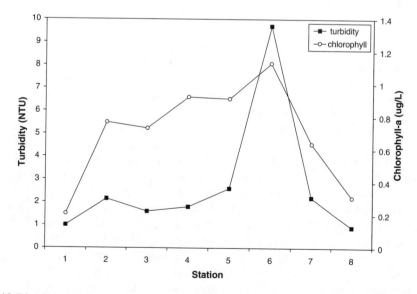

**FIGURE 13.54** Chlorophyll concentrations and turbidity (measured with a Lamotte 2008 turbidity meter) along the transect shown in Figure 13.53. The Looe Key reefs are located at stations 4 and 5.

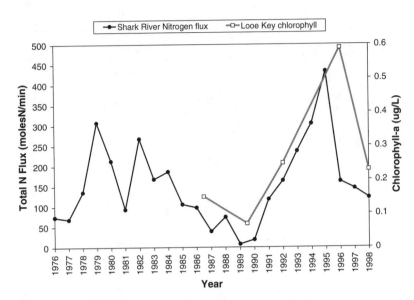

**FIGURE 13.55**   Annual nitrogen flux through the Shark River as measured at the S12A, S12B, S12C, and S12D structures (SFWMD data files) and average chlorophyll concentrations measured by Brian Lapointe near Looe Key in the lower Florida Keys by Lapointe et al. (2002). N flux is calculated monthly by multiplying N concentrations times water flow rates.

## POLICY IMPLICATIONS

It is ironic that South Florida has a high percentage of its area under environmental protection, yet major ecological degradation has been observed over the past two decades while under that "protection." Efforts are now underway to reverse some of this environmental decline.

### WATER QUANTITY

In the past century, large areas of South Florida have been drained in order to convert wetlands to agricultural and urban areas. This has resulted in a large decline in the amount of freshwater runoff into some coastal waters of South Florida and an alteration of those ecosystems. In an effort to reverse some of the ecological damage, efforts are now underway to restore some of the freshwater flow that existed a century ago (USACE/SFWMD, 1999).

In the case of Florida Bay, it has been hypothesized that the large seagrass dieoff that occurred in 1987 was a result of the reduced freshwater flow and resulting hypersalinity (Robblee et al., 1991). This has provided the rationale for pumping more freshwater from the Everglades system into Florida Bay. An examination of the data, however, does not support this hypothesis. While hypersaline water may generate physiological stress on seagrasses, there is very little temporal or spatial correlation between high salinity and the seagrass dieoff in 1987 in Florida Bay. McIvor et al. (1994) have documented that salinities up to 70 psu have occurred in Florida Bay in the past 50 years. The highest salinities observed in various studies conducted in Florida Bay, summarized by McIver et al. (1994), are presented in Table 13.2. The highest salinity observed during the time of the seagrass dieoff in 1987 was 46.6 psu, while no massive seagrass dieoffs were observed at previous times when salinities up to 70.0 psu were measured. The data of Fourqurean and Robblee (1999) show salinities in the 1980s and 1990s generally lower, and certainly not higher than those of the 1950s, 1960s, and 1970s.

Although the spatial distribution of salinity in 1989–1990 (Figure 13.56) has been used to explain the seagrass dieoff, freshwater flow through Taylor Slough and the South Dade Convey-ance System into Florida Bay was greatly reduced in 1989 and 1990 due to drought, compared

**TABLE 13.2**
**Highest Salinities Observed in Various Studies in Florida Bay**

| Years | Highest Salinity Observed (psu) |
|-------|--------------------------------|
| 1955–1957 | 70.0 |
| 1956–1958 | 57.8 |
| 1964–1965 | 65.3 |
| 1965–1966 | 67.7 |
| 1967–1968 | 44.8 |
| 1973–1976 | 66.6 |
| 1984–1985 | 44.0 |
| 1984–1986 | 50.0 |
| 1986–1987 | 46.6 |
| 1989–1990 | 59.0 |

*Source:* From McIvor, C.C. et al., in Davis, S.M. and Ogden, J.C., Eds., *Everglades: The Ecosystem and Its Restoration*, St. Lucie Press, Delray Beach, FL, 1994, pp. 117–146. With permission.

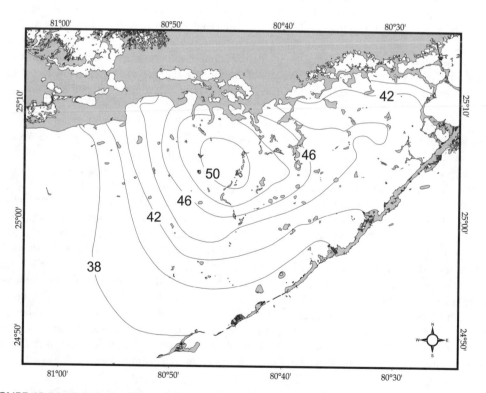

**FIGURE 13.56** Salinity isopleths redrawn from Fourqurean et al. (1992b).

to 1987 (a non-drought year) when the seagrass dieoff occurred (Figure 13.25). Apparently, no baywide salinity data for 1987 exist, but the bay was likely not as saline in 1987 as in 1989 and 1990 when the drought occurred and probably not much different from 1984 to 1986. There is no evidence that the seagrass died off in 1987 as a result of increasing salinity. The 1987 dieoff in fact occurred after freshwater runoff had increased for a number of years (Figure 13.25).

While some seagrass dieoff in 1987 did occur in areas of high salinity, much of the seagrass dieoff reported by Robblee et al (1991) occurred in areas (Figure 13.4) that had near-normal marine salinity even in the drought years of 1989 and 1990 (Figure 13.56). Using satellite imagery, Stumpf et al. (1999) have shown that even larger areas of seagrass to the west of Florida Bay (Figure 13.4) died off as well, as local fishers and other boaters had reported at the time (DeMaria, 1996). The fact that this area of seagrass dieoff has normal marine salinity and occasionally lower salinity because it is downstream of the Shark River outfall suggests that high salinity was not the dominant cause of the seagrass dieoff. The lack of significant spatial or temporal correlations between high salinity and seagrass dieoff suggests that simply pumping more freshwater into Florida Bay will not solve the ecological problem. Indeed, the data (Figure 13.25) show that SFWMD began pumping more N-rich freshwater into Florida Bay from the Everglades-agricultural system well before the seagrass dieoff.

## WATER QUALITY

Managers of a subtropical ecosystem such as South Florida are faced with a dilemma. South Florida provides an ideal climate for agriculture and living conditions for humans, but these activities inevitably generate nutrients. Subtropical marine ecosystems have strong thermoclines which lead to oligotrophic photic zones. The coral reef and seagrass ecosystems that develop under such oligotrophic conditions are very susceptible to elevated nutrient concentrations. Can agriculture and large human populations coexist with oligotrophic ecosystems? Current efforts are underway to partially restore the quantity of freshwater to coastal ecosystems of South Florida. Just as important is the quality of the water. As a result of various human activities, the freshwater today has much higher concentrations of nutrients than a century ago. In deciding how much freshwater should be pumped into the coastal waters of South Florida, the ecological benefits of more freshwater need to be balanced by the ecological costs of injecting more nutrients into these oligotrophic ecosystems.

The U.S. Army Corps of Engineers and SFWMD (1999) have proposed to pump even more freshwater from the Everglades agricultural system into Florida Bay as a way of ecologically "restoring" Florida Bay and reducing the algal blooms. This will increase the N load in the bay and it is hypothesized that this will increase the algal blooms even more. The spatial and temporal correlations between the freshwater runoff from agricultural lands in the Everglades and the algal blooms in Florida Bay support the hypothesis that the increased freshwater runoff is the cause of the algal blooms in Florida Bay. It has also been hypothesized that this increased N loading, which started in the early 1980s, led to the seagrass dieoff (Lapointe et al., 2002; Lapointe and Barile, 2001).

The importance of P in the P-limited Everglades has been recognized (Davis, 1994; Lodge, 1994), and efforts are underway to reduce P runoff from agricultural areas. Unfortunately, current restoration plans such as the Everglades Forever Act (State of Florida, 1994) and the Comprehensive Everglades Restoration Plan (USACE/SFWMD, 1999) ignore N as an important limiting nutrient in South Florida coastal waters. Much of the anthropogenic P is chemically scavenged before it makes it to the coastal waters of South Florida. Much of the N is not scavenged and does reach the coastal waters. This N does not generate large algal blooms in P-limited ecosystems in the east, such as south Biscayne Bay and eastern Florida Bay, but does in N-limited ecosystems in the west, such as central and western Florida Bay.

Any ecosystem restoration plans must consider the ecological effects of N in freshwater runoff. While much of the anthropogenic P does not make it to the coastal waters at the present time because of chemical scavenging, calcium carbonate can become saturated with P. The possibility that P scavenging will decline in the future as a result of continued P loading should be considered.

In conclusion, it is hypothesized that if more freshwater from the Everglades agricultural system is pumped into Florida Bay, as proposed, the algal blooms will increase and the ecological problems of Florida Bay will get worse, not better.

## The Reduction of Eutrophication

It is clear that eutrophication has drastically altered many South Florida coastal ecosystems. For ecosystem restoration to be successful, this eutrophication must be reduced. From a strictly ecological perspective, the best way to reduce eutrophication is to eliminate the nutrient source. Politically and economically, this may not be possible in South Florida because of the high human immigration rate and the importance of agriculture.

A second option is to reduce the flow of nutrient-rich freshwater into the coastal ecosystems. The benefits of reduced freshwater runoff must be balanced against the ecological benefits of lower salinity in estuaries for some organisms. Furthermore, as freshwater that accumulates on the South Florida peninsula during the rainy season must go somewhere, the requirement of flood control needs to be considered and balanced against the ecological harm of dumping excess nutrient-rich freshwater into the coastal ecosystems.

If we assume that the nutrient sources cannot be greatly reduced and freshwater runoff cannot or should not be reduced, other actions could be taken. In the case of south Biscayne Bay, which is P limited, a shift from direct canal injection to sheetflow of freshwater through marshes would be expected to reduce the P load and eutrophication there. In the case of Florida Bay, where N is the major problem, the current marshes between the SFWMD canals and Florida Bay do not appear to take out much of the N. A shift of freshwater runoff from Taylor Slough back to the east through the C111 canal would move the N-rich freshwater farther away from the P-rich seawater of western Florida Bay and reduce the algal blooms.

## ACKNOWLEDGMENTS

Funding for this research has been provided by the South Florida Ecosystem Restoration, Prediction, and Modeling Program of NOAA's Coastal Ocean Program, EPA's Florida Keys National Marine Sanctuary Water Quality Protection Program, and The Morris Family Foundation. Samples were collected under permits #19970016, 19970068, 19980013, 1999036, and 2000080 from Everglades National Park, permit #97S-072 from Florida Department of Environmental Protection, and Permit #FKNMS (UR)-02-96 from Florida Keys National Marine Sanctuary. I have been helped with much of the field sampling, sample processing, and data processing by Maiko Suzuki Ferro and Amie Gimon Davis. Fred Tooker has taken us throughout Florida Bay in his boat many times for sampling and has provided excellent insights into the bay and its history. Numerous others have helped with field sampling on various occasions, particularly Dave Forcucci. Discussions with Brian Lapointe over the years have provided many insights into South Florida ecosystems. The website maintained by Joe Boyer showing the data collected in the FIU water quality monitoring program has been extremely useful and stimulated some of the ideas presented here. The water flow, water level, and nutrient data were obtained from South Florida Water Management District hydrometeorological and water quality databases with help from Angela Chong. Some of the nutrient and chlorophyll data were collected by the Southeast Environmental Research Program at FIU with funding from the South Florida Water Management District through Everglades National Park (SFWMD/NPS Coop. Agreement C-7919 and NPS/SERP Coop. Agreement 5280-2-9017) and by the U.S. Environmental Protection Agency under Agreement #X994621-94-0. I would like to thank Bob Howarth, John Lamkin, Brian Lapointe, Charlie Yentsch, and Dick Zimmerman for reading and commenting on an earlier version of this manuscript.

## REFERENCES

Bacchus, S.T. 2002. The "ostrich" component of the multiple stressor model: undermining south Florida, in *The Everglades, Florida Bay, and Coral Reefs of the Florida Keys: An Ecosystem Sourcebook*, Porter, J.W. and Porter, K.G., Eds., CRC Press, Boca Raton, FL.

Bell, P.R.F. 1992. Eutrophication and coral reefs — some examples in the Great Barrier Reef lagoon, *Water Res.*, 26:553–568.

Bodeanu, N. 1992. Algal blooms and the development of the main phytoplanktonic species at the Romanian Black Sea littoral in conditions of intensification of the eutrophication process, pp. 891–906, in *Marine Coastal Eutrophication*, Vollenweider, R.A., Marchetti, R., and Viviani, R., Eds., Elsevier, Amsterdam.

Boesch, D.F., N.E. Armstrong, C.F. D'Elia, N.G. Maynard, H.N. Paerl, and S.L. Williams. 1993. *Deterioration of the Florida Bay Ecosystem: An Evaluation of the Scientific Evidence*, report to the Interagency Working Group on Florida Bay, U.S. Department of the Interior, National Park Service, Washington, D.C.

Bokuniewicz, H. and B. Pavlik. 1990. Groundwater seepage along a barrier island, *Biogeochemistry*, 10:257–276.

Bosence, D. 1989. Biogenic carbonate production in Florida Bay, *Bull. Mar. Sci.*, 44:419–433.

Boyer, J.N. and R.D. Jones. 1999. Effects of freshwater inputs and loading of phosphorus and nitrogen on the water quality of eastern Florida Bay, pp. 545–561, in *Phosphorus Biogeochemistry in Subtropical Ecosystems*, Reddy, K.R., O'Connor, G.A., and Schelske, C.L., Eds., Lewis Publishers, Boca Raton, FL.

Boyer, J.N., J.W. Fourqurean, and R.D. Jones. 1997. Spatial characterization of water quality in Florida Bay and Whitewater Bay by multivariate analyses: zones of similar influence, *Estuaries*, 20:743–758.

Brand, L.E. 1988. *Assessment of Plankton Resources and Their Environmental Interactions in Biscayne Bay, Florida*, report to Miami–Dade Department of Environmental Resources Management, 204 pp.

Brand, L.E. 1996. *The Onset, Persistence and Fate of Algal Blooms in Florida Bay*, presented at the 1996 Florida Bay Science Conference, Dec. 11, 1996, Key Largo, FL.

Brand, L.E., M.D. Gottfried, C.C. Baylon, and N.S. Romer. 1991. Spatial and temporal distribution of phytoplankton in Biscayne Bay, Florida, *Bull. Mar. Sci.*, 49:599–613.

Brand, L.E., M.S. Ferro, and A.G. Davis. 2001. Nutrient sources for algal blooms in Florida Bay, *Estuaries* (in review).

Burnison, B.K. 1980. Modified dimethyl sulfoxide (DMSO) extraction for chlorophyll analysis of phytoplankton, *Can. J. Fish. Aquatic Sci.*, 37:729–733.

Bush, P.W. and R.H. Johnston. 1988. Groundwater Hydraulics, Regional Flow and Groundwater Development of the Floridan Aquifer System in Florida and in parts of Georgia, South Carolina, and Alabama. U.S. Geological Survey Professional Paper 1403-C, Washington, D.C.

Butler, M.J., IV, J.H. Hunt, W.F. Herrnkind, M.J. Childress, R. Bertelsen, W. Sharp, T. Matthews, J.M. Field, and H.G. Marshall. 1995. Cascading disturbances in Florida Bay, USA: cyanobacteria blooms, sponge mortality, and implications for juvenile spiny lobsters *Panulirus argus*, *Mar. Ecol. Prog. Ser.*, 129:119–125.

Cable, J.E., G.C. Bugna, W.C. Burnett, and J.P. Chanton. 1996. Application of $^{222}$Rn and $CH_4$ for assessment of groundwater discharge to the coastal ocean, *Limnol. Oceanogr.*, 41:1347–1353.

Cambridge, J.L. and A.J. McComb. 1986. The loss of seagrass in Cockburn Sound, Western Australia. II. Possible causes of seagrass decline, *Aquatic Bot.*, 24:269–285.

Capone, D.G. and M.F. Bautista. 1985. A groundwater source of nitrate in nearshore marine sediments, *Nature*, 313:214–216.

Carlson, P.R., L.A. Yarbro, and T.R. Barber. 1994. Relationship of sediment sulfide to mortality of *Thalassia testudinum* in Florida Bay, *Bull. Mar. Sci.*, 54:733–746.

Cociasu, A., L. Dorogan, C. Humborg, and L. Popa. 1996. Long-term ecological changes in Romanian coastal waters of the Black Sea, *Mar. Poll. Bull.*, 32:32–38.

Compton, J.S. 1997. Origin and paleoceanographic significance of Florida's phosphorite deposits, pp. 195–216, in *The Geology of Florida*, Randazzo, A.F. and Jones, D.S., Eds., University Press of Florida, Gainesville.

Cooper, S.R. and G.S. Brush. 1991. Long-term history of Chesapeake Bay anoxia, *Science*, 254:992–995.

Corbett, D.R., J. Chanton, W. Burnett, K. Dillon, C. Rutkowski, and J. Fourqurean. 1999. Patterns of groundwater discharge into Florida Bay, *Limnol. Oceanogr.*, 44:1045–1055.

Corbett, D.R., K. Dillon, W. Burnett, and J. Chanton. 2000. Estimating the groundwater contribution into Florida Bay via natural tracers, $^{222}$Rn and $CH_4$, *Limnol. Oceanogr.*, 45:1546–1557.

Corredor, J.E., R.W. Howarth, R.R. Twilley, and J.M. Morell. 1999. Nitrogen cycling and anthropogenic impact in the tropical interamerican seas, *Biogeochemistry*, 46:163–178.

Cunningham, K.J., D.F. McNeill, L.A. Guertin, P.F. Ciesielski, T.M. Scott, and L. deVerteuil. 1998. New Tertiary stratigraphy for the Florida Keys and southern peninsula of Florida, *Geol. Soc. Am. Bull.*, 110:231–258.

Davis, R.W. 1997. Geology of the Florida coast, pp. 155–168, in *The Geology of Florida*, Randazzo, A.F. and Jones, D.S., Eds., University Press of Florida, Gainesville.

Davis, S.M. 1994. Phosphorus inputs and vegetation sensitivity in the Everglades, pp. 357–378, in *Everglades: The Ecosystem and Its Restoration*, Davis, S.M. and Ogden, J.C., Eds., St. Lucie Press, Delray Beach, FL.

Dawes, C.J., C. Uranowski, J. Andorfer, and B. Teasdale. 1999. Changes in the macroalgal taxa and zonation at the Content Keys, Florida, *Bull. Mar. Sci.*, 64:95–102.

DeGrove, J.M. 1984. History of water management in South Florida, pp. 22–27, in *Environments of South Florida: Present and Past II*, Gleason, P.J., Ed., Miami Geological Society.

DeKanel, J. and J.W. Morse. 1978. The chemistry of orthophosphate uptake from seawater onto calcite and aragonite, *Geochim. Cosmochim. Acta*, 42:1335–1340.

D'Elia, C.F., K.L. Webb, and J.W. Porter. 1981. Nitrate-rich groundwater inputs to Discovery Bay, Jamaica: a significant source of N to local coral reefs? *Bull. Mar. Sci.*, 31:903–910.

DeMaria, K. 1996. *Changes in the Florida Keys Marine Ecosystem Based Upon Interviews with Experienced Residents*, The Nature Conservancy and Center for Marine Conservation, Key West, FL. 105 pp.

Doering, P.H. and R.H. Chamberlain. 1998. Water quality in the Caloosahatchee estuary, San Carlos Bay and Pine Island Sound, Florida, pp. 229–240, in *Proc. of the Charlotte Harbor Public Conference and Technical Symposium*, Technical Report No. 98-02. South Florida Water Management District, West Palm Beach, FL.

Durako, M.J. 1994. Seagrass die-off in Florida Bay (USA): changes in shoot demographic characteristics and population dynamics in *Thalassia testudinum*, *Mar. Ecol. Prog. Ser.*, 110:59–66.

Dustan, P. and J.C. Halas. 1987. Changes in the reef-coral community of Carysfort Reef, Key Largo, Florida: 1974 to 1982, *Coral Reefs*, 6:91–106.

Enos, P., and R.D. Perkins. 1977. Quaternary Sedimentation in South Florida, *Geol. Soc. Am. Mem.*, 147, 198 pp.

EPA. 1992. *Water Quality Protection Program for the Florida Keys National Marine Sanctuary: Phase I Report*, final report submitted to the Environmental Protection Agency under Work Assignment 3-225, Contract No. 68-C8-0105. Battelle Ocean Sciences, Duxbury, MA, and Continental Shelf Associates, Inc., Jupiter, FL.

Fanning, K.A., R.H. Byrne, J.A. Breland II, P.R. Betzer, W.S. Moore, R.J. Elsinger, and T.E. Pyle. 1981. Geothermal springs of the west Florida continental shelf: evidence for dolomitization and radionuclide enrichment, *Earth Plant. Sci. Lett.*, 52:345–354.

Fanning, K.A., J.A. Breland, and R.H. Byrne. 1982. Radium-226 and radon-222 in the coastal waters of west Florida: high concentrations and atmospheric degassing, *Science*, 215:667–670.

Fourqurean, J.W. and M.B. Robblee. 1999. Florida Bay: a history of recent ecological changes, *Estuaries*, 22:345–357.

Fourqurean, J.W., G.V.N. Powell, and J.C. Zieman. 1992a. Relationships between porewater nutrients and seagrasses in a subtropical carbonate environment, *Mar. Biol.*, 114:57–65.

Fourqurean, J.W., J.C. Zieman and G.V.N. Powell. 1992b. Phosphorus limitation of primary production in Florida Bay: evidence from C:N:P ratios of the dominant seagrass *Thalassia testudinum*, *Limnol. Oceanogr.*, 37:162–171.

Fourqurean, J.W., R.D. Jones, and J.C. Zieman. 1993. Processes influencing water column nutrient characteristics and phosphorus limitation of phytoplankton biomass in Florida Bay, FL, USA: inferences from spatial distributions, *Estuar. Coast. Shelf Sci.*, 36:295–314.

Frankovich, T.A. and J.W. Fourqurean. 1997. Seagrass epiphyte loads along a nutrient availability gradient, Florida Bay, USA, *Mar. Ecol. Prog. Ser.*, 159:37–50.

Fraser, T.H. and W.H. Wilcox. 1981. Enrichment of a subtropical estuary with nitrogen, phosphorus and silica, pp. 481–498, in *Estuaries and Nutrients*, Neilson, B.J. and Cronin, L.E., Eds., Humana Press, Clifton, NJ.

Froelich, P.N., L.W. Kaul, J.T. Byrd, M.O. Andreae, and K.K. Roe. 1985. Arsenic, barium, germanium, tin, dimethylsulfide and nutrient biogeochemistry in Charlotte Harbor, Florida, a phosphorus-enriched estuary, *Estuar. Coast. Shelf Sci.*, 20:239–264.

Giblin, A.E. and A.G. Gaines. 1990. Nitrogen inputs to a marine embayment: the importance of groundwater, *Biogeochemistry*, 10:309–328.

Giesen, W.B.J.T., M.M. van Katwijk, and C. den Hartog. 1990. Eelgrass condition and turbidity in the Dutch Wadden Sea, *Aquatic Bot.*, 37:71–85.

Ginsburg, R.N. and E.A. Shinn. 1994. South Florida's environments are geological inheritances: the past is the key to the present, *Bull. Mar. Sci.*, 54:1075–1076.

Harding, L.W. and E.S. Perry. 1997. Long-term increase of phytoplankton biomass in Chesapeake Bay, 1950–1994, *Mar. Ecol. Prog. Ser.*, 157:39–52.

Havens, K.E., N.G. Aumen, R.T. James, and V.H. Smith. 1996. Rapid ecological changes in a large subtropical lake undergoing cultural eutrophication, *Ambio*, 25:150–155.

Hobbie, J. 1974. Ecosystems receiving phosphate wastes, pp. 252–270, in *Coastal Ecological Systems of the United States*, III, Odum, H.T., Copeland, B.J., and McMahan, E.A., Eds., The Conservation Foundation, Washington, D.C.

Hoffmeister, J.E. 1974. *Land from the Sea: The Geologic Story of South Florida*, University of Miami Press, 143 pp.

Howarth, R.W. 1988. Nutrient limitation of net primary production in marine ecosystems, *Ann. Rev. Ecol. Syst.*, 19:89–110.

Jensen, H.S., K.J. McGlathery, R. Marino, and R.W. Howarth. 1998. Forms and availability of sediment phosphorus in carbonate sand of Bermuda seagrass beds, *Limnol. Oceanogr.*, 43:799–810.

Justic, D., N.N. Rabalais, R.E. Turner, and Q. Dortch. 1995. Changes in nutrient structure of river-dominated coastal waters: stoichiometric nutrient balance and its consequences, *Estuar. Coast. Shelf Sci.*, 40:339–356.

Katz, B.G. 1992. *Hydrochemistry of the Upper Floridan Aquifer, Florida*, U.S. Geological Survey Water Resources Investigations Report 91-4196, Tallahassee, FL. 37 pp.

Kemp, W.M., R.R. Twilley, J.C. Stevenson, W.R. Boynton, and J.C. Means. 1983. The decline of submerged vascular plants in upper Chesapeake Bay: summary of results concerning possible causes, *Mar. Tech. Soc. J.*, 17:78–889.

Kitano, Y., M. Okumura, and M. Idogaki. 1978. Uptake of phosphate ions by calcium carbonate, *Geochem. J.*, 12:29–37.

Kohout, F.A. 1960. Cyclic flow of salt water in the Biscayne Aquifer of southeastern Florida, *J. Geophys. Res.*, 65:2133–2141.

Kohout, F.A. 1965. A hypothesis concerning cyclic flow of salt water related to geothermal heating in the Floridan Aquifer, *Trans. N.Y. Acad. Sci.*, 28:249–271.

Kohout, F.A. 1967. Groundwater flow and the geothermal regime of the Floridian Plateau, *Trans. Gulf Coast Assoc. Geol. Soc.*, 17:339–354.

Kohout, F.A., H.R. Henry, and J.E. Banks. 1977. Hydrogeology related to geothermal conditions of the Floridan Plateau, pp. 1–41, in *The Geothermal Nature of the Florida Plateau*, Smith, D.L. and Griffin, G.M., Eds., Florida Geological Survey Special Publ. No. 21.

Kreitman, A. and L.A. Wedderburn. 1984. Hydrogeology of South Florida, pp. 405–426, in *Environments of South Florida: Present and Past II*, Gleason, P.J., Ed., Miami Geological Society.

Kusek, K.M., G. Vargo, and K. Steidinger. 1999. *Gymnodinium breve* in the field, in the lab, and in the newspaper: a scientific and journalistic analysis of Florida red tides, *Contrib. Mar. Sci.*, 34:1–229.

Kuta, K.G. and L.L. Richardson. 1996. Abundance and distribution of black band disease on coral reefs in the northern Florida Keys, *Coral Reefs*, 15:219–223.

Lane, E. 1994. *Florida's Geological History and Geological Resources*, Florida Geological Survey Special Publ. No. 35, Tallahassee, FL. 64 pp.

Lapointe, B.E. 1987. Phosphorus and nitrogen-limited photosynthesis and growth of *Gracilaria tikvahiae* (Rhodophyceae) in the Florida Keys: an experimental field study, *Mar. Biol.*, 93:561–568.

Lapointe, B.E. 1989. Macroalgal production and nutrient relations in oligotrophic areas of Florida Bay, *Bull. Mar. Sci.*, 44:312–323.

Lapointe, B.E. and P.J. Barile. 2001. Seagrass die-off in Florida Bay: an alternative interpretation, *Estuaries* (in press).

Lapointe, B.E. and M.W. Clark. 1992. Nutrient inputs from the watershed and coastal eutrophication in the Florida Keys, *Estuaries*, 15:465–476.

Lapointe, B.E. and W.R. Matzie. 1997. *High Frequency Monitoring of Wastewater Nutrient Discharges and Their Ecological Effects in the Florida Keys National Marine Sanctuary*, final report of Special Study to the Water Quality Protection Program, U.S. Environmental Protection Agency, Marathon, FL, 64 pp.

Lapointe, B.E. and J.D. O'Connell. 1989. Nutrient-enhanced growth of *Cladophora prolifera* in Harrington Sound, Bermuda: eutrophication of a confined, phosphorus-limited marine ecosystem, *Estuar. Coast. Shelf Sci.*, 28:347–360.

Lapointe, B.E., M.M. Littler, and D.S. Littler. 1987. A comparison of nutrient-limited productivity in macroalgae from a Caribbean barrier reef and from a mangrove ecosystem, *Aquatic Bot.*, 28:243–255.

Lapointe, B.E., J.D. O'Connell, and G.S. Garrett. 1990. Nutrient couplings between on-site sewage disposal systems, groundwaters, and nearshore surface waters of the waters of the Florida Keys, *Biogeochemistry*, 10:289–307.

Lapointe, B.E., M.M. Littler, and D.S. Littler. 1992. Nutrient availability to marine macroalgae in siliciclastic versus carbonate-rich coastal waters, *Estuaries*, 15:75–82.

Lapointe, B.E., W.R. Matzie, and P.J. Barile. 2002. Biotic phase-shifts in Florida Bay and back reef communities of the Florida Keys: linkages with historical freshwater flows and nitrogen loading from Everglades runoff, in *The Everglades, Florida Bay, and Coral Reefs of the Florida Keys: An Ecosystem Sourcebook*, Porter, J.W. and Porter, K.G., Eds., CRC Press, Boca Raton, FL.

Lapointe, B.E., D.A. Tomasko, and W.R. Matzie. 1994. Eutrophication and trophic state classification of seagrass communities in the Florida Keys, *Bull. Mar. Sci.*, 54:696–717.

Larsson, U., R. Elmgren, and F. Wulff. 1985. Eutrophication and the Baltic Sea: causes and consequences, *Ambio*, 14:9–14.

Lee, T., E. Johns, D. Wilson, E. Williams, and N. Smith. 2002. Transport processes linking South Florida coastal ecosystems, in *The Everglades, Florida Bay, and Coral Reefs of the Florida Keys: An Ecosystem Sourcebook*, Porter, J.W. and Porter, K.G., Eds., CRC Press, Boca Raton, FL.

Ley, J. 1995. *C-111 Interim Construction Project*, South Florida Water Management District, West Palm Beach, FL.

Light, S.S. and J.W. Dineen. 1994. Water control in the Everglades: a historical perspective, pp. 47–84, in *Everglades: The Ecosystem and Its Restoration*, Davis, S.M. and Ogden, J.C., Eds., St. Lucie Press, Delray Beach, FL.

Littler, M.M., D.S. Littler, and B.E. Lapointe. 1988. A comparison of nutrient and light-limited photosynthesis in psammophytic versus epilithic forms of *Halimeda* (Caulerpales, Halimedaceae) from the Bahamas, *Coral Reefs*, 6:219–225.

Lodge, T.E. 1994. *The Everglades Handbook: Understanding the Ecosystem*, St. Lucie Press, Delray Beach, FL, 228 pp.

Maddox, G.L., J.M. Lloyd, T.M. Scott, S.B. Upchurch, and R. Copeland. 1992. *Florida's Ground Water Quality Monitoring Program Background Hydrogeochemistry*, Florida Geological Survey Special Publ. No. 34, Tallahassee, FL. 347 pp.

Mallinson, D.J., J.S. Compton, S.W. Snyder, and D.A. Hodell. 1994. Strontium isotopes and Miocene sequence stratigraphy across the northeast Florida Platform, *J. Sediment. Res.*, B64:392–407.

Maragos, J.E., C. Evans, and P. Holtbus. 1985. Reef corals in Kaneohe Bay six years before and after termination of sewage discharges, *Proc. Fifth Int. Coral Reef Congress*, 4:189–194.

McGlathery, K.J., R. Marino, and R.W. Howarth. 1994. Variable rates of phosphate uptake by shallow marine carbonate sediments: mechanisms and ecological significance, *Biogeochemistry*, 25:127–146.

McIvor, C.C., J.A. Ley, and R.D. Bjork. 1994. Changes in freshwater inflow from the Everglades to Florida Bay including effects on biota and biotic processes: a review, pp. 117–146, in *Everglades: The Ecosystem and Its Restoration*, Davis, S.M. and Ogden, J.C., Eds., St. Lucie Press, Delray Beach, FL.

McPherson, B.F. and R. Halley. 1997. *The South Florida Environment: A Region Under Stress*, U.S. Geological Survey Circular 1334, 61 pp.

Mee, L.D. 1992. The Black Sea in crisis: a need for concerted international action, *Ambio*, 21:278–285.

Meyers, J.B., P.K. Swart, and J.L. Meyers. 1993. Geochemical evidence for groundwater behavior in an unconfined aquifer, south Florida, *J. Hydrol.*, 148:249–272.

Millham, N.P. and B.L. Howes. 1994. Nutrient balance of a shallow coastal embayment: I. Patterns of groundwater discharge, *Mar. Ecol. Prog. Ser.*, 112:155–167.

Miller, J.A. 1997. Hydrogeology of Florida, pp. 69-88, in *The Geology of Florida*, Randazzo, A.F. and Jones, D.S., Eds., University Press of Florida, Gainesville.

Missimer, T.M. 1984. The geology of South Florida, pp. 385-404, in *Environments of South Florida: Present and Past II*, Gleason, P.J., Ed., Miami Geological Society.

Moffat, A.S. 1998. Global nitrogen overload problem grows critical, *Science*, 279:988–989.

Moore, W.S. 1996. Large groundwater inputs to coastal waters revealed by $^{226}$Ra enrichments, *Nature*, 380:612–614.

Moore, W.S. 1999. The subterranean estuary: a reaction zone of ground water and sea water, *Mar. Chem.*, 65:111–125.

Nehring, D. 1992. Eutrophication in the Baltic Sea, pp. 673-682, in *Marine Coastal Eutrophication*, Vollenweider, R.A., Marchetti, R., and Viviani, R., Eds., Elsevier, Amsterdam.

Nelson, J.E. and R.N. Ginsburg. 1986. Calcium carbonate production by epibionts on *Thalassia* in Florida Bay, *J. Sediment. Petrol.*, 56:622–628.

Nixon, S.W. 1995. Coastal marine eutrophication: a definition, social causes, and future concerns, *Ophelia*, 41:199–219.

Nordlie, F.G. 1990. Rivers and springs, pp. 392–425, in *Ecosystems of Florida*, Myers, R.L. and Ewel, J.J. Eds., University of Central Florida Press, Orlando.

Oberdorfer, J.A., M.A. Valentino, and S.V. Smith. 1990. Groundwater contribution to the nutrient budget of Tomales Bay, California, *Biogeochemistry*, 10:199–216.

Odum, H.T. 1953. *Dissolved Phosphorus in Florida Waters*, Florida Geological Survey Report of Investigations, pp. 1–40.

Ogden, J.C., J.W. Porter, N.P. Smith, A.M. Szmant, W.C. Jaap, and D. Forcucci. 1994. A long-term interdisciplinary study of the Florida Keys seascape, *Bull. Mar. Sci.*, 54:1059–1071.

Orth, R.J. and K.A. Moore. 1983. Chesapeake Bay: an unprecedented decline in submerged aquatic vegetation, *Science*, 222:51–53.

Parker, G. 1984. Hydrology of the pre-drainage system of the Everglades in southern Florida, pp. 28–37, in *Environments of South Florida: Present and Past II*, Gleason, P.J., Ed., Miami Geological Society.

Parkinson, R.W. 1989. Decelerating Holocene sea-level rise and its influence on southwest Florida coastal evolution: a transgressive/regressive stratigraphy, *J. Sediment. Petrol.*, 59:960–972.

Parsons, T.R., Y. Maita, and C.M. Lalli. 1984. *A Manual of Chemical and Biological Methods for Seawater Analysis*, Pergamon Press, New York, 173 pp.

Paul, J.H., J.B. Rose, J.Brown, E.A. Shinn, S. Miller, and S.R. Farrah. 1995a. Viral tracer studies indicate contamination of marine water by sewage disposal practices in Key Largo, Florida, *Appl. Environ. Microbiol.*, 61:2230–2234.

Paul, J.H., J.B. Rose, S. Jiang, C. Kellogg, and E. A. Shinn. 1995b. Occurrence of fecal indicator bacteria in surface waters and the subsurface aquifer in Key Largo, Florida, *Appl. Environ. Microbiol.*, 61:2235–2241.

Phlips, E.J. and S. Badylak. 1996. Spatial variability in phytoplankton standing crop and composition in a shallow inner-shelf lagoon, Florida Bay, Florida, *Bull. Mar. Sci.*, 58:203–216.

Phlips, E.J., T.C. Lynch, and S. Badylak. 1995. Chlorophyll *a*, tripton, color, and light availability in a shallow tropical inner-shelf lagoon, Florida Bay, USA, *Mar. Ecol. Prog. Ser.*, 127:223–234.

Phlips, E.J., S. Badylak, and T.C. Lynch. 1999. Blooms of the picoplanktonic cyanobacterium Synechococcus in Florida Bay, a subtropical inner-shelf lagoon, *Limnol. Oceanogr.*, 44:1166–1175.

Pitts, P.A. 1994. An investigation of near-bottom flow patterns along and across Hawk Channel Florida Keys, *Bull. Mar. Sci.*, 54:610–620.

Pitts, P.A. 1997. An investigation of tidal and nontidal current patterns in Western Hawk Channel, Florida Keys, *Cont. Shelf Res.*, 17:1679–1687.

Porter, J.W. and O.W. Meier. 1992. Quantification of loss and change in Floridian reef coral populations, *Am. Zool.*, 32:625–640.

Porter, J.W. et al. 2002. Detection of coral reef change by the Coral Reef Monitoring Program, *The Everglades, Florida Bay, and Coral Reefs of the Florida Keys: An Ecosystem Sourcebook*, Porter, J.W. and Porter, K.G., Eds., CRC Press, Boca Raton, FL.

Powell, G.V.N., W.J. Kenworthy, and J.W. Fourqurean. 1989. Experimental evidence for nutrient limitation of seagrass growth in a tropical estuary with restricted circulation, *Bull. Mar. Sci.*, 44:324–340.

Richardson, K. and B.B. Jorgensen. 1996. Eutrophication: definition, history and effects, pp. 1–19, in *Eutrophication in Coastal Marine Ecosystems: Coastal and Estuarine Studies*, Vol. 52, American Geophysical Union, Washington, D.C.

Richardson, L.L. 1997. Occurrence of the black band disease cyanobacterium on healthy corals of the Florida Keys, *Bull. Mar. Sci.*, 61:485–490.

Richardson, L.L., W.M. Goldberg, K.G. Kuta, R.B. Aronson, G.W. Smith, K.B. Ritchie, J.C. Halas, J.S. Feingold, and S.L. Miller. 1996. Florida's mystery coral-killer identified, *Nature*, 392:557–558.

Riggs, S.R. 1979. Phosphorite sedimentation in Florida: a model phosphogenic system, *Econ. Geol.*, 74:285–314.

Riggs, S.R. 1984. Paleoceanographic model of Neogene phosphorite deposition, U.S. Atlantic continental margin, *Science*, 223:123–131.

Robblee, M.B., T.B. Barber, P.R. Carlson, Jr., M.J. Durako, J.W. Fourqurean, L.M. Muehlstein, D. Porter, L.A. Yarbro, R.T. Zieman, and J.C. Zieman. 1991. Mass mortality of the tropical seagrass *Thalassia testudinum* in Florida Bay (USA), *Mar. Ecol. Prog. Ser.*, 71:297–299.

Roberts, H.H., T. Whelan, and W.G. Smith. 1977. Holocene sedimentation at Cape Sable, south Florida, *Sediment. Geol.*, 18:25–60.

Rudnick, D.T., Z. Chen, D.L. Childers, J.N. Boyer, and T.D. Fontaine. 1999. Phosphorus and nitrogen inputs to Florida Bay: the importance of the Everglades watershed, *Estuaries*, 22:398–416.

Ryther, J.H. and W.M. Dunstan. 1971. Nitrogen, phosphorus and eutrophication in the coastal marine environment, *Science*, 171:1008–1013.

Schomer, N.S. and R.D. Drew. 1982. *An Ecological Characterization of the Lower Everglades, Florida Bay and the Florida Keys*, U.S. Fish and Wildlife Service, Office of Biological Services, FWS/OBS-82/58.1, Washington, D.C., 246 pp.

Scoffin, T.P. 1970. The trapping and binding of subtidal carbonate sediments by marine vegetation in Bimini lagoon, Bahamas, *J. Sediment. Petrol.*, 40:249–273.

Scott, T.M. 1997. Miocene to Holocene history of Florida, pp. 57–67, in *The Geology of Florida*, Randazzo, A.F. and Jones, D.S., Eds., University Press of Florida, Gainesville.

SFWMD. 1992. *Surface Water Improvement and Management Plan for the Everglades*, South Florida Water Management District, West Palm Beach, FL, 471 pp.

SFWMD. 1995. *Florida Bay Adaptive Environmental Assessment and Management Workshop Report*, July 11–13, Duck Key, FL, South Florida Water Management District, 24 pp.

Shier, D.E. 1969. Vermetid reefs and coastal development in the Ten Thousand Islands, southwest Florida, *Geol. Soc. Am. Bull.*, 80:485–508.

Shinn, E.A., R.S. Reese, and C.D. Reich. 1994. *Fate and Pathways of Injection-Well Effluent in the Florida Keys*, U.S. Geological Survey Open File Report 94-276.

Short, F.T. 1987. Effects of sediment nutrients on seagrasses: literature review and mesocosm experiments, *Aquatic Bot.*, 27:41–57.

Short, F.T., W.C. Dennison, and D.G. Capone. 1990. Phosphorus-limited growth of the tropical seagrass *Syringodium filiforme* in carbonate sediments, *Mar. Ecol. Prog. Ser.*, 62:160–174.

Silberstein, K., A.W. Chiffings, and A.J. McComb. 1986. The loss of seagrass in Cockburn Sound, Western Australia. III. The effect of epiphytes on productivity of *Posidonia australis* Hook. F., *Aquatic Bot.*, 24:355–371.

Simmons, G.M. 1992. Importance of submarine groundwater discharge (SGWD) and seawater cycling to material flux across sediment/water interfaces in marine environments, *Mar. Ecol. Prog. Ser.*, 84:173–184.

Smith, D.L. and G.M. Griffin. 1977. *The Geothermal Nature of the Floridan Plateau*, Florida Bureau of Geology Special Publ. No. 21, 161 pp.

Smith, N.P. 1994. Long-term Gulf-to-Atlantic transport through tidal channels in the Florida Keys, *Bull. Mar. Sci.*, 54:602–609.

Smith, S.V., W.J. Kimmerer, E.A. Laws, R.E. Brock, and T. W. Walsh. 1981. Kaneohe Bay sewage diversion experiments: perspective on ecosystem responses to nutritional perturbation, *Pac. Sci.*, 35:279–385.

Smith, T.J., III, J.H. Hudson, M.B. Robblee, G.V.N. Powell, and P.J. Isdale. 1989. Freshwater flow from the Everglades to Florida Bay: a historical reconstruction based on fluorescent banding in the coral *Solenastrea bournoni*, *Bull. Mar. Sci.*, 44:274–282.

Snyder, G.H. and J.M. Davidson. 1994. Everglades agriculture: past, present, and future, pp. 85–115, in *Everglades: The Ecosystem and Its Restoration*, Davis, S.M. and Ogden, J.C., Eds., St. Lucie Press, Delray Beach, FL.

Sproul, C.R. 1977. Spatial distribution of groundwater temperature in South Florida, pp. 65-90, in *The Geothermal Nature of the Florida Plateau*, Smith, D.L. and Griffin, G.M., Eds., Florida Geological Survey Special Publ. No. 21. Tallahassee, FL.

Squires, A.P., H. Zarbock, and S. Janicki. 1998. Loadings of total nitrogen, total phosphorus and total suspended solids to Charlotte Harbor, pp. 187–200, in *Proc. of the Charlotte Harbor Public Conference and Technical Symposium*, Technical Report No. 98-02.

Stamm. D. 1994. *The Springs of Florida*, Pineapple Press, Sarasota, FL, 112 pp.

State of Florida, 1994. Everglades Forever Act, Florida Statutes Chapter 373.4592. Tallahassee, FL.

Stockman, K.W., E.A. Shinn, and R.N. Ginsburg. 1967. The production of lime mud by algae in south Florida, *J. Sediment. Petrol.*, 37:633–648.

Stumpf, R.P., M.L. Frayer, M.J. Durako, and J.C. Brock. 1999. Variations in water clarity and bottom albedo in Florida Bay from 1985 to 1997, *Estuaries*, 22:431–444.

Thayer, G.W., P.L. Murphey, and M.W. LaCroix. 1994. Responses of plant communities in western Florida Bay to the die-off of seagrasses, *Bull. Mar. Sci.*, 54:718–726.

Tomas, C.R., B. Bendis, and K. Johns. 1999. Role of nutrients in regulating plankton blooms in Florida Bay, pp. 323–337, in *The Gulf of Mexico Large Marine Ecosystem*, Kumpf, H., Steidinger, K., and Sherman, K., Eds., Blackwell Scientific, London.

Tomasko, D.A. and B.E. Lapointe. 1991. Productivity and biomass of *Thalassia testudinum* as related to water column nutrient availability and epiphyte levels: field observations and experimental studies, *Mar. Ecol. Prog. Ser.*, 75:9–17.

Tomasko, D.A. and B.E. Lapointe. 1994. An alternative hypothesis for the Florida Bay seagrass die-off, *Bull. Mar. Sci.*, 54:1086.

Top, Z., L.E. Brand, R.D. Corbett, W. Burnett, and J. Chanton. 2001. Helium as a tracer of groundwater input into Florida Bay, *J. Coastal Res.* (in press).

Turner, R.E. and N.N. Rabalais. 1991. Changes in Mississippi River water quality this century: implications for coastal food webs, *Bioscience*, 41:140–147.

Turner, R.E. and N.N. Rabalais. 1994. Coastal eutrophication near the Mississippi river delta, *Nature*, 368:619–621.

Upchurch, S.B. and A.F. Randazzo. 1997. Environmental geology of Florida, pp. 217–249, in *The Geology of Florida*, Randazzo, A.F. and Jones, D.S., Eds., University Press of Florida, Gainesville.

USACE and SFWMD. 1999. *Central and Southern Florida Project Comprehensive Review Study*, Vols. 1–10, United States Army Corps of Engineers and South Florida Water Management District.

Valiela, I, J. Costa, K. Foreman, J.M. Teal, B. Howes, and D. Aubrey. 1990. Transport of groundwater-borne nutrients from watersheds and their effects on coastal waters, *Biogeochemistry*, 10:177–197.

Vitousek, P.M. and R.W. Howarth. 1991. Nitrogen limitation on land and in the sea: how can it occur? *Biogeochemistry*, 13:87–115.

Vollenweider, R.A. 1992a. Coastal marine eutrophication: Principles and control, pp. 1–20, in *Marine Coastal Eutrophication*, Vollenweider, R.A. Marchetti, R., and Viviani, R., Eds., Elsevier, Amsterdam.

Vollenweider, R.A. 1992b. Eutrophication, structure and dynamics of a marine coastal system: results of a ten-year monitoring along the Emilia-Romagna coast (Northwest Adriatic Sea), pp. 63–106, in *Marine Coastal Eutrophication*, Vollenweider, R.A., Marchetti, R., and Viviani, R., Eds., Elsevier, Amsterdam.

Wang, J.D., J. van de Kreeke, N. Krishnan, and D.W. Smith. 1994. Wind and tide response in Florida Bay, *Bull. Mar. Sci.*, 54:579–601.

Wanless, H.R. and M.G. Tagett. 1989. Origin, growth and evolution of carbonate mudbanks in Florida Bay, *Bull. Mar. Sci.*, 44:454–489.

Wanless, H.R., R.W. Parkinson, and L.P. Tedesco. 1994. Sea level control on stability of Everglades wetlands, pp. 199–223, in *Everglades: The Ecosystem and Its Restoration*, Davis, S.M. and Ogden, J.C. Eds., St. Lucie Press, Delray Beach, FL.

Warzeski, E.R., K.J. Cunningham, R.N. Ginsburg, J.B. Anderson, and Z.-D. Ding. 1996. A Neogene mixed siliciclastic and carbonate foundation for the Quaternary carbonate shelf, Florida Keys, *J. Sed. Res.*, 66:788–800.

Whitaker, F.F. and P.L. Smart. 1990. Active circulation of saline ground waters in carbonate platforms: evidence from the Great Bahama Bank, *Geology*, 18:200–203.

Whitaker, F.F. and P.L. Smart. 1993. Circulation of saline ground water in carbonate platforms: a review and case study from the Bahamas, pp. 113–132, in *Diagenesis and Basin Development*, Horbury, A.D. and Robinson, A.G., Eds., American Association of Petroleum Geology, Tulsa, OK.

Whitaker, F.F. and P.L. Smart. 1997. Hydrogeology of the Bahamian Archipelago, pp. 183–216, in *Geology and Hydrogeology of Carbonate Islands*, Vacher, H.L. and Quinn, T.M., Eds., Elsevier, Amsterdam.

Yarbro, L.A. and P.R. Carlson. 1998. Seasonal and spatial variation of phosphorus, iron and sulfide in Florida Bay sediments. 1998 Florida Bay Science Conference, Miami, FL.

Zieman, J.C., J.W. Fourqurean, and R.L. Iverson. 1989. Distribution, abundance and productivity of seagrasses and macroalgae in Florida Bay, *Bull. Mar. Sci.*, 44:292–311.

Zieman, J.C., R. Davis, J.W. Fourqurean, and M.B. Robblee. 1994. The role of climate in the Florida Bay seagrass die-off, *Bull. Mar. Sci.*, 54:1088.

# 14 Linkages Between the South Florida Peninsula and Coastal Zone: A Sediment-Based History of Natural and Anthropogenic Influences

*Terry A. Nelsen, Ginger Garte, and Charles Featherstone*
NOAA-AOML-OCD

*Harold R. Wanless*
Department of Geological Sciences, University of Miami

*John H. Trefry, Woo-Jun Kang, and Simone Metz*
Division of Marine and Environmental Systems, Florida Institute of Technology

*Carlos Alvarez-Zarikian, Terri Hood, Peter Swart, and Geoffrey Ellis*
Division of Marine Geology and Geophysics, RSMAS, University of Miami

*Pat Blackwelder*
NOVA Southeastern University Oceanographic Center
Division of Marine Geology and Geophysics, RSMAS, University of Miami

*Leonore Tedesco, Catherine Slouch, Joseph F. Pachut, and Mike O'Neal*
Department of Geology, Indiana University–Purdue University at Indianapolis

## CONTENTS

# INTRODUCTION

Peninsular Florida is a broad carbonate platform (Parker and Cooke, 1944) of which South Florida is a partly inundated portion that is well removed from most continental influences (Wanless and Tagett, 1989). The subtle topographic features of South Florida that were formed during the Pliocene and Pleistocene include the Everglades depression which extends through the center of South Florida and carries freshwater flow southward from Lake Okeechobee to the sea via the Shark River Slough (SRS) complex (Figure 14.1). This depression is bounded by exposed limestone ridges to the west (Pliocene) and east (Pleistocene). To the south, earlier Holocene buildups define a complex of shallow bays, lakes, and channels, many of which are infilled with mangrove swamp deposits and fringed by mangrove forests. Within these mangrove forests tidal rivers serve as conduits of freshwater runoff from the central Everglades and as pathways of landward transport of tidal, winter storm, and hurricane water as well as suspended sediment. Holocene sediments have built coastlines and marine banks which define the present coastal complex, and a natural coastal dam of mangrove peat and storm-levee marl inhibits, but does not block, saline intrusion into the Everglades depression.

The transition of the South Florida landscape into the coastal zone is gradual with near-coastal elevations averaging <2 m while seaward of the shoreline Florida Bay has a mean depth of only ~2 m. Such low relief is conducive to bidirectional exchange of materials in contrast to more traditional unidirectional seaward transport experienced in most of the world's other coastal estuaries.

Seaward and south of peninsular Florida is the shallow, subtropical estuary known as Florida Bay. Covering ~2200 km$^2$, it forms an anastomozing complex of Holocene banks of carbonate mud sediments which subdivide Florida Bay into smaller bays or lakes. Sediment thickness varies from 3 to 4 m at the western margin to about 1 m in the northeastern portion. These sediments are a mix of carbonate, organic, and siliceous materials of biogenic origin which are forming and accumulating

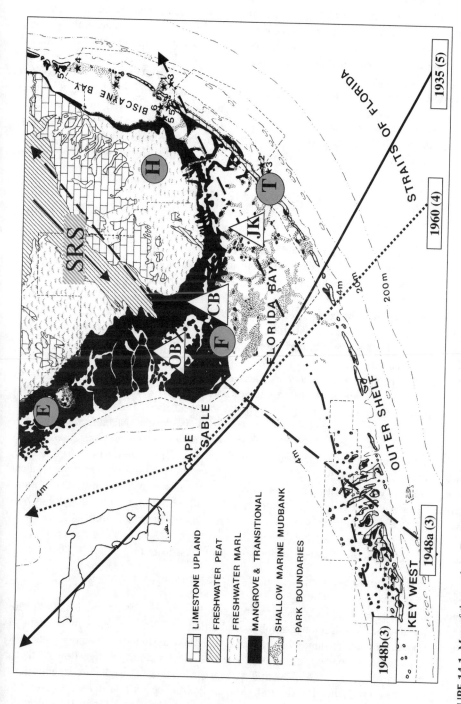

**FIGURE 14.1** Map of the southern Everglades and Florida Bay illustrating vegetative, rock, and soil types; Shark River Slough (SRS), Oyster Bay (OB), Coot Bay (CB), and Jimmy Key (JK) core sites; Everglades (E), Homestead (H), Flamingo (F), and Tavernier (T) rain gauge stations; and paths of the Labor Day Hurricane (1935), two unnamed 1948 hurricanes (1948a,b) and Hurricane Donna (1960) with category strengths in parentheses.

along with minor quartzose sands that are being supplied by southward littoral drift and by dissolution of local limestone (Wanless et al., 1989; Frederick, 1994). Whitewater Bay, behind Cape Sable, is 1 to 2 m deep and generally has little sediment cover. However, a few organic-rich carbonate banks occur adjacent to the lower Shark River in an area known as Oyster Bay.

In the absence of long-term monitoring studies and archived documentation, our objective was to examine the sedimentary record to decipher the area's history of natural and anthropogenic changes. To accomplish this, we chose three regions representative of the South Florida coastal zone: (1) Oyster Bay, adjacent to Shark River, the seaward outflow of the dominant regional freshwater source; (2) Jimmy Key, located in central Florida Bay; and (3) Coot Bay, an isolated bay neither directly influenced by Shark River nor central Florida Bay (Figure 14.1). Each coring site was selected because of rapid recent sediment accumulation and well-preserved, finely laminated stratification. The latter can be compared to the pages of a history book upon which nature and man potentially can record their influences. To deconvolute this record we must understand the story told by biological, chemical, and physical evidence contained within these sediments.

## SAMPLING AND METHODS

### Site Selection, Quality Control, and Sampling

An interpretable historical record requires undisturbed sediment sequences in which to establish a meaningful temporal context. Coring sites were selected after analysis of: (1) sequential aerial photographs (1927–present) which helped identify sites that had been free of macrovegetation since the time of the earliest aerial photography; and (2) burrowing activity and published information on the growth of shallow marine banks and deltas (Wanless and Tagett, 1989). From this, subsequent reconnaissance coring sought to verify that sites were areas of active sediment accumulation, had well-preserved stratification (minimal bioturbation activity), and maintained barebottom environment through the sequence (e.g., no periods of seagrass colonization). Cores opened in the field guided selection of unopened cores that were then screened in the laboratory for quality of stratification (X-radiography), with the most promising cores age dated ($^{210}$Pb) to determine whether an interpretable historical record was recovered. Of the many locations sampled, only locations satisfying these criteria were resampled by obtaining four to six closely spaced (<2 m diameter separation) cores at each site using 10-cm PVC and 7.6-cm aluminum tubes. Cores used in this study were X-radiographed, opened, described, and photographed for evaluation of sediment types and structures. Subsampling intervals were 1 to 3 cm and based on variations or breaks in sediment type and/or X-radiography boundaries and not simply by centimeter-by-centimeter intervals. Samples were collected and allocated for age dating, metals, pollen, isotopes, mineralogy, and biostratigraphy from the same stratigraphic interval to ensure subsequent comparability.

### Geochemical and Radiometric Sample Processing and Assessment

Sediments were analyzed for total organic C, $CaCO_3$, Al, Pb, and water content and were age dated using excess $^{210}$Pb and $^{137}$Cs techniques.

After inorganic carbon removal (HCl), sediment total organic C was determined using a Carlo-Erba NA1500 nitrogen/carbon/sulfur analyzer by combustion of 10-mg samples in tin cups at 1000°C. The sediment carbonate content was determined using the gasometric technique of Schink et al. (1978). Total sediment Al concentrations were determined, after complete digestion, by flame atomic absorption spectrometry as described in Trefry and Metz (1984), and Pb concentrations were obtained using a Perkin-Elmer ELAN 500 inductively coupled plasma mass spectrometer. Analytical precision was determined using replicate samples with levels better than 10% for each element. Accuracy was determined by analysis of certified standard reference materials available from the National Institute of Standards and Technology (NIST) and the National Research Council of Canada. Concentrations of all analytes were within certified limits.

Sediment accumulation rates and age determinations were measured from excess [210]Pb and [137]Cs geochronologies. The gamma activities of [210]Pb at 46.5 keV, [214]Pb at 351.8 keV, and [137]Cs at 661.6 keV were determined using a Princeton Gamma-Tech intrinsic germanium detector (Cutshall et al., 1983). Counting efficiencies and accuracy were determined by analyzing SRM 4350B, a NIST sediment sample calibrated for environmental radioactivity. The sediment accumulation rate was determined using the Constant Initial Concentration (CIC) model for excess [210]Pb as commonly used in marine and estuarine sediments (Koide et al., 1973; Goldberg et al., 1979; Finney and Huh, 1989). An age for each layer in each core was obtained by dividing cumulative dry mass of sediment by the sediment accumulation rate determined from data for excess [210]Pb. Sediment accumulation rates consider burial of only dry sediment and are expressed in $g/cm^2/yr$. These rates were normalized for varying water content in the sediment and, in some cases, enhanced differences among sites. The [137]Cs profile was used to help verify the dates derived via the CIC model with excess [210]Pb. Cesium-137 is a fission product introduced to the environment during nuclear weapons testing; the first appearance of [137]Cs occurred in about 1950 and maximum fallout during 1963 (Peirson, 1971). When sediments are not disturbed by biological or physical mixing, dates determined using excess [210]Pb should compare well with those determined using [137]Cs.

## STABLE ISOTOPES

Adult ostracod valves (*Malzella floridana, Peratocytheridea setipunctata*) and foraminifer tests (*Ammonia parkinsoniana typica*) were isolated from the >63 µm sediment fraction by wet-sieving and were dried. Well-preserved specimens were cleaned in a distilled water bath to remove adhering debris. Then 6 to 8 ostracod valves or foraminifer tests were immersed in methanol in "copper boats" and powdered. The carbonate samples were processed by an automated carbonate device attached to a Finnigan-MAT 251 mass spectrometer. The external precision (0.02‰ for $\delta^{13}C$) was calculated from replicate analyses of the internal laboratory calcite standard. The stable isotope composition of the organic C fraction of the sediment was analyzed for the Oyster Bay core. Whole sediment samples were air dried and acidified (5% HCl) to remove calcium carbonate. Approximately 1-mg samples were analyzed by an Automated Nitrogen and Carbon Analyzer interfaced to a 20/20 magnetic-sector stable-isotope mass spectrometer (ANCA, Europa Scientific, Ltd.) for $\delta^{13}C$ composition. The analytical precision of the method, as calculated from replicate analyses of standard reference materials, was 0.1‰ for $\delta^{13}C$ composition. All isotopic data are reported in per mil (‰) units relative to the Pee Dee Belemnite (PDB) reference and are corrected for the usual isobaric interferences.

## BENTHIC MICROFAUNA

Core sediment intervals were subsampled and wet-sieved using a 63-µm mesh, and the coarse fraction was reserved for microfaunal analysis (benthic ostracods and foraminifera). The Oyster Bay core was sampled at 1-cm intervals and the Jimmy Key core at 3-cm intervals. High organic content necessitated the use of ashing and 30% hydrogen peroxide treatments of the Oyster Bay sediments, while Jimmy Key sediments did not require this pretreatment. Subsample weights were determined to calculate total population abundances expressed as tests or valves per gram dry sediment weight. In cases where abundances were sufficiently high, 300 tests or valves were isolated from each interval and identified to species level when possible. Individuals of selected species were prepared for stable isotope measurements as noted above.

Biogenic community structure shifts were evaluated using relative and absolute abundance and the Shannon–Weiner Diversity Index (SWDI). The SWDI is defined as:

$$H = -\sum_{i=1}^{N} p(i) \ln p(i)$$

where $H$ is the measure of diversity, $N$ is the number of species counted, and $p(i)$ is the proportion of the total number of individuals belonging to the $i$th species (MacArthur, 1983). This index incorporates a measure of species evenness as well as number and is sensitive to the presence of rare species.

## POLLEN

Pollen subsamples from the Oyster Bay core were concentrated following standard procedures (Wood et al., 1996) that included removal of silica (HF) and plant material (200-µm screen) as well as acetolization and suspension in silicon oil before slide mounting for identification. All subsamples were spiked with *Lycopodium* yield tracer for calculation of absolute pollen concentrations (Stockmarr, 1971; Benninghoff,1962) which allowed determination of true temporal pollen changes, not relative changes caused by increases or decreases in other taxa or changes in sedimentation rate. Sediment bulk densities were calculated for subsequent count normalization.

Pollen grains were identified under light and fluorescence microscopy by comparison to a reference set of tropical pollen consisting of herbarium samples from Everglades National Park (ENP) and a computerized database system. The sparse pollen abundances found in this study were consistent with the findings of others for South Florida (Riegel, 1965; Willard and Holmes, 1997). This necessitated, in some cases, lower pollen counts (i.e., 100 grains) than the more ideal counts of 300 grains.

Riegel (1965) completed the first extensive palynological analysis of ENP to determine the relationship between pollen incorporated into surface sediments and the plant communities of the same area. His dataset consists of the percent frequency of 92 pollen and spore categories at 102 surface sample sites, representing six of the ten major vegetation communities in ENP. His data were presented as tables and abundance histograms (Riegel, 1965, Appendix II) that delineated trends in pollen distribution throughout the Everglades. These data (limited herein to his *in situ* depositional localities) and his conclusions are the basis for the modern analog pollen communities utilized in this study.

We have focused on modern analog samples taken from the mangrove swamp because the Oyster Bay site has been situated within the mangrove swamp throughout the time period of this study. To establish the local pollen assemblage for the mangrove swamp from Riegel's data, species composition and abundance data for the 31 sites within the mangrove swamp were analyzed to determine the average pollen frequency. For each site, Riegel's percent frequency data were summed for five important taxa associations. Mean and standard deviations were calculated and then used to define the reference local-pollen assemblage and its variability. Comparisons of downcore pollen assemblages were made to this reference assemblage.

## CLIMATOLOGICAL AND HYDROLOGICAL DATA

Salinity data were obtained from both the HisSal05 (Orlando et al., 1997) and the Florida International University/SERP (Boyer et al., 1997) databases. The former contains salinity data collected since 1955 and, although discontinuous, provides the best historical picture of regional salinity available. The FIU/SERP database provided continuous monthly data from September 1992 to March 1998. Both databases were sorted for stations proximal to our sediment coring sites. Freshwater flow data for SRS were obtained from the U.S. Geological Survey (USGS) which provided continuous monthly flow values from 1939 to present. Rainfall data were obtained from NOAA's National Climate Data Center for Everglades, Flamingo, Homestead, and Tavernier stations (Figure 14.1) with continuous monthly data extending from the present back to 1926, 1951, 1911, and 1936, respectively. Data analysis utilized measures of central tendency (i.e., mean, standard deviation), trend analysis (trend = $n - N/\sigma$, where $n$ = a single datum point from a dataset where $N$ and $\sigma$ = mean and standard deviation, respectively), and statistical correlations.

# RESULTS

## THE SEDIMENTS

The laminated sequences recovered were preserved in very shallow oxygenated waters with an extremely soft bottom, the latter being inhospitable to macrobenthic organisms which could cause bioturbation. These areas were not associated with seagrasses or other macrofloral communities throughout the history of the cores, although living seagrass communities sometimes occurred within 10 m of the sample sites.

Stratification occurs as irregular alternations of fine laminae dominated by differing amounts of fine quartz sand, fine-to-coarse carbonate silt derived from disaggregation of skeletal grains, organic detritus, and opaline silica diatom frustules and sponge spicules. In northwestern Whitewater Bay, at the Oyster Bay site, laminae are alternations of mangrove organics, carbonate silt, and fine quartz. The mangrove organic detritus is introduced from tidal release from mangrove swamps and eroding channel and bay flanks. The quartz and much of the carbonate silt at this site are derived as winter storm resuspensions from the offshore marine environment. Within central Florida Bay, laminae at the Jimmy Key site are dominantly carbonate with textural, organic content and carbonate grain-type differences producing a distinctly laminated sequence. the carbonate silt and clay at this site are derived from local surficial production and from recycling bank deposits on eroding northern and eastern flanks. Diatoms are derived from local benthic organic mats, settling from planktonic suspension, and recycling. Silic-clastic clays and detrital dolomite comprise a small percentage of the sediments. Although individual sub-millimeter-thick laminae differ greatly in texture and composition, the sequences, when sampled on a centimeter scale, display remarkably uniform sediment mineralogical composition and texture. This is important, as the different compositional elements represent different sources and different styles of sediment transport. That these parameters remain uniform reflects the uniformity of the combined processes through time and eliminates changing sediment regimes as a factor in observed parameters.

## GEOCHRONOLOGY

Vertical profiles for the activities of excess $^{210}$Pb and $^{137}$Cs were used to obtain sediment accumulation rates (cm/yr), to provide a useful time-vs.-depth perspective for identification of erosional/depositional events and for placing other data records into a temporal context.

In the Oyster Bay core, the sediment accumulation rate from the profile for excess $^{210}$Pb (Figure 14.2A) was approximately 1.1 cm/yr (0.6 g/cm$^2$/yr). The high-resolution profile in this core shows maximum $^{137}$Cs activity levels at 31 cm and no detectable $^{137}$Cs below 50 cm (Figure 14.2B). This pattern is consistent with peak inputs of $^{137}$Cs to the environment during 1963 and no inputs prior to 1950. When sediment ages, determined from data for excess $^{210}$Pb, are compared with sediment depth in the $^{137}$Cs profile, a good fit with the 1950 and 1963 periods was observed (Figure 14.2B). The Coot Bay core site also shows a well-defined profile for excess $^{210}$Pb that yields a sediment accumulation rate of 0.4 cm/yr (0.15 g/cm$^2$/yr) and good agreement with the vertical profile for $^{137}$Cs (Figure 14.2C,D). At Jimmy Key, the excess $^{210}$Pb profile for the upper 39 cm of the core yields a calculated sediment accumulation rate of about 1.0 cm/yr (~0.78 g/cm$^2$/ yr), again in good agreement with the $^{137}$Cs profile (Figure 14.2E,F). These relatively good fits (±2 to 3 years) for two separate geochronological methods support the validity of the sediment accumulation rates and associated ages for these cores and thus provide a solid temporal context for other data presented below.

In addition to providing a time line and accumulation-rate information, fine structural features were observed in the excess $^{210}$Pb and $^{137}$Cs profiles, including sediment erosion/deposition bands that are tied by year to hurricane events. These discontinuities appear as lower levels of excess $^{210}$Pb than would be expected from the overall trend. These lower levels are most likely a result of the combined effects of losing sediment from that site during these storms and input of sediment with lower levels of excess $^{210}$Pb (older sediment or organic-rich debris) from upland or other source sites. Specifically, the vertical profile for excess $^{210}$Pb for the Oyster Bay core shows three discontinuities that correspond (within ±3 yr) to 1960, 1948, and 1936 (Figure 14.2A). In the more open

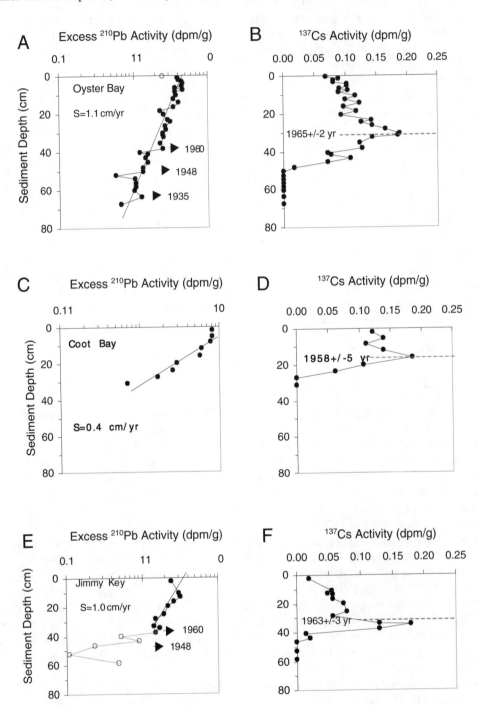

**FIGURE 14.2** Excess $^{210}$Pb and $^{137}$Cs profiles used to establish sediment geochronology for Oyster Bay (A, B), Coot Bay (C, D), and Jimmy Key (E, F) cores.

waters of Florida Bay at Jimmy Key, the discontinuities found at ~1960 and ~1948 (Figure 14.2E) are far more accentuated than at Oyster Bay. In contrast, in the sheltered environment of Coot Bay, no discontinuity events were observed in the excess $^{210}$Pb profile (Figure 14.2C). The dates of these discontinuities correspond to Hurricane Donna (1960), an unnamed storm in 1948, and the Great Labor Day Hurricane of 1935 (Figure 14.1). The lack of discernible large-scale disturbances of the sediment column since 1960 enables us to study a relatively uninterrupted sediment column over the past 40 years. A critical segment of the Oyster Bay core deposited during this period is shown in Figure 14.3A and contains both the discontinuity shown in Figure 14.2A and the subsequent strata.

## SEDIMENT COMPOSITION

In order to place the sedimentary sequence in context, it is important to discriminate between local production and sediment recycling. A large proportion of the sediment in this study area is transported into the site from seaward or landward sources, or derived from reworking of previous sediment deposits near the study sites. The Everglades drainage basin is not a source for clay or detrital dolomite; most of the clay and dolomite are derived from marine transport from the north (Manker and Griffin, 1969). The presence of a small percentage of quartz, clay, and detrital dolomite minerals in the silt- and clay-size fractions demonstrates such transport.

The three major components of sediments from the study areas are carbonate, aluminosilicates, and organic matter. Whole-core averages show an increasing $CaCO_3$ content from Oyster Bay ($57 \pm 5\%$) to Coot Bay ($71 \pm 7\%$) to Jimmy Key ($83 \pm 2\%$) with a concurrent inverse trend for both Al ($1.6 \pm 0.3\%$, $1.1 \pm 0.5\%$, $0.8 \pm 0.1\%$, respectively) and organic C ($7.1 \pm 1.3\%$, $4.5 \pm 2.2\%$, $1.8 \pm 0.3\%$, respectively). Al concentrations closely parallel that of organic C in the sediment and decrease by a factor of two from the Oyster Bay to the Coot Bay sites, and by a factor of four to

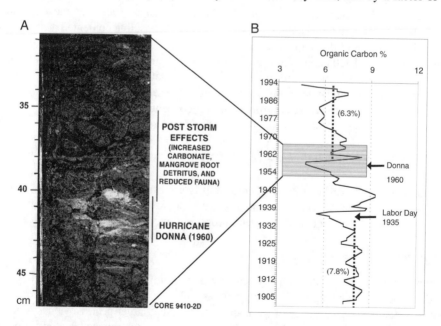

**FIGURE 14.3** Effects of Hurricane Donna on layering and organic C in Oyster Bay. (A) Photograph of a section of the Oyster Bay core including the interval dated at 1960. The light areas are carbonate-rich sediment and reflect onshore influx of carbonate sediment by Hurricane Donna. (B) Organic C profile of the Oyster Bay core. The Hurricane Donna interval has a core-wide low in organic C, followed by a post-storm peak in organic C. The Labor Day Hurricane (1935) intervals exhibit a similar pattern.

the Jimmy Key site. Correlation coefficients for Al vs. organic C were calculated for these three sites and indicated a progressive decline from Coot Bay (r = 0.91) to Oyster Bay (r = 0.86) to Jimmy Key (r = 0.42). The close covariance of Al with organic C in the Oyster and Coot Bay cores can best be explained by erosion of proximal coastal mangrove peat providing fine mangrove detritus. This is co-transported along with clay and quartz previously concentrated by dissolution of storm-derived carbonate-rich marine sediments. This correlation is weaker at the Jimmy Key site because it is more distal from the mangrove environments than either of the other sites.

The high-resolution organic C records from all three coring sites contain distinctly different trends. In the Oyster Bay core (Figure 14.3a), the largest excursions in organic C coincide with two of the major hurricane events to impact this site and have significantly lower values (at the 95% confidence level) after the mid-1950s than prior to the mid-1930s (Figure 14.3B). The later transition is more evident as a trend-analysis (Figure 14.4A). This trend demonstrates a clear decline in the sediment organic C content between approximately the first and second half of this century punctuated only by excursions associated with hurricane events. At Coot Bay organic C values increase upcore (Figure 14.4B) while at Jimmy Key the organic C record is more segmented (Figure 14.4C). Specifically, three segments exist at Jimmy Key: (1) a gradual increase from the bottom of the core (about the turn of the century) upward to about the early to mid-1930s; (2) gradual decrease from this point to about the mid-1970s; (3) thereafter to the present, a large increase to the top of the core in the mid-1990s. The significance of these trends, relative to natural and anthropogenic influences, will be discussed below.

The historical records for concentrations of total Pb at our three study sites are shown in Figure 14.5A–C. In sediments deposited prior to 1940 at Oyster Bay and Jimmy Key and prior to 1930 at Coot Bay, concentrations of Pb correlate well with concentrations of Al (r = 0.94). This relationship supports the importance of terrigenous inputs of Pb to the natural system and uniformity in the natural Pb content baseline (pre-anthropogenic) of the aluminosilicate clays (Pb/Al = $2.7 \times 10^{-4}$). After these dates, the Pb/Al ratio changes such that most points exhibit higher levels of Pb at a given Al content than the natural Pb baseline content. These observations document the presence of discernible levels of anthropogenic Pb in the sediments of Oyster Bay and Jimmy Key since 1940 and in the Coot Bay sediments since 1930. At Oyster Bay and Jimmy Key, increases in the Pb/Al ratio are most notable after 1950.

The strong relationship between Pb and Al in pre-1930 sediment supports use of the approach of Trefry et al. (1985) to calculate anthropogenic Pb concentrations in each core. Anthropogenic Pb concentrations can be calculated using the equation (see Trefry et al., 1985, for general format):

$$\text{Anthropogenic Pb } (\mu g/g) = (\text{Total Pb})_{\text{time t}} - ((\text{Pb/Al})\text{pre-1940} \times (\text{Al})_{\text{time t}})$$

In this calculation, the pre-1930 Pb/Al relationship is used to determine natural Pb levels in the sediment as a function of Al (clay) content. Then, the natural Pb level is subtracted from the total Pb level to calculate anthropogenic Pb concentrations presented in Table 14.1 and Figures 14.5A–C.

### BENTHIC MICROFAUNA

Total abundances and the SWDI for Oyster Bay and Jimmy Key sediment-derived populations exhibit dramatic shifts and significant temporal variability at both sites (Figure 14.6). The SWDI is sensitive to the presence/absence of trace species, such that larger positive values reflect a more diverse community, while lower numbers indicate a community dominated by fewer species.

Abundances at Oyster Bay range from <25 to >350 foraminifer tests and ostracod valves per gram of sediment, while at Jimmy Key they ranged from 0 to 180 and 0 to 230, respectively. At Oyster Bay a period of benthic foraminifer and ostracod high abundance extended from the early-1950s to late-1970 (Figure 14.6A). A similar, though less distinct, period also occurred at Jimmy

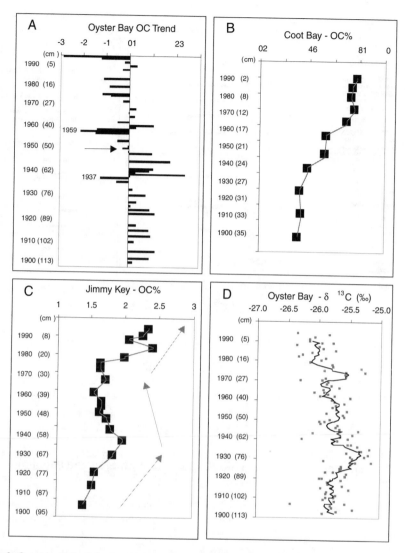

**FIGURE 14.4** Organic C (weight percent) profiles over the last century in the Oyster Bay, Coot Bay, and Jimmy Key cores. (A) Trend analysis of the Oyster Bay data. Arrow indicates a *ca.* 1950 transition period from higher than average to lower than average values for organic C. Intervals with dates indicate Hurricane Donna (upper) and the Labor Day Hurricane (lower) periods. (B) Coot Bay profile demonstrates increasing organic C. (C) Jimmy Key organic C data exhibit three trends: slightly increasing (pre-1930s), slightly decreasing (1930s to 1970s), and substantially increasing (post-1970s). (D) The δ¹³C data for the organic C fraction of the Oyster Bay core.

Key (Figure 14.6C). For convenience, a working definition of these abundance changes are as follows: (1) Zone B is a high-abundance period (~1950– late-1970s), and (2) Zones A and C are periods after and before Zone B, respectively. Two periods of reduced diversity are evident at Oyster Bay. The first constitutes a period of more than a decade, extending from the mid-1930s to late-1940s, while the second reduction starts in the late-1970s and extends to the time of core recovery in 1994 (Figure 14.6B). At Jimmy Key, a zone devoid of both microfauna was encountered which was centered at the mid- to late-1930s. As at Oyster Bay, a second decline in the SWDI starts in the late-1970s and extends upward to core recovery in 1997 (Figure 14.6D). Summary statistics of these data are presented in Table 14.2.

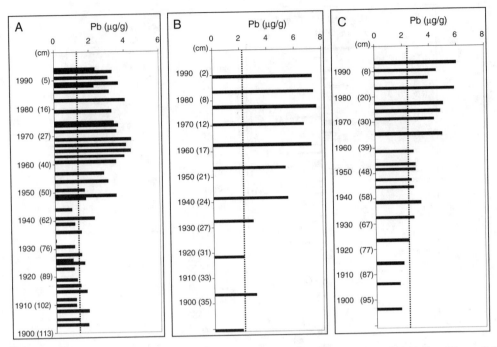

**FIGURE 14.5**  Curves for the total Pb record over the last century in (A) Oyster Bay, (B) Coot Bay, and (C) Jimmy Key cores. The vertical dashed line indicates estimated background Pb levels at each site, while values in excess of this may be interpreted as an anthropogenic signal.

**TABLE 14.1**
**Historical Accounting of Anthropogenic Pb (μg/g) at Study Sites and Population Growth in South Florida**

| Decade | Oyster Bay Pb | Coot Bay Pb | Jimmy Key Pb | Population (x10⁶) |
|---|---|---|---|---|
| 1990s | 2.2 | 3.5 | 2.8 | 2.17 |
| 1980s | 2.8 | 4.2 | 1.9–3.5 | 1.78 |
| 1970s | 2.5 | 4.4 | 2.3–3.0 | 1.36 |
| 1960s | 3.2 | 3.6 | 2.9 | 1.00 |
| 1950s | 2.1 | 3.9 | 0.8 | 0.53 |
| 1940s | 0.5–1.0 | 2.7 | 0.4 | 0.29 |
| 1930s | <0.5 | 1.6 | <0.2 | 0.16 |
| Mean | 1.60 | 2.99 | 0.72 | Dade, Broward, Monroe |

During the low-diversity period of the late-1970s to the present in Oyster Bay (Figure 14.7A), benthic foraminifer and ostracod communities exhibit dominance by a single species (*Ammonia parkinsoniana typica* and *Peratocytheridea setipunctata*, respectively). The combination of low absolute abundance and high relative abundance (Figures 14.7A vs. 14.7B) indicates a survivor-type dominance rather than opportunism.

Atypical assemblages of benthic foraminifer species were observed at specific cores depths at both Oyster Bay and Jimmy Key sites (Figure 14.8). At Oyster Bay, an abnormally high concentration (39% vs. 13% core-wide) of continental shelf foraminifer species was observed over a 3-cm interval estimated to occur in the early 1960s (Figure 14.8A). This temporally coincides with the passage of Hurricane Donna, a category 4 storm (Figure 14.1). Over the same time period a

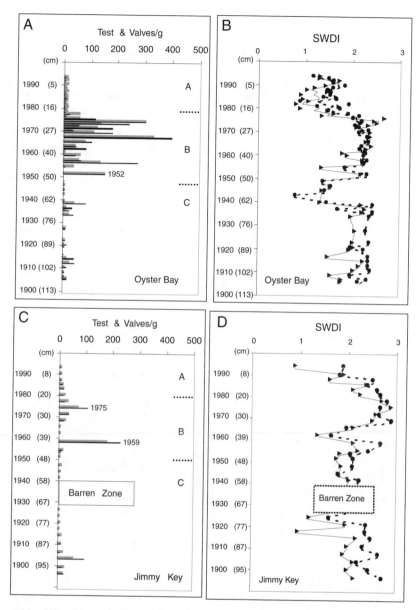

**FIGURE 14.6** Oyster Bay core (A) foraminifer and ostracod abundance and (B) Shannon–Weiner Diversity Index (SWDI) over the last 100 years. Jimmy Key core (C) foraminifer and ostracod abundance and (D) SWDI over the last 100 years. Zones A, B, and C as well as the "Barren Zone" are discussed in the text.

similar effect was also observed at Jimmy Key (Figure 14.8b). Foraminifera atypical to the Jimmy Key site occurred throughout the core in average abundance of ~6% except at one interval where nearly one fourth (~22%) of the foraminifer population were species atypical to this site. As at Oyster Bay, this interval was estimated to occur in the late 1950s to early 1960s and was also temporally coincident with Hurricane Donna. Above this horizon, a gradual increase and decrease in these atypical species reflect a long-term response to natural conditions at this site.

## STABLE ISOTOPIC COMPOSITIONS

The stable isotopic composition ($\delta^{18}O$ or $\delta^{13}C$) of calcareous organisms responds to three factors: water chemistry, temperature, and physiology. The important environmental forcing factors that

**TABLE 14.2**
**Microfaunal Shannon–Weiner Diversity Index and Abundances**

| | Oyster Bay | | | | | | Jimmy Key | | | | | |
| | Foraminifer | | | Ostracods | | | Foraminifer | | | Ostracods | | |
| Zone | Mean | σ | #/g | Mean | σ | #/g | Mean | σ | #/g | Mean | σ | #/g |
|---|---|---|---|---|---|---|---|---|---|---|---|---|
| A | 1.48 | 0.25 | 10 | 1.22 | 0.30 | 3 | 2.23 | 0.37 | 13 | 1.76 | 0.56 | 10 |
| B | 2.17 | 0.14 | 90 | 2.08 | 0.42 | 126 | 2.48 | 0.41 | 36 | 2.28 | 0.44 | 45 |
| C | 1.90 | 0.50 | 6 | 1.51 | 0.74 | 14 | 2.03 | 0.53 | 12 | 1.50 | 0.80 | 13 |
| t-test @ 95% | A ≠ B ≠ C | | | A ≠ B ≠ C | | | A = B, A = C, B ≠ C | | | A = B, A = C, B ≠ C | | |

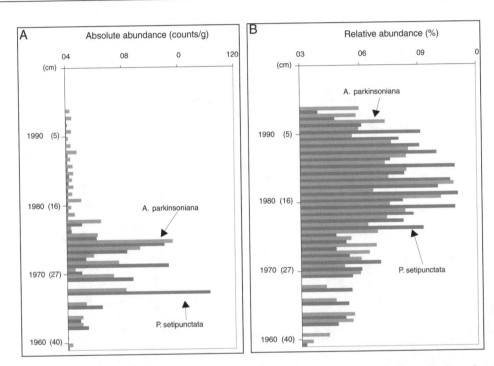

**FIGURE 14.7** The upper portion (1960–1995) of the Oyster Bay core distribution of the dominant benthic ostracod (*Peratocytheridea setipunctata*) and foraminifer species (*Ammonia parkinsoniana typica*). From the late-1970s on, both species exhibit (A) low absolute abundance and (B) high relative abundance, suggesting a survivor-type response to stressful conditions.

control the $\delta^{18}O$ and $\delta^{13}C$ values are water composition and temperature. Although the same species of foraminifera (*Ammonia parkinsoniana typica*) was analyzed at both sites, low abundances made the analysis of the ostracods *Peratocytheridea setipunctata* (Oyster Bay) *and Malzella floridana* (Jimmy Key) necessary. Differences in physiology give rise to a so-called vital effect, where different species inhabiting the same environment have similar responses to environmental forcing but are offset in absolute value from one another. Vital effects of these two species are currently under investigation.

For oxygen, the $\delta^{18}O$ of the water changes in response to evaporation and therefore in most instances is related to salinity. Increased evaporation preferentially removes $^{16}O$, leaving the water behind enriched in $^{18}O$. Under these circumstances the $\delta^{18}O$ value of the water becomes more positive as evaporation proceeds. If the water body being evaporated is freshwater, with a low initial salt content, then there is usually no correlation between salinity and $\delta^{18}O$. In contrast, in marine

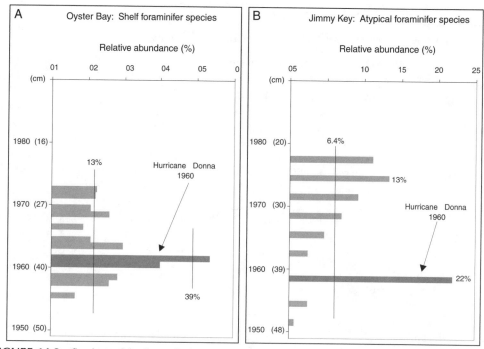

**FIGURE 14.8**   Sections of the Oyster Bay and Jimmy Key cores containing Hurricane Donna storm intervals. (A) The Hurricane Donna interval in Oyster Bay exhibits a core-wide high in continental shelf benthic foraminifer species. (B) In the Jimmy Key core, the Hurricane Donna interval contains a high percentage of benthic foraminifer species atypical for this site.

water, evaporation leads to a correlation between these two parameters. Usually marine waters have $\delta^{18}O$ values of between 0 and $+1‰$ Standard Mean Ocean Water (SMOW), while freshwaters are more negative. In South Florida, the $\delta^{18}O$ of the precipitation is approximately $-3‰$ (Meyers et al., 1993). Overall, three types of responses of the oxygen isotopic signal of calcareous organisms can be expected: (1) one response occurs during dry years when moderate changes in precipitation will mix freshwaters with relatively positive $\delta^{18}O$ values with marine waters; (2) during extremely dry years, evaporation of Everglades water produces enriched $\delta^{18}O$ values; and (3) during very wet periods, the freshwater will retain its original negative $\delta^{18}O$ value and when mixed with marine water produces a positive covariance between oxygen and salinity.

The carbon isotopic composition of calcareous organisms can be principally related to the $\delta^{13}C$ of the dissolved inorganic carbon (DIC) as well as the physiological state of the organism. Although under normal marine conditions physiological processes are probably the most important processes in controlling inter-annual variations in the $\delta^{13}C$ of the skeleton, in situations such as Oyster and Florida Bays variations in the DIC probably dominate. Simplistically, the $\delta^{13}C$ and $\delta^{18}O$ can be combined to provide an indication of the prevalent hydrological processes. During the wet season an area such as Oyster Bay will be inundated by freshwater which will be isotopically depleted in both carbon and oxygen isotopes. During the dry season, the area will be influenced more by marine water that will be heavier in both $\delta^{13}C$ and $\delta^{18}O$. Under dry to normal conditions, however, the freshwater will still retain its isotopically depleted carbon isotopic signature, but will be isotopically enriched in oxygen. Summary data for $\delta^{18}O$ and $\delta^{13}C$ from both Oyster Bay and Jimmy Key sites are listed in Table 14.3.

On a sample-by-sample basis, $\delta^{13}C$ and $\delta^{18}O$ data from Oyster Bay are generally positively correlated for both foraminifera and ostracods, suggesting a normal interaction between fresh and marine waters. When viewed on a longer time scale, a linear fit to the entire dataset of both $\delta^{13}C$ and $\delta^{18}O$ shows a consistent upcore trend to less fresh/more marine for ostracods and mixed results for foraminifera (Table 14.3). Ostracod $\delta^{18}O$ trends from Oyster Bay (Figure 14.9A) and Jimmy Key

**TABLE 14.3**
**Stable Isotopes for Microfauna from Oyster Bay and Jimmy Key**

| Oyster Bay | $\delta^{18}O$ – Ostracod[a] | $\delta^{18}O$ – Foraminifer[b] | $\delta^{13}C$ – Ostracod[a] | $\delta^{13}C$ – Foraminifer[b] |
|---|---|---|---|---|
| Mean | −1.56 | 0.02 | −9.67 | −7.89 |
| SD | 0.82 | 0.90 | 0.65 | 0.51 |
| Max | 0.08 | 2.37 | −7.59 | −7.14 |
| Min | −3.98 | −2.69 | −10.6 | −9.67 |
| | More marine | No change | More marine | No change |

| Jimmy Key | $\delta^{18}O$ – Ostracod[c] | $\delta^{18}O$ – Foraminifer[b] | $\delta^{13}C$ – Ostracod[c] | $\delta^{13}C$ – Foraminifer[b] |
|---|---|---|---|---|
| Mean | −0.80 | −0.38 | −4.39 | −2.52 |
| SD | 0.92 | 0.82 | 1.76 | 1.15 |
| Max | 0.47 | 0.74 | −1.00 | −0.92 |
| Min | −2.87 | −1.70 | −6.06 | −4.35 |
| | More marine | More marine | More marine | No change |

[a] *Peratocytheridea setipunctata.*
[b] *Ammonia parkinsoniana typica.*
[c] *Malzella floridana.*

(Figure 14.9B) are compatible with the trend (Figure 14.9C) of the low-salinity tolerant foraminifer species *Elphidium gunteri* (Parker et al., 1953; Phleger, 1966a,b; Poag, 1978; Murray, J. 1991), thus reinforcing this interpretation.

The normal interaction between fresh and marine waters, noted above, also allows evaluation of the relative importance of freshwater sources (rainfall vs. runoff) at Oyster Bay. Rainfall can account for input of freshwater to both the Oyster Bay and Jimmy Key sites while the influence of runoff from SRS can be assumed to be greater at Oyster Bay than at Jimmy Key (Figure 14.1). In order to evaluate which influence may play a dominant role at Oyster Bay, the $\delta^{18}O$ data for *Peratocytheridea setipunctata* (Figure 14.9A) were compared with historical local rainfall (Flamingo, Figure 14.1) and SRS flow data (Figure 14.10). The interval 1983–1995 (core top) was chosen because the sediment age-dating error estimates (±1 year) provide the most accurate comparison between core geochronology and absolute yearly values for flow and rainfall. Direct comparison of the $\delta^{18}O$ record indicates that the inverse covariance is best with rainfall.

## POLLEN

Pollen incorporated into wetland, estuarine, and coastal surface sediments reflects both the vegetation at a site and pollen contributed to that site from upstream and upwind. Pollen assemblages in vertical sequences are commonly assigned to parent vegetation communities and used to interpret environmental changes through time. Therefore, sites are usually selected to minimize the transport signal. In this study, the focus is different and we use the pollen record through time to track variations in the transport signal. While there are changes in vegetation, and thus pollen, on time scales of interest in this study, they are largely associated with high-magnitude events, such as hurricanes. The vegetation in the area is otherwise largely similar (Davis, 1943; Craighead and Gilbert, 1962; and Bancroft et al., 1994).

There are differences between vegetation communities and pollen zones. Vegetation communities in southwestern Florida, particularly in ENP, vary in complexity. Some communities are dominated by a few species, while others are composed of more complex mixtures with no truly dominant species (Davis,1943; Loveless,1959, Gunderson, 1994). Moreover, in southwestern Florida, the boundaries between vegetation communities are commonly well defined with distinct

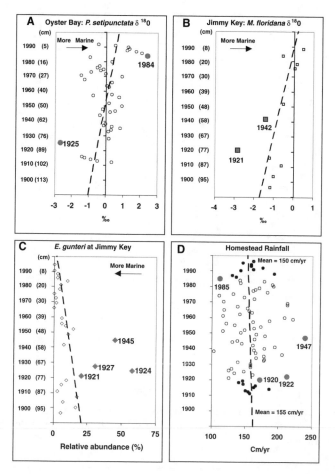

**FIGURE 14.9** Stable isotope values and faunal relative abundance in Oyster Bay and Jimmy Key cores and Homestead rainfall. (A) The $\delta^{18}O$ data for the ostracod *Peratocytheridea setipunctata* with linear trend indicating more marine conditions upcore. (B) At Jimmy Key a similar trend is shown for *Malzella floridana*. (C) The benthic foraminifera *Elphidium gunteri*, which prefers lower salinities, exhibits a marked decrease in relative abundance upcore at Jimmy Key. (D) Homestead rainfall exhibits marked excursions from the trend during years in which concomitant variability is seen in the microfaunal isotope and species relative abundance data (A, B, C). Specific dates are discussed in the text.

zonation that is strongly related to salinity, hydroperiod, elevation, and substrate (Riegel, 1965; Cohen, 1968). In contrast, pollen zones are blurred because of the mixing of pollen by fluvial, tidal, and atmospheric processes (Berglund, 1973; Bonny, 1976; Caratini et al., 1973; Cross et al., 1966). This is especially true in estuarine settings such as the Oyster Bay core site.

We have combined a series of approaches to read the pollen record in the finely laminated sequences at Oyster Bay. We have used a modern analog approach to establish the local pollen assemblage in the vicinity of the sediment record and have established key pollen taxa associations which are indicators of the different physiographic provinces in the Lower Everglades/Florida Bay ecosystem. These pollen associations are important because they are particularly useful in identifying both local pollen contributions, as well as those from upstream of the depositional site. This allows for assessment of both the source and the magnitude of pollen input. With the local pollen assemblage for these five taxa associations as a reference (Table 14.4), the downcore departure of pollen assemblages from the local pollen assemblage were interpreted as resulting from "exotic" pollen transported to the site. In this way, intervals of high vs. low flushing and high vs. low

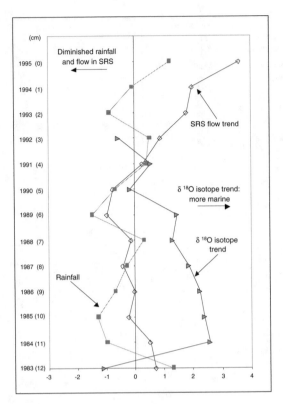

**FIGURE 14.10**   Comparison between regional rainfall, Shark River Slough flow, and $\delta^{18}O$ data for the ostracod *Peratocytheridea setipunctata* demonstrating the closest temporal correlation between changes in $\delta^{18}O$ data and regional rainfall.

**TABLE 14.4**
**Average Percentages of Pollen of Five Major Taxa Associations in the Local Pollen Assemblage**

| | Coastal → Upland | | | | |
| | Mangrove | Mangrove–Conocarpus | Brackish | Chenopods | Aquatics |
|---|---|---|---|---|---|
| Average % | 57 | 52 | 8 | 12 | 0.8 |
| 2 SD | 22 | 24 | 8 | 14 | 2 |
| Range % | 36–77 | 34–73 | 3–23 | 3–28 | 0–5 |

sedimentation rates can be identified and used in conjunction with other environmental proxies to interpret flow and discharge characteristics of the SRS over the past century.

The five pollen associations described below and their abundances in the local pollen assemblage are given in Table 14.4:

1. *Mangroves:* The *Rhizophora, Avicennia, Laguncularia, Conocarpus, Batis,* and *Salicornia* association represents the pollen from the mangrove ecosystem and accounts for ~57% of the local pollen assemblage. Significant negative departures from this represent "exotic" input.
2. *Mangroves–Conocarpus:* Conocarpus accounts for about 4% of the mangrove sum of the local mangrove swamp pollen assemblage and may be used as an indicator of brackish input. As a result, large positive departures in *Conocarpus* abundances may be indicative of non-local inputs.

3. *Brackish:* Includes triporate grains, primarily *Myrica* (wax myrtle) and *Chrysobalanus* (cocoplum); *Osmunda* and *Polypodium* (ferns); Cyperaceae (sedge family), Gramineae (grass family), *Typha* (cattails), *Taxodium* (bald cypress), and Compositae (includes the aster family). *Rhizophora, Conocarpus,* and chenopods are also important pollen contributors to the brackish marsh but they are not included here because they are tracked separately. Brackish taxa are taken to indicate pollen sourced from both the brackish and transitional marsh areas upstream of the mangrove swamp and influenced by freshwater input from the slough region as well as overland sheet flow. They are downstream of the slough, allowing differentiation of the relative degree of freshwater flushing down the SRS.

4. *Chenopods:* From the pigweed family, primarily freshwater flowering plants, these comprise more than 60% of samples from the headwater areas of the SRS and effectively separate the freshwater marsh–slough from the freshwater marsh with hammocks. Changes in their abundance are viewed as indicators of variation in the degree of freshwater flushing from the headwater area of the SRS and the slough environment. *Salicornia,* although a chenopod, is identifiable and included with the mangrove pollen sum.

5. *Aquatics: Nymphaea* (water lily), *Sagittaria* (arrowhead), *Utricularia* (bladderwort), *Umbelliferae* (hornwort), and cf. *Ovoidites* (the zygote of the green alga *Spirogyra* [van Geel, 1978]) are aquatic species whose pollen characterize freshwater marsh communities (Riegel, 1965) well upstream of the core site. Their overall absence from the local pollen assemblage (i.e., <1%; Table 14.4) makes aquatics useful indicators of flushing. Differences in the abundances of aquatics and chenopods can be used to independently assess flushing sources.

In the Oyster Bay core, pollen and spores were concentrated and counted from 24 intervals and plots (Figure 14.11A) of the five important taxa associations were made. Co-plotted for comparison with these data are "statistical envelopes" resulting from a two-standard-deviations range around the average abundance of the local assemblage for the five key taxa associations derived from the Riegel data (Table 14.4). Filled symbols indicate intervals that exceed this envelope. The absolute grains per gram of sample (Figure 14.11B) allows us to determine whether or not changes in pollen through time are just a relative effect caused by increases or decreases in other taxa and to determine changes in sedimentation rate.

Eight samples (0–1, 11–12, 26–27, 31–32, 39–40, 81–82, 87–88,103–104 cm; Figure 14.11; ~1994, 1984, 1970, 1966, 1960, 1925, 1921, and 1910, respectively) had grain counts of less than 100 grains and were excluded from the frequency percent plots (Figure 14.11A). These intervals are included in the plot of absolute pollen counts because they represent intervals with very little pollen (Figure 14.11B). These intervals all had extremely high counts of marker grains indicating either low input of pollen or high sedimentation rates diluting the pollen signal.

Table 14.5 summarizes the import of "exotic" pollen into the Oyster Bay site and delineates source areas of "exotic" pollen. This analysis provides a partial upland flushing history of the SRS area. Intervals possessing pollen from the slough represent times of high freshwater flushing; from the brackish marsh area, times of moderate freshwater flushing. Intervals of low to no flushing possess pollen derived from the local area and have very low amounts of transported pollen.

## HURRICANES

Hurricanes are capable of immediate and long-term environmental change (Tabb and Jones, 1962; Ball et al., 1967; Perkins and Enos, 1968; Meeder et al., 1994; Smith et al., 1994). Immediate damage includes the total destruction of mangrove forests (Wanless et al., 1994, Figure 8.10). Moreover, in low-relief areas such as this study area, storm-induced sediment transport can be

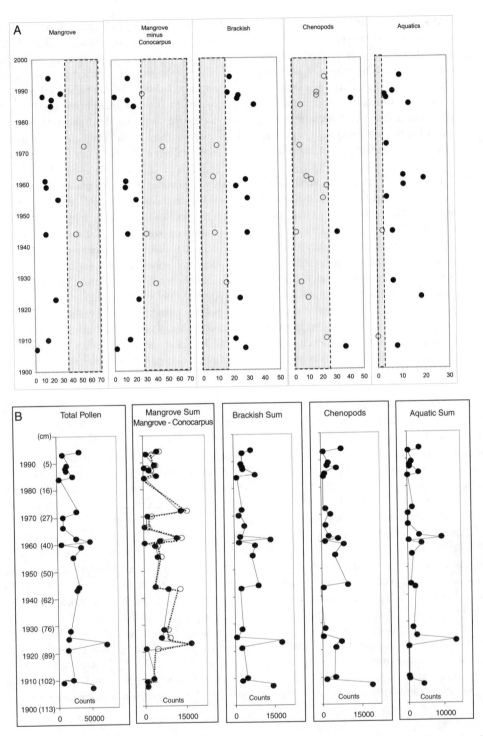

**FIGURE 14.11** (A) Percent abundance of pollen for five important taxa associations in 16 intervals of the Oyster Bay core. Dashed lines are 2 standard deviations about the mean of the local pollen assemblage. Open symbols are within the range of the local pollen assemblage. Filled symbols represent samples with a significant component of transported pollen. (B) Absolute pollen counts of five important taxa associations in 24 intervals of the Oyster Bay core. Note: Mangrove Sum is dashed line with open symbol and Mangrove–Conocarpus Sum is solid line and symbol.

**TABLE 14.5**
**Flushing Record from Shark River Slough to the Oyster Bay Core Site**

| Core Depth (cm) | Geochronology-Based Time Equivalent | Mangrove | Mangrove-Conocarpus | Brackish | Chenopods | Aquatics | Flushing Signal Source Area |
|---|---|---|---|---|---|---|---|
| 0 | 1994 | F | F | f | f | F | Slough |
| 5–6 | 1989 ± 1 | f | NF | F | NF | f | No/low Flushing |
| 7–8 | 1988 ± 1 | F | F | F | NF | NF | Brackish Marsh |
| 8–9 | 1987 ± 1 | F | F | F | F | f | Brackish Marsh |
| 10–11 | 1985 ± 1 | F | F | F | NF | F | Slough |
| 24–25 | 1972 ± 1 | NF | NF | NF | NF | f | No/low Flushing |
| 36–37 | 1962 ± 2 | NF | NF | NF | NF | F | No/low Flushing |
| 38–39 | 1961 ± 2 | F | F | F | NF | F | Slough |
| 40–41 | 1959 ± 2 | F | F | F | f | F | Slough |
| 44–45 | 1955 ± 2 | f | F | F | NF | F | Slough |
| 56–57 | 1945 ± 3 | F | F | F | F | NF | Brackish Marsh |
| 57–58 | 1944 ± 3 | NF | NF | NF | NF | F | Brackish Marsh |
| 77–78 | 1928 ± 4 | NF | NF | f | NF | F | No/low Flushing |
| 84–85 | 1923 ± 4 | F | F | F | NF | F | No/low Flushing |
| 101–102 | 1910 ± 5 | F | F | F | NF | F | Slough |
| 106–107 | 1907 ± 5 | F | F | F | F | NF | Brackish Marsh |
| | | | | | | | Slough |

*Note:* F = Taxa association indicates high degree of flushing; f = taxa association indicates some degree of flushing; NF = taxa association reflects local pollen assemblage and therefore no flushing contribution of transported pollen.

bidirectional with landward surge-transport of marine carbonates and seaward ebb-transport of terrestrial materials.

Since the turn of the century, 14 hurricanes have directly influenced the study area. Representative storm paths are plotted in Figure 14.1. The impact of windfields from Hurricane Donna (1960, category 4) on Jimmy Key, Coot Bay, and Oyster Bay locations was evaluated using the NOAA Spectral Application of Finite-Element Representation (SAFER) Model. The resultant windfields indicated that gale-force (≥18 m/sec) and greater winds impacted all core sites for more than one day, and hurricane-force winds (>33 m/sec) impacted each core site for ≥15 hours with primarily easterly and subsequent westerly winds. For the core at Oyster Bay this translated into both offshore and subsequent onshore winds. The sediment record at the Oyster Bay site confirms both onshore transport of shelf carbonates (Figure 14.3A) and continental shelf fauna (Figure 14.8A) as well as subtle erosion/deposition zones (Figure 14.2A) as immediate consequences of Hurricane Donna. Additionally, a time-lagged record for various parameters such as the amount (Figure 14.3B) and type of organic C (Figure 14.12) transported to this site were observed. At Jimmy Key, this resulted in westward, then eastward windfields.

## SALINITY

At each core site a record of historical salinity from 1957 to present allowed temporal analysis over decadal, yearly, and seasonal scales. At the decadal scale, the results from Oyster Bay and Jimmy Key indicate that each station had a large salinity range (Figure 14.13A). Oyster Bay salinity varied from nearly fresh (3.3) to marine (≥38), while Jimmy Key ranged from brackish (16.4) to hypersaline (>55), with Jimmy Key having both a higher mean salinity (33.5 vs. 22.9) and range (38.8 vs. 30.5). These data also reflect the decadal-scale temporal variation for periods of above (filled symbols) and below (open symbols) average salinities (Figure 14.13A). At the monthly level (Figure 14.13B) data for 1992–1998 show variability that is statistically similar at both sites. Trends in these data indicate: (1) broad-scale seasonal changes which correspond between sites, and (2) event-scale changes (e.g., June 1997; Figure 14.13B, arrow) from marine (~34) to fresh/brackish (~10) salinities in Oyster Bay which are not reflected at Jimmy Key.

To contrast how periods of above- and below-average salinity differed, on a monthly basis, all data used to construct Figure 14.13A were averaged into monthly groupings for the above ($n = 293$) and below ($n = 367$) average periods at both sites (Figures 14.13C,D). The shaded areas (Figure 14.13C,D) represent the differences between above- and below-average monthly salinities. Viewed in this manner, it is clear that:

1. During periods of elevated salinity, the average value at Jimmy Key is nearly double that at Oyster Bay (44.5 vs. 24.9).
2. Similarly, below-average salinity differs by more than a factor of three (29.8 vs. 9.3).
3. The monthly differences were nearly identical (15.6 vs. 14.7).
4. Nonuniformity in the salinity differences from month-to-month, however, suggest different influences on salinity during these periods.

A previous study in Florida Bay (Montague and Ley, 1993) suggested that variability is more important than mean salinity for changes in biomass of the benthic fauna and flora. In that study, for every increase of 3 in the standard deviation of the salinity, ($+\Delta S_\sigma \cong 3$), total benthic biomass decreased by a factor of 10. Changes of this magnitude were observed in our cores for both benthic foraminifer and ostracod abundances (Figure 14.6). Therefore, the relationship between benthic microfauna and known salinity data was examined in a manner that allowed for a limited evaluation of this hypothesis at two temporal levels: (1) Zones A vs. B at Oyster Bay, and (2) the periods 1975 ± 1 vs. 1979–1985 at Jimmy Key (Table 14.6). Due to the lack of salinity data prior to the mid-1950s, Zone C could not be evaluated. The precipitous drop in abundance in both benthic

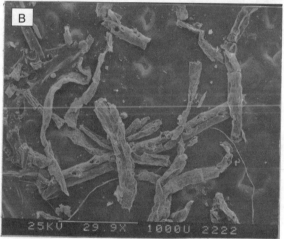

**FIGURE 14.12** Electron micrographs of mangrove macro-organic materials found in the Oyster Bay core (see Figure 14.3A). (A) A pre-Hurricane Donna interval dominated by mangrove leaf detritus. (B) A post-Hurricane Donna core interval dominated by mangrove root-hair material. See text for discussion.

foraminifera and ostracods in the late 1970s to early 1980s and a more interval-specific drop at Jimmy Key were examined. Both sites showed the following: (1) reductions in mean salinities were accompanied by increases in $S_\sigma$, and (2) all changes in both foraminifer and ostracod abundances were in a direction and approximate magnitude predicted by changes in $S_\sigma$ (Table 14.6). Oyster Bay exhibited the best example of this relationship, closely approximating the change ($+\Delta S_\sigma \cong 3$) observed by Montague and Ley (1993). The significance and environmental implications of this will be considered below.

## RAINFALL

Changes in salinity can be influenced directly by rainfall and indirectly by runoff to the coastal zone. Four spatially relevant rainfall stations, with sufficient data history to present (Figure 14.1, Flamingo (1951), Tavernier (1936), Everglades (1931), Homestead (1911)), were evaluated. Statistical comparisons, based on a 6-month running correlation of monthly data over common time periods (1951–1995; $n \approx 540$) established that rainfall at these stations was highly correlated ($r \geq 0.97$). Given this, Homestead rainfall records, back to 1911, were sorted and grouped to

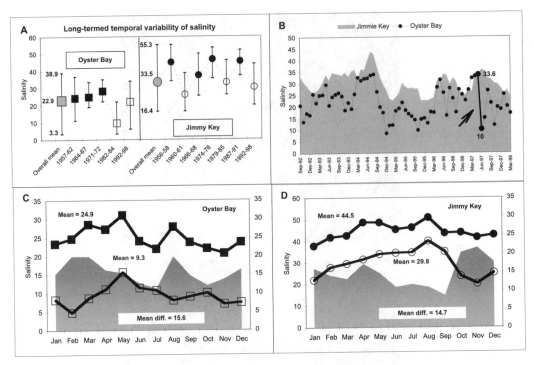

**FIGURE 14.13** Salinity variation in Oyster Bay and Jimmy Key over the last 40+ years. (A) Decadal scale (mid-1950s to 1998) temporal variability in salinity. The initial gray symbols are the means for the entire data set with ranges indicated by vertical lines. Average salinities for discrete time intervals are subsequently shown; values greater than the overall site mean are shown in solid symbols and those lesser than the mean are open symbols. (B) Monthly salinity data from Jimmy Key and Oyster Bay (1992 to 1998) highlighting (arrow) major rain-induced change. (C) Oyster Bay salinity data from part (A) above are combined for higher than average (solid symbols) and lower than average (open symbols) salinity periods, and mean monthly values were determined. The monthly differences between high- and low-salinity periods are indicated by the shaded area. (D) The same analysis performed for the Jimmy Key data. The right scale for parts (C) and (D) is for the mean difference salinity values.

## TABLE 14.6
## Standard Deviation of Salinity and Microfaunal Abundances

| Oyster Bay | | | | | Jimmy Key | | | | |
|---|---|---|---|---|---|---|---|---|---|
| Zone | n | $S_\sigma$ | F: #/g | O: #/g | | Zone | n | $S_\sigma$ | F: #/g | O: #/g |
| A | 76 | 8.8 | 10 | 3 | | 1979–85 | 33 | 5.5 | 18 | 13 |
| B | 214 | 6.1 | 90 | 126 | | 1975 ± 1 | 31 | 3.9 | 68 | 104 |
| ΔAB | | 2.7 | | | | Δ | | 1.7 | | |

*Note:* n = number of observations; #/g = valves or tests per gram of sediment.

corresponded exactly to the salinity periods shown in Figure 14.13A and were then co-plotted with salinity data for both Oyster Bay (Figure 14.14A) and Jimmy Key (Figure 14.14B). A clear anti-thetical relationship is shown indicating the strong influence of rainfall on salinity at each study site over the decadal time scale. On a much shorter time scale rainfall also has a profound effect on salinity. This is seen in Figure 14.13B (arrow) where the previously noted one-month change

(June 1997) in salinity from marine (33.6) to brackish/fresh (~10) directly resulted from a near-record one-month local rainfall (Flamingo) of 54.6 cm. Taken together, these data provide a cogent argument for strong direct rainfall control on coastal salinities in the study area from decadal to monthly scales.

## FRESHWATER FLOW

Coastal salinities can also be influenced by coastward overland freshwater flow. The largest single source of freshwater input from the South Florida peninsula to the adjacent coastal ocean is the SRS, which is of additional importance to this study for its proximity to the Oyster Bay core site (Figure 14.1). In order to quantify this flow and its seasonal patterns, flow data for the period of record (1940–present) were analyzed for periods of above- and below-average flow (i.e. >mean flow ± 1 σ) and the results are plotted in Figure 14.14C. From this it is clear that above-average periods are bimodal (~July and October), while below-average flow years are unimodal (October).

The relationship between regional rainfall and flow into SRS was investigated using trend analysis and then compared to water management construction projects and flow regulation (Figure 14.15A). Summarizing Light and Duneen (1994), these anthropogenic changes include:

- In ~1960 construction commenced on major modifications of the Everglades hydroscape in which the major construction period (~1960–1963) was followed by a series of short-term early water management schemes.
- In 1970, the Congressionally mandated water-flow plan known as the Minimum Allocation Plan (MAP, ~1970 to early 1980s) was implemented and followed by a regulated flow plan that was believed to be more natural, known as the Rainfall Plan (RFP).

The rainfall and flow records show the best correlations before 1960 (r = 0.84) in the pre-management period and after 1983 (r = 0.65) during the RFP. The balance of the time showed moderate (r = 0.40) correlation during the construction and early management phases with the lowest occurring during the MAP (r = −0.06) where essentially no correlation existed between rainfall and flow (Figure 14.15A). Given the high correlation between rainfall and flow (r = 0.84) during the unregulated flow period, regional rain at Homestead appears to be a good proxy for qualitatively estimating flow down SRS prior to historical flow records. During the period of flow records for SRS, flow volumes are shown in Figure 14.15B in the form of both cumulative flow and period averages for times of high and low flow, where the period averages are based on intervals between inflection points (rate changes) in the cumulative record.

## DISCUSSION

### NATURAL INFLUENCES

#### Hurricanes

Offsets in the excess $^{210}$Pb profiles at both Oyster Bay and Jimmy Key (Figure 14.2A,E) were coincident with the hurricanes of 1935, 1948, and 1960 (Donna), whereas the more sheltered coring site in Coot Bay showed no detectable disturbance. These offsets gave the first indication that normal sediment accumulation was interrupted at multiple levels and that each recorded storm events. The paleo-windfields for Hurricane Donna (1960), computed for each site by the NOAA SAFER Hurricane Model, confirmed hurricane-force winds for ≥15 hours with both easterly and westerly components. At Oyster Bay onshore sediment transport corresponding to Hurricane Donna was demonstrated by a distinct layer of continental-shelf-derived carbonate sediment (Figure 14.3A) and microfaunal species (Figure 14.8A). Downcore from this carbonate layer (Figure 14.3A) two

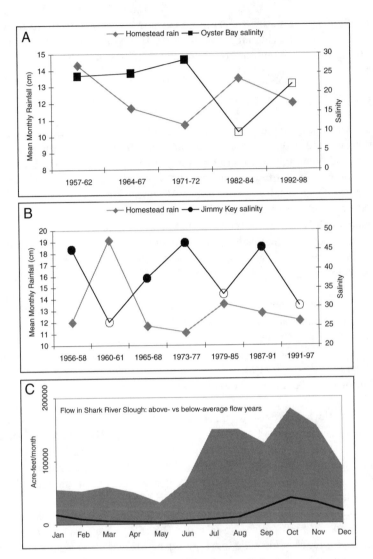

**FIGURE 14.14**   Rainfall, salinity, and Shark River Slough flow data. (A) Oyster Bay and (B) Jimmy Key salinities for time intervals shown in Figure 14.13A (symbol designations same) vs. regional (Homestead) mean monthly rainfall for those same time periods. Note strong inverse relationship. (C) Seasonal pattern of Shark River Slough flow for above- and below-average (>mean ± 1σ) flow years.

similar layers were coincident with the offsets in excess $^{210}$Pb (Figure 14.2A,E), which corresponded to the hurricanes of 1935 and 1948. This observation of landward transport not only conforms to post-Donna field observations (Ball et al., 1967; Perkins and Enos, 1968) but also validates the quality of sediment ages estimated by our geochronology.

The long-term consequences of Hurricane Donna show a complex overprint of natural and anthropogenic influences with reduced organic C followed by elevated organic C corresponding to and following Donna (Figure 14.3B, box) and the Labor Day Hurricane of 1935. Typical organic litter being released by mangrove forests is platy detritus dominantly composed of decayed leaves. Microscopic examination of the Oyster Bay core indicated this detrital fraction (>500 µm) dominated (Figure 14.12A) except just above hurricane layers. Following the Labor Day Hurricane of 1935 and Donna, the macro-organic detritus was entirely thin tubes — fragments of the red mangrove root–hair peat (Figure 14.12B) — which persisted upwards for several years (Figure 14.3B; from

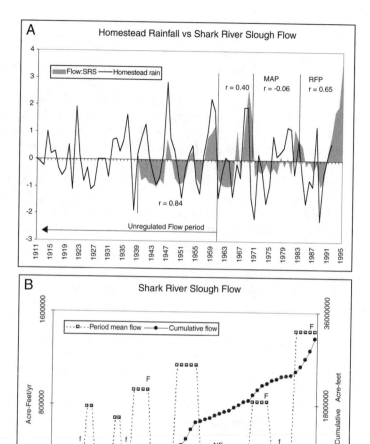

**FIGURE 14.15**  (A) Trend analysis of Homestead rainfall and Shark River Slough flow (1911–1995) where "0" represents average conditions for both. Time intervals representing different water management strategies (MAP = Monthly Allocation Plan; RFP = Rainfall Plan) are shown. The degree of correlation between rainfall and flow is calculated for each time interval. (B) Cumulative (right axis) and period mean (left axis) flow down Shark River Slough. Time intervals for calculation of period mean flow are based on inflection points of cumulative flow curve. The letters F, f, and NF represent periods of major flushing, flushing, and no flushing, respectively, for pollen from upland areas of Shark River Slough (see Table 14.5 and text for discussion).

39 to 35 cm). These intense category 4 and 5 storms defoliated and completely destroyed the mangrove forest community in the vicinity of Oyster Bay, eliminating the leaf litter source and initiating a period of erosion of the dead and decaying root-hair mangrove peat. Only category 4 and 5 storms decimate the mangrove forests (Wanless et al., 1994, Figure 8.10). Lesser storms that crossed the area did not initiate this change.

Within Florida Bay previous studies have suggested that hurricane activity promotes resuspension and flushing of organic C from the bay system, promoting a healthier ecosystem (Swart et al., 1996). Evidence from the long-term organic C record at Jimmy Key suggests that the frequency of hurricane activity may influence carbon loading in the sediments and supports the flushing concept. Based on hurricane records and estimated degree of direct impact on Florida Bay

(Sam Huston, NOAA Hurricane Research Div., pers. comm.), the two periods of organic C increase (~1900 to mid-1930s; post-mid-1970s) correspond to periods of four and one hurricane, respectively (Figure 14.4C). In contrast, the longer period of mildly declining organic C (mid-1930s to mid-1970s) corresponds to a period in which nine hurricanes were estimated to have had a direct impact on Florida Bay.

At Oyster Bay, continental shelf species averaged ~13% of the microfauna throughout the core except at the time of Hurricane Donna, when these species account for ~39% of the total faunal assemblage (Figure 14.8A). This immediate increase is likely the result of co-transport with hurricane-induced movement of continental shelf carbonates into this site (Figure 14.3A). Major hurricanes can also impact the diversity of microfaunal communities on longer time scales. At Oyster Bay, subsequent to the Labor Day Hurricane of 1935 (category 5), a decade-long reduction in the SWDI was observed for both foraminifer and ostracod communities (Figure 14.6B). The low diversities in the 1940s temporally correspond to the elevated organic C (Figure 14.3B) at this same period, and we postulate that resulting changes may have adversely affected these communities. A similar response to Hurricane Donna was not observed.

At Jimmy Key, the immediate hurricane impact resulted from physical processes and was different for the 1960 and 1935 hurricanes. The highest abundances for foraminifer tests and ostracod valves occurred at the Donna (category 4) level followed by a marked decrease in abundance and diversity in both groups (Figure 14.6C,D, 1959 ± 4). Additionally at this level, foraminifer tests and ostracod valves atypical to this site abruptly increased ~4× over the whole-core mean of ~6% (Figure 14.8B), suggesting concentration by physical processes resulting from bidirectional sediment transport and re-sedimentation. In contrast, the lowest foraminifer and ostracod abundance corresponded to an interval temporally coincident with the Labor Day Hurricane of 1935 (category 5). Abundances were so low (Figure 14.6C,D; Barren Zone) that community characterization was not possible. The core intervals immediately above the hurricane deposit indicate a decade-long decrease in both foraminifer and ostracod community diversities. These data are important in that they point to the need to recognize hurricane intervals within sedimentary sequences for accurate interpretation of microfaunal data.

### Salinity, Rainfall, and Flow from Shark River Slough

Based on long-term historical salinity data available from Oyster Bay and Jimmy Key, decadal to monthly time series showed high degrees of variability (Figure 14.13A,B). On the decadal scale, at both sites, this variability was linked to regional rainfall (Figure 14.14A,B) indicating that long-term salinity changes were strongly controlled by long-term, regional-scale precipitation patterns. On the monthly scale, salinity variability was synchronized between sites with short-term local rain events (Figure 14.13B, June, 1997: 54.6 cm rain $\Rightarrow \Delta S = 33.6$ to 10.0), accounting for intra-site divergence.

This rainfall-driven linkage implies that salinity patterns should reflect rainfall forcing, with the potential exception of sites adjacent to the outflow from SRS. When viewed on a seasonal basis for low- and high-salinity years (Figure 14.13C), Oyster Bay can be evaluated in conjunction with contemporary rainfall and related freshwater flow patterns (Figure 14.14C). The bimodal (July, September highs) nature of South Florida rainfall presented by Duever et al. (1994) correctly represents average conditions. During non-average conditions, such as low-salinity years at Oyster Bay (Figure 14.13C, open symbols), seasonal rainfall typically increases such that ~50% of these increases occur during June with ~95% of the total yearly increase occurring before August. This not only modifies freshwater outflow timing (Figure 14.14C), but also can account for difference patterns between high- and low-salinity periods in Oyster Bay (Figure 14.13C). Specifically, at Oyster Bay, the August and December–March salinity-difference maxima suggest a ≥1-month phase lag with summer and fall outflow modes (Figure 14.14C, July–August and October). These periods are separated by a flow reduction (September)/salinity increase (October). Although the data provide a less compelling argument at Jimmy Key, minimum salinity patterns during September–January (Figure 14.13D, open symbols) suggest delayed linkage to terrestrial freshwater outflow patterns which may influence central Florida Bay as well.

The relative influence of rainfall and freshwater outflow at Oyster Bay was evaluated with the $\delta^{18}O$ record of microfauna (Figure 14.10). The 1983–1995 interval was chosen due to its having: (1) minimal differences between sediment geochronology (±1 year) and absolute dates for rainfall and flow, and (2) RFP water management period, shown to be most representative of pre-management unregulated flow/rainfall relationships (Figure 14.15A). Direct comparison of these data show that the inverse covariance is best between $\delta^{18}O$ and rainfall (Figure 14.10), indicating that rainfall, rather than freshwater runoff, is the dominant driver of salinity changes, even in Oyster Bay adjacent to the mouth of SRS.

On a longer time scale, the microfaunal $\delta^{18}O$ record (Figure 14.9A,B) indicated a trend toward more marine conditions since about the turn of the century that was concordant with changes in salinity-sensitive foraminifera (Figure 14.9C, *Elphidium gunteri*). Regional rainfall during this period (Figure 14.9D) has declined from 155 cm/yr (mean of first decade of data) to 150 cm/yr (most recent decade). Although these rainfall means are not statistically different (95% confidence level), major departures from the microfaunal trends for both $\delta^{18}O$ and *E. gunteri* support a strong linkage to regional rainfall at essentially the yearly level.

The data in Figures 14.9A–C were re-examined for major departures from the long-term linear trends. Results indicate that these divergent data correlate with the regional rainfall record (Figure 14.9D). It is important to restate here that geochronology-based downcore ages have associated error estimates such that core dates in the 1940s and 1920s have a range of ±3 and ±4 years at Oyster Bay and ±6 and ±8 years at Jimmy Key, respectively. Within the limit of age estimates, the $\delta^{18}O$ and *Elphidium gunteri* records in the early to mid-1920s and in the mid-1940s (Figure 14.9A–C) indicated excursions toward less-marine conditions contemporaneous with increased regional rainfall during both periods (Figure 14.9D). Similarly, a mid-1980s extreme excursion to more marine conditions at Oyster Bay (Figure 14.9A) coincided with record low regional rainfall (Figure 14.9D). Although a point-by-point matchup of rainfall/microfaunal data cannot be accomplished with these data, we believe that when viewed together, $\delta^{18}O$ and *E. gunteri* patterns at both sites provide compelling evidence for short-term, essentially yearly coupling of microfaunal variations to regional rainfall and the attendant salinity changes. Accordingly, these observations support rainfall/microfaunal response at essentially yearly time scales and imply validity of overall longer-term trends.

Periods of elevated "exotic" pollen tracers found throughout the last century in Oyster Bay sediments (Figure 14.11) were semi-quantified (Table 14.5) for upland source region and thus flushing distance. The results of Table 14.5 indicate three levels of flushing: (1) major flushing and transport of pollen from the upland freshwater slough to the study site (F), (2) dominant flushing and transport from less distal brackish marsh areas (f), and (3) No significant flushing from either region (NF). Using the F, f, and NF conventions, pollen-based flushing estimates compared favorably with measured SRS flow for all periods of record (Figure 14.15B). Since a high degree of positive covariance has been demonstrated between flow in SRS and rainfall before 1960 (Figure 14.15A), estimates of SRS flushing should also compare well with regional rainfall data which extends back to 1911. From Table 14.5, no/low flushing for 1928 ± 4 and the highest category of flushing for 1923 ± 4 are predicted. Comparison with regional rainfall (Figure 14.15A) again indicates compatible evidence supporting both the use of palynology as an indicator of flushing from SRS as well as validating estimated age dates, even back to the turn of the century.

## Anthropogenic Influences

### Everglades Watershed Development and Management

A detailed history of the Everglades drainage basin is presented by Light and Dineen (1994), from which this summary was abstracted for anthropogenic changes relevant to this study:

- 1915–28: Construction of the trans-Everglades Tamiami Trail roadway linking Miami with Florida's west coast. Culverts under the roadway allowed unrestricted flow that was not monitored for about the first decade. Early impacts on the Everglades are still debated.
- 1954–59: Construction of Everglades Agricultural Area (EAA) partitioning 700,000 acres of rich muck lands to the north from the remainder of the Everglades to the south through a series of levees and water-diversion canals.
- 1960–63: Major construction period during which vast areas of sawgrass prairie north of Tamiami Trail were delimited with levees and borrow canals, and water flow into the SRS was controlled by a series of gated spillways in order to create the current system of Water Conservation Areas.
- 1963–70: A series of short-term water management plans for flow to SRS.
- 1970–early 1980s: Congressionally mandated Monthly Allocation Plan of water delivery to SRS.
- Post-early 1980s: Water delivery to SRS under the Rainfall Plan.

Two major and concurrent shifts occurred in the early 1950s that were recorded in the Oyster Bay sediments. The first was a decline in the organic C content of the sediment relative to pre-1940s sediments (Figures 14.3B, 14.4A), and the second was the onset of major increases in foraminifer and ostracod abundances (Figure 14.6A, Zone B). The statistically valid decrease of organic C (Figure 14.3B; 7.8 to 6.3%) commenced in the early 1950s and continued unabated until sediment recovery (1994), interrupted only by the Hurricane Donna event (~1959–62). Moreover, the transition from above- to below-average organic C temporally coincided with the onset on the late 1940s to early 1950s of a resurgence in watershed modification and impoundment of 700,000 acres of organic-rich soils during the creation of the EAA. Evidence from regional rainfall and freshwater flow data (Figure 14.15A) indicates that a high degree of coherence still existed between these parameters from before this transition period until 1960 and water flow trends appear unaltered. It is important to note that, despite these changes in the concentration of organic C at Oyster Bay, no co-occurring change in the $\delta^{13}$C (Figure 14.4D, $\Delta‰<1$) indicated that the source of the organic C had not changed.

Concurrent with this transition of decreased organic C and increased anthropogenic modification of the watershed was a dramatic increase in foraminifer and ostracod abundance at Oyster Bay (Figure 14.6A, Zone B). The abundance increase (*ca.* 1950) indicates only a biomass increase of the pre-1935 communities (similar diversities; Figure 14.6B) and implies the onset of conditions favorable to the existing communities of microfauna. The temporal correlation with decreased organic C and impoundment of the EAA suggests a possible water quality change that may have influenced these communities, but this cannot be verified with our current data.

The abundance increase was followed by an equally dramatic decline during the late 1970s to early 1980s (Figure 14.6A, Zone A). Existing data strongly indicate the ensuing abundance and diversity drop is related to the changing nature of the salinity field at Oyster Bay after ~1980. Specifically, Zones A and B (Figure 14.6A) were not only statistically different for both abundance and diversity but concurrently the $S_\sigma$ had changed by an amount observed by others to reduce macrobenthic biomass by nearly an order of magnitude (Tables 14.2 and 14.6). Montague and Ley (1993), in their study of seagrass and macrobenthic fauna in Florida Bay, showed that benthic biomass declined by an order of magnitude when the $S_\sigma$ increased by 3. Our data parallel this with a 9:1 decline in abundance associated with a $\Delta S_\sigma = 2.7$ between Zones A and B (Table 14.6). These observations provide a cogent argument for salinity-related changes as the driving factor for both decreased abundances and diversity between Zones A and B, but do not identify the cause of this salinity change. Although this abundance decline takes place at approximately a time of changing water-management strategy, more regional natural influences, such as regional rainfall, may also play a role, probably a dominant one, as noted earlier for Oyster Bay by ostracod $\delta^{18}O$ data (Figure 14.10).

Nearly concurrent with changes in Zones A to B at Oyster Bay (Figure 14.6A) are less abrupt but similar transitions at Jimmy Key (Figure 14.6C) that also link changing salinity fields with variability in microfaunal abundance. The Jimmy Key data indicate a statistically valid reduction in abundance which coincides with a rise in the $S_\sigma$ that parallels similar changes at Oyster Bay in both direction and approximate magnitude (Table 14.6). Moreover, the gradual decadal (Figure 14.6C, ~1970s+) abundance change observed for both foraminifera and ostracods at Jimmy Key appears to be a long-term, more natural transition also related to environmental changes such as salinity and rainfall. Comparison of these abundance changes to rainfall trends (Figure 14.15A) confirms this hypothesis. Specifically, rapid declines in rainfall during the late 1960s through early to mid-1970s (Figure 14.15A) correspond to above average salinities (Figure 14.13A) and the onset and peak of foraminifer and ostracod abundances at Jimmy Key. Increasing and above-average rainfall thereafter, until approximately the mid-1980s, corresponds to declining abundances (Figure 14.6C) and salinities (Figure 14.13A).

Embedded within the abundance data for Oyster Bay (Figure 14.6A) is a subfield of information derived from two critical species, the foraminifer *Ammonia parkinsoniana typica* and the ostracod *Peratocytheridea setipunctata*. Both field and culture studies (Phleger, 1956, 1966a,b; Keyser, 1976; Poag, 1978; Garbett and Maddocks, 1979; Murray, 1991) have shown these species to be tolerant of wide fluctuations in salinity which tends to reduce community diversity through stress. Specifically, between ~1980 and core recovery (1995), both microfaunal abundance and diversity are significantly lower than in the preceding two decades (Figure 14.6A, Zone A vs. B). Comparison of the absolute abundances for these two species (Figure 14.7A) mirrors the changes in abundance for the entire microfaunal community (Figure 14.6A). In contrast, their relative abundance, on the average, doubles upcore from Zone B to A, reaching intervals throughout the 1980s of >50% of the total population (Figure 14.7B). This indicates a stressed microfaunal community dominated by species which survived during prolonged periods of reduced flow and drought (Figure 14.15A). When viewed as a whole, microfaunal abundance and diversity, salinity, flow in SRS, and rainfall data for both Oyster Bay and Jimmy Key suggest that anthropogenic influences play a secondary role to natural influences such as regional rainfall.

## Anthropogenic Pb

In our study areas all anthropogenic Pb levels are <5 µg/g relative to a maximum of 20 µg/g for Mississippi Delta sediments (Trefry et al., 1985). The latter is a repository for the Mississippi River drainage basin with a watershed that encompasses ~41% of the continental United States. Lower levels of anthropogenic Pb in our Florida study areas result from a lower magnitude of Pb sources and lesser amounts of fine-grained clay to transport this Pb. Overall, the trends in anthropogenic Pb inputs vs. time show peak inputs during the 1960s to the 1980s with slight decreases during the 1990s (Table 14.1). In general, these trends result from the combined history of population growth in South Florida and the use of Pb additives in gasoline. Congressionally mandated regulations limiting Pb additives in gasoline were initiated during the 1970s and led to a continued decline in the use of Pb in gasoline since the late 1970s. Population growth in South Florida has continued to increase dramatically to the present (Table 14.1). This rapid growth of human activity in South Florida somewhat diminishes the effects of lower Pb additives in gasoline since the 1970s. While the overall impact of anthropogenic Pb in the sediments of Oyster, Coot, and Florida Bays is small, their record of inputs agree with established time lines for these cores.

# SUMMARY

Sediments from Oyster Bay adjacent to the outflow from the SRS, central Florida Bay near Jimmy Key, and Coot Bay account for nearly a century of natural and anthropogenic influences on the environment. When geochemical, biological, and physical data were interpreted in conjunction with historical regional rainfall, salinity, and freshwater flow from SRS, a pattern of natural and anthropogenic influences emerged. Our results indicated the following:

- *Hurricane Influences:* Paleo-hurricane influences in the study area, visible as light sediment layers and verified by offsets in $^{210}$Pb geochronology data, have imprinted the sediment record with signatures of immediate and long-term changes. Immediate consequences included abrupt changes in microfaunal abundance and community structure as well as quantitative organic C content. Long-term changes included the qualitative nature of macro-organic matter (mangrove leaf litter vs. root hairs) and microfaunal diversity. In central Florida Bay, decade-long trends of alternating sediment organic C buildup and reduction coincide respectively with periods of infrequent and more frequent hurricane storm activity. This supports major storm frequency as a potential controlling mechanism for carbon storage and removal in Florida Bay.
- *Regional rainfall and freshwater flow in Shark River Slough:* Rainfall at Homestead proved representative over the study area and when compared to SRS flow indicated a strong positive correlation during periods preceding construction of the present Water Conservation Areas. After construction, correlations between regional rainfall and flow in SRS varied with water management practices, ranging from essentially no correlation during the Monthly Allocation Plan to correlation approaching pre-management levels for the Rainfall Plan. Analysis of regional pollen identified taxa associations that allowed discrimination of pollen zonations from coastal mangrove to upland slough environments. This successfully allowed interpretation of the sediment record for historical periods of major to minor flushing from SRS. From the late 1940s to mid-1950s a dual transition occurred in the sediments near the outflow of the SRS. Organic C content permanently declined from above to below average and concurrent with this was the onset of major increases in foraminifer and ostracod abundances. These events temporally coincided with the construction of the Everglades Agricultural Area which impounded 700,000 acres of organic-rich swamp land and were not observed for sediments representing the same time period in central Florida Bay.
- *Salinity trends and microfaunal responses:* Changes in salinity, both near the outflow of the SRS and in central Florida Bay, showed a direct response to regional rainfall from the decadal to the monthly time scales. At both sites, foraminifer and ostracod data also indicate direct correlation to rainfall patterns for temporal scales ranging from decadal down to the limit-of-resolution of geochronology. At Oyster Bay, ostracod stable isotope ($\delta^{18}$O) trends correlated better with variations in regional rainfall than with freshwater outflow from the adjacent SRS. An order-of-magnitude drop in foraminifera and ostracod abundances in the late 1970s at Oyster Bay and a more gradual decrease in the mid-1970s at Jimmy Key did not correlate to changes in mean salinity, but rather to increases in the standard deviation of salinity ($S_\sigma$). Stable isotope ($\delta^{18}$O, $\delta^{13}$C) trends for ostracods and foraminifera at both sites showed mixed signals with most data suggesting upcore trends to more marine conditions since the turn of the century.
- *Anthropogenic Pb:* Anthropogenic Pb was observed at all core sites, although it was significantly lower than at other documented sites such as the Mississippi Delta. As at the Mississippi Delta, however, trends of anthropogenic Pb in the sediments covaried with the rise and fall of leaded gasoline use in the U.S.

## CONCLUSIONS

The South Florida ecosystem is characterized by a low-relief transition between peninsular Florida and the adjacent shallow coastal zone. Natural features such as the Shark River Slough form a major coastward conduit for freshwater and terrestrial materials, while major storms that impact the area, such as hurricanes, can provide energy for landward transport of marine sediments. Coastal sediments serve as reservoirs of historical information critical to understanding the natural and anthropogenic impacts on this ecosystem. To decipher this history we investigated the physical, biological, and chemical nature of these sediments at locations indicative of the continental–coastal transition as well as central Florida Bay. Interpretation of results was facilitated by comparison with existing regional salinity, rainfall, and freshwater flow data. Placed within the context of geochronology, sediment sequences were interpretable to about the turn of the century, thus making them ideal for contrasting time of minimal anthropogenic impact to present conditions.

Results indicated that major storms, such as the great Labor Day Hurricane of 1935 and Hurricane Donna (1960), accounted for both immediate and long-term impacts on the ecosystem. Immediate impacts were sediment erosion and redeposition, microfaunal changes in abundance and community diversity, and changes in the organic carbon content of sediments. On a decadal scale, changes in seaward transport of macro-organics resulted from the destruction of fringing mangrove forests and formed a distinct post-hurricane signature.

Salinity records varied on decadal to monthly time scales and correlated with changing patterns in regional rainfall. Moreover, regional rainfall, represented by the 80+ year record at Homestead, Florida, indicated high correlation with flow into Shark River Slough prior to major watershed construction which began in the early 1960s. During subsequent periods of water management strategies, enacted from the mid-1960s to present, results indicate essentially no correlation between regional rainfall and flow during the Monthly Allocation Plan. In contrast, correlations most closely paralleled pre-construction, apparently more natural conditions during the subsequent Rainfall Plan. A nontraditional use of pollen allowed evaluation of the degree of paleo-flushing from Shark River Slough that not only correlated well with existing flow and rainfall records but also suggested validity as a flushing proxy for pre-record times.

Temporally coherent with the early 1950s creation of the Everglades Agricultural Area, recovered sediments adjacent to the mouth of Shark River underwent a dual transition with a statistically significant upcore reduction in organic carbon content to present and a concurrent significant increase in microfaunal abundance. This was not observed in contemporaneous sediments from Florida Bay. Investigated parameters for the benthic microfaunal community (foraminifera and ostracods) population characteristics such as stable isotopic compositions, abundance, and community diversity exhibited changes and trends that more closely paralleled natural rather than anthropogenic influences. Changes in the stable isotopic values of the microfauna indicated, within the limits of our geochronology, direct responses to regional rainfall. Such responses more closely paralleled rainfall as opposed to freshwater runoff, even adjacent to the outflow of the Shark River Slough. Furthermore, long-term trends for both stable isotopes and relative abundance of salinity-sensitive species indicated a statistically valid upcore trend toward less fresh, more marine conditions at both study sites. This trend was contemporaneous with a weak decline in regional rainfall over the same time span. Moreover, crashes in microfaunal abundances at Oyster Bay and more gradual declines at Jimmy Key correlated not with changes in mean salinity, but rather with changes in the standard deviation of salinity ($\Delta S_\sigma$). An order of magnitude reduction in abundance corresponded with increased $\Delta S_\sigma \cong 3$ in Oyster Bay. This abundance drop was concurrent with an equally dramatic decline in community diversity characterized by survivor-type dominance by two microfaunal species and occurred over a period of drought at both sites as well as a period of reduced flow from Shark River Slough.

The heavy metal Pb, observed in all cores, was elevated above background by mid-century but the level of anthropogenic lead was minimal compared to major continental depocenters such as the Mississippi Delta. However, like the Mississippi Delta, which integrates the signal from 41% of the continental United States, the anthropogenic Pb observed in sediments recovered from Oyster Bay covaried with rises and declines in the use of leaded gasoline in the United States.

## ACKNOWLEDGMENTS

Salinity data were provided by Drs. J. Boyer and R. Jones at Florida International University/SERP as funded by the South Florida Water Management District and Everglades National Park. The HisSal05 dataset was provided by Dr. M. Robblee, T. Ross, of the National Climate Data Center, provided rainfall data; and the USGS provided flow data for the Shark River Slough. Dr. T. Schmidt provided useful historical information on Florida Bay and the Everglades during numerous conversations. A. Risi, B. Michaels, and J. Schull provided both field and laboratory assistance. S. Houston provided valuable advice on hurricanes and output data from the NOAA SAFER model. To the above we are grateful for their cooperation and assistance.

## REFERENCES

Ball, M.M., E.A. Shinn, and K.W. Stockman. 1967. The geologic effects of Hurricane Donna in south Florida, *J. Geol.*, 75:583.

Bancroft, G.T., A.M. Strong, R.J. Sawicki, W. Hoffman, and S.D. Jewell. 1994. Relationship among wading bird foraging patterns, colony locations, and hydrology of the Everglades, in *Everglades: The Ecosystem and Its Restoration*, Davis, S.M. and Ogden, J.C., Eds., St. Lucie Press, Delray Beach, FL, p. 615.

Benninghoff, W.S. 1962. Calculation of pollen and spore density in sediments by addition of exotic pollen in known quantities, *Pollen et Spores*, 4:332.

Berglund, B.E. 1973. Pollen dispersal and deposition in an area of southeast Sweden: some preliminary results, in *Quaternary Plant Ecology*, Birks, H.J.B. and West, R.G., Eds., Blackwell Scientific, Oxford, p.117.

Bonny, A.P. 1976. Recruitment of pollen to the seston and sediment of some Lake District lakes, *J. Ecology*, 64:859.

Boyer, J.N., J.W. Fourqurean, and R. D. Jones. 1997. Spatial characterization of water quality in Florida Bay and Whitewater Bay by multivariate analyses: zones of similar influence, *Estuaries*, 20:743.

Caratini, C., F. Blasco, and G. Thanikaimoni. 1973. Relation between the pollen spectra and the vegetation of a south Indian mangrove, *Pollen et Spores*, 15:281.

Cohen, A.D. 1968. The Petrology of Some Peats of Southern Florida (with Special Reference to the Origin of Coal), Doctoral dissertation, Pennsylvania State University, University Park.

Craighead, F.C. and V.C. Gilbert. 1962. The effects of Hurricane Donna on the vegetation of southern Florida, *Q. J. Florida Acad. Sci.*, 25:1.

Cross, A.T., G.G. Thompson, and J.B. Zaitzeff. 1966. Source and distribution of palynomorphs in bottom sediments, southern part of Gulf of California, *Mar. Geol.*, 4:467.

Cutshall, N., I.L. Larsen, and C.R. Olsen. 1983. Direct analysis of $^{210}$Pb in sediment samples: self-absorption corrections, *Nucl. Instr. Methods*, 206:309.

Davis, J.H. 1943. The Natural Features of Southern Florida, Especially the Vegetation and the Everglades, Bull. No. 25, Florida Geological Survey, Tallahassee, FL.

Duever, M.J., J.F. Meeder, L.C. Meeder, and J.M. McCollom. 1994. The climate of south Florida and its role in shaping the Everglades ecosystem, in *Everglades: The Ecosystem and Its Restoration*, Davis, S.M. and Ogden, J.C., Eds., St. Lucie Press, Delray Beach, FL, 225.

Finney, B.P. and C.A. Huh. 1989. History of metal pollution in the southern California Bight: an update, *Environ. Sci. Tech.*, 23:294.

Frederick, B.C. 1994. The Development of the Holocene Stratigraphic Sequence within the Broad-Lostman's River Region, Southwest Florida Coast, Masters' thesis, University of Miami, Coral Gables, FL.

Garbett, E.C. and R.F. Maddocks. 1979. Zoogeography of Holocene Cytheracean Ostracodes in the Bays of Texas, *J. Paleontol.*, 53:841.

Goldberg, E.D., J.J. Griffin, V. Hodge, M. Koide, and H. Windom. 1979. Pollution history of the Savannah River estuary, *Environ. Sci. Techol.*, 13:588.

Gunderson, L.H. 1994. Vegetation in the Everglades: determinants of community composition, in *Everglades: The Ecosystem and Its Restoration*, Davis, S.M. and Ogden, J.C., Eds., St. Lucie Press, Delray Beach, FL, 323.

Keyser, D. 1976. Ecology and zoogeography of recent brackish-water Ostracoda (Crustacea) from southwest Florida, Sixth International Ostracod Symposium, Saalfelden, 207.

Koide, M., K.W. Bruland, and E.D. Goldberg. 1973. Th-228/Th-232 and Pb-210 geochronologies in marine and lake sediments, *Geochim. Cosmochim. Acta*, 37:1171.

Light, S.S. and J.W. Duneen. 1994. Water control in the Everglades: a historical perspective, in *Everglades: The Ecosystem and Its Restoration*, Davis, S.M. and Ogden, J.C., Eds., St. Lucie Press, Delray Beach, FL, 47.

Loveless, C.M. 1959. A study of the vegetation of the Florida Everglades, *Ecology*, 40:1.

MacArthur, R.H. 1983. Patterns of species diversity, in *Diversity*, Patrick, R.P., Ed., Hutchinson Ross Publishing, Stroudsburg, 14.

Manker, J.P. and G.M. Griffin. 1969. Distribution of silicate minerals in Florida Bay: geology of the American Mediterranean, *Geol. Soc. Trans.*, 19:505.

Meeder, J.F., R. Jones, J.J. O'Brien, M.S. Ross, R.J. Sawicki, and A.M. Strong. 1994. Effects of Hurricane Andrew on Thalassia ecosystem dynamics and the stratigraphic record, *Bull. Mar. Sci.*, 54:1080.

Meyers, J., P.K. Swart, and J. Meyers. 1993. Geochemical evidence for groundwater behavior in an unconfined aquifer, south Florida, *J. Hydrol.*, 148:249.

Montague, C.L. and J.A. Ley. 1993. A possible effect of salinity fluctuation on abundance of benthic vegetation and associated fauna in northeastern Florida Bay, *Estuaries*, 16:703.

Murray, J.W. 1991. *Ecology and Paleoecology of Benthic Foraminifera*, John Wiley & Sons, New York.

Orlando, S.P., M.B. Robblee, and C.J. Klein. 1997. *Salinity characteristics of Florida Bay: A Review of the Archived Data Set (1955–1995)*, NOAA/Office of Ocean Resources, Conservation, and Assessment, 38 pp.

Parker, G.G. and C.W. Cooke. 1944. Late Cenozoic geology of southern Florida, with a discussion of ground water, *Florida Geol. Surv. Bull.*, 27:119.

Parker, F.L., F.B. Phleger, and J.F. Peirson. 1953. *Ecology of Foraminifera from San Antonio Bay and Environs, Southwest Texas*, Contributions for the Cushman Foundation for Foraminiferal Research, special paper 2:1.

Peirson, D.H. 1971. Worldwide deposition of long-lived fission products from nuclear explosions, *Nature*, 234:78-80.

Perkins, R.D. and P. Enos. 1968. Hurricane Betsy in the Florida–Bahamas area: Geologic effects and comparison with Hurricane Donna, *J. Geol.*, 76:710.

Phleger, F.B. 1956. *Significance of Living Foraminiferal Populations Along the Central Texas Coast*, Contributions for the Cushman Foundation for Foraminiferal Research, 7:106.

Phleger, F.B. 1966a. Living foraminifera from coastal marsh, southwestern Florida, *Boletin de la Sociedad Geologica Mexicana*, 28:45.

Phleger, F.B. 1966b. Patterns on living marsh foraminifera in south Texas coastal lagoons, *Boletin de la Sociedad Geologica Mexicana*, 28:1.

Poag, C.W. 1978. Paired foraminiferal ecophenotypes in Gulf coast estuaries: ecological and paleoecological implications, *Trans. Gulf Coast Assoc. Geol. Soc.*, 28:395.

Riegel, W.L. 1965. Palynology of Environments of Peat Formation in Southwestern Florida. Doctoral dissertation, Pennsylvania State University, University Park, 189 pp.

Schink, J.C., J.H. Stockwell, and R.A. Ellis. 1978. An improved device for gasometric determination of carbonate in sediment, *J. Sediment. Petrol.*, 48:651.

Smith, T.J., M.B. Robblee, H.R. Wanless, and T.W. Doyle. 1994. Mangroves, hurricanes, and lightning strikes, *BioScience*, 44:256.

Stockmarr, J. 1971. Tablets with spores used in absolute pollen analysis, *Pollen et Spores*, 8:615.

Swart, P.K., G.F. Healy, R.E. Dodge, P. Kramer, J.H. Hudson, R.B. Halley, and M.B. Robblee. 1996. The stable oxygen and carbon isotopic record from a coral growing in Florida Bay: a 160 year record of climatic and anthropogenic influence, *PALAEO*, 123:219.

Tabb, D.C. and A.C. Jones. 1962. Effect of Hurricane Donna on the aquatic fauna of North Florida Bay, *Trans. Am. Fish. Soc.*, 9:375.

Trefry, J.H. and S. Metz. 1984. Selective leaching of trace metals from sediments as a function of pH, *Anal. Chem.*, 56:745.

Trefry, J.H., S. Metz, R.P. Trocine, and T.A. Nelsen. 1985. A decline in lead transport by the Mississippi River, *Science*, 230:439.

van Geel, B. 1978. A paleoecological study of Holocene peat bog sections in Germany and the Netherlands, based on the analysis of pollen, spores, and macro- and microscopic remains of fungi, algae, cormophytes, and animals, *Rev. Paleobot. Palynol.*, 25:1.

Wanless, H.R. and M.G. Tagett. 1989. Origin, growth, and evolution of carbonate mudbanks in Florida Bay, *Bull. Mar. Sci.*, 44:454.

Wanless, H.R., L.P. Tedesco, V. Rossinsky, and J.J. Dravis. 1989. Carbonate environments and sequences of Caicos platform with an introductory evaluation of south Florida, in *28th International Geologic Congress Field Trip Guidebook*, T374, American Geophysical Union, 75 pp.

Wanless, H.R., R.W. Parkinson, and L.P. Tedesco. 1994. Sea level control on stability of Everglades Wetlands, in *Everglades: The Ecosystem and its Restoration*, Davis, S.M. and Ogden, J.C., Eds., St. Lucie Press, Delray Beach, Fl., 199–223.

Willard, D.A. and C.W. Holmes. 1997. *Pollen and Geochronological Data from South Florida: Taylor Creek Site 2*, Open-File Report 97-35, U.S. Geological Survey, Reston, VA.

Wood, G.D., A.M. Gabriel, and J.C. Lawson. 1996. Palynological techniques: processing and microscopy, in *Palynology: Principles and Applications*, Jansonius, J. and McGregor, D.C., Eds., American Association of Stratigraphic Palynologists Foundation, Salt Lake City, UT, 10, 29.

# 15 Rapid Remote Assessments of Salinity and Ocean Color in Florida Bay

*Eurico J. D'Sa and James B. Zaitzeff*
Office of Research and Applications, NOAA/NESDIS

*Charles S. Yentsch*
Bigelow Laboratory for Ocean Science

*Jerry L. Miller*
Naval Research Laboratory

*Russell Ives*
NOAA/NOS

## CONTENTS

## ABSTRACT

We investigated a multisensor approach using an airborne salinity mapper and a hyperspectral radiometer together with satellite ocean color data to map salinity and water quality indicators in Florida Bay in experiments conducted in October 1997 and February 1998. The airborne salinity mapper provided the first synoptic view of sea surface salinity distribution over large regions of the bay. A salinity map of the central interior bay reveals a low-salinity regime due to freshwater discharge from the Everglades. The intricate network of mudbanks and small islands in the region also appears to influence the salinity distribution. In the western region of the bay around Cape Sable, the salinity image shows the extent of freshening of the southwest Florida Shelf waters due to freshwater discharge from Shark River Slough and its indirect influence on bay salinity. Ocean color images obtained with an airborne hyperspectral radiometer that was flown along with the salinity mapper demonstrated its utility by optically recording the effects of a storm that passed through the region. Phytoplankton chlorophyll estimated from SeaWiFS ocean color satellite data generally indicated an increase in concentration with decreasing salinity due to possible effects of enhanced nutrients. We propose that synoptic airborne salinity mapping, together with monitoring of nutrients and ocean color, is an essential water management tool that can be used for estimating freshwater budgets to aid the restoration efforts in Florida Bay.

## INTRODUCTION

The Florida Bay estuary is a complex and dynamic environment that has experienced a variety of environmental changes that has affected its sensitive ecosystem (Robblee et al., 1991; Lapointe and Clark, 1992; Phlips and Badylak, 1996). Anthropogenic changes in water drainage patterns in South Florida have significantly reduced and altered the pattern of flow into Florida Bay (Light and Dineen, 1994) that may have contributed, along with drought conditions, to periods of prolonged hypersalinity in large areas of the bay (Fourqurean et al., 1992). A major indicator used in monitoring and predicting the health, hydrography, and habitat potential of the bay has been salinity (Montague and Ley, 1993). It plays a significant role in the biological and physical/chemical processes in coastal waters, and together with bio-optical indicators such as light attenuation (Stumpf et al. 1999), phytoplankton pigments, and nutrients (Boyer et al. 1999), is essential in studying water quality in the bay. Factors controlling salinity and optical properties in Florida Bay are quite complex. Rather than the large, open system it appears to be on maps, the bay is made up of many shallow basins that are partially separated by an intricate network of banks. Hydrologic and hydrodynamic models being developed to predict freshwater inflows and freshwater salinity relationships use existing time-series data from monitoring stations in the bay. Though invaluable in developing and calibrating these models, the data are restricted in time and space due to the small number of stations normally occupied. Given the complexity of bay topography and the spatial and temporal patterns of freshwater inflow, our understanding of salinity and ocean color properties would greatly benefit from synoptic, broad-scale, and dense measurements that could be obtained from remote sensing.

The ability to map salinity remotely is an important breakthrough in developing low and high-altitude platforms for coastal oceanic studies (Lagerloef et al., 1995; Le Vine et al., 2000). Recent research experiments with NOAA's scanning low-frequency microwave radiometer (SLFMR) conducted in the Chesapeake Bay and offshore, Florida Bay, and in the coastal waters of South Carolina (Miller et al., 1996; Goodberlet et al., 1997; D'Sa et al., 2000a) demonstrated the proof-of-concept and operational capability of airborne salinity data acquisition. Maps of sea surface salinity were produced with an accuracy of about 1 practical salinity unit (psu) when compared to *in situ* measurements. The passive microwave radiometer also measures sea surface temperature at nadir with an accuracy of about 0.5°C. This study describes the first remote monitoring of the sea surface salinity field in Florida Bay made in conjunction with ocean color and *in situ* measurements of salinity and bio-optical variables. The high resolution and synoptic coverage of the salinity field measured by the airborne microwave radiometer provides the capability to enhance the calibration and testing of the hydrodynamic models and reveals details of salinity contours that would be missed by *in situ* measurements.

Simultaneous remote measurements of salinity and ocean color extend our monitoring capability to important water quality indicators such as turbidity, phytoplankton pigment distribution, suspended sediment, and colored dissolved organic matter. In open ocean, or case 1 waters, reliable estimates of phytoplankton biomass have been obtained using ocean color algorithms (Gordon and Morel, 1983) mainly because optical properties in these waters are dominated by phytoplankton chlorophyll and associated and covarying detrital pigments. In coastal, or case 2 waters, correlations between phytoplankton, suspended sediments, and colored dissolved organic matter often exhibit large spatial and temporal variations (Tassan, 1994) due to local processes such as freshwater inflows, bottom resuspension, and tidal effects making it difficult to obtain accurate estimates using remote sensing data. Additionally, bottom effects and variable concentrations of suspended sediments in shallow coastal waters contribute to the water-leaving radiance and complicates satellite retrieval of marine constituents using ocean color remote sensing algorithms. However, recent algorithms used in processing SeaWiFS (Sea-Viewing Wide Field-of-View Sensor) satellite data (O'Reilly et al., 1998; Hu et al., 2000; Siegel et al., 2000) have shown improvements in SeaWiFS retrievals of chlorophyll pigment concentrations in coastal waters. To complement satellite data, we also used an airborne hyperspectral imaging radiometer (AISA) to obtain ocean color data at shorter scales of spatial and temporal resolution. This pilot research and applications study involving airborne salinity and ocean color mapping has been conducted to aid in designing a water quality monitoring program to guide water management decisions concerning Florida Bay.

## MATERIALS AND METHODS

Florida Bay is a shallow tropical bay about 2000 km$^2$ that is bounded by the mainland to the north and the islands of Florida Keys to the southeast. It has depths of 1 to 2 m with deeper waters occurring westwards from the bay into the Gulf of Mexico. Experiments were conducted over a one-week period in early October 1997 and January/February 1998 and were comprised of simultaneous shipborne and aircraft measurements of physical and bio-optical variables that included salinity, temperature, ocean color, and chlorophyll concentrations. Shipboard transects were made on alternate days that covered central-interior and the outer bay regions. Remote salinity and ocean color data were obtained using a single-engine De Havilland Beaver aircraft flying primarily at about 1 and 2.6 km altitude over the bay, with each survey run generally taking 3 hours or less. *In situ* salinity and bio-optical variables such as chlorophyll *a*, total suspended sediments, and colored dissolved organic matter (CDOM) were obtained at predetermined sampling stations (Figure 15.1, top panel).

A scanning low-frequency microwave radiometer operating at 1.413 GHz (Goodberlet et al., 1997) was located on a single-engine aircraft for remotely measuring sea surface salinity and temperature. Radiometric brightness temperature is sensitive to the electrolytic (salt) content of water through the influence of dielectric properties on the surface emissivity. Salinity is found in terms of emissivity $e(S, Ts, \theta)$, where the radiation temperature from the ocean surface is a function of salinity $S$, sea surface temperature $T_s$, and the incidence angle $\theta$ (Blume et al., 1978). Salinity measurements with the airborne microwave radiometer were compared and calibrated with *in situ* measurements made with a Hydrolab CTD system and had accuracies of 1 psu. Airborne transects were conducted over Florida Bay by flying east–west (Figure 15.1, bottom panel) or north–south transects that were spaced 2.5 nautical miles (at 2.6-km altitude) and consisted of several runs over a predetermined ship-track such that the aircraft passed over the ship several times as the ship progressed along the track. At a nominal altitude of 2.6 km and a speed of 100 knots, the pixel (ground spatial resolution) was about 1 km$^2$.

The AISA hyperspectral imaging spectrometer (Okkonen et al., 1997), mounted on the aircraft along with the salinity mapper, was programmed to acquire multispectral data at wavelengths restricted to the SeaWiFS satellite wavebands centered at 443, 490, 510, 555, 670, 765, and 865 nm. A SeaWiFS (1.1-km resolution) satellite image of October 4, 1997, acquired from NASA, was processed for the Florida Bay region using the NASA standard processing software SeaDAS V. 4.0.

## RESULTS AND DISCUSSION

Airborne microwave images of sea surface salinity of Florida Bay regions obtained with the salinity mapper are shown in Color Figures 15.1 and 15.2.* A salinity image obtained in the western region of the bay around Cape Sable (Color Figure 15.1) on October 3, 1997, reveals the influence of freshwater discharge from the Shark River Slough and the flow of low salinity waters along the

---

* Color figures follow page 648.

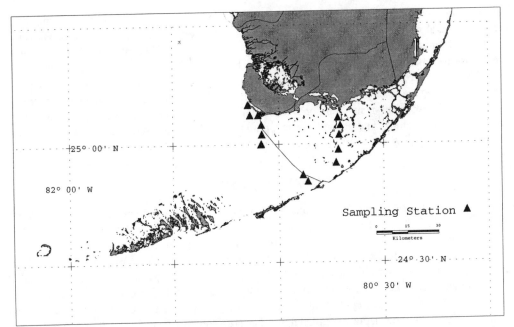

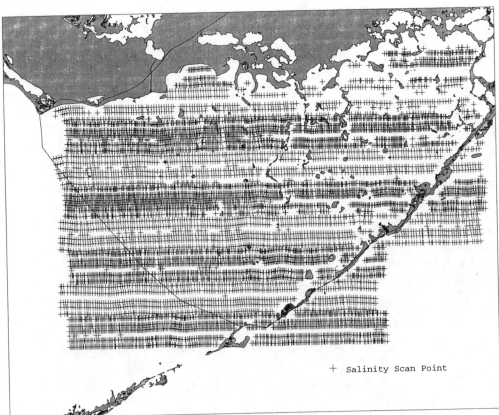

**FIGURE 15.1** Aircraft flight path at about 2.6-km altitude over Florida Bay showing the acquisition of salinity data (+) during a 3-hour survey to map the salinity field with the airborne microwave radiometer. Image pixel resolution is about 1 km². Top panel shows the sampling station locations (▲) in the central interior and outer Bay.

East Cape, thus indicating its indirect influence on the bay salinity. A high-resolution salinity image of a region in the central interior bay (Color Figure 15.2) obtained on October 4, 1997, with the aircraft flying at about 1-km altitude reveals many interesting features in the salinity field, such as lower salinities to the northeast of the bay, with the small islands in the region appearing to constrain water-mixing resulting in non uniform salinity distributions. Low-salinity values observed in some discrete spots in the image could be artifacts or local hotspots of freshwater sources that require further investigation. Overall, sea surface salinity in Florida Bay exhibited a typical salinity pattern with lower and variable salinity in the east and increasing to the west and south (D'Sa et al., 2000a). Along western and southern boundaries of the bay, influence of the Gulf of Mexico and Atlantic Ocean waters dominated as indicated by the higher salinity field in the region. Relationships between remotely measured salinity and total chlorophyll pigment concentration and CDOM absorption at 400 nm measured at the central interior and outer bay stations in October 1997 have been investigated (D'Sa et al., 2000a). Chlorophyll concentrations at stations in the inner bay generally increased northwards and correlated with increasing CDOM concentrations and decreasing salinity levels (indicating nutrient-enhanced waters to the north). Chlorophyll pigment estimates obtained from SeaWiFS data (Color Figure 15.3) were successful in retrieving chlorophyll from a large area of the bay and generally indicated an increase in chlorophyll concentrations with decreasing salinities. This suggests a possible influence of increased nutrients on the pigment distributions. However, the satellite retrieval algorithms were not able to retrieve chlorophyll from the interior regions of the bay, possibly due to the effects of bottom reflectance and atmospheric correction failure (pixels shown masked in Color Figure 15.3). Results using an airborne hyperspectral radiometer (AISA) are shown (in Figure 15.2 and Color Figure 15.4) to demonstrate its utility in detecting localized events at small spatial resolutions. An east–west transect (AB980131) taken with the airborne radiometer over the northern region of the bay on January 31 (Figure 15.2) was repeated on February 5, 1998, and the enlarged color composite images are shown in Color Figure 15.4. Features related to bottom topography, channel flows, and bottom vegetation are observed in the pre-storm color composite transect image (Color Figure 15.4, top panel). Water depth is also seen to vary and is reflected in the variable shades in the color image. Areas in white correspond to shallow regions with water depths around 0.3 m and high bottom reflectance, while areas in red correspond to small islands with vegetation. The effect of the passage of a storm front on February 2, 1998, that was accompanied by 3 to 4 inches of precipitation over a 2-day period is evident in the color composite image of the same region obtained on February 5 (Color Figure 15.4, bottom panel). This image appears to be without the many features seen on the pre-storm transect. High concentrations of suspended material in the water column, possibly resuspended from the bottom sediments by the frontal passage, appear to be the main reason for the absence of bottom reflectance in the shallow areas observed in the pre-storm transect. These images obtained with the airborne radiometer show the potential of using airborne ocean color remote sensing to monitor small-scale events over shorter time frames than possible with satellite ocean color images (Harding et al., 1992).

Major freshwater sources influencing the salinity field in Florida Bay include precipitation, the Biscayne Aquifer (groundwater), and surface water from Taylor River Slough and Shark River Slough, and sheetflow from the lower Florida Bay watershed (Orlando et al., 1997). The interaction of bay waters with the Atlantic Ocean and the Gulf of Mexico and their tidal effects are the other major factors determining salinity in the bay. Exchanges with the Atlantic are through several passes in the Florida Keys, while the Gulf waters directly influence salinity on the western and northwestern regions of the bay. The extensive mudbank system, however, dampens tidal effects and reduces its influence in the central and northeastern bay, resulting in nonuniform salinity distributions. Gulf waters entering the bay primarily from the northwest (near East Cape) entrain freshwater from the Shark River Slough and other smaller rivers in southwest Florida as indicated by the lower salinities observed in the vicinity of Cape Sable (Color Figure 15.1). A high-resolution

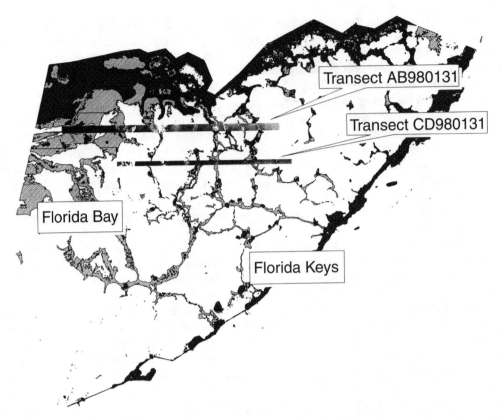

**FIGURE 15.2** An example of two flight lines (transects AB980131 and CD980131) showing regions of the bay imaged by the airborne hyperspectral radiometer.

salinity map (Color Figure 15.2) of the central interior bay reveals salinity features that are determined to a large extent by the intricate network of mudbanks and small islands in the region.

The reason for the presence of some localized freshwater pixels in the image is uncertain. It could be measurement artifacts of the scanning microwave radiometer, presence of small islands of sub-pixel resolution, or hotspots of localized freshwater sources. Previous studies (Orlando et al., 1997) have indicated that aquifers located to the north and separated by limestone and sand boundary along the northern boundary of Florida Bay may permit seepage along the boundary. However, the extent of the seepage is unknown. The location of these low-salinity spots on the image provides an opportunity to identify the presence of any groundwater sources in the bay. Salinity in Florida Bay is known to follow a seasonal pattern that is influenced by the wet season (June–October) and the dry season (January–May). Salinity distributions included near-ocean salinities along the western and south boundaries of the bay, lower salinity region in the northeastern bay, and in recent times the appearance of a hypersaline area in the north-central bay (Robblee, 1997). With the experiment being conducted towards the end of an unusually wet season influenced by the El Niño event, a lower than usual salinity field appears to have prevailed with no presence of hypersaline conditions being observed in the northcentral bay.

One of the major environmental problems of the decade concerns the so-called restoration of water flow through the Everglades ecosystem and a characteristic set of problems associated with 40 years of land development. The development involves land and water for recreational, retirement, growth, and agricultural interests. Most of this converges on the southern reaches of the Florida peninsula, which originally was a shallow river with sluggish flow into the headwaters of Florida Bay. Diversion of this flow into canals was a major part of land development. These canals diverted the normal flow roughly at right angles into the Atlantic and Gulf coastal regions. For two decades environmentalists believed the Everglades was running out of water and they argued that the flow must be increased to sustain the unique ecosystem. This was the crux of the restoration plan, with federal support being used to divert and pump waters through the system. A problem with the plan was that much of the water to be diverted to the South contains large amounts of nutrients. The advocates of the "watering" plan stressed that damage to local ocean waters, by eutrophication, would not occur because the nutrients would be absorbed by the dense vegetation in the Everglades, before entering Florida Bay.

In 1987, environmentalists and fisherman noted massive die-off of the resident seagrasses in Florida Bay (Robblee et al., 1991; Robblee and Daniels, 1997). This was associated with a decrease in water clarity and the disappearance of bottom fishes. Some argued that the decline of the Florida Bay ecosystem was due to increased salinity related to reduction of freshwater entering the system. Others believe that increasing nutrients are the cause and further increases in flow will further damage this tropical ecosystem. Florida Bay, once a shallow lagoonal system with turtle grass and solitary coral, is now dominated by blooms of micro- and macroalgae. Declining transparency of the waters is affecting the growth of fringing corals, which exhibit diseases that appear to be borne by freshwater. The nutrient chemistry of the freshwater flowing into Florida Bay is further evidence for eutrophication. Normal nitrogen to phosphorus ratios for ocean water is 10–5:1, waters flowing into Florida Bay are 100:1 or greater! (Lapointe et al., 1994). Clearly, the ability of the Everglades to absorb the agriculture nutrients has been overestimated.

Modeling the uptake of nutrients by the Everglades is a formidable task if the goal is to be an effective environmental management tool. To return Florida Bay to an original state prior to eutrophication requires that the waters entering Florida Bay have nearly unmeasurable concentrations of nitrogen and phosphate. The key to this is effective monitoring of nutrients, at selected sites, and large-scale measurements of salinity. A short-term view of the salinity patterns throughout Florida Bay is available through the efforts of the Florida Geological Survey. Their observations document the freshening of Florida Bay during the period of 1994–1996. During this period, the highest salinities recorded in the Bay were 30 psu, comparable to a large estuary such as Chesapeake Bay. Central regions of Florida Bay were about 15 psu; waters around the Keys adjacent to nearshore coral reefs were about 20 psu. The parent salinity of the ocean waters is somewhat greater than 3.8% (Orlando et al., 1997), which means that the volume of Florida Bay was made up of 10% of the freshwater inflow. During this period, the exact nitrogen and phosphate concentrations in the freshwater are unknown. Recent measurements show that the freshwater runoff is highly enriched with nitrogen, some 100 to 1000 times higher than the oceanic waters. It would seem likely that the entire bay waters had a minimal concentration of enrichment of 10%, whereas the northeastern areas contained 50 to 100% of the enriched freshwater.

The discrete measurement of salinity is time consuming and expensive. At the present time, it requires the efforts of three agencies and in some cases does not give a synoptic representation of the salinity patterns in Florida Bay. If the nutrient attenuation model by Everglade vegetation is to be followed, then it is essential that an extensive program of measurement of salinity, critical biochemical nutrients, and bio-optical variables be measured. We have demonstrated that this is possible and that agencies responsible for overseeing this environment could utilize airborne remote sensing of salinity as a means for estimating the freshwater budget entering Florida Bay. Such a program coupled with selected sites for nutrient analysis could provide the basis for the estimate of nutrient influx by freshwater into Florida Bay.

## CONCLUSION

This pilot field experiment using shipboard and aircraft measurements allowed the development and testing of sampling strategies and data analysis that could be incorporated in future monitoring efforts in Florida Bay. Results from this and other studies have indicated that the airborne salinity mapper can map sea surface salinity regimes at the rate of 100 km²/hr with salinity accuracies of 1 psu. A new salinity mapper currently under construction is anticipated to have an accuracy of 0.1 psu. The salinity mapper provided the first high-resolution description of surface salinity field in Florida Bay. Synoptic measurements over large areas of the bay revealed details of salinity distributions not possible with *in situ* measurements. Detection of salinity anomalies such as possible sources of freshwater in the bay is feasible, as revealed by the high-resolution image obtained of the central-interior bay. Inner bay salinity distributions were closely linked to colored dissolved organic matter absorption and provide the potential of using salinity measurements to determine dissolved organic matter concentrations for use in remote sensing algorithms. Recent remote sensing ocean color algorithms (Carder et al., 1999) that estimate CDOM distributions were used to derive salinity from SeaWiFS data based on a conservative salinity–CDOM mixing behavior (D'Sa et al., 2000b). Airborne salinity measurements could be used to verify these estimates over large spatial scales.

Even if small salinity fluctuations will not harm most marine organisms, salinity is still the best clue as to where a particular water mass originated. Specifically, any reduction or elevation in salinity, no matter how small, indicates that the water mass has interacted with terrestrial runoff (hyposalinity) or shallow, nearshore embayments (hypersalinity). This chapter demonstrates the utility of using the salinity signal to assess the origin of a water mass in South Florida. At present, specific applications of the salinity mapper are in regard to water management to restore the Florida Everglades and the environmental quality of Florida Bay and its fringing coral reefs.

## ACKNOWLEDGMENTS

The authors would like to thank the USGS, John Hunt from FMRI, and Nancy Diersling for their help in this project. We also thank J. Klein and B. Warner of NOAA and R. Steward from the University of South Florida for their support during field operations. Jim Zaitzeff passed away during the preparation of this manuscript. His vision and efforts were largely responsible for the success of this work.

## REFERENCES

Blume, H.-J., B.M. Kendall, and J.C. Fedors. 1978. Measurement of ocean temperature and salinity via microwave radiometry, *Bound. Layer Meteorol.*, 13:295–308.

Boyer, J.N., J.W. Fourqurean, and R.D. Jones. 1999. Seasonal and long-term trends in water quality of Florida Bay (1989–1997), *Estuaries*, 22:417–430.

Carder, K.L., F.R. Chen, Z.P. Lee, S.K. Hawes, and D. Kamykowski. 1999. Semianalytic moderate-resolution imaging spectrometer algorithms for chlorophyll *a* and absorption with bio-optical domains based on nitrate-depletion temperatures, *J. Geophys. Res.*, 104:5403–5421.

D'Sa, E.J., J. Zaitzeff, and R.G. Steward. 2000a. Monitoring water quality in Florida Bay with remotely sensed salinity and in situ bio-optical observations, *Int. J. Remote Sensing*, 21:811–816.

D'Sa, E.J., C. Hu, and F.E. Muller-Karger. 2000b. Estimation of colored dissolved organic matter and salinity fields in case 2 waters using SeaWiFS: examples from Florida Bay and Florida Shelf, pp. 34–38, in Proc. PORSEC 2000, Goa, India.

Fourqurean, J.W., J.C. Zieman, and G.V.N. Powell. 1992. Phosphorus limitation of primary production in Florida Bay: evidence from C:N:P ratios of the dominant seagrass *Thalassia testudinum*, *Limnol. Oceanogr.*, 37:162–171.

Goodberlet, M.A., C.T. Swift, K.P. Kiley, J.L. Miller, and J.B. Zaitzeff. 1997. Microwave remote sensing of coastal zone salinity, *J. Coastal Res.*, 13:363–372.

Gordon, H.R. and A. Morel. 1983. Remote assessment of ocean color for interpretation of satellite visible imagery: a review, in *Lecture Notes on Coastal and Estuarine Studies*, Bowman, M., Ed., Springer-Verlag, Berlin.

Harding, L.W., Jr., E.C. Itsweire, and W.E. Esaias. 1992. Determination of phytoplankton chlorophyll concentrations in the Chesapeake Bay with aircraft remote sensing, *Remote Sensing Environ.*, 40:79–100.

Hu, C., K.L. Carder, and F.E. Muller-Karger. 2000. Atmospheric correction of SeaWiFS Imagery over turbid coastal waters: a practical method, *Remote Sensing Environ.*, 74:195–206.

Lagerloef, G., C. Swift, and D.M. Le Vine. 1995. Sea surface salinity: the next remote sensing challenge, *Oceanography*, 8:44–50.

Lapointe, B.E. and M.W. Clark. 1992. Nutrient inputs from the watershed and coastal eutrophication in the Florida Keys, *Estuaries*, 15:465–476.

Lapointe, B.E., D.A. Tomasko, and W.R. Matzie. 1994. Eutrophication and trophic state classification of seagrass communities in the Florida Keys, *Bull. Mar. Sci.*, 54:696–717.

Le Vine, D.M., J.B. Zaitzeff, E.J. D'Sa, J.L. Miller, C. Swift, and M. Goodberlet. 2000. Sea surface salinity: Toward an operational remote sensing system, pp. 321–325, in *Satellites, Oceanography and Society*, Halpern, D., Ed., Elsevier Science, New York.

Light, S.S. and J.W. Dineen. 1994. Water control in the Everglades: a historical perspective, pp. 47–84, in *Everglades: The Ecosystem and Its Restoration*, Davis, S.M. and Ogden, J.C., Eds., St. Lucie Press, Delray Beach, FL.

Miller, J.L., M. Goodberlet, and J. Zaitzeff. 1996. Remote sensing of salinity in the littoral zone: application to Chesapeake Bay, *EOS Trans. Am. Geophys. Union*, 46:F342.

Montague, C.L. and J.A. Ley. 1993. A possible effect of salinity fluctuation on abundance of benthic vegetation and associated fauna in northeastern Florida Bay, *Estuaries*, 16:703–717.

Okkonen, J., T. Hyvarinen, and E. Herrala. 1997. AISA airborne imaging spectrometer — on its way from hyperspectral research to operative use, pp. 197–203, in Proc. Third Int. Airborne Rem. Sen. Conf. ERIM, Ann Harbor, MI.

O'Reilly, J.E., S. Maritorena, B.G. Mitchell, D.A. Siegel, K.L. Carder, S.A. Garver, M. Kahru, and C.R. Mcclain. 1998. Ocean color chlorophyll algorithms for SeaWiFS, *J. Geophys. Res.*, 103:24937–24953.

Orlando, S.P., M.B. Robblee, and C.J. Klein. 1997. Salinity characteristics of Florida Bay: a review of the archived data set (1955–1995). NOAA Rept., Silver Spring, MD.

Phlips, E.J. and S. Badylak. 1996. Spatial variability in phytoplankton standing stock and composition in a shallow tropical inner-shelf lagoon, Florida Bay, Florida, *Bull. Mar. Sci.*, 58:203–216.

Robblee, M.B. 1997. Historical salinity conditions in Florida Bay: qualitative and quantitative observations, *Proc. Tech. Symposium — U.S. Geological Survey Program on the South Florida Ecosystem*, 385:75–76.

Robblee, M.B. and A. Daniels. 1997. Temporal and spatial variation in seagrass associated fish and invertebrates in western Florida Bay: s decadal comparison, *Proc. Tech. Symposium — U.S. Geological Survey Program on the South Florida Ecosystem*, 385:77–78.

Robblee, M.B., T.B. Barber, P.R. Carlson, M.J. Durako, J.W. Fourqurean, L.K. Muehlstein, D. Porter, L.A. Yarbro, R.T. Zieman, and J.C. Zieman. 1991. Mass mortality of the tropical seagrass *Thalassia testudinum* in Florida Bay, *Mar. Ecol. Progr. Ser.*, 71:297–299.

Siegel, D.A., M. Wang, S. Maritorena, and W. Robinson. 2000. Atmospheric correction of satellite ocean color imagery: the black pixel assumption, *Appl. Opt.*, 21:3582–3591.

Stumpf, R.P., M.L. Frayer, M.J. Durako, and J.C. Brock. 1999. Variations in water clarity and bottom albedo in Florida Bay from 1985 to 1997, *Estuaries*, 22:431–444.

Tassan, S. 1994. Local algorithms using SeaWiFS data for the retrieval of phytoplankton, pigments, suspended sediment, and yellow substance in coastal waters, *Appl. Opt.*, 33:2369–2377.

# 16 Spatial and Temporal Patterns of Phytoplankton in Florida Bay: Utility of Algal Accessory Pigments and Remote Sensing to Assess Bloom Dynamics

*Laurie L. Richardson*
Department of Biological Sciences, Florida International University

*Paul V. Zimba*
U.S.D.A. Agricultural Research Service

## CONTENTS

## INTRODUCTION

The shallow lagoon system of Florida Bay receives water via drainage from the Everglades National Park wetlands of the South Florida Peninsula, circulation (current and wave activity) of the adjacent Gulf of Mexico, and tidal-controlled exchange of Atlantic Ocean waters transported between the Florida Keys. Florida Bay can thus be considered a nexus between South Florida aquatic ecosystems that include wetlands, coral reefs, and the shallow, turbid Gulf of Mexico.

Extensive phytoplankton blooms have occurred in Florida Bay since 1989. Numerous hypotheses have been formulated regarding the cause of the blooms; however, no conclusive results have been obtained and to this date the cause of the blooms is not known (Fourqurean and Robblee, 1999). Perturbation of the bay from nutrient-rich waters of the Gulf as well as an alteration of surface sheetflow of water (both quantitatively and qualitatively) through the Everglades wetlands system have been implicated (Fourqurean et al., 1993; Boyer et al., 1999; Phlips et al., 1999; Tomas et al., 1999). The bloom has initiated cascading biological disturbances that have resulted in sponge and lobster mortalities and has contributed to the massive dieoff of seagrasses within the bay (Butler et al., 1995). It is postulated that perturbed Florida Bay waters are contributing to the observed degradation of coral reefs of the Florida Keys (Porter et al., 1999). The potential long-term impacts of the bloom are not known (Fourqurean and Robblee, 1999).

Florida Bay consists of many different shallow basins (depth <3 m) that are hydrologically distinct due to separation by subtidal flats that are often only a few centimeters below the surface. Within these discrete basins, phytoplankton population composition differs in both species diversity and standing stock biomass (Phlips and Badylack, 1996). Among these mixed assemblages the dominant phytoplankton include cyanobacteria, diatoms, and dinoflagellates (Phlips and Badylack, 1996). Blooms of cyanobacteria, in particular, have been implicated in the biological degradation of the bay's aquatic ecosystem (Butler et al., 1995). Spatial and temporal heterogeneity of phytoplankton are typical in estuarine systems (Platt et al., 1970; Tomas et al., 1999). Consequently, extensive sampling that includes both numerous stations and high frequency of sample collection is required to assess phytoplankton in these highly variable systems.

From May 1994 to September 1996 we investigated the use of two relatively new approaches to assess phytoplankton community structure and dynamics in Florida Bay. These methods — detection of taxonomically significant algal accessory pigments in discrete water samples and hyperspectral remote sensing — were used to map phytoplankton distribution on a regional scale.

The algal accessory pigment approach is based on the fact that different algal groups possess, in addition to chlorophyll $a$, a suite of photosynthetic (and photoprotective) accessory pigments, many of which are taxonomically distinct (Rowan, 1989). The detection of these unique pigments, termed indicator pigments, can reveal the taxonomic identity of specific algal groups (Gieskes, 1991; Millie et al., 1993). This approach has been used successfully to determine phytoplankton composition in lakes (Leavitt et al., 1989; Wilhelm et al., 1991; Lizotte and Priscu, 1993; Soma et al., 1993), estuaries (Tester et al., 1995), and seas (Gieskes and Kraay, 1983).

All of the 80+ known algal pigments (a number that includes both indicator as well as more generally distributed pigments) are well characterized in terms of their chemical and optical properties (Foppen, 1971; Morton, 1975). These properties are the bases for analytical detection procedures. The preferred method for isolating and quantifying algal accessory pigments from natural water samples is high-performance liquid chromatography (HPLC), for which methodologies for extracting, identifying, and quantifying algal pigments are well established (Mantoura and Llewellyn, 1983; Wright and Shearer, 1984; Jeffrey et al., 1997). The major benefit of the pigment technique to assess phytoplankton is the elimination of cell counting, which normally takes about 3 hours per sample. For a review of accessory pigment applications, see Gieskes (1991).

Our second approach to studying Florida Bay phytoplankton dynamics is remote sensing. Remote sensing of aquatic ecosystems has historically focused on the measurement of chlorophyll $a$, a photosynthetic pigment common to all algal groups (Rowan, 1989). Chlorophyll $a$ is usually detected in remote sensing applications by an optical signal (blue absorbance maximum) at 433 nm (Clark, 1981), although recent research has focused on a second chlorophyll $a$ optical signal at 700 nm (Gitelson, 1992). Remote sensing measurements of chlorophyll $a$ have been used to estimate phytoplankton biomass, phytoplankton dynamics, and phytoplankton-associated primary productivity from the regional to the global scale (Smith and Baker, 1982).

Advances in remote sensing technology now offer enhanced capabilities for remote sensing of algal pigments. In particular, hyperspectral datasets are currently available that cover the entire visible to infrared areas of the electromagnetic spectrum by means of a large number of contiguous, narrow bands. Hyperspectral imaging sensor data further expand remote sensing capabilities for ecosystem studies by providing complete spectra within each pixel of an image (Goetz et al., 1985).

The ability to derive spectra from remote sensing imagery is a valuable tool for the detection of algal pigment signatures (Dekker et al., 1992a; Millie et al., 1993; Richardson et al., 1994; Richardson, 1996; Schalles et al., 1997). Laboratory-based identification of accessory pigments (first separated by HPLC) is based on comparison of the absorption spectrum of a pigment sample with the absorption spectra of pigment standards, an approach that is valid because each individual pigment is spectrally unique. Analysis of spectra derived from hyperspectral imagery can be considered a parallel approach to absorbance spectra-based pigment identification, as there is a direct relationship between the absorbance features of algal accessory pigments and spectral reflectance signatures of different types of phytoplankton (Dekker et al., 1992b; Richardson et al., 1994; Richardson, 1996). Reflectance spectra compose the optical signal that is detectable by remote sensing. In this way the spectra within hyperspectral image data can be analyzed to extract information that is diagnostic of spectrally based algal signatures. This approach has been successfully used by geologists who use hyperspectral imagery to detect and map geological features based on spectral signatures of minerals (Boardman and Kruse, 1994; Boardman et al., 1995; Farrand and Harsanyi, 1995; Kruse et al., 1996).

Our investigation into remote sensing of phytoplankton in Florida Bay used the airborne visible-infrared imaging spectrometer (AVIRIS) sensor, a hyperspectral imaging sensor flown on NASA's ER-2 high-altitude aircraft. AVIRIS is the first hyperspectral imaging sensor to be deployed at high altitude that is available for environmental research (Vane et al., 1993). It is a prototype for future generations of hyperspectral satellite-borne sensors. To date, aquatic applications of AVIRIS have been limited; in addition to our results on AVIRIS detection of algal accessory pigments in hypersaline aquatic environments (Richardson et al., 1994) and Florida Bay (Kruse et al., 1997; Richardson and Kruse, 1999), AVIRIS data have been used to estimate chlorophyll $a$ in an oligotrophic lake (Hamilton et al., 1993) and marine coastal waters (Carder et al., 1993).

To understand the impacts of algal blooms in Florida Bay, it is critical to assess both the spatial and temporal distributions of the highly variable phytoplankton community. The traditional method involves time-consuming and labor-intensive phytoplankton counting to identify the algal species present and cell densities, and supplementing this information with measurement of chlorophyll $a$ in water samples to determine total phytoplankton biomass (Edler, 1979). In this chapter we will demonstrate how measurements of Florida Bay algal accessory pigments, supplemented by enumerated samples, can be used to support interpretation of hyperspectral imaging (AVIRIS) data. We will then show how hyperspectral image data can be processed to map the spatial distribution of different phytoplankton assemblages in Florida Bay. Our results will illustrate the utility of using advanced remote sensing technology to study the synoptic distributions of mixed phytoplankton blooms on a regional scale. We will conclude with a discussion of the potential use of hyperspectral remote sensing to aid in understanding complex aquatic ecosystems, with a focus on the still unknown complexities of Florida Bay and its interactions with the other aquatic ecosystems of South Florida.

## MATERIALS AND METHODS

### ALGAL PIGMENTS

Florida Bay water samples (1 liter) were collected and filtered onto 4.25-cm GF/F filters and stored frozen at −70°C. Prior to analysis, the samples were thawed, placed in 100% acetone, sonicated in 5-sec bursts for approximately 30 sec, and allowed to extract for 1 hr. After extraction, the filtrate was separated using a needle-pierced conical polypropylene centrifuge tube placed in a slightly larger test tube and centrifuged at low rpm until the extract collected in the bottom tube following the methods of Wright and Shearer (1984). Samples were then washed two times (or until clear) with 100% acetone and evaporated (to concentrate the sample) using nitrogen gas, and the final volume was measured for each sample. Next, 20 μl of each sample solution (20% 0.5 $M$ ammonium acetate, 80% sample extract) was injected onto a Hewlett Packard (HP) 1090 HPLC equipped with a HP 200 × 2.1 mm column containing 5 μm hypersil ODS for pigment separation. An elution gradient similar to that of Mantoura and Llewellyn (1983) was used as follows: time zero, 80:20 methanol:0.5 $M$ ammonium acetate; 0 to 10 minutes, change to 80:20 methanol:acetonitrile; 10 to 20 minutes, isocratic hold with 80:20 methanol:acetonitrile; 20 to 25 minutes, return to 80:20 methanol:0.5 $M$ ammonium acetate; 25 to 30 minutes, re-equilibration with 80:20 methanol:0.5 $M$ ammonium acetate. The flow rate was 0.7 ml/min, and the oven temperature was 40°C. All peaks were detected at 440 nm with a 10-nm bandwidth with the exception of myxoxanthophyll, which was detected at 475 nm (also 10-nm bandwidth). Pigments in the resulting chromatogram were quantified and identified by HP chemstation software incorporated into the system. Samples were identified and quantified using the following pigment standards: chlorophylls $a$ and $b$, ß-carotene, and lutein (from Sigma); chlorophyll $c$ (from an extract of an *Isochrysis* culture); and fucoxanthin and diadinoxanthin (purchased from L. Van Heukelem). Myxoxanthophyll was quantified by using the standard curve for diadinoxanthin but ratioing the extinction coefficients of myxoxanthophyll and diadinoxanthin.

Samples for pigment analysis were collected over a 29-month period from May 1994 through September 1996. Intervals between sampling ranged from <1 to 4 weeks with 3 to 10 stations sampled on each sampling date. A total of 157 surface water samples were analyzed for the eight pigments listed above, plus chlorophyllide $a$, a natural degradation product of chlorophyll $a$. (The significance of each pigment will be discussed in the results section.) Samples for pigment analysis were collected in 17 individual basins throughout Florida Bay, with the sample location coordinates determined by a Trimble GPS.

## PHYTOPLANKTON CELL COUNTS

Six phytoplankton samples were collected for light microscopy-based taxonomic identification and cell counts for comparison with the pigment data. The samples were collected concurrently with samples for accessory pigment analysis from one sampling station (Rankin basin) on six dates (5/4/94, 7/7/94, 10/5/94, 11/2/94, 12/1/94, and 1/5/95). Phytoplankton samples (unconcentrated) were preserved immediately in Lugols solution and were later enumerated using Ütermohl sedimentation counting methodology. For each sample a minimum of ten random fields (to provide >300 cells) were identified at 400×. After counting, the sedimentation chamber was scanned at 100× to identify and enumerate larger, rarer cell forms. Biovolume equivalents were used to resolve the three orders of magnitude difference in cell size in the phytoplankton taxa identified. Biovolumes (as μm$^3$/l) were calculated using cell measures and standard formulae (Edler, 1979).

## REMOTE SENSING

Remote sensing data were used in support of this research. Two images were analyzed from the satellite (Landsat-5)-borne Thematic Mapper (TM), which provides broadband (70 to 80 nm) data in seven bands at a spatial resolution of 30 m. These data were used to reveal the spatial extent of the phytoplankton blooms throughout the bay by detection of the green reflectance of chlorophyll $a$. The two scenes were selected to represent the overall status of the bay before and after the onset of major phytoplankton blooms (1989). The second sensor used was the AVIRIS hyperspectral imaging sensor flown on NASA's high-altitude ER-2 aircraft at an altitude of 65,000 ft. The AVIRIS sensor (Vane et al., 1993) provides spectral data from 365 to 2500 nm via 224 contiguous bands, each of which is $ca.$ 10 nm wide. At 65,000 ft, the spatial resolution of the sensor is 20 m. AVIRIS data, collected on one day, were analyzed to discriminate and map different phytoplankton assemblages throughout Florida Bay by detecting the spectrally based optical signatures from taxonomically significant algal accessory pigments.

## IMAGE PROCESSING

Landsat-5 TM data for Florida Bay were collected on two dates: June 3, 1988, and July 19, 1993. The TM data were processed using ERDAS IMAGINE version 8.2 image analysis software housed on a SUN SPARC system 20SX workstation. Stepwise histogram modeling was used to discriminate spectral features in the water.

The AVIRIS data (11 scenes, each $10 \times 12$ km) were collected on March 23, 1996. The data were atmospherically corrected using ATREM (Atmospheric Removal Program) and georeferenced prior to image processing. Image analysis was performed using the Environment for Visualizing Imagery (ENVI) version 3.0 image analysis software installed on a Dell Inspiron 3000 computer. For image classification purposes, a spectral library was generated by extracting end-member spectra from the image data. End-member spectra were derived from ground-truth pixels (corresponding to the sampling station) located using GPS data obtained during sampling. Spectra utilized to build the spectral library were derived from a spectral subset of the AVIRIS data. The spectral subset consisted of 60 bands (bands 1 through 60) that provide data from 365 to 935 nm, corresponding to the visible and near-infrared areas of the spectrum. It is this area where phytoplankton optical signals are present. Whereas the complete AVIRIS dataset covers the spectrum from 365 nm to 2.5 μm, the longer wavelengths are absorbed by water and therefore are not useful for aquatic applications. For a detailed discussion of the approach used to process the AVIRIS imagery see Kruse et al. (1997) and Richardson and Kruse (1999).

## RESULTS AND DISCUSSION

### PIGMENTS

The range of pigments extracted from 1 liter of Florida Bay surface waters for each of the 157 samples is shown in Table 16.1. Values ranged from 0 for each pigment (either not present or undetectable in the 1-liter sample) to varying maxima. The pigment with the highest concentrations was chlorophyll $a$, which serves as both a light-harvesting pigment and the photosynthetic reaction center pigment and is present in all eukaryotic algae and cyanobacteria. The remainder of the pigments serve as either light-harvesting or photoprotective pigments and were present in lesser amounts. The significance of each individual pigment is discussed below.

**TABLE 16.1**

**Range of Pigments Measured Between May 5, 1994, and September 25, 1996, in 17 Basins in Florida Bay[a]**

| Pigment | Minimum Value | Maximum Value | Mean | Standard Deviation |
|---|---|---|---|---|
| Chlorophyll $a$ | 0.0 | 14.6 | 2.18 | 2.712 |
| Chlorophyllide $a$ | 0.0 | 1.4 | 0.09 | 0.225 |
| Chlorophyll $b$ | 0.0 | 0.3 | 0.003 | 0.026 |
| Chlorophyll $c_1/c_2$ | 0.0 | 6.4 | 0.32 | 0.674 |
| Fucoxanthin | 0.0 | 9.6 | 0.50 | 1.016 |
| Diadinoxanthin | 0.0 | 1.5 | 0.17 | 0.201 |
| Myxoxanthophyll | 0.0 | 1.2 | 0.22 | 0.254 |
| Lutein/zeaxanthin | 0.0 | 9.0 | 0.83 | 1.287 |
| β-Carotene | 0.0 | 3.1 | 0.35 | 0.465 |

[a] Pigment concentrations are in μg/l; $N = 157$.

Table 16.2 presents an example of the spatial variability of algal pigment concentrations within five different basins of Florida Bay on one date (locations of the basins are designated in Figure 16.1). The pigment data yield much information about the phytoplankton present in the individual basins. Chlorophyll *a* data from this date indicate that Calusa had the most dense phytoplankton population, whereas Porpoise had the least. The values for chlorophyllide *a* (a chlorophyll *a* degradation product) suggest the presence of healthy phytoplankton populations in Porpoise, Pollock, and Twin Key basins (concentration of zero), whereas degraded phytoplankton cells were abundant in Calusa and Whipray. The algal accessory pigment chlorophyll *b* is present only in the Chlorophyta (green algae). Thus, the chlorophyll *b* data reveal that on this date no green algae were present among the phytoplankton of any of the five basins sampled, or were present at a concentration too low to be detected in the 1- liter sample. Chlorophyll $c_1/c_2$ (a mixture of two types of chlorophyll *c* which coelute in our HPLC elution gradient) indicates the presence of members of the Chromophyta, which includes the diatoms and dinoflagellates. These comprise two of the three dominant phytoplankton groups found in Florida Bay, with diatoms normally the more dominant of the two (Phlips and Badylak, 1996). On this date, chromophytes were present in all basins except Porpoise. Myxoxanthophyll, another indicator pigment, is a carotenoid that is specific to cyanophytes (cyanobacteria, commonly called blue-green algae), the third dominant phytoplankton member found in the bay. Members of this algal group were present in all five basins on this date.

**TABLE 16.2**
**Spatial Variability of Algal Pigments Present (µg/l) in Five Basins of Florida Bay on March 22, 1995**

| Pigment | Calusa | Whipray | Porpoise | Pollock | Twin Key Basin |
|---|---|---|---|---|---|
| Chlorophyll *a* | 1.77 | 1.07 | 0.83 | 0.99 | 1.17 |
| Chlorophyllide *a* | 0.16 | 0.12 | 0.00 | 0.00 | 0.00 |
| Chlorophyll *b* | 0.00 | 0.00 | 0.00 | 0.00 | 0.00 |
| Chlorophyll $c_1/c_2$ | 0.16 | 0.14 | 0.00 | 0.19 | 0.24 |
| Fucoxanthin | 0.52 | 0.29 | 0.22 | 0.29 | 0.25 |
| Diadinoxanthin | 0.29 | 0.21 | 0.12 | 0.13 | 0.15 |
| Myxoxanthophyll | 0.13 | 0.13 | 0.12 | 0.09 | 0.12 |
| Lutein/zeaxanthin | 0.19 | 0.14 | 0.20 | 0.16 | 0.14 |
| β-Carotene | 0.22 | 0.14 | 0.19 | 0.17 | 0.19 |

The remaining pigments identified are not normally used for taxonomic determinations as they are widely distributed among major algal groups. ß-Carotene and zeaxanthin are found in most of the algal groups, while lutein, fucoxanthin, and diadinoxanthin are each present in several groups. Although not taxonomically specific, diadinoxanthin and fucoxanthin concentrations are useful for assessing phytoplankton in Florida Bay, as they are found in diatoms and dinoflagellates but not cyanobacteria. For a detailed treatment of the taxonomic distribution of algal pigments, see Rowan (1989). For the purposes of this chapter, we will focus on the indicator pigments.

In summary, based on the pigment data in Table 16.2 it can be seen that the bloom was healthy in Porpoise, Pollock, and Twin Key Basins, but exhibited some degradation (dead algal cells) in Calusa and Whipray basins. Calusa had the highest concentration of phytoplankton, while Porpoise and Pollock had the lowest. Cyanobacteria were present in all basins with the lowest concentrations in Pollock, diatoms/dinoflagellates were present in all basins but one, and green algae were not detectable in the pigment analyses.

Tables 16.3 and 16.4 consist of both pigment data (Table 16.3) and phytoplankton cell counts (Table 16.4) for six samples collected on six dates from one basin (Rankin). In Table 16.3 only the indicator pigment subset of the total pigment dataset is depicted (as discussed above) for comparison

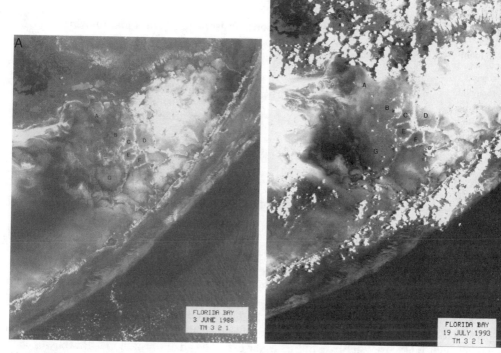

**FIGURE 16.1** (A) Landsat-5 Thematic Mapper image of Florida Bay on June 3, 1988, before phytoplankton bloom development. (B) Thematic Mapper image of Florida Bay on July 19, 1993, with extensive phytoplankton bloom present. Individual basins are indicated by letters: A = Rankin, B = Whipray, C = Calusa, D = Jimmie, E = Pollock, F = Porpoise, G = Twin Keys Basin. The remaining 10 basins sampled in this study (Peterson, Rabbit, Old Dan Bank, Lignum Vitae Key, Garfield, Johnson, Murray, Crockadile, Manatee, Near Eagle Pass, and Blackwater) are not depicted.

with the cell counts. In Table 16.4 both cell identifications and major algal group (subdivision, see table legend) are presented. It can be seen that the indicator accessory pigments are indicative of phytoplankton bloom composition in Florida Bay. For example, for samples collected on 5/4/94, the pigment data reveal that both chromophytes (chlorophyll $c_1/c_2$) and cyanobacteria (myxoxanthophyll) were present but not green algae (chlorophyll $b$). The cell count data support the pigment data with high counts of diatoms (chromophytes, as seen in subdivision codes 1 and 2) and by smaller amounts of dinoflagellates (also chromophytes, subdivision code 4); high counts of one unidentified cyanobacterial species (subdivision code 5); and no green algae (subdivision code 6). The cell count data also reveal information about the ecological status of the phytoplankton in the bay. For example, the samples from two dates, 7/7/94 and 10/5/94, contained many dead pennate diatoms that were probably resuspended from the sediments. The 10/5/94 sample also contained many flagellates that had lost their flagella.

Table 16.4 also includes biovolumes for each phytoplankton type counted. These data are summarized in Table 16.5, in which biovolumes for the three main phytoplankton groups were calculated from the data of Table 16.4. In Table 16.5 both total biovolumes and percentage of total biovolume are reported. Thus, again for 5/4/94, it can be seen that the bloom was composed of 82% (by biovolume) diatoms, 2% dinoflagellates, and 5% cyanobacteria. On 11/2/94, the bloom was dominated by cyanobacteria (47% biovolume), with both diatoms (9%) and dinoflagellates (38% present). This is visible both in the pigment data (high values for myxoxanthophyll) and the cell counts. It is important to note at this point that biovolume is a number that corresponds to an entire cell and is not limited to the pigment-containing part of the cell's interior (chloroplasts for

**TABLE 16.3**
**Temporal Variability of Algal Indicator Pigments Present (µg/l) in Rankin Basin[a]**

| Pigment | 5/4/94 | 7/7/94 | 10/5/94 | 11/2/94 | 12/1/94 | 1/5/95 |
|---|---|---|---|---|---|---|
| Chlorophyll $a$ | 7.96 | 14.59 | 11.93 | 4.63 | 3.08 | 5.09 |
| Chlorophyllide $a$ | 0.52 | 0.51 | 0.48 | 0.54 | 0.48 | 0.51 |
| Chlorophyll $b$ | 0.00 | 0.00 | 0.00 | 0.00 | 0.00 | 0.00 |
| Chlorophyll $c_1/c_2$ | 0.64 | 0.40 | 0.48 | 0.69 | 0.44 | 1.32 |
| Myxoxanthophyll | 0.50 | 0.68 | 0.92 | 0.93 | 0.78 | 0.53 |

[a] These pigments were analyzed from water samples that were collected concurrently with samples for phytoplankton cell counts.

the eukaryotic algae, thyllakoid membrane system for cyanobacteria). This and related points will be addressed in the conclusions.

Table 16.6 illustrates the utility of interpreting the presence vs. absence of diagnostic algal accessory pigments to reveal phytoplankton population dynamics over time. Data collected from one basin (Whipray) over a time span of the full 29 months show that the bloom varied in both quantity (total concentration of phytoplankton) and quality (bloom composition). It can be seen that the bloom is most dense in winter months, as previously reported by other investigators (Boyer et al., 1999). At times, there were no cyanobacteria present; at other times, no Chromophytes. In this particular basin, green algae were not detectable over the 29-month period. This table, in conjunction with Tables 16.2 and 16.3, illustrates the extreme variability of Florida Bay phytoplankton in terms of both spatial and temporal patterns.

The final pigment data subset presented here is shown in Table 16.7, which shows the concentrations of diagnostic pigments in water samples collected as ground-truth data during the AVIRIS overflight. On the date of the AVIRIS flight, pigment ground-truth data were collected and analyzed for seven stations. Four of the stations were imaged in one of the 11 AVIRIS scenes collected. This one scene (Figure 16.2) will be discussed in this chapter, and only pigment data from the four imaged stations will be presented. (The complete dataset, which includes all nine pigments from the seven stations, can be found in Richardson and Kruse, 1999).

The data in Table 16.7 reveal that on 3/23/96 the four basins sampled had three different bloom compositions: diatoms only (Calusa); diatoms and cyanobacteria (Pollock and Porpoise); and diatoms, cyanobacteria, and green algae (Jimmie). While Pollock and Porpoise contained two blooms with the same microalgae present (cyanobacteria and diatoms), they were present in different proportions, as seen by considering pigment ratios. Thus, Pollock has a ratio of myxoxanthophyll to chlorophyll $a$ of 1:7, whereas Porpoise has a corresponding ratio of 1:6. The chlorophyll $c$ to chlorophyll $a$ ratios in both basins were similar, suggesting that cyanobacteria constitute a relatively larger proportion of the mixed bloom in Porpoise than in Pollock.

## REMOTE SENSING

In addition to detection of algal accessory pigments as indicators of phytoplankton bloom dynamics, the purpose of this study was to use remote sensing to determine phytoplankton dynamics on a regional scale. Figure 16.1 presents two satellite (TM) images that clearly depict the condition of Florida Bay before and after the onset of phytoplankton blooms (1989). Figure 16.1A is a scene acquired before the bloom occurred (1988), and Figure 16.1B is a second scene of the same area after bloom development (1993). When comparing the two images it should be noted that in 1988 the water was optically clear, and the sediment bottom as well as benthic seagrass communities can be seen. In 1993, however, the bay was optically dense due to phytoplankton. In the second

## TABLE 16.4
## Phytoplankton Cell Counts from the Six Samples of Table 16.3

| Algal Species | Subdivision Code[a] | No. Cells/L | Biovolume ($\mu m^3$/L) |
|---|---|---|---|
| **May 4, 1994** | | | |
| 5- to 10-μm centric diatoms | 2 | 906,841 | 160,148,121 |
| 15- to 25-μm centric diatoms | 2 | 1,088,210 | 3,758,677,340 |
| *Amphidinium* c.f. *shroederi* | 4 | 90,684 | 76,739,521 |
| *Thalassiosira/Minidiscus* | 2 | 1,541,630 | 237,195,192 |
| *Amphora mexicana* | 1 | 90,684 | 1,865,369,880 |
| Bluegreen spheres | 5 | 16,685,881 | 353,657,248 |
| <10-μm flagellates | 3 | 11,426,201 | 879,017,643 |
| *Ceratium hircus* | 4 | 1000 | 6,600,000 |
| *Cocconeis* spp. | 1 | 1400 | 11,088,000 |
| 50- to 75-μm pennate diatoms | 1 | 2800 | 34,372,800 |
| Unknown dinoflagellates | 4 | 600 | 4,320,000 |
| *Coscinodiscus* spp. | 2 | 1000 | 139,887,000 |
| *Prorocentrum micans* | 4 | 800 | 13,727,600 |
| *Protoperidinium* c.f. *oblongum* | 4 | 200 | 12,434,400 |
| *Gonyaulax spinifera* | 4 | 200 | 30,520,800 |
| | | | |
| **July 7, 1994** | | | |
| *Gymnodinium* (or unknown small armored) | 4 | 453,421 | 720,230,000 |
| <3-μm bluegreen spheres | 5 | 17,320,670 | 286,500,000 |
| *Oscillatoria* spp. | 5 | 600 | 1,657,900 |
| *Prorocentrum micans* | 4 | 800 | 96,800,000 |
| *Ceratium hircus* | 4 | 2000 | 132,000,000 |
| <5 μm flagellates | 3 | 9,521,835 | 185,798,000 |
| 25- to 35-μm centric diatoms | 2 | 272,052 | 1,800,600,000 |
| *Coscinodiscus* spp. | 2 | 200 | 15,260,000 |
| *Nitzchia* spp. | 1 | 90,684 | 88,527,600 |
| *Gymnodinium splendens* | 4 | 200 | 1,584,000 |
| *Polykrikos hartmanii* | 4 | 600 | 56,050,000 |
| *Nitzchia sigma* | 1 | 200 | 3,400,000 |
| *Pleurosigma/Gyrsigma* spp. | 1 | 200 | 23,100,000 |
| *Dictyocha fibula* | 7 | 200 | 840,000 |
| *Gyrodinium* spp. | 4 | 800 | 115,480,000 |
| *Prorocentrum mexicanum* | 4 | 200 | 4,730,000 |
| *Prorocentrum minimum* | 4 | 90,684 | 391,750,000 |
| *Dinophysis* spp. | 4 | 200 | 4,056,000 |
| Unknown dinoflagellates | 4 | 400 | 864,000 |
| | | | |
| **October 5, 1994** | | | |
| <10-μm pennate diatoms | 1 | 226,710 | 7,118,700 |
| <25-μm pennate diatoms | 1 | 226,710 | 51,009,800 |
| 50- to 75-μm pennate diatoms | 1 | 1200 | 5,220,000 |
| *Cyclotella* sp. | 2 | 226,710 | 82,001,007 |
| 5- to 10-μm centric diatoms | 2 | 680,131 | 120,131,539 |
| *Amphora* spp. | 1 | 226,710 | 12,414,640 |
| Armored dinoflagellates | 4 | 151,140 | 581,889,000 |
| *Calycomonas wulffii* | 3 | 151,140 | 18,136,800 |

**TABLE 16.4 (CONTINUED)**
**Phytoplankton Cell Counts from the Six Samples of Table 16.3**

| Algal Species | Subdivision Code[a] | No. Cells/L | Biovolume ($\mu m^3$/L) |
|---|---|---|---|
| **October 5, 1994** | | | |
| Bluegreen spheres | 5 | 3,778,506 | 80,085,435 |
| *Synechococcus elongatus* | 5 | 1,209,122 | 102,509,363 |
| <5-$\mu$m flagellates | 3 | 6,725,740 | 329,981,619 |
| *Triploceras* sp. | 6 | 200 | 2,296,125 |
| *Prorocentrum micans* | 4 | 600 | 10,295,700 |
| *Gymnodinium splendens* | 4 | 200 | 3,168,000 |
| | | | |
| **November 2, 1994** | | | |
| *Gymnodinium* (or unknown small armored) | 4 | 90,684 | 38,540,700 |
| <3-$\mu$m bluegreen spheres | 5 | 6,074,318 | 128,750,000 |
| *Synechococcus elongatus* | 5 | 725,292 | 61,490,000 |
| *Oscillatoria* spp. | 5 | 272,052 | 1,420,000,000 |
| *Merismopedia glauca* | 5 | 6400 | 321,500 |
| *Prorocentrum micans* | 4 | 4200 | 72,070,000 |
| *Ceratium hircus* | 4 | 800 | 52,800,000 |
| <5-$\mu$m flagellates | 3 | 6,436,964 | 136,000,000 |
| *Cryptomonas* spp. | 3 | 90,684 | 71,180,000 |
| 25- to 50-$\mu$m pennate diatoms | 2 | 11,200 | 2,900,800 |
| <10-$\mu$m centric diatoms | 2 | 362,646 | 113,870,000 |
| *Amphora* spp. | 1 | 8000 | 43,200,000 |
| *Cocconeis* spp. | 1 | 600 | 2,304,000 |
| *Gonyaulax grindleyi* | 4 | 1000 | 900,000,000 |
| *Protoperidinium* spp. | 4 | 600 | 240,000,000 |
| *Mastogloia* spp. | 1 | 400 | 3,870,000 |
| *Pleurosigma/Gyrsigma* spp. | 1 | 2000 | 126,000,000 |
| *Hyalotheca* sp. | 6 | 2200 | 1,942,800 |
| | | | |
| **December 1, 1994** | | | |
| 10- to 20-$\mu$m centric diatoms | 2 | 634,789 | 896,960,000 |
| <35-$\mu$m centric diatoms | 2 | 90,684 | 987,510,000 |
| *Synechococcus elongatus* | 5 | 1,450,946 | 153,760,000 |
| <5-$\mu$m flagellates | 3 | 13,330,568 | 512,000,000 |
| *Prorocentrum micans* | 4 | 1400 | 9,240,000 |
| *Prorocentrum mexicanum* | 4 | 1000 | 8,850,000 |
| *Gymnodinium splendens* | 4 | 4600 | 72,864,000 |
| Unknown armored dinoflagellates | 4 | 2000 | 14,400,000 |
| *Ceratium hircus* | 4 | 200 | 13,200,000 |
| 25- to 50-$\mu$m pennate diatoms | 1 | 1200 | 5,328,000 |
| Bluegreen spheres | 5 | 6,075,837 | 128,780,000 |
| | | | |
| **January 5, 1995** | | | |
| *Cosmarium* sp. | 6 | 200 | 3,570,000 |
| <3-$\mu$m bluegreen spheres | 5 | 22,126,930 | 156,000,000 |
| *Merismopedia tenuissima* | 5 | 125,473 | 1,575,900 |
| *Synechococcus elongatus* | 5 | 4,624,891 | 32,674,855 |
| *Oscillatoria* spp. | 5 | 400 | 270,000 |

## TABLE 16.4 (CONTINUED)
## Phytoplankton Cell Counts from the Six Samples of Table 16.3

| Algal Species | Subdivision Code[a] | No. Cells/L | Biovolume ($\mu m^3/L$) |
|---|---|---|---|
| **January 5, 1995** | | | |
| *Merismopedia glauca* | 5 | 3200 | 160,800 |
| *Prorocentrum micans* | 4 | 600 | 9,517,200 |
| *Gyrodinium* spp. | 4 | 400 | 16,981,120 |
| <5-$\mu$m flagellates | 3 | 20,857,352 | 654,900,000 |
| *Cryptomonas* sp. | 3 | 90,684 | 51,254,600 |
| *Nitzchia marginulata* | 2 | 200 | 300,000 |
| <10-$\mu$m centric diatoms | 2 | 1,813,683 | 569,500,000 |
| *Coscinodiscus* spp. | 2 | 400 | 38,460,000 |
| *Rhizosolenia alata* f. *indica* | 2 | 600 | 47,100,000 |
| *Cyclotella* spp | 2 | 90,684 | 328,000,000 |
| *Rhizosolenia imbricata* v. *shrubsolei* | 2 | 362,737 | 2,090,000,000 |
| <20-$\mu$m pennate diatoms | 1 | 272,052 | 61,211,700 |
| *Synedra* spp. | 1 | 600 | 7,200,000 |
| *Diploneis* spp. | 1 | 1800 | 6,480,000 |
| *Achnanthes* sp. | 1 | 544,105 | 48,969,450 |
| *Amphora* sp. | 1 | 90,684 | 126,960,000 |
| *Cocconeis* spp. | 1 | 544,105 | 3,808,700,000 |
| *Nitzchia* spp. | 1 | 1600 | 360,000 |
| *Dimeroogramma* c.f. *minor* | 1 | 1600 | 360,000 |
| *Surirella* c.f. *comta* | 1 | 200 | 8,775,000 |
| *Pleurosigma/Gyrosigma* spp. | 1 | 6400 | 80,640,000 |

[a] Subdivision codes are as follows: 1 = pennate diatoms, 2 = centric diatoms, 3 = microflagellates, 4 = dinoflagellates, 5 = cyanobacteria, 6 = nonflagellated green algae, 7 = unknown.

## TABLE 16.5
## Relative Biovolumes of Cyanobacteria, Diatoms, Dinoflagellates, and Others[a]

| Algal Group | Biovolumes ($\mu m^3/L$) (%) | | | | | |
|---|---|---|---|---|---|---|
| | 5/4/94 | 7/7/94 | 10/5/94 | 11/2/94 | 12/1/94 | 1/5/95 |
| Cyanobacteria | $3.64 \times 10^8$ | $2.87 \times 10^8$ | $1.83 \times 10^8$ | $1.61 \times 10^9$ | $2.83 \times 10^8$ | $1.91 \times 10^8$ |
| | (5%) | (8%) | (13%) | (47%) | (10%) | (2%) |
| Diatoms | $6.21 \times 10^9$ | $1.93 \times 10^9$ | $2.65 \times 10^8$ | $2.92 \times 10^8$ | $1.89 \times 10^9$ | $7.16 \times 10^9$ |
| | (82%) | (52%) | (19%) | (9%) | (67%) | (89%) |
| Dinoflagellates | $1.44 \times 10^8$ | $1.50 \times 10^9$ | $5.85 \times 10^8$ | $1.30 \times 10^9$ | $1.19 \times 10^8$ | $1.70 \times 10^7$ |
| | (2%) | (40%) | (42%) | (38%) | (4%) | (<1%) |
| Other | $8.79 \times 10^8$ | $1.32 \times 10^7$ | $3.50 \times 10^8$ | $2.09 \times 10^8$ | $5.12 \times 10^8$ | $7.10 \times 10^8$ |
| | (12%) | (<1%) | (25%) | (6%) | (18%) | (9%) |
| Total biovolumes | $7.60 \times 10^9$ | $37.3 \times 10^9$ | $1.38 \times 10^9$ | $3.41 \times 10^9$ | $2.80 \times 10^9$ | $8.08 \times 10^9$ |
| | (100%) | (100%) | (100%) | (100%) | (100%) | (100%) |

[a] Data calculated from values in Table 16.4.

**TABLE 16.6**
**Temporal Variability of Diagnostic Pigments (µg/l) in Whipray Basin, Florida Bay**

| Date | Chlorophyll *a* | Chlorophyllide *a* | Chlorophyll *b* | Chlorophyll $c_1/c_2$ | Myxoxanthophyll |
|------|------|------|------|------|------|
| 5/4/94 | 3.74 | 0.21 | 0.00 | 0.16 | 0.34 |
| 6/8/94 | 3.37 | 0.13 | 0.00 | 0.00 | 0.30 |
| 7/7/94 | 4.41 | 0.37 | 0.00 | 0.40 | 0.38 |
| 8/3/94 | 3.24 | 0.14 | 0.00 | 0.19 | 0.28 |
| 10/5/94 | 5.32 | 0.00 | 0.00 | 0.26 | 0.21 |
| 11/2/94 | 5.07 | 0.00 | 0.00 | 0.49 | 0.48 |
| 12/1/94 | 5.39 | 0.00 | 0.00 | 0.60 | 0.38 |
| 1/5/95 | 10.11 | 0.85 | 0.00 | 1.73 | 0.93 |
| 3/22/95 | 1.07 | 0.12 | 0.00 | 0.14 | 0.13 |
| 5/4/95 | 1.90 | 0.00 | 0.00 | 0.28 | 0.24 |
| 7/4/95 | 1.78 | 0.00 | 0.00 | 0.42 | 0.33 |
| 10/31/95 | 2.40 | 0.00 | 0.00 | 0.25 | 0.22 |
| 12/5/95 | 5.00 | 0.00 | 0.00 | 0.00 | 0.82 |
| 1/11/96 | 3.19 | 0.00 | 0.00 | 0.23 | 0.22 |
| 2/1/96 | 0.70 | 0.00 | 0.00 | 0.12 | 0.08 |
| 4/4/96 | 0.46 | 0.00 | 0.00 | 0.10 | 0.00 |
| 5/14/96 | 0.53 | 0.00 | 0.00 | 0.09 | 0.09 |
| 6/13/96 | 1.30 | 0.00 | 0.00 | 0.23 | 0.00 |
| 9/25/96 | 0.23 | 0.00 | 0.00 | 0.06 | 0.06 |

**TABLE 16.7**
**Diagnostic Pigments Present in Four Basins Imaged in the AVIRIS Scene of Figures 16.2 and 16.3**

| Basin | Chlorophyll *a* | Chlorophyll *b* | Chlorophyll $c_1/c_2$ | Myxoxanthophyll | Bloom Composition[a] |
|------|------|------|------|------|------|
| Calusa | 0.39 | 0 | 0.14 | 0 | D |
| Jimmie | 0.80 | 0.13 | 0.13 | 0.08 | C, D, G |
| Pollock | 0.42 | 0 | 0.14 | 0.06 | C, D |
| Porpoise | 0.36 | 0 | 0.12 | 0.06 | C, D |

*Note:* Sampling stations are indicated in Figure 16.2; pigment values are in µg/l.

[a] C = cyanobacteria; D = diatoms; G = green microalgae.

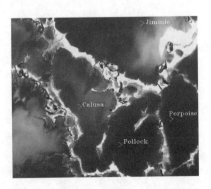

**FIGURE 16.2** AVIRIS image of Florida Bay on March 23, 1996. Pigment sampling stations are indicated by arrows.

**FIGURE 16.3** AVIRIS image classified using spectra corresponding to the pigment sampling stations. White = Calusa (diatoms only); light gray = Porpoise (diatoms and cyanobacteria); medium gray = Pollock (diatoms and cyanobacteria, with relatively less cyanobacteria present than the same mixed bloom in Porpoise); and dark gray = Jimmie (cyanobacteria, diatoms, and green algae). Black = unclassified (sediment banks).

scene, the bottoms of the basins cannot be seen, although the shallow mudbanks that separate the basins are evident. The stations sampled and referred to in Tables 16.1 to 16.7 are depicted in Figure 16.1.

Both Figures 16.1A and 16.1B are TM band 2 (520 to 600 nm), which is sensitive to the green area of the visible spectrum. Green reveals the presence of chlorophyll, indicating the presence of algae in aquatic ecosystems. It is well known that TM data are valuable in showing the spatial extent of phytoplankton blooms and the difference in the size of the blooms over time, via detection of the green reflectance of chlorophyll *a* (Galat and Verdin, 1989). Much more optical information is available from hyperspectral remote sensing data. Figures 16.2 and 16.3 show AVIRIS data of central Florida Bay. The AVIRIS scene in Figure 16.2 is a gray-scaled version of a natural color composite processed to enhance the phytoplankton optical signal. This figure is presented to show the location of the four sample stations where the pigment data of Table 16.7 were collected during the AVIRIS flight. (Station locations can be compared with those depicted in Figure 16.1.)

Figure 16.3 is a processed version of the same AVIRIS scene shown in Figure 16.2. Figure 16.3 is the result of a supervised classification in which the AVIRIS image data were classified according to phytoplankton assemblage using spectra derived from the imagery. To implement this, four spectra were extracted from the four image pixels that corresponded to the sample stations of Figure 16.2 using functions available within the ENVI image processing software. Each of the four classes (see figure legend) thus represent the four phytoplankton bloom types determined using the accessory pigment data (Table 16.7). It can be seen that each of the four basins contains at least two bloom assemblages and that each of the assemblages occurs as patches even within the basins of Florida Bay.

## CONCLUSIONS

Florida Bay is a complicated ecosystem that has, for unknown reasons, shifted in recent years from an optically clear lagoon to a turbid, phytoplankton-dominated system. The shift has negatively affected seagrass, fish, and invertebrate populations (Butler et al., 1995; Fourqurean and Robblee, 1999) and has potentially affected the health of the nearby coral reefs (Porter et al., 1999). Although a number of studies have been aimed at understanding the cause of the shift, all have, to date, been inconclusive (Fourqurean and Robblee, 1999). Some of the effects of the bloom are just beginning to be understood.

It is clear that a rapid assessment technique, such as provided by remote sensing and demonstrated in this chapter, is required to understand a large-scale ecosystem as complex as Florida Bay. In the absence of remote sensing, many of the studies aimed at elucidating the causation of Florida Bay phytoplankton blooms have used numerous sampling stations throughout the Bay that are sampled on a weekly to monthly basis. These large datasets have then been analyzed statistically in conjunction with similarly intensive water quality datasets also collected throughout the bay

(Boyer et al., 1999). Unfortunately, results remain inconclusive (Fourqurean and Robblee, 1999; Porter et al., 1999).

While it is known that Florida Bay phytoplankton are phosphorus limited (Fourqurean et al., 1993; Phlips et al., 1999; Tomas et al., 1999), there is no evidence of a mass phosphorus input that initiated the blooms in 1989 and sustained them in later years. An early conjecture focused on the observed hypersalinity of Florida Bay, resulting from decreased input of freshwater from Everglades wetlands, as the cause of phytoplankton blooms; this hypothesis has since been discounted (Fourqurean et al., 1993; Fourqurean and Robblee, 1999). However, it is now thought that the elevated salinity of Florida Bay is selecting for deleterious cyanobacteria within the phytoplankton bloom assemblage (Phlips et al., 1999) and may be contributing to the decline in health of the coral reefs of the Florida Keys. The latter connection is via a proposed model in which turbid, hypersaline water from Florida Bay flows out to the reef tract as part of the normal hydrologic cycle, but then sinks below less saline water and then negatively affects the health of coral on the reef (Porter et al., 1999).

Sources of nutrient enrichment of Florida Bay have been postulated to originate from several sources. Sewage outflows from the Florida Keys cities and septic tanks may release increased nutrients to this area (Corbett et al., 1999). Alternative hypotheses include delivery of nutrients via the Loop Current from western Florida into Florida Bay (Smith, 1994). Elevated phosphorus levels are common in waterways of south and western Florida, in part due to mining practices associated with the limestone bedrock formation that runs along the main axis of this region. Sheet flow from agricultural fields and surface hardening (massive use of concrete paving) of coastal regions may further contribute to elevated nutrient levels within Florida Bay and also contribute to salinity stress of submersed plants (Montague and Ley, 1993).

It has also been postulated that nutrients have been enriching Florida Bay from an underground hydrological link; however the carbonate mud sediments of Florida Bay appear to have very low hydraulic conductivity and are too impermeable to allow significant nutrient transport (Juster et al., 1997). Another water quality parameter that has been investigated in conjunction with the phytoplankton blooms is increased turbidity. While a dramatic increase in suspended sediments has been documented (Boyer et al., 1999) and implicated in the seagrass mortalities, it cannot be determined whether or not this increase is a cause or an effect of the phytoplankton blooms (Porter et al., 1999; Stumpf et al., 1999).

One model recognizes the complexity of Florida Bay by detailing the complicated resource limitation that exists among and between all members of the microbial plankton community (Lavrentyev et al., 1998). This model suggests an unraveling of the nutrient vs. resource requirements as a plausible approach to understanding the perturbed bay. While this approach has been carried out for phytoplankton (Phlips et al., 1995, 1999; Tomas et al., 1999) and each group has demonstrated phosphorus limitation of Florida Bay phytoplankton, results do not explain the massive bloom. Lavrentyev et al. (1998) suggest that all plankton must be included in this experiment, including nonphotosynthesizing heterotrophs. Another hypothesis that has recently been proposed is that iron may be the limiting nutrient that is responsible for the phytoplankton blooms (Fourqurean, pers. comm.). In any event, the cause of the blooms remains a mystery.

Use of remote sensing, in conjunction with measurement of taxonomically specific algal accessory pigments, may help determine what causes the Florida Bay phytoplankton blooms. The South Florida aquatic ecosystems are all perturbed on a regional scale. Since point sampling and statistical analysis of water quality factors vs. biological components has failed to reveal the cause and effect of the perturbations, perhaps a pattern could be discerned by studying each ecosystem as an entity. The only way to do this is via remote sensing.

As mentioned previously, many investigators have used the algal accessory pigment approach to assess phytoplankton community structure (Gieskes and Kraay, 1983; Leavitt et al., 1989; Wilhelm et al., 1991; Lizotte and Priscu, 1993; Millie et al., 1993; Soma et al., 1993; Tester et al., 1995). Others have used the pigment approach in conjunction with remote sensing. These investigators have used remote sensing of algal pigments to determine the population composition of phytoplankton (Dekker et al., 1992b; Millie et al., 1993; Richardson et al., 1994; Richardson, 1996), to detect harmful algal blooms (Carder and Steward, 1985), and to assess water quality by inferences from the types of bloom detected (Dekker et al., 1992a; Schalles et al., 1997).

While the approach works, remote sensing of phytoplankton accessory pigments is complicated by the fact that each individual phytoplankton cell contains several pigments. Additionally, there are almost always multiple types of phytoplankton present as a mixed suspension. While laboratory pigment analysis is simplified by the ability to physically separate individual pigments prior to identification, the reflectance spectral data used for remote sensing are influenced by the net sum of many overlapping pigments. While derivative spectra, particularly the fourth derivative, are useful in augmenting subtle spectral features (as in slope changes caused by absorbance and reflectance features within overlapping pigments), a very real criticism of this approach is that often there are wavelength shifts of individual pigment signals which could complicate interpretation of the subtle signals augmented by derivative analysis. Such shifts can occur due to the interplay between water scattering and reflectance maxima (Schalles et al., 1997), changes in the physiological status of phytoplankton (Heath et al., 1990), and, during the laboratory analysis of pigments that is commonly used for ground-truthing reflectance spectra, shifts in wavelength position of absorbance features due to solvents used in the extraction process (Millie et al., 1993). Investigators have found that cell counts are recommended to support interpretation of algal accessory pigment data (Tester et al., 1995), a finding with which we concur and which is evident from the results we have presented in this chapter. While biovolumes are also useful ground-truth data, interpretation of biovolume data vs. pigment data is complicated by the fact that algal cells are not uniformly pigmented and species–specific differences in pigment composition are common (Zimba et al., 1990, 1999). Thus, biovolume-to-pigment ratios should be interpreted with caution. For region-scale studies, however, it is often valid to determine the major algal groups that are present, and species composition is not necessary.

Our results show that in Florida Bay algal populations can be discriminated by AVIRIS based on the overall shape of reflectance spectra, which is a composite result of spectral pattern recognition of the different pigment spectra both within individual phytoplankton cells and in mixed populations. The spectral pattern approach alleviates the need to separate all of the subtle effects that can shift a pigment signature in terms of peak positions such as physiology, pigment packaging, etc. Hyperspectral sensors detect specific algal signatures in a manner analogous to remote sensing applications currently in use in geology, in which the spectral signatures of minerals, even those that are mixed with other signals in a reflectance spectrum, comprise the basis for image classification (Farrand and Harsanyi, 1995; Boardman and Kruse, 1994). Our approach, in fact, stems directly from the geological application (Kruse et al., 1997).

Because of the capability to discriminate between different algal populations, remote sensing is a potentially valuable tool to aid in not only understanding the complex phytoplankton community of Florida Bay but also determining interactions between blooms and (optically detectable) events in the surrounding aquatic ecosystems. These ecosystems are hydrologically linked, and disturbances of any individual ecosystem have an impact on the entire South Florida aquatic system. All of the major aquatic ecosystems of South Florida are seriously perturbed, and, while this chapter has focused on the problems of Florida Bay, it is well known that the coral reefs of the coastal Atlantic Ocean of the Florida Keys are suffering major degradation (Porter et al., 1999) and that the adjacent Gulf of Mexico is experiencing novel hypoxic "dead zones" (Rabalais et al., 1999) as well as increased red tides composed of toxin-producing dinoflagellates (Steidinger and Vargo, 1988). The major problems associated with alteration of water flow through South Florida's wetlands and from the Everglades Agricultural Area are under intense investigation.

The remote sensing approach has already been used in one study (Stumpf et al., 1999) that used Advanced Very High Resolution Radiometer (AVHRR) data to look at the relationships between water clarity and bottom reflectance (to map seagrass beds) in Florida Bay from 1985 to 1997. This study, however, was focused on detection of suspended sediments and not phytoplankton.

Archived TM (satellite) data are an existing dataset that could potentially provide quantitative data about the occurrence and extent of phytoplankton blooms in Florida Bay prior to 1989. As pointed out by Fourqurean and Robblee (1999), very little data about Florida Bay phytoplankton exists prior to the dramatic blooming in 1989. It is not known, for example, if this is a historically recurrent event. Analysis of TM band 2 radiance data for all available years, prior to 1989 to the present, would yield much needed information about the historical status of Florida Bay. Such data could be found in the archived TM data; however, these data are at risk of being destroyed in a NASA housecleaning effort and will be lost if the individual image tapes are not purchased in the near future. Each scene costs $400 ($600 if georeferenced and radiometrically corrected); information on retrieving these data can be obtained from NASA.

At this time the most promising remote sensing data that offer the most extensive potential information are hyperspectral imaging sensor data. While it is difficult and costly to obtain AVIRIS data, hyperspectral imaging sensors are scheduled to be launched and deployed on satellites in the current time frame. The first of these, NASA's hyperspectral Hyperion, was successfully launched in 2000. However, the data are only available to select data validation teams. A second highly relevant remote sensing technology is the new microwave radiometer with the capability of detecting and measuring the salinity of surface waters. Data from this sensor collected in conjunction with hyperspectral data analyzed to map phytoplankton communities would potentially answer many of the questions concerning the relationship between phytoplankton blooms and changing salinity.

In conclusion, remote sensing can detect broad-scale patterns and changes that are critically important to understand ecosystems at a regional scale. Such information would greatly benefit management of these ecosystems. For investigations of ecosystem-wide perturbation events, remote sensing offers the capability of quantitative measurements of large spatial areas acquired in synoptic datasets. Of special importance for Florida Bay, remote sensing instruments and data analysis capabilities are now available that can be used as a research tool to actively study this complex ecosystem.

## ACKNOWLEDGMENTS

We thank Fred Kruse for atmospheric correction of the AVIRIS data, Vince Ambrosia for processing the Thematic Mapper data, and an anonymous reviewer for comments on the manuscript. Accessory pigments were analyzed by Daniel Buisson. This research was supported by the National Aeronautics and Space Administration (grants NAGW-2092, NAG5-3124 and NAG5-8156 to LLR), and the Florida Marine Research Institute (contract No. MR031 to PVZ). This is contribution No. 35 from the Tropical Biology Program at Florida International University.

## REFERENCES

Bidigare, R.R., J.H. Morrow, and D.A. Kiefer. 1989. Derivative analysis of spectral absorption by photosynthetic pigments in the western Sargasso Sea, *J. Mar. Res.*, 47:323–341.

Bjørnland, T. and S. Liaaen-Jensen. 1989. Distribution patterns of carotenoids in relation to chromophyte phylogeny and systematics, pp. 37–60, in *The Chromophyte Algae: Problems and Perspectives*, Green, J.C., Leadbeater, B.S.C., and Diver, W.L., Eds., Clarendon Press, Oxford.

Boardman, J.W. and F.A. Kruse. 1994. Automated spectral analysis: a geologic example using AVIRIS data, north Grapevine Mountains, Nevada, *Proc. 10th Conf. Geol. Rem. Sensing*, I:407–418.

Boardman, J.W., F.A. Kruse, and R.O. Green. 1995. Mapping target signatures via partial unmixing of AVIRIS data, *Proc. 5th JPL Airborne Earth Science Workshop*, I:23–26.

Boyer, J.N., J.W. Fourqurean, and R.D. Jones. 1999. Seasonal and long-term trends in the water quality of Florida Bay (1989–97), *Estuaries*, 22:417–430.

Butler, IV, M.J., J.H. Hunt, W.F. Herrnkind, M.J. Childress, R. Bertelsen, W. Sharp, T. Matthews, J.M. Field, and H.G. Marshall. 1995. Cascading disturbances in Florida Bay, USA: cyanobacterial blooms, sponge mortality, and implications for juvenile spiny lobster *Panulirus argus*, *Mar. Ecol. Prog. Ser.*, 129:119–125.

Carder, K.L. and R.G. Steward. 1985. A remote-sensing reflectance model of a red tide dinoflagellate off west Florida, *Limnol. Oceanogr.*, 30:286–298.

Carder, K.L., P. Reinersman, R.F. Chen, F. Muller-Karger, C.O. Davis, and M. Hamilton. 1993. AVIRIS calibration and application in coastal oceanic environments, *Remote Sensing Environ.*, 44:205–216.

Clark, D. 1981. Phytoplankton pigment algorithms for the Nimbus-7 CZCS, pp. 227-237, in *Oceanography from Space*, Gower, J.F.R. Ed., Plenum, New York.

Corbett, D.R., J. Chanton, W. Burnett, K. Dillon, and C. Rutkowski. 1999. Patterns of groundwater discharge into Florida Bay, *Limnol. Oceanogr.*, 44:1044–1055.

Dekker, A.G., T.J. Malthus, M.M. Wijnen, and E. Seyhan. 1992a. Remote sensing as a tool for assessing water quality in Loosdrecht lakes, *Hydrobiology*, 233:137–159.

Dekker, A.G., T.J. Malthus, M.M. Wijnen, and E. Seyhan. 1992b. The effect of spectral bandwidth and positioning on the spectral signature analysis of inland waters, *Remote Sensing Environ.*, 41:211–225.

Demetriades-Shah, T.H., M.D. Steven, and J.A. Clark. 1990. High resolution derivative spectra in remote sensing, *Remote Sensing Environ.*, 33:55–64.

Edler, L. 1979. Recommendations for marine biological studies in the Baltic Sea: plankton and chlorophyll, *The Baltic Marine Biologists*, publ. 5, 38 pp.

Farrand, W.H. and J.C. Harsanyi. 1995. Discrimination of poorly exposed lithologies in imaging spectrometer data, *J. Geophys. Res.*, 100:1565–1578.

Foppen, J.H. 1971. Tables for the identification of carotenoid pigments, *Chromatogr. Rev.*, 14:133–298.

Fourqurean, J.W. and M.B. Robblee. 1999. Florida Bay: A history of recent ecological changes, *Estuaries*, 22:345–357.

Fourqurean, J.W., R.D. Jones, and J.C. Zieman. 1993. Processes influencing water column nutrient characteristics and phosphorus limitation of phytoplankton biomass in Florida Bay, FL, USA: inferences from spatial distributions, *Estuar. Coast. Shelf Sci.*, 36:295–314.

Galat, D. and J.P. Verdin. 1989. Patchiness, collapse and succession of a cyanobacterial bloom evaluated by synoptic sampling and remote sensing, *J. Plankton Res.*, 11:925–948.

Gieskes, W.W. 1991. Algal pigment fingerprints: clue to taxon specific abundance, productivity, and degradation of phytoplankton in seas and oceans, pp. 61-99, in *Particle Analysis in Oceanography*, Demers, S., Ed., Vol. G27, NATO ASI Series, Springer-Verlag, Berlin.

Gieskes, W.W. and G.W. Kraay. 1983. Dominance of Cryptophyceae during the phytoplankton spring bloom in the central North Sea detected by HPLC analysis of pigments, *Mar. Biol.*, 75:179–185.

Gitelson, A. 1992. The peak near 700 nm on radiance spectra of algae and water: relationships of its magnitude and position with chlorophyll concentration, *Int. J. Remote Sensing*, 13:3367–3373.

Goetz, A.F.H., G. Vane, J. Solomon, and B.N. Rock. 1985. Imaging spectrometry for Earth remote sensing, *Science*, 228:1147–1153.

Hamilton, M.K., C.O. Davis, W.J. Rhea, S.H. Pilorz, and K.L. Carder. 1993. Estimating chlorophyll content and bathymetry of Lake Tahoe using AVIRIS data, *Remote Sensing Environ.*, 44:217–230.

Heath, M.R., K. Richardson, and T. Kiørboe. 1990. Optical assessment of phytoplankton nutrient depletion, *J. Plankt. Res.*, 12:383–396.

Jeffrey, S.W., R.F.C. Mantoura, and S.W. Wright. 1997. *Phytoplankton Pigments in Oceanography: Guidelines to Modern Methods*, SCOR/UNESCO, Paris, 661 pp.

Juster, T., P.A. Kramer, H.L. Vacher, P.K. Swart, and M. Stewart. 1997. Groundwater flow beneath a hypersaline pond, Cluett Key, Florida Bay, Florida, *J. Hydrol.*, 197:339–369.

Kruse, F.A., A.B. Lefkoff, J.W. Boardman, K.B. Heidebrecht, A.T. Shapiro, P.J. Barloon, and A.F.H. Goetz. 1993. The spectral image processing system (SIPS): interactive visualization and analysis of imaging spectrometer data, *Remote Sensing Environ.*, 44:145–163.

Kruse, F.A., J.H. Huntington, and R.O. Green. 1996. Results from the 1995 AVIRIS Geology Group Shoot, *Proc. 2nd Int. Airborne Remote Sensing Conf.*, 1:211–220.

Kruse, F.A., L.L. Richardson, and V.G. Ambrosia. 1997. Techniques developed for geologic analysis of hyperspectral data applied to near-shore hyperspectral ocean data, *Proc. 4th Int. Conf. Remote Sensing Mar. Coast. Environ.*, 1:233–246.

Lavrentyev, P.J., H.A. Bootsma, T.H. Johengen, J.F. Cavaletto, and W.S. Gardner. 1998. Microbial plankton response to resource limitation: insights from the community structure and seston stoichiometry in Florida Bay, USA, *Mar. Ecol. Prog. Ser.*, 165:45–57.

Leavitt, P.R., S.R. Carpenter, and J.F. Kitchell. 1989. Whole lake experiments: the annual record of fossil pigments and zooplankton, *Limnol. Oceanogr.*, 34:700–717.

Lizotte, M.P. and J.C. Priscu. 1993. Algal pigments as markers for stratified phytoplankton populations in Lake Bonney (dry valleys), *Antarc. J. U.S.*, 27:259–260.

Mantoura, R.F.C. and Llewellyn, C.A. 1983. The rapid determination of algal chlorophyll and carotenoid pigments and their breakdown products in natural waters by reverse-phase high performance liquid chromatography, *Anal. Chim. Acta*, 151:297–314.

Millie, D.F., H.W. Paerl, J.P. Hurley, and G.J. Kirkpatrick. 1993. Algal pigment determinations in aquatic ecosystems: analytical evaluations, applications, and recommendations, *Cur. Top. Bot. Res.*, 1:1–13.

Montague, C.L. and J.A. Ley. 1993. A possible effect of salinity fluctuations on abundance of benthic vegetation and associated fauna in northeastern Florida Bay, *Estuaries*, 16:703–17.

Morton, A.M. 1975. *Biochemical Spectroscopy*, Vol. 1. Wiley, New York.

Phlips, E.J. and S. Badylak. 1996. Spatial variability in phytoplankton standing crop and composition in a shallow inner-shelf lagoon, Florida Bay, Florida, *Bull. Mar. Sci.*, 58:203–216.

Phlips, E.J., T.C. Lynch, and S. Badylak. 1995. Chlorophyll *a*, tripton, color and light availability in a shallow tropical inner-shelf lagoon, Florida Bay, USA, *Mar. Ecol. Prog. Ser.*, 127:223–234.

Phlips, E.J., S. Badylak, and T.C. Lynch. 1999. Blooms of the picoplanktonic cyanobacterium *Synechococcus* in Florida Bay, a subtropical inner-shelf lagoon, *Limnol. Oceanogr.*, 44:1166–1175.

Platt, T., L. Dickie, and R. Trites. 1970. Spatial heterogeneity of phytoplankton in a nearshore environments, *J. Fish. Res. Bd. Can.*, 27:1453–1473.

Porter, J.W., S.K. Lewis, and K.G. Porter. 1999. The effect of multiple stressors on the Florida Keys coral reef ecosystem: a landscape hypothesis and a physiological test, *Limnol. Oceanogr.*, 44:941–949.

Rabalais, N.N., R.E. Turner, and W.J. Wiseman, Jr. 1999. Hypoxia in the Northern Gulf of Mexico: linkages with the Mississippi River, pp. 323–337, in *The Gulf of Mexico Large Marine Ecosystem*, H. Kumpf et al., Eds., Blackwell Scientific, Malden, MA.

Richardson, L.L. 1996. Remote sensing of algal bloom dynamics, *BioScience*, 46:492-501.

Richardson, L.L. and F.A. Kruse. 1999. Identification and classification of mixed phytoplankton assemblages using AVIRIS image-derived spectra, *Proc. 8th Ann. JPL Airborne Earth Sci. Workshop*, Pasadena, CA, JPL Pub. 99–17, pp. 339–347.

Richardson, L.L., D. Buisson, C.J. Liu, and V. Ambrosia. 1994. The detection of algal photosynthetic accessory pigments using airborne visible-infrared imaging spectrometer (AVIRIS) spectral data, *Mar. Technol. Soc. J.*, 28:10–21.

Rowan, K.S. 1989. *Photosynthetic Pigments of Algae*, Cambridge University Press, Cambridge, U.K.

Schalles, J.F., F.R. Schiebe, P.J. Starks, and W.W. Troeger. 1997. Estimation of algal and suspended sediment loads (singly and combined) using hyperspectral sensors and integrated mesocosm experiments, *Proc. 4th Int. Conf. Remote Sensing Mar. Coast. Environ.*, I:247–258.

Smith, N. 1994. Long-term Gulf–Atlantic transport through tidal channels in the Florida Keys, *Bull. Mar. Sci.*, 54:602–609.

Smith, R.C. and K.S. Baker. 1982. Oceanic chlorophyll concentrations as determined by satellite (Nimbus-7 Coastal Zone Color Scanner), *Mar. Biol.*, 66:269–279.

Soma, Y., T. Imaizumi, K.-I. Yagi, and S.-I. Kasuga. 1993. Estimation of algal succession in lake water using HPLC analysis of pigments, *Can. J. Fish. Aquatic Sci.*, 50:1142–1146.

Steidinger, K.A. and G.A. Vargo. 1988. Marine dinoflagellate blooms: dynamics and impacts, pp. 373–401, in *Algae and Human Affairs*, Lembi, O.C.A. and Waaland, D.J.R., Eds., Cambridge University Press, Cambridge, U.K.

Stumpf, R.P., M.L. Frayer, M.J. Durako, and J.C. Brock. 1999. Variations in water clarity and bottom albedo in Florida Bay from 1985 to 1997, *Estuaries*, 22:431–444.

Tester, P.A., M.E. Geesey, C. Guo, H.W. Paerl, and D.F. Millie. 1995. Evaluating phytoplankton dynamics in the Newport River estuary (North Carolina, USA) by HPLC-derived pigment profiles, *Mar. Ecol. Prog. Ser.*, 124:237–245.

Tomas, C.R., B. Bendis, and K. Johns. 1999. Role of nutrients in regulating plankton blooms in Florida Bay, pp. 323–337, in *The Gulf of Mexico Large Marine Ecosystem*, Kumpf, H. et al., Eds., Blackwell Scientific, Malden, MA.

Vane, G., R.O. Green, T.G. Chrien, H.T. Enmark, E.G. Hansen, and W.M. Porter. 1993. The airborne visible/infrared imaging spectrometer (AVIRIS), *Remote Sensing Environ.*, 44:127–143.

Wilhelm, C., I. Rudolph, and W. Renner. 1991. A quantitative method based on HPLC-aided pigment analysis to monitor structure and dynamics of the phytoplankton assemblage: a study from Lake Meerfelder (Eifel, Germany), *Arch. fur Hydrobiol.*, 123:21–35.

Wright, S.W. and J.D. Shearer. 1984. Rapid extraction and high performance liquid chromatography of chlorophylls and carotenoids from marine phytoplankton, *J. Chromatogr.*, 294:281–295.

Zimba, P.V., M.J. Sullivan, and H.E. Glover. 1990. Carbon fixation in cultured marine benthic diatoms, *J. Phycol.*, 26:306–311.

Zimba, P.V., C.P. Dionigi, and D.F. Millie. 1999. Evaluating the relationship between photopigment synthesis and 2-methylisoborneol accumulation in cyanobacteria, *J. Phycol.*, 35:1422–1429.

# 17 Modern Diatom Distributions in Florida Bay: A Preliminary Analysis

*Jacqueline K. Huvane*
Duke University Wetland Center, Nicholas School of the Environment

## CONTENTS

## INTRODUCTION

Florida Bay has been influenced by changes in freshwater flow from the Everglades in the last half century as a result of water diversion practices.[1] This has prompted researchers to ask to what extent salinity levels in the bay have been affected by these changes. For instance, have fluctuations in salinity increased in recent decades? In addition, seagrass dieoffs in the late 1980s have been attributed in part to anthropogenic factors such as salinity changes and nutrient stress.[1,2] However, the effects of anthropogenic activities in South Florida on the Florida Bay ecosystem have not been fully investigated.

This study is part of a larger ecosystem history of South Florida project being conducted in conjunction with researchers at the U.S. Geological Survey, Reston.[3–6] The goal of the project is to document historical environmental changes in Florida Bay to evaluate the effects of anthropogenic disturbance in the context of the natural variability of the system. One portion of this project deals with the use of diatoms in sediment cores from Florida Bay to infer past changes in the environment. In order to make such ecological inferences, it is necessary to obtain information about the modern diatom flora of the bay. This kind of information aids the interpretation of diatom assemblages from Florida Bay sediment cores to make inferences about past environmental conditions. Knowledge of the long-term natural variability of the South Florida ecosystem is essential when considering management plans, and in the absence of long term monitoring data paleoecological data provide a useful proxy. Interpretation of the paleoecological data is greatly enhanced when regional autoecological data are available for the bioindicators under consideration. This chapter presents preliminary findings of an ongoing study of the modern diatom flora found in surface sediment samples collected in Florida Bay during February 1998.

The utility of diatoms as indicators of environmental change has been well documented.[7–9] Diatoms are unicellular algae that have a siliceous frustule, or shell, that is typically preserved in sedimentary environments. The various taxa are characteristic of different physical habitats, such as epiphytic taxa that grow attached to aquatic plants, planktonic taxa that grow in the open water, and epipelic taxa that grow on the bottom sediments. In addition, species distributions are closely linked to water quality. For instance, many taxa have known salinity and nutrient preferences.[10,11] Diatoms have been successfully used as environmental indicators in other estuarine ecosystems.[12–14]

Recent studies of the diatom distributions in Florida Bay are few. DeFelice[15] examined the benthic diatom assemblages in upper Florida Bay and associated sounds. Montgomery[16] examined diatom communities associated with coral reefs in the Florida Keys. Several studies of subtropical diatom floras in the region contain little ecological information. These include a study of the littoral diatoms of Cuba,[17–21] diatoms from the Indian River, Florida,[22–25] planktonic diatoms from Choctawhatchee Bay, FL,[26] epiphytic diatoms from the Gulf Coast of Florida,[27] and benthic diatoms of coastal mangroves of South Florida.[28]

The purpose of this study is to characterize the diatom assemblages from a suite of surface sediment samples from sites in Florida Bay (Figure 17.1). These sites are part of a 26-site U.S. Geological monitoring survey of Florida Bay.[29] The main goal of this project is to identify diatom species and assemblage indicators of salinity. To date, samples from 17 sites have been analyzed. Preliminary results show that certain diatom taxa may be used as indicators of salinity.

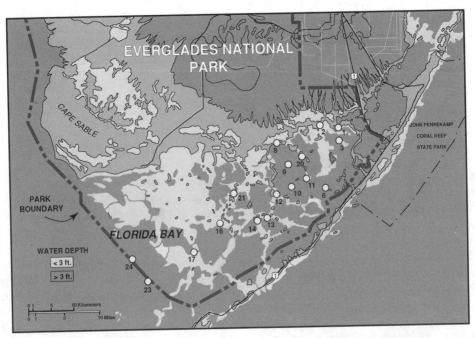

**FIGURE 17.1** Map of Florida Bay showing the 26 U.S. Geological Survey monitoring sites. Site numbers correspond to the numbers given in Table 17.1. This map is a modified version of a map provided by the U.S. Geological Survey.

# METHODS

## STUDY AREA

Florida Bay is a lagoonal estuary located south of the Everglades National Park and north and west of the Florida Keys. Approximately 80% of the bay is considered part of the Everglades National Park. In Florida Bay, seagrass, which provides habitat for epiphytic diatoms, covered about 80% of the bay prior to dieoffs in the late 1980s.[1,30,31] Mangrove isles cover less than 2% of the bay and specific diatom assemblages are known to occur on mangrove prop roots.[1, 22] Hard-bottom areas of calcium carbonate rock with a thin layer of carbonate sediment also occur. These areas contain sponges, corals, and macroalgae.[32] Diatoms are also found associated with corals,[16] macroalgae, and the hard bottom (field observation, 1999).

The U.S. Geological Survey has established 26 monitoring sites within the park boundary of Florida Bay.[3,5] Salinity, temperature, water depth, and other variables are recorded at each site.[29] Salinity data up to and including February 1998 were used to calculate the salinity averages for each of the 17 sites discussed in this paper. These data were used because the surface sediment diatom samples collected in February 1998 represent several previous years of deposition. There-fore, an average salinity based on salinity levels during the sampling years up to and including the surface sediment sampling year are thought to best reflect the salinity conditions under which those diatoms lived. The number of salinity samples used in the calculation for each site varied (see Table 17.1). Water depths at these sites in January 1998 ranged from 0.2 to 1.9 m at one of the midbasin sites (Table 17.1). Average salinity ranged between 7.7 and 33.7 ppt (Table 17.1). Detailed records of salinity, temperature, and field observations are available from the U.S. Geological Survey South Florida Ecosystem History Database.[29]

## SAMPLE COLLECTION AND PROCESSING

The sediment samples used in this study are from the top 1 to 2 cm of single shallow push cores (5-cm diameter) collected at the monitoring sites in January 1998 (Figure 17.1). Approximately 5 cc was taken from each surface sediment sample for diatom extractions. A modification of the diatom extraction procedure by Funkhauser and Evitt[33] was used to prepare samples. Briefly, samples were sequentially treated with hydrogen peroxide, hydrochloric acid, and finally nitric acid and potassium dichromate, with distilled water rinses performed between each treatment. A known volume of the resultant diatom slurry was permanently mounted on microscope slides with Naphrax®(Northern Biological Supply; Ipswich, U.K.) for enumeration and identification. Approx-imately 400 diatom valves per sample were counted.

## DIATOM IDENTIFICATIONS

Taxonomic identifications were aided by a number of resources including Peragallo and Peragallo,[34] Hustedt,[35,36] Patrick and Reimer,[37] Foged,[17] DeFelice,[15] Montgomery,[16] Stephens and Gibson,[23–25] Yohn and Gibson,[38,39] Foged,[40] Krammer and Lange-Bertalot,[41–44] Navarro et al.,[21] and Cooper.[45] However, some diatom taxa in the dataset have not been positively identified to species level. Some of these taxa may not yet be described in the literature.

## DATA ANALYSIS

A detrended correspondence analysis (DCA) was used to ordinate the diatom assemblages. This was done with the program PCORD.[46] DCA is an indirect ordination technique that disperses species data along axes that represent theoretical or latent variables.[47] Detrended correspondence analysis was developed to correct for the "arch affect," a mathematical artifact that occurs when the second axis shows a systematic, often quadratic relationship to the first axis.[47] The subset of

**TABLE 17.1**
**The Site Names and Numbers of the Samples Analyzed to Date[a]**

| Site Name | Site Number | Avg. Salinity[b] (ppt) | Standard Deviation (ppt) | Latitude | Longitude | Water Depth (m) |
|---|---|---|---|---|---|---|
| Trout Creek H$_2$O Station | 2 | 7.7 | 7.9 | 25.212 | 80.533 | 1.1 |
| Little Madeira Bay Mouth | 8 | 13.3 | 4.4 | 25.176 | 80.632 | 0.3 |
| Shell Creek Mouth | 1 | 15.0 | 6.2 | 25.207 | 80.487 | 0.9 |
| Duck Key H$_2$O Station | 3 | 19.0 | 4.9 | 25.180 | 80.489 | 0.5 |
| Mid-basin south of Little Madeira | 9 | 19.1 | 3.3 | 25.134 | 80.621 | 1.8 |
| Pass Key | 20 | 19.8 | 2.8 | 25.153 | 80.577 | 0.2–0.3 |
| Park Key southwest bank | 11 | 20.5 | 3.2 | 25.106 | 80.563 | 0.8 |
| Mid basin (Park and Russell Key) | 10 | 21.4 | 2.8 | 25.085 | 80.590 | 1.9 |
| Russell Key southeast bank | 12 | 23.7 | 3.0 | 25.072 | 80.638 | 0.5–0.6 |
| Butternut Key H$_2$O Station | 6 | 24.3 | 4.8 | 25.136 | 80.472 | 1.1 |
| Buttonwood Keys H$_2$O Station | 21 | 27.4 | 2.1 | 25.071 | 80.736 | 0.5 |
| Bob Allen Key Mudbank | 13 | 28.7 | 2.7 | 25.019 | 80.665 | 0.3 |
| Bob Allen Key H$_2$O Station | 14 | 29.6 | 3.1 | 25.026 | 80.664 | 0.8 |
| Mud Bank southeast of Sid Key | 16 | 30.7 | 2.9 | 25.012 | 80.764 | 0.5 |
| Rabbit Key H$_2$O Station | 17 | 32.8 | 1.6 | 24.981 | 80.825 | 0.2–0.3 |
| Schooner bank | 24 | 33.1 | 1.6 | 24.960 | 80.976 | 0.5 |
| Sprigger bank | 23 | 33.7 | 0.9 | 24.913 | 80.938 | 0.5 |

[a] The average salinity and standard deviation at each site are given, as well as the water depth from which the surface sediment samples were collected in 1998. Site location is specified by the latitude and longitude.
[b] The salinity averages are based on a various number of sampling dates. Sites 1–14: February and July 1995, 1996, 1997 and February 1998; site 16: July 1995, February and July 1996, 1997, and February 1998; site 17: February and July 1996 and 1997 and February 1998; site 20: February and July 1997 and February 1998; sites 23 and 24: July 1997 and February 1998.

diatom taxa used in this analysis included 47 taxa that were present at percentages of at least 2.5% in one sample.

Species diversity was calculated for each sample using the Shannon–Weiner diversity index ($H'$):[48]

$$H' = -\sum_{i=1}^{s} p_i(\ln p_i) \tag{17.1}$$

where $p_i$ = the proportion of the total number of valves belonging to the $i_{th}$ taxa in the assemblage, and $S$ = the number of species.

Pearson correlation coefficients were calculated to determine whether there were correlations between species diversity and average salinity and between species diversity and the percentage of unidentifiable valves.

Niche breadth ($B_i$) was calculated for each of the 47 taxa used in the DCA:[49]

$$B_i = \mathrm{Exp}\left[-\sum_{r=1}^{Q} n_{ir}/N_i \ln(n_{ir}/N_i)\right] \tag{17.2}$$

where $n_i$ is equal to the number of individuals (valves) of the $i$th taxon at the $r$th station, and $N_i$ is the total number of individuals found at all $Q$ sites. One way niche breadth can be defined is by the distribution of species over environmental gradients, whether those gradients are known or unknown.[50] In this case, one of the known environmental variables is salinity. Using the above equation, $B_i$ will range between 1 and $Q$ (the total number of sites). High values of $B_i$ suggest that the species has a fairly broad niche breadth or broad distribution with respect to the salinity range of the sites, while a smaller $B_i$ suggests a narrow niche breadth. It is important to remember that the data here are based on surface sediment samples and not live specimen samples.

Average salinity optima were calculated for each of the 47 taxa included in the DCA analysis based on a weighted average as follows:[47]

$$u_i = (y_1 x_1 + y_2 x_2 + \ldots y_n x_n)/(y_1 + y_2 + \ldots y_n) \qquad (17.3)$$

where $u_i$ = weighted average salinity optimum, $y_n$ = abundance of species $y$ at sites 1, 2, ..., $n$, and $x_n$ = the average salinity at sites 1, 2, ..., $n$. These optima give an idea of the salinity preference of each taxon. However, it should be noted that salinity optima presented here are only based on 17 samples and should be used with caution. Birks[51] gives a detailed discussion of weighted averaging methods.

## RESULTS

### SALINITY DATA

The average and standard deviation of the salinity data for the 17 sites are given in Table 17.1. Figure 17.2 shows a graphic representation of the salinity range of the 17 sites. Average salinity and the standard deviation are negatively correlated ($r^2 = 0.72$). The lowest average salinity sites — 2, 8, 1, and 3 — had the highest standard deviations, indicating greater seasonal and/or annual fluctuations. The most saline sites — 17, 23, and 24 — had the lowest variation. This may be the result of sampling error, as only two sampling dates were used to calculate the average and standard deviation for sites 23 and 24. However, salinity measurements taken after February 1998 indicate that these sites continue to have low salinity variation. Site 6, which had a fairly high salinity, was unusual in that it also had one of the higher standard deviations. Again, salinity measurements taken after February 1998 indicate the same trend of high salinity variation at this site.

### SURFACE SEDIMENT SAMPLES

To date, 260 taxa in 32 genera have been identified from the modern surface sediment samples. However, many of these taxa have been given only a generic designation and not a species designation (e.g., *Synedra* sp. A), so this number may change with further investigation. The five most common genera in the samples are *Navicula, Mastogloia, Amphora, Nitzschia,* and *Diploneis*. The ten most abundant taxa (based on average abundance across the 17 samples) are *Amphora coffeaeformis* (Agardh) Kutzing, *Cocconeis placentula* var. *euglypta* (Ehrenberg) Cleve, *Cyclotella* cf. *litoralis* Lange and Syversten, *Nitzschia granulata* Grunow, *Cyclotella* cf. *striata* (Kutzing) Grunow, *Amphora tenerrima* Hust. *Synedra* sp. A, *Cocconeis* cf. *placentula* var. *pseudolineata* Geitler, *Grammatophora* cf. *oceanica* var. *macilenta* (Wm. Smith) Grunow, and *Mastogloia crucicula* Grunow. Representative images of several taxa are shown in Figures 17.3 and 17.4.

### Ordination

The gradient length of the first DCA axis is relatively long at 3.0 standard deviation (s.d.) units. The second axis has a shorter gradient length of 2.0 s.d. units. In general, if two sites are 4 s.d.

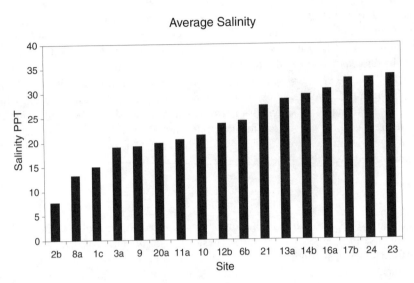

**FIGURE 17.2** Plot of the average salinity (ppt) by site. Sites are arranged from low to high salinity.

units apart, then they would not have species in common.[47] The eigenvalue (a measure of the importance of the axis) for axis 1 is 0.59. Values over 0.5 typically indicate a good separation of sites or samples along the axis.[47] The scatterplot of the DCA shows that the three sites with the highest salinity (17, 23, 24) have similar diatom assemblages that group together (Figure 17.5). These samples contain high abundances of epiphytic *Cocconeis* species. The samples from two of the less saline sites (2 and 3) have more distinct diatom assemblages and do not group with other sites in the analysis. In general, the higher salinity sites are in the upper right quadrant of the DCA diagram, while the lower salinity sites are in the lower half of the diagram. Interestingly, site 6, which has a higher average salinity and also a high standard deviation, has a DCA score that is more similar to the low salinity/high variation sites.

## Species Diversity

Diatom species diversity (H′) for the 17 samples ranges between 2 and 3.3. The average species diversity of the 17 samples is 2.8. Figure 17.6 shows the difference from the average diversity by site. Sites are arranged from the lowest to highest average salinity. Although preservation of diatom valves is poorer at some sites, (thereby increasing the number of unidentified taxa present), there is no obvious relationship between diversity and the percent unidentified taxa (Pearson correlation $r^2 = 0.06$). However, the sample with the lowest species diversity also has the highest percentage of unidentified taxa. Diversity is not clearly related to average salinity ($r^2 = 0.07$); however, the three sites with the highest average salinity (17, 23, 24) have diversity values above average.

## Individual Taxa

The number of samples in which each taxa occurred (N), niche breadth ($B_i$), and average salinity optima (S) are presented for the 47 common taxa in Table 17.2. Several of these taxa are discussed in greater detail in this section.

   Percentages of taxa arranged by site in order of increasing average can be found in Figures 17.7 and 17.8. *Amphora coffeaeformis* occurs in 15 of the 17 sites and has a niche breadth of 11.8 (Table 17.2, Figure 17.7). It has a salinity optimum of 23.5 ppt. This taxon has been noted in sediment samples from a tidal flat in Korea[52] and has also been noted as an epiphyte.[27]

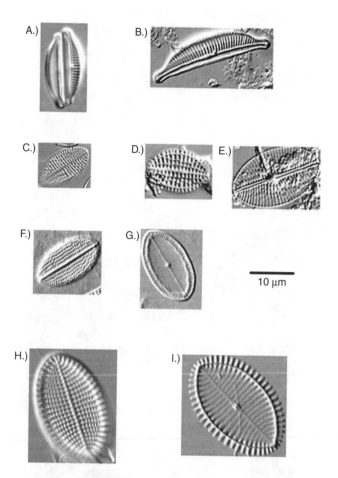

**FIGURE 17.3** Representative images of some of the common diatom taxa found in Florida Bay. A 10-μm bar is given for scale. (A) *Amphora tenerrima*; (B) *Amphora coffeaeformis*; (C) *Cocconies* cf. *placentula pseudolineata*; (D) *Cocconeis disculoides*, pseudoraphe valve; (E) *Cocconeis disculoides*, raphe valve; (F) *Cocconeis placentula* var. *euglypta*, pseudoraphe valve; (G) *Cocconeis placentula* var. *euglypta*, raphe valve; (H) *Cocconeis scutellum*, pseudoraphe valve; (I) *Cocconeis scutellum*, raphe valve.

*Amphora tenerrima* Hustedt, an epiphytic taxon,[27] occurs in 15 of the 17 samples and has a niche breadth of 10.1. It has a salinity optimum of 20.1 ppt (Table 17.2, Figure 17.7).

*Cocconeis placentula* var. *euglypta* occurs in all 17 samples; however, it has a fairly low niche breadth of 8.6 due to its scarcity at many of the sites (Table 17.2, Figure 17.7). This taxon is most common in samples from the less saline sites (2, 8, 1), although it occurs in one of the samples from a high salinity site (17). Accordingly, it has a low salinity optimum of 16.9 ppt. In North America, this taxon is typically associated with freshwater environments;[37] however, it is known to occur in brackish water[15,45] and in higher salinity environments.[53] This taxon has been identified as epiphytic in Florida Bay.[15] Sullivan[27] has found *Cocconeis* species growing on samples of seagrass from the Gulf Coast of Florida. In particular, *Cocconeis placentula* var. *euglypta* was common (approximately 14% of the total diatom valves counted) on a sample of *Thalassia testudinum* Konig.

*Cocconeis scutellum* Ehrenberg occurs in 13 of the samples and has a niche breadth of 6.8 (Table 17.2, Figure 17.7). This taxon is most common in a sample from site 21, which had an average salinity of 27.4 ppt. It is also fairly common in samples from sites 11, 17, and 24. It has a salinity optimum of 26.7 ppt. This taxon has been found on seagrass samples from Florida Bay[15]

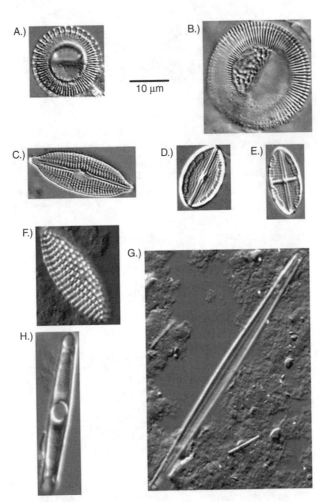

**FIGURE 17.4** Representative images of some of the common diatom taxa found in Florida Bay. (A) *Cyclotella* cf. *striata*; (B) *Cyclotella* cf. litoralis; (C) *Mastogloia corsicana*; (D) *Mastogloia ovalis*; (E) *Mastogloia crucicula*; (F) *Nitzschia granulata*; (G) *Synedra* sp. A; (H) *Grammatophora* cf. *oceanica* var. *macilenta*.

and on mangrove prop roots.[22] *Cocconeis scutellum* is also described as an epiphyte identified from Caja de Meuros Island, Puerto Rico,[21] and is common in sediment cores from the Chesapeake Bay.[45]

*Cocconeis disculoides* Hust. occurs in only seven samples and has a niche breadth of 4.6 (Table 17.2, Figure 17.7). This taxa is more common in samples from the less saline sites (2, 8, 1, 9) and is absent from samples from the more saline sites (17, 24, 23). It has a low salinity optimum of 13.8 ppt. DeFelice[15] describes this taxon as an epiphyte in Florida Bay.

*Cyclotella* cf. *litoralis* Lange & Syvertsen is common, occurring in 16 or the 17 samples examined to date (Figure 17.7). It has a niche breadth of 12 and a salinity optimum of 23.6 ppt (Table 17.2). This taxon is described as a marine planktonic species.[54]

*Cyclotella* cf. *striata* (Kutzing) Grunow occurs in 13 samples and has a niche breadth of 9.3 (Table 17.2, Figure 17.7). It is similar to *Cyclotella litoralis*, and the two may have been confused on occasion. Foged[17] describes this taxon as a mesohalobe to a polyhalobe. In this study it has an average salinity optimum of 22 ppt.

*Diploneis* cf. *didyma* is present in seven samples and has a niche breadth of 3.7 (Table 17.2, Figure 17.7). DeFelice[15] describes this taxon as mainly epipelic (growing on sediment). This taxon is only found at sites in this study with an average salinity above 20 ppt and has a salinity optimum of 30.4 ppt.

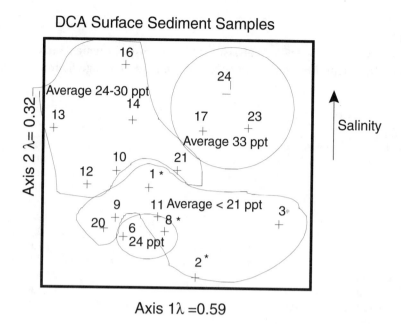

**FIGURE 17.5** Scatterplot of the detrended correspondence analysis (DCA) results. Numbers represent the site numbers given in Table 17.1. The length of the first DCA axis is 3.0 s.d., and the second axis 2.0 s.d.

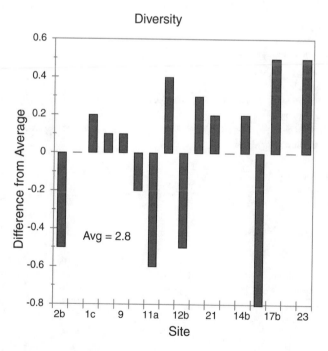

**FIGURE 17.6** A plot of the difference from the average diversity ($H'$ average = 2.8) for each site is shown.

*Grammatophora* cf. *oceanica* var. *macilenta* occurs in 15 samples, with a niche breadth of 5.4 (Table 17.2, Figure 17.7). This taxon shows a preference for higher salinities, as it is only common at the more saline sites (16, 17, and 24) and has a salinity optimum of 28.9 ppt. However, it is not abundant at the most saline site (24), and does occur at the less saline sites (8 and 1). This genera is more typical of marine littoral environments.[45]

**TABLE 17.2**

**The 47 Taxa Used in the DCA Analysis, Number of Sites in which Each Taxon Occurred ($N$), Niche Breadth Index ($B_i$), and Weighted Average Salinity Optima**

| Taxa | N | $B_i$ | S |
|---|---|---|---|
| *Achnanthes* cf. *amoena* Hust. | 4 | 1.6 | 10.3 |
| *Achnanthes* sp. 1 | 1 | 1 | 19 |
| *Amphora coffeaeformis* (Agardh) Kutz. | 15 | 11.6 | 23.5 |
| *Amphora tenerrima* Hust. | 15 | 10.1 | 21.8 |
| *Amphora* sp. 1 | 5 | 3.7 | 22.2 |
| *Amphora* sp. 2 | 1 | 1 | 19 |
| *Amphora* sp. 3 | 2 | 2 | 13.7 |
| *Amphora* cf. *granulata* Greg. | 10 | 9.2 | 25.2 |
| *Amphora* af. *acuta* Greg. | 4 | 2.8 | 32.4 |
| *Amphora* sp. 4 | 9 | 6.8 | 26.5 |
| *Amphora* sp. 5 | 5 | 3 | 25.1 |
| *Cocconeis disculoides* Hust. | 7 | 4.4 | 13.8 |
| *Cocconeis placentula* var. *euglypta* (Ehr.) Cleve | 17 | 8.6 | 16.9 |
| *Cocconeis placentula* var. *lineata* (Ehr.) V.H. | 5 | 2.4 | 30.7 |
| *Cocconeis placentula* var. *pseudolineata* Geitler | 4 | 3.4 | 32.4 |
| *Cocconeis scutellum* Ehr. | 13 | 6.6 | 26.7 |
| *Cymatosira belgica* Grunow | 10 | 8.1 | 27.7 |
| *Cyclotella* cf. *litoralis* Lange & Syversten | 16 | 12 | 23.6 |
| *Cyclotella* sp. 1 | 4 | 3 | 20.1 |
| *Cyclotella* cf. *striata* (Kutz.) Grunow | 13 | 9.3 | 22 |
| *Diploneis* cf. *didyma* (Ehr.) Ehr. | 7 | 3.7 | 30.4 |
| *Diploneis* cf. *suborbicularis* (Greg.) Cleve | 2 | 1.3 | 33.7 |
| *Diploneis oblongella* (Naeg. ex. Kutz.) Ross | 8 | 4.8 | 25.4 |
| *Diploneis ovalis* (Hilse) Cleve. | 3 | 2.2 | 30.9 |
| *Grammatophora* cf. *oceanica* var. *macilenta* (Wm. Smith) Grun. | 15 | 5.4 | 28.9 |
| *Mastogloia corsicana* Grunow | 8 | 4 | 28.3 |
| *Mastogloia crucicula* Grunow | 15 | 11.4 | 23.6 |
| *Mastogloia erythrea* Grunow | 11 | 7.7 | 24.1 |
| *Mastogloia ovalis* A. Schmidt | 13 | 9 | 21.6 |
| *Mastogloia subaffirmata* Hust. | 3 | 1.6 | 27.4 |
| *Navicula* cf. *eidrigiana* Carter | 5 | 3.5 | 20.7 |
| *Navicula scopuloroides* | 9 | 5.6 | 20.9 |
| *Navicula* sp. 1 | 9 | 6 | 20.9 |
| *Navicula* cf. *ramosisimma* (Ag.) Cleve. | 4 | 1.5 | 18.6 |
| *Fallacia* sp. 1 | 3 | 1.8 | 9.7 |
| *Navicula* sp. 2 | 1 | 1 | 19 |
| *Neodelphineas pelagica* Takano | 6 | 4 | 23.9 |
| *Nitzschia* cf. *panduriformis* Greg. | 10 | 6.8 | 28.6 |
| *Nitzschia granulata* Grunow | 9 | 6.2 | 20.1 |
| *Nitzschia marginulata* Grunow | 4 | 2.3 | 21.7 |
| *Nitzschia* sp. 1 | 1 | 1 | 19 |
| *Nitzschia* cf. *grossestriata* Hust. | 7 | 4 | 27.2 |
| *Rhopolodia acuminata* Krammer | 10 | 7.4 | 21.1 |
| *Rhopolodia* cf. *gibberula* (Ehr.) O. Mull. | 10 | 7 | 24.1 |
| *Rhopolodia* cf. *musculus* (Kutz.) O. Mull. | 6 | 3.2 | 23.5 |
| *Synedra* sp. A | 12 | 7.4 | 26.8 |
| *Thalassiosira* sp. 1 | 1 | 1 | 33.1 |

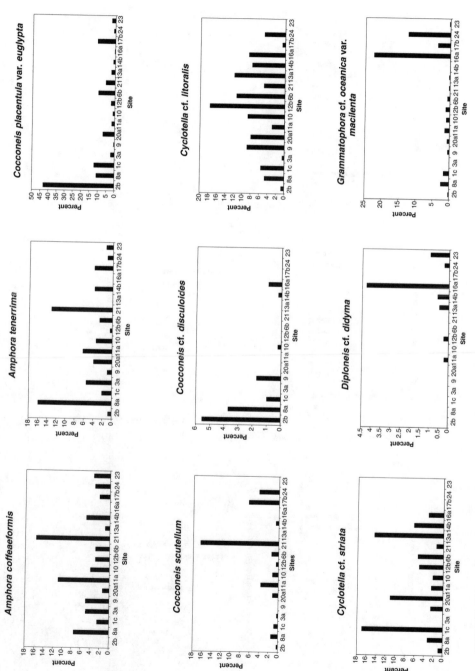

**FIGURE 17.7** Plots of individual taxa percent abundances by site. Sites are arranged from lowest to highest average salinity.

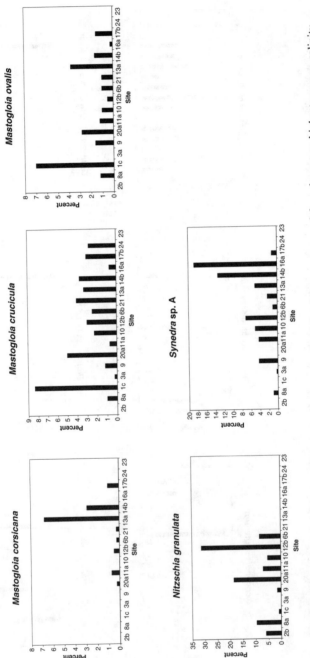

**FIGURE 17.8**  Plots of individual taxa percent abundances by site. Sites are arranged from lowest to highest average salinity.

*Mastogloia crucicula* occurs in 15 of the 17 samples, with a niche breadth of 11.4 (Table 17.2, Figure 17.8). This taxon is most common at site 1 but generally does not appear to have any salinity preference within the range of these sites. Stephens and Gibson[23] report that this taxon has a salinity range of 16 to 42 ppt. In this study, it has a salinity optimum of 23.6 ppt. This taxon is a fairly common epiphyte.[15,22,23,27]

Another species of *Mastogloia, M. corsicana* Grunow, shows a preference for the more saline sites and has a high salinity optimum of 28.3 ppt (Table 17.2, Figure 17.8). It occurs in only eight of the samples and has a niche breadth of 4.

*Mastogloia ovalis* A. Schmidt is present in 13 of the samples and has a niche breadth of 9 (Table 17.2, Figure 17.8). It is most abundant from the site 1 sample and absent in samples from sites 2, 3, 23, and 24. This species is epiphytic and has been found growing on seagrass samples.[27] Stephens and Gibson[23] report that this taxon has a salinity range of 16 to 42 ppt. In this study, it has a salinity optimum of 21.6 ppt.

*Nitzschia granulata* Grunow is restricted to sites that had an average salinity below 24 ppt. It is present in nine samples and has a niche breadth of 6.2 and a fairly low salinity optimum of 20.1 ppt (Table 17.2, Figure 17.8). DeFelice[15] describes this taxon as epipelic. Krammer and Lange-Bertalot[42] report that this taxon is common on sandflats.

*Synedra* sp. A occurs in 12 samples and has a niche breadth of 7.4 (Table 17.2, Figure 17.8). This taxon is more common at sites with an average salinity over 19 ppt, but it does not occur at the two most saline sites (23 and 24). It has a high salinity optimum of 26.8 ppt, making it useful as an indicator of intermediate to high salinity conditions.

## DISCUSSION

The ordination analysis of the diatom assemblage data indicates that the diatom flora of Florida Bay is fairly diverse. This is probably due in part to the restricted circulation in the bay that is a result of the network of mudbanks.[1] However, some taxa, such as *Amphora coffeaeformis*, occur at most sites. This particular taxon had a fairly wide niche breadth, suggesting that it may not be a useful indicator of salinity. However, a recent study by Sala et al.[55] has shown that this taxon may be indistinguishable from other taxa when viewed under a light microscope. Therefore, without the aid of electron microscopy, what is identified as *A. coffeaeformis* may actually consist of several taxa. *Cyclotella* cf. *litoralis* (Figure 17.7) was also found at many of the sites, although it has been described as a marine taxon.[54] This diatom may have been confused with the similar *C. striata*, thereby increasing its apparent niche breadth. Because this taxon is planktonic, it may move through the bay more easily (via wind or current) than a benthic taxon. This may also account for its presence in many of the sediment samples. Another planktonic diatom taxon, *Rhizosolenia*, has been reported to bloom in Florida Bay;[56] However, this taxon is lightly silicified and does not preserve well in sediment samples.[45] This may explain why *Rhizosolenia* frustules were not noted in any samples from this dataset.

The samples from mid-range salinity sites (9, 20, 11, 10, 12, and 6) group together on the ordination diagram. This is not unexpected, as these are less likely to contain either freshwater or marine taxa. Unlike the high salinity sites (17, 23, and 24), some of the samples from sites at the lower end of the salinity range (1, 2 and 3) do not group together on the diagram. This indicates that these samples are relatively dissimilar in terms of their diatom assemblages. Sites 1 and 2 are situated near the mouths of Shell Creek and Trout Creek, respectively. Outflow from these creeks may contribute freshwater, as well as transport of diatoms from the creeks to these sites. Thus, the resulting fluctuations in salinity as well as the input of diatoms from the different creeks may explain the distinct diatom assemblages found at these sites.

Species diversity was expected to be higher at the more marine sites (closer to the edge of the park boundary). DeFelice[15] found that diatom species diversity in Florida Bay generally increased from land towards more open areas of water. At one of these sites (23) diversity was fairly high.

Site 23 is located in the southwestern portion of the bay along the western park boundary. However, site 1 has a diatom species diversity higher than the average and is located in the northeast portion of the bay, relatively close to the land margin. The higher diversity at both these sites may be due in part to their location along an ecotone. Site 6 displays a relatively high species diversity value. This site had an unusually high average salinity coupled with a higher salinity variance. The higher diversity at this site is probably related to this higher degree of salinity variation.

Four of the five sites with lower than average diversity had salinities below 24 ppt. The exception was site 16, which had the lowest diversity ($H' = 2$) but one of the highest average salinities (30.7 ppt). Analysis of samples from the other sites that are part of the U.S. Geological Survey monitoring program may indicate whether there is a significant relationship between diatom species diversity and salinity.

Several diatom species showed a salinity preference for lower salinity sites. *Cocconeis disculoides* had a relatively low niche breadth ($B_i = 4.6$) with a preference for low salinity sites. *Cocconeis placentula* var. *euglypta* also showed a slight preference for the low salinity sites. However, this species had a higher niche breadth ($B_i = 8.6$) and may be less useful as an indicator of low salinity. Another potential indicator of lower salinity is *Nitzschia granulata*. The absence of this taxon from the higher salinity sites strongly suggests a restricted distribution associated with salinity. Examination of samples from the other monitoring sites should help to better define the distribution of this taxon in Florida Bay.

Several taxa display a preference for the higher salinity sites. *Cocconeis scutellum* is more common in samples from the high salinity sites and has a fairly low niche breadth ($B_i = 6.8$). *Grammatophora* cf. *oceanica* var. *macilenta*, a marine littoral species,[45] shows a preference for samples from the high salinity sites and has a fairly narrow niche breadth ($B_i = 5.4$). Another indicator of higher salinities is *Mastogloia corsicana*, which also displays a narrow niche breadth ($B_i = 4$). *Synedra* sp. A may also be useful as an indicator of higher salinity. Although this taxon is not present in samples from the two most saline sites, it has a high average salinity optimum and is uncommon or absent from samples taken from the lower salinity sites.

Not all diatom taxa display a relationship to average salinity. This may be the result of several factors, one being methodological. The samples analyzed here represent a relatively large temporal and spatial scale. For example, the top 1 to 2 cm of the pushcore likely represent several years of deposition. Salinity in Florida Bay fluctuates both seasonally and annually; therefore, salinity conditions over a few years may change significantly.[1] Diatom populations also vary seasonally. For example, *Mastogloia crucicula* was present in *Thalassia* samples collected from the Florida Keys during March, August, and December, but December samples contained higher percentages (approximately 3 to 4× higher) than those collected in spring and summer.[16] Averaging several years of both diatom deposition and salinity data may obscure the relationship between salinity and diatom taxa abundance. In addition, the surface sediment samples integrate diatoms from various habitats in the area (e.g., planktonic, epiphytic, and epipelic). Examination of diatoms from a variety of habitats collected periodically over the course of several years may provide better information concerning salinity preferences. The advantage of using surface sediment samples, however, is that the collection and analysis are quicker *because* there is an integration of time and space. Given constraints of time and money, a dual approach that combines surface sediment samples and fewer collections from different habitats may be preferable.

Another possibility for the lack of a clear relationship between some diatom taxa and salinity is that some of these taxa may actually have relatively broad salinity ranges. For instance, *Mastogloia crucicula* is present at many sites and is reported in the literature as having a relatively broad salinity range of 16 to 42 ppt.[23] Many of the brackish water species may be adapted to salinity fluctuations like those seen in Florida Bay.

Finally, other factors besides salinity most likely influence diatom distributions in Florida Bay. For example, nutrient concentrations are known to affect diatom composition in estuarine environments.[13] Another potential variable influencing diatom distributions is light availability. For example, light may be attenuated by factors such as turbulence or planktonic algal blooms, thus limiting benthic algal growth. Other factors that may influence diatom distributions are subaquatic vegetation cover, water temperature, and water depth.

Several of the taxa discussed in this chapter have been identified as epiphytic, that is, growing on submerged aquatic vegetation. The author is currently conducting another study to examine the diatoms from seagrass samples collected in Florida Bay. This study will help define which diatom taxa may be used as indicators of subaquatic vegetation. For instance, DeFelice[15] found that *Cocconeis placentula* var. *euglypta* was the dominant taxa found growing on the seagrass *Thalassia*. Preliminary data collected to date suggest that this taxon is still a common epiphyte on *Thalassia* (Huvane, unpubl. data). In addition, several epiphytic species of *Cocconeis* identified in this study have different salinity preferences. For instance, *Cocconeis scutellum* has a higher salinity optimum than either *Cocconeis placentula* var. *euglypta* or *Cocconeis disculoides*. These different salinity preferences would allow researchers to distinguish salinity changes from changes in subaquatic vegetation cover based on diatom indicator species.

## CONCLUSIONS

The diatom flora of Florida Bay is fairly diverse, and preliminary investigations suggest that there are new species that have not been described in the literature. Further taxonomic and ecological studies of the diatom flora of this region are warranted. In this study, the less saline sites appear to be more distinct in terms of their diatom assemblages. This may be the result of the larger fluctuations in salinity at these sites and the transport of diatoms from nearby creeks. In general, the sites with the highest salinity displayed the highest species diversity. This may be due to their location at the edge of the bay, at the ecotone between the bay and the marine environment.

Although Florida Bay experiences variation in salinity on both a seasonal and annual basis, several taxa have been identified as indicators of either low or high salinity conditions. This study suggests that diatoms can be effectively used as proxies for salinity conditions in Florida Bay in sediment core studies. The abundance of epiphytic diatom taxa in the surface sediments suggests that diatoms may also be used to infer the abundance of subaquatic vegetation in Florida Bay. A study is currently underway to identify these indicators from seagrass samples collected from Florida Bay.

## ACKNOWLEDGMENTS

Funding for this study was provided by the U.S. Geological Survey, Department of the Interior, under assistance award No. 1434-HQ-00-AG-00000. The views and conclusions contained in this document are those of the author and should not be interpreted as necessarily representing the official policies, either expressed or implied, of the U.S. Government.

The Duke University Wetland Center, Nicholas School of the Environment, provided additional support. I would like to thank G. Lynn Brewster-Wingard of the U.S. Geological Survey, Reston, VA, who provided sediment samples for diatom analyses. The Keys Marine Lab, Florida Institute of Oceanography, Long Key, FL provided field assistance. Michael Sullivan at Mississippi State University and Andrzej Witkowski at the University Szczecinski in Poland provided help with diatom taxonomy. Finally, I would like to thank Sherri Cooper of the Duke University Wetland Center for providing assistance with diatom taxonomy and a critical review of this paper.

## REFERENCES

1. McIvor, C.C., Ley, J.A., and Bjork, R.D., Changes in freshwater inflow from the Everglades to Florida Bay including effects on biota and biotic processes: a review, in *Everglades: The Ecosystem and Its Restoration*, Davis, S.M. and Ogden J.C., Eds., St. Lucie Press, Delray Beach, FL, 1994, p. 117.

2.  Lapointe, B.E., Tomasko, D.A., and Matzie, W.R., Eutrophication and trophic state classification of seagrass communities in the Florida Keys, *Bull. Mar. Sci.,* 54, 696, 1994.

3.  Brewster-Wingard, G.L., Ishman, S.E., Edwards, L.E., and Willard, D.A., Preliminary report on the distribution of modern fauna and flora at selected sites in northcentral and northeastern Florida Bay, U.S. Geological Survey Open-File Report, 96–732, 1996.

4.  Brewster-Wingard, G.L., Ishman, S.E., and Holmes, C.W., Environmental impacts on the Southern Florida coastal waters: a history of change in Florida Bay, *J. Coast. Res.,* 26, 162, 1998.

5.  Scott, T.M., Means, G.H., and Brewster-Wingard, G.L., Progress report on sediment analyses at selected faunal monitoring sites in northcentral and northeastern Florida Bay, U.S. Geological Survey Open-File Report, 97–534, 1997.

6.  Ishman, S.E., Cronin, T.M., Brewster-Wingard, G.L., Willard, D.A., and Verado, D.J., A record of ecosystem change, Manatee Bay, Barnes Sound, Florida, *J. Coast. Res.,* 26, 125, 1998.

7.  Battarbee, R.W., Charles, D.F., Dixit, S.S., and Renberg, I., Diatoms as indicators of surface water acidity, in *The Diatoms — Applications for the Environmental and Earth Sciences,* Stoermer, E.F. and Smol, J.P., Eds., Cambridge University Press, New York, 1999, p. 85.

8.  Hall, R.I. and Smol, J.P., Diatoms as indicators of lake eutrophication, in *The Diatoms — Applications for the Environmental and Earth Sciences,* Stoermer, E.F. and Smol, J.P., Eds., Cambridge University Press, New York, 1999, p. 128.

9.  Smol, J.P., Paleolimnological approaches to the evaluation and monitoring of ecosystem health: providing a history for environmental damage and recovery, in *Evaluating and Monitoring the Health of Large-Scale Ecosystems,* Rapport, D.J., Gaudet, C.L., and Calow, P., Eds., NATO ASI Series Vol. 128, Springer-Verlag, Berlin, 1995, p. 301.

10. Cooper, S.R., Huvane J.K., Vaithiyanathan, P., and Richardson C.J., Calibration of diatoms along a nutrient gradient in Florida Everglades Water Conservation Area-2A, U.S.A., *J. Paleolimnol.,* 22, 413, 1999.

11. Fritz, S.C., Juggins, S., Battarbee, R.W., and Engstrom, D.R., Reconstruction of past changes in salinity and climate using a diatom-based transfer function, *Nature,* 352, 706, 1991.

12. Cooper, S.R., Estuarine paleoenvironmental reconstructions using diatoms, in *The Diatoms — Applications for the Environmental and Earth Sciences,* Stoermer, E.F. and Smol, J.P., Eds., Cambridge University Press, New York, 1999, p. 352.

13. Cooper, S.R., Chesapeake Bay watershed historical land use: impact on water quality and diatom communities, *Ecol. Appl.,* 5, 3, 703, 1995.

14. Juggins, S., *Bibliotheca Diatomologica,* Vol. 25, *Diatoms in the Thames Estuary, England: Ecology, Palaeoecology, and Salinity Transfer Function,* Lange-Bertalot, H., Ed., Gerbruder Borntraeger, Berlin.

15. DeFelice, D.R., Model Studies of Epiphytic and Epipelic Diatoms of Upper Florida Bay and Associated Sounds, M.S. thesis, Duke University, Durham, NC, 1976.

16. Montgomery, R.T., Environmental and Ecological Studies of the Diatom Communities Associated with the Coral Reefs of the Florida Keys (Vols. I and II), Ph.D. dissertation, Florida State University, Tallahassee, FL, 1978.

17. Foged, N., *Bibliotheca Diatomologica,* Vol. 5, Freshwater *and Littoral Diatoms from Cuba,* J. Cramer, Herschberg, 1984.

18. Navarro, J.N., A survey of the marine diatoms of Puerto Rico, VI: suborder Raphidineae: family Naviculaceae (Genera *Haslea, Mastogloia,* and *Navicula*), *Bot. Marina,* 26, 119, 1983.

19. Navarro, J.N., A survey of the marine diatoms of Puerto Rico, VII: suborder Raphidineae: families Auriculaceae, Epithemiaceae, Nitzschiaceae, and Surirellaceae, *Bot. Marina,* 26, 393, 1983.

20. Navarro, J.N., A survey of the marine diatoms of Puerto Rico, V: suborder Raphidineae: families Achnanthaceae and Naviculaceae (excluding *Navicula* and *Mastogloia*), *Bot. Marina,* 25, 321, 1982.

21. Navarro, J.N., Perez, C., Arce, N., and Arroyo, B., Benthic marine diatoms of Caja de Muertos Island, Puerto Rico, *Nova Hedwigia,* 49, 3-4, 333, 1989.

22. Navarro, J.N., *Bibliotheca Phycologica,* Vol. 61, *Marine Diatoms Associated with Mangrove Prop Roots in the Indian River, Florida, U.S.A.,* J. Cramer, Hirschberg, 1982.

23. Stephens, F.C. and Gibson, R.A., Ultrastructural studies on some *Mastogloia* (Bacillariophyceae) species belonging to the group Ellipticae, *Bot. Marina,* 22, 499, 1979.

24. Stephens, F.C. and Gibson, R.A., Ultrastructural studies of some *Mastogloia* (Bacillariophyceae) species belonging to the groups Undulatae, Apiculatae, Lanceolatae, and Paradoxae, *Phycologia,* 19, 2, 143, 1980.

25. Stephens, F.C. and Gibson, R.A., Ultrastructural studies on some *Mastogloia* species of the group *Inaequales* (Bacillariphyceae), *J. Phycology,* 16, 354, 1980.

26. Prasad, A.K. S.K., Nienow, J.A., and Livingston, R.J., The genus *Cyclotella* (Bacillariophyta) in Choctawhatchee Bay, Florida, with special reference to *C. striata* and *C. choctawhatcheeana* sp. nov., *Phycologia,* 29, 418, 1990.

27. Sullivan, M.J., Community structure of epiphytic diatoms from the Gulf Coast of Florida, U.S.A., in *Proceedings of the Seventh International Diatom Symposium,* Mann, D.G., Ed., Otto Koeltz Science Publishers, Koenigstein, 1982, p. 373.

28. Ross, M.S., Gaiser, E.E., Meeder, J.F., and Lewin, M.T., Multi-taxon analysis of the "white zone," a common ecotonal feature of South Florida coastal wetlands, in *The Everglades, Florida Bay, and Coral Reefs of the Florida Keys: An Ecosystem Sourcebook*, Porter, J.W. and Porter, K.G., Eds., CRC Press, Boca Raton, FL, 2002.

29. U.S. Geological Survey, South Florida Ecosystem History Database Web Site, http://flaecohist.er.usgs.gov/database/, 1999.

30. Robblee, M.B., Barber, R., Carlson, P.R., Durako, M.J., Fourqurean, J.W., Muehlstein L.K., Porter D., Yarbro, L.A., Zieman, R.T., and Zieman, J.C., Mass mortality of the tropical seagrass *Thalassia testudinum* in Florida Bay (USA), *Mar. Ecol. Prog. Ser.*, 71, 297, 1991.

31. Zieman, J.C., Fourqurean, J.W., and Iverson, R.L., Distribution, abundance and productivity of seagrasses and macroalgae in Florida Bay, *Bull. Mar. Sci.*, 44, 1, 292, 1989.

32. Butler, M.J., Hunt J.H., Herrnkind, W.F., Childress, M.J., Bertelsen, R., Sherp, W., Matthews, T., Field, J.M., and Marshall, H.G., Cascading disturbances in Florida Bay, U.S.A.: cyanobacterial blooms, sponge mortality, and implications for juvenile spiny lobsters *Panulirus argus*, *Mar. Ecol. Prog. Ser.*, 129, 119, 1995.

33. Funkhauser, J.W. and Evitt, W.R., Preparation techniques for acid insoluble microfossils, *Micropaleontology*, 5, 3, 369, 1959.

34. Peragallo, H. and Peragallo, M., *Diatomées marines de France et de districts maritimes voisins*, M. J. Tempère, Grez-sur-Loing, 1897–1908.

35. Hustedt, F., *Die Kieselalgen Deutschlands, Österreichs und der Schweiz unter Berücksichtigung der übrigen Länden Europas sowie der angrenzenden Meeresgebiete*, 3 vols., Dr. L. Rabenhorst's Kryptogamen-flora von Deutschland, Österreich und der Schweiz, Akademische Verlagsgesellschaft, Leipzig, 1927–1964.

36. Hustedt, F., *Marine Littoral Diatoms of Beaufort, North Carolina*. Duke University Press, Durham, NC, 1955.

37. Patrick, R. and Reimer, C.W., *The Diatoms of the United States Exclusive of Alaska and Hawaii*, Vol. 1, The Academy of the Natural Sciences of Philadelphia, Philadelphia, PA, 1966.

38. Yohn, T.A. and Gibson, R.A., Marine diatoms of the Bahamas. II. *Mastogloia* Thw. *ex* Wm. Sm. species of the groups Decussatae and Ellipticae, *Bot. Marina*, 25, 41, 1982.

39. Yohn, T.A. and Gibson, R.A., Marine diatoms of the Bahamas. III. *Mastogloia* Thw. *ex* Wm. Sm. species of the groups Inaequales, Lanceolatae, Sulcatae, and Undulatae. *Bot. Marina*, 25, 277, 1982.

40. Foged, N., *Bibliotheca Diatomologica*, Vol. 12, *Diatoms in the Volo Bay, Greece*, J. Cramer, Berlin, 1986.

41. Krammer, K. and Lange-Bertalot, H., *Bacillariophyceae 2/1. Naviculaceae, Süßwasserflora von Mitteleuropa*, Fischer, Stuttgart, 1986.

42. Krammer, K. and Lange-Bertalot, H., *Bacillariophyceae 2/2. Bacillariaceae, Epithemiaceae, Surirellaceae, Süßwasserflora von Mitteleuropa*, Fischer, Stuttgart, 1988.

43. Krammer, K. and Lange-Bertalot, H., *Bacillariophyceae 2/3. Centrales, Fragilariaceae, Eunotiaceae, Süßwasserflora von Mitteleuropa*, Fischer, Stuttgart, 1991.

44. Krammer, K. and Lange-Bertalot, H., *Bacillariophyceae 2/4. Achnanthaceae, Süßwasserflora von Mitteleuropa*, Fischer, Stuttgart, 1991.

45. Cooper, S.R., Diatoms in sediment cores from the mesohaline Chesapeake Bay, U.S.A., *Diatom Res.*, 10, 1,39,1995.

46. McCune, B. and Mefford, M.J., Multivariate analysis of ecological data, version 2.0, MjM Software, Gleneden Beach, OR, 1997.

47. Ter Braak, C.J.F., Ordination, in *Data Analysis in Community and Landscape Ecology*, Jongman, R.H.G., Ter Braak, C.J.F., and Van Tongeren, O.F.R., Eds., Cambridge University Press, Cambridge, U.K., 1995 p. 91.

48. Wetzel, R.G., *Limnology*, CBS College Publishing, New York, 1983.

49. McIntire, C.D. and Overton, W.S., Distributional patterns in assemblages of attached diatoms from Yaquina Estuary, Oregon, *Ecology*, 52, 758, 1971.

50. Levins, R., *Evolution in Changing Environments*, Princeton University Press, Princeton, NJ, 1968.

51. Birks, H.J.B., Quantitative palaeoenvironmental reconstructions, in *Statistical Modeling of Quaternary Science Data*, Technical Guide 5, Maddy, D. and. Brew, J.S., Eds., Quaternary Research Association, Cambridge, MA, 1995, p. 161.

52. Oh, S.H. and Koh, C.H., Distribution of diatoms in the surficial sediments of the Mangyung-Dongjin tidal flat, west coast of Korea (Eastern Yellow Sea), *Mar. Biol.*, 12, 487, 1995.

53. Noel, D., Diatoms from two Mediterranean salt works: Salin-de-Giraud (S.E. France) and Santa Pola (Eastern Spain) similarities and differences, in *Proceedings of the Eighth International Diatom Symposium*, Ricard, M., Ed., Koeltz Scientific Books, Koenigstein, Germany, 1984, p. 655.

54. Lange, C.B. and Syversten, E.E., *Cyclotella litoralis* sp. nov. (Bacillariophyceae) and its relationships to *C. striata* and *C. stylorum*, *Nova Hedwigia*, 48, 341, 1989.

55. Sala, S.E., Sar, E.A., and Ferrario, M.E., Review of materials reported as containing *Amphora coffeaeformis* (Agardh) Kutzing in Argentina, *Diatom Res.*, 13, 323, 1998.

56. Steidinger, K.A., Lukas-Black, S., Richards, S., Richardson, B., and McRae, G., Florida Bay Microalgae, Proc. 1998 Florida Bay Science Conf., May 12–14, 1998, Miami, FL.

# 18 Seagrass Distribution in South Florida: A Multi-Agency Coordinated Monitoring Program

*James W. Fourqurean*
Department of Biological Sciences and Southeast Environmental
Research Center, Florida International University

*Michael J. Durako*
Department of Biology and Center for Marine Science Research,
The University of North Carolina at Wilmington

*Margaret O. Hall*
Florida Marine Research Institute, Florida Fish and Wildlife
Conservation Commission

*Lee N. Hefty*
Miami–Dade Department of Environmental Resources Management

## CONTENTS

## INTRODUCTION

Seagrass beds are a vitally important component of the nearshore marine environment. Seagrasses provide habitat for commercially and economically important fish and invertebrates and feeding grounds for wading and diving birds, as well as enhance sediment stability, decrease wave energy, and increase water clarity (see reviews by McRoy and Helfferich, 1977; Phillips and McRoy, 1980). Seagrass beds are very sensitive to changes in their environment and are particularly vulnerable to any decrease in the transmission of light through the water column and dredging of the sandy and muddy bottoms on which they grow. Much human activity in the coastal zone has the potential to deleteriously affect seagrasses. Dredging and filling of coastal areas for navigation and development can directly remove potential seagrass habitat (Zieman and Zieman, 1989), alter hydrological conditions that lead to erosion (Giesen et al., 1990; Larkum and West, 1990), and cause a reduction in light available to seagrasses by increasing turbidity (Onuf, 1994). Increasing human population density in coastal regions has often led to eutrophication, which can reduce light available for seagrasses; eutrophication has been implicated in the loss of seagrasses from many areas of the world (e.g., Orth and Moore, 1983; Cambridge et al., 1986). Recreational and commercial use of seagrass beds also can damage them. For example, contact of the bottom by outboard motors can cause scars that can take years to recover (Zieman, 1976); the cumulative impacts of such frequent events can lead to complete loss of seagrass beds from heavily trafficked areas (Sargent et al., 1995). Commercial harvesting of shellfish can also have severe effects on seagrass beds (Thayer et al., 1984).

Seagrasses are a dominant component of the hydroscape of South Florida, and they occupy the position between the freshwater environments of the mainland and the deep ocean. Seagrass communities are found from the mangrove-lined estuaries of Florida Bay, the Shark River drainage, and the Ten Thousand Islands out to back-reef environments and open continental shelf waters. Six species of rooted aquatic vascular plants, or seagrasses, are commonly found in south Florida: *Thalassia testudinum* Banks ex. König (turtle grass), *Syringodium filiforme* Kützing (manatee grass), *Halodule wrightii* Ascherson (shoal grass), *Halophila decipiens* Ostenfeld, *Halophila engelmanni* Ascherson, and *Ruppia maritima* L. (widgeon grass). One additional species, *Halophila johnsonii* Eiseman, occurs in Florida, but its distribution is limited to the Indian River Lagoon and extreme northern Biscayne Bay (Eiseman and McMillan, 1980), which is outside of the geographic scope of this paper. The general patterns of the distribution and relative abundance of these species are described (see Zieman, 1982; Zieman and Zieman, 1989, for review), but specific information on the areal extent of seagrass species in south Florida is incomplete (Iverson and Bittaker, 1986). In general, *R. maritima* is restricted to areas near freshwater sources. In areas of stable salinity, stable sediments, and high light availability, *T. testudinum* is often dominant. In slightly deeper or more frequently disturbed areas, *H. wrightii* and/or *S. filiforme* are often found. The *Halophila* species generally are restricted to low-light environments such as deep waters where <15% of surface light penetrates to the bottom, or to shallow turbid waters.

Previous surveys have documented the widespread occurrence of seagrasses in the South Florida region. In the area of Florida Bay within Everglades National Park, there are *ca.* 2000 km$^2$ of seagrasses, mostly dominated by *Thalassia testudinum* (Zieman et al., 1989). Using diver surveys, Iverson and Bittaker (1986) estimated that an additional 2900 km$^2$ of seagrass beds can be found in outer Florida Bay (defined as water depths >2 m); these beds were a mixture of *T. testudinum*, *Syringodium filiforme*, *Halodule wrightii*, and *Halophila decipiens*. A more intensive *in situ* and aerial survey of the entire southeastern Gulf of Mexico region documented 16,600 km$^2$ of seagrass beds in the area north of the Florida Keys and south of Cape Romano (Continental Shelf Associates, 1991). By far the most common seagrass encountered in this large area was *H. decipiens*. On the Atlantic Ocean side of the Florida Keys, at least an additional 1029 km$^2$ of seagrass beds has been reported (Klein and Orlando, 1994); this brings the estimate of total seagrass habitat in the South Florida region to at least 17,629 km$^2$ of semicontinuous beds.

The nearshore marine and estuarine habitats of South Florida are managed by a diverse group of governmental agencies at local, state, and federal levels (Figure 18.1). At the local level, county agencies are charged with protection of biotic resources; three counties occupy the shoreline of our study area: Monroe, Miami–Dade, and Collier. The State of Florida's Department of Environmental Protection (FDEP) has jurisdiction on biotic resources in state waters (i.e., within 3 nautical miles of the shoreline). Some of the marine area controlled by the state is further managed by subagencies of FDEP. For example, John Pennekamp Coral Reef State Park occupies a sizeable portion of the potential seagrass habitat in South Florida; the state parks are administered by their own agency (FDEP Division of Parks and Recreation). The South Florida Water Management District, a Florida state agency, is charged with environmental protection of state waters in addition to its primary goals of flood control and water supply. Many agencies of the federal government also exercise control over marine waters of the area. Within the Department of the Interior, the National Park Service (NPS) and the Fish and Wildlife Service (FWS) each control large areas in south Florida. Everglades National Park and Dry Tortugas National Park are largely marine parks. The FWS operates a number of wildlife sanctuaries in the region that have large areas of seagrass habitat within their boundaries. The U.S. Department of Commerce is also involved in management of the region; the Key Largo, Looe Key, and Florida Keys National Marine Sanctuaries are operated by the National Oceanographic and Atmospheric Administration, an agency of the Department of Commerce. The U.S. Environmental Protection Agency (EPA) also has regulatory authority over the marine waters in South Florida. Each agency that has some administrative authority over the marine environment has its own mission; these missions sometimes conflict. This myriad of overlapping agencies is also a regulatory gauntlet for people who wish to exploit the resource (e.g., tourism operators, fishermen) as well as for scientists doing research in the area.

While the details of each agency's mission vary, they all have the same goal: a healthy, stable, and sustainable environment. All of the agencies have also recognized the need for proper resource assessment and monitoring of the seagrass communities of South Florida. The critical role of seagrasses in South Florida has recently been demonstrated. A poorly understood dieoff of dense stands of *Thalassia testudinum* in Florida Bay began in 1987 (Fourqurean and Robblee, 1999). In the initial stages of this dieoff, *ca.* 4000 ha of dense *T. testudinum* beds in western and central Florida Bay died suddenly (Robblee et al., 1991). While this area of seagrass was small when compared to the total amount of seagrass habitat in South Florida, the ramifications of the loss were great. Turbidity in the water column and algal blooms followed the loss of seagrasses (Phlips et al., 1995), leading to a dieoff of sponges (Butler et al., 1995) and a general decline in seagrass beds that had survived the initial dieoff over an area of *ca.* 1000 km$^2$ (Hall et al., 1999). While deterioration of the seagrass beds across the entire region has yet to occur in South Florida, the fact that the western half of Florida Bay continues to respond to changes wrought by the catastrophic loss of a relatively small area of seagrass (Durako et al., 2002) underscores the importance of healthy seagrass beds to a sustainable marine environment in South Florida. Regulatory agencies in South Florida have taken the opportunity to act in a coordinated effort before region-wide degradation, in the hopes that we will be able to detect, and possibly avert, regional-scale seagrass loss.

Monitoring programs have been implemented in response to three major seagrass-related concerns in South Florida: the relationship of seagrass communities to water quality in the Florida Keys National Marine Sanctuary (FKNMS), changing freshwater runoff in northeast Florida Bay, and the poorly understood seagrass dieoff event that began in Florida Bay in 1987. Communication among scientists and resource managers in South Florida has led to the complementary design of these three monitoring programs. The programs not only are providing data to address the original question of concern, but are also providing data that can be combined to give a comprehensive view of the distribution and status of seagrass communities in the region as a whole. The goal of this paper is not to address any of the questions that led to the original creation of the monitoring efforts, but to use the data to develop an integrated description of the distribution, relative abundance, and species composition of the seagrass communities from the entire South Florida region.

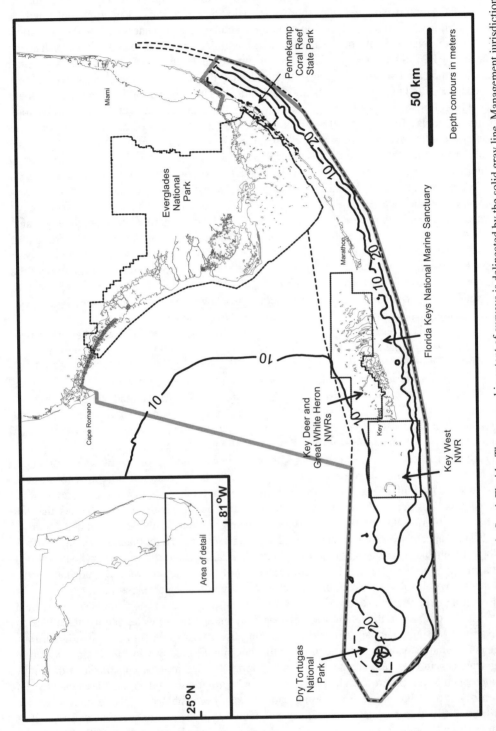

**FIGURE 18.1** Area of seagrass surveys in South Florida. The geographic extent of surveys is delineated by the solid gray line. Management jurisdictional boundaries are given for the major management areas in the region.

# THREE SEAGRASS MONITORING PROGRAMS
# IN SOUTH FLORIDA

## SEAGRASS COMMUNITIES AS INDICATORS OF WATER QUALITY IN THE FLORIDA KEYS NATIONAL MARINE SANCTUARY

The FKNMS was established by the Florida Keys National Marine Sanctuary and Protection Act of 1990 to "preserve and protect the physical and biological components of the South Florida estuarine and marine ecosystem to ensure its viability for the use and enjoyment of present and future generations" (NOAA, 1996). Seagrasses are an important biological component of the FKNMS. Water quality and the health of seagrass communities have been linked in many locations around the world; as water quality has deteriorated, seagrass communities have been lost (e.g., Orth and Moore, 1983; Cambridge et al., 1986). Concern has been raised over the role of eutrophication and its relation to the status of seagrass communities in the waters of the FKNMS (Lapoint et al., 1990; Tomasko and Lapointe, 1991; Lapointe and Clark, 1992; Lapointe et al., 1994). Because of these concerns, the U.S. EPA established a monitoring program in 1995 designed to define the status and trends of seagrass communities as a part of its comprehensive Water Quality Protection Plan for the FKNMS (Figure 18.2). This program was designed to determine regional-scale gradients in the status of the seagrass communities of the sanctuary.

## GAUGING THE EFFECTS OF CHANGING FRESHWATER FLOW ON BENTHIC COMMUNITIES OF FLORIDA BAY

Much of the historic freshwater inflow to Florida Bay has been severely altered by canal dredging and dike building in the Everglades ecosystem directly to the north, altering the pattern of salinity in Florida Bay (Smith et al., 1989; Light and Dineen, 1994; McIvor et al., 1994). The present system is one in which hypersalinity is common (Tabb et al., 1962; Fourqurean et al., 1993). It has been hypothesized that changes in the freshwater flow into Florida Bay have led to changes in benthic communities, such that *Thalassia testudinum* is more prevalent in northeast Florida Bay today than historically, when *Halodule wrightii* was more common (Zieman, 1982). Salinity plays a very important role in controlling benthic plant communities in the upper estuaries of Florida Bay; areas of high variability in salinity have low biomass of submerged plants (Montague and Ley, 1993). Currently, water managers are attempting to restore much of the historic flow of freshwater to the northeastern part of Florida Bay by engineering manipulations of the C-111 canal system. If these changes have an effect on salinity in Florida Bay, it is probable that benthic communities in Florida Bay will respond to the hydrologic changes. The South Florida Water Management District (SFWMD) and Miami–Dade County Department of Environmental Resources Management (DERM) began a monitoring program in 1993 to assess the effects of changing freshwater flows on the macrophyte communities of northeast Florida Bay (Figure 18.3). It should be obvious that both the quality and the quantity of the water flowing into Florida Bay are critical to the success of this management program.

## DETERMINING THE CAUSES AND EXTENT OF SEAGRASS DIEOFF IN FLORIDA BAY

Florida Bay is currently undergoing an unprecedented modification of its ecosystems (Fourqurean and Robblee, 1999). The mass mortality of seagrasses within Florida Bay (Robblee et al., 1991) and the more recent widespread algal blooms (Butler et al., 1995; Phlips et al., 1995; Phlips and Badylak, 1996) may have far-reaching consequences on the habitat quality and restoration potential of this important ecosystem. Causes of the mortality of seagrasses have yet to be fully described, but it is clear that a pathogen (Durako and Kuss, 1994), sulfide toxicity (Carlson et al., 1994), and salinity (Zieman et al., 1999) all play some role in the mortality of the dominant seagrass in Florida Bay, *Thalassia testudinum*. In 1995, the FDEP initiated a monitoring and research program designed

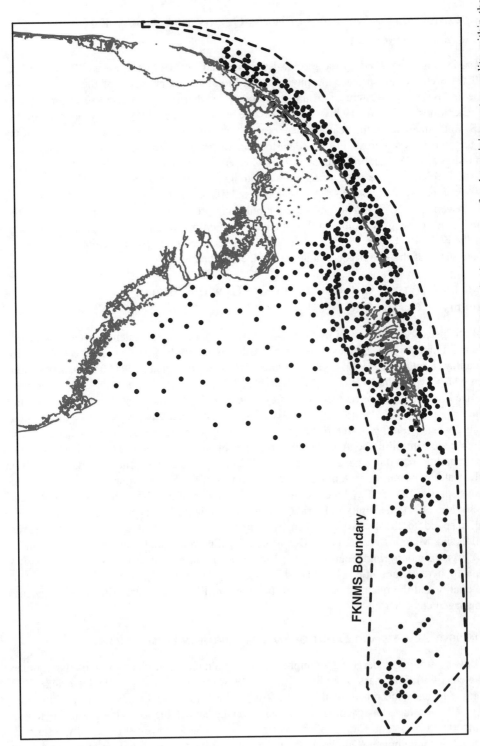

**FIGURE 18.2**  Station locations for the U.S. Environmental Protection Agency-funded monitoring program for determining water quality within the Florida Keys National Marine Sanctuary (FKNMS).

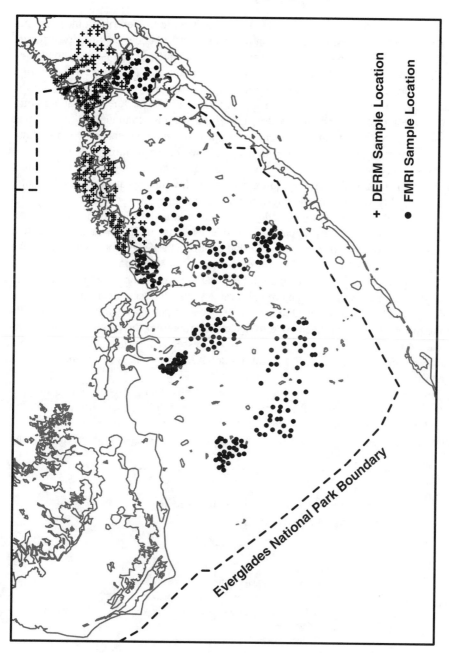

**FIGURE 18.3**  Station locations in Florida Bay within Everglades National Park. Filled circles are sites sampled in the Florida Department of Environmental Protection's Fish-Habitat Assessment Program (FHAP), funded by Florida DEP, the U.S. Geological Survey–Biological Resources Division, and Everglades National Park, designed to investigate the causes and consequences of seagrass dieoff in Florida Bay. Crosses indicate Miami–Dade County Department of Environmental Resources Management (DERM) sampling locations in the South Florida Water Management District/DERM-funded project investigating the consequences of changing freshwater discharge into Florida Bay on benthic communities.

to provide spatially comprehensive status and trends information on the benthic communities of Florida Bay (Figure 18.3). Trend data from this monitoring program are reported elsewhere in this volume (Durako et al., 2002).

## METHODS

Plant ecologists have worked for many years to devise the best metric for describing the structural characteristics of plant communities. Each question that may be asked about community structure has its own optimal method for assessment. Moreover, the scale at which a study is being conducted also influences the sampling methods. The prime questions motivating the seagrass monitoring programs in south Florida are: (1) What species make up the seagrass beds? (2) What are the relative abundances of the species? (3) Are there spatial trends in the structure of seagrass communities? (4) Are there temporal trends in the structure of the seagrass communities? Given that the area to be assessed is *ca.* 19,000 km$^2$, the methods adopted for these projects required rapidity and precision, sometimes at the expense of detail. Hence, we chose to utilize a rapid, visual assessment technique developed early in the 20th century by the plant sociologist Braun–Blanquet (Braun–Blanquet, 1972). This method is very quick, requiring only minutes at each sampling site; yet, it is robust and highly repeatable, thereby minimizing among-observer differences. In this method, a series of quadrats are randomly placed on the bottom at a given location. Each quadrat is examined by a scientist using SCUBA apparatus. All species occurring in the quadrat are listed, and a ranking based on abundance of the species in that quadrat is assigned for each species. We have adopted a modified Braun–Blanquet scale for our work in south Florida (Table 18.1). Cover, as defined for this purpose, is the fraction of the total quadrat area that is obscured by a particular taxon when viewed from directly above. The only allowable scores for each taxon in each quadrat are listed in Table 18.1. The choice of quadrat size is also very important for this technique; it is important that the quadrats be of sufficient size to accurately represent the make-up of the community, yet small enough so that they may be rapidly assessed, sometime under very turbid conditions. We have found that quadrats 0.5 m on a side (0.25 m$^2$) work well in South Florida seagrass communities.

Slightly different methods are used to ensure an unbiased placement of sampling quadrats in the three monitoring programs. In the FKNMS program, 10 quadrats are placed at each site by locating the quadrats at predetermined random distances along a 50-m transect placed in a north–south direction at each site. In the water management and seagrass dieoff monitoring

---

**TABLE 18.1**
**Braun–Blanquet Abundance Scale Used to Assess Seagrass Density**[a]

| Cover Class | Description |
|---|---|
| 0 | Absent |
| 0.1 | Solitary individual ramet, less than 5% cover |
| 0.5 | Few individual ramets, less than 5% cover |
| 1 | Many individual ramets, less than 5% cover |
| 2 | 5–25% cover |
| 3 | 25–50% cover |
| 4 | 50–75% cover |
| 5 | 75–100% cover |

[a] Cover is defined as the fraction of the bottom that is obscured by the species when viewed by a diver from directly above.

programs, four sample quadrats are haphazardly placed at each site. In the SFWMD/DERM, the quadrats are placed off of the port, starboard, bow, and stern of the small boat used as a research vessel, resulting in a spacing of about 5 m between quadrats. In the seagrass dieoff program, the quadrats are placed a few meters north, south, east, and west of the site location, resulting in a similar layout of quadrats as the water management program.

From the raw observations of species cover in each quadrat at a site, a single density estimate is calculated for each plant taxon encountered in the quadrats at a site. Density is calculated as $D_i = \Sigma S_{ij}/N$, where $D_i$ = density of taxon $i$; $j$ = quadrat number from 1 to $N$, the total number of quadrats sampled at a site; and $S_{ij}$ = the Braun–Blanquet score for taxon $i$ in quadrat $j$. For any taxon, $D$ can range between 0 and 5, the maximum Braun–Blanquet score. At a site, however, the sum of all taxa $D$ values can actually be greater than 5. This results from the relatively broad cover ranges for each Braun–Blanquet value and the fact that seagrass canopies are three dimensional. It should also be noted that a taxon may be observed at a site by the sample collector, but unless the taxa falls within one of the randomly placed observation quadrats, the taxon receives a $D = 0$. For this reason, our methods underestimate the true areal distribution of rare taxa by defining a lower density limit for inclusion in the survey. In addition, species richness $S$ is calculated for each site by summing the number of taxa for which $D > 0$.

When attempting to describe the distribution of habitat types in a landscape, it is important to sample in a way that allows for unbiased interpolation of the actual sample points to produce the distribution maps. This means that all points within the landscape must have an equal probability of being sampled, and that sampling effort be quasi-evenly distributed across the landscape. Yet, pure random distribution of sampling points often leads to clumped and nonuniformly distributed data points. To meet both of these requirements, we have used the stratified random method of hexagonal tesselation, developed by the U.S. EPA's EMAP program, to locate our sampling locations. The entire region to be sampled was defined and, based on the number of samples to be collected, the region was divided into hexagonal subunits. One random location was then chosen as a sample site from within each hexagonal subunit. These randomly-chosen sites are located in the field using differential global positioning systems (DGPS) which is accurate to ±5 m in South Florida.

Sites within the boundaries of the FKNMS (Figure 18.2) were sampled during the summer months of 1996 and 1997. Additionally, 100 sites were sampled in a roughly triangular area north of the FKNMS defined by Cape Romano, Key West, and Florida Bay during August 1998 as part of the FKNMS program. Data within Florida Bay were all collected in the summer of 1998 (Figure 18.3): the seagrass dieoff program sampled 378 sites, and the SFWMD/DERM program sampled 228 sites within Florida Bay.

Point data on species density were used to produce continuous maps of the density of seagrass species, as well as maps of species richness. A krigging algorithm (Watson, 1992) was used to interpolate between the random point data. A spatial analysis program (SURFER, Golden Software; Golden, CO) was used to compute areas of seagrass coverage from these interpolated surfaces.

Since no species density data were normally distributed, correlations between densities of species, and between species densities and depth were tested using the nonparametric Spearman's $\rho$; significances of correlations were assessed using two-tailed tests.

# RESULTS

We assessed the seagrass species composition and density of 1207 sites distributed across 19,402 km$^2$ of nearshore marine and estuarine environments in South Florida (Figure 18.1). At these sites, a total of 8434 quadrats (0.25 m$^2$) were sampled, covering an area of 2108.5 m$^2$. At least one species of seagrass was common enough to be counted in our quadrats at 1056 of the 1207 sites, or 87.5% of all sampling sites (Table 18.2). *Thalassia testudinum*, found at 898 sites, was the most commonly encountered species. *Halodule wrightii* was the second most commonly encountered species, occurring at 459 sites, followed by *Syringodium filiforme* (239 sites), *Halophila decipiens* (96 sites), *Ruppia maritima* (41 sites) and *Halophila engelmanni* (28 sites).

Differing morphology and life-history characteristics are apparent in the comparison of the relative densities of the species (Table 18.2). With two exceptions, only *Thalassia testudinum* and *Syringodium filiforme* were found to occur at very high density ($D > 4$; $4 \equiv 50–75\%$ cover; Table 18.1); 6.0% of all 1207 sites sampled had very dense cover of *T. testudinum*, and 1.8% of all sites had very dense beds of *S. filiforme*. Because seagrass beds in the region often contain more than one seagrass species, very dense beds of total seagrass cover were found at 18.1% of the sites sampled. Density greater than 4 was very rare for the seagrass species of smaller stature than *T. testudinum* and *S. filiforme*. Even the two larger species were most often found to have moderate density at most sites.

Species-specific differences in density tendency were found (Table 18.2). Restricting the analysis to only those sites where a species was found, *Thalassia testudinum* and *Syringodium filiforme* were most frequently encountered at $D$ between 1 and 2, although $D$ was higher and lower than this mode at a significant number of sites. The other species were almost always found at lower $D$: *Halodule wrightii*, *Halophila decipiens*, *Halophila engelmanni*, and *Ruppia maritima* were most commonly found to have D between 0.1 and 0.5. This lower mean D may have multiple causes. Some species, such as *H. wrightii* and *H. engelmanni*, are often found as understory plants beneath a canopy of *T. testudinum* or *S. filiforme*. Other species, like *H. decipiens* and *R. maritima*, tend to occur at the extremes of the available habitat for seagrasses, and their $D$ may be limited by the environment. Because more than one species may contribute to the overall seagrass $D$, sites with seagrass were most frequently observed in the 2 to 3 density class.

The density of one species was frequently correlated with densities of other seagrasses (Table 18.3). No relationship between the density of *Thalassia testudinum* and *Syringodium filiforme* was observed, but *T. testudinum* density was positively correlated to *Halodule wrightii* density and negatively correlated to the densities of *Halophila engelmanni*, *Halophila decipiens*, and *Ruppia maritima*. *Syringodium filiforme* density was not correlated to the densities of *H. wrightii* or *H. decipiens* but was positively correlated to *H. engelmanni* density and negatively correlated to *R. maritima* density. *Halodule wrightii* density was negatively correlated with *H. decipiens* density and positively correlated with the density of *H. engelmanni* and *R. maritima*. *Halophila decipiens* and *R. maritima* densities were negatively correlated, while no significant relationship between the densities of the two congeners of *Halophila* was found. No significant relationship was present between *H. engelmanni* and *R. maritima*, most likely due to the small number of stations where either species occurred.

Water depth was significantly related to the densities of all seagrass species except for *Halophila engelmanni* (Table 18.3). Densities of *Thalassia testudinum*, *Halodule wrightii*, and *Ruppia maritima* were higher in shallow water, while *Syringodium filiforme* and *Halophila decipiens* densities were higher in deeper water. Owing to the coastal nature of the region surveyed, shallow sites were much more common than deep sites. 43% of all of the sites fell within the depth range of 0 to 2 m; fewer than 10% of the sites were deeper than 10 m (Table 18.4). The likelihood of finding *T. testudinum* at a site decreased as site depth increased. More than 80% of sites shallower than 4 m supported *T. testudinum*. While *H. wrightii* was most likely to be encountered at the shallowest sites, a significant number of relatively deep stations also supported this species. *Ruppia maritima* was restricted to only those sites shallower than 2 m. *Syringodium filiforme* was much less common at the shallowest sites than at mid-depth sites; it was particularly common in the depth range 6 to 8 m; 45.6% of all sites in this depth class supported *S. filiforme*. *Halophila decipiens*, in contrast, was absent from the shallowest sites, but was found at over 50% of all sites sampled that were deeper than 18 m. *Halophila engelmanni* presence showed no clear relationship with water depth.

With the exception of *Ruppia maritima*, the seagrass species had similar ranges of depth of occurrence, but clear differences existed in the median depth at which each species was recorded (Table 18.5). *Ruppia maritima* was never found deeper than 1.4 m, with a median depth of 0.9 m.

**TABLE 18.2**
**Distribution of Seagrass Density (D) at the 1207 Sampling Sites**

| Species | Density Class (D) | | | | | | | | |
|---|---|---|---|---|---|---|---|---|---|
| | 0 | 0 < D ≤ 0.1 | 0.1 < D ≤ 0.5 | 0.5 < D ≤ 1 | 1 < D ≤ 2 | 2 < D ≤ 3 | 3 < D ≤ 4 | 4 < D ≤ 5 | D > 5 |
| **Number of Sites** | | | | | | | | | |
| Thalassia testudinum | 309 | 33 | 116 | 111 | 240 | 201 | 124 | 73 | 0 |
| Syringodium filiforme | 968 | 24 | 32 | 44 | 70 | 25 | 22 | 22 | 0 |
| Halodule wrightii | 748 | 56 | 162 | 76 | 109 | 43 | 12 | 1 | 0 |
| Halophila decipiens | 1111 | 27 | 26 | 11 | 13 | 11 | 7 | 1 | 0 |
| Halophila engelmanni | 1179 | 11 | 12 | 3 | 2 | 0 | 0 | 0 | 0 |
| Ruppia maritima | 1166 | 2 | 18 | 10 | 9 | 2 | 0 | 0 | 0 |
| Σ D for all seagrasses | 151 | 32 | 80 | 79 | 223 | 233 | 190 | 155 | 64 |
| **Fraction of All Sites Sampled (%)** | | | | | | | | | |
| Thalassia testudinum | 25.6 | 2.7 | 9.6 | 9.2 | 19.9 | 16.7 | 10.3 | 6.0 | 0.0 |
| Syringodium filiforme | 80.2 | 2.0 | 2.7 | 3.6 | 5.8 | 2.1 | 1.8 | 1.8 | 0.0 |
| Halodule wrightii | 62.0 | 4.6 | 13.4 | 6.3 | 9.0 | 3.6 | 1.0 | 0.1 | 0.0 |
| Halophila decipiens | 92.0 | 2.2 | 2.2 | 0.9 | 1.1 | 0.9 | 0.6 | 0.1 | 0.0 |
| Halophila engelmanni | 97.7 | 0.9 | 1.0 | 0.2 | 0.2 | 0.0 | 0.0 | 0.0 | 0.0 |
| Ruppia maritima | 96.6 | 0.2 | 1.5 | 0.8 | 0.7 | 0.2 | 0.0 | 0.0 | 0.0 |
| Σ D for all seagrasses | 12.5 | 2.7 | 6.6 | 6.5 | 18.5 | 19.3 | 15.7 | 12.8 | 5.3 |
| **Fraction of Sites Where Species Occurs (%)** | | | | | | | | | |
| Thalassia testudinum | | 3.7 | 12.9 | 12.4 | 26.7 | 22.4 | 13.8 | 8.1 | 0.0 |
| Syringodium filiforme | | 10.0 | 13.4 | 18.4 | 29.3 | 10.5 | 9.2 | 9.2 | 0.0 |
| Halodule wrightii | | 12.2 | 35.3 | 16.6 | 23.7 | 9.4 | 2.6 | 0.2 | 0.0 |
| Halophila decipiens | | 28.1 | 27.1 | 11.5 | 13.5 | 11.5 | 7.3 | 1.0 | 0.0 |
| Halophila engelmanni | | 39.3 | 42.9 | 10.7 | 7.1 | 0.0 | 0.0 | 0.0 | 0.0 |
| Ruppia maritima | | 4.9 | 43.9 | 24.4 | 22.0 | 4.9 | 0.0 | 0.0 | 0.0 |
| Σ D for all seagrasses | | 3.0 | 7.6 | 7.5 | 21.1 | 22.1 | 18.0 | 14.7 | 6.1 |

**TABLE 18.3**
**Correlations (Nonparametric Spearman's ρ) Between Densities of Seagrass Species and Both Seagrass Species Density and Water Depth from the 1207 Seagrass Sampling Sites[a]**

| | Depth | T. testudinum | S. filiforme | H. wrightii | H. decipiens | H. engelmanni | R. maritima |
|---|---|---|---|---|---|---|---|
| Water depth | — | **< 0.001** | **< 0.001** | **< 0.001** | **< 0.001** | 0.734 | **< 0.001** |
| Thalassia testudinum | −0.350 | — | 0.580 | **0.005** | **< 0.001** | **0.014** | **< 0.001** |
| Syringodium filiforme | 0.291 | 0.016 | — | 0.092 | 0.953 | **0.034** | **0.001** |
| Halodule wrightii | −0.451 | 0.080 | −0.049 | — | **0.006** | **< 0.001** | **< 0.001** |
| Halophila decipiens | 0.317 | −0.314 | −0.002 | −0.079 | — | .0251 | **0.050** |
| Halophila engelmanni | 0.010 | −0.071 | 0.061 | 0.176 | 0.033 | — | 0.316 |
| Ruppia maritima | −0.262 | −0.129 | −0.092 | 0.226 | −0.058 | −0.029 | — |

[a] Correlation coefficients are below the diagonal; (—) 2-tailed significances are above the diagonal. Significant correlations ($p \leq 0.05$) are in boldface type.

**TABLE 18.4**
**Frequency of Encountering Seagrass Species as a Function of the Depth of the Sample Site**

| Depth Interval (m) | Number of Sites | Thalassia testudinum | Syringodium filiforme | Halodule wrightii | Halophila decipiens | Halophila engelmanni | Ruppia maritima | Any Species |
|---|---|---|---|---|---|---|---|---|
| | | | | | Percent of Sites Occupied by | | | |
| 0–2 | 518 | 83.0 | 6.4 | 60.2 | 0.0 | 2.9 | 7.1 | 95.0 |
| 2–4 | 301 | 89.0 | 24.3 | 29.6 | 6.0 | 2.7 | 0.0 | 96.0 |
| 4–6 | 121 | 61.2 | 37.2 | 17.4 | 21.5 | 0.8 | 0.0 | 81.8 |
| 6–8 | 114 | 68.4 | 45.6 | 21.1 | 11.4 | 0.0 | 0.0 | 78.9 |
| 8–10 | 64 | 50.0 | 31.3 | 6.3 | 14.1 | 0.0 | 0.0 | 64.1 |
| 10–12 | 36 | 27.8 | 27.8 | 16.7 | 25.0 | 5.6 | 0.0 | 55.6 |
| 12–14 | 18 | 11.1 | 11.1 | 0.0 | 16.7 | 0.0 | 0.0 | 27.8 |
| 14–16 | 12 | 25.0 | 16.7 | 8.3 | 41.7 | 8.3 | 0.0 | 58.3 |
| 16–18 | 10 | 10.0 | 20.0 | 10.0 | 40.0 | 0.0 | 0.0 | 40.0 |
| 18–20 | 7 | 0.0 | 0.0 | 14.3 | 57.1 | 14.3 | 0.0 | 57.1 |
| 20–22 | 1 | 0.0 | 0.0 | 0.0 | 100.0 | 0.0 | 0.0 | 100.0 |
| 22–24 | 0 | nd | nd | nd | nd | nd | nd | nd |
| 24–26 | 4 | 0.0 | 0.0 | 0.0 | 75.0 | 0.0 | 0.0 | 75.0 |
| 26–28 | 1 | 0.0 | 0.0 | 0.0 | 100.0 | 0.0 | 0.0 | 100.0 |

*Note:* nd = no data.

**TABLE 18.5**
**Depth Range of Sample Sites Where the Six Seagrass Species Were Collected**

| Species | n | Min. Depth | Max. Depth | Mean Depth | Median Depth |
|---|---|---|---|---|---|
| Thalassia testudinum | 898 | 0.2 | 18.0 | 3.0 | 2.1 |
| Syringodium filiforme | 239 | 0.9 | 18.0 | 5.1 | 4.6 |
| Halodule wrightii | 460 | 0.2 | 18.6 | 2.3 | 1.4 |
| Halophila decipiens | 96 | 2.4 | 26.5 | 8.7 | 6.2 |
| Halophila engelmannii | 28 | 1.4 | 18.3 | 3.9 | 1.9 |
| Ruppia maritima | 41 | 0.4 | 1.4 | 0.9 | 0.9 |

*Thalassia testudinum* and *Syringodium filiforme* were found to have the same maximum depth of 18.0 m, but the median depth for *T. testudinum*, 2.1 m, was shallower than the median depth for *S. filiforme*, 4.6 m. *Halodule wrightii* penetrated slightly deeper in the water column, with a maximum depth of 18.6 m, but the median depth of 1.4 m illustrates the fact that it was most commonly found in shallow water. *Halophila engelmanni* was similar to *H. wrightii* in maximum and median depth. *Halophila decipiens* showed a much different pattern with respect to depth; it was found as deep as 26.5 m, with a median depth of 6.2 m.

Many (47.6%) of the 1207 sampled sites supported more than one species of seagrass (Figure 18.4). Even though it was relatively common for seagrass species to co-occur, a slim plurality (40.0%) of the 1207 sites supported only 1 seagrass species. Two seagrasses were found at 37.8% of all sites. Higher species richness was uncommon; three species were found at 8.6% of sites, and only 1.2% of sites had four or more species. No clear spatial pattern in species richness was apparent; relatively diverse (>3 species) seagrass beds were found on both the Atlantic Ocean and Gulf of Mexico sides of the Florida Keys (Figure 18.5). The only 2 sites with five species (*Thalassia testudinum*, *Syringodium filiforme*, *Halodule wrightii*, *Halophila decipiens*, and *Halophila engelmanni*) were found within the Dry Tortugas National Park.

Because sampling intensity varied spatially due to different goals of the three monitoring programs, frequency of occurrence data (Table 18.2) cannot be used directly to calculate the relative importance, in terms of area, of the six seagrass species in south Florida. Instead, maps of the occurrence of each species were analyzed for their areal extent. *Thalassia testudinum* was the most common seagrass in the sampling region. Density of *T. testudinum* was highest in Florida Bay, in the area between the upper Florida Keys and the reef tract, and in the shallow, protected waters north and west of Key West (Figure 18.6). In all, 8482 km$^2$ of *T. testudinum* beds were mapped, which was 43.7% of the 19,402 km$^2$ survey area. Roughly half of this total area was made up of very sparse *T. testudinum* cover: 3927 km$^2$ of the *T. testudinum* area had $D < 1$ (Table 18.6).

Second to *Thalassia testudinum* in terms of areal extent was *Halophila decipiens*, which was found to cover 7410 km$^2$, or 38.2% of the survey area (Table 18.6). In contrast to *T. testudinum*, however, *H. decipiens* was found predominantly in the waters of the southwest Florida Shelf, to the west of the Florida Mainland and to the north of the FKNMS (Figure 18.7). Most of this coverage consists of low-density seagrass beds: Of the 7410 km$^2$ of total area, 4652 km$^2$ consisted of areas where $D < 1$. Only rarely did *H. decipiens* form very dense beds; the area for which $D > 3$ was less than 1% of the total area surveyed.

*Syringodium filiforme* was also commonly encountered and was found to cover 4879 km$^2$. While *Thalassia testudinum* had the highest density immediately adjacent to the Florida Keys and in Florida Bay (Figure 18.6), *S. filiforme* density generally increased in an offshore direction until reaching the reef tract (Figure 18.8). A very dense bed of *S. filiforme* dominated the area to the north of the middle Florida Keys, north of Marathon and west of Florida Bay, encompassing about 350 km$^2$. Most of the area that supported *S. filiforme* had sparse cover; 3537 km$^2$ of the total area of *S. filiforme* had $D < 1$ (Table 18.6).

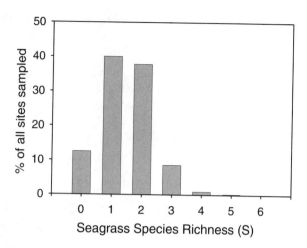

**FIGURE 18.4** Frequency histogram of the Species Richness, $S$, at sampling locations. $S$ is defined as the number of seagrass species occurring at a station (see text).

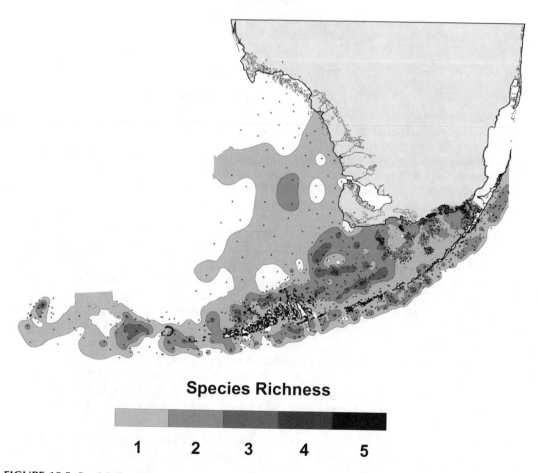

**FIGURE 18.5** Spatial distribution of species richness of seagrass beds across the South Florida hydroscape. Small crosses indicate sampling points.

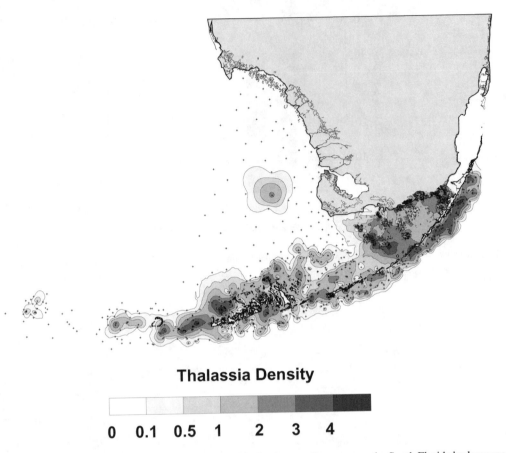

## Thalassia Density

0   0.1   0.5   1   2   3   4

**FIGURE 18.6** Spatial distribution of the density of *Thalassia testudinum* across the South Florida hydroscape. Small crosses indicate sampling points. Density scale is in Braun–Blanquet density units (see text and Table 18.1).

The only other species of seagrass that covered a large proportion of the surveyed area was *Halodule wrightii*; it occupied 3540 km², or 18.2% of the surveyed area. While *H. wrightii* was found sporadically throughout the region, it was most common in Florida Bay, on the Gulf of Mexico side of the Florida Keys, and in an area west of Key West known as the Quicksands (Figure 18.9). Of all of the area supporting *H. wrightii*, 83% had $D < 1$. The other two species encountered, *Halophila engelmanni* and *Ruppia maritima*, were found to be very limited in spatial extent. In the extreme upper estuaries of Florida Bay, *R. maritima* occupied 73 km² (Figure 18.10). *Halophila engelmanni* was occasionally observed, found in 143 km² scattered around the survey area (Figure 18.10).

The individual species distributions combine to produce a very large area of almost continual seagrass cover (Figure 18.11). 75.4% of the total surveyed area supported seagrasses, resulting in a total area of seagrass beds in the region of 14,622 km² (Table 18.6). Of this total area, 5197 km² was very sparse, with $D < 1$. Most of these sparse areas were dominated by *Halophila decipiens*, such as the southwest Florida Shelf area north of Key West and the relatively deep water between the Quicksands and Dry Tortugas National Park. The densest areas of seagrass were generally on the Gulf of Mexico side of the Florida Keys. On the Atlantic Ocean side of the Keys, seagrass beds were more dense in the Upper Keys than farther west.

**TABLE 18.6**
**Area Inventory of Seagrass Species in the Surveyed Region[a]**

| Species | Density Class (D) | | | | | | | |
|---|---|---|---|---|---|---|---|---|
| | $0 \leq D \leq 0.1$ | $0.1 < D \leq 0.5$ | $0.5 < D \leq 1$ | $1 < D \leq 2$ | $2 < D \leq 3$ | $3 < D \leq 4$ | $D > 4$ | $D > 0.1$ |
| **Area in a Density Class (km²)** | | | | | | | | |
| Thalassia testudinum | 10920 | 2193 | 1734 | 2657 | 1370 | 472 | 55 | 8482 |
| Syringodium filiforme | 14523 | 2421 | 1116 | 718 | 249 | 196 | 179 | 4879 |
| Halodule wrightii | 15862 | 2163 | 772 | 554 | 49 | 1 | 0 | 3540 |
| Halophila decipiens | 11992 | 2984 | 1668 | 1838 | 780 | 138 | 2 | 7410 |
| Halophila engelmanni | 19259 | 132 | 10 | 1 | 0 | 0 | 0 | 143 |
| Ruppia maritima | 19329 | 43 | 20 | 10 | 0 | 0 | 0 | 73 |
| Σ D for all seagrasses | 4780 | 3052 | 2145 | 4183 | 3112 | 1473 | 657 | 14622 |
| **Fraction of Surveyed Area (%)** | | | | | | | | |
| Thalassia testudinum | 56.3 | 11.3 | 8.9 | 13.7 | 7.1 | 2.4 | 0.3 | 43.7 |
| Syringodium filiforme | 74.9 | 12.5 | 5.8 | 3.7 | 1.3 | 1.0 | 0.9 | 25.1 |
| Halodule wrightii | 81.8 | 11.1 | 4.0 | 2.9 | 0.3 | 0.0 | 0.0 | 18.2 |
| Halophila decipiens | 61.8 | 15.4 | 8.6 | 9.5 | 4.0 | 0.7 | 0.0 | 38.2 |
| Halophila engelmanni | 99.3 | 0.7 | 0.0 | 0.0 | 0.0 | 0.0 | 0.0 | 0.7 |
| Ruppia maritima | 99.6 | 0.2 | 0.1 | 0.1 | 0.0 | 0.0 | 0.0 | 0.4 |
| Σ D for all seagrasses | 24.6 | 15.7 | 11.1 | 21.6 | 16.0 | 7.6 | 3.4 | 75.4 |

[a] Total area of the survey was 19,402 km² (Figure 11.1).

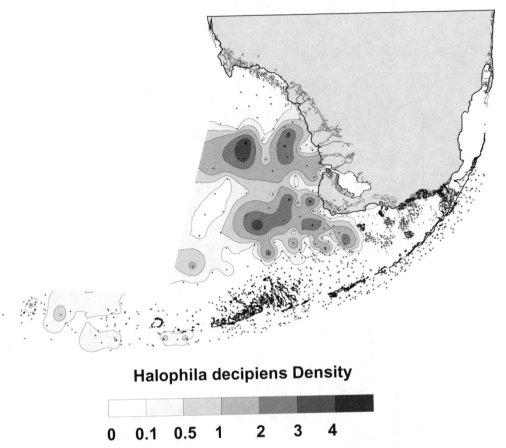

## Halophila decipiens Density

0    0.1   0.5   1    2    3    4

**FIGURE 18.7** Spatial distribution of the density of *Halophila decipiens* across the South Florida hydroscape. Small crosses indicate sampling points. Density scale is in Braun–Blanquet density units (see text and Table 18.1).

## DISCUSSION

The 14,622 km² of seagrasses in South Florida ranks this area among the most expansive documented seagrass beds on Earth, comparable to the back-reef environment of the Great Barrier Reef in Australia (Lee Long et al., 1996) and the Miskito Bank of Nicaragua (Phillips et al., 1982). Accordingly, the economic impact and ecological importance of the South Florida seagrass beds are significant (Zieman, 1982). Fisheries landings in the Florida Keys total over $12 \cdot 10^6$ kg annually of mostly seagrass-associated organisms (Bohnsack et al., 1994), and over half of all employment in the Florida Keys is dependent on outdoor recreation (NOAA, 1996). For the larger part, these outdoor activities are reliant on the clear waters and healthy marine habitats of the marine environment.

Proper environmental stewardship requires accurate data on the present state of resources. Prior to the initiation of the three monitoring programs that supplied data for this chapter, there was only a general understanding of the magnitude and composition of the seagrass beds of South Florida. Our work has provided baseline data that will be required for assessing the efficacy of management of the marine environment in South Florida. In terms of areal extent, seagrasses are, by far, the most commonly encountered habitat type in the survey area. At 87.5% of randomly selected stations, at least one species of seagrass was present; on an areal basis, this translated to seagrass present over 75.4% of the surveyed area. The remaining area was predominantly unvegetated soft-bottom

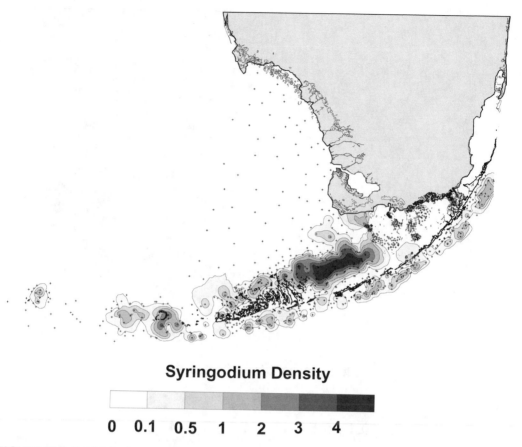

**Syringodium Density**

0  0.1  0.5  1  2  3  4

**FIGURE 18.8** Spatial distribution of the density of *Syringodium filiforme* across the South Florida hydroscape. Small crosses indicate sampling points. Density scale is in Braun–Blanquet density units (see text and Table 18.1).

communities. Coral reef communities, while in many respects the most valued and visible benthic habitat type in the region, make up only a small percentage of the total bottom cover in the survey area (Porter, 2002).

Analyses of the spatial scope required for this assessment are often impossible because of the magnitude of the task of collecting the data and because of overlapping jurisdictional boundaries. Careful coordination between management agencies and research groups ensured that data collected by different principle investigators, for different goals funded by different agencies, could be pooled and analyzed as a whole. This type of cooperation should serve as a model to other groups embarking on the assessment of resources over large geographic ranges.

In the nearshore environments of the survey area, *Thalassia testudinum* was the dominant seagrass. *T. testudinum* may be limited to shallow water because of its high light requirement. This requirement is a consequence of its relatively low proportion of leaves to roots and rhizomes compared to the other seagrass species found in the area (Fourqurean and Zieman, 1991). Nutrient availability also plays a role in *T. testudinum* distribution. This species is the competitive dominant in the high-light, low-nutrient environment of Florida Bay (Fourqurean et al., 1995). Phosphorus availability, which limits the biomass of *Thalassia testudinum*, increases from east to west in Florida Bay (Fourqurean et al., 1992; Fourqurean et al., 1993); it also increases from onshore to offshore on the ocean side of the Florida Keys (Szmant and Forrester, 1996). Experimental increases in phosphorus availability have resulted in other seagrasses outcompeting *T. testudinum* and become

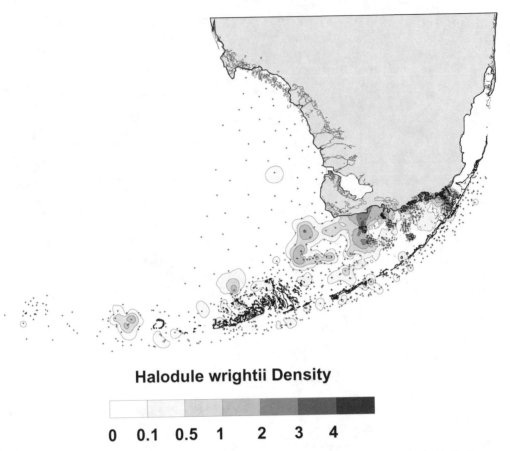

## Halodule wrightii Density

0    0.1    0.5    1    2    3    4

**FIGURE 18.9** Spatial distribution of the density of *Halodule wrightii* across the South Florida hydroscape. Small crosses indicate sampling points. Density scale is in Braun–Blanquet density units (see text and Table 18.1).

dominant (Fourqurean et al., 1995). We hypothesize that the increase in the abundance of *Syringodium filiforme* with distance from shore, as well as the very dense bed of *S. filiforme* north of Marathon, may be partially a response to relatively high phosphorus availability. Only in areas of relatively high phosphorus availability can *S. filiforme* outcompete *T. testudinum*. This hypothesis remains to be confirmed by experimental manipulation.

Interspecific differences in light requirements allow some species of seagrasses to grow in deeper water than others. Most seagrass genera have a minimum light requirement of >10% of surface irradiance (Duarte, 1991). Species in the genus *Halophila*, however, are often found in waters deeper than species of other genera (e.g., Lee Long et al., 1996), suggesting that *Halophila* spp. have lower light requirements. The median depth of sites that supported *H. decipiens* was 6.2 m, compared to 4.6 m for *S. filiforme* and 2.1 m for *Thalassia testudinum*. This lower light requirement of *Halophila* spp. is probably the factor responsible for the expansive beds of *H. decipiens* that we documented in the deeper water areas of our survey area. These areas are deep enough to prevent adequate light from reaching the bottom to support the larger species *Thalassia testudinum* and *Syringodium filiforme*. Of interest is the observation that *H. decipiens* was completely absent from shallow (<2.4 m) areas. Without experimental evidence, we can only hypothesize that *H. decipiens* is competitively displaced from higher light environments by other seagrass species. In contrast to *H. decipiens*, median depth for *H. engelmanni* was a relatively shallow 1.9 m. We never found extensive meadows dominated by *H. engelmanni*; instead, it was

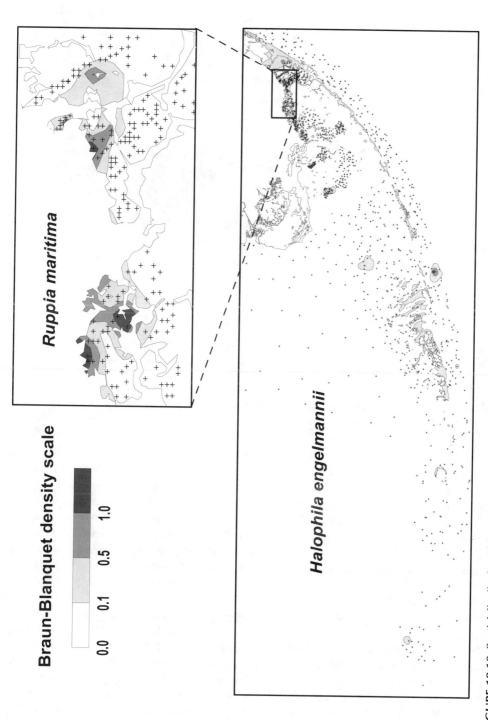

**FIGURE 18.10**  Spatial distribution of the density of *Halophila engelmanni* (main map) and *Ruppia maritima* (inset) across the South Florida hydroscape. Small crosses indicate sampling points. Density scale is in Braun–Blanquet density units (see text and Table 18.1).

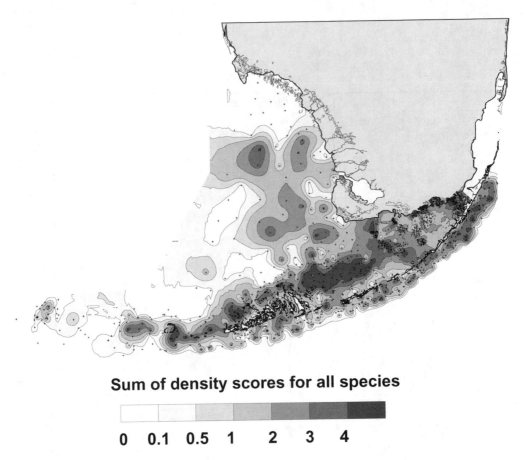

## Sum of density scores for all species

0   0.1   0.5   1   2   3   4

**FIGURE 18.11** Spatial distribution of the sum of the density scores for all seagrass species across the South Florida hydroscape. Small crosses indicate sampling points. Density scale is in Braun–Blanquet density units (see text and Table 18.1).

encountered as a sparse understory species, generally associated with denser beds of *Syringodium filiforme* and *Halodule wrightii*. It is probable that the generally low light requirements of *Halophila* spp. allow *H. engelmanni* to exist as an understory plant, but what is not clear are the life-history differences between *H. decipiens* and *H. engelmanni* that allow *H. engelmanni* to be a successful understory species, while its congener *H. decipiens* rarely occurs as an understory. Also, the minimum light requirements for *H. engelmanni* do not appear to be any greater than those for *H. decipiens*, as *H. engelmanni* has been documented growing at 90-m depth within the study area (den Hartog, 1970). Thus, it is unclear why *H. decipiens* is a meadow-former in deep water, while *H. engelmanni* is not.

While *Halophila* species were restricted to areas of truly marine, near-constant salinity, the other seagrass species were also found in Florida Bay, where salinity is strongly influenced by runoff from mainland Florida and by exchange of oceanic water with the Gulf of Mexico. Florida Bay can be either hypo- or hypersaline, depending on location, season, and year. Deviations from normal seawater salinity are deleterious to most seagrasses, but there is apparently a range in tolerances of species to salinity variation. Of the non-*Halophila* species, *Ruppia maritima* is the most tolerant of hyposalinity events; it is so tolerant of freshwater that it is often found growing in completely freshwater. This fact has led some authors (e.g., den Hartog, 1970) to exclude *R. maritima* from membership within the polyphyletic group of seagrasses. Of the remaining species encountered in our surveys, *Halodule wrightii* is the most tolerant of salinity fluctuation, *Thalassia testudinum* has intermediate tolerance, and *Syringodium filiforme* is the least tolerant (McMillan and Moseley, 1967). The extreme northeastern portions of Florida Bay are subject to very large salinity variability; the salinity range for the period 1991–1994 for northeast Florida Bay was 50‰ (Frankovich and Fourqurean, 1997). It is likely that this salinity variation limits the ability of all species but *R. maritima* to flourish in the extreme northeastern parts of Florida Bay. It is not clear, however, why *R. maritima* is not often found in other parts of the survey area. From distributional evidence around point sources of nutrients in Florida Bay, nutrient availability may have a role in determining *R. maritima* distribution. Adjacent to point sources of phosphorus, *R. maritima* dominates the benthic flora; farther from the point sources, *H. wrightii* and *T. testudinum* dominate (Powell et al., 1991). These authors interpreted these observations as evidence that *R. maritima* can only compete with other seagrass species in high-nutrient areas or where salinity variability limits the other species.

It has been suggested that changing water management practices on mainland Florida have led to changes in distribution of seagrasses in Florida Bay. Surveys of Florida Bay from the mid-1970s recorded large areas in central and eastern Florida Bay that were dominated by *Halodule wrightii* (Schmidt, 1979), yet these areas were reported to be dominated by *Thalassia testudinum* in the 1980s (Zieman et al., 1989), and were dominated by *T. testudinum* in our surveys. Zieman (1982) speculated that these changes were the result of changes in timing and amount of freshwater runoff.

Concerns for the state of the seagrass beds of South Florida are well founded. While currently the seagrass beds are nearly continuous and apparently healthy, there is cause for alarm. Localized cases of coastal eutrophication have led to loss of seagrasses in the study area (Lapointe et al., 1990; Tomasko and Lapointe, 1991; Lapointe and Clark, 1992; Lapointe et al., 1994). Seagrass dieoff in Florida Bay is still poorly understood (Fourqurean and Robblee, 1999), and the increase in turbidity that followed the dieoff continues to effect change in western Florida Bay (Hall et al., 1999; Durako et al., 2002). We now have the baseline data against which to measure future changes in these communities.

The present distribution and species composition of seagrasses in South Florida is a result of the interaction of many factors, the most important being water depth, water clarity, and nutrient availability. Changes in the movement and quality of water in the region, whether natural or anthropogenic, are likely to cause changes in the large-scale patterns in abundance and composition of these seagrass beds. Because nearshore oceanic water quality is determined by the interaction of coastal influences, marine influences, and human activities, it is clear that proper management of seagrass beds in South Florida requires holistic knowledge of the entire hydroscape of south Florida. Timing and amounts of freshwater runoff can change coastal salinity. Degradation of water quality of the freshwater runoff can directly effect nutrient availability and water clarity. Restriction of water exchange with the open ocean can alter salinity patterns and nutrient availability. Anthropogenic actions both in the marine and mainland realms can change nutrient availability and water clarity. Because any of these actions has the potential to alter the seagrasses of South Florida, all of these activities must be managed to ensure the continued existence of the seagrass communities in their current state. It is also likely that the first symptoms of a changing coastal environment will be a change in species composition of seagrass beds, not a wholesale loss of seagrass cover (e.g., Hall et al., 1999; Durako et al., 2002). For this reason, accurate data on the species composition of the seagrass communities must be collected periodically as a measure of the state of the coastal environment.

## CONCLUSIONS

Seagrass beds are an important, often dominant component in many coastal marine environments; however, there are few locations in the world where seagrasses are as dominant in the hydroscape as in South Florida. Because of the close proximity between human activities and seagrass communities, seagrass beds are being increasingly threatened in many locations worldwide. Seagrass beds are being lost due to the combined effects of dredging, filling, and water quality degradation throughout their range. Often, habitat degradation is only recognized after a vital resource is lost or severely altered. In South Florida, the importance of seagrasses to the economic vitality and ecological integrity of the region has long been recognized; this recognition has led to the development of coordinated seagrass monitoring programs involving government agencies from federal, state, and local levels; academic institutions; and private-sector environmental groups. While smaller scale seagrass declines have been documented, these monitoring programs have been largely implemented *before* regional-scale habitat degradation has severely affected the distribution of seagrasses. The data from these monitoring efforts provide a baseline view of the distribution and abundance of seagrasses of the region that is without precedent.

Clear jurisdictional boundaries in the seagrass-supporting marine areas of South Florida provide both a help and a hindrance to the development of an integrated seagrass monitoring effort. These jurisdictional boundaries — National Marine Sanctuaries, National Parks, National Wildlife Refuges, state waters, state parks, county parks, etc. — clearly define the entity in government that is responsible for proper environmental stewardship and set up clear areas of responsibility. Delineation can also be to the detriment of a coordinated effort, as governmental agencies have independent staffs and differing mandates, often leading to disparities among science and monitoring programs. Because the components of the hydroscape do not respect political boundaries, many resources occur across multiple jurisdictions. Further, the environmental factors controlling the distribution of resources also do not respect jurisdictional boundaries. The regional, cross-ecosystem nature of environmental phenomena make a coordinated effort paramount if proper data are to be collected to address questions of environmental sustainability.

Funding agencies, management groups, and university scientists in South Florida have recognized the need for complementary monitoring of seagrass ecosystems. Three major seagrass monitoring efforts are ongoing: a U.S. EPA-funded program addressing status and trends of seagrasses within the Florida Keys National Marine Sanctuary; a State of Florida–U.S. Department of Interior (U.S. Geological Survey and Park Service)-funded program assessing the seagrass communities of Florida Bay; and a program funded by the South Florida Water Management District and Miami-Dade County that concentrates on seagrass distribution in the upper estuaries of Florida Bay. Together, these programs are producing regional scale maps of the distribution of benthic marine habitats over a 19,402-km$^2$ area. Seagrasses were found to occur in 75.4% of this total area, or 14,622 km$^2$.

## ACKNOWLEDGMENTS

Monitoring projects of this size would not be possible if not for the hard field and laboratory work of a legion of people. On the FKNMS project, Braxton Davis, Cassie Furst, Leanne Rutten, Brad Peterson, Craig Rose, and Alan Willsie did the lion's share of the field work. The project was greatly facilitated by Captain Dave Ward and the *R/V Magic*. The U.S. Environmental Protection Agency provided funding for the FKNMS through cooperative agreement X994620-94 with Florida International University. The Florida Bay Fish-Habitat Assessment Program (FHAP) has benefited from field and lab assistance from Donna Berns, Scott Daeschner, Nancy Diersing, Scott Fears, Jeff Hall, Manuel Merello, Erica Moulton, and Leanne Rutten. Financial and logistical support for FHAP was provided by the Florida Department of Environmental Protection (#3720-2060-204 D1), U.S. Geological Survey (#98HQAG2186), and Everglades National Park. For the SFWMD/DERM program, Jason Bacon, Forrest Shaw, Kenneth Liddell, Susan Kim, and Martin Roch conduct all of the field work and assist with data management. They are recognized for their commitment and hard work in conducting this program. Cecelia Weaver is SFWMD's project manager and is recognized for her continued guidance and support of this program. This is contribution number 154 of the Southeast Environmental Research program at Florida International University.

# REFERENCES

Bohnsack, J.A., D.E. Harper, and D.B. McClellan. 1994. Fisheries trends from Monroe County, Florida, *Bull. Mar. Sci.*, 54:982–1018.

Braun–Blanquet, J. 1972. Plant Sociology: The Study of Plant Communities, Hafner Publishing, New York.

Butler, M.J., IV, J.H. Hunt, W.F. Herrnkind, M.J. Childress, R. Bertelsen, W. Sharp, T. Matthews, J.M. Field, and H.G. Marshall. 1995. Cascading disturbances in Florida Bay, USA: cyanobacterial blooms, sponge mortality, and implications for juvenile spiny lobsters *Panulirus argus*, *Mar. Ecol. Prog. Ser.*, 129:119–125.

Cambridge, M.L., A.W. Chiffings, C. Brittan, L. Moore, and A.J. McComb. 1986. The loss of seagrass in Cockburn Sound, Western Australia. II. Possible causes of seagrass decline, *Aquatic Bot.*, 24:269–285.

Carlson, P.R., L.A. Yarbro, and T.R. Barber. 1994. Relationship of sediment sulfide to mortality of *Thalassia testudinum* in Florida Bay, *Bull. Mar. Sci.*, 54:733–746.

Continental Shelf Associates. 1991 Southwest Florida nearshore benthic habitat study narrative report. U.S. Department of the Interior, Minerals Management Service, Gulf of Mexico Regional Office 89-0080.

den Hartog, C. 1970. *Sea-Grasses of the World*, North-Holland Publishing Company, Amsterdam.

Duarte, C.M. 1991. Seagrass depth limits, *Aquatic Bot.*, 40:363–377.

Durako, M.D. and K.M. Kuss. 1994. Effects of *Labyrinthula* infection on the photosynthetic capacity of *Thalassia testudinum*, *Bull. Mar. Sci.*, 54:727–732.

Durako, M.J., M.O. Hall, and M. Merello. 2002. Patterns of change in the seagrass-dominated Florida Bay hydroscape, in *The Everglades, Florida Bay, and Coral Reefs of the Florida Keys: An Ecosystem Sourcebook*, Porter, J.W. and Porter, K.G., Eds., CRC Press, Boca Raton, FL.

Eiseman, N.J. and C. McMillan. 1980. A new species of seagrass, *Halophila johnsonii*, from the Atlantic coast of Florida, *Aquatic Bot.*, 9:15–19.

Fourqurean, J.W. and M.B. Robblee. 1999. Florida Bay: a history of recent ecological changes, *Estuaries*, 22:345–357.

Fourqurean, J.W. and J.C. Zieman. 1991. Photosynthesis, respiration and whole plant carbon budget of the seagrass *Thalassia testudinum*, *Mar. Ecol. Prog. Ser.*, 69:161–170.

Fourqurean, J.W., J.C. Zieman, and G.V.N. Powell. 1992. Phosphorus limitation of primary production in Florida Bay: evidence from the C:N:P ratios of the dominant seagrass *Thalassia testudinum*, *Limnol. Oceanogr.*, 37:162–171.

Fourqurean, J.W., R.D. Jones, and J.C. Zieman. 1993. Processes influencing water column nutrient characteristics and phosphorus limitation of phytoplankton biomass in Florida Bay, FL, USA: inferences from spatial distributions, *Estuar. Coast. Shelf Sci.*, 36:295–314.

Fourqurean, J.W., G.V.N. Powell, W.J. Kenworthy, and J.C. Zieman. 1995. The effects of long-term manipulation of nutrient supply on competition between the seagrasses *Thalassia testudinum* and *Halodule wrightii* in Florida Bay, *Oikos*, 72:349–358.

Frankovich, T.A. and J.W. Fourqurean. 1997. Seagrass epiphyte loads along a nutrient availability gradient, Florida Bay, USA, *Mar. Ecol. Prog. Ser.*, 159:37–50.

Giesen, W.B.J.T., M.M. van Katwijk, and C. Den Hartog. 1990. Eelgrass condition and the turbidity in the Dutch Wadden Sea, *Aquatic Bot.*, 37:71–85.

Hall, M.O., M.D. Durako, J.W. Fourqurean, and J.C. Zieman. 1999. Decadal changes in seagrass distribution and abundance in Florida Bay, *Estuaries*, 22:445–459.

Iverson, R.L. and H.F. Bittaker. 1986. Seagrass distribution and abundance in eastern Gulf of Mexico coastal waters, *Estuar. Coast. Shelf Sci.*, 22:577–602.

Klein, C.J.I. and S.P.J. Orlando. 1994. A spatial framework for water-quality management in the Florida Keys National Marine Sanctuary, *Bull. Mar. Sci.*, 54:1036–1044.

Lapointe, B.E. and M.W. Clark. 1992. Nutrient inputs from the watershed and coastal eutrophication in the Florida Keys, *Estuaries*, 15:465–476.

Lapointe, B.E., J.D. O'Connell, and G.S. Garrett. 1990. Nutrient couplings between on-site sewage disposal systems, groundwaters, and nearshore surface waters of the Florida Keys, *Biogeochemistry*, 10:289–307.

Lapointe, B.E., D.A. Tomasko, and W.R. Matzie. 1994. Eutrophication and trophic state classification of seagrass communities in the Florida Keys, *Bull. Mar. Sci.*, 54:696–717.

Larkum, A.W.D. and R.J. West. 1990. Long-term changes of seagrass meadows in Botany Bay, Australia, *Aquatic Bot.*, 37:55–70.

Lee Long, W.J., R.G. Coles, and L.J. McKenzie. 1996. Deepwater seagrasses in northeastern Australia - how deep, how meaningful? pp. 41–50, in *Seagrass Biology: Proceedings of an International Workshop*, Kuo, J., Phillips, R.C., Walker, D.I., and Kirkman, H., Eds., Faculty of Sciences, University of Western Australia, Nedlands.

Light, S.S. and J.W. Dineen. 1994. Water control in the Everglades: a historical perspective, pp. 47–84, in *Everglades: The Ecosystem and Its Restoration*, Davis, S.M. and Ogden, J.C., Eds., St. Lucie Press, Delray Beach, FL.

McIvor, C.C., J.A. Ley, and R.D. Bjork. 1994. Changes in freshwater inflow from the Everglades to Florida Bay including effects on biota and biotic processes: a review, pp. 117–146, in *Everglades: The Ecosystem and Its Restoration*, Davis, S.M. and Ogden, J.C., Eds., St. Lucie Press, Delray Beach, FL.

McMillan, C. and F.N. Moseley. 1967. Salinity tolerances of five marine spermatophytes of Redfish Bay, Texas, *Ecology*, 48:503–506.

McRoy, C.P. and C. Helfferich. 1977. *Seagrass Ecosystems: A Scientific Perspective*, Marcel Dekker, New York.

Montague, C.L. and J.A. Ley. 1993. A possible effect of salinity fluctuation on abundance of benthic vegetation and associated fauna in northeastern Florida Bay, *Estuaries*, 16:703–717.

NOAA. 1996. *Florida Keys National Marine Sanctuary Final Management Plan/Environmental Impact Statement*, Vol. 1: *Management Plan*, National Oceanographic and Atmospheric Administration, U.S. Department of Commerce, Washington, D.C.

Onuf, C.P. 1994. Seagrasses, dredging and light in Laguna Madre, Texas, USA, *Estuar. Coast. Shelf Sci.*, 39:75–91.

Orth, R.J. and K.A. Moore. 1983. Chesapeake Bay: an unprecedented decline in submerged aquatic vegetation, *Science*, 222:51–53.

Phillips, R.C. and C.P. McRoy. 1980. *Handbook of Seagrass Biology: An Ecosystem Perspective*, Garland STPM Press, New York.

Phillips, R.C., R.L. Vadas, and N. Ogden. 1982. The marine algae and seagrasses of the Miskito Bank, Nicaragua, *Aquatic Bot.*, 13:187–195.

Phlips, E.J. and S. Badylak. 1996. Spatial variability in phytoplankton standing crop and composition in a shallow inner-shelf lagoon, Florida Bay, Florida, *Bull. Mar. Sci.*, 58:203–216.

Phlips, E.J., T.C. Lynch, and S. Badylak. 1995. Chlorophyll a, tripton, color, and light availability in a shallow tropical inner-shelf lagoon, Florida Bay, USA, *Mar. Ecol. Prog. Ser.*, 127:223–234.

Porter, J.W. 2002. Coral reef monitoring in South Florida, in *The Everglades, Florida Bay, and Coral Reefs of the Florida Keys: An Ecosystem Sourcebook*, Porter, J.W. and Porter, K.G., Eds., CRC Press, Boca Raton, FL.

Powell, G.V.N., J.W. Fourqurean, W.J. Kenworthy, and J.C. Zieman. 1991. Bird colonies cause seagrass enrichment in a subtropical estuary: observational and experimental evidence, *Estuar. Coast. Shelf Sci.*, 32:567–579.

Robblee, M.B., T.R. Barber, P.R. Carlson, M.J. Durako, J.W. Fourqurean, L.K. Muehlstein, D. Porter, L.A. Yarbro, R.T. Zieman, and J.C. Zieman. 1991. Mass mortality of the tropical seagrass *Thalassia testudinum* in Florida Bay (USA), *Mar. Ecol. Prog. Ser.*, 71:297–299.

Sargent, F.J., T.J. Leary, D.W. Crewz, and C.R. Kruer. 1995. *Scarring of Florida's Seagrasses: Assessment and Management Options*, Florida Marine Research Institute TR-1. St. Petersburg, FL.

Schmidt, T.W. 1979. *Ecological Study of the Fishes and Water Quality Characteristics of Florida Bay, Everglades National Park, Florida*, South Florida Research Center, Everglades National Park RSP-EVER N-36. Homestead, FL.

Smith, T.J., III, J.H. Hudson, M.B. Robblee, G.V.N. Powell, and P.J. Isdale. 1989. Freshwater flow from the Everglades to Florida Bay: a historical reconstruction based on fluorescent banding in the coral *Solenastrea bournoni*, *Bull. Mar. Sci.*, 44:274–282.

Szmant, A.M. and A. Forrester. 1996. Water column and sediment nitrogen and phosphorus distribution patterns in the Florida Keys, USA, *Coral Reefs*, 15:21–41.

Tabb, D.C., D.L. Dubrow, and R.B. Manning. 1962. *The Ecology of Northern Florida Bay and Adjacent Estuaries*, Technical Series #39, State of Florida Board of Conservation, Miami, FL, 81 pp.

Thayer, G.W., W.J. Kenworthy, and M.S. Fonseca. 1984. *The Ecology of Eelgrass Meadows of the Atlantic Coast: A Community Profile*, U.S. Fish and Wildlife Service 84/02. Washington, D.C.

Tomasko, D.A. and B.E. Lapointe. 1991. Productivity and biomass of *Thalassia testudinum* as related to water column nutrient availability and epiphyte levels: field observations and experimental studies, *Mar. Ecol. Prog. Ser.*, 75:9–17.

Watson, D.F. 1992. *Contouring: A Guide to the Analysis and Display of Spatial Data*, Vol. 10. Pergamon Press, Elmsford, NY.

Zieman, J.C. 1976. The ecological effects of physical damage from motor boats on turtle grass beds in southern Florida, *Aquatic Bot.*, 2:127–139.

Zieman, J.C. 1982. *The Ecology of the Seagrasses of South Florida: A Community Profile*, U.S. Fish and Wildlife Service 82/25. Washington, D.C.

Zieman, J.C., J.W. Fourqurean, and R.L. Iverson. 1989. Distribution, abundance and productivity of seagrasses and macroalgae in Florida Bay, *Bull. Mar. Sci.*, 44:292–311.

Zieman, J.C. and R.T. Zieman, 1989. *The Ecology of the Seagrass Meadows of the West Coast of Florida: A Community Profile*, U.S. Fish and Wildlife Service. 85/725. Washington, D.C.

Zieman, J.C., J.W. Fourqurean, and T.A. Frankovich. 1999. Seagrass die-off in Florida Bay: long-term trends in abundance and growth of turtlegrass, *Thalassia testudinum*, *Estuaries*, 22:460–470.

# 19 Patterns of Change in the Seagrass Dominated Florida Bay Hydroscape

*Michael J. Durako*
Center for Marine Science Research, The University of North Carolina at Wilmington

*Margaret O. Hall and Manuel Merello*
Florida Marine Research Institute

## CONTENTS

## INTRODUCTION

Seagrasses are a dominant component of many of the world's estuaries and shallow, coastal waters (Phillips and Meñez, 1988). The shallow distribution of seagrasses places them in close proximity to the land/sea interface, and they are subjected to the stresses and disturbances associated with this boundary. This distribution also places seagrass communities at the end of the watershed pipe, and their status may reflect larger, landscape-scale problems. Because most seagrasses are benthic perennial plants and are sedentary, they are continuously subject to stresses and disturbances that are associated with intra- and interannual changes in water quality. Thus, seagrasses act as integrators of net changes in water quality variables which tend to exhibit rapid and wide fluctuations when measured directly, and their status may reflect larger, landscape-scale problems. For these reasons, seagrasses may be one of the best indicators of changes in the condition of South Florida's coastal hydroscape.

Seagrasses provide critical habitat for many important fisheries species (Zieman, 1982; Thayer et al., 1984; Zieman and Zieman, 1989). It is this role of vegetative fisheries habitat that largely drives our interest in these plants. Loss or deterioration of seagrass habitat has generally been linked with loss of fisheries and a decline in habitat quality (Orth and Moore, 1983; Lombardo and Lewis, 1985; Dennison et al., 1993). Increasing development and use of coastal and estuarine systems have resulted in dramatic alterations of many seagrass beds (Orth and Moore, 1983; Lewis et al., 1985; Cambridge et al., 1986). Consequently, concern regarding widespread losses of seagrasses over the past several decades is increasing. Management actions have recently been directed towards instituting monitoring programs to assess the status and trends of seagrasses and linking these to changes in water quality (EPA, 1990) and even more recently, developing strategies to stop or reverse the losses of seagrasses (Neckles, 1994).

Seagrasses are the dominant biological community of the Florida Bay hydroscape, and they act as an intermediate link between the Everglades/South Florida and the Florida Keys hydroscapes. Seagrass beds, dominated by *Thalassia testudinum* Banks ex König (Turtle grass), historically covered over 90% of the 180,000 ha of subtidal mudbanks and basins within Florida Bay (Zieman et al., 1989). By comparison, mangrove islands cover only about 7% of the Bay. Because of the shallow nature of Florida Bay (mean depth <2 m; Schomer and Drew, 1982), seagrasses are also the dominant physical feature of Florida Bay, and their presence greatly affects physical, chemical, geological, and biological processes in this system. Seagrasses are important to the economy of South Florida because they provide food and shelter to numerous fish and invertebrate species, many of commercial importance within the region (Powell et al., 1989; Thayer and Chester, 1989; Tilmant, 1989; Chester and Thayer, 1990). Seagrass abundance, particularly the abundance of *T. testudinum*, is a valuable ecological indicator and to a large extent determines public perception regarding the "health" of the South Florida hydroscape (Goerte, 1994; Boesch et al., 1995). Thus, any changes in the distribution or abundance of seagrass within Florida may be perceived as a change in the health of the Bay.

The widespread dieoff of seagrasses, first observed in 1987, was one of the first and most conspicuous of the recent dramatic ecological changes that have occurred in the South Florida hydroscape (Robblee et al., 1991; Durako, 1994, 1995). Extensive areas of *Thalassia testudinum* began dying rapidly (4000 ha completely lost, 24,000 ha affected) during the summer of 1987, particularly in central and western Florida Bay (Robblee et al., 1991). Physiological stressors such as elevated water temperature, prolonged hypersalinity, excessive seagrass biomass leading to increased respiratory demands, hypoxia and sulfide toxicity, and disease are some of the factors thought to have contributed to *T. testudinum* dieoff; however, the causative mechanisms responsible for initiating the dieoff remain incompletely understood (Robblee et al., 1991; Carlson et al., 1994; Durako, 1994; Durako and Kuss, 1994).

The loss of seagrasses in Florida Bay has undergone at least two phases. The first phase was the initial dieoff which occurred during the relatively dry and clear period of 1987 to early 1991 (Robblee et al., 1991). Following the initial seagrass dieoff, which appeared to affect only *T. testudinum*, the Florida Bay hydroscape began exhibiting widespread declines in water clarity which may affect all seagrass and benthic macroalgae species in the Bay (Boesch et al., 1993; Phlips et al., 1995; Stumpf et al., 1999). The increased light attenuation, which began during the fall of 1991, is principally due to microalgal blooms and resuspended sediments associated with the loss of seagrasses on the western banks, and it has been most severe in the western and central Bay (Phlips and Badylak, 1996). Recent data from the U.S. Geological Survey indicate that loss of seagrass cover is a major factor in increased sediment resuspension in the Bay (Prager, 1998). The second phase of seagrass loss within Florida Bay coincided with a return to normal rainfall patterns. A comparison of seagrass distribution in Florida Bay between 1984 and 1994 indicated that the chronically turbid regions exhibited the most significant recent losses of *T. testudinum* and of the less common seagrasses *Halodule wrightii* Ascherson (Shoal grass), and *Syringodium fili-forme* Kützing (Manatee grass) (Hall et al., 1999).

The recent turbid conditions in western Florida Bay have complicated measurements and interpretations of seagrass losses and changes in species' distributions due to dieoff vs. changes attributable to light limitation. The turbidity has also precluded the use of a remote sensing approach to species-specific monitoring of seagrass distribution and abundance. Because of the continuing concern regarding the extent of seagrass changes within Florida Bay and the need to monitor the effects on seagrass communities of proposed water management alterations for the restoration of the Everglades/Florida Bay ecosystem, we initiated the Fisheries Habitat Assessment Program (FHAP) during spring 1995.

The approach utilized in the Florida Bay FHAP program for assessing status and trends in benthic macrophytes in Florida Bay is similar to the approaches being applied in the Florida Keys National Marine Sanctuary seagrass monitoring program and the Dade Environmental Resource Management/South Florida Water Management District monitoring program in northeast Florida Bay (Fourqurean et al., 2002). The establishment of compatible sampling protocols has provided a mechanism for the establishment of a regional management-oriented database of unprecedented scale (Fourqurean et al., 2002). The goal of FHAP is to provide information for spatial assessment and resolution of both intra- and interannual variability in the seagrass–macroalgae communities and, along with the aforementioned companion programs, FHAP will provide baseline data to monitor responses of these important benthic communities to possible future water management alterations or other restoration activities in the South Florida hydroscape. Here, we present a summary of the recent baywide distribution patterns of the two dominant seagrasses within Florida Bay, *Thalassia testudinum* Banks ex König (Turtle grass) and *Halodule wrightii* Ascherson (Shoal grass), and compare changes that have occurred between spring 1995 and spring 1998. We also present time-series maps illustrating detailed spatial and temporal variations in distribution and abundance of seagrasses within several basins that have exhibited the most dramatic changes in their hydroscape.

## METHODS

### FHAP Sampling Design

Sampling for FHAP is conducted twice per year, during spring (April–May) and fall (October–November). Each of ten basins, representing a range of conditions and gradients in Florida Bay, were partitioned into approximately 30 to 35 tessellated hexagonal grid cells. Sampling-station locations are randomly chosen from within each cell, for a total of about 330 stations per sample period (Figure 19.1). Sampling grids and station locations were generated using algorithms developed by the U.S. Environmental Protection Agency's Environmental Monitoring and Assessment Program (EMAP) and were provided by Dr. Kevin Summers (EPA, Gulf Breeze, FL). This type of sampling design results in systematic random sampling, scales the sampling effort to the size of the basin, and is well suited for interpolation (i.e., kriging) and mapping of the data. Stations were located using a handheld GPS (Magellan 5000).

At each station, seagrass cover was visually quantified within each of four, haphazardly located 0.25-$m^2$ quadrats using a modified Braun–Blanquet frequency/abundance scale (Mueller-Dombois and Ellenberg, 1974). This semiquantitative method requires relatively little time per sample (5 to 10 minutes for four quadrats), can be used for most plant communities, and has been shown to closely approximate quantitative characteristics of shoot density and standing crop for seagrasses (Durako et al., unpublished data). For a particular sample quadrat, the observer first lists all the species or plant groupings which are observed. A cover-abundance rating is then assigned using the following scale: 5, any number with cover of more than 75% of the quadrat; 4, any number with 50 to 75% cover; 3, any number with 25 to 50% cover; 2, any number with 5 to 25% cover; 1, numerous, but less than 5% cover or scattered with up to 5% cover; 0.5, few, with small cover; 0.1, solitary with small cover. The upper four scale values (5, 4, 3, 2) refer only to cover. The lower three scale values are primarily estimates of abundance (i.e., number of individuals per species). Sampling of replicate quadrats (4) at each sample point allows assessment of within- vs. among-station variability.

Upon completion of the visual assessment, ten short-shoots of *Thalassia* are collected and assayed for the presence of the marine slime mold *Labyrinthula*, within 24 hours of collection.

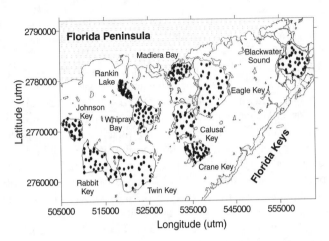

**FIGURE 19.1**  Location of Florida Bay stations and basin names for the spring 1995 Fish-Habitat Assessment Program (FHAP) sampling.

This organism is thought to be a possible cause of the seagrass dieoff (Durako and Kuss, 1994). The shoots are collected along a haphazard transect if *Thalassia* is abundant; if distribution is patchy, an area of approximately 10-m diameter area around the boat is surveyed.

## SEAGRASS DISTRIBUTION AND ABUNDANCE

Maps of seagrass distribution and abundance and changes in abundance were produced using a contouring and three-dimensional mapping program (Surfer version 6). The geostatistical gridding method of kriging was used to express the trends in the Braun–Blanquet data. A linear variogram model with no drift was used to calculate all grid node values. Changes in distribution and abundance were estimated by subtracting grid node values from differing sampling periods. Grid node values and difference values were visualized using contour maps. Planar areas for each cover class were calculated by the area differences among cut (positive) and fill (negative) volumes of the kriged grids using the grid volume command in Surfer. An estimate of the total abundance of a species within a basin was then obtained by multiplying the planar area for each cover class by the cover-class midpoint and adding the resulting values together.

## RESULTS

Despite seagrass die-off and the presence of widespread turbidity, *Thalassia testudinum* has remained the most widespread and abundant seagrass species in Florida Bay (Figure 19.2). However, the relative total abundance of *Thalassia* in the ten basins sampled by FHAP dropped from being over five times that of *Halodule wrightii* in spring 1995 to being just over three times more abundant in spring 1998 (Table 19.1). Most of this change in relative abundance has been due to widespread increases in *Halodule* distribution and abundance over this time period (Figure 19.3). total *Thalassia* abundance exhibited very little change (±8%) from spring 1995 to spring 1998, although there were some substantial changes in *Thalassia* abundance within individual basins (Table 19.1, Figure 19.4). This is in dramatic contrast to the almost doubling in total *Halodule* abundance over this same time period, with the greatest increases occurring in the western basins and in Blackwater Sound in northeast Florida Bay (Table 19.1, Figure 19.5). Estimated planar areas for the kriged Braun–Blanquet data for the spring 1995 sampling indicated that only about 1% (2.7 km²) of the area of the ten basins (273.3 km²) sampled by FHAP had no *Thalassia* cover. At this time, 39% (107.1 km²) of the area of the ten basins was without *Halodule*. The zero-abundance *Thalassia* area exhibited a small, but steady increase through spring 1998 (2.9, 3.4, and 3.5 km²

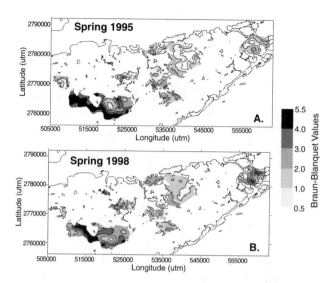

**FIGURE 19.2** Distribution and abundance of *Thalassia testudinum* in Florida Bay during spring 1995 (A) and spring 1998 (B) FHAP sampling. See text for description of Braun–Blanquet scale.

for spring 1996, 1997, and 1998, respectively). However, the zero-abundance area for *Halodule* dropped by over half over this period as this species became more widely distributed (95.3, 68.9, and 48.6 km² for spring 1996, 1997, and 1998, respectively).

Geographic patterns of *T. testudinum* abundance in spring 1995 indicate that this species was least abundant in the northcentral basins and increased in abundance towards the southwest where large areas of both Rabbit Key (19.2 km²) and Twin Key (12.4 km²) Basins had 75 to 100% cover of this species (Figure 19.2A). However, the western basins of Florida Bay exhibited the greatest losses of *Thalassia* from 1995 to 1998 (Table 19.1, Figure 19.6). Most of the central and northeastern basins exhibited little net change or increases in *Thalassia* over this time period. Rabbit Key basin lost over 35% of its *Thalassia* from spring 1995 to spring 1998 (Table 19.1); the loss pattern progressed as a front from the western bank and extended eastward (Figure 19.7). Twin Key Basin exhibited a 50% decline in *Thalassia* abundance from spring 1995 to 1997 but then exhibited an increase in abundance in spring 1998 to a level only about 12% lower than spring 1995 (Table 19.1). In Johnson Key Basin, *Thalassia* abundance doubled from spring 1995 to spring 1996 but has steadily declined since (Table 19.1, Figure 19.8). The losses in Johnson Key Basin correspond to the areas that have experienced highest turbidities along the edge of Sandy Key Bank. In contrast, total abundance of *Halodule* has increased over 400% in Johnson Key Basin over this time period and, since spring 1997, *Halodule* has become the most abundant seagrass within this basin. *Halodule* seems to have recruited from the banks surrounding Johnson Key Basin (Figure 19.9). *Halodule* abundance has also more than doubled in Rabbit Key Basin, and this species has increased about 80% in Twin Key Basin. During fall 1996, the small-bodied, low-light adapted species, *Halophila engelmannii* Ascherson, was observed at one station in Johnson Key Basin; by spring 1998, this species was present at 15 of the 32 stations (Figure 19.10).

## DISCUSSION

The Florida Bay hydroscape has recently exhibited environmental disturbances of unprecedented scale, including seagrass dieoff (Robblee et al., 1991), dense phytoplankton blooms (Phlips et al., 1995; Phlips and Badylak, 1996), sponge mortality, and changes in juvenile lobster population dynamics (Butler et al., 1995) and there are indications of cascading effects on plant and animal communities in adjacent systems (e.g., urchin population explosions and unbalanced growth of

**TABLE 19.1**
**Relative Total Abundances of Seagrasses in Ten Florida Bay Basins from Spring 1995 to Spring 1998**

| Basin | Area (km²) | 1995 | | 1996 | | 1997 | | 1998 | |
|---|---|---|---|---|---|---|---|---|---|
| | | Thalassia | Halodule | Thalassia | Halodule | Thalassia | Halodule | Thalassia | Halodule |
| Blackwater | 28.7 | 30.0 | 11.8 | 35.5 | 9.2 | 52.1 | 18.8 | 44.3 | 28.6 |
| Calusa Key | 26.4 | 38.7 | 0.0 | 44.7 | 5.1 | 28.5 | 4.5 | 38.5 | 6.3 |
| Crane Key | 15.3 | 26.0 | 2.9 | 27.3 | 2.4 | 32.9 | 4.3 | 26.8 | 5.6 |
| Eagle Key | 62.4 | 57.3 | 28.2 | 72.9 | 26.2 | 50.0 | 21.7 | 85.8 | 29.6 |
| Johnson Key | 14.3 | 12.8 | 6.1 | 23.5 | 13.1 | 17.2 | 20.5 | 10.1 | 25.0 |
| Madiera Bay | 12.4 | 16.1 | 6.3 | 22.4 | 4.1 | 23.0 | 6.0 | 27.2 | 4.2 |
| Rabbit Key | 31.8 | 128.9 | 10.0 | 125.0 | 10.7 | 105.3 | 11.8 | 94.4 | 24.7 |
| Rankin Lake | 5.8 | 3.9 | 3.9 | 3.8 | 3.6 | 3.6 | 8.7 | 20.7 | 7.5 |
| Twin Key | 54.3 | 159.8 | 11.3 | 122.7 | 17.7 | 105.8 | 25.3 | 142.7 | 20.7 |
| Whipray Bay | 21.9 | 16.8 | 6.2 | 29.0 | 6.6 | 31.5 | 8.0 | 30.6 | 12.6 |
| Total | 273.3 | 490.3 | 86.8 | 506.8 | 98.8 | 449.8 | 129.6 | 521.2 | 164.9 |

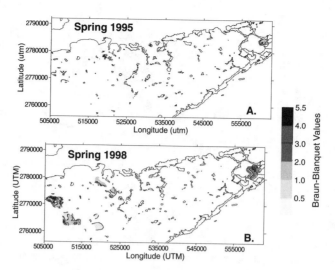

**FIGURE 19.3** Distribution and abundance of *Halodule wrightii* in Florida Bay during spring 1995 (A) and spring 1998 (B) FHAP sampling. See text for description of Braun–Blanquet scale.

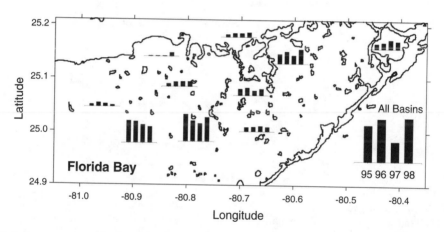

**FIGURE 19.4** Changes in relative abundance of *Thalassia testudinum* from spring 1995 to spring 1998 by basin and at the Bay scale. See text for description of methods used to calculate relative abundance.

*Syringodium filiforme*), (Rose et al., 1999; Kenworthy et al., 1998). Despite significant decreases in abundance in western Florida Bay from spring 1995 to spring 1998, *Thalassia testudinum* continues to be the dominant seagrass species in the bay. At the bay scale, abundance of *Thalassia* increased from the northeastern to southwestern bay, following patterns in sediment depth (Zieman et al., 1989) and phosphorus availability (Fourqurean et al., 1992). Only in Johnson Key Basin has species dominance shifted from *Thalassia* to *Halodule wrightii*. Likewise, Thayer et al., (1994) observed shifts from *Thalassia* dominance to *Halodule* dominance and back to *Thalassia* dominance, in terms of short-shoot densities in Johnson Key Basin from July 1990 to August 1992. They hypothesized that *Halodule* is more prone to rapid mortality induced by perturbations in water transparency than is *Thalassia* because of the former species' limited storage reserves. Hall et al., (1999) reported that *T. testudinum* shoot density and biomass fell significantly over the decade from 1984 to 1994 in western and central Florida Bay. The patterns of little change or an increase in abundance in the central basins, Rankin Lake and Whipray Bay, that we observed from 1995 to

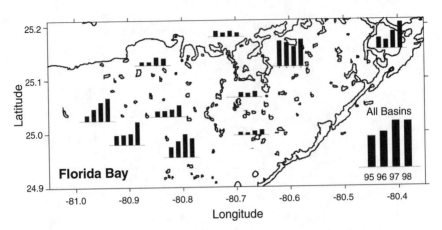

**FIGURE 19.5**  Changes in relative abundance of *Halodule wrightii* from spring 1995 to spring 1998 by basin and at the Bay scale. See text for description of methods used to calculate relative abundance.

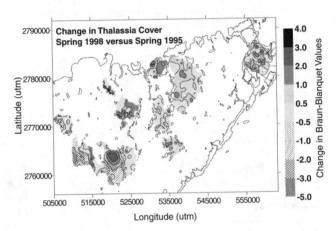

**FIGURE 19.6**  Net changes in distribution and abundance of *Thalassia testudinum* in Florida Bay between spring 1995 and spring 1998 FHAP sampling. See text for description of Braun-Blanquet scale and methods for calculating cover-abundance changes.

1998 reflect the fact that these basins had already lost much of their *Thalassia* in the original dieoff from 1988 to 1991 (Robblee et al., 1991; Durako, 1994; Thayer et al., 1994).

Johnson Key Basin was also unique in being the only basin sampled by FHAP to exhibit the recruitment of the low-light-adapted, small-bodied seagrass *Halophila engelmannii*. This species was first observed during the fall 1996 sampling, and it exhibited a continuous areal expansion through spring 1998. *Halophila engelmannii* has not previously been reported to occur within Florida Bay (Schomer and Drew, 1982; Zieman et al., 1989; Hall et al., 1999), although Iverson and Bittaker (1986) observed this species southwest of Sandy Key, beyond the Everglades National Park boundaries. These authors indicated that *H. engelmannii* only inhabited low-energy environments west of the Bay. Fourqurean et al., (1997) reported that *Halophila* distribution was restricted to the western portion of the Florida Keys National Marine Sanctuary in water too deep to support *Thalassia*. The occurrence and spread of *Halophila* within the shallow waters of Johnson Key Basin, along with the increasing abundance and recent dominance of *Halodule*, suggest a change to a shade-adapted seagrass community within this basin, and it may reflect the influence of the recent chronic turbidity. The chronic turbidity in this region may also have been responsible for

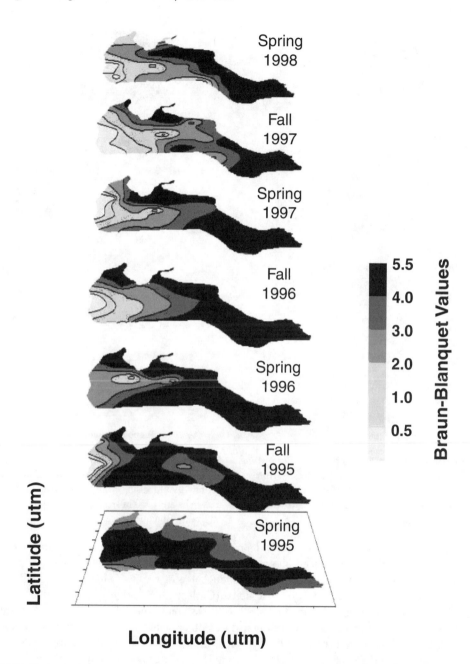

**FIGURE 19.7** Distribution and abundance of *Thalassia testudinum* in Rabbit Key Basin from spring 1995 to spring 1998. Latitude and longitude axis ticks are 1000 m apart. See text for description of Braun–Blanquet scale.

changes in the photosynthetic characteristics of the phytoplankton communities from 1994 to 1996; the changes suggested an adaptation to lower light conditions (i.e., increase in $\alpha$ and a decrease in $P_{max}$; Tomas et al., 1998).

Environmental stresses may result in a change in seagrass species composition and can weaken some species, making them vulnerable to disease (den Hartog, 1987; Short et al., 1988; Meuhlstein, 1989). Durako and Kuss (1994) found that infection by *Labyrinthula* significantly reduced

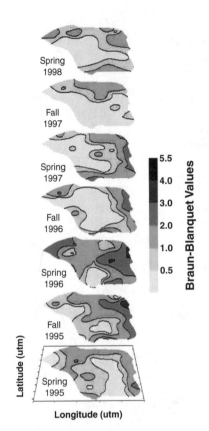

**FIGURE 19.8** Distribution and abundance of *Thalassia testudinum* in Johnson Key Basin from spring 1995 to spring 1998. Latitude and longitude axis ticks are 1000 m apart. See text for description of Braun–Blanquet scale.

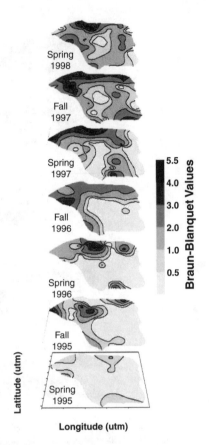

**FIGURE 19.9** Distribution and abundance of *Halodule wrightii* in Johnson Key Basin from spring 1995 to spring 1998. Latitude and longitude axis ticks are 1000 m apart. See text for description of Braun–Blanquet scale.

photosynthetic rates in *Thalassia*, resulting in a negative carbon balance (respiration rate > photosynthetic rate). In addition to reduced photosynthetic capacity and increased respiration, infected *Thalassia* short-shoots also exhibited reduced conductance of oxygen which would make them more susceptible to hypoxia and sulfide toxicity (Carlson et al., 1990; Durako et al., 1992). *Thalassia testudinum* is characteristic of stable, low-stress conditions (Neckles, 1994). A decrease in *Thalassia* abundance and an increase in density of other seagrass species in areas previously dominated by *Thalassia* are usually indicative of declining environmental conditions (Neckles, 1994). Powell et al., (1989) observed a change from *Thalassia* to *Halodule wrightii* after 3 years of nutrient enrichment from birds utilizing specially designed perches in eastern Florida Bay. This species change is consistent with the view that *Halodule* is a pioneering species that is more common in disturbed sites. The mean short-shoot density for *Thalassia* in western Florida Bay in 1989 (300 short-shoots $m^{-2}$; Durako, 1995) was almost 40% lower than the pre-dieoff mean value of 500 short-shoots $m^{-2}$ for the period from 1974 to 1980 reported by Iverson and Bittaker (1986). This decrease in density reflected the losses of this species in Florida Bay following the initial dieoff event, but before initiation of the widespread turbidity. The mean short-shoot density of *Thalassia* in the ten basins sampled by FHAP in spring 1998 was 275 short-shoots $m^{-2}$, reflecting a continuing decline in the abundance of this species. In contrast, the relative abundance of *Halodule* doubled from 1995 to 1998.

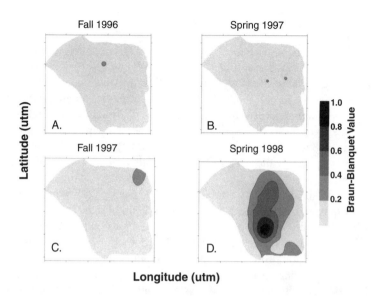

**FIGURE 19.10** Distribution and abundance of *Halophila engelmannii* in Johnson Key Basin during fall 1996 (A), spring 1997 (B), fall 1997 (C), and spring 1998 (D). Latitude and longitude axis ticks are 1000 m apart. See text for description of Braun–Blanquet scale.

While dieoff appears to affect only *Thalassia*, a change in the Florida Bay hydroscape that may affect all seagrass species is the widespread decline in water clarity that began in 1991 (Boesch et al., 1993). This turbidity, due to microalgal blooms and resuspended sediments, has been most severe in the western and central Bay (Phlips and Badylak, 1996). The microalgal blooms, consisting largely of cyanobacteria (possibly *Synechococcus elongatus*; Phlips and Badylak, 1996; Karen Steidinger, Florida Marine Research Institute, pers. comm.), may have been initiated by the nutrients liberated from the seagrass dieoff (Butler et al., 1995). Phlips et al., (1995) suggested that these turbid conditions would negatively affect all seagrasses in Florida Bay. Declining environmental conditions that result in reductions in available light have been implicated in seagrass declines worldwide (Cambridge and McComb, 1984; Dennison et al., 1993; Orth and Moore, 1993). The losses of *Thalassia* we observed in western Rabbit Key Basin occurred in the region of the basin that is relatively deep (>2 m), in addition to being chronically turbid. Thus, the losses here may be the result of light-stress-induced mortality, in addition to dieoff.

As noted by Hall et al., (1999), establishing the relative contribution of dieoff vs. light-stress-induced mortality to the losses of *Thalassia* in western Florida Bay is clearly problematic. This is mainly due to a high spatial coincidence among the distribution patterns of seagrass loss, *Labyrinthula* abundance (Blakesley et al., 1998), and turbidity (Phlips et al., 1995; Stumpf et al., 1999). Abundance of *Thalassia* decreased substantially in areas where both dieoff (Robblee et al., 1991) and light reduction (Phlips et al., 1995; Stumpf et al., 1999) have occurred. The increase in *Halodule* in the western basins (i.e., Rabbit Key and Johnson Key) may reflect its relatively lower light requirements (Williams and McRoy, 1982; Dunton and Tomasko, 1994), ability to rapidly spread into areas where the *Thalassia* canopy has been removed (Thayer et al., 1994), or resistance to disease. It should be noted that there was little change in *Thalassia* and *Halodule* abundance in basins which are periodically subjected to low salinities. This could reflect the existence of a low-salinity refugia from disease; *Labyrinthula* is never found in Florida Bay at salinities below 15 ppt (Blakesley et al., 1998). Thus, the patterns of seagrass loss in Florida Bay are partially consistent not only with predictions associated with dieoff, but also with those expected from chronic light reduction.

## CONCLUSIONS

Seagrasses, dominated by *Thalassia testudinum* (Turtle grass), are the principal biological community of the Florida Bay hydroscape and they act as an intermediate link between the Everglades and the Florida Keys ecosystems. Because of their shallow distribution and perennial growth habit, seagrasses integrate changes in nearshore water quality variables which tend to exhibit rapid and wide fluctuations when measured directly. Thus, seagrass distribution and abundance can be used as ecoindicators of the condition of South Florida's coastal hydroscape and to a large extent determines public perception regarding the "health" of the bay. Any changes in the distribution or abundance of seagrasses within Florida may also affect the integrity of the linkages between the adjacent ecosystems.

The hydroscape of Florida Bay has exhibited dramatic changes following the onset of seagrass dieoff in the late 1980s and the initiation of widespread and chronic turbidity in late 1991. This recent turbidity has precluded the use of a remote sensing approach to species-specific monitoring of seagrass distribution and abundance. Because of the continuing concern regarding the extent of seagrass changes within Florida Bay and the need to monitor the effects on seagrass communities of water management alterations being proposed for the restoration of the Everglades/Florida Bay ecosystem, the Florida Bay Fisheries Habitat Assessment Program (FHAP) was initiated during spring 1995.

Analyses of seagrass cover/abundance changes since spring 1995 indicate that *Thalassia testudinum* (Turtle grass) remains widely distributed and is still the dominant seagrass within Florida Bay. However, this species has exhibited significant declines in abundance along the western margin of the bay, a region far removed from the Everglades/Florida Bay land–sea interface. Estimated planar areas for the kriged cover/abundance data for the spring 1995 sampling indicated that only about 1% (2.7 km$^2$) of the area of the ten basins sampled by FHAP (273.3 km$^2$) had no *Thalassia*. At this time, 39% (107.1 km$^2$) of the area of the ten basins was without *Halodule*. The zero-abundance *Thalassia* area has exhibited a small, but steady increase through spring 1998 (2.9, 3.4, and 3.5 km$^2$ for spring 1996, 1997, and 1998, respectively). In contrast, the zero-abundance area for *Halodule* has dropped by more than half over this period (95.3, 68.9, and 48.6 km$^2$ for spring 1996, 1997, and 1998, respectively), as this species became much more widespread and abundant.

The recent losses of *Thalassia* have corresponded with areas where turbidity has been most persistent, and the losses may be the result of light-stress-induced mortality, in addition to dieoff. Increases in the distribution and abundance of the lower light-adapted seagrass *Halodule wrightii* (Shoal grass) and the recent appearance (fall 1996) and spread of the small-bodied, low-light-adapted *Halophila engelmannii* (Star grass) in western Florida Bay further indicate a change to a shade-adapted hydroscape. These patterns of changes, occurring in the benthic-perennial seagrasses, raise significant questions regarding the efficacy of hydrologic modifications being proposed as part of the South Florida Restoration program.

The spatial patterns of change in the seagrass-dominated hydroscape of Florida Bay from 1995 to 1998 suggest that, currently, the most perturbed environment is along the western Bay margin, bordering the open waters of the Gulf of Mexico. In contrast, much of the focus of management and restoration efforts in the South Florida hydroscape have been directed toward landscape-scale modifications to the existing flood-control system to increase the quantity of freshwater delivered to northeast Florida Bay via C-111, Taylor River Slough, and along the Everglades/Florida Bay land–sea interface. If the metrics by which the public will judge the success of these restoration efforts are increases in seagrass abundance and water clarity, then resource managers must consider the system as it is now and focus their efforts accordingly. Although the initial dieoff is generally thought to have been initiated in the interior basins of the Bay (Robblee et al., 1991), the greatest changes in the present system are occurring far from the Everglades/Florida Bay land–sea interface. A recent analysis of satellite Advanced Very High Resolution Radiometer (AVHRR) reflectance data suggested the possibility that there was a massive loss of benthic macrophytes west of Florida

Bay just prior to the observation of dieoff within the Bay (Stumpf et al., 1999). Additionally, the population explosion of sea urchins (*Lytechinus variegatus*) in the *Syringodium filiforme*-dominated seagrass bed southwest of Florida Bay in 1997, which has resulted in substantial losses of *Syringodium*, also reflects the perturbed status of this western boundary (Rose et al., 1999). Western Florida Bay and these adjacent hydroscapes form a hydrodynamic link between the Everglades and the coastal waters of the southwestern Florida peninsula/eastern Gulf of Mexico to the north, and the Florida Keys reef tract and the Atlantic Ocean to the south (Schomer and Drew, 1982). The seagrass communities of this region form an important buffer by intercepting the flow of water along this region and reducing nutrient and particulate loads in the waters reaching the reef tract (Kenworthy et al., 1998). Current restoration plans have not considered the possible effects of the planned water delivery modifications on these outer Florida Bay benthic communities. If losses of seagrasses continue along this margin, the increased water flows could result in increase fluxes of material out of Florida Bay and into the reef tract (Kenworthy et al., 1998). Resource managers need to consider actions that might aid in the re-establishment of seagrass cover along the western basins and banks. This would be an important step in reducing sediment-resuspension-induced turbidity along this boundary (Prager, 1998), and these actions could reverse the cascading declines that characterize the present system.

## ACKNOWLEDGMENTS

Financial support was provided by the Florida Department of Environmental Protection (#3720-2060-204 D1), U.S. Geological Survey (#98HQAG2186), and Everglades National Park. We would like to thank Leanne Rutten, Jeff R. Hall, Scott Fears, Donna Berns, and Nancy Diersing (Florida Marine Research Institute) for assistance in field sampling.

## REFERENCES

Blakesley, B.A., J.H. Landsberg, B.B. Ackerman, R.O. Reese, J.R. Styer, C.O. Obordo, and S.E. Lucas-Black. 1998. Slime mold, salinity, and statistics: implications from laboratory experimentation with turtlegrass facilitate interpretation of field results from Florida Bay, Florida Bay Science Conference, University of Miami, May 12–14, 1998.

Boesch, D.F., N.E. Armstrong, C F. D'Elia, N.G. Maynard, H.W. Paerl, and S.L. Williams, 1993. *Deterioration of the Florida Bay Ecosystem: An Evaluation of the Scientific Evidence*, Report to the Interagency Working Group on Florida Bay, National Fish and Wildlife Foundation, Washington, D.C., 27 p.

Boesch, D.F., N.E. Armstrong, J.E. Cloern, L.A. Deegan, R.D. Perkins, and S.L. Williams. 1995. *Report of the Florida Bay Science Review Panel on Florida Bay Science Conference: A Report by Principal Investigators*, Program Management Committee, Florida Bay Research Program, Miami, FL, 19 p.

Butler, M.J., J.H. Hunt, W.F. Herrnkind, M.J. Childress, R. Bertelsen, W. Sharp, T. Matthews, J.M. Field, and H.G. Marshall. 1995. Cascading disturbances in Florida Bay, USA: cyanobacterial blooms, sponge mortality, and implications for juvenile spiny lobsters, *Panulirus argus*, *Mar. Ecol. Prog. Ser.*, 129:119–125.

Cambridge, M.L., A.W. Chiffings, C. Brittan, L. Moore, and A.J. McComb 1986. The loss of seagrass in Cockburn Sound, Western Australia. II. Possible causes of seagrass decline, *Aquatic Bot.*, 24:269–285.

Carlson, P.R., M.J. Durako, T.R. Barber, L.A. Yarbro, Y. deLama, and B. Hedin. 1990. Catastrophic mortality of the seagrass *Thalassia testudinum* in Florida Bay. Ann. Rept., Florida Dept. Environmental Regulation, Office of Coastal Zone Management, Tallahassee, FL, 54 pp.

Carlson, P.R., L.A. Yarbro, and T.R. Barber. 1994. Relationship of sediment sulfide to mortality of *Thalassia testudinum* in Florida Bay, *Bull. Mar. Sci.*, 54:733–746.

Chester, A.J. and G.W. Thayer. 1990. Distribution of spotted seatrout (*Cynoscion nebulosus*) and gray snapper (*Lutjanus griseus*) juveniles in seagrass habitats in western Florida Bay, *Bull. Mar. Sci.*, 46:345–357.

Dennison, W.C., R.J. Orth, K.A. Moore, J.C. Stevenson, V. Carter, S. Kollar, P.W. Bergstrom, and R.A. Batuik. 1993. Assessing water quality with submersed vegetation, *BioScience*, 43:86–94.

den Hartog, C. 1987. "Wasting disease" and other dynamic phenomena in *Zostera* beds, *Aquatic Bot.*, 27:3–14.

Dunton, K H. and D.A. Tomasko. 1994. *In situ* photosynthesis in the seagrass *Halodule wrightii* in a hypersaline subtropical lagoon, *Mar. Ecol. Prog. Ser.*, 107:281–293.

Durako, M.J. 1995. Indicators of seagrass ecological condition: An assessment based on spatial and temporal changes associated with the mass mortality of the tropical seagrass *Thalassia testudinum*, in *Changes in Fluxes in Estuaries: Implications For Science to Management*, Dyer, K.R. and Orth, R.J., Eds., pp. 261–266 Olsen & Olsen, Fredensborg, Denmark.

Durako, M.J. 1994. Seagrass die-off in Florida Bay (USA): changes in shoot demography and populations dynamics, *Mar. Ecol. Prog. Ser.*, 110:59–66.

Durako, M.J. and K.M. Kuss. 1994. Effects of *Labyrinthula* infection on the photosynthetic capacity of *Thalassia testudinum*, *Bull. Mar. Sci.*, 54(3):727–732.

Durako, M.J., T.R. Barber, J.B.C. Bugden, P.R. Carlson, J.W. Fourqurean, R.D. Jones, D. Porter, M.B. Robblee, L.A. Yarbro, R.T. Zieman, and J.C. Zieman. 1992. Seagrass die-off in Florida Bay. Pp. 14–15. In J.D. Jacobsen, Ed., Proc. 1992 Symp. Gulf of Mexico, US EPA, Tarpon Springs, FL.

EPA. 1990. *Environmental Monitoring and Assessment Program Overview*, EPA/600/9-90/001, Environmental Protection Agency, 5 pp.

Fourqurean, J.W., J.C. Zieman, and G.V.N. Powell. 1992. Relationships between porewater nutrients and seagrasses in a subtropical carbonate environment, *Mar. Biol.*, 114:57–65.

Fourqurean, J.W., M.J. Durako, and J.C. Zieman. 1997. *Seagrass Status and Trends Monitoring: Annual Report*, FY1996. Florida Keys National Marine Sanctuary Water Quality Protection Program, U.S. Environmental Protection Agency, Washington, D.C.

Fourqurean, J.W., M.J. Durako, M.O. Hall, and L.N. Hefty. 2002. Seagrass distribution in South Florida: a multi-agency coordinated monitoring program, in *The Everglades, Florida Bay, and Coral Reefs of the Florida Keys: An Ecosystem Sourcebook*, Porter, J.W. and Porter, K.G., Eds., CRC Press, Boca Raton, FL.

Goerte, R.W. 1994. *The Florida Bay Economy and Changing Environmental Conditions*, U.S. Library of Congress, Congressional Research Service Report No. 94-435 ENR. Washington, D.C., 19 pp.

Hall, M.O., M.J. Durako, J.W. Fourqurean, and J.C. Zieman. 1999. Decadal-scale changes in seagrass distribution and abundance in Florida Bay, *Estuaries*, 22(2B):445–459.

Iverson, R.l. and H.F. Bittaker. 1986. Seagrass distribution and abundance in Eastern Gulf of Mexico coastal waters, *Estuar. Coast. Shelf Sci.*, 22:577–602.

Kenworthy, W.J., A.C. Swartzschild, M.S. Fonseca, D. Woodruff, M.J. Durako, and M.O. Hall. 1998. Ecological and optical characteristics of a large *Syringodium filiforme* meadow. I. The southeastern Gulf of Mexico. Florida Bay Science Conference, University of Miami, May 12–14, 1998.

Lewis, R.R., M.J. Durako, M.D. Moffler, and R.C. Philips. 1985. Seagrass meadows of Tampa Bay: a review, pp. 210–246, in *Proceedings, Tampa Bay Area Scientific Information Symposium*, Treat, S.F., Simon, J.L., Lewis, R.R., and Whitman, R.L., Eds., Florida Sea Grant Publ. 65. Burgess Publishing, Minneapolis, MN, 663 pp.

Lombardo, R. and R.R. Lewis. 1985. Commercial fisheries data: Tampa Bay, pp. 614– 634, in *Proceedings, Tampa Bay Area Scientific Information Symposium*, Treat, S.F., Simon, J.L., Lewis, R.R., and Whitman, R.L., Eds., Florida Sea Grant Publ. 65, Burgess Publishing, Minneapolis, MN, 663 pp.

Mueller-Dombois, D. and H. Ellenberg. 1974. *Aims and Methods of Vegetation Ecology*, John Wiley & Sons, New York, 547 pp.

Muehlstein, L.K. 1989. Perspectives on the wasting disease of eelgrass *Zostera marina*, *Dis. Aquatic Org.*, 7:211–221.

Neckles, H. A. 1994. *Indicator Development: Seagrass Monitoring and Research in the Gulf of Mexico*, U.S. Environmental Protection Agency, EPA/620/R-94/029, Gulf Breeze, FL, 62 pp.

Orth, R.J. and K.A. Moore. 1983. Chesapeake Bay: an unprecedented decline in submerged aquatic vegetation, *Science*, 222: 51–53.

Phillips, R.C. and E.G. Meñez. 1988. Seagrasses, *Smithsonian Cont. Mar. Sci.*, No. 14, 104 pp.

Phlips, E.J., T.C. Lynch, and S. Badylak. 1995. Chlorophyll *a*, tripton, color, and light availability in a shallow tropical inner-shelf lagoon, Florida Bay, USA, *Mar. Ecol. Prog. Ser.*, 127:223–234.

Phlips, E.J. and S. Badylak. 1996. Spatial variability in phytoplankton standing crop and composition in a shallow inner-shelf lagoon, Florida Bay, Florida, *Bull. Mar. Sci.*, 58:203–216.

Powell, A.B., D.E. Hoss, W.F. Hettler, D.S. Peters, and S. Wagner 1989. Abundance and distribution of ichthyoplankton in Florida Bay and adjacent waters, *Bull. Mar. Sci.*, 44:35–48.

Powell, G.V.N., W.J. Kenworthy, and J.W. Fourqurean. 1989. Experimental evidence for nutrient limitation of seagrass growth in a tropical estuary with restricted circulation, *Bull. Mar. Sci.*, 44:324–340.

Prager, E. 1998. Sediment resuspension in Florida Bay. Proc. Florida Bay Science Conference, University of Miami, May 12–14, 1998.

Robblee, M.B., T.R. Barber, P.R. Carlson, M.J. Durako, J.W. Fourqurean, L.K. Muehlstein, D. Porter, L.A. Yarbro, R.T. Zieman, and J.C. Zieman. 1991. Mass mortality of the tropical seagrass *Thalassia testudinum* in Florida Bay (USA), *Mar. Ecol. Prog. Ser.*, 71:297–299.

Rose, C.D., W.C. Sharp, W.J. Kenworthy, J.H. Hunt, W.G. Lyons, E.J. Prager, J.F. Valentine, M.O. Hall, P. Whitfield, and J.W. Fourqurean. 1999. Sea urchin overgrazing of a large seagrass bed in outer Florida Bay, *Mar. Ecol. Prog. Ser.*, 190:211–222.

Schomer, N.S. and R.D. Drew. 1982. *An Ecological Characterization of the Lower Everglades, Florida Bay, and the Florida Keys*, U.S. Fish and Wildlife, FWS/OBS-82/58.1, Washington, D.C., 246 pp.

Short, F.T., B.W. Ibelings, and C. den Hartog. 1988. Comparison of a current eelgrass disease to the wasting disease in the 1930s, *Aquatic Bot.*, 30:295–304.

Stumpf, R.P., M.L. Frayer, M.J. Durako, and J.C. Brock. 1999. Variations in water clarity and bottom albedo in Florida Bay from 1985 to 1997, *Estuaries*, 22(2B):431–444.

Thayer, G.W. and A.J. Chester. 1989. Distribution and abundance of fishes among basin and channel habitats in Florida Bay, *Bull. Mar. Sci.*, 44:200–219.

Thayer, G.W., W.J. Kenworthy, and M.S. Fonseca. 1984. *The Ecology of Eelgrass Meadows of the Atlantic Coast: A Community Profile*, U.S. Fish and Wildlife, FWS/OBS-84/02, Washington, D.C., 147 pp.

Thayer, G.W., P.L. Murphy, and M.W. Lacroix. 1994. Responses of plant communities in western Florida Bay to the die-off of seagrasses, *Bull. Mar. Sci.*, 54: 718–726.

Tilmant, J.T. 1989. A history and an overview of recent trends in the fisheries of Florida Bay, *Bull. Mar. Sci.*, 44:3-22.

Tomas, C.R., B. Bendis, and L. Houchin. 1998. Florida Bay algal blooms: Spatial and temporal variations in primary production. Florida Bay Science Conference, University of Miami, May 12–14, 1998.

Williams, S.L. and C.P. McRoy. 1982. Seagrass productivity: the effect of light on carbon uptake, *Aquatic Bot.*, 12:321–344.

Zieman, J.C. 1982. *The Ecology of the Seagrasses of South Florida: A Community Profile*, U.S. Fish and Wildlife, FWS/OBS-82/25, Washington, D.C., 123 pp.

Zieman, J.C. and R.T. Zieman. 1989. *The Ecology of the Seagrasses of West Florida: A Community Profile*, U.S. Fish and Wildlife, Biological Report 85(7.25), Washington, D.C., 155 pp.

Zieman, J.C., J.W. Fourqurean, and R.L. Iverson. 1989. Distribution, abundance and productivity of seagrasses and macroalgae in Florida Bay, *Bull. Mar. Sci.*, 44:292–311.

# 20 Linkages Between Estuarine and Reef Fish Assemblages: Enhancement by the Presence of Well-Developed Mangrove Shorelines

*Janet A. Ley*
Faculty of Fisheries and Maritime Environment, Australian Maritime College

*Carole C. McIvor*
USGS–Biological Resources Division, Center for Coastal Geology

## CONTENTS

## INTRODUCTION

Florida Bay is an important habitat for species of commercial and recreational interest in southern Florida. Fishing effort within Florida Bay targets mainly spotted sea trout (*Cynoscion nebulosus*) and subadult gray snapper (*Lutjanus griseus*) (Tilmant, 1989). Over 200 mangrove islands and extensive mangrove shorelines provide a potentially important habitat for species of fishery interest in Florida Bay, but relevant information is limited by the lack of fish surveys targeting mangrove fishes. Fishery species were not abundant in enclosure net surveys conducted in mangroves of central and western Florida Bay, possibly due to the small size of the area sampled (Thayer et al., 1987). A sampling approach targeting a larger area of mangrove habitat may provide more information on the value of this habitat in supporting populations of important fishery species.

Mangrove habitats have functional value to fishery species through enhancement of food resources and through provision of structure useful in reducing predation risk (Odum et al., 1982). The role of mangroves in support of estuarine food webs derives from enhancement of nearshore secondary production via detrital-based food chains (Odum and Heald, 1975). More recently, details of alternative trophic pathways and processes have been identified for mangroves within estuarine food webs (Robertson and Blaber, 1992); however, little definitive work has been done on the role of mangroves in providing shelter from predation. In one Australian study, mangroves fringing tidal creeks (mainly *Rhizophora*) were used by larger fishes, whereas smaller fishes penetrated deeply into the mangrove forest (mainly *Ceriops*) (Vance et al., 1996). However, structural features such as prop-root density and distribution did not correlate with spatial variation in fish distribution among small overwash islands in Tampa Bay (Mullin 1995). In shallow waters of Florida Bay, spatial variation in the distribution of small epibenthic fishes correlated with structural features of seagrass beds (Sogard et al., 1987). Furthermore, juvenile *L. griseus* were most abundant in seagrass beds with relatively greater seagrass biomass (Chester and Thayer, 1990). Outside of Florida, relationships between spatial variation in habitat structure and fish assemblages have been intensively investigated on coral reefs. On a broad scale, variation in fish diversity, density, and species composition correlates with the distribution of physiographic zones on coral reefs (i.e., reef crest, fore-reef), or with general types of coralline structure (i.e., massive vs. branching) (Sale, 1991). At a localized scale, measures of three-dimensional structure (i.e., aspects of holes in the coral substrate) were important in determining fish density and diversity on Hawaiian coral reefs (Friedlander and Parrish, 1998). Evidence from coral reef studies suggests that fishes tend to use shelter appropriate to their body size, and that predation-induced competition may result in shelter of appropriate size being a limiting resource for reef fishes (Friedlander and Parrish, 1998; Hixon, 1991). Similarly, red mangrove prop-root habitats having more massive structural features may harbor larger sized fishes. Thus, for this study we sought to quantify assemblages of larger fishes concentrated in the mangrove prop-root habitats and to identify the degree to which variation in mangrove structural features contributed to spatial variation in the distribution of fishes.

Spatial variation in estuarine fish distribution may also correspond with variation in physicochemical features among locations. The distribution and abundance of fishes have been correlated with differences in turbidity across an estuarine gradient in northeastern Australia (Cyrus and Blaber, 1992). Lower catches of major species upstream corresponded with greater upstream variation in salinity when compared with sites nearer the sea in a northeastern Australian creek (Sheaves, 1998). However, even in the absence of differences in salinity or turbidity, fish distribution was distinctly different upstream compared with downstream sites in an Australian mangrove creek system (Bell et al., 1988). Differences in fish distribution in the latter instance were attributed to limitations on dispersal processes in the early life-history phases of fishes. This finding implies that longer term trends in physicochemical features, such as a flush of freshwater coincident with the period of larval migration, may explain why fishes are not present in some locations (Sheaves, 1998). Thus, among the suite of variables included in the analysis of physicochemical factors influencing fish distribution, annual (or longer term) variation may be just as important as mean salinity and turbidity.

To examine the influence of both structural and physicochemical features on mangrove fish assemblages, a range of conditions must exist in a relatively small area. Northeastern Florida Bay provides an appropriate area for such an endeavor due to the range of conditions found in the six subbasins that comprise this portion of the Bay (Boyer et al., 1997; Ley et al., 1999).

The fish fauna of mangrove shoreline habitats can be subdivided into at least three categories dependent on fish size, position in the water column, and behavior. Benthic forage fish (e.g., Gobiidae, Cyprinodontidae) are small (<15 cm) and are primarily found close to the substrate (Ley et al., 1999). Water-column forage fish (e.g., Engraulidae, Atherinidae) are also small but typically occupy the middle and upper water column in large schools. Large roving fish (e.g. Lutjanidae, Mugilidae) are larger (>15 cm when fully grown) and occur throughout the mid- and lower water column. Because mangrove habitats are highly structured and strongly three-dimensional in character, larger fishes are more likely to distinguish among mangrove sites than smaller species. In addition, the ability of a researcher to directly observe the location of fish relative to specific sites is necessary to establish relationships between habitat and fish. The scale of the mangrove prop-root habitat coupled with the pragmatic challenges of quantitatively sampling fish in mangroves and observational factors make the use of visual surveys (e.g., Jones and Thompson, 1978) appropriate for this type of investigation. However, small forage species are not well sampled using visual surveys in the mangrove habitats (Ley et al., 1999). Thus, the current investigation concentrated on the large roving fish group, which also includes the species of greatest fishery interest.

We addressed the following objectives: (1) Characterize the fish assemblages inhabiting mangrove shoreline habitats and examine variation across the estuarine gradient, (2) determine the variability in species composition and density across the gradient and assess the distinctiveness of mangrove fish assemblages in the subbasins, (3) analyze the relative contribution of a suite of physicochemical and structural parameters in determining the densities of fish in the mangrove shorelines, and (4) look for linkages between adjacent ecosystems in the South Florida hydroscape as reflected in the structure of mangrove-associated fish assemblages.

## MATERIALS AND METHODS

### STUDY AREA

Florida Bay is a lagoonal estuary (2200 km²) located at the southern tip of Florida, U.S. (25°10′ N latitude). Mudbanks and mangrove islands divide Florida Bay into over 30 shallow (<2 m) subbasins interconnected by deeper (2- to 4-m) passes. This study focuses on six of these subbasins, located in northeastern Florida Bay (250 km²). Freshwater flows through numerous small creeks and via overland sheetflow from the eastern Everglades into northeastern Florida Bay along its entire northern border. The influence of freshwater decreases southward in northeastern Florida Bay, as exchange with the central bay, the Atlantic Ocean, and Barnes Sound increases (Figure 20.1); however, tidal influence is negligible and evaporation rates are high. Salinity in Florida Bay often exceeds 35, especially during the winter/spring dry season.

The Florida Keys archipelago forms the southern and eastern boundary of Florida Bay. This chain of islands extends from Soldier Key in the north to Dry Tortugas in the southwest. The Florida Keys Reef Tract on the Atlantic Ocean side of the Keys exhibits a diverse pattern of hardgrounds, patch reefs, and bank reefs from 25 m to 13 km offshore (Jaap, 1984). Seaward of the reefs are the Straits of Florida and the Florida current, a subsystem of the Gulf Stream. The Keys act as barriers to water exchange between Florida Bay and the reef tract. Such exchange is limited to the passes occurring between the Keys in the chain; however, the boundary of northeastern Florida Bay is Key Largo, one of the largest Keys, with extensive reef development offshore.

The type of mangrove habitat surveyed herein is "mangrove-fringing forest" (Odum et al., 1982). In northeastern Florida Bay, these common communities were dominated by *Rhizophora mangle*. Two other mangrove species (*Avicennia germinans* and *Laguncularia racemosa*) were

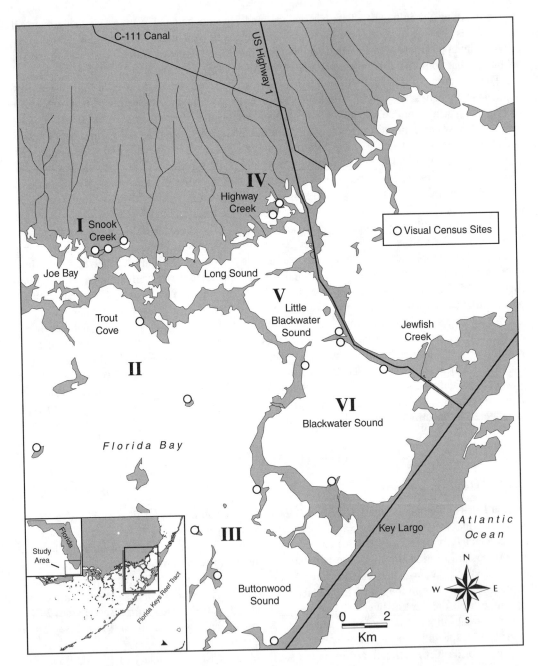

**FIGURE 20.1** Map showing northeastern Florida Bay study sites. Roman numerals indicate subbasins as defined in the analysis: (I) upstream-west (uw); (II) midstream-west (mw); (III) downstream-west (dw); (IV) upstream-east (ue); (V) midstream-east (me); (VI) downstream-east (de).

restricted to the higher intertidal portion of the mangrove fringe at the mid- and downstream (but not upstream) sites. A natural berm roughly 30 cm high by 30 cm wide separates the *Rhizophora* and *Avicennia* forests at most of the sites surveyed. If viewed from above, the mangrove fringed shorelines appear to be scalloped, with points extending into deeper (>1 m) water connected by shallower arcs. The distance between points is generally from 80 to 100 m.

Seagrass and algal species differ along the up- to downstream gradient (Montague and Ley, 1993). Widgeon grass (*Ruppia maritima*) and shoal grass (*Halodule wrightii*) occur upstream; turtle grass (*Thalassia testudinum*) and shoal grass occur at mid- and downstream sites. Algae including *Lyngbya*, *Batophora*, *Chara*, *Laurencia*, *Sargassum*, *Penicillus*, and *Udotea* sporadically occur throughout the study area. Storm surges and currents deposit clumps of dead algae and seagrass blades in various stages of decomposition along the mangrove prop-root edges.

Two sets of three subbasins were selected along the gradient (Figure 20.1). The western system consisted of Snook Creek, Trout Cove, and Buttonwood Sound (subbasins I to III); the eastern, Highway Creek, Little Blackwater Sound, and Blackwater Sound (subbasins IV to VI).

## Fish Sampling Regime

Fishes were censused visually along the mangrove fringe. Seventeen transects 50 to 70 m in length were selected for repeated sampling. Each transect generally encompassed a deeper (approx. 1.5 m) mangrove point and the adjacent shallower (0.5 m) shoreline. Permanent stations were designated at 10-m intervals along each transect. A snorkeler remained at a station for 30 seconds, an adequate period for recording the number of fish by species and approximate total length of each. Four repetitive swims were made at each site per survey. Only unique species and size classes of fish were added to the underwater data sheets after the first swim to avoid counting same fishes more than once. Densities for each visual census sample were determined by dividing the abundance of fish observed at a station by the estimated area of observation (i.e., area of a half-circle with horizontal visibility or fringe-width as the radius).

## Habitat Variables

For each monthly collection, salinity was measured with a handheld, temperature-compensated refractometer. From the monthly data, mean and standard deviation of salinity were determined. To determine relative water clarity and range of visibility, a line was extended to the maximum horizontal distance at which a snorkeler in the water could discern a white stationary pole (5-cm diameter). Each visual census site consisted of five to eight substations individually marked with a piece of flagging tape tied to an outer mangrove branch. At each substation, habitat features were measured along a transect perpendicular to the shoreline and passing under the flag. Starting at the inside berm, a 1.0-m² frame, internally divided into quarters, was set at 3-m intervals along the transects to 6 m off the mangrove fringe. The following data were collected:

- *Water depth:* Average of four measurements taken with a meter stick, one in each quarter of the frame
- *Fringe width:* Linear distance from the berm edge to the outer boundary of the mangrove fringe
- *Tree height:* Average of four measurements made with calibrated extension poles, one in each quarter of the frame
- *Prop root density:* Total number per frame
- *Seagrass, algae, and litter (SAV):* Sum of the separate visual estimates of the volume (depth × % cover) of seagrass, algae, and litter; estimates made in each quarter of the frame; depth of vegetation determined with a meter stick

## ANALYSIS

### Transformations

Skewness and kurtosis of graphed distribution patterns were examined to determine the appropriateness of transformation of data. Natural log transformation ($\ln(x + 1)$) effectively normalized the data for the density of fish. Following similar analysis, prop root density and volume of seagrass, algae, and litter were square-root transformed.

## ANOVA

To test the hypothesis that a subbasin was a source of variation in habitat variables and overall densities by fish category, one-way analyses of variance were conducted. *Post hoc* comparisons of means were tested using the least significant difference (LSD) test.

### SPECIES ASSEMBLAGES

Cluster analysis was conducted to organize species composition data into an ecologically mean-ingful structure. A clustering algorithm joined samples into successively larger clusters using a measure of similarity based on transformed densities for all species for each subbasin, resulting in a hierarchical tree diagram. To determine when two clusters were sufficiently similar to be linked together, average geometric (Euclidean) distance was calculated between all pairs of objects in two clusters when graphed in multidimensional space (i.e., unweighted pair-group average).

### ASSEMBLAGES VS. VARIABLES

Average fish density, number of species, and the Shannon–Weaver Diversity Index — $H' = \Sigma(p_i \ln p_i)$ where $p_i$ is the proportion of all individuals counted that were of species $i$ (Ludwig and Reynolds, 1988) — for each site ($n = 131$) were each used as dependent variables in three separate linear multiple regression analyses. Independent variables were the set of eight habitat variables. Using stepwise regression, the habitat variables that best predicted the fish assemblage parameter were analyzed. The forward stepwise procedure evaluated the independent variables at each step, adding or deleting them from the model based on specified criteria. In the first step, the variable that had the highest $F$ value (above 1.0) was entered into the regression equation. At each step, habitat variables already in the equation were removed if the $p$ value became greater than 0.10 due to the entry of other factors.

To examine associations between fish densities and environmental characteristics, data were examined using canonical correlation analysis (CCA), which generates linear combinations of variables (canonical roots) that maximize correlations among two sets of variables while it mini-mizes correlations within sets (Tabachnick and Fidell, 1989). Using several metrics, the method quantifies associations between densities of fish by species (fish dataset) and environmental con-ditions (environmental dataset) by sample. Canonical $R$ measures the overall association between the two datasets. The variance extracted measures the percent variance explained by each root within each individual dataset. The redundancy coefficient measures the amount of overall variation in one dataset as predicted by the other. Chi-squared (Bartlett's) tests measure the significance of each canonical root. To assist in ecological interpretation, each of the original variables is correlated to each root; correlations over 0.40 are considered ecologically meaningful (Stein et al., 1992). The first CCA included all sites and all variables in an examination of trends across the entire study area. A second CCA was conducted using the downstream sites and a subset of the variables to specifically focus on the influence of habitat structure.

## RESULTS

### SPECIES COMPOSITION

A total of 26,510 individuals from 24 species of large roving fish (with four additional taxa identified to family) were censused (Table 20.1). Of these, 90.7% of the abundance (20 taxa) were estuarine transients, 9.0% (four species) were residents, and 0.1% (four taxa) were occasional visitors. The majority (87.5% of abundance) of the large roving fishes were estuarine transients that spawn offshore and exist as adults on coral reefs of the Florida Keys Reef Tract. These fishes apparently recruit to mangroves in northeastern Florida Bay mangroves late in their first year of life (Ley et al., 1999). Smallest individuals were found in the downstream sites year-round, but a peak in recruitment of *Lutjanus griseus* juveniles occurred in February and March (Figure 20.2).

## TRENDS ACROSS THE GRADIENT

Number of species was greater downstream, with midstream sites intermediate, and upstream sites having lowest values (Table 20.2). Density of all species combined was significantly greater at the downstream-east subbasin (VI). The other subbasins were similar to each other in terms of fish densities (Table 20.2, Figure 20.3). The Diversity Index, combining the effects of both number of species and density, did not vary systematically along the gradient. Cluster analysis separated the fish assemblages into two distinct clusters: upstream sites vs. a combination of mid- and downstream sites (Figure 20.4). This separation occurred because mid- and downstream sites included numerous reef-oriented estuarine transients such as Lutjanidae and Haemulidae, families that seldom occurred upstream (Table 20.1).

Distribution of six of the eight most abundant species varied systematically across the estuarine gradient (not *Mugil cephalus* or *Strongylura notata*) (Table 20.3). For each of these six species, upstream sites had lower densities than mid- and downstream sites (Figure 20.5).

Systematic trends were evident for each habitat variable across the gradient, with upstream sites consistently differing from mid- and downstream (Table 20.4). Upstream sites had lower, more variable salinity; reduced water clarity; narrower fringe width; shorter trees; shallower water depth; and reduced vegetation but a greater density of mangrove prop roots (Figure 20.6).

## FISH AND HABITAT ASSOCIATIONS

Multiple regression analyses using all sites and all taxa indicated the strong influence of salinity regime on number of species (Table 20.5). As salinity standard deviation increased, the number of species of large roving fish present at a site decreased dramatically. This relationship accounted for 56% of the variation in species number among sites. Density of all fish combined (total density) was related (19.1%) to the occurrence of taller trees, deeper water, and wider mangrove fringe at a site. Multiple regression results for the Diversity Index ($H'$) indicated relatively high diversity at low salinity sites.

Three significant roots were derived in a CCA, which included seven habitat variables and fish densities for a subset of the eight most abundant fish species at all 131 sites throughout the study area (Table 20.6). A total of 41% of the variation in fish density was attributable to the significant habitat variables (redundancy); however, most of this variation was accounted for by the first root alone (36%). Habitat factors most well correlated with Root 1 were less salinity variation, greater water clarity, wider fringe, deeper water, and greater volume of submersed vegetation. Sites with these characteristics supported greater densities of *Haemulon sciurus*, *Lutjanus griseus*, and *Sphyraena barracuda*. Roots 2 and 3 each represented very little of the variation and had low correlations with the species or variables.

Two significant roots were derived from a second CCA using only downstream sites ($n = 64$), five habitat structural variables, and densities of *L. griseus*, *H. sciurus*, and *S. barracuda* (Table 20.7). For Root 1, highly correlated habitat variables accounted for 8% of the variation in fishes. Thus, greater densities of *H. sciurus* occurred at sites with deeper water and taller trees. Root 2 (14% redundancy) related greater densities of *L. griseus* and *S. barracuda* to sites with wider fringe, deeper water, taller trees, and greater volume of SAV.

**TABLE 20.1**
**Species of Large Roving Fish in Mangrove Prop-Root Habitats of Northeastern Florida Bay**

| Family/Species | Category | | Gradient Position | | | Total Abundance | Occurrence (No. of Sites) |
|---|---|---|---|---|---|---|---|
| | Age[a] | Residency[b] | Up | Mid | Down | | |
| Orectolobidae (nurse sharks) | | | | | | | |
| Ginglymostoma cirratum | J/A | R | | | 2 | 2 | 1 |
| Carcharhinidae (requiem sharks) | | | | | | | |
| Carcharhinus leucas | J | T | 2 | | | 2 | 2 |
| Dasyatidae (stingrays) | | | | | | | |
| Dasyatis americana | A | R | 1 | | 1 | 2 | 2 |
| Lepisosteidae (gars) | | | | | | | |
| Lepisosteus platyrhincus | J/A | F | 1 | | | 1 | 1 |
| Megalopidae (tarpon) | | | | | | | |
| Megalops atlanticus | J/A | T | | | 6 | 6 | 2 |
| Ariidae (sea catfish) | | | | | | | |
| Arius felis | A | R | | | 554 | 554 | 10 |
| Belonidae (needlefish) | | | | | | | |
| Strongylura notata | J/A | R | 148 | 589 | 1079 | 1816 | 124 |
| Centropomidae (snook) | | | | | | | |
| Centropomus sp. | J/A | T | 2 | 2 | 49 | 53 | 21 |
| Carangidae (jacks) | | | | | | | |
| Carangidae | A | T | | | 4 | 4 | 2 |
| Caranx hippos | J/A | T | 6 | 1 | 4 | 11 | 7 |
| Trachinotus goodei | J | T | | | 1 | 1 | 1 |
| Lutjanidae (snapper) | | | | | | | |
| Lutjanus apodus | J/A | T | | 234 | 198 | 432 | 68 |
| Lutjanus griseus | J/A | T | 85 | 5625 | 12,482 | 18,192 | 104 |
| Lutjanus jocu | J/A | T | | 32 | 39 | 71 | 35 |

| | a | b | | | | | |
|---|---|---|---|---|---|---|---|
| Haemulidae (grunts) | | | | | | | |
| *Haemulidae* | J | T | | | 1 | 1 | 1 |
| *Haemulon parrai* | J/A | T | 26 | | 113 | 139 | 12 |
| *Haemulon sciurus* | J/A | T | 72 | | 2696 | 2768 | 62 |
| Sparidae (porgies) | | | | | | | |
| *Archosargus probatocephalus* | J/A | T | 41 | | 122 | 163 | 36 |
| *Archosargus rhomboidalis* | J/A | T | | | 52 | 52 | 7 |
| *Lagodon rhomboides* | J/A | T | 2 | | 26 | 28 | 12 |
| Ephippidae (spadefish) | | | | | | | |
| *Chaetodipterus faber* | J/A | T | | | 18 | 18 | 5 |
| Scaridae (parrotfish) | | | | | | | |
| *Scaridae* | J/A | O | | | 4 | 4 | 4 |
| *Sparisoma radians* | J/A | O | 2 | | 1 | 3 | 3 |
| Mugillidae (mullet) | | | | | | | |
| *Mugil cephalus* | J/A | T | 93 | 12 | 1256 | 1361 | 41 |
| Sphyraenidae (barracuda) | | | | | | | |
| *Sphyraena barracuda* | J/A | T | 262 | 2 | 537 | 801 | 96 |
| Acanthuridae (surgeon fish) | | | | | | | |
| *Acanthurus chirurgus* | A | O | | | 9 | 9 | 3 |
| Balistidae (filefish) | | | | | | | |
| *Aluterus scriptus* | A | O | | | 1 | 1 | 1 |
| Diodontidae (spiny puffers) | | | | | | | |
| *Diodontidae* | J/A | T | 1 | | 14 | 15 | 7 |
| Totals | | | 340 | 6902 | 19,268 | 26,510 | 131 |

[a] Age categories: J = juvenile; A = adult.

[b] Residency categories: R = resident; T = transient; O = occasional visitor.

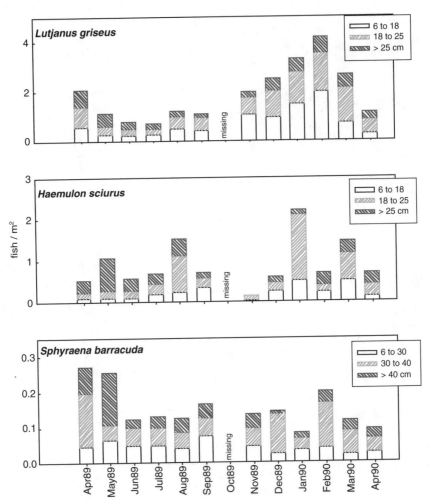

**FIGURE 20.2** Density by size class for top three species of estuarine transients in mangrove prop-root habitats of northeastern Florida Bay. Smallest size classes (bottom-most part of bars) shown in solid white; middle size classes (middle part of bars), upward cross-hatch; largest size classes (upper part of bars), downward cross-hatch. Data are for downstream sites ($n = 64$).

## TABLE 20.2
### Results of Analyses of Variance To Test for Significant Differences in Community Parameters Among Subbasins

| Parameter | SS Effect | F | p | Subbasin[a,b] |
|---|---|---|---|---|
| Number of species | 566.30 | 39.55 | <0.0000 | ue uw  mw me  dw de |
| Density | 18.59 | 3.95 | 0.0023 | ue uw dw me mw  de |
| Index of species diversity (H′) | 0.87 | 11.34 | <0.0000 | dw mw me  de  ue uw |

*Note:* Data are the transformed densities $((\ln(x + 1))$ of all 28 taxa of large roving fish censused in mangrove prop-root habitats throughout northeastern Florida Bay. Multiple comparisons were made using least significance difference (LSD) tests.

[a] Subbasins are listed in order of increasing fish density; underlined groups of subbasins are not significantly different.

[b] ue = upstream-east, uw = upstream-west, mw = midstream-west, me = midstream-east, dw = downstream west, de = downstream east.

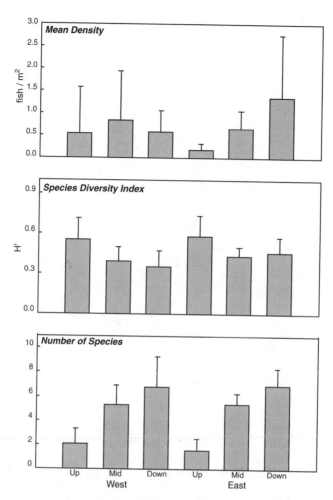

**FIGURE 20.3** Comparison of community variables among subbasins. Data are for all taxa (28) of large roving fish observed at all sites throughout northeastern Florida Bay ($n = 131$). Means (bars) and standard deviations (error bars) represented for all sites within a subbasin.

# DISCUSSION

## OVERALL PATTERNS ACROSS THE GRADIENT

Position along the estuarine gradient was a strong influence on the spatial distribution patterns of large roving fish assemblages and mangrove habitat features in the northeastern Florida Bay system. Number of fish species, species composition, and densities of six of the eight most abundant species varied systematically along the gradient. Additionally, values of each of the eight habitat variables varied in accordance with gradient location. In general, upstream sites had fewer fish and less development of the mangrove habitat. In this study, as in others investigating the influence of both physicochemical and habitat structural variables on fish assemblages along a longitudinal estuarine gradient (e.g., Loneragan et al., 1986; Bell et al., 1988; Sheaves, 1998), upstream habitats were characterized by lower numbers of species and unique fish assemblages. Such patterns are often attributed to a combination of biotic and abiotic factors and processes that vary across the gradient, but quantitative relationships have been difficult to test experimentally due to the scale and complexity of the habitat.

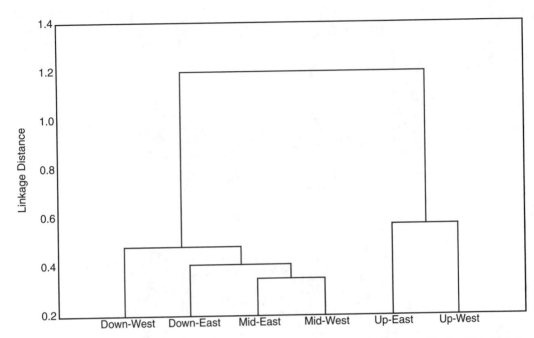

**FIGURE 20.4** Results of cluster analysis to compare community composition among subbasins. Data include all taxa of large roving fish.

**TABLE 20.3**
**Results of Analyses of Variance To Test for Significant Differences Among Subbasins**

| Species | SS Effect | F | p | Subbasin[a,b] |
|---|---|---|---|---|
| *Strongylura notata* | 1.76 | 5.73 | 0.0001 | uw dw  de   me mw ue |
| *Lutjanus apodus* | 0.86 | 10.32 | <0.0001 | ue uw me de dw  mw |
| *Lutjanus griseus* | 45.12 | 17.43 | <0.0001 | ue uw  dw de me mw |
| *Lutjanus jocu* | 0.01 | 3.75 | 0.0034 | uw ue dw  de mw me |
| *Haemulon sciurus* | 5.6 | 7.64 | <0.0001 | ue uw mw  me  dw de |
| *Archosargus probatocephalus* | 0.19 | 6.95 | <0.0001 | ue uw me de  dw mw |
| *Mugil cephalus* | 3.77 | 4.56 | 0.0007 | mw me ue dw  de uw |
| *Sphyraena barracuda* | 1.05 | 16.32 | <0.0001 | ue uw  dw me de mw |

*Note:* Data are transformed densities ($\ln (x + 1)$) of the eight most abundant species censused in mangrove prop-root habitats. Multiple comparisons were made using least significance difference (LSD) tests.

[a] Subbasins are listed in order of increasing fish density; underlined groups of subbasins are not significantly different.
[b] ue = upstream-east, uw = upstream-west, mw = midstream-west, me = midstream-east, dw = downstream west, de = downstream east.

### Systemwide Relationships

Considering the northeastern Florida Bay system as a whole, among the variables tested, salinity variation contributed most strongly to differences in numbers of species across the estuarine gradient. Because fish species are known to differ in their physiological capacity to tolerate changes in salinity (Moyle and Cech, 1988), this factor is often identified as important in explaining differences in species occurrence across an estuarine gradient (e.g., Cyrus and Blaber, 1992;

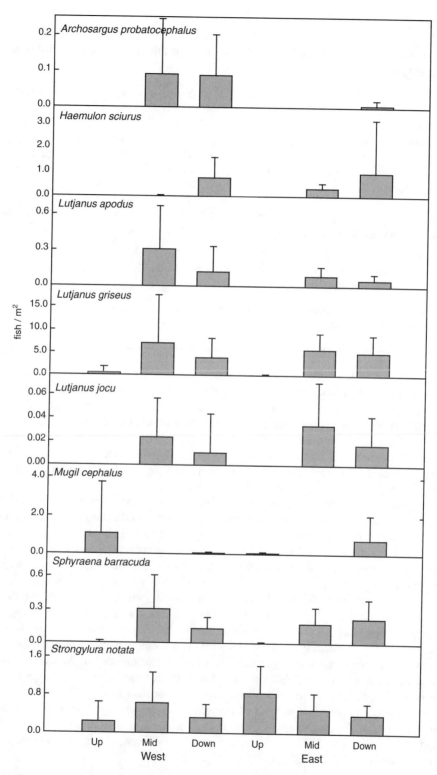

**FIGURE 20.5** Comparison of average fish densities among subbasins for the eight most abundant species of large roving fish in mangrove prop-root habitats. Means (bars) and standard deviations (error bars) represented for all sites within a subbasin.

**TABLE 20.4**
**Results of Analyses of Variance To Test for Significant Differences Among Subbasins in Habitat Variables**

| Parameter | SS Effect | F | p | Subbasin[a,b] |
|---|---|---|---|---|
| Salinity mean | 9861.95 | 1712.49 | <0.0001 | ue  uw  me  mw de  dw |
| Salinity standard deviation | 3034.69 | 7260.92 | <0.0001 | de  dw  me  mw  ue  uw |
| Water clarity | 863.08 | 61.82 | <0.0001 | ue uw  mw me  dw  de |
| Mangrove fringe width | 2105.20 | 20.67 | <0.0001 | ue uw  mw mevdw de |
| Mangrove tree height | 299,940.10 | 7.00 | <0.0001 | ue uw  dw me  de  mw |
| Prop-root density | 66.57 | 6.72 | <0.0001 | de dw mw  me ue uw |
| Water depth | 24,305.38 | 19.41 | <0.0001 | ue  uw  mw de me dw |
| Submersed vegetation and plant litter (SAV) | 496,887.30 | 43.52 | <0.0001 | ue uw  me  dw de mw |

*Note:* Data are measured values from 131 sites throughout northeastern Florida Bay. Multiple comparisons were made using least significance difference (LSD) tests.

[a] Subbasins are listed in order of increasing fish density; underlined groups of subbasins are not significantly different.
[b] ue = upstream-east, uw = upstream-west, mw = midstream-west, me = midstream-east, dw = downstream west, de = downstream east.

Peterson and Ross, 1991). However, for most species of large roving fishes no tests of osmoregulatory capacity have been conducted. In one related series of tests, however, juveniles of two species of large roving fish common in warm, temperate estuaries of mid-Atlantic U.S., differences in salinity preferences between the species could not be attributed to differences in physiological tolerances (Moser and Gerry, 1989). Instead, the investigators hypothesized that critical habitat factors correlated with salinity change (e.g., secondary factors such as species and density of submersed vegetation) may be responsible for preferences displayed in experimental tests and for resultant distribution differences between the two species.

The hypothesis that fishes respond to secondary factors in the process of habitat selection among different salinity regimes may be evident in comparison of the diets of benthic forage fishes in northeastern Florida Bay. Algae formed a larger part of the diets of fishes living upstream, whereas fishes living mid- and downstream consumed proportionally greater amounts of benthic invertebrates, a higher quality food source (Ley et al., 1994). This difference in food quality along the gradient may have been a function of the reduced submersed vegetation upstream, in turn a probable result of wide and often rapid variations in upstream salinity (Montague and Ley, 1993). In addition to reduced food resources, the finding that many species of large roving fishes in northeastern Florida Bay may have avoided upstream habitats could also be due to a lack of adequate shelter. Six abundant species tested tended to prefer more well-developed mangrove sites. Factors measured at upstream sites generally had lesser values related to habitat structure (e.g., mangrove fringe width, tree height, volume of submersed vegetation). As suggested for low development of submersed vegetation upstream (Montague and Ley, 1993), salinity conditions may also contribute to low development of mangroves in the upstream subbasins (Smith, 1992; McIvor et al., 1994). Thus, although the habitat variables measured strongly influenced the distribution of several abundant fish species, the nature of the influences (i.e., physiological, trophic, behavioral) remains to be determined.

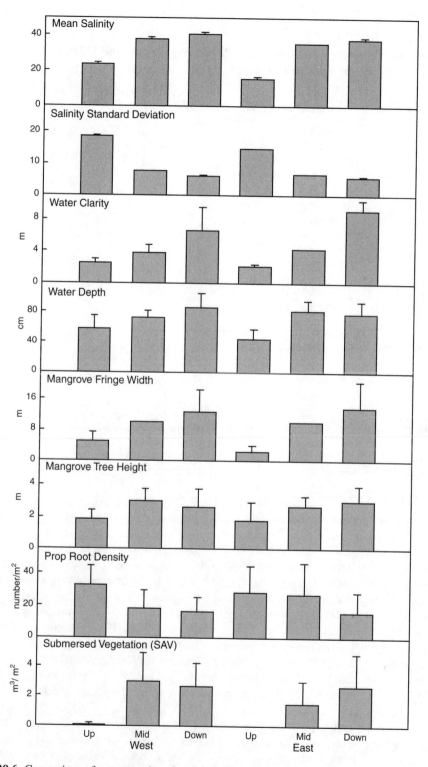

**FIGURE 20.6** Comparison of average values for eight habitat variables among subbasins. Means (bars) and standard deviations (error bars) represented for all sites within a subbasin.

**TABLE 20.5**
**Results of Three Separate Multiple Linear Regressions Testing the Relationships Between Habitat Parameters and Measures of Fish Community Structure**

| Dependent Variable[a] | Independent Variables | Sign | Partial $F$ | Partial $R^2$ | Model $R^2$ |
|---|---|---|---|---|---|
| Total density[***] | Mangrove tree height | + | 9.80[**] | 7.10 | 7.10 |
| | Water depth | + | 3.97 | 2.80 | 9.86 |
| | Mangrove fringe width | – | 7.56[*] | 5.07 | 14.93 |
| | Salinity standard deviation | – | 2.37 | 1.57 | 16.50 |
| | SAV (seagrass, algae, and litter) | – | 4.01 | 2.60 | 19.10 |
| No. of species[***] | Salinity standard deviation | – | 165.24[***] | 56.00 | 56.00 |
| | Water depth | + | 56.83[***] | 13.30 | 69.64 |
| | Water clarity | + | 10.30[*] | 2.30 | 71.92 |
| | Mangrove tree height | + | 5.50 | 1.20 | 73.09 |
| | Salinity mean | + | 3.57 | 0.74 | 73.84 |
| | Mangrove fringe width | – | 2.26 | 0.47 | 74.31 |
| | Prop-root density | – | 1.12 | 0.23 | 74.54 |
| Diversity Index ($H'$)[***] | Salinity mean | – | 48.94[***] | 27.00 | 27.50 |
| | Prop-root density | + | 8.52[*] | 4.50 | 32.00 |
| | Mangrove tree height | – | 5.47 | 2.80 | 34.80 |
| | Mangrove fringe width | + | 4.18 | 2.10 | 36.90 |
| | Water depth | – | 4.78 | 2.33 | 39.20 |
| | Salinity standard deviation | – | 3.13 | 1.49 | 40.70 |
| | SAV (seagrass, algae, and litter) | – | 1.34 | 0.64 | 41.30 |

*Note:* Fish data are transformed densities ($\ln(x + 1)$) of all 28 taxa of large roving fish censused in mangrove prop-root habitats. Habitat parameters are measured values for 131 sites throughout northeastern Florida Bay.

[a] Significance level: [***] $p < 0.0001$; [**] $p < 0.001$; [*] $p < 0.01$.

## DOWNSTREAM RELATIONSHIPS

In the analysis of downstream basins, where salinity variation was not a factor that varied among the sites, the dominant species of large roving fish differentiated among the mangrove habitats based on structural features. *Lutjanus griseus* and *Sphyraena barracuda* were more abundant at sites having more widely fringed shorelines, and *Haemulon sciurus* preferred sites with deeper water and taller trees. In general, greater densities coincided with sites having greater mangrove habitat development (i.e., prop-roots extending deeply into the water, wider fringe); however, quantitative relationships between fish densities and habitat variables were not strong (<20 %) and the spatial distribution of the species largely overlapped. Most of the mangrove shoreline sites downstream in northeastern Florida Bay were occupied by large schools of *L. griseus* and *H. sciurus*. By occupying these highly developed mangrove sites, risk of predation may be reduced. Potential predators in northeastern Florida Bay, mainly sharks (Carcharhinidae), were often seen patrolling the fringe but were seldom observed within the mangrove fringe. Nighttime numbers of Lutjanidae significantly increased on seagrass-covered mudbanks (Sogard et al., 1989), possibly due to feeding migrations from the mangrove fringes. Similarly, in mangrove shorelines of Puerto Rico, well-ordered migrations away from the mangrove prop-root habitat occurred at dusk for subadult *Haemulon sciurus* (Rooker and Dennis, 1991). Although not a routine part of our study, we occasionally observed twilight migrations of *L. griseus* and *H. sciurus* from the mangrove fringes in northeastern Florida Bay. The subadults in the mangrove habitats apparently seek the shelter of well-developed mangrove fringe habitats during the daytime, migrating to nearby seagrass beds to

**TABLE 20.6**
**Results of Canonical Correlation Analysis for Eight Species of Large Roving Fish and Seven Environmental Variables Over All Sampling Sites ($n$ = 131)**

| | Total | Correlation with Canonical Roots | | |
| | | First Root | Second Root | Third Root |
|---|---|---|---|---|
| Fish dataset | | | | |
| *Archosargus probatocephalus* | | 0.30 | **–0.68** | 0.15 |
| *Haemulon sciurus* | | **0.59** | **0.41** | **0.58** |
| *Lutjanus apodus* | | 0.31 | **–0.70** | 0.33 |
| *Lutjanus griseus* | | **0.87** | –0.25 | –0.07 |
| *Lutjanus jocu* | | 0.32 | –0.29 | –0.08 |
| *Mugil cephalus* | | –0.08 | **0.42** | **0.51** |
| *Sphyraena barracuda* | | **0.61** | –0.14 | –0.06 |
| *Strongylura notata* | | –0.08 | –0.31 | 0.26 |
| Variance extracted | 52% | 22% | 20% | 10% |
| Environmental dataset | | | | |
| Salinity variation | | **–0.83** | 0.12 | –0.05 |
| Water clarity | | **0.68** | **0.52** | 0.03 |
| Mangrove fringe width | | **0.79** | –0.08 | **–0.52** |
| Mangrove tree height | | **0.56** | –0.15 | **0.46** |
| Prop-root density | | **–0.50** | 0.12 | –0.09 |
| Water depth | | **0.89** | 0.10 | 0.07 |
| Submersed vegetation | | **0.75** | **–0.46** | 0.07 |
| Variance extracted | 68% | 53% | 8% | 7% |
| Redundancy | 41% | 36% | 3% | 2% |
| Chi-square tests | | | | |
| Canonical R | | 0.82 | 0.66 | 0.56 |
| Canonical R-square | | 0.68 | 0.44 | 0.32 |
| Chi-square | | 271.20 | 133.72 | 63.08 |
| *p* | | **0.0001** | **0.0001** | **0.0004** |

*Note:* Bold type indicates ecologically significant correlations.

feed at night. This pattern parallels the behavior of adults that shelter among the branching corals on the reef during the day and migrate to nearby seagrass beds to feed at night (Meyer and Schultz, 1985; Starck and Schroeder, 1971).

## LINKAGES: BAY AND REEF

*Lutjanus griseus*, *Haemulon sciurus*, and *Sphyraena barracuda*, among the most abundant large roving fishes in the northeastern Florida Bay mangroves, are species that occupy coral reefs on the Atlantic Ocean side of the Florida Keys as adults. Individuals of these species were mainly present in the downstream mangrove shorelines of northeastern Florida Bay as larger juveniles (year 1 to 3 size classes), recruiting year-round to the mangrove habitats (Ley et al., 1999). As larvae, these recruits would most likely have originated on the ocean side of the Florida Keys. Post-larval *L. griseus* enter Florida Bay via inlets through the Keys and settle in nearby seagrass beds (Chester and Thayer, 1990). When they reach larger sizes, these species apparently seek daytime shelter in the mangroves. Thus, the mangroves may serve as a staging habitat for certain species of estuarine transients that are resident on coral reefs as adults. The reproductive mode of Lutjanidae, Haemulidae, and Sphyraenidae is typical of the group of tropical fishes termed "migrating spawners"

**TABLE 20.7**
**Results of Canonical Correlation Analysis for Three Species**
**of Large Roving Fish and Five Environmental Variables Over Downstream**
**Sampling Sites ($n = 64$)**

| | Total | Correlation with Canonical Roots | | |
| --- | --- | --- | --- | --- |
| | | First Root | Second Root | Third Root |
| Fish dataset | | | | |
| *Haemulon sciurus* | | **−0.99** | −0.14 | 0.05 |
| *Lutjanus griseus* | | −0.30 | **0.88** | 0.36 |
| *Sphyraena barracuda* | | −0.01 | **0.43** | −0.90 |
| Variance extracted | 100% | 36% | 33% | 32% |
| | | | | |
| Environmental dataset | | | | |
| Mangrove fringe width | | 0.15 | **0.92** | −0.35 |
| Mangrove tree height | | **−0.73** | **0.40** | 0.11 |
| Prop-root density | | 0.24 | **−0.43** | −0.16 |
| Water depth | | **−0.62** | **0.44** | −0.19 |
| Submersed vegetation | | −0.05 | **0.55** | 0.75 |
| Variance extracted | 69% | 20% | 34% | 15% |
| Redundancy | 22.4% | 8.2% | 13.9% | 0.3% |
| | | | | |
| Chi-square tests | | | | |
| Canonical R | | 0.65 | 0.64 | 0.13 |
| Canonical R-square | | 0.42 | 0.41 | 0.02 |
| Chi-square | | 63.23 | 31.76 | 1.03 |
| *p* | | **<0.0001** | **0.0001** | 0.7943 |

*Note:* Bold type indicates ecologically significant correlations.

(Johannes, 1978). Species in this group have been characterized as growing into large roving adults (>25 cm TL), having mass offshore spawning migrations with consequent wide dispersal of young. This strategy may provide a selective advantage for these fishes by reducing losses of eggs and larvae from predators that tend to be highly abundant on coral reefs (Johannes, 1978). In northeastern Florida Bay, the nearest inlet through the Keys to any of the mangrove study sites was Tavernier Creek, a long, sinuous passage 18 km away. A small patch reef is near the creek mouth, but the main body of the extensive Florida Keys reef tract occurs a further 13 km offshore. Sites with closer proximity to the coral reef (e.g., mangrove shorelines on the ocean side of the Keys) may have more species of reef-oriented migrating spawners using the mangrove shorelines as subadult staging areas (e.g., *Haemulon flavolineatum* and *Ocyrus chrysurus*, as observed in Puerto Rico by Rooker and Dennis, 1991), but surveys are needed to confirm this hypothesis.

In the mangrove shoreline habitats of northeastern Florida Bay, assemblages of large roving fishes would be depauperate if not for the presence of reef-oriented estuarine transients. Does use of these habitats as a nursery enhance the Keys reef populations? In a review of interactions among fish assemblages in tropical shallow-water habitats, the question of possible linkages between reef populations and neighboring mangrove habitats could not be resolved because so few surveys of mangrove fish assemblages had been conducted (Parrish, 1989). Since that review was published, 12 quantitative studies documenting the direct use of mangrove habitats by fish have been published (Table 20.8). Lutjanids, haemulids, or sphyraenids were included among the dominant large roving fishes in six of these mangrove investigations. Two of these studies were in tropical Australia (Robertson and Duke, 1990a,b; Sheaves, 1998), three in southern Florida (Sheridan, 1991; Thayer et al., 1987, this study), and one in Puerto Rico (Rooker and Dennis, 1991). In four of the surveys

**TABLE 20.8**
**Summary of Published Literature Quantifying Fish Assemblages in Mangrove Shoreline Habitats**

| Author | Location | Methods | Frequency | Mangrove Species | Habitat Features | Five Most Abundant Species of Large Roving Fish |
|---|---|---|---|---|---|---|
| Morton (1990) | Australia; Moreton Bay; 27 31′ S | Enclosure net: Set against the bank within a path cleared of roots (1 site). High tide set, low tide fish removal. Net: 240 × 2 m; mesh: 18 mm. | 14 times, 1987–88 | *Avicennia marina* | Tide range: Spring, 2.3 m; Neap, 1.1 m. Salinity: 24–36. Temperature: 16–26°C | *Mugil georgi* (Mugilidae), *Sillago analis* (Sillaginidae), *Acanthopagrus australis* (Sparidae), *Girella tricuspidata* (Kyphosidae), *Selenotoca* (Scatophagidae) |
| Laegsgaard and Johnson (1995) | Australia; Moreton Bay; 27 31′ S | Trap nets: Set in small natural gutters with wings extended along mangrove fringe (4 nets, 2 sites). High tide set, low tide fish removal. Net: 8 × 2 m; mesh: 1 mm. | 24 times, 1991–93 | *Avicennia marina* | Tide range: Spring, 2.5 m; Neap, 0.1. Salinity: 27–37. Temperature: 17–28°C | *Platycephalus fuscus* (Platycephalidae), *Acanthopagrus australis* (Sparidae), *Mugil georgii* (Mugilidae), *Liza argentea* (Mugilidae), *Sillago* spp. (Sillaginidae) |
| Halliday and Young (1995) | Australia; Tin Can Bay; 25 54′ S | Enclosure nets: Set against the bank within a path cleared of roots (4 nets, 2 consecutive nights). High tide set, low tide fish removal. Size: 40 × 25 × 1.2 m; mesh: 18 mm. | 12 times, 1991–93 | *Rhizophora stylosa*, *Avicennia marina*, *Bruguiera gymnorhiza* | Water level range: 0.7–1.2 m. Salinity: 22–36. Temperature: 15–32°C | *Sillago* spp. (Sillaginidae), *Mugil georgii* (Mugilidae), *Acanthopagrus australis* (Sparidae), *Mugil cephalus* (Mugilidae), *Hyporhamphus ardelio* (Hemiramphidae) |
| Robertson and Duke (1990a,b) | Australia; Alligator Creek; 19 21′ S | Trap net: Set in small natural gutters with wings extended along mangrove fringe (1 net, 3 consecutive days). High tide set, low tide fish removal. Net: 13 × 1 m; mesh: 3 mm. | 7 times, 1985–86 | *Rhizophora stylosa*, *Avicennia marina*, *Ceriops tagal* | Tide range: minimum: 2.2 m; maximum: 3.5 m. Salinity: 30–38. Temperature: 21–31°C | *Acanthopagrus berda* (Sparidae), *Pomadasys kaakan* (Haemulidae), *Liza subviridis* (Mugilidae), *Zenarchopterus buffonis* (Hemiramphidae), *Selenotoca multifasciata* (Scatophagidae) |

**TABLE 20.8 (CONTINUED)**
**Summary of Published Literature Quantifying Fish Assemblages in Mangrove Shoreline Habitats**

| Author | Location | Methods | Frequency | Mangrove Species | Habitat Features | Five Most Abundant Species of Large Roving Fish |
|---|---|---|---|---|---|---|
| Sheaves (1998) | Australia; Alligator, Cattle, and Barramundi Creeks; 18 15' to 19 25' S | Antillean Z-traps; Set among subtidal snags in mangrove creeks (12 traps). Trap: 1.8 × 1.1 × 0.6 m; mesh: 12.5 mm. | 8 times, 1991–93 | *Rhizophora* spp., *Ceriops* spp., *Bruguiera* spp., *Xylocarpus* spp., *Avicennia marina* | Tide range: NA Salinity range: 0–50 Temperature: 21–32°C | *Lutjanus russelli* (Lutjanidae) *Acanthopagrus berda* (Sparidae) *Acanthopagrus australis* (Sparidae) *Arius argyropleuron* (Ariidae) *Epinephalus coioides* (Serranidae) |
| Vance et al. (1996) | Australia; Embley River; 12 40' S | Enclosure nets: Set within a path cleared of roots (3 sites, 2 consecutive days). High tide set, low tide fish removal. Net: 100 m circumference × 2 m; mesh: 2 mm. | 2 times, Nov. 1992 and Mar. 1993 | *Rhizophora mangle, Ceriops tagal* | Tide range: 2.4–2.8 m Salinity: 9.5–36.9 Temperature: 29–31°C | *Zenarchopterus buffonis* (Hemiramphidae) *Acanthopagrus berda* (Sparidae) *Toxotes chatareus* (Toxotidae) *Liza subviridis* (Mugilidae) *Scatophagus argus* (Scatophagidae) |
| Mullin (1995) | Tampa Bay, FL; 27 49' N | Enclosure nets: Encircling isolated mangrove islands (6 sites). High tide set, low tide fish removal. Net: 40-m circumference; mesh: 3.2 mm. | 6 times, summer, 1990 | *Rhizophora mangle* | Tide range: Spring, 0.7; Neap, 0.4 Salinity: mean 30 (SE.4) Temperature: NA | *Sarotherodon melanotheron* (Cichlidae) *Chloroscombrus chrysosura* (Carangidae) *Archosargus probatocephalus* (Sparidae) *Lagodon rhomboides* (Sparidae) *Leiostomus xanthurus* (Sciaenidae) |

| Study | Location | Mangrove | Sampling | Methods | Environmental parameters | Fish species |
|---|---|---|---|---|---|---|
| Sheridan (1991) | Rookery Bay, FL; 26 01′ N | *Rhizophora mangle* | 4 times, 1988–89 | Enclosure traps: Dropped within a path cleared of roots (8 sites). Fish collected by pumping trap contents through a 1-mm mesh plankton net. Solid circular trap: 1.8-m diameter. | Water level: 15–56 cm; Salinity: 15–37; Temperature: 19–31°C | *Lagodon rhomboides* (Sparidae), *Lutjanus griseus* (Lutjanidae), *Orthopristis chrysoptera* (Haemulidae), Sciaenidae (larval), Haemulidae (larval) |
| Ley et al. (This study) | Florida Bay, FL; northeast; 25 10′ N | *Rhizophora mangle* | 13 times, 1988–89 | Visual census: Snorkeling along mangrove shorelines (4 repetitions, 17 sites). Transects of 70 m. | Water level: 59–82 cm (monthly mean); Salinity range: 0.1–60.0; Temperature range: 20–34°C | *Lutjanus griseus* (Lutjanidae), *Haemulon sciurus* (Haemulidae), *Strongylura notata* (Belonidae), *Mugil cephalus* (Mugilidae), *Sphyraena barracuda* (Sphyraenidae) |
| Ley et al. (1999) | Florida Bay, FL; northeast; 25 10′ N | *Rhizophora mangle* | 13 times, 1988–89 | Enclosure nets: Set against the bank within a path cleared of roots; rotenone applied; fish retrieved with handheld dip-nets (3 nets at each of 6 sites). Net: 32 × 2 m; mesh: 3 mm. | Water level (monthly mean): 35–58 cm; Salinity range: 0.2–58.0; Temperature range: 17–35°C | *Strongylura notota* (Belonidae), *Sphyraena barracuda* (Sphyraenidae), *Lutjanus griseus* (Lutjanidae), *Mugil curema* (Mugilidae), *Mugil cephalus* (Mugilidae) |
| Thayer et al. (1987) | Florida Bay, FL; central and western; 25 10′ N | *Rhizophora mangle* | 8 times, 1984–85 | Enclosure net: Set against the bank within a path cleared of roots; rotenone applied; fish retrieved with handheld dip-nets (2 nets at each of 4 sites). Net: 32 × 2 m; mesh: 3 mm. | Water level (site mean): 0.8–0.9 m; Salinity range: 5.5–42; Temperature: NA | *Strongylura* spp. (Belonidae), *Mugil curema* (Mugilidae), *Sphyraena barracuda* (Sphyraenidae), *Mugil cephalus* (Mugilidae) |
| Rooker and Dennis (1991) | Puerto Rico; 18 00′ N | *Rhizophora mangle* | 7 times, 1986–87 | Visual census: Snorkeling along mangrove shorelines (8 sites, 3 times per day, 1 time at night). Transects of 15 m. | Water level: 1.0–1.5 m; Salinity: 34–37; Temperature 27–30°C | *Lutjanus griseus* (Lutjanidae), *Haemulon* spp. (juveniles), *Lutjanus apodus* (Lutjanidae), *Haemulon flavolineatum* (Haemulidae), *Haemulon sciurus* (Haemulidae), *Ocyurus chrysurus* (Lutjanidae) |

where Lutjanidae, Haemulidae, or Sphyraenidae were not among the dominant families, coral reefs were absent from the more subtropical study regions (Morton Bay, Australia; Tampa Bay, FL). Thus, in a wide variety of mangrove sites and in studies employing a diversity of sampling techniques, reef-oriented transients have been found to utilize mangrove shoreline habitats. The current investigation indicates that individuals of these species selected habitats with characteristics similar to the coral reef, having greater habitat size and clear water of near-marine salinity. Direct evidence is needed to determine if individuals that develop in Florida Bay migrate out to the reef as adults and enhance reef fish populations. However, in a comparison of fish communities among Florida reefs (Pennekamp, Dry Tortugas, Biscayne Bay areas), assemblages at the Pennekamp reefs (located nearest northeastern Florida Bay) were most diverse and fishes most abundant (Tilmant, 1984). Furthermore, a family of migratory spawners (Haemulidae) was the top-ranking family at Pennekamp in contrast to the other sites where nonmigrating spawners (Pomacentridae and Scaridae) ranked first. Thus, the proximity of well-developed mangrove shorelines functioning as staging habitats for juveniles of migrating spawners potentially enhances both reef and estuarine fish assemblage.

## SUMMARY AND CONCLUSIONS

Mangroves are being destroyed worldwide despite their known value for many important fishery species harvested in coastal habitats. Identifying linkages with other habitats and conditions that contribute to greater habitat quality will provide management information on sustaining biodiversity and fish production. We quantified fish assemblages in 131 estuarine mangrove sites over 13 consecutive months using visual census surveys. We also measured corresponding features of the sites spanning a complex estuarine gradient from upstream near sources of freshwater inflow to downstream nearer inlets from the Florida Keys Reef Tract. Among the assemblage of fishes observed, the large roving fish guild included several species of fishery interest. This guild consisted of 26,510 individuals from 28 taxa. Number of species and densities of six of the eight most abundant species increased downstream along the gradient. In a multivariate analysis, habitat features of the sites explained 36% of the variation in spatial distribution of a subset of the fishes comprised of the top eight species. This multivariate factor was most strongly correlated with the distributions of three fish species and several of the habitat variables. Greater densities of juvenile blue-striped grunts (*Haemulon sciurus*), gray snapper (*Lutjanus griseus*), and great barracuda (*Sphyraena barracuda*) occurred at sites where (1) water was deeper, clearer, and more marine in salinity regime; (2) mangrove structure was more highly developed; and (3) submersed vegetation was more abundant. Accounting for 82.1% of the individuals surveyed, these three species of estuarine transients share a complex life-history pattern that includes major habitat shifts with life-history stage. As young juveniles, they occupy shallow nearshore habitats, and as adults they commonly occupy coral reef habitats. In northeastern Florida Bay, they are highly abundant in well-developed mangrove prop-root habitats as later stage juveniles. Well-developed mangrove sites near the reef may function as staging habitats for juveniles of barracuda, grunts, and snappers, potentially enhancing both reef and estuarine fish assemblages.

## REFERENCES

Bell, J.D., A.S. Steffe, and M. Westoby. 1988. Location of seagrass beds in estuaries: effects on associated fish and decapods, *J. Exp. Mar. Biol. Ecol.*, 122, 127–146.

Boyer, J.N., J.W. Fourqurean, and R.D. Jones. 1997. Spatial characterization of water quality in Florida Bay and Whitewater Bay by multivariate analyses: zones of similar influence, *Estuaries*, 20(4), 743–758.

Chester, A.J. and G.W. Thayer. 1990. Distribution of spotted seatrout (*Cynoscion nebulosus*) and gray snapper (*Lutjanus griseus*) juveniles in seagrass habitats of western Florida Bay, *Bull. Mar. Sci.*, 46(2), 345–357.

Cyrus, D.P. and S.J.M. Blaber. 1992. Turbidity and salinity in a tropical northern Australian estuary and their influence on fish distribution, *Est. Coast. Shelf Sci.*, 35, 545–563.

Friedlander, A.M. and J.D. Parrish. 1998. Habitat characteristics affected fish assemblages on a Hawaiian coral reef, *J. Exp. Mar. Biol. Ecol.*, 224, 1–30.

Halliday, I.A. and W.R. Young. 1996. Density, biomass and species composition of fish in a subtropical *Rhizophora stylosa* mangrove forest, *Mar. Freshwater Res.*, 47, 609–615.

Hixon, M.A. 1991. Predation as a process structuring coral reef fish communities, in *The Ecology of Fishes on Coral Reefs*, Sale, P.F., Ed., Academic Press, San Diego. CA, pp. 475–505.

Jaap, W.C. 1984. *The Ecology of the South Florida Coral Reefs: A Community Profile*, FWS/OBS-82/08, U.S. Fish and Wildlife Service, Washington, D.C., 138 pp.

Johannes, R.E. 1978. Reproductive strategies of coastal marine fishes in the tropics, *Environ. Biol. Fish.*, 3(1), 65–84.

Jones, R.S. and M J. Thompson. 1978. Comparison of Florida reef fish assemblages using a rapid visual technique, *Bull. Mar. Sci.*, 28(1), 159–172.

Laegdsgaard, P. and C.R. Johnson. 1995. Mangrove habitats as nurseries: unique assemblages of juvenile fish in subtropical mangroves in eastern Australia, *Mar. Ecol. Prog. Ser.*, 126, 67–81.

Ley, J.A., C.C. McIvor, and C.L. Montague. 1999. Fishes in mangrove prop-root habitats of northeastern Florida Bay: distinct assemblages across an estuarine gradient, *Estuar. Coast. Shelf Sci.*, 48(6), 701–723.

Ley, J.A., C.L. Montague, and C.C. McIvor. 1994. Food habits of mangrove fishes: a comparison along estuarine gradients in northeastern Florida Bay, *Bull. Mar. Sci.*, 54(3), 881–899.

Loneragan, N.R., I.C. Potter, R.C. Lenanton, and N. Caputi. 1986. Spatial and seasonal differences in the fish fauna in the shallows of a large Australian estuary, *Mar. Biol.*, 92, 575–586.

Ludwig, J.A. and J.F. Reynolds. 1988. *Statistical Ecology*, John Wiley & Sons, New York.

McIvor, C.C., J.A. Ley, and R.D. Bjork. 1994. Changes in freshwater inflow from the Everglades to Florida Bay including effects on biota and biotic processes: a review, in *Everglades: The Ecosystem and Its Restoration*, Davis, S.M. and Ogden, J.C., Eds., St. Lucie Press, Delray Beach, FL, pp. 117–146.

Meyer, J.L. and E.T. Schultz. 1985. Migrating of haemulid fishes as a source of nutrients and organic matter on coral reefs, *Limnol. Oceanogr.*, 30(1), 146–156

Montague, C.L. and J.A. Ley 1993. A possible effect of salinity fluctuation on abundance of benthic vegetation and associated fauna in northeastern Florida Bay, *Estuaries*, 16(4), 703–717.

Morton, R.M. 1990. Community structure, density and standing crop of fishes in a subtropical Australian mangrove area, *Mar. Biol.*, 105, 385–394.

Moser, M.L. and L.R. Gerry. 1989. Differential effects of salinity changes on two estuarine fishes, *Leiostomus xanthurus* and *Micropogonias undulatus*, *Estuaries*, 12(1), 35–41.

Moyle, P.B. and T.J. Cech. 1988. *Fishes: An Introduction to Ichthyology*. Prentice Hall, Englewood Cliffs, NJ.

Mullin, S.J. 1995. Estuarine fish populations among red mangrove prop roots of small overwash islands, *Wetlands*, 15(4), 567–573.

Odum, W.E. and E.J. Heald. 1975. The detritus-based food wed of an estuarine mangrove community, in *Estuar. Research*, Cronin, L.E., Ed., Academic Press, New York, pp. 265–286.

Odum, W.E., C.C. McIvor, and T.J. Smith, III. 1982. *The Ecology of the Mangroves of South Florida: A Community Profile*, FWS/OBS 81/24, U.S. Fish and Wildlife Service, Washington, D.C., 144 pp.

Parrish, J.D. 1989. Fish communities of interacting shallow-water habitats in tropical oceanic regions, *Mar. Ecol. Prog. Ser.*, 58, 143–160.

Peterson, M.S. and S.T. Ross. 1991. Dynamics of littoral fishes and decapods along a coastal river-estuarine gradient, *Estuar. Coast. Shelf Sci.*, 33, 467–483.

Robertson, A.I. and N.C. Duke. 1990a. Recruitment, growth and residence times of fishes in a tropical Australian mangrove system, *Estuar. Coast. Shelf Sci.*, 31, 723–743.

Robertson, A.I. and N.C. Duke. 1990b. Mangrove fish-communities in tropical Queensland, Australia: spatial and temporal patterns in densities, biomass and community structure, *Mar. Biol.*, 104, 369–379.

Robertson, A.I. and S.J.M. Blaber. 1992. Plankton, epibenthos, and fish communities, in *Tropical Mangrove Ecosystems*, Robertson, A.I. and Alongi, D.M. Eds., American Geophysical Union, Washington, D.C., pp. 293–326.

Robins, C.R. and G.C. Ray. 1986. *A Field Guide to Atlantic Coast Fishes of North America*, Houghton Mifflin, Boston.

Rooker, J.R. and G.D. Dennis. 1991. Diel, lunar and seasonal changes in a mangrove fish assemblage off southwestern Puerto Rico, *Bull. Mar. Sci.*, 9(3), 684–698.

Sale, P.F. 1991. Habitat structure and recruitment in coral reef fishes, in *The Ecology of Fishes on Coral Reefs*, Sale, P.F., Ed., Academic Press, San Diego, CA, pp. 197–210.

Sheaves, M. 1998. Spatial patterns in estuarine fish faunas in tropical Queensland: a reflection of interaction between long-term physical and biological processes? *Mar. Freshwater Res.*, 49, 31–40.

Sheridan, P.F. 1992. Comparative habitat utilization by estuarine macrofauna within the mangrove ecosystem of Rookery Bay, Florida, *Bull. Mar. Sci.*, 50(1), 21–39.

Smith, III, T.J. Forest structure. 1992, in *Tropical Mangrove Ecosystems*, Robertson, A.I. and Alongi, D.M. Eds., American Geophysical Union, Washington, D.C., pp 101–136.

Sogard, S.M., G.V.N. Powell, and J.G. Holmquist. 1987. Epibenthic fish communities on Florida Bay banks: relations with physical parameters and seagrass cover, *Mar. Ecol. Prog. Ser.*, 40, 25–39.

Sogard, S.M., G.V.N. Powell, and J.G. Holmquist. 1989. Utilization by fishes of shallow, seagrass-covered banks in Florida Bay: 2. Diel and tidal patterns, *Environ. Biol. Fish.*, 24(2), 81–92.

Starck, II, W.A. and R.E. Schroeder. 1971. Investigations on the gray snapper, *Lutjanus griseus*, *Stud. Trop. Oceanogr.*, 10, 1–224.

Stein, D.L., B.N. Tissot, M.A. Hixon, and W. Barss. 1992. Fish-habitat associations on a deep reef at the edge of the Oregon continental shelf, *Fish. Bull.*, 90, 540–551.

Tabachnick, B.G. and L.S. Fidell. 1989. *Using Multivariate Statistics*, Harper Collins, New York.

Thayer, G.W., D.R. Colby, and W.F. Hettler. 1987. Utilization of the red mangrove prop root habitat by fishes in south Florida, *Mar. Ecol. Prog. Ser.*, 35, 25–38.

Tilmant, J.T. 1984. Reef fish, in *The Ecology of the South Florida Coral Reefs: A Community Profile*, Jaap, W.C., Ed., FWS/OBS-82/08, U.S. Fish and Wildlife Service, Washington, D.C.

Tilmant, J.T. 1989. A history and overview of recent trends in the fisheries of Florida Bay, *Bull. Mar. Sci.*, 44(1), 3–22.

Vance, D.J., M.D.E. Hayward, D.S. Heales, R.A. Kenyon, N.R. Loneragan, and R.C. Pendry. 1996. How far do prawns and fish move into mangroves? Distribution of juvenile banana prawns *Penaeus merguiensis* and fish in a tropical mangrove forest in northern Australia, *Mar. Ecol. Prog. Ser.*, 131, 115–124.

# 21 Nesting Patterns of Roseate Spoonbills in Florida Bay 1935–1999: Implications of Landscape Scale Anthropogenic Impacts

*Jerome J. Lorenz*
National Audubon Society Tavernier Science Center

*John C. Ogden*
South Florida Water Management District

*Robin D. Bjork*
Wildlife Conservation Society and Department of Fisheries and Wildlife, Oregon State University

*George V. N. Powell*
World Wildlife Fund

## CONTENTS

0-8493-2026-7/02/$0.00+$1.50
© 2002 by CRC Press LLC

# INTRODUCTION

Florida Bay is a subtropical estuary that lies between the tip of the south Florida mainland and the Florida Keys (Figure 21.1). The bay is dependent on freshwater runoff from adjacent uplands that mixes with marine water from the Atlantic Ocean and the Gulf of Mexico to produce habitat conditions that historically supported a highly productive flora and fauna. South Florida's unique ecosystems face increasing pressures from a rapidly expanding human population. The ecological decline of Florida Bay exemplifies this pressure (Van Lent et al., 1993; Fourqurean and Robblee, 1999). Increasingly, attention is being focused on the problems of the Bay and what can be done to ameliorate them.

Historically, freshwater entered the Florida Bay estuary from the Everglades via Taylor Slough and numerous associated creeks to the east of the slough (Figure 21.1). However, drainage of the Everglades for urban and agricultural uses and flood-control measures substantially reduced much of this freshwater source (Fourqurean and Robblee, 1999). Prior to the mid-1960s, water-management efforts were focused primarily in the northern and central parts of the Everglades (Light and Dineen, 1994). Although these early activities altered hydropatterns in the southeastern part of the Everglades (Van Lent et al., 1993), the impact on direct freshwater inflows to Florida Bay was relatively small (Johnson and Fennema, 1989).

Beginning in 1964, flow patterns toward Florida Bay were significantly altered by the construction and operation of canals in the extreme southeastern Everglades basin (Johnson and Fennema, 1989). Concerns about the environmental impacts of this canal system on Everglades National Park (ENP) resulted in the U.S. Congress authorizing the South Dade Conveyance System (SDCS) in 1968 (Light and Dineen, 1994). The SDCS was designed to provide Taylor Slough with a predictable freshwater supply from upstream components of the Everglades system via an array of pumps and canals added to the existing canals in the C-111 basin (Light and Dineen, 1994). In 1970, Congress also legislated implementation of the Minimum Schedule of Water Deliveries to ENP which dictated that the canal system be operated such that Shark River Slough, Taylor Slough and the southeastern basin receive a predetermined minimum amount of freshwater per year (Light and Dineen, 1994). It was believed that the SDCS and the Minimum Schedule would provide Florida Bay with an adequate supply of freshwater; however, the SDCS was not completed until June 1982 (Van Lent et al., 1993), and, as a result, the Minimum Schedule was never fully implemented (Johnson and Fennema, 1989; Van Lent et al., 1993; Light and Dineen, 1994). During the dry season of 1983, unseasonably heavy rains resulted in emergency regulatory water releases into ENP which resulted in ecologically undesirable conditions (Light and Dineen, 1994). In March 1983, an environmental emergency was declared and the Minimum Schedule of Water Deliveries was suspended (Light and Dineen, 1994). In 1984 the Minimum Schedule was officially terminated by congress. Johnson and Fennema (1989) showed that between 1970 and 1982 water levels in Taylor Slough were lower

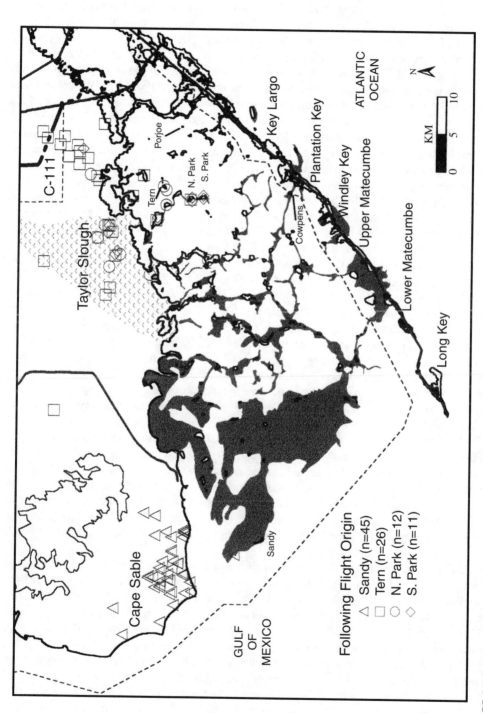

**FIGURE 21.1** Map of Florida Bay, southern mainland, and mainline Keys. Figure 21.1 also indicates final foraging location of 94 Roseate Spoonbills followed by fixed-wing aircraft from four colonies in 1989.

than in prior years and, consequently, less water was reaching Florida Bay. Their findings indicated that the Minimum Schedule failed in its design to provide Florida Bay with an adequate water supply.

Beginning with the March 1983 emergency actions, the recently completed SDCS was used to implement a new schedule for canal operations (Light and Dineen, 1994), ultimately called the Experimental Program of Water Deliveries (the Experimental Program did not officially began until 1984 when Congress ended the Minimum Schedule; U.S. Army Corps of Engineers, 1995). Although designed to deliver an adequate water supply to Taylor Slough and Florida Bay, the SDCS was primarily used for flood-control measures under the Experimental Program of Water Deliveries (Johnson and Fennema, 1989). Johnson and Fennema (1989) compared hydrologic conditions from 1982 to 1988 to conditions from 1956 to 1964 and found that water levels in the Taylor Slough headwaters were lowered by an estimated 45 to 60 cm as a result of the operation of SDCS under the Experimental Program. Van Lent et al. (1993) indicated that operation of the SDCS under the Experimental Program resulted in significant water loss from the system through groundwater seepage. These conditions ultimately resulted in further lowering of water levels throughout eastern ENP (Van Lent et al., 1993) with even less freshwater reaching Florida Bay than during the Minimum Schedule of Water Deliveries period (Johnson and Fennema, 1989). The Experimental Program went through several iterations (called "tests") but continued as the operational schedule through the spring of 1999 (U.S. Army Corps of Engineers, 1995).

Hydrologic models indicate that the amount of water flowing through the Everglades toward Florida Bay has been greatly reduced compared to historic flows (Fennema et al., 1994). McIvor et al. (1994) developed a model of historic salinities for the euryhaline area between the freshwater Everglades and saltwater Florida Bay. The model indicated that marine conditions have occurred more frequently and that brackish and marine periods have been longer in duration as a result of water management practices of the last several decades. Meeder et al. (1996) used paleoecological techniques to demonstrate that the marine environment had intruded into the historically freshwater wetlands north of Florida Bay at a rate faster than could be accounted for by sea-level rise. They also showed that changes in the rate of marine intrusion were correlated with the construction and changes in operation of the SDCS. Ross et al. (2000) performed vegetation surveys between the C-111 canal and Florida Bay and compared the results to similar surveys of the same area performed by Egler in 1948 (Egler, 1952). The comparisons indicated that the mangrove-dominated area has expanded inland by as much as 3.3 km since the 1948 study, supplanting historic freshwater sawgrass marshes. Cumulatively, these studies indicate that the wetlands north of Florida Bay have experienced higher salinities, longer periods of saline intrusion, and shorter hydroperiods due to anthropogenic manipulation of water resources than would have occurred in a non-managed system. Many researchers have concluded that the reduction in freshwater flow from the Everglades was at least partially responsible for recent ecological changes in the Florida Bay ecosystem (Boesch et al., 1993; Fourqurean and Robblee, 1999).

Florida Bay's problems are not limited to the diversion of freshwater. The construction of roadbeds between the Florida Keys has reduced connectivity between the estuary and the marine environment (Hudson et al., 1989). Furthermore, urban and agricultural pollutants from both mainland and Keys sources may be contributing factors in the recent ecological degradation of Florida Bay (Lapointe and Clark, 1992; Lapointe and Matzie, 1996; Brand, 2000). Between 1940 and 1990, the number of people residing in the Florida Keys increased from 14,000 to 78,000 (Lott et al., 1996). This more than fivefold increase in population was also accompanied by a boom in the tourism industry. An estimated 1.9 million people visited the Keys in 1990, with an average of more than 35,000 visitors staying in the Keys on a daily basis (Lott et al., 1996). Between 1955 and 1972, the desire for waterfront housing resulted in large scale dredge-and-fill projects which converted mangrove wetlands into canal-lined subdivisions with easy ocean or bay access (Lott et al., 1996). Between Lake Surprise on Key Largo and the southern end of Lower Matecumbe Key, an estimated 39.2% of the original mangrove wetlands were filled during this period (Strong and Bancroft, 1994). Lower Matecumbe Key, Plantation Key, and Southern Key Largo were most strongly impacted by urban development with wetland losses of 65.3, 49.8, and 43.8%, respectively (Strong and Bancroft, 1994). The impact of the Keys' human population on the Florida Bay ecosystem is currently a contentious issue (Lott et al., 1996), and is the subject of a large-scale, multi-agency planning process (Keys Carrying Capacity Study). However, the impact of the loss of wetland habitat on animal populations has largely been overlooked.

Federal, state, and local authorities are currently involved in the process of redesigning the human-controlled features of the southern Florida landscape, to a large extent aimed at reversing the ecological degradation of Florida Bay. So that restoration will be an iterative process, indicator species will be selected to evaluate the progress of these redesign activities. Roseate Spoonbills (*Ajaia ajaja*) have been selected as a potential indicator species for the Florida Bay ecosystem; however, little information about the responses of spoonbills to anthropogenic perturbations in southern Florida has been published. This chapter presents spoonbill nesting data collected since 1935 and examines spoonbill responses to various impacts to the system over the last 65 years. The conclusions drawn are used to develop potential responses of spoonbills to future restoration efforts.

## REVIEW

Roseate Spoonbills feed by tactolocation (Allen, 1942; Dumas, 2000). While feeding, a spoonbill swing its head so that the partially open bill is swept through the water in a semicircular motion. This action is performed while walking forward, giving the feeding spoonbill a somewhat inebriated appearance while feeding. When the highly sensitive "spoon" encounters a prey item, the bill snaps shut (Allen, 1942). With little cessation in forward progress, the spoonbill tosses the prey to the back of the throat by suddenly jerking the head backward out of the water.

This highly specialized feeding behavior requires that Roseate Spoonbills forage in extremely shallow water environments (less than 20 cm; Powell, 1987), but it also allows them to take advantage of an abundant food source in the form of small demersal fishes and crustaceans found in coastal wetlands (Allen, 1942). A variety of prey items have been described from spoonbill dietary investigations; however, small fishes of the orders Cyprinodontidae and Poeciliidae appear to be the primary dietary items (Allen, 1942; Dumas, 2000). Sheepshead minnows (*Cyprinodon variegatus*) and sailfin mollies (*Poecilia latipinna*) were found in virtually every analysis of spoonbill diet reviewed by Dumas (2000). Crop samples from 25 nestlings from two colonies in different parts of Florida Bay contained 85 to 90% small fish (primarily Cyprinodontidae and Poeciliidae), with the remainder being mostly shrimp of the genus *Palaemonetes* (Powell and Bjork, 1990).

Like many wading bird species, spoonbills appear to depend on the wet/dry seasonal cycle of southern Florida to successfully reproduce (Kushlan, 1978; Frederick and Collopy, 1989a; Bjork and Powell, 1994). During the wet season, the prey base exploits vast, seasonally inundated wetlands, thereby increasing in number and biomass (Kushlan, 1978; Loftus and Eklund, 1994; DeAngelis et al., 1997; Lorenz, 2000). High freshwater flows toward the coastal wetlands of Florida Bay may promote high secondary production as well (Lorenz, 1999, 2000). During the dry season, the abundant prey organisms are concentrated in the drying wetlands or are forced into the remaining pools, creeks and sloughs (Kushlan, 1978; Loftus and Kushlan, 1987; Loftus and Eklund, 1994; DeAngelis et al., 1997; Lorenz, 2000). Spoonbill breeding in Florida Bay is timed with seasonally low-water depths (Allen, 1942; Bjork and Powell, 1994), presumably because the adults depend on this concentration effect to capture enough food to raise young (Allen, 1942; Frederick and Collopy, 1989a; Bjork and Powell 1994, 1996; Dumas, 2000). Gradual and consistent declining water levels throughout the nesting cycle may be critical for successful nesting in spoonbills (Bjork and Powell, 1996). Evidence of high reproductive success during breeding seasons with gradual drydowns and poor reproduction during seasons of a reversed drying trend on the foraging grounds supports this hypothesis (Bjork and Powell, 1994). Therefore, ideal conditions for nesting spoonbills would result from high water levels in the headwaters during the wet season and low water levels in the coastal habitats during the dry season. Because of their highly specialized foraging behavior, spoonbills may, admittedly, be more sensitive to disruptions in the wet/dry cycle than other predatory species dependent on the coastal wetlands of Florida Bay. However, it is this sensitivity that makes the Roseate Spoonbill a good candidate for biological monitoring in this system.

Of the Roseate Spoonbills breeding in Florida, 90% nest on mangrove islands in Florida Bay within the boundaries of Everglades National Park (ENP) (Bjork and Powell, 1996). At present, they are a Species of Special Concern in Florida (Wood, 1997). Historical accounts from the 1800s indicate that a large population of Roseate Spoonbills once existed in the state, including a large presence in Florida Bay (summarized in Allen, 1942; Powell et al., 1989). During the last half of the 1800s, and continuing through the early 1900s, the Roseate Spoonbill population was greatly reduced by direct human disturbance. During this period, plume hunting was a highly lucrative profession in southern Florida (Graham, 1990) and spoonbills were heavily affected (Allen, 1942). Spoonbills may have been targeted by plume, subsistence, and sport hunters or may have been incidentally impacted by these activities (Allen, 1942; Graham, 1990). Regardless of intention, hunting practices had greatly reduced spoonbill numbers by the time prohibition of plume hunting and protection of nesting colonies began in the early 1900s. Illegal sport and subsistence hunting probably continued the decline, and, by the early 1930s, fewer than 200 pairs were thought to nest in Florida (Allen, 1942). By 1935, continued human depredation on adults and eggs had probably eliminated all colonies except the Bottle Key colony, which had been reduced to 15 pairs (Allen, 1963). By 1941, it was thought to be the only remaining active colony in Florida (Allen, 1942). After 1940, no spoonbill counts were made until Allen resumed his Florida Bay surveys in 1950. At that time he found larger numbers of nesting birds and a greater number of active colonies (64 pairs at seven colonies in 1950–51) compared to his surveys in the 1930s (Allen, 1963).

Powell et al. (1989) recounted the history of Roseate Spoonbill nesting in Florida Bay from 1935 to 1989. Table 21.1 and Figure 21.2 summarizes their findings. Approximately every 10 years from 1955 through 1978, the nesting population doubled (Figure 21.2). The 1978–79 season experienced a dramatic increase in the number of breeding spoonbills over the previous two years, with the population exceeding 1200 breeding pairs (Figure 21.2). No information is available for the 5 years subsequent to 1978–79, but by 1984 (when population surveys were re-initiated) the nesting population had declined to about 400 nests. With the exception of the 1978–79 peak, spoonbill nesting effort in Florida Bay was approximately 400 to 800 nests (Figure 21.2) in about 15 to 20 colonies (Table 21.1) in all years surveyed since 1975–76 (Figure 21.2).

A general temporal pattern in nesting effort was observed at most colonies. Nest numbers were initially low for the first year(s), followed by a period of increase, a peak of variable duration, and finally by a period of decrease (Table 21.1; Powell et al., 1989). This "boom and bust" temporal sequence in nesting effort can loosely be described as a bell-shaped pattern. The record at Cowpens, Tern, and Porjoe Keys are good examples of the roughly bell-shaped pattern observed at many colonies through time (Figure 21.3). Several of the more recently initiated colonies presumably are at various stages within this pattern. For example, nesting at Sandy Key appears to be in a stabilized peak period (Figure 21.3) and may decline at some future time, if the pattern holds.

Colonies in close spatial proximity tend to follow the same approximate temporal sequence in nesting effort resulting in a similar bell-shaped pattern when colonies are combined on a regional scale (Figure 21.2). This trend may indicate that the observed pattern is a result of environmental fluctuations on a smaller than baywide scale. Our hypothesis is that, initially, environmental conditions favor nesting, and effort presumably increases to some level. On varying time scales, conditions apparently change and the colonies within a given region of the bay decline. The pattern may also be repeating itself on a baywide scale (with current conditions at a stable peak stage or in the early stages of decline), which may give reason for concern. Understanding the driving mechanism(s) behind the observed patterns is critical if spoonbills are to be used as a gauge by which the general condition of Florida Bay is measured and by which management and restoration decisions are made.

**TABLE 21.1**
**Number of Roseate Spoonbill Nests by Colony and Year**

| Year | Southeast | | | | | | | | | | | | Northwest | | |
|---|---|---|---|---|---|---|---|---|---|---|---|---|---|---|---|
| | Bottle | Stake | Cowpens | Cotton | West | Low | Pigeon | Crab | East | Crane | East Butternut | Middle Butternut | Frank | Oyster | Sandy |
| 1935–36 | 15 | ○ | ○ | ○ | ○ | ○ | ○ | ○ | ○ | ○ | ○ | ○ | ○ | ○ | ○ |
| 1936–37 | 5 | ○ | ○ | ○ | ○ | ○ | ○ | ○ | ○ | ○ | ○ | ○ | ○ | ○ | ○ |
| 1937–38 | 6 | ○ | ○ | ○ | ○ | ○ | ○ | ○ | ○ | ○ | ○ | ○ | ○ | ○ | ○ |
| 1938–39 | 16 | ○ | ○ | ○ | ○ | ○ | ○ | ○ | ○ | ○ | ○ | ○ | ○ | ○ | ○ |
| 1939–40 | 3 | ○ | ○ | ○ | ○ | ○ | ○ | ○ | ○ | ○ | ○ | ○ | ○ | ○ | ○ |
| 1942–43 | | | | | | | | | | | | | | | |
| 1943–44 | | | | | | | | | | | | | | | |
| 1950–51 | 5 | 12 | 3 | 18 | ○ | ○ | ○ | ○ | ○ | ○ | ○ | ○ | ○ | ○ | ○ |
| 1951–52 | 5 | 26 | 10 | 10 | ○ | ○ | ○ | ○ | ○ | ○ | ○ | 2 | ○ | ○ | ○ |
| 1952–53 | 0 | 34 | 8 | 0 | ○ | ○ | ○ | ○ | ○ | ○ | ○ | 4 | ○ | ○ | ○ |
| 1953–54 | 0 | 30 | 28 | ○ | ○ | ○ | ○ | ○ | ○ | ○ | ○ | ○ | ○ | ○ | ○ |
| 1954–55 | 13 | 47 | 39 | 0 | 32 | 25 | ○ | ○ | ○ | ○ | ○ | ○ | ○ | ○ | ○ |
| 1955–56 | 4 | 55 | 70 | 27 | 5 | 18 | ○ | ○ | ○ | ○ | ○ | ○ | ○ | ○ | ○ |
| 1956–57 | 0 | 45 | 70 | 25 | 0 | 22 | ○ | ○ | ○ | ○ | ○ | ○ | ○ | ○ | ○ |
| 1957–58 | | | 40[a] | ○ | ○ | ○ | ○ | ○ | ○ | ○ | ○ | ○ | ○ | ○ | ○ |
| 1958–59 | 5 | 5 | 75 | 20 | ○ | ○ | ○ | ○ | ○ | ○ | ○ | ○ | ○ | ○ | ○ |
| 1959–60 | 0 | 8 | 65 | 3 | ○ | ○ | ○ | ○ | ○ | ○ | ○ | ○ | ○ | ○ | ○ |
| 1960–61 | 0 | ○ | 70 | ○ | ○ | ○ | ○ | ○ | ○ | ○ | ○ | ○ | ○ | ○ | ○ |
| 1961–62 | 0 | 35 | 50 | 3 | ○ | ○ | ○ | ○ | ○ | ○ | ○ | ○ | ○ | ○ | ○ |
| 1962–63 | 0 | 10 | 75 | 15 | ○ | ○ | ○ | ○ | ○ | 45 | ○ | ○ | ○ | ○ | ○ |
| 1966–67 | 0 | 15 | 0 | 0 | ○ | 20 | ○ | 0 | ○ | 24 | ○ | ○ | ○ | ○ | ○ |
| 1968–69 | 5 | A | A | 0 | ○ | 2 | ○ | 3 | ○ | A | ○ | ○ | ○ | ○ | ○ |
| 1969–70 | 5 | ○ | ○ | 0 | 0 | 0 | ○ | ○ | ○ | A | ○ | ○ | 0 | ○ | ○ |
| 1972–73 | | ○ | ○ | ○ | ○ | ○ | ○ | ○ | ○ | ○ | ○ | ○ | A[b] | 0 | 0[b] |
| 1973–74 | 14[b] | ○ | ○ | ○ | ○ | ○ | ○ | ○ | ○ | ○ | ○ | ○ | ○ | 20[b] | ○ |
| 1974–75 | 25 | ○ | ○ | ○ | ○ | ○ | ○ | ○ | ○ | A | ○ | ○ | 24[b] | 57[b] | ○ |
| 1975–76 | | ○ | 30 | ○ | ○ | ○ | ○ | ○ | ○ | ○ | ○ | ○ | 47 | 105 | 4[b] |
| 1976–77 | | ○ | ○ | 0 | ○ | 0 | A | ○ | 0 | A | 0 | A | ○ | ○ | ○ |
| 1977–78 | 0 | 41 | 13 | 0 | A | 0 | A | A | 0 | ○ | ○ | ○ | ○ | 81[b] | 25 |
| 1978–79 | 23 | 77 | 69 | 0 | 0 | 4 | 30 | 0 | 0 | 31 | 8 | 0 | 36 | 56 | 80 |
| 1979–80 | | ○ | 7[c] | ○ | ○ | ○ | ○ | ○ | ○ | A | 4 | 0 | 62 | 64 | ○ |
| 1980–81 | | | 6[c] | | | | | | | | | | | | |
| 1984–85 | 0 | 0 | 12 | 0 | ○ | 0 | 0 | 0 | 12 | 14 | 0 | 26 | 0 | 3 | 62 |
| 1985–86 | 0 | 0 | 0 | ○ | 7 | 0 | 0 | 8 | 6 | 17 | 0 | 66 | 0 | 6 | 139 |

**TABLE 21.1 (CONTINUED)**
**Number of Roseate Spoonbill Nests by Colony and Year**

| Year | Southeast | | | | | | | | | | | | Northwest | | |
|---|---|---|---|---|---|---|---|---|---|---|---|---|---|---|---|
| | Bottle | Stake | Cowpens | Cotton | West | Low | Pigeon | Crab | East | Crane | East Butternut | Middle Butternut | Frank | Oyster | Sandy |
| 1986–87 | 0 | 0 | 15 | 0 | 2 | 0 | 0 | 0 | 7 | 27 | 0 | 42 | 0 | 0 | 107 |
| 1987–88 | 0 | 0 | 14 | 0 | 9 | 0 | 12 | 0 | 0 | 12 | 8 | 14 | 4 | 0 | 151 |
| 1988–89 | 6 | 0 | 14 | 0 | 8 | 0 | 0 | 0 | 0 | 21 | 10 | 10 | 13 | 4 | 223 |
| 1989–90 | 19 | 8 | 9 | 0 | 1 | 0 | 3 | 0 | 0 | 17 | 6 | 12 | 22 | 9 | 163 |
| 1990–91 | 30 | 19 | 8 | 0 | 8 | 0 | 8 | 0 | 0 | 13 | 11 | 18 | 23 | 8 | 221 |
| 1991–92 | 40 | 18 | 8 | 0 | 6 | 0 | 5 | 0 | 0 | 13 | 4 | 17 | 46 | 45 | 232 |
| 1992–93 | | | | | | | | | | | | | | | 200 |
| 1994–95 | | | | | | | | | | | | | | | 250 |
| 1995–96 | | | | | | | | | | | | | | | 130 |
| 1996–97 | | | | | | | | | | | | | | | 196 |
| 1997–98 | | | | | | | | | | | | | | | 165 |
| 1998–99 | 28 | 0 | 10 | 0 | 0 | 0 | 8 | 7 | 2 | 11 | 4 | 47 | 125 | 23 | 177 |

| Year | Central | | | | Northeast | | | | | | | Southwest | | | |
|---|---|---|---|---|---|---|---|---|---|---|---|---|---|---|---|
| | Bob Allen | Manatee | Jimmie Channel | South Park | North Park | Tern | North Nest | South Nest | Porjoe | Duck | Pass | East Buchanon | West Buchanon | Barnes | Twin |
| 1935-36 | 0 | 0 | 0 | 0 | 0 | 0 | 0 | 0 | 0 | 0 | 0 | 0 | 0 | 0 | 0 |
| 1936-37 | 0 | 0 | 0 | 0 | 0 | 0 | 0 | 0 | 0 | 0 | 0 | 0 | 0 | 0 | 0 |
| 1937-38 | 0 | 0 | 0 | 0 | 0 | 0 | 0 | 0 | 0 | 0 | 0 | 0 | 0 | 0 | 0 |
| 1938-39 | 0 | 12 | 0 | 0 | 0 | 0 | 0 | 0 | 0 | 0 | 0 | 0 | 0 | 0 | 0 |
| 1939-40 | 0 | 0 | 0 | 0 | 0 | 0 | 0 | 0 | 0 | 0 | 0 | 0 | 0 | 0 | 0 |
| 1942-43 | 0 | 0 | 0 | 0 | 0 | 0 | 0 | 0 | A[d] | 0 | 0 | 0 | 0 | 0 | 0 |
| 1943-44 | 0 | 0 | 0 | 0 | 0 | 0 | 0 | 0 | A[d] | 0 | 0 | 0 | 0 | 0 | 0 |
| 1950-51 | 0 | 0 | 0 | 0 | 0 | 10[e] | 5 | 0 | 11 | 0 | 0 | 0 | 0 | 0 | 0 |
| 1951-52 | 0 | 0 | 0 | 0 | 0 | 9[e] | 5 | 0 | 14 | 0 | 0 | 0 | 0 | 0 | 0 |
| 1952-53 | 0 | 0 | 0 | 0 | 0 | 0 | 8 | 0 | 12 | 0 | 0 | 0 | 0 | 0 | 0 |
| 1953-54 | 0 | 0 | 0 | 0 | 0 | 0 | 23 | 0 | 20[f] | 0 | 0 | 0 | 0 | 0 | 0 |
| 1954-55 | 0 | 0 | 0 | 0 | 0 | 0 | 9 | 0 | 9 | 0 | 0 | 0 | 0 | 0 | 0 |
| 1955-56 | 0 | 0 | 0 | 0 | 0 | 0 | 26 | 0 | 9 | 0 | 0 | 0 | 0 | 0 | 0 |
| 1956-57 | 0 | 0 | 0 | 0 | 0 | 0 | 1 | 0 | 20 | 0 | 0 | 0 | 0 | 0 | 0 |
| 1957-58 | 0 | 0 | 0 | 0 | 0 | 0 | 0 | 0 | 0 | 0 | 0 | 0 | 0 | 0 | 0 |
| 1958-59 | 0 | 0 | 0 | 0 | 0 | 0 | 20 | 0 | 20 | 0 | 0 | 0 | 0 | 0 | 0 |
| 1959-60 | 0 | 0 | 0 | 0 | 0 | 0 | 24 | 0 | 16 | 0 | 0 | 3 | 0 | 0 | 0 |
| 1960-61 | 0 | 0 | 0 | 0 | 0 | 0 | 10 | 0 | 20 | 0 | 0 | 25 | 0 | 0 | 0 |

| Year | | | | | | | | | | | | | |
|---|---|---|---|---|---|---|---|---|---|---|---|---|---|
| 1961-62 | 0 | | | | 0 | 0 | 0 | 20 | 0 | 25 | 0 | 0 | 0 |
| 1962-63 | 0 | 4 | | | 50 | 0 | 0 | 15 | 0 | 10 | 100 | 3 | 0 |
| 1966-67 | 20 | | | | 181 | 0 | 0 | | | | | 2 | 0 |
| 1968-69 | | | | | 175 | 0 | 0 | 0 | | | 75 | 0 | 0 |
| 1969-70 | | | | | 150 | A | A | 0 | | | 100 | 0 | 0 |
| 1972-73 | | | | | 115[b] | | | | | | | | |
| 1973-74 | | | | | 200[b] | | | | | | | | |
| 1974-75 | | | | | 180[b] | | | | | | | | |
| 1975-76 | | 0 | 0 | | 495 | 0 | 30 | 20 | 0 | 25 | 25 | 0 | 0 |
| 1976-77 | | | | | | | | | | | | | |
| 1977-78 | 1 | 0 | | | 272 | 0 | 10 | 64 | 0 | 20 | 42 | 0 | 0 |
| 1978-79 | 5 | 0 | A | | 591 | 0 | 33 | 64 | 0 | 68 | 85 | A | 0 |
| 1979-80 | | | | | 300[g] | | | | | | | | |
| 1980-81 | | | | | | | | | | | | | |
| 1984-85 | | 12 | 3 | 20 | 170 | 0 | | 77 | 0 | 27 | 8 | 0 | 0 |
| 1985-86 | 8 | 19 | 5 | 4 | 184 | 0 | | 118 | 0 | 0 | 7 | 0 | 0 |
| 1986-87 | 9 | 6 | 6 | 0 | 158 | 0 | 21 | 99 | 0 | 8 | 9 | 0 | 0 |
| 1987-88 | 15 | 28 | 10 | 9 | 110 | 0 | 21 | 75 | 0 | 8 | 0 | 0 | 0 |
| 1988-89 | 34 | 13 | 39 | 50 | 153 | 0 | 43 | 25 | 0 | 19 | 6 | 0 | 0 |
| 1989-90 | 35 | 18 | 17 | 36 | 113 | 0 | 39 | 28 | 0 | 1 | 8 | 0 | 5 |
| 1990-91 | 24 | 19 | 25 | 30 | 113 | 0 | 39 | 60 | 0 | 8 | 4 | 0 | 8 |
| 1991-92 | | 47 | 25 | 36 | 172 | 0 | 59 | 66 | 0 | 7 | 9 | 0 | 8 |
| 1992-93 | | | | | 102 | | | | | 0 | 0 | | |
| 1994-95 | | | | | 100 | | | 4 | | | | | |
| 1995-96 | | | | | 100 | | | 0 | 100 | | | | |
| 1996-97 | | | | | 92 | | | 0 | 54 | | | | |
| 1997-98 | | | | | 95 | | | 0 | 8 | 26 | | | |
| 1998-99 | 11 | 23 | 7 | 39 | 60 | 0 | 19 | 9 | 9 | 0 | 0 | 0 | 2 |

*Note:* Data from 1935–1988 as per Powell et al. (1989) unless otherwise noted; 1989–1999 from this study. "A" indicates colony was active with a small number of nests but no count was made. Empty cells indicate no data collected for that colony/year. Years without colony checks were omitted.

[a] Allen (1963) reported 52 nests at other unspecified colonies and that colony checks were incomplete.
[b] Ogden (field notes).
[c] Sprunt (field notes).
[d] Eifler field notes.
[e] Nests on Little Tern Key. R. P. Allen. The present status of the Roseate Spoonbill, a summary of events 1943–63. Unpublished report, National Audubon Society Research Center, Tavernier, FL.
[f] Porjoe colony disrupted by humans and abandoned early in nesting. R. P. Allen. The present status of the Roseate Spoonbill, a summary of events 1943–63. Unpublished report, National Audubon Society Research Center, Tavernier, FL.
[g] William B. Robertson notes.

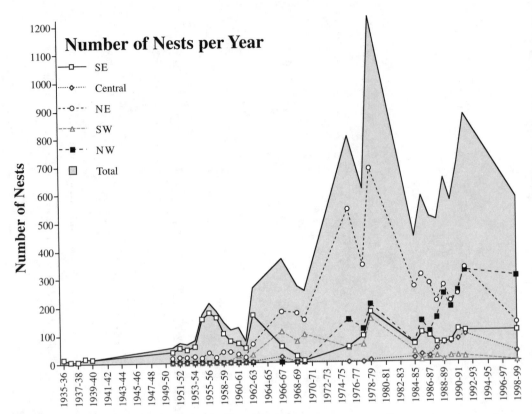

**FIGURE 21.2**  Total number of Roseate Spoonbill nests in Florida Bay from 1935 to 1999. Number of nests per colony group are also indicated for each year.

In his landmark work on Roseate Spoonbills, Allen (1942) identified a variety of factors that play a role in the selection and suitability of nesting sites. Some factors identified by Allen do not lend themselves to explaining the observed bell-shaped temporal pattern in nesting effort. These include suitable nesting site (semi-protected red mangroves are found on virtually every bay key), minimal tidal range (most of the nesting colonies are in areas that have minimal or no diurnal tide), reasonable water quality (water quality throughout the bay was excellent until the late 1980s and the pattern was observed prior to this change), and climatic factors (would be expected to act on a baywide scale and therefore would not likely impact colonies individually or on a regional scale). Allen (1942) also identified four factors that could potentially explain the observed pattern: (1) parasites and disease may impact nesting success and result in colony abandonment; (2) direct human disturbance (hunting and general colony disruption) may result in colony abandonment; (3) disruption of nesting by raccoon invasion and subsequent nest predation may disband a colony; and (4) availability and quality of food resources for a colony may change through time resulting in variable usage of the colony. The following sections review existing evidence that might support or refute these factors as explanations for the observed pattern in colony size through time.

## DISEASES, PARASITES, AND CONTAMINANTS

In 1940, Allen (1942) found evidence of an ectoparasite infestation in a Roseate Spoonbill colony on the Texas coast. The colony was abandoned by adults and chick mortality soon followed. Allen (1942) proposed that the parasites weakened the nestlings so they could not perform the necessary begging behavior needed to elicit a feeding response by parents. Bjork and Powell (1994) also indicated that disease and parasites were a source of nesting mortality in Florida Bay Roseate

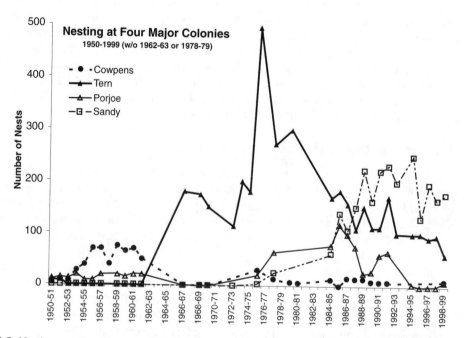

**FIGURE 21.3** Nesting record at Cowpens, Tern, Porjoe, and Sandy Key colonies for the period 1950–1999.

Spoonbill colonies between 1987 and 1992. M. G. Spalding (pers. comm.) performed necropsies on chicks in Florida Bay colonies between 1987 and 1992. Results indicated that viral, bacterial, and endoparasitic infections were present and may have played a role in nestling mortality; however, prevalences were low. Spalding also found high prevalences of ectoparasitic infestations, with some chicks having sufficient enough numbers to cause health problems.

These results indicate that disease and parasites could result in colony abandonment or lowered nesting success rates. In the case of the 1940 Texas colony, Allen (1942) found no evidence of the ectoparasites within the colony the following year. The colony formed normally and successfully fledged young. Bjork and Powell (1994) indicated that disease and parasites did not appear to be a factor in either colony formation or overall colony nesting success; rather, the impacts appeared to be on the level of individual nests. Spalding (pers. comm.) indicated that disease-related mortality had a minor impact on the overall nesting success in Roseate Spoonbills. Furthermore, Allen (1942), Bjork and Powell (1994), and Spalding (pers. comm.) all indicated that other factors may have acted synergistically with the disease/parasitic agents in causing observed nest mortality. Based on these findings, it seems unlikely that disease/parasitic agents explain the observed boom-and-bust patterns in nesting.

Spalding (pers. comm.) also examined the effects of environmental contaminants on wading birds. Spoonbill eggshells, nestlings, and adults showed only low concentrations of organochloride compounds and there was little evidence supporting poisoning by organophosphates and carbamates. Sundlof et al. (1994) found mercury concentrations in tissues of some spoonbills nestlings at levels associated with emaciation, disease, and weakness in other species; however, in this case, the condition was not directly implicated in nestling mortality. Interestingly, mercury concentrations were much higher in nestlings from colonies in northeastern Florida Bay when compared to nestlings from the northwestern bay (Sundlof et al., 1994). These differences in mercury concentrations are pertinent to observed differences in spoonbill nesting patterns between these regions of the bay (see Results and Discussion section).

In reference to tactile feeding wading birds, Spalding (pers. comm.) concluded that disease/contamination-related mortality was minor compared to nest abandonment caused by human disturbance, inclement weather, and inadequate food resources. This conclusion is supported by observations of Florida Bay spoonbill nesting success between 1995 and 1999. During this period, success rates were extremely low and adults were observed to abandon colonies following inclement weather events (see also Bjork and Powell, 1994). Dead chicks were collected and found to be emaciated (JJL, pers. obs.). Live chicks in a weakened state were collected and found to be starving. Several of these birds recovered when fed a diet of frozen bait fish and were ultimately released to the wild (Laura Quinn, Florida Keys Wild Bird Center; pers. comm.).

## DIRECT HUMAN DISTURBANCE

According to Allen (1942), direct human disturbance (hunting and general colony disruption) resulted in the relocation of nesting effort. At the time of Allen's initial studies, spoonbill colonies remained largely unprotected and human transgressions were still a threat to the small population of spoonbills left in Florida Bay. However, in 1947, most of the spoonbill colonies were incorporated into ENP, and boat landings on colony islands were restricted. Since then, remarkably few incidents of humans disturbing spoonbill colonies in Florida Bay have been reported. Allen (1963) indicated that photographers caused the collapse of the Porjoe colony in 1953; however, the colony formed in subsequent years, indicating that the impact of direct human disturbance was not persistent. Bjork and Powell (1994) speculated that the decline at Porjoe Key between 1988 and 1990 was a result of human visitation to the island. The recovery in 1990 coincided with ENP posting a 30-m minimum approach distance around the island. Bjork and Powell (1994) also reported that, in 1991, someone had cut the mangroves into the East Buchanan colony and the cleared path was reportedly being used as a sightseeing destination. The colony was abandoned at that time and ENP posted this site as well. When next surveyed in 1998–99, evidence of recent spoonbill nesting was found within the colony site, though there were no active nests that year. Although detrimental to nesting activity, direct human disturbance of Florida Bay colonies has been temporally and spatially isolated and is not likely to have played a role in the observed bell-shaped pattern in nesting distribution through time.

Allen (1942) suggested that even mild human disturbance at any time in the nesting cycle could lead to failure of the colony, indicating that the collection of scientific data could alter nesting patterns. However, a study of nesting Roseate Spoonbills in Texas (White et al., 1982) found high reproductive success and no significant difference in success between colonies visited weekly throughout the nesting period and one that was visited only twice. We have seen no evidence that our research presence in spoonbill colonies negatively affected spoonbill nesting. For example, reproduction at Sandy Key was close to 100% in 3 of the 5 years that intensive observations were being made there (1988–1992). Activities during this period included monitoring nesting success at 4- to 6-day intervals and radio-tagging juveniles. Nest success was 88% at Tern Key during the 1990–91 nesting period, even though we visited the colony at 4- to 6-day intervals to monitor nests and tag adult and nestlings. Finally, nest counts at many colonies that followed the bell-shaped temporal pattern were made only after fledging of young occurred. These observations indicate that visits by researchers do not explain the temporal pattern in nesting.

## RACCOONS

After discovering evidence of raccoon predation and subsequent abandonment of the Manatee Key colony in 1940, Allen (1942) formulated a hypothesis that the natural limiting factor determining colony location and nest success in Florida Bay was nest predation by raccoons (*Procyon lotor*).

Raccoons have been implicated as efficient nest predators in various studies (Lapinot, 1951; Frederick and Collopy, 1989b). By surveying various bay keys for raccoon habitation, Allen concluded that raccoons had little difficulty in moving along the banks from the mainland to the keys and from one key to another. Keys in contact with a raccoon source (e.g., the mainland or the mainline Florida Keys) via a mudbank were most likely to have raccoons, while those more distant along a bank were less likely to have raccoons. Isolated keys (i.e., those not connected by banks) tended not to have raccoons.

In a study of White-Crowned Pigeon (*Columba leucocephala*) nesting patterns in Florida Bay, Strong et al. (1991) examined the distribution of potential nest predators on bay keys. Based on surveys conducted between 1987 and 1989, they found the raccoon distribution in Florida Bay to be nearly parapatric with nesting pigeons. The impact of other potential predators was found to be minimal. They concluded that the presence of raccoons made potential nesting sites unsuitable. Furthermore, their data supported Allen's (1942) assessment that raccoons appear to migrate from island to island along the mud banks.

Figure 21.4 shows that the distribution of raccoons from 1987 to 1989 (Strong et al., 1991) was parapatric with Roseate Spoonbill nesting colonies from 1935 to 1999. Strong et al. (unpubl. data) also assembled a list of bay keys reported to have had raccoon activity prior to and after the 1987–1989 survey (Figure 21.4). Of the 30 spoonbill nesting colonies, only four were located on keys that had ever had reports of raccoons: Pass, East Bob Allen, Manatee, and Cowpens Keys. The complete spoonbill nesting record at Pass Key included only eight nests over three consecutive years (Table 21.1), suggesting that conditions at Pass Key were never ideal for spoonbill nesting. Although raccoons were observed on East Bob Allen Key in the 1970s, no signs of raccoons were observed during the 1987–1989 survey. The record of spoonbill nesting on this key began in 1986–87 (Table 21.1), so it appears that spoonbill nesting and raccoon presence were not contemporaneous. Allen (1942) reported the disruption of the Manatee Key nesting colony by raccoons in 1940. Since then, nesting effort at Manatee has been sparse and intermittent (Table 21.1). Cowpens Key had been the subject of intense scrutiny by National Audubon Society researchers between 1950 and 1980 because of a large mixed-species wading bird colony (Allen, notes; Alexander Sprunt IV, pers. comm.). In 1978, a single raccoon appeared on the island in late spring and was removed prior to the following nesting season (Alexander Sprunt IV, pers. comm.). This is the only record of a raccoon at Cowpens.

These data and our observations support Allen's hypothesis that raccoon presence is a primary determinant in colony site selection and nesting success of spoonbills. Most large colonies formed on isolated islands that were not connected to the mainland or the mainline keys by banks (e.g., Tern, Porjoe, Pigeon, Duck, Butternut Keys; Nest Keys; and Buchanan Keys). Several large colonies were located along a bank but were isolated by the presence of deep channels that cut through the bank (e.g., Sandy, Frank, Cowpens, and Cotton Keys) thereby cutting off the colony from any potential raccoon source. However, most of these colonies exhibited the bell-shaped temporal pattern (Table 21.1) even though the evidence indicates that no raccoon perturbations ever occurred. Clearly, spoonbills select colony locations that do not have raccoons, but this still does not explain the temporal pattern in question.

## Availability and Quality of Food Resources

By the time Allen (1942) completed his monograph, drainage of the Everglades was well underway and, even at this early date, the scientific community was beginning to express concerns over the impact of human development on southern Florida ecosystems. Allen addressed the possibility that human activities could be affecting spoonbills through alteration of their prey base. However, he dismissed the possibility, citing the spoonbill's "maritime nature" (a reference to their fidelity to using Florida Bay for nesting over the freshwater wetlands of the Everglades proper) and the lack of evidence that drainage of the Everglades was impacting the Florida Bay ecosystem. Specifically

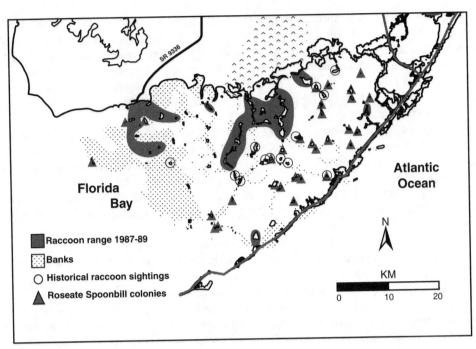

**FIGURE 21.4** Location of all spoonbill nesting sites from 1935 through 1999 and the distribution of raccoons in Florida Bay between 1987 and 1989. None of the spoonbill nesting sites was located within the 1987–1989 raccoon range. Keys reported to have had resident raccoons prior to 1987 and after 1989 are also indicated. (Adapted from Strong et al., 1991.)

he stated that "… there does not appear to be any obvious connection between desertion of spoonbill colonies and drainage of the Everglades." However, by the late 1950s and early 1960s, Allen (1963) recognized that human impacts to the spoonbill's environment were altering nesting patterns. Specifically, he speculated that declines in nesting over time at colonies near the upper Florida Keys might be explained by habitat loss of primary foraging locations to urban development (Table 21.2).

We hypothesize that land and water management practices in southern Florida have reduced the quantity and availability of food resources for Roseate Spoonbills and, as a result, are the major factors that determine nesting patterns in Florida Bay. The primary anthropogenic perturbations to spoonbill foraging grounds have been the filling of wetlands for urban development in the upper Florida Keys (Strong and Bancroft, 1994; Lott et al., 1996) and the alteration of wetland type and function along the northeast coast of Florida Bay and southern Biscayne Bay through water management practices (McIvor et al., 1994; Lorenz, 2000; Ross et al., 2000). We will attempt to show through an examination of data collected over the last 65 years that anthropogenic alterations to spoonbill foraging habitat have resulted in the observed patterns of nesting distribution and success in Florida Bay. More specifically, anthropogenic alterations to spoonbill foraging habitat have resulted in the observed patterns, indicating that spoonbills are sensitive to changes in the abundance, availability, and quality of their prey.

Although most spoonbills nest in the bay, during recent years most of these nesting birds have made daily flights to foraging grounds in the mainland wetlands north of Florida Bay. The salinity, water depths, and drying patterns of these wetlands have been altered by water management practices over the last 40 years (McIvor et al., 1994), thereby altering the spoonbill prey base (Lorenz, 1999). Further, we hypothesize that plans to restore more natural surface and groundwater flows into the mainland estuaries north of Florida Bay should positively influence spoonbill nesting patterns in the bay. This perceived sensitivity of spoonbills, particularly to the effects of water management practices

**TABLE 21.2**
**Summary of Spoonbill Foraging Habitat Lost to Dredge-and-Fill Urban Development in the Upper Florida Keys[a]**

| Decade | Spoonbill Colonies Impacted | % Loss of 1955 Foraging Habitat | Name of Development | Location | Approximate Construction (Nesting Year Impacted) | Estimated Foraging Acres Filled |
|---|---|---|---|---|---|---|
| 1955–1965 | Southeast | 28 | Boatman's Colony | Northern Plantation Key | 1955 | 7 |
| | | | Indian Waterways[b] | Northern Plantation Key | 1958 (1958–59) | 65 |
| | | | Plantation Key Colony[b] | Northern Plantation Key | 1959 (1959–60) | 90 |
| | | | Tropical Atlantic Shores[b] | Northern Plantation Key | 1959 (1959–60) | 18 |
| | | | Holiday Isle Resort | Southern Windley Key | 1958–1959 | 15 |
| | | | Venetian Shores (phase 1) | Southern Plantation Key | 1959–1960 | 12 |
| | | | Key Heights | Southern Plantation Key | 1960 | 5 |
| | Southwest | 9 | Fiesta Key Campground | Fiesta Key | 1955–56 | 11 |
| | | | Layton | Long Key | 1955–56 | 10 |
| | | | Tollgate Shores/Safety Harbor | Southern Lower Matecumbe Key | 1955 | 77 |
| | | | Matecumbe Sand Beach/White Marlin Beach[b] | Southern Lower Matecumbe Key | 1958–59 | 44 |
| 1965–1975 | Southeast | 7 | Davis Shores | Northern Lower Matecumbe | 1959–60 | 9 |
| | | | Venetian Shores (phase 2) | Southern Plantation Key | 1965–72 | 13 |
| | | | Hammer Point[b] | Southern Key Largo | 1971–72 (1971–72) | 71 |
| | | | Tavernier Town Shopping Center[b] | Southern Key Largo | 1971–72 (1971–72) | 10 |
| | Southwest | 6 | Tavernier Creek/Tavernaero | Northern Plantation Key | 1971–72 | 20 |
| | | | Port Antigua (central region)[b] | Southern Lower Matecumbe Key | 1971–72 (1971–72) | 134 |

[a] Only developments that resulted in a loss of 5 acres or more are included.
[b] Important Roseate Spoonbill foraging sites as reported by Allen (notes), Watson (notes), and Sprunt (pers. comm.).

on northeastern Florida Bay, makes them a highly appropriate indicator species for biological monitoring. However, an understanding of the driving mechanism(s) behind the observed nesting patterns is critical if spoonbills are to be used as a gauge by which the general condition of Florida Bay is measured or by which management and restoration decisions and success are assessed.

## MATERIALS AND METHODS

The 65 years of information on nesting and foraging patterns of Roseate Spoonbills in Florida Bay was principally gathered during four distinct study periods. Robert Porter Allen (RPA) led research efforts from 1935 until his death in 1963. Information attributed to RPA comes from his two books on the subject (Allen, 1942; Allen, 1947), an incomplete and unpublished report (Allen, 1963), and detailed field notes and correspondences housed at National Audubon Society's Field Research Office in Tavernier, FL. Between 1966 and 1980, spoonbill data were principally collected through a coordinated effort among John C. Ogden (JCO), William B. Robertson, Jr., Barbara W. Patty, and Alexander Sprunt IV. George V.N. Powell (GVNP) and Robin D. Bjork (RDB) led a comprehensive study on spoonbill nesting and foraging from 1984 to 1993. Beginning in 1994, Jerome J. Lorenz (JJL) collected nesting information. Each of these research efforts utilized different techniques and variable effort in assessing spoonbill nesting and foraging patterns in Florida Bay.

### ASSESSMENT OF FORAGING LOCATION

In order to identify the direction of foraging grounds from nesting colonies, flight-line counts similar to those described by Dusi and Dusi (1978) were made by RPA, JCO et al., and JJL. Initially, the colony island was circumnavigated by boat in order to determine the least number of positions needed to observe birds arriving and leaving the colony from any direction. Once identified, the boat was anchored at each position for a predetermined amount of time and the number and direction (compass heading) of Roseate Spoonbills arriving and leaving the colony were then counted. In order to analyze the data, flights were grouped based on eight major compass points (i.e., all flights with compass headings between 22.5 and 67.5 were categorized as northeast, from 67.5 to 112.5 were east, etc.).

GVNP and RDB used following flights (Bjork and Powell, 1994) to better characterize foraging flights made by nesting spoonbills. Individual birds were followed using a fixed-wing aircraft from their nesting colonies to the first foraging location. During the flight, a distance of at least 200 m was maintained between the aircraft and the birds to avoid influencing their flight path. When a flock was followed from a colony, one individual was selected as the target individual to be followed if the flock separated in flight. During the 1989–90 nesting season, 94 spoonbills were followed from four colonies (Sandy = 45, Tern = 26, N. Park = 12, S. Park =11). Mean and standard deviation in flight distance were calculated from these data. Also, initial flight compass heading and the compass heading from the colony to the destination site were recorded. The initial compass heading and final compass direction were compared using product-moment correlation coefficient ($r^2$) to determine if flight-line observations were a good indicator of the direction of foraging location.

### COLONY SURVEYS

RPA surveyed Florida Bay by boat investigating locations of Roseate Spoonbill activity for nesting sites (Allen, 1942). Subsequent investigators continued this practice while also investigating all islands that were previously reported to have spoonbill colonies. The techniques and timing of nest count surveys varied according to the researcher. RPA believed that spoonbills would abandon a colony in response to human disturbance within the colony (Allen, 1942) and therefore strongly avoided entering active colonies. However, he observed nesting spoonbill activity from a blind on the periphery of a colony and found that one member of a nesting pair will relieve the other around dawn and dusk (Allen, 1942). This information allowed him to estimate the number of nests within the colony by making crepuscular counts of the number of spoonbills entering and leaving a colony. Half of the total number of flights was approximately equal to the number of nests and averaging over a period of several days strengthened this approximation. He tested the accuracy of this method on several occasions by entering the colony following fledging of young and counting the empty nests.

Subsequent researchers generally entered the colonies and physically counted the number of nests. This method was more accurate and less time consuming but was also potentially disruptive to nesting activity. In an attempt to minimize any impact, visits to colonies did not begin until late in the incubation period, well after the birds were settled into the nesting cycle. Furthermore, researchers minimized the time in the colony to less than 1 hour whenever possible, and active colonies were entered only during mild climatic conditions. Many surveys were not completed until after chicks fledged so that there was no chance of colony disruption. Such surveys ran the risk of underestimating the number of nests, as vacant nests were often destroyed by other wading birds in need of nesting material. Therefore, post-fledging surveys were made immediately after young left the island. JCO et al., also used aerial surveys to estimate the number of nests within certain colonies. Since spoonbills generally nest in dense mangrove understory, nest counts made from the air are not considered as accurate as the other methods. Powell et al. (1989) annotated the results of Florida Bay Roseate Spoonbill nesting surveys from 1935 to 1988 with the type of survey used.

Starting about 1978, spoonbills began to occasionally nest a second time per season. Initially, these efforts were small in number (on the order of 10 nests) but have increased in effort in recent years (several hundred pairs re-nested in 1998–99). Second nestings were not included as part of this analysis.

The latitude and longitude (units measured in fractions of a degree rather than minutes and seconds) of each colony were multiplied by the number of nests within the colony to generate mean nest location for all Florida Bay nests. Standard deviations were also calculated for both longitude and latitude. These calculations were made for the first survey (1935) and at approximately 10-year intervals thereafter, except during the 1940s when no surveys were performed. No survey was made within a year of 1964-65 so both the 1962–63 and 1966–67 means were substituted. Likewise, no survey was made within a year of 1994–95 so both the 1990–91 and 1998–99 mean were substituted.

## Nesting Success

The most accurate estimates of spoonbill nesting success were collected by GVNP and RDB from 1987–88 through 1991–92. Reproductive data were collected from the three largest spoonbill colonies (Sandy, Tern, and Porjoe Keys) except during 1989–90 and 1990–91, when only Sandy and Tern Key colonies were monitored. Transects were established in each colony so that representative samples of nests could be monitored. Nests were individually marked with numbered plastic tags; each colony was visited every 4 to 6 days and the contents of each nest were recorded. Information on clutch size, duration of nesting, reproductive success, and cause of nest failure was collected for each nest. Estimates of reproductive success for each colony were calculated using the Mayfield (1975) method which prorates survival of repeatedly visited nests for the entire reproductive cycle and allows the sample to be extrapolated to estimate success of the entire colony. Survival rates were estimated through 21 days of age.

JJL used a variation of this mark/revisit technique between 1994 and 1999. Nests were marked at Tern, Sandy, and Duck Keys and revisited at approximately 14-day intervals. On the initial visit, between 25 and 50 nests were marked (unless the colony had fewer than 25 nests, in which case all nests were marked) and the number of eggs per nest counted. Nests were monitored until failure or until all surviving chicks reached >21 days of age. All chicks >21 days of age were considered a successful nesting attempt. Subsequent counts of flying young of the year at the colonies corroborated this method.

Young-of-the-year spoonbills are easily distinguishable from adults and become highly conspicuous from outside the colony during the fledging process (Allen, 1942). The total production of a colony can be estimated by counting the number of visible fledglings just prior to their leaving the colony. An estimate of nest success can be calculated by dividing the fledglings count by the number of nests within the colony. RPA, JCO et al., and JJL used this method to estimate nest success, and estimates were corroborated with the biologists' impressions of success of the colony from their field notes. Furthermore, each researcher supplemented their success estimates with additional data. RPA strengthened the estimate by making multiple counts of the fledglings from the time they first became conspicuous until the time they left the colony. The largest of these counts was then used as the estimate for colony production (Richard T. Paul, pers. comm., has validated this technique). JCO et al., would first count the number of visible fledglings from outside the colony and then would confirm the count upon entering the colony. Furthermore, they would make note of the number of chicks found dead within the colony, thereby supporting their estimate of nest success (also see Robertson et al., 1983). JJL used this method in 1998–99 while performing a baywide nesting survey. The results were supported by estimates made at nearby focal colonies using mark/revisit technique.

Nesting success estimates are reported for each region within the bay (from now on referred to as colony groups). Frequently, accurate nesting success estimates were collected at only one colony per colony group. In such cases, the individual colony estimate was used for the entire colony group. In cases where more than one colony per group had accurate nest success estimates, the total production of the combined colonies with estimates was divided by the total number of nests within those colonies to produce a success estimate for the colony group. Although accurate nest success rates were not made for all years, field notes frequently allowed for a qualitative assessment as to whether the colony succeeded (more than one chick per nest) or failed (less than one chick per nest). These were included as part of the analysis of nesting success.

## FORAGING HABITAT INFORMATION

Strong and Bancroft (1994) determined that almost 40% of the mangrove habitat on the mainline Keys between Lake Surprise and the southern end of Lower Matecumbe Key was filled by 1989. However, not all mangrove forest types (Lugo and Snedaker, 1974) are equally valuable as foraging habitats to spoonbills. The highly specialized feeding behavior requires that they feed in open, shallow-water habitats such as mangrove ponds and mudbanks (Allen, 1942) in water depths less than 20 cm (Powell, 1987). In reference to mangrove-forest types, dwarf mangrove forests provide large expanses of suitable foraging habitat. Along the mainline Keys this habitat type generally occurs as part of the transitional zone between Florida Bay and the hardwood hammocks that grow on the uplands (Strong and Bancroft, 1994). The amount of dwarf-mangrove forest, open-mangrove slough, and mangrove-lined ponds lost to development on the mainline Keys needed to be estimated so as to understand spoonbill foraging patterns on the Keys.

Photographic evidence was used to evaluate the loss of spoonbill foraging habitat in the upper Florida Keys (Figure 21.5). Low-level aerial photographs of the area from Lake Surprise on Key Largo through Long Key were used to quantify suitable spoonbill foraging habitat (i.e., open-canopy mangrove wetlands). Unfortunately, only a partial set of photographs was available from 1945 so these could not be used in this analysis. However, a comparison of the 1955 photographs to the partial set from 1945 revealed relatively little change in wetland habitat during the intervening decade. Based on this comparison, we felt that photographs taken in March 1955 portrayed the Upper Keys wetlands in a relatively pristine state, even though a small amount of wetlands had been filled (Figure 21.5A). Photographs from 1965 and 1975 were used to reflect the amount of spoonbill foraging habitat lost to dredge-and-fill development on a decadal scale (Figure 21.5B). All wetland areas that remained intact through present were visited to verify that habitat types were correctly identified from the aerial photographs. Image analysis software was used to compare the relative amount (percentage) of spoonbill foraging habitat left in 1964 and 1975 compared to the more pristine conditions of 1955. By visually inspecting sets of aerial photographs taken in 1955, 1959, 1964, 1971, and 1975, the location and approximate timing of fill projects were identified. Where possible, survey records and government documents were used to corroborate the timing of dredge-and-fill projects; however, government regulations were relatively lax prior to 1972, and accurate records for many projects were not available.

Seasonal rainfall and deviation from mean seasonal rainfall from 1963 to present were calculated from monthly rainfall data acquired from the South Florida Water Management District's Surface Water Conditions Detail Report. Monthly averages were from approximately 150 rainfall collection stations located from the Kissimmee region southward; therefore, these data reflect conditions for the entire southern Florida landscape.

## RESULTS AND DISCUSSION

### LOCATION OF FORAGING GROUNDS BY COLONY GROUP

The mean foraging flight distance by nesting spoonbills was calculated to be 12.4 ± 5.8 km (mean ± standard deviation) based on the 1989–90 following flights. The data were skewed toward shorter flights, with 83.3% of the flights less than 16 km; however, several flights greater than 30 km were recorded. Figure 21.6 shows the location of all spoonbill colonies observed in Florida Bay since 1950 and the mean foraging circumference (based on the 1989–90 estimate) around each colony. Bancroft et al. (1994) demonstrated that wading birds nesting in the Everglades selected colony locations based on proximity to primary foraging grounds. They argued that a given colony's primary foraging area would be located within the mean foraging flight radius at the time of colony formation (Bancroft et al., 1994). Based on these conclusions, Florida Bay's spoonbill colonies were divided into five geographical colony groups based on primary foraging ground location as predicted by the mean foraging flight radius. The 12 colonies within the southeastern colony group have a mean flight radius that includes southern Key Largo and Plantation Key (Figure 21.6A), areas that had extensive mangrove forests prior to urban development on those keys (Strong and Bancroft, 1994). Spoonbills from these colonies probably used these wetlands as primary foraging grounds. The four southwestern colonies encompass Lower Matecumbe Key, an area that was also rich in mangrove swamps prior to urban development and was likely the primary foraging area for birds nesting in this colony group (Figure 21.6B). Flight radii around the seven northeastern colonies predict that these birds would forage in the mainland mangroves from Taylor Slough to Manatee Bay (Figure 21.6B). The three colonies in the northwestern colony group would be predicted to forage among the extensive saltmarshes and mangrove swamps of Cape Sable (Figure 21.6C). The central colony group is represented by four colonies that fringe both eastern mainland and Plantation Key habitats within their mean foraging flight circumference (Figure 21.6C). Birds from these colonies probably would feed in a variety of habitats. The 1989–90 mean flight distance is used here as a tool to subdivide the colonies into functional groups and is not meant to indicate that the birds feed exclusively within the areas depicted in Figure 21.6. Indeed, the relatively large standard deviation (5.8 km) indicates that birds foraged well beyond the area indicated in Figure 21.6. However, the division into colony groups is justified based on the observation that wading birds nest in relative proximity to their primary foraging grounds (Bancroft et al., 1994).

The following of foraging spoonbills in fixed-wing aircraft in 1989–90 indicated that the initial compass heading of a bird leaving a colony and the final foraging location of that bird (compass heading from the colony of origin to the final foraging location) were strongly correlated ($r^2 = 0.78$). These results indicate that the direction of flight lines from a colony are good predictors of the direction of the final foraging location; therefore, flight-line observations can be used to support the predicted location of primary foraging grounds based on mean foraging flight radii (Table 21.3).

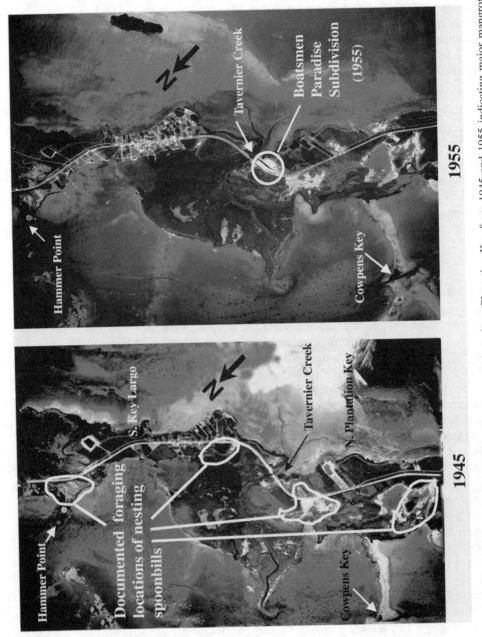

**FIGURE 21.5** (A) Aerial photographs of southern Key Largo and northern Plantation Key from 1945 and 1955 indicating major mangrove wetlands documented as spoonbill foraging locations.

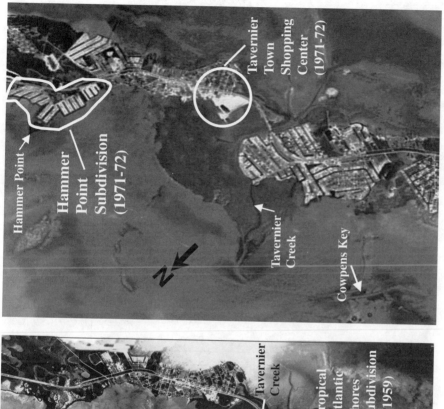

**1975**

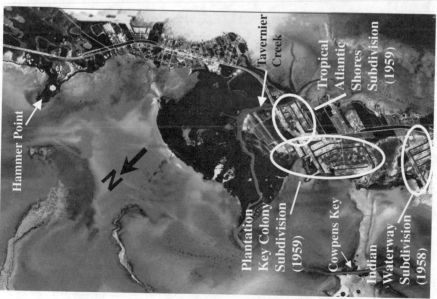

**1964**

(B) Aerial photographs of southern Key Largo and northern Plantation Key from 1964 and 1975 indicated the major dredge-and-fill subdivisions that destroyed wetlands known to be spoonbill foraging locations.

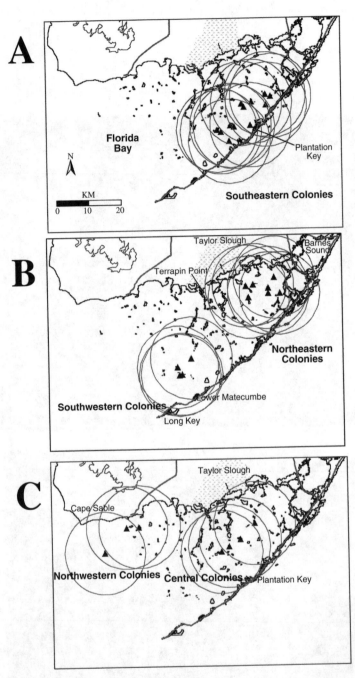

**FIGURE 21.6** Spoonbill colonies (triangles) identified by colony group (i.e., region). Rings around each spoonbill colony indicate the mean flight distance (12.4 km) of foraging spoonbills based on 1989 following flights. Flight distance suggests that colonies can be separated into colony groups based on likely foraging locations. (A) The southeastern colony group foraged on Plantation Key and adjacent areas. (B) The southwestern colony group likely foraged in the vicinity of Lower Matecumbe Key and the northeastern colony group foraged on the mainland between Taylor Slough and Barnes Sound. (C) The northwestern colony group was in proximity to foraging grounds on Cape Sable and the central colony group may have foraged widely within the bay, including both mainland and mainline Keys habitats.

**TABLE 21.3**
Results of Flight Line Surveys for the Southeast, Northeast, and Northwest Colony Groups (1957–1999)

| Year | Colony | Dates Observed | Hours Observed | Total Nests | Total Flights | Flight Direction (Percent Total Flights) | | | | | | | |
|---|---|---|---|---|---|---|---|---|---|---|---|---|---|
| | | | | | | N | NE | E | SE | S | SW | W | NW |
| **Southeast[a]** | | | | | | | | | | | | | |
| 1957–58 | Cowpens | 6 | 8.0 | 40 | 260 | 3 | 7 | 48 | 30 | 2 | 1 | 0 | 9 |
| 1957–58 | Stake | 1 | 1.0 | ? | 19 | 11 | 0 | 58 | 0 | 32 | 0 | 0 | 0 |
| 1958–59 | Bottle | 2 | 2.0 | 5 | 18 | 22 | 6 | 11 | 22 | 11 | 22 | 6 | 0 |
| 1958–59 | Cowpens | 6 | 9.3 | 75 | 660 | 7 | 24 | 52 | 5 | 2 | 3 | 3 | 3 |
| 1958–59 | Stake | 1 | 1.0 | 5 | 11 | 9 | 0 | 45 | 9 | 0 | 0 | 18 | 18 |
| 1960–61 | Cowpens | 6 | 8.8 | 70 | 804 | 17 | 36 | 25 | 8 | 3 | 1 | 8 | 2 |
| 1961–62 | Cowpens | 7 | 6.8 | 50 | 417 | 24 | 15 | 27 | 6 | 9 | 1 | 13 | 6 |
| 1962–63 | Cowpens | 8 | 8.0 | 75 | 678 | 10 | 20 | 22 | 9 | 0 | 5 | 23 | 12 |
| Total 5 | 3 | 37 | 44.9 | 320 | 2867 | 13 | 23 | 33 | 9 | 3 | 3 | 10 | 6 |
| **Northeast[b]** | | | | | | | | | | | | | |
| 1958–59[a] | N. Nest | 1 | 1.0 | 20 | 35 | 31 | 31 | 23 | 14 | 0 | 0 | 0 | 0 |
| 1973–74 | Tern | 1 | 1.2 | 200 | 203 | 0 | 90 | 5 | 0 | 0 | 0 | 5 | 0 |
| 1975–76 | Tern | 1 | 1.0 | 495 | 212 | 33 | 33 | 0 | 0 | 0 | 0 | 0 | 33 |
| 1976–77 | Porjoe | 1 | 1.0 | ? | 62 | 3 | 30 | 41 | 0 | 0 | 0 | 25 | 0 |
| 1976–77 | Tern | 2 | 2.0 | ? | 245 | 25 | 50 | 14 | 0 | 0 | 0 | 5 | 7 |
| 1977–78 | Porjoe | 3 | 3.3 | 64 | 101 | 27 | 42 | 20 | 5 | 0 | 0 | 4 | 3 |
| 1977–78 | Tern | 4 | 4.0 | 272 | 422 | 18 | 37 | 3 | 0 | 3 | 1 | 6 | 32 |
| 1978–79 | Porjoe | 4 | 8.0 | 64 | 241 | 22 | 43 | 17 | 7 | 0 | 3 | 2 | 5 |
| 1978–79 | Tern | 6 | 9.0 | 591 | 1545 | 12 | 58 | 8 | 0 | 1 | 1 | 7 | 14 |
| 1998–99[c] | Tern | 2 | 1.5 | 60 | 62 | 0 | 0 | 0 | 0 | 0 | 0 | 0 | 100 |
| Total 7 | 3 | 25 | 32.0 | 1766 | 3129 | 15 | 51 | 11 | 1 | 1 | 1 | 6 | 14 |
| **Northwest[b]** | | | | | | | | | | | | | |
| 1977–78 | Frank | 1 | 1.0 | 36 | 12 | 33 | 0 | 0 | 0 | 0 | 0 | 0 | 67 |
| 1977–78 | Oyster | 1 | 1.0 | 56 | 6 | 52 | 30 | 0 | 0 | 0 | 0 | 0 | 18 |
| 1978–79 | Frank | 2 | 4.5 | 62 | 49 | 38 | 13 | 2 | 0 | 0 | 0 | 4 | 42 |
| 1978–79 | Oyster | 4 | 7.8 | 64 | 213 | 21 | 19 | 5 | 0 | 0 | 1 | 3 | 50 |
| 1978–79 | Sandy | 1 | 2.0 | 80 | 66 | 41 | 33 | 0 | 0 | 0 | 0 | 0 | 26 |
| Total 2 | 3 | 9 | 16.3 | 298 | 345 | 28 | 21 | 3 | 0 | 0 | 1 | 2 | 44 |

a Allen (field notes).
b Robertson (field notes) and Ogden (field notes), unless otherwise noted.
c Lorenz (field notes).

Flight-line surveys taken from 1957 to 1963 at southeastern colonies indicated that birds flew predominantly east and northeast toward Plantation Key and southern Key Largo as was predicted (Table 21.3). More than 55% of the almost 3000 flights observed from the southeastern colonies between 1957 and 1963 were directed towards Plantation Key and southern Key Largo (Table 21.3). This was particularly evident prior to 1960 (Table 21.3). Field notes from several researchers confirmed that Roseate Spoonbills were foraging heavily within the wetlands on southern Key Largo and Plantation Keys during this period (Figure 21.5A; Robert P. Allen, field notes; Jack Watson, field notes; Alexander Sprunt IV, pers. comm.). Collectively, these observations indicate that birds nesting in the southeast region heavily depended on mainline Keys wetland habitat as principal feeding areas.

Most of the flight-line observations from northeastern colonies were made between 1973 and 1979. The results strongly indicate that these colonies focused foraging effort toward the mainland wetlands north of eastern Florida Bay (Table 21.3). Of more than 3000 observed flights, 80% were toward the mainland. Following flights made in 1989–90 confirmed that birds from the northeastern colonies predominantly foraged in the mangrove wetlands north of the bay (Figure 21.1).

Although the number of flight-line counts in the northwestern colony group was relatively small compared to the northeastern and southeastern groups, the results strongly indicate that birds from these colonies foraged on Cape Sable (Table 21.3). From 1977–79, 94% of the 345 flights observed were directed toward the Cape. The 1989–90 following flights confirmed that the wetlands of Cape Sable were the primary foraging ground for these colonies (Figure 21.1).

No flight line counts were made for colonies in the central colony group; however, 11 of the 1989–90 following flights originated at South Park Key in the central colony group. Six of the followed birds foraged on the mainland north of the colony. Five others foraged on bay keys (Figure 21.1). Although limited, these results suggest that birds from central colonies utilize both the mainland wetlands and wetlands on bay keys for foraging.

No flight line nor following flight data were collected for colonies in the southwest colony group prior to 1975; however, field notes (William B. Robertson, 1978; JCO, pers. obs.; Alexander Sprunt IV, pers. comm.) reported that foraging spoonbills utilized the wetlands on Lower Matecumbe and Long keys during the period when the southwestern colonies were most active (1966–1971; Figure 21.7A). These reports support our prediction based on mean foraging radii that spoonbills in the southwestern colonies foraged principally in wetlands on Lower Matecumbe Key (Figure 21.6B).

## TEMPORAL CHANGES IN RELATIVE NESTING EFFORT BY COLONY GROUP

Between 1935 and 1940, Roseate Spoonbills nested exclusively in the southeastern colony group, principally on Bottle Key (Table 21.1). When surveys were resumed in 1950, spoonbills had expanded their nesting range throughout the southeastern colony group and into the northeastern colony group (Table 21.1, Figure 21.7A). Through the 1950s, 70 to 90% of the nesting activity occurred in the southeastern colonies, with the remaining 10 to 30% in the northeast (Figure 21.7A). Between 1963–64 and 1968–69, the number of spoonbills nesting in the southeast colony group dropped to about 10%. Since that time, southeastern colonies have only accounted for 2 to 18% of the nests found in Florida Bay.

The decline in the southeast colonies was concurrent with increases in nesting effort in both the southwestern and northeastern colonies (Figure 21.7A). The southwestern colonies peaked at about 30 to 40% of nesting effort in the late 1960s and early 1970s, after which they declined to between 8 and 12% from mid-1970s through mid-1980s. Since 1984–85, the southwestern colonies have made up less than 5% of the total nesting effort in Florida Bay (Figure 21.7A).

The northeastern colonies included about 60% of the nesting effort by 1968–69 and remained at this level through 1984–85 (Figure 21.7A). After the 1984–85 nesting season, the number of nests in northeastern colonies began to decline, and by 1991–92 less than 40% of the nests were located in the northeastern colony group. This decline continued through the 1990s, and by the 1998–99 survey only 22% of the nests were found in the northeastern bay.

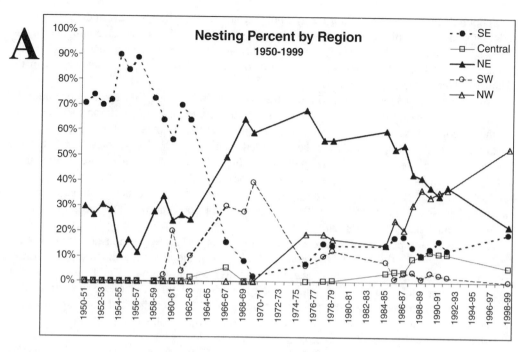

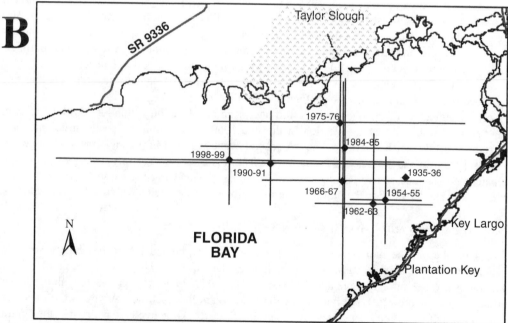

**FIGURE 21.7** (A) Percent of total spoonbill nests found in each colony group between 1950 and 1999. (B) Mean nest location within Florida Bay. Diamond indicates mean, and error bars indicate the standard deviation on both the longitudinal and latitudinal axes. Intervals are approximately 10 years. In 1935–36, all nests were located at the Bottle Key colony with no variance in nesting location. No surveys were performed in the 1940s. No survey was made within a year of 1964–65 so both the 1962–63 and 1966–67 means were plotted. No survey was made within a year of 1994–95 so both the 1990–91 and 1998–99 mean were plotted.

Spoonbills first nested in the northwest colony group in 1974. Between the mid-1970s and the mid-1980s about 20% of the nests occurred there (Figure 21.7A). The percentage in the northwest increased to about 40% by the early 1990s, roughly equal to the number of nests found in the northeast at that time. By 1998–99, the majority of spoonbill nests (53%) were located in the northwestern colony group (Figure 21.7A).

Low numbers of spoonbills nested intermittently at the four central bay colonies beginning in the early 1960s. During the 1980s, spoonbills began to nest more consistently on these colonies and they accounted for more than 10% of the nests from 1987–88 to 1991–92.

Three general nesting periods can be identified by the examination of mean (and standard deviation) nest location through time (Figure 21.7B). From 1935 to about 1963, Roseate Spoonbills nested predominantly in the southeastern part of the bay as indicated by the relative compact standard deviations (Figure 21.7B). During this period the general direction in range expansion was toward the southwest; however, in 1962–63, range expansion occurred to the north as indicated by the latitudinal standard deviation.

A relatively abrupt change in the direction of nesting range expansion occurred between 1962–63 and 1966–67 (Figure 21.7B), indicating the start of the second nesting period. Between 1966–67 and 1984–85, the mean nest location was in the northeastern bay. The standard deviation around mean latitude became more compact, indicating the pronounced decrease in nesting effort in the two southern colony groups (Figure 21.2; Figure 21.7). Longitudinal standard deviation increased during this period owing to the colonization in the northwestern colonies and to the subsequent increase in effort at those colonies (Figure 21.7A).

The third nesting period (about 1984–85 through 1998–99) was characterized by the increase in nesting in the northwestern colony group and the decline in the northeastern effort. Figure 21.7 shows that by the mid-1980s spoonbills began to select nesting locations in the northwest colonies over those in the northeast. This westward movement was apparent by 1990–91 and continued through 1998–99 (Figure 21.7B). Low variance in the latitudinal direction indicates that effort in the southern bay was relatively low while the wide east–west variation indicates significant nesting still occurred in the northeastern bay (Figure 21.7B).

The examination of relative nesting effort and mean nest location within Florida Bay indicates that spoonbills shifted nesting location in the mid-1960s away from the southern colonies and moved toward the northeastern colonies. A second event occurred in the mid-1980s when preferences shifted from the northeast to the northwest. The changes were relatively swift, having become apparent in just four or five nesting cycles in both cases. Knowing the causes of these shifts is paramount to understanding the nesting patterns exhibited by spoonbills within individual colonies.

## NESTING PATTERNS AND EXTERNAL INFLUENCES BY COLONY GROUP

Two anomalies in the number of nesting pairs must be addressed before proceeding with the identification of potential causal agents leading to the observed changes in spoonbill nesting patterns. The observation of more than 1200 nests in the 1978–79 nesting season was unprecedented in the record (Figure 21.2). All but two of the other years in which nest surveys were done had fewer than half of the number of nests counted in 1978–79. All active colony groups (the central colony group was not active at the time) showed sharp increases in the number of nests indicating that the increase was a baywide phenomena (Figure 21.2). A similar situation occurred during the 1962–63 nesting season, although the magnitude of the increase was much smaller than that of 1978–79. In 1962–63, the total number of nests reached an unprecedented high to that point in time and the nest numbers nearly doubled in all active colony groups over the previous nesting year (Figure 21.2, Table 21.1).

Allen (1942) proposed that spoonbills from Caribbean islands (most likely Cuba) periodically immigrate to Florida Bay, thereby providing a possible explanation for the sharp increases in the spoonbill population between 1961–62 and 1962–63 and again between 1977–78 and 1978–79.

The anecdotal evidence presented by Allen (1942) and his subsequent field notes indicate that such immigrations might be episodic. Robertson et al. (1983) indicated that, although some communication between Cuba and Florida Bay probably occurs, the frequency of such migrations are unknown and are probably not part of an annual migration pattern. Allen's notes indicate that an immigration episode occurred following the 1961–62 breeding cycle (February 1962). Three flocks totaling more than 250 Roseate Spoonbills (many, if not all, of them adults) were observed arriving in the Key Largo area from offshore. One of the flocks was observed over the Gulf Stream heading toward southern Key Largo. The following year, the nesting population of Florida Bay increased by almost 400 birds (200 nests). Although no such observations were reported prior to the 1978–79 nesting season, it is possible that an influx of adult spoonbills from the Caribbean could account for the observed increase.

Subsequent results and discussion focus on trends within the Florida Bay population by colony group. Because the baywide increases in the spoonbill nesting population in 1962–63 and 1978–79 obfuscate trends on a regional level, these two years have been removed from Figure 21.8 and will not be addressed further.

## Southeastern and Southwestern Colony Groups

The record of nesting in the southeastern colonies indicates that the number of nests increased exponentially from the original 15 nests on one key (Bottle Key) in 1935–36 to a peak of almost 200 nests in 1955–56 (Table 21.1, Figure 21.8). This period of growth was followed by an almost complete collapse of the southeastern nesting colonies to a low of five nests in 1969–70 (Figure 21.8). The colonies recovered during the 1970s, and by the mid-1980s they were supporting between 75 and 100 nests annually (Figure 21.8).

At the time of the 1960s collapse, Cowpens Key (Figure 21.5) was the largest spoonbill colony in Florida Bay (Table 21.1) and, as such, was the subject of frequent monitoring. Records indicate that Cowpens supported 50 to 75 nests annually between the years 1955–56 and 1961–62 (Figure 21.3). The next survey (performed in 1966–67) found the colony deserted, and subsequent surveys revealed only low level nesting efforts (Figure 21.3). Flight-line counts taken during the peak nesting years at Cowpens indicated a dramatic change in foraging patterns by breeding spoonbills (Figure 21.9; Allen, field notes). In 1957–58, almost all foraging flights were directed toward the east and southeast (Figure 21.9). Large numbers of spoonbills in breeding plumage were observed foraging in two large wetlands on the northern end of Plantation Key less than 2 km to the east and southeast of the Cowpens colony (Figure 21.5). These observation strongly indicate that spoonbills nesting at the Cowpens colony were primarily foraging in these wetlands.

Between, the 1957–58 and 1958–59 breeding cycles, the wetland to the southeast was filled for urban development (Indian Waterways subdivision; Figure 21.5C, Table 21.2). Flight-line counts from 1958–59 indicated that spoonbills no longer directed flights toward the southeast, presumably because there were no longer any substantive foraging grounds in that direction (Figure 21.9). Nesting spoonbills appeared to respond to this perturbation by redirecting flight intensity toward the east and northeast (Figure 21.9). Presumably, these birds were foraging in the wetland due east of Cowpens and in the large wetland near Hammer Point on southern Key Largo approximately 5 km to the northeast (Figure 21.5). Again, Allen's notes and the recollections of Alexander Sprunt IV (pers. comm.) indicate that these locations were heavily frequented by spoonbills during this approximate time period (Table 21.2). No observations of the Cowpens colony are available for 1959–60, but by 1960–61 most of the large wetland to the east of the colony had also been filled for urban development (Plantation Key Colony and Tropical Atlantic Shores subdivisions; Figure 21.5C, Table 21.2) with the predictable result of greatly reduced foraging effort in that direction (Figure 21.9).

Prior to the filling of these two wetlands on Plantation Key, spoonbills rarely flew from the colony toward the north, west, or south (Figure 21.9). Following the destruction of feeding sites, spoonbill flight directions from Cowpens moved increasingly in these directions and by the 1961–62 and 1962–63 breeding seasons, there was no apparent preference in flight direction from the colony (Figure 21.9). Sometime between the 1962–63 and the 1966–67 breeding cycles, Cowpens Key was abandoned by nesting spoonbills (Figure 21.3). Similarly, the number of nests in the southeastern colonies declined sharply following urban development in the late 1950s (Table 21.2; Figure 21.8).

Roseate Spoonbills first initiated colonies in the southwest colony group in the early 1960s; however, large numbers were not reported until 1966–67 (Table 21.1, Figure 21.8). Between 1966–67 and 1970–71, the southwestern colonies supported approximately 100 nests, or about 35% of Florida Bay's nesting spoonbill population (Figure 21.8). This increase coincided with the decline in the southeast colonies (Figure 21.8), suggesting that birds that abandoned colonies in the southeastern colony group may have relocated in the southwestern colonies. Between 1970–71 and 1975–76 the number of nests in the southwest declined by about 50%. The downward trend has persisted at a more gradual rate during the remaining period of record (Figure 21.8).

The southwestern colonies were the least studied of the five colony groups. No information regarding flight lines or foraging locations was collected prior to 1975. However, anecdotal evidence suggests that the birds nesting in this region foraged in the expansive wetlands that existed on Lower Matecumbe and Long Keys prior to 1972 (Table 21.2). Large portions of these wetlands were filled in the early 1970s (Table 21.2). Based on the observed impact of dredge-and-fill projects on the Cowpens Key colony, it is likely that the filling of approximately 134 acres of wetlands on Lower Matecumbe Key in 1972 (Port Antigua subdivision; Table 21.2) resulted in the observed rapid decline of nests in the southwestern colonies between 1971 and 1976.

Between 1950 and 1972, spoonbills nesting in the southern colony groups averaged more than two offspring per nest and succeeded in all of the 17 years that data were collected (Table 21.4). Following dredge-and-fill development, spoonbills average 0.22 young per nest and had successful nestings in only 2 of 10 years that nesting success data were collected for the southeastern colonies and 2 of 8 years for the southwestern colonies (Table 21.4). Observations from the southern colony groups indicate that development in the Keys resulted in loss of primary foraging habitat. As a result, nesting effort in individual colonies apparently declined in relation to amount of nearby primary foraging habitat lost to development. Spoonbills may have responded by switching to nesting sites in other regions. Although a 1972 law stopped large-scale dredge-and-fill projects in the Keys (Lott et al., 1996), developers rushed to fill massive amounts of wetlands in the early 1970s so as to circumvent the intent of the law (Table 21.2). During this push, the last known important spoonbill foraging habitats were filled on the mainline Keys. Breeding spoonbills in the southern part of the bay no longer had access to prime foraging grounds and were probably no longer able to efficiently gather enough food for young. As a result, the remnant nesting population in these colonies has been less successful in fledging young since 1972 (Table 21.4). Prior to the final development push, spoonbills in both colony groups were still producing young at the highest rates observed in Florida Bay for the period of record.

The relative lack of hurricanes between 1950 and 1999 may have also played a role in the declining use of the southern colonies. Moderate to large hurricanes affect the upper Keys and Florida Bay at approximately 25-year intervals (Gentry, 1974). These events destroy mangroves and open up expansive areas to spoonbills for foraging. Over time, mangroves fill these areas back in (Bjork and Powell, 1990). The lack of a major hurricane impact on the upper Keys and Florida Bay since 1965 has resulted in loss of spoonbill foraging habitat and may have played a role in the decline of spoonbills in the southern bay (Bjork and Powell, 1990). However, the impact of hurricanes on spoonbill foraging is sporadic and probably minor compared to the impact of urban development. That hurricane impacts were minor compared to urban development was indicated by the fact that spoonbills did not recolonized the Cowpens colony following Hurricane Betsy in 1965, even though Betsy substantially opened the mangrove canopy in the Upper Keys and eastern Florida Bay.

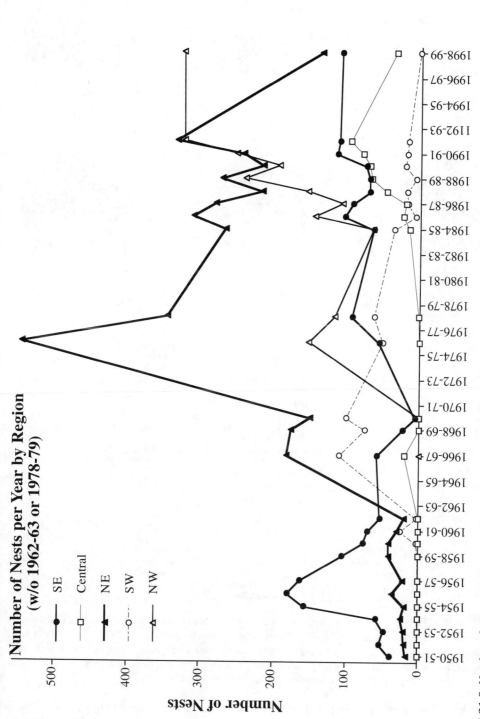

**FIGURE 21.8** Number of nests per year by colony group from 1950 to 1999. The 1962–63 and 1978–79 nesting years have been removed so that regional trends are more visually clear (see text for justification).

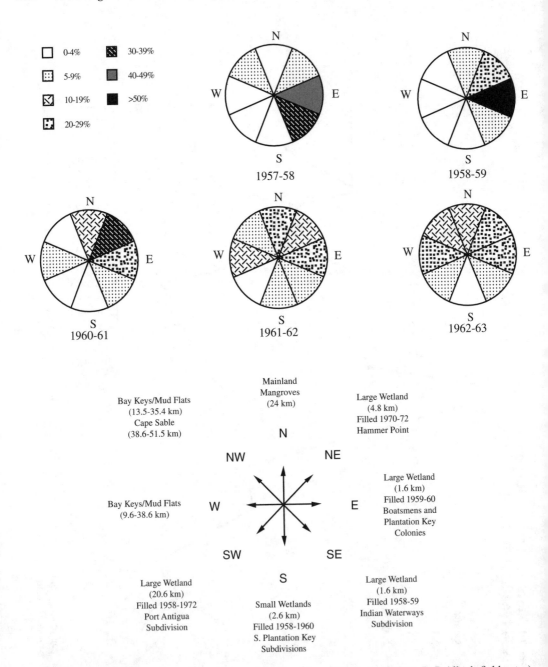

**FIGURE 21.9** Flight line surveys for Cowpens Key colony from 1957 to 1963 (from R. P. Allen's field notes). Pie charts summarize the percentage of flights toward eight compass headings (see Table 21.3 for details about flight-line surveys). Rosetta indicates nearest foraging location and approximate distance to that location in parentheses; also indicates timing of dredge-and-fill urban development at specific locations on the mainline Keys (see Table 21.2 for specifics about dredge-and-fill developments).

Sea-level rise has changed the topography of many of the bay keys. For example; in 1935 when Spoonbills initially recolonized Florida Bay, Bottle Key (site of the founding colony) had internal lagoons that were largely covered by a canopy of scrubby mangrove (Alexander Sprunt IV, photograph) reminiscent of the current colony site at Tern Key. These mangroves were destroyed in Hurricane Donna and did not recover. Wanless et al. (1994) demonstrated that loss of wetland

**TABLE 21.4**
**Estimated Annual Roseate Spoonbill Production
(Number of Chicks per Nest) for Southeastern (SE)
and Southwestern (SW) Nesting Subregions Before
and After Major Development of the Upper Florida Keys**

| Year | Estimated Chicks per Nest | |
|------|------|------|
| | SE | SW |
| **Pre-Development** | | |
| 1950–51 | 2.7 | |
| 1951–52 | 2.5 | |
| 1952–53 | 1.9 | |
| 1953–54 | 2.0 | |
| 1954–55 | 1.7 | |
| 1955–56 | 2.5 | |
| 1956–57 | 1.0 | |
| 1957–58 | 3.0 | |
| 1958–59 | 2.7 | |
| 1959–60 | 2.5 | |
| 1960–61 | 2.5 | |
| 1961–62 | 2.0 | |
| 1962–63 | 2.0 | |
| 1964–65 | | S |
| 1965–66 | S | |
| 1966–67 | 2.6 | 2.1 |
| 1968–69 | | 2.2 |
| Mean | 2.25 | 2.15 |
| % success | 100 | 100 |
| **Post-Development** | | |
| 1974–75 | S | S |
| 1975–76 | | F |
| 1977–78 | F | 0.21 |
| 1978–79 | 0.3 | 0.24 |
| 1979–80 | 0 | |
| 1981–82 | S | S |
| 1982–83 | F | F |
| 1983–84 | F | F |
| 1986–87 | F | F |
| 1998–99 | 0.4 | |
| Mean | 0.22 | 0.23 |
| % success | 22 | 25 |

*Note:* For some years, numerical estimates were not calculated but observations indicated success or failure in nesting. In such cases, S indicates successful nesting (greater than 1 chick per nest) and F indicates failed nesting (less than 1 chick per nest).

habitat to sea-level rise is manifested during catastrophic events. Similar changes in habitat to that of Bottle Key were demonstrated to have occurred on Cape Sable when Hurricane Donna passed over the area (Wanless et al., 1994). The change in habitat on the bay keys as a result of sea-level rise as manifested during Hurricane Donna in 1960 and Hurricane Betsy in 1965 could have resulted in the destruction of important nesting and foraging sites.

## Northeastern Colony Group

Roseate Spoonbills began nesting in the northeastern colony group in the early 1940s (Table 21.1). From 1950 to 1960, the number of nests gradually increased (Figure 21.8), accounting for as much as 30% of the total nests in Florida Bay (Figure 21.7A); however, the total number of nests never exceeded 50 (Table 21.1; Figure 21.8). Between the 1961–62 and the 1966–67 nesting seasons, the number of nests in the northeast increased to about 200 (Figure 21.8). This increase roughly coincided with the decline in the southeast colonies (Figure 21.8), suggesting that some of the birds that abandoned colonies in the southeastern colony group relocated to the northeastern colonies. The number of nests increased to a peak in the late 1970s (Figure 21.8; Table 21.1).

No data were collected in the early 1980s; however, by 1984–85 the number of nests in the northeast had noticeably declined (Figure 21.8). With the exception of the 1992–93 nesting season, the number of nests continued a decline from 1984–85 to 1998–99 (Figure 21.8). Although complete colony surveys were not made between 1992–93 and 1998–99, nest counts at the two largest colonies in the northeastern colony group (Tern and Porjoe Keys) provide evidence of a regional decline in nesting during the intervening years (Figure 21.3; Table 21.1).

Following flights (Figure 21.1), flight-line counts (Table 21.3) and tracking of radio-tagged spoonbills (Bjork and Powell, 1994) demonstrated that spoonbills nesting in the northeastern colonies foraged almost entirely in the coastal wetlands associated with Taylor Slough on the mainland north and northeast of eastern Florida Bay. Johnson and Fennema (1989) examined data from water-level recorders in the Taylor Slough watershed from 1956 to 1988. They found that water levels in the headwaters region of Taylor Slough were correlated with the amount of freshwater reaching Florida Bay. By including intra-annual variation in rainfall in their analyses they showed that water management practices had profoundly impacted the Taylor Slough/northeastern Florida Bay watershed. Subsequent studies strongly supported this conclusion (Van Lent et al., 1993; McIvor et al., 1994; Meeder et al., 1996; Lorenz, 2000; Ross et al., 2000).

Johnson and Fennema (1989) identified three distinct periods of water management that affected this region. During period I (1956–1964), the impact of water management on the wetlands was relatively small. Period II (1970–82) was characterized by the completion of the C-111 and associated canals and operation of those canals under the Minimum Schedule of Water Deliveries. Water management during period II resulted in much lower wet season water levels and slightly lower dry season water levels in the headwaters of Taylor Slough (Figure 21.10). The above-mentioned correlation between water levels in the Taylor Slough headwaters and freshwater flows toward Florida Bay indicates that much less freshwater was reaching the coastal areas during the wet season and that dry seasons may have been relatively drier (e.g., lower water levels, shorter hydroperiods) during period II. Period III (1983–1988) was characterized by the completion of the SDCS and its operation under the Experimental Program of Water Deliveries. During period III, wet season water levels remained about the same as period II (i.e., lower than the relatively unaltered condition of period I); however, dry season water levels were significantly higher than in either of the two preceding periods (Figure 21.10). This indicates that during the dry season, period III water management practices artificially increased water levels in the coastal wetlands, thereby reducing intra-annual variation in water depth (Figure 21.10; Johnson and Fennema, 1989).

To examine the effect of water management on spoonbill nesting patterns, we included the years 1964 through 1969 as part of period II. Johnson and Fennema (1989) did not include these years in period II because the C-111 canal was being constructed during this period and therefore the impact of the canal was not fully realized. However, they state that water levels were artificially lowered during construction of the canal system, thereby impacting the coastal wetlands and justifying the inclusion of these years in period II (i.e., period II is defined from 1964 to 1982). Likewise, the SDCS and the Experimental Program of Water Deliveries has remained largely unchanged in structure and operation since 1988 (Ley et al., 1995; U.S. Army Corps of Engineers,

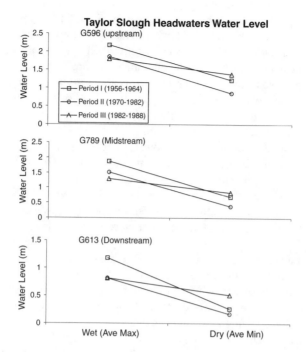

**FIGURE 21.10** Changes in water level in the Taylor Slough headwaters. Period I was prior to significant water management impacts in the Taylor Slough region; period II, completion of C-111 and associated canals and Minimum Schedule of Water Deliveries management regime; period III, completion of the South Dade Conveyance System (1982) and Experimental Program of Water Deliveries management regime (1984). (Adapted from Johnson and Fennema, 1989.)

1995); therefore, Johnson and Fennema's (1989) period III will be extended to include 1989 through 1999 (i.e., period III is defined from 1982 to 1999).

During period I, between 11 and 34% of Florida Bay's Roseate Spoonbill population nested in the northeastern colonies (Figure 21.7A) and foraged in the mainland coastal wetlands (Table 21.3). Only two estimates of nest production were made at northeastern colonies during period I. In 1955–56, the north Nest Key colony was estimated to have produced 2.3 chicks per active nest, and in 1961–62 the Porjoe colony was estimated to have produced 3.0 chicks per active nest.

Period II corresponds to an era of rapid growth in the number of Roseate Spoonbills nesting in the northeastern colonies (Figure 21.8). Initially, period II had high success rates and colonies succeeded in all years surveyed between 1964 and 1975 (Table 21.5); however, between 1975 and 1982, 4 out of 5 years surveyed resulted in nest failure (Table 21.5). All years that failed had lower than average rainfall during the wet season, while 8 of 10 successful years had above-average rainfall during the wet season (Table 21.5). Overall, the percentage of successful years was relatively high (71%) during period II, but the estimated production rates were much smaller than those found in the southeastern colonies during the 1960s (1.38 vs. 2.25 mean chicks per nest; Tables 21.4 and 21.5).

At spoonbill foraging grounds in northeastern Florida Bay, water management practices would have resulted in greatly reduced freshwater flow in the wet season during periods II and III (Johnson and Fennema, 1989; McIvor et al., 1994). Several studies (McIvor et al., 1994; Meeder et al., 1996; Ross et al., this volume) indicated that salinity in the coastal wetlands increased and became more variable during this period, supporting the findings of Johnson and Fennema (1989). Lorenz (1999, 2000) found that higher and more variable salinity in the same coastal wetlands resulted in lower prey-base biomass for foraging spoonbills. All 4 years that spoonbill nesting failed during period II had lower than average rainfall during the wet season (Table 21.5), possibly exacerbating the impacts of water management practices and further reducing the prey production through increased variability in salinity. Likewise, 8 of the 10 successful years during this period occurred when wet season rainfall was above average, possibly ameliorating the impact of water management resulting in relatively higher prey abundance. During these 8 years, dry-season rainfall ranged from 62% above average to 48% below average indicating that dry-season rainfall was not likely a factor in determining success so long as sufficient rainfall occurred during the wet season.

During period II, artificially lower dry-season water levels in the Taylor Slough headwaters may have resulted in shorter hydroperiods and lower water levels in the coastal wetlands (Johnson and Fennema, 1989; Lorenz, 2000). In turn, this may have enhanced the concentration effect on prey, making them more available to predators (Kushlan, 1978; DeAngelis et al., 1997); that is, the wetlands as a whole may have suffered a reduction in secondary production due to increased salinity variability, but the fish that were produced were presumably more available to predators such as spoonbills. This may help explain the relatively high success rates during period II relative to period III (Table 21.5).

Both years that succeeded with lower than average rainfall during the wet season also experienced drought conditions during the dry season (Table 21.5, 1964–65 and 1981–82). During drought years, low rainfall in the wet season would likely reduce the swings in salinity, making the ecotone higher in mean salinity but lower in variance. Even though mean salinity would be higher, the more constant salinity environment might be more favorable to primary and secondary production as compared to a fluctuating salinity environment (Montague and Ley, 1993; Ley et al., 1994; Lorenz, 1999). Furthermore, drought conditions would result in extreme low water events during the dry season thereby enhancing the prey concentration effect (DeAngelis et al., 1997; Lorenz, 2000). Cumulatively, these drought conditions would likely result in a more abundant and more available prey base for nesting spoonbills when compared to years with low water levels during the wet season and high water levels during the dry season. As previously stated, the best conditions for nesting spoonbills presumably would be above-average water levels in the wet season followed by below-average water levels during the dry. Nest success estimates from period II indicate that the worst conditions for nesting spoonbills would be the opposite: low-water wet seasons and high-water dry seasons. Drought conditions for the entire annual cycle appear to be intermediate.

Water management during Period III caused the mainland coastal wetlands to lose much of the seasonal nature of the hydrologic cycle (Figure 21.10; Johnson and Fennema, 1989), resulting in low-water wet seasons followed by high-water dry seasons (i.e., the worst case conditions for nesting spoonbills as per period II results). These conditions would likely result in intermittent periods of flood and drydown throughout the year, making prey concentration events less predictable (Bjork and Powell, 1994; Lorenz, 2000). Furthermore, these conditions would also allow bay waters to encroach on the mainland coastal wetlands more frequently, resulting in higher mean salinity overall and more pronounced pulses in salinity (Van Lent,1993; McIvor, 1994). Increased freshwater input during the dry season would do little to alleviate the artificially high salinity, as bay levels would be naturally low (Powell, 1987), thus allowing the freshwater to run off quickly and mix with marine waters (Baratta and Fennema, 1994). Mixing would be exacerbated by the frequent cold fronts that impact the bay during the dry season (Holmquist et al., 1989; Baratta and Fennema, 1994) and would increase variability in salinity. Higher means and greater fluxes in salinity would likely result in lower primary and secondary production (Montague and Ley, 1993; Lorenz, 1999, 2000) in the coastal wetlands during period III compared to periods I or II. The decline in prey production through changes in the salinity regime combined with the unpredictability in prey availability through reversals in dry season drydowns would make it more difficult for spoonbills to forage successfully on a consistent basis.

These conditions predictably resulted in low nesting success (Table 21.5) and colony abandonment (Figure 21.8) in the northeastern colonies during period III. Nesting failed in most of the years recorded (only 36% of the years successful) while the estimated production was only 0.67

**TABLE 21.5**
**Estimated Annual Roseate Spoonbill Production (Number of Chicks per Nest) for the Northeastern (NE) and Northwestern (NW) Subregions Before and After the Major Structural and Operational Changes in the South Dade Conveyance System that Occurred in 1982**

| Year | Estimated Chicks Per Nest | | % Deviation from Mean Rainfall | |
| | NE | NW | Wet | Dry |
|---|---|---|---|---|
| | **Period II** | | | |
| 1964–65 | S | | –0.03 | –0.27 |
| 1965–66 | S | | **0.07** | 0.07 |
| 1966–67 | **1.8** | | **0.13** | –0.48 |
| 1967–68 | **1.4** | | **0.08** | –0.03 |
| 1968–69 | **2.2** | | **0.27** | 0.07 |
| 1969–70 | **2.0** | | **0.24** | 0.62 |
| 1971–72 | **2.0** | | **0.05** | 0.13 |
| 1973–74 | **1.6** | S | **0.02** | –0.34 |
| 1974–75 | **2.2** | S | **0.12** | –0.25 |
| 1975–76 | 0.8 | 0.5 | –0.02 | –0.07 |
| 1977–78 | 0 | F | –0.04 | 0.12 |
| 1978–79 | 0.6 | 0.5 | –0.05 | 0.27 |
| 1979–80 | 0.5 | | –0.05 | 0.15 |
| 1981–82 | S | S | –0.21 | –0.44 |
| Mean | 1.38 | 0.48 | 0.04 | –0.03 |
| % success | 71 | 50 | | |
| | **Period III** | | | |
| 1982–83 | 0.0 | S | 0.14 | 0.42 |
| 1983–84 | 0.5 | | 0.05 | 0.32 |
| 1986–87 | 0.0 | 0.0 | –0.02 | 0.14 |
| 1987–88 | **1.2** | 2.5 | **–0.02** | **–0.21** |
| 1988–89 | **1.9** | 1.9 | **–0.16** | **–0.29** |
| 1989–90 | **2.2** | 2.4 | **–0.16** | **–0.18** |
| 1990–91 | **1.8** | 2.1 | **–0.10** | **0.40** |
| 1991–92 | **1.3** | 2.0 | **–0.01** | **–0.29** |
| 1992–93 | 0.0 | 0.0 | 0.16 | 0.22 |
| 1994–95 | 0.0 | 1.6 | 0.21 | 0.16 |
| 1995–96 | 0.1 | 0.1 | 0.33 | 0.08 |
| 1996–97 | 0.2 | 0.4 | –0.15 | 0.03 |
| 1997–98 | 0.0 | 0.4 | 0.00 | 0.48 |
| 1998–99 | 0.3 | 1.4 | 0.02 | –0.38 |
| Mean | 0.67 | 1.24 | 0.02 | 0.06 |
| % success | 36 | 62 | | |

*Note:* For some years, numerical estimates were not calculated but observations indicated success or failure in nesting. In such cases, S indicates successful nesting (greater than 1 chick per nest) and F indicates failed nesting (less than 1 chick per nest).

chicks per active nest (Table 21.5). The only years with successful nesting were 5 consecutive years during a prolonged drought (1987–1992), as would also have been predictable from period II results. The apparent exception during this drought period was in 1990–91 when most of the above-average dry season rainfall occurred in April and May, after the nesting cycle was completed.

## Central Colony Group

Significant nesting did not take place at the central colonies until the mid-1980s, although some nesting occurred in the 1960s, (Table 21.1, Figure 21.7). The increase in nesting effort in the central colonies coincided with the decline in nesting effort in the northeast, possibly indicating that birds relocated to this colony group. To date, little attention has been paid the central colonies, and estimates of nest success are sparse. In 1966–67, nests at Manatee Key produced 2.3 chicks per nest. In 1978–79, the central colonies failed entirely, and in 1998–99, they produced an estimated 0.4 chicks per nest.

## Northwestern Colony Group

Roseate Spoonbills began nesting in the northwestern colony group in the early 1970s (Table 21.1) concurrent with the decline of the southwestern colonies (Figure 21.8). Following the destruction of wetlands on Lower Matecumbe Key in 1972 (Table 21.2), the direction of foraging flights from the Buchanan Keys was predominantly toward Cape Sable (JCO, pers. obs.). Robertson (1978) indicated that the majority of spoonbills nesting in the southwestern bay foraged on Cape Sable although they continued to supplement their diet by feeding on Lower Matecumbe and Long Keys. These observations suggest that destruction of wetlands on the mainline Keys resulted in a redirection in foraging effort that ultimately led to a change in nesting location from the southwest to the northwest in the 1970s.

Nesting effort in the northwest began an exponential increase starting in 1984–85 (Figure 21.8). This increase coincided with the collapse of colonies in the northeastern region and with the continued decline of the southwestern colonies. Initially, nesting success in the northwest and the northeast were comparable. From 1973–74 to 1981–82, nest success was documented in both regions for 6 years (Table 21.5), and comparisons of the individual years show that the two colony groups had very similar success rates. However, following the completion and operation of the SDCS in 1982 (period III in the northeastern region), the northwestern colonies exhibited much higher nest production and had a higher annual success rate than the northeastern colony group (Table 21.5).

## NESTING PATTERNS ON INTER-REGIONAL AND BAYWIDE SCALES

A comparison of nesting effort and nest success in the northeast and northwest colony groups suggests that operation of the SDCS under the Experimental Program of Water Deliveries has had a negative impact on Roseate Spoonbills. During period II similar success rates in the northeast and northwest were probably due to the effects of systemwide conditions that either allowed for successful nesting or caused failure throughout the spoonbill nesting range in any given year. During period III success rates in the northeast were much lower than the northwest. While climatological conditions affected both regions equally, water management practices in the northeast caused a deterioration of foraging conditions for spoonbills in that colony group. This conclusion is most evident from the decline in nesting effort in the northeast during the same period. An increase in the number of nests in the northwestern and central colonies over the same time period suggests that many of the birds abandoned colonies in the northeast in favor of nesting sites in other colony groups.

The most striking implication of these findings is that current water management practices in the southern Everglades have resulted in the ecological degradation of the coastal wetlands in northeastern Florida Bay. The most likely difference within the northeastern spoonbill niche between periods II and III was the changes in the hydrology and hydrography caused by the SDCS and the Experimental Program of Water Deliveries. These hydrologic and hydrographic changes presumably impacted spoonbills by altering the availability of prey (Lorenz, 1999, 2000; Dumas, 2000). The severity of this wetland degradation was such that the spoonbill response (decline in nesting success and abandonment of colonies) was similar to the complete destruction of more than half of the suitable foraging wetlands on the mainline Florida Keys. Because a variety of other vertebrate predators (piscivorous fishes, reptiles including crocodilians, wading birds, and other piscivorous birds) depend on this prey base during various stages of their life cycles (Lorenz, 2000), the impacts of the SDCS probably extend well beyond Roseate Spoonbill nesting patterns.

Although no data suggest that mercury poisoning has been a factor in spoonbill nesting patterns in Florida Bay, mercury concentrations were higher in nestlings from northeastern colonies when compared to nestlings from the northwestern colony group (Sundlof et al., 1994). These nestlings probably obtained the mercury from the small fish and aquatic invertebrates brought in by parents from the eastern mangrove region. Mercury contamination has a number of effects on birds (Eisler, 1987), many of which may affect both the initiation and success of reproduction. There is some evidence that these effects can lead to decreased nesting success (Eisler, 1987). Therefore, high mercury levels in the mangrove ecosystem north of eastern Florida Bay cannot be dismissed as a causal agent in the observed changes in spoonbill nesting patterns in the northeastern colonies. However, the large differences in the thresholds of various effects among species (Eisler, 1987; Thompson et al., 1991) prevent any meaningful assessment regarding the implications of mercury in Florida Bay spoonbill chicks or adults without species-specific effect studies. The most likely explanation for the patterns observed remains the environmental degradation of foraging grounds for the northeastern colonies.

The evidence presented indicates that Roseate Spoonbills are sensitive to anthropogenic alterations to their foraging grounds. Spoonbills appear to have responded to the destruction or degradation of primary foraging grounds by relocating nesting effort to other regions of the bay in proximity to other suitable foraging habitats. This switch to secondary foraging locations resulted in higher nest productivity compared to that found in the compromised colony group; however, overall production was reduced when compared to years prior to the given perturbation. For example, during the period that nesting predominantly occurred in the southeastern colonies (1950 to the mid-1960s), mean nesting success was estimated at 2.25 chicks per nest and success occurred in all years surveyed (Table 21.4). Following the destruction of foraging grounds on the mainline Keys, spoonbills began to shift their nesting effort toward the northeastern colonies (Figure 21.7). Nest success in the southeastern colonies dropped to 0.22 birds per nest with success in only 22% of the nesting years (Table 21.4). During the succeeding period when spoonbills predominantly nested in the northeastern colonies (1964–1982), mean nesting success was estimated at 1.38 chicks per nest and success in 71% of the years (Table 21.5). Although this is much higher than that found in the southeastern colonies after Keys urban development, it is a substantial drop in success compared to the pre-development period.

Another change in nesting pattern occurred starting in 1982 with the completion of the SDCS and the beginning of the Experimental Program of Water Deliveries. In subsequent years, operation of the SDCS presumably degraded the primary foraging ground for the northeastern colonies, namely the coastal wetlands of Taylor Slough and the C-111 Basin. Mean nesting success in the northeastern colonies fell to 0.67 chicks per nest and success in only 36% of the years for the period 1982–1999. Similar to the shift from the southeastern colonies to the northeastern colonies following Keys development, spoonbills began to shift their nesting effort toward the northwest after the completion of the SDCS in 1982 (Figure 21.7). From 1982–1999, northwestern colonies had a mean estimated success rate of 1.24 and succeeded in 62% of the nesting years (Table 21.5). Although reproductive success was much higher than the northeastern colonies for the same period, it is a telling decline from success rates during period II in the northeastern colonies.

The abandonment of the southeastern colonies between 1956 and 1961 (Figure 21.8) occurred while nesting success was still high (Table 21.4). It appears that spoonbills can continue to nest successfully when nearby foraging grounds are lost, so long as suitable foraging grounds are available within flight range. However, spoonbills do not maintain this strategy indefinitely, particularly when new colony sites are available closer to preferred foraging locations. For example, following the dredge-and-fill development of the late 1950s, most of the remaining foraging areas on the Keys were on Lower Matecumbe, possibly leading to the increase in nesting effort in the southwest colonies. Furthermore, the last major foraging ground in the Keys outside of Lower Matecumbe was Hammer Point on southern Key Largo (Figure 21.5). This wetland was sufficiently far north that it fell within the mean flight radius of several of the northeastern colonies. Increased nesting at these colonies may have been triggered by easy access to both the mainland coastal wetlands and the remaining Keys wetland. Similarly, following loss of habitat on Lower Matecumbe Key, spoonbills in the southwestern colonies continued to successfully nest by making foraging flights toward Cape Sable. Shortly thereafter, nesting began in the northwestern colony group, closer to the Cape. Therefore, it appears that proximity to a readily available prey base would be the major reason for relocating. Bancroft et al. (1994) found that increased flight distance to feeding grounds corresponded with nest abandonment in Everglades wading bird colonies. In addition, they reasoned that changes in habitat quality resulted in changes in nesting colony location, and that the drop in habitat quality was likely due to changes in the prey base dynamics (Bancroft et al., 1994; Loftus and Eklund, 1994). These findings may apply to spoonbills as well. Proximity to prime foraging locations would result in less energy expenditure in gaining access to the food resource. Furthermore, less flight time between the nest and the food resource allows for greater foraging time, thereby increasing the likelihood of success.

Although spoonbills may locate their nests in proximity to their primary foraging ground, they are not restricted to feeding exclusively at that location. While flight-line and following-flight data indicated that most foraging flights were directed toward the closest suitable foraging site, several of the following flights were in excess of 20 km (longest was 38 km) toward more distant resources (Bjork and Powell, 1994). This indicates that spoonbills, like other wading birds, will forage more widely, presumably when nearby resources become scarce (Bancroft et al., 1994). It follows that the high success rates in the 1950s and 1960s may, therefore, be attributable not only to the high quality of foraging grounds on the mainline Keys but also to the variety of foraging habitats within reasonable flight distance (e.g., mainland coastal wetlands, interior wetlands on bay keys, mudflats). As an example, nesting success at Cowpens in 1961–62 and 1962–63 was relatively high (about 2 chicks per nest in both years), even though the primary foraging grounds for this colony were already destroyed. Flight-line counts indicated that these birds were making a significant number of flights toward the north (mainland coastal wetlands), northwest (Cape Sable), and west (bay keys and mudbanks) as well as toward the remaining Keys wetlands to the east (Figure 21.9). Even though their proximal foraging ground was gone, they were able to achieve a high degree of success by using the mosaic of habitats available throughout the Florida Bay landscape (Figure 21.9).

Wading bird feeding behavior requires that a dependable prey source be available during the nesting season within a reasonable flight distance from the colony (Bancroft et al., 1994). Spoonbill feeding behavior requires similar conditions (Bjork and Powell, 1996; Dumas, 2000). However, the nature of the prey base itself is somewhat ephemeral. During a typical dry season, wind-driven tides in Florida Bay can reflood foraging grounds for periods ranging from hours to weeks (Holmquist et al., 1989; Baratta and Fennema, 1994; Lorenz, 2000), thereby dispersing the prey base (Loftus and Eklund, 1994; DeAngelis et al., 1997; Lorenz, 2000). However, if several foraging habitats exist within flight range, then it is highly unlikely that all of them would be flooded at the same time by wind-driven tides. The ability to be wide ranging has allowed spoonbills to forage in remote areas when primary foraging grounds were flooded, but the loss of foraging habitat diversity in the Florida Bay landscape (e.g., destruction of wetlands in the Keys and ecological degradation of mainland coastal habitats in northeastern Florida Bay) has reduced the likelihood that suitable foraging habitat will be found on a consistent basis. Low foraging success over just a period of days could result in a failed nesting attempt through abandonment and starvation. In short, the unaltered Florida Bay had a mosaic of foraging habitats that resulted in successful reproduction in almost every year. The degradation of many of these wetlands may have resulted in undependable and inconsistent food resources and increasing nesting failures. Based on these observations, it seems unlikely that the described anthropogenic disturbances simply impacted discrete colonies; rather, the impacts were probably baywide and cumulative in effect. The relatively recent trend toward nesting in the central colony group may be an attempt by spoonbills to nest in locations that have equal access to a variety of foraging habitat types.

## ECOSYSTEM RESTORATION AND SPOONBILLS AS INDICATORS OF SUCCESS

Current plans for restoring the Everglades watershed include the resumption of more natural freshwater flows through Taylor Slough and the C-111 Basin. The evidence presented suggests that nesting spoonbills foraging in these wetlands respond to changes in prey availability. In turn, prey availability in the wetland is a function of hydrologically related conditions on at least three levels, all of which will be improved by a resumption of natural fresh water flows. First, length of wet-season hydroperiod is related to overall prey-base abundance by determining the time that fish can exploit seasonally inundated portions of the wetland (Loftus and Eklund, 1994; DeAngelis et al., 1997; Lorenz, 1999). This is presumably related both to generation time and individual growth rates of the various fish species (DeAngelis et al., 1997). Second, the biomass of the prey base is impacted by hydrographic conditions; specifically, fish production is related to the salinity regime such that increased freshwater flow is correlated with increased prey-base fish stock (Montague and Ley, 1993; Ley et al., 1994; Lorenz, 1999). Third, the prey base is impacted by dry-season water levels in that the density of fish per unit area (availability) at any given time is dictated by the severity of drying conditions and resulting concentration effect (Kushlan, 1978; Loftus and Eklund, 1994; DeAngelis et al., 1997; Lorenz, 2000). Gradual and constant declines in water levels over the course of the nesting period result in consistent concentrations of fish while reversals in water level declines disrupt the concentration effect (Kushlan, 1978; Bjork and Powell, 1994; DeAngelis et al., 1997; Lorenz, 2000). All three of these components must be taken into account when evaluating the response of spoonbills to restoration efforts.

No hydrologic nor hydrographic data are available for the Taylor Slough and C-111 regions prior to inception of water management in the Everglades. Although this area was moderately impacted by water management prior to the 1960s, it was the construction of the canals in south Dade County that significantly altered the hydrology, hydrography, and ecology of this region (Johnson and Fennema, 1989; Van Lent, 1993; McIvor et al., 1994). Therefore, restoration of more natural conditions would be expected to most closely resemble the conditions that occurred during period I (1956–1964). Compared to current conditions (period III), this would result in longer wet-season hydroperiods, lower and less variable salinity, and greater intra-annual (i.e., cycle of wet and dry seasons) variation in water depth (Johnson and Fennema, 1989; Van Lent et al., 1993; McIvor et al., 1994). We predict that these conditions would be beneficial to spoonbills nesting in the northeast colony group because they would result in a more robust and available prey base on the foraging grounds and would increase nesting numbers and nesting success in northeastern Florida Bay over the current condition. Therefore, any improvement in nesting effort and nesting success in the northeastern colonies should be considered positive in evaluating restoration efforts.

During period I, Florida Bay's spoonbill population was still suffering from near extirpation in the 1930s and was relatively small (Figure 21.2). Spoonbill nesting effort in the northeastern colony group was minimal and received little scientific attention (Allen, 1963). As a result, the response of spoonbills to pre-1960s conditions is difficult to gauge. A return to nesting numbers and success rates of the 1970s is a possibility, but hydrologic conditions during period II (1964–1982) were very different than period I (Johnson and Fennema, 1989; Van Lent et al., 1993). Although arguments can be made that period II conditions were both better and worse than period I, the paucity of empirical data almost precludes a meaningful prediction of the actual response. An examination of two response scenarios is instructive and sets the stage for evaluating restoration efforts.

### Restoration Scenario One: Wetland Conditions During Period II Were Better Suited to Successful Foraging than Period I

The northeastern colonies may not have been suitable for nesting during period I and became suitable in period II because of anthropogenic manipulations of hydrologic parameters. During the dry season, the Taylor Slough headwaters were, on average, 10 to 35 cm higher during period I compared to period II (Johnson and Fennema, 1989). The implication is that, during the dry season, the mainland coastal wetlands had higher water levels during period I compared to period II (Johnson and Fennema, 1989). If such conditions occurred they may not have been conducive to successful foraging by spoonbills, as the concentration effects of drydowns would have been less pronounced during period I than period II. Furthermore, if water was consistently deeper than 20 cm, then spoonbills may have been prevented from foraging in the area all together (Powell, 1987). Anthropogenic lowering of dry season water levels during period II may have created opportunities for spoonbills to forage in areas that were previously too deep, thereby explaining at least some of the increase in the spoonbill population during period II. If this scenario is accurate, then the pre-1960s system may have been too wet for spoonbills. These conditions would have resulted in lower prey availability and lower foraging success in spoonbills than in period II. Therefore, a return to peak northeastern nesting levels of the late 1970s would not be expected.

### Restoration Scenario Two: Wetland Conditions During Period I Were as Good or Better Suited to Successful Foraging than Period II

It is possible that foraging conditions in the mainland coastal wetlands were better in period I than period II. Wet-season freshwater flow to the coastal wetlands was greater during period I than period II (Johnson and Fennema, 1989; Fennema et al., 1994), thereby increasing wet-season hydroperiod and lowering salinity and salinity variation. Consequently, the prey base would have been more robust overall. Although dry-season water levels were, on average, higher in the Taylor Slough headwaters during period I compared to period II, the differences were less pronounced downstream; therefore, the most downstream water-level station (G613) was the best indicator of coastal wetland water level (Johnson and Fennema, 1989). At G613, dry-season low water averages for periods I and II were very similar (Figure 21.10), indicating that the concentration effect would have occurred similarly in both periods. Given these conditions, suitable fish concentrations may have been achieved at higher water levels and for longer periods, thereby providing spoonbills a reliable prey source earlier in the dry season and prey concentrations that were consistent throughout the nesting cycle. In this scenario, the quality of foraging habitat for spoonbills in the mainland coastal wetlands would have been as high, if not higher, during period I compared to period II, and a return to these conditions would result in increased nesting effort and success in the northeastern colonies. Restoration could result in better quality of foraging habitat than that which occurred from 1964 to 1982, and the spoonbill population in the northeastern colonies might meet or exceed the peak densities of the late 1970s.

### Empirical Assessment of Two Scenarios

Some empirical data can be used to support the second scenario. Flight-line data from the southeastern colony group during period I indicated that the mainland was being significantly used as a supplemental foraging area (Table 21.3). Initial expansion of nesting into the northeastern colonies began prior to 1950 and increased through 1960, indicating that birds were choosing to nest in the northeast prior to loss of Keys foraging habitat. Although only one hour of flight-line data was collected for a northeastern colony during period I, these limited observations show that the birds were primarily foraging on the mainland. During this period a sizable percentage of the Florida Bay spoonbill population successfully nested in the northeastern colony group during period I

(Figure 21.7A). Estimates of nesting success were made in only 2 years during period I (1955–56 and 1961–62); however, these success estimates were about twice the average (2.3 and 3.0 chicks per nest, respectively) of nesting success in the northeastern colonies during period II (Table 21.5). These observations indicate that the foraging quality of the mainland wetlands was high enough during period I to attract foraging birds nesting in other regions of Florida Bay. Furthermore, the foraging quality was high enough to stimulate some spoonbills to nest in the northeastern colonies over the southeastern colonies.

A possible reason for the under-utilization of northeastern colonies during period I may simply have been that the quality of wetlands on the mainline Keys were superior to those on the mainland. Mean spoonbill nesting effort moved southwestward (Figure 21.7B) from the founding colony (Bottle Key), suggesting that spoonbills preferred nesting locations closer to the Keys foraging grounds during period I (Figure 21.7B). Florida Bay's spoonbill population was so small as a result of near extirpation in the 1930s that the entire nesting population could succeed by exploiting these "preferred" food resources. If this were the case, then the northeastern colonies may have been suitable for nesting during period I but were not used because nesting sites in the southeastern colony group were readily available. Following destruction of Keys foraging sites, rapid expansion into the northeastern and southwestern colonies co-occurred. The expansion to the southwest cannot be explained by human activities, as there was no similar quality "enhancement" of primary foraging grounds (such as was postulated for the northeast in scenario one). This suggests that the impetus for movements to the northeast and southwest was most likely the loss of proximal foraging habitat in the southeast and not an enhancement of the northeastern foraging grounds.

As a counterpoint to this scenario, the very high success rates for the 2 years during period I do not necessarily indicate that the quality of foraging habitat in northeastern Florida Bay was higher in period I vs. period II, as wetlands on the mainline keys were still largely unaltered. The northeastern colonies may have been successful because adult birds were supplementing their diets by foraging in Keys wetlands in addition to the coastal wetlands on the mainland (as indicated by north Nest Key flight-line counts in 1958–59; Table 21.3).

The fate of the Roseate Spoonbill in Florida Bay is not clear. The insular nature of Cape Sable should protect the foraging grounds of the northwestern colonies from being degraded by future detrimental land and water management plans. Therefore, nesting should continue in the north-western colonies with high enough success to sustain the population in that area. However, sea-level rise has impacted Cape Sable wetlands and will continue to do into the foreseeable future (Wanless et al., 1994). As a result, the capacity of the Cape Sable foraging grounds to sustain a nesting population is not well understood and prevents a definitive prediction of future population stability in the region. Therefore, improving conditions in the northeastern colony foraging grounds through restoration is essential to stabilizing the Florida Bay Roseate Spoonbill population and ensuring viability into the future. Regardless of the fate of the northeastern and northwestern colonies, we predict that spoonbills will never exhibit the high nesting success rates of the 1950s and 1960s simply because the mosaic of foraging habitats across the Florida Bay landscape has been permanently altered by urbanization of the Florida Keys.

## CONCLUSIONS

The preponderance of Roseate Spoonbill information from Florida Bay indicates that land and water management practices in and around the Florida Bay landscape have affected nesting patterns. More specifically, these data support the hypothesis that anthropogenic alterations to the landscape explain the "boom-and-bust" temporal and spatial pattern in spoonbill nesting in the bay.

Observations made in the 1950s and 1960s indicated that spoonbills nesting in the southeastern and southwestern regions of Florida Bay depended on mainline Keys wetland habitat as principal feeding areas. Destruction of these habitats as a result of urban development in the Keys resulted in a dramatic decline of nesting success followed by the abandonment of these regions by nesting spoonbills. Following urbanization of the Keys, spoonbill nesting increased in the northeastern region of the bay. The principle foraging grounds for these colonies was the mainland wetlands to the north. These wetlands were heavily impacted by changes in water management practices in the early 1980s, resulting in further spoonbill nesting failure. Subsequently, spoonbills abandoned nesting sites in the northeast region in favor of sites in the northwest region of Florida Bay. This sequential decline in nesting success followed by a relocation of nesting effort as a result of anthropogenic alterations to primary foraging grounds indicates that spoonbill responses were related to food abundance, availability, and quality. The overall effect on the spoonbill population was a baywide decline in nesting success.

Bancroft et al. (1994) demonstrated that wading birds nesting in the Everglades selected colony locations based on proximity to primary foraging grounds. This leads to the conclusion that wading birds will abandon one nesting area in favor of another if conditions on the primary foraging grounds become too degraded. Presumably, this is why spoonbills serially abandoned various nesting regions in Florida Bay. The destruction of wetlands on the mainline Keys prompted the transition to northern nesting locations, while the westward movement in the early 1980s was prompted by changes in water management practices affecting this region. The most striking implication of these findings is that water management practices in the southern Everglades resulted in an ecological degradation of the coastal wetlands in northeastern Florida Bay to such a degree that the response by spoonbills was similar to that caused by the total destruction of wetlands on the Keys.

Accomplishing the restoration target of a return to pre-1960s hydrologic conditions in the mainland coastal wetlands would be a substantial improvement over current conditions. Longer wet season hydroperiods, lower and less variable salinity, and greater intra-annual variation in water depth would likely increase prey abundance and make prey more available to spoonbills. A return to the high nesting success rates of the 1950s and 1960s may be unlikely given that foraging habitat heterogeneity within the Florida Bay landscape has been permanently altered. The limited empirical evidence indicates that a return to nesting effort and success rates of the 1970s is possible, although the degree of recovery is debatable.

## ACKNOWLEDGMENTS

Foremost, we wish to acknowledge the pioneering work of Robert Porter Allen and William B. Robertson, Jr. In addition to their contributions to this work, their vigilance and dedication have greatly facilitated restoration efforts in the Everglades and Florida Bay. Alexander Sprunt IV, Peter C. Frederick, Marilyn G. Spalding, and Richard J. Sawicki have made valuable contributions to this manuscript. We would like to thank David Barrow (Barrow Surveying and Mapping) and John Kipp for contributing their expert knowledge of the Florida Keys and Florida Bay. This project could not have been done without the dedicated field staff at NAS who not only endured but persevered under the most difficult of field conditions. We would especially like to thank Lori Oberhofer, Robin Corcoran, Ginny Oshaben, and Rich Paul. We would also like to thank all the staff members at the SFWMD, ENP, USGS-BRD, and FIU for their support and assistance over the years. This project was partially funded by the South Florida Water Management District, the U.S. Army Corps of Engineers, Everglades National Park, the John D. and Catherine T. MacArthur Foundation, the Elizabeth Ordway Dunn Foundation, and John and Florence Schumann Foundation.

## REFERENCES

Allen, R.P. 1942. *The Roseate Spoonbill*, Dover Publications, New York.

Allen, R.P. 1947. *The Flame Birds*, Dodd, Mead, New York.

Allen, R.P. 1963. The present status of the roseate spoonbill, a summary of events 1943–1963. Unpublished report of the National Audubon Society Research Center, Tavernier, FL.

Audubon, J.J. 1960. *Audubon and His Journals*, Vol. 2, Dover Publications, New York.

Bancroft, G.T., A.M. Strong, R.J. Sawicki, W. Hoffman, and S.D. Jewell. 1994. Relationships among wading bird forage patterns, colony locations, and hydrology in the Everglades, pp. 615–658, in *Everglades: The Ecosystem and Its Restoration*, Davis, S.M. and Ogden, J.C., Eds., St. Lucie Press, Delray Beach, FL.

Baratta, A.M. and R.J. Fennema. 1994. The effects of wind, rain, and water releases on the water depth and salinity of northeast Florida Bay, *Bull. Mar. Sci.*, 54:1072.

Bjork, R.D. and G.V.N. Powell. 1994. *Relationships Between Hydrologic Conditions and Quality and Quantity of Foraging Habitat for Roseate Spoonbills and Other Wading Birds in the C-111 Basin*, final report to the South Florida Research Center, Everglades National Park, Homestead, FL.

Bjork, R.D. and G.V.N. Powell. 1996. Roseate Spoonbill, pp. 295–308, in Rare and endangered biota of Florida. Vol. 5: Birds, Rodgers, J.A., Kale, H.W., and Smith, H.T., Eds., University of Florida Press, Gainesville.

Boesch, D.F., N.E. Armstrong, C.F. D'Elia, N.G. Maynard, H.W. Paerl, and S.L. Williams. 1993. *Deterioration of the Florida Bay Ecosystem: An Evaluation of the Scientific Evidence*, report to the interagency working group on Florida Bay, South Florida Water Management District, West Palm Beach, FL.

Brand, L.E. 2000. An evaluation of the scientific basis for "restoring" Florida Bay by increasing freshwater runoff from the Everglades. Rosenstiel School of Marine and Atmospheric Science, University of Miami.

DeAngelis, D.L., W.F. Loftus, J.C. Trexler, and R.E. Ulanowicz. 1997. Modeling fish dynamics and effects of stress in a hydrologically pulsed ecosystem, *J. Aquatic Ecosys. Stress Rec.*, 6:1–13.

Dumas, J. 2000. Roseate Spoonbill (*Ajaia ajaja*), in *The Birds of North America*, Poole, A. and Gill, F., Eds., No. 490, The Birds of North America, Inc., Philadelphia, PA.

Dusi, J.L. and R.D. Dusi. 1978. Survey methods used for wading birds studies in Alabama, pp. 207–212, in *Wading Birds: Research Report No. 7 of the National Audubon Society*, Sprunt, IV, A., Ogden, J.C., and Winckler, S., Eds., National Audubon Society, New York.

Egler, F.E. 1952. Southeast saline Everglades vegetation, *Vegetatio*, 3:213–265.

Eisler, R. 1987. *Mercury Hazards to Fish, Wildlife, and Invertebrates: A Synoptic Review*, Biol. Rep. 85, U.S. Fish and Wildlife Service, Washington, D.C.

Fennema, R.J., C.J. Neidrauer, R.A. Johnson, T.K. MacVicar, and W.A. Perkins. 1994. A computer model to simulate natural everglades hydrology, pp. 249–289, in *Everglades: The Ecosystem and Its Restoration*, Davis, S.M. and Ogden, J.C., Eds., St. Lucie Press, Delray Beach, FL.

Fourqurean, J.W. and M.B. Robblee. 1999. Florida Bay: a history of recent ecological changes, *Estuaries*, 22:345–357.

Frederick, P.C. and M.W. Collopy. 1989a. Nesting Success of five ciconiiform species in relation to water conditions in the Florida Everglades, *Auk*, 106:625–634.

Frederick, P.C. and M.W. Collopy. 1989b. The role of predation in determining reproductive success of colonially nesting wading birds in the Florida Everglades, *The Condor*, 91:860–867.

Gentry, R.C. 1974. Hurricanes in south Florida, pp. 73–81, in *Environments of South Florida: Present and Past*, Gleason, P.J., Ed., Miami Geologic Society.

Graham, Jr., F. 1990. *The Audubon Ark: A History of the National Audubon Society*, Knopf, New York.

Holmquist, J.G., G.V.N. Powell, and S.M. Sogard. 1989. Sediment, water level and water temperature characteristics of Florida Bay's grass-covered mud banks, *Bull. Mar. Sci.*, 44:348–364.

Hudson, J.H., G.V.N. Powell, M.B. Robblee, and T.J. Smith, III. 1989. A 107-year-old coral from Florida Bay: barometer of natural and man induced catastrophes, *Bull. Mar. Sci.*, 44:283–291.

Johnson, R.A. and R.A. Fennema. 1989. Conflicts over flood control and wetland preservation in the Taylor Slough and eastern panhandle basins of Everglades National Park, pp. 451–462, in *Wetlands: Concerns and Successes*, Fisk, D.W., Ed., American Water Resources Association, Bethesda, MD.

Kushlan, J.A. 1978. Feeding ecology of wading birds, pp. 249–248, in *Wading Birds: Research Report No. 7 of the National Audubon Society*, Sprunt, IV, A., Ogden, J.C., and Winckler, S., Eds., National Audubon Society.

Lapinot, A.C. 1951. Raccoon predation on the Great Blue Heron, *Ardea herodias*, *The Auk*, 68:235–236.

Lapointe, B.E. and W.E. Clark. 1992. Nutrient inputs from the watershed and coastal eutrophication in the Florida Keys, *Estuaries*, 15:465–476.

Lapointe, B.E. and W.R. Matzie. 1996. Effects of stormwater nutrient discharges on the eutrophication processes in nearshore waters of the Florida Keys, *Estuaries*, 19:422–435.

Ley, J.A., C.L. Montague, and C.C. McIvor. 1994. Food habits of mangrove fishes: a comparison along estuarine gradients in northeastern Florida Bay, *Bull. Mar. Sci.*, 54:881–889.

Ley, J., L. Gulick, J. Branscome, M. Ostrovsky, S. Traver, and B. Kacvinsky. 1995. C-111 interim construction project first evaluation report: Nov. 1987–Nov. 1993 (FDEP Permit No. 131654749). South Florida Water Management District, West Palm Beach, FL.

Light, S.S. and J.W. Dineen. 1994. Water control in the Everglades: a historical perspective, pp. 47–84, in *Everglades: The Ecosystem and Its Restoration*, Davis, S.M. and Ogden, J.C. Eds., St. Lucie Press, Delray Beach, FL.

Loftus, W.F. and A.-M. Eklund. 1994. Long-term dynamics off an Everglades small-fish assemblage, p. 461-483, in *Everglades: The Ecosystem and Its Restoration*, Davis, S.M. and Ogden, J.C. Eds., St. Lucie Press, Delray Beach, FL.

Loftus, W.F. and J.A. Kushlan. 1987. Freshwater fishes of southern Florida, *Bull. Florida State Mus. Biol. Sci.*, 31:137–344.

Lorenz, J.J. 1999. The response of fishes to physicochemical changes in the mangroves of northeast Florida Bay, *Estuaries*, 22:500–517.

Lorenz, J.J. 2000. Impacts of Water Management on Roseate Spoonbills and Their Piscine Prey in the Coastal Wetland of Florida Bay, Ph.D. dissertation, University of Miami.

Lott, C., R. Dye, and K.M. Sullivan. 1996. *Historical Overview of Development and Natural History of the Florida Keys*. Vol. 3: *Site Characterization for the Florida Keys National Marine Sanctuary and Environs*, The Preserver of the Farley Court of Publishers for the Nature Conservancy, Zenda, WI.

Lugo, A.E. and S.C. Snedaker. 1974. The Ecology of Mangroves, *Ann. Rev. Ecol. Syst.*, 5:39–63.

Mayfield, H.F. 1975. Suggestions for calculating nest success, *Wilson Bull.*, 73:84–99.

McIvor, C.C., J.A. Ley, and R.D. Bjork. 1994. Changes in freshwater inflow from the Everglades to Florida Bay including effects on the biota and biotic processes: a review, pp. 117–146, in *Everglades: The Ecosystem and Its Restoration*, Davis, S.M. and Ogden, J.C. Eds., St. Lucie Press, Delray Beach, FL.

Meeder, J.F., M.S. Ross, G. Telesnicki, and P.L. Ruiz. 1996. *Vegetation Analysis in the C-111/Taylor River Slough Basin: Marine Transgression in the Southeast Saline Everglades, Florida, USA — Rates, Causes and Plant and Sediment Responses*, final report to the South Florida Water Management District (Contract C-4244).

Montague, C.L. and J.A. Ley. 1993. A possible effect of salinity fluctuation on abundance of benthic vegetation and associated fauna in northeastern Florida Bay, *Estuaries*, 16:703–717.

Powell, G.V.N. 1987. Habitat use by wading birds in a subtropical estuary: implications of hydrography, *Auk*, 104:740–749.

Powell, G.V.N., R.D. Bjork, J.C. Ogden, R.T. Paul, A.H. Powell, and W.B. Robertson. 1989. Population trends in some Florida Bay wading birds, *Wilson Bull.*, 101:436–457.

Powell, G.V.N. and R.D. Bjork. 1990. Relationships between hydrologic conditions and quality and quantity of foraging habitat for Roseate Spoonbills and other wading birds in the C-111 basin. Second annual report to the South Florida Research Center, Everglades National Park, Homestead, FL.

Robertson, Jr., W.B. 1978. Roseate Spoonbill Biology Progress Report. Everglades National Park, Homestead, FL.

Robertson, Jr., W.B. 1979. *1979 Annual Report*, Everglades National Park, Homestead FL.

Robertson, Jr., W.B., L.L. Breen, and B.W. Patty. 1983. Movement of marked Roseate Spoonbills in Florida with a review of present distribution, *J. Field Ornithol.*, 54:225–236.

Ross, M.S., J.F. Meeder, J.P. Sah, P.L. Ruiz, and G. Telesnicki. 2000. The southeast saline Everglades revisited: a half-century of coastal vegetation change, *J. Veg. Sci.*, 11:101–112.

Strong, A.M. and G.T. Bancroft. 1994. Patterns of deforestation and fragmentation of mangrove and deciduous seasonal forests in the upper Florida Keys, *Bull. Mar. Sci.*, 54:795–804.

Strong, A.M., R.J. Sawicki, and G.T. Bancroft. 1991. Effects of predator presence on the nesting distribution of White Crowned Pigeons in Florida Bay, *Wilson Bull.*, 103:415–425.

Sundlof, S.F., M.G. Spalding, J.D. Wentworth, and C.K. Steible. 1994. Mercury in livers of wading birds (Ciconiiformes) in southern Florida, *Arch. Environ. Contam. Toxicol.*, 27:299–305.

Thompson, D.R., K.C. Hamer, and R.W. Furness. 1991. Mercury accumulation in Great Skuas (*Catharacta skua*) of known age and sex, and its effects upon breeding and survival, *J. Appl. Ecol.*, 28:672–84.

U.S. Army Corps of Engineers. 1995. Environmental assessment and finding of no significant impact, test iteration 7 of experimental program of water deliveries to Everglades National Park, Central and Southern Florida project. Jacksonville District, South Atlantic Division.

Van Lent, T. J., R. Johnson, and R. Fennema. 1993. *Water Management in Taylor Slough and the Effects on Florida Bay*, report to South Florida Research Center, Everglades National Park, Homestead FL.

Wanless, H.R., R.W. Parkinson, and L.P. Tedesco. 1994. Sea level control on stability of Everglades wetlands, pp. 199–223, in *Everglades: The Ecosystem and Its Restoration*, Davis, S.M. and Ogden, J.C. Eds., St. Lucie Press, Delray Beach, FL.

White, D.H., C.A. Mitchell, and E. Cromartie. 1982. Nesting ecology of Roseate Spoonbills at Nueces Bay, Texas, *Auk*, 99:275–284.

Wood D.A. 1997. *Official Lists of Endangered and Potentially Endangered Fauna and Flora in Florida*, Florida Game and Freshwater Fish Commission, Tallahassee, FL.

# Section III

## Florida Reef Tract

*Little Butternut Key #1*
by
Clyde Butcher

# 22 A View from the Bridge: External and Internal Forces Affecting the Ambient Water Quality of the Florida Keys National Marine Sanctuary (FKNMS)

*Joseph N. Boyer and Ronald D. Jones*
Southeast Environmental Research Center

## CONTENTS

## INTRODUCTION

The Florida Keys are an archipelago of subtropical islands of Pleistocene origin which extend in a northeast to southwest direction from Miami to Key West and out to the Dry Tortugas (Figure 22.1). In 1990, President Bush signed into law the Florida Keys National Sanctuary and Protection Act (HR5909) which designated a boundary encompassing more than 2800 square nautical miles of islands, coastal waters, and coral reef tract as the Florida Keys National Marine Sanctuary (FKNMS). The Comprehensive Management Plan (NOAA, 1995) required the FKNMS

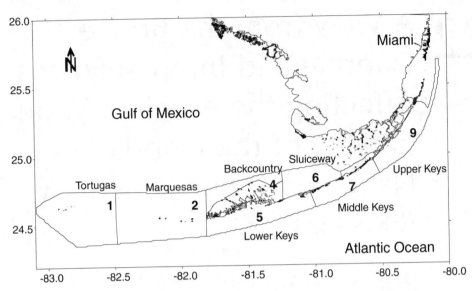

**FIGURE 22.1** Map of South Florida showing the FKNMS boundary, segments, and common names of segments.

to have a Water Quality Protection Plan (WQPP), thereafter developed by the Environmental Protection Agency (EPA) and the State of Florida (EPA, 1995). The contract for the water quality monitoring component of the WQPP was subsequently awarded to the Southeast Environmental Research Program at Florida International University and the field sampling program began in March 1995.

The waters of the FKNMS are characterized by complex water circulation patterns over both spatial and temporal scales with much of this variability due to seasonal influence in regional circulation regimes. The FKNMS is directly influenced by the Florida Current, the Gulf of Mexico Loop Current, inshore currents of the southwest Florida Shelf (Shelf), and tidal exchange with both Florida Bay and Biscayne Bay. Advection from these external sources has significant effects on the physical, chemical, and biological composition of waters within the FKNMS, as does internal nutrient loading and freshwater runoff from the Keys themselves. Water quality of the FKNMS may be directly affected both by external nutrient transport and internal nutrient loading sources; therefore, the geographical boundary of the FKNMS must not be thought of as enclosing a distinct ecosystem but rather as being one of political/regulatory definition.

Ongoing quarterly sampling of more than 200 stations in the FKNMS and Shelf, as well as monthly sampling of 100 stations in Florida Bay, Biscayne Bay, and the mangrove estuaries of the southwest coast, has provided us with a unique opportunity to explore the spatial component of water quality variability. By stratifying the sampling stations according to depth, regional geography, distance from shore, proximity to tidal passes, and influence of Shelf waters we are able to report some preliminary conclusions as to the relative importance of external vs. internal factors on the ambient water quality within the FKNMS.

## METHODS

### SITE CHARACTERISTICS AND SAMPLING DESIGN

A spatial framework for FKNMS water quality management was proposed on the basis of geographical variation of regional circulation (Klein and Orlando, 1994). The final implementation plan (EPA, 1995) partitioned the FKNMS into nine segments which were collapsed to seven for routine sampling (Figure 22.1). Station locations were developed using a stratified random design along onshore/offshore transects in segments 5, 7, and 9 or within the EPA Monitoring and Assessment Program (EMAP) segments grid cells in Seg. 1, 2, 4, and 6.

Segment 1 (Tortugas) includes the Dry Tortugas National Park and surrounding waters and is most influenced by the Loop Current and Dry Tortugas Gyre. Segment 2 (Marquesas) includes the Marquesas Keys and a shallow sandy area between the Marquesas and Tortugas called the Quick-sands; Segment 4 (Backcountry) contains the shallow, hard-bottomed waters of the gulfside Lower Keys. Segments 2 and 4 are both influenced by water moving south from the Shelf. Segment 6 can be considered as part of western Florida Bay. This area is referred to as the Sluiceway as it heavily influenced by transport from Florida Bay and the Shelf (Smith, 1994). Segment 5 (Lower Keys), 7 (Middle Keys), and 9 (Upper Keys) include the inshore, Hawk Channel, and reef tract of the Atlantic side of the Florida Keys. The Lower Keys are most influenced by cyclonic gyres spun off of the Florida Current; the Middle Keys, by exchange with Florida Bay; and the Upper Keys, by the Florida Current frontal eddies and to a certain extent by exchange with Biscayne Bay. All three oceanside segments are also influenced by wind and tidally driven lateral Hawk Channel transport (Pitts, 1997).

## FIELD SAMPLING

The period of record of this study was from March 1995 to April 1998 and included 12 quarterly sampling events. For each event, field measurements and grab samples were collected from 150 fixed stations within the FKNMS boundary (Figure 22.1). Depth profiles of temperature (°C), salinity (psu, practical salinity scale), dissolved oxygen (DO, mg $l^{-1}$), photosynthetically active radiation (PAR, $\mu E\ m^{-2}\ sec^{-1}$), *in situ* chlorophyll *a* specific fluorescence (FSU), optical backscatterance (OBS) turbidity, depth as measured by pressure transducer (m), and density ($\sigma_t$, kg $m^{-3}$) were measured by CTD casts (Seabird SBE 19). The CTD was equipped with internal RAM and operated in stand-alone mode at a sampling rate of 0.5 sec. The vertical light attenuation coefficient ($K_d$, $m^{-1}$) was calculated at 0.5-m intervals from PAR and depth using the standard exponential equation (Kirk, 1994) and averaged over the station depth. This was necessary due to periodic occurrence of optically distinct layers within the water column. During these events, $K_d$ was reported for the upper layer.

In the Backcountry area (segment 4; Figure 22.1) where it was too shallow to use a CTD, surface salinity and temperature were measured using a combination salinity–conductivity–temperature probe (Orion model 140). DO was measured using an oxygen electrode (Orion model 840) corrected for salinity and temperature. PAR was measured using a Li-Cor irradiance meter equipped with two $4\pi$ spherical sensors (LI-193SB) separated by 0.5 m in depth and oriented at 90° to each other. The light meter measured instantaneous differences between sensors which were then used to calculate $K_d$ from in-air surface irradiance.

Water was collected from approximately 0.25 m below the surface and at approximately 1 m from the bottom with a Teflon-lined Niskin bottle (General Oceanics), except in the Backcountry where it was collected directly into sample bottles. Duplicate, unfiltered water samples were collected using 3× sample-rinsed 120-ml HDPE bottles for analysis of total constituents. Duplicate water samples for dissolved nutrients were collected using 3× sample-rinsed 150-ml syringes which were then filtered by hand through 25-mm glass fiber filters (Whatman GF/F) into 3× sample-rinsed 60-ml HDPE bottles. The wet filters used for chlorophyll *a* (Chl *a*) analysis were placed in 1.8-ml plastic centrifuge tubes to which 1.5 ml of 90% acetone/water was added (Strickland and Parsons, 1972).

Unfiltered samples were kept at ambient temperature in the dark during transport to the laboratory. During shipboard collection in the Tortugas/Marquesas and overnight stays in the Keys, unfiltered samples were analyzed for alkaline phosphatase activity (APA) and turbidity (see Analytical section) prior to refrigeration. Filtered samples and filters were kept on ice in the dark during transport. During shipboard collection in the Tortugas/Marquesas and overnight stays in the lower Keys, filtrates and filters were frozen until analysis.

## LABORATORY ANALYSIS

Unfiltered water samples were analyzed for total organic carbon (TOC), total nitrogen (TN), total phosphorus (TP), silicate (Si(OH)$_4$), alkaline phosphatase activity (APA), and turbidity. TOC was measured by direct injection onto hot platinum catalyst in a Shimadzu TOC-5000 after first acidifying to pH < 2 and purging with CO$_2$-free air. TN was measured using an ANTEK 7000N Nitrogen Analyzer using O$_2$ as carrier gas to promote complete recovery of the nitrogen in the water samples (Frankovich and Jones, 1998). TP was determined using a dry ashing, acid hydrolysis technique (Solórzano and Sharp, 1980). Si(OH)$_4$ was measured using the molybdosilicate method (Strickland and Parsons, 1972). The APA assay measures the activity of alkaline phosphatase, an enzyme used by bacteria to mineralize orthophosphate from organic compounds. The assay is performed by adding a known concentration of an organic phosphate compound (methylfluorescein phosphate) to an unfiltered water sample. Alkaline phosphate in the water sample cleaves the orthophosphate, leaving methylfluorescein, a highly fluorescent compound. Fluorescence at initial and after 2-hr incubation were measured using a Gilford Fluoro IV Spectrofluorometer (excitation = 430 nm, emission = 507 nm) and subtracted to give APA in $\mu M$ h$^{-1}$ (Jones, 1996). Turbidity was measured using an HF Scientific model DRT-15C turbidimeter and reported in NTU.

Filtrates were analyzed for soluble reactive phosphorus (SRP), nitrate + nitrite (NO$_x^-$), nitrite (NO$_2^-$), and total ammonia (NH$_4^+$) on a four-channel autoanalyzer (Alpkem model RFA 300). Filters for Chl *a* content ($\mu$g l$^{-1}$) were allowed to extract for a minimum of 2 days at –20°C before analysis. Extracts were analyzed using a Gilford Fluoro IV Spectrofluorometer (excitation = 435 nm, emission = 667 nm). All analyses were completed within 1 month after collection in accordance with SERC laboratory quality control guidelines.

Some parameters were not measured directly but were calculated by difference. Nitrate (NO$_3^-$) was calculated as NO$_x^-$ – NO$_2^-$ and dissolved inorganic nitrogen (DIN) as NO$_x^-$ + NH$_4^+$; total organic nitrogen (TON) was defined as TN–DIN. All concentrations are reported as $\mu M$ unless noted. All elemental ratios discussed were calculated on a molar basis. Percent DO saturation (DO$_{sat}$ as %) was calculated using the equations of Garcia and Gordon (1992).

## STATISTICAL ANALYSIS

Stations were grouped four different ways for statistical analysis: by surface or bottom samples, surface by segment, surface by transect distance, and surface by shore type. These groupings were subjectively defined using best available knowledge in an effort to provide information as to source, transport, and fate of water quality components. For the first grouping, stations were selected as being >3 m depth where both surface and bottom samples were collected and stratified by depth. The second grouping included surface samples stratified by segment (Figure 22.1) in accordance with the implementation plan (EPA, 1995). The third grouping consisted of those surface stations situated on oceanside transects aggregated according to their distance from shore: Alongshore, Hawk Channel, or Reef Tract. In addition, we initiated a transect of stations in the Tortugas off Loggerhead Key to serve as a reference. Because sampling at these locations in the Tortugas was only recently set up to address this question, the data are more sparse. Also, having only two "channel" stations in the Tortugas made the data more susceptible to outlier conditions.

One of the concerns of this program is to determine the contribution of water movement through the passes of the Keys to the water quality of the reef. To this end we decided to characterize a last grouping of transects as to shore type: those that were adjacent to land off Biscayne Bay — Old Rhodes Key, Elliot Key, and the Safety Valve (BISC); those that abutted land in Key Largo, Middle, and Lower Keys (LAND); and those transects which are aligned along an open channel or pass through the Keys (PASS). These grouping strategies may be changed when enough data are collected (*ca.* 5 to 7 yr) to be analyzed using a statistically objective, multivariate approach as has been done previously for Florida Bay and Ten Thousand Islands (Boyer et al., 1997; Boyer and Jones 1998).

Typical water quality variables are usually skewed to the left, resulting in non-normal distributions; therefore it is more appropriate to use the median as the measure of central tendency. Data distributions of selected water quality variables are reported as box-and-whiskers plots. The box-and-whisker plot is a powerful statistic as it shows the median, range, and data distribution, as well as serving as a graphical, nonparametric analysis of variance (ANOVA). The center horizontal line of the box is the median of the data, the top and bottom of the box are the 25th and 75th percentiles (quartiles), respectively; and the ends of the whiskers are the 5th and 95th percentiles, respectively. The notch in the box is the 95% confidence interval of the median. When notches between boxes do not overlap, the medians are considered significantly different. Outliers (<5th and >95th percentiles) were excluded from the graphs to reduce visual compression. Differences in variables were also tested between groups using the Wilcoxon Ranked Sign test (comparable to $t$-test) and among groups by the Kruskall–Wallace test (ANOVA) with significance set at $p < 0.05$.

In an effort to elucidate the contribution of external factors to the water quality of the FKNMS and to visualize gradients in water quality over the region, we combined data from other portions of our water quality monitoring network: Florida Bay, Biscayne Bay, Whitewater Bay, Ten Thousand Islands, and the Shelf (Figure 22.1). Data from these 153 additional stations were collected during the same month as the FKNMS surveys and analyzed by SERC personnel using similar methodology and quality control procedures as previously described. Spatial contour maps of median water quality variables (of 12 surveys) were generated in SURFER (Golden Software) using the kriging algorithm.

# RESULTS

## OVERALL WATER QUALITY

Summary statistics for all surface water quality variables including data from all 12 sampling events and all stations for the period of record are shown in Table 22.1 as median value, minimum value, maximum value, and number of samples measured. Overall, the region was warm and euhaline with a median temperature of 26.7°C and salinity of 36.2. The median DO was 6.2 mg l$^{-1}$, or ~93% DO$_{sat}$. On this coarse scale, the FKNMS exhibited very good water quality with median NO$_3^-$, NH$_4^+$, and TP concentrations of 0.10, 0.32, and 0.18 $\mu M$, respectively. NH$_4^+$ was the dominant DIN species in almost all of the samples (~70 %). DIN comprised a small fraction (4%) of the TN pool, with TON making up the bulk (11.4 $\mu M$). SRP concentrations were very low (median 0.01 $\mu M$) and was only 6% of the TP pool. Chlorophyll $a$ concentrations were also very low overall, 0.25 $\mu$g l$^{-1}$, but ranged from 0.01 to 6.8 $\mu$g l$^{-1}$. Median turbidity was low (0.6 NTU) as reflected in a low Kd (0.206 m$^{-1}$). This resulted in a median photic depth (to 1% incident PAR) of ~22 m. Molar ratios of N to P suggested a general P limitation of the water column; median TN:TP = 67 and median DIN:SRP = 49.

## STATIONS GROUPED BY DEPTH

Some general differences were observed for those stations at a depth of >3 m when both surface and bottom concentrations were measured. Temperature, DO, TOC, and TON were significantly higher at the surface, while salinity, NO$_3^-$, NO$_2^-$, NH$_4^+$, TP, and turbidity were significantly higher in bottom waters. No significant differences in SRP, APA, or Si(OH)$_4$ with depth were found.

## STATIONS GROUPED BY SEGMENT

NO$_3^-$ was highest in the Backcountry (0.18 $\mu M$), followed by the Lower and Middle Keys. Interestingly, NO$_3^-$ concentrations in the Sluiceway (0.08 $\mu M$) were not significantly different than the Upper Keys, Tortugas, or Marquesas. Median NO$_2^-$ concentrations in the Tortugas and Marquesas (0.03 $\mu M$) were significantly lower than for the other Keys segments. NH$_4^+$ was highest in the

**TABLE 22.1**
**Summary Statistics for Each Water Quality Variable in the FKNMS**

|  | Median | Minimum | Maximum | $n$ |
|---|---|---|---|---|
| $NO_3^-$ | 0.08 | 0.00 | 1.61 | 862 |
| $NO_2^-$ | 0.04 | 0.00 | 0.22 | 1180 |
| $NH_4^+$ | 0.27 | 0.01 | 2.44 | 1191 |
| TON | 9.65 | 3.64 | 42.67 | 1180 |
| TP | 0.17 | 0.01 | 0.66 | 1183 |
| SRP | 0.01 | 0.00 | 0.12 | 915 |
| APA | 0.04 | 0.01 | 0.84 | 1029 |
| Chlorophyll $a$ | 0.25 | 0.01 | 6.81 | 1188 |
| TOC | 198.70 | 86.98 | 1054.79 | 1186 |
| $Si(OH)_4$ | 0.59 | 0.00 | 37.36 | 902 |
| Turbidity | 0.46 | 0.01 | 12.97 | 1083 |
| Salinity | 36.20 | 31.80 | 38.40 | 1140 |
| Temperature | 26.45 | 17.50 | 39.60 | 1146 |
| DO | 6.20 | 4.60 | 10.40 | 1123 |
| $DO_{sat}$ | 92.50 | 68.28 | 150.03 | 1123 |
| $K_d$ | 0.16 | 0.01 | 1.12 | 1138 |
| TN:TP | 62.10 | 15.89 | 1356.21 | 1181 |
| DIN:SRP | 43.83 | 1.40 | 935.33 | 906 |
| $\%I_o$ | 17.21 | 0.00 | 97.03 | 1138 |

Backcountry (0.46 $\mu M$) and lowest in the Tortugas, Marquesas, and Upper Keys (~0.25 $\mu M$). The Middle Keys had the most variability in DIN for any of the oceanside segments.

Total phosphorus was highest in the Backcountry and Sluiceway (~0.21 $\mu M$) and lowest in the Upper Keys (0.15 $\mu M$), with the remaining segments being intermediate (Figure 22.2). SRP was very low (~0.01 $\mu M$) for all areas but was slightly elevated and most variable in the Marquesas and Backcountry. Median $Si(OH)_4$ concentrations were highest in the Sluiceway (4.9 $\mu M$); lowest in the Tortugas, Marquesas, and Upper Keys (~0.45 $\mu M$); and intermediate in the Backcountry, Lower, and Middle Keys (~1.4 $\mu M$). Consistently higher chlorophyll $a$ concentrations were observed in the Marquesas (0.36 $\mu$g l$^{-1}$) than for any other area of the FKNMS (Figure 22.2). Lowest chlorophyll $a$ concentrations were found in the Upper Keys (0.20 $\mu$g l$^{-1}$).

The organic C and N pools as well as APA showed remarkable similarity in relative concentration among segments (Figure 22.3). Highest TOC (~260 $\mu M$), TON (~15.5 $\mu M$), and APA (~0.08 $\mu M$ hr$^{-1}$) were observed in the Backcountry and Sluiceway which declined southwest towards the Tortugas and northeast towards the Upper Keys. Median $DO_{sat}$ was relatively similar among segments but was significantly higher and more variable in the Backcountry and Sluiceway (~95%). This result would not have been evident had we only reported DO in mg l$^{-1}$ as it was not significant across segments.

Salinity was comparable for most segments but was slightly lower in the Sluiceway (Figure 22.3). Salinity in the Sluiceway and Backcountry were highly variable and precluded any statistical discrimination from the other segments. Turbidity was highest in the Backcountry (1.0 NTU), Sluiceway (0.85 NTU), and Marquesas (0.83 NTU), with lowest turbidity occurring in the Tortugas segment (0.35 NTU; Figure 22.3). The shallow Quicksands area in the Marquesas probably accounted for the elevated turbidity in this segment. The Middle Keys showed high variability in turbidity although the overall median was low.

The Tortugas, Marquesas, and Lower Keys had significantly higher water temperature (~27.3°C) than the other segments (Figure 22.4). Light attenuation showed a pattern similar to turbidity, with highest $K_d$ in Sluiceway (0.36 m$^{-1}$) and Backcountry stations (0.31 m$^{-1}$) and lowest $K_d$ in the

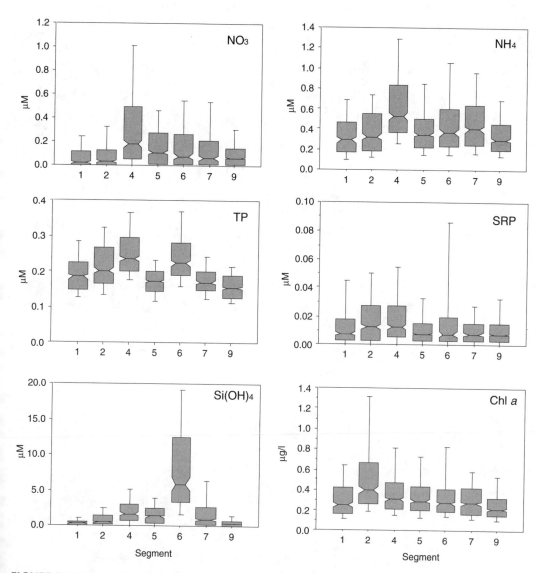

**FIGURE 22.2** Box-and-whisker plots of $NO_3^-$, $NH_4^+$, TP, SRP, $Si(OH)_4$, and chlorophyll *a* stratified by FKNMS segment: 1, Tortugas; 2, Marquesas; 4, Backcountry; 5, Lower Keys; 6, Sluiceway; 7, Middle Keys; and 9, Upper Keys.

Tortugas (0.12 m$^{-1}$). This works out to respective median photic depths of 13, 15, and 38 m. Median TN:TP ratios in the Tortugas (55) and Marquesas (52) were significantly lower than in the other segments (Figure 22.4). Much of this difference was due to decreased TON concentrations in these areas rather than increased in TP. Lowest DIN:SRP ratios were found in the Marquesas (29) followed by the Tortugas (39). Decreased DIN as well as elevated SRP in the Marquesas relative to the Tortugas were responsible for these differences.

## STATIONS GROUPED BY DISTANCE ALONG TRANSECT

Median concentrations of $NO_3^-$ in the Middle and Lower Keys were significantly higher in alongshore stations than those of Hawk Channel and Reef Tract (Figure 22.5). Alongshore $NO_3^-$ in the Upper Keys and Tortugas (~0.1 µ*M*) was not nearly as high as found in the Middle and Lower

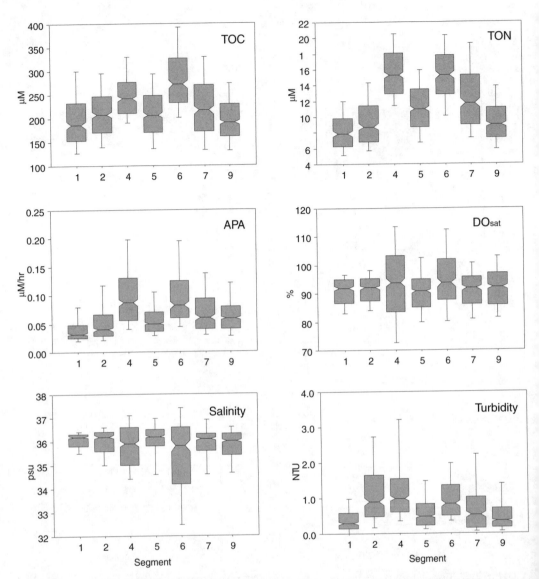

**FIGURE 22.3** Box-and-whisker plots of TOC, TON, APA, $DO_{sat}$, salinity, and turbidity by FKNMS segment.

Keys (~2.5 $\mu M$). $NO_3^-$ concentrations on the Reef Tract and offshore in the Tortugas and Upper and Middle Keys were comparable (~0.05 $\mu M$) and were all significantly lower than for the Lower Keys. $NH_4^+$ concentrations followed trends similar to $NO_3^-$, being higher in Alongshore stations in the Middle and Lower Keys and declining with distance offshore (Figure 22.5). Alongshore $NH_4^+$ was highest in the Middle Keys (~0.5 $\mu M$). No significant differences in $NH_4^+$ were seen among Hawk Channel, Reef Tract, and Tortugas groups (~0.3 $\mu M$).

Alongshore TP concentrations (~0.18 $\mu M$) were significantly higher than the Reef Tract only in the Lower Keys (Figure 22.5). The Middle Keys and Tortugas showed no offshore trend, while the Upper Keys showed a slight increasing trend in TP from shore to reef. TP concentrations in the Tortugas and Lower and Middle Keys were comparable (~0.18 $\mu M$), while the Upper Keys were lowest overall (~0.16 $\mu M$). The major trends in TP were mirrored by SRP but were not statistically significant. Median $Si(OH)_4$ concentrations dropped dramatically with distance offshore in the Middle Keys (Figure 22.5). In the Lower Keys, $Si(OH)_4$ was significantly lower only in the

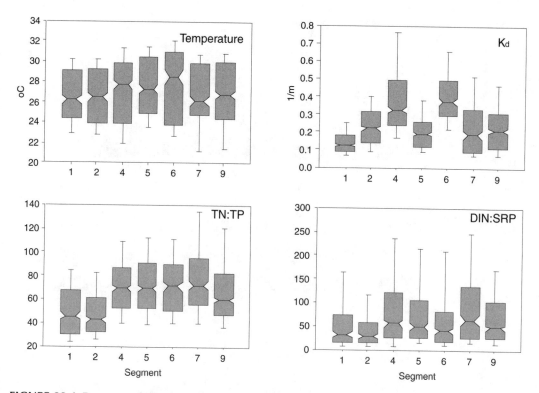

**FIGURE 22.4** Box-and-whisker plots of temperature, $K_d$, TN:TP, and DIN:SRP by FKNMS segment.

Reef Tract stations. There was no difference in $Si(OH)_4$ concentrations in the Upper Keys or Tortugas transects. Alongshore $Si(OH)_4$ concentrations were highest in the Middle Keys (~3 $\mu M$), while Reef Tract concentrations were highest in the Lower Keys (~0.5 $\mu M$).

There was no significant trend in chlorophyll *a* with distance from land in the Lower Keys (Figure 22.6), although there was a slight decline in the Middle Keys and a small increase in the Upper Keys. Chlorophyll *a* in the Offshore Tortugas sites was significantly lower than Alongshore and Hawk Channel sites and was comparable to levels in the Upper Keys (0.2 $\mu g$ $^{-1}$). TOC in the Lower, Middle, and Upper Keys was elevated at Alongshore stations and declined sequentially through Hawk Channel to the Reef Tract (Figure 22.6). No significant difference in TOC within Tortugas groups (~170 $\mu M$) was found, and it was similar to Reef Tract concentrations in the Keys. Highest TOC in Alongshore stations occurred in the Middle Keys (~250 $\mu M$). TON concentrations exhibited patterns similar to those for TOC (data not shown).

Turbidity in all segments declined significantly with distance from land (Figure 22.6). All Reef Tract and offshore Tortugas sites had comparably low turbidity levels (~0.2 NTU). Highest Alongshore turbidity was found in the Middle Keys (1.3 NTU). No significant differences in Alongshore turbidity in the Tortugas and Lower and Upper Keys were observed (~0.6 NTU). Trends in median salinity with distance offshore were small; trends in salinity variability were large (Figure 22.6). Salinity from Alongshore to Reef Tract increased significantly in the Upper Keys, whereas in the Lower Keys salinity actually decreased offshore. No significant change in salinity was observed along the Middle Keys, Marquesas, or Tortugas transects. In all segments, the Alongshore station salinities were much more variable than those of Reef Tract and offshore. Reef Tract and Tortugas offshore salinities were not significantly different; therefore, Alongshore salinity in the Lower Keys was higher than local seawater values, while Alongshore station salinity in the Upper Keys was depressed relative to local seawater values.

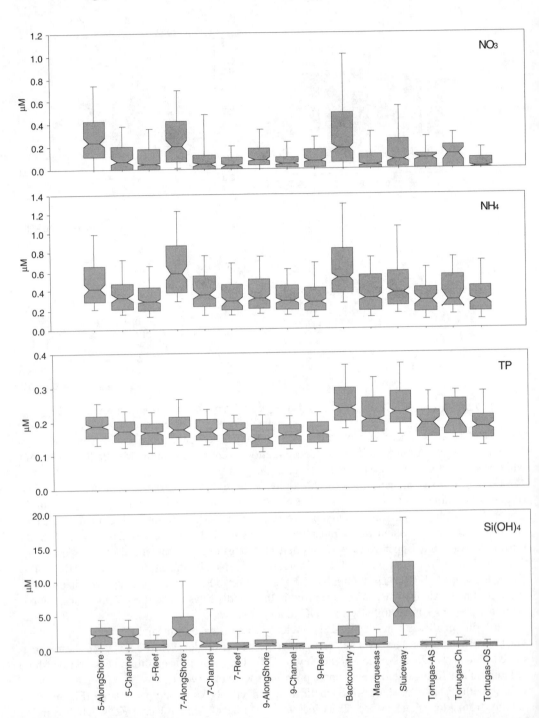

**FIGURE 22.5** Box-and-whisker plots of $NO_3^-$, $NH_4^+$, TP, $Si(OH)_4$ by FKNMS segment and location on transect from land. In the Tortugas segment, AS = alongshore, CH = channel, and OS = offshore.

## STATIONS GROUPED BY SHORE TYPE

Oceanside transects showed marked differences in water quality when grouped by shore type (Figure 22.7). Transects situated on open channels (PASS) through the Keys were elevated in $NO_3^-$, $NH_4^+$, TP, $Si(OH)_4$, chlorophyll $a$, TOC, and turbidity relative to those against the island chain

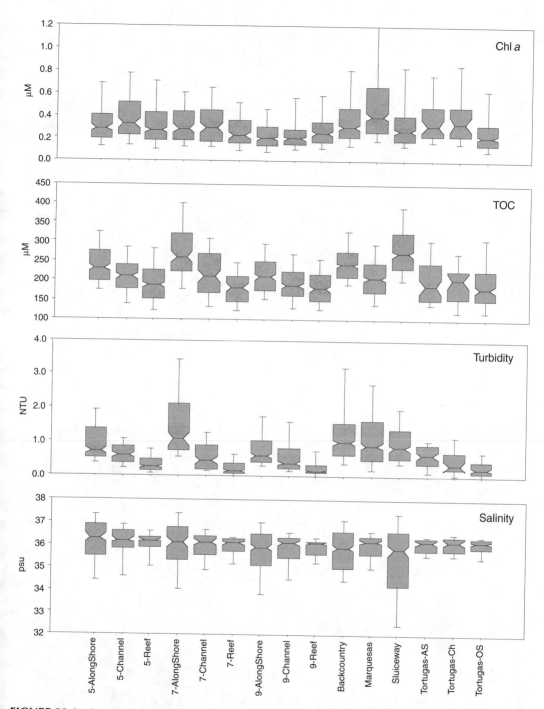

**FIGURE 22.6** Box-and-whisker plots of chlorophyll *a*, TOC, turbidity, and salinity by FKNMS segment and location on transect from land.

(LAND). Both salinity and temperature were significantly lower in PASS transects than for LAND. Although these differences were statistically significant, the absolute differences were very small, being only fractional. We also found that these effects diminished rapidly with distance offshore (data not shown). Interestingly, those transects located along Biscayne National Park (Old Rhodes Key, Elliot Key, the Safety Valve) were lowest of all for $NO_3^-$, $NH_4^+$, TP, $Si(OH)_4$, and chlorophyll *a*.

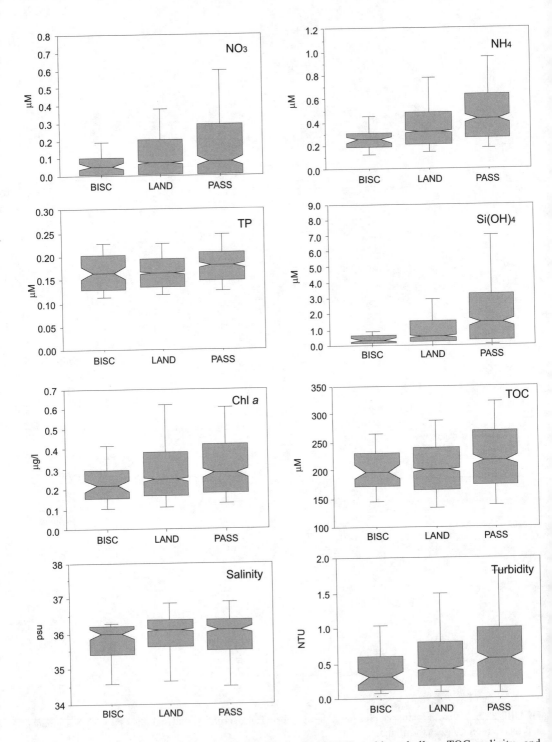

**FIGURE 22.7** Box-and-whisker plots of $NO_3^-$, $NH_4^+$, TP, $Si(OH)_4$, chlorophyll $a$, TOC, salinity, and turbidity for oceanside Keys transects stratified as being located along Biscayne Bay National Park (BISC), against land in Key Largo, Middle and Lower Keys (LAND), or in line with major passes in the Keys (PASS).

# DISCUSSION

Distinguishing internal from external sources of nutrients in the FKNMS is a difficult task. The finer discrimination of internal sources into natural and anthropogenic inputs is even more difficult. Most of the important anthropogenic inputs are regulated and most likely controlled by management activities; however, recent studies have shown that nutrients from shallow sewage injection wells may be leaking into nearshore surface waters (Corbett et al., 1999). Advective transport of nutrients through the FKNMS was not measured by the existing fixed sampling plan; however, nutrient distribution patterns may be compared to the regional circulation regimes in an effort to visualize the contribution of external sources and advective transport to internal water quality of the FKNMS.

Circulation in coastal South Florida is dominated by regional currents such as the Loop Current, Florida Current, and Tortugas Gyre and by local transport via Hawk Channel and alongshore Shelf movements (Klein and Orlando, 1994). Regional currents may influence water quality over large areas by the advection of external surface water masses into and through the FKNMS (Lee et al., 1994) and by the intrusion of deep offshore ocean waters onto the reef tract as internal bores (Leichter et al., 1996). Local currents become more important in the mixing and transport of freshwater and nutrients from terrestrial sources (Smith, 1994; Pitts, 1997).

Spatial patterns of salinity in coastal South Florida show these major sources of freshwater to have more than just local impacts (Color Figure 22.1*). In Biscayne Bay, freshwater is released through the canal system operated by SFWMD; the impact is clearly seen to affect northern Key Largo by causing a depression in median salinity coupled with high variability in alongshore sites (Figures 22.6 and 22.7). Freshwater entering northeast Florida Bay via overland flow from Taylor Slough and C-111 basin in ENP can be seen to mix in a southwest direction. The extent of influence of freshwater from Florida Bay on alongshore salinity in the Keys is less than that of Biscayne Bay but it is more episodic. Transport of low-salinity water from Florida Bay does not affect the Middle Keys sites enough to depress the median salinity in this region but is manifested as increased variability (Figure 22.6). On the west coast, the large influence of the Shark River Slough, which drains the bulk of the Everglades and exits through the Whitewater Bay/Ten Thousand Islands mangrove complex, is clearly seen as impacting the Shelf waters (Color Figure 22.1). The mixing of Shelf waters with the Gulf of Mexico produces a salinity gradient in a southwest direction that extends out to Key West. This freshwater source does not seem to impact the Backcountry because of its shallow nature but instead follows a trajectory of entering western Florida Bay and exiting out through the channels in the Middle Keys (Smith, 1994). This net transport of lower salinity water from mainland to reef in open channels through the Keys is observed more as an increase in the range and variability of salinity than as a large depression in salinity (Figure 22.7).

In addition to surface currents, internal tidal bores regularly impact the Key Largo reef tract (Leichter et al., 1996). Internal bores are episodes of higher density, deep water intrusion onto the shallower shelf or reef tract. Depending on their energy, internal tidal bores can promote stratification of the water column or cause complete vertical mixing as a breaking internal wave of subthermocline water. To determine the extent of stratification, we calculated the difference between surface and bottom density, delta sigma-t ($\Delta\sigma_t$), where positive values denoted greater density of bottom water relative to the surface. The resulting graph of $\Delta\sigma_t$ (Color Figure 22.2), shows that the southwest area of the Tortugas segment tends to experience the greatest frequency of stratification events. The decreased temperature and increased salinity in bottom waters from intrusion of deeper denser oceanic waters to this region may also account for increases in $NO_3^-$, TP, and SRP in these bottom waters. For example, in April 1998 a mass of colder, nutrient-laden water from the Gulf of Mexico moved up onto the Tortugas reefs and fueled a large benthic macroalgae bloom (J. Porter, pers. comm.). This event was observed throughout most of the eastern Gulf as far north as Pensacola. At the two most southwest stations (Table 22.2), temperatures dropped ~4°C, $NO_3^-$ increased 3

---

* Color figures follow page 648.

**TABLE 22.2**
**Physical and Chemical Differences Between Surface and Bottom Waters During April 25, 1998, Stratification Event at Two Sites in the Southwest Tortugas (Segment 1) of the FKNMS**

|  | Station #345 | | Station #350 | |
|---|---|---|---|---|
|  | Surface | Bottom | Surface | Bottom |
| Salinity | 36.00 | 36.00 | 36.10 | 36.00 |
| Temperature | 24.00 | 19.70 | 23.60 | 20.20 |
| $NO_3^-$ | 0.05 | 3.66 | 0.00 | 3.23 |
| $NO_2^-$ | 0.01 | 0.81 | 0.04 | 0.35 |
| $NH_4^+$ | 0.94 | 1.14 | 0.32 | 0.57 |
| TP | 0.23 | 0.39 | 0.19 | 0.40 |
| SRP | 0.02 | 0.22 | 0.01 | 0.20 |
| $Si(OH)_4$ | 0.09 | 2.07 | 0.29 | 2.26 |
| TOC | 170.10 | 155.48 | 171.42 | 168.35 |
| TON | 4.86 | 2.25 | 5.45 | 4.01 |
| Chlorophyll $a$ | 0.94 | 1.62 | 1.10 | 1.82 |
| Turbidity | 0.17 | 0.50 | 0.43 | 1.11 |
| $DO_{sat}$ | 94.28 | 80.91 | 93.90 | 78.74 |

orders of magnitude, SRP and $Si(OH)_4$ increased by a factor of 100, and TP, turbidity, and *in vivo* chlorophyll $a$ specific fluorescence (measured via CTD) all doubled. As there was only a small increase in $NH_4^+$ during this event, we believe the general case of elevated $NH_4^+$ and turbidity found in bottom waters throughout the FKNMS is most probably due to benthic flux and resuspension and not to subthermocline advection.

Surface $Si(OH)_4$ concentrations exhibited a pattern similar to salinity (Color Figure 22.3). The source of $Si(OH)_4$ in this geologic area of carbonate rock and sediments is from siliceous periphyton (diatoms) growing in the Shark River Slough, Taylor River Slough, and C-111 basin watersheds. Unlike the Mississippi River plume with chlorophyll $a$ concentrations of 76 µg l$^{-1}$ (Nelson and Dortch, 1996), phytoplankton biomass on the Shelf (1 to 2 µg l$^{-1}$ chlorophyll $a$) was not sufficient to account for the depletion of $Si(OH)_4$ in this area. Therefore, $Si(OH)_4$ concentrations on the Shelf were rapidly depleted by mixing alone, allowing us to use $Si(OH)_4$ as a semi-conservative tracer of freshwater in this system (Ryther et al., 1967; Moore et al., 1986). Unlike Florida Bay and the west coast, very little $Si(OH)_4$ loading to southern Biscayne Bay was found, mostly because the source of freshwater to this system is from canals which drain agricultural and urban areas of Dade County.

In the Lower and Middle Keys, it is clear that the source of $Si(OH)_4$ to the nearshore Atlantic waters is through the Sluiceway and Backcountry. $Si(OH)_4$ concentrations near the coast were elevated relative to the reef tract, with much higher concentrations occurring in the Lower and Middle Keys than in the Upper Keys (Figure 22.5). An interesting peak in $Si(OH)_4$ concentration can be seen in an area of the Sluiceway that is densely covered with the seagrass *Syringodium*. We are unsure as to the source but postulate that it may be due to benthic flux.

Visualization of spatial patterns of $NO_3^-$ concentration over South Florida waters provide an extended view of source gradients over the region (Color Figure 22.4). Biscayne Bay, Florida Bay, and the Shark River area of the west coast exhibited high $NO_3^-$ concentrations relative to the FKNMS and Shelf. Elevated $NO_3^-$ in Biscayne Bay is the result of loading from both the canal drainage system and from inshore groundwater (Alleman et al., 1995; Meeder et al., 1997). The source of $NO_3^-$ to Florida Bay is the Taylor River Slough and C-111 basin (Boyer and Jones, 1999; Rudnick et al., 1999), while the Shark River Slough impacts the west coast mangrove rivers and out onto the Shelf (Rudnick et al., 1999). We speculate that in both cases elevated $NO_3^-$ concentrations are the result of $N_2$ fixation/nitrification within the mangroves (Pelegri and Twilley, 1998). The oceanside transects off the uninhabited Upper Keys (off Biscayne Bay in segment 9) exhibited the lowest alongshore $NO_3^-$ compared to the Middle and Lower Keys. A similar pattern was observed in a previous transect survey from these areas (Szmant and Forrester, 1996). The authors also showed an inshore elevation of $NO_3^-$ relative to Hawk Channel and the reef tract which is also demonstrated in our analysis (Figure 22.5). Interestingly, $NO_3^-$ concentrations in all stations in the Tortugas transect were similar to those of reef tract sites in the mainland Keys; no inshore elevation of $NO_3^-$ on the transect off uninhabited Loggerhead Key was found. We suggest this difference is the due to the lack of human shoreline development in this area.

Color Figure 22.3 also shows that a distinct intensification of $NO_3^-$ occurs in the Backcountry region. Part of this increase may due to a local sources of $NO_3^-$, (septic systems and stormwater runoff around Big Pine Key) (Lapointe and Clark, 1992). However, another area, the Snipe Keys, also exhibits high $NO_3^-$ but is uninhabited by humans. This rules out the premise of septic systems being the only source of $NO_3^-$ in this area. It is important to note that the Backcountry area is very shallow (~0.5 m) and hydraulically isolated from the Shelf and Atlantic which results in its having a relatively long water residence time. Elevated $NO_3^-$ concentrations may be partially due to simple evaporative concentration as is seen in locally elevated salinity values (Color Figure 22.1). Another possibility is significant contribution of benthic $N_2$ fixation/nitrification in this very shallow area.

Total dissolved $NH_4^+$ concentrations were distributed in a manner similar to $NO_3^-$ with highest concentrations occurring in Florida Bay, the Ten Thousand Islands, and the Backcountry (data not shown). $NH_4^+$ concentrations were very low in Biscayne Bay because it is not a major component of loading from the canal drainage system. $NH_4^+$ also showed similarities with $NO_3^-$ in its spatial distribution, being lowest in the Upper Keys and highest inshore relative to offshore. No alongshore elevation of $NH_4^+$ concentrations was seen in the Tortugas, where levels were similar to those of reef tract sites in the mainland Keys. That the least developed portion of the Upper Keys in Biscayne National Park and uninhabited Loggerhead Key (Tortugas) exhibited lowest $NO_3^-$ and $NH_4^+$ concentrations is evidence of a local anthropogenic source for both of these variables along the oceanside of the Upper, Middle, and Lower Keys. This pattern of decline implies an onshore nitrogen source which is diluted with distance from land by low nutrient Atlantic Ocean waters.

Elevated DIN concentrations in the Backcountry, on the other hand, are not so easily explained. We postulate that the high concentrations found there are due to a combination of anthropogenic loading, physical entrapment, and benthic $N_2$ fixation. The relative contribution of these potential sources is unknown. Lapointe and Matzie (1996) have shown that stormwater and septic systems are responsible for increased DIN loading in and around Big Pine Key. The effect of increased water residence time in DIN concentration is probably small. Salinities in this area were only 1 to 2 psu higher than local seawater which resulted in a concentration effect of only 5 to 6%. Benthic $N_2$ fixation may potentially be very important in the nitrogen budget of the Backcountry. Measured rates of $N_2$ fixation in a *Thalassia* bed in Biscayne Bay, having very similar physical and chemical conditions, were 540 $\mu$mol N m$^{-2}$ d$^{-1}$ (Capone and Taylor, 1980). Without the plant community nitrogen demand, one day of $N_2$ fixation has the potential to generate a water column concentration of >1 $\mu M$ $NH_4^+$ (0.5 m deep). Much of this $NH_4^+$ is probably nitrified and may help account for the elevated $NO_3^-$ concentrations observed in this area as well. Clearly, $N_2$ fixation may be a significant component of the nitrogen budget in the Backcountry and it may be a exported as DIN to the FKNMS in general.

Spatial patterns in TP in South Florida coastal waters were strongly driven by the west coast sources (Color Figure 22.5). A small gradient in TP extended from the inshore waters of the Whitewater Bay/Ten Thousand Islands mangrove complex out onto the Shelf and Tortugas. A weak gradient also extended from northcentral Florida Bay to the Middle Keys. Brand (1998) has postulated that groundwater from a subterranean Miocene quartz sand channel, "the river of sand," containing high levels of phosphorus is the source of TP in this region. However, little evidence of this source exists to date and field data from Florida Bay do not indicate a subterranean source (Corbett et al., 1999; Boyer and Jones, unpubl. data). Finally, there was no evidence of a significant terrestrial source of TP to Biscayne Bay.

In the Keys, evidence of elevated TP in alongshore stations of the Middle and Lower Keys could be seen, but the differences were very small (Figure 22.5). The Upper Keys actually showed higher TP concentrations on the reef tract than inshore, implying an offshore source. Interestingly, the Tortugas area had higher TP concentrations than the Upper Keys as a result of Shelf water advection.

In South Florida coastal waters, very little TP is found in the inorganic form (SRP, $PO_4^-$); most is organic P (TOP). The distribution of SRP on the west coast and Shelf was similar to that of TP with the general gradient from the west coast to Tortugas remaining (Color Figure 22.6); however, the SRP distribution was distinctly different from that of TP in Florida Bay, Whitewater Bay, and Biscayne Bay. In central Florida Bay, the north–south gradient previously observed for TP was highly diminished for SRP, indicating that almost all the TP in central Florida Bay was in the form of TOP. It is unlikely that the source of TOP to this region is from overland flow or groundwater as this is also the region that expresses highest salinity. Alternately, we hypothesize that the presence of the Flamingo channel, running parallel to the southern coastline of Cape Sable, acts as a tidal conduit for episodic advection of inshore Shelf water to enter northcentral Florida Bay. Subsequent trapping and evaporation then may act to concentrate TOP in this region. The second difference in phosphorus distributions was the significant SRP gradient present in northeast Florida Bay that was not observed for TP. The sources of SRP to this area are the Taylor River Slough and C-111 basin (W. Walker, pers. comm.; Boyer and Jones, 1999; Rudnick et al., 1999).

Whitewater Bay displayed an east–west gradient in SRP concentrations which increased with salinity, leading us to conclude that the freshwater inputs from the Everglades were not a source of SRP to this area. Finally, there was evidence of a significant onshore–offshore SRP gradient in southern Biscayne Bay, most probably as a direct result of canal loading and groundwater seepage to this region (A. Lietz, pers. comm.; Meeder et al., 1997).

Concentrations of TOC (Color Figure 22.7) and TON (not shown) are remarkably similar in pattern of distribution across the South Florida coastal hydroscape. The decreasing gradient from west coast to Tortugas was very similar to that of TP. A steep gradient with distance from land was observed in Biscayne Bay. Both these gradients were due to terrestrial loading. On the west coast, the source of TOC and TON was from the mangrove forests. Our data from this area show that concentrations of TOC and TON increased from Everglades headwaters through the mangrove zone and then decreased with distance offshore. In Biscayne Bay, much of the TOC and TON is from agricultural land use. The high concentrations of TOC and TON found in Florida Bay were due to a combination of terrestrial loading (Boyer and Jones, 1999), *in situ* production by seagrass and phytoplankton, and evaporative concentration (Fourqurean et al., 1993).

Advection of Shelf and Florida Bay waters through the Sluiceway and passes accounted for this region and the inshore area of the Middle Keys having highest TOC and TON of the FKNMS (Figure 22.7). Strong offshore gradients in TOC and TON existed for all mainland Keys segments (Figure 22.6) but not for the Tortugas transect. Part of this difference may be explained by the absence of mangroves in the single Tortugas transect. The higher concentrations of TOC and TON in the inshore waters of the Keys implies a terrestrial source rather than simply benthic production and sediment resuspension. Main Keys reef tract concentrations of TOC and TON were similar to those found in the Tortugas.

Much emphasis has been placed on assessing the impact of episodic phytoplankton blooms in Florida Bay on the offshore reef tract environment. Spatial patterns of chlorophyll *a* concentrations (Color Figure 22.8) showed that northwest Florida Bay, Whitewater Bay, and the Ten Thousand Islands exhibited high levels of chlorophyll *a* relative to Biscayne Bay, the Shelf, and FKNMS. The highest chlorophyll *a* concentrations were found in west coast mangrove estuaries (up to 45 µg l$^{-1}$ in Alligator Bay, TTI). Chlorophyll *a* is also routinely high (~2 µg l$^{-1}$) in northwest Florida Bay along the channel connecting the Shelf to Flamingo Key. It is interesting that chlorophyll *a* concentrations are higher in the Marquesas (0.36 µg l$^{-1}$) than in other areas of the FKNMS (Figure 22.2). When examined in context with the entire South Florida ecosystem, it is obvious that the Marquesas zone should be considered a continuum of the Shelf rather than a separate management entity. This shallow sandy area (often called the Quicksands) acts as a physical mixing zone between the Shelf and the Atlantic Ocean and is a highly productive area for other biota, as it encompasses the historically rich Tortugas shrimping grounds. A chlorophyll *a* concentration of 1 µg l$^{-1}$ in the water column of a reef tract might be considered a problem as it indicates eutrophication. Conversely, a similar chlorophyll *a* level in the Quicksands indicates a productive ecosystem that feeds a valuable shrimp fishery.

The oceanside transects in the Upper Keys (segment 9) exhibited the lowest overall chlorophyll *a* concentrations of any zone in the FKNMS. Ocean transects showed a slight increase in chlorophyll *a* on the reef tract in this area (Figure 22.6). Transects off the Middle and Lower Keys showed that a drop in chlorophyll *a* occurred at reef tract sites with no linear decline with distance from shore (data not shown). Interestingly, chlorophyll *a* concentrations in the Tortugas transect showed a pattern similar to the mainland Keys. Inshore and Hawk Channel chlorophyll *a* concentrations among Middle Keys, Lower Keys, and Tortugas sites were not significantly different. As inshore chlorophyll *a* concentrations in the Tortugas were similar to those in the Middle and Lower Keys, we see no evidence of persistent phytoplankton bloom transport from Florida Bay; however, we did see some evidence of increased chlorophyll *a* in those stations situated along the major passes in the Keys relative to those abutting land (Figure 22.7). The differences between these two groupings were very small (0.25 vs. 0.20 µg l$^{-1}$), suggesting that although significant bloom transport events do occur, they are not routinely observed with a quarterly sampling program.

Along with phosphorus concentration, turbidity is probably the second most important determinant of local ecosystem health. The fine, low-density carbonate sediments in this area are easily resuspended, rapidly transported, and have high light-scattering potential per gram of material. High-water column turbidity and transport directly affect filter-feeding organisms by clogging their feeding apparatus and by increasing local sedimentation rate. Sustained high turbidity of the water column indirectly affects benthic community structure by decreasing light penetration and promoting seagrasses extinction. Large-scale observations of turbidity clearly show patterns of onshore–offshore gradients that extend out onto the Shelf to the Marquesas (Color Figure 22.9). In the last 7 years, turbidities in Florida Bay have increased dramatically in the northeast and central regions (Boyer et al., 1999), potentially as a consequence of destabilization of the sediment from seagrass dieoff (Robblee et al., 1991).

Strong turbidity gradients were observed for all Keys transects (Figure 22.6), but reef tract levels were remarkably similar regardless of inshore levels. High alongshore turbidity is most probably due to the shallow water column being easily resuspended by wind and wave action. Inshore stations in the Middle Keys had higher turbidity than other segments. Transects aligned with major passes had slightly greater turbidity than those against land but the difference was not statistically significant (Figure 22.7). Light extinction ($K_d$) was highest alongshore and improved with distance from land (data not shown). This trend was expected, as light extinction is directly related to water turbidity.

Using the DIN:SRP ratio is a relatively simple method of determining phytoplankton nutrient limitation status of the water column (Redfield, 1958). Most of the South Florida hydroscape was shown to have DIN:SRP values >> 16:1, indicating the potential for phytoplankton to be limited by phosphorus at these sites (Color Figure 22.10). The bulk of Florida Bay and both southern and northern Biscayne Bay were severely P limited, mostly as a result of high DIN concentrations. All of the FKNMS is routinely P limited using this metric. Interestingly, the Marquesas/Quicksands area was the least P limited of all zones and exhibited a significant regression between SRP and chlorophyll *a*. Only in the northern Ten Thousand Islands and Shelf did nitrogen become the limiting nutrient. The south–north shift from P to N limitation observed in the west coast estuaries has been ascribed to changes in land use and bedrock geochemistry of the watersheds (Boyer and Jones,

1998). The west coast south of 25.4 N latitude is influenced by overland freshwater flow from the Everglades and Shark River Slough having very low phosphorus concentrations relative to nitrogen. Above 25.7 N latitude, the bedrock geology of the watershed changes from carbonate- to silicate-based and land use changes from relatively undeveloped wetland (Big Cypress Basin) to a highly urban/agricultural mix (Naples, FL).

The large scale of this monitoring program has allowed us to assemble a much more holistic view of broad physical/chemical/biological interactions occurring over the South Florida hydroscape. Much information has been gained by inference from this type of data collection program: Major nutrient sources have been confirmed, relative differences in geographical determinants of water quality have been demonstrated, and large-scale transport via circulation pathways has been elucidated. In addition, we have shown the importance of looking "outside the box" for questions asked within. Rather than thinking of water quality monitoring as being a static, nonscientific pursuit, it should be viewed as a tool for answering management questions and developing new scientific hypotheses. One of the more important management questions to be answered is "Is the water quality better or worse than it used to be?" As it stands, this monitoring program based on quarterly sample intervals may require up to 10 years before small trends may be detected because of seasonal variability and background noise.

## CONCLUSIONS

Results of an ongoing, spatially extensive water quality monitoring program in the FKNMS have been reported. Stratification of sites according to depth showed that temperature, dissolved oxygen, total organic carbon, and total organic nitrogen were significantly higher at the surface, while salinity, nitrate, nitrite, ammonium, total phosphorus, and turbidity were greater in bottom waters. Stratification according to geographical region showed that the Upper Keys generally had lower nutrient concentrations than the Middle or Lower Keys. In the Lower Keys, inorganic nitrogen and total phosphorus were elevated in the Backcountry area. The offshore Marquesas/Quicksands area exhibited the highest phytoplankton biomass (chlorophyll a) for any segment of the FKNMS. Declining inshore to offshore trends were observed for $NO_3^-$, $NH_4^+$, $Si(OH)_4$, TOC, TON, and turbidity for all oceanside transects. TP concentrations in the Lower Keys transects decreased with distance offshore but increased along transects in the Upper Keys, mostly because of low concentrations alongshore. Stations stratified as being off land or channel/pass showed that those stations situated along channels/passes through the Keys had higher nutrient concentrations, phytoplankton biomass, and turbidity than those stations off land. The differences were statistically significant but the absolute differences were very small and not likely to be biologically important. The water quality of the FKNMS is put in perspective with data from 150 stations sampled in the southwest Florida Shelf Florida Bay, Whitewater Bay, Ten Thousand Islands, and Biscayne Bay. It becomes clear from this analysis that the ambient water quality in the Lower Keys and Marquesas is most strongly influenced by water quality of the southwest Florida Shelf; the Middle Keys, by southwest Florida Shelf and Florida Bay transport; the Backcountry, by internal nutrient sources; and the Upper Keys, by Florida Current intrusion.

## ACKNOWLEDGMENTS

We thank all the field and laboratory technicians involved with this project including: J. Absten, O. Beceiro, N. Black, S. Bolanos, C. Covas, D. Diaz, T. Frankovich, B. Gilhooly, S. Kaczynski, E. Kotler, S. Perez, P. Sterling, F. Tam, and especially Pete Lorenzo. We want to thank the captains and crew of the *R/V Bellows* and *R/V Suncoaster* of the Florida Institute of Oceanography for their professional support of the monitoring program. We also appreciate the laboratory and lodging support supplied by the Florida Program of the National Underwater Research Center,

UNC-Wilmington. This project was possible due to continued funding by the U.S. EPA under Agreement #X994621-94-0. Funding for other portions of the SERC Water Quality Monitoring Network was provided by the South Florida Water Management District and Everglades National Park (SFWMD/NPS Coop. Agreement C-7919 and NPS/SERC Coop. Agreement 5280-2-9017). This is contribution #155 of the Southeast Environmental Research Center at Florida International University.

This publication is dedicated to the memory of Cristina Menendez.


# REFERENCES

Alleman, R.W. et al., 1995. *Biscayne Bay Surface Water Improvement and Management*, technical supporting document, South Florida Water Management District, West Palm Beach, FL.

Boyer, J.N. and R.D. Jones. 1998. Influence of coastal morphology and watershed characteristics on the water quality of mangrove estuaries in the Ten Thousand Islands–Whitewater Bay complex. Proceedings of the 1998 Florida Bay Science Conference, University of Florida Sea Grant.

Boyer, J.N. and R.D. Jones. 1999. Effects of freshwater inputs and loading of phosphorus and nitrogen on the water quality of Eastern Florida Bay, pp. 545–561, in Phosphorus Biogeochemistry in Sub-Tropical Ecosystems, Reddy, K.R., O'Connor, G.A., and Schelske, C.L., Eds., Lewis Publishers, Boca Raton, FL.

Boyer, J.N., J.W. Fourqurean, and R.D. Jones. 1997. Spatial characterization of water quality in Florida Bay and Whitewater Bay by multivariate analysis: zones of similar influence (ZSI), *Estuaries*, 20:743–758.

Boyer, J.N., J.W. Fourqurean, and R.D. Jones. 1999. Seasonal and long term trends in the water quality of Florida Bay (1989–1997), *Estuaries*, 22:417–430.

Brand, L. 1998. The role of groundwater in the Florida Bay ecosystem. Proceedings of the 1998 Florida Bay Science Conference, University of Florida Sea Grant.

Capone, D.G. and B.F. Taylor. 1980. Microbial nitrogen cycling in a seagrass community, pp. 153-161, in *Estuarine Perspectives*, Kennedy, V.S., Ed., Academic Press, New York.

Corbett, D.R., J. Chanton, W. Burnett, K. Dillon, and C. Rutowski. 1999. Patterns of groundwater discharge into Florida Bay, *Limnol. Oceanogr.*, 44:1045–1055.

Department of Commerce. 1995. Florida Keys National Marine Sanctuary Draft Management Plan/Environmental Impact Statement, National Oceanic and Atmospheric Administration, Sanctuary Programs Division, Washington, D.C., 245 pp.

EPA. 1995. *Water Quality Protection Program for the Florida Keys National Marine Sanctuary: Phase III Report*, final report submitted to the Environmental Protection Agency under Work Assignment 1, Contract No. 68-C2-0134, Battelle Ocean Sciences, Duxbury, MA, and Continental Shelf Associates, Inc., Jupiter FL.

Fourqurean, J.W., R.D. Jones, and J.C. Zieman. 1993. Processes influencing water column nutrient characteristics and phosphorus limitation of phytoplankton biomass in Florida Bay, FL, USA: inferences from spatial distributions, *Estuar. Coast. Shelf Sci.*, 36:295–314.

Frankovich, T.A. and R.D. Jones. 1998. A rapid, precise, and sensitive method for the determination of total nitrogen in natural waters, *Mar. Chem.*, 60:227–234.

Froelich, P.N., Jr., D.K. Atwood, and G.S. Giese. 1978. Influence of Amazon River discharge on surface salinity and dissolved silicate concentrations in the Caribbean Sea, *Deep Sea Res.*, 25:735–744.

Garcia, H.E. and L.I. Gordon. 1992. Oxygen solubility in seawater: better fitting equations, *Limnol. Oceanogr.*, 37:1307–1312.

Jones, R.D. 1996. Phosphorus cycling, pp. 343–348, in *Manual of Environmental Microbiology*, Hurst, G.J., Knudsen, G.A., McInerney, M.J., Stetzenbach, L.D., and Walter, M.V., Eds., ASM Press, Materials Park, OH.

Kirk, J.T.O. 1994. *Light and Photosynthesis in Aquatic Systems*, 2nd ed., Cambridge University Press, Cambridge, U.K.

Klein, C.J. and S.P. Orlando, Jr. 1994. A spatial framework for water-quality management in the Florida Keys National Marine Sanctuary, *Bull. Mar. Sci.*, 54:1036–1044.

Lapointe, B.E. and M.W. Clark. 1992. Nutrient inputs from the watershed and coastal eutrophication in the Florida Keys, *Estuaries*, 15:465-476.

Lapointe, B.E. and W.R. Matzie. 1996. Effects of stormwater nutrient discharges on eutrophication processes in nearshore waters of the Florida Keys, *Estuaries*, 19:422–435.

Lee, T.N., M.E. Clarke, E. Williams, A.F. Szmant, and T. Berger. 1994. Evolution of the Tortugas gyre and its influence on recruitment in the Florida Keys, *Bull. Mar. Sci.*, 54:621–646.

Leichter, J.J., S.R. Wing, S.L. Miller, and M.W. Denny. 1996. Pulsed delivery of subthermocline water to Conch Reef (Florida Keys) by internal tidal bores, *Limnol. Oceanogr.*, 41:1490–1501.

Meeder, J.F., J. Alvord, M. Byrns, M.S. Ross, and A. Renshaw. 1997. *Distribution of Benthic Nearshore Communities and Their Relationship to Groundwater Nutrient Loading*, final report to Biscayne National Park.

Moore, W.S., J.L. Sarmiento, and R.M. Key. 1986. Tracing the Amazon component of surface Atlantic water using [228]Ra, salinity, and silica, *J. Geophys. Res.*, 91:2574–2580.

Nelson, D.M. and Q. Dortch. 1996. Silicic acid depletion and silicon limitation in the plume of the Mississippi River: evidence from kinetic studies in spring and summer, *Mar. Ecol. Prog. Ser.*, 136:163–178.

NOAA. 1995. *Florida Keys National Marine Sanctuary Draft Management Plan/Environmental Impact Statement*, National Oceanic and Atmospheric Administration, Washington, D.C.

Pelegri, S.P. and R.R. Twilley. 1998. Heterotrophic nitrogen fixation (acetylene reduction) during leaf litter decomposition of two mangrove species from South Florida, USA, *Mar. Biol.*, 131:53–61.

Pitts, P.A. 1997. An investigation of tidal and nontidal current patterns in Western Hawk Channel, Florida Keys, *Cont. Shelf Res.*, 17:1679–1687.

Redfield, A.C. 1958. The biological control of chemical factors in the environment, *Am. Sci.*, 46:205–222.

Robblee, M.B., T.B. Barber, P.R. Carlson, Jr., M.J. Durako, J.W. Fourqurean, L.M. Muehlstein, D. Porter, L.A. Yarbro, R.T. Zieman, and J.C. Zieman. 1991. Mass mortality of the tropical seagrass *Thalassia testudinum* in Florida Bay (USA), *Mar. Ecol. Prog. Ser.*, 71:297–299.

Rudnick, D., Z. Chen, D. Childers, T. Fontaine, and J.N. Boyer. 1999. Phosphorus and nitrogen inputs to Florida Bay: the importance of the Everglades watershed, *Estuaries*, 22:398–416.

Ryther, J.H., D.W. Menze, and N. Corwin. 1967. Influence of the Amazon River outflow on the ecology of the western tropical Atlantic. I. Hydrography and nutrient chemistry, *J. Mar. Res.*, 25:69–83.

Solorzano, L. and J.H. Sharp. 1980. Determination of total dissolved phosphorus and particulate phosphorus in natural waters, *Limnol. Oceanogr.*, 25:754–758.

Smith, N.P. 1994. Long-term Gulf-to-Atlantic transport through tidal channels in the Florida Keys, *Bull. Mar. Sci.*, 54:602–609.

Strickland, J.D.H. and T.R. Parsons. 1972. A practical handbook of seawater analysis, *Bull. Fish. Res. Bd. Can.*, 167:107–112.

Szmant, A.M. and A. Forrester. 1996. Water column and sediment nitrogen and phosphorus distribution patterns in the Florida Keys, USA, *Coral Reefs*, 15:21–41.

# 23 Biotic Phase-Shifts in Florida Bay and Fore Reef Communities of the Florida Keys: Linkages with Historical Freshwater Flows and Nitrogen Loading from Everglades Runoff

*Brian E. Lapointe and William R. Matzie*
Harbor Branch Oceanographic Institution, Inc.

*Peter J. Barile*
Florida Tech, Division of Marine & Environmental Systems

## CONTENTS

### BACKGROUND

This chapter describes how human alteration of the Everglades watersheds in South Florida have increased nutrient loads to "downstream" seagrass communities in Florida Bay and coral reefs in the Florida Keys. We also describe how these hydrological and biogeochemical modifications have resulted in nutrient enrichment of coastal waters, initiating harmful algal blooms and seagrass dieoff in Florida Bay and loss of coral communities on the Florida reef tract. Historically, nutrient concentrations were very low in the upland freshwater ecosystems in South Florida and controlled largely by nutrient concentrations in rainfall. Sheetflow drained southward through the Everglades marsh system and fringing mangroves where natural biogeochemical nutrient cycling prevented the buildup of elevated nutrient concentrations.[1] The extensive mangrove forests of the estuarine portion of the Everglades, combined with the oligotrophic seagrass and coral reef communities of Florida Bay and the Florida Keys, represented, collectively, the most biologically diverse ecosystem within the continental U.S.

Human alteration of the south Florida watershed (Kissimmee–Okeechobee–Everglades drainage basin) began around the turn of the century and was designed to make South Florida hospitable for human settlement.[2] During the 1930s, drainage of some 2800 km² of Everglades wetlands south of Lake Okeechobee was initiated for large-scale agriculture (sugarcane, sod, and winter vegetables) in the Everglades Agricultural Area (EAA), where sugarcane cultivation increased considerably after 1960. Today, the Everglades hydrologic system has been greatly modified by 2240 km of canals, 19 pumping stations, and 125 control structures. To the east and south of the EAA are the Water Conservation Areas (WCAs), remnant portions of the Everglades that were diked and used for flood control and water supply (Figure 23.1). Inflows of fresh surface water from the Everglades to Florida Bay occur from 20 creek systems fed by Taylor Slough and the C-111 canal in the eastern bay and by Shark River Slough, the largest drainage of the Everglades, which flows into coastal waters around Cape Sable, in turn, which flow through Conchie Channel into western Florida Bay and southward towards the Middle and Lower Keys.

Beginning in the late 1970s, additional structural modifications to the human-controlled hydrologic system were initiated to increase the conveyance of water south to Everglades National Park.[2] These included the Interim Action Plan, a regulatory effort intended to ameliorate the adverse impacts of backpumping nutrient-rich stormwater runoff from the EAA into Lake Okeechobee. Instead, the stormwater runoff was diverted southward into the WCAs to provide increased quantities of water to Everglades National Park. This plan became the long-term water management strategy in South Florida. The increased nutrient loads associated with the Interim Action Plan markedly changed the sawgrass plains and diverse aquatic sloughs of the Everglades to more heterotrophic and monotypic cattail stands.[3,4] These flows were further augmented by the "flow-through plan" and "rainfall plan" (1983 to present) that linked water deliveries to Shark River Slough to weather conditions and hydraulic gradients between Everglades National Park and WCA-3A.[5]

Phosphorus was identified as the nutrient of primary concern to expansion of the cattails in the Everglades as a result of increased discharges,[4] and the high nitrogen concentrations associated with these discharges were not considered with respect to their effects on downstream seagrass and coral reef communities. We explore this issue in detail here and suggest that the regional water quality deterioration in Florida Bay and the Florida Keys represents an important and timely case study of the global nitrogen-overload problem that has resulted from human alteration of the nitrogen cycle[6,7] in South Florida.

## NITROGEN LOADING FROM THE EVERGLADES TO DOWNSTREAM COASTAL WATERS

The relatively conservative behavior of nitrogen (N) compared to phosphorus (P) in the freshwaters of the Everglades indicates that the increased water deliveries (~1 million acre ft/yr) to Shark River Slough in the early 1980s (1982–1984) increased nitrogen loads to downstream coastal waters, including those in western Florida Bay and the Middle and Lower Keys (Figure 23.2). The long-term average total nitrogen (N) and phosphorus (P) concentrations for the spatial gradient from the EAA southward through Shark River Slough,[8] compared with the first nutrient data for Florida Bay,[9] show that the Everglades have very limited capacity for removal of N compared to P and represent an external source of N to Florida Bay (Figure 23.3). As the freshwater flows south from Lake Okeechobee and the EAA through the Everglades increased in the early 1980s, P would have been selectively removed by biogeochemical processes, indicated by an increase in the N:P ratio

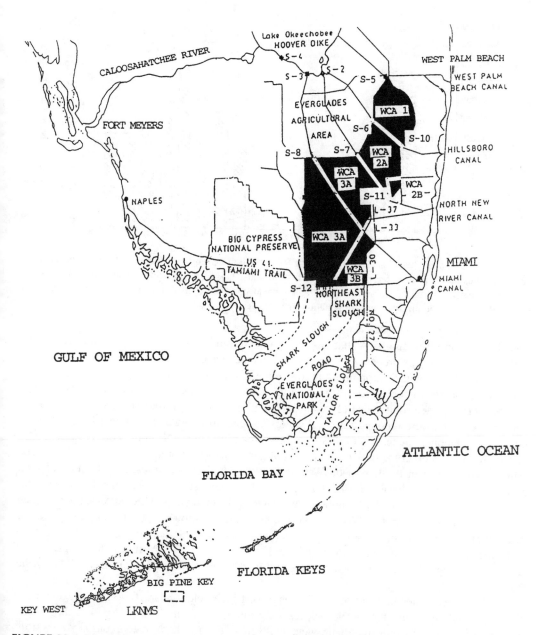

**FIGURE 23.1** Map of South Florida showing Florida Bay and the Florida Keys. Note the Everglades Agricultural Area (EAA) south of Lake Okeechobee, the Water Conservation Areas (WCAs), Shark River Slough, Taylor Slough, Everglades National Park, and Looe Key National Marine Sanctuary (LKNMS) south of Big Pine Key.

from ~80:1 at the EAA pump stations to ~200:1 at the S-12 structures at the headwaters of Shark River Slough. The greater uptake of P compared to N along this large-scale Everglades nutrient gradient is consistent with a smaller scale (10-km) nutrient gradient analysis in WCA-2A that found rapid uptake of P but relatively little uptake of N.[10]

The increase in the N:P ratio between the EAA and Shark River Slough is diagnostic of nitrogen contamination in carbonate systems[11] and indicative of non-point-source nitrogen enrichment associated with agricultural stormwater runoff. A major nitrogen source, in addition to fertilizers, is the

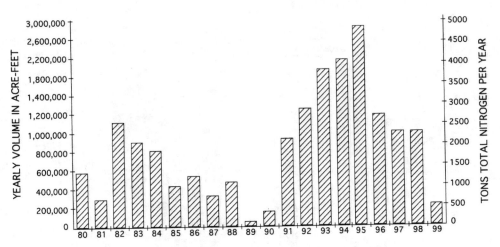

**FIGURE 23.2** Annual flows (on left) into Shark River Slough (Everglades National Park) from the S-12 structures plus flows from levee 30 to 67 between 1980 and 1994. Corresponding annual nitrogen loads (on right) were calculated based on annual flows and a mean concentration of 1.51 mg/l for total nitrogen at station P35, the southernmost monitoring station in Shark River Slough.

organic peat topsoil of the EAA that has undergone subsidence due to drainage, mineralization, and stormwater runoff, a process that has resulted in an average loss of ~2.5 cm of topsoil per year since the 1920s.[12] Denitrification rates are low in the Everglades,[13] which, together with a reduced ability for N accretion compared to P in the Everglades sediments,[14] limits the uptake or removal of N being transported south toward Florida Bay. In the early 1980s, nitrate concentrations 30 times above background were reported for canal waters adjacent to Everglades National Park.[13] Ammonium is the predominant species of dissolved inorganic nitrogen (DIN) in the northern Everglades surface waters, although the bulk of the total nitrogen is dissolved organic nitrogen (DON).[14] The limited uptake of N in the Everglades WCAs results in high concentrations (~108 $\mu M$) of total nitrogen throughout Shark River Slough (Figure 23.3), a value sixfold higher than the mean event concentration of total N for an unpolluted wetland (~18 $\mu M$ total N; U.S. EPA National Urban Runoff Program). The threshold total N and P concentrations shown in Figure 23.3 represent the mean water-column concentrations at which dieoff of *Thalassia testudinum* occurred from nutrient enrichment in sewage-impacted waters of the Florida Keys.[15] The N threshold, in particular, illustrates the high degree of nitrogen enrichment of both the Everglades and Florida Bay by the years 1989–1990.

As DON-rich water flows through the mangrove fringe below Shark River Slough, microbes mineralize and ammonify the DON pool, resulting in enhanced ammonium export to the coastal marine environment.[16] Compared to nitrogen, relatively little phosphorus enters coastal waters from Everglades runoff because of the high N:P ratios of surface waters in the Everglades. However, inner-shelf coastal waters off southwest Florida entering Florida Bay contain high concentrations of phosphorus (~1.0 $\mu M$ total phosphorus)[9,18] and represent a major regional source of P enrichment to Florida Bay[9,17,18] and the Middle and Lower Keys.[18] Rivers along the entire southwest coast of Florida have P concentrations substantially higher than most rivers in North America as a consequence of the distribution of phosphatic-rock formations in their watershed and additional burdens from the phosphate mining industry.[19] The long-term advective inputs of phosphorus from the southwest Florida shelf into western Florida Bay, combined with nitrogen-rich runoff from the Everglades, mix in the vicinity of Cape Sable where water-column N:P ratios of ~20:1 occur.[9,18] These N:P ratios are close to the Redfield Ratio (16:1 by atoms) and indicate N-limited growth of both macroalgae and phytoplankton, not P limitation as assumed by some scientists.[9,20]

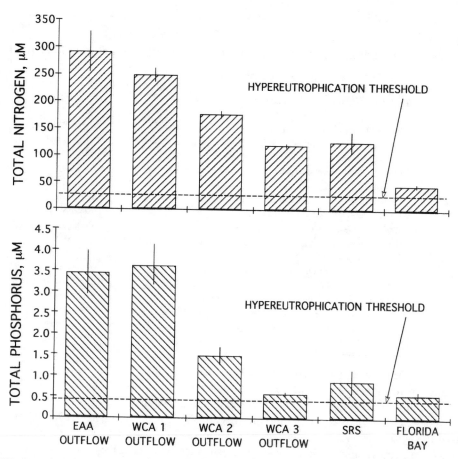

**FIGURE 23.3** Mean concentrations of ±1 S.D. total nitrogen and total phosphorus for the period of record from 1978 to 1994 for the Everglades Agricultural Area (EAA) outflows (S-5A, S-6, S-7, and S-8), outflows from WCA-1 (S-10A, B, C, D, E), WCA-2 (S-11A, B, C), WCA-3 (S-12 A, B, C, D), and the mean of four interior Shark River Slough (SRS) monitoring stations (P33, 34, 35, 36), and Florida Bay. (Data from References 8 and 9.) Also shown are the empirically derived total nitrogen and total phosphorus threshold concentrations at which *Thalassia testudinum* died-off from nutrient enrichment in wastewater-impacted nearshore waters of the Florida Keys.[15]

## NITROGEN ENRICHMENT INITIATES BIOTIC PHASE SHIFTS IN FLORIDA BAY

Following implementation of the Interim Action Plan in the late 1970s and early 1980s, Florida Bay experienced a series of ecological changes (Table 23.1) that were unprecedented during the period of human observation.[21] Historically, the shallow (<2 m), clear waters of Florida Bay supported a diverse community dominated by the seagrass *Thalassia testudinum* (Turtle grass).[22] The extensive *T. testudinum* meadows played a key role not only in the overall primary production in Florida Bay, but other functions as well, including stabilization of bottom sediments and provision of nursery grounds for fishes and invertebrates. den Hartog [23,24] developed a succession scheme illustrating the development of the *T. testudinum* climax communities characteristic of Florida Bay (Figure 23.4A); he specifically noted that "pollution leads to quite aberrant developments as the original vegetation becomes degraded and replaced by a vegetation consisting of species which are not normally involved in the succession series, or occur only as rare companion species." A very similar model was later proposed by Zieman[25] (Figure 23.4B), but that model lacked the "regressive

**TABLE 23.1**
**Chronology of Biotic Phase Shifts in Florida Bay and Downstream Florida Reef Tract**

| Date | Biotic event sequence | Ref. |
|---|---|---|
| Pre-1977 | *Thalassia* and corals are dominant biota | Dustan[51] <br> Tabb[22] |
| 1979 | Blooms of *Ulva*, *Laurencia* in western Florida Bay region | SFWMD[27] |
| 1983 | *Laurencia* most abundant macroalga in Florida Bay; heavy epiphytes on seagrasses of the mainland fringe of Florida Bay | Zieman and Fourqurean[29] |
| 1983–1985 | *Laurencia* bloom covers northeast bay region and expands to the south and west | SFWMD[27] |
| 1984–1991 | 44% decrease in coral cover at Looe Key | Porter and Meier[53] |
| 1987 | Massive dieoff of *Thalassia* in Florida Bay | Robblee et al. (1987) |
| 1987 | Large scale bleaching of corals on reef tract | Porter and Meier 1992 |
| 1991 | Change in finfish and other fisheries species in composition in central Florida Bay region | SFWMD[27] |
| 1991–1996 | Expansive blooms of phytoplankton (*Synechococcus*) in central and western Florida Bay | Phlips and Badylak[20] <br> Lapointe and Clark[61] |
| 1991–1992 | Dieoff of sponges in central Florida Bay region | Butler et al.[83] |
| 1996 | Unprecedented incidence of coral diseases and mortality coincide with peak SRS flows and chl-a at Looe Key | EPA (1997) <br> Lapointe and Matzie 1997 <br> Jaap et al. 2000 |

development" due to excessive macroalgae that was an element of the den Hartog model. Importantly, nitrogen fixation by cyanobacteria in the anaerobic root–rhizome layer is considered the major source of nitrogen supporting *T. testudinum* communities in oligotrophic tropical waters.[26]

The progression of habitat deterioration in western Florida Bay in the late 1970s followed the increased freshwater flows and associated nitrogen loads associated with implementation of the Interim Action Plan when commercial and sport fishermen first noted the appearance of seagrass epiphytes and seasonal blooms of macroalgae.[27] These phenomena are well known ecological indicators of elevated water-column nutrient concentrations in seagrass meadows of South Florida.[15,28] Land-based nutrient enrichment was first evident from observations of heavy epiphytization of seagrass communities along the entire mainland seagrass community at the interface between the Everglades and Florida Bay in 1983.[29] Controlled microcosm studies have demonstrated that low-level DIN enrichment significantly increases attached-blade epiphytes on *T. testudinum*, reducing light availability, decreasing productivity, and ultimately causing seagrass dieoff.[15,28]

Unlike seagrasses that have extensive root-rhizome systems and can access sediment pore waters as nutrient sources to support growth, frondose macroalgae must acquire their nutrition directly from the water column and are therefore excellent indicators of water-column nutrient availability. In Bermuda, experimental nutrient enrichment of eutrophic *T. testudinum* meadows led to significant and dramatic reductions in above-ground biomass and increases in percent cover of the red macroalga *Spyridia hypnoides*.[30] By the early 1980s, the red macroalga *Laurencia* and the green macroalga *Ulva* formed extensive blooms to the west of Cape Sable (plume of Shark River Slough) and southward to Sandy Key in western Florida Bay.[27] In the summer of 1983, *Laurencia* was the second most abundant marine plant in Florida Bay (behind *T. testudinum*) but, because of its highly aggregated (contagious) distribution, *Laurencia* also had the highest localized biomass of any submerged macrophyte in Florida Bay (664 g dry wt m$^{-2}$).[29] Increased biomass of *Laurencia* was a predictable response to increased nitrogen inputs to Florida Bay because experimental nutrient-enrichment bioassays have shown that its productivity is DIN limited in Florida Bay.[31] Controlled, continuous culture studies have shown that low-level DIN enrichment to concentrations of 0.5 to 1.0 $\mu M$ can sustain maximum growth rates of macroalgae, which lead to blooms of eutrophic indicator species such as *Ulva*.[32] Quantitative sampling in Florida Bay during the summer of 1983

**A**

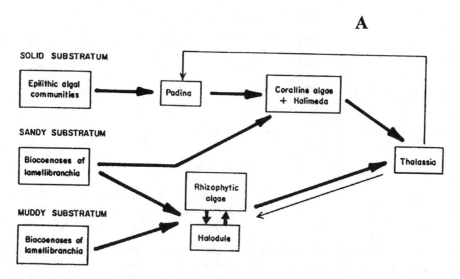

Scheme of the succession series leading to the *Thalassia testudinum* association. Thick arrows = progressive developments; thin arrows = regressive developments. After den Hartog (1973).

**B**

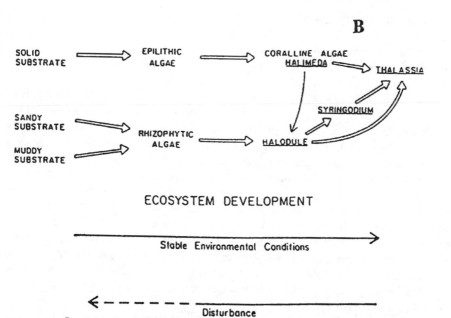

ECOSYSTEM DEVELOPMENT

Ecosystem development patterns in south Florida marine waters. This is a generalized pattern, and all stages may not be present. Note that in the absence of disturbance that the tendency is to a Thalassia climax.

**FIGURE 23.4** (A) Schematic of succession series leading to the *Thalassia testudinum* association. (Adapted from den Hartog. With permission.[23,24]) (B) Ecosystem development patterns for *T. testudinum* in South Florida marine waters. (From Zieman, J.C., *The Ecology of the Seagrasses of South Florida: A Community Profile*, FWS/OBS-82/25, U.S. Fish and Wildlife Service, Washington, D.C.)

indicated a baywide average biomass of 74.5 g m$^{-2}$ for seagrasses and 40.5 g m$^{-2}$ for macroalgae.[29] This high macroalgae-to-seagrass biomass ratio for Florida Bay in 1983 (Figure 23.5) is similar to that reported for mixed *T. testudinum*/macroalgae communities in sewage-enriched eutrophic waters of the Florida Keys,[15] indicating a comparable baywide eutrophic condition in Florida Bay at this

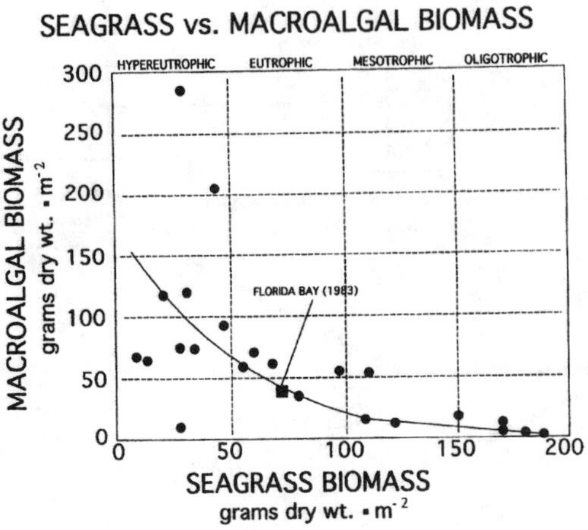

**FIGURE 23.5** Relationship between macroalgal biomass and seagrass biomass for *Thalassia testudinum* communities along a eutrophication gradient in the Florida Keys.[15] Also shown is the average value for Florida Bay, based on stations quantitatively sampled in July and August, 1983. (From Zieman, J.C. and Fourqurean, J.W., *The Distribution and Abundance of Benthic Vegetation in Florida Bay, Everglades National Park*, final report to South Florida Research Center.)

time. A recent survey of the macroalgal diversity by Dawes et al.[33] in Florida Bay near the Content Keys (northwest of Big Pine Key) indicates that there has been a dramatic decrease in species diversity of 73% between 1964 and 1995. This work suggests the role of nutrification in causing this crash in biodiversity, while also promoting the invasion of several nutrient pollution indicator species, including *Ulva lactuca* and *Chaetomorpha branchygona*.

By 1987, large-scale dieoff of *T. testudinum* began in central and western Florida Bay, and to date over 100,000 acres have been lost.[34] The explanation promoted by some scientists[35-37] and national environmental groups to explain the dieoff is the "hypersalinity hypothesis." According to this model (see Figure 6.12 in McIvor et al.[37]), drainage of South Florida for human development reduced freshwater flows from the Everglades and resulted in elevated salinities in Florida Bay which, combined with a lack of hurricanes and high summer temperatures, resulted in "stress" and dieoff of *T. testudinum*. The model suggested that following seagrass dieoff the decomposing seagrass detritus released nutrients that increased internal nutrient cycling, which supported the escalating algal blooms in the region. We believe, however, that this widely accepted model for seagrass dieoff in Florida Bay is flawed in at least four aspects: (1) freshwater deliveries to Florida Bay had increased, not decreased, in the years prior to dieoff; (2) elevated salinities in Florida Bay were not a recent phenomenon; (3) *T. testudinum* grows in numerous hypersaline environments; and (4) hurricane frequencies in Florida Bay had not decreased in recent years. Below we provide details of the basis for our opinions.

Most importantly, the overall trend from the 1960s to present is one of increased freshwater flows,[2] not decreased flows as stated in the model.[37] Modifications designed to increase deliveries of freshwater to Everglades National Park (ENP) by the South Florida Water Management District (SFWMD) included the construction of various spillways, canals, and levees during projects from 1965–1973, 1974–1979, and 1980–1985.[2] As shown in Figure 23.2, freshwater flows had increased substantially from Shark River Slough (>1 million acre ft/yr) in the early 1980s, just prior to dieoff of *T. testudinum*. Historical flows from Shark River Slough have been estimated at ~1 million acre ft/yr.[38]

Furthermore, as summarized by McIvor et al.,[37] hypersalinity had characterized Florida Bay for over 40 years, and the highest salinity of 70 ppt was recorded in 1956. No dieoff of *T. testudinum* was reported during that period. In fact, Tabb et al.,[22] who made detailed observations of Florida

Bay seagrasses following this peak in salinity, concluded that *T. testudinum* thrived under hyper-saline conditions by stating: "With marked reduction in salinity beginning in the winter of 1957 and ending in 1960, the *T. testudinum* underwent decline in size and abundance ... they did not return in abundance until the drought of 1961–2, reaching peak growth and coverage in the spring of 1962. Thus, it appears that long periods of near or slightly above normal salinities are a requirement for maximum growth of *T. testudinum*." These observations are supported by physio-logical growth studies of *T. testudinum* in response to varying salinity. In Laguna Madre, TX, McMillan and Moseley[39] measured seagrass productivity over a wide range of salinities by deter-mining how fast clipped blades grew to their original length. Despite the limitations imposed by this technique, *T. testudinum* continued to grow up to at least 60 ppt.

Finally, an obvious problem with the hurricane element of the model is that it ignored the storm track of Hurricane Floyd, which crossed Florida Bay on a southwest to northeast track in 1987, the year large-scale dieoff began. The model also ignored Hurricane Andrew, which crossed the southern Everglades from east to west in 1992. In reviewing the storm tracks and dates for all hurricanes and tropical storms that have affected Florida from the 1870s to present,[40] the frequency of hurricanes crossing Florida Bay (11 in 110 years) has not changed from decade to decade. The 21-year interval between hurricanes from 1966 (Inez) to 1987 (Floyd) was exceeded by a span of 23 years between 1905 and 1928. Extensive seagrass meadows found in low-energy sites in Western Australia show no symptoms of dieoff, despite not having experienced a major storm event in over 60 years.[41] Furthermore, neither water quality nor seagrass productivity has improved as a result of the direct impact of Hurricane Georges, which struck the Florida Keys and Florida Bay in September 1998.

Alternatively, direct evidence of the eutrophic status of Florida Bay (Table 23.2.) was apparent in the drought years of 1989–1990 when the first baywide water quality measurements revealed a mean DIN concentration of 2.68 μM (maximum was ~11 μM).[9] The ecological significance of these elevated concentrations was not recognized by Fourqurean et al.,[9] as they stated, "Our measurements do not indicate any baywide nutrient enrichment." Water column DIN concentrations are typically very low or undetectable (<0.5 μM) in unpolluted tropical *T. testudinum* meadows that rely on cyanobacterial N fixation in the rhizosphere.[26] For example, in Shark Bay, Western Australia, hypersalinity (55 ppt) and very low DIN (0.35 μM) and SRP (0.03 μM) concentrations supports healthy seagrass communities.[42] Fourqurean et al.[9] reported a mean baywide salinity of 41.1 ppt for Florida Bay in the drought years of 1989–1990, a value that was much lower than the upper salinity tolerance of *T. testudinum* (>60 ppt[39]) and the maximum recorded salinities of ~70 ppt reported for Florida Bay in the 1950s.[37] Fourqurean et al.[9] also reported a baywide mean chlorophyll *a* concentration of 1.05 μg/l (maximum was 4.86 μg/l) for Florida Bay in this period, a value much higher than that reported for productive *T. testudinum* communities behind the forereef at Looe Key National Marine Sanctuary[15,62] (Table 23.2). Alternatively, the high mean chlorophyll a con-centrations reported for western Florida Bay,[9] an area that was not influenced by hypersalinity, correlated with extensive dieoff of *T. testudinum*.[34,35]

In an attempt to reverse the ongoing seagrass dieoff and water quality degradation assumed to have resulted from hypersalinity,[37] water managers increased Everglades flows to values well above historic levels,[38] beginning in 1991 from both Shark River Slough (Figure 23.2) and Taylor Slough (Figure 23.6). Following the increased flows and nitrogen loads from Shark River Slough and Taylor Slough between 1991 and 1995, dense phytoplankton blooms and turbidity developed in the downstream receiving waters of central and western Florida Bay as salinity decreased. During this 4-year period, salinity decreased significantly (Tukey HSD test, $p < 0.0001$) in the eastern, central and western sections of the bay (Figure 23.7D) with an overall baywide decrease of 44%. As the flows initially increased between 1991 and 1992 (Figures 23.2 and 23.6), ammonium concentrations increased significantly (Tukey HSD test, $p = 0.003$) in the central bay (Figure 23.7C) to an annual mean concentration of ~15 μM (and maximum values to 120 μM) in 1992.

**TABLE 23.2**
**Annual Mean Concentrations of Salinity (ppt.), Ammonium, Nitrate + Nitrite, Dissolved Inorganic Nitrogen (DIN), Soluble Reactive Phosphorus (SRP), or Total Phosphorus (TP)**

| Year | n | Salinity | Ammonium | Nitrate + Nitrite | DIN | SRP or TP | Chl a | Ref. |
|---|---|---|---|---|---|---|---|---|
| | | | | Florida Bay | | | | |
| 1989–90 | 187 | 41.4 ± 0.4 | 1.89 ± 0.15 | 0.72 ± 0.07 | 2.68 ± 0.21 | 0.03 ± 0.01 | 1.05 ± 0.07 | Fourqurean et al.[9] |
| 1991 | 433 | 38.1 ± 1.0 | 4.84 ± 0.82 | 0.87 ± 0.23 | 5.81 ± 1.05 | 0.65 ± 0.08[a] | 1.33 ± 0.04 | SFWMD[27] |
| 1992 | 433 | 31.4 ± 9.1 | 14.6 ± 0.16 | 1.84 ± 0.49 | 18.09 ± 8.12 | 0.64 ± 0.16[a] | 1.71 ± 1.40 | SFWMD[27] |
| 1993 | 263 | 31.3 ± 0.8 | 4.47 ± 1.06 | 1.68 ± 0.99 | 6.15 ± 2.05 | 1.02 ± 0.44[a] | 2.41 ± 0.18 | SFWMD[27] |
| 1994 | 484 | 30.5 ± 0.7 | 6.60 ± 1.16 | 1.98 ± 0.94 | 8.59 ± 2.10 | 0.34 ± 0.01[a] | 2.52 ± 0.22 | SFWMD[27] |
| 1995 | 523 | 23.8 ± 0.5 | 8.02 ± 1.43 | 2.11 ± 1.20 | 10.14 ± 2.63 | 1.01 ± 0.48[a] | 1.75 ± 0.21 | SFWMD[27] |
| | | | | Looe Key | | | | |
| 1984 | 258 | — | 0.34 ± 0.29 | 0.15 ± 0.16 | 0.49 ± 0.32 | 0.05 ± 0.07 | NA | Lapointe and Smith[59] |
| 1986–87 | 48 | — | 0.23 ± 0.18 | 0.39 ± 0.30 | 0.49 ± 0.32 | 0.05 ± 0.07 | 0.15 ± 0.08 | Lapointe et al.[60] |
| 1989–90 | 16 | — | 0.10 ± 0.02 | 0.20 ± 0.29 | 0.31 ± 0.28 | 0.04 ± 0.03 | 0.07 ± 0.02 | Lapointe and Clark[61] |
| 1992 | 12 | — | 1.96 ± 0.99 | 0.76 ± 0.52 | 2.72 ± 1.35 | 0.22 ± 0.09 | 0.25 ± 0.11 | Lapointe and Matzie[62] |
| 1996 | 90 | — | 0.39 ± 0.54 | 0.55 ± 0.41 | 0.88 ± 0.68 | 0.04 ± 0.03 | 0.59 ± 0.46 | Lapointe and Matzie[64] |
| 1998 | 52 | — | 0.64 ± 0.47 | 0.54 ± 0.36 | 1.18 ± 0.69 | 0.09 ± 0.02 | 0.23 ± 0.15 | Lapointe (unpubl. data) |

*Note:*  All values expressed as means ± SE (μM concentrations) in Florida Bay and Looe Key. Chlorophyll *a* expressed as μg/L. n = number of samples.

[a] Total phosphorus.

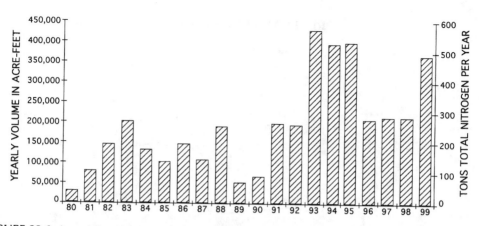

**FIGURE 23.6** Annual flows (on left) into Taylor Slough (Everglades National Park) from the S-175 combined with flows from the Taylor Slough bridge constitute total flows to Taylor Slough. Corresponding annual nitrogen loads (on right) were calculated based on annual flows and a mean concentration of 1.40 mg/l for total nitrogen at station P37, the southernmost monitoring station in Taylor Slough.

The ammonium enrichment fueled the ongoing bloom of the small (~1 μm) cyanobacterium *Synechococcus* in the central bay, where its growth was limited by DIN.[43] The cyanobacterial bloom resulted in significant (Tukey HSD test, $p < 0.0001$) increases in chlorophyll *a* concentrations between 1992 and 1994 (Figure 23.7B), which decreased ammonium concentrations following the spike in 1992. Chlorophyll *a* concentrations also increased significantly (*post hoc* Tukey HSD test, $p < 0.0001$) between 1991 and 1993 in western Florida Bay (Figure 23.7B), where phytoplankton growth was also limited by nitrogen.[43] (See also Chapter 13, Figures 13.12 and 13.13.) The increased phytoplankton biomass resulted in severe light attenuation and further seagrass dieoff in central and western Florida Bay,[35] which, together with increased precipitation/flocculation of dissolved organic matter (DOM) from the increased freshwater inputs, detritus, and cascading sediment resuspension associated with wind events following seagrass dieoff, led to significant (Tukey HSD test, $p < 0.0001$) and dramatic increases in turbidity between 1991 and 1995 (Figure 23.7A). The increase in turbidity in the central and western bay following increased nitrogen loads since 1991 is evident in AVHRR satellite imagery that showed increased reflectance from the central and western bay in this timeframe.[44]

Our interpretation of nitrogen-fueled eutrophication rather than hypersalinity as the more significant causative factor underpinning the deterioration of Florida Bay is consistent with other recent reviews of Florida Bay. Our model for the nutrient-mediated biotic phase shifts in Florida Bay is presented in Figure 23.8. Boesch et al.[21] concluded that "nutrient inputs stimulating (algal) blooms in the western bay could result from increased nutrient loading in the Shark River Slough discharge" and that "the western bay blooms may be characteristic of a troublesome and growing trend of coastal eutrophication." A NOAA National Estuarine Eutrophication Assessment recently ranked Florida Bay as having a "high" level of expression of eutrophic conditions, which was based upon chlorophyll *a* levels, macroalgal abundance problems, epiphyte abundance problems, low dissolved oxygen, nuisance and toxic algal blooms, and loss of submerged aquatic vegetation.[45]

## NUTRIENT ENRICHMENT AND DECLINE OF CORAL REEFS IN THE FLORIDA KEYS

The development of the modern Florida Reef Tract, the third largest bank–barrier coral reef ecosystem in the world (~2800 km²), has occurred during the past 4000 years of sea-level rise. The hermatypic (reef-forming) corals primarily responsible for reef accretion are *Montastraea annularis, M. cavernosa, Acropora palmata, A. cervicornis, Diploria* spp., *Siderastrea siderea,*

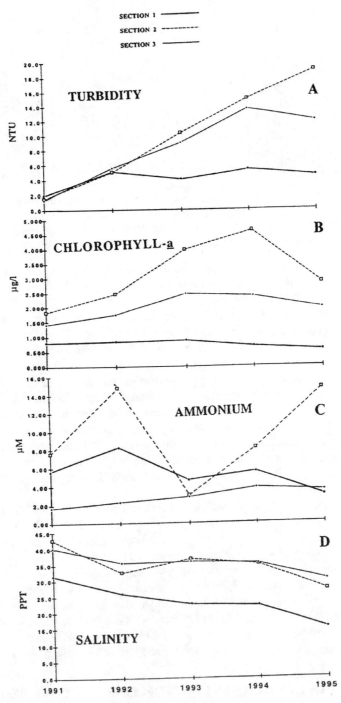

**FIGURE 23.7** Annual mean values of turbidity (A), chlorophyll *a* (B), ammonium (C), and salinity (D) in three regions of Florida Bay between 1991 and 1995. Section 1 represents mean values for stations in the eastern bay; section 2, the central bay; and section 3, the western bay. (Data are from the Florida International University water quality monitoring program for Florida Bay and were provided by the South Florida Water Management District.)

**Biotic phase-shift:**

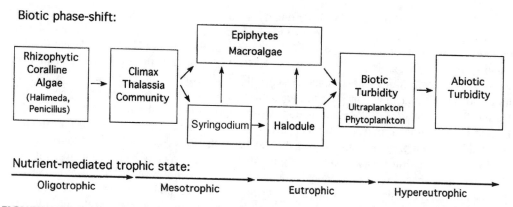

**Nutrient-mediated trophic state:**

FIGURE 23.8 Nutrient-mediated biotic phase shift model for Florida Bay. The benthic community undergoes biological transition under nutrient enrichment until phytoplankton blooms dominate pelagic and benthic production.

and *Colpophyllia* sp.[46] Circulation studies in the Florida Bay–Florida Keys region using moored current meters have shown long-term net flow of water from the inner southwest Florida shelf southeastward into Florida Bay and toward the tidal channels of the Middle and Lower Keys.[47] (See also Chapter 12, Figure 12.7.) Florida Bay water flows out into the Atlantic Ocean through the tidal channels, where it turns in a southwesterly direction and flows westward along Hawk Channel toward the Dry Tortugas. A significant offshore deflection directs flow in a vector toward the outer forereef.[48] Geologists have suggested that the environmental stress associated with the "inimical" waters of Florida Bay prevented coral reef growth and accretion in areas offshore of the major tidal passes and downstream of these outflows.[49,50]

Coral reef stress and degradation were observed on offshore forereefs in the Florida Keys concurrent with the decline of water quality in Florida Bay in the late 1970s.[46] The first report of coral disease ("plague") and expansion of algal turf was in 1977 at Carysfort Reef, one of the best-developed forereefs in the northern Keys, which had >70% coral cover.[51] In the mid-1980s, infections of black-band disease (*Phormidium corallyticum*) increased cover of macroalgae (*Halimeda* spp., *Dictyota* spp.), and the cyanobacterium *Lyngbya* spp. were reported for offshore forereefs.[52,53] In the summer of 1987, water temperatures reached 31.5°C in the "core area" of Looe Key National Marine Sanctuary (LKNMS), which correlated with the first massive coral bleaching event on the Florida Reef Tract. Between 1984 and 1991, the offshore forereefs lost between 24 and 44% of their hard (hermatypic) coral cover, with the highest loss occurring at LKNMS downstream of the largest tidal channel (Moser Channel, Seven-Mile Bridge).[52] During this period water visibility also decreased at LKNMS as reported by the Underseas Dive Shop on Big Pine Key (Mary Ann Rockett, pers. comm.).

In response to the deteriorating water quality and loss of coral cover in the Florida Keys, over 50 scientists and managers attended a NOAA-sponsored research and management workshop in 1988.[46] Attendees ranked the major threats and problems facing coral reefs of the Florida Keys. Although anchor damage had been previously considered the most widespread problem facing the reefs in the 1970s,[51] attendees at this meeting identified excessive nutrient loading as the top priority issue to be addressed to protect the reefs by concluding, "Excessive nutrients invading the Florida Reef Tract from the Keys and Florida Bay are a serious and widespread problem."[46] Global attention to the declining status of Florida's coral reefs heightened in 1990 when *National Geographic* featured the article "Florida's Imperiled Coral Reefs,"[54] which specifically noted the problems of nutrient pollution, algal blooms, and coral diseases. This increased awareness helped spur the creation of the Florida Keys National Marine Sanctuary (FKNMS) and Protection Act on November 5, 1990, to "provide long term protection to the coral reef resources of the Florida Keys."

Recognizing the critical role of water quality to sustaining coral reefs in the Florida Keys, the U.S. Congress directed the U.S. Environmental Protection Agency (EPA) and the State of Florida, represented by the Florida Department of Environmental Protection (FDEP), to develop a Water Quality Protection Program (WQPP) for the FKNMS. The purpose of the WQPP was to recommend priority corrective actions and compliance schedules addressing point- and non-point-source pollution to restore and maintain the chemical, physical, and biological integrity of the FKNMS. During Phase I of the WQPP, information was compiled and synthesized on the status of the FKNMS. Whereas the problem of nutrient enrichment from Florida Bay had been specifically noted by attendees at the previous NOAA workshop in 1988, the EPA *Phase I Report*[55] concluded that, "Even though the ultimate cause of the seagrass dieoff has yet to be proven conclusively, it seems clear that anthropogenic nutrient input to the surface water is not responsible." Alternatively, the hypersalinity hypothesis was advanced by scientists[35] and federal agencies (see Redfield[56]) to explain the deterioration of Florida Bay as well as downstream coral reefs in the Florida Keys. Specifically, coral stress and dieoff were hypothesized to result from the offshore movement of near-bottom hypersaline (37 to 38 ppt) water from Florida Bay to the offshore reefs.[57,58] Accordingly, increased flows from the Everglades were identified as a "priority strategy" for improving water quality in the FKNMS Management Plan, and this water management policy was implemented by simultaneously increasing flows from both Shark River Slough and Taylor Slough into Florida Bay.[37]

A long-term water quality monitoring database for LKNMS shows that the deterioration of water quality in Florida Bay following increased Everglades runoff since 1991 has had a significant impact on water quality of this downstream offshore forereef. Beginning in 1984, a series of studies included monitoring of low-level DIN, SRP, and chlorophyll *a* concentrations in the "core area" of LKNMS. In the first study from October 1984 to October 1985, annual mean DIN concentrations were relatively low (DIN = 0.49 $\mu M$; Table 23.2)[59] but were likely above historic values, especially those prior to increased Everglades flows that began in the 1960s. The next two studies, conducted in 1986–1987[60] and 1989–1990,[61] collectively showed a trend of decreasing annual mean DIN concentrations as Everglades flows decreased with the drought of the late 1980s (Table 23.2). In 1992, flows from the Everglades increased (Figures 23.2 and 23.6) following implementation of the new "restoration policy," and annual mean DIN concentrations at LKNMS spiked to very high concentrations (>2 $\mu M$; Table 23.2)[62] concurrent with a DIN (ammonium) spike in central Florida Bay (see Figure 23.7C), indicating a linkage between DIN in Florida Bay and downstream reefs at LKNMS. The 1992 DIN concentrations at LKNMS exceeded the threshold value of ~1.0 $\mu M$ noted for the demise of coral reefs from eutrophication[63] and the development of macroalgal blooms.[32] Annual mean DIN concentrations at LKNMS in 1996[64] and 1998[65] were lower than in 1992, but still averaged ~1.0 $\mu M$, a value indicative of eutrophic coral reefs. Eight out of nine statistical comparisons (Tukey HSD test) of annual mean DIN concentrations for 1992, 1996, and 1998 with those of 1984–1985, 1986–1987, and 1989–1990 are significantly higher (seven comparisons are $p < 0.00002$, one is $p = 0.0183$, and one is $p = 0.1353$), indicating significant DIN enrichment in the decade of the 1990s compared to that of the 1980s.

Increased annual chlorophyll *a* concentrations at LKNMS also followed the DIN enrichment spike in Florida Bay, indicating that the phytoplankton bloom in central and western Florida Bay expanded downstream into a larger scale regional bloom between 1993 and 1996 following the increased nitrogen loads from the Everglades. The annual mean chlorophyll *a* concentrations averaged 0.15 $\mu g/l$ in 1986–1987 and decreased to minimal annual values of <0.10 $\mu g/l$ in the drought years of 1989–1990 (Table 23.2). Coincident with the DIN spike at LKNMS in 1992, the annual mean chlorophyll-a increased to 0.25 $\mu g/l$ followed by a further increase to the maximum annual mean of 0.59 $\mu g/l$ in 1996 (Table 23.2). The 1996 annual mean chlorophyll *a* is significantly (Tukey HSD test, $p < 0.000028$) higher than any of the other years on record, and followed peak freshwater flows and nitrogen loads from the Everglades in 1994–1995. Chlorophyll *a* is considered one of the best long-term indicators of eutrophication on coral reefs, and values >0.3 $\mu g/l$ are considered typical of eutrophic reefs, such as those in Kaneohe Bay, Hawaii, which were impacted by nutrients derived from a sewage outfall in the 1960s.[63] While there is evidence that nutrients derived from sewage are contributing to the eutrophication problem in the Florida Keys,[61,62,64,66] the rapid deterioration of water quality at LKNMS following the decrease in salinity in Florida Bay points to land-based nitrogen sources in the Everglades as a major contributor to the overall regional eutrophication problem. This evidence is contrary to the conclusion of Szmant and Forrester[67] which ascribes nutrient enrichment of the reef tract to offshore upwellings and high nutrient concentrations in the middle Keys to:

> … a recent change caused by the seagrass dieoff [whereby] phytoplankton blooms supported by the decay products of the dead plant matter have been spreading slowly throughout the northern and western parts of the Bay and may have begun to reach the [tidal] passes in the Long Key area by 1990.

The effects of nutrients, particularly nitrogen, on coral reefs, are predictable and multifaceted and include profound physiological and ecological consequences for the delicate symbiosis between zooxanthellae and the host coral polyp. Increased DIN and SRP concentrations at LKNMS after 1991 (Table 23.2) correlated with decreased coral cover, which we ascribe to multiple mechanisms driven by nutrient enrichment. Elevated DIN reduces calcification in reef corals,[68] which contributes to a reduction in overall "reef growth." Falkowski et al.[69] and Muscatine et al.[70] have shown that ammonium enrichment of corals in the Red Sea resulted in significant increases in population density and chlorophyll $a$ per cell of symbiotic zooxanthellae. Dubinsky and Stambler[71] described the physiological basis by which such nutrient-mediated increases in zooxanthellae density lead to decreased photosynthesis and increased respiration per cell. The overall effect is a reduction in the amount of photosynthate translocated to the symbiotic coral animal which leads to "stress" and the lack of control of zooxanthellae. Over a 6-year period, the zooxanthellae density of corals in Mauritius varied considerably, correlating most significantly with nitrate concentration of the water column.[72] This supports a consensus opinion at a coral reef workshop[73] that nutrient availability is a key factor related to "coral bleaching" (i.e., loss of zooxanthellae).

Indeed, the Florida reef tract experienced the first massive bleaching event in the world in the summer of 1987, an event that followed increased nitrogen loading from the Everglades in the early and mid-1980s (Table 23.1). Nitrogen enrichment and associated increases in phytoplankton biomass alter the heat budget of coastal waters through the "black-body effect," which could result in slight increases in water temperature. Although elevated temperatures have been identified as a cause of some of the massive coral bleaching phenomena, the global nitrogen overload problem that now affects the entire planet[6,7] may also contribute to these events not only through increased warming (algal alteration of heat budgets of the water column) but also by increased respiration and stress in the corals. The recent global expansion of coral bleaching events may directly involve nitrogen enrichment as a contributing factor to localized elevated sea-surface temperature, including antecedent physiological stress upon the coral–algal symbiosis.

The increased chlorophyll $a$ concentrations we measured at LKNMS in the 1990s would also stress reef corals by reducing the quality and quantity of submarine irradiance while increasing sedimentation rates. The peak chlorophyll $a$ concentrations in 1996 correlated directly with outbreaks of multiple coral diseases (250% increase in disease[52]) which were followed by a 40% loss of coral cover between 1996 and 1999 throughout the FKNMS.[74] This may have resulted, in part, from increased light attenuation and sedimentation associated with the nitrogen-fueled regional eutrophication. In addition to reduced photosynthetically active radiation due to absorption of light by chlorophyll $a$ and dissolved organic matter, simultaneous increases in sedimentation rates were associated with the high turbidity plumes flowing from Florida Bay through the tidal passes to the outer reefs as evident in AVHRR imagery.[44] Increased sedimentation rates are well known to decrease the viability of hermatypic reef corals.[75] Because reef corals have photosynthesis-to-respiration ratios close to unity (1),[76] even a slight reduction in light quality or quantity, coupled with increased sedimentation rates, would predictably increase physiological stress to the hermatypic corals that are keystone species of these offshore forereef communities.

As reef coral cover declined on the offshore forereefs of the FKNMS over the past two decades, the cover of macroalgae and turf algae has simultaneously increased. Turf algae (<2 cm high) and macroalgae (>2 cm high) now dominate the offshore reefs, collectively accounting for 48 to 84% of the cover.[77] Increased mean DIN and SRP concentrations would increase the growth and reproduction of macroalgae,[32] which would include endolithic filamentous algae, which represent the majority of algae on coral reefs.[78] Long-term monitoring at LKNMS has shown that the significant increase in annual mean DIN concentrations from <0.49 $\mu M$ in 1986 to >1.0 $\mu M$ in the 1990s was accompanied by shifts from coral to algal domination (see Tables 23.1 and 23.2). According to the Relative Dominance Model (RDM),[79] a shift away from corals and towards turf algae would indicate decreased grazing under low nutrient levels; however, a shift from corals to macroalgae is predicted from increased nutrient levels. Because LKNMS has extremely abundant schools of migrating herbivorous fishes (protected by federal law) that graze macroalgae and turf algae at very high rates,[80] the increased DIN and SRP concentrations that we measured in the 1990s would appear to be the primary mechanism mediating the shift from coral to algal domination. Furthermore, coralline red algal crusts are becoming increasingly common on dead coral substrate in the core area of LKNMS, which the RDM predicts to occur under conditions of high nutrient levels and high grazing.

## SUMMARY AND CONCLUSIONS

The nutrient transport phenomena and biotic shifts described in this chapter demonstrate the hydrological and biogeochemical linkages between the Everglades, Florida Bay, and downstream coral reefs. We believe the series of ecological events over the past several decades in Florida Bay and the Florida reef tract follows, with few exceptions, the predictable responses of shallow subtropical seagrass and coral reef ecosystems to escalating nutrient loading. This was particularly evident between 1991 and 1995 when increased flows and associated nitrogen loads[81] correlated with trends of increasing DIN, phytoplankton blooms, and turbidity in the central and western bay,[82] initiating ecological disturbances that included a sponge dieoff,[83] loss of macroalgal biodiversity in Florida Bay,[33] and increased nutrient concentrations, algal blooms, and coral dieoff on downstream reefs of the FKNMS.[57,62,64] The biotic changes that occurred on a regional scale in South Florida are similar to eutrophication processes described for other coral reef regions around the world, such as Kaneohe Bay, Hawaii.[84] However, in the case of South Florida, these impacts have occurred on unprecedented temporal and spatial scales.

The impacts of nitrogen enrichment on eutrophication of coastal waters have been recognized for at least 50 years.[85,86] Globally, coastal ecosystems have been enriched with increasing loads of nitrogen from anthropogenic sources;[87,88] in recent decades[6], this has resulted in unprecedented alteration of these ecosystems.[7,88,89] Oligotrophic tropical seagrass and coral reef communities, such as those that occurred historically in South Florida, are particularly vulnerable to even slight increases in nutrient concentrations. Unfortunately, scientists and water managers accepted the hypersalinity hypothesis readily with little attempt to experimentally verify this model or examine alternative hypotheses (see Redfield[56]). "Getting the science right" and "getting the right science" are the two critical aspects of any environmental risk assessment[90] and the omission of these two steps in the development of the Everglades Restoration Plan was a key factor contributing to the water quality degradation of Florida Bay and Florida Keys as described in this chapter. We hope this case study of regional water quality degradation will serve to inspire water resource managers in other coral reef regions around the world to implement policies based on sound science to moderate the pervasive and escalating trend of nutrient pollution in coastal ecosystems.

## ACKNOWLEDGMENTS

The authors thank Leslie Wedderburn and David Rudnick of the South Florida Water Management District for providing water quality and flow data. Ms. Colleen Murphy, a HBOI volunteer, assisted with data analysis and graphics. The comments of David Tomasko, Dick Zimmerman, John Ryther, Charles and Clarice Yentsch, Larry Brand, Peter Bell, and Steve Howe are gratefully acknowledged. This research was supported by funds from the Herbert W. Hoover Foundation, National Oceanic and Atmospheric Administration (Marine Sanctuaries Division), and the Coastal Ocean Program (NOAA). This is Contribution No. 1403 from the Harbor Branch Oceanographic Institution, Inc.

# REFERENCES

1. Davis, J.H., The natural features of southern Florida, *Florida Geol. Surv. Bull.*, 25, 1, 1943.
2. Light, S.S. and Dineen, J.W., Water control and the Everglades: a historical perspective, in *Everglades: The Ecosystem and Its Restoration*, Davis, S.M. and Ogden, J.C., Eds., St. Lucie Press, Delray Beach, FL, 1994.
3. Belanger T.V. and Scheidt, D.J., Effects of nutrient enrichment on the Florida Everglades, *Lake Reservoir Manage.*, 5(1), 101, 1989.
4. Davis, S.M., Phosphorus inputs and vegetation sensitivity in the Everglades, in *Everglades: The Ecosystem and Its Restoration*, Davis, S.M. and Ogden, J.C., Eds., St. Lucie Press, Delray Beach, FL, 1994.
5. Gunderson, L.H. et al., Lessons from the Everglades, *Bioscience* (suppl.), S, 225, 1995.
6. Moffat, A.S., Global nitrogen overload problem grows critical, *Science*, 279, 988, 1998.
7. Vitousek, P. et al., Human alteration of the global nitrogen cycle: sources and consequences, *Ecol. Appl.*, 7(3), 737, 1997.
8. Germain, G.J., *Surface Water Quality Network Database*, South Florida Water Management District, technical memorandum, West Palm Beach, FL, 1994, 236 pp.
9. Fourqurean, J.W., Jones, R.D., and Zieman, J.C., Processes influencing water column nutrient characteristics and phosphorus limitation of phytoplankton in Florida Bay, FL, *Estuar. Coast. Sci.*, 36, 295, 1993.
10. Urban, N.H., Davis, S.M., and Aumen, N.G., Fluctuations in sawgrass and cattail densities in Everglades WCA2A under varying nutrient, hydrologic and fire regimes, *Aquatic Bot.*, 46, 203, 1993.
11. Murray, J.P. et al., Groundwater contamination by sanitary landfill leachate and domestic wastewater in carbonate terrain: principal source diagnosis, chemical transport characteristics, and design characteristics, *Water Res.*, 15, 745, 1981.
12. Snyder, G.H. and Davidson J.M., Everglades agriculture: past, present, and future, in *Everglades: The Ecosystem and Its Restoration*, Davis, S.M. and Ogden, J.C., Eds., St. Lucie Press, Delray Beach, FL, 1994.
13. Gordon, A.S. et al., Denitrification in marl peat sediments in the Florida Everglades, *Appl. Environ. Microbiol.*, 52(5), 987, 1986.
14. Reddy, K.R. et al., Long-term nutrient accumulation rates in the Everglades, *Soil Sci. Soc. Am. J.*, 57, 1147, 1993.
15. Lapointe, B.E., Tomasko, D.A., and Matzie, W.R., Eutrophication and trophic state classification of seagrass communities in the Florida Keys, *Bull. Mar. Sci.*, 54, 696, 1994.
16. Twilley, R.R., Chen, R., and Koch, M., Biogeochemistry and forest development of mangrove wetlands in southwest Florida: implications to nutrient dynamics of Florida Bay, *Florida Bay Sci. Conf.* (abstr.), 1996, 93.
17. Fourqurean, J.W., Zieman, J.C., and Powell, G.V.N., Phosphorus limitation of primary production in Florida Bay: evidence from C:N:P ratios of the dominant seagrass *Thalassia testudinum*, *Limnol. Oceanogr.*, 37(1), 162, 1992.
18. Lapointe, B. E., Matzie, W.R., and Clark, M.W., Phosphorus inputs and eutrophication on the Florida Reef Tract, in *Proceedings of the Colloquium on Global Aspects of Coral Reefs: Health, Hazards, and History*, Ginsburg, R., Ed., University of Miami, 1993, 106.
19. Odum, H.T., Dissolved phosphorus in Florida waters, *Florida Geol. Surv.*, 1, 1, 1953.
20. Phlips, E.J. and Badylak, S., Spatial variability in phytoplankton standing crops and composition of Florida Bay, *Bull Mar. Sci.*, 58, 203, 1996.
21. Boesch, D.F. et al., *Deterioration of the Florida Bay Ecosystem: An Evaluation of the Scientific Evidence*, FWS/PMC–Florida Bay Research Program, Miami, FL, 1993.
22. Tabb, D.C., Dubrow, D.L. and Manning, R.B., *The Ecology of Northern Florida Bay and Adjacent Estuaries*, State of Florida Board of Conservation, Tech. Ser. No. 39, The Marine Laboratory, University of Miami, FL, 1962.
23. den Hartog, C., The dynamic aspect of the ecology of seagrass communities, *Thalassia Jugoslavica*, 7, 101, 1973.
24. den Hartog, C., Structure, function, and classification in seagrass communities, in *Seagrass Ecosystems*, McRoy, P. and Helfferich, C., Eds., Marcel Dekker, New York, 1977, p. 89.

25. Zieman, J.C., *The Ecology of the Seagrasses of South Florida: A Community Profile*, FWS/OBS-82/25, U.S. Fish and Wildlife Service, Washington, D.C., 1982, p. 123.

26. Patriquin, D.G., The origin of nitrogen and phosphorus for the growth of the marine angiosperms *Thalassia testudinum*, *Mar. Biol.*, 15, 35, 1972.

27. South Florida Water Management District, Proceedings of the Florida Bay adaptive environmental assessment and management workshop, July 11–13, 1995.

28. Tomasko, D.A. and Lapointe, B.E., Productivity and biomass of *Thalassia testudinum* as related to nutrient availability and epiphyte levels, *Mar. Ecol. Prog. Ser.*, 75, 9, 1991.

29. Zieman, J.C. and Fourqurean, J.W., *The Distribution and Abundance of Benthic Vegetation in Florida Bay*, Everglades National Park, final report to South Florida Research Center, 1985, p. 63.

30. McGlathery, K.J., Physiological controls on the distribution of the macroalgae *Spyridea hypnoides*: patterns along a eutrophication gradient in Bermuda, *Mar. Ecol. Prog. Ser.*, 87, 173, 1995.

31. Delgado, O. and Lapointe, B.E., Nutrient-limited productivity of calcareous versus fleshy macroalgae in a eutrophic, carbonate-rich tropical marine environment, *Coral Reefs*, 13, 151, 1995.

32. Lapointe, B.E., Nutrient thresholds for bottom-up control of macroalgal blooms on coral reefs in Jamaica and southeast Florida, *Limnol. Oceanog.*, 42(5-2), 1119, 1997.

33. Dawes, C.J., Uranowski, C., Andorfer, J., and Teasdale, B., Changes in the macroalga taxa and zonation at the Content Keys, Florida, *Bull. Mar. Sci.*, 64(1), 95, 1999.

34. Robblee, M.B. et al., Mass mortality of the tropical seagrass *Thalassia testudinum* in Florida Bay, *Mar. Ecol. Prog. Ser.*, 71, 297, 1991.

35. Zieman, J.C., Fourqurean, J.W., and Frankovich, T.A., Seagrass die-off in Florida Bay: long-term trends in abundance and growth of turtle grass *Thalassia testudinum*, *Estuaries*, 22(2b), 460, 1999.

36. Durako, M.J., Hall, M.O., and Merello, M., Patterns of change in the seagrass dominated Florida Bay hydroscape, in *The Everglades, Florida Bay, and Coral Reefs of the Florida Keys: An Ecosystem Sourcebook*, Porter, J.W. and Porter, K.G., Eds., CRC Press, Boca Raton, FL, 2002.

37. McIvor, C.C., Ley, J.A., and Bjork, R.D., Changes in freshwater inflow from the Everglades to Florida Bay, including the effects on biota and biotic responses, in *Everglades: The Ecosystem and Its Restoration*, Davis, S.M. and Ogden, J.C., Eds., St. Lucie Press, Delray Beach, FL, 1994.

38. Smith, T.J., Hudson, J.H., Robblee, M.B., Powell, G.V.N., and Isdale, P.J., Freshwater flow from the Everglades to Florida Bay: a historical reconstruction based on fluorescent banding in the coral *Solenastrea bournoni*, *Bull. Mar. Sci.*, 44, 274, 1989.

39. McMillan, C. and Moseley, F.N., Salinity tolerances of five marine spermatophytes of Redfish Bay, Texas, *Ecology*, 48, 503, 1967.

40. Doehring, F. et al., *Florida Hurricanes and Tropical Storms 1871–1993: An Historical Survey*, Florida Sea Grant Tech. Paper-71, 1994.

41. Kirkman H. and Kuo, J., Pattern and process in southern Western Australian seagrasses, *Aquatic Bot.*, 37, 369, 1990.

42. Atkinson, M.J., Low phosphorus sediments in a hypersaline marine bay, *Estuar. Coast. Shelf Sci.*, 24, 335, 1987.

43. Larentyev, P.J., Bootsma, H.A., Johengen, T.H., Cavaletto, J.F., and Gardner, W.S., Microbial plankton resource limitation and insights from community structure and seston stoichiometry in Florida Bay, *Mar. Ecol. Prog. Ser.*, 165, 43, 1998.

44. Stumpf, R.P., Frayer, M.L., Durako, M.J., and Brock, J.C., Variations in water clarity and bottom albedo in Florida Bay from 1985–1997, *Estuaries*, 22(2b), 431, 1999.

45. Bricker, S.B., Clement C.C., Pirhalla, D.E., Orlando, S.P., and Farrow, D.R.G., *National Estuarine Eutrophication Assessment: Effects of Nutrient Enrichment in the Nation's Estuaries*, National Oceans Studies Special Projects Office, National Oceanic and Atmospheric Administration, Silver Springs, MD, 1999.

46. NOAA, *Results of a Workshop on Coral Reef Research and Management in the Florida Keys: A Blueprint for Action*, Tech. Rep., National Undersea Research Program, Washington, D.C., 1988.

47. Smith, N.P. and Pitts, P.A., Regional-scale and long-term transport patterns in the Florida Keys, in *The Everglades, Florida Bay, and Coral Reefs of the Florida Keys: An Ecosystem Sourcebook*, Porter, J.W. and Porter, K.G., Eds., CRC Press, Boca Raton, FL, 2002.

48. Pitts, P.A., An investigation of near-bottom flow patterns along and across Hawk Channel, Florida Keys, *Bull. Mar. Sci.*, 54(3), 610, 1994.

49. Ginsberg, R.N., Compiler, *Proceedings of the Colloquium on Global Hazards to Coral Reefs: Health, Hazards, and History*, Rosenstiel School of Marine and Atmospheric Science, University of Miami, 1994.

50. Lidz, B.H. and Hallock, P., Sedimentary petrology of a declining reef ecosystem, Florida Reef Tract (U.S.A.), *J. Coast. Res.*, 16(3), 675, 2000.

51. Dustan, P., Vitality of coral populations off Key Largo, FL: recruitment and mortality, *Environ. Geol.*, 2, 51, 1977.

52. Harvell, C.D., Kim, K., Burkholder, J.M., Colwell, R.R., Epstein, P.R., Grimes, D.J., Hofmann, E.E., Lipp, E.K., Osterhaus, A.D.M.E., Overstreet, R.M., Porter, J.M., Smith, G.W., and Vasta, G.R., Emerging marine diseases: climate links and anthropogenic factors, *Science*, 285, 1505, 1999.

53. Porter, J.W. and Meier, O., Quantification of loss and change in Floridian reef coral populations, *Am. Zool.*, 32, 625, 1992.

54. Ward, F., Florida's coral reefs are imperiled, *National Geographic*, 178, 115, 1990.

55. EPA. *Water Quality Protection Program for the Florida Keys National Marine Sanctuary: Phase I Report*, Office of Wetlands, Oceans, and Watersheds, U.S. Environmental Protection Agency, Washington, D.C., 1991.

56. Redfield, G.W., Ecological research for aquatic science and environmental restoration in south Florida, *Ecol. Appl.*, 10(4), 990, 2000.

57. Porter, J.W., Meier, O.W., Tougas, J.I., and Lewis, S.K., Modification of the south Florida hydroscape and its effect on coral reef survival in the Florida Keys, *Bull. Ecol. Soc. Am.*, 75(2, part 2), 184, 1994.

58. Porter, J.W., Lewis, S.K., and Porter, K.G., The effect of multiple stressors on the Florida Keys coral reef ecosystem: a landscape hypothesis and a physiological test, *Limnol. Oceanogr.*, 44(3, part 2), 941, 1999.

59. Lapointe, B.E. and Smith, N.P., *A Preliminary Investigation of Upwelling as a Source of Nutrients to Looe Key National Marine Sanctuary*, Tech. Rep. No. NA84AAA04157, National Oceanic and Atmospheric Administration, Marine Sanctuaries Division, Washington, D.C., 1987.

60. Lapointe, B.E., Smith, N.P., Pitts, P.A., and Clark, M.W., *Baseline Characterization of Chemical and Hydrographic Processes in the Water Column of Looe Key National Marine Sanctuary*, final report to the National Oceanic and Atmospheric Administration, Office of Ocean and Coastal Resource Management, Washington, D.C., 1992.

61. Lapointe, B.E. and Clark, M.W., Nutrient inputs from the watershed and coastal eutrophication in the Florida Keys, *Estuaries*, 15, 465, 1992.

62. Lapointe, B.E. and Matzie, W.R., Effects of stormwater nutrient discharges on eutrophication processes in nearshore waters of the Florida Keys, *Estuaries*, 19, 422, 1996.

63. Bell, P.R.F., Eutrophication and coral reefs: some examples in the Great Barrier Reef lagoon, *Water Res.*, 26, 553, 1992.

64. Lapointe, B.E. and Matzie, W.R., *High Frequency Monitoring of Wastewater Nutrient Discharges and Their Ecological Effects in the Florida Keys Marine Sanctuary*, final report to U.S. Environmental Protection Agency for Florida Keys National Marine Sanctuary Water Quality Protection Program, Special Studies grant, 1997.

65. Lapointe, B.E., Brand, L., and Barile, P.J., *A Comparative Study of Water Quality and Coral Reef Communities in the Florida Keys National Marine Sanctuary and Biscayne National Park*, final report the Herbert W. Hoover Foundation, 1999.

66. Lapointe, B.E., O'Connell, J.D., and Garrett, G.S., Nutrient couplings between on-site sewage disposal, ground-waters and surface waters of the Florida Keys, *Biogeochemistry*, 10, 289, 1990.

67. Szmant, A.M. and Forrester, A., Water column and sediment nitrogen and phosphorus distribution patterns in the Florida Keys, USA, *Coral Reefs*, 15, 21, 1996.

68. Marubini, F. and Davies, P.S., Nitrate increases zooxanthellae density and reduces skeletogenesis in corals, *Mar. Biol.*, 127, 319, 1996.

69. Falkowski, P.G., Dubinsky, Z., Muscatine, L., and McLoskey, L., Population control in symbiotic corals, *Bio-Science*, 43, 606, 1989.

70. Muscatine, L., Falkowski, P.G., Dubinsky, Z., Cook, P.A., and McClosky, L.R., The effect of external nutrient resources on the population dynamics of zooxanthellae in a reef coral, *Proc. R. Soc. Lond. B.*, 236, 311, 1989.

71. Dubinsky, Z. and Stambler, N., Marine pollution and coral reefs, *Global Change Biol.*, 2, 511, 1996.

72. Fagoonee, I., Wilson, H.B., Hassell, M.P., and Turner, J.R., The dynamics of zooxanthellae populations: a long-term study in the field, *Science*, 283, 843, 1999.

73. D'Elia, C.F., Buddemeier, R.W., Smith, S.V., *Workshop on Coral Bleaching, Coral Reef Ecosystems, and Global Change: Report of Proceedings*, Maryland Sea Grant Publication UM-SG-TS-91-03, 1991.

74. Jaap, W.C., Porter, J.W., Wheaton J., Hackett, K., Lybolt, M., Callahan, M., Tsokos, C., Yanev, G., and Dustan, P., Coral Reef Monitoring Project Executive Summary, Presented to the U.S. Environmental Protection Agency Science Advisory Panel, December 2000, 15 pp.

75. Rogers, C.S., Sublethal and lethal effects of sediments applied to common Caribbean corals in the field, *Mar. Poll. Bull.*, 14, 378, 1983.

76. Smith, S.V. and Buddemeier, R.W., Global change and coral reef ecosystems, *Ann. Rev. Ecol. Syst.*, 23, 89, 1992.

77. Chiappone, M. and Sullivan, K.M., Rapid assessment of reefs in the Florida Keys: results from a synoptic survey, *Proc. 8th Int. Coral Reef Symp.*, 2, 1509, 1997.

78. Odum, H.T. and Odum, E.P., Trophic structure and productivity of the windward coral reef community on Eniwetok Atoll, *Ecol. Monogr.*, 25, 291, 1955.

79. Littler, M.M. and Littler, D.S., Models of tropical reef biogenesis: the contribution of algae, *Prog. Phycol. Res.*, 3, 323, 1984.

80. Littler, M.M., Littler, D.S., and Lapointe, B.E., Baseline studies of herbivory and eutrophication on dominant reef communities of Looe Key national Marine Sanctuary. National Oceanic and Atmospheric Administration, U.S. Department of Commerce, Washington, D.C., 1986.

81. Rudnick, D.T., Chen, Z., Childers, D.L., Boyer, J.N., and Fontaine, T.D., Phosphorus and nitrogen inputs to Florida Bay: the importance of the Everglades watershed, *Estuaries*, 22(2b), 398, 1999.

82. Boyer, J.N., Fourqurean, J.W., and Jones, R.D., Seasonal and long-term trends in the water quality of Florida Bay (1989–1997). *Estuaries*, 22(2b), 412, 1999.

83. Butler, M.J., Hunt, J.H., Herrnkind, W.F., Childress, M.J., Bertelsen, R., Sharp, W., Matthews, T., Field, J.M., and Marshall, G., Cascading disturbances in Florida Bay, USA: cyanobacterial blooms, sponge mortality, and implications for juvenile spiny lobsters *Panularus argus*, *Mar. Ecol. Prog. Ser.*, 129, 119, 1995.

84. Banner A.H., Kaneohe Bay, Hawaii: urban pollution and a coral reef ecosystem, *Proc. 2nd Int. Coral Reef Symp.*, *Brisbane*, 2, 685, 1974.

85. Ryther, J.H., The ecology of phytoplankton blooms in Moriches Bay and the Great South Bay, Long Island, New York, *Biol. Bull.*, 106, 198, 1954.

86. Ryther, J.H. and Dustan, W.M., Nitrogen, Phosphorus, and eutrophication in the coastal marine environment, *Science*, 171, 1008, 1971.

87. Howarth, R.W. et al., Regional nitrogen budgets and riverine N and P fluxes for the drainages to the north Atlantic ocean: natural and human influences, *Biogeochemistry*, 35, 75, 1996.

88. National Research Council, *Clean Coastal Waters: Understanding and Reducing the Effects of Nutrient Pollution*, Ocean Studies Board, Water Science and Technology Board, 2000, p. 391.

89. Howarth, R., Anderson, D., Cloern, J., Elfring, C., Hopkinson, C., Lapointe, B., Malone, T., Marcus, N., McGlathery, K., Sharpley, A., and Walker, D., Nutrient pollution of coastal rivers, bays, and seas, in *Issues in Ecology*, No. 7, Ecological Society of America, Washington, D.C., 2000.

90. National Academy of Sciences, *Understanding Risk: Informing Decisions in a Democratic Society*, Stern, P.C. and Fineberg, H.V., Eds., National Research Council, Washington, D.C., 1996.

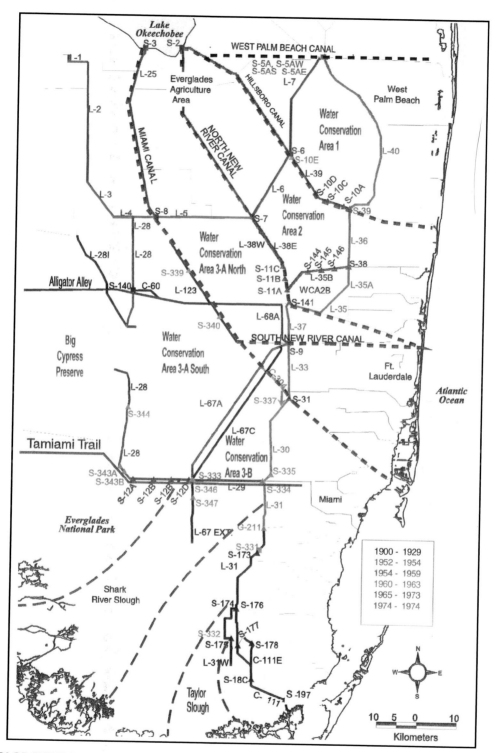

**COLOR FIGURE 2.1** Construction sequence of canals, levees, and water control structures associated with the C&SF Project and the South Dade Conveyance System. (From Light, S. S. and J. W. Dineen. 1994. Water control in the Everglades: a historical perspective, in *Everglades: The Ecosystem and Its Restoration*. S. M. Davis and J. C. Ogden (Eds.). St. Lucie Press, Delray Beach, FL. With permission.)

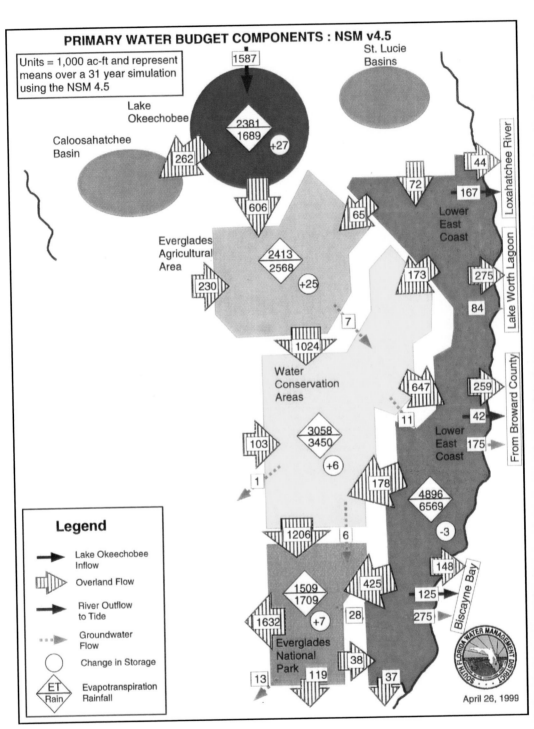

**COLOR FIGURE 2.2** Primary water budget (V1.0) components that resulted from a 31-year simulation of the Natural Systems Model v4.5. Data are aggregated by major basins. All values are annual averages in thousand acre-feet.

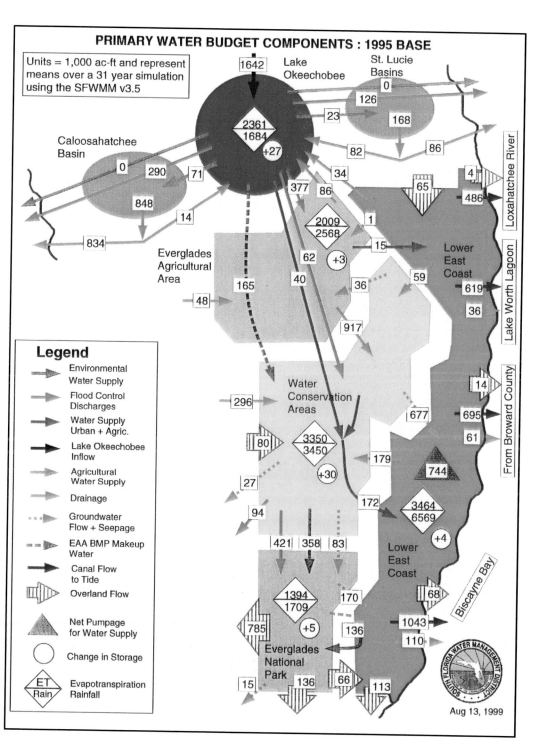

**COLOR FIGURE 2.3** Primary water budget (V1.0) components that resulted from a 31-year simulation of the South Florida Water Management Model v3.5. Data are aggregated by major basins. All values are annual averages in thousand acre-feet.

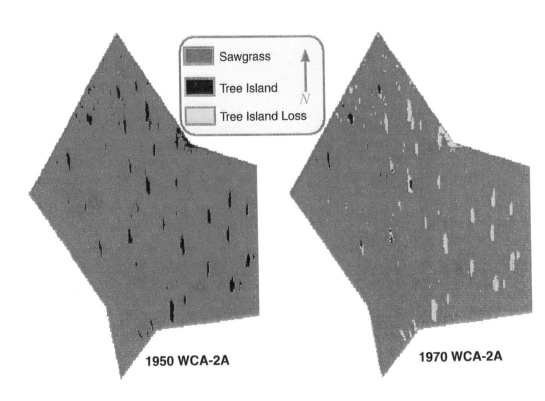

**COLOR FIGURE 2.4** The loss of tree islands in WCA-2A between 1950 and 1970, due to conversion to marsh habitat. This 20-year period had relatively prolonged high water levels, and followed a 30-year period of prolonged low water levels and high soil oxidation. The 1950 map was digitized by Worth from a Loveless survey. (From Worth, D.F. 1988. Environmental response of WCA-2A to reduction in regulation schedule and marsh draw-down. South Florida Water Management District, West Palm Beach, FL; Loveless, C. M. 1959. A study of vegetation of the Florida Everglades. *Ecology* 40(1): 1-9.) The 1970 map was taken from Dineen (Dineen, J.W. 1972. Life in the tenacious Everglades. Central and South Florida Flood Control District. In Depth Report. Vol. 1, No. 5, 112 pp.; Dineen, J. W. 1974. Examination of water management alternatives in Conservation Area 2A. In Depth Report 2(3):1-11, Central & Southern Flood Control District, West Palm Beach, FL, USA.)

**FIGURE 3.1** Typical distribution of periphyton (whitish floating material) within an oligotrophic slough in the northern Everglades (reference station U3, WCA-2A) during the summer wet season. Thick mats of periphyton are typically associated with the submerged macrophyte *Utricularia purpurea* and can completely cover the water surface in oligotrophic, open-water areas of the Everglades during the summer months, providing food and habitat for invertebrates and small fish. A benthic layer of periphyton (not shown in picture) is maintained in these habitats throughout the year. Sparse vegetation includes *Nymphaea odorata* (floating leaves) and *Eleocharis cellulosa* (erect stems).

**FIGURE 4.1** Photographs of the flume setup. (a) The Loxahatchee NWR flume from the air, showing the five parallel 100-m floating walkways, the four experimental channels, the 10-m mixing areas (lower right), and the instrumentation platform. Water flows from lower right to upper left in this photo. (b) Closeup of the instrumentation platform showing solar panels, the weather station tower, precipitation collectors, the P reservoir and pump array, and computer center (beige box, center). (c) Closeup of the pump array that injects P from the reservoir (back) to each channel, based on calculations and signals from the flow sensors and computer. (d) Example of how mid-channel sampling is conducted with walkway bridges.

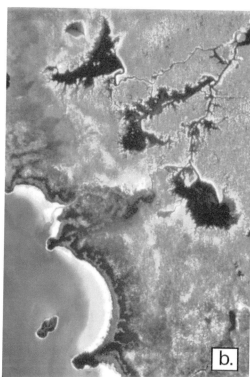

**FIGURE 9.1**   Color infrared aerial photographs depicting (a) healthy mangrove vegetation (red) before Hurricane Andrew, and (b) dead and damaged mangroves (gray) after the hurricane.

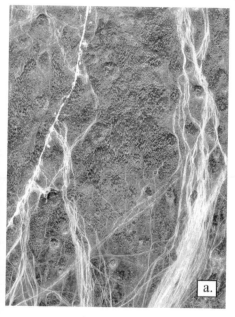

0            50 m

**FIGURE 9.2**   (a) Color infrared aerial photograph; and (b) a photograph taken from a helicopter of ORV trails in Big Cypress National Preserve.

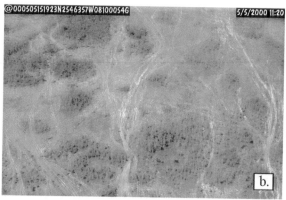

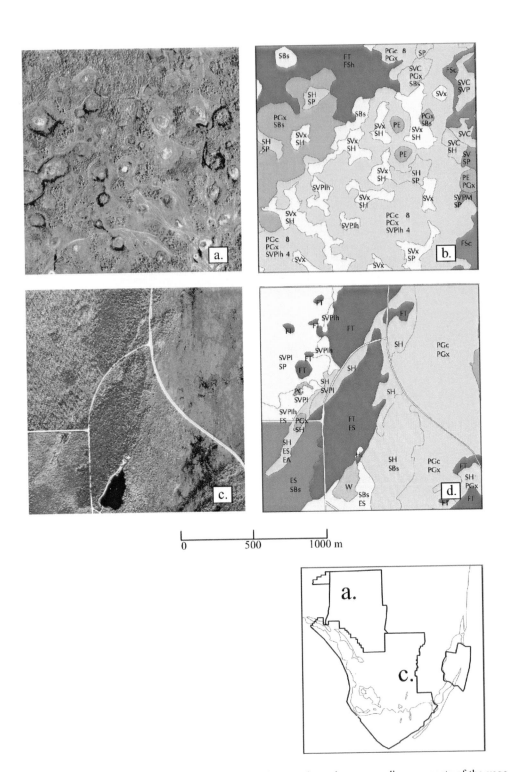

**FIGURE 9.3** Selected sections of color infrared aerial photographs and corresponding segments of the vegetation map/database reveal the diversity of vegetation patterns in Big Cypress National Preserve (a) and (b); and in Everglades National Park (c) and (d). Plant communities depicted in the map segments include: subtropical hardwood forest (FT), sawgrass prairie (PGc), and slash pine savanna with sabal palms (SVx) in Big Cypress National Preserve; and exotic Brazilian pepper (ES), slash pine savanna with hardwoods (SVPIh), shrubland (SH), and sawgrass prairie (PGc) in Everglades National Park.

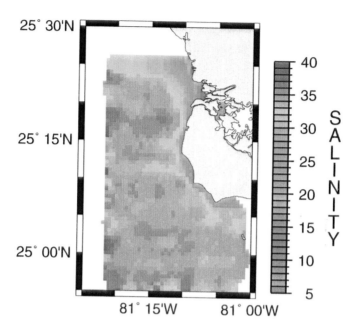

**FIGURE 15.1** Sea surface salinity image of the northwest boundary of Florida Bay near Cape Sable (3 October 1997) showing variable salinity and the influence on salinity due to the flow from the Shark River Slough. Remote salinity values were compared and intercalibrated with *in situ* measurements obtained during the over-flights on the ship track.

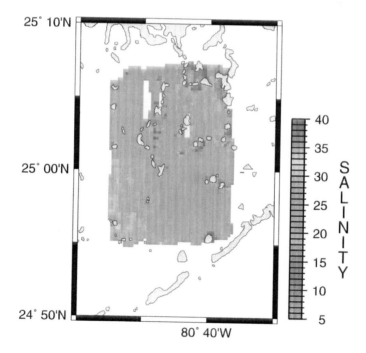

**FIGURE 15.2** High-resolution sea surface salinity image of the central interior bay taken on 4 October 1997 with the airborne salinity mapper flying at an altitude of about 1 km.

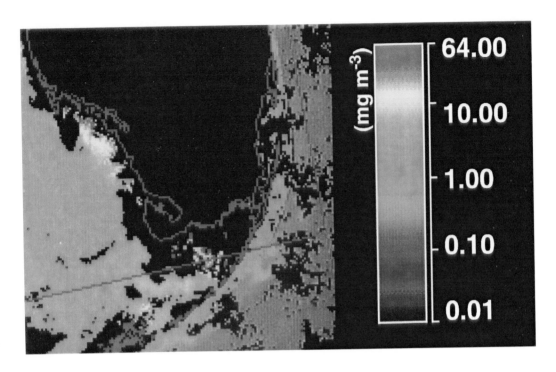

**FIGURE 15.3**  Chlorophyll-*a* (mg m⁻³) derived from SeaWiFS data obtained on 4 October 1997 and processed using the NASA standard software processing package SEADAS 4.0.

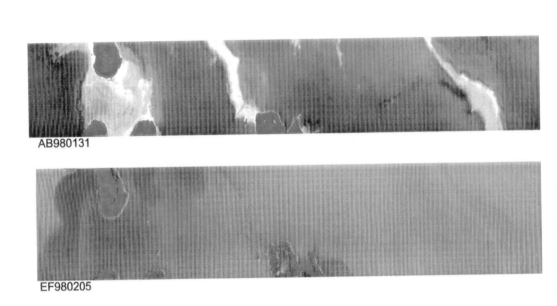

AB980131

EF980205

**FIGURE 15.4**  (Top) An example of two flight lines (transects AB980131 and EF980205) showing regions of the bay imaged by the airborne hyperspectral radiometer; (bottom) a color composite image of a section of the transect AB980131 taken on 31 January 1998 (top panel). The same sectional transect  (bottom panel) imaged on 5 February after a storm passage on 2 February 1998.

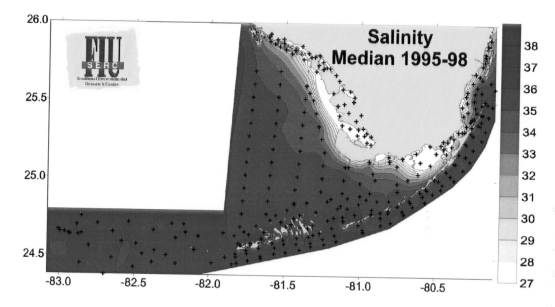

**FIGURE 22.1** Contour map of median salinity generated from fixed stations (+) in South Florida coastal waters.

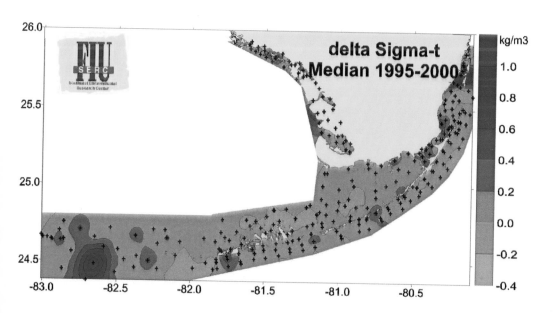

**FIGURE 22.2** Contour map of median delta sigma-t ($\Delta\sigma_t$ in kg m$^{-3}$) generated from fixed stations (+) in South Florida coastal waters reported as the median for 12 sampling events. $\Delta\sigma_t$ is the difference in density between surface and bottom waters where positive values mean the bottom is more dense than the surface.

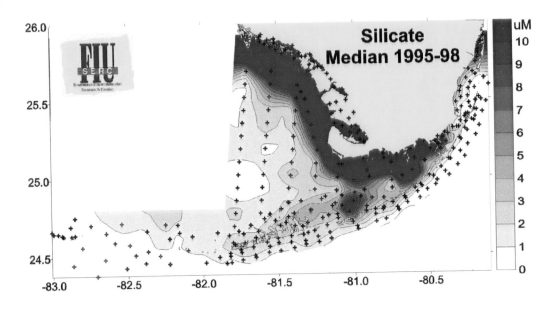

**FIGURE 22.3**  Contour map of median Si(OH)$_4$ ($\mu M$) generated from fixed stations (+) in South Florida coastal waters.

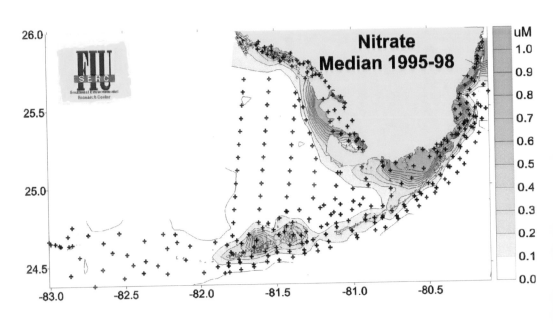

**FIGURE 22.4**  Contour map of median NO$_3^-$($\mu M$) generated from fixed stations (+) in South Florida coastal waters.

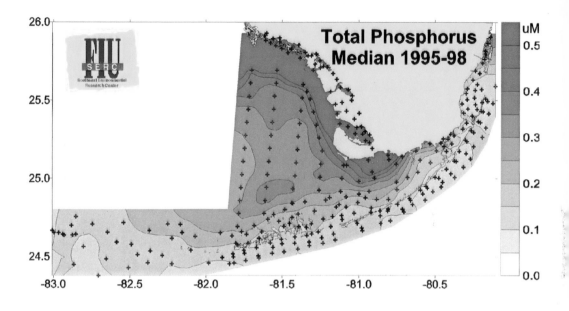

**FIGURE 22.5** Contour map of median TP (μM) generated from fixed stations (+) in South Florida coastal waters.

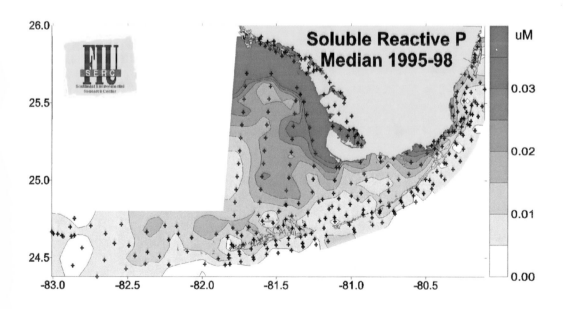

**FIGURE 22.6** Contour map of median SRP (μM) generated from fixed stations (+) in South Florida coastal waters.

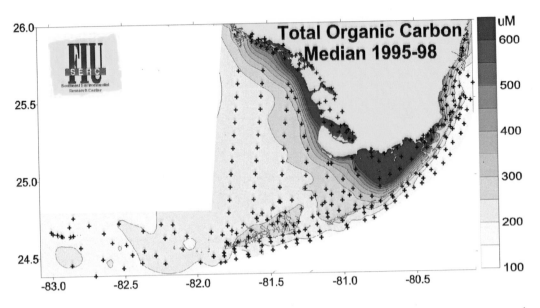

**FIGURE 22.7** Contour map of median TOC ($\mu M$) generated from fixed stations (+) in South Florida coastal waters.

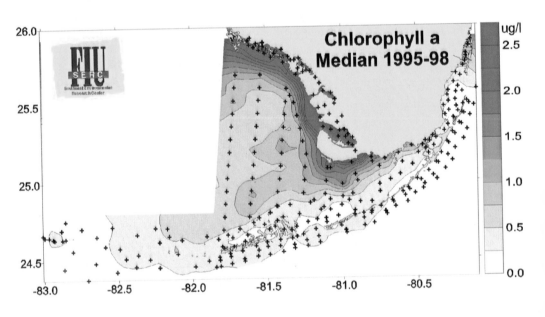

**FIGURE 22.8** Contour map of median Chl $a$ ($\mu$g l$^{-1}$) generated from fixed stations (+) in South Florida coastal waters.

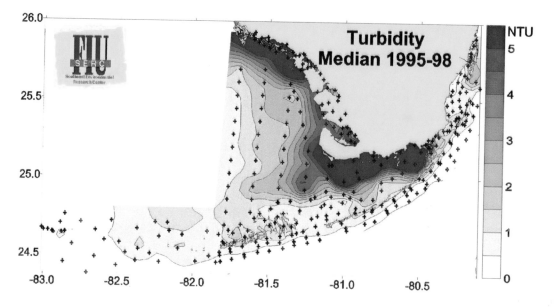

**FIGURE 22.9** Contour map of median turbidity (NTU) generated from fixed stations (+) in South Florida coastal waters.

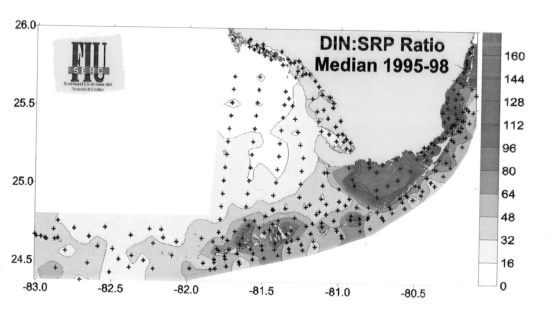

**FIGURE 22.10** Contour map of median DIN:SRP ratio generated from fixed stations (+) in South Florida coastal waters.

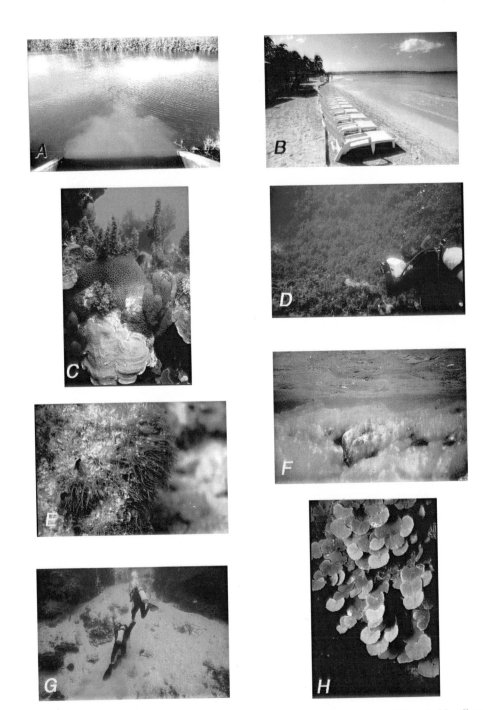

**FIGURE 35.1** A. Phytoplankton bloom (green plume) in the effluent being discharged from the Negril sewage treatment ponds into the South Negril River. B. Receding calcium carbonate beach in Long Bay. C. Hard corals and zoanthids being overtopped by the phaeophyte *Sargassum polyceratium* on shallow reefs at Davis Cove. D. Meadows of *S. polyceratium* on the shallow reef at Orange Bay (North Negril) in December 1997. E. The turf-like chlorophyte *Cladophora fuliginosa* on shallow reefs at Davis Cove, December 1997. F. Blooms of the chlorophyte *Chaetomorpha linum* and blue-green algae smothering hard corals on the shallow reefs of Orange Bay (North Negril) following the onset of the summer wet season and increased concentrations of soluble reactive phosphorus — August 1998. G. "*Halimeda* hash" talus slope produced by the calcifying chlorophyte *Halimeda copiosa* and other species on the deep "spur-and-groove" zone (25 m) off Green Island, October 1998. H. Closeup of the calcifying chlorophyte *Halimeda copiosa* — an abundant calcifying macroalga on deep reefs of the Negril Marine Park.

# 24 Shoreline Nutrients and Chlorophyll *a* in the Florida Keys, 1994–1997: A Preliminary Analysis

*Brian D. Keller*
Florida Keys National Marine Sanctuary

*Arthur Itkin*
Statistical Consultant

## CONTENTS

## INTRODUCTION

Ecological changes in Florida Bay during the past decade have raised concerns about water quality of the bay (Fourqurean and Robblee, 1999). These changes included large-scale seagrass dieoffs starting in 1987 (Robblee et al., 1991; Carlson et al., 1994; Zieman et al., 1994, 1999), prolonged and extensive phytoplankton blooms (Phlips and Badylak, 1996; Boyer et al., 1997, 1999), and an extensive dieoff of sponges associated with a cyanobacteria bloom in 1991 (Butler et al., 1995). Water quality issues associated with these events include: (1) periods of hypersalinity (Fourqurean et al., 1993; Zieman et al., 1999) and high water temperatures (Zieman et al., 1994) that may have been exacerbated by water management practices in southern Florida, and (2) inputs of nutrients as a result of the seagrass dieoffs (Fourqurean et al., 1993) or from human activities such as agriculture (Tomasko and Lapointe, 1994).

The condition of coral reefs along the Florida reef tract is also of concern because of declining coral populations and increasing growth of benthic macroalgae (Porter and Meier, 1992; Szmant and Forrester, 1996). Between the reef tract and Florida Bay lie the Florida Keys, which have a resident population of approximately 80,000, are visited by millions of tourists annually, and are a source of nutrients to nearshore waters (Lapointe and Clark, 1992). In the Keys, Lapointe et al. (1990) showed that groundwater concentrations of nitrogen and phosphorus near septic tanks were higher than background levels. Subsurface groundwater movements transport these nutrients at greatly reduced concentrations with increasing distance from the source. This flux provides nutrients to "venetian" canals, which are widespread in the Keys, and surface waters (Lapointe et al., 1990). Anthropogenic inputs of nutrients to nearshore waters occur throughout the Florida Keys, with concentrations decreasing to low (oceanic) values within approximately 1 to 3 km from land oceanside of the Keys (Lapointe and Clark, 1992). Szmant and Forrester (1996) found elevated water-column concentrations of nitrogen and chlorophyll *a* in inshore waters oceanside of the Upper Keys, which were at oligotrophic levels within 0.5 km of shore. In contrast, phosphorus concentrations often were greater offshore than nearshore, possibly because of periodic upwelling along the shelf edge.

An ongoing program of water quality monitoring in Florida Bay (Boyer et al., 1999) and along the Florida Keys (Boyer and Jones, 2002) does not include sampling of shoreline waters, where anthropogenic effects may be strongest. The purpose of this study is to present water quality monitoring data for Keys shorelines and to contrast these findings with those of nearshore studies.

## MATERIALS AND METHODS

Since 1994, The Nature Conservancy (TNC) has trained volunteers to collect seawater samples and environmental data. In order to ensure data integrity, the training includes instruction on completing data forms, calibrating field equipment, and careful handling of water samples. To help ensure consistency in data collection, TNC staff conducted periodic evaluations of data- and sample-collection techniques.

Sampling stations were located at homes and workplaces of the volunteers and were distributed from Key Largo to Cudjoe Key (Figure 24.1). They were classified in three regions: Upper (Mile Marker [MM] 106–80 = distance in miles from Key West), Middle (MM <80–40), and Lower Keys (MM <40–0). Some stations faced Florida Bay or the Gulf of Mexico ("bayside") and others faced the Atlantic Ocean ("oceanside"). The sites included both developed (residential canals and boat basins) and natural, unobstructed shorelines. Developed shorelines included various kinds of "venetian" canals (e.g., dead-end, open-ended, aerated) and boat basins. Water samples were collected from docks, sea walls, or the shoreline. A total of 18 stations were categorized according to these three characteristics (Table 24.1).

Volunteers were instructed to collect seawater samples at their stations each week during a low tide. During the first year of the program, sampling was conducted during a high and a low tide each week. Analysis of these data showed no significant difference in water quality parameters between tidal levels. After the first year, low tide was selected for weekly sampling; only the low-tide data from the first year of the program were analyzed here. In practice, not all stations were sampled each week. The actual number of samples at a given station during a given month varied from 0 to 6.

Volunteers collected and froze a water sample for laboratory analysis of total nitrogen (TN, $\mu$M) and total phosphorus (TP, $\mu$M) concentrations. For determination of chlorophyll *a* (Chl *a*, $\mu$g/L) concentrations, two 60-mL aliquots of seawater were drawn into a syringe and then squirted through a filter unit containing a Whatman glass microfiber filter (GF/F, 25-mm diameter). The filter was placed in a vial, which was closed within a brown opaque bottle and frozen. The water and filter

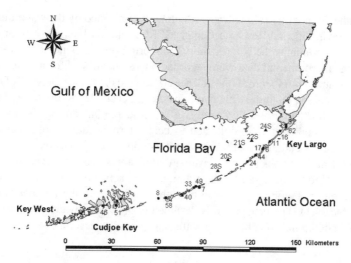

**FIGURE 24.1** Map of the Florida Keys showing station locations identified by number (see Table 24.1). Filled circles are locations of the eight stations sampled for three years (3, 6, 7, 8, 11, 16, 17, and 24). Filled squares are locations of the additional ten stations sampled during November 1996–October 1997. Filled triangles are locations of the five offshore stations in Florida Bay sampled by SERC.

**TABLE 24.1**
**Distribution Among Site Characteristics of the 18 Stations Included in this Analysis**

| Region | Shoreline | Side | N | Station No. |
|---|---|---|---|---|
| Upper Keys | Developed | Bayside | 5 | 6, 17, 24, 32, 44 |
| Middle Keys | Developed | Bayside | 0 | — |
| Lower Keys | Developed | Bayside | 2 | 3, 43 |
| Upper Keys | Natural | Bayside | 2 | 11, 16 |
| Middle Keys | Natural | Bayside | 3 | 8, 33, 49 |
| Lower Keys | Natural | Bayside | 0 | — |
| Upper Keys | Developed | Oceanside | 1 | 62 |
| Middle Keys | Developed | Oceanside | 2 | 7, 58 |
| Lower Keys | Developed | Oceanside | 1 | 51 |
| Upper Keys | Natural | Oceanside | 0 | — |
| Middle Keys | Natural | Oceanside | 1 | 40 |
| Lower Keys | Natural | Oceanside | 1 | 46 |
| Upper Keys | | | 8 | |
| Middle Keys | | | 6 | |
| Lower Keys | | | 4 | |
| | Developed | | 11 | |
| | Natural | | 7 | |
| | | Bayside | 12 | |
| | | Oceanside | 6 | |

*Note:* Upper Keys had 8 stations; Middle Keys, 6; Lower Keys, 4. The shoreline of 11 stations was developed; the shoreline of 7 stations was natural. Of the 18 stations, 12 were bayside and 6 were oceanside.

samples were gathered monthly for analyses, which were conducted by the water quality laboratory of the Southeast Environmental Research Center (SERC), Florida International University. This is the same facility that conducts the water quality monitoring programs noted above.

The following environmental data were recorded on a standardized form: date, time, tidal stage from tide tables, Beaufort number for wind and sea state, wind direction, category of current strength and current direction based on fluorescein dye movement, Secchi depth (m), time of Secchi reading, sea-surface temperature (°C), specific gravity (g/kg), and rainfall during the previous 24 hours (mm). The environmental data were collected to help interpret patterns of variation in the three water quality parameters and are not presented in this preliminary analysis.

The addition of new stations and the termination of others occurred over the 3 years reported here (November 1994 to October 1997); eight stations were active for the entire period. The varying number and location of stations explains, in part, the imbalance in the number of stations among station types (Table 24.1). The program was designed to take full advantage of sampling opportunities, rather than attempting to adhere to a statistical sampling design.

## DATA ANALYSIS

As described earlier, 18 sampling stations were categorized according to three characteristics: *region* (Upper, Middle, and Lower Keys), *shoreline* (developed vs. natural), and *side* (bayside or oceanside), for a total of $3 \times 2 \times 2 = 12$ possible spatial groupings. Table 24.1 gives the number of stations for each of the groupings. There was considerable imbalance of replication among the categories, and no stations were available for three of the groupings.

Two periods were selected for this analysis. Data were available from a maximum of 18 stations for the 1-year period from November 1996 to October 1997. The other 2 years had less complete coverage of the 12 possible spatial groupings and are not included in this analysis. To examine longer term variation, data were analyzed for eight stations that were monitored from November 1994 to October 1997.

Because of the imbalance, effects due to each of the major variables were evaluated separately; comparisons were made among regions, between shorelines, and between sides. Evaluation of possible interactions among these factors was not considered to be practical at this time, so conclusions drawn should be considered preliminary. Comparisons were also made between the dry season (November to April) and the wet season (May to October). Only three response variables were evaluated: TN, TP, and Chl *a* concentrations. Because the number of samples taken per month was variable, all individual determinations were averaged to give a monthly mean. Months in which no samples were available were considered to be missing. Seasonal means (dry vs. wet season) were calculated as the average of the available monthly means for that period. Yearly means were calculated as the average of the two seasonal means.

All data were analyzed by analysis of variance procedures. Natural logarithms of the data were used in the analysis in order to satisfy the assumptions underlying the statistical procedures. Accordingly, the means reported are geometric means. Comparisons among the 3 regions (Upper vs. Middle vs. Lower Keys) and comparisons among the three years (for the longer-term data) were based on least significant differences (LSDs); they were not adjusted for multiple comparisons. Statistical significance was set at $p < 0.05$; if $0.05 < p < 0.10$, a trend was noted.

Because the number and distribution of sampling stations depended on volunteers, the amount of data collected was not as extensive as we might wish. More definitive conclusions would require larger numbers of stations, which we are trying to obtain. It should be recognized that a comparison that is declared to be not statistically significant does not necessarily mean that no difference exists. An effect may have been present, but the sample size was too small to be able to statistically detect that difference. As already noted, our conclusions should be considered to be preliminary.

We also examined the correlation between the three response variables of interest. Over the 1-year period (November 1996 to October 1997) common to the 18 stations, correlation coefficients *within each station* were calculated based only on complete datasets (i.e., samples with all three values). In addition, the 1-year means were calculated for each variable within each station and used in the calculation of correlation coefficients *across stations*. All data were converted to natural logarithms prior to analysis.

A monitoring program to describe spatial and temporal patterns of water quality in Florida Bay was initiated in 1989 (Boyer et al., 1997, 1999). This study has 28 sampling stations in the bay, five of which are within several kilometers of the Upper Keys (Figure 24.1). Joe Boyer (SERC) provided data for these stations for the period November 1996 to October 1997. We compared geometric means of the five SERC stations with geometric means of the bayside stations sampled in the Upper Keys over this same 1-year period ($N = 5$ for developed shorelines; $N = 2$ for natural shorelines).

## RESULTS AND DISCUSSION

### SPATIAL GROUPINGS, NOVEMBER 1996 TO OCTOBER 1997

#### Total Nitrogen

Monthly mean values of TN ranged from 13.6 to 177.0 µM. The annual mean was greatest in the Upper Keys followed by the Lower and Middle Keys (Table 24.2), but only the Upper and the Middle Keys were significantly different. Comparing shoreline data for the Upper Keys with SERC's nearshore stations in Florida Bay, TN was elevated at natural shorelines (52.3 vs. 39.4 µM). Surprisingly, TN was lower at developed stations (41.3 µM) and similar to the nearshore value. The decrease in TN between the Upper and the Middle Keys (Table 24.2) was consistent with the spatial trend in total organic nitrogen reported in Florida Bay by Fourqurean et al. (1993).

The bayside mean for TN was significantly greater than the oceanside mean (Table 24.2). The oceanside shoreline concentration (28.0 µM) was higher than that reported by Szmant and Forrester (1996) at inshore stations off Key Largo and Long Key sampled in March–July 1990 (19.8 ± 7.0 [SD] and 21.8 ± 15.5 µM, respectively). Values at developed and natural shorelines were not significantly different. TN levels were significantly higher during the wet season (Table 24.2).

#### Total Phosphorus

Monthly mean values of TP ranged from 0.17 to 5.25 µM. The annual mean value showed no statistically significant differences with regard to region, shoreline, side, and season, even though the value for developed shorelines was nearly 40% greater than the value for natural shorelines (Table 24.2). As was the case for TN, TP was elevated at natural shorelines compared to nearshore concentrations in Florida Bay (0.31 vs. 0.17 µM); the nearshore mean equaled the minimum measurement at shoreline stations. Unlike TN, TP also was elevated at developed shorelines (0.49 µM, nearly three times the nearshore value). As reported here (Table 24.2), Fourqurean et al. (1993) measured little variation in TP across this region of the bay. The oceanside level of TP (0.36 µM) was substantially higher than concentrations reported by Szmant and Forrester (1996) off Key Largo and Long Key (0.21 ± 0.12 and 0.21 ± 0.09 µM, respectively).

#### Chlorophyll *a*

Monthly mean values of Chl *a* ranged from 0.04 to 12.69 µg/L. The annual mean value showed some suggestive differences: The mean value for the Lower Keys was greater by two- to threefold than the mean values at the Middle and Upper Keys, and the mean value for developed shorelines showed a trend toward being greater than the mean value at natural shorelines (Table 24.2); however, no statistically significant differences were found. Shallow "backcountry" waters north of the Lower Keys may support nitrogen fixers such as epiphytic cyanobacteria that contribute to the productivity of this region (J. Fourqurean, pers. comm.). Dry and wet season results were almost identical.

**TABLE 24.2**
**Geometric Means of Water Quality Parameters for the Three Categories of Sampling Stations**

|  | Upper Keys (8) | Middle Keys (6) | Lower Keys (4) | Seasons (18) |
|---|---|---|---|---|
| Total nitrogen ($\mu M$) |  |  |  |  |
| Dry Season | 37.9 | 28.9 | 32.0 | 33.4 |
| Wet Season | 48.3 a | 28.5 b | 41.4 a,b | 39.1* |
| Nov. 96–Oct. 97 | 42.8 a | 28.7 b | 36.4 a,b |  |
| Total phosphorus ($\mu M$) |  |  |  |  |
| Dry Season | 0.37 | 0.51 | 0.50 | 0.44 |
| Wet Season | 0.45 | 0.44 | 0.73 | 0.50 |
| Nov. 96–Oct. 97 | 0.41 | 0.47 | 0.60 |  |
| Chlorophyll $a$ ($\mu g/L$) |  |  |  |  |
| Dry Season | 0.56 | 0.40 | 0.95 | 0.57 |
| Wet Season | 0.50 | 0.40 | 1.28 | 0.57 |
| Nov. 96–Oct. 97 | 0.53 | 0.40 | 1.11 |  |

|  | Developed (11) | Natural (7) | Seasons (18) |
|---|---|---|---|
| Total nitrogen ($\mu M$) |  |  |  |
| Dry Season | 31.9 | 35.7 | 33.4 |
| Wet Season | 39.9 | 37.9 | 39.1* |
| Nov. 96–Oct. 97 | 35.7 | 36.8 |  |
| Total phosphorus ($\mu M$) |  |  |  |
| Dry Season | 0.47 | 0.39 | 0.44 |
| Wet Season | 0.59 | 0.38 | 0.50 |
| Nov. 96–Oct. 97 | 0.53 | 0.38 |  |
| Chlorophyll $a$ ($\mu g/L$) |  |  |  |
| Dry Season | 0.82 + | 0.32 | 0.57 |
| Wet Season | 0.77 + | 0.36 | 0.57 |
| Nov. 96–Oct. 97 | 0.80 + | 0.34 |  |

|  | Bayside (12) | Oceanside (6) | Seasons (18) |
|---|---|---|---|
| Total nitrogen ($\mu M$) |  |  |  |
| Dry Season | 38.0* | 25.8 | 33.4 |
| Wet Season | 44.3* | 30.4 | 39.1* |
| Nov. 96–Oct. 97 | 41.0* | 28.0 |  |
| Total phosphorus ($\mu M$) |  |  |  |
| Dry Season | 0.49 | 0.36 | 0.44 |
| Wet Season | 0.58 | 0.37 | 0.50 |
| Nov. 96–Oct. 97 | 0.53 | 0.36 |  |
| Chlorophyll $a$ ($\mu g/L$) |  |  |  |
| Dry Season | 0.71 | 0.36 | 0.57 |
| Wet Season | 0.64 | 0.47 | 0.57 |
| Nov. 96–Oct. 97 | 0.67 | 0.41 |  |

*Note:* Sample sizes are in parentheses. Significantly different means are designated by an asterisk (*; $p < 0.05$, ANOVA); letters indicate the LSD results, $p < 0.05$), for the three regions of the Keys. A plus sign (+) indicates a trend toward significance ($0.05 < p < 0.10$).

Unlike TN and TP, Chl *a* was lower at natural shorelines than nearshore in Florida Bay (0.20 vs. 0.33 µg/L). However, the value of Chl *a* at developed, bayside shorelines in the Upper Keys (0.86 µg/L) was more than twice the nearshore concentration. As reported here (Table 24.2), Fourqurean et al. (1993) measured little variation in Chl *a* across this region of the bay. As was seen for TN and TP, oceanside Chl *a* (0.41 µg/L) was higher than concentrations reported by Szmant and Forrester (1996) off Key Largo and Long Key (0.24 ± 0.25 and 0.31 ± 0.09 µg/L, respectively).

## TEMPORAL VARIATION, NOVEMBER 1994 TO OCTOBER 1997

### Total Nitrogen

Monthly mean values of TN ranged from 11.5 to 90.6 µ*M* for the eight stations that were monitored for 3 years. The annual mean values of TN were essentially the same (Table 24.3). Over the entire period, the mean wet season value was significantly greater than the dry season value. Within each of the 3 years, the wet season mean was numerically greater than the dry season mean. This mirrors the seasonal difference documented in the spatial comparisons. The level of TN reported here was substantially greater than the level reported by Boyer and Jones (2002) for the Florida Keys National Marine Sanctuary over much of the same 3-year period (range of 3.65 to 46.94 µ*M*).

### TABLE 24.3
### Geometric Means of Water Quality Parameters
### for the Three Years of the Study

|  | Dry Season | Wet Season | Full Year |
|---|---|---|---|
| Total nitrogen (µ*M*) |  |  |  |
| Nov. 94–Oct. 95 | 38.1 | 39.7 | 38.9 |
| Nov. 95–Oct. 96 | 36.0 | 41.0 | 38.4 |
| Nov. 96–Oct. 97 | 37.3 | 44.6 | 40.8 |
| Overall | 37.1 | 41.7* |  |
| Total phosphorus (µ*M*) | 0.34 | 0.38 | 0.36 |
| Nov. 94–Oct. 95 | 0.40 | 0.36 | 0.38 |
| Nov. 95–Oct. 96 | 0.37 | 0.53 | 0.44 |
| Nov. 96–Oct. 97 | 0.37 | 0.41 |  |
| Overall |  |  |  |
| Chlorophyll *a* (µg/L) | 0.53 | 0.75 | 0.63 a |
| Nov. 94–Oct. 95 | 0.56 | 0.58 | 0.57 a,b |
| Nov. 95–Oct. 96 | 0.43 | 0.41 | 0.42 b |
| Nov. 96–Oct. 97 | 0.50 | 0.56 |  |
| Overall |  |  |  |

*Note:* $N = 8$ stations. Significantly different means are designated by an asterisk (*; $p < 0.05$, ANOVA); letters indicate the LSD results ($p < 0.05$) for the 3 years of the study.

### Total Phosphorus

Monthly mean values of TP ranged from 0.15 to 4.21 µ*M*. No statistically significant differences were found among years or between seasons over the 3 years of the study (Table 24.3). Interestingly, the year-to-year increases paralleled the temporal trend of increasing TP measured by Jones and Boyer (2000) for the Florida Keys National Marine Sanctuary. As was the case with TN, the level of TP reported here was much higher than the level reported by Boyer and Jones (2002; 0.01 to 0.66 µ*M*).

## Chlorophyll *a*

As reported by Boyer and Jones (2002), Chl *a* was highly variable; monthly mean values of Chl *a* ranged from 0.04 to 9.84 µg/L). A consistent decrease in the annual mean value of Chl *a* was seen from the first year on (Table 24.3). The difference between the first and third years was statistically significant. The drop in Chl *a* during the third year of the study (Table 24.3) coincided with declining phytoplankton blooms in the north-central and western regions of Florida Bay (Boyer and Jones, 1999; J. Hunt, pers. comm.). As was the case with TN and TP, the level of Chl *a* reported here for inshore waters was much higher than the Sanctuary-wide level, which ranged from 0.01 to 6.81 µg/L (Boyer and Jones, 2002).

### CORRELATIONS BETWEEN VARIABLES

As stated above, correlation coefficients between the three variables were calculated *within stations* based on complete sets of water quality samples collected over a 1-year period (Table 24.4). Note that only the correlations between TN and TP were consistently positive, with more than half (11 of 18) statistically significant. Correlations between TN and Chl *a* and between TP and Chl *a* were not consistent and tended to be of smaller magnitude.

Based on the annual means for each of the 18 stations, the *across-station* correlation coefficients were all positive, but only the correlation between TP and Chl *a* was statistically significant (Table 24.4).

The positive correlations between TN and TP in samples collected from each station (Table 24.4) may be indicative of a common source. The largest components of TN and TP are organic rather

**TABLE 24.4**
**Correlation Coefficients Between Water Quality Parameters**

| Station No. | N | TN vs. TP | TN vs. Chl *a* | TP vs. Chl *a* |
|---|---|---|---|---|
| 3 | 47 | 0.76* | 0.10 | −0.17 |
| 6 | 40 | 0.63* | −0.09 | 0.18 |
| 7 | 52 | 0.22 | 0.11 | −0.46* |
| 8 | 26 | 0.51* | 0.10 | 0.08 |
| 11 | 31 | 0.56* | 0.37* | 0.35 |
| 16 | 42 | 0.21 | −0.08 | −0.26 |
| 17 | 45 | 0.64* | 0.25 | 0.22 |
| 24 | 32 | 0.71* | 0.39* | 0.28 |
| 32 | 42 | 0.82* | −0.30 | −0.06 |
| 33 | 33 | 0.32 | 0.38* | 0.43* |
| 40 | 39 | 0.28 | 0.09 | −0.19 |
| 43 | 41 | 0.29 | 0.12 | 0.39* |
| 44 | 51 | 0.36* | 0.08 | −0.19 |
| 46 | 40 | 0.71* | 0.46* | 0.36* |
| 49 | 33 | 0.84* | 0.42* | 0.52* |
| 51 | 22 | 0.41 | −0.02 | 0.53* |
| 58 | 24 | 0.23 | 0.30 | 0.33 |
| 62 | 48 | 0.59* | −0.17 | 0.14 |
| Annual mean | 18 | 0.22 | 0.25 | 0.79* |

*Note:* $N$ = sample size; TN = concentration of total nitrogen ($\mu M$); TP = concentration of total phosphorus ($\mu M$); Chl *a* = concentration of chlorophyll *a* (µg/L). An asterisk (*) indicates $p < 0.05$.

than dissolved inorganic (e.g., Boyer et al., 1997), and each measurement of TN and TP for a particular sampling event comes from the same small volume of water; therefore, this correlation may stem from variations in the amount of surface-water organic material.

The strong positive correlation between TP and Chl *a* in the across-station calculation (Table 24.4) may be indicative of a time-averaged association, in this case of year-long phosphorus loading and the concentration of Chl *a*. A positive association of TP and Chl *a* concentrations in eastern Florida Bay (Boyer and Jones, 1999) indicates that this region of the bay is phosphorus limited (Fourqurean et al., 1993). The results of this study suggest that shoreline waters of the Keys may also be phosphorus limited.

## CONCLUSION

The concentrations of TP and Chl *a* did not differ significantly in shoreline waters among the three regions of the Keys. The concentration of TN was greater in the Upper than in the Middle Keys. TN and TP did not differ significantly between developed and natural shorelines, even though the concentration of total phosphorus was elevated by nearly 40% at developed shorelines relative to natural shorelines. A trend could be seen, however, for greater Chl *a* at developed than at natural shorelines. TN was significantly greater at bayside than at oceanside stations, but TP and Chl *a* did not differ significantly. A strong, positive correlation existed between TP and TN at many of the stations, but not between TN and Chl *a* or between TP and Chl *a*. Using annual station means to calculate across-station correlation coefficients, a significant positive correlation was present only between TP and Chl *a*, suggesting phosphorus limitation.

The concentrations of TN and TP did not vary significantly over the 3 years of this study (November 1994 to October 1997), although TP increased somewhat from year to year as reported for the Florida Keys National Marine Sanctuary (Jones and Boyer, 2000). Chl *a* declined significantly between the first and the last year of the study. TN was significantly greater during the wet season (May to October) than during the dry season (November to April), but there was no significant seasonal difference for TP and Chl *a*.

The results of this study indicate inputs of nitrogen and phosphorus to marine waters along the shoreline and an associated increase in the concentration of phytoplankton. Possible connections between these shoreline effects, nearshore ecosystems, and the reef tract may be episodic and require further investigation.

## ACKNOWLEDGMENTS

This study would not have been possible without the thousands of hours of effort contributed by Florida Bay Watch volunteers over the years. Ron Jones (Southeast Environmental Research Center, Florida International University) provided advice about the sampling protocol, and the water quality laboratory he directs analyzed the water samples; Joe Boyer provided water quality data for selected SERC stations. Paul Dye was Program Manager of Florida Bay Watch until 1998. Fran Decker coordinated the program until 1997; Mary Enstrom and Sherry Dawson coordinated volunteer involvement. Bill Miller and Julie Overing maintained the database and produced data reports from 1994 until 1997. Funding was provided by the U.S. Environmental Protection Agency–Region IV, Everglades National Park, U.S. Fish and Wildlife Service, South Florida Water Management District, Orvis Company, John Smale, Johnny Morris Foundation, Yamaha Contender Miami Billfish Tournament, Perkins Charitable Foundation, and Curtis and Edith Munson Foundation. The Florida Keys National Marine Sanctuary and Florida Marine Research Institute contributed office space and other in-kind support for the program. We thank Joe Boyer and an anonymous reviewer for commenting on the manuscript.

# REFERENCES

Boyer, J.N. and R.D. Jones. 1999. Effects of freshwater inputs and loading of phosphorus and nitrogen on the water quality of eastern Florida Bay, pp. 545–561, in *Phosphorus Biogeochemistry in Sub-Tropical Ecosystems*, Reddy, K.R., Ed., Lewis Publishers, Boca Raton, FL.

Boyer, J.N. and R.D. Jones. 2002. A view from the bridge: external and internal forces affecting the ambient water quality of the Florida Keys National Marine Sanctuary, in *The Everglades, Florida Bay, and Coral Reefs of the Florida Keys: An Ecosystem Sourcebook*, Porter, J.W. and Porter, K.G., Eds., CRC Press, Boca Raton, FL, 2001.

Boyer, J.N., J.W. Fourqurean, and R.D. Jones. 1997. Spatial characterization of water quality in Florida Bay and Whitewater Bay by multivariate analysis: zones of similar influence, *Estuaries*, 20:743–758.

Boyer, J.N., J.W. Fourqurean, and R.D. Jones. 1999. Seasonal and long-term trends in the water quality of Florida Bay (1989–1997), *Estuaries*, 22:417–430.

Butler, IV, M.J., J.H. Hunt, W.F. Herrnkind, M.J. Childress, R. Bertelsen, W. Sharp, T. Matthews, J.M. Field, and H.G. Marshall. 1995. Cascading disturbances in Florida Bay, USA: cyanobacteria blooms, sponge mortality, and implications for juvenile spiny lobsters *Panulirus argus*, *Mar. Ecol. Prog. Ser.*, 129:119–125.

Carlson, P.R., L.A. Yarbro, and T.R. Barber. 1994. Relationship of sediment sulfide to mortality of *Thalassia testudinum* in Florida Bay, *Bull. Mar. Sci.*, 54:733–746.

Fourqurean, J.W. and M.B. Robblee. 1999. Florida Bay: a history of recent ecological changes, *Estuaries*, 22: 345–357.

Fourqurean, J.W., R.D. Jones, and J.C. Zieman. 1993. Processes influencing water column nutrient characteristics and phosphorus limitation of phytoplankton biomass in Florida Bay, FL, USA: inferences from spatial distributions, *Estuar. Coast. Shelf Sci.*, 36:295–314.

Jones, R.D. and J.N. Boyer. 2000. *Florida Keys National Marine Sanctuary Water Quality Monitoring Project: 1999 Annual Report*, Tech. Rep. #T121, Southeast Environmental Research Center, Florida International University, Miami.

Lapointe, B.E. and M.W. Clark. 1992. Nutrient inputs from the watershed and coastal eutrophication in the Florida Keys, *Estuaries*, 15: 465–476.

Lapointe, B.E., J.D. O'Connell, and G.S. Garrett. 1990. Nutrient couplings between on-site sewage disposal systems, groundwaters, and nearshore surface waters of the Florida Keys, *Biogeochemistry*, 10:289–307.

Phlips, E.J. and S. Badylak. 1996. Spatial variability in phytoplankton standing crop and composition in a shallow inner-shelf lagoon, Florida Bay, Florida, *Bull. Mar. Sci.*, 58:203–216.

Porter, J.W. and O.W. Meier. 1992. Quantification of loss and change in Floridian reef coral populations, *Am. Zool.*, 32: 625–640.

Robblee, M.B., T.R. Barber, P.R. Carlson, Jr., M.J. Durako, J.W. Fourqurean, L.K. Muehlstein, D. Porter, L.A. Yarbro, R.T. Zieman, and J.C. Zieman. 1991. Mass mortality of the seagrass *Thalassia testudinum* in Florida Bay (USA), *Mar. Ecol. Prog. Ser.*, 71:297–299.

Szmant, A.M. and A. Forrester. 1996. Water column and sediment nitrogen and phosphorus distribution patterns in the Florida Keys, USA, *Coral Reefs*, 15:21–41.

Tomasko, D.A. and B.E. LaPointe. 1994. An alternative hypothesis for the Florida Bay seagrass die-off, *Bull. Mar. Sci.*, 54:1086.

Zieman, J.C., R. Davis, J.W. Fourqurean, and M.B. Robblee. 1994. The role of climate in the Florida Bay seagrass die-off, *Bull. Mar. Sci.*, 54:1088.

Zieman, J.C., J.W. Fourqurean, and T.A. Frankovich. 1999. Seagrass die-off in Florida Bay: long-term trends in abundance and growth of turtle grass, *Thalassia testudinum*, *Estuaries*, 22:460–470.

# 25 Tidal and Meteorological Influences on Shallow Marine Groundwater Flow in the Upper Florida Keys

*Christopher D. Reich, Eugene A. Shinn, and Todd D. Hickey*
U.S. Geological Survey, Center for Coastal Geology

*Ann B. Tihansky*
U.S. Geological Survey, Water Resources Division

## CONTENTS

## INTRODUCTION

Historically, groundwater-management issues in the Florida Keys have not been a major concern of local residents; however, in recent years changes in the water quality of Florida Bay and the health of neighboring coral reefs have focused attention on the possibility that groundwater may play a vital role in the health of marine ecosystems both nearshore and along the reef tract. Groundwater contamination from on-site disposal systems (i.e., septic tanks and cesspools) and wastewater injection has been targeted as a likely source of nutrients and pathogens leading to eutrophication/nutrification of nearshore surface waters (Lapointe et al., 1990; Shinn et al., 1994).

In addition to nearshore impacts, it was also believed that contaminants were being transported from onshore areas to the reef tract, as much as 8 km away, by surface- and/or groundwater movement (Shinn et al., 1994).

Creation of the Florida Keys National Marine Sanctuary in 1990 prompted the U.S. Environmental Protection Agency (EPA) to develop a Water Quality Protection Program (WQPP) for the Florida Keys region (EPA, 1991, 1992, 1996). This region encompasses all of the nearshore environments of the Keys, the reef tract, and portions of Florida Bay. The coordination of federal and non-federal agencies through the WQPP has led to the investigation of several groundwater-movement related issues: (1) groundwater flow around wastewater injection wells (Paul et al., 1995a,b; Monaghan, 1996; Corbett et al., 1999); (2) fate and transport of wastewater contaminants to the offshore reef environments (Shinn et al., 1994); (3) septic tank discharge and its impacts on canal systems (Lapointe et al., 1990; Paul et al., 1995a,b); (4) quality and quantity of submarine groundwater discharge (SGD) as well as nutrient loading of surface waters in response to groundwater seepage (Corbett et al., 1999; Shinn et al., in press); and (5) natural groundwater-flow conditions in the Florida Keys (Halley et al., 1994; this study).

In this study, the installation of a circular cluster of paired, multidepth wells on opposing sides of Key Largo and the use of Fluorescein and Rhodamine dyes as groundwater tracers verified that marine groundwater flow occurs from Florida Bay beneath Key Largo and ultimately into the Atlantic Ocean. Weather, bay level, and tidal fluctuations play dynamic roles in controlling groundwater movement in the upper Florida Keys area. This particular area was selected to represent natural gradient-driven flow of groundwater in the absence of any extraneous sources, such as artificial recharge from sewage injection wells.

## DESCRIPTION OF STUDY AREA

### SITE LOCATION

Key Largo, FL, lies within a subtropical climate with mean January temperatures of 21°C and mean July temperatures of 28°C. Annual rainfall for the upper Keys averages about 140 cm, with almost two thirds of the rain occurring between May and October (Halley et al., 1997). The study site is shown in Figure 25.1. The tracer experiments were conducted using pairs of nested wells (two per borehole) at two offshore well cluster sites. The bayside well cluster (BSWC) is located in Florida Bay (25°04.252′ N × 80° 28.120′ W), while the oceanside well cluster (OSWC) is located directly across Key Largo in the Atlantic Ocean (25°03.990′ N × 80°27.941′ W). The circular clusters of wells are located approximately 60 m and 80 m from the shoreline in 1.1-m and 1.8-m water depth (mean high tide), respectively (Figure 25.1). Well depths are 13.6 m and 6.1 m below rock surface at both sites. These sites were chosen because they straddle a relatively narrow (500-m) unpopulated portion of Key Largo and are unaffected by artificial recharge from shallow sewage injection wells. Four onshore transect wells (18-m deep) were core-drilled to obtain water-table profiles across the island between the well clusters.

### HYDROGEOLOGIC SETTING

The Florida Keys region is underlain by one of the most permeable limestones (Key Largo Formation) in the world. The islands have little vertical relief (5 m maximum at Windley Key) and function as discontinuous, semipermeable dams separating the large lagoon of Florida Bay from the Atlantic Ocean. The low relief of the islands and the high permeability of the limestone prevent the development of a well-defined freshwater lens. When combined with meteorological events, these unique conditions result in rapid multidirectional flow patterns within the groundwater system.

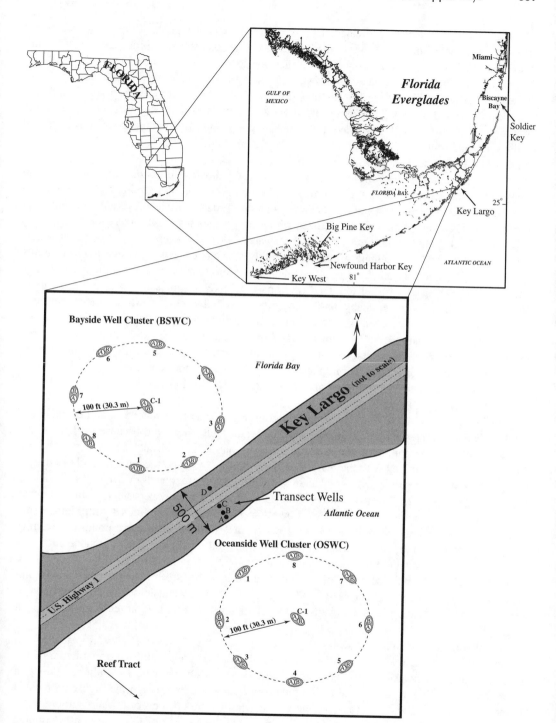

**FIGURE 25.1** Site locations of the dye tracer experiments in Key Largo, FL. Layout and position of wells at each well cluster in relation to Key Largo, Florida Bay, and the reef tract are shown for orientation. The central wells, C-1A,B at both Bayside Well Cluster (BSWC) and Oceanside Well Cluster (OSWC), are the sites where dye was injected. The transects of wells located on the island are also shown.

The Florida Keys can be divided naturally into two distinct geologic units of Pleistocene age — a coralline boundstone facies of the Key Largo Limestone and the oolitic facies of the Miami Limestone (Miami oolite). The Key Largo Limestone extends from Soldier Key in the northeast to Newfound Harbor Keys to the southwest, a distance of approximately 176 km, and westward beneath the Miami oolite of the Lower Keys (Figure 25.1). The Lower Keys consist of cemented spherical oöids and peloids that begin at Big Pine Key and extend past Key West and beneath the Marquesas Keys in the Gulf of Mexico. Both the Key Largo and Miami oolite formations dip down to the south and southwest (Hoffmeister and Multer, 1968; Perkins, 1977; Shinn et al., 1989; Davis et al., 1992; Halley et al., 1997).

Core data from both well cluster sites depict a vuggy grainstone limestone, typical of reef and back-reef deposits of the Key Largo Formation (Figure 25.2). Core description terminology used in Figure 25.2 and throughout the text is after Dunham (1962). Estimated hydraulic conductivity for this lithologic unit within the Key Largo Limestone ranges from 1400 meters per day (m/d) (Vacher et al., 1992) to 12,000 m/d (Fish and Stewart, 1990). However, horizons of impermeable calcrete, also referred to as caliche, often cap the highly permeable Key Largo Limestone. Generally speaking, five major horizons have been detected within the Key Largo Formation, each horizon being formed during a sea-level lowstand (Perkins, 1977). Where present and continuous, these caliche units create an impermeable barrier to vertical groundwater flow. These calcrete surfaces can form up to 6-cm-thick coatings and can be assumed to have a permeability of several orders of magnitude less than that of the limestone above and below. Core C-1 obtained at BSWC (Figures 25.1 and 25.2) contained the only significant laminated caliche crust horizon in all cores obtained. Lithoclasts or fragments of caliche were observed in other cores, but the presence of a continuous laminated caliche crust was absent (Figure 25.2). This indicates that vertical movement of groundwater is not impeded at these sites.

Owing to a high volume of precipitation and highly porous limestone, a large percentage of rainfall likely infiltrates on land and recharges the water table. However, because of the extreme permeability, tidal pumping, and presence of shallow marine groundwater, the newly recharged freshwater is rapidly mixed with the marine groundwater or is transported laterally and discharged along the coast. Profiles of specific conductance measured at transect wells drilled across the island of Key Largo (Figure 25.1) between the well clusters show a relatively persistent (1- to 2-m thick) brackish lens between the two study sites. Lowest specific conductance of 21.4 milliSiemens per centimeter (mS/cm) was measured in the top 1 m of transect Well C. This corresponds to a salinity of approximately 13 parts per thousand (ppt). At a 5-m well depth, the specific conductance was similar to that of seawater (55.0 mS/cm; 35 ppt), and below 5 m water was hypersaline (59.2 mS/cm; 38 ppt). However, geophysical techniques have revealed localized freshwater lenses around highly populated areas of Key Largo (Ciriello, 1997). Pockets of freshwater lenses were determined to originate from wastewater injection wells and septic tanks, not from natural recharge events such as rainfall.

In addition to the high permeability of the Key Largo Limestone and the role it plays in dispersing freshwater, tides of the Atlantic Ocean also have a great influence on structure of the brackish-water lens beneath the island. The Atlantic Ocean exhibits a typical semidiurnal tide with a range of ~1 m (Figure 25.3). The tidal signal propagates rapidly through the highly permeable island (Halley et al., 1994). Because of the rise and fall of the tides every 6 hours, groundwater dynamics beneath the island are extremely complex. On the bayside of Key Largo, astronomical tides are nearly absent and bay tides instead reflect, for the most part, meteorological (wind) conditions (Figure 25.3) (Smith, 1994).

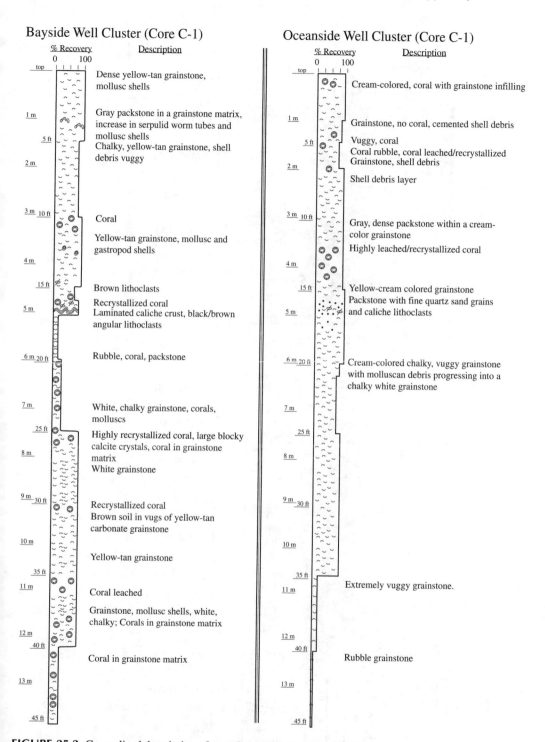

**FIGURE 25.2** Generalized description of core C-1 (OSWC and BSWC). Core logs depict typical Key Largo limestone with skeletal material comprising grainstones and packstones along with corals and molluscan shells. Core C-1 at BSWC contains a thick caliche crust (~3 to 5 cm) not present in other cores taken at OSWC or BSWC.

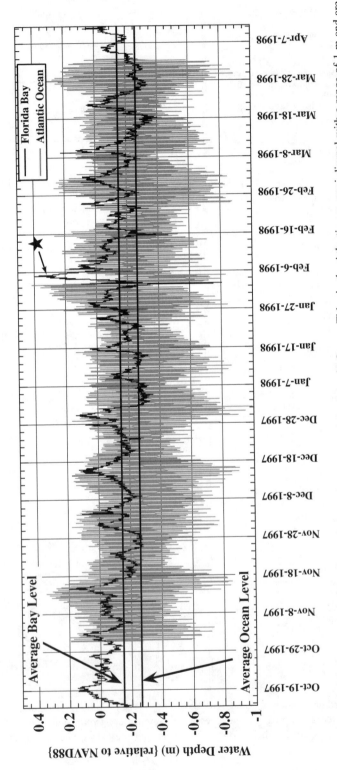

**FIGURE 25.3** Fluctuations of water levels for tides for Florida Bay and the Atlantic Ocean. Tides in the Atlantic are semi-diurnal with a range of 1 m and are characteristic of astronomical tides, whereas tides in Florida Bay do not fluctuate with any regularity but vary according to meteorological conditions. The star on February 3 shows peak effects of the Groundhog Day storm that was associated with a severe cold front and winds up to 100 mph. The severe cold front created a drastic swing in bay water level.

# METHODS

## WELL INSTALLATION

Well location was critical for this study, as natural-gradient conditions were essential in obtaining valid groundwater flow data. Absence of artificial influences (septic tank leachate or disposal well effluents) to the groundwater system were key factors for choosing this unpopulated region of Key Largo for the installation of the well clusters (Figure 25.1). BSWC well nests were installed in July 1995; OSWC wells, in February 1996. Ten 7.6-cm holes were core-drilled to a depth of 13.6 m at each site using a hydraulic-powered rotary drill. All drilling and underwater well installations were performed from a 7.6-m-long barge. All wells were constructed of 2.54-cm threaded-PVC pipe and a 1.5-m-long well screen (0.010-inch slot) (Figure 25.4). The deep well screens were set between 12.1 and 13.6 m below the rock surface. In order to complete the deep wells, coarse quartz sand was poured into the borehole, creating a sand pack surrounding the well screen, followed by a slurry of cement forming a plug capping the sampling interval. The shallow wells were inserted into the borehole, alongside the deep well, to a depth of 6.1 m below the rock surface. A plug of quick-setting cement around the well heads and the opening of the hole sealed the borehole from the surface water (Figures 25.4 and 25.5). The central hole was drilled first and completed with a nest of two wells. The eight peripheral well nests were placed in a circular array around the central well nest to create a 60.6-m-diameter well cluster (Figure 25.1).

## DYE INJECTION AND SAMPLING

Both Rhodamine WT and Fluorescein were used as groundwater tracers. The decision to use them was based on previous work in karst and limestone environments (e.g., Atkinson et al., 1973; Smart and Laidlaw, 1977; Davis et al., 1980; Sabatini and Austin, 1991). Both Rhodamine and Fluorescein are semiconservative groundwater tracers that travel well through limestone because they are not adsorbed onto the limestone and do not degrade in high salinity water (Smart and Laidlaw, 1977). Dye was injected into the central wells of the well cluster and sampling occurred in the surrounding wells. For consistency, at both well clusters the shallow central wells (C-1B) received Fluorescein dye and the deep central wells (C-1A) received Rhodamine WT dye. The initial injection of Fluorescein at OSWC occurred in March 1996. Rhodamine was injected in June 1996. A second injection of both Fluorescein and Rhodamine was conducted in February 1997. The initial dye injection of Fluorescein and Rhodamine at BSWC was in August 1996, and a second injection of both dyes was made in February 1997. In addition to the second injection at BSWC, a known concentration of a sulfur hexafluoride ($SF_6$) solution was also injected along with the Fluorescein in the shallow well (C-1B). Samples for $SF_6$ were collected concurrently with the water for dye analyses and sent to Florida State University's Department of Oceanography Lab, where they were analyzed on a liquid gas chromatograph (LGC). $SF_6$ is detectable in the picomolar range, making it an ideal groundwater tracer. Care must be taken during sample collection, however, because $SF_6$ is adsorbed onto and contaminates plastic tubing, as it is highly volatile when in contact with the atmosphere. Copper tubing and dedicated silicon tubing were used during the collection of $SF_6$ to eliminate or reduce the opportunity for cross-contamination.

Two 55-gal drums were filled with surface seawater for preparing dye solutions. Fluorescein powder (500 g) was added to one drum and 3 L of a 20% liquid Rhodamine solution was added to the second. A small 5-gpm, 12-volt D.C. pump and dye-dedicated tubing were used to inject both dyes into the respective wells. A chaser volume (~50 gal) of ambient seawater was pumped into each well to ensure dye dispersal in the rock. Fluorescein dye issued from small solution holes up to 3 m away at both OSWC and BSWC sites during injection, demonstrating the connectivity between the shallow groundwater and the overlying surface water. Rhodamine dye injected deep (13.6 m), however, was not observed to leak into the overlying surface water.

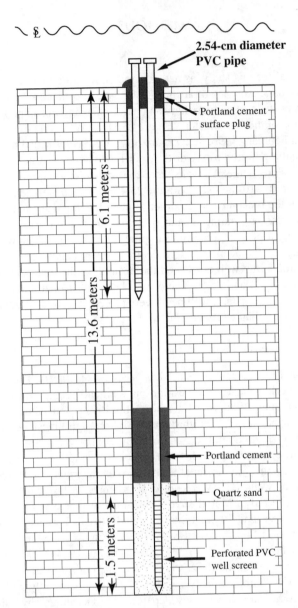

**FIGURE 25.4** Installation design for all wells at BSWC and OSWC.

Subsequent sampling of water from the circular array of wells was accomplished using a portable 12-volt D.C. pump, tubing, and a coupler with O-rings to ensure a tight seal on the underwater wellhead. Collecting water from each well required anchoring a small boat over each site. A diver connected the coupler and tubing to the well head. Each well was purged for at least three well volumes. To eliminate any photodecay of the dyes water samples were collected in amber HDPE bottles. Samples were analyzed in the lab on a Turner Designs Fluorometer following fluorometric procedures according to Wilson (1968) and Aley and Fletcher (1971).

## Tidal Measurements

Surface water levels at the OSWC and BSWC were recorded with two pressure transducers attached to the sea floor. The pressure transducers were surveyed in to the nearest benchmarks such that

**FIGURE 25.5** Photograph of a completed underwater well nest. Each pair of well heads is sufficiently exposed above the sea floor that a coupler and tubing can be attached to sample the groundwater.

data could be processed relative to a common datum (e.g., North American Vertical Datum of 1988, NAVD88). Measurements of the surface water levels and groundwater pressure allowed some insight into understanding what controls groundwater flow in the Florida Keys. Tidally induced groundwater pressure changes were periodically monitored with an underwater manometer device (Reich, 1996).

## RESULTS AND DISCUSSION

Flow velocities obtained under natural-gradient conditions were calculated from breakthrough curves using the first-arrival time rather than the $C_{0.5}$ point along the advective front (Domenico and Schwartz, 1990). The $C_{0.5}$ point takes into account both the dispersion and advection of a tracer in a groundwater flow field. The first-arrival time for obtaining velocity was employed because not all breakthrough curves were sigmoid shaped (i.e., bell-shaped; Parnell, 1986). BSWC-2B (Figure 25.7) is a good example of a sigmoid-shaped curve showing an increase in dye concentration, maximum concentration peak, and then a decrease to near background levels. This indicates that advective flow is strong through these well sites. However, non sigmoid-shaped curves (BSWC-5A; Figure 25.7) are suggestive of low advective flow and higher dispersion of tracer resulting in a slow movement of dye through those well sites.

### OCEANSIDE WELL CLUSTER (OSWC)

Data were collected from March 23, 1996, to July 31, 1997. Results after the first injection of Fluorescein in March 1996 and Rhodamine in June 1996 are shown in Figure 25.6. Flow velocities ranged from 0.1 to 1.68 m/d for Fluorescein (shallow injection) and 0.22 to 0.76 m/d for Rhodamine (deep injection). The first arrival of dye (indicated by ◆) at wells 5B and 6A suggests that net

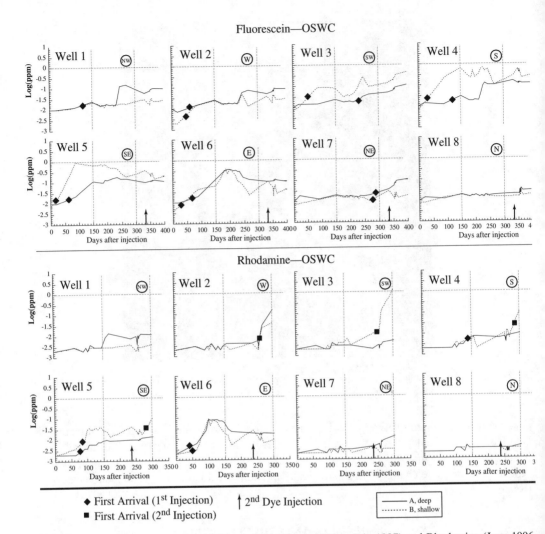

**FIGURE 25.6** Breakthrough curves for Fluorescein (March 1996 to July 1997) and Rhodamine (June 1996 to July 1997) at the OSWC. Well numbers are indicated on each chart along with direction from the injection point. Arrival time of the first injection (◆) is used to calculate groundwater-flow velocity. Arrows depict the time of second dye injection (February 1997). A solid box (■) shows first arrival of dye after the second injection. Dye concentrations have been log-normalized so that all data can be displayed on the same scale.

groundwater flow was easterly and southeasterly, or in the offshore direction. Dye eventually dispersed in the subsurface and appeared at other wells in the cluster.

Both dyes were injected a second time in February 1997 (see days 238 and 332 in Figure 25.6 for Rhodamine and Fluorescein, respectively). Interestingly, after the second injection, the flow rates increased and direction of flow was to the west. First arrival (indicated by a solid box) of Rhodamine was within 16 days at well 3B and 20 days at wells 2A and 2B, resulting in a calculated velocity of 1.9 m/d. No obvious peak of Fluorescein was detected after the second injection.

## BAYSIDE WELL CLUSTER (BSWC)

Data at BSWC were collected from August 21, 1996, to July 31, 1997. Breakthrough curves at BSWC are shown in Figure 25.7. Horizontal flow velocity based on the first arrival of dye (indicated by a diamond) ranged from 0.36 to 2.52 m/d for both Fluorescein and Rhodamine. The flow velocity

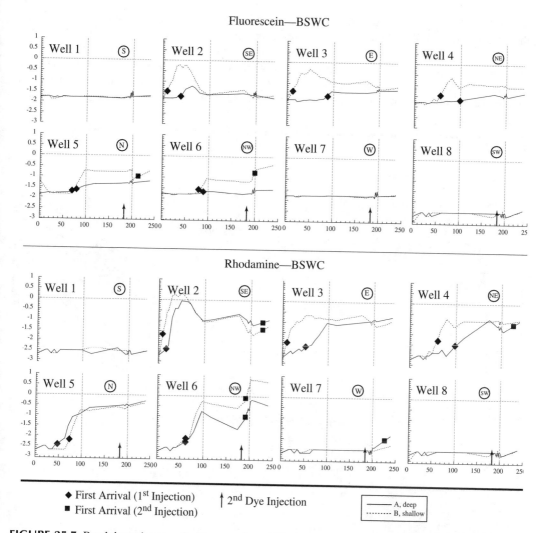

**FIGURE 25.7** Breakthrough curves for Fluorescein and Rhodamine (August 1996 to July 1997) at the BSWC. Well numbers are indicated on each chart along with direction from the injection point. Arrival time of the first injection (◆) is used to calculate groundwater-flow velocity. Arrows depict the time of second dye injection (February 1997) of dyes occurred. A solid box (■) shows first arrival after the second injection. Dye concentrations have been log-normalized so that all data can be displayed on the same scale.

was 1.5 times greater than that observed at OSWC, and the primary direction of flow was in a southeasterly direction.

The arrival of Rhodamine at the shallow wells prior to detection in the deep wells suggests that the vertical-flow component on the bayside of Key Largo is greater than was observed at OSWC. However, the opposite was not true; for example, Fluorescein dye did not appear to move vertically downward at either OSWC or BSWC. The greater upward vertical flow of groundwater at BSWC suggests a greater potential for seepage of nutrient-rich or contaminated groundwater into the surface water than at OSWC. In addition, based on preliminary results from seepage meters, measured seepage rates are greatest at BSWC (Shinn et al., in press). For example, seepage rates collected at BSWC on June 6, 1996, were 38 to 50 L/m²/day compared to 8 to 20 L/m²/d obtained on the same date and under the same conditions at OSWC.

Fluorescein and Rhodamine were injected for the second time into the central wells at BSWC, one day after OSWC injection. $SF_6$ was injected along with Fluorescein. Dye injection occurred at day 182 in Figure 25.7 (indicated by arrow). Patterns in flow rate and direction were similar to those observed at OSWC during the second injection. Flow rates exceeded 3.0 m/d (wells 6A and 6B) and the flow direction had reversed to the northwest. $SF_6$ results showed similar trends with a northwesterly direction and rates of 3.0 m/d. Groundwater flow velocity data from OSWC and BSWC are summarized in Figure 25.8 as a vector diagram. This illustrates the net direction of groundwater flow for both the first and second injection periods.

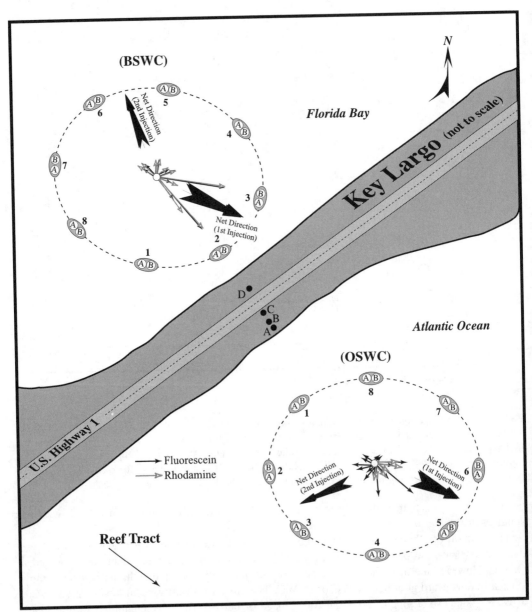

**FIGURE 25.8** Vector diagram summarizing the direction and velocity from the breakthrough curves for the first injection experiment. Fluorescein (solid black arrow) and Rhodamine (grey hollow arrow) are to the same scale and represent magnitude of velocity. Note that arrows are longer at BSWC than at OSWC, indicating greater flow rates. Large arrows indicate net direction of groundwater flow observed during both the first and second injection experiment.

## TIDES AND TIDAL PUMPING

The variability in results from one location to another and from one dye-injection period to another are the result of dynamic controlling factors, such as tides, tidal pumping, and storm events. The result of tidal interaction between bay and ocean create a daily fluctuation in the potential for movement of groundwater beneath the Keys (Shinn et al., 1999). In addition to normal tides, storm winds depress bay or ocean water levels for extended periods of time, resulting in potentially higher flow rates and reverse flow directions.

The most dynamic forcing factor in the Florida Keys is the Atlantic Ocean tides. The Atlantic Ocean water level fluctuates on a semidiurnal tidal cycle: two high and two low tides per day with a tidal range of approximately 1 m. The level of Florida Bay, however, does not fluctuate on a daily tidal cycle; instead, bay water-level fluctuation is driven primarily by meteorological events such as wind speed and direction. The plot in Figure 25.3 displays both Atlantic Ocean and Florida Bay water-level fluctuation over a 6-month period and the average water level at both BSWC and OSWC sites.

It was imperative that water level be compared to a common datum because previous data collected by Halley et al. (1994) and Smith (1994) indicated that Florida Bay water level was, on average, higher than the Atlantic Ocean water level. Pressure transducer data in this study show that water level in this part of Florida Bay is consistently 12 to 15 cm higher than that on the Atlantic side of Key Largo two thirds of the time. The head difference between OSWC and BSWC surface water sets up the potential for groundwater flow to occur from Florida Bay toward the Atlantic. The velocity of bay-to-ocean groundwater flow can be calculated based on Darcy's Equation:

$$v = (K/n)(dh/dl)$$

where $v$ is the velocity (m/d), $K$ is the hydraulic conductivity (m/d), $n$ is the effective porosity (dimensionless), and $dh/dl$ is the hydraulic head gradient (dimensionless). Assuming a conservative hydraulic conductivity of 1000 m/d, an effective porosity of 0.10, and an average difference in head of 0.13 m over a distance of 500 m, a flow velocity of 2.6 m/d is obtained. This flow rate is equivalent to velocity observed from the tracer experiments. This approach was used to calculate flow rates at Davies Reef, Australia, as well as at Eniwetok Atoll, Marshall Islands in the Pacific (Buddemeier and Oberdorfer, 1986).

Advective flow is the primary result of the surface water head gradient, but the greatest influence on dispersive groundwater flow is the direct result of tidal pumping. Atlantic tides fluctuate every 6 hours, four times each day, creating a "pump" that reverses the potential subsurface flow (Parnell, 1986; Serfes, 1991; Underwood et al., 1992). Tidal pumping drives both lateral and vertical groundwater flow (Figure 25.9) and is primarily a nearshore phenomenon that dissipates in magnitude offshore in both the ocean and bay directions (pers. obs.). Observations of tidal pumping and the decrease in amplitude offshore were measured using the underwater manometer (Reich, 1996). Well-head pressures at OSWC and BSWC were measured with the underwater manometer during a rising ocean tide (Figure 25.10). It is apparent that pressures in wells at BSWC are directly linked to tidal fluctuations on the ocean side. Generally, slower horizontal and vertical flow rates were observed on the Atlantic side. Differences between OSWC and BSWC are probably the result of: (1) frictional resistance of flow through the island, and (2) presence of a less saline groundwater lens beneath the island. Groundwater flowing from the bay toward the island is forced to rise and seep into the overlying water as it approaches the island.

## STORM EVENTS

Weather events such as winter frontal systems or tropical storms have a dramatic impact on bay and ocean water elevations and also influence the pattern of groundwater flow. Surface water levels

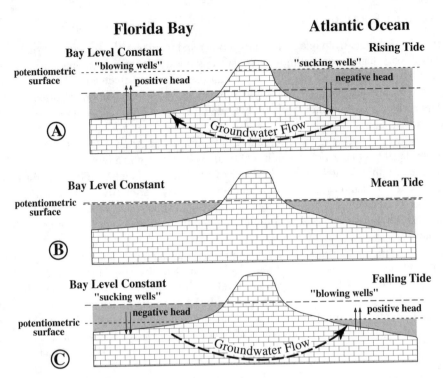

**FIGURE 25.9** Fluctuating tidal conditions of the Atlantic Ocean with groundwater flow resulting from tidal pumping. Florida Bay water level remains constant while the Atlantic Ocean fluctuates. Atlantic Ocean tides rise (A), creating a head gradient in the direction of Florida Bay. This is observed in the wells as negative pressure or "sucking" wells on the ocean side and positive pressure or "blowing" wells on the bayside. Between high and low tides (B), surface-water levels are the same on both sides of Key Largo, resulting in no head gradient across the island. As the tide falls on the Atlantic side (C), the head gradient shifts to the Atlantic, providing the potential for ground-water flow in that direction. Monitoring wells display negative pressure or "sucking" on the bayside and positive pressure or "blowing" on the Atlantic side. Groundwater has the potential to flow toward the Atlantic two thirds of the time.

at BSWC and OSWC recorded during the passing of Hurricane Georges September 25–27, 1998, show how water levels respond to a shift in wind direction and speed (Figure 25.11). The impact of wind speed and direction on water level is seen most dramatically in Florida Bay. During Hurricane Georges, bay levels dropped ~1 m and the ocean rose ~1 m, which created a head gradient of ~2 m between BSWC and OSWC (Figure 25.11). The cause for such a drop in the bay is its shallow depth — mean water depth is only ~1 m. In addition to its shallow depth, the bay is semi-enclosed, open on the west to the Gulf of Mexico and on the south to the Atlantic through small tidal channels running between the keys.

During the second dye injection experiment in February, a cold front passed through the area with 15- to 25-knot winds from the east to northeast. Because of the cold front, dye was detected first in wells OSWC 2A,B and 3A,B and BSWC 5B and 6A,B, opposite to what was observed after the first injection experiment. The groundwater flow direction was to the northwest, and the rate had increased from 1.7 to 1.9 m/d at OSWC and from 2.5 to 3.0 m/d at BSWC. Flow reversal was a direct result of lowering of the bay water level and raising of the ocean water level. The magnitude of Hurricane Georges, a minor Category 2 hurricane, likely represents extreme conditions for measured effects of storms. Events such as strong hurricanes (Category 3 and greater) have a significantly greater impact on groundwater flow than the less severe cold front winds that occurred after the second dye experiment.

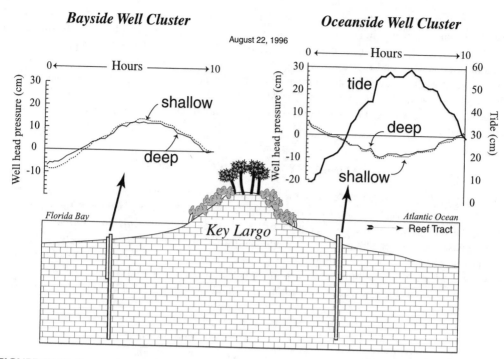

**FIGURE 25.10** Pressure in OSWC and BSWC wells during a rising ocean tide. Pressure data were collected with the underwater manometer every 15 minutes for 10 hours. Tide in the bay varied by only 3 cm and therefore is not included in the figure. Notice that the shallow well generally exhibits more positive and negative values than the deep well. This is a result of the limestone dampening the tidal signal with increasing depth.

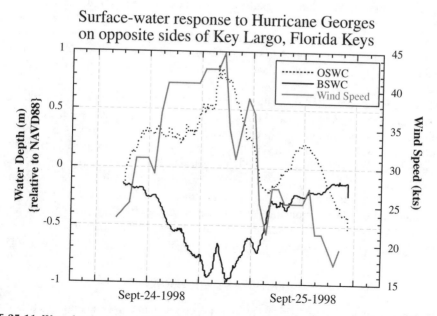

**FIGURE 25.11** Water-level response in Florida Bay and Atlantic Ocean during Hurricane Georges. Plotted wind speed shows a direct correlation with height of water level in both the bay and ocean. Wind direction was from the east to southeast according to C-MAN weather station at Molasses Reef.

Natural groundwater flow can be altered locally where businesses or other facilities have installed shallow sewage injection wells. These systems can inject large volumes of freshwater to depths up to 18 to 27 m (60 to 90 ft). The freshwater is injected into a completely marine groundwater system, and the result is a rapid upward ascent of the treated sewage waters due to the differences between the density of fresh and saltwater. These systems have been shown to create their own freshwater lenses (pers. obs.; Ciriello, 1997). Once the treated sewage waters reach the water table, they are subject to lateral movement and eventually may leak or seep into the nearshore surface waters.

## SUMMARY AND CONCLUSIONS

The results presented here demonstrate the significant influences that bay level, daily tidal cycles, and meteorology have on shallow (<14 m) marine groundwater flow in the Upper Keys. The dye tracer analyses, based on the breakthrough curves, demonstrate both horizontal and vertical flow. Horizontal groundwater movement is predominately toward the reef tract (southeasterly flow). A component of vertical flow likely contributes groundwater seepage up into overlying surface waters and along with it the possibility of transporting nutrients or contaminants into coastal nearshore waters. In addition, this study shows that normal groundwater flow rates (2.5 m/d) toward the reef tract are punctuated by higher flow events (3.0 m/d in the reverse direction) during winter cold fronts.

Based on the data from this study, the greatest forcing factor in the Upper Keys is the difference in water level between Florida Bay and the Atlantic Ocean. Data collected from pressure transducers on either side of Key Largo consistently indicate that the upper Florida Bay water level (averaged over 6 months) is 12 to 15 cm higher than the surface of the Atlantic Ocean. This head gradient over a distance of 500 m would be sufficient to drive groundwater movement at rates comparable to those observed in the tracer experiments. The higher water level on the bayside correlates two thirds of the time with a net direction of flow toward the reef tract. However, during storm events, easterly winds can depress the bay water level while raising the ocean water level, creating a head gradient opposite to the normal gradient. In most cases, the reversed gradient is higher than the average 12 to 15 cm and has the potential to result in a temporarily higher groundwater flow rate. The bay-directed flow rates observed in the tracer experiment provide evidence that reverse flow occurs. Storm events that might push water into the bay, however, would increase the water level in the bay and cause higher flow rates toward the ocean, mimicking the normal tidal pumping flow patterns.

With tourism and residential population rising in the Florida Keys, local and regional concerns over water quality changes of the nearshore and coral reef environments have increased. Influx of anthropogenic nutrients from sewage injection wells, septic tanks, and cesspools increases as the population increases. Evidence provided in this paper as well as other preliminary results from age dating of groundwater (Böhlke et al., 1999) indicates that groundwater seepage occurs in the nearshore. It seems highly likely that once contaminated groundwater discharges into nearshore surface waters, pollutants can also be transported offshore to the reef tract.

In summary, these data demonstrate that, whatever sewage treatment systems are installed in the Florida Keys, what must be taken into account is that both surface discharge and shallow subsurface injection will almost certainly find their way into the nearshore marine waters.

## GLOSSARY OF TERMS

**Boundstone.**  A rock unit for which the original components were bound together during deposition.

**Calcrete.**  A dense and often laminated carbonate which can coat the surfaces of limestones. It is formed during exposure of limestone to rain water, resulting in a precipitation of calcium carbonate. Several successions of this precipitation result in a laminated calcrete. Calcrete formation typically takes place beneath a thin soil which can give the calcrete a brownish color. Synonymous with caliche or soilstone crust.

**Grainstone.** Carbonate rocks that are mud free and grain supported.

**Hydraulic conductivity.** A measurement describing the rate at which water can move through a permeable unit.

**Oöids.** A spherical carbonate grain formed by precipitating calcium carbonate from seawater; most typically formed in high-energy environments. A rock unit composed primarily of oöids is an oolitic limestone.

**Peloids.** An oblong-shaped, fine-grained carbonate that may have been produced as fecal matter (pellets) by living animals such as crustacea. Peloids can also form via other mechanisms in warm, shallow seas.

**Pleistocene.** Period of time in the geologic history of Earth designated as the ice age. It spans a period of time from about 2 million years ago to 10,000 years ago. Large fluctuations in sea level occurred as the ice sheets advanced and retreated.

**Vug.** A small cavity in limestone; an underground chamber or cavity.

## ACKNOWLEDGMENTS

The authors wish to thank Jeff Chanton and Kevin Dillon at the Florida State University for supplying and analyzing water samples for $SF_6$. Much appreciation goes to staff members at the U.S. Geological Survey laboratory in Ocala, FL, for allowing us to borrow their fluorometer for the duration of the project. The authors also wish to acknowledge Barbara Lidz and one anonymous reviewer for their diligence and keen eyes in helping to fine-tune this chapter.

## REFERENCES

Aley, T. and M.W. Fletcher. 1971. The water tracer's cookbook, *Missouri Speleol.*, 16(3):1–32.

Atkinson, T.C., D.I. Smith, J.J. Lavis, and R.J. Whitaker. 1973. Experiments in tracing underground waters in limestone, *J. Hydrol.*, 19:323–349.

Böhlke, J.K., L.N. Plummer, E. Busenberg, T.B. Coplen, E.A. Shinn, and P. Schlosser. 1999. Origins, residence times, and nutrient sources of marine ground water beneath the Florida Keys and nearby offshore areas. Proceedings of South Florida Restoration Science Forum, May 17–19, 1999, Boca Raton, FL, Geological Survey Open-File Report 99-181:2–3.

Buddemeier, R.W. and J.A. Oberdorfer. 1986. Internal hydrology and geochemistry of coral reefs and atoll islands: key to diagenetic variations, in *Reef Diagenesis*, Schroeder, J.H. and Purser, B.H., Eds., Springer-Verlag, Berlin, pp. 91–111.

Ciriello, D. 1997. Geophysical Analysis of the Effects of Artificial Recharge on the Ground Water Lenses of Key Largo, Florida, M.S. thesis, University of South Florida, Tampa, 97 pp.

Corbett, R.D., J. Chanton, W. Burnett, K. Dillon, C. Rutkowski, and J.W. Fourqurean. 1999. Patterns of groundwater discharge into Florida Bay, *Limnol. Oceanogr.*, 44(4):1045–1055.

Davis, R.A. Jr., A.C. Hine, and E.A. Shinn. 1992. Holocene development on the Florida Peninsula, in *Quaternary Coasts of the United States: Marine and Lacustrine Systems*, SEPM Special Publication 48: 193–212.

Davis, S.N., G.M. Thompson, H.W. Bentley, and G. Stiles. 1980. Ground-water tracers — a short review, *Ground Water*, 18(1):14–23.

Domenico, P.A. and F.W. Schwartz. 1990. *Physical and Chemical Hydrogeology*, John Wiley & Sons, New York.

Dunham, R.J. 1962. Classification of carbonate rocks according to depositional texture, in *Classification of Carbonate Rocks*, Ham, W.E., Ed., American Association of Petroleum Geologists Memoir, Houston, TX. pp. 108–121.

EPA. 1991. *Water Quality Protection Program for the Florida Keys National Marine Sanctuary: Phase I Report*, Continental Shelf Associates and Battelle Ocean Sciences, Washington, D.C.

EPA. 1992. *Water Quality Protection Program for the Florida Keys National Marine Sanctuary: Phase II Report*, Continental Shelf Associates and Battelle Ocean Sciences, Washington, D.C.

EPA. 1996. *Water Quality Protection Program for the Florida Keys National Marine Sanctuary*, First Biennial Report to Congress, Washington, D.C. 32 pp.

Fish, J.E. and M. Stewart. 1990. *Hydrogeology of the Surficial Aquifer System, Dade County, Florida*, U.S. Geological Survey Water-Resources Investigation Report 90–4108, Tallahassee, FL. 50 pp.

Halley, R.B., H.L. Vacher, E.A. Shinn, and J.W. Haines. 1994. Marine geohydrology: dynamics of subsurface sea water around Key Largo, Florida. Abstracts with Programs, Geological Society of America Annual Meeting, Seattle, WA, p. A-411.

Halley, R.B., H.L. Vacher, and E.A. Shinn. 1997. Geology and hydrogeology of the Florida Keys, in *Geology and Hydrogeology of Carbonate Islands*, Vacher, H.L. and Quinn, T., Eds., Developments in Sedimentology 54, Elsevier Science. pp. 271–248.

Hoffmeister, J.E. and H.G. Multer. 1968. Geology and origin of the Florida Keys, *Geol. Soc. Am. Bull.*, 79:1487–1502.

Lapointe, B.E., J.E. O'Connell, and G.S. Garrett. 1990. Nutrient couplings between on-site sewage disposal systems, groundwaters, and nearshore surface waters of the Florida Keys, *Biogeochemistry*, 10:289–307.

Monaghan, L.B. 1996. The Hydrogeochemical Behavior of Wastewater-Derived Nutrient Elements in the Groundwaters of Long Key, Florida, M.S. thesis, Pennsylvania State University, State College, PA. 197 pp.

Parnell, K.E. 1986. Water movement within a fringing reef flat, Orpheus Island, North Queensland, Australia, *Coral Reefs*, 5:1–6.

Paul, J.H., J.B. Rose, J. Brown, E.A. Shinn, S. Miller, and S.R. Farrah. 1995a. Viral tracer studies indicate contamination of marine waters by sewage disposal practices in Key Largo, Florida, *Appl. Environ. Microbiol.*, 61(6):2230–2234.

Paul, J.H., J.B. Rose, S. Jiang, C. Kellogg, and E.A. Shinn. 1995b. Occurrence of fecal indicator bacteria in surface waters and the subsurface aquifer in Key Largo, Florida, *Appl. Environ. Microbiol.*, 61(6):2235–2241.

Perkins, R.D. 1977. Depositional framework of Pleistocene rocks in south Florida, in *Quaternary Sedimentation in South Florida, Part II*, Enos, P. and Perkins, R.D., Eds., Geological Society of America Memoir 147, Boulder, CO. pp. 131–198.

Reich, C.D. 1996. Diver-operated manometer: a simple device for measuring hydraulic head in underwater wells, *J. Sediment. Res.*, 66(5):1032–1034.

Sabatini, D.A. and T. A. Austin. 1991. Characteristics of Rhodamine WT and Fluorescein as adsorbing ground-water tracers. *Ground Water*, 29(3):341–349.

Serfes, M.E. 1991. Determining the mean hydraulic gradient of ground water affected by tidal fluctuations, *Ground Water*, 29(4):549–555.

Shinn, E.A., B.H. Lidz, R.B. Halley, J.H. Hudson, and J.L. Kindinger. 1989. *Reefs of Florida and the Dry Tortugas: International Geological Congress*, Field Trip Guidebook T176, American Geophysical Union, Washington, D.C., 53 pp.

Shinn, E.A., R.S. Reese, and C.D. Reich. 1994. *Fate and Pathways of Injection-Well Effluent in the Florida Keys*, U.S. Geological Survey Open-File Report 94-276, 116 pp.

Shinn, E.A., C.D. Reich, and T.D. Hickey. 1999. Tidal pumping as a diagenetic agent. Program with Abstracts, American Association of Petroleum Geologists Annual Meeting, San Antonio, Texas: A-129.

Shinn, E.A., C.D. Reich, and T.D. Hickey. In press. Seepage meters and Bernoulli's revenge, *Estuaries*.

Smart, P.L. and I.M.S. Laidlaw. 1977. An evaluation of some fluorescent dyes for water tracing, *Water Resour. Res.*, 12(1):15–33.

Smith, N.P. 1994. Long-term Gulf-to-Atlantic transport through tidal channels in the Florida Keys, *Bull. Mar. Sci.*, 54(3):602–609.

Underwood, M.R., F.L. Peterson, and C.I. Voss. 1992. Groundwater lens dynamics of atoll islands, *Water Resour. Res.*, 28(11):2889–2902.

Vacher, H.L., M.J. Wightman, and M.T. Stewart. 1992. Hydrology of meteoric diagenesis: effect of Pleistocene stratigraphy on freshwater lenses of Big Pine Key, Florida in *Quaternary Coasts of the Unites States: Marine and Lacustrine Systems*, SEPM Special Publication, 48, pp. 213–219.

Wilson, J.F., Jr. 1968. Fluorometric procedures for dye tracing, in *Techniques of Water-Resources Investigations of the United States Geological Survey*, 3, 33 pp.

# 26

# The "Ostrich" Component of the Multiple Stressor Model: Undermining South Florida

*Sydney T. Bacchus*
Applied Environmental Services

## CONTENTS

## THE STATE OF THE STATE

### TWENTIETH CENTURY TOLL

The plight of coral reef ecosystems in South Florida has been summarized by Lidz et al. (1997) and Dustan (1999). Increasing coral mortality and extensive proliferation of algae occurred during the 1980s (Dustan, 1985). In 1987, corals throughout Florida expelled the symbiotic algae (zooxanthellae) that provide color to corals and are required for growth. Minor bleaching had occurred in Florida prior to 1987, but no previous reports of such a severe case existed (Jaap, 1979, 1985). Bleaching in the summer of 1990 was more extensive, resulting in severe mortality of corals, including the hydrocoral *Millepora*. During the 1980s, massive corals (e.g., *Montastraea annularis*) suffered severe mortality caused by black-band disease (Richardson and Carlton, 1993; Rutzler and Santavy, 1983; Rutzler et al., 1983). Other coral maladies include white pox disease and ring bleaching. An historic and pictorial synopsis of coral diseases and decline in South Florida is provided by the U. S. Geological Survey (USGS, 1999a) as follows: 1973 to 1974, black-band disease makes its first appearance in the Keys; 1978, *Acropora* corals experience dieoff and black-band disease is present; 1983, the sea urchin *Diadema* experiences dieoff and algal infestation of dead corals occurs; 1985 to 1986, black-band disease is rampant and dieoff of *Acropora* corals occurs; 1987, bleaching of corals in the Florida Keys occurs; 1994, bleaching of corals occurs in the Dry Tortugas (USGS, 1999b).

The reef-building corals are not alone in the catastrophic decline and dieoff in the vicinity of the Florida Keys. Other reef-dwelling organisms, such as the soft coral Caribbean sea fans (*Gorgonia* spp.), also are exhibiting severe signs of stress and falling prey to disease (refer to the following discussion on predisposition to disease).

A similar event has occurred with the mass mortality of seagrass (*Thalassia testudinum*) in Florida Bay. The mass dieoff was initiated in 1987, with more than 4000 ha (9880 ac) of turtle grass lost, and an additional 23,000 ha (56,810 ac) in decline by 1994 (Durako and Kuss, 1994; Robblee et al., 1991). This mass mortality of seagrass reportedly was due to hypersaline conditions in Florida Bay during a period of low rainfall, and the subsequent decay of the seagrass reportedly resulted in eutrophication of the Bay (J. Zieman, pers. comm.). However, Brand (2000) has refuted the hypothesized influence of hypersalinity and seagrass mortality in the eutrophication of Florida Bay, based on "highest salinity" data for a period of record beginning in 1955. The referenced dataset showed that the highest salinities recorded during six of eight sample periods prior to the dieoff in 1987 were greater than the highest salinity recorded during the 1986–1987 period.

It is important to note that organisms respond to both acute and chronic stresses. The "highest salinity" dataset evaluated by Brand (2000) would address the acute stress of an organism exposed to a single (or few) extreme high-temperature event(s). The referenced dataset, however, would not address the impact to organisms under chronic stress of prolonged increased salinity (even slight increases) that may have been occurring for many years. It is also important to note that groundwater mining of the Biscayne aquifer, and subsequent diversion of submarine groundwater discharge (SGD) had been occurring for approximately half a century, as is discussed in the subsequent section describing groundwater perturbations.

The described distinction between acute and chronic stresses is not meant to infer that Brand's proposed influence of external sources of nutrients is unfounded, nor that his argument that nutrients released by the dying/decaying seagrass could not be the source of eutrophication of Florida Bay is unfounded. To explore the role of external sources of nutrients in the demise of seagrasses, Cockburn Sound, Western Australia can be used as an example. At Cockburn Sound, increased epiphyte load on seagrasses, in addition to increased levels of chlorophyll and phosphate in the water column, were attributed to the addition of nutrients from discharged effluent and injected wastewater surfacing in the Sound. The subsequent deterioration of the large seagrass meadow in the area receiving those external nutrients was attributed to the 2- to 8-times increase in epiphytes covering the blades of the seagrass at that site, as compared with another portion of the seagrass meadow where external nutrients were not reaching the surface waters (Silberstein et al., 1986).

The increase in epiphytes calculated in that Cockburn Sound study was conservative because only the epiphytes shading the seagrass blades were considered. The larger epiphytes that extended beyond the blades were not included in their estimates. For example, the macrophytes *Ulva*, *Enteromorpha*, and *Calothrix* were found only at the site with the effluent discharge. Thus, in that case, the addition of external nutrients from effluent appeared to be the sole factor initiating phytoplankton blooms, increased epiphyte loads, and subsequent seagrass decline (Silberstein et al. 1986). The high concentrations of epiphytes and blanketing layers of filamentous algae associated with the deteriorating seagrass meadows in that study mirror the conditions of the distressed seagrass in Florida Bay and the declining coral reefs associated with the Keys (Bacchus, unpubl. data; D. DeMaria, pers. comm.).

The foraminifera *Amphistegina gibbosa* is another example of a coastal organism in distress. This foraminifera has a fossil record of approximately 50 million years, and its shells make a significant contribution of sand-sized sediments in nearshore zones. In the Florida Keys, new disease symptoms have been prevalent in *A. gibbosa* since the summer of 1991, with a decline in population densities of approximately 95% in 1992. Disease symptoms continued through the 1996 sampling period and included mottling to bleaching, abnormal calcification, lesions on the shell surface that permit invasion by epiphytic and boring organisms, and damage to asexual reproduction. The cause of the new, unnamed disease is unknown, but various theories have been posed (Hallock et al., 1992, 1995; Toler and Hallock, 1998).

The health of coral reefs is a primary concern because of their ecological significance to many other organisms, their importance to the tourist industry, and their protection of the coast from storm surge. Studies designed to monitor or assess the conditions of coral reefs (e.g., Aronson et al., 1994) have not considered the potential influence of SGD perturbations. Those perturbations can include decreases in, or elimination of, the discharge of pristine, low-salinity, low-nutrient ground-water of constant temperature and the concomitant loss of benefits derived by coastal organisms associated with those historic discharges. A second type of SGD perturbation is a shift in water quality to groundwater transport of anthropogenic contaminants such as excessive nutrients, microbial pathogens, and other pollutants (e.g., endocrine disruptors). One source of those types of contaminants is "treated" effluent injected into the aquifers.

As an example of how SGD may be linked to environmental problems in South Florida, increased noncalcareous, macroalgal growth and the decline and mortality of the coral reefs initially were attributed to the decline of the herbivorous sea urchin *Diadema*. However, in 1996, the U.S. Environmental Protection Agency (USEPA, 1996) indicated that eutrophication appeared to be a significant factor in the loss of coral in South Florida. Nutrient sources identified by Lidz et al. (1997) mirrored those identified by the USEPA, and included surface runoff, live-aboard boats, cess pits, sewage outfalls, package plants (injecting treated sewage effluent into the shallow, porous limestone beneath the Florida Keys), and septic tanks. Those external nutrient sources (excluding surface "runoff") are listed as ranked by the USEPA, in increasing order, based on the "pounds/day of total nitrogen" contributed (USEPA, 1996).

The volume of water associated with the nutrient loadings was not provided in the 1996 USEPA report, although the volume of water involved is critical with respect to the transport of those contaminants into the surrounding surface waters. A second critical factor is recognizing that injection wells generally concentrate the waste of many individual units (many people) into one narrow bore hole (well) in the extremely permeable limestone. This greatly increases the mobility of the contaminants, in terms of both the speed and distance of travel. Some studies in the Keys have documented that tracers introduced into septic systems reach surface water adjacent to the test units (Dillon et al., 1999). However, in at least three cases where tracer solutions were flushed into septic tanks in the Keys, the tracer was never detected in surface waters (Paul et al., 2000; J. Paul, pers. comm.). In one of the experiments where the tracer from the septic tank was not detected in surface water, the same type of tracer introduced into a shallow injection well in the Lower Keys (Saddlebunch) traveled in a southeasterly direction, toward the reef tract on the ocean side, at rates up to 141 m h$^{-1}$ (463 ft h$^{-1}$). The flow rate of effluent injected into the shallow carbonate aquifer of the Keys for a 24-h period, based on the data from Paul et al. (2000), would be 3.4 km d$^{-1}$ (2.1 mi d$^{-1}$).

For comparative purposes, Lapointe et al. (1990) determined lateral flow rates in the shallow Miami Oolite aquifer near septic tanks in the Lower Keys. Flow rates under those conditions ranged from 0 to 0.2 m d$^{-1}$ with a mean of 0.1 m d$^{-1}$ (0.3 ft d$^{-1}$), while flow rates in the underlying Key Largo Limestone aquifer in the same location were approximately 1.0 to 1.3 m d$^{-1}$, with a mean of 1.14 m d$^{-1}$ (3.7 ft d$^{-1}$). A subsequent analysis of groundwater flow associated with septic tanks, also in the Lower Keys (Big Pine Key), was conducted by Dillon et al. (1999) using sulfur hexafluoride as an artificial tracer of the septic tank effluent. Their results suggested that transport rates were approximately 0.1 to 1.9 m h$^{-1}$ (0.4 to 6.1 ft h$^{-1}$). Conversion of those rates to flow for a 24-h period yields rates of approximately 2.4 to 45.6 m d$^{-1}$ (9.6 to 146.4 ft d$^{-1}$). More rapid rates of transport (3.7 m h$^{-1}$) were observed for the shallow upper portion of the Key Largo Limestone aquifer, but plumes simply moved back and forth due to tidal pumping, without significant dispersal from the site.

Several important points are illustrated in those studies regarding the movement of effluent from septic tanks vs. the movement of effluent injected into shallow wells. The first and most critical point is that the potential migration rates can be in the range of three orders of magnitude greater for effluent injected into shallow aquifers in the Keys than for effluent associated with septic tanks in the same area. This determination is based on a comparison of the 3.4 km d$^{-1}$ injection-well discharge rate observed by Paul et al. (2000) and the 0.004 km h$^{-1}$ rate, as the most rapid rate calculated by Dillon et al. (1999) for effluent discharged from septic tanks in the same vicinity. Second, the effluent in the shallow injection wells has been shown to travel considerable distances from the point of injection, appearing in surface waters near sensitive coral reefs. Similar observations have not been made for effluent from septic tanks. Recall the observation that plumes associated with septic tanks simply move back and forth due to tidal pumping, without significant dispersal from the site, as described above.

Equally important is the fact that all disinfection treatments used for treated effluent (chlorine, chlorine dioxide, ozone, peracetic acid, and ultraviolet radiation), including the treated effluent that is injected, produce bacterial mutagenicity (Monarca et al., 2000). Septic tanks do not involve artificial treatment of the wastes; therefore, the induced mutagenicity is avoided. Finally, numerous types of septic systems are in operation throughout the Keys, due to changes in permitting requirements that have occurred over time (S. Borrero, pers. comm.). However, none of the tracer studies that have been conducted in the Keys septic systems have provided crucial information regarding the specific systems tested (e.g., construction/design details, date constructed, period of operation). Therefore, no scientific basis exists for the widespread conclusions that all (or even most) septic tank systems in the Keys are introducing a significant nutrient load (or other pollutants) into the surface waters. Likewise, the similar presumption that cesspits are a significant factor in the water quality degradation of Florida Bay and surface waters surrounding the coral reefs is without scientific evidence.

The tracer studies that have been conducted do not represent a definitive body of research regarding dispersal of effluent from shallow injection wells and septic tanks (or lack of dispersal in the latter case). They do, however, illustrate the potential for significant environmental damage associated with injection wells, the approach that is perceived as the more technologically advanced and desirable means of sewage disposal in the Keys. To date, no studies appear to have been conducted to determine the flow rates and discharge locations of effluent injected into the underlying Floridan aquifer system via deep wells. Consequently, much of the nearshore eutrophication that has been attributed to septic systems and other, less "technologically advanced" means of sewage disposal (cesspits) may, in fact, be due to injected effluent (both shallow and deep) resurfacing as induced discharge.

The flow associated with the injected effluent that was described by Paul et al. (2000) is presumed to be occurring through preferential flow paths in the Key Largo Limestone formation. A preferential flow path is any area in the formation (aquifer) that has characteristics that differ from the surrounding area and through which groundwater can move more easily (and rapidly). Preferential flow paths can facilitate water movement in any direction, from horizontal to vertical. Those pathways may be in the form of interbedded material with greater permeability than the surrounding area (e.g., unconsolidated sand or shell; interbedded peat) or void spaces in the rock (e.g., fractures, dissolution cavities).

The rate of flow observed by Paul et al. (2000) is slightly slower than the 7 to 8 km d$^{-1}$ rates of groundwater flow measured by Beck (1989) and by Patten and Klein (1989) in the exposed carbonate platform portion of the upper Floridan aquifer. Those sites were in an adjacent groundwater basin and the same groundwater basin (Bush and Johnston, 1988), respectively, as the Florida Keys. More rapid flow through the upper Floridan aquifer system would not be unusual based on the larger, more numerous preferential flow paths in the aquifer; likewise, a considerably more rapid rate of flow should be expected for effluent injected into highly cavernous formations, as is reported for the lower Floridan aquifer.

The 3.4 km d$^{-1}$ rate of flow documented by Paul et al. (2000) in the submerged portion of the carbonate aquifer of the Florida Keys was similar to the 1.0 km d$^{-1}$ hydraulic conductivities of the Pleistocene carbonate aquifer estimated by Oberdorfer and Buddemeier (1985) in the coral reefs of Enjebi Island, Eniwetok Atoll. The extensive injection of fluid wastes at the Florida site may have accounted for the higher flow rates. Oberdorfer and Buddemeier (1985) stressed the need for a clear understanding of water movement to predict contaminant pathways and migration associated with the coral reefs. Their concern was based on the knowledge that the reef system was hydraulically very heterogeneous, resulting in the bulk flow occurring through the high permeability zones characterized by voids and rubble at the Enjebi Island sites, as well as sites at Davies Reef in the central Australian Great Barrier Reef (Oberdorfer and Buddemeier, 1986). The presence of geologic heterogeneity at their sites (including cavernous voids) was similar to that of the Florida coral reefs.

When the USEPA originally adopted the Underground Injection "Control" rule (40 CFR Part 146), regulating the injection of minimally treated effluent and other wastes into both shallow and deep wells, the basis was that this action would "relieve stress to surface water environments" and "reduce impacts to surface ecosystems" (National Archives and Records Administration, 2000). However, the scientific basis for those statements could not be substantiated by the USEPA (H. Beard, pers. comm.). Furthermore, the USEPA indicated that the waste injection rule was implemented and has been maintained without information regarding the fate (discharge location) or environmental impact of the injected wastes (W. Diamond, pers. comm.). Despite the failure of the federal and state agencies to conduct the essential research to determine that the injected wastes were not violating the federal Clean Water Act and the federal Endangered Species Act, sufficient knowledge exists to determine that wastes injected into karstic carbonate aquifers, such as those in Florida, cannot be "controlled." Consequently, the use of the term "control" in the federal rule and related state rules that regulate the injection of waste into aquifers is misleading.

The number of shallow (Class V) effluent-injection wells in the Florida Keys reportedly has increased from 700 in 1992 (USEPA, 1992) to approximately 1000 (USGS, 2001). Repeated attempts to obtain specific information from the Florida Department of Environmental Protection (FDEP), however, regarding the exact number and location of all waste-injection wells in the Keys (and South Florida) were unsuccessful. The FDEP is the state agency to which implementation of the USEPA's injection rule in Florida was delegated.

The oceanside waters adjacent to the Keys are incorporated into the Florida Keys National Marine Sanctuary (FKNMS) where no discharges are allowed. Florida also has designated those waters as Outstanding Florida Waters. Despite the recently documented rapid migration and discharge of shallow-well injected effluent into surface waters, FDEP continues to issue permits for the injection of minimally treated effluent. This has been the case even when the injection sites are immediately adjacent to sensitive areas such as nesting beaches for sea turtles, where driller's logs of the injection wells show zones for preferential flow (Division of Administrative Hearings, 2000).

Sea turtles are federally listed species. In the permit involving the sea turtle nesting beach (as is the case typically), no specific permit conditions were required for determining the presence or concentrations of pollutants (including excessive nutrients) in the injected effluent that discharges to surface waters, despite the proximity of the injection site to critical nesting habitat for sea turtles and the fact that manatees (also federally listed) use those waters as habitat. Ironically, FDEP is the same agency that is converting to waterless toilet systems at numerous state parks in the Keys.

A recent outbreak of sea turtle deaths and diseases was reported on Christmas Eve 2000 in Marathon, just southwest of the site of the referenced contested injection well (Fuss, 2000). Gruesome tumors have increased in prevalence in all sea turtle species. Those tumors can cover the turtles' eyes (incapacitating their sight), cover the bases of their flippers (preventing them from swimming), and engulf their internal organs to the point of erupting through their shells. Many of the tumor-ridden sea turtles brought to the Turtle Hospital in Marathon were from the Indian River area, while others were transported to the Turtle Hospital from the Upper Keys. Necropsies have revealed extensive tumors engulfing the kidneys of those sea turtles (S. Schaf, pers. comm.).

One significant potential source of nutrients (and other pollutants), omitted from the USEPA list of contributors referenced above, is sewage effluent injected into the Boulder Zone of the lower Floridan aquifer. That minimally treated effluent is injected under a discontinuous lower permeability layer, theoretically at depths of approximately 610 m (2000 ft). Minimally treated effluent has been injected into this karst aquifer system via deep wells for many years in the Indian River area, while effluent is being injected into both deep wells and shallow wells in the vicinity of the Upper Keys. Approximately 400 million gallons per day (MGD) of minimally treated municipal effluent reportedly is injected into the lower Floridan aquifer throughout South Florida's sensitive coastal area, including the Miami area. In fact, approximately one fourth (110 MGD) of all effluent permitted for injection into deep wells in Florida is injected at the Miami–Dade Blackpoint facility. Those injection wells are in close proximity to Biscayne Bay, at the northern end of Florida's coral reef tract (National Archives and Records Administration, 2000; USEPA, 2000).

The Boulder Zone of the Lower Floridan is composed of highly permeable carbonate rocks underlying South Florida and reportedly contains cavernous voids, cavities, and extensive fractures (Kaufman, 1973). This zone is considerably more permeable than the Key Largo Limestone formation. The formation water in the lower Floridan aquifer is saline and much more dense than the low-salinity injected effluent. Vertical discontinuities, including fractures, sinkholes, and solution pipes, have been documented extending from the lower Floridan aquifer through the "confining" zone to the overlying aquifers, including throughout the east coast of Florida (see Bacchus, 2000a, for a summarized discussion, with emphasis on Maslia and Prowell, 1990; Snyder et al., 1989; Spechler and Wilson, 1997). An example of the types of submerged vertical discontinuities in South Florida that may be facilitating upward migration of the less dense injected wastes is provided in Figure 26.1 (Meyer, 1989). As summarized by Bacchus (2000a), vertical migration of groundwater has been documented in the Floridan aquifer system in response to anthropogenic groundwater perturbations. The widespread nature of this phenomenon has been noted recently in England, where it is associated with fluid extractions several hundreds of meters (approximately 1000 ft) deep (Goerres et al., 1999; B. Goerres, pers. comm.).

Lapointe (1997, 1999) presented a convincing case against the exclusive top–down approach (loss of herbivores) proposed in past attempts to explain increasing proliferation of noncalcareous, macroalgal growth (macroalgal blooms) in South Florida and other areas. As an alternative hypothesis, he proposed a controlling bottom–up model (where herbivore influences are linear/minimal and nutrient influences are exponential/maximal) to explain the profuse growth of frondose macroalgae associated with the ailing coral reefs. His experimental data support the conclusion that in the absence of anthropogenic nutrient sources, noncalcareous macroalgal growth remains in check due to extremely limited nutrients (either nitrogen or phosphorus), even in the absence of herbivorous grazers.

In addition to the specific, nutrient-response experiments he conducted, Lapointe (1997) evaluated four reef sites in southeast Florida at approximately equal intervals from Jupiter Island to just north of the Palm Beach Inlet, where prolific growth of the chlorophyte *Codium isthmocladum* was occurring. His investigations at those sites suggested that SGD of dissolved inorganic nitrogen (N) and soluble reactive phosphorus (P) appear to be an important route of nutrient enrichment supporting dense growths of the frondose macroalgae *C. isthmocladum*.

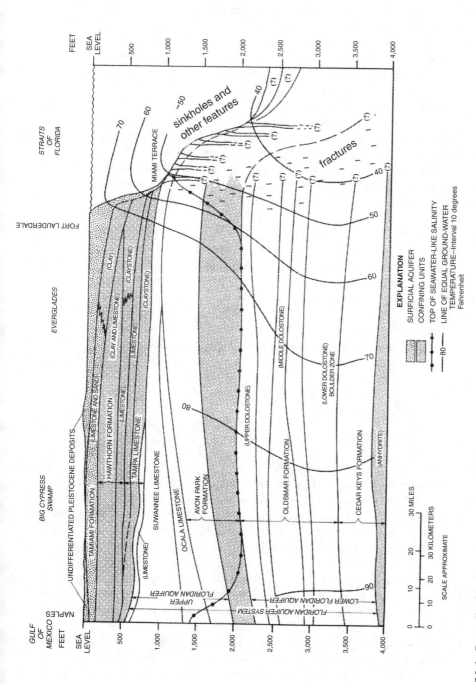

**FIGURE 26.1**  Generalized hydrogeologic section through southern Florida showing submarine sinkholes, fractures, and other discontinuity features, in addition to isotherms and upper limit of groundwater with salinity equivalent to seawater in the Floridan aquifer system. (From Meyer, F.W. *Hydrogeology, Ground-Water Movement, and Subsurface Storage in the Floridan Aquifer System in South Florida*, U.S. Geological Survey Professional Paper 1403-G, 1989.)

Although Lapointe (1997) acknowledged inputs of nutrients from surface water sources, a Class I injection well also is operational in the vicinity of the reefs he studied. Sewage effluent is being injected into the Floridan aquifer system at that deep-well injection site at depths greater than 900 m (2953 ft). Pitt and Meyer (1976) described the general hydrologic conditions at that deep-well injection site in West Palm Beach, including construction of the injection and monitoring wells. Subsequently, the problems associated with disposing of the saline water from the receiving aquifer were described (Pitt et al., 1977). It is important to note that no monitoring wells, in conjunction with those deep-well injection activities, were established in the sensitive coastal areas of the submerged carbonate platform (used synonymously with the terms *shelf* and *plateau*), including the vicinity of the reefs that Lapointe studied.

Pitt and Meyer (1976) revealed that the principal areas of concern in evaluating the proposed site were (1) supply wells in the unconfined aquifer within 1.6 km (1 mi) of the injection well, (2) artesian wells within 4.8 km (3 mi) of the injection well, and (3) observation wells in the unconfined aquifer at the site. The ultimate capacity of effluent generated by the secondary waste-water treatment plant in West Palm Beach was expected to be 64 MGD, with 40 MGD as a peak flow for the first phase (Pitt and Meyer, 1976). This 64-MGD injection rate is the equivalent of approximately 352 football fields of water 20 cm (8 in) deep injected into the aquifer adjacent to these reefs every day.

Converting that volume of injected effluent into one example of hypothetical effluent travel through the reportedly cavernous Boulder Zone of the lower Floridan aquifer, for a single day's injection of effluent, would yield a plume of effluent, laden with nutrients and other pollutants, approximately 20 cm deep and 48 m (150 ft) wide that extends for 32 km (20 mi). The Grand Bahamas lie approximately 100 km (62 mi) east of West Palm Beach; therefore, effluent injected into the Boulder Zone in West Palm Beach on a Monday hypothetically could reach the Grand Bahamas by approximately Wednesday of the same week. This example does not consider the potential impediment to flow that the plume of effluent injected into the same geologic zone in the Bahamas (Weech, 1997) may have on the West Palm Beach plume, as both plumes emanate outward and upward.

## MODIFICATION OF THE MULTIPLE STRESSOR MODEL

In an attempt to provide additional protection to South Florida's ecological resources, the Florida Keys National Marine Sanctuary and Protection Act (Public Law 101-605) was signed on November 16, 1990. This Act created the FKNMS, to be managed by the National Oceanic and Atmospheric Administration (NOAA). The FKNMS encompasses 4820 km² (2800 mi²) of nearshore waters extending south from Miami to the Dry Tortugas. The sanctuary includes the entire 280-km length of coral reef tract ecosystems located on the Atlantic Ocean side of the Florida Keys, in addition to the sea bottom out to a distance of 17 km (10.5 mi) on the Gulf side. The FKNMS also incorporates John Pennekamp Coral Reef State Park and the Key Largo National Marine Sanctuary (Lidz and Shinn, 1991; Shinn et al., 1989; USEPA, 1996).

As a requirement of the Act, the USEPA and the State of Florida were to initiate a comprehensive water quality monitoring and research/special studies program "for the successful restoration and maintenance of the water quality and other resources of the Sanctuary." This approach was summarized in the First Biennial Report to Congress, 1996 (USEPA, 1996).

Although that USEPA report addressed several important components (e.g., water temperature fluctuations, anthropogenic increases in nutrient levels and other contaminants, reductions in transparency), deep-well injection of treated sewage was not identified as one of the sources of those problems. Likewise, there was no discussion in the report regarding any serious threats associated with effluent injected into shallow wells, such as widespread eutrophication or the large-scale introduction of pathogens, endocrine disruptors, and other pollutants. Neither the "Domestic Waste-water" nor the "Hazardous Materials" sections of the report make any reference to those threats.

Under the "Water Quality Monitoring Program" section of that USEPA report, a long-term, comprehensive water quality monitoring program was proposed; however, no monitoring appears to have been conducted to determine the fate of deep-well injection of minimally treated sewage effluent into the Boulder Zone of the Floridan aquifer system at any wells in Florida. The closest deep-well injection site at the time of the USEPA report was the Miami–Dade facility, located approximately 0.8 km (0.5 mi) from the shore at Biscayne Bay, at the northern extent of the FKNMS and reef tract. In fact, one paragraph of the USEPA report simultaneously recommends investigating the possibility of effluent migrating through the Boulder Zone into Sanctuary waters, and implementing deep-well injection of sewage in the Keys.

The current, quarterly collection of water quality data is grossly inadequate to document potential direct or diffuse discharges of injected effluent at various points associated with coral reefs, particularly when those discharges may occur in pulses. In fact, 4 years after the release of the referenced USEPA report, no comprehensive investigation had been conducted to determine where the effluent injected into those deep wells was going or even what pollutants, other than excess nutrients, were contained in the injected effluent. Unfortunately, deep-well injection of municipal effluent was scheduled to begin in Key West by spring 2001 (City of Key West, 2000), despite the lack of data regarding the fate and impact of contaminants that currently are being injected via deep-wells at the Miami–Dade site near Biscayne Bay and elsewhere in Florida. Deep-well injection of effluent (and other wastes) is the inverse equivalent of constructing taller smoke stacks to spew pollutants higher in the atmosphere and may simply ensure that the contaminants and associated adverse impacts are more widespread and more difficult to control.

The design capacity for the Key West Wastewater Treatment Plant is 10 MGD, with a current capacity of 8 MGD (City of Key West, 2000). Converting this design capacity volume to the hypothetical plume described for injected effluent in West Palm Beach, it would take only about one month for this effluent to flow from Key West to the Dry Tortugas and associated Riley's Hump, Tortugas Bank, and Sherwood Forest. Coincidentally, that time frame is comparable to the time frame determined (29 days) using the flow rate calculated by Paul et al. (2000) in their second injection well study in the Lower Keys. In that study, tracers injected into a shallow injection-well appeared in surface waters at the most distant sample site within 27 hours after injection, yielding a migration rate of approximately 141 m h$^{-1}$ (463 ft h$^{-1}$). At that rate, the effluent injected at Key West could reach the Marquesas Keys in less than 9 days, but the transport of the Key West municipal effluent is predicted to be considerably more rapid. This prediction is based on two factors. First, the Boulder Zone, where the deep-well injection would be permitted in Key West, is reported to be considerably more permeable than the aquifer receiving the shallow-well injected effluent in the Lower Keys experiment by Paul et al. (2000). The second factor is that the focused volume of the discharge will be more than 10 times the volume of effluent injected at the sites where the Paul et al., experiment was conducted.

The Dry Tortugas and associated environmentally sensitive areas recently were designated as the Tortugas Ecological Reserve because of the high-quality coral reefs and fish-spawning areas such as Riley's Hump, the Tortugas Bank, and Sherwood Forest. Ironically, a final Environmental Impact Statement (EIS) in excess of 300 pages was prepared for the proposed designation of this National Reserve. Conversely, no EIS appears to have been prepared or even considered for the deep-well injection of effluent at Key West, or any other location along South Florida's sensitive coast.

It is important to note that many marine species associated with the deep reef habitats and limestone ledges at depths extending to approximately 550 m (1800 ft) were observed by submersibles at Tortugas South during the 2000 Sustainable Seas Expedition (NOAA, 2000). This is comparable to the dense colonies of organisms dominated by the giant clam *Calyptogena soyoae* and associated with the freshwater discharges at depths of approximately 1170 m (3510 ft) along the Sagami Trough, at the northern end of the Philippine Sea plate (Tsunogai et al., 1996; T. Gamo, pers. comm.). The assemblages of organisms associated with the limestone ledges at Tortugas South coincide with the target depth for the municipal effluent-injection wells scheduled for the Key West facility, where no environmental assessment of the potential impacts from deep-well injection of effluent has been completed.

Porter et al. (1999) and Brand (2000) expanded the government's focus on altered flow of surface water via the extensive canal system excavated in South Florida (USEPA, 1996). Porter et al. (1999) proposed a hypothetical scenario to describe a complex series of land-sea linkages in Florida Bay. Their Florida Bay hypothesis depicts environmental responses in Florida Bay under conditions of "natural water flow" and "reduced water flow." In both cases, the "water flow" is synonymous with surface water. In their hypothetical scenario, the source of the clean (nutrient-poor, low-turbidity, hyposaline) surface water is the historic, unaltered flow of surface water from the Everglades. The reduced surface water flow condition includes nutrient-laden, turbid water discharging from the Everglades.

As indicated previously, Brand (2000) presented a convincing argument to refute a dominant hypothesis (Durako, 1994; Carlson et al., 1994) regarding the seagrass dieoff in Florida Bay. That seagrass dieoff, reportedly initiated in 1987, in conjunction with subsequent decomposition of that organic matter and sediment resuspension, released nutrients that generated the persistent algal blooms in Florida Bay according to that dominant hypothesis. Data were presented by Brand (2000) to show that those persistent blooms were not and are not spatially consistent with areas of seagrass dieoff but instead coincided temporally with periods of increased releases of nutrient-laden surface water from agricultural areas in the Everglades. Estimated annual flow of surface water into Florida Bay increased from 1981 through 1983 and remained high through 1988. Average annual rainfall data at Flamingo showed a period of low rainfall from 1984 through 1986 and again in 1989, the latter also coinciding with a significant reduction in estimated surface water flow into Florida Bay. Brand also emphasized the value of the long-term, historic knowledge of the local residents (e.g., fishermen, boaters), who began observing a systematic decline in water quality of Florida Bay around 1981 (DeMaria, 1996), before extensive water quality monitoring was initiated.

Also as indicated previously, one aspect of the seagrass dieoff hypothesis referenced above is that hypersalinity in Florida Bay was a significant factor in the death of the seagrass. Consequently, federal and regional regulatory agencies have proposed to continue and increase discharges of nutrient-laden surface water from the Everglades agricultural area into Florida Bay (USACOE/SFWMD, 1999) without determining the contaminant load of such discharges. The water management policy decision to implement coastal restoration based on the presumption that large-scale discharges of surface water were required to control hypersalinity resulted in increased discharges of surface water from the Everglades from 1991 through 1995, with concomitant increases in ammonium, chlorophyll $a$, and turbidity (see Lapointe et al., 2002, for a more detailed discussion).

Certainly past alterations in the flow and quality of surface water in South Florida are sufficient to cause environmental degradation in Florida Bay. Unfortunately, cessation/alteration of the subsurface flow of pristine, low-salinity, low-nutrient SGD of constant temperature to Florida Bay and the coral reef ecosystems on the ocean side could magnify the damage considerably. Reducing natural, historical SGD may influence vertical mixing in the saline water column and, in conjunction with pulsed releases of freshwater from canals, result in or exacerbate the stratified, hypersaline conditions in Florida Bay during certain periods of the year. Such conditions have been reported by Porter et al. (1999) to be contributing ecosystem stressors and markers inferring other stressors. Those conditions would be expected to intensify during periods of low rainfall, when the volume of water removed from the Biscayne aquifer via groundwater mining tends to increase and the discharges of surface water from the agricultural areas tend to decrease.

As a worst-case (but highly probable) scenario, the same subsurface flow-ways that once delivered pristine, low-salinity, low-nutrient SGD of constant temperature, may now be delivering effluent from shallow and deep injection wells. While the introduction of excessive nutrients and other pollutants via surface water (e.g., canals) would decrease during periods of low rainfall, the same pollutants contained in injected effluent, and transported via preferential flow paths, would be more likely to remain relatively constant. In this manner, anthropogenic perturbations of SGD conditions could result in highly localized, yet widespread changes in salinity, temperature, and nutrients as multiple physical and chemical stressors. Additionally, effluent-contaminated SGD could introduce biological pollutants, including microbes lethal to present and future generations of coastal organisms. Any one or a combination of those changes could predispose coastal organisms such as corals, sea turtles, manatees, and seagrasses to a variety of fungal, bacterial, and viral diseases to which they would be immune otherwise.

The time has come to take the ostrich's approach and get a view below the ground. This chapter expands on the published literature and unpublished observations describing evidence of historic SGD (summarized by Bacchus, 2000a), with emphasis on South Florida's coastal ecosystems. Additional stressors, not addressed by or investigated under the federal/state/regional initiative, are proposed in this chapter. With the addition of groundwater stressors, a modification of the hypotheses by Porter et al. (1999) and Brand (2000) is proposed to incorporate the impact of anthropogenic alterations of groundwater flow and quality as critical components of the multiple stressor model (Figure 26.2A and B, respectively). Although the alternative models proposed in this chapter are a form of the complex series of land–sea linkages referenced by Porter et al. (1999), they can be described more specifically as surface–groundwater linkages. Finally, examples of testable hypotheses are provided to generate scientifically based answers to some of the critical questions, as required before the environmental problems of coastal South Florida can be resolved.

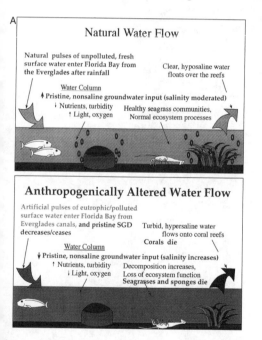
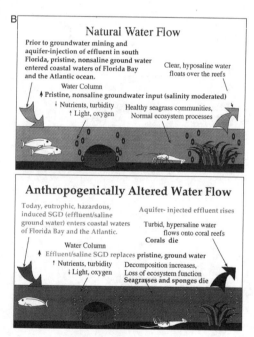

**FIGURE 26.2** Modified Florida Bay hypothesis, including perturbations of groundwater flow (A) and quality (B) as components of the multiple stressor model for Florida Bay and the Florida Keys National Marine Sanctuary (coral reefs). (Modified from Porter et al., 1999.)

## SWISS CHEESE

### DOWN AND OUT

The Floridan aquifer system is a regional, karstic limestone and dolomite groundwater system that extends throughout Florida and the Coastal Plain portions of Georgia, South Carolina, and Alabama (Figure 26.3). This aquifer system was divided into six subregions in conjunction with the Regional Aquifer-System Analysis (RASA) program initiated in 1978 by the USGS. The area within the South Florida subregion is shown in Figure 26.3, and is designated area "G" (Krause and Randolph, 1989).

The karst features of this aquifer system include characteristic breaches in the low permeability zones that separate the upper and lower Floridan aquifer, and the upper Floridan from overlying surficial aquifers (e.g., the Hawthorn Formation). Therefore, the Floridan aquifer system, including

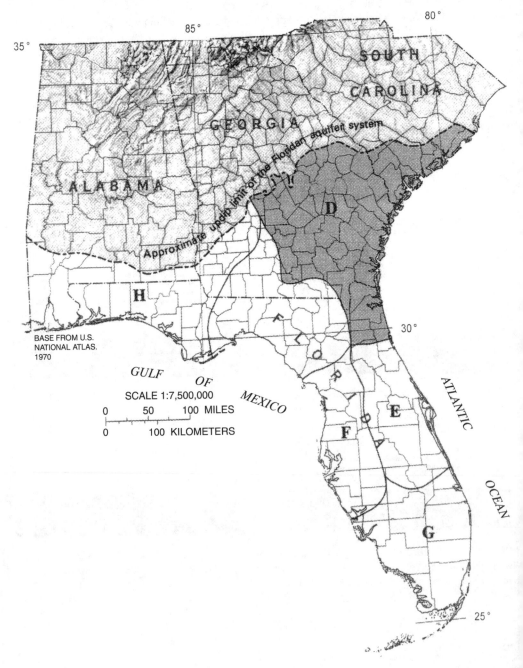

**FIGURE 26.3** Approximate extent of the Floridan aquifer system and designated subregions. (From Krause, R.E. and Randolph, R.B., *Hydrology of the Floridan Aquifer System in Southeast Georgia and Adjacent Parts of Florida and South Carolina*, U.S. Geological Survey Professional Paper 1403-D, 1989).

overlying zones of lower permeability, can be compared to a block of Swiss cheese. The solid parts of the cheese represent areas where zones of lower permeability are intact, while the holes in the cheese represent breaches in these barriers and the carbonate aquifers, such as vertical fractures and solution/collapse features (e.g., sinkholes). Where the lower permeability zones are breached, interaction between overlying and underlying aquifers is more pronounced.

Bacchus (2000a) provides a synopsis of past and present conditions of this highly permeable aquifer system, including examples of responses to anthropogenic groundwater perturbations such as groundwater mining. The general hydrologic characteristics prior to and following exploitation of the aquifer system in South Florida are described most succinctly in the following excerpts from Fish and Stewart (1991, p. 44), which provide insight into the lateral dynamism of South Florida's groundwater system:

> Predevelopment hydrologic conditions in Dade County have been described by Parker and others (1955, pp. 580–584). During that time, ground-water and surface-water levels were higher than at present, many springs discharged along the shoreline and on the bottom of Biscayne Bay, and freshwater wells flowed near the mouth of the Miami River. In the Coconut Grove area, water filled The Everglades and ponded behind the Atlantic Coastal Ridge (sometimes termed a "reef" in historical reports) to within 3 mi of Biscayne Bay. Dry-season water levels were about 10 ft above sea level. During the wet season, water rose sufficiently high to flow the "reef" through low spots in the Atlantic Coastal Ridge and the Miami River.

> The first serious efforts to change the natural hydrologic conditions began in 1907 in the New River Basin in Ft. Lauderdale (outside the study area) and in 1909 in the Miami River. Dredging operations deepened the Miami River through the Atlantic Coastal Ridge and extended a channel (canal) into The Everglades. As a result, the water level near the canal was lowered from land surface to 6 ft below land surface (Parker and others, 1955, p. 584). Water formerly stored in The Everglades was allowed to flow freely to the ocean. By 1913, both the New River Canal and the Miami Canal had been completed to Lake Okeechobee. The result was drainage of The Everglades and a general lowering of water levels by several feet.

> The lowering of water levels in The Everglades had several major hydrologic consequences. *The natural flow system was disturbed so that the coastal springs and artesian wells no longer flowed.* More importantly, the freshwater, saltwater interface, originally along the coast because of the high water table, began to encroach landward. Gradually, many private supply wells near the coast and the Miami public well field had to be moved farther inland. Other uncontrolled coastal drainage canals were installed, and the lowest water levels of record occurred in May and June 1945 (Klein and others, 1975) at the end of a prolonged drought (fig. 21). In northern Dade County, the hydraulic gradient was seaward, but only 1.5 ft in 18 mi, and in southern Dade County water levels were below sea level because of evapotranspiration. After this drought, control structures were placed in the canals near the coast to prevent overdrainage, and canals were gradually extended inland to provide better stormwater drainage.

> The reduction or elimination of a seasonal or temporary ground-water mound by construction of the present drainage system, as shown in figures 18 and 20, has had a substantial effect on the ground-water flow system. Under predevelopment, wet-season conditions, groundwater flowed away from the mound in all directions, including westward away from the coast. Parker and others (1955, p. 211) stated that this condition commonly occurred in this area before the construction of the drainage canals. *Some of the westward flow was discharged to The Everglades by springs.* The mound formed a temporary ground-water divide or barrier to flow from the interior toward eastern or southeastern coastal Dade County. It was approximately coincident with the natural drainage (fig. 5). [emphasis added]

Between 1955 and 1990, total freshwater withdrawals in Florida increased 245% (Marella, 1995), with agricultural water withdrawals in Dade and Collier Counties ranging from 100 to 400 MGD by 1990 (Marella, 1992). The environmental impacts of those agricultural withdrawals do not appear to have been accessed. The period of 1955 to 1957, however, represented the highest salinity reported for Florida Bay for the period from 1955 to 1990. Note that massive seagrass dieoff was not reported for the period from 1955 to 1957, when the highest salinity was reported.

The environmental impacts of municipal groundwater withdrawals also rarely are assessed or reported. However, Hofstetter and Sonenshein (1990) and Sonenshein and Hofstetter (1990) describe the impacts of the Northwest Well Field in Dade County constructed in the eastern portion of the wetland that has been designated Everglades National Park and Wildlife Management Area/Water Conservation Area No. 3 (west of the Miami Canal, east of Levee 30 and the L-30 Canal, and north of the Tamiami Canal).

Groundwater withdrawals from the unconfined Biscayne aquifer began at this well field in May 1983, just prior to initiation of seagrass dieoff in Florida Bay. At the conclusion of their study, water levels had declined in 30% of the 168 km$^2$ (65 mi$^2$) study area since the well field began operation. Water levels were lowered below land surface in 15% of the wellfield study area, resulting in total dewatering of associated wetlands. The wetlands formerly were dominated by native herbs (primarily sawgrass), shrubs, or a combination of herbs and shrubs. Ten years after pumping was initiated, composition had shifted to woody, upland plants dominated by the alien pest species melaleuca (*Melaleuca quinquenervia*). The authors concluded that the adverse environmental impacts documented in the area surrounding this well field were a direct result of the groundwater mining (Hofstetter and Sonenshein, 1990; Sonenshein and Hofstetter, 1990).

Tree islands in the Everglades are considered to be key indicators of the health of the Everglades ecosystem. Trees on tree islands in the Everglades Water Conservation Areas have experienced considerable decline over the past few decades. The cause of the decline of the tree islands in the Everglades has not been identified (Orem et al., 2001; Willard et al., 2001; William Orem, pers. comm.). The association of those tree islands with the municipal groundwater supply wells for Miami (e.g., the Northwest Well Field) and areas of other withdrawals from that aquifer, however, strongly suggests that groundwater mining is a critical factor.

Withdrawals from the Northwest Well Field have ranged from 64 to 134 MGD, approximately one fourth of the agricultural withdrawals reported for 1990 from the same aquifer referenced above. Ironically, the lower range of groundwater withdrawals at this well field is the same as the amount of effluent injected at the West Palm Beach facility described previously. Recall that 64 MGD would be the equivalent removal of water approximately 20 cm deep and 48 m (150 ft) wide extending for 32 km (20 mi) daily from the Biscayne aquifer at this location alone. That distance suggests the diversion of water reaching to the center of the Everglades and including diversion of water from Indian Reservations in the Everglades. To approximate the 134 MGD withdrawals at that site, the magnitude in the example would more than double. A recent study by Fitterman and Deszca-Pan (1999) documented that the salt/fresh interface in Everglades National Park is 8 to 20 km inland from the coast, indicating saltwater saturation of the aquifer. That suggests that groundwater mining of the Biscayne aquifer may be having profound significance, not only for the proposed "Comprehensive Restoration Plan" for the Everglades, but also for the health of Florida Bay.

Annually, withdrawals from that well field are slightly greater from March through August (Sonenshein and Hofstetter, 1990). Insufficient information is available to determine what portion of that groundwater would have been discharged to and utilized by wetlands in the Everglades, as well as what portion would have been SGD utilized by other sensitive coastal ecosystems, including Florida Bay. Increased withdrawals during the summer months (peak growing season) particularly intensifies adverse impacts on wetlands and trees, and may have comparable adverse impacts on sensitive submerged coastal ecosystems deprived of historic SGD. As indicated previously, the cause of the decline of the tree islands in the Everglades has not been identified; however, high levels of phosphorus are associated with those tree islands, suggesting the discharge of groundwater near those tree islands (Orem et al., 2001; Willard et al., 2001).

As suggested in Figure 26.1, the Floridan aquifer system does not stop abruptly at the water's edge along the coast. The general lateral extent of the aquifer system is shown in Figure 26.4, as the geologic cross-section through the Florida–Bahama Platform (southern Florida) that extends

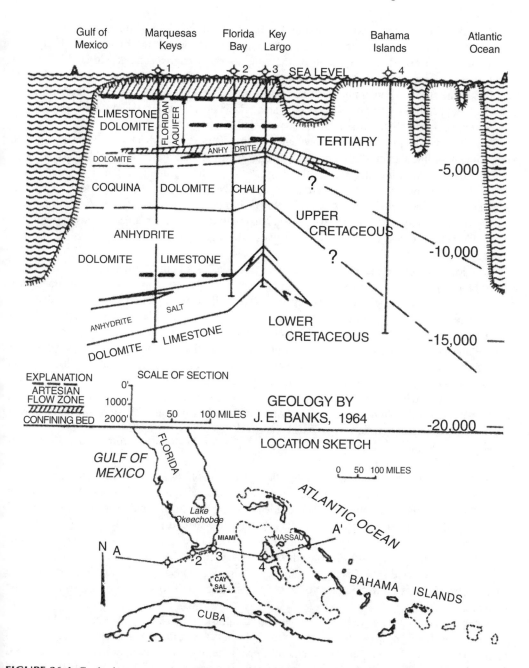

**FIGURE 26.4** Geologic cross-section through the Florida–Bahama Platform (southern Florida) showing the extent of the Floridan aquifer system (upper and lower) from the Gulf of Mexico to the Atlantic Ocean. (From Kohout, F.A., *Trans. Gulf Coast Assoc. Geol. Soc.*, 27:339–354, 1967. With permission.)

from the Gulf of Mexico, through the Marquesas, Florida Bay, the Florida Keys, and Bahama Islands, to the Atlantic Ocean. Meyer (1989) suggests that vertical discontinuity features in the form of fractures, sinkholes, and related breaches are common near the seaward margin of the upper and lower zones of the Floridan aquifer system. Such vertical features are depicted in his generalized hydrogeologic section off the east coast of South Florida, and extending more than 1300 m (4000 ft) below sea level (Figure 26.1).

The well field described above, draining the Biscayne aquifer, is located approximately 25 km (16 mi) northwest of the northern boundary of Biscayne National Park. Kohout and Kolipinski (1967) provide an excellent discussion of the zone of freshwater discharge along the coast of Biscayne Bay during the middle of the century, prior to initiation of pumping at the North West Well Field. The chloride content in the zone of diffusion of the Biscayne aquifer in this area ranged from 16 parts per million (ppm), which is characteristic of freshwater, to approximately 19,000 ppm, which is characteristic of seawater.

Photographs have provided visual evidence of the historic groundwater seepage through the bay sediments in the Cutler study area in Biscayne Bay, FL and comparable subsurface seepage at Jekyll Island, GA, at low tide (Kohout and Kolipinski, 1967). Similar evidence of groundwater seepage during low tides was observed along the shore of New Smyrna Beach, FL (between Daytona Beach and Cape Canaveral) throughout the 1960s (Bacchus, unpubl. data). Dense beds of living bivalves known locally as "coquina" were associated with those seepage areas in the exposed sand at low tide. Those dense populations of bivalves no longer can be found along those beaches (Bacchus, unpubl. data).

In the lower margin of the intertidal zone along the beach south of Marineland (north of Flagler Beach, FL), aggregations of cemented coquina "rocks" with dense populations of sea anemones (*Bunodosoma* sp.) and associated marine organisms were observed through the late 1980s (Bacchus, unpubl. data). Within the next decade, the sea anemones in the intertidal zone were replaced by a thick growth of the green macroalga, *Ulva lactuca*, which is an indicator of coastal eutrophication (Harline and Thorne-Miller, 1981). The coastline at this location was undeveloped throughout the period of both observations, with the dunes and dense native dune vegetation intact. This suggests that local surface runoff was not a factor in the prolific growth of this eutrophication indicator (Bacchus, unpubl. data). Dense populations of a similar sea anemone were observed (Bacchus, unpubl. data) along the shoreline immediately north of Cockburn Sound, Western Australia, during an international intercalibration experiment that evaluated SGD in the Sound. At that location, discharge of low-salinity groundwater is known to occur (Appleyard, 1994; Johannes, 1980; Johannes and Hearn, 1985).

Freshwater springs are known to have flowed freely off South Florida's coast, including Biscayne Bay, prior to extensive groundwater mining and related perturbations (Kohout, 1966; Parker, 1984; Schomer and Drew, 1982). Evidence of freshwater discharge from the submerged platform, farther from shore in Biscayne Bay, had been provided previously by the presence of artesian flow from wells, demonstrating that there was sufficient pressure in the Biscayne aquifer for water to discharge to Biscayne Bay through bay sediments (Kohout and Kolipinski, 1967). Kohout (1965) described the Biscayne aquifer as "highly permeable, solution-riddled limestone that thins from a maximum thickness of 200 feet near the coast to a featheredge 40 miles west of Miami."

In conjunction with the study by Kohout and Kolipinski (1967), the distribution of biological components of Biscayne Bay also was estimated. The flora and fauna evaluated included species of algae, seagrasses, porifera, cnidaria, annelida, echinodermata, mollusca, arthropoda, and fishes. Five of the 42 species captured in trawl samples were associated with hydrological factors, and their spatial distribution was related to salinity more closely than to other environmental factors. The spatial arrangements of 17 other species also suggested the influence of salinity. The strongest inference of zonation with respect to salinity gradients was exhibited by attached plants and sessile animals (Kohout and Kolipinski, 1967). Their early work is one of the first to suggest the ecological importance of the now-overlooked influence of SGD. Other chemical components of the discharge, such as nutrients, were not considered.

Although it is intuitive that the extensive groundwater mining, and injection of effluent and other waste in South Florida have had significant impacts on both the quantity and quality of SGD, those impacts (and subsequent ecological impacts) remain unassessed. Likewise, previous equations defining SGD lack recognition of those anthropogenic aspects. The following equation (modified from Li et al., 1999, and Zektser et al., 1983), incorporates a component that addresses those anthropogenic impacts:

$$D_{SGD} = D_n + D_a + D_w + D_t \qquad (1)$$

where $D_{SGD}$ is the total submarine groundwater discharge; $D_n$ is the natural, historic aquifer discharge; $D_a$ is the anthropogenic aquifer discharge; $D_w$ is the outflow circulation (recirculation) from wave action; and $D_t$ is the outflow circulation (recirculation) from tide action. If groundwater is extracted (e.g., for agricultural, industrial, potable municipal use), then $D_a$ is negative. If water (or liquid waste such as municipal effluent) is injected into the aquifer and not "recovered," then $D_a$ is positive. If fluid injections and extractions are coupled (e.g., aquifer storage and recovery), then $D_a$ pulses between positive and negative, respectively. Those pulses cannot be averaged to infer no net change in the aquifer or groundwater conditions because of the significant adverse impacts that may occur to coastal and inland ecosystems as the result of these pulsed groundwater perturbations.

The first term in Equation (1) initially was defined by Zektser et al. (1983) as net groundwater discharge. Later, this term was defined by Moore (1996) and Church (1996) as net groundwater discharge plus recirculation. Most recently, Li et al. (1999) subdivided the two components in that term, redefining $D_n$ as net groundwater discharge estimated using aquifer recharge data and establishing two new terms for wave-induced and tide-induced recirculation.

Due to the extensive anthropogenic groundwater perturbations in South Florida (and many other locations in the U.S.), as well as the subsequent ecological implications, $D_n$ now is proposed to represent the natural, historic groundwater discharge. The proposed use of that term includes both direct discharge (e.g., via vents in the sea floor) and diffuse discharge (e.g., via seepage) components throughout the submerged carbonate platform and along the margin of the platform. The new term proposed in this chapter ($D_a$) represents anthropogenic increases/decreases in the natural discharge (including associated chemical and physical changes in the water), as indicated above.

Both the volume and the characteristics of aquifer-injected fluids may differ substantially from that subsequently discharged to surface waters as SGD. Consequently, the volume of groundwater withdrawn from a karst aquifer cannot be used to calculate $D_a$ reductions in SGD because of potential induced recharge from terrestrial systems and delayed aquifer responses. Likewise, combining the volume of injected fluids and withdrawn fluids to determine "net $D_a$" also is misleading and invalid because it ignores the ecological impacts of both the physical and chemical changes associated with the injected fluids and the withdrawn fluids. Determining "net $D_a$" also could provide highly inaccurate estimates of nutrient loading to surface waters due to induced discharge of eutrophic wastewater.

Clearly the anthropogenic discharge component of the equation is the most complex. The primary focus of SGD research and quantification (e.g., most seepage studies) currently is recirculation ($D_w$ and $D_t$). Because of the intimate linkage between groundwater and surface water in South Florida, the extensive anthropogenic groundwater perturbations occurring there, and the significant ecological implications of those perturbations, South Florida is the ideal location to initiate an in-depth evaluation of the anthropogenic component of this equation. The ecosystems in the Florida Keys are in a crisis state of decline and the role of anthropogenic perturbations to SGD in the Florida Keys requires immediate attention.

Investigations of SGD should include direct discharge components, as well as diffuse discharge components, throughout the submerged carbonate platform and along the margin of the platform. This approach is required because freshwater SGD has been reported as diffuse seepage from submerged land forms and more direct discharge along the margins of these landforms, which are deep and in some cases distant from shore. The most far-reaching of the reported freshwater SGD may be that of the "shimmering curtain" at depths of 1170 m (3510 ft) along the Sagami Trough, at the northern end of the Philippine Sea plate. Here, the seepage water had a seeping flux of 310 to 500 $m^3 d^{-1}$ (Tsunogai et al., 1996; T. Gamo, pers. comm.).

Highlights of historical development of the physical concepts leading to an awareness of SGD are provided in Table 26.1. Simmons and Netherton (1987) discussed the documentation of localized SGD along the southeast Atlantic coast since the late 1960s, but indicated that prior focus had been on salinity measurements for verification of the presence of freshwater, rather than on the presence of nutrients or pollutants that could be of ecological importance. The data they collected in the vicinity of Key Largo between November 1984 and September 1985 provided additional evidence of SGD, based on measurements of the quantity and quality of submarine seepage associated with a deep coral reef ecosystem.

Both seepage meters and submarine piezometers (pipes open at the bottom and installed to designated depths in the substrate to determine groundwater pressure at the designated depth) were used at the following four sites in that study: French Reef (Key Largo National Marine Sanctuary); Ajax Reef (Biscayne National Park, north of the Sanctuary); Alligator Reef (south of the sanctuary); and Conch Reef (also south of the sanctuary). The sites south of the sanctuary were considered reference sites. Samples were collected from shallow (4.6 and 9.1 m) and deep (30.5 and 36.5 m) depths associated with the reefs, during both high and low tides.

Data from both sampling devices verified differences in chemical composition between the seepage water and ambient sea water, and documented the presence of a positive head (pressure). Although differences in salinity between SGD water and ambient ocean water were small, Simmons and Netherton (1987) concluded that the differences probably were ecologically significant. This conclusion was based on findings in Perth, Australia, that a decrease of 1% (0.35 ppt) corresponded to a "nitrate increase of approximately 2 to 4 $\mu$g L$^{-1}$, which was several times the mean nitrate concentration of undiluted sea water in the immediate vicinity." They also documented higher

---

**TABLE 26.1**
**Historical Development of Some of the Physical Concepts Leading to an Awareness of Submarine Groundwater Discharge**

| Researchers | Contribution |
|---|---|
| Newell et al. (1953) Adams and Rhodes (1960) Simms (1984) | Described history of quantitative developments; important conclusion from ecological perspective was that seawater can and does cycle in advective way through sediments due to fresh groundwater hydraulic head. |
| Cooper (1959) | Hypothesized explanation of mixing zone, or zone of dispersion; continuous circulation of seawater was observed in various field studies; attempted to quantify amount of mixing due to tidal fluctuations. |
| Henry (1959, 1964) | Quantitatively corroborated Cooper's hypothesis and used advection–diffusion equation to account for hydrodynamic dispersion. |
| Kohout (1960, 1964) | One of first to suggest/quantify continuous cycling of seawater as a result of hydrodynamic dispersion. |
| Kohout (1965, 1967) Kohout et al. (1977) | Postulated thermal convection as a possible mechanism for seawater circulation in the Floridean plateau (also known as the Florida platform). |
| Manheim (1967) | Identified localized SGD along the southeastern Atlantic coast. |
| Riedl (1971) Riedl and Machan (1972) | Demonstrated significance of surf action on movement of seawater through marine sediments in intertidal and subtidal zone of sandy beach. |
| Riedl et al. (1972) | Developed mathematical equation to explain seawater circulation through marine sediments by means of subtidal pumping; concluded that the presence of oxidized sediments indicates that circulation is occurring. |
| Nixon et al. (1984) | Described relatively new concept of benthic–pelagic coupling. |
| Simms (1984) | Proposed reflux, defined as a type of saline fluid convection, as a mechanism of seawater cycling, and more fully described Kohout convection, reflux, and cycling along coastal mixing zones. |
| Reilly and Goodman (1985) | Described history of quantitative development: advective cycling of seawater through sediments due to fresh groundwater head. |

concentrations of major cations and various trace metals, in addition to temperatures up to 1°C less in the SGD.

Based on these results, Simmons and Netherton concluded that the SGD represents a mechanism for moving nutrients and trace minerals across the sediment–water interface that may be "global in its nature." They further suggested that the longshore current of the Gulf Stream interacts with the zone of diffusion along the southeastern coast of the U.S., carrying SGD water toward shore along the bottom, then upwelling and transporting the water offshore for a specified distance, before it falls again in the water column.

An extension of this concept may have been reflected in a phenomenon observed by local Fisheries Researcher Don DeMaria during one of his dives. As summarized in his letter dated February 10, 1994, to the Chairman of the FKNMS, DeMaria described a layer of cold, dirty water approximately 9 to 12 m (30 to 40 ft) thick moving along the bottom at the outer edge of a reef. The total depth was approximately 30 m (100 feet) and the thick layer of turbid water reportedly was "pushing" the fish into deep water as the mass of turbid water moved offshore. From approximately the same area to a few kilometers east (24°26.009′ N, 81°55.880′ W) of where the "wall" of turbid water was observed, DeMaria also observed purple-colored algae covering the bottom and much of the reef (DeMaria, unpubl. data; pers. comm.).

Those "cold, dirty" layers of water along the sea floor were reported previously by DeMaria in letter to the Chairman dated July 10, 1992. In that initial letter, DeMaria noted that he often encounters a layer of cold, dirty water along the sea floor while diving on wrecks north of Key West, and that the "bottom current" runs counter to the surface current. Additionally, he indicated that farther north of Key West many small holes (~15-cm diameter), fissures, and sinkholes can be seen in the "porous limestone" bottom and that very cold "freshwater is constantly flowing out of this bottom." Finally, he stated that the algal blooms and poor visibility along the reef line on the south side of the Florida Keys was disturbing, and stressed that he had never seen algae blanket the bottom of the deeper areas of the reef in that manner before. He further indicated that some areas were "so heavily blanketed that the underlying reef is no longer visible." He concluded by stating that he did not know of anyone else who had ever encountered the total blanketing of a reef by algae in the manner that he had observed, and that it "appears that the deep water algae stretches from at least Key West to the Tortugas, a distance of at least 60 nautical miles."

The FKNMS had been notified previously by DeMaria regarding the proliferation of algae associated with reefs in a letter to the Chairman dated June 2, 1992. In that letter, DeMaria reported observing a thick layer ("like a blanket") of "green and hairlike" algae covering everything (including seafans and sponges), during a dive on May 31, 1992, southwest of Key West, at a depth of 34 m (110 ft). DeMaria further reported that, in the past, this area had a large population of small ornamental reef fish but now no small fish could be found and few other fish were observed in the area. DeMaria also indicated that this was the first time in his diving career that he had observed algal growth of that nature associated with a reef in deep water.

Hunt and Herrnkind (1993) described the role that the proliferation of algae may have played in the dieoff of the bay's sponge population. Dustan (1999) indicated that macroalgae not only can shade coral tissue, causing bleaching and eventually death, but also can abrade the soft coral tissue as the algae waves in the surge. Finally, microalgal filaments at the edge of corals form "sediment dams" which prevent corals from clearing sediment off their surface, thus suffocating the live coral tissue. This functional disease is known as "algal-sediment encroachment" (Dustan, 1999). More than 10 years ago, warnings were published that algae was covering large areas of the reefs, and in many cases covering entire coral colonies throughout Florida and the Bahamas (Dustan, 1987). Now, the algae on the outer reefs of Key Largo is so abundant it has been termed "metastatic," and the algae reportedly is so thick on Molasses Reef that it is like a rug (Dustan, 1999).

The general response from agency and academic personnel who were queried regarding those observations was that it simply represented a "typical upwelling event." However, another explanation is that some of the injected effluent is discharging at the platform margins, either continuously or in large, periodic pulses (belches) from the underlying aquifers, creating polluted layers of cold water that blanket the coastal floor due to temperature differences. Those cold, polluted walls of water would force mobile organisms offshore to an uncertain fate, while engulfing nonmobile organisms such as sensitive coral reefs. Flowing groundwater that is focused in preferential flow paths, with fluxes and flow trajectories that vary in both time and space, is not a new concept. Those responses were illustrated most recently and lucidly by Dahan et al. (2000).

Based on our knowledge of the hydrogeology of South Florida, including documented karstic features throughout the submerged platform (summarized in Bacchus, 2000a, with emphasis on Kohout, 1967; Meyer, 1989; and Swayze and Miller, 1984), we also would expect a significant portion of the minimally treated effluent injected into the lower Floridan aquifer at the Miami–Dade site (and Key West site, if permitted) to resurface along the margins of the Florida Plateau, in close proximity to the east coast of Florida (in the Straits of Florida), as well as along the western margin, at a greater distance from the shore (Figure 26.5; see Enos and Perkins, 1977). Discharges of that nature would provide a significant source of excess nutrients, in addition to an assortment of yet-to-be-identified contaminants, that would "slosh" up onto the shelf during "typical upwelling events." Those scenarios could explain phenomena similar to those described in DeMaria's letter, as well as the increasing occurrence of filamentous green algae observed in more recent years in the Dry Tortugas and other deep-water coral reefs in the Florida Keys (D. DeMaria, pers. comm.; J. Porter, pers. comm.).

Numerous residents in the Florida Keys have observed equally numerous direct-flow SGD points throughout the submerged shelf (platform) on both the bay and ocean side (e.g., D. DeMaria, pers. comm.; K. Simpson, pers. comm.; D. Telewicz, pers. comm.). Extensive cold water was flowing from all but one of the discharge points observed during a reconnaissance of some of these areas late in 2000 (Bacchus, unpubl. data). The existing water quality monitoring program for the FKNMS does not appear to have monitoring stations established at these locations (or to even acknowledge their existence), and water quality samples appear to be collected primarily on a quarterly schedule. Clearly that approach is not conducive to identifying or evaluating potential direct discharges of injected effluent in areas throughout and along the margins of the platform that may be having a significant adverse impact on sensitive coastal organisms such as corals and sea turtles.

With regard to diffuse discharges, which are the typical focus of SGD studies, the measurements reported by Simmons and Netherton (1987) indicated that the rates of discharge in their study were independent of tidal action and depth. That observation suggests that the water involved in the SGD at the reefs that Simmons and Netherton evaluated may be linked to a source with a greater hydraulic gradient. The Floridan aquifer system is a source with a greater hydraulic gradient.

Subsequent research by Simmons (1992) at French Reef revealed that SGD was a common occurrence, at least to water depths of 30 and 35 m (90 and 105 ft). Discharge from October 1984 through July 1986 at French Reef in the Florida Keys ranged from 8.9 L m$^{-2}$ d$^{-1}$ (for depths <27 m) to 5.4 L m$^{-2}$ d$^{-1}$ (for depths 27 to 29 m). Diel tidal pulses associated with the reef crest, back reef, fore reef, and deep reef of French Reef during July and August 1986 were inconsistent within reef position for two sample periods of three replicates each, except for a weak inverse relationship between discharge rate and tidal height at the deep reef location.

Results from the transect of piezometers installed perpendicular to French Reef, at a depth of 30.3 m (90.9 ft) and where measurements were taken at 3-hour intervals from July 31 to August 1, 1986, may have been of greatest significance. The three piezometers were placed approximately 1.0 m (3.0 ft) into the sediment and approximately 10 m (30 ft) apart (the distance of the piezometers from the reef was not provided). Consistently negative head measurements (–4 to –8 mm) were associated with the piezometer farthest from the reef (#1), while the greatest positive measurements (1 to 4 mm) were associated with the center piezometer (#2). The piezometer closest to the reef (#3) provided measurements similar to, but lower than, those for the center piezometer (0 to 2 mm). These data document that even in areas of the sea floor that may appear similar or homogeneous

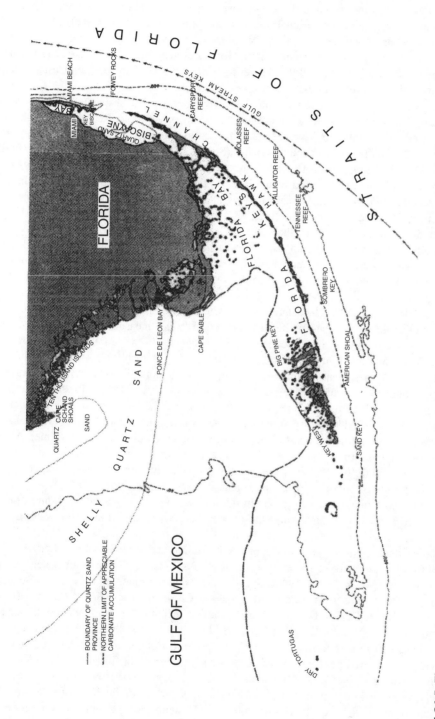

**FIGURE 26.5** The approximate east–west extent of the Florida Plateau and alignment of major coral reef systems in South Florida. (From Enos, P. and Perkins, R.D., *Quaternary Sedimentation in South Florida*, Memoir 147, The Geological Society of America, Boulder, CO, 1977. With permission.)

on the surface (due to masking by overlying sediments), SGD may vary significantly. A logical explanation for those observed variations is that underlying geological formations contain preferential flow paths that are capable of discharging groundwater nonuniformly, at discrete locations. Those results also suggest that discharge of groundwater can be associated with coral reefs.

Similar attempts by other researchers to measure seepage in the Keys using seepage meters produced variable results, which were interpreted as equipment-related problems (E. Shinn, pers. comm.). Understandably, seepage meters are better suited for measuring seepage in areas with thick, unconsolidated substrate. Areas with rock outcrops at or near the surface make proper seating of the equipment difficult. Simmons (1992), however, attributed the differences he measured to the high degree of spatial variability in that system.

Different SGD responses at different locations overlying an eroded carbonate platform should be expected when the underlying strata are not homogeneous or isotropic. Reef ridges and karstic features associated with coral reefs have been documented in numerous locations including Belize (Macintyre et al., 2000), the Cocos (Keeling) Islands (Searlel, 1994), and Papua New Guinea (Purdy and Bertram, 1993; Williams, 1972). Those findings further suggest that localized SGD may occur in the vicinity of coral reefs. In fact, Purdy and Bertram (1993) hypothesized that preferential reef colonization occurred in Papua New Guinea on polygonal karst relief that was formed on limestone surfaces during Pleistocene subaerial exposure.

Based on the information above, the selection of sample site locations for placement of instruments and collection of SGD data is a critical factor, particularly in areas with coral reefs. Specifically, in areas underlain by a karst aquifer, a study design using data collection points and transects that are not linked to underlying hydrogeological features and potential preferential flow paths cannot provide an accurate evaluation of the nature of or total SGD contribution.

Additional research is needed to determine whether zones of preferential flow are associated with specific reef positions or whether the varied range in values measured in piezometers along the transects in previous studies simply reflect variable responses of the sampling devices or point differences in flow from underlying aquifers. The results of Simmons (1992) suggest that links to different aquifers, or different zones within the same aquifer (e.g., upper Floridan vs. lower Floridan) could be responsible for those different responses. The presence of fractures or dissolution features have been shown to result in vertical flow links between various aquifers and the surface water.

A second site location evaluated by Simmons (1992) was on the southeastern continental shelf in Onslow and Long Bays (off the coast of the South Carolina/North Carolina state line). At that location, discharge ranged from approximately 6 to 20 L m$^{-2}$ d$^{-1}$ during the summer of 1987. At one sample location, 20 m (60 ft) deep, a persistent negative hydraulic head was observed, with a mean influx of seawater of 10.8 L m$^{-2}$ d$^{-1}$. No explanation was proposed for this negative hydraulic gradient, but the most likely cause is groundwater mining on the mainland. Bacchus (2000b) described some of the nearshore environmental impacts that may be linked to groundwater mining on the mainland.

One detailed description of such links between aquifer and surface water is provided by Manheim (1967). He summarized the knowledge of local submarine discharge of water near the southeastern Atlantic coast in the late 1960s. Specific sites were identified off the coast of Beaufort, SC, and Florida. The Florida sites were off the coast of Crescent Beach, South Daytona Beach, northeast of Cape Kennedy, Eau Gallie, and Biscayne Bay/Miami. Results of the Joint Oceanographic Institutions' Deep Earth Sampling (JOIDES) Program previously had documented freshwater in sedimentary strata as far as 120 km (75 mi) seaward of Florida's Atlantic coast in the vicinity of Jacksonville. Manheim's summary includes a photograph of freshwater flowing from a drill pipe on a drilling ship at a depth of approximately 130 m (390 ft) below the sea floor at approximately 60 km (37 mi) from shore. Discharges occurring in the vicinity of mid-shelf and beyond appear to be of Pleistocene origin or older (F. Maheim, pers. comm.). Outcrops on the Blake Escarpment also included pre-Eocene strata.

The Blake Platform in the vicinity of the documented freshwater strata also included a number of large depressions. Manheim (1967) discussed several factors that may have contributed to the formation of those depressions. Subsidence due to submarine discharge was one explanation proposed for the formation of those abrupt, local depressions in an otherwise "tightly cemented and erosion-resistant sediment bottom." Those depressions frequently occur in areas of probable outcrops or lightly buried exposures of Eocene rock. As an example, the submarine spring approximately 3.2 km (2 mi) seaward of Crescent Beach, FL, is a hole 37 m (111 ft) below sea level that is eroded in a flat regional bottom of about 17 m (56 ft).

Another factor discussed by Manheim (1967) was the possible initial breaching of the Blake Platform strata at the site of their research submarine's "lost buoyancy" and at similar locations during the partial emergence of the present continental margin during the Pleistocene. When the platform was exposed, infiltration of groundwater may have created the "preferred channels of communication," which could have served as sites of preferential leakage when the platform was resubmerged. This process is consistent with the description of terrestrial erosion during earlier periods of low sea level (Shinn, 1988, 1999).

The final contributing factor described by Manheim (1967) for the formation of the depressions was the possible presence of stronger bottom currents beneath the Gulf Stream during times of lower sea levels. He indicated, however, that it is difficult to visualize how bottom currents could have scoured or otherwise formed the characteristically karst-like rounded depressions that are several tens of meters deep. Chemical and flow characteristics of water discharging from the Crescent Beach Spring, off Florida's east coast, suggest that both the Floridan and surficial aquifers contribute to the flow (P. Swarzenski, pers. comm.).

In a written communication from V.T. Stringfield to R.O. Vernon in 1965, the former calculated the potential for SGD in a submarine outcrop to a depth of approximately 500 m (1500 ft). That site was located approximately 100 km (62 mi) from shore. His calculation was based on the assumption of an original potentiometric head of 18 m (54 ft) in the Floridan aquifer and a gradient of 0.06 m km$^{-1}$. However, the intensity of the flow implied by the observations of their research submarine *Aluminaut* at 510 m (1673 ft) depth and more than 200 km (124 mi) from shore, suggested that the potentiometric heads assumed by Stringfield either were very conservative or that fluid flow was following different pathways than those generally accepted (Manheim, 1967).

Those submerged features are not confined to the northeastern coast of Florida. Extensive sinkhole-like depressions on the Pourtales Terrace off the coast of South Florida at depths of 300 to 350 m (900 to 2800 ft) were described by Jordan (1954) and Jordan et al. (1964). More recently, Lidz et al. (1991) generated a series of high-resolution seismic-reflection profiles off the Lower Keys in Florida during an investigation of a multiple outlier-reef tract system approximately 0.5 to 1.5 km (0.3 to 0.9 mi) seaward of the bank margin. Signatures suggesting karstic subsidence and dissolution features were observed in those profiles.

The evidence described by Manheim (1967) led to his conclusion that fluids were moving from the continent to offshore areas via the Tertiary strata. He included a simplified illustration of numerous "minor leakage" channels, as vertical conveyances from the Tertiary formation to the surface, with "major discharge" from the Tertiary emanating from the outcrop of the strata at the margin of the Florida Platform. Previously, Stringfield (1966) had predicted the discharge of freshwater along the continental margin (see Figure 26.4), based on the fact that the strata bearing the freshwater (the Floridan aquifer system) appeared to be exposed in underwater outcrops along the Florida–Hatteras Slope. The geologic formation of that portion of the aquifer was described as chiefly late Eocene in age.

Manheim concluded that the considerable anisotropy (nonuniformity in all directions) of Tertiary sediments of the southeastern Atlantic continental margin is demonstrated by the apparent ability of a layer approximately 200 m (656 ft) thick transmitting freshwater at least 200 km (124 mi) from shore. He indicated that those conditions suggest a horizontal/vertical permeability ratio of more than 1000. Communication between pore fluids of underlying Tertiary and Mesozoic strata and bottom waters was deemed plausible (Manheim, 1967). In simple terms, he indicated that water from the deep aquifers was discharging to the overlying surface waters prior to injection of effluent into the deep aquifers. Injection of large volumes of wastes into those aquifers will merely ensure that more water is discharged to the overlying surface waters as induced discharge.

The early work by Manheim, described above, focused on SGD near the outer extent of the submerged aquifer of the east coast. Possibly the most recent and convincing documentation of significant SGD in Florida Bay was conducted via a geochemical approach, using helium isotopes ($^4$He, the more abundant isotope, and $^3$He, tritium) and $^{222}$radon ($^{222}$Rn, a conservative noble gas) as tracers of groundwater inflow (Top et al., 2001). Large quantities of both of these isotopes are dissolved in water while it is underground, in contact with various aquifers in that area. Stable radioactive decay products, such as helium, accumulate over geological time, while short-lived $^{222}$Rn generally reflects instantaneous shallow-aquifer characteristics. Surface water injected into carbonate aquifers (e.g., injected effluent) could acquire the He signature immediately (Z. Top, pers. comm.). For example, if surface water injected into the deep Floridan aquifer mixes with an equal volume of aquifer water, then the helium excess in the mix could be higher than 200%. Surface water samples throughout Florida Bay and at selected locations on the ocean side of the Keys showed helium and radon anomalies during both the summer and winter of 1998 and 1999 (Top et al., 2001).

Significant helium anomalies persisted in the shallow (approximately 1.5 m deep) waters of Florida Bay. This finding supports the presence of a groundwater source, with helium flux increasing significantly in the winter, based on the 2-year study period. The highest excesses of $^4$He were observed in the northern sector of Florida Bay in a band parallel to the coast. The observed pattern of localized clusters may be due to discrete groundwater sources, or it may be a sampling artifact. The greatest anomalous areas of $^4$He were south of Flamingo (where persistent high levels of phosphorus occur), along the southern shore of Cape Sable, in the vicinity of Rankin Bight and Whipray Basin (areas of initial seagrass dieoff), in the vicinity of Sandy Key Basin and Rabbit Key Basin (areas of subsequent seagrass dieoff), near the northeast sector near Porjoe Key (where Corbett et al., 1999, measured reduced salinity). Anomalous levels of $^4$He also were associated with both the bay side and ocean side of Tavernier and Key Largo. Rankin Bight and Porjoe Key exhibited high groundwater flux, with the latter exhibiting low salinity (suggesting low salinity SGD in that area). Moderate excesses of $^4$He were associated with shallow, nearshore reefs from Largo to Upper Matecumbe, in the vicinity of Cheeca Rocks.

The concentration anomalies of both $^3$He and $^4$He isotopes further suggest the presence of both "young" and "old" groundwater discharging into Florida Bay. The source of the young groundwater could be the shallow aquifer, while the source of the old groundwater most logically would be the Floridan aquifer system. The former increased during the summer and the latter increased in the winter (Top et al., 2001). Coincidentally, isolated higher concentrations of chlorophyll a that were measured quarterly and averaged (see Figure 6 of Brand, 2000; Figure 2 of Brand, 2002), coincide with the areas of greatest $^4$He excesses identified by Top et al. (2001) in both Florida Bay and along the coral reef tract in the FKNMS.

Radon excesses ($\delta$ values, which are inversely related to wind speed) increased during the summer, as did $^3$He, suggesting a rain-enhanced shallow groundwater source for the radon. Although the deep aquifer also could be a source for $^{222}$Rn, the short half life of 3.8 days theoretically decreases the probability that the source of $^{222}$Rn in Florida Bay is from the deeper aquifers. No radon excesses were associated with Lower Matecumbe (Top et al., 2001), where USEPA identified a water quality "hot spot" reportedly attributed to groundwater-laden nutrients from septic tanks. The sample used to designate that location as a "hot spot," however, was collected from a channel with live-aboard boats (P. Yananton, pers. comm.), suggesting that those boats were a more likely source of the pollutants identified in the sample.

Clusters of low-level radon excess anomalies were documented around Long Key, where shallow-well injection of effluent occurs. Similar clusters were located in the vicinity of former seagrass dieoffs in Florida Bay and throughout the northern half of Florida Bay, northern Upper Matecumbe, and southern Plantation Key. Numerous shallow-well injection sites are operating on the latter two Keys. The highest level of radon excesses occurred at Tavernier and Largo (bay side), which are additional locations where numerous shallow-well injection sites are operating. The >50% increase in estimated radon flux from summer to winter during the period of this study suggests a trend of strong seasonality.

Corbett et al. (1999) made similar findings using $^{222}$Rn and methane ($CH_4$, another natural chemical tracer of groundwater discharge). Additionally, they evaluated the abundance of nitrogen isotopes associated with macroalgae and seagrasses in Florida Bay and the coral reef tract on the ocean side of the Keys. They concluded that both macroalgae and seagrasses showed significant $^{15}$N enrichment (a potential indicator of groundwater-derived nutrients) in areas of high groundwater discharge. Finally, they evaluated nutrient concentrations in interstitial porewater, concluding that groundwater discharge may provide as much N and P to the eastern portion of Florida Bay as surface inputs from the Everglades (e.g., the C-111 canal). They note, however, that the inputs clearly are not uniform, and that areas near solution holes or tidal springs may have a substantially greater nutrient flux into surface waters than suggested by their data.

What is the significance of the documentation that both Florida Bay and the ocean side of the Keys are receiving considerable groundwater discharge? The ratios of total N to total P and inorganic N to inorganic P throughout Florida Bay have been determined to be greater than the Redfield ratio of 16 (the ratio required for algal growth), and the data suggest P limitation in the east and N limitation in the west of Florida Bay. Based on those data, Brand (2000, 2002) hypothesized that much of the N entering the Everglades comes from discharges of eutrophic agricultural runoff from the Everglades (including pulsed discharges from canals), suggesting that the proposed approach to increase the input of freshwater to Florida bay by increasing discharges of eutrophic surface water from the Everglades (USACOE/SFWMD, 1999) is fatally flawed and will perpetuate algal blooms in Florida Bay. Finally, he proposed that the source of P may be via groundwater discharging in areas of coarse-grained quartz sand deposits in Florida Bay.

The distribution of coarse-grained sand deposits that Brand has shown would not account for the patchy distribution of chlorophyll *a* "hot spots" in Florida Bay, nor the isolated areas of higher chlorophyll *a* concentrations associated with the reef tracts. The co-occurrence of these sand deposits with areas of preferential flow in underlying karstic carbonate rock, however, could account for the patchy distribution of chlorophyll *a* "hot spots" in Florida Bay. Conversely, karstic preferential flow paths on the ocean side of the Keys could account for the isolated areas of high chlorophyll *a* concentration associated with the coral reefs.

Top et al. (2001) have addressed the significant impacts that groundwater-transported pollutants can have on coastal ecology and public health, describing finfish and shellfish kills related in part to groundwater-laden nutrients and providing examples of Atlantic coastal areas where groundwater accounted for half of the nitrogen loading of the sediments. They described the implication of presumed leakage of effluent injected into the Boulder Zone via deep wells throughout South Florida. Also referenced are results of a recent tritium/helium isotope and chlorofluorocarbon tracer survey in Everglades National Park which show evidence of deep aquifer water in some shallow wells (R. Price, manu. in prep.). That tracer survey further supports the need for a more comprehensive study of injection wells. Direct measurements from wells in the deep aquifer were recommended by Top et al. (2001), based on their estimates that groundwater from both deep and shallow aquifers appears to be a significant fraction of the circulating volume of Florida Bay.

## Mainlining

Although limited information is available regarding the ecological impacts of groundwater mining in South Florida, including the impacts on historic SGD, considerably less is known about the responses of the aquifer system to the injection of large volumes of fluid wastes such as minimally treated sewage effluent. A total of 81 deep wells (Class I) throughout Florida inject large volumes of wastes, primarily minimally treated municipal sewage, into the same Tertiary strata described above by Manheim (Figure 26.6). All but two of those deep-well injection facilities are located in the southern half of the state, with the majority injecting wastes along the coast (FDEP, 1999).

In December 1969, the Secretary of the Interior directed the USGS to begin a research program to evaluate the "effects of underground waste disposal on the Nation's subsurface environment, with particular attention to ground-water supplies" (Kaufman, 1973). Admittedly, this was no small undertaking; however, the directive overlooked the potential resurfacing of this waste into coastal environments via SGD and the subsequent environmental implications.

During the 1960s, a federal policy statement on wastewater injection indicated that subsurface injection should be used as a waste-disposal method only as a last alternative — and then only with great caution and for a limited period of time. The potential for use of aquifer zones to "store" water and dispose of waste was suggested by Vernon (1970). In 1970, with the creation of the USEPA, a policy statement similar to the 1960s policy statement was issued, emphasizing the temporary aspect of authorized deep-well injection of wastewater (Hickey and Vecchioli, 1986).

By mid-1971, more than 7 billion gallons of waste had been injected into the Boulder Zone of Florida's karst regional aquifer. The Boulder Zone is composed of highly permeable carbonate rocks and reportedly contains cavernous voids and cold, dense saline water. In South Florida, the Boulder Zone underlies the coastal areas at depths from approximately 870 to 1430 m (2600 to 4300 ft). Approximately the upper half of this zone (where the wastes are injected) particularly is characterized by cavities, extensive fractures, and solution porosity. The hydraulic head of the seawater in the Boulder Zone was determined to be at approximately sea level in the early 1970s (Kaufman, 1973). This suggests a high probability for rapid vertical discharge of any nonsaline fluids injected into that zone.

During the initial stages of consideration, Kaufman (1973) expressed numerous concerns regarding the uncertainty of injecting wastes into the karst aquifer. Those uncertainties included: (1) whether hydraulic separation of the various aquifers would remain effective under the anthropogenic perturbations of injecting large volumes of waste, (2) the apparent hydraulic connection with the Straits of Florida, and (3) the regional hydrodynamic circulation pattern and direction, rate of movement, and ultimate fate of injected liquid waste. He indicated that an effective assessment of long-term environmental impact of waste-disposal feasibility would include considerations of the "regional circulation patterns and other complexities of the sub-surface environment" in South Florida. Kaufman's conclusions included the following:

> The deep subsurface aquifer systems are complex, and knowledge of the regional hydraulic gradients and circulation patterns, recharge and discharge boundaries, and permeability distributions within these saline zones remains scant. In South Florida, the effectiveness of hydraulic separation under man-induced perturbations remains to be verified. ... The long-term environmental impact of subsurface waste injection in Florida is only conjectural at this time. An adequate assessment would appear to require comprehensive evaluation and monitoring of the subsurface environment on a regional basis, in addition to intensive localized studies.

Approximately 30 years later, our "knowledge of the regional hydraulic gradients and circulation patterns, recharge and discharge boundaries, and permeability distributions within these saline zones" still remains scant. No evidence could be found that intensive localized studies have been designed or implemented to assess the fate and long-term environmental impact of subsurface waste injection in South Florida or any other area of the state where such disposal is being conducted. All of the information available, however, suggests that extensive contamination of associated sensitive coastal ecosystems can occur from the injected waste. In fact, by the early 1970s numerous hydraulic and geochemical effects of the waste injections already had been documented. Following are examples of some of those effects: (1) reversal of natural hydraulic gradients, with establishment

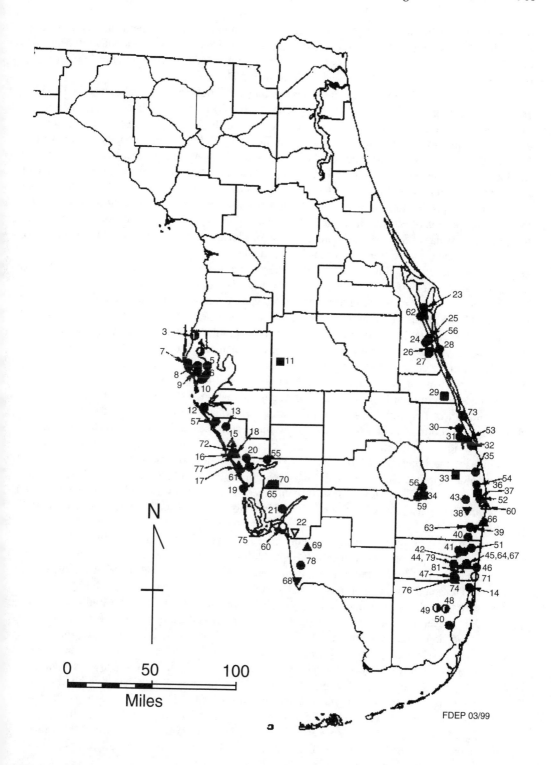

**FIGURE 26.6** The location of Class I (deep-well) injection facilities in South Florida, with the Miami–Dade Black Point effluent-injection facility shown as the southeasternmost facility as of 2000. (From Florida Department of Environmental Protection, Class I injection facilities map, Tallahassee, FL, 1999.)

of potential for updip migration of fluids; (2) increased localized aquifer permeability and transmissivity; (3) dissolution of the carbonate aquifer and localized cavern development beneath the injection sites; (4) upward movement of wastes; (5) evolution of high concentrations of hydrogen sulfide, nitrogen, methane, and other gases (Kaufman, 1973).

Some of the available information indicates that the highly transmissive Eocene has been divided arbitrarily into upper (Eo-1), mid- (Eo-2), and lower (Eo-3) units. The highest transmissivity of the upper zone includes a large area north of Collier and Broward Counties, a smaller zone in northeast Monroe County in the vicinity of the oil fields, and a band approximately 24 km (15 mi) wide that extends from the vicinity of the injection wells in Miami generally southwest under Florida Bay, the Lower Keys, and the Dry Tortugas National Park (the focus of attempts to adopt stronger "protection" measures). The highest transmissivity of the Eo-2 unit extends throughout the southern peninsular north of Lee, Hendry, and Broward Counties, with an additional zone under the northeast portion of Florida Bay. The highest transmissivity of the Eo-3 unit extends throughout the entire peninsular of South Florida, including all of Florida Bay and the Keys (Puri et al., 1973).

In Belle Glade, FL, upward migration of hot acidic waste (pH 2.5 to 4.5) associated with a sugar mill occurred approximately 27 months after waste injection began in 1966. The waste was injected at a depth of approximately 500 m (1500 ft). The upward migration re-occurred again, within 15 months after resumption of waste injection, in a well extending to approximately 700 m (2000 ft). That waste is partially neutralized almost immediately to a pH of 5.5 by dissolution of limestone. That upwelling reconfirms the reality of hydraulic connections between the injection zone and overlying zones (Kaufman et al., 1973; Kaufman and McKenzie, 1975). Belle Glade is located in Palm Beach County, approximately 80 km (50 mi) northwest of Miami and approximately 64 km (40 mi) west of West Palm Beach/Palm Beach, in the vicinity of where Lapointe (1997) documented the proliferation of frondose macroalgal growth associated with four reefs.

The impact of injecting low-pH substances in a carbonate system was elucidated in the conclusion that injection of acidic liquid waste into the permeable carbonate zone of the late Eocene in north Florida "improved" the permeability of the injection zone near the injection well by dissolution of the limestone (Puri et al., 1973). The mere flow of injected liquids will increase dissolution in a carbonate channel, in the absence of chemical dissolution from low-pH substances. Increasing the permeability of the subsurface unit receiving injected waste only ensures more rapid movement of the waste away from the point of injection and potentially toward environmentally sensitive locations.

Secondary sewage-plant effluent reportedly injected into the highly cavernous limestone of early Eocene age at a depth of approximately 1000 m (3000 ft) in the Miami area since approximately 1971 exhibited only "slight bottom-hole pressure increase during injection" (Puri et al., 1973). The most logical reason for this response is the rapid movement (laterally, vertically, or both) of injected wastes away from the area of injection. Meyer (1984) provided a modified illustration (originally presented by Meyer in 1974) of the hydrogeologic section from the Everglades to Bimini, including a stylized injection well extending into the Boulder Zone. Under a best-case scenario, the injected waste would move laterally, into the Straits of Florida. Temperature gradients and data from acoustic televiewer surveys, however, have suggested that the wastes may move vertically via high-angle fractures through the overlying low-permeability "confining zones" into sensitive coastal areas. Because the density of the injected waste is approximately 1000 g $L^{-1}$ and the density of the native fluid (saline water) is approximately 1025 g $L^{-1}$, the native saline water is displaced by the more buoyant injected waste and could experience preferential upward flow (Meyer, 1984). This suggests that the source of the "old" groundwater that Top et al. (2001) identified isotopically may be induced discharge from the effluent injected into the lower Floridan aquifer.

Vecchioli (1979) reiterated the historic single focus of monitoring on potential adverse impacts of injection of "waste liquids" exclusively on "potable-water resources." The "effective aquifer-system monitoring program" he described included no monitoring of sensitive coastal ecosystems to ensure that the wastes are not surfacing in those sensitive areas via preferential flow paths.

Likewise, his monitoring program made no attempt to ensure that the hypersaline water was not being displaced into sensitive coastal areas with sensitive coastal ecosystems. Finally, his monitoring program also failed to include any monitoring of ecologically sensitive "indicator" organisms or ecosystems that could be harmed by such discharges.

Hickey and Vecchioli (1986) summarized the status of the carbonate aquifer system in South Florida as one of the most permeable rock masses in the world, indicating that it was first used as a receiving zone for treated sewage in 1971 (Miami), reportedly at a depth of approximately 1000 m (3000 ft). Since that time, the majority of the Class I (hazardous waste) wells in South Florida use the Boulder Zone; however, the first use of a pressure well for subsurface injection of treated sewage into a saline subsurface zone occurred in Broward County in 1959, at a depth of approximately 400 m (1200 ft).

Records compiled by the Florida Department of Environmental Regulation (now FDEP) in 1981 reported that 52 Class I injection wells and 80 Class II injection wells were in operation in Florida. Class II wells are used for discharge of fluids brought to the surface in connection with conventional oil or natural gas production, for enhanced recovery of oil or natural gas, and for storage of hydrocarbons which are liquid at standard temperature and pressure (Hickey and Vecchioli, 1986). The majority of the Class I injection wells are in Dade, Broward, and Palm Beach Counties, along the southeast coast of Florida just north of the ailing Keys and Biscayne Bay, and in Pinellas County on the west-central coast (Hickey and Vecchioli, 1986). As indicated previously, currently a total of 81 Class I wells inject large volumes of wastes (primarily minimally treated municipal sewage) throughout Florida, with all but two located along the southern coast of Florida (FDEP, 1999). Approximately half of those 81 wells reportedly are violating, or have been predicted to violate federal laws prohibiting migration of the injected wastes (National Archives and Records Administration, 2000).

In what may be the first geological investigation of a deep-well injection site in Florida funded by a nongovernmental organization, 10 of the 17 effluent-injection wells at the Miami–Dade deep-well injection facility were determined to have been constructed improperly, with the lowermost casing failing to extend below the semiconfining unit, into the Boulder Zone. Consequently, the minimally treated effluent injected into those 10 wells actually had been discharging into the upper Floridan aquifer since the time the wells became operational (McNeill, 2000).

That investigation also verified that, in addition to the construction problems, there were no data to verify the actual subsurface migration path(s) beyond the site of the effluent injected into the Boulder Zone (the permitted zone) or the effluent injected above the semiconfining zone (in violation of the permit). The semiconfining unit becomes progressively more shallow to the west, however, which would facilitate westward movement of the more buoyant effluent injected into the saline Boulder Zone. The monitor well in the northwest corner of the site of the active injection wells was the first to detect ammonia in the upper monitor zone. More importantly, the ammonia (a compound contained in effluent) was detected in the northwest corner of the site before the adjacent injection well began receiving effluent. Those findings suggest that considerable vertical and westward lateral migration of the injected effluent had occurred since injection in the initial wells at the Miami–Dade site (McNeill, 2000).

As part of that investigation, the hydraulic conductivity values collected from several studies conducted at the injection site by CH2M Hill were reviewed. That review revealed that the "confining" zone separating the intended injection zone (Boulder Zone) from the upper Floridan aquifer was not nearly as "impermeable" as earlier proposed by CH2M Hill when they assessed the feasibility of injecting effluent in that area. That investigation further documented that CH2M Hill originally estimated a travel time of 343 years for movement of injected effluent upward into the upper Floridan aquifer; however, recalculation of the data by McNeill (2000) revealed upward travel times of 3.2 to 17.5 years, within the range of the time of first detection (11 years). CH2M Hill is the same firm responsible for the deep-well injection facility proposed to begin injecting effluent in Key West in spring 2001, while the wastewater treatment facility processing the effluent for injection is the "sister" organization of CH2M Hill, based on the information available on the city's website (City of Key West, 2000).

The "time of first detection" referenced by McNeill should not be misinterpreted to mean that the effluent had not migrated laterally (beyond the site) or vertically (into the Upper Floridan) prior to the "time of first detection," but simply that the migrating effluent had not been detected in the grossly inadequate monitoring wells prior to that time. If monitoring wells are not in contact with preferential flow paths or the path of the plume, the probability is low that the migrating effluent plume will be detected. There is no indication that the monitoring wells were associated with preferential flow paths or the path of the plume. The fact that there were no reported significant increases in "pressure" at the injection site and the fact that both the lower and upper portions of the Floridan aquifer are highly permeable suggests that there has been rapid migration of the injected effluent both westward (toward the Everglades and Everglades National Park) and eastward (toward Biscayne National Park and the FKNMS), as suggested by McNeill (2000).

The Everglades, the focus of an approximately $8 billion "restoration" effort, is located to the west of the Miami–Dade effluent injection site. Consequently, at least a portion of the severe eutrophication of the Everglades could be due to the injected effluent that has been migrating in a westward direction from the Miami–Dade site since initiation of the injections in those wells in 1983. That updip migration of injected effluent from the Miami–Dade site also may be a significant factor in the decline of trees on tree islands in the Everglades the past few decades and the high nutrient levels and other pollutants associated with the tree islands reported by Orem et al. (2001) and Willard et al. (2001). Neither that potential source of contamination of the Everglades nor the extensive removal of groundwater (e.g., agricultural and municipal withdrawals, including approximately 134 MGD municipal withdrawals for Miami and 7,812,171,000,000 gallons per year withdrawn and piped to Monroe County/Florida Keys) that historically supported the Everglades ecosystems has been recognized or addressed in the Comprehensive Restoration Plan that the $8 billion is funding. The Everglades also includes Everglades National Park and Reservations for Native Americans.

As indicated in Figure 26.4, the Floridan aquifer extends westward to the Gulf of Mexico. It is not unrealistic to presume that a portion of this westwardly migrating effluent also is discharging along the submerged carbonate platform in the Gulf of Mexico. Support for this presumption is provided by DeMaria's observations of "walls" of cold turbid water moving across the sea floor and cold turbid freshwater constantly flowing from many small holes, fissures, and sinkholes in the "porous limestone" bottom north of Key West, as described in the previous section. Additional support is provided by the fact that harmful algal blooms (including red tide) and hypoxia (reduced oxygen conditions) repeatedly occur in the Gulf of Mexico off the coast of South Florida.

Based on Congressional findings, harmful algal blooms have resulted in fish kills, the deaths of numerous endangered West Indian manatees, beach and shellfish bed closures, threats to public health and safety, concern among the public about the safety of seafood, and possibly an estimated $1,000,000,000 in economic losses during the past decade. Harmful algal blooms and blooms of nontoxic algal species may lead to other damaging marine conditions such as hypoxia, which are harmful or fatal to fish, shellfish, and benthic organisms. Excessive nutrient loading to coastal waters is a factor believed to cause hypoxia (Public Law 105-383, Title VI, Section 605, Authorization of Appropriations).

Burkholder and Glasgow (1997), Burkholder et al. (1995), and Glasgow et al. (1995) provided considerable evidence that outbreaks of toxic ambush-predator dinoflagellates in coastal areas of the southern U.S. are linked to anthropogenic eutrophication of coastal waters. Additionally, Glasgow et al. (1995) described responses of humans exposed to the aerosols from the ichthyotoxins associated with blooms of these dinoflagellates (e.g., narcosis, respiratory distress with asthma-like symptoms, severe stomach cramping, nausea, vomiting, and eye irritation with reddening and blurred vision, autonomic nervous system dysfunction, central nervous system dysfunction, cognitive impairment and short-term memory loss, asthma-like symptoms, exercise fatigue, sensory symptoms, elevated hepatic enzyme levels and high phosphorus excretion suggesting hepatic and renal dysfunction, easy infection, and low counts of several T-cell types which may indicate immune system suppression).

Because of the magnitude of this problem in the Gulf of Mexico, a total of $52,250,000 was appropriated for disbursement by NOAA during 1999–2001 for research, education, and monitoring activities related to the prevention, reduction, and control of harmful algal blooms and hypoxia (Public Law 105-383, Title VI, Section 602, Findings). It does not appear that NOAA has considered the inevitable discharge of injected effluent in South Florida throughout and along the margins of the submerged Florida Platform as a factor in harmful algal blooms and hypoxia in the Gulf of Mexico. More specifically, no funding has been provided by NOAA to investigate the role of injected effluent in the algal blooms occurring in NOAA's FKNMS and Florida Bay. Additional potential impacts of anthropogenic nutrient loading in oligotrophic (nutrient limited) coastal waters are described below.

## TOO MUCH OF A GOOD THING

The surface waters surrounding coral reef ecosystems previously have been characterized as clear and low in dissolved and particulate nutrients (Lewis, 1977; Odum, 1990; Odum and Odum, 1955, 1956, 1957). Consequently, coral reef ecosystems have evolved very efficient mechanisms for trapping and retaining nutrients (Muscatine and Porter, 1977). Therefore, even concentrations that are below the level of detection can be ecologically significant to corals (Dustan, 1999). For example, in the coral reef nutrient threshold model demonstrated by Bell (1992), nutrient concentrations as low as 0.006 mg $L^{-1}$ of dissolved inorganic P and 0.014 mg $L^{-1}$ of dissolved inorganic N were found to have adverse effects on coral reefs (Bell, 1992). Similar thresholds for soluble reactive P (0.009 to 0.189 mg $L^{-1}$) and dissolved inorganic N (0.01 mg $L^{-1}$) were reported for macroalgal overgrowth of coral and seagrass habitats over a broad geographic range (Bell, 1992; Lapointe et al., 1992). Likewise, Lapointe (1999) revealed that macroalgal growth was maximal and exponential at very low levels of dissolved inorganic N (0.007 to 0.014 mg $L^{-1}$), summarizing data for macroalgal species from six different genera (see also D'Elia and DeBoer, 1978; Duarte, 1995).

As a comparison, levels of P and N in sewage effluent discharged after receiving advanced wastewater treatment (AWT) and "polishing" by man-made wetlands are 1.88 mg $L^{-1}$ and 6.16 mg $L^{-1}$, respectively (FDEP, unpubl. data). These AWT discharge levels are more than two orders of magnitude above the "adverse effects" levels of P and N for corals.

The research of D'Elia et al. (1981) documented submarine springs and seeps near the reef at the mouth and along the southern and western shorelines of Discovery Bay, Jamaica. They also documented concentrations of 80 µg/L N (primarily as nitrate) in undiluted springwater that was virtually devoid of P. The strong inverse correlation between salinity and N concentration documented at the reef suggested that point-source SGD was contributing to eutrophication of the oligotrophic surface waters of the back reef of Caribbean coral reefs. The flow of the SGD was not measured, but was reported as sufficient to reduce salinity significantly (D'Elia et al., 1981). Kohout (1966) described the lack of consideration given to the influence of submarine springs on coastal hydrology. More than 40 years later, the same lack of consideration appears to be widespread.

A similar phenomenon of SGD associated with coral reefs was documented by Lewis (1987), using seepage meters and miniature piezometers for direct, *in situ* measurements of groundwater seepage flux onto coral reefs at Barbados, West Indies. His measurements confirmed both spatial and temporal variations in SGD. Seepage flux was approximately twice as high in the wet season as in the dry season, with measured fluxes comparable to groundwater discharge estimates based on aquifer models. Nutrient loads, however, were not estimated in that study.

During the twentieth century, a multitude of anthropogenic inputs to the coastal waters in South Florida have tipped the delicate nutrient balance and resulted in too much of a good thing, with respect to available nutrients. Surface water sources of excess nutrients abound, including the extensive network of drainage canals constructed throughout the Everglades and South Florida, that transport stormwater runoff and agricultural drainage to sensitive coastal surface waters. The research referenced above in other areas of the Caribbean basin, however, suggests that nutrients also are entering South Florida's coastal waters via another (subsurface) anthropogenic source. This subsurface connection (the linkage of groundwater and surface water) appears to be resulting in point-source eutrophication and transport of other pollutants at reef sites via SGD. In addition to increasing the biomass of macroalgae around and on the reefs, eutrophication reduces the reproductive capacity of hermatypic reef corals (Tomascik, 1991).

Dillon et al. (2000) noted that the practice of disposing of sewage via shallow injection wells introduces extraordinary amounts of nutrients into the groundwaters of the Florida Keys. They indicated further that if the wastewater plume reaches surface waters rapidly, with little dilution or nutrient uptake, that human and ecosystem health could be at risk and different wastewater disposal methods would be needed in the Florida Keys. One of the objectives of their study was to attempt to determine the amount of dilution that occurred before the contaminated groundwaters reached nearby surface waters. A companion study conducted at the same site sought to determine the fate of the nutrients in the minimally treated effluent injected under the low-flow conditions at that site (Corbett et al., 2000).

The studies were conducted in the Middle Keys (Keys Marine Laboratory on Long Key) to determine the rate of transport, using conservative artificial tracers that were added to a shallow (10 to 30 m) sewage injection well. Dillon et al. (2000) detected what they termed rapid flow (0.20 to 2.20 m h$^{-1}$), reportedly representing flow through solution channels in the karst aquifer. They also observed slow flow (<0.003 to 0.14 m h$^{-1}$), reportedly representing diffusive flow through the limestone's primary porosity.

The "rapid" rate of flow detected at their Middle Keys injection-well site was comparable to the flow rates reported by Dillon et al. (1999) for the septic tank study in the Lower Keys. Another similarity between those two injection well and septic tank studies was the finding at the Middle Keys site that the tracers injected in the shallow wells reportedly remained in the vicinity for months, due to limited circulation in the area. The findings of the Dillon et al. (2000) study have been interpreted to infer that effluent injected into the shallow wells in the Keys is not migrating significant distances from the site and that the majority of nutrients in the injected effluent are adsorbed to/taken up by the carbonate aquifer. Those interpretations are not well founded.

Several design constraints were associated with those two studies, which attempted to determine directions and rates of groundwater transport and the level of dilution/uptake of the contaminated water from effluent injected into the shallow well at that Long Key site. The first constraint was the fact that the samples were collected from observational wells that had not been constructed with the specific purpose of intersecting preferential flow paths that may be present on the site. Consequently, the primary path of the plume from the injected effluent may not have been coincident with the monitoring wells.

The second constraint was that the only location for samples collected from the surface water was a "canal" ("boat basin"?) adjacent to Long Key. Consequently, the injected tracers (and effluent) could have been migrating, via preferential flow paths, and discharging to surface waters at greater distances or in other locations from their single sample point in the canal (as in the study by Paul et al., 2000, described previously in the introduction).

The third constraint was that the volume of effluent injected into the well at that site (2600 L d$^{-1}$) was considerably less than the volume of effluent injected into typical shallow injection wells in the Keys. For example, the shallow injection wells permitted by FDEP at a site proposed for a condominium complex in Islamorada, would (if it becomes operational) introduce approximately 22 times that amount (56,775 L d$^{-1}$) of minimally treated sewage effluent to the highly permeable shallow aquifer. Ironically, the shallow, highly permeable aquifer where the effluent is permitted to be injected extends under the last remaining sea turtle nesting beach in the Upper Keys, located on the opposite side of U.S. 1 from those injection wells. Based on the three constraints described above, the flow conditions at the Long Key effluent-injection site would not be applicable to other effluent-injection sites in the Keys.

A similar study by Dillon et al. (2001) was conducted in the Middle Keys in an attempt to determine nutrient removal due to adsorption from effluent injected under "high discharge conditions." That study, however, was conducted on Key Colony Beach, a man–made island composed entirely of fill material. Responses in artificially created islands cannot be presumed to represent responses in natural geologic formations. Furthermore, samples were collected from wells located without any knowledge of preferential flow paths.

The overriding constraint of those studies was the presumption that dispersion (flow) of the tracers (and effluent) was homogenous and isotropic. That suggests that their tracers radiated outward evenly, in all directions from the point of discharge, with respect to rates and levels of dilution/uptake. In reality, it is known that this is not the case for any site in the Florida Keys (or for any karst aquifer). In fact, Corbett et al. (2000) identified an additional constraint on the dispersal and dilution of the injected effluent. Their salinity profiles of the injection well documented that the injected effluent "moves out laterally at the base of the casing (18 m deep), with little penetration to greater depths." They attributed this lack of downward flow/mixing to the "buoyant nature of the wastewater with respect to the ambient groundwater." Therefore, their inferences that dilution of the tracers (and uptake of nutrients in the effluent) was occurring may have been the result of samples collected along the periphery of the tracer/effluent plume(s). The bulk of the plume may have been flowing through a preferential flow path not intersected by a monitoring well, with little or no dilution/uptake of the tracer (or nutrients and other pollutants in the effluent).

Even if it is presumed that the monitoring wells intersected the flow-path in which the plume was traveling, larger volumes of injected effluent (as in the small facility across U.S. 1 from the sea turtle nesting beach) will increase the rate of flow. Increased flow will increase the diameter and length of the preferential flow paths (recall the documented findings by Kaufman, 1973, described previously). The increasing dimensions of the preferential flow paths will decrease the probability that the injected effluent will come into contact with the aquifer matrix (the carbonate rock of the aquifer that provides the theoretical means by which large amounts of the nutrients are taken up/removed from the effluent). Consequently, the conclusions of those studies (at Key Colony Beach and Long Key), and in particular the conclusion by Corbett et al. (2000) that their "results suggest that nutrients injected in the subsurface are removed rapidly from solution and thus may not have a significant impact on surface waters," are not applicable to the widespread, large-scale aquifer injection of effluent that is occurring throughout the Keys and South Florida.

Additional constraints of the "nutrient uptake" experiment by Corbett et al. (2000) included use of an inorganic phosphate tracer at a concentration of approximately three times the salt content of the groundwater. It would have been extremely difficult for a plume with a tracer of that density to have "risen" in the water column to the point where it could have been accurately detected by their shallow monitoring wells, in close proximity to the injection well, for a monitoring period of less than 5 days (rather than the previous monitoring period of 70 days). Their subsequent experiment, with a nonhypersaline nutrient tracer, was conducted using a single monitoring well. Clearly, the bulk of the plume (with associated nutrients) could have by-passed that single monitoring well. If the carbonate aquifer was involved in active uptake of nutrients from the injected effluent, it would be logical that more nutrient uptake would occur with septic tanks, where the focused volume of effluent is less and the contact time is greater than with injection wells.

Recent findings regarding a similar approach in Western Australia, of injecting nutrient-rich wastewater into the Barst carbonate aquifer of that region, can be used as a model for the Keys. In that case, the wastewater was injected into the aquifer approximately 1 km (0.6 mi) from Cockburn Sound to prevent eutrophication of the Sound associated with openwater discharge. Unfortunately, the groundwater plume discharged to the Sound with minimal reduction of nutrients, contributing to algal blooms and seagrass dieoff. Consequently, efforts now are underway to attempt to extract the nutrient-laden groundwater via pumping wells (Appleyard, 1994; Department of Commerce and Trade, 2000; A. Smith and J. Turner, pers. comm.).

To increase the validity of related research in the future, travel/surface water discharge times should be determined based on methods used by Paul et al. (2000). Then, preferential discharge points should be identified and the sample collection locations should focus on those preferential discharge points to determine if the groundwater discharge points are directing effluent-related pollutants to coral reefs and other areas in the FKNMS. Furthermore, inferences about large-volume effluent injection systems should not be made based on responses of low volume effluent-injection systems. Despite those constraints, Corbett et al. (2000) is correct in noting that the secondary treatment that the sewage is subjected to prior to injection does not remove significant amounts of dissolved nitrogen, phosphorus, heavy metals, or many pathogenic bacteria and viruses (Davis and Cornwell, 1991).

Based on the faulty presumptions described above (including the scientifically unsubstantiated belief that the primary sources of sewage contamination to the surface waters of the Keys are cess pits and septic tank leachate), the FDEP continues to issue permits for injection of minimally treated sewage effluent into the highly permeable karst aquifer system via shallow and deep wells. The well-documented findings of the Cockburn Sound studies, referenced above, suggest that the Keys represent a groundwater remediation site of Super Fund proportion.

As indicated previously, the deep injection wells, known as "Class I" wells, are located primarily along the coast of South Florida. On the east coast, the Class I wells extend from St. Lucie County, south to Dade County (Figure 26.6). The wastes injected into the Class I wells primarily are minimally treated municipal sewage effluent and what is termed "nontoxic liquid wastes." No tests appear to have been conducted to determine the actual toxicity, or other hazardous impact, of any of the deep-well injected wastes to sensitive coastal ecosystems in South Florida. Those wastes theoretically are injected into the Boulder Zone, the lower zone of the Floridan aquifer system. Repeated attempts to obtain information from the FDEP regarding the total number and locations of the shallow injection wells (Class V) in South Florida, and more specifically the Florida Keys, were unsuccessful. Those wells primarily receive minimally treated sewage effluent from nonmunicipal sources.

The injection of so-called "nontoxic wastes" into the Floridan aquifer system began in 1943, and injection of treated municipal sewage began in 1959 in Broward County. Injection of treated municipal sewage into the Boulder Zone of the Floridan aquifer system began in Dade County (Figure 26.6, site 50) in 1971 (Meyer, 1989). Currently, treated sewage is injected into the Boulder Zone of the Floridan aquifer system at the Black Point facility in Miami via 25 wells (E. Shinn, pers. comm.). As indicated previously, this injection site is near the northern boundary of the Florida Keys National Marine Sanctuary.

The sole regulatory focus governing those injected wastes has been the potential contamination of current sources of potable groundwater from the upward migration of the injected effluent. No consideration has been given to potential impacts on coastal ecosystems that may be linked to submarine discharge of these substances. As indicated previously, however, the saline Boulder Zone extends under Biscayne Bay, the Straits of Florida, the Bahama Islands, the Everglades, the Florida Keys, Marquesas Key, and the Dry Tortugas complex, terminating in the Atlantic Ocean to the east and in the Gulf of Mexico to the west (Kohout, 1965). Consequently, the low-salinity sewage effluent that is injected into Boulder Zone will be less likely to mix than to stratify in a concentrated layer on top of the saline water. That concentrated layer then could be forced upward into sensitive coastal ecosystems such as coral reefs and the Everglades via SGD, as well as displacing saline groundwater into those sensitive coastal ecosystems.

The term "Boulder Zone" is a misnomer, suggesting that large stones are present. In reality, the Boulder Zone reportedly is characterized by cavernous dolomite and limestone (Kohout, 1965). The highly fractured dolomite breaks off in boulder-sized chunks that complicate the drilling process. Because of the large cavities in this zone, permeability is considerably greater than in the overlying upper Floridan aquifer and the Key Largo Limestone. Kohout (1965) reports that this increase in permeability generally:

> ... occurs 900 feet below the top of the Floridan aquifer at an average depth of 1900 feet below sea level in southern Florida. ... Borehole photographs of this zone show many caverns 2 to 6 feet in height below 1900 feet, and one large 90-foot cavern at a depth of 2600 feet below sea level. The Boulder Zone caverns are developed through a vertical range of fully 1000 feet from 1900 to 2900 feet below sea level in southern Florida.

The saline water in the lower Floridan aquifer is under hydrostatic pressure. This hydrostatic pressure results in artesian flow at the Continental shelf "from depths more than 1000 feet deeper than those reported in inland areas where the piezometric surface is higher" (Kohout, 1965). In northeast Florida, reductions in hydrostatic pressure in the overlying upper Floridan aquifer, due to extraction of fresh groundwater for municipal and other uses, have resulted in upward leakage of saline water from underlying zones, including the lower Floridan aquifer. The avenues for this upward leakage were vertical fractures that interconnect the upper and lower Floridan aquifer with underlying zones, and penetrate the intermediate "confining" unit that lies between the upper Floridan and surficial aquifers (Krause and Randolph, 1989; Spechler, 1994; Spechler and Phelps, 1997).

It has not been determined how deep these vertical fractures extend or how widespread similar vertical fractures are throughout Florida. However, recent seismic data and natural gamma logs from wells on the mainland and along the submerged platform of the Atlantic Coast in the Indian River Lagoon and barrier island near Vero Beach revealed a southeast dipping of the Hawthorn Group in response to subsidence or dissolution in the underlying Floridan aquifer system. The Hawthorn Group is regarded as the "confining" layer that separates the upper Floridan aquifer from the surficial aquifers and surface waters. Fluid migration, rock movement, and dissolution along a deeper fault zone were identified as possible mechanisms for the subsidence (Kindinger et al., 1997). That study confirmed the presence of structural features along the submerged platform that could serve as "point-recharge" locations for induced recharge. Those same features could serve as preferential flow paths for upward movement of injected wastes to equalize pressure differentials between overlying aquifers. This concept of induced discharge of injected wastes (including effluent) has not been addressed by the agencies that issue permits for the wells and "protected" water quality in Florida's coastal waters, including the FKNMS.

It is reasonable to assume that similar vertical fractures or vertical dissolution and collapse features exist in South Florida. In fact, a large (~600-m wide) sediment-filled depression (with characteristics similar to Red Snapper Sink east of Crescent Beach, FL) was discovered in 1991, between Garden Cove and Elbow Reef near Key Largo. Although the total depth of the Elbow Reef sinkhole has not been determined, the lime mud that has filled in the large cavity has been confirmed to extend to a depth of at least 55 m (180 ft) below the sea floor (Lidz et al., 1997; Shinn et al., 1996). High-resolution seismic-reflection profiles across Hawk Channel and White Bank suggest that numerous other small vertical dissolution features or shafts occur in association with the reef tracts. Many of the reefs in that area no longer are growing vigorously and are in a state of decline. The declining reefs are located both interior of the platform margin and along the outer edge of the platform margin (Lidz et al., 1997).

It is realistic to hypothesize that prior to groundwater mining along the southeastern coast of Florida, pristine, low-salinity, low-nutrient groundwater of constant temperature flowed from the submerged platform (e.g., Hawk Channel, White Bank, and Florida Bay) as both diffuse and direct discharge. It also is realistic to hypothesize that this flow provided essential elements, including limited quantities of nutrients capable of sustaining thriving coral reefs, but insufficient to allow rampant growth of macroalgae and phytoplankton.

That hypothesis can be expanded to envision the same diffusion and flow paths introducing excess nutrients and toxic or otherwise hazardous compounds from sewage effluent (based in part on the reduction in hydrostatic pressure in the freshwater aquifer zone(s) and the increase in hydrostatic pressure in the saline aquifer zone(s)). As indicated previously, the low-salinity effluent may form a distinct layer along the top of the more dense, saline water in the lower Floridan aquifer. This would promote preferential expelling of the contaminated layer through vertical structures, such as fractures and dissolution/collapse shafts. Even if the injected effluent was saline, eliminating the density difference, the mere volume of the injected effluent and Swiss cheese nature of the aquifer would ensure that the injected effluent would reach surface waters as induced discharge via displacement.

The conceptual model of the large sinkhole associated with White Bank near Key Largo, shown in Figure 9 of Lidz et al. (1997) illustrates the typical horizontal dissolution channels that are connected to these dissolution/collapse features. The authors indicate that the sinkhole probably is part of an underground cave system that formed as freshwater dissolved the limestone, possibly during a period of lower sea level. Those horizontal channels may continue for considerable distances, providing preferential flow from the coast to reef systems in the bay.

Previous authors have identified antecedent topography and the position of paleosea level as the primary factors controlling initiation and development of coral reefs in the Florida Keys (Shinn et al., 1989; Lidz and Shinn, 1991). Pristine, low-salinity, low-nutrient SGD of constant temperature should be added to the list of primary factors. The historic SGD may have been a critical source of essential elements, including nutrients, in limited amounts capable of sustaining coral reef health and growth but insufficient to promote proliferation of macroalgae and phytoplankton.

As an example, the question has been posed as to why reef-building corals do not grow near the shore in South Florida (Shinn, 1999). Under present-day conditions, surface discharge of stormwater with associated turbidity and pollutants probably is a major factor precluding the establishment of reef-building corals. However, from an historic standpoint, the answer may be the absence of sufficient localized discharge of pristine, low-salinity, low-nutrient groundwater of constant temperature in those sensitive coastal areas to sustain reef-building corals.

Reefs inshore of the outer barrier in the Florida Keys (e.g., Grecian Rocks, Key Largo Dry Rocks) reportedly exhibit approximately 15 m (45 ft) of vertical growth during the past 6000 years, to keep pace with the rising sea level. Deeper reefs exhibit only a thin veneer of living reef organisms covering an 80,000-year-old reef ridge (Shinn, 1999). Limited available light and lower temperatures probably are major factors influencing the growth of that coral. Both light availability and temperature, however, can be influenced adversely if historic pristine, low-salinity, low-nutrient SGD of constant temperature is replaced by nutrient-laden SGD derived primarily from injected effluent, or if the historic SGD ceases because of groundwater mining on the mainland. Consequently, the historic preferential distribution of essential elements to the referenced reefs inshore may have been another factor favoring their growth. Lidz et al. (1991) report that thick Holocene reefs also have been found in similar off-bank settings, including the northern upper Florida Keys (Carysfort Reef), the reef located farther north off Ft. Lauderdale, FL, and the outer reef off Belize at Carrie Bow Cay.

If coral reefs are associated with areas of historic erosion and dissolution in the underlying carbonate platform, where historic preferential flow of pristine SGD was occurring, those areas could have been highly localized sources of both essential (although limited) elements and vertical mixing due to the discharge of fresh water from beneath the denser, saline coastal waters. Unfortunately, groundwater mining and excavation of ditches and canals on the mainland of South Florida have diverted that original subsurface influx of pristine, low-salinity, low-nutrient groundwater of constant temperature and replaced it with pulses of contaminated fresh water entering the system as surface water from the Everglades (Brand, 2000). Therefore, the mixing mechanisms are dramatically different from historical conditions and stratification is promoted. Furthermore, if underlying areas of carbonate erosion and dissolution are associated with coral reefs, the localized SGD now may be conveying the contaminants that are being injected via deep wells on the mainland. More diffuse SGD could be spreading those wastes throughout the Florida Platform and emitting the remainder from the margin of the platform to become components of upwellings, as described by Simmons and Netherton (1987).

Simmons (1992) suggested the possibility of nutrients reaching reefs through groundwater movement and seepage, based on his discovery of low-salinity water seeping from bottom sediment in water 40 m (131 ft) deep off Key Largo. The outlier reefs associated with the lower reefs described by Lidz et al. (1991) appeared to be pre-Holocene in origin and associated with Pleistocene beaches that formed terrace-like features. They were unable to hypothesize why the "south-facing environment of the Florida Platform was optimum for development of outlier reefs," although they did suggest that paleodrainage patterns across and through the porous platform limestone may have influenced local outlier-reef growth.

In November/December 2000, the SGD Working Group of the Scientific Committee on Oceanic Research (SCOR) and Land-Ocean Interactions in the Coastal Zone (LOICZ) conducted a SGD Intercalibration Experiment in Cockburn Sound, Western Australia. That area is underlain by the Tamala Limestone formation, a karstic carbonate aquifer similar to the Floridan aquifer system. At the onset of the experiment, subsidence features were predicted to be associated with terrace features landward and seaward of Garden Island (Bacchus, unpubl. data). Garden Island (an outcrop) forms the western boundary of Cockburn Sound and is located approximately 6 to 9 km (4 to 6 mi) west of the mainland. Although SGD of low-salinity water near the vicinity of the Intercalibration Experiment had been documented at least 15 years before (Johannes and Hearn, 1985) and the ecological significance of such discharge had been described even earlier (Johannes, 1980), the role of karstic preferential flow paths in the vertical discharge of low-salinity groundwater does not appear to have been addressed.

During the Intercalibration Experiment, seismic transects were run normal to the shoreline from the mainland, with one transect extending to Garden Island, using equipment operated by Stieglitz (see Stieglitz and Ridd, 2000, for a description of the sonar equipment used in the Intercalibration Experiment). At the point where the Garden Island transect intersected the approximate 20-m (65-ft) contour (presumed to be a terrace), a subsurface feature was detected that was similar to subsidence features that have been identified in Florida, including South Florida, on both exposed (the mainland) and submerged areas of the carbonate platform (SCOR Working Group, unpubl. data). Time and weather constraints during the experiment precluded a more extensive investigation of the subsidence-like feature during the experiment. However, the subsurface feature that was identified near Garden Island could have represented a potential preferential flow path for historic vertical discharge of pristine, low-salinity, low-nutrient groundwater of constant temperature.

Outcrops of subsurface strata appear to occur along terrace margins. Historically, those outcrops could have provided preferential flow of essential elements, including low concentrations of essential nutrients for the reefs (via SGD, as suggested by Simmons 1992) in an otherwise oligotrophic sea surrounding the reefs. For example, extensive "vents" occur in the vicinity of Tavernier Key along the same contour where the shallow reefs are aligned in the Florida Keys. A thick layer of organic material is associated with those vents, suggesting that contour line represented a densely vegetated paleoshoreline. Recent discharge of groundwater from these "vents" has been observed at least since 1996 (D. Trelewicz, pers. comm.) and was confirmed in 2000 (Bacchus, unpubl. data). There is no indication that a systematic attempt has been made to identify those types of preferential flow paths or to monitor the frequency, volume, and chemical characteristics of the discharges.

It is important to note that the injected effluent typically has been treated with chlorine, generally with no dechlorination. When chlorine comes into contact with organic matter it can form compounds known as trihalomethanes (chloroform, bromoform, dibromochloromethane, and bromodichloromethane). Those compounds have been classified as "probable human carcinogens." People who drink water with trihalomethanes have an increased risk of getting various types of cancer, including liver and kidney cancer. Approximately half of the SGD areas of direct discharge observed during a recent reconnaissance had thick layers of organic material associated with them (Bacchus, unpubl. data). Previous studies also have documented organic layers within living coral reefs associated with the Florida Keys. If direct discharge of injected effluent is occurring via preferential flow paths, such as those with associated organic layers, organisms exposed to this water could experience significant adverse impacts (e.g., the kidney tumors in sea turtles referenced previously).

The live coral at Cheeca Rocks has exhibited a rapid decline over the past 10 years (Bacchus, unpubl. data; P. Yananton, pers. comm.). Cheeca Rocks lies approximately 2 km southeast of the midpoint of Upper Matecumbe Key. It is one of several shallow, inner reefs established along the same bathymetric contour (approximately 4 m), but southwest of the Tavernier Key seafloor vents.

In conjunction with the decline of coral reefs at Cheeca Rocks, a proliferation of macroalgae has become established on both the declining and the recently dead coral throughout Cheeca Rocks. It does not appear that an extensive monitoring program has been established at Cheeca Rocks (or other declining reefs in the NOAA Sanctuary) that would permit a meaningful evaluation of any changing water quality conditions that may be associated with the death of the corals. Indicator species, however, can provide a wealth of information as "integrators" of environmentally significant occurrences that have not been monitored or are difficult to detect.

Certain species of algae have been used as indicators of waters polluted with treated or untreated domestic sewage for more than 40 years. It is well documented that the number and species of algae (and other organisms) present in a body of water change once the water becomes contaminated with domestic sewage or effluent (Palmer, 1959). The original focus regarding algae as indicators of nutrient contamination of our water bodies was on surface water. However, for at least 20 years, the potential has been recognized for anthropogenic nutrient loading of coastal areas to occur via groundwater, with concomitant responses (blooms) by algae (Sewell, 1982).

Many of the species indicative of the types of polluted waters referenced above are blue-green algae, with species of *Oscillatoria* being one of the common pollution indicators (Palmer, 1959). Ironically, the mass mortality of the sea urchin *Diadema antillarum* in the Caribbean was associated with a prevalence of *Oscillatoria*, based on reports from St. Croix for the period of December 1983 through September 1984 (Carpenter, 1985). That coincidence raises the question of whether domestic sewage effluent may have been a factor in both the dieoff of the sea urchins and the composition (and increased growth) of the algae. An extensive discussion of the responses of submerged aquatic vegetation to nutrient loading is provided by Duarte (1995). Carpenter (1985) also noted that the red macroalga *Ceramium bysoideum* was another species that increased in abundance at St. Croix as the sea urchin dieoff progressed.

The largest and most widespread of the macroalgae covering the Cheeca Rocks reef in late September 2000 was identified as *Ceramium corniculatum*. Samples of that filamentous red alga examined by staff at the Smithsonian Institution in Washington, D.C., were reported to be notably longer than other specimens of that species that routinely are encountered by Smithsonian algal specialists. They have collected and identified algae throughout the Florida Keys, as well as the rest of the world, for many years. The robust growth form of that macroalgal species at Cheeca Rocks implied the availability of nutrients in excess of those typically occurring under natural conditions (B. Brooks, pers. comm.).

Another of the macroalgae growing at Cheeca Rocks during that time period when the dense stands of *C. corniculatum* were observed was the calcareous green alga identified as *Halimeda simulans*. That macroalgal species often is associated with nutrient-rich areas (B. Brooks, pers. comm.).

A final algal indicator proliferating throughout Cheeca Rocks during that time period was the cyanobacteria (primitive blue-green algae) *Schizothrix*. Several undetermined species of *Schizothrix*, in addition to *S. calcicola*, occurred within those blue-green algal masses (B. Brooks, pers. comm.). Both the color and texture of those algal masses resembled soggy matzo balls. Specimens ranged in size from slightly larger than a golf ball to approximately the size of a baseball. With few exceptions, those cyanobacterial masses were growing in the center of patches of dead coral.

Filaments of cyanobacteria, like the *Schizothrix* proliferating at Cheeca Rocks, reportedly trap sediments (B. Brooks, pers. comm.) which can increase damage to living corals, as discussed previously. The dead zone of coral at Cheeca Rocks surrounding most of the *Schizothrix* masses was large (approximately 0.5-m diameter), with respect to the size of the cyanobacteria. That observation suggested that some factor other than sediment trapped by the cyanobacterial masses was killing the coral. Cyanobacteria, in addition to other turf algae, also reportedly repopulate any surface that results from the dieoff of coral or coralline algae on a reef system (B. Brooks, pers. comm.). Some of the cyanobacterial masses, however, appeared to be growing on coral with no dead tissue visible at the immediate point of contact with the cyanobacterial mass.

An additional noteworthy observation was that a considerable amount of exposed carbonate substrate was present at Cheeca Rocks that appeared to be available for colonization by cyanobacteria. The cyanobacterial masses, however, appeared to be confined to the dying corals. Furthermore, the blue-green algal masses were located exclusively in the center of the dead coral area, suggesting the potential for chemotactic interaction between the cyanobacterium and the dying coral. Finally, cyanobacteria reportedly are "early colonizers" of stressed reefs and are resistant to chlorine (B. Brooks, pers. comm.). The minimally treated effluent that is injected, via shallow wells, into the Key Largo Limestone formation throughout the Keys (including Upper Matecumbe Key) is dosed with chlorine immediately prior to injection. The chlorinated effluent reportedly does not undergo dechlorination prior to injection into the shallow (or deep) injection wells.

The influence on the reefs of injected effluent (laden with nutrients, carbon, chlorine, and other pollutants) discharging preferentially from the sea floor, potentially in close proximity to the coral reefs, does not appear to have been evaluated in the FKNMS. Miller et al. (1999), however, conducted an *in situ* experiment that may have provided inadvertent insight into aspects of those responses.

The intended purpose of their study was to evaluate the effects of nutrients and large herbivores on reef algae. Their source of nutrients, however, was slow-release fertilizer (tree) spikes anchored in cinder blocks under bare carbonate substrate (quarried Pleistocene coral rock). Those treatments were compared with control substrate slabs without spikes. Although not addressed in the paper by Miller et al., the fertilizer spikes reportedly contain chlorine to suppress the growth of algae on the fertilizer spikes, so that release of the nutrients to the target trees will not be hampered. An independent experiment conducted in a small pool, using the same brand of fertilizer spikes used by Miller et al. (1999), revealed the release of chlorine from the fertilizer spikes at concentrations comparable to those used to control algae in swimming pools (B. Brooks, unpubl. data). That finding suggests that the Miller et al. (1999) study was, in fact, an evaluation of the effects of chlorine on reef algae.

Not surprisingly, Miller et al. (1999) documented some results that they had not predicted, including lower values for most categories of algae on the "nutrient-enriched" substrates. In fact, the filamentous cyanobacteria was the only category of algae in their experiment that was significantly stimulated by "nutrient" enrichment (Miller et al., 1999). This is consistent with the reputation of cyanobacteria of being more tolerant of stressors, including chlorine, than other macroalgae (B. Brooks, pers. comm.). Those results also suggest that preferential discharge of chlorine-laden injected effluent may be occurring in the immediate vicinity of Cheeca Rocks, and may be responsible not only for the death of the corals, but also for the proliferation of the blue-green algal masses.

The *in situ* study by Miller et al. (1999) had additional complications, including their selection of an offshore reef as the site for their experiment. Pickles Reef, one of the offshore reefs at Key Largo, was selected based on the presumption that offshore reefs were least likely to be exposed to anthropogenic nutrients from the mainland. Unfortunately, their presumption considered only surfacewater transport of the nutrients, and ignored groundwater transport and induced discharge of injected effluent (e.g., from the Miami–Dade facility and throughout the Keys). Based on published literature describing SGD along the platform margin (addressed previously) and observations of local residents (e.g., D. DeMaria, pers. comm.), offshore reefs can exhibit symptoms of more severe eutrophication than reefs closer to shore.

An additional problem with the study by Miller et al. (1999) was the fact that ambient nutrient levels were not monitored continuously throughout the experiment at their offshore reef site. Only periodic ambient data from Molasses Reef and Conch Reef (4 km NE and 6 km SW of Pickles Reef, respectively) were available to supplement ambient nutrient samples at the site that they collected on days 1, 9, 24, and 41 of the experiment. Clearly, periodic pulsed releases of nutrients from induced discharge of effluent could have occurred at the site of their experiment without their knowledge. In fact, for day 24 ambient samples from their experiment, both total inorganic N and soluble reactive P exceeded the threshold levels posed by Bell (1992) and Lapointe (1997) for nutrient saturation of frondose macroalgae.

One of the most valuable findings from the study by Miller et al. (1999) may have been the documentation that growth of cyanobacteria (vs. other categories of macroalgae) apparently was favored significantly in the presence of chlorine, even though the additives were detectable only within 1 cm above the experimental substrate. The growth advantage of cyanobacteria also may have been related to the presence of nutrients that exceeded the saturation levels for other macroalgae (see Lapointe, 1999). Those findings further suggest that induced discharge of injected effluent in the immediate vicinity of the reefs could disperse through the reefs and result in severe adverse impacts, after which concentrations of contaminants could then be reduced to levels that resembled ambient conditions.

Averaged chlorophyll $a$ concentrations, from samples collected quarterly beginning in 1996, show a "hot spot" of high chlorophyll $a$ in the vicinity of Cheeca Rocks. Those data also show concentrations of chlorophyll $a$ that are higher than in surrounding areas in the vicinity of outer reef tracts (see Figure 6 of Brand, 2000; and Figure 2 of Brand, 2002). Ironically, the largest concentrations (25 and 15%) of "drifters" (simulating current-transported organisms) released at Riley's Hump in the Tortugas were recovered in the immediate vicinity of Cheeca Rocks. Those drifters were released in conjunction with the spawning (the release of the early stage) of mutton snapper (NOAA, 2000).

Those drift data suggest that the largest concentration of the early life stage for at least one commercially and recreationally important fish species in the Florida Keys is transported to a location where coral decline and death are severe, a chlorophyll $a$ "hot spot" has been identified, and indicators of pollutants, including chlorine, are present. The current conditions in the vicinity of Cheeca Rocks suggest a potential life-threatening situation for the larval fish that are being transported now. The destination of the majority of the "drifters" referenced above (simulating transport of young mutton snapper from the Tortugas Marine Reserve) also coincided with the area where operation permits recently were issued by FDEP for numerous new injections of sewage effluent into the shallow aquifer. Finally, the data suggest that the provisions of the Tortugas Marine Reserve spawning locations will not protect this species, particularly if at least 40% of its young are transported to a potential pollution "hot spot" near Cheeca Rocks.

Riley's Hump is known for its richness of fish and other marine life, in addition to being the location of the adult snapper producing the larval fish. Riley's Hump and associated Dry Tortugas and Tortugas Bank are located west of Key West and appear to be a relict reef (NOAA, 2000). Due to the relict reef status, this entire environmentally critical area is predicted to have preferential flow paths similar to those observed and inferred at other coral reefs. In fact, SGD recently was documented from a submerged carbonate outcrop (e.g., relict reef) off the coast of North Carolina, providing further evidence for present-day movement of groundwater from the mainland to, and discharge from, submerged areas of the platform (S. Skrabal, pers. comm.). Those observations fortify the concerns regarding preferential flow to and discharge from Riley's Hump and other areas of the Dry Tortugas of municipal effluent injected into deep wells in Key West.

The observation in 1996 of a plume of foul turbid water emanating from the base of Carysfort Reef on the ocean side and rising to the surface (C. Mitchell, pers. comm.) provides additional evidence suggesting that groundwater, laden with contaminants and transported from the mainland, may be discharging preferentially at various "protected" reef tracts in the Florida Keys. The source of the foul turbid water could have been the 110 MGD of minimally treated effluent injected into the Boulder Zone at the Miami–Dade Black Point injection facility near Biscayne Bay; however, numerous shallow-well effluent-injection sites occur throughout Key Largo and the Upper Keys. Those shallow wells also could be the source. Such discharges may be highly irregular and unpredictable at any specific location due to the following factors: (1) the potentially large numbers of preferential flow paths present, (2) the tendency for new flow paths to develop or become interlinked as prolonged injection increases dissolution of the karstic aquifers, and, (3) pulsed or other pressure changes prior to and following discharges of injected effluent into coastal waters via preferential flow paths. It is important to note that Carysfort Reef was the first reef where Dustan encountered the "white plague" coral disease (discussed previously). That disease was rare in 1974, but has reached epidemic proportions now (Dustan, 1999).

Swart et al. (2000) attempted to use coprostanol (a sterol associated with sewage) to detect the presence of sewage from septic tanks and outfalls in sediments of the FKNMS, as well as nitrogen isotopic composition ($\delta^{15}N$ levels), to determine the origin (e.g., recycled municipal waste) of nitrogen in sedimentary organic material as an indicator of the source of nitrogen in the corals and other biota of the FKNMS. Their study was based on the presumption that septic systems were a "major source of nutrients in the Florida Keys," in addition to runoff, outfalls, and live-aboard boats (no citation provided). Based on their hypothesis that nutrients from those sources were being transported to coral reefs via surface water, they established a series of transects along which bottom sediment and suspended particulate samples were collected throughout the FKNMS. There is no indication that their sample sites were established in conjunction with any preferential discharge points for groundwater in the sea floor.

They concluded that the nitrogen cycle is not understood sufficiently in coral reef systems to draw any conclusions regarding $\delta^{15}N$ levels and the source of nitrogen, and that the use of $\delta^{15}N$ to indicate levels of pollution in coral reef environments is not recommended. They further concluded that, although sewage sterols (e.g., coprostanol) and other domestic pollution markers (e.g., trialkylamines) are useful in assessing surface transport of associated pollutants to open sea, the instability of coprostanol in oxic environments may jeopardize the validity of the use of this tracer in coral reef environments (Swart et al., 2000).

It is important to note that, of the 50 locations in the Keys sampled in their study, Anne's Beach exhibited one of the highest concentrations of coprostanol and cholesterol (indicators of human sewage). The levels at Anne's Beach were similar to, but higher than levels of those compounds in the Doctor's Arm Canal. Ironically, Anne's Beach is an extensive undeveloped stretch of naturally vegetated beach lacking septic systems, cess pits, and related sources of sewage that were the focus of the Swart et al. (2000) study. The source of those indicators of human sewage identified in that study was not known (P. Swart, pers. comm.); however, those contaminants may represent a threat to sea turtles at the last remaining sea turtle nesting beach in the Upper Keys, north of Anne's Beach, as well as to humans.

An intimately related point is the fact that their study was designed to test the hypothesis that effluent was being transported, via surface water, from the sources of effluent referenced in their study. Their hypothesis disregarded the potential role of effluent that is injected into both shallow and deep wells and transported/discharged to the coral reefs as groundwater via preferential flow paths. Any effluent from sources such as septic systems, runoff, outfalls, and live-aboard boats (anchored at shore) would experience significant mixing, dispersion, and dilution in surface waters prior to reaching coral reefs (if effluent from such sources are transported to coral reefs).

Conversely, injection wells concentrate large volumes of effluent into a small vertical pipe, which focuses the effluent through existing preferential flow paths. The diameter and extent of these flow paths increase with extended use (as documented by Kaufman, 1973, and described previously). Those ever-enlarging preferential flow paths function as natural pipelines that can transport injected effluent rapidly, with virtually no mixing, dispersion and dilution, to sensitive areas such as coral reefs. For example, an operational shallow injection well is located on the opposite side of U.S. 1, in proximity to Anne's Beach. The fate of effluent injected into the shallow well near Anne's Beach should be a logical starting point for future investigations regarding the source of the high levels of the human sewage markers at Anne's Beach.

Previously, the constraints associated with some of the studies that have been conducted in the Keys (e.g., Corbett et al., 2000; Dillon et al., 2000, 2001) were described. The applicability of that type of research could be improved in the future by identifying preferential discharge points and collecting samples from those locations to determine if the identified groundwater discharge points are directing effluent-related pollutants to coral reefs and other areas in the FKNMS. In fact, such documentation already exists for the Lower, Middle, and Upper Keys (e.g., Saddlebunch Keys, Marathon, Long Key, and Key Largo), indicating that the primary direction of flow from shallow effluent-injection wells is in a southeasterly direction, toward the coral reefs of the Keys (Paul et al., 1995, 1997, 2000; Reich et al., 2002). Additionally, tracers added to shallow effluent-injection wells in the Lower Keys in one of the studies by Paul et al. (2000) verified that injected tracers entered surface waters only at the most remote sample site, which was a creek channel. In a subsequent experiment at the same injection well, the injected tracers surfaced again from the creek channel, then from a submerged sinkhole (Paul et al., 2000).

The findings by Paul et al. (2000) are relevant to the observations by Reich et al. (2002) that Rhodamine WT dye injected into wells 6.1 m (20 ft) deep (in both the bay and the ocean sides of the Keys) was dispersed from small solution holes up to 3 m (10 ft) away during injection, while dye injected into wells 13.6 m (44.6 ft) deep did not show a similar response. This should not be interpreted to mean that fluids injected at those deeper wells (or effluent-injection wells) would not reach surface waters. In fact, those results suggest that either the deeper wells where they injected the dye were not connected to preferential flow paths or that the dispersal of the injected dye was delayed beyond the observation period or the location of observation. Apparently no geophysical analyses were conducted in the vicinity of the two study areas prior to the construction of the central (injection) wells and the surrounding observation wells.

The relationship between karst features (including natural channels and sinkholes) on the exposed (terrestrial) portion of the platform and submerged portion of the platform recently was described by Faught and Donoghue (1997). They used those submerged terrestrial features in the Gulf of Mexico to identify potential sites of concentrated archeological relicts. Submerged pale-ochannels are not unique to the southeast Atlantic coast. More recently, Stieglitz and Ridd (2000) identified seafloor depressions that appear to be paleochannels on the central Great Barrier Reef Shelf near Townsville, in Australia. Information compiled from interviews with local fishermen indicated those features are (1) approximately 10 to 20 m (33 to 66 ft) in diameter, (2) areas of enhanced shrimp and fish activity, (3) regarded as freshwater springs, (4) located almost exclusively off the coast from relatively large estuaries, and (5) at depths of approximately 20 m (66 ft). Clusters of approximately 100 of these depressional features were verified using side-scan sonar, seismic monitors, and divers. The depressions were aligned along the 20 m isobath (the seaward extension of the Holocene sediment wedge), parallel to the mainland shore and associated islands. Those seafloor depressions exhibited numerous horizontal "caverns" extending from the perimeter at a depth of approximately 4 m (13 ft) below the sea floor (Stieglitz and Ridd, 2000), similar to the caverns associated with the submerged Key Largo sinkhole that was described previously.

It is logical to presume that paleochannels serving as preferential flow paths, discharging ground-water as described above, are associated with the Shark River and Taylor River Slough. It also is logical to presume that such paleochannels could account for the discharge of groundwater contam-inated with effluent from injection wells, as well as agricultural pollutants from the Everglades similar to those noted by Brand (2000, 2002) that enter Florida Bay via discharges of surface water. Furthermore, it is logical to presume that the "many small holes (~15-cm diameter), fissures, and sinkholes" that DeMaria observed in the "porous limestone" bottom north of Key West, where very cold "freshwater is constantly flowing" may be associated with paleochannels of the Shark River.

Finally, it also is logical to presume that Black Creek, immediately adjacent to the Miami–Dade municipal effluent injection facility, is associated with geologic features, such as fractures (south-west-trending) or paleochannels, that also may function as a preferential flow path. If such preferential flow paths occur at that location, the minimally treated effluent could be transported to the adjacent sensitive coastal ecosystems of the FKNMS. Unfortunately, videotapes of open-ocean effluent discharge substantiate the fact that large numbers of marine organisms are attracted to the nutrient/contaminant-laden source of freshwater, even in the absence of any topographic relief of the sea floor (Broward County, unpubl. data).

Discussions have occurred within the federal and state regulatory agencies regarding future intentions to "upgrade" the injected sewage to advanced wastewater treatment (AWT) standards, including what has been termed "drinking water standards." The most obvious problems associated with the continued aquifer injection of effluent in Florida (even at AWT or "drinking water standards") have been described previously and include: (1) the documented adverse impacts of nutrients at AWT levels (even "polished" AWT) on coral reef ecosystems; (2) the failure of drinking water standards to require monitoring, regulation, or preclusion of compounds known to be (or potentially) harmful to organisms (e.g., nonylphenol and its breakdown products), including South Florida's sensitive coastal ecosystems; (3) the fact that additional injections merely will force all of the residual, more nutrient-laden effluent out of the aquifers; and (4) the continued dissolution of the carbonate aquifers receiving the injectate.

## IN THE HEAT OF THE MOMENT

Porter et al. (1999) describe warm, hypersaline conditions that persisted in Florida Bay beyond the recent period of below-average rainfall in South Florida. They also described the transport of this warm, hypersaline water through channels between the Florida Keys and the Long Key viaduct to Florida Keys coral reefs. The adverse impacts to corals exposed to elevated temperatures are summarized by Porter et al. (1999), in conjunction with the results of their experiments. Specifically, they found that the exposure of coral to a 3°C elevation in temperature (with no change in salinity) resulted in significant reductions in both net photosynthesis and the ratio of net photosynthesis to respiration within 6 hours of exposure. Under those conditions, all experimental organisms died within 12 hours. Their experiments were conducted without introduced pathogens, and the rapid death of the corals suggested that pathogens were not a factor in the rapid death of the corals.

Kushmaro et al. (1996) have shown that elevated water temperatures enhance the growth and pathogenicity of the *Vibrio* bacterium recently identified as a causal agent in bleaching of some coral species. Several species of *Vibrio* are enteric pathogens associated with human sewage. Although large-scale climate events can elevate coastal water temperatures, it is conceivable that perturbations of SGD also are capable of resulting in small-scale, discrete areas of elevated water temperatures that may be significant physiologically and pathologically.

The highest infection frequency of corals at Carysfort Reef off Key Largo occurred in the late summer and early fall of 1975, when water temperatures reached the yearly maximum between 29 and 30°C. Overgrowth by algae also was observed, and the declining vitality of the corals was compared to that observed at the once pristine reefs of Kaneohe Bay, Hawaii (Dustan, 1977). It is important to note that in Hawaii aquifer injection of sewage (generally only secondarily-treated) began about the time the practice was initiated in Florida. However, regulation of injection well activities in Hawaii did not begin until July 1984 (C. Hew, pers. comm.).

During a 1987 study by Lapointe et al. (1990), less variation in the annual ranges of temperature (23 to 29°C) and salinity (0.0 to 0.1 ppt) was observed in the shallow groundwater of the Key Largo Limestone aquifer at the Keys Deer National Wildlife Refuge (KDNWR) control site than in adjacent Florida Bay surface water (19 to 34°C and 33 to 45 ppt, respectively) for the same period. The cooler temperatures of this groundwater (with high values 5°C lower than high values in the surface water), suggest that historically SGD could have ameliorated high temperatures in Florida Bay that may occur during periods such as La Niña events.

Further evidence of the influence of SGD on surrounding marine temperatures is provided by Manheim (1967). During subsurface investigations off the coast of northeast Florida in 1966, the 80-ton, deep submersible *Aluminaut* was passing over a depression approximately 50 m (150 ft) deeper than the surrounding sea floor and suddenly lost an estimated half ton of buoyancy. The general bottom depth in the area was approximately 510 m (1530 ft) and the water temperature was approximately 2°C cooler than the surrounding bottom temperature in that area. The loss of buoyancy was explained by the outflow of water considerably less saline (less dense) than the surrounding seawater. Similar depressions reportedly were present in the area but could not be investigated.

Another factor in coral decline and retarded growth in the Middle and Lower Keys reportedly is exposure to cold Gulf of Mexico waters in the winter months (Shinn, 1999). The colder temperatures experienced by the more exposed and deeper reefs may have been ameliorated during the winters in the past by pristine, low-salinity, low-nutrient, groundwater of constant temperature from the Floridan aquifer system, preferentially bathing the reefs with water warmer than ambient surface water in the winter and cooler than ambient surface water in the summer.

Stringfield (1966) described temperature differences between the water discharged from the submarine spring approximately 4 km (2.5 mi) east of Crescent Beach, FL. That spring apparently was reported first in 1913 and was described as approximately 18 m (55 ft) below sea level. Stringfield and Cooper (1951) stated that the discharge occurs through a sinkhole similar to the large springs on land. Recent seismic evaluations of the area, however, suggested that, although the Hawthorn formation is breached, the depression more closely resembles a "blowout" maintained by artesian flow (J. Kindinger, pers. comm.).

The floor of the spring, approximately 42 m (126 ft) below sea level, is pitted with conical depressions as much as 4 m (12 ft) in diameter and 2 m (6 ft) deep, but no limestone outcrops were observed. Water samples from the submarine spring in the middle of the century had chloride contents of 7680 and 8720 ppm, comparable to artesian wells along the coast at that time. Discharge from the spring at that time was estimated at 40 ft$^3$ s$^{-1}$(cfs), creating a boil approximately 25 m (75 ft) in diameter at the surface of the Atlantic Ocean. The temperature of the water discharging from the submarine spring was 21°C (71°F) and 22°C (72°F) on November 3 and 16, respectively, during the mid-century investigation. This corresponded with the 21.8°C (71.25°F) temperature recorded at a depth of approximately 40 m (121 ft) in the submarine spring in 1925.

The global occurrence of SGD and the localized magnitude of its influence on temperature is evident in the temperature differences between fresh SGD and deep (~1170 m) ambient seawater documented along the Sagami Trough, at the northern end of the Philippine Sea plate. The temperature of seepage water at that site (where dense populations of organisms thrived, as previously described) was 9°C (48.2°F) higher than ambient seawater (Tsunogai et al., 1996). Certainly it is conceivable that organisms adapted to the consistent temperatures of shallow and deep areas with significant SGD could experience adverse impacts if those discharges were diminished or eliminated.

## PRESERVES IN PERIL

### Is Money the Solution … or the Problem?

Billions of tax dollars have been devoted to identifying the cause(s) of, seeking solutions for, and attempting to correct the environmental problems of South Florida. Those problems include widespread decline and death of corals and other marine organisms in various preserves associated with the Florida Keys, in addition to the demise of the Everglades. Some of the proposed causes not described in the Biennial Report for the Florida Keys (USEPA, 1996), as discussed above, but receiving considerable state and federal funding for research include El Niño/La Niña events, global warming, and African soil settling as dust in the southeastern U.S. (USGS, 1999a). As indicated previously, no inferences have been made regarding possible impacts of, or eliminated sources of historic (natural) SGD, due to dramatic increases in groundwater mining of the regional karst aquifer and overlying surficial aquifers in Florida. Likewise, no mention has been made regarding the potential role of minimally treated sewage effluent and associated pollutants injected into the Floridan aquifer system that may be surfacing (via flow-ways that transported historic SGD) in coastal ecosystems.

Despite that lack of information, vast sums of federal, state, and local funding are resulting in the massive increase in injected waste in South Florida. Permits continue to be issued for construction and operation of new effluent-injection wells throughout South Florida. The impending deep-well injection facility in Key West, the most environmentally sensitive location for a deep-injection well, reportedly will cost $39 million to convert the existing ocean outfall facility to a system that will inject effluent into the Floridan aquifer system. The irony is that the Key West ocean outfall has been discharging secondarily treated effluent since 1954, with the FDEP claiming no adverse environmental impacts from that ocean outfall facility (City of Key West, 2000). Discharges via ocean outfall reportedly are exposed to considerable mixing and dilution in the ocean currents, contrary to what would be expected for injected effluent flowing rapidly through karst channels, and discharging into surface waters near sensitive coastal ecosystems.

Apparently neither the federal nor the state regulatory agencies permitting the effluent injections in South Florida have calculated the tons/year of nutrients (and other pollutants) that are transported by and discharged from those karstic pipelines. To provide perspective, a recent study conducted in Western Australia determined that as much as "73 tonnes/year" of nitrogen is discharged into Jervoise Bay, Cockburn Sound, via groundwater, along 3.9 km (2.4 mi) of coastline. The Tamala Limestone formation, which extends under Cockburn Sound, is a semiconfined karstic aquifer similar to the Floridan aquifer system, with respect to the presence of numerous preferential flow paths. In that study, the highest concentrations of ammonia nitrogen (>15 mg $L^{-1}$) were found downgradient from the injection of wastewater identified as the "southern plume" point source. Slightly lower levels (3-15 mg $L^{-1}$) were found downgradient from former sludge beds at the wastewater treatment plant ("northern plume" point source). Plumes from these two point sources constituted the greatest contribution of nitrogen to Jervoise Bay. The conclusion of that study was that a Management Plan should be prepared so that the amount of nitrogen entering the groundwater, and ultimately the sensitive coastal waters, could be reduced (Department of Commerce and Trade, 2000).

That study was prompted in 1994, when the Australian government decided that a comprehensive sampling program should be undertaken to improve the current estimates of the groundwater nutrient fluxes. An early study determined that most of the nitrogen in the groundwater was in the form of ammonia, which is readily available for use by algae after discharging to Cockburn Sound. That study also determined that approximately 80% of the nitrogen discharge was taking place from the karst limestone aquifer. In addition to nearshore discharges of groundwater (including discharge from springs and seeps where the limestone aquifer is "unconfined"), it was determined that discharges of groundwater from the cavernous Tamala Limestone aquifer also may occur several hundreds of meters offshore. Finally, it was concluded that: (1) despite the progressively decreasing amount of nutrients discharging to Cockburn Sound via outfalls, groundwater discharge eventually will become the major source of nutrients; and (2) even if the sources of groundwater contamination were immediately halted, the discharge of contaminated groundwater to Cockburn Sound could continue for many years (Appleyard, 1994).

Ironically, the nutrient-laden wastewater was injected into the deeper, carbonate aquifer, approximately 1 km (0.6 mi) from Cockburn Sound, to prevent eutrophication of the Sound associated with openwater discharge. That scenario is identical to the shift to aquifer injection along South Florida's coast, including Miami–Dade County and throughout the Keys, as well as the impending municipal effluent injections in Key West. Based on the documented magnitude of the nutrient-laden groundwater plume discharging to Cockburn Sound, however, and the impacts of eutrophication (e.g., algal blooms, seagrass dieoffs), efforts now are under way in Western Australia to attempt to extract the nutrient-laden groundwater, via pumping wells (A. Smith and J. Turner, pers. comm.). Similar, but significantly more costly, groundwater remediation is predicted to be the next course of action required to "restore" South Florida's ecosystems. In fact, South Florida (including the Keys) is predicted to be the focus of the largest, most costly groundwater remediation effort in the history of Super Fund projects, in an effort to reverse the environmental damage due to aquifer disposal of wastes.

One of many studies receiving continued funding to evaluate possible causes of the plight of the corals in the Keys is the African dust hypothesis. As support for the African dust hypothesis, Shinn et al. (2000) noted that the initial reports of Atlantic coral diseases in Florida and the eastern Bahamas (San Salvador), in addition to the decline of corals in the Caribbean in the 1970s, are coincidental with large increases in transatlantic dust transport. They further proposed that the "hundreds of millions of tons/year of soil dust" crossing the Atlantic during the last 25 years could be a significant factor in the decline of coral reefs and other ecosystems. The African dust is proposed to be a source of atmospheric nutrient enrichment and an intermittent supply of fungus spores and bacterial cysts. For example, they indicated that African dust is an efficient substrate for delivering *Aspergillus* spores and that spores are absent from air samples when the air is clear.

It is important to note that water also is an efficient substrate for delivering fungus spores, including *Aspergillus* spores (as well as bacteria and viruses). This suggests that spores of that common fungus (as well as other microbes) could be transported via injected effluent and discharged at various coral reef sites. The decline of coral reefs and other coastal ecosystems also is coincidental with the initiation of deep-well injection of minimally treated municipal sewage in South Florida (Puri et al., 1973) and the Bahamas (Weech, 1997). This fact suggests that effluent injection may be transporting countless pathogens and other pollutants (including excess nutrients) directly to coral reef ecosystems, as well as to the Everglades and Florida Bay.

Shinn et al. (2000) noted that one of the flaws in their hypothesis (that dust, transported aerially from Africa, is a significant contributor in the decline of coral reefs and other ecosystems) is the fact that existing oceanographic charts do not verify currents capable of transporting pathogens to remote islands in the eastern Bahamas, such as San Salvador. Although the general circulation patterns for surface water in the coastal areas of South Florida and the Caribbean are known (as illustrated in Shinn et al., 2000), the flow patterns and discharge points for the injected effluent have not been identified or mapped. It is realistic to presume, however, that injected effluent flowing through preferential channels in the karstic aquifer is not constrained by circulation patterns of surface water and could be discharged at remote islands, including San Salvador. In fact, aquifer-injection of secondarily treated sewage effluent occurs on the island of San Salvador (J. Bowleg, pers. comm.).

A second flaw in the African dust hypothesis is the patchiness of the areas of coral reef decline and disease in the Florida Keys. If that area is being blanketed with pathogen/nutrient-laden dust from Africa (which may be dispersed more widely via surface currents), then more uniform rather than discrete areas of algal blooms and disease outbreaks should be occurring in the vicinity of the Keys. Direct, induced discharge of injected effluent from preferential flow paths (at certain reefs or portions of reefs and in Florida Bay) could account for the patchy distribution of coral reef decline and disease in the Florida Keys and Florida Bay, where algal blooms also may be patchy. To mirror the conclusion by Shinn et al. (2000), that pathogen transport and/or nutrient enrichment by dust is an emerging unexplored research area pertaining to corals and marine organisms, the same can be said for injected effluent as an emerging unexplored research area.

The most fitting example may be the new bacterial species of *Sphingomonas* which has been identified as the "cause" of advanced "white plague." White plague is one of several fatal coral diseases now occurring throughout the Florida Keys (Richardson et al., 1998). The term "white plague" was chosen based on the discoloration of the coral after its living tissue dies. That disease reportedly was not known prior to the 1970s, when injection of effluent was initiated at the Miami–Dade facility. Perhaps the name "white plague" more appropriately should be associated with the source of the most recent approach for disposing of human sewage along South Florida's coast via injection wells.

The notion that increasingly more "high-tech" and expensive solutions (e.g., centralized sewage treatment facilities with aquifer-injection of AWT-quality effluent) will solve the environmental problems of South Florida is without factual basis. The bottom line is that groundwater mining of the Biscayne aquifer (that supported the Everglades, Florida Bay, and oceanside ecosystems) and aquifer injection of wastes must be curtailed. Numerous alternatives are available that are less costly and scientifically sound. For example, with the arrival of the 21st century we should be able to comprehend the extreme lack of logic in using potable groundwater as a medium for human excrement, while simultaneously proposing to spend approximately $4 million federal tax dollars to construct hundreds of new deep-injection wells to compensate for depletion of the aquifer (as evaluated by the National Research Council Committee on Restoration of the Greater Everglades Ecosystem, 2001).

Since approximately 50% of municipal water reportedly is used for toilets, eliminating potable water as a receiving medium for human excrement would result in a significant reduction in groundwater diverted from the Everglades, Florida Bay, and sensitive oceanside ecosystems. That action also could eliminate aquifer injection in the Keys. The following discussion lends additional support for the need of such action. A summary is provided of responses to anthropogenic groundwater perturbations by terrestrial species that are linked to deeper aquifers by discontinuities in "confining" layers in the southeastern Coastal Plain. The knowledge of those responses can be applied to the recent declines of various coastal organisms and ecosystems in South Florida.

## CANARIES IN THE MINE SHAFT

Responses of species in terrestrial ecosystems of the southeastern Coastal Plain parallel those of the coral reef ecosystems on the submerged platform in South Florida. Pond cypress (*Taxodium ascendens*), the dominant tree species in forested depressional wetlands throughout the southeastern Coastal Plain, has been identified as a hydroecological indicator species of anthropogenic groundwater perturbations (Bacchus, 1999b). Pond cypress exhibits symptoms of stress associated with groundwater mining of the Floridan aquifer system more rapidly than associated species; therefore, it can be compared to the canaries used in mine shafts in the past to warn miners that conditions were unsafe, so that immediate action could be taken.

The approach developed to detect unsustainable aquifer yield using pond cypress has been applied to the interior of the Wilderness Area in Cumberland Island National Seashore. Results suggest that significant and permanent ecological damage has occurred on this barrier island due to anthropogenic groundwater perturbations (Bacchus, 1999a, 2000b). Federally designated Wilderness Areas theoretically are providing the highest degree of protection of any lands held in trust for the public. There is a general unawareness, however, of the magnitude of damage that can occur to these and other theoretically "protected" public resources from anthropogenic groundwater perturbations such as extraction of groundwater and oil, aquifer "storage" and "recovery" (ASR), and injection of fluid wastes into the aquifers.

Cumberland Island National Seashore is contained within the lateral extent of impacts from groundwater mining along the east coast of Georgia and northern Florida. That area of impact is represented by a massive coalesced cone of depression, extending from north of Savannah, GA, to the Jacksonville, FL area and west to the Okefenokee Swamp, including the Okefenokee National Wildlife area (Bush and Johnston, 1988). Groundwater withdrawals from the Floridan aquifer system were initiated in the Savannah area in 1885. The cone of depression from prolonged groundwater mining represents a decline in the potentiometric surface of the Floridan aquifer from premining conditions of approximately 9 m (30 ft) at Cumberland Island and a decline of more than 6 m (20 ft) at the center of the Okefenokee Swamp (Johnston et al., 1980, 1981).

In reality, the declines in the potentiometric surface from groundwater withdrawals were considerably greater than those reflected in the 1980 example, for several reasons. First, the potentiometric surface of the aquifer was significantly lower during part of the 1970s. Second, there are many withdrawals from the aquifer that are not monitored and therefore are not included in the estimated levels used to compile the 1980 map. Additionally, extensive "predevelopment" data are not available to determine what the actual conditions were in the aquifer prior to the initiation of groundwater mining.

Krause and Randolph (1989) estimated that the volume of water extracted for the coastal counties of Georgia was approximately 350 MGD. The coastal counties are responsible for the majority of the groundwater mining that has resulted in the referenced cones of depression. Peak withdrawals for the Northwest Well Field (supplying potable water for Miami–Dade County), described earlier in this chapter, are approximately 134 MGD (Sonenshein and Hofstetter, 1990). The groundwater withdrawals for municipal use at the Northwest Well Field in Dade County alone represent almost half of the volume of water extracted for the coastal counties of Georgia.

The location of the study referenced previously by Kindinger et al. (1997) was in the vicinity of the Indian River Aquatic Reserve and Pelican National Wildlife Refuge and confirmed the presence of structural features along the submerged platform that could serve as "point-recharge" locations for induced recharge. Those same features could serve as preferential flow paths for upward movement of injected wastes, including the equalization of any pressure differentials between overlying aquifers. Such "point-discharges" could account for the "slight bottom-hole pressure increase during injection" of effluent at the Miami–Dade facility that was reported by Puri et al. (1973). No studies have been conducted to confirm that similar breaches in the theoretical "confining" zones do not exist in the vicinity of the Miami–Dade and impending Key West deep-well injection facilities.

Discussion of the uncalculated impacts of unsustainable aquifer yield (including impacts to associated areas that theoretically are "protected") was provided by Bacchus (1999b, 2000b). Some of the more significant areas in South Florida that theoretically are "protected" are listed in Table 26.2, in alphabetical order. In general, those areas were designated for protection because of their unique and sensitive flora and fauna. No "protection" of those systems has been provided from subsurface assaults associated with the underlying aquifers.

As an example of impacts that can occur to organisms in a theoretically "protected" area, the cambium of sapling pond cypress trees was penetrated by nonaggressive, opportunistic fungi (*Botryosphaeria rhodina* and *Fusarium solani*) during a prolonged water-stress experiment conducted under controlled conditions. The same response occurred in mature pond cypress trees *in situ*, at sites where groundwater mining was occurring (Bacchus et al., 2000, unpubl. data). Both of the opportunistic fungus species from those experiments exhibit widespread occurrence, but apparently do not affect pond cypress in the absence of anthropogenic stressors. Those findings provide additional support for the case that anthropogenic stressors can increase the susceptibility of organisms to commonly occurring microbes such as fungi.

Dustan (1999) described various sources of coral reef mortality, adding carbon loading to the existing concerns over nitrogen and phosphorus loading. He suggested that although the anthropogenic loadings (including carbon) may be diluted, those compounds still may affect reef health adversely and be ecologically significant. Brand (2000) described the importance of the N/P ratio in Florida Bay in promoting algal blooms that influence the coral reefs. It does not appear that anyone has considered or investigated the role of carbon/nitrogen (C/N) ratio perturbations in the rapidly declining health of the coral reef ecosystems. The importance of the C/N ratio in biocontrol of fungal pathogens is well known, with early work having been done approximately 30 to 40 years ago (Ko and Lockwood, 1967; Phipps and Barnett, 1975; Shigo et al., 1961). The following section describes components of South Florida's coastal ecosystems that are under attack by fungal pathogens, with possible implications to anthropogenic shifts in the natural C/N ratio.

**TABLE 26.2**
**Examples of Some of the Theoretically Protected Areas in South Florida**
**That May Be Influenced by Withdrawals From, or Injections of Effluent**
**and Other Wastes into the Underlying Aquifer System**

Bahia Honda State Park
Big Cypress National Preserve[a]
Biscayne Bay and Card Sound Aquatic Preserve
Biscayne National Park
Coupon Bight Aquatic Preserve
Crocodile Lake National Wildlife Refuge
Dry Tortugas National Park[a]
Everglades National Park
Florida Keys National Marine Sanctuary (including Key Largo and Looe Key National Marine Sanctuaries)
Fort Zachary Taylor National Marine Sanctuary
Great White Heron National Wildlife Refuge
John Pennekamp Coral Reef State Park
Key Largo Hammocks Botanical Site[a]
Key West National Wildlife Refuge[a]
Lignumvitae Aquatic Preserve
Lignumvitae Key State Botanical Site[a]
National Key Deer Refuge[a]
Shell Key State Preserve
San Pedro State Underwater Archeological Site

[a] Areas that include land-based natural resources.

What might suggest similarities between structural preferences of the depressional pond cypress wetlands (terrestrial canaries) and coral reefs? Evidence suggests that the present-day reefs in Florida formed after the carbonate platform flooded, approximately 6000 years ago, and that those reefs were associated with former depressional wetlands that were drowned by the rising sea. As indicated previously, prior to the most recent rise in sea level approximately 6000 years ago, much of the submerged carbonate platform associated with Florida (the shallow Gulf of Mexico and Atlantic coastal bays) was exposed during lower sea levels. In addition to the ability of mildly acidic rainwater to erode an exposed carbonate platform (see previous discussion), plants also generate acidic substances, which further promote karstic dissolution (Shinn, 1988, 1999).

As described by Lidz et al. (1997), the Pleistocene rock surface of the southeast Florida Platform had been exposed and vegetated by terrestrial plants prior to formation of the South Florida reefs on the drowned platform. The area along the platform margin that supports the extensive reef system from Elbow Reef to French Reef was the equivalent of a barrier island system thousands of years ago when sea level was approximately 10 to 12 m (30 to 36 ft) lower (Figure 5 from Lidz et al., 1997). This again suggests that subsurface discontinuities responsible for hydrogeological phenomenon in terrestrial ecosystems (e.g., depressional pond cypress wetlands and herbaceous depressional wetlands on the barrier islands) are similar to those associated with the drowned features where coral reefs became established.

Observations of reefs in Biscayne National Park confirm that corals did not begin growing everywhere after the last rise in sea level. Topography has been identified as a significant factor in the "gross distribution" of coral reefs; however, local factors have been less clear (Shinn, 1988). One factor that may have contributed to localized distribution and establishment of coral reefs is SGD. Additional support for this conclusion is provided by the observation that several small terraces control the distribution of today's living reefs (Shinn, 1988, 1999). For example, the Grecian Rocks Reef was initiated at a break in slope (Shinn, 1988). Lidz et al. (1991) found that Holocene reefs also formed on the slope leading down to the top of the sea cliff in the Bahamas.

Shinn (1988) reported that for every reef drilled off Florida (for core analysis), the top or thickest part of the reef was found to overlie sand. For example, at the spur and groove reefs of Looe Key, the spurs are composed of elkhorn coral underlain by approximately 3 m (10 ft) of sand (which can serve as a preferential flow path). A similar association was documented approximately 1.6 km (1 mi) from the shore of the northern end of Miami Beach, where a north–south trending reef was found to be localized over a relic beach ridge. In the latter case, the role of topography was stressed in elevating the reef above lower surrounding areas that are more susceptible to sedimentation. The conclusion was that corals first become established on topographically high areas (Shinn, 1988).

Another factor that may be involved in those topographically high areas is the prehistoric drainage pattern developed when the beach ridges were exposed. Beach ridges are thick deposits of highly permeable sands, where rapid percolation of rainwater is focused. This focused percolation of naturally acidic water could magnify the dissolution of underlying carbonate strata in an area surrounding the base of these ridges. After submergence of those ridges, the dissolution channels may have provided preferential flow paths for seepage of groundwater with essential elements for promoting the growth and maintaining the health of coral reef ecosystems.

Precht (1994) addressed various uses and misuses of the term guild when applied to coral reefs, as summarized below:

> The term "guild" was originally defined by Root (1967) as a "group of species that exploit the same type of environmental resources in a similar way." This term is useful in ecology because it groups together species, without regard to taxonomic position, that overlap significantly in their resource use. Root (1967) considered "guild" to be "the most evocative and succinct term for groups of species having similar exploitation patterns." The essential points of Root's definition are that guild members use the same resources, and that they use them in a similar way. These resources can be food, shelter, or a host of other requisites that living organisms need in order to survive and reproduce.

> In spite of the original definition of the term "guild" Fagerstrom (1987, 1988a, 1991) has applied the term in an altogether different way. He has argued that reef communities can be subdivided into "guilds" based on the distribution of reef taxa along a spatial resource (related to reef surface). His rationale for selecting the guild concept is based on his interpretation that the unifying environmental resource is substratum.

Although Precht's reasoning seems to be supported for past applications of this term to coral reefs, a new use is proposed in this chapter. If coral reefs historically have been associated with areas of enhanced interaction between groundwater and surface water, as are depressional freshwater wetlands, perhaps these sensitive ecosystems are members of the same regional guild as depressional freshwater wetlands. Both appear to have been "exploiting" freshwater discharge from carbonate aquifers.

## PREDISPOSITION TO DISEASE, AND BEYOND

The first factor contributing to the use of pond cypress as a terrestrial indicator of anthropogenic perturbations of groundwater is the structural component of its ecosystem (as described above). The second factor is its predisposition to disease and subsequent premature decline that result from those perturbations. Opportunistic fungal pathogens such as *Botryospheria* and *Fusarium* have been identified from the interior wood of pond cypress trees associated with areas of groundwater mining. Those opportunistic pathogens were not found in the interior wood of pond cypress trees located in areas not associated with groundwater mining (Bacchus et al., 2000, unpubl. data).

*Botryospheria* and *Fusarium* are opportunistic fungi that generally are widespread on tree bark and in soil, respectively. Abnormal water stress enables those fungi to penetrate and infect trees that are capable of remaining infection free under unstressed conditions (Bacchus et al., 2000; Brown and Britton, 1986; Kerry Britton, pers. comm.; Ocamb and Juzwik, 1995; C. Ocamb, pers.

comm.). Trees in depressional wetlands and associated uplands in groundwater mining areas of the southeastern Coastal Plain may exhibit signs of root decay and basal decay characteristic of fungal infections (e.g., windthrow) within a year after initiation of the perturbation. Unfortunately, the death of those trees may be delayed for 15 to 20 years after the onset of the anthropogenic stress, as the premature decline slowly intensifies. Pond cypress can be used as a hydroecological indicator of anthropogenic groundwater perturbations because signs of premature decline in pond cypress appear earlier and are more readily visible than signs of premature decline in other species (Bacchus, 1999b; Bacchus et al., 2000). Thus, pond cypress can be used as the terrestrial canary for detecting anthropogenic groundwater perturbations.

Sea fans (*Gorgonia*, a soft coral) appear to be succumbing to a fungus identified as *Aspergillus sydowii*, which has been implicated in mass mortalities of the sea fans. That sea fan pathogen is considered to be a common, cosmopolitan saprobic fungus, and has been isolated from many types of terrestrial environments, including soils from Alaska to the tropics. That fungus also has been cultured from subtropical marine waters near the Bahamas and the Straits of Florida, and has been found in both eulittoral zones and oceanic zones, including isolations from waters collected as deep as 4450 m (13,350 ft). Prior to the sea fan mass mortalities, that fungal species had not been recognized as the cause of widespread disease in plants or animals. However, several species of *Aspergillus* are opportunistic animal pathogens, generally infecting individuals with compromised immune systems. Likewise, the infection of sea fans by *A. sydowii* may be the result of opportunistic pathogenicity due to weakening of the host from stressors, such as water pollution or environmental factors (Geiser et al., 1998; Nagelkerken et al., 1997; Roth et al., 1964; Smith et al., 1996).

Rinaldi (1983) described the role of the compromised host in the invasive fungal disease of humans by species of the genus *Aspergillus*. The significance of host vigor in avoiding infection by this fungus was emphasized. Other factors that may influence the ability of a pathogen to infect its host include competition between the pathogen and competing antagonists. For example, *Trichoderma*, another fungus commonly found in soils, is regarded as an antagonist of fungal pathogens. *Trichoderma* exhibited reduced competitive ability in laboratory experiments when higher concentrations of C were present, relative to available N. That finding suggested that a delicately balanced C/N ratio is required for maximum competition (Overmier, 1977). During the same experiments, the fungal pathogen *Gliocladium virens* required high levels of C for successful invasion of *Diplodia* colonies. Therefore, the C/N ratio may influence pathogenicity by affecting the competitive ability of antagonists or by increasing the ability of fungal pathogens to invade their host, independent of any increased susceptibility of the host organism due to other stressors.

Despite the fact that disruption of the C/N ratio can facilitate infection by opportunistic fungi, not all organisms exhibit equal susceptibility to infection. Certain organisms are more sensitive to environmental perturbations than other organisms, and those organisms are considered the "canaries" that are issuing early warnings, as discussed previously. Those organisms may be hypersensitive (e.g., sea fans, certain species of corals) responding more rapidly and/or more severely to perturbations, including succumbing to infection by fungi, bacteria, or viruses.

Hydrologic perturbations such as the interception and diversion of pristine, low-salinity, low-nutrient SGD of constant temperature, and the possible replacement of that resource with treated effluent containing excessive nutrients and other contaminants could induce a state of physiological distress in ecosystems of the same "guild" (e.g., coral reefs), on the submerged platform. A chronic state of physiological distress could promote predisposition to disease caused by pathogens that may have been present historically or introduced recently (e.g., via injected effluent or transported African dust). It also could render organisms more susceptible to other stressors such as changes in solar irradiance.

An event similar to the mass mortality of sea fans has occurred with mass mortality of seagrasses in Florida Bay. The seagrasses also became the victim of a fungus that is considered to be a nonaggressive (opportunistic) species on seagrasses throughout South Florida (J. Zieman, pers. comm.). That fungus is a marine slime mold (*Labyrinthula* sp.) that has been identified as endemic to the South Florida area, meaning that it does not occur naturally in any other areas. The fungus implicated in the mass seagrass mortality in Florida Bay does not occur in Africa (the origin of the aerially dispersed dust in the African dust theory). Furthermore, that marine slime mold does not produce the type of resistant structures that would allow long-range aerial distribution (D. Porter, pers. comm.). Therefore, it is unlikely that the African dust dispersed across the ocean is the source of the pathogen implicated in the mass mortality of the seagrasses in Florida Bay.

Evidence is mounting that seagrasses, like other organisms, can be weakened by environmental stressors and made more vulnerable to disease (Den Hartog, 1987; Muehlstein, 1989; Short et al., 1988). Although the mass dieoff of seagrasses in Florida Bay was reported in 1987 (reportedly due to hypersaline conditions in Florida Bay during a period of low rainfall), our current state of knowledge suggests that the hypersaline event was not due to low rainfall alone and was not the sole or possibly even the most significant stressor (as described previously). The similar dieoff of seagrasses in Cockburn Sound, Western Australia (also described previously), suggested that induced discharge of injected waste water may have played a significant role in predisposing the seagrasses in Florida Bay to the opportunistic fungal disease.

Durako and Kuss (1994) suggested that the dieoff of turtle grass (*Thalassia testudinum*) in Florida Bay was density dependent because it was observed only in areas that previously supported very dense populations of turtle grass. They also noted that the lower density stands that were less affected by the dieoff also were in areas of lower salinity. However, it is conceivable that the denser stands of turtle grass were associated with areas of historic SGD that originally provided more favorable growing conditions for the turtle grass. It also is conceivable that those areas of SGD subsequently became areas where saline water from deeper aquifers was surfacing (supported by the data of Top et al., 2001) as induced discharge, due to deep-well injected effluent. In that case, point discharges of anthropogenic nutrients and contaminants could have reduced the vigor of the seagrass and increased the vigor of pathogens, such as the undescribed marine slime mold.

Duarte (1995) provided extensive insight into the feedback mechanisms leading to the "domino effect" that occurs as coastal eutrophication results in a shift from ecosystems dominated by relatively slow-growing, nutrient-conserving macrophytes such as seagrasses to systems dominated by rapidly growing phytoplankton and macroalgae. In the latter case, greater amounts of dissolved organic carbon are released and available for recycling. *Ceramium corniculatum*, the red alga that was reported covering the coral reefs at Cheeca Rocks, is an example of the thin, finely-textured macroalgae described by Duarte that results in the "domino effect" with respect to the release and rapid recycling of C.

The sensitivity of corals to increases in C was demonstrated by Mitchell and Chet (1975). They exposed coral heads to low concentrations of various substances, including crude oil (100 ppm) and organic matter (1000 ppm, in the form of dextrose), under controlled conditions in laboratory aquaria. Many of the colonies of coral died after 24 hours of exposure at those low concentrations. The addition of the organic matter (dextrose) resulted in the same level of increased mucus production that was associated with the exposure of the coral heads to crude oil. In fact, the corals died within 24 hours after dextrose was added to the water. Concomitantly, the bacterial population associated with the coral heads "reached an extraordinarily high peak" of $10^7$ cells ml$^{-1}$ within 24 hours after addition of the dextrose.

Their analysis of bacterial isolates from the coral surface in the presence of the crude oil indicated that 15 to 25% of the bacteria isolated were capable of growing on coral tissue extract as the sole C and N source. That finding suggests those bacteria could co-occur with corals in low numbers under natural, oligotrophic conditions, without external sources of nutrients. Of equal significance, Mitchell and Chet (1975) discovered that approximately 60% of the bacteria identified in those experiments were motile (capable of moving through the water), Gram-negative rods. They further noted that more than 50% of the motile bacteria displayed chemotaxis to the coral mucus, meaning that they were attracted to the chemical signature of the coral mucus. The majority of those bacteria also were capable of growing on coral tissue extract as the sole C and N source.

Corals produce mucus in response to both chemical (e.g., crude oil) and physical (e.g., sand) stressors. Coral mucus is composed of polysaccharides, molecules containing many sugars (which are composed of carbon atoms). Dextrose is a "hexose," which simply is a sugar molecule composed of six C atoms. Although dextrose is not a substance that would be considered toxic, it is a source of C, as is the mucus that is produced by the corals when they are under stress.

Microbes, such as bacteria and fungi, use C as a food source. In the experiment conducted by Mitchell and Chet (1975), the bacterial population associated with the coral heads increased at the same rate as the production of mucus by the coral. By additional experimentation with antibiotics (penicillin and streptomycin), they also were able to show that the coral death was due to the bacteria (including two predatory bacteria), rather than the actual stressors. The results were the same, even with an order of magnitude increase in the concentration of crude oil. Their research was critical in showing that even when concentrations of pollutants were insufficient to kill the corals directly, the increased stimulation of omnipresent microbes and microbial processes was sufficient to cause coral death.

The role of increased C in the coral deaths documented in the experiments by Mitchell and Chet (1975) support Dustan's concerns regarding the implication of carbon loading and coral death. That may be the case with the injection wells in South Florida, particularly in the Keys. Their experiments also are important with respect to the concerns raised in this chapter regarding disruption of the C/N ratio.

The additional C entering the water near the coral reefs, via induced discharge of injected effluent, may be a factor in the assault of corals by *Sphingomonas* (white plague). Recall that white plague initially was reported from coral reefs near Key Largo at approximately the time that the Miami–Dade facility began injecting effluent near Biscayne Bay. *Sphingomonas* is representative of the ultramicrobacteria in oligotrophic marine waters (Fegatella and Cavicchioli, 2000). Edwards (2000) reiterates the differences in responses of microorganisms under artificially nutrient-rich conditions and their natural, oligotrophic environments where they are exposed to starvation conditions and grow slowly, or not at all. The presence of organic matter in the water column also has been shown to increase the survival time of bacterial pathogens in the water, such as *Vibrio cholerae*, the human pathogen that causes cholera (Joseph and Bhat, 2000). More chilling is the mounting evidence that symbiotic organisms (e.g., bacteria, fungi) can become pathogenic towards their hosts under abnormal conditions (Bacchus, unpubl. data; Hentschel et al., 2000).

The implications described above, coupled with the increasing incidence of unexplained (and often unidentifiable) diseases and mortality in other marine organisms (including federally listed sea turtles and manatees) should provide sufficient impetus for a comprehensive investigation of the potential impact of injected wastes on coastal ecosystems and associated organisms.

The preceding discussion addresses the ability of environmental conditions that have been subjected to anthropogenic alterations to increase the susceptibility of (predispose) organisms such as corals and seagrasses to infection by commonly occurring, opportunistic pathogens. Also addressed is the potential for those altered environmental conditions to increase the virulence of commonly occurring pathogens. A related scenario would be the introduction of pathogens such as viruses, fungi, and bacteria into environments foreign to those in which they evolved. The relevant example of this scenario is the injection of large volumes of effluent containing human pathogens into the marine environment.

There are at least two significant differences to consider between the dispersal and discharge of those pathogens via groundwater flow channels and disposal via ocean outfall pipes. The first is the cooler, more stable temperature of groundwater transport of effluent. The second is the potential for longer periods of incubation in the absence of light. Both of those conditions can extend the period of viability for at least some pathogens (J. Paul, pers. comm.). Finally, consideration must be given to the possibility that the introduction of microbes into a new environment or exposure of existing organisms to those altered environmental conditions may be facilitating the evolution of new organisms, as is suggested by recent findings.

Evidence also is mounting that organisms with rapid regeneration times (e.g., microbes) are evolving equally rapidly in the severely disturbed environments we are creating. This particularly is true for sewage-laden, eutrophic coastal waters. For example, Parveen et al. (1997) used multiple-antibiotic-resistance profile homology to determine that *E. coli* isolates from sewage discharge point sources were markedly more diverse than isolates from non-point sources (e.g., stormwater runoff). Those findings provide additional evidence that our natural resistance, as well as our medical resistance to those organisms are under serious threat.

Responses are similar at the ecosystem level. Burkholder and Glasgow (1997), Burkholder et al. (1995), and Glasgow et al. (1995) provide disturbing details regarding the increasing frequency, magnitude, severity, and range of outbreaks of toxic ambush-predator dinoflagellates in coastal areas of the southern U.S. Those organisms were undescribed and unknown to science until recently. The marine slime mold implicated in the mass mortality of seagrasses in Florida Bay may represent another example of microbes evolving rapidly in coastal areas that are experiencing significant anthropogenic eutrophication and contamination by other pollutants.

Likewise, differences in population composition of bacteria were noted in pond cypress trees associated with groundwater mining that were dying and those not associated with groundwater mining (healthy trees) in the *in situ* experiment described previously. Fatty acid analysis was used to identify bacteria cultured from sterilized samples of internal woody tissue in that *in situ* experiment. Samples from the pond cypress trees associated with groundwater mining areas not only contained unidentifiable bacteria, but also bacteria that were composed primarily of fatty acids that were unidentified and undescribed (Bacchus, unpubl. data).

Shinn (1988) concluded that corals have survived the natural disasters and sea-level fluctuations for the last 500 million years of Earth's history, and predicted that they will "survive the vagaries of sea-level oscillations well into the future." Unfortunately, that statement ignores the geologic time scale at which those changes were occurring, as compared to the rapid time scale of the anthropogenic changes occurring now. The corals (and other coastal organisms) did not have to contend with the magnitude of anthropogenic changes that have occurred within the past century and have accelerated at record pace within the past several decades (as described above), while simultaneously attempting to adjust to the natural changes in their environment. Whether the corals, the interdependent reef species, and other coastal organisms can survive these recent, anthropogenic assaults remains to be seen.

## THE NEED TO RESCALE

### LANDSCAPE VS. REGIONAL SCALE

A truly comprehensive scientific evaluation of the cause of coral reef decline and other environmental problems in South Florida's coastal waters requires a rescaling of our thought processes and research approaches in two opposing directions. First, we must increase the scale of the perceived sources of the problem from a landscape scale to a regional scale. Rescaling in this direction has been initiated with the recognition of adverse impacts to water quality and water quantity in coastal waters of South Florida from the alteration of historic surface water contributions (e.g., ditching and diking throughout the Everglades and other areas of South Florida). Unfortunately, no recognition has been given to the impacts of similar or greater magnitude due to anthropogenic alterations of the groundwater system.

Available data suggest that groundwater perturbations may be more far reaching than surface water perturbations. For example, Krause and Randolph (1989) determined that the natural groundwater divide along the southwestern boundary of the Floridan aquifer subregion in southeast Georgia and adjacent parts of Florida and South Carolina had been breached due to groundwater mining that was occurring primarily along the Atlantic coast. The artificial gradient created by groundwater withdrawals along the coast in that subregion produced a flow of approximately 201 ft$^3$ s$^{-1}$ across that boundary into their study area (Krause and Randolph, 1989).

Based on their findings, the volume of water migrating from the adjacent subregion due to coastal groundwater mining would be approximately 17,366,400 ft$^3$ d$^{-1}$. Converting that daily volume (that is breaching the groundwater "divide") into a horizontal column of water with a width and height of 1 ft would create a horizontal column of water extending approximately 5292 km (3289 mi). As perspective, the distance from the east coast of Florida at Biscayne Bay, due west, to the west coast of Florida is less than 100 km (70 mi). In the opposite direction, the distance to the Bahamas would represent only about 200 km (140 mi).

Groundwater withdrawals from the west-central Florida subregion and the subregion described by Krause and Randolph (1989) are similar (985 MGD and 645 MGD, respectively), based on simulated pumpage for 1980 (Bush and Johnston, 1988). This suggests that groundwater mining in the west-central Florida subregion likewise may be resulting in the pirating of groundwater from the adjacent South Florida subregion. That would reduce groundwater available for submarine discharge as diffuse seepage and flow from submarine springs in the South Florida subregion. The amount of diffuse upward leakage for the South Florida subregion in 1980 was 76% less than the amount of diffuse upward leakage in that subregion approximately 100 years earlier, before initiation of groundwater mining (Bush and Johnston, 1988). Those estimates, however, did not include reductions in diffuse upward leakage of spring discharge from the submerged platform.

The land-based source of approximately half of the amount of water supplied for the simulated 1980 pumpage in the South Florida region was increased recharge (Bush and Johnston, 1988). Anthropogenic increases in recharge due to groundwater mining is known as "induced recharge." The source of this induced recharge is from surface water, such as streams and wetlands, and the surficial aquifer that supports these systems, as well as the upland habitats that tap the water table.

The report by Sonenshein and Hofstetter (1990) regarding hydrologic effects of the Northwest Well Field in Dade County did not address the long-term fate of the portions of the wetland in the Everglades National Park and Wildlife Management Area/Water Conservation Area No. 3. Evaluation of delayed and long-term wetland impacts throughout the remainder of the 65-square mile study area and protected wetlands in Everglades National Park and Wildlife Management Area/Water Conservation Area No. 3 is not possible, considering the relatively short period of time between initiation of pumping at the well field and release of their report.

Both Everglades National Park and Wildlife Management Area/Water Conservation Area No. 3 would be expected donors of groundwater in response to withdrawals at that well field. Likewise, freshwater SGD to Biscayne National Park and other "protected" areas in South Florida, such as those listed in Table 26.1, may have been reduced beyond the land-based reductions reported by for 1980 (Bush and Johnston, 1988). Neither of the two reports addressed potential impacts to SGD of the groundwater mining at that well field.

Initiation of groundwater mining from the Northwest Well Field coincides with the first major hypersaline event recorded for Florida Bay in the mid 1980s. Although that period also coincided with reduced rainfall, the reduction or cessation of pristine, low-salinity, low-nutrient groundwater discharge to Florida Bay during that period (due to extensive groundwater mining in the surrounding area) may have been the additional stressor that dealt the fatal blow to some of South Florida's coastal ecosystem components (e.g., seagrass beds in Florida Bay). Support for this hypothesis comes from the fact that alterations in surface flow were initiated many years ago and numerous periods of low rainfall have occurred since those alterations. That response is consistent with the destruction of sensitive ecosystems that has occurred in the west-central Florida subregion of the Floridan aquifer system, as indicated by the following statement from the Southwest Florida Water Management District (SFWMD, 1996, p. 6-4):

Environmental impacts to lakes and wetlands observed in the Northern Tampa Bay WRAP area are variable, and include wetland species changes, intrusion of upland species, ground subsidence, rapid and severe desiccation and oxidation of soils, loss of overstory, severe fire damage, wildlife loss, and complete loss of habitat. The spatial magnitude and severity of these impacts can not be attributed to variations in rainfall.

Although the Northwest Well Field is closer to the northern boundaries of the FKNMS and Biscayne National Park than to Florida Bay, it is possible that the reduction or cessation of pristine, low-salinity, low-nutrient SGD, due groundwater mining at the Northwest Well Field, was significant in Florida Bay. Ample evidence exists to support the concept of enhanced subsurface connections between the Floridan aquifer system and the Everglades National Park wetlands, based on knowledge of other depressional wetlands associated with this regional karst aquifer system (summarized in Bacchus, 2000b). The Northwest Well Field is located in the eastern boundary of the wetland that has been designated as Everglades National Park. Additional decreases in freshwater via SGD could have resulted in more significant increases in salinities in Florida Bay.

This scaled concept recently was described by Dustan (1999), with respect to the decline of the corals. A series of nested stresses were addressed, as a hierarchic structuring of multiple stresses based on local, regional, and global levels. Using that concept, a shallow-well injection of effluent could be considered local, while a deep-well injection of effluent could be considered regional, because deep-well injections have a more far-reaching potential for environmental harm. Dustan stressed the need to ensure that implementation of "management" practices will not create a new ecological stress, as may be the case with the shift from open-ocean outfalls to injection of effluent.

## Parts per Million vs. Parts per Billion and Trillion

In addition to increasing the scale of the perceived sources of perturbations contributing to the environmental problems in South Florida, we also must decrease the scale with respect to potential water quality problems. For example, no monitoring currently is being conducted to evaluate the introduction and escalation of endocrine disruptors in South Florida's coastal waters. Environmental monitoring focuses on the toxic impacts of pollutants (often in the range of parts per million), rather than impacts of pollutants that disrupt the normal functioning of hormones (usually in the range of parts per billion or parts per trillion). The former can lead to death of the organism that is exposed, while the latter can result in the loss of future generations of exposed organisms.

A general comparison of nonlethal (low) levels of compounds that cause toxic responses to organisms and the orders of magnitude lower levels of exposure to endocrine-disruptive compounds that may be "lethal" to all future generations (after exposure of the initial generation) is shown in Figure 26.7 (right and left, respectively). In some cases, hazardous levels of endocrine disruptors may be below current detection limits of sampling equipment.

Synthetic chemicals such as nonylphenol and its breakdown products are capable of disrupting hormonal function and are becoming more widespread in the environment. Alkylphenol polyethoxylate commonly is used in non-ionic surfactants (soaps and wetting solutions). For example, these compounds are biotransformed to several stable metabolic products, including nonylphenol and its breakdown products. Many of those compounds, including nonylphenol, are lipophilic; therefore, they are stored in fatty tissue and are considerably more toxic than the parent compound (Reinhard et al., 1982; Giger et al., 1984, 1987; Granmo et al., 1989; Holt et al., 1992; Li and Schroder, 2000).

Those compounds can function as estrogen mimics. Estrogen mimics can cause deformities in penises and testicles, in addition to reducing sperm counts in males exposed to those chemicals. Exposure to very small concentrations of these estrogen mimic compounds also can cause breast cancer cells to proliferate under laboratory conditions (Colborn et al., 1996). Marine mammals bioaccumulate nonylphenol (Ekelund et al., 1990).

In addition to being components of household and industrial detergents, those compounds are added to products such as polystyrene and polyvinyl chloride (PVC), as an antioxidant to make the plastics more stable and less brittle, and to pesticides, contraceptive creams, and personal care products. Those chemicals can leach into water. Bacteria in animals' bodies, in the environment, and in sewage treatment plants (STPs) degrade those compounds into nonylphenol and other chemicals that can mimic estrogens (Colborn et al., 1996).

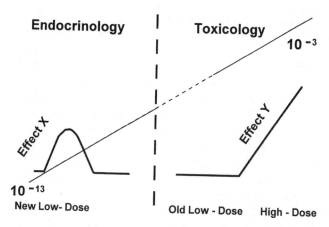

**FIGURE 26.7** Comparison of scales for (left) new low-dose, nonreactive levels of endocrine disruptor chemicals and (right) old low-dose, sublethal levels of toxic chemicals (Theo Colborn, World Wildlife Fund, previously unpublished).

Approximately 40% (molar concentration) of the total nonylphenol in STPs can reach surface waters via secondary effluents (Ahel et al., 1994). During the 1980s, fish in streams exposed to discharge from STPs in England appeared to be sexually confused, containing characteristics of both sexes. Controlled tests were conducted with caged fish exposed to water flowing from the STP. After 3 weeks of exposure, levels of the estrogen marker in the fish exposed to the STP water in the stream had increased 1000 times more than levels in control fish. In the summer of 1988, a nationwide study was conducted at 28 sites in England and Wales. Dramatic increases in the estrogen marker were found in all cases where the fish remained alive through the tests. In one case the increase in the marker for the exposed fish was 100,000 times the level in the control (Colborn et al., 1996; Jobling et al., 1998).

The use of alkylphenols, the breakdown product of alkylphenol polyethoxylates contained in detergents, was a strong suspect. Extended research tested fish exposed to alkylphenols to determine if: (1) those compounds caused estrogenic responses in fish similar to responses in breast cancer cells in the laboratory, and (2) levels in the environment were high enough to cause the estrogenic responses in the fish. The answer to both questions was yes. However, a range of chemicals could be contributing to the response observed in the fish exposed to streams with STP discharge (Colborn et al., 1996; Jobling et al., 1998). Toxicity of nonylphenol and its breakdown products to other organisms, including coastal organisms, also have been documented (Granmo et al., 1989).

Use of alkylphenol polyethoxylates has increased since the 1940s. Because of concern expressed during the past decade regarding the persistence and toxicity of those chemicals to aquatic life, several European countries banned the use of the most common of those chemicals in household cleaners by the late 1980s. Additionally, 14 European and Scandinavian countries phased out their use by 2000 (Colborn et al., 1996).

In 1995, nonylphenol and its ethoxylates (NPEs) were added to Canada's list of substances that are toxic or are capable of becoming toxic. Partial justification for that concern was the presence of nonylphenol and NPEs in the effluent from municipal wastewater treatment plants and the knowledge that nonylphenol and NPEs have been shown to produce endocrine-disrupting effects in fish and other organisms. An environmental risk assessment conducted recently in Canada determined that nonylphenol also should be considered toxic, based on the Canadian Environmental Protection Act (Davidson et al., 2000).

Currently, neither the USEPA nor FDEP have required or initiated monitoring to determine if estrogen mimic compounds (or any other environmentally harmful chemicals) are present in the treated effluent that is being injected into the highly permeable aquifers in South Florida. In fact, the USEPA currently does not even regulate or require monitoring for nonylphenol and related compounds in drinking water in the U.S. (USEPA Hot Line/Labat-Anderson, pers. comm.). Consequently, even effluent reportedly treated to "drinking water standards" could contain hazardous and toxic levels of nonylphenol and its breakdown products.

Results of research described previously clearly show that effluent injected into shallow (Class V) injection wells in the Florida Keys rapidly enters surface waters, including areas near sensitive coastal ecosystems. Studies similar to those conducted for shallow injection wells have not been conducted for deep injection wells (primarily Class I). Therefore, no information is available regarding the degree to which the treated effluent (even at theoretical upgraded "drinking water standards") injected into deep wells may be transporting hazardous and toxic compounds, such as endocrine disruptors, to the sensitive ecosystems.

Nutrients have been the sole focus of the effluent-related studies in the Keys, in part because the rapidly increasing eutrophication (e.g., proliferation of algal growth) in areas such as Florida Bay is obvious even to the casual observer. If reliable scientific evidence was available to support the belief that significant amounts of excess nutrients in the injected effluent were being adsorbed or taken up by the carbonate aquifers, that information would not suggest that other pollutants (e.g., endocrine disruptors) were being adsorbed or diluted similarly, prior to discharging into those environmentally sensitive areas. Unfortunately, there is insufficient evidence even to suggest that adequate nutrient removal is occurring.

Even if the endocrine disruptor chemicals were subjected to the same dilution processes as the phosphates and dissolved nitrogenous compounds, their environmental damage occurs at concentrations orders of magnitudes lower than nutrient pollutants. At highly diluted concentrations they could present a significant environmental hazard to organisms in Florida Bay and associated reefs on the Atlantic side of the Florida Keys.

Endocrine disruptors are not the only chemicals of concern in discharging groundwater. A study conducted in the current boundaries of the "Sanctuary" (FKNMS) in 1983 documented numerous pesticide peaks and heavy metal concentrations ranging from 100 to 10,000 times greater than mean seawater values in groundwater discharging from the bases of both Carysfort and French Reefs. Heavy metals identified in the groundwater discharging from those reefs included cadmium, chromium, copper, iron, lead, mercury, and zinc. Salinities of the groundwater were as low as 10 ppt at Carysfort Reef, while ambient salinity was 33 ppt (Simmons and Love, 1987). National Park Service data from 1994 also documented pulses of freshwater that were sufficient to lower ambient salinity from approximately 35 ppt to 28 ppt in groundwater discharging from the base of another deep reef near Carysfort, in Biscayne National Park (R. Curry, pers. comm.).

Those freshwater discharges verify surface/groundwater connections associated with the reefs and preferential discharges of groundwater. Such connections strongly suggest that injected fluids, such as minimally treated effluent and the proposed injection of 1.7 billion gallons of contaminated surface water (as proposed by the Florida Legislature in 2001 for "restoration" of the Everglades), ultimately would be discharging at the reef tracts.

Simmons and Love (1987) emphasized the potential adverse impacts to benthic organisms exposed to groundwater discharges laden with these hazardous substances. They further predicted that those contaminated groundwater discharges ultimately could result in the demise of the coral reefs. Later, Top et al. (2001) addressed the significant impacts that groundwater-transported pollutants can have on coastal ecology and public health. They described kills of finfish and shellfish that were related to, in part, groundwater-laden nutrients. They also provided examples of Atlantic coastal areas where groundwater accounted for half of the nitrogen loading of the sediments. Finally, they described the implication of presumed leakage of effluent injected into the Boulder Zone throughout South Florida, via deep wells.

## QUICK FIX: SHORT-TERM USES VS. LONG-TERM ABUSES OF UNDERMINING FLORIDA

The discussions above have addressed some of the long-term problems associated with groundwater mining (e.g., reductions in SGD and potential loss of natural buffer against hypersaline conditions during periods of drought, damage to the structure of the aquifer, destruction of sensitive terrestrial ecosystems including wetlands). Another form of mining with similar long-term damage resulting from relatively short-term extraction is excavation of the solid component of the aquifer (e.g., limerock and sand). Limerock and sand are used to construct roads and buildings. Mining for phosphate in rock produces comparable damage.

Groundwater occupies void spaces in the substrate. In a sandy surficial aquifer this void (pore) space increases from approximately 20% in an unmined state to 100% following mining. The void space for South Florida limestone is approximately 40%. When limerock is dredged, the limestone matrix of the aquifer that contains the groundwater is removed permanently. The resulting pit (100% pore space for the entire dimension of the pit) fills with aquifer water that previously occupied smaller spaces in the limestone matrix, relative to the size of the pit. Consequently, the water that flows into the pit due to gravity results in a reduction of groundwater levels for a considerable distance in all directions around the pit. This reduction in groundwater levels is permanent and irreversible.

In addition to the lowering of groundwater levels by removing the substrate that contained the water, water levels are reduced further by rock (and sand) mining as groundwater becomes "surface water." Under those conditions, the aquifer is exposed to the atmosphere, and an unspecified volume of aquifer water is lost to the atmosphere through evaporation. Evaporation of surface water is controlled by physical forces; therefore, as long as there is wind or a temperature or humidity gradient, additional loss of water will occur that would not take place if the water was under the ground.

In Florida, roots of native plants generally are shallow because the water table is (or was) high. With native plants there is an additional factor that is not present with evaporation of surface water. It is a biological factor. This biological factor can provide shade to reduce temperature and air movement at the surface of the substrate and increase humidity, all of which can reduce the loss of water to the atmosphere. Furthermore, when water table levels are low, many native plants have highly evolved mechanisms to reduce or halt transpiration and loss of water. Those biological factors that ameliorate loss of water are removed when the structure of the aquifer is mined.

A third form of mining in South Florida is the extraction of crude oil. Approximately 2,270,000 barrels of oil were extracted annually from 16 of approximately 100 oil wells in South Florida (Lloyd and Tootle, 1992). Like groundwater mining, extraction of oil is a fluid removal process that can result in land mass subsidence, in addition to other environmental problems. Ultimately, changes in surface elevation can be detrimental to upland species, as well as aquatic species, including coastal species. Extractions of groundwater, rock and mineral resources, and oil are forms of relatively short-term exploitation of natural resources in South Florida that can result in long-term or permanent adverse environmental impacts, including hydrologic alterations, on both the exposed and submerged areas of the carbonate platform.

## ADJUSTING THE COURSE

In a recent report released by the USEPA's Scientific Advisory Board (1999), "hydrologic alteration" was assigned the highest ecological risk rank of 33 major environmental stressors in the U.S. (Figure 26.8). Climate change, another stressor in the "highest risks" category, results in hydrologic alterations, among other changes. Hydrologic alterations that are anthropogenic in origin, however, can result in changes that are orders of magnitude greater than those resulting from climate change, in a relatively instantaneous period of time compared to that of global climate change.

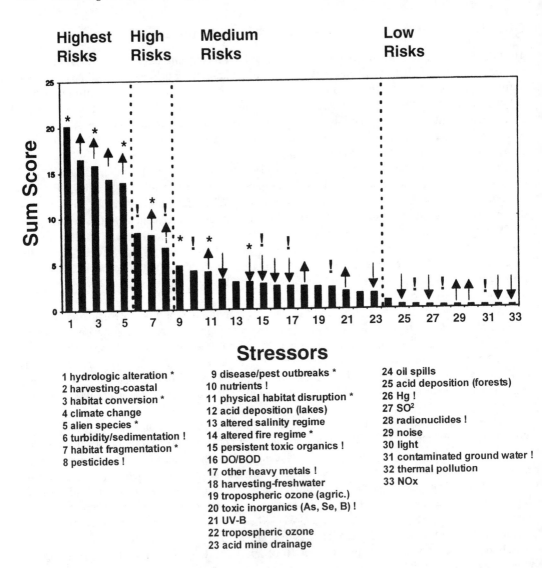

**FIGURE 26.8** Ecological risk ranking scores for 33 major environmental stressors by the USEPA Scientific Advisory Board. Asterisks added denote risks to ecosystems linked to groundwater quantity alterations (e.g., groundwater mining; aquifer storage and recovery; and aquifer injection of wastes) in karst aquifer systems. Exclamation marks added denote risks to ecosystems that are linked to water quality alterations (e.g., aquifer-injection of wastes) in karst aquifer systems. (Modified from Scientific Advisory Board, 1999).

Ecological systems are much more resilient to change that occurs slowly, such as that associated with climate change, than to rapid change, such as that associated with hydrologic alterations that are anthropogenic in origin. Likewise, those hydrologic alterations ultimately may result in many of the other stressors included in Figure 26.8 as secondary impacts. Examples can include habitat conversion, proliferation of introduced alien species, turbidity/sedimentation, habitat fragmentation, increased concentrations of pesticides, disease/pest outbreaks, critical changes in nutrient availability (either increases or decreases), altered salinity regimes (either increases or decreases), altered fire regime, changes in dissolved oxygen (DO)/biological oxygen demand (BOD), increases in heavy metals, and increases in toxic inorganics.

Currently, no federal laws specifically address hydrologic alterations of ecological systems via anthropogenic groundwater perturbations. The environmental problems related to those anthropogenic groundwater perturbations (hydrologic alterations) can be addressed under existing federal laws. The application of those laws historically has not included aspects of environmental damage due to anthropogenic groundwater perturbations.

Recommendations of the First Biennial Report to Congress, 1996 include restoring historical freshwater flow to Florida Bay and conducting research to understand the effect of water transport from Florida Bay on water quality and resources in the Sanctuary (USEPA, 1996). The new millennium has arrived, however, and still no studies have been initiated or proposed to evaluate the effect that diversion of historic groundwater discharge from this sensitive area has had on the ecosystems of the FKNMS or Florida Bay. The only reference to deep well injection in the report is that it is to be evaluated and implemented by the City of Key West. Although the deep-well injection of effluent in Key West was scheduled to begin in spring 2001 (City of Key West, 2000), no EIS or other scientific investigation has been conducted to determine the potential impacts of that proposed action. Until the roles of the additional stressors described above are evaluated thoroughly, it will be difficult for the goals of the Monitoring and Special Studies Program, stated in the First Biennial Report to Congress, to be met.

In general, the large-scale, long-term, adverse environmental impacts of hydrologic alterations have received relatively little attention. Consequently, the recent recognition of hydrologic alterations as a major environmental stressor is of great significance (Scientific Advisory Board, 1999). The lack of attention probably is due to three primary factors. The first factor is the difficulty in measuring the physical and ecological changes accurately. For example, Bacchus (1995, 1998) described the difficulties of measuring ecological impacts due to groundwater mining in terrestrial ecosystems.

The second factor is the considerable lapse in time (e.g., 20 years) that may transpire between the onset of the physical perturbation and the blatant expression of ecological distress (e.g., ecosystem collapse), as the rebound ability of elastic response mechanisms is exceeded. The problem is exacerbated when the connection between the physical changes and potential ecological response is not recognized at the onset of the physical perturbation (see Bacchus, 1998).

The final factor is poor dissemination of available information. For example, results of scientific studies often are published in scientific journals without conveying the information to the people who have the ability to support the necessary research and resolve the problem (e.g., general public, elected government officials, staff of regulatory agencies, members of the judicial system).

Examples of some of the testable hypotheses to investigate the environmental impacts of groundwater perturbations in South Florida are provided in Table 26.3. Comprehensive scientific studies testing those hypotheses should have been conducted prior to initiation and during the early implementation stages of the USEPA rule permitting the injection of wastes into the aquifers via shallow and deep wells, particularly in South Florida's sensitive environment. As indicated previously, such comprehensive studies still have not been conducted (W. Diamond, pers. comm.).

Despite the fact that the essential comprehensive studies have not been conducted, the existing body of scientific knowledge supports each of the testable hypotheses listed in Table 26.3 (as described in this chapter). Of particular relevance is the realization that a similar scenario of groundwater contamination on a much smaller scale in Western Australia (Cockburn Sound) led to eutrophication of the coastal waters and subsequent abandonment of aquifer-injected wastewater. That case refutes the presumption held by some that significant adsorption of nutrients occurs in carbonate aquifers under conditions of significant anthropogenic perturbations. That realization and the current attempts to control, halt, or reverse the nutrient-laden groundwater plume discharging into those coastal waters specifically provide support for testable hypothesis $H_4$. As in the early attempts to "improve" conditions in the Everglades (via ditches and canals), the aquifer injection of wastes throughout South Florida's coast soon will be recognized as an environmental nightmare, out of a magnitude that we have not yet witnessed in South Florida or elsewhere.

**TABLE 26.3**
**Examples of Some Testable Hypotheses To Investigate the Environmental Impacts of Groundwater Perturbations in South Florida**

$H_1$:   An extensive geophysical survey of South Florida's nearshore areas will identify fractures, dissolution features, collapse features, and other types of discontinuities in the carbonate aquifer system that can serve as preferential flow paths to the coral reefs and Florida Bay for wastes (e.g., effluent) injected into shallow and deep aquifers on the mainland (e.g., the Miami–Dade facility, and the Florida Keys).

$H_2$:   An extensive geophysical survey of the Everglades will identify fractures, dissolution features, collapse features, and other types of discontinuities in the carbonate aquifers that can serve as preferential flow paths for deep-well-injected municipal effluent to migrate from the Miami–Dade facility into the Everglades.

$H_3$:   Comprehensive tracer studies in all shallow and deep waste-injection wells in South Florida, including the Florida Keys, will demonstrate that the injected wastes are discharged into coastal waters in proximity to coral reefs and other sensitive ecosystems, thus constituting "open-water" disposal of those wastes, in violation of Chapter 403 of the Florida Statutes and the federal Clean Water Act.

$H_4$:   Hydrogeologic/isotopic studies will demonstrate that the discharge to nearshore surface waters of wastes injected into wells on the mainland of South Florida, including the Keys, cannot be controlled, halted, or reversed.

$H_5$:   Detailed chemical analyses of the wastes injected into shallow and deep wells in South Florida, including the Keys, will demonstrate that in addition to excessive nutrients, the wastes contain pathogens, endocrine disruptors, and other pollutants that are being discharged to coastal surface waters in violation of Chapter 403 of the Florida Statutes and the federal Clean Water Act.

$H_6$:   Hydroecological studies will demonstrate that the large-scale groundwater alterations of the Biscayne aquifer (e.g., agricultural withdrawals; municipal withdrawals for Miami–Dade County and the Florida Keys; structural mining of the aquifer) are significant factors in the decline of South Florida ecosystems, including the Everglades, Florida Bay, and the coral reefs.

*Note:* Includes the potential role of submarine groundwater discharge in the transport of excessive nutrients, pathogens, endocrine disruptors, and other pollutants to coral reef ecosystems, Florida Bay, and the Everglades.

## CONCLUSIONS AND RECOMMENDATIONS

### SUMMARY

The regional karst aquifer system underlying South Florida is a dynamic rather than static system, changing both spatially and temporally in response to anthropogenic perturbations. The historic SGD in South Florida occurred throughout the submerged carbonate platform (including along the margin of the platforms, and at outcrops in terraces, and associated karst dissolution/subsidence features) in two forms: (1) large-scale direct preferential discharge, and (2) small-scale diffuse preferential seepage. Published and unpublished data suggest that the historic discharge of pristine, low-salinity, low-nutrient groundwater of constant temperature in coastal areas was significant in maintaining the coral reefs and other oceanside ecosystems, Florida Bay, and the Everglades.

The hypothesis presented in this chapter suggests that the areas of preferential groundwater discharge throughout the submerged carbonate platform represented microniches with essential elements (e.g., low levels of nutrients, minerals) and moderated temperature and salinity regimes that were lacking in the surrounding, historically oligotrophic waters. Consequently, this hypothesis further suggests that coral reefs in South Florida, and possibly throughout their entire range, developed in areas that were receiving significant contributions of historic SGD. This hypothesis also suggests that the living reefs are members of a "regional guild" of ecosystems, possibly including seagrass beds, that are dependent on that natural groundwater resource and that mobile organisms, such as sea turtles and fish, are attracted to areas of freshwater SGD. The proposed hypothesis also suggests that temperature and salinities in Florida Bay were moderated by historic discharges of pristine, low-salinity, low-nutrient groundwater of constant temperature, and that the hydroperiods of the Everglades wetlands were influenced significantly by the historic SGD.

Thus ecosystems in Florida Bay and the Everglades also may be members of the "regional groundwater guild."

The historic SGD in South Florida, including discharges to the oceanside coral reef ecosystems Florida Bay, and the Everglades, has experienced two types of extensive anthropogenic perturbations. First, groundwater mining (e.g., Biscayne aquifer), channelization, canalization (e.g., Everglades), and other excavations (e.g., limerock mining) on the mainland have resulted in the reduction or cessation of that historic SGD, in addition to severely disrupting the hydroperiod of wetlands such as the Everglades. Second, South Florida's highly permeable karst aquifer system has been receiving large volumes of fluid wastes (including minimally treated effluent) via deep and shallow injection wells since the early 1970s. This waste presumably contains microbial pathogens and other pollutants (e.g., endocrine disruptors), in addition to excessive nutrients. The necessary research and monitoring to determine (1) all of the contaminants present in the waste, (2) where the waste is discharging, and (3) the environmental impacts of the wastes are not required or conducted. Published and unpublished data support the conclusion that injected effluent displaces water in the carbonate aquifers (including the injected effluent), and that induced discharge to surface waters occurs.

It is plausible to assume that the same subsurface flow paths that supplied pristine, low-salinity, low-nutrient groundwater of constant temperature in coastal areas historically, now are serving as points of preferential discharge of fluid wastes injected into wells along South Florida's coast. It also is plausible to assume that those contaminants are entering surface waters as both direct and diffuse discharges (including induced discharges). The hypothesis presented in this chapter suggests that at least a portion of the injected waste is being transported directly to coral reefs, Florida Bay, and the Everglades via preferential flow paths in the aquifer, with minimal dilution, dispersion, or adsorption due to the preferential flow paths and rapid rate of flow. The hypothesis presented in this chapter also suggests that SGD containing these injected wastes are entering the Gulf of Mexico, Straits of Florida, Gulf Stream, and Atlantic Ocean as induced discharges to surface waters.

This hypothesis proposes that the injected effluent and induced discharges: (1) contribute to the eutrophication/hypoxia of South Florida's sensitive coastal waters; (2) predispose South Florida's coastal organisms to mass mortalities and diseases (e.g., via reduced host resistance, increased pathogen virility, decreased antagonist vigor, introduction of alien microbes); and (3) contribute to hydroperiod disruption of the Everglades at a level capable of preventing restoration of that wetland.

The increased anthropogenic nutrients and other contaminants from deep-well injection of sewage effluent, such as that injected at the Miami–Dade Black Point facility, may exceed the localized eutrophication/pollution currently generated in the Keys. Thus, expensive federal initiatives such as the $213 million designated to support the "Florida Keys Water Quality Improvement Act" and to increase the volume of injected effluent in the Keys ironically would exacerbate the problem rather than ameliorate it. Likewise, other costly federal initiatives such as the $7.8 billion Everglades Restoration Project and the Harmful Algal Blooms and Hypoxia Research and Control Act of 1998 (Public Law 105-383), which designated more than $52 million to address the problem of harmful algal blooms and hypoxia in the Gulf of Mexico, are being circumvented by the continued and expanded groundwater mining and aquifer injection of fluid wastes in South Florida, proposed and implemented by federal, state, regional, and local government agencies.

## INITIAL RECOMMENDATIONS

Currently, wells drilled into the highly permeable karst carbonate aquifer system in Key West have been permitted to inject municipal sewage effluent, while no data are available regarding where or how rapidly this effluent will reach the surface waters of the sensitive South Florida ecosystems. No disposal of effluent into aquifer-injection wells not currently in operation (disposing of effluent) should occur until a comprehensive EIS has been completed for all operational aquifer-injection wells in Dade and Monroe Counties, addressing all of the potential environmental impacts of the injected effluent, including impacts to endangered and threatened species. The potential impacts evaluated should include impacts to the Florida Everglades, Florida Bay, and the coral reefs due to the continued diversion of formerly pristine groundwater discharges to those ecosystems for the current use as a receiving fluid for human excrement.

The available literature supports the conclusion that effluent injected into the karst carbonate aquifer system in South Florida (at any depth and any treatment level) cannot be discounted as a potential source of eutrophication and contamination of South Florida's ecosystems (including the theoretically protected coral reefs, Florida Bay, and the Everglades). Therefore, the "ostrich component" (anthropogenic groundwater perturbations, including induced recharge and discharge) should be incorporated into any multiple stressor model developed for testing in South Florida ecosystems.

Anthropogenic alterations of groundwater such as those associated with large-scale subsurface fluid extraction and well injections (e.g., aquifer "storage" and "recovery," "disposal" of treated effluent, "nontoxic" waste, and brine wastes) are examples of the adage "knowledge before wisdom." Technological advances provided the ability for large-scale exploitation of the Floridan aquifer system (including overlying aquifers) before we were able to comprehend the ramifications of those exploitations. Because all large-scale subsurface perturbations in South Florida are capable of resulting in significant adverse impacts to federal, as well as state, regional, and local protected areas (e.g., FKNMS, Everglades National Park), a comprehensive EIS should be completed, addressing all of the potential environmental impacts prior to initiating any additional activities of this nature in South Florida. A comprehensive EIS also should be conducted for all existing activities of this nature in South Florida. The EIS should determine the impact of those activities on protected areas in South Florida (specifically, impacts on the proposed restoration of the Everglades) and their role in initiating or perpetuating harmful algal blooms and hypoxia, including those in the Gulf of Mexico, pursuant to Public Law 105-383.

Historically, the agencies regulating and attempting to maintain natural resources have been compartmentalized into aspects of either physical sciences (e.g., geology, hydrology, and geochemistry) or life sciences (e.g., fish and wildlife). This compartmentalization has perpetuated the lack of awareness and understanding of the critical interrelationships between the organisms and ecosystems those agencies are charged with maintaining and the reliance of those organisms and ecosystems on groundwater. This historic compartmentalization was a primary factor in the need to produce this book. Although the multidisciplinary approach is gaining recognition, recent agency decisions and actions in South Florida suggest a severely limited understanding of those critical interrelationships. All students earning an undergraduate degree from institutes of higher learning in the state of Georgia are required to complete an environmental literacy program. A similar program should be initiated in Florida, emphasizing the ecological importance of the surfacewater/groundwater linkage associated with Florida's regional karst aquifer system. All agency personnel involved in regulating or maintaining Florida's natural resources also should be required to complete the environmental literacy program.

Scientific monitoring programs generally are not designed to provide information about the ecological components most intricately linked to perturbations of the regional karst carbonate aquifer system in Florida. For example, groundwater data are not collected, analyzed, or reported in a manner conducive to detecting anthropogenic changes that may be detrimental to associated organisms. More specifically, "average annual" data may be useful for hydrologic purposes, but it is of limited use for living components of ecosystems with seasonal requirements and adverse reactions to induced extreme events masked in reported "average annual" conditions. Likewise, water quality data collected quarterly or even monthly also have severe limitations for detecting pulsed events that may be life threatening for sensitive organisms. Even when the data are available on a finer scale, the locations where the data are collected (chosen randomly or systematically without consideration for preferential discharge from the underlying aquifer) are not likely to coincide with the locations where the greatest (or any) impact may be occurring. The magnitude and full impact of anthropogenic perturbations of the Floridan aquifer system cannot be realized or corrected until basic changes are made in the collection of hydrologic data. An extensive multidisciplinary approach needs to be developed and adopted for the new millennium. Areas of characteristic discontinuities in lower permeability (semiconfining) zones, both land-based and submarine, must become the new targets for intensive, long-term monitoring. Reference sites should be established for comparable intensive monitoring in subregions where minimal anthropogenic groundwater perturbations have occurred. For example, the selection of sample site locations for placement of instruments and collection of SGD data is a critical factor, particularly when associated with coral reefs. Specifically, in areas underlain by a karst aquifer, a study design using data collection points and transects that are not linked to underlying hydrogeological features and potential preferential flow paths cannot provide an accurate evaluation of the SGD contribution.

## ACKNOWLEDGMENTS

Comments on the manuscript were contributed by Barrett Brooks, Bill Burnett, Jaye Cable, Don and Karen DeMaria, Randall Denker, Jack Kindinger, Frank Manheim, June Oberdorfer, James Porter, Chris Reich, Don Rosenberry, Eugene Shinn, Ian Thomas, Zafer Top, and Patrick Yananton. Additional contributions were made by the members and activities of the SCOR/LOICZ Working Group on SGD. Support was provided by Ellen Hemmert, Maurice Spitz, Greg Thompson, and James Thompson. Gratitude also is expressed to Pat Mixson for her assistance in gaining access to numerous USGS publications reviewed for this chapter, to the UGA Reference Librarians for assistance in securing copies of various documents, and to Thelma Richardson for graphics assistance. Kerry Britton provided insight into aspects of biocontrol of fungal pathogens. Finally, I am indebted to George Brook, Todd Rasmussen, and Mark Stewart for providing the ability to visualize the subsurface world of karst aquifers, and to the UGA Institute of Ecology for providing exposure to a myriad ecological concepts, problems, and solutions.

## REFERENCES

Adams, J.E. and M.L. Rhodes. 1960. Dolomitization by seepage refluxion, *Bull. Am. Ass. Petrol. Geol.*, 44:1912–1920.

Ahel, M., W. Giger, and M. Koch. 1994. Behaviour of alkylphenol polyethoxylate surfactants in the aquatic environment. I. Occurrence and transformation in sewage treatment, *Water Res.*, 28:1131–1142.

Appleyard, S.J. 1994. *The Discharge of Nitrogen and Phosphorus from Groundwater into Cockburn Sound, Perth Metropolitan Region*, Hydrogeology Report No. 1994/39, 8 p. + app.

Aronson, R.B., P.J. Edmunds, W.F. Precht, D.W. Swanson, and D.R. Levitan. 1994. Large-scale, long-term monitoring of Caribbean coral reefs: simple, quick inexpensive techniques, *Atoll Res. Bull.*, No. 421, 19 pp.

Bacchus, S.T. 1995. Improved assessment of baseline conditions and change in wetlands associated with groundwater withdrawal and diversion, pp. 158–167, in Hatcher, K.J., Ed., Proc. of the 1995 Georgia Water Resources Conference, April 11–12, 1995, The University of Georgia, Athens, GA.

Bacchus, S.T. 1998. Determining sustainable yield in the southeastern Coastal Plain, pp. 503–519, in *Land Subsidence Case Studies and Current Research*, Borchers, J.W., Ed., Proc. of the Dr. Joseph F. Poland Symposium, October 2–8, 1995, Sacramento, CA, Association of Engineering Geologists Special Publ. No. 8.

Bacchus, S.T. 1999a. Cumberland Island National Seashore: linking offshore impacts to mainland withdrawals from a regional karst aquifer, pp. 463–472, in Hatcher, K.J., Ed., Proc. of the 1999 Georgia Water Resources Management Conference, March 30–31, 1999, The University of Georgia, Athens, GA.

Bacchus, S.T. 1999b. New Approaches for Determining Sustainable Yield from the Regional Karst Aquifer of the Southeastern Coastal Plain, Ph.D. dissertation, University of Georgia, Athens, GA, 172 pp.

Bacchus, S.T. 2000a. Predicting nearshore environmental impacts from onshore anthropogenic perturbations of ground water in the southeastern Coastal Plain, USA, pp. 609–614, in Interactive Hydrology: 3rd International Hydrology and Water Resources Symposium of the Institution of Engineers, Australia, November 20–23, 2000, Perth, Western Australia.

Bacchus, S.T. 2000b. Uncalculated impacts of unsustainable aquifer yield and evidence of subsurface interbasin flow, *J. Am. Water Resour. Assoc.*, 36(3):457–481.

Bacchus, S.T, T. Hamazaki, K.O. Britton, and B.L. Haines. 2000. Soluble sugar composition of pond-cypress: a potential hydroecological indicator of groundwater perturbations, *J. Am. Water Resour. Assoc.*, 36(1):1–11.

Banks, J.E. 1964. Petroleum in Comanche (Cretaceous) section, Bend area, Florida, *Bull. Am. Assoc. Petrol. Geol.*, 44:1737–1748.

Beck, B.F. 1989. *Engineering and Environmental Impacts of Sinkholes and Karst*, Proc. of the Third Multidisciplinary Conference on Sinkholes and the Engineering and Environmental Impacts of Karst, St. Petersburg Beach, FL, October 2–4, 1989, A.A. Balkema Publishers, Brookfield, VT, 384 pp.

Bell, P.R.F. 1992. Eutrophication and coral reefs: some examples in the Great Barrier Reef Lagoon, *Water Res.*, 26:553–568.

Brand, L.E. 2000. *An Evaluation of the Scientific Basis for "Restoring" Florida Bay by Increasing Freshwater Runoff from the Everglades*, report submitted to the EPA Florida Keys National Marine Sanctuary Water Quality Protection Program, www.aoml.noaa.gov/ocd/sferpm/brand/FLBreport.html.

Brand, L.E. 2002. The transport of terrestrial nutrients to South Florida coastal waters, in *The Everglades, Florida Bay, and Coral Reefs of the Florida Keys: An Ecosystem Sourcebook*, Porter, J.W. and Porter, K.G., Eds., CRC Press, Boca Raton, FL.

Brown, E.A. and K.O. Britton. 1986. *Botryosphaeria* diseases of apple and peach in the southeastern United States, *Plant Dis.*, 70:480–484.

Burkholder, J.M. and H.B. Glasgow, Jr. 1997. *Pfiesteria piscicida* and other *Pfiesteria*-like dinoflagellates: behavior, impacts and environmental controls, *Limnol. Oceanogr.*, 42(5):1052–1075.

Burkholder, J.M., H.B. Glasgow, Jr., and C.W. Hobbs. 1995. Fish kills linked to a toxic ambush-predator dinoflagellate: distribution and environmental conditions, *Mar. Ecolo. Prog. Ser.*, 124(43):43–61.

Bush, P.W. and R.H. Johnston. 1988. *Ground-Water Hydraulics, Regional Flow, and Ground-Water Development of the Floridan Aquifer System in Florida and in Parts of Georgia, South Carolina, and Alabama: Regional Aquifer-System Analysis*, U.S. Geological Survey Professional Paper 1403-C, 80 pp. + maps.

Carlson, P.R., L.A. Yarbro, and T.R. Barber. 1994. Relationship of sediment sulfide to mortality of *Thalassia testudinum* in Florida Bay, *Bull. Mar. Sci.*, 54:733–746.

Carpenter, R.C. 1985. Sea urchin mass-mortality: effects on reef algal abundance, species composition, and metabolism and other coral reef herbivores, pp. 53–60, in Gabrie, C. and Salvat, B., Eds., Proc. of the Fifth International Coral Reef Congress, Tahiti, May 27–June 1, 1985.

Church, T.M. 1996. An underground route for the water cycle, *Nature*, 380:579-580.

City of Key West. 2000. Department, Utilities, Sewer, OMI, www.keywestcity.com.

Colborn, T., D. Dumanoski, and J.P. Myers. 1996. Our Stolen Future: Are We Threatening Our Fertility, Intelligence, and Survival? — A Scientific Detective Story, Penguin Books, New York, 306 pp.

Cooper, H.H. 1959. A hypothesis concerning the dynamic balance of freshwater and saltwater in a coastal aquifer, *J. Geophys. Res.*, 64:461–467.

Corbett, D.R., J. Chanton, W. Burnett, K. Dillon, C. Rutkowski, and J.W. Fourqurean. 1999. Patterns of groundwater discharge into Florida Bay, *Limnol. Oceanogr.*, 44:1045–1055.

Corbett, D.R., L. Kump, K. Dillon, W. Burnett, and J. Chanton. 2000. Fate of wastewater-borne nutrients under low discharge conditions in the subsurface of the Florida Keys, USA, *Mar. Chem.*, 69:99–115.

Dahan, O., R. Nativ, E.M. Adar, B. Berkowitz, and N. Weisbrod. 2000. On fracture structure and preferential flow in unsaturated chalk, *Ground Water*, 38(3):444–451.

Davidson, N., M. Servos, J. Maguire, D. Bennie, B. Lee, P. Cureton, R. Sutcliffe, and T. Rawn. 2000. The environmental risk assessment of nonylphenol and its ethoxylates in the Canadian aquatic environment, p. 147, in *Earth Science in the 21st Century: Paradigms, Opportunities and Challenges*, SETACs 21st Annual Meeting in North America, November 12–16, Nashville, TN.

Davis, M.L. and D.A. Cornwell. 1991. *Introduction to Environmental Engineering*, McGraw-Hill, New York, 822 pp.

D'Elia, C.E. and J.A. DeBoer. 1978. Nutritional studies of two red algae. 2. Kinetics of ammonium and nitrate uptake, *J. Phytocol.*, 14:266–272.

D'Elia, C.E., K.L. Webb, and J.W. Porter. 1981. Nitrate-rich groundwater inputs to Discovery Bay, Jamaica: a significant source of N to local reefs? *Bull. Mar. Sci.*, 31:903–910.

Den Hartog, C. 1987. "Wasting disease" and other dynamic phenomena in *Zostera* beds, *Aquatic Bot.*, 27:3–14.

DeMaria, K. 1996. *Changes in the Florida Keys Marine Ecosystem Based on Interviews with Experienced Residents*, The Nature Conservancy and Center for Marine Conservation, Washington, D.C., 105 pp.

Department of Commerce and Trade. 2000. *Groundwater Monitoring of Nitrogen Discharges into Jervoise Bay — March 2000*, report prepared and submitted by PPK Environment & Infrastructure Pty, Ltd., 32 pp. + app.

Dillon, K.S., D.R. Corbett, J.P. Chanton, W.C. Burnett, and D.J. Furbish. 1999. The use of sulfur hexafluoride ($SF_6$) as tracer of septic tank effluent in the Florida Keys, *J. Hydrol.*, 220:129–140.

Dillon, K.S., D.R. Corbett, J.P. Chanton, W.C. Burnett and L. Kump. 2000. Bimodal transport of a waste water plume injected into saline ground water of the Florida Keys, *Ground Water*, 38(4):624–634.

Dillon, K.S., W. Burnett, J. Chanton, G. Kim, D.R. Corbett, L. Kump, and K. Elliott. 2001. Fate of wastewater-borne phosphate under high discharge conditions into karst Key Largo limestone. Final Report to the U.S. Environmental Protection Agency from the Department of Oceanography, Florida State University #1368-732-28, Tallahassee, FL, 79 pp.

Division of Administrative Hearings. 2000. *Port Antigua Townhouse Association, Inc., and Port Antigua Property Owners Association* v. *Seanic Corporation and Department of Environmental Protection*, State of Florida Case Nos. 00-00137 and 00-0139, Tallahassee, FL.

Duarte, C.M. 1995. Submerged aquatic vegetation in relation to different nutrient regimes, *Ophelia*, 41:87–112.

Durako, M.J. 1994. Seagrass die-off in Florida Bay (USA): changes in shoot demographic characteristics and population dynamics in *Thalassia testudinum*, *Mar. Ecol. Prog. Ser.*, 110:59–66.

Durako, M.J. and K.M. Kuss. 1994. Effects of *Labyrinthula* infection on the photosynthetic capacity of *Thalassia testudinum*, *Bull. Mar. Sci.*, 54(3):727–732.

Dustan, P. 1977. Vitality of reef coral populations off Key Largo, Florida: Recruitment and mortality, *Environ. Geol.*, 2:51–58.

Dustan, P. 1985. Community structure of reef-building corals in the Florida Keys: Carysfort Reef, Key Largo and Long Key, Dry Tortugas, I, 288:1–27.

Dustan, P. 1987. Preliminary observations on the vitality of reef corals in San Salvador, Bahamas, pp. 57–65, in Proc. of the Third Symposium on the Geology of The Bahamas, Fort Lauderdale, FL, CCFL Bahamian Field Station.

Dustan, P. 1999. Coral reefs under stress: sources of mortality in the Florida Keys, *Natural Resour. Forum*, 19990500, 23(2):147–155.

Edwards, C. 2000. Problems posed by natural environments for monitoring microorganisms, *Molec. Biotechnol.*, 15(3):211–223.

Ekelund, R. Å Bergman, Å. Ganmo, and M. Berggren. 1990. Bioaccumulation of 4-nonylphenol in marine animals: a re-evaluation, *Environ. Pollut.*, 64:107–20.

Enos, P. and R.D. Perkins. 1977. *Quaternary Sedimentation in South Florida*, Memoir 147, The Geological Society of America, Boulder, CO.

Faught, M.K. and J.F. Donoghue. 1997. Marine inundated archaeological sites and paleofluvial systems: examples from a karst-controlled continental shelf setting in Apalachee Bay, Northeastern Gulf of Mexico, *Geoarcheol. Int. J.*, 12:417–458.

FDEP. 1999. Class I injection facilities map, Florida Department of Environmental Protection, Tallahassee, FL.

Fegatella, F. and R. Cavicchioli. 2000. Physiological responses to starvation in the marine oligotrophic ultramicrobacterium *Sphingomonas* sp. strain RB2256, *Appl. Environ. Microbiol.*, 665:2037–2044.

Fish, J.E. and M. Stewart. 1991. *Hydrogeology of the Surficial Aquifer System, Dade County, Florida*, U.S. Geological Survey Water-Resources Investigations Report 90-4108, Tallahassee, FL, 50 pp. + maps.

Fitterman, D.V. and M. Deszcz-Pan. 1999. Geophysical mapping of saltwater intrusion in Everglades National Park, pp. 16–17, in U.S. Geological Survey Program on the South Florida Ecosystem: Proc. of South Florida Restoration Science Forum, May 17–19, 1999, Boca Raton, FL, U.S. Geological Survey Open-File Report 99-181.

Fuss, L. 2000. Endangered turtles' illness still a mystery: 11 loggerheads are found near death — experts fear many more unreported. Miami Herald, Dec. 24, 2000, www.herald.com/content/today/news/keys/digdos/049046.htm.

Geiser, D.M., J.W. Taylor, K.B. Ritchie, and G.W. Smith. 1998. Cause of sea fan death in the West Indies, *Nature*, 394:137–138.

Giger, W., P.H. Brunner, and C. Schaffner. 1984. 4-Nonylphenol in sewage sludge: accumulation of toxic metabolites from nonionic surfactants, *Science*, 225:623–625.

Giger, W., M. Ahel, M. Koch, H.U. Laubscher, C. Schaffner, and J. Schneider. 1987. Behaviour of alkylphenol polyethoxylate surfactants and of nitrilotriacetate in sewage treatment, *Water Sci. Technol.*, 19:449–460.

Glasgow, H.B., Jr., J.M. Burkholder, D.E. Schmechel, P.A. Tester, P.A. Rublee. 1995. Insidious effects of a toxic estuarine dinoflagellate on fish survival and human health, *J. Toxicol. Environ. Health*, 46:501–522.

Goerres, B., J. Campbell, and H. Kotthoff. 1999. *Vertical Ground Movements in the Lower Rhenish Embayment*, International Union of Geodesy and Geophysics, Birmingham, England, p. A59.

Granmo, Å., R. Ekelund, K. Magnusson, and M. Berggren. 1989. Lethal and sublethal toxicity of 4-nonylphenol to the common mussel (*Mytilus edulis*), *Environ. Pollut. Ser. A.*, 59:115–127.

Hallock, P., H.K. Talge, K. Smith, and E.M. Cockey. 1992. Bleaching in a reef-dwelling foraminifer, *Amphistegina gibbosa*, pp. 44–49, in Proceedings of the Seventh International Coral Reef Symposium, Guam, 1992, Vol. 1.

Hallock, P., H.K. Talge, E.M. Cockey, and R.G. Muller. 1995. A new disease in reef-dwelling foraminifer: implications for coastal sedimentation, *J. Foraminiferal Res.*, 25:280–286.

Harline, M.M. and B. Thorne-Miller. 1981. Nutrient enrichment of seagrass beds in a Rhode Island coastal lagoon, *Mar. Biol.*, 65:221–229.

Henry, H.R. 1959. Salt intrusion into fresh-water aquifers, *J. Geophys. Res.*, 64:1911–1919.

Henry, H.R. 1964. Effects of dispersion on salt encroachment in coastal aquifers, pp. 70–84, in *Sea Water in Coastal Aquifers*, Cooper, H.H. et al., Eds., U.S. Geological Service Water-Supply Paper 1613–C, Washington, D.C.

Hentschel, U., M. Steinert, and J. Hacker. 2000. Common molecular mechanisms of symbiosis and pathogenesis, *Trends Microbiol.*, 8(5):226–231.

Hickey, J.J. and J. Vecchioli. 1986. *Subsurface Injection of Liquid Waste with Emphasis on Injection Practices in Florida*, U.S. Geological Survey Water-Supply Paper 2281, Restin, VA, 25 pp.

Hofstetter, R.H. and R.S. Sonenshein. 1990. *Vegetative Changes in a Wetland in the Vicinity of a Well Field, Dade County, Florida*, U.S. Geological Survey Water-Resources Investigations Report 89–4155, Tallahassee, FL, 16 pp.

Holt, M.S., G.C. Mitchell, and R.J. Watkinson. 1992. The environmental chemistry, fate and effects of nonionic surfactants, pp. 89–144, in *The Handbook of Environmental Chemistry*, Vol. 3, Part F: *Anthropogenic Compounds, Detergents*, de Oude, N.T., Ed., Springer-Verlag, Berlin.

Hunt, J. and W. Herrnkind. 1993. *Sponge and Lobster Research*, final report to the Florida Department of Natural Resources, Contract C-8077.

Jaap, W.C. 1979. Observations on zooxanthellae expulsion at Middle Sambo Reef, Florida Keys, *Bull. Mar. Sci.*, 29: 414–422.

Jaap, W.C. 1985. An epidemic zooxanthellae expulsion during 1983 in the lower Florida Keys coral reefs: hyperthermic etiology, *Int. Coral Reef Congr. Proc.*, 6:143–148.

Jobling, S., M. Nolan, C.R. Tyler, G. Brighty and J.P. Sumpter. 1998. Widespread sexual disruption in wild fish, *Environ. Sci. Technol.*, 32:2498–2506.

Johannes, R.E. 1980. The ecological significance of submarine discharge of groundwater, *Mar. Ecol. Prog. Ser.*, 3:365–373.

Johannes, R.E. and C.J. Hearn. 1985. The effect of submarine groundwater discharge on nutrient and salinity regimes in a coastal lagoon off Perth, Western Australia, *Estuar. Coast. Shelf Sci.*, 21:789–800.

Johnston, R.H., R.E. Krause, F.W. Meyer, P.D. Ryder, C.H. Tibbals and J.D. Hunn. 1980. *Estimated Potentiometric Surface for the Tertiary Limestone Aquifer System, Southeastern United States, Prior to Development*, U.S. Geological Survey Open-File Report 80–406, Restin, VA, map.

Johnston, R.H., H.G. Healy, and L.R. Hayes. 1981. *Potentiometric Surface for the Tertiary Limestone Aquifer System, Southeastern United States, May 1980*, U.S. Geological Survey Open-File Report 81–486, Restin, VA, map.

Jordan, G.F. 1954. Large sink holes in Straits of Florida, *Bull. Am. Assoc. Petrol. Geol.*, 38:1810–1817.

Jordan, G.F., R.H. Malloy, and J.W. Kofoed. 1964. Bathymetry and geology of Pourtales Terrace, Florida, *Mar. Geol.*, 1:259–287.

Joseph, S. and K.G. Bhat. 2000. Effect of iron on the survival of *Vibrio cholerae* in water, *Indian J. Med. Res.*, 111:115–117.

Kaufman, M.I. 1973. Subsurface wastewater injection, Florida. Am. Civil Engineers Proc. Paper 9598, *J. Irrigation Drainage Div.*, 99:53–70.

Kaufman, M.I., D.A. Goolsby, and G.L. Faulkner. 1973. Injection of acidic industrial waste into a saline carbonate aquifer: geochemical aspects, pp. 526–551, in Braunstein, J., Ed., Second International Symposium on Underground Waste Management and Artificial Recharge, New Orleans, Louisiana, September 26–30, 1973, Vol. 1, American Association of Petroleum Geologists, New Orleans, LA.

Kaufman, M.I. and D.J. McKenzie. 1975. Upward migration of deep-well waste injection fluids in Floridan Aquifer, South Florida, *USGS J. Res.*, 3:261–271.

Kindinger, J.L., J.B. Davis, and J.G. Flocks. 1997. *Seismic Stratigraphy of the Central Indian River Region, Indian River County, Florida*, Indian River Region, U.S. Geological Survey Open-File Report 97–723, Tallahassee, FL.

Ko, W.-H. and J.L. Lockwood. 1967. Soil fungi-stasis: relation to fungal spore nutrition, *Phytopathology*, 57:894–901.

Kohout, F.A. 1960. Cyclic flow of saltwater in the Biscayne aquifer of southeastern Florida, *J. Geophys. Res.*, 65(7):2133–2141.

Kohout, F.A. 1964. Flow of fresh water and saltwater of the Biscayne aquifer of the Miami area, Florida U.S. Geological Survey Water Supply Paper 1613-C, Tallahassee, FL.

Kohout, F.A. 1965. A Hypothesis Concerning Cyclic Flow of Salt Water Related to Geothermal Heating in the Floridan Aquifer, *N.Y. Acad. Sci. Trans. Ser. 2*, 28:249–271.

Kohout, F.A. 1966. Submarine springs: a neglected phenomenon of coastal hydrology. Central Treaty Organization, Symp. on Hydrology and Water Research, U.S. Dept. of Interior, Washington, D.C., pp. 391–413.

Kohout, F.A. 1967. Ground-water flow and the geothermal regime of the Floridian Plateau. *Trans.: Gulf Coast Assoc. Geol. Soc.*, 27:339–354.

Kohout, F.A. and M.C. Kolipinski. 1967. Biological zonation related to groundwater discharge along the shore of Biscayne Bay, Miami, Florida, pp. 488–499, in *Estuaries*, Lauff, G.H., Ed., Conf. on Estuaries, Jekyll Island, GA, Publ. No. 83, American Association of Advanced Science, Washington, D.C.

Kohout, F.A., H.R. Henry, and J.E. Banks. 1977. Hydrogeology related to geothermal conditions of the Floridan Plateau, pp. 1–41, in *The Geothermal Nature of the Floridan Plateau*, Smith, D.L. and Griffin, G.M., Eds., Florida Bureau of Geology Special Publ. No. 21, Tallahassee, FL.

Krause, R.E. and R.B. Randolph. 1989. *Hydrology of the Floridan Aquifer System in Southeast Georgia and Adjacent Parts of Florida and South Carolina*, U.S. Geological Survey Professional Paper 1403-D, Restin, VA, 65 pp. + plates.

Kushmaro, A., Y. Loya, M. Fine, and E. Rosenburg. 1996. Bacterial infection causes bleaching of the coral *Oculina patagonica*, *Nature*, 380:396.

Lapointe, B.E. 1997. Nutrient thresholds for bottom-up control of macroalgal blooms on coral reefs in Jamaica and southeast Florida, *Limnol. Oceanogr.*, 42:1119–1131.

Lapointe, B.E. 1999. Simultaneous top-down and bottom-up forces control macroalgal blooms on coral reef (reply to the comment by Hughes et al.), *Limnol. Oceanogr.*, 44(6):1586–1592.

Lapointe, B.E., J.D. O'Connell, and G.S. Garrett. 1990. Nutrient couplings between on-site sewage disposal systems, groundwaters, and nearshore surface waters of the Florida Keys, *Biogeochemistry*, 10:289–307.

Lapointe, B.E., M.M. Littler, and D.S. Littler. 1992. Modification of benthic community structure by natural eutrophication: the Belize Barrier Reef, *Proc. 7th Int. Coral Reef Symp.*, 1:323–334.

Lapointe, B.E., W.R. Matzie, and P.J. Barile. 2002. Biotic phase shifts in Florida Bay and back reef communities of the Florida Keys: linkages with historical freshwater flows and nitrogen loading from Everglades runoff, in *The Everglades, Florida Bay, and Coral Reefs of the Florida Keys: An Ecosystem Sourcebook*, Porter, J.W. and Porter, K.G., Eds., CRC Press, Boca Raton, FL.

Lewis. J.B. 1977. Processes of organic production on coral reefs, *Biol. Res.*, 52:305–347.

Lewis. J.B. 1987. Measurements of groundwater seepage flux onto a coral reef: Spatial and temporal variations, *Limnol. Oceanogr.*, 32:1165–1169.

Li, L., D.A. Barry, F. Stagnitti, and J.-Y. Parlange. 1999. Submarine groundwater discharge and associated chemical input to a coastal sea, *Water Resour. Res.*, 35(11):3253–3259.

Li, L., H.Q. and H.F. Schroder. 2000. Surfactants: standard determination methods in comparison with substance specific mass spectrometric methods and toxicity testing by *Daphnia magna* and *Vibrio fischeri*, *Water Sci. Technol.*, 42(7-8):391–398.

Lidz, B.H. and E.A. Shinn. 1991. Paleoshores, reefs and a rising sea: South Florida, USA, *J. Coast. Res.*, 7:204–229.

Lidz, B.H., A.C. Hine, E.A. Shinn, and J.L. Kindinger. 1991. Multiple outer-reef tracts along the south Florida bank margin: outlier reefs, a new windward-margin model, *Geology*, 19:115–118.

Lidz, B.H., E.A. Shinn, M.E. Hansen, R.B. Halley, M.W. Harris, S.D. Locker, and A.C. Hine. 1997. Maps Showing Sedimentary and Biological Environments, Depth to Pleistocene Bedrock, and Holocene Sediment and Reef Thickness from Molasses Reef to Elbow Reef, Key Largo, South Florida. Miscellaneous Investigations Series, U.S. Geological Survey Map I-2505.

Lloyd, J.M. and C.H. Tootle. 1992. *1990 and 1991 Florida Petroleum Production and Exploration, Including Florida Petroleum Reserve Estimates*, Florida Geological Survey Information Circular No. 108, Tallahassee, FL, 31 pp.

Macintyre, I.G., W.F. Precht, and R.B. Aronson. 2000. Origin of the Pelican Cays Ponds, Belize, *Atoll Res. Bull.* No. 466, 11 pp.

Manheim, F.T. 1967. Evidence for submarine discharge of water on the Atlantic continental slope of the southern United States, and suggestions for further research, *N.Y. Acad. Sci. Trans. Ser.*, 2, 29:839–853.

Marella, R.L. 1992. *Water Withdrawals, Use and Trends in Florida, 1990*, U.S. Geological Survey Water-Resources Investigations Report 92-4140, Tallahassee, FL, 38 pp.

Marella, R.L. 1995. *Water Use Data by Category, County, and Water Management District in Florida, 1950–90*, U.S. Geological Survey Open-File Report 94-521, Tallahassee, FL, 84 pp.

Maslia, M.L. and D.C. Prowell. 1990. Effect of faults on fluid flow and chloride contamination in a carbonate aquifer system.

McNeill, D.F. 2000. *A Review of Upward Migration of Effluent Related to Subsurface Injection at Miami–Dade Water and Sewer South District Plant*, report prepared for the Sierra Club, Miami Group, 30 pp.

Meyer, F.W. 1984. Disposal of liquid wastes in cavernous dolostones beneath southeastern Florida, in hydrogeology of karstic terranes, *Int. Assoc. Hydrogeol.*, 1:211–216.

Meyer, F.W. 1989. *Hydrogeology, Ground-Water Movement, and Subsurface Storage in the Floridan Aquifer System in Southern Florida*, U.S. Geological Survey Professional Paper 1403-G, 59 pp.

Miller, M.W., M.E. Hay, S.L. Miller, D. Malone, E.E. Sotka, and A.M. Szmant. 1999. Effects of nutrients versus herbivores on reef algae: a new method for manipulating nutrients on coral reefs, *Limnol. Oceanogr.*, 44(8):1847–1861.

Mitchell, R. and I. Chet. 1975. Bacterial attack of corals in polluted seawater, *Microb. Ecol.*, 2:227–233.

Monarca, S., D. Feretti, C. Collivignarelli, L. Guzzella, I. Zerbina, G. Bertanza, and R. Pedrazzani. 2000. The influence of different disinfectants on mutagenicity and toxicity of urban wastewater, *Water Res.*, 34(17):4261–4269.

Moore, W.S. 1996. Large groundwater inputs to coastal waters revealed by $^{226}$Ra enrichment, *Nature*, 380:612–614.

Muehlstein, L.K. 1989. Perspectives on the wasting disease of eelgrass *Zostera marina*, *Dis. Aquatic Org.*, 7:211–221.

Muscatine, L. and J.W. Porter. 1977. Reef corals: Mutualistic symbioses adapted to nutrient-poor environments, *BioScience*, 27:454–460.

Nagelkerken, I.K. Buchan, G.W. Smith, K. Bonair, P. Bush, J. Garzon-Ferreira, L. Botero, P. Gayle, C.D. Harvell, C. Heberer, K. Kim, C. Petrovic, L. Pors, and P. Yoshioka. 1997. Widespread disease in Caribbean sea fans. II. Patterns of infection and tissue loss, *Mar. Ecol. Prog. Ser.*, 160:255–263.

National Archives and Records Administration. 2000. Revision to the Federal Underground Injection Control (UIC) Requirements for Class I - Municipal Wells in Florida; Proposed Rule, *Federal Register*, 65(131):42234–42245.

National Research Council Committee on Restoration of the Greater Everglades Ecosystem. 2001. Aquifer Storage and Recovery in the Comprehensive Everglades Restoration Plan: A Critique of the Pilot Projects and Related Plan for ASR in the Lake Okeechobee and Western Hillsboro Areas. National Academy Press, Washington, D.C., 61 pp.

NOAA. 2000. *Final Supplemental Environmental Impact Statement/Final Supplemental Management Plan, Florida Keys National Marine Sanctuary*, National Oceanic and Atmospheric Administration, Washington, D.C., 310 pp.

Newell, N.D., J.K. Rigby, A.G. Fischer, A.J. Whiteman, J.E. Hilcox, and J.S. Bradley. 1953. *The Permian Reef Complex of the Guadalupe Mountains Region, Texas and New Mexico*, W. H. Freeman, Salt Lake City, UT.

Nixon, S.W., M.E.Q. Pilson, C.A. Oviatt, P. Donaghay, B. Sullivan, S. Seitzinger, D. Rudnick, and J. Frithsen. 1984. Eutrophication of a coastal marine ecosystem: an experimental study using the MERL Microcosms, pp. 135–195, in *Flows of Energy and Materials in Marine Ecosystems: Theory and Practice*, Fasham, M.J., Ed., Plenum Press, New York.

Oberdorfer, J.A. and R.W. Buddemeier. 1985. Coral reef hydrogeology, pp. 307–312, in Proc. of the Fifth International Coral Reef Congress, Tahiti, May 27–June 1, 1985, Vol. 3: Symposia and Seminars (A).

Oberdorfer, J.A. and R.W. Buddemeier. 1986. Coral-reef hydrology: Field studies of water movement within a barrier reef, *Coral Reefs*, 5:7–12.

Ocamb, C.M. and J. Juzwik. 1995. *Fusarium* species associated with rhizosphere soil and diseased roots of eastern white pine seedlings and associated nursery soil, *Can. J. Plant Pathol.*, 17:325–330.

Odum, E.P. 1990. *How To Prosper in a World of Limited Resources*, The 1990 Ferdinand Phinizy Lecture, the University of Georgia, Athens, GA, 18 pp.

Odum, E.P. and H.T. Odum. 1957. Zonation of corals on Japtan Reef, Eniwetok Atoll, *Atoll Res. Bull.*, 52.

Odum, H.T. and E.P. Odum. 1955. Trophic structure and productivity of a windward coral reef community on Eniwetok Atoll, *Ecol. Monogr.*, 25:291–320.

Odum, H.T. and E.P. Odum. 1956. Corals as producers, herbivores, carnivores, and possibly decomposers, *Ecology*, 37(2):385.

Orem, W.H., H.E. Lerch, A.L. Bates, A. Boylan, and M. Corum. 2001. Nutrient geochemistry of sediments from two tree islands in Water Conservation Area 3B, The Everglades, Florida, in *Tree Islands of the Everglades*, Sklar, F. and van der Val, A., Eds., Kluwer Academic, Dordrecht/Norwell, MA.

Overmier, K.A. 1977. Antagonism of *Gliocladium virens* and *Trichoderma harzianum* Toward *Diplodia gossypina*, Master of Science thesis, University of Georgia, Athens, GA, 52 pp.

Palmer, C.M. 1959. *Algae in Water Supplies*, U.S. Dept. of Health, Education, and Welfare Public Health Service, Washington, D.C., 88 pp.

Parker, G.G. 1984. Hydrology of the pre-drainage system of the Everglades in Southern Florida, pp. 28–37, in *Environments of South Florida: Present and Past*, Gleason, P.J., Ed., Miami Geological Society, Coral Gables, FL.

Parveen, S., R.L. Murphree, L. Edmiston, C.W. Kaspar, K.M. Portier, and M. Tamplin. 1997. Association of Multiple-Antibiotic-Resistance Profiles with Point and Nonpoint Sources of *Escherichia coli* in Apalachicola Bay, *Appl. Environ. Microbiol.*, 63(7):2607–2612.

Patten, T.H. and J.-G. Klein. 1989. Sinkhole formation and its effect on Peace River hydrology, pp. 25–31, in Beck, B.F., Ed., Proc. of the Third Multidisciplinary Conference on Sinkholes and the Engineering and Environmental Impacts Karst, St. Petersburg Beach, FL, October 2–4, 1989, A.A. Balkema Publishers, Brookfield, VT.

Paul, J.H., J.B. Rose, J.K. Brown, E.A. Shinn, S. Miller, and S.R. Farrah. 1995. Viral tracer studies indicate contamination of marine waters by sewage disposal practices in Key Largo, FL, *Appl. Environ. Microbiol.*, 61:2230–2234.

Paul, J.H., J.B. Rose, S.C. Jiang, X. Zhou, P. Cochran, C.A. Kellog, J.B. Kang, D.W. Griffin, S.R. Farrah, and J. Lukasik. 1997. Evidence for groundwater and surface marine water contamination by waste disposal wells in the Florida Keys, *Water Res.*, 31:1448–1454.

Paul, J.H., M.R. McLaughlin, D.W. Griffin, E.K. Lipp, R. Stokes, and J.B. Rose. 2000. Rapid movement of wastewater from on-site disposal systems into surface waters in the Lower Florida Keys, *Estuaries*, 23(5):662–668.

Phipps, P.M. and H.L. Barnett. 1975. Effect of nitrogen nutrition on the free amino acid pool of *Choanephora cucurbitarum* and parasitism by two haustorial mycoparasites, *Mycologia*, 67:1128–1142.

Pitt, W.A., Jr. and F.W. Meyer. 1976. *Ground-Water Quality at the Site of a Proposed Deep-Well Injection System for Treated Wastewater, West Palm Beach, FL*, U.S. Geological Survey Open-File Report 76-91, Tallahassee, FL, 43 pp.

Pitt, W.A., Jr., F.W. Meyer, and J.E. Hull. 1977. Disposal of salt water during well construction: problems and solutions, *Ground Water*, 15:276–283.

Popenoe, P., F.A. Kohout, and F.T. Manheim. 1984. Seismic-reflection studies of sinkholes and limestone dissolution features on the northeastern Florida shelf, pp. 43–57, in Beck, B.F., Ed., Proc. of First Multidisciplinary Conference on Sinkholes: Orlando, FL, A.A. Balkema Publishers, Accord, MA.

Porter, J.W., S.K. Lewis, and K.G. Porter. 1999. The effect of multiple stressors on the Florida Keys coral reef ecosystem: a landscape hypothesis and a physiological test, *Limnol. Oceanogr.*, 44:941–949.

Precht, W.F. 1994. The use of the term guild in coral reef ecology and paleoecology: a critical evaluation, *Coral Reefs*, 13:135–136.

Precht, W.F. and R.B. Aronson. 1997. White band disease in the Florida Keys: a continuing concern, *Reef Encounter*, 22:14–16.

Purdy, E.G. and G.T. Bertram. 1993. *Carbonate Concepts from the Maldives, Indian Ocean*, American Association of Petroleum Geologists, Studies in Geology No. 34, Tulsa, OK, 56 pp.

Puri, H.S., G.L. Faulkner, and G.O. Wilson. 1973. Hydrogeology of subsurface liquid waste storage in Florida, pp. 825–850, in Braunstein, J., Ed., Second International Symposium on Underground Waste Management and Artificial Recharge, New Orleans, LA, Sept. 26–30, 1973, Vol. 2. American Association of Petroleum Geologists, New Orleans, LA.

Reich, C.D., E.A. Shinn, T.D. Hickey, and A.B. Tihansky. 2002. Tidal and meteorological influences on shallow marine groundwater flow in the upper Florida Keys, in *The Everglades, Florida Bay, and Coral Reefs of the Florida Keys: An Ecosystem Sourcebook*, Porter, J.W. and Porter, K.G., Eds., CRC Press, Boca Raton, FL.

Reilly, T.E. and A.S. Goodman. 1985. Quantitative analysis of saltwater-freshwater relationships in groundwater systems: — a historical perspective, *J. Hydrol.*, 80:125–160.

Reinhard, M., N. Goodman, and K.E. Mortelmans. 1982. Occurrence of brominated alkylphenol polyethoxy carboxylates in mutagenic wastewater concentrates, *Environ. Sci. Technol.*, 16:351–362.

Richardson, L.L. and R.G. Carlton. 1993. Behavioral and chemical aspects of black band disease of corals: an *in situ* field and laboratory study, *Am. Acad. Underwater Sci. Proc.*, 11:107–116.

Richardson, L.L., W.M. Goldberg, K.G. Kuta, R.B. Aronson, G.W. Smith, K.B. Ritchie, J.C. Halas, J.S. Feingold, and S.M. Miller. 1998. Florida's mystery coral-killer identified, *Nature*, 392:557–558.

Riedl, R. 1971. How much seawater passes through sandy beaches? *Int. Rev. Ges. Hydrobiol.*, 56:923–946.

Riedl, R. and R. Machan. 1972. Hydrodynamic patters in lotic intertidal sands and their bioclimatological implications, *Mar. Biol.*, 13:179–209.

Riedl, R., N. Huang, and R. Machan. 1972. The subtidal pump: a mechanism of interstitial water exchange by wave action, *Mar. Biol.*, 13:210–221.

Rinaldi, M.G. 1983. Invasive aspergillosis, *Rev. Infect. Dis.*, 5:1061–1077.

Robblee, M.B., T.R. Barber, P.R. Carlson, M.J. Curako, J.W. Fourqurean, L.K. Muehlstein, D. Porter, L.A. Yarbro, R.T. Zieman, and J.C. Zieman. 1991. Mass mortality of the tropical seagrass *Thalassia testudinum* in Florida Bay (USA), *Mar. Ecol. Prog. Ser.*, 71:297–299.

Roth, F.J., Jr., P.A. Orpurt, and D.G. Ahearn. 1964. Occurrence and distribution of fungi in a subtropical marine environment, *Can. J. Bot.*, 42:375–383.

Rutzler, K. and D.L. Santavy. 1983. The black band disease of Atlantic reef corals, Part I: Description of the cyanophyte pathogen P.S.Z.N.I., *Mar. Ecol.*, 4:301–319.

Rutzler, K., D.L. Santavy, and A. Antonius. 1983. The black band disease of Atlantic reef corals. Part III. Distribution, ecology, and development: P.S.Z.N.I., *Mar. Ecol.*, 4:329–358.

Schomer, N.S. and R.D. Drew. 1982. *An Ecological Characterization of the Lower Everglades and the Florida Keys*, U.S. Fish and Wildlife Service Report #14-16-009-80-999, Office of Biological Services, Washington, D.C. FWS/OBS 82/58.1, 236 pp.

Scientific Advisory Board. 1999. Integrated Environmental Decision-Making in the 21st Century: Peer Review Draft, May 3, 1999. A Report from the EPA Science Advisory Board's Integrated Risk Project, http://www.epa.gov/sab/drrep.htm.

Searlel, D.E. 1994. Late Quaternary morphology of the Cocos (Keeling) Islands, *Atoll Res. Bull.* No. 401, 13 pp.

Sewell, P.L. 1982. Urban groundwater as a possible nutrient source for an estuarine benthic algal bloom, *Estuar. Coast. Shelf Sci.*, 15:577–580.

SFWMD. 1996. *Northern Tampa Bay Water Resource Assessment Project*, Vol. 1, *Surface-Water/Ground-Water Interrelationships*, Southwest Florida Water Management District Brooksville, FL, 351 pp. + app.

Shigo, A.L., C.D. Anderson, and H.L. Barnett. 1961. Effect of concentration of host nutrients on parasitism of *Piptocephalis xenophila* and *P. virginiana*, *Phytopathology*, 51:616–620.

Shinn, E.A. 1988. The Geology of the Florida Keys, *Oceanus*, 31:47–53.

Shinn, E.A. 1999. *Geologic History of Florida Reefs: Many Unanswered Questions*, http://www.fknms.nos.noaa.gov.

Shinn, E.A., B.H. Lidz, J.L. Kindinger, J.H. Hudson, and R.B. Halley. 1989. *Reefs of Florida and the Dry Tortugas: A Guide to the Modern Carbonate Environments of the Florida Keys and the Dry Tortugas*, Field Trip Guide Book T176 for the 28th International Geological Congress, Washington, D.C., American Geophysical Union, 55 pp.

Shinn, E.A., C.D. Reich, S.D. Locker, and A.C. Hine. 1996. A giant sediment trap in the Florida Keys, *J. Coast. Res.*, 12:953–959.

Shinn, E.A., G.W. Smith, J.M. Prospero, P. Betzer, M.L. Hayes, V. Garrison, and R.T. Barber. 2000. African dust and the demise of Caribbean coral reefs, *Geophys. Res. Lett.*, 27(19):3029–3032.

Short, F.T., B.W. Ibelings, and C. Den Hartog. 1988. Comparison of a current eelgrass disease to the wasting disease in the 1930s, *Aquatic Bot.*, 30:295–304.

Silberstein, K., A.W. Chiffings, and A.J. McComb. 1986. The loss of seagrass in Cockburn Sound, Western Australia. III. The effect of epiphytes on productivity of *Posidonia australis* Hook. F, *Aquatic Bot.*, 24:355–371.

Simmons, G.M., Jr. 1992. Importance of submarine groundwater discharge (SGWD) and seawater cycling to material flux across sediment/water interfaces in marine environments, *Mar. Ecol. Prog. Ser.*, 84:173–184.

Simmons, G.M., Jr. and F.G. Love. 1987. Water quality of newly discovered submarine groundwater discharge into a deep coral habitat, p. 155–163, in *Scientific Applications of Current Diving Technology on the U.S. Continental Shelf: Results of a Symposium Sponsored by the National Undersea Research Program*, Cooper, R.A. and Shepard, A.N., Eds., University of Connecticut at Avery Point, Groton, CT, May 1984, Symposium Series for Undersea Research, NOAA's Undersea Research Program Vol. 2, No. 2, Washington, D.C.

Simmons, G.M., Jr. and J. Netherton. 1987. Groundwater discharge in a deep coral reef habitat: evidence for a new biogeochemical cycle?, pp. 1–12, in *Diving for Science '86*, Mitchell, C.T., Ed., Proc. of the American Academy of Underwater Science Sixth Annual Scientific Diving Symposium, Oct. 31–Nov. 3, 1986, Tallahassee, FL.

Simms, M. 1984. Dolomitization by groundwater-flow systems in carbonate platforms, *Trans. Gulf Coast Assoc. Geol. Soc.*, 34:411–420.

Smith, G.W., L.D. Ives, I.A. Nagelkerken, and K.B. Ritchie. 1996. Caribbean sea-fan mortalities, *Nature*, 383:487.

Snyder, S.W., M.E. Evans, A.C. Hine, and J.S. Compton. 1989. Seismic expression of solution collapse features from the Florida Platform, pp. 281–297, in Third Multidisciplinary Conference on Sinkholes, St. Petersburg Beach, FL.

Sonenshein, R.S. and R.H. Hofstetter. 1990. *Hydrologic Effects of Well-Field Operations in a Wetland, Dade County, Florida*, U.S. Geological Survey Water-Resources Investigations Report 90–4143, Tallahassee, FL, 16 pp.

Spechler, R.M. 1994. *Saltwater Intrusion and the Quality of Water in the Floridan Aquifer System, Northeastern Florida*, U.S. Geological Survey Water-Resources Investigations Report 92–4174, 76 pp.

Spechler, R.M. and G.G. Phelps. 1997. Saltwater intrusion in the Floridan aquifer system, northeastern Florida, pp. 398–400, in Hatcher, K.J., Ed., Proc. of the 1997 Georgia Water Resources Conference, March 20–22, 1997, The University of Georgia, Athens, GA.

Spechler, R.M. and W.L. Wilson. 1997. Stratigraphy and hydrogeology of a submarine collapse sinkhole on the continental shelf, northeastern Florida, pp. 61–66, in Beck, B.F., Stephenson, J.B., and Herring, J.G., Eds., The Engineering Geology and Hydrogeology of Karst Terranes: Proc. of the 6th Multidisciplinary Conference on Sinkholes and the Engineering and Environmental Impacts of Karst, Springfield, MO, April 6–9, Balkema Publishers, Brookfield, VT.

Stieglitz, T. and P. Ridd. 2000. Submarine groundwater discharge from paleochannels? "Wonky holes" on the Inner Shelf of the Great Barrier Reef, Australia, pp. 189–194, in Interactive Hydrology: 3rd International Hydrology and Water Resources Symposium of the Institution of Engineers, Australia, November 20–23, 2000, Perth, Western Australia.

Stringfield, V.T. 1966. *Artesian Water in Tertiary Limestone in the Southeastern States.* U.S. Professional Paper 517, Washington, D.C., 226 pp.

Stringfield, V.T. and Cooper., H.H., Jr. 1951. Geologic and hydrologic features of an artesian submarine spring east of Florida, *Florida Geol. Surv. Rep. Invest.*, 7(2):61–72.

Swart, P.K., G. Ellis, and P. Milne. 2000. *The Impact of Anthropogenic Waste on the Florida Reef Tract*, final report prepared by the University of Miami for the U.S. Environmental Protection Agency, 27 pp. + app.

Swayze, L.J. and W.L. Miller. 1984. *Hydrogeology of a Zone of Secondary Permeability in the Surficial Aquifer of Eastern Palm Beach County, Florida*, U.S. Geological Survey Water-Resources Investigations Report, 83-4249, Tallahassee, FL, 38 pp.

Toler, S.K. and P. Hallock. 1998. Shell malformation in stressed *Amphistegina* populations: relation to biomineralization and paleoenvironmental potential, *Mar. Micropaleontol.*, 34:107–115.

Tomascik, T. 1991. Settlement patterns of Caribbean scleractinian corals on artificial substrata along an eutrophication gradient, Barbados, West Indies, *Mar. Ecol. Prog. Ser.*, 77:261–269.

Top, Z., L.E. Brand, R.D. Corbett, W. Burnett, and J. Chanton. 2001. Helium and radon as tracers of groundwater input into Florida Bay, *J. Coast. Res.*, 17:00–00 (in press).

Tsunogai, U., J. Ishibashi, H. Wakita, T. Gamo, T. Masuzawa, T. Nakatsuka, Y. Nojiri, and T. Nakamura. 1996. Fresh water seepage and pore water recycling on the seafloor: Sagami Trough subduction zone, Japan, *Earth Planetary Sci. Lett.*, 138:157–168.

USACOE/SFWMD. 1999. *Central and Southern Florida Project Comprehensive Review Study*, Vols. 1–10, U.S. Army Corps of Engineers/South Florida Water Management District.

U.S. Army Corps of Engineers/South Florida Water Management District. 2000. Final Programmatic Environmental Impact Statement: Rock Mining - Freshwater Lakebelt Plan, Miami-Dade County, FL, USCOE, Jacksonville, FL, 105 pp. + app.

USEPA. 1991. *Water Quality Protection Program for the Florida Keys National Marine Sanctuary: Phase 1 Report*, Contract No. 68-C8-0105, Continental Shelf Associates and Battelle Ocean Sciences, Duxbury, MA.

USEPA. 1992. *Water Quality Protection Program for the Florida Keys National Marine Sanctuary: Phase II*, Contract No. 68-C8-0105, Continental Shelf Associates and Battelle Ocean Sciences, Duxbury, MA, 329 pp.

USEPA. 1996. *Water Quality Protection Program for the Florida Keys National Marine Sanctuary*, First Biennial Report to Congress.

USEPA. 2000. *EPA Proposes a New Rule To Protect Underground Sources of Drinking Water from Wastewater Disposal in South Florida*, EPA 816-F-00-022, U.S. Environmental Protection Agency, Washington, D.C.

USGS. 1999a. Center for Coastal Geology: http://coastal.er.usgs.gov/african_dust

USGS. 1999b. South Florida Restoration Science Forum: http://sofia.usgs.gov/sfrsf/rooms/coastal/flbay/.

USGS. 2001. South Florida Restoration Science Forum: http://sofia.usgs.gov/projects/groundwtr_flow/.

Vecchioli, J. 1979. Monitoring of subsurface injection of wastes, Florida, *Ground Water*, 17:244–249.

Vernon, R.O. 1970. *The Beneficial Uses of Zones of High Transmissivities in the Florida Subsurface for Water Storage and Waste Disposal*, Information Circular No. 70, Florida Bureau of Geology, Tallahassee, FL.

Weech, P. 1997. Deep-well disposal in the Bahamas, *Bahamas J. Sci.*, 4(3):6–13.

Willard, D.A., C.W. Holmes, J.B. Murray, W.H. Orem, and L.M. Weimer. 2001. Evolution of fixed tree islands in the Florida Everglades: environmental controls, in *Tree Islands of the Everglades*, Sklar, F. and van der Val, A., Eds., Kluwer Academic, Dordrecht/Norwell, MA.

Williams, P.W. 1972. Morphometric analysis of polygonal karst in New Guinea, *Geol. Soc. Am. Bull.*, 83:761–796.

Zektser, I.S., R.G. Dzhamalov, and T.I. Safronova. 1983. Role of submarine groundwater discharge in the water balance of Australia, *IAHS-AISH Publ.*, 1:209–219.

# 27 Detection of Coral Reef Change by the Florida Keys Coral Reef Monitoring Project

*James W. Porter, Vladimir Kosmynin, Kathryn L. Patterson, and Karen G. Porter*
Institute of Ecology, University of Georgia

*Walter C. Jaap, Jennifer L. Wheaton, Keith Hackett, and Matt Lybolt*
Florida Marine Research Institute

*Chris P. Tsokos*
Department of Mathematics, University of South Florida

*George Yanev*
University of South Florida

*Douglas M. Marcinek*
Department of Fisheries and Aquatic Sciences, University of Florida

*John Dotten*
Florida Keys National Marine Sanctuary

*David Eaken*
South Florida Regional Laboratory, Florida Fish and Wildlife Conservation Commission

*Matt Patterson*
Biscayne National Park

*Ouida W. Meier*
Department of Biology, Western Kentucky University

*Mike Brill and Phillip Dustan*
Department of Biology, University of Charleston

-8493-2026-7/02/$0.00+$1.50
© 2002 by CRC Press LLC

## CONTENTS

## INTRODUCTION

The decline of coral reefs is widely perceived both on a global scale (Richmond, 1993; Birkeland, 1997; Bryant and Burke, 1998; Wilkinson et al., 1999; Porter and Tougas, 2001) and on a regional scale, as in the Florida Keys (Dustan and Halas, 1987; Porter and Meier, 1992; Dustan, 1999). While coral reef scientists have, in principle, embraced the concept that coral reefs are threatened, they have also impugned the few quantitative studies that actually support this contention. Current studies are criticized as being either (1) too short in duration (Connell, 1997) or (2) too limited in geographic extent (Murdoch and Aronson, 1999). This dilemma has contributed to uncertainty in the conservation community (Risk, 1999), with coral reef scientists and managers suspecting that there is a problem but not believing that they have the statistical certainty needed to push for the strongest possible conservation measures. The Florida Keys Coral Reef Monitoring Project (CRMP) was initiated for three reasons: (1) to overcome the spatial and temporal criticisms of previous studies, (2) to rigorously determine change in coral species richness and relative benthic cover, and (3) to provide the baseline data necessary to evaluate the success of future management actions in the Florida Keys.

The Florida Keys National Marine Sanctuary Protection Act (HR5909) designated over 2,800 square nautical miles of coastal waters as the Florida Keys National Marine Sanctuary (FKNMS). In cooperation with NOAA, the U.S. Environmental Protection Agency (EPA) and the State of Florida implemented a Water Quality Protection Program (WQPP) to monitor seagrass habitats, coral reefs, hardbottom communities, and water quality (Hankinson and Conklin, 1996). The WQPP explicitly acknowledged that, in the absence of high-quality monitoring data, neither the present situation nor the efficacy of any future management actions could be evaluated. The specific objective of the monitoring program was to provide the data needed to make unbiased, statistically rigorous statements about the status and trends of benthic marine communities in the Florida Keys. Goals of the monitoring program also included identifying, when possible, causes for, and spatial distribution of, ecosystem change.

Every aspect of the design and implementation of the CRMP was derived from these goals. As part of defining the monitoring program, a series of *a priori,* spatially explicit, mutually exclusive, testable hypotheses was developed to evaluate the data (Table 27.1). Testing these

**TABLE 27.1**
**Spatially Explicit Testable Hypotheses for Detecting Coral Reef Change Through Time**

| Hypotheses | Emergent Pattern |
|---|---|
| $H_1$  Null hypothesis (dynamic equilibrium) | No net change |
| $H_2$  Global increase in coral reefs | Overall net increase |
| $H_3$  Global decrease in coral reefs | Overall net decline |
| $H_4$  Regional changes | Localized changes |
|    (a)  $H_{4a}$ Florida Bay water impact |    (a)  Reefs near passes decline |
|    (b)  $H_{4b}$ Florida Keys pollution |    (b)  Reefs near population centers decline and/or reefs near shore decline |
|    (c)  $H_{4c}$ Visitor impact |    (c)  Reefs visited by divers/fishermen decline |

hypotheses was also meant to elucidate linkages among ecosystems in the South Florida hydroscape. The sampling strategy and methods were developed in conjunction with the EPA, FKNMS, Continental Shelf Associates, and the principal investigators in 1994. The major criteria for coral reef monitoring included sanctuary-wide spatial coverage, repeated sampling, and statistically valid findings to document status and trends of the coral communities. The results were also intended to assist managers in understanding, protecting, and restoring the living marine resources of the FKNMS (Norse, 1993; Klein and Orlando, 1994). The CRMP allows us to detect problems, identify trends, and evaluate solutions.

## METHODS

Sampling sites were chosen using stratified random EPA E-MAP procedures (Overton et al., 1991). In 1994, 37 reef sites located within five of the nine EPA Water Quality Segments (Figure 27.1) were randomly selected. Based upon the existence of previous monitoring activity, three additional sites (Carysfort Reef, Looe Key, and Western Sambo) were added, bringing the list to a total of 40 reef monitoring sites within the boundaries of the FKNMS. Permanent station markers delineating four sampling stations within each of the 40 sites were installed in 1995. The first station for each site was located by going to the randomly generated latitude and longitude and choosing the closest appropriate reef type (offshore shallow reefs, offshore deep reefs, or patch reefs). For hardbottoms, the first station was chosen by going to the randomly located latitude and then swimming to the closest suitable benthic habitat. The remaining three stations were placed on adjacent suitable habitat at a minimum distance of 5 m apart. For uniform hardbottom environments, this frequently resulted in a parallel arrangement of the four stations (Figure 27.2); for offshore shallow reefs, this process frequently resulted in the selection of adjacent coral reef spurs (Figure 27.3). Sampling was initiated in 1996, and 160 stations among 40 sites were sampled annually through 2000. Three additional sites were installed and sampled in the Dry Tortugas beginning in 1999 (Figure 27.1). The project's 43 sampling sites included 7 hardbottom, 11 patch, 12 offshore shallow, and 13 offshore deep reef sites. Field sampling consisted of station species inventories and video transects conducted at all four stations at each site (Figure 27.3).

### STATION SPECIES INVENTORY

Counts of stony coral species (Milleporina and Scleractinia) present within each station provided data on stony coral species richness (S). Two observers conducted simultaneous timed (15-min) inventories within the approximately $22 \times 2$-m stations (Figure 27.3) and entered data on underwater data sheets. Each observer recorded all stony coral taxa, including fire corals, and enumerated long-spined urchins (*Diadema antillarum*) within the station boundaries. After recording the data,

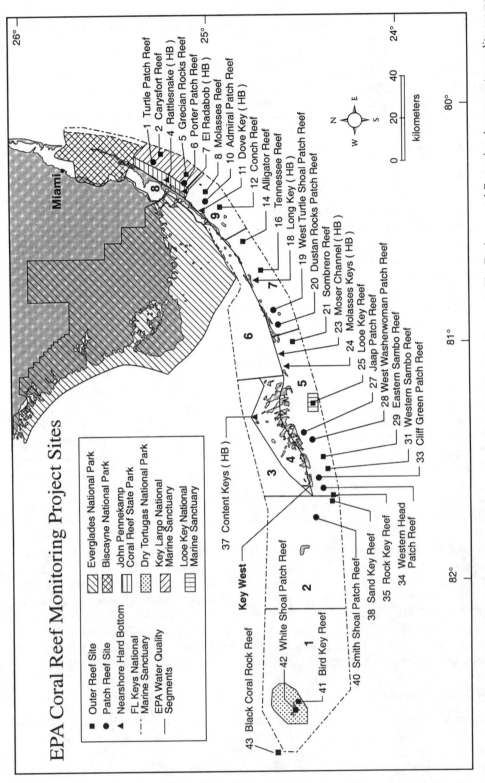

**FIGURE 27.1** Map of the Florida Keys Coral Reef Monitoring Project Sites (including Dry Tortugas). The Environmental Protection Agency water quality segments (1 to 9) are also shown on this map.

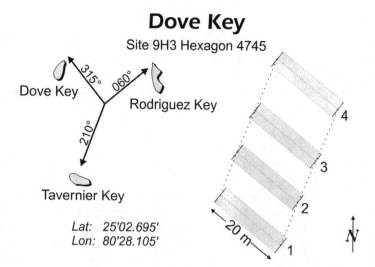

**FIGURE 27.2** At each of the randomly selected Coral Reef Monitoring Project sites, four individual stations are laid out on the coral reef for sampling. This example from the Dove Key Hardbottom site shows station placement on a topographically uniform substrate.

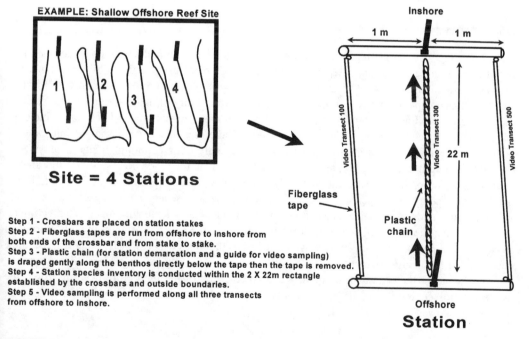

**FIGURE 27.3** Schematic diagram of the setup for a shallow-water, offshore coral reef site. Placement of the permanent stainless-steel survey pins and the removable material for station installation are also shown.

observers spent 5 min underwater confirming the presence of species recorded by only one observer during the 15-min search period. Taxonomic differences were addressed during this period. Data sheets were verified aboard the vessel and forwarded to the Florida Marine Research Institute for data entry and processing. This method facilitated data collection with broad spatial coverage and with an optimal expenditure of time and effort.

## Diseases/Conditions

During the timed species count, if a colony of any coral species within the station exhibited specific signs of either bleaching or disease, the appropriate code letter (H, bleaching; B, black-band disease; W, white disease; or O, other disease) was entered for that species on the data form. This method scored the presence or absence of these conditions for each species; it did not produce an estimate of the percent of all colonies with each disease.

## Videography

All video sampling through 1999 was conducted with a Sony CCD-VX3 Hi 8-mm analogue video camera with full automatic settings utilizing two 50-watt artificial lights. Beginning in 2000, the project upgraded to digital video, filming all sites with a Sony TRV 900 4-mm digital video camera in an Amphibico underwater housing. A chain was laid on the surface of the reef directly underneath each of the three transect lines (Figure 27.3). The camera was pointed straight down and maintained at 40 cm above the reef surface. A convergent laser light system indicated distance from the reef surface for filming. This position and height generally allowed the camera to record a swath of reef surface that was 40 cm wide and 22 m long. The videographer filmed a clapper board prior to beginning each transect. The clapper board displayed a record of date and location of each film segment. Filming was conducted at a constant swim speed of about 4 m/min, yielding approximately 9000 video frames per transect. Representative images for all transects were frame-grabbed, written to, and archived on a CD-ROM.

## Image Analysis

During filming, camera settings were optimized (progressive scan and sport mode) to maximize the quality of individual frames. Because there is considerable overlap in the field of view of video images captured at a rate of 30 frames per second, it was necessary to digitize only 120 frames for more than complete coverage of the sea floor. From this library of images, frames were selected that abutted but did not overlap by more than 15% with the previous image. Each image included an area approximately 30 cm high by 40 cm wide. Therefore, of the 120 captured frames, only about 60 abutting frames were selected by a trained analyst for counting. Image analysis was conducted using a custom software application, PointCount for Coral Reefs. When the analyst opened each image, the software automatically inserted ten random points over the image. Selected benthic taxa (stony coral, octocoral, zoanthid, sponge, seagrass, and macroalgae) and substrate were identified under each point. The software has a "point and click" feature that creates a comma-separated value file (*.csv). Data were formatted as a csv file to simplify current and future data storage and retrieval. After all images were analyzed, the analyst performed a quality assurance check of the file using Microsoft Excel before forwarding the file to the project's data manager. The data manager then ran another quality assurance check of the file before entering it into the master Microsoft Access database. The software allowed comments to be associated with each point or image and added to the file.

## Statistical Analyses

In addition to the descriptive methods of organizing and summarizing the data, hypothesis testing was performed to analyze the percent cover, species richness, and disease/condition data. The decision to reject or not to reject the null hypothesis, that there is no significant difference in the data for certain years, was based on the minimum detectable difference for different significance levels and powers. Six combinations for significance level ($\alpha$) and power ($1-\beta$) were considered: $\alpha = 0.05$, $1-\beta = 0.75$; $\alpha = 0.10$, $1-\beta = 0.75$; $\alpha = 0.20$, $1-\beta = 0.75$; $\alpha = 0.05$, $1-\beta = 0.80$; $\alpha = 0.10$, $1-\beta = 0.80$; $\alpha = 0.20$, $1-\beta = 0.80$. When the one-sided alternative was tested, the above values for

$\alpha$ must be divided by two. The output consists of the minimum detectable difference for a certain pair ($\alpha$, $1-\beta$), which was used to construct a $(1-\alpha)\%$ confidence interval and provide a measure of the test accuracy.

## VIDEO DATA

To ensure at least 90% similarity between point assessments of the observers, the Bray–Curtis dissimilarity measure was utilized. Each species was assigned a coefficient of dissimilarity based on the inter-observer data files. In addition to the three principal investigators, the point count assessments of three additional observers were included. The list of species in a station was reviewed, and, if necessary, rare (below 5% contribution) species were combined into the category "other coral" until the sum of the dissimilarity coefficients of the species present became less than 10%. Hypotheses for the difference between species proportions on the station level were tested: (1) within stony coral (i.e., a particular species vs. total stony coral), and (2) total stony coral vs. total coral coverage.

If the conditions for normality were met, the hypothesis testing was completed for the difference between two proportions for all pairwise comparisons between the years: 1996 vs. 1997, 1997 vs. 1998, and 1996 vs. 1998. The output was the minimum detectable difference between the two proportions for all combinations of three significance levels (0.05, 0.10, and 0.20) and two power levels (0.75 and 0.80).

Tests were run for $H_o$: $p_1-p_2 = 0$ vs. $H_1$: $p_1-p_2 \neq 0$. If either one of the one-sided alternatives were of interest, $H_o$: $p_1-p_2 = 0$ vs. $H_1$: $p_1-p_2 > 0$ or $H_1$: $p_1-p_2 < 0$, then the above significance level must be divided by two. To perform the above hypothesis testing, an S-PLUS code was written.

Results presented herein were for the significance level of 0.20 (0.10 for a one-sided alternative) and a power of 0.75. The results for change of one particular stony coral species were conditional on the total stony corals. For instance, if the total stony coral category decreased, then the contribution of one species within stony coral could actually increase, even if the presence of this species remained unchanged.

## SPECIES INVENTORY AND DISEASE CONDITION DATA

To study species richness, the hypothesis of whether there was a significant difference in the proportion of the number of stations where each species was present was tested. Results were calculated sanctuary-wide on data for the period 1996 to 2000. Species richness was highly influenced by the presence or absence of relatively rare species. By contrast, the percent coral cover analysis was most influenced by the population dynamics of the most common coral species.

Pairwise tests were run for total stony coral and for individual species with adequate data to determine whether there was significant change in number of stations with the presence of a certain disease/condition from 1996 to 2000.

## RESULTS

Results were reported by regions defined as follows: Upper Keys (north Key Largo to Conch Reef), Middle Keys (Alligator Reef to Molasses Keys), Lower Keys (Looe Key to Smith Shoal), and Tortugas (Dry Tortugas to Tortugas Banks) (Figure 27.1).

### STONY CORAL SPECIES RICHNESS

Figure 27.4 shows coral species loss from the Florida Keys CRMP Stations between 1996 and 2000 (all reporting in this section is relative to the time period 1996 to 2000). Coral species were more likely to disappear from stations than to appear in them. For instance, 34 of 43 coral species (79.1%) were found in fewer CRMP stations, and only nine species (20.9%) were found in more stations. Further, the magnitude of the losses greatly exceeded the magnitude of the gains (Figure 27.4).

In the Upper Keys, losses of stony coral species were observed at 32 of 52 stations (61.5%); 11 stations had an increase; and at nine stations, the presence of stony coral species was unchanged (Figure 27.5). In the Middle Keys, 32 of 44 stations (72.7%) had losses of stony coral species; six

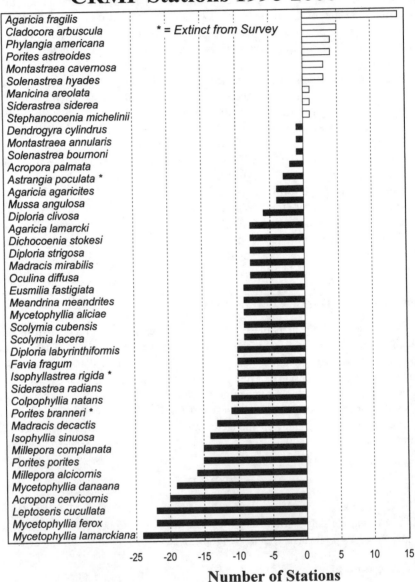

**FIGURE 27.4** Coral species loss from the Florida keys CRMP Stations, 1996 to 2000, is shown for species encountered in the survey. Species were more likely to disappear from stations than appear in them; in addition, the magnitude of the losses exceeds the magnitude of gains.

stations had an increase; and six stations had no change (Figure 27.6). In the lower Keys, 43 of 64 stations (67.2%) experienced losses in stony coral species; 15 stations had gains; and six stations were unchanged (Figure 27.7).

Sanctuary-wide, 16 of 28 (57.1%) hardbottom stations had stony coral species losses; five hardbottom stations gained; and seven stations were unchanged. Of the patch reef stations, 29 (72.5%) had stony coral species losses; six stations gained; and five stations were unchanged. For shallow reef stations, 28 of 48 (58.3%) showed stony coral losses; 14 stations gained; and six were unchanged. Of 44 deep reef stations, 34 (77.3%) had stony coral species losses; seven stations gained; and three stations were unchanged (Table 27.2).

Sanctuary-wide, Smith Shoal Station 2 had the maximum of five stony coral species gained. Greatest losses in stony coral species richness were at Carysfort Deep, Station 4; Cliff Green, Station 2; and Conch Deep, Station 1, where each lost nine species.

For the one-year intervals between 1996, 1997, and 1998, no significant change was seen in the number of stations where the majority of stony coral species were present, based on the statistical confidence level chosen. However, between 1998 and 1999, significant losses did occur in the number of stations where four stony coral species were found (*Millepora alcicornis, Mycetophyllia danaana, Mycetophyllia ferox,* and *Porites porites*), but significant gains were recorded for *Oculina diffusa* and *Scolymia lacera*. Again between 1999 and 2000, for the majority of stony coral species, no significant change in their presence sanctuary-wide was observed. However, when the first and last year's data were compared, eight stony coral species (*Acropora cervicornis, Leptoseris cucullata, Millepora alcicornis, Millepora complanata, Mycetophyllia danaana, Mycetophyllia ferox, Mycetophyllia lamarckiana,* and *Porites porites*) exhibited a significant loss in the number of stations

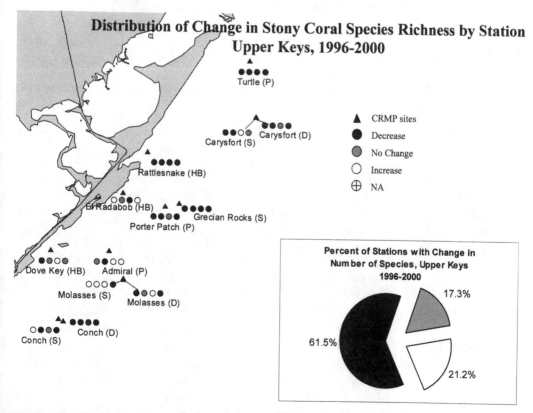

**FIGURE 27.5** The distribution of change in stony coral species richness by station is shown for the Upper Keys, 1996 to 2000.

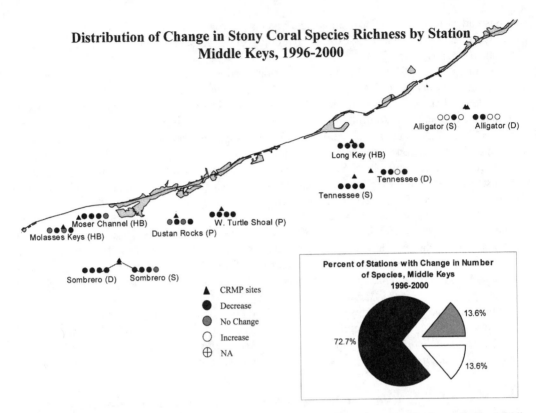

**FIGURE 27.6** The distribution of change in stony coral species richness by station is shown for the Middle Keys, 1996 to 2000.

where they were present. Only *Agaricia fragilis* showed a significant gain over the time span sampled (Table 27.3).

## Coral Condition

Scleractinian corals in CRMP stations sanctuary-wide experienced a significant increase in disease infections from 1996 to 2000 (Table 27.4). This increase occurred among white diseases, other diseases, and total diseases. All disease conditions exhibited increasing frequency of coral infection from 1996 to 1997 and 1997 to 1998. The incidence of black-band disease also increased significantly between 1996 and 1998, but, because of significant decreases between 1998 and 1999, the frequency of this disease actually was unchanged between 1996 and 2000. From 1999 to 2000, no significant change in the distribution of disease infections was evident. Over the 5 years, there was a significant increase in white diseases and other diseases (Table 27.4). Overall, increases were found in the number of stations containing diseased coral, the number of coral species with disease, and the number of different types of diseases that were observed (Porter et al., in press).

## Stony Coral Cover

For 1996 through 1998, of the 143 stations with sufficient coral cover to analyze, 64 stations (44.8%) had significant coral cover loss, 73 stations (51.0%) showed no significant change, and only six stations (4.2%) exhibited a significant gain in coral cover (Figure 27.8). Our survey has documented a highly significant loss of living coral throughout the Florida Keys. A general sanctuary-wide trend of decline in stony coral cover from 1996 to 1999 is presented in Figure 27.9. For all years except 1996 to 1997, the decrease in mean percent coral cover was significant, with

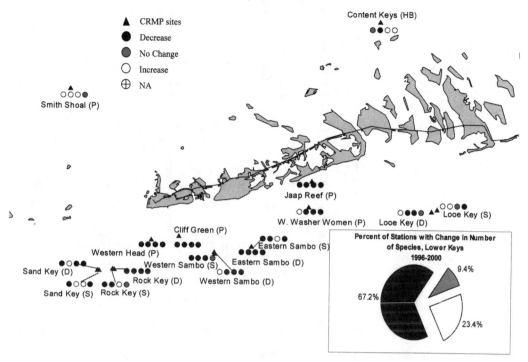

**FIGURE 27.7** The distribution of change in stony coral species richness by station is shown for the Lower Keys, 1996 to 2000.

a *p* value of 0.03 or less for the Wilcoxon rank-sum test. Between 1996 and 1999, a 38% reduction in living coral in the Florida Keys has occurred (Figure 27.9).

The Upper Keys experienced a greater relative change in stony coral cover, as illustrated in Figure 27.10. A greater percent of Upper Keys stations had significant loss of coral compared with the Lower and Middle Keys stations. In the Upper Keys, 25 stations (62.5%) experienced significant loss of coral cover, 13 (32.5%) had no significant change, and only two stations showed a significant gain in coral cover (Figure 27.11). In the Middle Keys, nine stations (22.5%) experienced significant coral losses, 30 stations (75.0%) had no significant change, and only one station showed a significant gain in coral cover (Figure 27.12). In the Lower Keys, an equal number of stations — 30 (47.6%) — lost a significant amount of coral cover or had no significant change, whereas only three stations gained coral cover (Figure 27.13).

By habitat type, for 1996 to 1998, 13 patch reef stations (32.5%) lost significant cover; of the remaining 27 stations (67.5%), cover was statistically unchanged. No patch reef station gained coral cover. For offshore shallow sites, 26 stations (54.2%) had significant loss of coral cover; 19 stations (39.6%) had no significant change in coral cover; and only three stations showed significant gains. In the offshore deep reef habitat, 23 stations (52.3%) lost significant coral cover; 19 stations (43.2%) had no significant change in cover; and only two gained significant cover (Table 27.5). Because coral cover is very sparse on most hardbottom habitats (Chiappone and Sullivan, 1994), hypothesis testing was not possible for all hardbottom stations. For the 11 hardbottom stations with sufficient coral cover, two stations (18.2%) showed significant coral cover loss; one station (9.1%) had a significant increase; and eight stations (72.7%) showed no significant change (Table 27.5).

Overall, between 1996 and 1998, we documented significant losses in coral cover for almost half of all coral reefs in the Florida Keys (Figure 27.8).

**TABLE 27.2**
**The Number of Coral Reef Monitoring Project Stations With Change in the Number of Species of Stony Corals, by Habitat Type, 1996–2000**

Habitat Type

| | Hardbottom Communities | | | Patch Reefs | | | Shallow Offshore Reefs | | | Deep Offshore Reefs | | |
|---|---|---|---|---|---|---|---|---|---|---|---|---|
| | Lost | Gained | No Change | Lost | Gained | No Change | Lost | Gained | No Change | Lost | Gained | No Change |
| Lower Keys | 1 | 2 | 1 | 15 | 4 | 1 | 11 | 6 | 3 | 16 | 3 | 1 |
| Middle Keys | 9 | 0 | 3 | 6 | 0 | 2 | 8 | 3 | 1 | 9 | 3 | 0 |
| Upper Keys | 6 | 3 | 3 | 8 | 2 | 2 | 9 | 5 | 2 | 9 | 1 | 2 |
| Total | 16 | 5 | 7 | 29 | 6 | 5 | 28 | 14 | 6 | 34 | 7 | 3 |
| Percent | 57.1% | 17.9% | 25.0% | 72.5% | 15.0% | 12.5% | 58.3% | 29.2% | 12.5% | 77.3% | 15.9% | 6.8% |

**TABLE 27.3**
**Significant Change in the Number of Stations Where a Stony Coral Species was Present, Sanctuary-Wide, 1996–2000**

| Species | 1996–1997 | 1997–1998 | 1998–1999 | 1999–2000 | 1996–2000 |
|---|---|---|---|---|---|
| *Acropora cervicornis* | 0 | 0 | 0 | 0 | Loss |
| *Acropora palmata* | 0 | 0 | 0 | 0 | 0 |
| *Agaricia agaricites* | 0 | 0 | 0 | 0 | 0 |
| *Agaricia fragilis* | Gain | 0 | 0 | 0 | Gain |
| *Agaricia lamarcki* | 0 | 0 | 0 | 0 | 0 |
| *Astrangia poculata* | — | — | — | — | — |
| *Astrangia solitaria* | — | — | — | — | — |
| *Cladocora arbuscula* | 0 | 0 | 0 | 0 | 0 |
| *Colpophyllia natans* | 0 | 0 | 0 | 0 | 0 |
| *Dendrogyra cylindrus* | — | — | — | — | — |
| *Dichocoenia stokesi* | 0 | 0 | 0 | 0 | 0 |
| *Diploria clivosa* | 0 | 0 | 0 | 0 | 0 |
| *Diploria labrynthiformis* | 0 | 0 | 0 | 0 | 0 |
| *Diploria strigosa* | 0 | 0 | 0 | 0 | 0 |
| *Eusmilia fastigiata* | 0 | 0 | 0 | 0 | 0 |
| *Favia fragum* | 0 | 0 | 0 | 0 | 0 |
| *Isophyllastrea rigida* | — | — | — | — | — |
| *Isophyllia sinuosa* | — | — | — | — | — |
| *Leptoseris cucullata* | 0 | 0 | 0 | 0 | Loss |
| *Madracis decactis* | 0 | 0 | 0 | 0 | 0 |
| *Madracis mirabilis* | 0 | 0 | 0 | 0 | 0 |
| *Madracis pharensis* | — | — | — | — | — |
| *Manicina areolata* | 0 | 0 | 0 | 0 | 0 |
| *Meandrina meandrites* | 0 | 0 | 0 | 0 | 0 |
| *Millepora alcicornis* | 0 | Loss | Loss | 0 | Loss |
| *Millepora complanata* | 0 | 0 | 0 | 0 | Loss |
| *Montastraea annularis* | 0 | 0 | 0 | 0 | Loss |
| *Montastraea cavernosa* | 0 | 0 | 0 | 0 | 0 |
| *Mussa angulosa* | 0 | 0 | 0 | 0 | 0 |
| *Mycetophyllia aliciae* | 0 | 0 | 0 | 0 | 0 |
| *Mycetophyllia danaana* | 0 | 0 | Loss | 0 | Loss |
| *Mycetophyllia ferox* | 0 | 0 | Loss | 0 | Loss |
| *Mycetophyllia lamarckiana* | 0 | 0 | 0 | 0 | Loss |
| *Oculina diffusa* | 0 | 0 | Gain | 0 | 0 |
| *Oculina robusta* | — | — | — | — | — |
| *Phyllangia americana* | 0 | 0 | 0 | 0 | 0 |
| *Porites astreoides* | 0 | 0 | 0 | 0 | 0 |
| *Porites braneri* | — | — | — | — | — |
| *Porites porites* | 0 | 0 | Loss | 0 | Loss |
| *Scolymia cubensis* | 0 | 0 | 0 | 0 | 0 |
| *Scolymia lacera* | 0 | 0 | Gain | Loss | 0 |
| *Siderastrea radians* | 0 | 0 | 0 | 0 | 0 |
| *Siderastrea siderea* | 0 | 0 | 0 | 0 | 0 |
| *Solenastrea bournoni* | 0 | 0 | 0 | 0 | 0 |
| *Solenastrea hyades* | — | — | — | — | — |
| *Stephanocoenia michelinii* | 0 | 0 | 0 | 0 | 0 |

*Note:* $\alpha = 0.1$; $1-\beta = 0.75$; 0 indicates no significant change; — indicates insufficient sample size to make a determination.

**TABLE 27.4**
**Significance of Change in Disease Conditions Recorded at 160 CRMP Stations in the Florida Keys National Marine Sanctuary Between 1996 and 2000**

| Time | 1996–1997 | 1997–1998 | 1998–1999 | 1999–2000 | 1996–2000 |
|---|---|---|---|---|---|
| Black-band disease | Significant increase | Significant increase | Significant decrease | No significant change | No significant change |
| White diseases | Significant increase | Significant increase | No significant change | No significant change | Significant increase |
| Other diseases | Significant increase | Significant increase | Significant increase | No significant change | Significant increase |

*Note:* $\alpha = 0.10$; $1-\beta = 0.75$.

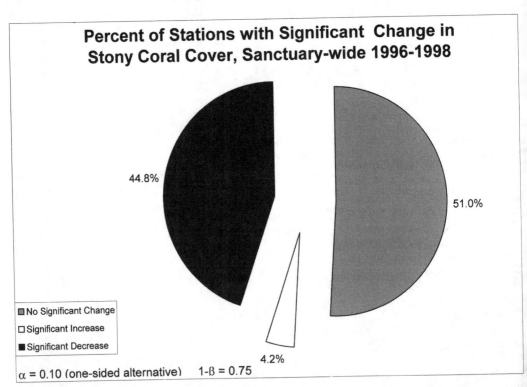

**FIGURE 27.8** The percent of stations with significant change in stony coral cover sanctuary-wide is shown for 1996 to 1998.

## DISCUSSION

In his survey on the health of coral reefs worldwide, Connell (1997) reviewed quantitative studies of coral abundance from 65 monitoring studies that were at least 4 years long. The data he reviewed demonstrated that coral cover did not decline in 29% of the studies, declined and recovered in another 29%, and declined and did not fully recover in the remaining 42%. His conclusion was that "coral assemblages are relatively stable over ecological time scales." He did add one caveat that, "The Western Atlantic region was more unstable than the Indo-Pacific; declines without subsequent recovery occurred in 57% of Western Atlantic examples, but only 29% of those in the

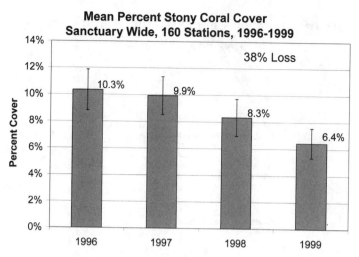

**FIGURE 27.9** The mean percent stony coral cover sanctuary-wide is shown for 1996 to 1999. During this 3-year period, the Florida Reef Tract lost 38% of its coral cover.

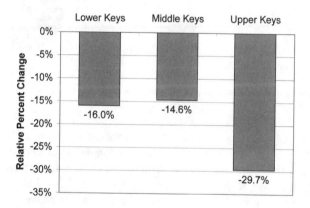

**FIGURE 27.10** The relative change in percent stony coral cover by region is shown for 1996 to 1998.

Indo-Pacific." Our data from the Florida Keys are in striking contrast to his general conclusion, but do show the overall trend for Florida that he infers for the Western Atlantic. In our study, 44.8% of all stations showed significant decline without recovery between 1996 and 1998 (Figure 27.8) and, given the overall 38% coral loss demonstrated between 1996 and 1999 (Figure 27.9), we expect that the percent of stations showing significant decline will continue to increase as the full statistical analyses are completed for 2000 and 2001.

Because we used a stratified random design, our data can be used to test several spatially explicit models (Table 27.1). Our data immediately falsify both Hypothesis 1 (the null hypothesis, which suggests a dynamic equilibrium and no net change in coral cover) and Hypothesis 2, global increase in coral cover, which predicts an overall net increase. Almost half of our stations are losing coral cover (Figure 27.8) and substantially more than half are losing coral species (Table 27.2). Elements of our data alternatively support both Hypothesis 3 (global decline in coral reefs) and Hypothesis 4 (regional declines in coral reefs). Hypothesis 3 is supported because we did record declines everywhere (Figures 27.5 to 27.7, 27.11 to 27.13). This could easily be interpreted as a general non-site-specific decline in living coral; however, Figure 27.10 and Table 27.2 demonstrate

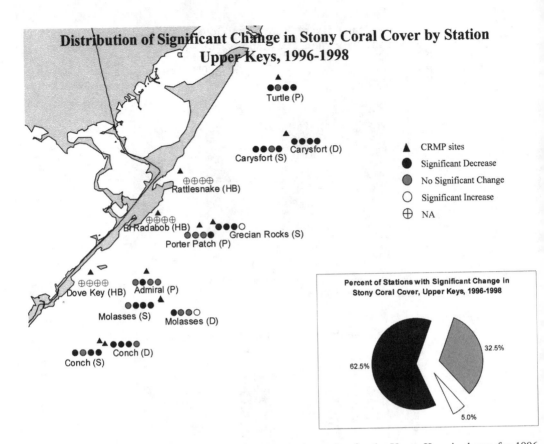

**FIGURE 27.11** Significant change in stony coral cover by station for the Upper Keys is shown for 1996 to 1998.

that declines are highest in the Upper Keys. This observation may support Hypotheses 4b, which posits that coral reef decline will be highest near human population centers (Pastorok and Bilyard, 1985; Lapointe et al., 1990; Richmond, 1993). With this in mind, future research and data analysis will be focused on the Lower Keys to see if the trend identified for the Key Largo area repeats itself for coral reefs near Key West. Our data do not support either of the two remaining regional hypotheses. Hypothesis 4a suggests that reefs near passes in the Middle Keys will decline fastest due to their proximity to water emerging from Florida Bay (Porter et al., 1999). While historical evidence exists to prove the negative influence of Florida Bay water on the associated coral reefs offshore (Ginsburg and Shinn, 1994; Shinn et al., 1994; Pitts, 1994; Smith, 1994; Porter et al., 1999; Cook et al., this volume), our survey data for the years 1996 to 1998 (Figure 27.10) definitely do not show this. Further, our data also do not support the contention (Hypothesis 4c) that visitation by divers and snorkelers will accelerate coral reef decline, as some of the reefs in the Middle Keys, where coral declines are more modest, are more frequently visited than some reefs in the Upper Keys, where coral decline is much higher (Figures 27.5 through 27.7). Further, even within a region, declines occurred on some reefs that are infrequently visited, and, finally, increases occurred on some reefs that are visited by divers.

It is highly desirable to identify the causes for coral reef decline in the Florida Keys. Although single stresses are more easily understood and mitigated than multiple stresses, a simplistic view is almost certainly incorrect. We suspect that coral decline in the Florida Keys is caused by a series of multiple stressors, some of which are strongly influenced by global climate change (Smith and Buddemeier, 1992; Fitt et al., 1993; Hoegh-Guldberg, 1999; Kleypas et al., 1999), and some of which are more strongly influenced by regional processes (Porter et al., 1999). While hurricane

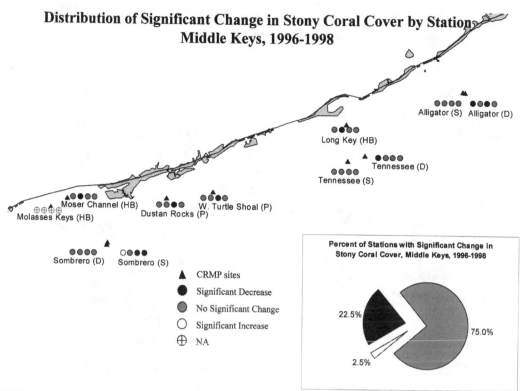

**FIGURE 27.12** Significant change in stony coral cover by station for the Middle Keys is shown for 1996 to 1998.

activity contributed to reef decline (Hurricane Georges hit the Florida Keys in September 1998), hurricanes did not start the downward trend (Figure 27.9). By the same token, elevated late-summer sea-surface temperatures made 1998 a bleaching year, and, while this may also have contributed to the decline, it definitely did not initiate it (Figure 27.9). Water quality measurements in the Florida Keys suggest that several water quality parameters are deteriorating in the FKNMS. For instance, sanctuary-wide, phosphorous has increased significantly, as have chlorophyll *a* concentrations and water turbidity in several places (Lapointe et al., 1993; Boyer and Jones, 2002). Optimal coral recruitment, growth, and survival are achieved under conditions of low nutrient concentrations and low turbidity (Lapointe, 1997, 1999; Hughes et al., 1999), and we are examining both the coral population data and the water quality data (Ogden et al., 1994) to see if adult coral mortality is linked to deteriorations in any of these factors.

One contributing factor that we are quite certain has caused coral decline in the Florida Keys is coral disease (Holden, 1996; Harvell et al., 1999; Richardson et al., 1998). Disease has caused substantial coral loss, especially on the deep fore-reef slope of Carysfort Reef in the Upper Keys (Porter et al., in press). At present, we do not know if the elevated incidence of disease identified during our project (Table 27.4) is itself influenced by regional factors, global factors, or, more likely, by both. Finally, we strongly suspect that the species losses identified in Table 27.3 are at least in part also causally related to coral disease. Species in the genera *Acropora*, *Montastraea*, and *Mycetophyllia* are the hardest hit by disease, and these species are also the ones to show significant decline (Table 27.3).

Between 1996 and 1999, the average rate of coral loss was 12.6% per year (Figure 27.9). These losses confirm both the direction of change and the magnitude of change identified in previous studies (Porter and Meier, 1992). Even over ecological time scales, let alone geological time scales, these loss rates are unsustainable.

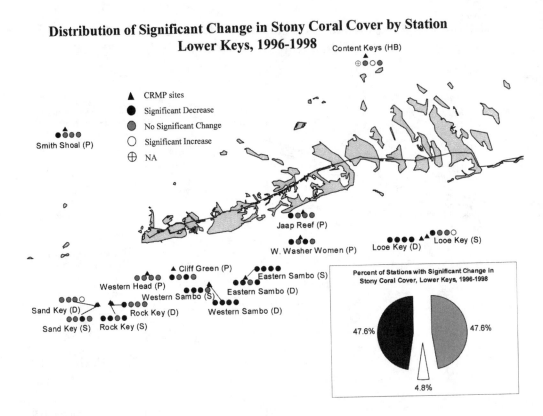

**FIGURE 27.13**   Significant change in stony coral cover by station for the Lower Keys is shown for 1996 to 1998.

## CONCLUSIONS

1. A significant loss of coral species richness occurred in the Florida Keys National Marine Sanctuary; between 1996 and 2000, 67% of 160 stations lost species.
2. By the end of the survey (2000), 80% of all coral species were found in fewer stations than at the beginning (1996).
3. From 1996 to 1998, stony coral cover significantly declined in 45% of the 143 stations that could be analyzed.
4. From 1996 to 1999, coral cover in the Florida Keys declined by 38%. Heaviest losses were in the Upper and Lower Keys.
5. Hurricane Georges (which hit the Florida Keys in September 1998) continued a downward trend in live coral cover seen in the previous two years, but the hurricane did not start this decline.
6. The number of stations where diseases occur has increased, as has the number of coral species affected. The number of new disease phenomena observed has also increased.
7. The survey documented an average rate of coral loss of 13% per year in the Florida Keys; these loss rates are high and unsustainable.

**TABLE 27.5**
**Number of Coral Reef Monitoring Project Stations With Significant Change in Coral Cover, by Habitat Type, 1996–1998**

| | Habitat Type | | | | | | | | | | | |
|---|---|---|---|---|---|---|---|---|---|---|---|---|
| | Hardbottom Communities | | | Patch Reefs | | | Shallow Offshore Reefs | | | Deep Offshore Reefs | | |
| | Lost | Gained | No Change | Lost | Gained | No Change | Lost | Gained | No Change | Lost | Gained | No Change |
| Lower Keys | 0 | 1 | 2 | 6 | 0 | 14 | 12 | 1 | 7 | 12 | 1 | 7 |
| Middle Keys | 2 | 0 | 6 | 2 | 0 | 6 | 2 | 1 | 9 | 3 | 0 | 9 |
| Upper Keys | 0 | 0 | 0 | 5 | 0 | 7 | 12 | 1 | 3 | 8 | 1 | 3 |
| Total | 2 | 1 | 8 | 13 | 0 | 27 | 26 | 3 | 19 | 23 | 2 | 19 |
| Percent | 18.2% | 9.1% | 72.7% | 32.5% | 0.0% | 67.5% | 54.2% | 6.3% | 39.6% | 52.3% | 4.5% | 43.2% |

*Note:* $\alpha = 0.1$; $\beta = 0.75$.

## ACKNOWLEDGMENTS

We thank the 1996–2000 CRMP Sampling Team (alphabetically): Lonny Anderson, Bobby Barratachea, Mike Callahan, Katie Fitzsimmons, Jeff Jones, Jim Kidney, James Leard, Sarah Lewis, Laurie Maclaughlin, Leanne Miller, Jamie O'Brien, and Tom Trice. We also thank Mel Parsons, Gary Collins, Phillip Murphy, Ken Potts, Ed McLean, and the crew of the U.S. EPA's research vessel, *OSV Peter W. Anderson* for logistical support. John Halas, Ben Haskill, Billy Causey, G.P. Schmahl, and Steve Bumbgardiner from the Florida Keys National Marine Sanctuary helped with station installation.

## REFERENCES

Birkeland, C., Ed. 1997. *Life and Death of Coral Reefs*, Chapman & Hall, New York.

Boyer, J. and Jones, R. 2002 A view from the bridge: external and internal forces affecting the ambient water quality of the Florida Keys National Marine Sanctuary, in *The Everglades, Florida Bay, and Coral Reefs of the Florida Keys: An Ecosystem Sourcebook*, Porter, J. and Porter, K., Eds., CRC Press, Boca Raton.

Bryant, D. and L. Burke. 1998. *Reefs at Risk: A Map-Based Indicator of Threats to the World's Coral Reefs*, World Resources Institute, Washington, D.C., 56 pp.

Chiappone, M. and K.M. Sullivan. 1994. Patterns of coral abundance defining nearshore hardbottom communities of the Florida Keys, *Florida Sci.*, 57:108–125.

Connell, J.H. 1997. Disturbance and recovery of coral assemblages, *Proc. Eighth Int. Coral Reef Symp.* (*Panama, R. de Panama*), 1:9–22.

Dustan, P. 1999. Coral reefs under stress: sources of mortality in the Florida Keys, *Nat. Resour. Forum*, 23:147–155.

Dustan, P. and J.C. Halas. 1987. Changes in the reef coral community of Carysfort Reef, Key Largo, Florida: 1974 to 1982. *Coral Reefs*, 16:91–106.

Fitt, W.K., H.J. Spero, J. Halas, M.W. White, and J.W. Porter. 1993. Recovery of the coral *Montastraea annularis* in the Florida Keys after the 1987 Caribbean "bleaching event," *Coral Reefs*, 12:57–64.

Ginsburg, R.N. and E.A. Shinn. 1994. Preferential distribution of reefs in the Florida Reef Tract: The past is the key to the present, in P*roceedings of the Colloquium on Global Aspects of Coral Reefs: Health, Hazards, and History, 1993*, Ginsburg, R.N., Ed., Rosenstiel School of Marine and Atmospheric Science, University of Miami, pp. 21–26.

Hankinson, J.H. and E.J. Conklin. 1996. *Water Quality Protection Program for the Florida Keys National Marine Sanctuary: First Biennial Report to Congress*, U.S. Environmental Protection Agency Office of Water, Washington, D.C.

Harvell, C.D., K. Kim, J.M. Burkholder, R.R. Colwell, P.R. Epstein, J. Grimes, E.E. Hofmann, E.K. Lipp, A.D.M.E. Osterhaus, R.M. Overstreet, J.W. Porter, G.W. Smith, and G.R. Vasta. 1999. Emerging marine diseases: climate links and anthropogenic factors, *Science*, 285:1505–1510.

Hoegh-Guldberg, O. 1999. Climate change, coral bleaching and the future of the world's coral reefs, *Mar. Freshwater Res.*, 50:839–866.

Holden, C. 1996. Coral disease hot spot in the Florida Keys, *Science*, 274:2017.

Hughes, T.P., A.M. Szmant, R. Steneck, R. Carpenter, and S. Miller. 1999. Algal blooms on coral reefs: What are the causes? *Limnol. Oceanogr.*, 44:1583–1586.

Klein, C.J. and S.P. Orlando. 1994. A spatial framework for water quality management in the Florida Keys National Marine Sanctuary, *Bull. Mar. Sci.*, 54:1036–1044.

Kleypas, J.A., R.W. Buddemeier, D. Archer, J.-P. Gattuso, C. Langdon, and B.N. Opdyke. 1999. Geochemical consequences of increased atmospheric carbon dioxide on coral reefs, *Science*, 284:118–120.

Lapointe, B.E. 1997. Nutrient thresholds for bottom-up control of macroalgal blooms on coral reefs in Jamaica and southeast Florida, *Limnol. Oceanogr.*, 42:1119–1131.

Lapointe, B.E. 1999. Simultaneous top-down and bottom-up forces control macroalgal blooms on coral reefs (reply to the comment by Hughes et al.), *Limnol. Oceanogr.*, 44:1586–1592.

Lapointe, B.E., J.D. O'Connell, and G.S. Garrett. 1990. Nutrient couplings between on-site sewage disposal systems, groundwaters, and nearshore waters of the Florida Keys, *Biogeochemistry*, 10:289–307.

Lapointe, B.E., W.R. Matzie, and M.W. Clark. 1994. Phosphorous inputs and eutrophication on the Florida reef tract, in *Proceedings of the Colloquium on Global Aspects of Coral Reefs: Health, Hazards, and History, 1993*, Ginsburg, R.N., Ed., Rosenstiel School of Marine and Atmospheric Sciences, University of Miami, pp. 106–112.

Murdoch, T.J.T. and R.B. Aronson. 1999. Scale dependent spatial variability of coral assemblages along the Florida Reef Tract, *Coral Reefs*, 18:341–351.

Norse, E. 1993. *Global Marine Biological Diversity: A Strategy for Building Conservation into Decision Making*, Island Press, Washington, D.C., 383 pp.

Ogden, J.C., J.W. Porter, N.P. Smith, A.M. Szmant, W.C. Jaap, and D. Forcucci. 1994. A long-term interdisciplinary study of the Florida Keys seascape, *Bull. Mar. Sci.*, 54:1059–1071.

Overton, W.S, D. White, and D.L. Stevens. 1991. *Design Report for the Environmental Monitoring Assessment Program (EMAP)*, EPA/600/3-91/053, U.S. Environmental Protection Agency, Washington, D.C.

Pastorok, R.A. and G.R. Bilyard. 1985. Effects of sewage pollution on coral-reef communities, *Mar. Ecol. Prog. Ser.*, 21:175–189.

Pitts, P.A. 1994. An investigation of near-bottom flow patterns along and across Hawk Channel, Florida Keys, *Bull. Mar. Sci.*, 54:610–620.

Porter, J.W. and O.W. Meier. 1992. Quantification of loss and change in Floridian reef coral populations, *Am. Zool.*, 32:625–640.

Porter, J.W. and J.I. Tougas. 2001. Reef ecosystems: threats to their biodiversity, in *Encyclopedia of Biodiversity*, Vol. 5, Levin, S., Ed., Academic Press, New York, pp. 73–95.

Porter, J.W., S.K. Lewis, and K.G. Porter. 1999. The effect of multiple stressors on the Florida Keys coral reef ecosystem: a landscape hypothesis and a physiological test, *Limnol. Oceanogr.*, 44:941–949.

Porter, J.W., P. Dustan, W.C. Jaap, K.L. Patterson, V. Kosmynin, O.W. Meier, M.E. Patterson, M. Parsons, and S.L. Rathbun. (in press) Patterns of spread of coral disease in the Florida Keys, *Hydrobiologia.*

Richardson, L.L., R.B. Aronson, W.M. Goldberg, G.W. Smith, K.B. Ritchie, J.C. Halas, J.S. Feingold, and S.L. Miller. 1998. Florida's mystery coral-killer identified, *Nature*, 392:557–558.

Richmond, R.H. 1993. Coral reefs: present problems and future concerns resulting from anthropogenic disturbance, *Am. Zool.*, 33:524–536.

Risk, M.J. 1999. Paradise lost: how marine science failed the world's coral reefs, *Mar. Freshwater Res.*, 50:381–387.

Shinn, E.A., B.H. Lidz, and M.W. Harris. 1994. Factors controlling the distribution of Florida Keys reefs, *Bull. Mar. Sci.*, 54:1084.

Smith, N.P. 1994. Long-term Gulf-to-Atlantic transport through tidal channels in the Florida Keys, *Bull. Mar. Sci.*, 54:602–609.

Smith, S.V. and R.W. Buddemeier. 1992. Global change and coral reef ecosystems, *Ann. Rev. Ecol. Syst.*, 23:89–118.

Wilkinson, C., O. Linden, H. Cesar, G. Hodgson, J. Rubens, and A.E. Strong. 1999. Ecological and socioeconomic impacts of 1998 coral mortality in the Indian Ocean: an ENSO impact and a warning of future change? *Ambio*, 28:188–196.

# 28 The Influence of Nearshore Waters on Corals of the Florida Reef Tract

*Clayton B. Cook*
Harbor Branch Oceanographic Institution

*Erich M. Mueller*
Center for Tropical Research

*M. Drew Ferrier*
Department of Biology, Hood College

*Eric Annis*
Darling Marine Center/University of Maine

## CONTENTS

0-8493-2026-7/02/$0.00+$1.50
© 2002 by CRC Press LLC

## INTRODUCTION

At the distal end of water flow through the South Florida hydroscape lies the Florida Reef Tract. This discontinuous group of reefs is at the interface between the nearshore environment of Hawk Channel and the oceanic conditions of the Florida Straits. Depending upon their location along the tract and temporally variable oceanographic conditions, the reefs are subject to various combinations of nearshore or oceanic water. Florida Bay has exerted an increasing influence on the waters of Hawk Channel since sea level rise allowed its waters to pass through the Keys archipelago (Ginsburg and Shinn, 1964; Shinn et al., 1994). (Influences that affect Florida Bay are discussed elsewhere in this volume.) In turn, the effects of Florida Bay water on the reefs have also increased, particularly in the Middle Keys (Lower Matecumbe to Big Pine Key) where broad channels permit considerable exchange of Florida Bay waters with those of Hawk Channel. Reefs are poorly developed offshore of these channels, although geological evidence indicates that Holocene reef development was considerable in these areas until broaching of the Keys approximately 4000 years BP (Shinn, 1963).

Summarizing previous literature, Chiappone (1996) stated that "...water exchange between Florida Bay and the Atlantic Ocean significantly impeded coral growth in certain areas, particularly the middle Florida Keys." Reefs found offshore of the Middle Keys are limited in development and are usually found where islands provide a barrier to direct Florida Bay influence (e.g., Sombrero Reef near Marathon). Tidal currents flow in and out of Florida Bay via channels through the Keys but with a net flow outward from Florida Bay (Smith, 1994). Within Hawk Channel, between the Keys proper and the Reef Tract, flows are generally southwest to westward (Pitts, 1994; cf. Figure 28.1). How much of the Florida Bay water actually reaches the Florida reef tract is uncertain and certainly varies with wind and current conditions; however, it appears clear that waters of Florida Bay have influence on the reef tract, particularly in the Middle and Lower Keys. Extensive development of reefs in the Upper Keys is likely due to low Florida Bay influence because of the barrier provided by the Keys in this region and the westward movement of bay water after entering Hawk Channel (Shinn et al., 1994).

So, why is Florida Bay water deleterious to reef development, especially in the Middle Keys? High turbidity (Roberts et al., 1982), variable temperature (Shinn, 1966) and salinity (Shinn et al., 1989), and elevated nutrients (Szmant and Forrester, 1994) of Florida Bay waters have been suggested as possible reasons for these "inimical effects" (Ginsburg and Shinn, 1964). Porter et al. (1999) have emphasized that these stressors are likely to act in concert in impacting the Florida Keys coral reef ecosystem. In this chapter, we consider these stressors in the context of known effects on reef corals and our own preliminary transplantation experiments in the Middle Keys.

Florida Bay has been documented as a source and sink for fine carbonate sediments, and these sediments are one cause of elevated turbidity in the Bay. Calcareous green algae, particularly *Penicillus* spp., produce fine carbonate particles which are normally trapped and stabilized as sediments by seagrass meadows. However, the extensive loss of *Thalassia testudinum* during the late 1980s and early 1990s (Robblee et al., 1991) has been largely responsible for increasing the turbidity of Florida Bay by allowing wind-driven resuspension of these unstabilized carbonate sediments and seagrass detritus (see Thayer et al., 1994). Extensive algal blooms (Smith and Robblee, 1994) have also increased turbidity in the bay in recent years.

Because of their shallow depth (<10 m), the waters of Florida Bay and the southeastern Gulf of Mexico that overlie the Southwest Florida Shelf are subject to large temperature and salinity fluctuations. Polar cold fronts during the winter can rapidly lower water temperatures to as low as 9°C during extreme events (Hudson, 1981; Porter et al., 1982). Summer insolation during periods of low wind velocities can raise temperatures up to 40°C (Schmidt and Davis, 1978). These extremes are well beyond the optimal range for reef corals (20 to 30°). Salinities can vary considerably, as well, particularly during the wet season in summer when evaporation can produce high salinities that can be quickly lowered by rainfall and runoff from the South Florida mainland. Management of upstream water supply has had considerable effects on the salinity of Florida Bay (Light and

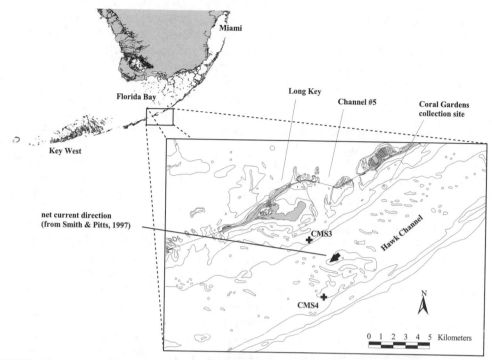

**FIGURE 28.1** Collection site (Coral Gardens) and array sites at Long Key (CMS3 and CMS4) used in study. Broad arrow near Tennessee Reef shows location of current meter (deployed by N. Smith, Harbor Branch Oceanographic Institution) and net current direction over the period.

Dineen, 1994). Reductions in water flow during the 1970s and 1980s generally raised salinity to levels where the bay had become significantly hypersaline (40 to 50‰; Boesch et al., 1993). This problem was addressed by the South Florida Water Management District in 1993 by increasing freshwater flow into the Everglades and thus Florida Bay. However, the effects of these salinity fluctuations on corals in the Florida Reef Tract are not clear. In one of the few published studies on the effects of salinity on coral physiology, Muthiga and Szmant (1987) found that changes of ±10‰ had little effect on photosynthesis and respiration of the coral *Siderastrea siderea*, although at 42‰ photosynthesis decreased by 25%. Porter et al. (1999) have demonstrated that coral productivity is reduced by the interactions of elevated salinity and temperature.

The role of elevated nutrient levels as a factor in the decline of coral reefs has been a subject of recent debate (Lapointe, 1997; Szmant, 1997; Hughes et al., 1999). Florida Bay has seen an increase in nutrient levels from the Gulf of Mexico and the South Florida watershed and the release of nutrients from sediments following seagrass dieoffs (Boesch et al., 1993). These processes have resulted in elevated nitrogen concentrations, so that phosphorus can be limiting in the bay, especially in the northeast portions (e.g., Fourqurean et al., 1993). The sources and effects of nutrients on coral reefs in the Florida Reef Tract have been subjects of considerable concern and debate. In the Middle Keys can be found a gradient of decreasing nitrogen concentrations from the bay to the reef tract (Szmant and Forrester, 1996), so that bay waters reaching the reefs during ebbing tides could deliver nutrients to the reef. Anthropogenic sources in the Keys proper increase dissolved nutrient levels in the inshore waters (Lapointe and Matzie, 1996), but the extent to which this input affects the offshore reef tract in the Keys is unclear (Szmant and Forrester, 1996). Other possible nutrient inputs to reef waters include upwelling, tidal bore events, and regeneration from sediments (Leichter et al., 1996; Szmant and Forrester, 1996). Typically, coral reef development is greatest in oligotrophic waters and is adversely affected by elevated nutrients.

Nutrient-related effects on reef corals may be indirect, as when macroalgae proliferate and outcompete corals under eutrophic conditions (Lapointe, 1997, but see Hughes et al., 1999), or direct. However, studies of the direct effects of dissolved nutrients on corals have yielded contradictory results. Ferrier-Pagès et al. (2000) have summarized the effects of nutrient addition on coral growth under laboratory and field conditions. In general, their review indicates that ammonium inhibits linear growth and calcification at concentrations between 2 and 15 $\mu M$, while nitrate reduces total calcification between 1 and 20 $\mu M$ (Marubini and Davies, 1996). It should be noted that these concentrations are generally higher than ambient reef levels, and other studies (Meyer and Schultz, 1985; Atkinson et al., 1995) indicate that lower concentrations of dissolved inorganic nitrogen are associated with increased coral growth. Phosphate acts as a crystal poison in calcification (Simkiss, 1964), and most studies indicate that elevated phosphate depresses coral growth (Kinsey and Davies, 1979; Stambler et al., 1991; Ferrier-Pagès et al., 2000). In contrast, Steven and Broadbent (1997) found increased calcification of corals on the Great Barrier Reef when phosphate was added to a final concentration of 4 $\mu M$ phosphate.

Despite the widespread impression that Florida Bay and other nearshore waters are detrimental to corals growing on the Florida Reef Tract, surprisingly few studies directly address this question. Hudson (1981) measured the linear growth of corals on a transect from the Keys to the Reef Tract starting at Snake Creek and found that linear growth rate increased with distance from shore. We examined the possibility that nearshore waters, particularly those emanating from Florida Bay, are responsible for decreased coral vigor (thus, reef development) in the Middle Keys by measuring the growth and nutrient exposure of *Montastraea faveolata* explants over one year at two locations near Long Key, a site of maximal impact from inshore waters. Growth was assessed by measuring changes in buoyant weight, areal extension, number of polyps, and radial extension. Nutrient exposure was assessed by physiological signals in the coral zooxanthellae (*Symbiodinium* sp.) that are commonly used to assess the nutrient sufficiency of marine algae and other plants (Flynn, 1990). These signals (elemental ratios, free amino acid content, and ammonium enhancement of dark carbon fixation) integrate both long-term nutrient history and nutrient inputs to corals from both dissolved and particulate (e.g., host-feeding) sources (Cook et al., 1997). We also examined the photosynthetic capability of the zooxanthellae, as calcification in corals is enhanced by symbiont photosynthesis (Vandermeulen and Muscatine, 1974).

## METHODS

We used a pneumatic drill fitted with a diamond coring bit to obtain 24 cores (5.1-cm diameter ~2.5 cm deep) from each of four colonies of *Montastraea faveolata* (Knowlton et al., 1992) at an inshore patch reef near Lower Matecumbe Key (Coral Gardens, 24°50.154′ N; 80°43.751′ W Figure 28.1). Core holes in the donor colonies were later filled with pre-cast cement plugs and Portland Type 2 cement. The cores were imaged, stained with 10 mg l$^{-1}$ Alizarin Red S for 24 hours secured in PVC collars, and weighed using the buoyant weight technique (Mettler AT400 balance 0.1-mg resolution). The explants were then deployed on two arrays such that all of the colonie were equally represented. The inshore array (CMS3; 4-m depth) was located near the shorewar edge of Hawk Channel (24°47.868′ N; 80°47.093′ W) and received direct flow from Florida Ba via Channel #5 on ebbing tides. The offshore array (CMS4; 5-m depth) was located on the oute edge of Hawk Channel near Tennessee Reef (24°45.475′ N; 80°46.370′ W). Based on informatio on net current speed and direction (Figure 28.1), it received less water flow from Florida Bay.

At approximately quarterly intervals over one year, 12 cores were retrieved from each of th arrays for assessment of growth and the nutrient status of the symbiotic zooxanthellae. Core numbe for collection at each interval were randomly selected at the start of the study so that each collectic consisted of matched pairs from each site from the same colonies located at corresponding positio on the arrays. After thorough cleaning of the PVC collars to remove fouling organisms, buoya weights were measured to determine total calcification (Jokiel et al., 1978). Tissues were remove from the skeletons using 0.22-μm-filtered seawater (FSW) in a recirculating water-jet system (Annis, 1998) based on a Water-Pik™ design; the extracts were used for biomass and nutrient exposure assays. Surface areas of the cleaned skeletons were determined using the foil method (Marsh, 1970), as were counts of the number of polyps. For measurements of linear extension, cores were bisected transversely using a diamond sawblade. One half was roughly polished with 600 grit silicon carbide sandpaper, and images of the polished face were recorded using an Olympus SZH10 stereomicroscope with a Sony video camera and SVHS recorder. For each core section, five points within intercalyx regions were selected for vertical growth measurements by making measurements of the distance from the alizarin stain line to the uppermost dissepiment (Lamberts, 1978) directly from the monitor screen.

The nutrient exposure of the corals was assessed by determining the nutrient status of freshly isolated zooxanthellae. Physiological parameters such as elemental ratios (C:N:P), free amino acid pools, and the ammonium enhancement of dark carbon fixation is commonly used to assess the nutrient status of marine plants such as microalgae (Flynn, 1990) and seagrasses (e.g., Fourqurean et al., 1992). These parameters integrate nutrient exposure from all sources over time, as has been demonstrated for zooxanthellae in host tissue (Cook et al., 1997). In addition, biomass characteristics of coral tissue such as numbers of zooxanthellae and protein content also reflect the environmental history of the host (Muller-Parker et al., 1994b; Fitt et al., 2000). These integrating measurements are particularly important for multitrophic organisms such as reef corals that receive nutrient inputs from a variety of sources.

Host tissues and zooxanthellae were separated and prepared according to the protocol developed by Muller-Parker et al. (1994b). Measurements of the zooxanthellae included cell counts, chlorophyll $a$ and $c_2$ content (Jeffrey and Humphrey, 1975), and elemental ratios (CNP). For the latter, frozen samples on pre-combusted GF/F filters were sent to the Analytical Services Laboratory of the University of Maryland (Chesapeake Biological Laboratory). C and N content was determined with an Exeter Analytical Model CE-440 elemental analyzer, and P content was determined by the method of Aspila et al. (1976). For analysis of the free amino acid (FAA) pools, zooxanthellae were collected on sterile 1.2-μm nylon syringe filters, rinsed with FSW and frozen. FAA were extracted from the filters in 2 ml of HPLC-grade distilled water for 1 hr at 70°C, derivatized with $o$-phthaldialdehyde (OPA) and separated by HPLC with a reversed-phase C18 column (Lindroth and Mopper, 1979; Ferrier, 1992). Protein content of animal supernatants was determined by the Lowry procedure (Muller-Parker et al., 1994b). In addition, suspensions of freshly isolated zooxanthellae were used for determination of photosynthetic rates and ammonium enhancement of dark carbon fixation (20 μ$M$ NH$_4$Cl) with $^{14}$CO$_2$ (Cook et al., 1992). For photosynthesis, the algae were incubated with NaH$^{14}$CO$_3$ for 30 min under fluorescent lamps producing an irradiance of 200 μmol photon m$^{-2}$ sec$^{-1}$.

For each quarterly sampling, we made between-sites comparisons of matched pairs (i.e., cores from the same colony at the same position on each array) with paired $t$-tests assuming unequal variances. For other statistical analyses (seasonal, colony, and overall effects), multi-way ANOVA and correlation analyses were used. *Post hoc* comparisons between groups were performed with Tukey's HSD procedure to examine seasonal and intercolony differences at each site. All datasets were normalized with the graphical procedures of Systat (SPSS, Inc.) prior to analysis.

# RESULTS

## GROWTH AND SURVIVAL OF CORAL EXPLANTS

### Coral Explant Survival

All 96 explants appeared to recover from the coring, handling, and alizarin treatment; however, over the one-year duration of the experiment, several explants suffered physical damage (breakage or tissue abraded by buoy lines) and one lost tissue, possibly due to hydrogen sulfide-rich water emanating from under the core through a gap in the epoxy. We saw no evidence of disease. Damaged cores were excluded from the datasets, so that the final sample sizes were 43 at the inshore site and 47 at the offshore array. Incomplete pairs were eliminated from between-site, paired $t$-test comparisons, but healthy unpaired cores were included in overall comparisons. For the 12 explants on each array deployed for the entire experiment, the survival rates were 83.3% inshore (385 to 390 days) and 91.7% offshore (361 days).

## Mass Accretion

Deposition of total $CaCO_3$ was greater at the offshore site (CMS4) during each of the quarterly samplings (Figure 28.2; paired $t$-tests; $p < 0.05$ for each interval). Corals at this site had accretion rates that were 25 to 46% greater than those at the inshore site (CMS3). Combining the rates for all sampling periods, the mean accretion rate of the offshore corals was 34.9 ± 11.9 mg $CaCO_3$ day$^{-1}$ (mean ± SD), while that of corals at CMS3 was 21.1 ± 7.7 mg $CaCO_3$ day$^{-1}$ ($p < 0.001$; two-sample $t$-test). The mean $CaCO_3$ accretion rate by all inshore corals was 39.7% lower than those offshore. The data in Figure 28.1 are not corrected for surface area. The area-corrected rates for offshore corals also exceeded those of the inshore corals during every sampling period (Table 28.1). Overall, the accretion rate per square centimeter of the offshore corals was 37.4% greater than that of the inshore corals.

Our measurements integrate growth from the time of deployment to the time of collection, so that any seasonal differences in calcification (e.g., decreased growth during winter months) may not be evident. A three-way ANOVA of the entire dataset for Figure 28.1 showed a strong effect of site ($p < 0.001$; cf. Figure 28.2) but no effects of sampling date or colony, We examined the data within each site for effects of sampling time and donor colony. Explants at the inshore site showed significant differences between dates ($p < 0.05$), with the October 1997 samples having slightly increased rates over the other samples. No effects of either sampling date or donor colony were found at the offshore site. The area-corrected data in Table 28.1 suggest decreased calcification rates of the offshore corals after the last sample (March 1997, after the winter period); however, this was more likely due to the increase in surface area of these corals during this period (Figure 28.3A), as no suggestion of decreased accretion rates of these corals can be seen in Figure 28.2.

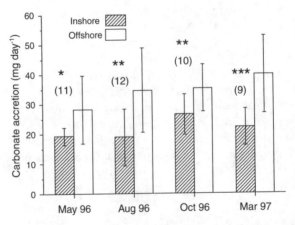

**FIGURE 28.2** Rates of buoyant weight increase (carbonate accretion) between paired coral cores at the inshore (CMS3) and offshore (CMS4) sites during each sampling period. Rates are integrated over the time between the deployment and collection of cores. Significance levels are from paired $t$-tests of log-transformed data. The variances for the March 1996 dataset were not homogeneous following this transformation. Error bars represent ±1 SD; $n$ value in parentheses. *, $p < 0.05$; **, $p < 0.01$; ***, $p < 0.001$.

**TABLE 28.1**
**Rates of Buoyant Weight Increase (Accretion Rates) from Figure 28.2**
**Normalized to Surface Area During Each of the Measurement Intervals**

| Collection Date | Accretion Rate (mg CaCO$_3$ day$^{-1}$ cm$^{-2}$) | | $n$ | $p$ |
|---|---|---|---|---|
| | CMS3 | CMS4 | | |
| May 1996 | 0.93 ± 0.17 | 1.45 ± 0.58 | 11 | <0.05 |
| August 1996 | 0.64 ± 0.19 | 1.29 ± 0.71 | 12 | <0.01 |
| October 1996 | 0.77 ± 0.15 | 1.17 ± 0.19 | 10 | <0.01 |
| March 1997 | 0.75 ± 0.24 | 0.92 ± 0.19 | 10 | <0.001 |

*Note:* The data were transformed with an inverse square root transformation prior to comparisons between sites by paired *t*-tests. Variances were homogenous except for the May 1996 dataset. Data reported as mean ± SD.

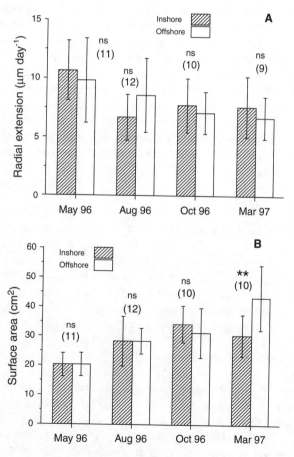

**FIGURE 28.3** (A) Rates of linear increase (radial extension) of the coral cores in Figure 28.2, as determined from alizarin staining. Significance levels are from paired *t*-tests of log-transformed data (all variances homogeneous). Statistical conventions are the same as in Figure 28.2. (B) Surface areas of the corals shown in Figure 28.2. Significance levels from paired *t*-tests of log-transformed data (all variances homogeneous). Statistical conventions are the same as in Figure 28.2.

## Radial Growth

In contrast to the mass accretion data, no differences were found in radial growth (linear extension) of corals at the two sites during any sampling period, as indicated by the deposition of new skeleton above the alizarin stain lines (Figure 28.3A). The mean rates of extension ranged from 7.0 to 10.7 µm day$^{-1}$ (inshore) and 6.6 to 9.8 at the offshore site. Averaged over all corals in the study, these values correspond to extrapolated yearly growth rates of $3.02 \pm 0.86$ mm yr$^{-1}$ at the inshore site and $2.92 \pm 0.95$ mm yr$^{-1}$ at the offshore site (range 2.4 to 3.9 mm yr$^{-1}$). As expected from the paired comparisons, the ANOVA did not reveal any between-site differences. Taken together with the mass accretion and surface areal data (below), these results indicate that the inshore corals were producing less dense skeletons. For each sample we calculated the density of newly deposited skeleton by dividing the weight of CaCO$_3$ added during each interval by the extension of the skeleton (mg CaCO$_3$ µm$^{-1}$ extension). For the offshore corals this value was $4.92 \pm 2.18$ ($n = 47$); for the inshore corals, $3.04 \pm 2.00$ ($n = 43$; $p < 0.001$). The analysis indicates a 38% decrease in the density of skeletal material added at the inshore site.

## Areal Growth

Coral explants showed evident skeletal and tissue growth over the epoxy and PVC collars during the experiment, and this growth is evident as increased surface areas of the samples over time at both sites (inshore: $r = 0.460$, $p < 0.01$; offshore: $r = 0.713$, $p < 0.001$) No between-site differences were observed in the surface areas of the explants during the first three collections (Figure 28.3A). However, during the last sampling (March 1997) corals at the offshore site had 30% more surface area than those at the inshore site ($p < 0.05$; Table 28.2). The offshore corals more than doubled their surface area (a 100% increase) between May 1996 and March 1997, while the inshore corals at CMS3 showed only a 50% increase. We noticed that the inshore corals routinely had greater algal growth and other bio-fouling at the margins of the live tissue than did those at the offshore site. Whether this increased fouling resulted in inhibition or regression of coral tissue is not clear.

## Zooxanthellae Biomasses

Overall, the densities of zooxanthellae normalized to coral surface area or per-milligram host protein were similar at both sites (Table 28.2). There were no between-site differences at each measurement interval except during August 1996, when corals at the inshore site (CMS3) had more zooxanthellae per square centimeter than those at the offshore site. ANOVA showed no seasonal differences in zooxanthellae per unit area, in contrast to the findings of Fitt et al. (2000), who reported a decline in zooxanthellae densities of corals in the Florida Keys during late summer and autumn. Our values for zooxanthellae per host protein for corals in October 1996 were lower than those in May 1996 or March 1997 ($p < 0.05$ for both; Tukey *post hoc* comparisons), suggesting that there might have been a fall decline in these values. Unfortunately, we had no protein samples for the summer of 1996.

Our values for the numbers of zooxanthellae per square centimeter of coral surface are generally lower than those reported from other species of "normal" (i.e., unbleached) corals in the *Montastraea annularis* complex (Szmant and Gassman, 1990; Fitt et al., 1993, 2000; Cook et al., 1994). Sectioned skeletons of these corals revealed that significant amounts of brown coral tissue remained, indicating that the removal of tissue from these relatively large (<20 cm$^2$) pieces of coral was less than 100%. While our values for the total numbers of zooxanthellae (particularly per unit area) are underestimates, subsequent data normalized to cell numbers in this chapter refer to determination made on final cell suspensions and do not depend on the efficiency of tissue recovery.

**TABLE 28.2**
**Between-Site Comparisons for All Parameters of Zooxanthellae Measured in this Study, Without Regard to Time of Sampling**

| Parameter | CMS3 | | CMS4 | | |
|---|---|---|---|---|---|
| | Mean ± SD | *n* | Mean ± SD | *n* | *p* |
| $10^6$ zooxanthellae cm$^{-2}$ | 0.80 ± 0.50 | 39 | 0.67 ± 0.40 | 41 | ns |
| $10^6$ zooxanthellae mg protein$^{-1}$ | 0.60 ± 0.31 | 28 | 0.50 ± 0.24 | 31 | ns |
| Chlorophyll *a* per cell[a] | 4.17 ± 1.67 | 41 | 2.98 ± 1.49 | 45 | <0.001 |
| Chlorophyll *c* per cell | 1.31 ± 0.61 | 41 | 1.02 ± 0.55 | 43 | ns |
| Chlorophyll *a* / chlorophyll $c_2$ | 3.45 ± 0.84 | 41 | 3.19 ± 0.83 | 43 | ns |
| C:N ratios | 6.50 ± 0.51 | 36 | 6.81 ± 0.49 | 47 | <0.01 |
| N:P ratios | 33.52 ± 5.58 | 36 | 28.71 ± 4.75 | 47 | <0.001 |
| C:P ratios | 220.57 ± 45.07 | 36 | 194.99 ± 31.89 | 47 | <0.001 |
| gln:glu | 1.15 ± 0.81 | 37 | 1.17 ± 0.69 | 46 | ns |
| Basic FAA/total FAA[b] | 0.22 ± 0.09 | 37 | 0.22 ± 0.08 | 46 | ns |
| Ammonium enhancement[c] | 1.17 ± 0.18 | 23 | 1.07 ± 0.16 | 28 | < 0.05 |
| Photosynthesis per cell (pg C cell$^{-1}$ h$^{-1}$) | 5.64 ± 3.86 | 31 | 5.296 ± 3.86 | 33 | ns |
| Photosynthesis per chlorophyll *a* (pg C µg chlorophyll $a^{-1}$ h$^{-1}$) | 1.21 ± 0.68 | 31 | 1.641 ± 0.77 | 32 | <0.05 |

*Note:* Comparisons by two-sample *t*-tests assuming unequal variances; data were log-transformed except where indicated; ns = not significant ($p > 0.05$).

[a] Data transformed as $y = $ chlorophyll $a^{0.2}$.
[b] Data transformed as $y = $ (basic/total)$^{0.6}$.
[c] Data transformed as $y = $ (NH$_4$ enh)$^{0.5}$.

## INDICES OF NUTRIENT SUFFICIENCY OF ZOOXANTHELLAE

### Elemental Ratios

C:N ratios of zooxanthellae from both sites were consistently low during every sampling, clustering around the N-sufficient Redfield ratio of 6.6:1 (Figure 28.4A). No between-site differences in C:N were observed, except in the October samples, for which ratios of zooxanthellae from the offshore corals were slightly higher than those from CMS3 ($p < 0.05$). The N:P ratios of zooxanthellae from both sites were generally elevated, roughly double the Redfield ratio of 16:1. N:P ratios of zooxanthellae from the inshore corals were greater than those of the offshore corals in the summer samples ($p < 0.01$), and possibly in the autumn ($p < 0.06$; Figure 28.4B). The winter and spring samples showed no between-site differences. As with the N:P ratios, C:P ratios for zooxanthellae from both sites were generally twice the Redfield ratio for P-sufficient phytoplankton (Figure 28.4C). C:P values were higher at the inshore site than offshore in the August samples (Figure 28.4C; $p < 0.01$); no other between-site differences were observed.

The overall pattern of the elemental data indicates that zooxanthellae at both sites were N sufficient throughout the year and probably were P limited. P limitation was greater at the inshore site in the summer, and a seasonal analysis revealed that both N:P and C:P ratios in the August samples were greater than other samples at this site (Tukey, $p < 0.05$). P limitation in Florida Bay waters also is most pronounced during the summer months (Fourqurean et al., 1993).

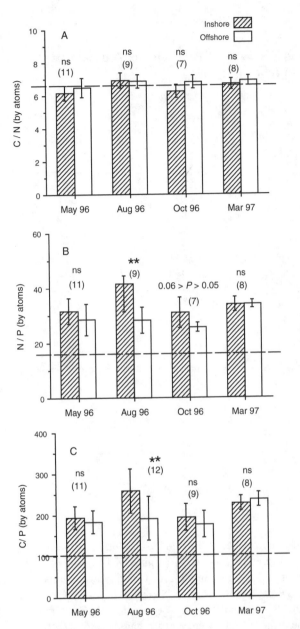

**FIGURE 28.4** Elemental ratios of zooxanthellae isolated from the corals in Figure 28.2. (A) C:N; (B) N:P; (C) C:P. Significance levels are from paired $t$-tests of log-transformed data (all variances homogeneous).

## FAA Ratios of Zooxanthellae

Typically nitrogen-replete algae store N as free basic amino acids with high N content (lysine, arginine, glutamine). As with the C:N ratios, neither ratios of glutamine to glutamate (gln:glu) nor basic to total FAAs showed any between-site differences in our samples; overall values are given in Table 28.2. Gln:glu ratios greater than 0.5 generally indicate N-sufficient microalgae (Flynn, 1990), as do gln-to-total ratios over 0.25. Our gln:glu data were well in excess of 0.5, with the overall mean for both sites greater than 1 (more glutamine than glutamate in the free amino acid pool), and the basic-to-total ratios approached 0.25. These ratios generally indicated N sufficiency at both sites throughout the study.

## Ammonium Enhancement of Dark Carbon Fixation

The addition of ammonium typically increases dark carbon fixation by nitrogen-limited microalgae, while it has little effect on N-sufficient ones (Flynn, 1990; Cook et al., 1992). We only assayed ammonium enhancement in three samples (May, August and October), but as with the C:N and FAA data, no differences in ammonium enhancement ratios (ammonium dark rates/seawater dark rates) between the two sites could be found at these times, although the overall dark enhancement ratio at the inshore site was slightly higher than the offshore site. In general, ratios at both sites were very low (Table 28.2), and there actually was no overall effect of ammonium addition on dark carbon fixation in these samples ($t$-test). These data complement those for C:N and FAA ratios in demonstrating N sufficiency for zooxanthellae at both sites in our Long Key study.

## Chlorophyll Content

Figure 28.5 summarizes the data on chlorophyll content of the zooxanthellae. During every sampling, zooxanthellae from corals at the inshore site contained more chlorophyll $a$ than did those at the offshore site (Figure 28.5A). Overall, zooxanthellae from the inshore corals had 25% more chlorophyll $a$ ($p < 0.001$; Table 28.2). Chlorophyll $c_2$ content at the inshore site was also greater during two of the sampling periods (Figure 28.5B). The ratios of the two pigments (chlorophyll $a$/chlorophyll $c_2$) generally ranged between 3 and 4 throughout the study, and showed no between-site differences (Figure 28.5C). The pigment content of zooxanthellae in corals can be influenced both by nitrogen supply and light conditions (Hoegh-Guldberg and Smith, 1989; Muscatine et al., 1989). Given the apparent similarity in nitrogen exposure of corals at the two sites, the differences in chlorophyll $a$ content were probably the result of differing light conditions at the two sites.

Fitt et al. (2000) have demonstrated a seasonal pattern of pigment content of zooxanthellae from corals in the Keys, with samples taken in the summer having consistently less chlorophyll than winter samples. Our samples did not show this trend; chlorophyll $a$ values at both sites were highest in October 1996 and lowest in March of 1997 (Tukey HSD, $p < 0.05$). The pigment ratios showed no effects of sampling date.

## Photosynthetic Rates of Isolated Zooxanthellae

Photosynthetic rates were only compared for the May, August, and October samples. Pooling data for all of the sites revealed no differences in per-cell photosynthetic rates, but photosynthesis per unit chlorophyll $a$ was 26% greater at the offshore site ($p < 0.05$; Table 28.2). Comparisons of matched pairs of corals at each sampling time showed this pattern only in the August samples ($p < 0.05$), with no between-site differences in the other samples.

## DISCUSSION

We chose the Long Key area as a site for this work in part because the influence of Florida Bay on the inshore waters (and possibly the Reef Tract) is greatest in this part of the Middle Keys (e.g., Szmant and Forrester, 1996). Without replicated sites, our results are inconclusive with respect to the general influence of Florida Bay on reef corals. However, our study produced four significant findings comparing these inshore and offshore sites: (1) total calcification, but not skeletal extension, was reduced in inshore corals; (2) corals at both sites were exposed to sufficient, if not excessive, sources of nitrogen; (3) corals at both sites were probably phosphorus limited, and this limitation was probably greater at the inshore site in the summer; (4) zooxanthellae from the inshore site had higher chlorophyll $a$ content and lower photosynthetic rates (per μg chlorophyll $a$) than those from the offshore site.

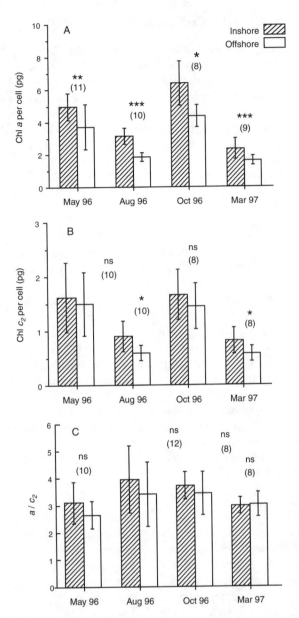

**FIGURE 28.5** Chlorophyll *a* content of zooxanthellae isolated from the corals in Figure 28.2. Significance levels are from paired *t*-tests of log-transformed data (all variances homogeneous). Statistical conventions are the same as in Figure 28.2.

## CORAL GROWTH: LINEAR EXTENSION

Two other studies of the growth of corals transplanted between inshore sites and offshore sites in the Florida Keys have been conducted. Shinn (1966) transplanted colonies of *Acropora cervicornis* from the Reef Tract off Key Largo to two locations: (1) on a pinnacle in Hawk Channel, and (2) in shallow water 300 ft from Key Largo Island. Linear extension, as determined by a banding technique, was twice as great at the donor site as at the two other sites. While the sites differed in a number of respects, temperature extremes were considered to be a major reason for the decreased growth (and increased mortality) at the inshore sites. However, it is not clear that the inshore corals were exposed to water from Florida Bay, as there are no major passes in this area of Key Largo.

Using alizarin staining techniques, Hudson (1981) measured skeletal extension rates of *Montastraea annularis* transplants along a transect from Snake Creek Channel (inshore) to Crocker Reef (offshore) in the Upper Keys. These are sites that would be affected by flow through Snake Creek and Whale Harbor Channels. He found a gradient in the rate of skeletal extension with nearshore corals growing more slowly than those offshore, and be correlated this result with increasing temperature stress near shore. We made no temperature measurements in our study. While our inshore corals very likely experienced a wider range of temperatures than the offshore corals, these differences in temperature exposure were not expressed as effects on skeletal extension rates. We did find evident differences in coral growth in regard to total carbonate deposition, which is also temperature dependent (Clausen and Roth, 1975). For reasons discussed below, we believe that water clarity was an important factor affecting total calcification at Long Key.

Comparison of our extension data with those of Hudson (1981) and other studies suggests that inshore waters affected corals at both of our sites. Corals at Hudson's most inshore site had mean extension rates of 2 mm yr$^{-1}$, while transplants on offshore reefs added 8 to 9 mm yr$^{-1}$. The extension rates that we measured at Long Key were similar to his inshore rates at both the inshore and offshore sites (2.4 to 3.9 mm yr$^{-1}$) and were lower than those reported for the *M. annularis* species complex in shallow waters elsewhere in the Florida Keys (Vaughan, 1915; Hudson, 1981) or from the Caribbean (Dustan, 1975; Graus and Macintyre, 1982; Tomascik and Sander, 1985). As temperature has been implicated in the reduction of extension rates of corals in the Keys, this may be a stress experienced by corals on reefs in the Middle Keys due to tidal exchange through the passes.

## CORAL GROWTH: TOTAL CALCIFICATION

Our finding that offshore corals deposited more total CaCO3 than inshore corals while exhibiting no differences in extension rate indicates that the inshore corals had produced a less dense skeleton (37% reduction). We are aware of no comparable studies for reef corals in the Florida Keys. Risk and Sammarco (1991) found increasing skeletal density in *Porites lobata* on the Great Barrier Reef with greater distance from shore and suggested that this was due either to increased light penetration offshore or to the inhibition of calcification inshore due to elevated nutrients. Foster (1979) examined the density of skeletons of *M. annularis* from a variety of habitats in Jamaica and found that corals from lagoonal environments had more porous skeletons than did those from patch or offshore reefs. Reciprocal transplants between lagoonal and other environments resulted in the skeletons of transplants taking on the morphology characteristic of the new environment. The lagoonal environment was characterized as having higher sedimentation rates and lower light levels than the other sites, and Foster (1980) suggested that differences in light levels were largely responsible for the growth differences in *M. annularis* at these sites. Similar results were reported by Dodge and Brass (1984), who measured both extension rates and total mass accretion of *M. annularis* at various sites in St. Croix. They found that corals from Christiansted Harbor, historically exposed to dredging activities and sewage effluent, exhibited reduced mass accretion compared to corals at more pristine sites, although extension rates were similar.

At Long Key there is a decreasing gradient of both dissolved nutrients and suspended material from Florida Bay to the Reef Tract (Szmant and Forrester, 1994, 1996). While we made no irradiance measurements at our sites, the patterns of chlorophyll *a* content of zooxanthellae in our study indicated lower light levels at the inshore site. Zooxanthellae in corals typically respond to lower light levels by increasing pigment content (e.g., Falkowski and Dubinsky, 1981). Because these corals were at comparable depths and showed no differences in zooxanthellae density (thus eliminating self-shading as a factor), reduced light penetration of the water column at the inshore site was most likely responsible. Increased turbidity inshore was supported by our own qualitative observations during dives and by subsequent field measurements (Jones and Boyer, 1999; Szmant, pers. comm.). Thus, the effects of nearshore waters in decreasing bulk calcium deposition by reef corals, but not skeletal extension, appear to be related to reduced water clarity. In South Florida this appears to be due to the increasing turbidity of Florida Bay, associated with seagrass dieoffs (Robblee et al., 1991).

It appears that calcification by reef-building corals involves at least two phases. Skeletal extension (constructing the "flimsy scaffolding" of Barnes and Crossland, 1980) occurs largely at night and does not require light (Barnes and Crossland, 1980; Gladfelter, 1983; Vago et al., 1997). Bulk deposition of calcium carbonate as an "infilling" process takes place during the day and may represent the light-enhanced calcification that is typical of zooxanthellate corals (Gladfelter, 1983). We suggest that increased turbidity of inshore waters inhibits this light-enhanced bulk calcification and has a lesser effect (or none) on skeletal extension. In contrast, the wider temperature ranges of inshore waters are likely to affect both processes, perhaps linked through the interacting effects of salinity and temperature on coral productivity (Porter et al., 1999).

## Effects of Nutrients

Another cause of decreased calcification at inshore reef sites could be elevated nutrients, especially nitrate, which can reduce calcification by corals at concentrations as low as 1 $\mu M$ (Marubini and Davies, 1996). Nitrate levels at Long Key decrease along a gradient from Florida Bay to the reef tract (Szmant and Forrester, 1996). However, nitrate concentrations both inshore and offshore at Long Key typically are less than 0.5 $\mu M$ (Szmant and Forrester, 1996; Jones and Boyer, 1999), and it not clear that nitrate concentrations in this range affect coral calcification.

It is generally thought symbiotic dinoflagellates in reef corals are limited by the availability of inorganic nutrients, especially nitrogen (e.g., Muscatine et al., 1989). All of the parameters of nitrogen sufficiency that we measured indicated that zooxanthellae from corals at both Long Key sites were nitrogen sufficient throughout the period of study. These included C:N ratios between 6.5 and 7.0, high levels of basic amino acids in the free amino acid pool, and the lack of significant enhancement of dark carbon fixation by ammonium. Our C:N values are lower than those found for coral zooxanthellae from the Red Sea (Muscatine et al., 1989), Hawaii (Muller-Parker et al., 1994a), Bermuda (Muller-Parker and Cook, unpub. data), and corals from the northern end of the Florida Reef Tract (McGuire and Szmant, 1997). Zooxanthellae from *M. annularis* in Bermuda typically exhibited significant enhancement of dark carbon fixation by ammonium with higher ammonium enhancement ratios, (Cook et al., 1994) in addition to higher C:N values. These comparisons all indicate that the corals at both of our Long Key sites were exposed to higher levels of nitrogen than those from other reef sites.

The sources of this elevated nitrogen are not clear, particularly for corals located at our offshore site. Dissolved and particulate nitrogen is clearly higher in Florida Bay (Szmant and Forrester, 1996), and dissolved N is elevated in the canals and other inshore waters in the Keys (Lapointe and Matzie, 1996). How much of this nutrient-enriched water actually impinges on the Florida Reef Tract is not clear (Porter et al., 1999). Current meter studies have shown a net transport along Hawk Channel, such that at least some of the tidal flow from inshore is diverted to the southwest (Pitts, 1994; Smith and Pitts, 1998). Our findings of P limitation of coral zooxanthellae, particularly in the summer at the inshore site indicate that some Florida Bay water reached this site, and possibly the offshore site as well. The waters of the eastern part of the bay typically are P limited in the summer (Fourqurean et al., 1992), and the P signals in the algae appear to reflect this. Other sources of dissolved nitrogen for corals on the Reef Tract could be upwelling events (Lee et al., 1992, 1994; Leichter et al., 1996) and particulate sources supplying nitrogen to coral zooxanthellae via host feeding (e.g., Cook et al., 1994). Regardless of the sources of these nutrients, it does not appear as though any evidence exists to date that elevated nutrients have a direct effect upon corals in the Florida Reef Tract.

## Productivity

The calcification of reef corals is intimately associated with symbiont productivity, although the linkages between them are still unclear and controversial (Gattuso et al., 1999). Numerous authors have noted that gross productivity of corals is stimulated by the addition of inorganic nutrients, primarily through the increase in symbiont numbers (Hoegh-Guldberg and Smith, 1989; Muscatine et al., 1989; Marubini and Davies, 1996; Ferrier-Pagès et al., 2000). In one of the few studies examining both the effects of long-term nutrient exposure on calcification and productivity of corals, Ferrier-Pagès et al. (2000) found an inverse relationship between the stimulation of productivity (P/R ratios) by nutrients and the inhibition of calcification. We measured photosynthesis, but not respiration, by isolated zooxanthellae from our corals. Carbon fixation per cell did not differ between inshore and offshore sites, although photosynthesis per unit chlorophyll *a* was greater at the offshore site (Table 28.2). The similar symbiont densities (Table 28.2) suggest that total photosynthesis by corals at the two sites was comparable under identical light conditions. As noted above, we believe that increased turbidity at the inshore site depressed productivity by reducing irradiance, thus total calcification by these corals, despite the elevated chlorophyll *a* content of the zooxanthellae. A related effect of turbidity on coral productivity would be increased respiratory rates (Telesnicki and Goldberg, 1995), which would further divert host energy from calcification processes. These possibilities should be examined by the use of *in situ* coral respirometry.

## OTHER EFFECTS OF INSHORE WATERS ON CORALS

We found that the areal growth of corals at the inshore site was depressed in the latter stages of our study (Figure 28.3B). We noted increased bio-fouling of the coral maintenance structure at the inshore site during every sampling period. The attached epibionts included algae, hydroids, and other organisms that grew at the periphery of the growing edges of the coral explants. This bio-fouling may have inhibited both areal increase and polyp formation at this site. Longer-term effects on coral growth are suggested by our finding that the inshore corals produced skeletons of lower density. Coral skeletons with decreased density are structurally weaker and would be more susceptible to mechanical breakage from storm waves and other physical forces. A related consequence of weaker skeletons might be increased susceptibility to bio-erosion. Sammarco and Risk (1990) reported increased bio-erosion of *Porites lobata* at inshore sites on the Great Barrier Reef, where skeletal density was low (Risk and Sammarco, 1991). However, bio-erosion has also been reported to increase in coral skeletons of higher density (Highsmith, 1981).

## CONCLUSIONS

Coral reefs of the Florida Reef Tract receive variable inputs of inshore waters and those of Florida Bay. Reef development is typically greatest where islands of the Florida Keys impede exchange of these waters and is least where passes permit such exchange. Temperature, salinity, turbidity, and elevated nutrient levels have been cited as "inimical effects" of inshore waters that inhibit coral growth. Previous work has indicated that temperature extremes reduce linear growth by corals in the Florida Reef Tract and operate with elevated salinity to reduce productivity. We performed transplant experiments using the major reef-building coral *Montastraea faveolata* at Long Key in the Middle Keys, where reefs are maximally affected by tidal exchange through passes. Corals at inshore and offshore sites exhibited no differences in linear growth rates, but skeletal deposition of $CaCO_3$ was 40% greater in offshore corals. We ascribe this effect to increased turbidity at the inshore site, as reported in other studies. Assays of nutrient exposure of coral zooxanthellae (elemental ratios, free amino acid pools, dark carbon) indicated no differences in nitrogen exposure at the two sites: symbionts were N sufficient, if not saturated, at both sites. Zooxanthellae from both sites had elevated N:P and C:P ratios, indicative of P limitation. Turbidity and temperature appear to be the major characteristics of inshore waters that affect corals on the Florida Reef Tract; to date, little evidence exists that elevated nutrients have a direct effect.

## ACKNOWLEDGMENTS

We thank Diane Silvia, Simon Davy, Paget Graham, Lori Hrdlicka, Chad McNutt, and Gisèle Muller-Parker for assistance, and the staff of the Keys Marine Laboratory for providing boat and lab facilities. Elemental analyses were provided by Analytical Services of the Chesapeake Biological Laboratory. We thank Ned Smith, Patrick Pitts, and Ron Jones for sharing data from unpublished reports, and Harold Hudson for his loan of a coral drilling bit. Funding was provided by the U.S. EPA Region IV Special Studies of the Florida Keys National Marine Sanctuary Water Quality Protection Program. Our coral field work was conducted under permits issued to CBC: FKNMS–(UR)–53–95, FDEP #95S-0412, 96S-046. This is Contribution #1398 of the Harbor Branch Oceanographic Institution, Inc., and #001 of the Center for Tropical Research.

## REFERENCES

Annis, E. 1998. Phosphatase Activity in the Zooxanthellae From the Sea Anemone *Aiptasia pallida*, M.S. thesis, Florida Institute of Technology, Melbourne, FL.

Aspila, I., H. Agemian, and A.S. Chau. 1976. A semi-automated method for the determination of inorganic, organic and total phosphate in sediments, *Analyst*, 101:187–197.

Atkinson, M.J., B. Carlson, and G.L. Crow. 1995. Coral growth in high-nutrient, low-pH seawater: a case study of corals cultured at the Waikiki Aquarium, Honolulu, Hawaii, *Coral Reefs*, 14:215–223.

Barnes, D.J. and C.J. Crossland. 1980. Diurnal and seasonal variations in the growth of a staghorn coral measured by time-lapse photography, *Limnol. Oceanogr.*, 25:1113–1117.

Boesch, D.F., D.E. Armstrong, C.F. D'Elia, N.G. Maynard, H.W. Paerl, and S.L. Williams. 1993. *Deterioration of the Florida Bay Ecosystem: An Evaluation of the Scientific Evidence*, final report. Florida Bay Interagency Working Group on Florida Bay (National Fish and Wildlife Service, National Park Service, South Florida Water Management District.)

Chiappone, M. 1996. *Geology and Paleontology of the Florida Keys and Florida Bay*, Farley, Zenda, WI.

Clausen, C.D. and A.A. Roth. 1975. Effect of temperature and temperature adaptation on calcification rate in the hermatypic coral *Pocillopora damicornis*, *Mar. Biol.*, 33:93–100.

Cook, C.B., G. Muller-Parker, and C.F. D'Elia. 1992. Ammonium enhancement of dark carbon fixation and nitrogen limitation in symbiotic zooxanthellae: effects of feeding and starvation of the sea anemone *Aiptasia pallida*, *Limnol. Oceanogr.*, 37:131–139.

Cook, C.B., G. Muller-Parker, and M.D. Ferrier. 1997. An assessment of nutrient sufficiency in symbiotic dinoflagellates, in *Proc. 8th Int. Coral Reef Symp.*, Vol. 1, Lessios, H.A., Ed., pp. 903-908, Panama City.

Cook, C.B., G. Muller-Parker, and C.D. Orlandini. 1994. Ammonium enhancement of dark carbon fixation and nitrogen limitation in zooxanthellae symbiotic with the reef corals *Madracis mirabilis* and *Montastraea annularis*, *Mar. Biol.*, 118:157–165.

Dodge, R.E. and G.W. Brass. 1984. Skeletal extension, density and calcification of the reef coral, *Montastraea annularis*: St. Croix, U.S. Virgin Islands, *Bull. Mar. Sci.*, 34:288–307.

Dustan, P. 1975. Growth and form in the reef-building coral *Montastraea annularis*, *Mar. Biol.*, 33:101–107.

Falkowski, P.G. and Z. Dubinsky. 1981. Light-shade adaptation of *Stylophora pistillata*, a hermatypic coral from the Gulf of Eilat, *Nature*, 289:172–174.

Ferrier, M.D. 1992. Fluxes and Metabolic Pools of Amino Acids in Algal-cnidarian Symbioses, Ph.D. thesis, University of Maryland, College Park.

Ferrier-Pagès, C., J.-P. Gattuso, S. Dallot, and J. Jaubert. 2000. Effect of nutrient enrichment on growth and photosynthesis of the zooxanthellate coral *Stylophora pistillata*, *Coral Reefs*, 19:103–113.

Fitt, W.K., H.J. Spero, J. Halas, M.W. White, and J.W. Porter. 1993. Recovery patterns of the coral *Montastraea annularis* after the 1987 "bleaching event" in the Florida Keys, *Coral Reefs*, 12:57–64.

Fitt, W.K., F.K. McFarland, M.E. Warner, and G.C. Chilcoat. 2000. Seasonal patterns of tissue biomass and densities of symbiotic dinoflagellates in reef corals and relation to coral bleaching, *Limnol. Oceanogr.*, 45:677–685.

Flynn, K.J. 1990. The determination of nitrogen status in microalgae, *Mar. Ecol. Prog. Ser.*, 61:297–307.

Foster, A.B. 1979. Phenotypic plasticity in the reef corals *Montastraea annularis* and *Siderastrea siderea*, *J. Exp. Mar. Biol. Ecol.*, 39:25–54.

Foster, A.B. 1980. Environmental variation in skeletal morphology within the Caribbean reef corals *Montastraea annularis* and *Siderastrea siderea*, *Bull. Mar. Sci.*, 30:678–709.

Fourqurean, J.W., J.C. Zieman, and G.V.N. Powell. 1992. Phosphorus limitation of primary production in Florida Bay: evidence from C:N:P ratios of the dominant seagrass *Thalassia testudinum*, *Limnol. Oceanogr.*, 37:162–171.

Fourqurean, J.W., R.D. Jones, and J.C. Zieman. 1993. Processes influencing water column nutrient characteristics and phosphorus limitation of phytoplankton biomass in Florida Bay, FL, USA: inferences from spatial distribution, *Estuar. Coast. Shelf Sci.*, 36:295–314.

Gattuso, J.-P., D. Allemand, and M. Frankignolle. 1999. Photosynthesis and calcification at cellular, organismal and community levels in coral reefs: a review on interactions and control by carbonate chemistry, *Am. Zool.*, 39:160–183.

Ginsburg, R.N. and E.A. Shinn. 1964. Distribution of the reef-building community in Florida and the Bahamas, *Am. Assoc. Petroleum Geol. Bull.*, 48:527.

Gladfelter, E.H. 1983. Skeletal development in *Acropora cervicornis*: II. Diel patterns of calcium carbonate accretion, *Coral Reefs*, 2:91–100.

Graus, R.R. and I.G. Macintyre. 1982. Variation in growth forms of the reef coral *Montastraea annularis* (Ellis and Solander): a quantitative evaluation of growth response to light distribution using computer simulation, *Smithsonian Cont. Mar. Sci.*, 12:441–464.

Highsmith, R.C. 1981. Coral bioerosion: damage relative to skeletal density, *Am. Nat.*, 117:193–198.

Hoegh-Guldberg, O. and G.J. Smith. 1989. Influence of the population density of zooxanthellae and supply of ammonium on the biomass and metabolic characteristics of the reef corals *Seriatopora hystrix* and *Stylophora pistillata*, *Mar. Ecol. Prog. Ser.*, 57:173–186.

Hudson, J.H. 1981. Response of *Montastraea annularis* to environmental change in the Florida Keys, in *Proc. 4th Int. Coral Reef Symp.*, Vol. 2, pp. 233–240, Manila.

Hughes, T., A. Szmant, R. Steneck, R. Carpenter, and S. Miller. 1999. Algal blooms on coral reefs: what are the causes? *Limnol. Oceanogr.*, 44:1583–1586.

Jeffrey, S.W. and G.W. Humphrey. 1975. New spectrophotometric equations for determining chlorophylls $a$, $b$, $c_1$, and $c_2$ in higher plants, algae, and natural phytoplankton, *Biochem. Physiol. Pflanz.*, 167:191–194.

Jokiel, P.L., J.E. Maragos, and L. Franzisket. 1978. Coral growth: buoyant weight technique, pp. 529–541, in *Coral Reefs: Research Methods*, Stoddart, D.R. and Johannes, R.E., Eds., Monogr. Oceanog. Methodol., Vol. 5, SCOR/UNESCO, Paris.

Jones, R.D. and J.N. Boyer. 1999. *1999 Annual Report of the Water Quality Monitoring Project for the Florida Keys National Marine Sanctuary*, final report, SERC Technical Report #T121, Florida International University, under EPA Agreement No. X994621-94-0.

Kinsey, D.W. and P.J. Davies. 1979. Effects of elevated nitrogen and phosphorus on coral reef growth, *Limnol. Oceanogr.*, 24:935–940.

Knowlton, N.K., E. Weil, A. Weight, and H.M. Guzmán. 1992. Sibling species in *Montastraea annularis*, coral bleaching and the coral climate record, *Science*, 255:330–333.

Lamberts, A.E. 1978. Coral growth: alizarin method, pp. 523–527, in *Coral Reefs: Research Methods*, Stoddart, D.R. and Johannes, R.E., Eds., Monogr. Oceanog. Methodol., Vol. 5, SCOR/UNESCO, Paris.

Lapointe, B.E. 1997. Nutrient thresholds for bottom-up control of macroalgal blooms on coral reefs in Jamaica and southeast Florida, *Limnol. Oceanogr.*, 42:1119–1131.

Lapointe, B.E. and W.R. Matzie. 1996. Effects of stormwater nutrient discharges on eutrophication processes in nearshore waters of the Florida Keys, *Estuaries*, 19:422–435.

Lee, T.N., C. Rooth, E. Williams, M. McGowan, A.F. Szmant, and M.E. Clarke. 1992. Influence of Florida Current, gyres and wind-driven circulation on transport of larvae and recruitment in the Florida Keys coral reefs, *Cont. Shelf Res.*, 12.

Lee, T.N., M.E. Clarke, E. Williams, A.F. Szmant, and T. Berger. 1994. Evolution of the Tortugas gyre and its influence on recruitment in the Florida Keys, *Bull. Mar. Sci.*, 54:621–646.

Leichter, J.J., S.R. Wing, S.L. Miller, and M.W. Denny. 1996. Pulsed delivery of subthermocline water to Conch Reef (Florida Keys) by internal tidal bores, *Limnol. Oceanogr.*, 41:1490–1501.

Light, S.S. and J.W. Dineen. 1994. Water control in the Everglades: a historical perspective, pp. 47–84, in *Everglades: The Ecosystem and Its Restoration*, Davis, S.M. and Ogden, J.C., Eds., St. Lucie Press, Delray Beach, FL.

Lindroth, P. and K. Mopper. 1979. High performance liquid chromatographic determination of subpicomole amounts of amino acids by precolumn fluorescence derivatization with *o*-phthalaldehyde, *Anal. Chem.*, 51:1167–1174.

Marsh, J.A., Jr. 1970. Primary production of reef-building calcareous red algae, *Ecology*, 51:255–263.

Marubini, F. and P.S. Davies. 1996. Nitrate increases zooxanthellae population density and reduces skeletogenesis in corals, *Mar. Biol.*, 127:319–328.

McGuire, M.P. and A.M. Szmant. 1997. Time course of responses to $NH_4$ enrichment by a coral-zooxanthellae symbiosis, in *Proc. 8th Int. Coral Reef Symp.*, Vol. 1, Lessios, H.A., Ed., pp. 909–914, Panama City.

Meyer, J.L. and E.T. Schultz. 1985. Migrating haemulid fishes as a source of nutrients and organic matter on coral reefs, *Limnol. Oceaongr.*, 30:146–156.

Muller-Parker, G., C.B. Cook, and C.F. D'Elia. 1994a. Elemental composition of the coral *Pocillopora damicornis* exposed to elevated seawater ammonium, *Pacific Sci.*, 48:234–246.

Muller-Parker, G., L.R. McCloskey, O. Hoegh-Guldberg, and P.J. McAuley. 1994b. The effect of ammonium enrichment on animal and algal biomass of the coral *Pocillopora damicornis*, *Pacific Sci.*, 48:273–283.

Muscatine, L., P.G. Falkowski, Z. Dubinsky, P.A. Cook, and L.R. McCloskey. 1989. The effect of external nutrient resources on the population dynamics of zooxanthellae in a reef coral, *Proc. R. Soc. Lond. B*, 236:311–324.

Muthiga, N.A. and A.M. Szmant. 1987. The effects of salinity stress on the rates of aerobic respiration and photosynthesis in the hermatypic coral *Siderastrea siderea*, *Biol. Bull.*, 173:539–551.

Pitts, P.A. 1994. An investigation of near-bottom flow patterns along and across Hawk Channel, Florida Keys, *Bull. Mar. Sci.*, 54:610–620.

Porter, J.W., J.F. Battey, and G.J. Smith. 1982. Perturbation and change in coral reef communities, *Proc. Natl. Acad. Sci. USA*, 79:1678–1681.

Porter, J.W., S.K. Lewis, and K.G. Porter. 1999. The effect of multiple stressors on the Florida Keys coral reef ecosystem: a landscape hypothesis and a physiological test, *Limnol. Oceanogr.*, 44:941–949.

Risk, M.J. and P.W. Sammarco. 1991. Cross-shelf trends in skeletal density of the massive coral *Porites lobata* from the Great Barrier Reef, *Mar. Ecol. Progr. Ser.*, 69:195–200.

Robblee, M.B., T.R. Barber, M.J. Durako, J.W. Fourqurean, L.K. Muehlstein, D. Porter, L.A. Yarbro, R.T. Zieman, and J.C. Zieman. 1991. Mass mortality of the tropical seagrass *Thalassia testudinum* in Florida Bay (USA), *Mar. Ecol. Progr. Ser.*, 71:297–299.

Roberts, H.H., J.L.J. Rouse, and N.D. Walker. 1982. Evolution of cold-water stress conditions in high latitude reef systems: Florida Reef tract and the Bahama Banks, *Carib. J. Sci.*, 19:55–60.

Sammarco, P.W. and M.J. Risk. 1990. Large-scale patterns in internal bioerosion of *Porites*: cross continental shelf trends on the Great Barrier Reef, *Mar. Ecol. Progr. Ser.*, 59:145–156.

Schmidt, T.W. and G.E. Davis. 1978. *A Summary of Estuarine and Marine Water Quality Information Collected in Everglades National Park, Biscayne National Monument and Adjacent Estuaries from 1879 to 1977*, final report, South Florida Research Center Report Series No. T-519.

Shinn, E.A. 1963. Spur and groove formation on the Florida Reef Tract, *J. Sediment. Petrol.*, 33:291–303.

Shinn, E.A. 1966. Coral growth-rate, an environmental indicator, *J. Palaeont.*, 40:233–240.

Shinn, E.A., B.H. Lidz, J.L. Kindinger, J.H. Hudson, and R.B. Halley. 1989. *A Field Guide: Reefs of Florida and the Dry Tortugas*, 28th International Geological Congress-Field Trip T176, Washington, D.C.

Shinn, E.A., B.H. Lidz, and M.W. Harris. 1994. Factors controlling distribution of Florida Keys reefs. Bull. Mar. Sci. 54:1084.

Simkiss, K. 1964. Phosphates as crystal poisons of calcification. *Biol. Rev.*, 39:487–505.

Smith, N.P. 1994. Long-term Gulf-to-Atlantic transport through tidal channels in the Florida Keys, *Bull. Mar. Sci.*, 54:602–609.

Smith, N.P. and P.A. Pitts. 1998. *Hawk Channel Transport Study: Pathways and Processes*, final report, South Florida Water Management District Contract No. C-6627-A1.

Smith, T.J. and M.B. Robblee. 1994. Relationships of sport fisheries catches in Florida Bay to freshwater inflow from the Everglades (abstr.), *Bull. Mar. Sci.*, 54:1084.

Stambler, N., N. Popper, Z. Dubinsky, and J. Stimson. 1991. Effects of nutrient enrichment and water motion on the coral *Pocillopora damicornis*, *Pacific Sci.*, 45(3):299–307.

Steven, A.D.L. and A.D. Broadbent. 1997. Growth and metabolic responses of *Acropora prolifera* to long term nutrient enrichment, in *Proc. 8th Int. Coral Reef Symp.*, Vol. 1, Lessios, H.A., Ed., pp. 867–872, Panama City.

Szmant, A.M. 1997. Nutrient effects on coral reefs: a hypothesis on the importance of topographic and trophic complexity to reef nutrient dynamics, in *8th Int. Coral Reef Symp.*, Vol. 2, Lessios, H.A., Ed., pp. 1527–1532, Panama City.

Szmant, A.M. and A. Forrester. 1994. Sediment and water column nitrogen and phosphorus distribution in the Florida Keys: SEAKEYS (abstr.), *Bull. Mar. Sci.*, 54:1085–1086.

Szmant, A.M. and A. Forrester. 1996. Water column and sediment nitrogen and phosphorus distribution patterns in the Florida Keys, USA, *Coral Reefs*, 15:21–41.

Szmant, A.M. and N.J. Gassman. 1990. The effects of prolonged "bleaching" on the tissue biomass and reproduction of the reef coral *Montastraea annularis*, *Coral Reefs*, 8:217–224.

Telesnicki, G.J. and W.M. Goldberg. 1995. Effects of turbidity on the photosynthesis and respiration of two south Florida reef coral species, Bull. Mar. Sci. 57: 527–539.

Thayer, G.W., P.L. Murphey, and M.W. LaCroix. 1994. Responses of plant communities in western Florida Bay to the die-off of seagrasses. *Bull. Mar. Sci.*, 54:718–726.

Tomascik, T. and F. Sander. 1985. Effects of eutrophication on reef-building corals. I. Growth rate of the reef-building coral *Montastraea annularis*, *Mar. Biol.*, 87:143–155.

Vago, R., E. Gill, and J.C. Collingwood. 1997. Laser measurements of coral growth, *Nature*, 386:30–31.

Vandermeulen, J.H. and L. Muscatine. 1974. Influence of symbiotic algae on calcification in reef corals: critique and progress report, pp. 1–19, in *Symbiosis in the Sea*, Vol. 2, Vernberg, W.B., Ed., The Belle W. Baruch Library in Marine Science, University of South Carolina Press.

Vaughan, T.W. 1915. The geologic significance of the growth-rate of the Floridian and Bahaman shoal-water corals, *J. Wash. Acad. Sci.*, 5:591–600.

# 29 Differential Coral Recruitment Patterns in the Florida Keys

*Jennifer I. Tougas and James W. Porter*
Institute of Ecology, University of Georgia

## CONTENTS

## INTRODUCTION

Because adult corals are sessile, emigration and immigration are dependent upon the mobile larval phase of the life cycle (Fautin et al., 1989). Coral larvae are produced through sexual reproduction (Richmond and Hunter, 1990). Successful cross-fertilization is a challenge for sessile marine invertebrates due to the diluting effects of the surrounding ocean (Oliver and Babcock, 1992). Corals have evolved numerous strategies to overcome challenges associated with fertilization in the open sea (Oliver and Babcock, 1992). Aggregation of parent colonies, synchronized spawning, and high reproductive output during reproductive events all increase the chance of fertilization (Levitan, 1995).

As an order of organisms, scleractinians are surprisingly flexible in their reproductive modes. Even within a given genus, some species may be hermaphroditic, producing both sperm and eggs within an individual colony, or gonochoric, producing only sperm or eggs within an individual colony (Szmant, 1991). Fertilization can take place externally during synchronous mass spawning events for certain species, or internally as for certain brooding species. In the case of the former,

0-8493-2026-7/02/$0.00+$1.50
© 2002 by CRC Press LLC

larval development takes place in the water column and several days can pass before the larvae are capable of settling. In the case of the latter, larval development takes place within the parent colony, and when later released the larvae are capable of settling within a matter of hours. Further, depending on the species, reproduction may occur once a year or many times throughout the year (Richmond and Hunter, 1990). Regardless of the mode of reproduction, the end result of sexual reproduction is mobile larvae.

Population dynamics of sessile invertebrates are partially estimated from death of adult colonies and recruitment of new colonies from mobile larvae (Hughes and Jackson, 1985; Roughgarden et al., 1985). Depending on the reproductive strategy of the parent, the behavior of the larvae, and the oceanographic features of the reef, coral larvae will either settle quickly and recruit to local populations or disperse and recruit to distant populations (Carlon and Olson, 1993; Sammarco, 1994). Asexual reproduction through fragmentation also increases the density of local populations, though new individuals are genetically identical to parent colonies (Neigel and Avise, 1983). Adult population decline has been well documented on selected reefs in the Florida Keys (Dustan and Halas, 1987; Ogden et al., 1994; Porter and Meier, 1992; Porter et al., 2002); however, little is known about larval recruitment on these reefs. Therefore, monitoring recruitment will provide needed insight into coral population dynamics and future prospects of these reefs.

Coral recruitment patterns, which have been studied extensively in the Pacific and in the Caribbean, vary with biological, physical, and environmental factors. Biological factors that influence coral recruitment include the reproductive strategy of the species (Bak and Engel, 1979; Hall and Hughes, 1996; Smith, 1992), larval densities (Fitzhardinge, 1985; Gay and Andrews, 1994; Smith, 1992), spawning location (Gay and Andrews, 1994), competition for space (Sammarco, 1980), and larval behaviors (Carlon and Olson, 1993; Lewis, 1974; Morse et al., 1988; Young, 1995). While the horizontal direction larvae travel is determined principally by oceanographic currents (Gay and Andrews, 1994), coral larvae respond to physical cues such as light (Maida et al., 1994; Morse et al., 1988; Young, 1995), depth (Harriott, 1985b; Rogers et al., 1984; Sammarco, 1994; Wallace, 1985; Young, 1995), and substrate orientation (Fisk and Harriott, 1990; Fitzhardinge, 1985; Harriott, 1985b; Harriott and Fisk, 1987; Maida et al., 1994). Finally, recruitment patterns are altered by the environmental conditions of the reef, including sedimentation rates (Rogers, 1990) and degree of eutrophication (Tomascik, 1991).

The Florida Keys span a wide geographical range. Reef development, as characterized by the composition of the adult communities, varies along latitudinal and bathymetric gradients. The development of these reefs is influenced by seasonal changes in light and temperature, as well as level of exposure to waters flowing from Florida Bay or waters carrying contaminants from land-based development (Porter et al., 1999). The purpose of the current study is to examine coral recruitment patterns along latitudinal and bathymetric gradients in the Florida Keys.

In this study, annual scleractinian and hydrocoral recruitment patterns were monitored between 1991 and 1994 in the Florida Keys. Like scleractinian corals, hydrocorals build limestone skeletons, harbor symbiotic zooxanthellae, and compete for light and space on reefs (Lewis, 1989). Recruitment rates were examined in regard to orientation of the substrate, height in the water column, depth, latitude, and year. In addition, the family compositions of scleractinian coral recruits were compared to resident adult populations on the reefs.

## MATERIALS AND METHODS

### Location Description

Figure 29.1 shows the location and Table 29.1 describes the principle characteristics of each reef within this study. The high latitude reefs within Biscayne National Park (BNP) represent the northernmost extent of reef development in the Florida Keys (Burns, 1985; Glynn et al., 1989; Ogden et al., 1994; Porter, 1987). Triumph Reef (TR), located near the northern boundary of the

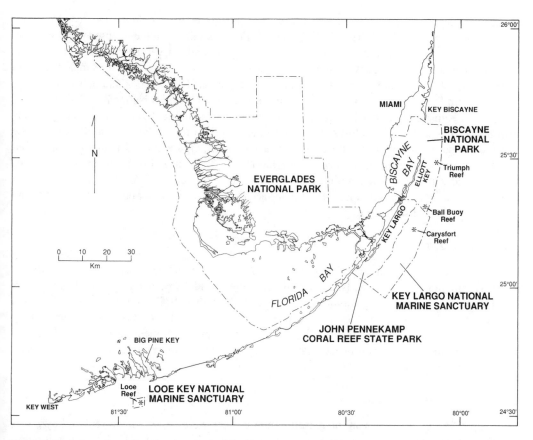

**FIGURE 29.1** Location of study sites in the Florida Keys. All reefs were studied in year 1; only Triumph and Ball Buoy Reefs were studied in years 2 and 3.

park, has a very low percent coral cover and poorly developed vertical relief. TR averages 5 m in depth and is dominated by the hydrocoral *Millepora alcicornis* and small encrusting or head corals such as *Agaricia agaricites* and *Porites astreoides*. In contrast, the shallower Ball Buoy Reef (BB) has a 30.6% coral cover and well-developed vertical relief. Located at the southern boundary of BNP, BB is dominated by the branching coral *Acropora palmata* (Meier, 1996).

Like Ball Buoy Reef, *Acropora palmata* dominates the shallow site of Carysfort Reef (CR). However, the deep site at CR, the deepest site in this study, is composed of massive corals including *Montastraea annularis* and *Colpophyllia natans* (Meier, 1996; White and Porter, 1985). Looe Reef (LR) is a well-developed spur and groove reef at the southernmost point of this study. Once dominated by *Acropora palmata* (Porter and Meier, 1992; White and Porter, 1985), the shallow site at LR is now composed primarily of *Montastraea annularis* and *M. cavernosa*. The deeper site is also dominated by these massive corals (Meier, 1996). It is important to note that, with the exception of Triumph Reef, the known reproductive mode of the dominant adult corals on these reefs is annual mass spawning of gametes (Richmond and Hunter, 1990).

## TREE DESIGN

For the first 2 years of the study, "settling trees" were constructed of 2-inch-diameter schedule 80 PVC trunks. Twelve ceramic tiles (11 cm × 11 cm) were attached to the trunk (design A, Figure 29.2). Ceramic tiles were chosen to monitor coral recruitment because they are readily available, require minimal preparation, have standardized surfaces within and between batches, are readily colonized by corals, and are easy to analyze (Harriott and Fisk, 1987). These tiles were arranged in pairs with their corrugated surfaces exposed to the water column. On each tree, four tiles were oriented in each of the following ways: (1) horizontally with the exposed surface facing down, (2) horizontally with the exposed surface facing up, and (3) vertically. These orientations were distributed among three levels on the tree. Levels 1, 2, and 3 were 0.5 m, 0.33 m, and 0.16 m, respectively, off of the sea floor. Vertical tiles were only located on levels 2 and 3. This design is modified from that used by W. Jaap (pers. comm.) in the Dry Tortugas National Park.

Results from the first year indicated that scleractinian recruitment occurred primarily on the downfacing tiles and on tiles closest to the substrate. For this reason, and in an effort to reduce cost, the tree design was modified for the third year (design B, Figure 29.2). The tree was simplified to hold five tiles that were distributed among levels 2 (0.33 m) and 3 (0.16 m) only. Three tiles were oriented with the grooved surface facing down and were present on levels 2 and 3. Vertically oriented tiles were paired with their grooved surface exposed to the water column and were found either on level 2 or 3 on each tree.

## Tree Deployment

Deployment details are summarized in Table 29.2. In year 1 of the study, annual recruitment patterns were monitored on TR, BB, CR, and LR in the Florida Keys (Figure 29.1). During years 2 and 3, recruitment was only monitored on TR and BB within BNP. Settling trees were bolted to stainless-steel stakes that had been permanently cemented on each reef (Porter and Meier, 1992). Because the trees were deployed on existing permanent stakes, recruitment rates measured through time are from the same location each year. The deployment depths were grouped into four categories: A (0–3 m), B (4–6 m), C (7–9 m), and D (10+ m) (Table 29.1).

## Collection and Processing

After approximately one year on the reef, the settling trees were collected and replaced by a new set (Table 29.2). The tiles were brought to the surface in individually marked bags and photographed using a 2:1 macro lens on a Minolta 3000i 35-mm camera with Ektachrome ASA 200 slide film. The tiles were then sealed in plastic bags containing 75% ethanol and a unique identification tag.

An intact tile measures 11 cm × 11 cm, or 0.0121 m²; however, many tiles fragmented over the course of the study. Therefore, surface areas of the recovered tiles were estimated by placing them on a tile-sized centimeter grid. The number of unoccupied squares was subtracted from 121 cm². The tiles were then analyzed under a dissecting microscope at 10× magnification. Colonies were identified as either Order Hydrocorallina or Order Scleractinia. For scleractinians, colonies were further identified to family using the guidelines of Vaughan and Wells (1943). Spat surface areas were estimated using length and width measurements for oblong spat or diameter measurements from fairly round spat.

## Statistics

Data were statistically analyzed using the procedure `proc genmod` in SAS Release 6.11 for Windows software. Since the data collected are count frequencies, the data are assumed to be Poisson distributed with a Chi-square distributed deviance. Parameters were estimated using the maximum log likelihood method in Equation (29.1):

$$\lambda = \exp \{\beta_0 + \beta_1 X_1 + \ldots + \beta_n X_n\} \tag{29.1}$$

where $\beta_0$ is the intercept term, $\beta_1 \ldots, \beta_n$ are partial slopes, and $X_1, \ldots, X_n$ are explanatory variables. Significance was determined by comparing the log likelihood ratio of reduced to full models with the Chi-square distribution (Crawley, 1993). For the hydrocorals, the ratio of variance to degrees of freedom was consistently greater than 2.0, indicating significance over dispersion of the data

## TABLE 29.1
### Characteristics and Locations of Coral Reefs Included in This Coral Settlement Study

| | Triumph Reef (TR) | | Ball Buoy Reef (BB) | | Carysfort Reef (CR) | | | | Looe Reef (LR) | | | |
| --- | --- | --- | --- | --- | --- | --- | --- | --- | --- | --- | --- | --- |
| | | | | | CR01 | | CR02 | | LR01 | | LR02 | |
| Latitude | 25°28.7' N | | 25°18.3' N | | 25°12.8' N | | 25°12.7' N | | 24°32.7' N | | 25°12.8' N | |
| Longitude | 80°06.7' W | | 80°11.2' W | | 80°13.3' W | | 80°13.2' W | | 81°24.5' W | | 80°13.3' W | |
| Depths | 4.6 to 6.5 m | | 2.3 to 5.4 m | | 2.4 to 3.5 m | | 13.7 to 15.5 m | | 5.4 to 6.7 m | | 6.65 to 8.6 m | |
| Depth codes | B, C | | A, B | | A, B | | D | | B, C | | C | |
| Percent cover of dominant corals on each reef (%)[a] | M. alcicornis[b] | 0.32 | A. palmata[d] | 25.6 | A. palmata[d] | 8.8 | M. annularis[d] | 20.4 | M. annularis[d] | 9.8 | M. cavernosa[d] | 16.6 |
| | A. agaricites[c] | 0.17 | P. astreoides[c] | 2.0 | A. agaricites[c] | 1.2 | C. natans[e] | 6.3 | M. cavernosa[d] | 4.7 | M. annularis[d] | 7.9 |
| | P. astreoides[c] | 0.11 | M. alcicornis[b] | 1.0 | M. complanata[b] | 1.1 | S. siderea[d] | 1.7 | C. natans[e] | 2.5 | S. siderea[d] | 0.7 |
| | S. michellini | 0.09 | M. complanata[b] | 0.9 | P. astreoides[c] | 0.8 | | | S. siderea[d] | 0.4 | | |
| Total percent live coral cover (year) | (1994) | 0.77 | (1994) | 30.6 | (1991) | 12.2 | (1991) | 30.8 | (1991) | 18.5 | (1991) | 26.7 |

*Note:* Settling tree deployment depths are divided into four categories: A (0 - 3 m), B (4 - 6 m), C (7 - 9 m) and D (10+ m).

[a] Data summarized from Meier (1996).

[b] Hydrocorals release dioecious medusae which carry gametes (Lewis, 1989).

[c] Species that brood their larvae and planulate (Richmond and Hunter, 1990).

[d] Species that are broadcast spawners (Richmond and Hunter, 1990).

[e] Unknown reproductive mode.

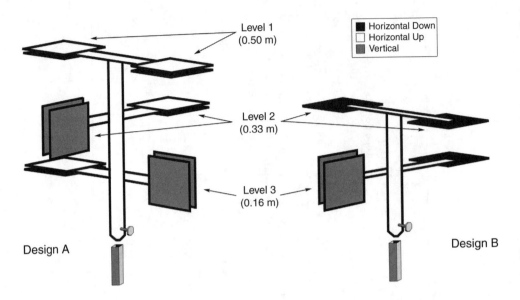

**FIGURE 29.2** Settling tree design A was used in year 1 and year 2. Based on results from year 1, the more efficient design B was used in year 3.

**TABLE 29.2**
**Coral Settlement Deployment Details for Each Reef and Year**

|  | Triumph Reef (TR) | Ball Buoy Reef (BB) | Carysfort Reef (CR) | Looe Reef (LR) |
|---|---|---|---|---|
| **Year 1** |  |  |  |  |
| Date deployed | July 30, 1991 | July 26, 1991 | July 12, 1991 | August 13, 1991 |
| Date retrieved | September 15, 1992 | September 16, 1992 | September 10, 1992 | August 16, 1992 |
| Days on reef | 413 | 418 | 425 | 369 |
| No. of trees deployed | 6 | 6 | 4 | 4 |
| No. of tiles deployed | 72 | 72 | 48 | 48 |
| No. of tiles retrieved | 20[a] | 61 | 37 | 37 |
| **Year 2** |  |  |  |  |
| Date deployed | September 18, 1992 | September 17, 1992 |  |  |
| Date retrieved | October 12, 1993 | October 12, 1993 |  |  |
| Days on reef | 389 | 390 |  |  |
| No. of trees deployed | 6 | 6 |  |  |
| No. of tiles deployed | 72 | 72 |  |  |
| No. of tiles retrieved | 59 | 62 |  |  |
| **Year 3[b]** |  |  |  |  |
| Date deployed | October 12, 1993 | October 12, 1993 |  |  |
| Date retrieved | November 30, 1994 | December 2, 1994 |  |  |
| Days on reef | 414 | 416 |  |  |
| No. of trees deployed | 6 | 6 |  |  |
| No. of tiles deployed | 30 | 30 |  |  |
| No. of tiles retrieved | 18 | 22 |  |  |

[a] Hurricane Andrew removed four trees from Triumph Reef in August, 1992.
[b] The modified tree design was used during year 3.

(Crawley, 1993). This was corrected using the ratio of deviance to degrees of freedom to estimate standard error. Recruitment densities within a treatment were calculated as in Equation (29.2):

$$\hat{\lambda}_i = \frac{\Sigma x_i}{\Sigma a_i} \qquad (29.2)$$

where $a_i$ is the area of tile $i$, and $x_i$ is the number of recruits on tile $i$. The variance was calculated as in Equation (29.3):

$$\hat{var}(\lambda) = \frac{1}{\bar{a}^2}\{s_x^2 + \hat{\lambda}^2 s_a^2 - 2\hat{\lambda}s_{xia} \qquad (29.3)$$

where $\bar{a} = \frac{1}{n}\Sigma a_i$, $s_x^2$ is the sample variance of $x_i$, $s_a^2$ is the sample variance of $a_i$, and $s_{xia}$ is the sample covariance of $x_i$ and $a_i$ defined as in Equation (29.4):

$$S_{xia} = \frac{\Sigma a_i x_i - (\Sigma a_i)(\Sigma x_i)/n}{n-1} \qquad (29.4)$$

The standard error was taken as the square root of the variance, and the 95% confidence interval was calculated as 1.96 times the standard error.

## RESULTS

### GENERAL OBSERVATIONS

Hydrocoral recruitment rates were consistently higher than scleractinian recruitment rates. A total of 381 colonies of *Millepora* spp. were found, while only 45 scleractinian recruits were found over the 3-year period (Table 29.3). In addition, averaging 1.19 cm², hydrocoral spat were consistently larger than the scleractinian spat, which averaged 0.18 cm². All of the scleractinian recruits were distributed among the families Agariciidae (35.6%), Poritidae (35.6%), and Faviidae (26.7%), all of which brood their larvae (Figure 29.3). These distributions were consistent through time and between reefs. None of the major reef building corals, which dominate these reefs as adults, recruited to any tiles at any time during the entire course of the study.

---

**TABLE 29.3**
**Composition of Recruits**

|  | Family Agariciidae | Family Faviidae | Family Poritidae | Scleractinia Totals | Hydrocorallina |
|---|---|---|---|---|---|
| No. of spat | 16 | 12 | 16 | 45[a] | 381 |
| Average size (cm²) (± std. dev.) | 0.38 (±0.64) | 0.06 (±0.05) | 0.08 (±0.13) | 0.18 (±0.41) | 1.19 (±3.40) |
| Largest spat (cm²) | 2.54 | 0.20 | 0.56 | 2.54 | 29.11 |

*Note:* Recruitment of hydrocorals was nearly eight times that of scleractinians; the average and largest sizes of the hydrocorals were larger than the scleractinians.

[a] One colony was too small to be assigned with confidence to a scleractinian family.

---

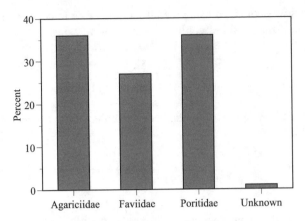

**FIGURE 29.3** All of the scleractinian recruits were from corals that brood their larvae. None of the recruits was from the major reef-building corals which dominate the adult communities of these reefs. This trend was consistent through time and between reefs.

## TREE EFFECTS

Table 29.4 summarizes the results of all statistical tests on the effects of settling plate orientation, height in the water column, reef location, and year of recruitment. The effects of the tree design were tested during year 1 of the study. In both scleractinians and hydrocorals, recruitment was significantly higher ($p < 0.01$) on downfacing tiles than on the other tile orientations (Table 29.5, Figure 29.4). For the scleractinian corals, recruitment rates ranged from 1.9 colonies/m$^2$ on horizontal tiles facing up to 42.8 colonies/m$^2$ on horizontal tiles facing down. There was a slight shift from downfacing tiles to vertical tiles with increasing depth (Figure 29.5). For the hydrocorals, recruitment rates ranged from 59.6 colonies/m$^2$ on upfacing tiles to 275.6 colonies/m$^2$ on downfacing tiles (Table 29.5, Figure 29.4).

The distance above the substrate of the settling tile also influenced recruitment rates. Scleractinian corals had significantly higher ($p < 0.01$) recruitment densities on level 3 (0.16 m) tiles, whereas hydrocorals had significantly higher ($p < 0.05$) recruitment rates on level 1 (0.5 m) tiles (Table 29.6, Figure 29.6).

**TABLE 29.4**
**Statistical Tests of the Influence of Settling Plate Orientation, Height, Depth, Latitude, and Year on Scleractinian and Hydrocoral Recruitment**

| | | Likelihood Ratio Test Statistic | |
| --- | --- | --- | --- |
| Effect | df | Scleractinia | Hydrocorallina |
| 1992 only | | | |
| Orientation | 2 | 31.67** | 21.68** |
| Height | 2 | 31.75** | 6.45* |
| Depth | 3 | 40.09** | 10.88* |
| Latitude (w/o CR02) | 3 | 2.95 | 11.36** |
| Biscayne National Park only, 1992–1994 | | | |
| Year | 2 | 14.52** | 24.81** |
| Latitude | 1 | 1.61 | 7.11** |

\*  Significant at $X^2_{df,0.05}$.
\*\* Significant at $X^2_{df,0.01}$.

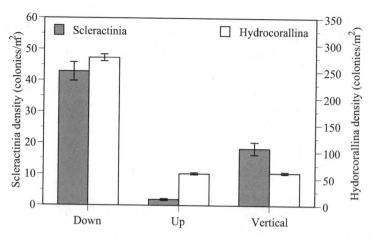

**FIGURE 29.4** Tree effects for tile orientation. Recruitment was significantly higher ($p < 0.01$) on downfacing tiles for all corals. Data are summarized from all reefs in 1992.

**TABLE 29.5**
**Effect of Tile Orientation on Recruitment**

| Orientation | Density (no./m² ± 95% confidence interval) | |
| --- | --- | --- |
| | Scleractinia | Hydrocorallina |
| Down | 42.8 ± 3.0 | 275.6 ± 6.6 |
| Up | 1.9 ± 0.3 | 59.6 ± 1.9 |
| Vertical | 18.3 ± 2.0 | 61.0 ± 2.1 |

*Note:* Data are grouped from all reefs in year 1.

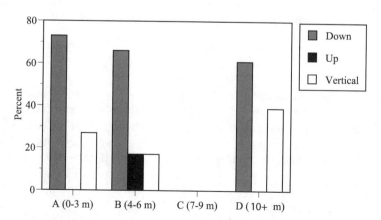

**FIGURE 29.5** Scleractinian recruits were consistently found on downfacing surfaces (gray) over all depths. While there is a slight trend toward vertical surfaces (white) with depth, the depth range of this study was insufficient to produce a distinguishable pattern. Recruits were rarely found on upfacing tiles (black).

**TABLE 29.6**
**Effect of Tile Distance above Substrate on Recruitment**

| | Density (no./m² ± 95% confidence interval) | |
| --- | --- | --- |
| Height Above Substrate | Scleractinia | Hydrocorallina |
| Level 1 (0.50 m) | 7.0 ± 0.5 | 245.7 ± 5.9 |
| Level 2 (0.33 m) | 6.5 ± 0.8 | 118.1 ± 5.0 |
| Level 3 (0.16 m) | 42.4 ± 3.2 | 75.2 ± 2.7 |

*Note:* In scleractinian corals, recruitment densities were significantly higher ($p < 0.01$) on the tiles closest to the reef; in hydrocorals, recruitment densities were significantly higher ($p < 0.05$) on tiles farthest away from the reef. Data are grouped from all reefs in year 1.

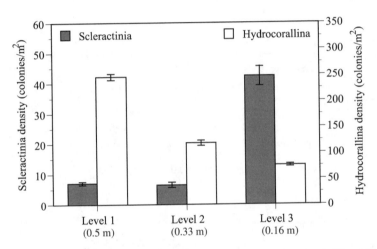

**FIGURE 29.6** In scleractinia, recruitment densities were significantly higher ($p < 0.01$) on tiles closest to the benthos (0.16 m). In contrast, significantly higher ($p < 0.05$) densities of hydrocorals were found on tiles that were farther away from the reef (0.50 m). Data are summarized from all reefs in 1992.

## REEF EFFECTS

The depth at which the settling trees were deployed affected coral recruitment. At 144.85 colonies/m², the deepest site (depth code D, 10+ m) had significantly higher ($p < 0.01$) scleractinian recruitment densities than any of the other depths (Table 29.7, Figure 29.7). The family composition of scleractinia spat also shifted with depth. While faviid corals were the commonest recruits at the shallowest depths, agariciid corals dominated the deepest sites (Figure 29.8). Hydrocoral recruitment was significantly higher ($p < 0.05$) at intermediate depths (depth code B, 4–6 m) than at the other depths (Table 29.7, Figure 29.7).

Because of the high recruitment rates at depth D (10+ m), which was only found on CR02, this reef was removed from the year 1 dataset to examine the effect of latitude. Latitude did not have a significant effect on scleractinian recruitment rates ($p > 0.05$); however, hydrocoral recruitment was highest ($p < 0.01$) at the northernmost reef (Table 29.8, Figure 29.9).

**TABLE 29.7**
**Effect of Depth on Recruitment**

| | Density (no./m² ± 95% confidence interval) | |
|---|---|---|
| Depth | Scleractinia | Hydrocorallina |
| A (0–3 m) | 8.2 ± 0.7 | 98.3 ± 3.8 |
| B (4–6 m) | 8.4 ± 0.7 | 222.4 ± 6.3 |
| C (7–9 m) | 0.0 ± 0.0 | 89.5 ± 2.4 |
| D (10+ m) | 144.9 ± 5.6 | 36.2 ± 1.6 |

*Note:* For scleractinians, recruitment was significantly higher ($p < 0.01$) at the deepest site in year 1; for hydrocorals, recruitment was significantly higher ($p < 0.05$) at intermediate depths and lowest at the deepest site. Data are grouped from all reefs in year 1.

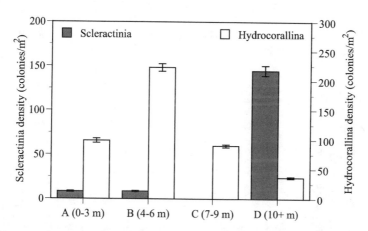

**FIGURE 29.7** The effects of depth on coral recruitment. Scleractinian recruitment was significantly higher ($p < 0.01$) at greater depths. Hydrocoral recruitment was highest at intermediate depths ($p < 0.05$). Data are summarized from all reefs in year 1.

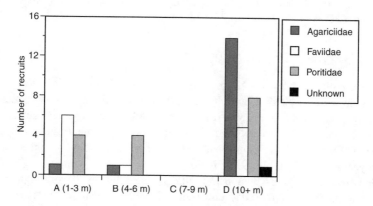

**FIGURE 29.8** Family composition of scleractinian spat at each depth. Faviid recruits dominate the shallow site, while agariciid corals dominate the deepest sites.

**TABLE 29.8**
**Effect of Latitude on Recruitment**

| | Density (no./m² ± 95% confidence interval) | |
|---|---|---|
| Depth | Scleractinia | Hydrocorallina |
| 25.29 N (TR) | 9.2 ± 0.9 | 224.9 ± 5.2 |
| 25.18 N (BB) | 5.7 ± 0.5 | 183.7 ± 6.1 |
| 25.13 N (CR) | 8.5 ± 0.9 | 72.2 ± 2.7 |
| 24.33 N (LR) | 4.7 ± 0.5 | 72.4 ± 1.9 |

*Note:* To test the effect of latitude without the confounding effect of high recruitment rates at the deepest site, the deepest site (CR02) was removed from the year 1 dataset. Latitude did not significantly affect scleractinian coral recruitment ($p > 0.05$); hydrocoral recruitment was highest ($p < 0.01$) at Triumph Reef, the northernmost reef.

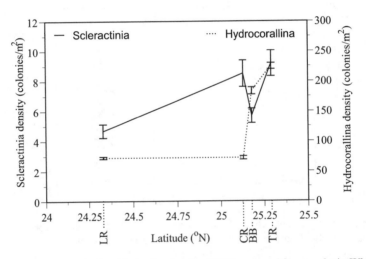

**FIGURE 29.9** Reef effects by latitude with the deepest site, CR02, removed from analysis. While scleractinian recruitment rates were not affected by latitude ($p > 0.05$), hydrocoral recruitment increased in the northern reefs ($p < 0.01$). Data are summarized from 1992 only.

## YEAR AND LATITUDE EFFECTS WITHIN BNP

Because data were collected only in BNP for 3 years, TR and BB data were used to test the effects of year and latitude within BNP. For both the scleractinians and the hydrocorals, recruitment rates were significantly lower ($p < 0.01$) in year 2, the 12 months following Hurricane Andrew. During year 2, no scleractinian recruits were found and hydrocoral recruitment rates dropped by 75% to 42.45 colonies/m² (Figure 29.10). Recruitment rates were comparable within the scleractinians and hydrocorals in years 1 and 3 (Table 29.9, Figure 29.10).

Data pooled from all 3 years demonstrate that latitude did not have a significant effect on scleractinian recruitment rates ($p > 0.05$); however, hydrocoral recruitment rates were significantly higher ($p < 0.01$) on the northern TR (Table 29.10, Figure 29.11) reefs in BNP.

## TABLE 29.9
### Effect of Year on Recruitment within Biscayne National Park

| Depth | Density (no./m² ± 95% confidence interval) | |
| | Scleractinia | Hydrocorallina |
|---|---|---|
| Year 1 (1991–1992) | 6.5 ± 0.7 | 193.4 ± 5.9 |
| Year 2 (1992–1993) | 0.0 ± 0.0 | 42.5 ± 2.7 |
| Year 3 (1993–1994) | 9.8 ± 0.9 | 148.7 ± 6.8 |

*Note:* To test the effect of year, data were compared for years 1 through 3 from BNP reefs only. Recruitment rates in both scleractinians and hydrocorals were significantly lower ($p < 0.01$) during year 2, the 12-month period after Hurricane Andrew passed directly over these reefs in August 1992.

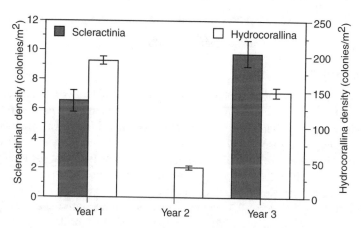

**FIGURE 29.10**  For all corals, recruitment rates decreased significantly ($p < 0.01$) in year 2, the 12-month period following Hurricane Andrew.

## TABLE 29.10
### Effect of Latitude on Recruitment within Biscayne National Park

| Latitude | Density (no./m² ± 95% confidence interval) | |
| | Scleractinia | Hydrocorallina |
|---|---|---|
| 25.29 N (TR) | 1.8 ± 0.4 | 143.4 ± 6.2 |
| 25.18 N (BB) | 5.6 ± 0.7 | 94.7 ± 4.4 |

*Note:* The effect of latitude within BNP was also tested over the three-year period. For scleractinian corals, latitude did not have a significant affect on recruitment ($p > 0.05$); however, for hydrocorals, recruitment rates were significantly higher ($p < 0.01$) on Triumph Reef at the northern boundary of BNP.

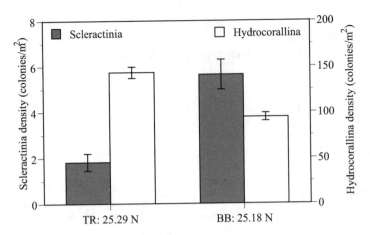

**FIGURE 29.11**    While latitude did not affect scleractinian recruitment rates ($p > 0.05$), hydrocoral recruitment rates were significantly higher on the northern reef (Triumph Reef, $p < 0.01$).

## DISCUSSION

### GENERAL OBSERVATIONS

Contrary to other studies in the Caribbean (Rogers et al., 1984), hydrocoral recruitment in the Florida Keys was higher than scleractinian recruitment. In the study by Rogers (1984), *Millepora* spp. consistently were less than 7% of the relative abundance of larval recruits. In our study, hydrocorals commonly comprised greater than 81% of the relative abundance. Only on the deepest reef, CR02, was scleractinian relative abundance (80%) greater than that for hydrocorals (20%).

*Millepora* spp. flourish and show some competitive advantage over corals in disturbed habitats (Lewis, 1989). In the Lower Keys, adult scleractinian populations have declined on reefs studied by Porter and Meier (1992). In BNP, reefs are subject to temperature fluctuations from Biscayne Bay (Hudson et al., 1994), pesticides from agriculture in South Florida (Glynn et al., 1989), and sedimentation from dredging operations and urbanization around Miami (Hudson et al., 1994). Along the Lower Keys, reefs are subject to temperature and salinity fluctuations from Florida Bay water and sedimentation from development (Ogden et al., 1994; Pitts, 1994). Diseases are also common along these reefs, and black-band disease was cited as one source of mortality of *Montastraea annularis* and *M. cavernosa* colonies in southern reefs of the Florida Keys (Porter and Meier, 1992). The high recruitment rates of hydrocorals relative to scleractinian corals may reflect differences in response to environmental stress.

Total scleractinian recruitment in this study in Florida (only 45 colonies over 3 years) was far lower than observed in similar studies elsewhere in the Caribbean (Rogers et al., 1984), Gulf of Mexico (Baggett and Bright, 1985), or the Pacific (Harriott and Banks, 1995; Harriott and Fisk 1987; Wallace, 1985). On Carysfort Reef, Dustan and Halas (1987) suggested that the recruitment of juvenile scleractinians may have decreased between 1975 and 1983. This corresponded with a decline in live coral cover attributed to damage from boat groundings and storms as well as increases in diseases, algal cover, and sedimentation. Increases in sedimentation, eutrophication water turbidity, and pollution can reduce recruitment success (Richmond, 1996) and are all associated with increased human populations in coastal areas, as is the case in the Florida Keys (Porter and Tougas, 2001).

In a study of high-latitude reefs in Eastern Australia, Harriott and Banks (1995) found lower recruitment rates than in other studies in the Great Barrier Reef. This low recruitment was attributed to a combination of low temperatures and the dominant current patterns which would quickly carry spawned gametes away from the reef. Whether the observed low recruitment rates in this study are due to the limitations associated with high latitude reefs or environmental stresses as discussed above, or both, is unclear.

The largest scleractinian recruit was an agariciid coral (cf. *Agaricia agaricites*) with a diameter of 2.54 cm². Assuming that this specimen settled soon after the tiles were deployed, and therefore grew for a year, this size is consistent with observed growth rates for this coral (Birkeland, 1977). Because the average size of the corals was so small (0.18 cm²), many of the corals present on the tiles probably settled shortly before they were collected. Average hydrocoral diameters were nearly 10 times that of scleractinian diameters, indicating a much higher growth rate in the hydrocorals. At 29.11 cm², the largest hydrocoral was far bigger than the largest scleractinian. This equates to a monthly lateral growth rate of 3.9 mm and is comparable to those found elsewhere for hydrocorals (Lewis, 1989).

Chiappone and Sullivan (1996) found a significant correlation between juvenile and adult abundance for spawning and brooding species in the Florida Keys. Other studies in the Pacific have also shown that the composition of the adult community is a good predictor of which corals would recruit (Harriott, 1985b). This was definitely not the case in the present study.

Consistent with observations of recruitment in photostations on these reefs by Porter and Meier (1992), none of the major reef building coral species common as adults, *Acropora*, *Montastraea*, *Diploira*, *Colpophyllia*, *Solenastrea*, or *Mycetophyllia*, recruited to the tiles during the 3 years of the study. Several researchers have found a discrepancy between the abundance of adult corals and their successful recruitment (Fitzhardinge, 1985; Harriott and Banks, 1995; Rogers et al., 1984). In the Caribbean, broadcast spawners, such as *Montastraea* spp. or *Acropora* spp., tend to dominate the adult community in terms of percent cover, yet tend to have few recruits and juveniles (Bak and Engel, 1979; Birkeland, 1977; Baggett and Bright, 1985). The absence of recruitment by a taxon could indicate the lack of larvae in the recruitment pool or inadequate experimental conditions (Wallace, 1985). This does not appear to be the case here as *Montastraea* spp. were observed spawning in August of 1992 and 1993 in the Florida Keys (pers. observ.). Furthermore, ceramic tiles, the artificial substrate used in this study, attracted the highest number and diversity of corals in a study of potential settling surfaces conducted by Harriott and Fisk (1987).

Thorson (1950) predicted that brooding would be more common in colder climates. This was supported by Harriott and Banks (1995), who found that the recruitment of hard corals on high-latitude reefs in eastern Australia was dominated by planulating species. This was in contrast to other studies in the Pacific and was true even though the dominant coral on the high-latitude reef was a broadcast spawner. In our study, brooding corals of the families Poritidae, Agariciidae, and Faviidae were the dominant corals to recruit. This has also been observed for many low-latitude reefs throughout the Caribbean (Bak and Engel, 1979; Birkeland, 1977; Rogers et al., 1984; Tomascik, 1991). Therefore, it is unclear if the dominance of recruits from planulating species in our study is the result of these reefs being high-latitude reefs or if it is the result of a larger pattern throughout the Caribbean.

Bak and Engel (1979) have attributed relative recruitment success between brooding and broadcasting species to differences in life-history strategies in the face of environmental conditions in the adult corals. Many brooding corals are successful at recruiting in locations that are frequently disturbed (Bak and Luckhurst, 1980). In these environments, even though adult mortality is relatively high, recruitment and maturation of larvae are successful, thus favoring settlement in these regions (Babcock, 1991). Because the larvae of brooding corals are released several times a year, are well developed when released, and settle relatively quickly, their chances of successfully settling on a suitable reef and recruiting to the population are relatively high (Babcock, 1991; Fadlallah, 1983; Kojis and Quinn, 1981).

Also common in shallow, frequently disturbed habitats are branching corals, such as *Acropora palmata*, which dominate two shallow reefs studied here (Table 29.2). While this coral is a broadcast spawner, it may rely more heavily on fragmentation than on sexual recruitment as a means of propagation (Babcock, 1991; Bak and Engel, 1979; Neigel and Avise, 1983), which could partially explain the lack of recruits of *A. palmata*. This has also been found in the Great Barrier Reef (Kojis and Quinn, 1981) for the branching coral *Porites andrewssii*, also a broadcast spawning species common in shallow waters.

The *Montastraea* spp. dominant in many reefs of this study (Table 29.1) are broadcast spawners (Szmant, 1991). Because fertilization and development take place in the water column over a period of days, the larvae are subject to higher mortality due to predation or by being transported to locations unsuitable for settlement (Babcock, 1991; Kojis and Quinn, 1981). Tomascik (1991) found a decrease in successful recruitment of spawning species along an increasing eutrophication gradient in Barbados; therefore, the absence of recruits from spawning species could reflect a stressed environment. This is also indicated by the observed decline of the adult populations of these reefs (Dustan and Halas, 1987; Porter and Meier, 1992).

When recruitment is successful for these species, survivorship tends to be relatively high and the adults are long lived, particularly in deeper, more stable environments (Bak and Luckhurst, 1980). Indeed, in the study by Porter and Meier (1992), CR02 declined the least, losing only 7.3% coral cover over a 7-year period. This reef is dominated by massive, long-lived colonies of *Montastraea annularis*, *Colpophyllia natans*, and *Siderastrea siderea* (Meier, 1996).

## TREE EFFECTS

Recruitment rates were highest on downfacing tiles. This result has been seen in many other studies in both the Caribbean and Pacific (Fisk and Harriott, 1990; Fitzhardinge, 1985; Harriott, 1985a; Harriott and Fisk, 1987; Maida et al., 1994). Settlement preferences seem to involve trade-offs between sedimentation rates, light availability, and competition from algae and other fast-growing benthic organisms.

Lack of recruitment on upper surfaces has been attributed to high sedimentation rates in several studies (Fisk and Harriott, 1990; Fitzhardinge, 1985; Harriott, 1985a; Maida et al., 1994). Coral larvae are unable to settle on shifting sediments (Rogers, 1990), and Babcock and Davies (1991) have shown that even slight increases in sedimentation rates dramatically reduces settlement on the upper surface of horizontal tiles for the Pacific coral *Acropora millepora*. While not quantified in the present study, we noticed a fair amount of sediment trapped in filamentous algae on upfacing tiles, which could have prevented coral settlement or smothered those that did settle.

Harriott and Banks (1995) found higher recruitment on upper surfaces on high latitude reefs and hypothesized that this may be a response to reduced light due to latitudinal increases in turbidity or to latitudinal decreases in light availability. Maida et al. (1994) found that while sedimentation influences settlement patterns, light availability determines growth rate and survivorship patterns of hermatypic corals. Birkeland (1977) also found that growth rates were higher on upper surfaces exposed to full sunlight; however, survivorship was lower on the upper surfaces due to competition with algae, bryozoans, and barnacles.

Settlement and survivorship on upper surfaces increase with depth (up to 37 m) as competition with other heliophilic species decreases (Bak and Engel, 1979; Birkeland, 1977; Wallace, 1985). A depth-mediated shift towards settlement on upper surfaces was not observed in this study. This may indicate that in the Florida Keys, sedimentation may have a stronger influence on settlement patterns than light availability, or that the depths in this study (17 m) were not great enough to reveal this trend. There was, however, a very slight trend towards increased settlement on vertical surfaces with depth in this study (Figure 29.5). A shift from lower surfaces to vertical tiles with depth has also been observed in St. Croix (Rogers et al., 1984), the East Flower Gardens (Baggett and Bright, 1985), and Geoffrey Bay (Babcock and Mundy, 1996). Other studies have shown a shift towards vertical tiles in turbid environments (Fisk and Harriott, 1990) and on eutrophic reefs (Tomascik, 1991). In each of these studies, light availability was cited as the limiting factor causing this shift.

The height above the substrate at which the tiles were deployed was a significant factor for both scleractinians and hydrocorals; however, the trends were opposite to each other. Scleractinian densities were highest on tiles closest to the reef, while hydrocoral densities were highest on tiles that were highest in the water column. One possible explanation for this is the reproductive and settlement modes of these corals. As discussed above, all of the scleractinian corals that recruited during this study are brooders. Well-developed larvae are released from the parent colony and swim up into the water column a short distance before settling (Carlon and Olson, 1993). The first settlement tree surface these corals would encounter would be the tiles closest to the substrate on the array. Hydrocorals, on the other hand, release gamete-bearing medusae that swim up into the water column (Lewis, 1989). Fertilization and larval development consequently take place in the water column (Fautin et al., 1989). As the larvae descend towards the benthos, they would first encounter the tiles farthest above the substrate on the settling trees.

## REEF EFFECTS

In addition to fixed variables such as depth and geographic location, reefs vary temporally in nutrient concentrations (Tomascik, 1991), light intensities (Falkowski et al., 1984), and community stability (Bak and Luckhurst, 1980). Coral species vary in life history strategies (Bak and Engel, 1979), competitive abilities (Lang, 1973; Sammarco, 1980) and settlement behavior (Carlon and Olson, 1993; Morse et al., 1988). Therefore, the effect of year, latitude, and depth on coral recruitment will also depend on specific environmental conditions of each reef and on the biological constraints of each coral species. In our study, depth had the greatest effect on both the density of recruitment and the family composition of dominant taxa.

By far the highest scleractinian recruitment densities were at 13 to 15 m, the greatest depths in this study. Depths from 11 to 18 m have also yielded the highest recruitment in St. Croix (Rogers et al., 1984) and parts of the Great Barrier Reef (Sammarco, 1994). Below these depths, recruitment and survivorship drop off dramatically (Rogers et al., 1984; Sammarco, 1994). For juvenile corals in the Florida Keys, Dustan (1977) found the highest densities at an intermediate depth of 9 m, and Chiappone and Sullivan (1996) found a general increase in densities with depth, though this trend was not statistically significant.

In contrast, recruitment in other areas of the Pacific seems to be higher on shallower reefs. Wallace (1985) found the highest recruitment rates at the reef shoulder (1 to 2 m), and Harriott (1985b) found fewer spat on deeper plates. The dominant recruits in these studies are from the families Acroporidae, Poritidae, and Pocilloporidae. In the Pacific, acroporids and poritids are generally branching, mass-spawning species common in shallow waters (Richmond and Hunter, 1990). On the other hand, pocilloporids are brooding species which have been shown to have a limited gene flow, indicative of high fragmentation and local recruitment (Stoddart, 1984). Based on these features, one would expect these corals to recruit heavily to shallow-water environments.

In Barbados, the effect of depth depended on the reef (Tomascik, 1991). On the more pristine reef, recruitment was higher in the deeper spur and groove formations than on the reef crest. However, on a reef in close proximity to terrestrial urban and agricultural runoff, recruitment was higher on the shallow reef crest than on the spur and groove. Tomascik (1991) suggested that decreases in light intensity due to turbidity from suspended sediments caused the shift in recruitment to the shallower depths.

Bak and Engel (1979) found that the effect of depth was heavily dependent on the species in question. Agariciid corals recruited most heavily on shallow reefs while poritid corals recruited fairly equally throughout the depth zones (Bak and Engel, 1979). In contrast, faviid corals dominated the shallow sites in our study. This is expected considering the distribution, life-history strategy, and competitive ability of these corals. A brooding coral, *Favia fragum* is common in frequently disturbed shallow reef zones, where it begins reproduction at an early age of less than 2 years and continues reproducing several times throughout the year (Szmant-Froelich et al., 1985). *Favia fragum* is also a poor competitor with algae and other fouling organisms (Sammarco, 1980).

While faviid recruits dominated shallow sites, agariciid and poritid corals dominated the deeper sites in the present study (Figure 29.8). Rogers et al. (1984) also found a species shift with depth in St. Croix where *Agaricia agaricites* comprised 60% of the recruits at 18 m on the East Wall. At 37 m, the deepest site of the study, *Madracis decactis* was the most abundant coral, followed by *Stephanocoenia michellini*, *A. lamarcki*, and *A. agaricites* (Rogers et al., 1984). Due to the presence of spawning species at greater depths, a shift in species composition with depth was also seen in a survey of juvenile abundance patterns in the Florida Keys (Chiappone and Sullivan, 1996). In contrast, Harriott (1985b) found that the relative abundance of different taxa was relatively constant across a depth gradient up to 9 m in the Great Barrier Reef.

Hydrocoral recruitment was highest at an intermediate depth range of 4 to 6 m (Figure 29.7). Rogers et al. (1984) also found higher densities of hydrocorals at shallower sites in St. Croix. On the East Wall, the relative abundance of *Millepora* spp. dropped from 11.3% at 9 m to 0.3% at 18 m (Rogers et al., 1984). Most of the *Millepora* spp. reviewed by Lewis (1989) are found at depths less than 10 m. Three plausible explanations for this pattern are light availability, feeding morphology, and life-history phenomena.

Hydrocorals may have a physiological limit to depth distribution, probably due to diminished light intensities that would reduce photosynthesis (Lewis, 1989). Second, platy species, such as *Millepora complanata*, may rely extensively on strong water movement common in shallow waters to deliver planktonic food. By orienting the broad side of the blade to prevailing currents (Lewis, 1989), micro-eddies may form behind the blade which may help in feeding. This phenomenon has been demonstrated in an *Agaricia* species (Helmuth and Sebens, 1993). Finally, hydrocoral life history strategies may play an important role in determining their distribution. Hydrocorals tend to be opportunists that are common in frequently disturbed areas (Lewis, 1989). In Curacao, Bak and Luckhurst (1980) have shown that environmental stability increases with depth on reefs. In more stable environments, long-lived competitive species, such as *Montastraea annularis* (Bak and Engel, 1979), are favored over short-lived opportunistic species (Skelton, 1993).

## LATITUDE

Due to the greater number of species in the southern Florida Keys (Porter, pers. comm.) and consistently warmer temperatures there (Ogden et al., 1994), one could expect higher recruitment rates due to an increase in the pool of available larvae and favorable growing conditions. Fisk and Harriott (1990) found a trend for increasing recruitment densities from south to north in the Great Barrier Reef, and Harriott and Banks (1995) found that reef location affects scleractinian recruitment. This was not the case in our study as recruitment rates were consistently low from north to south.

These contrasting patterns in the effect of latitude could reflect differences in the condition of the adult communities. Porter and Meier (1992) studied changes in coral populations in the Florida Keys between 1984 and 1991. LR, the southernmost reef, lost up to three coral species and had the greatest loss in percent cover (43.9%) of any of the reefs studied. The potential increase in the number of species contributing to the larval pool at this location could be offset by the decline in percent cover this reef has suffered; therefore, the health of the adult community may directly influence the abundance of coral recruits.

For hydrocorals, recruitment rates were higher at northern latitudes. This may be directly related to relative abundance of hydrocorals in the adult population at the different reefs. *Millepora* spp are one of the dominant coral species in the shallow sites of TR, BB, and CR but are absent as one of the dominant species at the southern LR (Meier, 1996).

## Year and Latitude Within BNP

As has been found in other studies (Baggett and Bright, 1985; Fisk and Harriott, 1990; Wallace, 1985), year was a significant factor in determining recruitment rates for both scleractinians and hydrocorals. During the second year of this study (1992–1993), scleractinian recruitment rates dropped to zero, and hydrocoral recruitment dropped to 75% of their previous levels. Fitzhardinge (1985) also found high variations in recruitment rates between years and attributed this either to variations in planulae abundance or to differential mortality of newly settled larvae, particularly when faced with competition from rapidly growing algae. General variability in larval supply or recruitment success may explain the drop in recruitment; however, the dramatic drop in both scleractinian and hydrocoral recruitment suggests otherwise. In independent studies, Ward (1995) and Hall and Hughes (1996) demonstrated that damage to adult colonies reduces fecundity; therefore, the damaging effects of Hurricane Andrew, which passed over BNP in August 1992, may have reduced local reproduction and gives an insight into coral population dynamics within BNP.

On August 22, 1992, Hurricane Andrew passed through BNP, directly between BB and TR reefs. A category 4 storm, Hurricane Andrew had sustained winds of 145 mph and recorded gusts of up to 175 mph. The effects of this major storm event were highly variable between reefs. The damage was concentrated in the northeast corner of the storm (TR), and the majority of the reefs in the Florida Keys were spared. Between 10 and 38% of the coral cover and 40 to 90% of algal cover were lost either to scouring or physical removal during the storm (Blair et al., 1994; Meier, 1996).

The Florida Current delivers water from the south to the north along the reef tract and farther offshore (Lee et al., 1995). Because the Lower Keys were spared damage from the storm, there is a potential source of larvae from upstream reefs to the south. If BNP coral populations are completely open to recruitment from other reefs, then recruitment rates should not drop significantly after the storm. If, however, the coral populations are self-seeding, then the damage from Hurricane Andrew could cause a significant shift in resource allocation toward tissue repair and away from reproduction so that recruitment would drop (Rogers, 1993). Because recruitment rates were lower during year 2 following Hurricane Andrew, the latter hypothesis is supported. However, planulation by colonies of *Porites astreoides* collected from within BNP was observed in May 1993, and the quantity of larvae produced from these colonies was not significantly different from other time periods (McGuire, pers. comm.). This suggests that factors other than planula production rates diminished post-hurricane recruitment and calls into question the hypothesis that synchronous, late-summer spawning events are timed to take advantage of newly bared substrates produced by hurricanes.

Evidence of self-seeding populations have been found elsewhere (Baggett and Bright, 1985; Chiappone and Sullivan, 1996) and is supported by the genetic distribution of other brooding species and the larval behavior of the corals present. In a study of the genetical structure of the brooding coral *Pocillopora damicornis*, Stoddart (1984) found that 84% of all sexual recruits were directly related to adults within the population. In a study of larval dispersal, Carlon and Olson (1993) followed the swimming behavior of *Favia fragum* and *Agaricia agaricites* coral larvae. Because larvae of brooding corals are well developed when released, they are capable of settling quickly. The distance larvae travel depends a great deal on the length of time the larvae remains in the water column (Carlon and Olson, 1993). Because *F. fragum* consistently settled within 10 minutes of release during the experiment and was a generalist in terms of settlement substrata, so Carlon and Olson (1993) concluded that gregarious distributions of adult *F. fragum* colonies was a direct result of localized recruitment from released larvae. *A. agaricites* remained in the water column for a longer period of time (Carlon and Olson, 1993). Sammarco (1994) also observed that larvae from brooding species tended to recruit to the reef within a few meters of the parent colonies. This evidence suggests that populations of brooding corals within BNP are self-seeding.

Latitude was not a significant factor for scleractinian recruitment within BNP. Hydrocoral recruitment, on the other hand, was higher at the northern TR reef than the southern BB reef over the 3-year period. This corresponds to the relative abundance of adult hydrocoral colonies on these reefs.

## CONCLUSIONS

This study has two main conclusions: (1) There is a disturbing parallel between the lack of juvenile coral recruitment measured in this study and the high death rate of adult corals documented in the Florida Keys (Porter et al., 2002) — with a high death rate and a low recruitment rate, the overall coral community in Florida is in trouble, and (2) a striking contrast exists between the low coral recruitment rates measured in our study and the much higher rates measured elsewhere in the Caribbean using the same techniques. This suggests that Floridian coral reefs have a reduced capacity to recover from stress or catastrophe, and this further emphasizes their status as an area of critical concern.

Adult populations on the Floridian reefs studied have been declining for several years. The relative success of hydrocoral to scleractinian recruitment could reflect differences in the ways in which these organisms are responding to the stresses which are causing the decline of the adult populations. Hydrocorals consistently recruited more often and grew faster than their scleractinian counterparts.

As has been found elsewhere in the Caribbean, there is a discrepancy between the composition of the adult community and the composition of juvenile scleractinian recruits. The lack of recruitment of mass-spawning species may be the result of reliance on fragmentation for propagation by branching species in shallow water or from detrimental effects of natural or anthropogenic stresses. It is also possible that the dominance of recruitment by brooding corals in this study is a function of high latitude reefs.

Recruits of both scleractinians and hydrocorals were consistently found on the lower surfaces of tiles. These patterns are consistent with a model that suggests that recruitment is influenced by high sedimentation on upper surfaces. It is interesting to observe that scleractinians and hydrocorals recruited differently to tiles that varied in height above the substrate. This phenomenon may be the result of the different reproductive modes of these corals: Fully developed larvae released from scleractinian corals may encounter the lower tiles more often than the higher tiles, while larvae produced in the water column by floating hydrocoral medusae would encounter the higher tiles more frequently.

Three trends were observed with depth for the scleractinian corals: recruitment was highest at the greatest depth, the preferred orientation shifted slightly from lower surfaces to vertical surfaces, and the family composition of the corals shifted from faviids to agariciids and poritids. The shift toward vertical surfaces is most likely in response to light, while the shift in species composition may reflect differences in life-history phenomena as faviid corals are poor competitors in the relatively stable environments found at depth.

Surprisingly, latitude was not a significant source of variation in scleractinian recruitment in our study as it is in other studies. This could reflect the relatively high decline of adult corals at the southern-most reef. Hydrocoral recruitment, on the other hand, was higher on northern reefs where *Millepora* spp. also comprise a much greater portion of the adult community.

Finally, recruitment rates of both scleractinians and hydrocorals were dramatically lower during the second year of the study, the year immediately following Hurricane Andrew. While this could be due to interannual variation in recruitment, as has been reported in other studies, it is also possible that Hurricane Andrew depressed recruitment directly. In our study, late-summer spawning events did not cause an increase in successful larval settlement. Because well-developed reefs upstream were not affected by the hurricane, this decline also suggests that reefs within BNP are self-seeding.

The low coral recruitment rates identified in our study are especially worrisome in the face of the high adult coral mortalities measured at these same places. Because circumstantial evidence suggests recruitment is locally restricted within reefs, conservation actions must be applied Keys-wide to be effective. Only the presence of healthy, reproductively active adult communities throughout the Keys will ensure successful juvenile recruitment regionwide.

# REFERENCES

Babcock, R.C. 1991. Comparative demography of three species of scleractinian corals using age- and size-dependent classifications, *Ecol. Monogr.*, 61:225–244.

Babcock, R.C. and P. Davies. 1991. Effects of sedimentation on settlement of *Acropora millepora*, *Coral Reefs*, 9:205–208.

Babcock, R.C. and C. Mundy. 1996. Coral recruitment: consequences of settlement choice for early growth and survivorship in two scleractinians, *J. Exp. Mar. Biol. Ecol.*, 206:179–201.

Baggett, L.S. and T.J. Bright. 1985. Coral recruitment at the East Flower Garden Reef (Northwestern Gulf of Mexico), *Proc. Fifth Int. Coral Reef Congr., Tahiti*, 4:379–384.

Bak, R.P.M. and M.S. Engel. 1979. Distribution, abundance and survival of juvenile hermatypic corals (scleractinia) and the importance of life history strategies in the parent coral community, *Mar. Biol.*, 54:341–352.

Bak, R.P.M. and B.E. Luckhurst. 1980. Constancy and change in coral reef habitats along depth gradients at Curacao, *Oecologia*, 47:145–155.

Birkeland, C. 1977. The importance of rate of biomass accumulation in early successional stages of benthic communities to the survival of coral recruits, in *Proc. Third Int. Coral Reef Symp., Miami, FL*, pp. 15–21.

Blair, S.M., T L. McIntosh, and B.J. Mostkoff. 1994. Impacts of Hurricane Andrew on the offshore reef systems of central and northern Dade County, Florida, *Bull. Mar. Sci.*, 54:961–973.

Burns, T.P. 1985. Hard-coral distribution and cold-water disturbances in south Florida: variation with depth and location, *Coral Reefs*, 4:117–124.

Carlon, D.B. and R.R. Olson. 1993. Larval dispersal distance as an explanation for adult spatial patterns in two Caribbean reef corals, *J. Exp. Mar. Biol. Ecol.*, 173:247–263.

Chiappone, M. and K.M. Sullivan. 1996. Distribution, abundance and species composition of juvenile scleractinian corals the Florida reef tract, *Bull. Mar. Sci.*, 58:555–569.

Crawley, M.J. 1993. *GLIM for Ecologists*, Blackwell Scientific, Oxford.

Dustan, P. 1977. Vitality of reef coral populations off Key Largo, Florida: recruitment and mortality, *Environ. Geol.*, 2:51–58.

Dustan, P. and J.C. Halas. 1987. Changes in the reef-coral community of Carysfort Reef, Key Largo, Florida: 1974 to 1982, *Coral Reefs*, 6:91–106.

Fadlallah, Y.H. 1983. Sexual reproduction, development and larval biology in scleractinian corals, *Coral Reefs*, 2:129–150.

Falkowski, P.G., Z. Dubinsky, L. Muscatine, and J. W. Porter. 1984. Light and the bioenergetics of a symbiotic coral, *Bioscience*, 34:705–709.

Fautin, D.G., J.G. Spaulding, and F.-S. Chia. 1989. Cnidaria, p. 463, in *Fertilization, Development and Parental Care*, Vol. IV, Part A, Adiyodi, K.G. and Adiyodi, R.G., Eds., John Wiley & Sons, Chichester.

Fisk, D.A. and V.J. Harriott. 1990. Spatial and temporal variation in coral recruitment on the Great Barrier Reef: implications for dispersal hypotheses, *Mar. Biol.*, 107:485–490.

Fitzhardinge, R. 1985. Spatial and temporal variability in coral recruitment in Kaneohe Bay (Oahu, Hawaii), *Proc. Fifth Int. Coral Reef Congr., Tahiti*, 4:373–378.

Gay, S.L. and J.C. Andrews. 1994. The effects of recruitment strategies on coral larvae settlement distributions at Helix Reef, in *The Biophysics of Marine Larval Dispersal*, Vol. 45, Sammarco, P.W. and Heron, M.L., Eds., American Geophysical Union, Washington, D.C.

Glynn, P.W., A.M. Szmant, E.F. Corcoran, and S.V. Cofer-Shabica. 1989. Condition of coral reef cnidarians from the northern Florida reef tract: pesticides, heavy metals and histopathological examination, *Mar. Pollut. Bull.*, 20:568–576.

Hall, V.R. and T.P. Hughes. 1996. Reproductive strategies of modular organisms: comparative studies of reef building corals, *Ecology*, 77:950–963.

Harriott, V.J. 1985a. Mortality rates of scleractinian corals before and during a mass bleaching event, *Mar. Ecol. Prog. Ser.*, 21:81–88.

Harriott, V.J. 1985b. Recruitment patterns of scleractinian corals at Lizard Island, Great Barrier Reef, *Proc. Fifth Int. Coral Reef Congr., Tahiti*, 4:367–372.

Harriott, V.J. and S.A. Banks. 1995. Recruitment of scleractinian corals in the Solitary Islands Marine Reserve, a high latitude coral-dominated community in Eastern Australia, *Mar. Ecol. Prog. Ser.*, 123:155–161.

Harriott, V.J. and D.A. Fisk. 1987. A comparison of settlement plate types for experiments on the recruitment of scleractinian corals, *Mar. Ecol. Prog. Ser.*, 37:201–208.

Helmuth, B. and K. Sebens. 1993. The influence of colony morphology and orientation to flow on particle capture by the scleractinian coral *Agaricia agaricites*, *J. Exp. Mar. Biol. Ecol.*, 165:251–278.

Hudson, H.H., K.J. Hanson, R.B. Halley, and J.L. Kindinger. 1994. Environmental implications of growth rate changes in *Montastraea annularis*: Biscayne National Park, *Bull. Mar. Sci.*, 54:647–669.

Hughes, T. and J. Jackson. 1985. Population dynamics and life histories of foliaceous corals, *Ecol. Monogr.*, 55:141–166.

Kojis, B.L. and N.J. Quinn. 1981. Reproductive strategies in four species of *Porites* (scleractinia), *Proc. Fourth Int. Coral Reef Symp., Manila*, 2:145–151.

Lang, J. 1973. Interspecific aggression by scleractinian corals. 2. Why the race is not only to the swift, *Bull. Mar. Sci.*, 23:260–279.

Lee, T.N., K. Leaman, and E. Williams. 1995. Florida Current meanders and gyre formation in the southern Straits of Florida, *J. Geophys. Res.*, 100:8607–8620.

Levitan, D.R. 1995. The ecology of fertilization in free-spawning invertebrates, pp. 464, in *Ecology of Marine Invertebrate Larvae*, McEdward, L., Ed., CRC Press, Boca Raton, FL.

Lewis, J.B. 1974. The settlement behavior of planulae larvae of the hermatypic coral *Favia fragum* (Esper), *J. Exp. Mar. Biol. Ecol.*, 15:165–172.

Lewis, J.B. 1989. The ecology of *Millepora*, *Coral Reefs*, 8:99–107.

Maida, M., J.C. Coll, and P.W. Sammarco. 1994. Shedding new light on scleractinian coral recruitment, *J. Exp. Mar. Biol. Ecol.*, 180:189–202.

Meier, O.W. 1996. *A Long-Term Study of Reef Coral Dynamics in the Florida Keys*, Institute of Ecology, University of Georgia, Athens, pp. vi - 285.

Morse, D.E., N. Hooker, A.N.C. Morse, and R.A. Jensen. 1988. Control of larval metamorphosis and recruitment in sympatric agariciid corals, *J. Exp. Mar. Biol. Ecol.*, 116:193–217.

Neigel, J.E. and J.C. Avise. 1983. Clonal diversity and population structure in a reef-building coral, *Acropora cervicornis*: self-recognition analysis and demographic interpretation, *Evolution*, 37:437-453.

Ogden, J.C., J.W. Porter, N.P. Smith, A.M. Szmant, W.C. Jaap, and D. Forcucci. 1994. A long-term interdisciplinary study of the Florida Keys seascape, *Bull. Mar. Sci.*, 54:1059–1071.

Oliver, J. and R. Babcock. 1992. Aspects of the fertilization ecology of broadcast spawning corals: sperm dilution effects and *in situ* measurements of fertilization, *Biol. Bull.*, 183:409–417.

Pitts, P.A. 1994. An investigation of near bottom flow patterns along and across Hawk Channel, Florida Keys, *Bull. Mar. Sci.*, 54:610-620.

Porter, J.W. 1987. *Species Profiles: Life Histories and Environmental Requirements of Coastal Fishes and Invertebrates (South Florida) — Reef-Building Corals*, U.S. Fish & Wildlife Service Biological Report.

Porter, J.W. et al. 2002. Detection of coral reef change by the Florida Keys Coral Reef Monitoring Project. In *The Everglades, Florida Bay, and Coral Reefs of the Florida Keys: An Ecosystem Sourcebook*, Porter, J.W. and Porter, Karen G., Eds., CRC Press, Boca Raton, FL.

Porter, J.W. and O.W. Meier. 1992. Quantification of loss and change in Floridian reef coral populations, *Am. Zool.*, 32:625–640.

Porter, J.W. and J.I. Tougas. 2001. *Reef Ecosystems: Threats to Their Biodiversity*. In *Encyclopedia of Biodiversity*, Volume 5, Levin, S.A., Ed., Academic Press, San Diego, CA.

Porter, J.W., S.K. Lewis, and K.G. Porter. 1999. The effect of multiple stressors on the Florida Keys coral reef ecosystem: a landscape hypothesis and a physiological test, *Limnol. Oceanogr.*, 44:941–949.

Richmond, R. 1996. Reproduction and recruitment in corals: critical links in the persistence of reefs, pp. 175–197, in *Life and Death of Coral Reefs*, Birkeland, C., Ed., Chapman & Hall, New York.

Richmond, R.H. and C.L. Hunter. 1990. Reproduction and recruitment of corals: comparisons among the Caribbean, the Tropical Pacific, and the Red Sea, *Mar. Ecol. Prog. Ser.*, 60:185–203.

Rogers, C.S. 1990. Responses of coral reefs and reef organisms to sedimentation, *Mar. Ecol. Prog. Ser.*, 62:185–202.

Rogers, C.S. 1993. Hurricanes and coral reefs: the intermediate disturbance hypothesis revisited, *Coral Reefs*, 12:127–137.

Rogers, C.S., H.C. Fitz, III, M. Gilnack, J. Beets, and J. Hardin. 1984. Scleractinian coral recruitment patterns at Salt River Canyon, St. Croix, U.S. Virgin Islands, *Coral Reefs*, 3:69–76.

Roughgarden, J., Y. Iwasa, and C. Baxter. 1985. Demographic theory for an open marine population with space-limited recruitment, *Ecology*, 66:54–67.

Sammarco, P.W. 1980. *Diadema* and its relationship to coral spat mortality: grazing, competition, and biological disturbance, *J. Exp. Mar. Biol. Ecol.*, 45:245–272.

Sammarco, P.W. 1994. Larval dispersal and recruitment processes in Great Barrier Reef corals: analysis and synthesis, in *The Biophysics of Marine Larval Dispersal*, Vol. 45, Sammarco, P.W. and Heron, M.L., Eds., American Geophysical Union, Washington, D.C.

Skelton, P. 1993. *Evolution: A Biological and Palaeontological Approach*, Addison-Wesley, Milton Keynes, Wokingham, England.

Smith, S.R. 1992. Patterns of coral recruitment and post-settlement mortality on Bermuda's reefs: comparisons to Caribbean and Pacific reefs, *Am. Zool.*, 32:663–673.

Stoddart, J.A. 1984. Genetic structure within populations of the coral *Pocillopora damicornis*, *Mar. Biol.*, 81:19–30.

Szmant, A.M. 1991. Sexual reproduction by the Caribbean reef corals *Montastraea annularis* and *M. cavernosa*, *Mar. Ecol. Prog. Ser.*, 74:13–25.

Szmant-Froelich, A., M. Reutter, and L. Riggs. 1985. Sexual reproduction of *Favia fragum* (Esper): lunar patterns of gametogenesis, embryogenesis and planulation in Puerto Rico, *Bull. Mar. Sci.*, 37:880–892.

Thorson. 1950. Reproductive and larval ecology of marine bottom invertebrates, *Biol. Rev.*, 25:1–45.

Tomascik, T. 1991. Settlement patterns of Caribbean scleractinian corals on artificial substrata along a eutrophication gradient, Barbados, West Indies, *Mar. Ecol. Prog. Ser.*, 77:261–269.

Vaughan, T.W. and J.W. Wells. 1943. *Revision of the Scleractinia*, Geological Society of America, Waverly Press, Inc., Baltimore, MD.

Wallace, C.C. 1985. Seasonal peaks and annual fluctuations in recruitment of juvenile scleractinian corals, *Mar. Ecol. Prog. Ser.*, 21:289–298.

Ward, S. 1995. The effect of damage on the growth, reproduction and storage of lipids in the scleractinian coral *Pocillopora damicornis* (Linneaus), *J. Exp. Mar. Biol. Ecol.*, 187:193–206.

White, M.W. and J.W. Porter. 1985. The establishment and monitoring of two permanent photograph transects in Looe Key and Key largo national Marine Sanctuaries (Florida Keys), *Proc. Fifth Int. Coral Reef Congr., Tahiti*, 6: 531–537.

Young, C.M. 1995. Behavior and locomotion during the dispersal phase of larval life, pp. 464, in *Ecology of Marine Invertebrate Larvae*, McEdward, L., Ed., CRC Press, Boca Raton, FL.

# 30 Aspergillosis of Sea Fan Corals: Disease Dynamics in the Florida Keys

*Kiho Kim*
Biology Department, American University

*C. Drew Harvell*
Ecology and Evolutionary Biology, Cornell University

## CONTENTS

## INTRODUCTION

A dominant feature of the South Florida hydroscape is the complex of patch and fringing coral reefs that dominate the shallow coastal waters along the Florida Keys. These reefs not only provide physical protection for the Keys and their coastal environments, but also support immense biological diversity which in turn sustains productive fisheries and a thriving tourism industry. However, intense use and concomitant degradation of coastal environments have threatened the health of these coral reefs. Increased inputs of nutrients have resulted in the eutrophication of nearshore waters (Lapointe et al., 1990; Lapointe and Clark, 1992; Lapointe and Matzie, 1996), and have been correlated with a decline in coral cover (Porter and Meier, 1992; Ogden et al., 1994). Several causal mechanisms have been posited for this decline, including the suggestion that eutrophic conditions favor both macroalgae, which outcompete corals for space, and phytoplankton blooms, which reduce light availability to corals (Pastorok and Bilyard, 1985; Done et al., 1996). In addition, there is growing awareness that diseases are contributing significantly to the loss of corals in the

0-8493-2026-7/02/$0.00+$1.50
© 2002 by CRC Press LLC

Florida Keys (Porter and Meier, 1992; Richardson et al., 1998, Harvell et al., in press) and elsewhere around the world (Epstein et al., 1998; Goreau et al., 1998). The virtual elimination of benthic dominants such as the sea urchin, *Diadema antillarum* (reviewed in Lessios, 1988) and acroporid corals (Gladfelter, 1982; Aronson and Precht, 1997; Greenstein et al., 1998; McClanahan and Muthiga, 1998) throughout the Caribbean as a result of epidemics clearly illustrates the importance of diseases in the ecology of coral reefs.

Since the discovery of a "plague" in the late 1970s (Dustan, 1977), approximately a dozen more pathologies and diseases affecting scleractinian and gorgonian corals have been described (reviewed in Richardson, 1998). In spite of the apparent increase in disease outbreaks, not only in frequency but also in their impact on marine populations (Harvell et al., 1999), quantitative assessments or epidemiological studies of coral diseases have been rare (e.g., Edmunds, 1991; Richardson et al., 1998). Such studies are important for assessing the impact of disease on coral reefs and elucidating the potential factors that contribute to or exacerbate disease outbreaks.

Sea fans and other gorgonian corals are common and frequently dominant members of shallow water reefs in the Florida Keys. An outbreak of fungal disease (aspergillosis) affecting sea fan corals (Nagelkerken et al., 1996, 1997; Smith et al., 1996) may have significant and as yet unknown consequences for the ecology of this already stressed ecosystem. Aspergillosis is of particular interest in the context of ecosystem linkages because it is caused by the terrestrial fungus *Aspergillus sydowii* (Smith et al., 1996; Geiser et al., 1998). Although *A. sydowii* has been previously detected in marine samples, its discovery as a new pathogen has led to the hypothesis that it crossed the land–sea boundary, perhaps as part of soil runoff, to emerge as a pathogen in the ocean. Following a preliminary survey (data from two reefs near Key West, FL, October 1996), which revealed that aspergillosis was highly prevalent among sea fan populations in the Florida Keys (Nagelkerken et al., 1997), we designed a study using standard epidemiological measures of disease (age-specific prevalence and severity) to assess modes of transmission and disease impact. An additional goal of the study was to ascertain possible links between disease and water quality in the Florida Keys.

## BACKGROUND

Aspergillosis of sea fans is an infectious, widespread fungal disease affecting Caribbean gorgonian corals, *Gorgonia* spp. (Nagelkerken et al., 1996, 1997). A primary symptom* of aspergillosis is one or more necrotic patches (i.e., lesions) surrounded by a narrow margin of darkly pigmented tissue (Figure 30.1). The necrosis exposes the underlying axial skeleton, which quickly becomes fouled and eventually lost during heavy surges. Aspergillosis of sea fans was first noted in 1995 in Saba, Netherland Antilles (K. Buchan, pers. comm.); subsequent surveys performed by the Caribbean Coastal Marine Productivity Program (CARICOMP) network indicated that the disease was widespread, extending from Trinidad to Colombia in the south, and from Mexico to the Florida Keys in the north (Nagelkerken et al., 1996).

The pathogen of sea fan aspergillosis was identified by its partial 18s ribosomal RNA sequence and by morphological and metabolic characteristics as a fungus of the genus *Aspergillus* (Smith et al., 1996), and subsequently as *A. sydowii* (Geiser et al., 1998). The role of *A. sydowii* as a pathogen of sea fans was confirmed by isolating hyphae from sick fans, inoculating healthy fans to produce disease symptoms, and re-isolating the agent from inoculated fans, thus fulfilling Koch's Postulates (Smith et al., 1996). Although the actual mechanism of spread is not yet known, the disease can be transmitted from one colony to another by grafting infected tissue onto healthy tissue and by inoculating healthy fans with pure cultures of *A. sydowii;* in both cases, disease symptoms develop within a few days (Smith et al., 1996).

---

\* In the disease literature, it is sometimes suggested that the term "sign" rather than "symptom" is more appropriate for nonhuman subjects; however, we use the term symptom (as well as "epidemic," "epidemiological," etc.) more broadly to include both human and nonhuman hosts (see, for example, Terminology Sub-Committee, 1973).

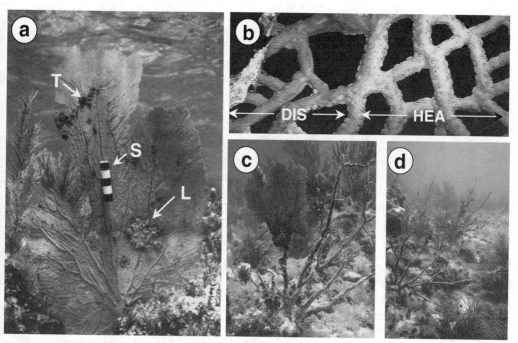

**FIGURE 30.1** Aspergillosis of sea fan corals. (a) Symptoms of aspergillosis include lesions (L) surrounded by a darkly pigmented margin. The lesions expose the underlying axial skeleton which becomes fouled. Tumors (T) also may result from aspergillosis infections. (b) Close-up of a lesion on *Gorgonia ventalina* showing the darkly pigmented margin (DIS) which is typically purple. The healthy tissue (HEA) in this photograph is lighter in color. Note the absence of coral polyps in the dark margin. (c) Advanced stage of disease. (d) An area of Western Dry Rocks reef (Key West, FL) with a large number of dead colonies. Scale bar (S) = 10 cm in total.

Several ecological and environmental factors have been associated with incidence of the disease (Nagelkerken et al., 1997). For instance, disease damage increased with colony size, decreased with wave exposure, and increased with depth on shallow reefs. Nagelkerken et al. (1997) also found that the sea fan predator *Cyphoma gibbosum* and tumors were more abundant on diseased than on healthy sea fans.

Caribbean sea fans were subject to a devastating epidemic during the early 1980s that wiped out entire populations along the southern coasts of the Caribbean basin (Guzmán and Cortés, 1984; Garzón-Ferreira and Zea, 1992). In spite of the ecological scale of this event, no additional information was gathered regarding the nature of the pathogen or even the nature of the symptoms, thereby preventing a comparison of the earlier sea fan epidemic with the current one.

## MATERIALS AND METHODS

### DISEASE SURVEY METHODS

Monitoring of sea fans was carried out in June 1997 at permanent transects established at five sites in the Florida Keys (Figure 30.2). Site selection was based on the presence of sea fan corals, the need for adequate spatial coverage of the Keys, and the proximity of these sites to environmental (i.e., water quality) monitoring locations (see below). At each location, transects were haphazardly positioned in gorgonian beds without reference to disease state. At some sites (Molasses, Western Dry Rocks, and Western Sambo Reefs), existing mooring buoys were used as site markers; at other sites (Alligator and Tennessee), 30-cm metal spikes were installed adjacent to sea fan beds. The site marker served as the origin for three haphazardly sited linear transects which radiated outward. To prevent re-sampling near the transect origin, the angles between transects were maximized and surveys were carried out starting from the 2-m mark for a total of 25 m. Additional spikes marked the ends of all transects. At each transect, all sea fans with bases within 1 m of either side of the transect line were examined. For each colony, the maximum height and the number and sizes of lesions (i.e., percent disease damage) were recorded. Aspergillosis was diagnosed by the presence of distinctive lesions accompanied by purpling of the surrounding tissue (Figure 30.1). From the transect data, estimates of disease prevalence (percent of individuals infected) and severity (percent of colony area affected by disease) were derived. In addition, the number of dead sea fan colonies (i.e., remains of axial skeleton) within the transects was recorded.

To estimate the contribution of sea fans to coral cover, the surface area of individual sea fan colonies was estimated using a power function derived from image analyses (NIH Image 1.61, U.S. National Institutes of Health) of photographs (taken April 1997) of 40 sea fan colonies ranging in height from 17 to 90 cm at Western Dry Rocks reef. The analysis yielded the following equation: surface area $(cm^2) = 0.624 \times$ height $(cm)^{1.967}$, $r = 0.821$.

## WATER QUALITY DATA

Water quality data presented here were provided by J. Boyer (Southeast Environmental Research Program, Florida International University). The monitoring program collects water quality data on a quarterly basis for a number of parameters including the following which were used in this study: dissolved inorganic nitrogen (DIN), total phosphorous (TP), chlorophyll $a$ (CHLA), and turbidity (TURB). Details of methods are given in Jones and Boyer (1999). These data were collected on June 7, 1997, which was approximately 1 week prior to our sea fan surveys. Because water quality data were unavailable for two of the sea fan monitoring sites (Western Dry Rocks and Western Sambo), data for nearby sites (Eastern Dry Rocks and Eastern Sambo, respectively) were used in their place. These water quality monitoring locations are located approximately 10 km east of their corresponding sea fan survey sites.

## STATISTICAL ANALYSIS

Disease prevalence and severity data were first arcsine transformed and tested for normality and homogeneity of variances using the Kolmogorov and Levene tests, respectively. The results of these tests are as follows: prevalence — Kolmogorov, $p = 0.147$; Levene, $F_{4,10} = 0.803$, $p = 0.550$; severity — Kolmogorov, $p = 0.583$; Levene, $F_{4,10} = 1.00$, $p = 0.450$. The transformed data were analyzed using ANOVA. Mortality data could not be adequately transformed to normality and homogeneity due to the abundance of zeros; therefore, these data were analyzed using the Kruskal–Wallis ANOVA by ranks.

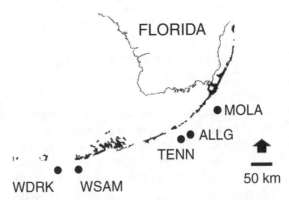

**FIGURE 30.2** Survey sites. Map of the Florida Keys showing the five sea fan monitoring sites established June 1997. At each site, three 25-m transects were marked using metal spikes. Abbreviations: MOLA = Molasses, ALLG = Alligator, TENN = Tennessee, WSAM = Western Sambo, WDRK = Western Dry Rocks

# RESULTS

## DESCRIPTION OF *GORGONIA VENTALINA* POPULATIONS

In total, we examined 822 colonies within 15 transects at five sites (an area of 750 m²). Within individual transects, the number of *Gorgonia ventalina* colonies ranged from 34 to 99, which was equivalent to densities of 0.79 to 1.62 colonies m⁻² (Figure 30.3). Colony size, in terms of colony height, ranged from 3 to 152 cm with mean and standard error of 44 ± 0.67 cm. All sea fans combined made up 91 m² of living coral surface area, or 182 m² if both sides of the colony are considered. If *G. ventalina* grew horizontally on the substratum, the combined contribution of all sea fans would account for a theoretical 12% of the total "coral cover" within our survey sites (or 24% if both sides are considered). On one transect at Alligator where sea fans are particularly abundant, they accounted for 23 (or 46%) of coral cover.

## DISEASE DYNAMICS

In our surveys, we documented 387 healthy sea fans, diagnosed 352 colonies as diseased, and noted 83 as dead. When averaged across sites, prevalence of aspergillosis was 42.6% (±3.19 SE) in the Florida Keys. Prevalence of aspergillosis was size specific and showed a monotonic increase with colony size (Figure 30.4). Among infected individuals, severity (how much of the colony is affected by the disease) ranged upward to slightly less than 100% (Figure 30.5) with a Keys-wide mean (± SE) of 23.2 ± 6.00%. However, half of the infected colonies had less than 5% of their colony area affected by the disease. For approximately 8% of the infected colonies, the disease caused significant damage, affecting more than 75% of the colony.

Disease prevalence ranged from 33 to 53% (Figure 30.6). However, we did not detect a significant site-to-site variation in an ANOVA ($F_{4,10}$ = 1.023, $p$ = 0.441). In contrast, severity varied across sites ($F_{4,10}$ = 5.07, $p$ = 0.017). Severity ranged from 11 to 43% and was higher at the two sites near Key West. The Upper/Lower Keys gradient was even more striking in the number of dead colonies (Kruskal–Wallis, $H$ = 11.6, $df$ = 4, $p$ = 0.020). Although there were, on average, 11.1 skeletal remains (= dead) per transect across all sites, we did not find any at Molasses; we found 28 skeletal remains at Western Sambo and 31 at Western Dry Rocks. Associated with the high numbers of dead colonies at these sites were high numbers of colonies that were severely affected by aspergillosis (i.e., severity of 76 to 99%; see Figure 30.5). Of the 29 colonies in total that were severely affected, 14 and 11 were found on transects at Western Sambo and Western Dry Rocks, respectively. These results suggest that mortality of sea fans at these Key West sites was likely due to aspergillosis.

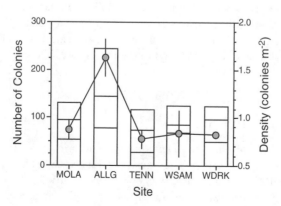

**FIGURE 30.3** Density of *Gorgonia ventalina*. For each site, the number of colonies within each of three 25 × 2-m transects (stacked bars) and mean density (circles; ± SE) are shown (total number of colonies = 822). Site abbreviations are as in Figure 30.2.

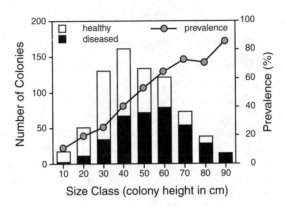

**FIGURE 30.4** Demography and disease of *Gorgonia ventalina*. Proportions of healthy (open bars) and diseased colonies (filled bars) and prevalence (percent according to colony size class; circles). Size class is based on colony height.

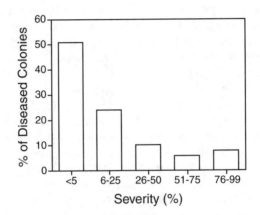

**FIGURE 30.5** Severity class. Frequency histogram of disease severity (percent of colony area affected by aspergillosis). Total number of diseased colonies = 352.

## WATER QUALITY DATA

Variations in the availability of two major nutrients (N and P), primary productivity (chlorophyll *a*), and water clarity (turbidity) illustrate the differences in habitat quality among our survey sites (Figure 30.7). These data indicate that coral reefs near Key West are characterized by higher nutrient loads and reduced light availability compared to the Upper Keys sites.

## DISCUSSION

### ASPERGILLOSIS IN THE FLORIDA KEYS

Prevalence of aspergillosis of *Gorgonia ventalina* was uniform in the Florida Keys (Keys-wide mean ± SE = 43 ± 3.54%, n = 823). On one transect at Tennessee Reef, more than 61% of the sea fans were infected. In 1995, when Nagelkerken et al. (1997) carried out a Caribbean-wide survey, more than 90% of the sea fans were diagnosed with aspergillosis at some sites (e.g., Saba). In comparison, most other coral diseases documented to date have been much less prevalent. For instance, cyanobacterial *(Phormidium)* infections of the gorgonian coral *Pseudopterogorgia acerosa* were found on less than 8% of the population (Feingold, 1988). Similarly, black-band disease

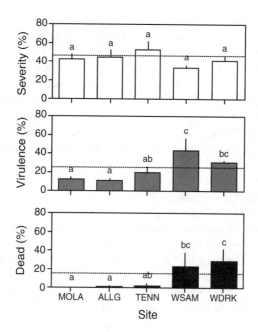

**FIGURE 30.6** Disease dynamics in the Florida Keys. Means (± SE) are given for disease prevalence (percent of infected individuals), severity (percent of colony affected by disease), and number of dead colonies at the monitoring sites. Dotted lines indicate Keys-wide means. Significant site-to-site differences in disease parameters, as noted in *post hoc* comparisons, are indicated by letter groupings (i.e., sites with the same letter are not significantly different). Site abbreviations are as in Figure 30.2.

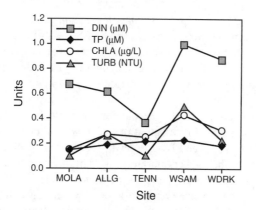

**FIGURE 30.7** Water quality. Variability in water quality across sea fan survey sites (June 1997). Abbreviations: DIN = dissolved inorganic nitrogen; TP = total phosphorous; CHLA = chlorophyll *a*; and TURB = turbidity. Site abbreviations are as in Figure 30.2. (Data provided by Southeast Environmental Research Program.)

(BBD), which is among the best characterized of all coral diseases (see Richardson, 1998), occurs at very low frequencies. Edmunds (1991) found that ≤5.5% of any given species in St. John, USVI, was affected by BBD. In the Florida Keys, Kuta and Richardson (1996) detected BBD on only 10 of 1397 (0.72%) scleractinian coral colonies surveyed. More recently, Richardson et al. (1998) showed that a bacterial disease referred to as "plague type II" affected a maximum of 38% of *Dichocoenia stokesi* colonies in the Keys.

Disease prevalence, which is a population-level parameter, is insufficient for fully assessing the impact of a disease on the host population. Critical to an assessment of impact is information on how the disease is affecting the host at the individual level; however, such detailed data are rare for coral diseases. In one report, Edmunds (1991) estimated that partial mortality in *Diploria strigosa* due to BBD was less than 4% per year. In comparison, the impact of aspergillosis on *Gorgonia ventalina* appears to be more substantial, with a Keys-wide mean disease severity of more than 23%. Furthermore, if we include the number of dead colonies in our assessment, the impact is more substantial. Indeed, a longitudinal photomonitoring study carried out at Conch Reef showed that between May 1998 and September 1999, rate of tissue loss due to aspergillosis was $31.9 \pm 8.81\%$ (mean $\pm$ SE, $n = 13$) per year among infected sea fans (Kim and Harvell, unpubl. data).

Our surveys indicate a monotonically increasing prevalence of aspergillosis with colony size (Figure 30.4; see also Nagelkerken et al., 1997), similar to the prevalence of algal infections of the gorgonian coral *Pseudopterogorgia* spp. (Goldberg and Makemson, 1981). Three explanations are likely for this pattern. First, an increasing prevalence with age is common in many epidemiological studies, because older individuals tend to accumulate infections over time (Anderson and May, 1998). Although age and size are not perfectly correlated for most colonial animals (Hughes, 1984), it is likely that large coral colonies are much older than small ones. Second, larger colonies may represent larger "targets" for waterborne pathogen transmission. Third, large colonies may be chemically more susceptible than small colonies. Similar to within-plant variation in chemical defense (Karban and Baldwin, 1997), there is within-colony variation in resistance to disease, with highest resistance occurring along the colony edge (i.e., new tissue) (Kim et al., 2000). Thus, larger colonies may have greater areas of the colony that are poorly defended chemically in comparison to smaller sea fans. Regardless of the mechanisms producing this pattern, the fact that larger colonies bear the brunt of the disease has important demographic implications for sea fans. To illustrate this point, consider the impact of aspergillosis at the time of our survey for two size groups of sea fans: small (height <50 cm) and large (>50 cm). Categorized this way and using the data summarized in Figure 30.4, there are approximately twice as many small colonies as large colonies (455 vs. 285), with a disease prevalence of 35% for small colonies and 62% for large colonies. Severity is approximately equal between the two size groups (both ~21%; data not shown); however, the actual surface area represented by the severity value is quite different for the two size groups. For instance, for two colonies, 25 and 75 cm tall, with 21% of colony area affected by disease, the actual surface area affected by disease is 0.007 and 0.064 m², respectively. Combined total surface area of all sea fans surveyed was 91 m², and of that 11 m² were affected by aspergillosis. Of the affected area, nearly 80% (or 9.1 m²) occurred on large colonies. Because larger coral colonies contribute proportionately more to the reproductive output of a reef even when accounting for differences in size (e.g., Karlson, 1986; Brazeau and Lasker, 1990; Coma et al., 1995; Hall and Hughes, 1996; Beiring and Lasker, 2000), the higher mortality — partial or complete — among the larger *G. ventalina* is likely to reduce fecundity and thus recruitment and to slow recovery from the current disease outbreak and future disturbance events. Moreover, by serving as substrates for the pathogen, larger colonies may also contribute disproportionately to the transmission of the disease.

### ENVIRONMENTAL CORRELATES

Although disease prevalence did not vary by site, the impact of the disease (severity) was greater among sea fans near Key West. Why is this area an apparent disease "hotspot"? Although the finding of this study must be taken as preliminary, given the limited amount of data available for analysis, one possibility is that poor water quality at these sites, as indicated by higher N concentrations and slightly lower water clarity (Figure 30.6), exacerbated disease severity. Growing evidence suggests that declines in coral cover (i.e., mortality), diversity, and general coral health are strongly linked to decreased water quality, primarily as a result of terrestrial inputs of nutrients and pollutants (e.g., Walker and Ormond, 1982; Maragos et al., 1985; Tomacik and Sanders, 1985,

1987a,b; Jackson et al., 1989; Tomacik, 1991; Guzmán and Holst, 1993; Edinger et al., 1998). However, the mechanism by which poor water quality affects coral host-pathogen interactions is not clearly understood. Done et al. (1996) suggested that degradation of coastal waters can favor macroalgae and other non-coral species (which outcompete corals for space and other resources) and promote phytoplankton blooms (which reduce light availability for photoautotrophy by corals) (reviewed in Pastorok and Bilyard, 1985). In the latter case, the net effect of poor habitat quality may be that corals become stressed and less able to resist infections (Kim et al., in press). Although there is no direct evidence of this for corals, studies of other marine species have shown that polluted waters can have immuno-suppressive effects on shrimp (Couch and Courtney, 1977), bivalves (Kim et al., 1998; Fisher et al., 1999), fish (Arkoosh et al., 1998), and marine mammals (Ross et al., 1996).

Clearly, poor water quality is not a prerequisite for high disease impact. According to Nagelkerken et al. (1997), aspergillosis was Caribbean-wide, occurring at remote locations that were considerably more pristine than the Florida Keys. In particular, they found higher disease prevalence among sea fan populations in San Salvador, Bahamas, than in the Florida Keys. Subsequent surveys in San Salvador have revealed significant partial and complete mortality among *Gorgonia ventalina* (Kim, Harvell, and Smith, unpubl. data). Similarly, McClanahan and Muthiga (1998) concluded that the virtual elimination of acroporids from a remote atoll in Belize, presumably due to white-band disease, was not related to eutrophication or other human influences. It is likely that in some instances (e.g., when a pathogen has newly emerged), the disease may be highly virulent and/or hosts highly susceptible, leading to widespread outbreak irrespective of environmental conditions. Mass mortality of the sea urchin *Diadema antillarum* in the early 1980s is widely believed to have resulted from a highly virulent pathogen (as yet unidentified) that rapidly swept throughout the Caribbean basin (reviewed in Lessios, 1988).

Sea fans and other gorgonian corals are prominent members of many Caribbean reef communities (e.g., Kinzie, 1973; Lasker and Coffroth, 1983; Yoshioka and Yoshioka, 1989), including those along Florida's coasts (Goldberg, 1973; Wheaton and Jaap, 1988; Chiappone and Sullivan, 1994). Indeed, *Gorgonia ventalina* alone theoretically represented 12% of the coral coverage in our survey areas (or 24% if both sides of the sea fans were considered). This level of coverage is comparable to all scleractinian corals combined at some of these sites (Dustan and Halas, 1987; Porter and Meier, 1992; Chiappone and Sullivan, 1994; Murdoch and Aronson, 1999).

Although gorgonian corals, unlike hermatypic corals, are not important contributors to the physical and structural aspects of reef development, their great abundance and dominant role in providing reef habitat make them important in the reef ecology. Given the recent widespread loss of scleractinian corals to hurricanes, bleaching events, and diseases (reviewed in Williams and Bunkley-Williams, 1990), the importance of sea fans as habitat providers is likely to be magnified and so, too, the impact of aspergillosis on the productivity of Florida Keys reefs.

## CONCLUSIONS

Diseases and bleaching are posing increasing threats to the health and stability of coral reefs worldwide (Harvell et al., 1999; Hoegh-Guldberg, 1999). In the Florida Keys, several disease outbreaks have accelerated a decline of corals which has been exacerbated by the degradation of coastal waters. *Aspergillus sydowii*, a terrestrial fungus thought to have been introduced into the ocean in runoff or African Dust (Shinn et al., 2000), was recently described as the causative agent of a disease (aspergillosis) affecting sea fan corals *(Gorgonia* spp.) Caribbean-wide. Because of the terrestrial source of this pathogen, we hypothesized that reefs most intensely affected by terrestrial inputs would be the most impacted by the disease. A census of sea fan populations in the Florida Keys revealed that disease prevalence was similar across five sites of varying terrestrial input, affecting ~47% of the colonies surveyed. Severity, as indicated by the percent of a colony affected by the disease, averaged 23% but ranged upward to slightly less than complete mortality.

Severity data indicated that the impact of aspergillosis was site specific and was highest on reefs near Key West. These sites also had higher concentrations of major nutrients and reduced water clarity, suggesting that the impact of the disease was related to water quality. The prevalence and severity of aspergillosis were concentrated in the larger size classes of sea fans; thus, the demographic impact of this disease should be substantial and have significant consequences for reef communities in the Florida Keys.

## ACKNOWLEDGMENTS

This work was supported by funds from NYS Hatch 183-6414, NSF OCE-9614004 and 9818830, NSF IBN-9408228, NURC-UNCW9821, and New England BioLabs Foundation. Portions of this work were carried out under permits issued by the Florida Department of Environmental Protection (98S-334), and the Florida Keys National Marine Sanctuary (FKNMS-268-97). We thank J. Boyer (SERP-FIU) for providing the water quality data. Support from the staff at the UNCW NOAA/NURC facility in Key Largo and from D. and C. Quirolo of Reef Relief have been invaluable to our work in the Florida Keys. Comments from A. P. Alker and E. A. Beiring are gratefully acknowledged.

## REFERENCES

Anderson, R.M. and May, R.M., *Infectious Diseases of Humans: Dynamics and Control*, Oxford University Press, London, 1998, 757.
Arkoosh, M.R., Casillas, E., Clemons, E., Kagley, A.N., Olson, R., Reno, P., and Stein, J.E., Effect of pollution on fish diseases: potential impacts on salmonid populations, *Journal of Aquatic Animal Health*, 10, 182–190, 1998.
Aronson, R.B. and Precht, W.F., Stasis, biological disturbance, and community structure of a Holocene coral reef, *Paleobiology*, 23, 326–346, 1997.
Beiring, E.A. and Lasker, H.R., Egg production by colonies of a gorgonian coral, Marine Ecology Progress Series, 196, 169–177, 2000.
Brazeau, D.A. and Lasker, H.R., Sexual reproduction and external brooding by the Caribbean gorgonian *Briareum asbestinum*, *Marine Biology*, 104, 465–474, 1990.
Chiappone, M. and Sullivan, K.M., Ecological structure and dynamics of nearshore hard-bottom communities in the Florida Keys, *Bulletin of Marine Science*, 54, 747–756, 1994.
Coma, R., Zabala, M. and Gili, J.-M., Sexual reproductive effort in the Mediterranean gorgonian *Paramuricea clavata*, *Marine Ecology Progress Series*, 117, 185–192, 1995.
Couch, J.A. and Courtney, L., Interaction of chemical pollutants and virus in a crustacean: a novel bioassay system, *Annals of the New York Academy of Sciences*, 79, 497–504, 1977.
Done, T.J., Ogden, J.C., Wieve, W., and Rosen, B.R., Biodiversity and ecosystem function of coral reefs, in *Functional Roles of Biodiversity: A Global Perspective*, Mooney, H., Cushman, J., Medina, E., Sala, O. and Schulze, E.-D., Eds., John Wiley & Sons, Chichester, 1996.
Dustan, P., Vitality of reef coral populations off Key Largo, Florida: recruitment and mortality, *Environmental Geology*, 2, 51–58, 1977.
Dustan, P. and Halas, J., Changes in the reef-coral community of Carysfort Reef, Key Largo, Florida: 1974–1982, *Coral Reefs*, 6, 91–106, 1987.
Edinger, E.N., Jompa, J., Limmon, G., Widjatmoko, W., and Risk, M.J., Reed degradation and coral biodiversity in Indonesia: effects of land-based pollution, destructive fishing practices and changes over time, *Marine Pollution Bulletin*, 8, 617–630, 1998.
Edmunds, P.J., Extent and effect of black band disease on a Caribbean reef, *Coral Reefs*, 10, 161–165, 1991.
Epstein, P.R., Sherman, B., Spanger-Siegfried, E., Langston, A., Prasad, S., and McKay, B., Marine Ecosystems: Emerging Diseases as Indicators of Change, Harvard Medical School, Boston, 1998, 85.
Feingold, J.S., Ecological studies of a cyanobacterial infection the Caribbean sea plume *Pseudopterogorgia acerosa* (Coelenterata: Octocorallia), *International Coral Reef Symposium*, 3, 157-162, 1988.
Fisher, W.S., Oliver, L.M., Walker, W.W., Manning, C.S., and Lytle, T.F., Decreased resistance of eastern oysters *(Crassostrea virginica)* to a protozoan pathogen *(Perkinsus marinus)* after sublethal exposure to tributyltin oxide, *Marine Environmental Research*, 47, 185–201, 1999.
Garzón-Ferreira, J., and Zea, S., A mass mortality of *Gorgonia ventalina* (Cnidaria: Gorgonidae) in the Santa Marta area, Caribbean coast of Colombia, *Bulletin of Marince Science*, 50, 522–526, 1992.

Geiser, D.M., Taylor, J.W., Ritchie, K.B., and Smith, G.W., Cause of sea fan death in the West Indies, *Nature*, 394, 137–138, 1998.

Gladfelter, W.B., White-band disease in Acropoa palmataL implications for the structure and growth of shallow reefs. *Bulletin of Marine Science* 32, 639–643, 1982.

Goldberg, W.M., The ecology of the coral-octocoral communities off the southeast Florida coast: geomorphology, species composition, and zonation, *Bulletin of Marine Science*, 23, 465–488, 1973.

Goldberg, W.M. and Makemson, J.C., Description of a tumorous condition in a gorgonian coral associated with a filamentous green alga. *International Coral Reef Symposium*, 2, 685–697, 1981.

Goreau, T., Cervino, J., Goreau, M., Smith, G., Richardson, L.L., Williams, E., Bruckner, A., Nagelkerken, I., Porter, J., Porter, K., Garzon-Ferreira, J., Hayes, R., Santavy, D., Peters, E., Littler, M., and Littler, D., Rapid spread of disease in Caribbean coral reefs, *Revista Biologica Tropical*, 46(suppl. 5), 157–171, 1998.

Greenstein, B.J., Curran, H.A. and Pandolfi, J.M., Shifting ecological baselines and the demise of *Acropora cervicornis* in the western North Atlantic and Caribbean Province: a Pleistocene perspective, *Coral Reefs*, 17, 249-261, 1998.

Guzmán, H.M. and Cortéz, J., Mortand de *Gorgonia flabellum* Linnaeus (Octocorallia: Gorgoniidae) en la Costa Caribe de Costa Rica, *Revista Biologica Tropical*, 32, 305–308, 1984.

Guzmán, H.M. and Holst, I., Effects of chronic oil-sediment pollution on the reproduction of the Caribbean reef coral *Siderastrea siderea*, *Marine Pollution Bulletin*, 26, 276–282, 1993.

Hall, V.R., and Hughes, T.P., Reproductive strategies of modular organisms: comparative studies of reef-building corals, *Ecology*, 77, 950–963, 1996.

Harvell, C.D., Kim, K., Burkholder, J., Colwell, R.R., Epstein, P.R., Grimes, J.J., Hofmann, E., Lipp, E.K., Osterhaus, A.D.M.E., Overstreet, R., Porter, J.W., Smith, G.W., and Vasta, G.R., Emerging marine diseases — climate links and anthropogenic factors, *Science*, 282, 1505–1510, 1999.

Harvell, C.D., Kim, K., Quirolo, C., Smith, G.W., and Weir, J., Mass mortality of *Briareum asbestinum* associated with the 1998 Caribbean coral bleaching, *Hydrobiologia*, in press.

Hoegh-Guldberg, O., Climate change, coral bleaching and the future of the world's coral reefs, *Marine and Freshwater Research*, 50, 839–866, 1999.

Hughes, T.P., Population dynamics based on individual size rather than age: a general model with a reef coral example, *American Naturalist*, 123, 778–795, 1984.

Jackson, J.B.C., Cubit, J.D., Keller, B.D., Batista, V., Burns, K., Caffey, H.M., Caldwell, R.L., Garrity, S.D., Getter, C.D., Gonzalez, C., Guzmán, H.M., Kaufmann, K.W., Knapp, A.H., Levings, S.C., J., M.M., Steger, R., Thompson, R.C., and Weil, E., Ecological effects of a major oil spill on Panamanian coastal marine communities, *Science*, 243, 37–44, 1989.

Jones, R.D. and Boyer, J.N., *An Integrated Surface Water Quality Monitoring Program for South Florida Coastal Waters*, Florida International University, Miami, 1999, 37 (http://serc.fiu.edu/wqmnetwork/)

Karban, R. and Baldwin, I.T., *Induced Responses to Herbivory*, University of Chicago Press, Chicago, 1997, 319.

Karlson, R., Disturbance, colonial fragmentation, and size-dependent life history variation in two reef cnidarians, *Marine Ecology Progress Series*, 28, 245–249, 1986.

Kim, Y., Powell, E.N, Wade, T., Presley, B., and Sericano, J., Parasites of sentinel bivalves in the NOAA status and trends program: distribution and relationship to contaminant body burden, *Marine Pollution Bulletin*, 37, 45–55, 1998.

Kim, K., Dobson, A.P., Gulland, F.M.D., and Harvell, C.D., Diseases in the conservation of marine diversity, in *Marine Conservation Biology: The Science of Maintaining the Sea's Biodiversity*, Norse, E.A., and Crowder, L.B., Eds., Island Press, Washington, D.C., in press.

Kim, K., Harvell, C.D., Kim, P.D., Smith, G.W., and Merkel, S.M., Fungal disease resistance of Caribbean sea fan corals (*Gorgonia* spp.), *Marine Biology*, 136, 259–267, 2000.

Kinzie, R.A., The zonation of West Indian gorgonians, *Bulletin of Marince Science*, 23, 93–155, 1973.

Kuta, K.G. and Richardson, L.L., Abundance and distribution of black band disease on coral reefs in the northern Florida Keys, *Coral Reefs*, 15, 219–223, 1996.

Lapointe, B.E. and Clark, M.W., Nutrient inputs from the watershed and coastal eutrophication in the Florida Keys, *Estuaries*, 15, 465–476, 1992.

Lapointe, B.E. and Matzie, W.R., Effects of stormwater nutrient discharges on eutrophication processes in nearshore waters of the Florida Keys, *Estuaries*, 19, 422–436, 1996.

Lapointe, B.E., O'Connell, J.D., and Garrett, G.S., Nutrient coupling between on-site sewage disposal systems, groundwaters and nearshore surface waters of the Florida Keys, *Biogeochemistry*, 10, 289–307, 1990.

Lasker, H.R. and Coffroth, M.A., Octocoral distribution at Carrie Bow Cay, Belize, *Marine Ecology Progress Series*, 13, 21–28, 1983.

Lessios, H.A., Mass mortality of *Diadema antillarum* in the Caribbean: what have we learned?, *Annual Review of Ecology and Systematics*, 19, 371–393, 1988.

Maragos, J.E., Evans, C., and Holthus, P., Reef corals in Haneohe Bay six years before and after termination of sewage discharges, *International Coral Reef Symposium*, 4, 189–194, 1985.

McClanahan, T.R. and Muthiga, N.A., An ecological shift in a remote coral atoll of Belize over 25 years, *Environmental Conservation*, 25, 122–130, 1998.

Murdoch, T.J.T. and Aronson, R.B., Scale-dependent spatial variability of coral assemblages along the Florida Reef Tract, *Coral Reefs*, 18, 341–351, 1999.

Nagelkerken, I., Buchan, K., Smith, G.W., Bonair, K., Bush, P., Garzón-Ferreira, J., Botero, L., Gayle, P., Heberer, C., Petrovic, C., Pors, L., and Yoshioka, P., Widespread disease in Caribbean sea fans: I. Spreading and general characteristics, *International Coral Reef Symposium*, 1, 679–682, 1996.

Nagelkerken, I., Buchan, K., Smith, G.W., Bonair, K., Bush, P., Garzón-Ferreira, J., Botero, L., Gayle, P., Harvell, C.D., Heberer, C., Kim, K., Petrovic, C., Pors, L., and Yoshioka, P., Widespread disease in Caribbean sea fans: II. Patterns of infection and tissue loss, *Marine Ecology Progress Series*, 160, 255–263, 1997.

Ogden, J.C., Porter, J. W., Smith, N.P., Szmant, A.M., Jaap, W.C., and Forcucci, D., A long-term interdisciplinary study of the Florida Keys seascape. *Bulletin of Marine Science*, 54, 1059–1071, 1994.

Pastorok, R.A. and Bilyard, G.R., Effects of sewage pollution on coral-reef communities, *Marine Ecology Progress Series*, 21, 175–189, 1985.

Porter, J.W. and Meier, O.W., Quantification of loss and change in Floridian reef coral populations, *American Zoologist*, 32, 625–640, 1992.

Richardson, L.L., Coral diseases: What is really known?, *Trends in Ecology and Evolution*, 13, 438–443, 1998.

Richardson, L.L., Goldberg, W.M., Kuta, K., Aronson, R.B., Smith, G.W., Ritchie, K.B., Halas, J.C., Feingold, J.S., and Miller, S.L., Florida's mystery coral-killer identified, *Nature*, 392, 557–558, 1998.

Ross, P., Swart, R.D., Loveren, H.V., Osterhaus, A., and Vost, J., The immunotoxicity of environmental contaminants to marine wildlife: a review, *Annual Review of Fish Diseases*, 6, 151–165, 1996.

Shinn, E.A., Smith, G.W., Propero, J.M., Betzer, P., Hayes, M.L., Garrison, V., and Barber, R.T., African dust and the demise of Caribbean reef corals, *Geophysical Research Letters*, 27, 3029–3032, 2000.

Smith, G.W., Ives, L.D., Nagelkerken, I., and Ritchie, K.B., Caribbean sea-fan mortalities, *Nature*, 383, 487, 1996.

Terminology Sub-Committee, A guide to the use of terms in plant pathology. *Phytopathological Papers*, 17, 1–53, 1973.

Tomascik, T., Settlement patterns of Caribbean scleractinian corals on artificial substrata along a eutrophication gradient, Barbados, West Indies, *Marine Ecology Progress Series*, 77, 261–269, 1991.

Tomascik, T. and Sanders, F., Effects of eutrophication on reef-building corals: I. Growth rate of the reef building coral *Montastraea annularis*, *Marine Biology*, 87, 143–155, 1985.

Tomascik, T. and Sanders, F., Effects of eutrophication on reef-building corals: II. Structure of scleractinian coral communities on fringing reefs, Barbados, West Indies, *Marine Biology*, 94, 53–75, 1987a.

Tomascik, T. and Sanders, F., Effects of eutrophication on reef-building corals: III. Reproduction of the reef-building coral *Porites porites*, *Marine Biology*, 94, 77–94, 1987b.

Walker, D.I. and Ormond, R.F.G., Coral death from sewage and phosphate pollution at Aqba, Red Sea, *Marine Pollution Bulletin*, 13, 21–25, 1982.

Wheaton, J.L. and Jaap, W.C., *Corals and Other Prominent Benthic Cnidaria of Looe Key, National Marine Sanctuary, Florida*, Florida Department of Natural Resources, St. Petersburg, 1988, 25.

Williams, E.H. and Bunkley-Williams, L.B., The world-wide coral reef bleaching cycle and related sources of coral mortality. *Atoll Research Bulletin*, 335, 1–63, 1990.

Yoshioka, P.M. and Yoskioka, B.B., Effects of wave energy, topographic relief and sediment transport on the distribution of shallow-water gorgonians of Puerto Rico, *Coral Reefs*, 8, 145–152, 1989.

# Section IV

## Policy, Management, and Conservation

*Three Snorkelers*
by
Don Kincaid

# 31  Water Quality Concerns in the Florida Keys: Sources, Effects, and Solutions

*William L. Kruczynski and Fred McManus*
U.S. Environmental Protection Agency

## CONTENTS

## INTRODUCTION

The Florida Keys are a chain of tropical islands surrounded by clear ocean waters teeming with sea life. The uniqueness and diversity of natural communities combine to make the Florida Keys ecosystem one of the "crown jewels" of our nation's natural treasure chest.

The Keys ecosystem is composed of several interdependent community types, including tropical hardwood forests, fringing mangrove wetlands, seagrass meadows, hard and soft bottoms, and coral reefs. This ecological diversity has made the Keys a popular place to live and an important vacation destination.

The current population of the Keys is approximately 78,000 permanent, year-round residents (1990 census). The population increases by about 25,000 during peak tourist season (winter months). Approximately 70% of Keys residents regularly participate in water-based activities, such as fishing (48%), snorkeling (45%), beach activities (38%), and observing wildlife and nature (36%) (Leeworthy and Wiley, 1997). Maintenance of the integrity and ecological health of marine and terrestrial environments is critical to the economy of the Keys. Approximately 3 million visitor trips annually are made to the Keys, totaling over 16 million person days. Visitors generate over $1.3 billion in direct output and tourism supports over 21,800 jobs in the Keys (English et al., 1996). Tourists come to the Keys for a variety of reasons: snorkeling (28%), scuba diving (8%), fishing (21%), wildlife observation (28%), beach activities (34%), and sightseeing (55%) (Leeworthy and Wiley, 1997).

Shallow water environments surrounding the Keys constitute extensive nursery areas and fishing grounds for a variety of commercially and recreationally important marine species. Monroe County ranks first in Florida in total volume of seafood landed (10% of state landings). In 1990, 19.7 million pounds of fin fish, shellfish, and other aquatic organisms were landed in Monroe County with a dockside value of $48.4 million (Adams, 1992). The spiny lobster is the most valuable harvest (>$20 million annually). Monroe County accounts for 91% of the total spiny lobster harvest and 44% of total harvest of pink shrimp and stone crab (Adams, 1992).

The natural communities that make up the Florida Keys ecosystem exist in a dynamic equilibrium. Changes to the physical–chemical conditions that result in a direct impact to one community type can have profound effects on adjacent community types. For example, coastal fringing wetlands filter upland runoff, stabilize sediments, and absorb some nutrients. Thus, wetlands help maintain clear, relatively nutrient-poor waters that facilitate luxuriant growth of seagrasses in adjacent waters. Upsetting this balance by removing wetland vegetation can result in a localized increase in nutrients and turbidity in nearshore waters that may reduce seagrass coverage. Coastal wetlands and seagrasses are important habitats for juvenile fishes, and a reduction in spatial coverage of these habitats can result in decreased fish populations that can further upset ecosystem functions. Loss of wetlands and seagrasses can increase water turbidity due to resuspension of sediments previously bound by their root systems which can have additional negative impacts on adjacent communities. Thus, subtle, single changes can have profound, cascading effects throughout the entire ecosystem.

Human activities have negatively impacted the ecological balance of the Florida Keys ecosystem (Voss, 1988). Cumulative, large-scale physical impacts, such as construction of barriers to tidal flushing, dredging and filling of seagrass beds and wetlands, and nutrient addition to waters surrounding the Keys have profoundly influenced the physical appearance of the Keys, as well as the balance of ecosystem functions. The impacts of many human activities are obvious, such as the approximately 30,000 acres of seagrasses that have been propeller scarred by boaters in the Keys (Sargent et al., 1995). Other impacts, such as water quality degradation, may not be immediately obvious to the casual observer. However, nutrient loading is a widespread factor that alters structure and function of aquatic ecosystems in coastal watersheds (Valiela et al., 1992).

The survival of the existing Florida Keys marine ecosystem is dependent upon clear, low-nutrient waters. This chapter is a summary of available information of nearshore water quality (canals, basins, and waters immediately adjacent to the Keys). The data demonstrate that the cumulative effects of continued discharges of nutrient-rich wastewater and stormwater into confined and some other adjacent nearshore waters have degraded the water quality of those waters (Barada and Partington, 1972; USEPA, 1975; Florida Department of Environmental Regulation, 1985, 1987, 1990; Lapointe et al., 1990; Lapointe and Clark, 1992). There is evidence that the degraded water quality has adversely impacted other nearshore communities (Lapointe and Clark, 1992; Lapointe et al., 1994; Lapointe and Matzie, 1996). If sources of nutrient enrichment continue unabated, it is likely that the ecological balance of nearshore communities of the Keys will be changed. Changes in nearshore community structure and function could result in stresses to other components of the Keys ecosystem. Because the tourist-based economy of the Keys is directly linked to a healthy Keys ecosystem, it is prudent to work diligently toward eliminating sources of excessive nutrients to this ecosystem.

Restoration of degraded portions of the Keys aquatic ecosystem may be possible, but it will require the combined effort of the entire community of the Florida Keys, with help from federal and state governments. Collectively, we are the stewards of this unique national treasure, and restoring and maintaining this ecosystem is a national goal. In recognition of the warning signals of degraded water quality, the U.S. Environmental Protection Agency (EPA) and the State of Florida, in conjunction with the National Oceanic and Atmospheric Administration (NOAA), have, at the direction of Congress, prepared a Water Quality Protection Program (WQPP) for the Florida Keys National Marine Sanctuary. It is hoped that full implementation of the WQPP will reverse the trend of environmental degradation and restore and maintain the Florida Keys marine ecosystem.

## HISTORY AND PHYSICAL SETTING

The Florida Keys are a chain of limestone islands that extend from the southern tip of the Florida mainland southwest to the Dry Tortugas, a distance of approximately 220 miles. The Keys are island remnants of ancient coral reefs (Upper Keys) and sand bars (Lower Keys) that flourished during a period of higher sea levels about 125,000 years ago (Pleistocene) (Hoffmeister and Multer, 1968; Shinn, 1988; Lidz and Shinn, 1991). During the last ice age, which started about 100,000 years ago, sea level dropped and exposed the ancient coral reefs and sand bars that form the present Keys. At that time of lower sea level, the Florida land mass was much larger than it is today and Florida Bay was forested. The sea level began to rise as polar ice caps started melting about 15,000 years ago, which resulted in flooding of some of the exposed land and led to our present-day geography. The existing outer coral reef tract that parallels the Florida Keys on the Atlantic Ocean side began forming between 6000 and 10,000 years ago. Reef growth rate ranges from 0.61 to 4.85 m (2 to 16 ft) per 1000 years (Shinn et al., 1977).

A continued rise of sea level resulted in flooding what we now call Florida Bay about 4000 years ago. At that time, coral communities thrived along the entire seaward edge of the Keys. As sea level rose further, it resulted in the establishment of tidal passes between the Keys. This was a significant event, as it resulted in the export of terrestrial material, sediments, and organic matter from Florida Bay to the Atlantic through the tidal passes. The export of that material resulted in conditions that no longer favored lush coral reef development in the regions of the major tidal passes (Middle Keys) (Ginsburg and Shinn, 1964; Shinn et al., 1989, 1994a; Lidz and Shinn, 1991; Ogden et al., 1994).

Florida Bay is a shallow embayment composed of basins separated by mudbanks and mangrove islands. Water quality in Florida Bay is highly variable. Discharges of either hot or cold water, with very high or low salinity, from Florida Bay through the tidal passes further limited development of the outer coral reefs. To the north and west of the Middle Keys, where the Reef Tract is more sheltered by the Keys from waters discharged from Florida Bay, vigorous coral reef growth continued (Lidz and Shinn, 1991; Shinn et al., 1989). Thus, prior to human impacts in South Florida, water exchange between Florida Bay and the Atlantic Ocean significantly impeded coral growth in the areas of major tidal passes as well as offshore.

Today, the Florida Keys outer reefs are a disjunct series of bank reefs located at the northern zoogeographic boundary of tropical waters. Because it is at the northern limit of coral reef development, the Keys Reef Tract regularly experiences natural stresses, such as winter temperatures below those normally associated with vigorous coral reef development. Also, the reef experiences higher summer temperature extremes than many other reefs in the Caribbean basin (Vaughn, 1918).

The Keys themselves consist of limestone rock formations. In the Upper Keys, these rock formations are composed of the Key Largo Limestone, which is the skeletal remains of the ancient Pleistocene reef. The Lower Keys, Big Pine Key and west, were formed by deposition and consolidation of sand bars (Miami oolite) over the underlying limestone. Over time, vegetation began growing on the exposed surfaces of the limestone and thin veneers of soils formed in some areas from weathering of the limestone and accumulation of organic matter from plants.

The Keys were vegetated from seeds, propagules, and uprooted or detached plant material carried from the Florida mainland and from Caribbean islands. This has resulted in a curious mix of tropical and subtropical vegetation in this unique geographic setting. Prior to the arrival of Europeans, the Keys consisted of diverse West Indian tropical hardwood forests on high ground, pine rocklands and freshwater wetlands on the interiors of larger islands (e.g., Big Pine Key), and vast expanses of mangrove wetlands that surrounded the islands and extended into tidal waters.

The waters surrounding the Keys were clear and supported an abundance and diversity of plant and animal life. Shallow areas were vegetated by acres of lush seagrasses in areas where sediments accumulated. Hard and soft corals thrived where limestone was exposed under the water. Large populations of queen conchs, sea turtles, and many species of sealife were supported by the productivity of this diverse, shallow water ecosystem. The shallow water and coral reef communities evolved in a low-nutrient subtropical sea environment, and the continued existence of this ecosystem is dependent upon maintenance of relatively low sediment and nutrient conditions.

## CHANGING TIMES

Although known to exist, the Florida Keys were largely uninhabited during the sixteenth, seventeenth, and eighteenth centuries, even though waters just offshore provided a major shipping route to and from Europe. During that time, the islands were occupied by Keys Indians, some settlers, and pirates who preyed on sea traffic. Scarcity of freshwater and the lack of a vast expanse of fertile soil prevented the populous settlement of the Keys. Undoubtedly, mosquitoes and diseases also played major roles in the lack of development in the Keys.

After Florida was ceded by Spain to the U.S. in 1821, Key West became an important military post, and island trade began to grow. Trading, fishing, cigar making, recovering goods from shipwrecks, and a limited agriculture base provided livelihoods for Keys residents (Viele, 1996). The Overseas Railway and Overseas Highway, completed in 1912 and 1938, respectively, connected the Keys to the mainland through a series of filled causeways and bridges. This transportation system, together with a water pipeline from the mainland built to supply the military in Key West during World War II, set the stage for post-war development of the Keys (Halley et al., 1997). The attractive climate, inexpensive land, beauty of the coral reefs, clear waters with abundant fishes, diversity of wildlife, and mosquito control all combined to make the Keys a very popular place to live and vacation.

Much of the physical alteration of the Keys to support the growing human population occurred from the 1950s through the 1970s. During that period, many acres of tropical hardwood hammocks were cleared to provide land for housing and commercial development. The attractiveness of waterfront development prompted the creation of "fastland" through dredging and filling of mangrove forests and seagrass beds to construct networks of finger-fill residential canals. More than 200 canals and access channels were dredged during that period (Florida Department of Environmental Regulation, 1987). Turbidity from the dredging and filling operations smothered adjacent areas of hard-bottom and seagrass habitats. Many canals were dug 10- to 20 ft deep to maximize the production of fill material excavated from the canal, and most canal systems were designed as long, dead-end networks with little or no tidal flushing at their upper ends. In general, water quality of newly dug canals was the same as areas of adjacent nearshore waters due to lack of input of nutrients from runoff and development.

## WATER QUALITY

The concept of what constitutes "good" water quality is complex. The definition of acceptable water quality is based upon several interrelated parameters, including how the water will be used (e.g., drinking, swimming, fishing), concentrations of materials in the water above natural background levels that could have a deleterious effect on plants or animals (pollution), and the presence of compounds not usually found in the water (contamination). Parameters typically measured during routine water quality studies are salinity, dissolved oxygen, turbidity, biochemical oxygen demand ($BOD_5$), chlorophyll, fecal coliform, and nutrient concentrations, predominantly nitrogen and phosphorus (Table 31.1). Contaminants include heavy metals, pesticides, herbicides, and other chemicals.

**TABLE 31.1**
**Parameters Measured in Water Quality Monitoring and Examples of Analyses**

| Water Quality Parameter | Examples of Methods of Analyses |
| --- | --- |
| Physical-Chemical Parameters | |
| Temperature | Thermistor or mercury thermometer |
| Conductivity/salinity | Electrometric |
| Dissolved oxygen | Winkler titration or polarographic sensor |
| pH | Electrometric |
| Light attenuation | PAR attenuation |
| Turbidity | Secchi disk or nephelometric |
| Depth | Measured line or pressure transducer |
| Nutrients | |
| Ammonia | Indophenol |
| Nitrate and nitrite | Cd reduction and diazo |
| Nitrite | Diazo |
| Total nitrogen | Combustion nitrous oxide chemoluminescence |
| Soluble reactive phosphorus | Molybdate |
| Total phosphorus | Digestion and molybdate |
| Organic carbon | Combustion and infrared detection |
| Biological Parameters | |
| Chlorophyll *a* | Fluorometric |
| Alkaline phosphatase activity | Fluorometric |
| Fecal coliform bacteria | Incubation and plate count |
| Biochemical oxygen demand | Incubation and oxygen analysis |

Water quality standards are acceptable limits for materials found in water and are defined in regulations. State of Florida water quality criteria are contained in Chapter 62-302 Florida Administrative Code. Rule 62-302.530 includes standards for Class III marine waters. Water quality standards for drinking water include acceptable levels (i.e., numeric limits of odor, taste, color, pollutants, and contaminants). These standards are aimed at reducing or eliminating compounds that are displeasing or potentially hazardous to people who drink the water.

Defining environmental water quality standards is more complex than drinking water standards and must be evaluated in an ecological and aesthetic context. Water quality standards are based on conditions that may result in a change in the quantity or health of the organisms that live in the water. However, because even pristine natural ecosystems undergo changes in response to natural variations and all ecosystems gradually change over time, it can be difficult to determine the exact point when changes in water quality parameters begin to cause degradation of the ecosystem.

The waters surrounding the Keys have been declared as "Outstanding Florida Waters" (OFW) by the State of Florida (Florida Department of Environmental Regulation, 1985). By regulation, input of materials that could be considered pollutants to open surface waters cannot exceed the concentration of those materials that naturally occur in water. However, ambient background conditions can change seasonally or at different phases of a tidal cycle. From a scientific standpoint, the declaration of OFW status for the waters of the Florida Keys does not solve the problem of defining acceptable limits of pollution. The range of water quality parameters measured throughout the Keys during a survey to support designation of the Florida Keys as OFW is given in Table 31.2 (Florida Department of Environmental Regulation, 1985). Because of the OFW designation, direct surface water discharges of pollutants have been eliminated or are being phased out.

**TABLE 31.2**
**Ranges of Water Quality Parameters Measured During a Survey to Support Designation of Waters Surrounding the Florida Keys as Outstanding Florida Waters**

| Water Quality Parameter | Ambient Stations (mg/l, except pH) | Canals (mg/l except pH) |
|---|---|---|
| Dissolved oxygen | 6.0–9.4 | 0.0–9.6 |
| pH | 7.0–8.4 | 7.3–8.3 |
| Total phosphorus | 0.001–0.054 | 0.005–0.083 |
| Total Kjeldahl nitrogen | 0.128–0.693 | 0.196–1.150 |
| Ammonia nitrogen | 0.051–0.160 | 0.057–0.239 |
| Organic nitrogen | 0.019–0.580 | 0.066–0.850 |
| Nitrate and nitrite | 0.000–0.027 | 0.002–0.054 |

*Source:* From Florida Department of Environmental Regulation (1985).

In order to establish pollutant standards, the effects of the pollutants on biological communities must be determined. Pollutant (or contaminant) levels become unacceptable when they result in detrimental changes to an organism or the biological community. This concept is easy to understand when the pollutant or contaminant results in loss or replacement of a community or a species; no one can argue against the fact that concentrations that cause death are unacceptable to the community or species that died! Measurements must be sufficiently sensitive to detect the subtle nonlethal changes that can slowly result in shifts in species dominance and community structure. These changes are signs that pollutants have reached concentrations that are resulting in unacceptable changes to the natural ecosystem. This threshold is called a non-numeric or "narrative" water quality standard.

In the Keys, two main problems are associated with wastewater pollution: fecal contamination (health risk) and nutrient enrichment (eutrophication). One important water quality standard concerns the presence of fecal coliform bacteria in the water. Birds and mammals excrete fecal coliform bacteria in fecal matter. The presence of fecal coliform bacteria in the water column is used as a measure of possible wastewater contamination of the water. Although fecal coliform bacteria are not a major health risk, they are easy to measure and can indicate the presence of other enteric (intestinal), disease-producing microbes. Presence of fecal coliform bacteria above the state standard of 800 colonies/100 ml of water (monthly average) is indicative of contamination by untreated sewage and is a public health concern. This bacteriological standard was developed for freshwater. Fecal coliform bacteria normally die when exposed to marine waters. However, fecal coliforms sometimes are present in tropical environments in the absence of any source of fecal contamination (Hazen, 1988). Therefore, it is questionable whether the existing standard is meaningful for marine systems (Dutka et al., 1974; Goodfellow et al., 1977; Loh et al., 1979). Normally, when fecal coliform bacteria are present in marine systems, it is an indicator of very recent fecal contamination. Low concentrations of fecal coliform bacteria should not necessarily be equated to low abundance of bacterial or viral pathogens in marine waters.

Coprostanol is a chemical that is produced during the digestion process and is a product of cholesterol decomposition. It is a better indicator of discharge of untreated sewage because unlike fecal coliform bacteria which are relatively short-lived in the marine environment, coprostanol accumulates in sediments and provides a long-term record of sewage pollution. However, measurement of coprostanol is impractical for routine monitoring because it requires sophisticated, expensive analysis and, at the present time, there is no regulatory standard for coprostanol.

## EUTROPHICATION

Nutrients, such as carbon, nitrogen, and phosphorus, are essential for the normal healthy functioning of all living cells. They are used in biosynthesis in all living matter. These nutrients and others, such as potassium and magnesium, which are present in very small amounts, are recycled in the ecosystem. When organisms excrete waste products or die, the nutrients present in the wastes or carcasses are made available through the decomposition process. The growth of plants is generally limited by the lack of one or more of these nutrients. New plant growth is dependent upon this recycling of bound nutrients. In this manner, an ecosystem maintains a "balance."

Ecosystems can utilize a certain amount of "new" nutrients. New nutrients may come from other adjacent (upstream) natural systems or may be introduced by human activities. Domestic wastewater is one major source of new nutrients to the aquatic environment. If nutrients are released into the environment in excessive amounts (eutrophication), they become pollutants because they disrupt the natural nutrient balance and result in unacceptable changes of community structure. Dramatic changes in community structure can result in a catastrophic collapse of an ecosystem.

Eutrophication often progresses through a sequence of stages. A typical progression involves: (1) enhanced primary productivity; (2) changes in plant species composition; (3) very dense phytoplankton blooms, often toxic; (4) anoxic conditions; (5) adverse effects on fish and invertebrates; and (6) changes in structure of benthic communities (GESMAP, 1990).

There are many documented examples of the collapse of an ecosystem due to nutrient enrichment. For the sake of simplicity, consider a simple pond ecosystem that is rarely visited by people. The pond ecosystem is in balance because the aquatic vegetation (grassbeds) that grows on the bottom of the pond supports a population of shrimp, which in turn supports a population of fish. As shrimp and fish grow and defecate, and eventually die, they return nutrients to the water which are taken up by the grasses and support grassbed growth. If you (or a raccoon) defecate or urinate into the pond, the nutrients that are added may result in increased grass growth which may cover more area of the pond bottom. Increased grassbeds will result in increased numbers of shrimp and bigger and more abundant fish. Thus, the pond ecosystem can assimilate some additional new nutrients without a significant negative change in structure or function.

However, if the area surrounding the pond becomes a popular campground, and all the campers dump their wastewater directly into the pond, the structure of the pond ecosystem will change drastically. Microscopic algae which were always present in the pond, but were held in check by low amounts of available nutrients, will grow, divide, and result in an algal bloom that will change the water color from clear to green. The green water (high chlorophyll) will absorb most of the sunlight that strikes the pond and will result in the death of the aquatic grasses living on the bottom of the pond. Death of the benthic grasses will result in death of the shrimp that are dependent upon them. The fish that eat shrimp will also starve, as they are not physically able to eat algae. Death of the benthic grasses, shrimp, and fish will result in the release of more nutrients into the water that will further fuel the algal bloom. The small number of fish in the pond that can eat algae can now explode in population size because of a seemingly unlimited amount of algae. Ultimately, the blooms of algae and fish will cause the collapse of the ecosystem when they respire at night, utilize all the dissolved oxygen, and die. This hypothetical, catastrophic collapse of a pond ecosystem is exactly the scenario that resulted in the ecological collapse of Lake Erie, portions of Tampa Bay, and many other bodies of water that received unacceptably high levels of nutrients. Addition of high levels of nutrients result in major changes in ecosystem structure and function and can lead to the eventual collapse of the ecosystem.

Generally, it is the total amount of nutrients, including micronutrients, entering a water body that can result in overloading of the system, not necessarily their concentration. It matters little whether nutrient addition comes from a single or a few concentrated sources of nutrients discharging into a water body or from many sources discharging lower concentrations of nutrients. The effect of the total loading to the receiving water body will be the same. When the system can no longer absorb increased levels of new nutrients without significantly changing ecosystem structure and function, the threshold of nutrient assimilative capacity of the system is reached.

A principal objective of wastewater treatment processes is to remove nutrients and other pollutants and dispose of them in a manner that does not cause unacceptable changes to the environment. Indeed, re-use of wastewater in suitable areas may cause desirable changes in the productivity of a cultivated field or forest (e.g., land application of wastewater).

Tropical marine hard-bottom and seagrass communities have evolved and thrive in relatively low nutrient (oligotrophic) conditions. Species in these communities efficiently take up nutrients and out-compete other, less-adapted species in low-nutrient environments. They cannot successfully compete with organisms that have evolved to take advantage of elevated nutrient loads. Therefore, nutrients added to oligotrophic systems are very quickly taken up by opportunistic species. Because of rapid uptake, nutrient concentrations in the water can be quite low and may not be detectable using traditional water quality sampling methods. Changes in the structure of the biological community (species abundance and composition) are important signs of nutrient enrichment in oligotrophic systems.

Nutrients are found in the foods, drinks, fertilizers, drinking water, and the like that are imported into the Keys every day. If these new nutrients get into the surface waters, they become available for use by the marine ecosystem. Small additions of nutrients may cause inconsequential changes, but if continued or increased over time, they can cause drastic shifts in the numbers and kinds of plants and animals. The change to ecosystems due to excess nutrients is called *eutrophication*, which means "too much food."

## SOURCES OF WATER QUALITY CONCERNS

### STORMWATER RUNOFF

Pollutants can be conveyed into surface waters when stormwater accumulates on land surfaces and runs off. Stormwater is considered a major source of pollutants to surface waters nationally. Runoff typically contains substances such as organic debris, silt, nitrogen, phosphorus, metals, and oils. The amount or load of pollutants is largely a function of rainfall quantity, imperviousness (i.e., the degree to which rainwater cannot soak into soil), and land use. In residential areas, for example, nutrients are a major part of the load. Pollutants from roadways include oils and metals. Soil characteristics can also play a major role in the types and quantities of pollutants that are retained on land.

In Florida, the Water Management Districts and local governments now impose a minimum level of stormwater treatment for all new developments. The criteria are intended to protect surface waters according to their use classification. Much of the development in the Florida Keys occurred prior to the existence of these criteria. Similar to other parts of the state at the time, stormwater was considered a nuisance as it resulted in flooding. Therefore, if stormwater systems were employed at all, they were typically designed to efficiently convey water off land surfaces as quickly as possible. These old systems are considered to be the most likely to cause water pollution; therefore, policies now in place seek to retrofit them whenever possible.

In most areas of the Keys, there was no stormwater management. Uncontrolled runoff can cause pollution of surface waters. Stormwater runoff from roadways, bridges, driveways and yards, roof tops, and shopping center parking lots contribute stormwater loading to surface waters. The amount of pollutant load caused by stormwater runoff can be estimated mathematically from the factors given above. Estimates of total loadings of nitrogen and phosphorus from wastewater and stormwater were summarized in the Phase II Report of the WQPP (USEPA, 1993). Assumptions used to generate those figures were recently reevaluated and the numbers have been revised (Table 31.3). These recent estimates attribute about 20% of the nearshore nitrogen load and about 45% of the phosphorus to stormwater (Table 31.3). These estimates, however, can vary widely depending on the magnitude of each factor. No estimate should be considered absolute, but should be viewed only in relationship to its potential impact.

**TABLE 31.3**
**Estimated Nutrient Loading (pounds per day) from Wastewater and Stormwater Sources in the Florida Keys**

| Source | Nitrogen | | Phosphorus | |
|---|---|---|---|---|
| | (lb/day) | (%) | (lb/day) | (%) |
| *Wastewater* | | | | |
| OSDS | 932 | 30.9 | 226 | 23.0 |
| Cesspits | 283 | 9.4 | 100 | 10.2 |
| Package plants | 758 | 25.3 | 152 | 15.5 |
| Central plants | 320 | 10.6 | 36 | 3.7 |
| Live-aboards | 84 | 2.8 | 30 | 3.0 |
| Subtotal | 2377 | 78.9 | 544 | 55.4 |
| *Stormwater* | | | | |
| Developed areas | 401 | 13.3 | 364 | 37.0 |
| Undeveloped areas | 234 | 7.8 | 75 | 7.6 |
| Subtotal | 635 | 21.1 | 439 | 44.6 |
| Total | 3012 | 100 | 983 | 100 |

*Source:* Modified from USEPA (1993).

# WASTEWATER

As is true for all animal life, humans derive nutrients and energy from the food we eat. We are not totally efficient in removing nutrients from our food, so human waste contains nutrients, such as carbon, nitrogen, and phosphorus. Typical residential wastewater flow is approximately 45 gal per person per day. Of that, approximately 35% (16 gal) is from the toilet (black water) and 65% (29 gal) is from sinks, bathtubs, and appliances (gray water) (Harkins, 1996). Nutrient concentrations of pollutants in black water and gray water are summarized in Table 31.4. Wastewater can enter canals and other nearshore waters from cesspits (4000 estimated), septic tanks (approximately 20,000), injection wells (750), ocean outfalls (1), and live-aboard vessels.

Based upon current best estimates (Table 31.3), approximately 80% of nitrogen loadings comes from wastewater. Onsite disposal systems (septic tanks and aerobic treatment systems) and cesspits account for 40.3% of nitrogen loadings. Approximately 55% of phosphorus loadings are from wastewater. Onsite disposal systems and cesspits account for 33.2% of total phosphorus loadings.

Disposal of wastewater from live-aboard vessels is a significant localized problem because of the low level of treatment, the tendency for live-aboard vessels to congregate in certain marinas or anchorages, and potential adverse health effects of discharging untreated wastewater. Many live-aboard vessels are permanently anchored, and mobile pumpout facilities are required to service those vessels, but there are few mobile pumpout facilities in the Keys. Overall, live-aboard vessels account for approximately 2.7% of total nitrogen and 2.9% of total phosphorus loading to the region's surface waters.

**TABLE 31.4**
**Nutrient Loadings from Residential Wastewater[a]**

| | Total Nutrient Loading by Fixture (gm/person/day) | | | |
|---|---|---|---|---|
| Nutrient | Toilets | Sinks/Showers | Total without Garbage Disposal | Total with Garbage Disposal |
| Carbon | 18.0 | 30.0 | 48.0 | 59.0 |
| Nitrogen | 6.5 | 1.5 | 8.0 | 9.0 |
| Phosphorus | 1.2 | 2.8 | 4.0 | 4.0 |
| | Concentration of Nutrients in Wastewater (mg/l) | | | |
| Nutrient | Without Garbage Disposal | | With Garbage Disposal | |
| Carbon | 280 | | 350 | |
| Nitrogen | 47 | | 53 | |
| Phosphorus | 24 | | 24 | |

[a] Typical residential wastewater flow is 45 gal/capita/day.

*Source:* From Harkins, J., *Nutrients in Wastewater and Nutrient Removal*, U.S. Environmental Agency, Region 4, Atlanta, GA, Washington, D.C., 1996.

## OTHER SOURCES

Nutrients come from a variety of other sources. Loadings to the waters of the Keys from most other sources, such as Florida Bay, Gulf of Mexico, oceanic upwelling, and atmospheric deposition, have not been quantified. Nutrient inputs from those sources external to the Keys may be greater than anthropogenic loadings from wastewater or stormwater emanating from the Keys. However, that does not diminish the importance of focusing on anthropogenic nutrient loadings and their effects on water quality and biological resources. Because maintenance of healthy, natural communities of the Keys is dependent on low-nutrient environments, localized sources of nutrients can have immediate negative impacts that can result in cascading effects throughout the ecosystem. Nutrient loadings from atmospheric sources are diffuse and evenly distributed over the Florida Keys. Wastewater nutrient loadings emanate from the land–water boundary and may cause concentration increases in canals and confined nearshore waters well above those caused from atmospheric or other sources. Similarly, upwelling of deep ocean waters can provide nutrients, particularly to the outer reef tract and areas seaward of the Keys. Although the concentration of nutrients in upwelled oceanic waters is low, the total loading to the reef system can be significant because of the high volume of water. External advective nutrient inputs are more diffuse than land-based, human-induced sources. Very little data are available on the physical processes driving advective and atmospheric loadings and their effects on water quality of the Florida Keys. This is a topic that requires further research.

Florida Bay has represented a source of nutrient-rich and turbid waters to the Florida Keys for approximately the last 4000 years. The discharge of Florida Bay waters through the major tidal passes between the Keys has arrested development of the outer reefs near those locations. In 1987, a significant decline of seagrasses began in Florida Bay. Although the cause of that dieoff is still debated, it was probably related to manipulation of historic delivery of freshwater to the Everglades. Several very dry years immediately preceded the initiation of the dieoff, and salinities in some parts of Florida Bay were approximately twice seawater strength (70 ppt). The dead seagrasses decomposed and their stored nutrients became available for phytoplankton algae. Also, sediments that the seagrasses bound with roots and rhizomes became waterborne with wind events and resulted in highly turbid water. Because corals thrive in clear, low-nutrient waters, the discharge of turbid, nutrient-rich Florida Bay water is probably having a detrimental effect on coral reef communities seaward of the tidal passes. Cook (1997) demonstrated that effect by measuring growth of coral transplants in the discharge from Florida Bay. Corals exposed to Florida Bay water grew slower and were less dense than corals transplanted to a reference site (Tennessee Reef). Corals within the influence of Florida Bay water also had a significantly higher concentration of symbiotic algae in their tissues, presumably in response to the more turbid conditions. Brand (1997) and others have tracked Florida Bay water out to the reef tract using chlorophyll concentrations or satellite imagery reflectance as a fingerprint of the water mass.

Several studies have analyzed sediments, primary producers, and/or consumers for trace metals and pesticides (Glynn et al., 1989; Manker, 1975; Skinner and Jaap, 1986; Strom et al., 1992). In general, the results are consistent with a relatively clean environment with some localized anthropogenic effects. For example, Strom et al. (1992) found relatively high cadmium at stations near the Seven-Mile Bridge and Newfound Harbor Key. Highest metal concentrations were found in consumers (sponges) which is indicative of bioaccumulation.

Marinas have the potential for polluting water or sediments from boat scraping and painting operations, fueling, and engine repair. Data are not available to quantify loadings of pollutants from marina operations.

Pesticides are a potential threat to marine life. Chemicals used in mosquito control are known to be toxic to aquatic crustaceans, such as lobsters, shrimp, and crabs. Pesticide levels in samples from the Keys have been historically low (Strom et al., 1992). Although the amounts of pesticides currently used by the Monroe County Mosquito Control Program are known, no information is available on the amount of pesticides that reach marine waters. Also, nothing is known about the environmental concentrations or effects of residual pesticides in marine waters. That is an area of research that has yet to be examined.

Other "natural" sources of pollutants include animal wastes, runoff from natural environments, and weed wrack. Although bird droppings can be a significant source of nutrients locally (for example, around breeding or roosting islands), they represent a redistribution and recycling of nutrients currently in the system and are generally not considered pollutants.

Weed wrack consists of detached blades of benthic seagrasses and algae that became wind-driven into large floating mats. These mats can become trapped along shorelines and in canal systems along the windward side of the Keys. Decomposition of the weed wrack removes oxygen from the water, releases nutrients, and forms toxic hydrogen sulfide gas. With wind shifts, weeds trapped along shorelines move offshore; however, weed wracks trapped in canal systems result in the buildup of organic debris. Decomposition of organic matter quickly strips all oxygen from stagnant canal waters. Mobile life forms (e.g., fish) may be able to leave the canal before succumbing to low oxygen concentrations. Other relatively non-mobile life forms that require oxygen (e.g., corals, benthic worms, and mollusks) cannot survive.

Many nearshore waters are very shallow with bottoms consisting of fine sediments. Fine sediments can be resuspended in the water column by disturbances, such as boat traffic. High-use areas are experiencing chronic turbidity generated by the growing number of recreational and commercial vessels that transit those waters. Turbid waters could detrimentally affect seagrass (shading) and adjacent hard-bottom communities (smothering). Research and monitoring are needed to quantify the effects of chronic turbidity on biological communities.

## CANALS AND OTHER CONFINED WATERS

Much variability exists in the design and physical characteristics of canal systems in the Keys. Differences in length, depth, slope, geometry, and underlying geology of canal systems, as well as the population density, affect the impacts of nutrient loading, flushing rates, and the water quality in the canals. The following summary of information on water quality findings in canals is based on studies for particular canal systems; however, many generalities about canal systems can be gleaned from this information.

Much of the pre-1970 information on canal systems in Florida was summarized by Barada and Partington (1972), who reviewed the literature and conducted a survey of environmental officials. Based on water quality data and the personal experience of the individuals surveyed, Barada and Partington concluded that excavating artificial canals causes serious environmental degradation within the canals themselves and in waters adjacent to canals. Deep, narrow, box-cut canals with dead-end configurations gradually accumulate oxygen-demanding and toxic sediments and organic wastes, causing low dissolved oxygen, objectionable odors (hydrogen sulfide gas), floating sludge, fish kills, and anaerobic and putrid conditions. Eutrophication of canals with poor circulation is accelerated by a heavy pollution load that is related to population density and shoreline length. Sources of pollution into the canals investigated include stormwater runoff, septic tanks, sewage effluent, and live-aboard houseboats.

Citing Smith, Milo, and Associates (1970), Barada and Partington concluded that no soils in Monroe County are suitable for septic tanks. The combination of high water table and extremely porous soils "nullifies the filtering capacity and virtually raw sewage is leached into the waterways." The report also recommended against the discharge of effluent from package plants into canals. Package plants with secondary treatment remove most of the organic material and bacteria but do not effectively remove dissolved contaminants, such as phosphates, nitrates, and other chemicals that contribute significantly to the degradation of water quality.

Chesher (1973) performed an environmental study of canals and quarries in the Lower Keys and concluded that the flow-through canal system at Summerland Key Cove had excellent water quality. Construction of that canal system was begun in 1957 and completed in 1971. At the time of Chesher's study, 69 houses were constructed on the 614-lot subdivision. The total population of the subdivision was 207, which included winter-only residents. All houses utilized septic tanks. Chesher generally found low levels of nutrients in the canals, relatively high oxygen, and no evidence of stratification. Mean nitrate concentration was about 0.03 mg/l (ppm) and mean phosphate was 0.06 mg/l. Fecal coliform bacteria ranged from 0 to 37 colonies/100 ml. The canal system configuration and orientation prevented any algae or seagrass from accumulating in the canal. Chesher also observed a diverse and numerous biotic community living in the canal system, including seagrasses, fish, lobsters, and many other species.

Chesher's results are atypical of other canal studies for a number of reasons. The Summerland Key Cove canal system was only 11% developed at the time of sampling. Also, nutrients were measured with a HACH kit, which is not as sensitive as standard analytical methods. There is no indication that oxygen measurements were made in early morning when daily minimums are expected.

It would be interesting to revisit the Summerland Key Cove canal system today. Chesher's findings of lush marine life is typical of newly dug canals. Barada and Partington (1972) reported that it is a common fallacy that finger canals provide a haven in which fish thrive. That condition may occur in the very early stages after canal excavation. A typical pattern is that in the first few months of spring, bottom animals and fish are abundant in newly dug canals. However, with the advent of summer and hot weather, dissolved oxygen in deeper waters of the canals drops to zero, or nearly so. Mortality of benthic organisms is heavy, and fish are absent. When cooler weather returns, benthic animals and fish may recolonize. But, as dead and decaying organic materials gradually build up in the canal bottom, the number and diversity of marine creatures declines and eventually there is virtually no desirable biological production in the canal. Taylor and Saloman (1968) found very little benthic life and half as many species of fish in a 10-year old, box-cut canal near St. Petersburg as in surrounding areas. They concluded that the accumulation of organic material and low dissolved oxygen in canals has a permanent adverse affect on fish and other marine life.

In 1972, during the peak of finger-fill canal construction in the Keys, the Florida Department of Pollution Control issued a dredge-and-fill moratorium halting all canal construction in the Keys until completion of a study to assess the effects of canal development on the marine habitats, plants, and animals. One important reason for that study was the apparent drop in average underwater visibility at the outer reefs from approximately 175 feet in 1968 to approximately 35 feet in 1973. They found that major turbidity problems persisted up to 2 years after the completion of a canal dredging project due to slow settling of very fine particles. Also, the repopulation by seagrasses in areas dredged for access channels was very slow; dredged grassbeds showed no signs of new growth after 10 years.

Ten canal systems were studied in the Florida Department of Pollution Control (1973) study. Depressed dissolved oxygen levels were frequently encountered in all canals. The average bottom concentration was less than 4.0 mg/l (the state standard) and often less than 1.0 mg/l. Surface and mid-water levels of dissolved oxygen of less than 4.0 mg/l were frequent. Long-term conditions of low oxygen concentrations resulted in the growth of anaerobic bacteria which produce hydrogen sulfide which is toxic to most other organisms. Most canal systems studied had reduced number of animal species and densities compared to reference sites. At the conclusion of the study, the moratorium on dredge-and-fill operations was lifted, provided strong enforcement measures were taken for violators of turbidity and other water quality parameters. In addition, water exchange and circulation of future canal systems would be critically examined. The FDER study and its recommendations effectively stopped construction of additional finger fill canal systems in the Keys.

The U.S. Environmental Protection Agency (USEPA, 1975) conducted a study of finger-fill canals in Florida and North Carolina and came to the same conclusions as the Florida Department of Pollution Control (1973) study. EPA concluded that poorly designed canals result in poor flushing, which, coupled with a seasonal inflow of freshwater, produced extensive salinity stratification in the canals. The bottom layer of high salinity water resulted in stagnation, putrification, and extensive nutrient enrichment of the water column. Canals greater than 4- to 5-feet deep regularly experienced violations of state water quality standards for dissolved oxygen (<4 mg/l).

The EPA (USEPA, 1975) compared the water quality of two canals on Big Pine Key at Doctor's Arm Subdivision. At that time, one of the canals was recently constructed and undeveloped and the other was sparsely developed with septic tank systems in Miami oolite substrate. Even though the canal was sparsely developed, they found reduced oxygen concentrations, increased biochemical oxygen demand, and increased fecal coliform bacteria compared to the undeveloped canal. The water quality in both the developed and undeveloped canals was poorer (higher nutrients and lower dissolved oxygen) than ambient conditions in a well-flushed adjacent area, Bogie Channel.

Other canal systems tested during the EPA study were in Punta Gorda, FL, and several locations in North Carolina. Those systems had greater nutrient levels in developed canals than the Big Pine site, probably because the canals systems at those locations were more densely developed. Total nitrogen and organic carbon were the most salient chemical constituents characterizing water quality differences between developed and undeveloped canal systems. In nearly every case, concentrations of those two nutrients were significantly greater in the developed waterways.

At all canals studied by EPA, a dye tracer was flushed down toilets to measure the time septic tank leachate reached adjacent waters. At Punta Gorda, the dye appeared in the canal within 25 hours at two sites. In North Carolina, the dye appeared after 60 hours in one test and 4 hours in a second test. Septic tanks at those locations were approximately 50 feet from the adjacent canals. Dye introduced into two septic systems on Big Pine Key did not appear in the canal within 150 hours, the duration of the study. The reason was thought to be due to a period of sustained high tides. Septic systems were installed in porous Miami oolite that has a high percolation rate (2 min per inch). However, during the time of the dye tracer study, the water surface in the canal was kept high due to natural tidal amplitude (spring tides) and wind-driven waters. During the time frame of the study at Doctor's Arm Subdivision, the observed high tides were higher than normal and the low tides were not low enough to effect a hydraulic gradient that would flush the leachate from the seepage field and disperse it to the canal. Subsequent to the EPA study, other studies in the Keys have demonstrated the rapid transmissivity of Keys substrates to wastewater and the influence of tides on the movement. Those studies are discussed below.

In 1985, the Florida Department of Environmental Regulation (FDER) studied the water quality of the waters surrounding the Florida Keys in preparation for the proposed designation of the waters of the Florida Keys as Outstanding Florida Waters (OFW). That study concluded that the majority of the Florida Keys met the criteria for designation as OFW but that certain areas, including canals and the vicinity of the Key West outfall, should not be included. Many of the canal systems tested exhibited low values in dissolved oxygen, high nutrient values, and violations of the fecal coliform standard. Ranges of some water quality parameters from canals and other ambient stations are given in Table 31.2.

Canals and other confined water bodies that demonstrated signs of eutrophication during the OFW study were listed as "hot spots" in the Phase II Report of the Water Quality Protection Program (Table 31.5; USEPA, 1993). That hot spot list was revised (Table 31.6) at an interagency workshop sponsored by the South Florida Water Management District (April 16, 1996). The revised list includes a relative priority ranking of the top 19 canal systems and other waters that demonstrate poor water quality based upon the literature and the collective experience of participants of the workshop. It also includes a brief description of potential solutions to the water quality problems for each prioritized hot spot. Three recommendations were made for all high priority, poorly designed canal systems: install best available technology (BAT) sewage treatment, collect and treat stormwater runoff, and improve canal circulation. Installation of pumpout facilities was added to the list of recommended solutions for hot spots that included live-aboard vessels. Improved circulation to canal systems is an essential component of restoration because water quality of even undeveloped canals generally deteriorates due to cumulative, long term loading of fine organic matter (high BOD), salinity stratification, and long residence time (USEPA, 1975). However, construction of flushing channels or installation of culverts to improve circulation may not be practicable at all locations due to physical constraints and quantity and quality of natural resources that would be impacted during or after construction.

The Florida Department of Environmental Regulation (1987) measured 32 water quality parameters at twelve nearshore sites in Marathon for one year (1984). Primary sampling sites were in canals and marina basins at Faro Blanco Marina, City Fish Market, Winn Dixie Shopping Center, Key Colony Beach Sewage Treatment Plant, and the 89th to 91st Street canal system. High levels of nutrients (0.14 mg/l ammonia) and fecal coliform bacteria (3400 colonies/100 ml) were found at Faro Blanco Marina during the tourist season (November to May) due to discharge of raw sewage from live-aboard vessels. Total Kjeldahl nitrogen, total phosphorus, and biochemical oxygen demand were significantly higher in the marina than in adjacent waters.

**TABLE 31.5**
**Florida Keys Water Quality Hot Spots[a]**

| Number | Site | Location |
|---|---|---|
| 1 | Ocean Reef Marina | Key Largo |
| 2 | Phase I and Dispatch Creek | Key Largo |
| 3 | C-111 Canal | Mainland |
| 4 | Sexton Cove and Lake Surprise Subdivisions | Key Largo |
| 5 | Cross Key Waterways Subdivision | Key Largo |
| 6 | Port Largo | Key Largo |
| 7 | Key Largo Fishery Marina | Key Largo |
| 8 | Marian Park and Rock Harbor Estates | Key Largo |
| 9 | Pirate Cove Subdivision | Key Largo |
| 10 | Winken, Blynken, and Nod | Key Largo |
| 11 | Blue Water Trailer Park | Key Largo |
| 12 | Hammer Point | Key Largo |
| 13 | Campbell's Marina | Plantation Key |
| 14 | Tropical Atlantic Shores Subdivision | Plantation Key |
| 15 | Plantation Key Colony* | Plantation Key |
| 16 | Indian Waterways | Plantation Key |
| 17 | Plantation Yacht Harbor | Plantation Key |
| 18 | Treasure Harbor | Plantation Key |
| 19 | Venetian Shores | Plantation Key |
| 20 | Holiday Isle Resort | Windley Key |
| 21 | Islamorada Fish House | Upper Matecumbe |
| 22 | Lorelei Restaurant | Upper Matecumbe |
| 23 | Stratton's Subdivision | Upper Matecumbe |
| 24 | Port Antigua | Lower Matecumbe |
| 25 | White Marlin Beach | Lower Matecumbe |
| 26 | Lower Matecumbe Beach | Lower Matecumbe |
| 27 | Caloosa Cove Marina* | Lower Matecumbe |
| 28 | Kampgrounds of America Marina | Fiesta Key |
| 29 | Long Key Estates and City of Layton* | Long Key |
| 30 | Outdoor Resorts of America | Long Key |
| 31 | Conch Key | Conch Key |
| 32 | Coco Plum Beach area* | Fat Deer Key |
| 33 | Bonefish Towers Marina* | Fat Deer Key |
| 34 | Coco Plum Causeway | Fat Deer Key |
| 35 | Key Colony Subdivision* | Vaca Key |
| 36 | Sea-Air Estates | Vaca Key (Marathon) |
| 37 | 90th Street Canal | Vaca Key |
| 38 | Winner Docks | Vaca Key |
| 39 | National Fish Market | Vaca Key |
| 40 | Faro Blanco Marina | Vaca Key |
| 41 | Boot Key Harbor | Vaca Key |
| 42 | Boot Key Harbor drainage area | Vaca Key |
| 43 | Marathon Seafood | Vaca Key |
| 44 | Little Venice | Vaca Key |
| 45 | Knight Key Campground | Knight Key |
| 46 | Sunshine Key Marina | Ohio Key |
| 47 | Bahia Shores | No Name Key |
| 48 | Doctors Arm | Big Pine Key |
| 49 | Tropical Bay | Big Pine Key |

**TABLE 31.5 (CONTINUED)**
**Florida Keys Water Quality Hot Spots[a]**

| Number | Site | Location |
|---|---|---|
| 50 | Whispering Pines Subdivision | Big Pine Key |
| 51 | Sands Subdivision area | Big Pine Key |
| 52 | Eden Pines Colony | Big Pine Key |
| 53 | Pine Channel Estates | Big Pine Key |
| 54 | Cahill Pines and Palms | Big Pine Key |
| 55 | Port Pine Heights | Big Pine Key |
| 56 | Sea Camp* | Big Pine Key |
| 57 | Coral Shores Estates | Little Torch Key |
| 58 | Jolly Roger Estates | Little Torch Key |
| 59 | Breezeswept Beach Estates* | Ramrod Key |
| 60 | Summerland Key Fisheries | Summerland Key |
| 61 | Summerland Key Cove | Summerland Key |
| 62 | Cudjoe Ocean Shore | Cudjoe Key |
| 63 | Venture Out Trailer Park | Cudjoe Key |
| 64 | Cutthroat Harbor Estates* | Cudjoe Key |
| 65 | Cudjoe Gardens Subdivision* | Cudjoe Key |
| 66 | Orchid Park Subdivision | Lower Sugarloaf Key |
| 67 | Sugar Loaf Shore Subdivision | Lower Sugarloaf Key |
| 68 | Sugar Loaf Lodge Marina* | Lower Sugarloaf Key |
| 69 | Bay Point Subdivision | Saddlebunch Keys |
| 70 | Porpoise Point* | Big Coppitt Key |
| 71 | Seaside Resort | Big Coppitt Key |
| 72 | Gulfrest Park* | Big Coppitt Key |
| 73 | Boca Chica Ocean Shores | Geiger Key |
| 74 | Tamarac Park | Geiger Key |
| 75 | Boca Chica Naval Air Station | Boca Chica Key |
| 76 | Boyd's Trailer Park | Stock Island |
| 77 | Alex's Junkyard | Stock Island |
| 78 | Ming Seafood | Cow Key |
| 79 | Oceanside Marina | Cow Key |
| 80 | Safe Harbor | Cow Key |
| 81 | Key West Landfill | Key West |
| 82 | House Boat Row | Key West |
| 83 | Garrison Bight Marina | Key West |
| 84 | Navy/Coast Guard Marina and Trumbo Point Fuel Storage Facility | Key West |
| 85 | Truman Annex Marina | Key West |
| 86 | Key West sewage treatment plant outfall | Key West |
| 87 | Key West Bight | Key West |
| 88. | Key West stormwater discharge | Key West |

[a] Areas with known or suspected (*; no data available) degraded water quality.

*Source:* Modified from USEPA (1993).

The 90th Street canal station was selected to monitor leachate from septic tanks and cesspits. FDER consistently found violations of dissolved oxygen (<4 mg/l) at the head of the deadend canal. With a single exception, mean monthly fecal coliform bacteria were higher at the end of the canal (3 to 37 colonies/100 ml) than mean concentrations at the canal mouth (1 to 6 colonies/100 ml). Fecal coliform concentrations were highest during Thanksgiving, Christmas, and New Year's holiday periods. The maximum reading was 1220 colonies/100 ml. Orthophosphate (0.04 mg/l) and mean chlorophyll concentrations (29 µg/l) were also significantly higher in the canal than at the reference site, indicating eutrophication and algal blooms. High levels of mercury, lead, zinc, copper, and hydrocarbons were found in the canal sediments, presumably from boats. Iron levels were significantly higher in the canal which is indicative of stormwater runoff (USEPA, 1975).

The Florida Department of Environmental Regulation (1987) measured coprostanol, a degradation product of cholesterol, which is excreted in human waste. The presence of coprostanol in marine sediments provides a historic record of sewage contamination. Coprostanol levels at Faro Blanco Marina were 2 to 50 times higher in marina sediments than in reference sediments. Coprostanol levels (mean = 256; maximum = 1645 ng/g) were highest in sediments directly below boat slips, indicating that the primary source of fecal contamination was from discharge of sewage from vessels. Reference stations averaged 34 ng/g coprostanol.

Concentration of coprostanol at the outfall of the Key Colony Beach sewage treatment plant (secondary treatment) was 294 ng/g. Of the three locations in which coprostanol was measured in that study, the area surrounding the Key Colony Beach outfall was the least impacted by sewage. That is not surprising because secondary treatment plants remove between 85 and 95% of total suspended solids (TSS) in raw sewage, and coprostanol is normally associated with TSS.

Coprostanol was found in sediments from the 89th, 90th, and 91st Street canals and exhibited spatial and temporal variability. Sediments from the 90th Street canal contained the highest coprostanol concentrations found in the study (2206 ng/g). All three canals sampled contained high levels of coprostanol and were heavily impacted by sewage-derived materials. Mean coprostanol concentration ranged from a maximum at the head of the 91st Street canal (1363 ng/g) to a minimum at the middle of the 90th Street canal (160 ng/g). In general, coprostanol levels decreased from the end of each canal toward the canal mouth, probably reflecting a flushing gradient within each canal. Substantial amounts of sewage-associated, fine-grained material appeared to be transported out of the canals by tidal exchange and deposited in the nearshore access channel, where coprostanol was measured at 681 ng/g. Coprostanol was undetectable (<10 ng/g) in four out of five sampling events at a station located approximately one mile offshore; at one sampling event, coprostanol was detected at that reference site in very low concentration (28 ng/g).

The Florida Department of Environmental Regulation (1987) measured water quality in a canal system that received stormwater drainage from the Marathon Winn Dixie shopping center. An occluded effluent pipe and inefficient drainage of the parking lot reduced the amount of stormwater discharged to the canal and the impact of stormwater runoff at that location could not be definitively evaluated due to the low discharge volume. Regardless, the study found significant gradients in the canal that could be the result of septic tank seepage and stormwater. Dissolved oxygen levels were significantly depressed at the head of the canal. Mean monthly levels ranged from 3.06 to 4.93 mg/l, whereas those at the mouth of the canal fell below 5.0 mg/l only once during the study. On 76% of the days sampled, the dissolved oxygen at the head of the canal was below the state minimum criterion.

At the Winn Dixie site, monthly concentrations of total nitrogen and ammonia nitrogen were statistically indistinguishable between canal waters and ambient waters. However, phosphorus concentrations at the stormwater discharge site (maximum = 0.04 mg/l) were significantly higher than those measured at the mouth of the canal (0.01 mg/l) and offshore (0.01 mg/l). Orthophosphate levels peaked during July and autumn (rainy season) at the canal head and averaged two to three times above those measured at the canal mouth.

Later, the Florida Department of Environmental Regulation (1990) conducted an intensive, one-year study to assess the water quality in Boot Key Harbor. Boot Key Harbor has approximately 400 live-aboard vessels during winter months. Stations were located in canals, the Harbor basin, and a reference site. Annual mean dissolved oxygen concentrations were lowest in canals and the basin (4.2 mg/l) compared to the reference stations (6.1 mg/l). Low dissolved oxygen levels in the canals and basin were due to poor flushing characteristics that resulted in the canals serving as sinks for organic matter. Regular violations of the state standard for dissolved oxygen were observed in the canals.

**TABLE 31.6**
**Priority Water Quality Hot Spots**

| Hot Spot | Location | Priority | Cause of Water Quality Problem | Potential Solutions |
|---|---|---|---|---|
| Sexton Cove and Lake Surprise Subdivisions | Key Largo | H | Poorly designed canals, use of septic tanks or cesspits, untreated runoff | Install best available technology OSDS or WWTP, install surface water system, improve canal circulation |
| Cross Keys Waterways | Key Largo | H | Poorly designed canals, use of septic tanks and cesspits, untreated runoff | Install best available technology OSDS or WWTP, install surface water system, improve canal circulation |
| Winken, Blynken, and Nod | Key Largo | H | High density, poorly designed canals, septic tanks or cesspits, marina, live-aboards, untreated runoff | Install best available technology OSDS or WWTP, improve canal circulation, install or improve marina surface water system, install pumpout, install surface water system |
| Conch Key | Conch Key | H | Poorly designed canals, septic tanks or cesspits, marina, untreated runoff | Install best available technology OSDS or WWTP, improve canal circulation, install or improve marina surface water system, install pumpout, install surface water system |
| Boot Key Harbor area | Vaca Key | H | Poorly designed canals, use of septic tanks or cesspits, marinas, live-aboards, seafood processing, untreated runoff | Install best available technology OSDS or WWTP, improve canal circulation, install surface water treatment, install pumpout |
| Little Venice Subdivision | Vaca Key | H | Poorly designed canals, use of septic tanks or cesspits, untreated runoff | Install best available technology OSDS or WWTP, improve canal circulation, install surface water system |
| Knight Key Campground | Knight Key | H | High density, poorly designed canals, use of septic tanks or cesspits, untreated runoff | Install best available technology OSDS or WWTP, improve canal circulation, install surface water system |
| Doctors Arm | Big Pine Key | H | High density, poorly designed canals, use of septic tanks or cesspits, untreated runoff | Install best available technology OSDS or WWTP, improve canal circulation, install surface water system |

| Location | Key | Priority | Sources | Recommendations |
|---|---|---|---|---|
| Eden Pines Colony | Big Pine Key | H | High density, poorly designed canals, use of septic tanks or cesspits, untreated runoff | Install best available technology OSDS or WWTP, improve canal circulation, install surface water system |
| Bay Point Subdivision | Saddlebunch Key | H | High density, poorly designed canals, use of septic tanks or cesspits, untreated runoff | Install best available technology OSDS or WWTP, improve canal circulation, install surface water system |
| Boca Chica Naval Air station | Boca Chica Key | H | Discharge of wastewater effluent to surface waters | Install injection well |
| Key West sewage treatment plant outfall | Key West | H | Discharge of wastewater effluent to surface waters | Install injection well |
| Key West stormwater outfalls | Key West | H | Discharge of untreated stormwater to nearshore waters | Retrofit surface water system to provide treatment |
| Hammer Point | Key Largo | M | Poorly designed canals, use of septic tanks or cesspits, untreated runoff | Install best available technology OSDS or WWTP, improve canal circulation, install surface water system |
| Tropical Bay | Big Pine Key | M | Poorly designed canals, use of septic tanks or cesspits, untreated runoff | Install best available technology OSDS or WWTP, improve canal circulation, install surface water system |
| Sands Subdivision including Whispering Pines | Big Pine Key | M | Poorly designed canals, use of septic tanks or cesspits, untreated runoff | Install best available technology OSDS or WWTP, improve canal circulation, install surface water system |
| Port Pine Heights | Big Pine Key | M | High density, poorly designed canals, use of septic tanks or cesspits, untreated runoff | Install best available technology OSDS or WWTP, improve canal circulation, install surface water system |
| Cudjoe Gardens | Cudjoe Key | L | High density, poorly designed canals use of septic tanks or cesspits, untreated runoff, | Install best available technology OSDS or WWTP, improve canal circulation, install surface water system |
| Gulfrest Park | Big Coppitt Key | L | High density, poorly designed canals, use of septic tanks or cesspits, untreated runoff | Install best available technology OSDS or WWTP, improve canal circulation, install surface water system |

*Source:* Modified from Florida Keys Water Quality Hot Spot Workshop, March 19, 1996.

Fecal coliform bacteria concentrations were highest at canal stations and were practically absent at reference stations. Highest fecal coliform levels were observed at stations with onsite disposal systems after a heavy rainfall. Fecal coliform levels in Boot Key Harbor basin stations were highest during winter months at stations in close proximity to live-aboard vessels. Violations of the state standard for fecal coliform bacteria were common.

Florida Bay Watch is a volunteer program to collect water quality data in Florida Bay and the Florida Keys. Bay Watch volunteers take water quality data that augment ongoing studies by agencies and institutions. Between July 1995 and June 1996, Bay Watch volunteers sampled 38 fixed nearshore stations, of which 16 were in residential canals, one in a boat basin, and 21 at natural shorelines (Florida Bay Watch Program, 1996). Immediately apparent is the variability of the data, both at any station and between stations. This may be due in part to varying climatological differences between sampling intervals; however, some basic generalities appear from this dataset. Twenty-one of the stations had enough data to determine seasonal trends. Of those, five of the canal sites had higher nitrogen during the wet season; others showed no seasonal variation. No significant trends, spatially or seasonally, in total phosphorus with station location were observed. Highest chlorophyll levels occurred in bayside canal sites.

In 1997, Bay Watch volunteers sampled 36 fixed nearshore stations for water quality. Nutrient data varied among stations because of the many differences between sampling sites, such as flushing rates, density and number of residences, proximity to injection wells or other discharges, and stormwater controls. However, these data are very useful in comparing and ranking nearshore waters. For example, the canal system on Duck Key is very well flushed due to its flow-through design and proximity to open waters of the Atlantic Ocean. Also, the density of residences is comparatively low on Duck Key. In contrast, The Eden Pines (Big Pine Key) and Ramrod Key canal systems are long, with many deadend fingers and relatively dense development. Differences in water quality parameters from Duck Key and the Eden Pines and Ramrod Key canals are striking (Table 31.7). For example, in 1997 mean total nitrogen was approximately twice as high in Eden Pines (40.5 $\mu M$) and Ramrod (35.8 $\mu M$) compared to Duck Key (19.8 $\mu M$). Mean total phosphorus and mean total chlorophyll $a$ showed similar trends. A natural, unobstructed shoreline at Grassy Key (bayside) is included in Table 31.7 for comparison. These data document the degraded water quality in poorly flushed, long, deadend canal systems (Florida Bay Watch Program, 1997).

**TABLE 31.7**
**Water Quality Data from Selected Florida Bay Watch Program Monitoring Sites**

| Location | Total Nitrogen ($\mu M$) | Total P ($\mu M$) | Chlorophyll a ($\mu g/l$) |
|---|---|---|---|
| Grassy Key (open) | 19.8 | 0.34 | 0.57 |
| Duck Key Canal | 19.8 | 0.21 | 0.28 |
| Eden Pines Canal | 40.5 | 1.04 | 2.78 |
| Ramrod Key Canal | 35.8 | 0.64 | 2.27 |

*Source:* From Florida Bay Watch Program, *Annual Report on Water Quality Data Collected by the Florida Bay Watch Volunteer Program, July 1996–June 1997*, 1997.

## OTHER NEARSHORE WATERS

Because the Florida Keys ecosystem is an "open" system and receives water from many sources, defining the causes of changes in the community structure becomes more difficult farther from the shore. Several studies have been performed to investigate the extent of impacts from land-based nutrient loading to nearshore habitats.

Lapointe and Clark (1992) measured water quality parameters at 30 stations during summer and winter to characterize seasonal extremes of measured variables. Sampling at each site was performed along an onshore–offshore transect. They found a gradient in nutrients from inshore to offshore. Man-made canal systems had significantly elevated concentrations of soluble reactive phosphorus (0.3 $\mu M$) compared to seagrass meadows (0.1 $\mu M$), patch reefs (0.05 $\mu M$), and offshore reef banks (0.05 $\mu M$). Ammonia was highest in canal systems and seagrass meadows (>1 $\mu M$) compared to patch and bank reef stations (<0.3 $\mu M$). Chlorophyll and turbidity were highest in canal systems and seagrass meadows and reached peak levels during summer months. Chlorophyll was >1 $\mu g/l$ at canal and <0.3 $\mu g/l$ at back reef stations. They concluded that widespread use of septic tanks increases the nutrient contamination of groundwaters that discharge into shallow nearshore waters, resulting in coastal eutrophication.

Seagrasses and other community components integrate the effects of nutrients in the water column over time. Growth of benthic algae, increased chlorophyll (phytoplankton) in the water column, as well as increased nutrient concentrations have been used to gauge the onset of eutrophication in tropical marine ecosystems (Bell, 1992). Lapointe et al. (1994) assessed how nutrient enrichment affects algal growth on seagrass blades (epiphytes) and the productivity and structure of the shallow-water turtle grass (*Thalassia testudinum*) community in the Keys. A stratified random sampling technique was utilized along three onshore–offshore transects perpendicular to shore in the Middle and Lower Keys. Inshore stations (hypereutrophic) were selected in areas receiving direct impacts of wastewater nutrient discharges and included a canal mouth with septic tanks and cesspits (Doctor's Arm Subdivision, Big Pine Key), live-aboard vessels (Houseboat Row, Key West), and a package sewage treatment plant (Fiesta Key Campground). Eutrophic and mesotrophic stations were located within approximately 1 km from land. Oligotrophic stations were located along the back reef at Alligator Reef, Looe Key, and Sand Key.

Total nitrogen and total phosphorus decreased linearly from inshore stations to offshore stations. Offshore stations had the highest shoot densities, areal biomass, and areal production rates and the lowest epiphyte levels. Nearshore seagrass meadows had greater diversity of primary producers, including macroalgae, attached seagrass epiphytes, high phytoplankton concentration (green water), and jellyfish (*Cassiopeia* spp.) Lapointe et al. (1994) concluded that nutrient-enhanced productivity of macroalgae and attached epiphytes not only leads to decreased productivity of turtle grass, but also may reduce dissolved oxygen levels, resulting in significant habitat damage prior to actual dieoff. Eutrophic seagrass meadows in the Florida Keys were found to have pre-dawn hypoxia (<2.0 mg/l dissolved oxygen) or anoxia (<0.1 mg/l) during warm, rainy periods. McClanahan (1992) reported that low pre-dawn oxygen concentrations were negatively correlated with species richness and diversity of mollusks in Florida Bay compared to waters with higher oxygen found offshore Key Largo. Lapointe et al. (1994) found that at concentrations of approximately 25 $\mu M$ nitrogen and 0.45 $\mu M$ phosphorus, turtle grass is replaced by shoalgrass (*Halodule wrightii*), an opportunistic seagrass. They concluded that nutrient enrichment from land-based sewage inputs can have significant effects on seagrass productivity for considerable distances from shore.

## OUTER CORAL REEFS

Szmant and Forrester (1996) measured distribution patterns of nutrients to determine whether anthropogenic nutrients from land-based sources may be reaching the outer reef tract. Samples were collected along seven transects oriented perpendicular to the shoreline and located from Biscayne National Park to Looe Key. Samples were taken along transects at stations located in tidal passes and canal mouths to approximately 0.5 km seaward of the outermost reef. Water column and sediment concentrations of nitrogen and phosphorus were measured.

In the Upper Keys, water column nitrogen (1 $\mu M$ $NO_3$) and chlorophyll (1 $\mu g/l$ chlorophyll *a*) were elevated near marinas and canals but returned to oligotrophic concentrations within 0.5 km of shore. Phosphorus concentrations were higher at offshore stations (>0.2 $\mu M$ $PO_4$) and were attributed to upwelling of deep water along the shelf edge at the time of sampling. Sediment interstitial nitrogen concentrations decreased from inshore to offshore stations which is indicative of an onshore source of nitrogen. There was some indication of a reverse trend for phosphorus that may be indicative of upwelling of deep oceanic waters as a source.

In the Middle Keys, both water column nutrients and chlorophyll concentrations were higher than observed in the Upper Keys, and there was less of an inshore–offshore gradient than noted in the Upper Keys. Sediment nutrients were also higher, and no differences were found in nutrient concentrations at nearshore and offshore areas. These observations may be explained by the mixing of Florida Bay waters with the waters adjacent to the Keys.

These data support the conclusion that outer reef areas in the Upper Keys are not accumulating elevated loads of land-derived nutrients via surface water flow and document moderately elevated nutrient and chlorophyll levels in many developed nearshore areas. The authors concluded that most of the anthropogenic and natural nutrients entering the coastal waters from shore appear to be taken up by nearshore algal and seagrass communities before they reach patch reef areas (about 0.5 to 1 km from shore). Further work is needed to determine whether nutrient-enriched ground-waters reach the reefs; however, these groundwaters would be expected to cause an enrichment of reef sediments, which was not observed.

Lapointe and Matzie (1996) used high-frequency sampling to track effects of periodic rainfall events on a transect that included stations in a canal on Big Pine Key (Port Pine Heights), a seagrass meadow (Pine Channel), a patch reef (Newfound Harbor), and an offshore reef (Looe Key). Lowest dissolved oxygen (<0.1 mg/l), maximum concentration of $NH_4^+$, total dissolved phosphorus, and chlorophyll, and minimum salinities were measured in the canal during a rain event. Concentrations of total dissolved phosphorus also increased at the seagrass, patch reef, and offshore reef stations after the initial rainfall event. Concentrations of NH4+ and chlorophyll increased at offshore stations approximately 1 to 3 weeks following the rain event. The authors suggested that rainfall events can rapidly flush nutrients into canals and adjacent nearshore waters. These nutrients may have the potential of impacting water quality for considerable distances from land; however, more research is required to substantiate these findings and define the area of impact. Lapointe and Matzie (1996) concluded that the effects of increased concentrations of nutrients in nearshore waters justifies that special precautions be taken in the treatment and discharge of wastewaters and stormwater runoff.

Lapointe and Matzie (1997) measured water quality parameters along a transect from the eastern shoreline of Big Pine Key to Looe Key from January to October 1996. The inshore station was located off Avenue J Canal and was down gradient of approximately 1000 septic tanks and cesspits. A patch reef station was located off Munson Island, and an offshore station was located along the back reef at Looe Key. Monthly samples were taken, along with high-frequency sampling prior to, during, and following selected rainfall and wind events. Lapointe and Matzie also measured nitrogen isotope ratios in macroalgae and seagrass blades to determine the source of the nitrogen; a higher ratio of $^{15}N/^{14}N$ in wastewater was found.

Highest levels of dissolved inorganic nitrogen, soluble reactive phosphorus, and chlorophyll occurred during periods of high winds, low tides, and rain events. The highest nitrogen and phosphorus concentrations were measured at the inshore station at low tide when tidal ranges were highest. Low tides allow rapid drainage of nutrient-enriched groundwater to adjacent surface waters.

Ratios of nitrogen isotopes were highest in a benthic algae at the nearshore station (5.0‰), intermediate at the patch reef station (3.5‰), and lowest at Looe Key (3.0‰). These data may indicate increasing wastewater nitrogen contributions to algae with increasing proximity to shore; however, there are many sources of nitrogen, only one isotopic indicator was used in that study, and more denitrification may occur inshore than offshore. Thus, additional research is required to quantitatively define the sources of nitrogen.

Lapointe and Matzie (1997) observed that a large area of seagrasses located near the mouth of the Avenue J Canal was covered by a heavy growth of attached and benthic algae and that approximately 2.5 acres of seagrasses had been replaced by benthic algae at that location. They also documented blooms of benthic algae and epiphytes on seagrass blades at Looe Key.

It is very difficult to quantify all sources of nutrients and their effects at offshore areas. For example, no quantitative information is available on the impacts of increased numbers of charter boats or other vessels that flush their heads and holding tanks at offshore areas. Reduction of predators and grazers is another confounding factor affecting the community composition of the outer reef. Preparation of a detailed nutrient budget for nearshore and offshore areas in the Florida Keys is a topic that requires further research.

## GROUNDWATER

Information on the geology and hydrogeology of the Florida Keys is summarized by Halley et al. (1997). Several studies have been performed that demonstrate the transmissivity of the substrates of the Florida Keys and the rapid exchange of wastewater from onsite systems or injection wells to surface waters.

Lapointe et al. (1990) measured significant nutrient enrichment of groundwaters contiguous to onsite disposal systems at several sites. Mean dissolved inorganic nitrogen (987 $\mu M$) was 400 times higher and mean soluble reactive phosphorus (9.77 $\mu M$) was 70 times higher in groundwater adjacent to a septic tank seepage field compared to a reference site. Concentrations of nitrogen and phosphorus decreased in the groundwater away from the septic tank toward the adjacent canal, presumably due to dilution by groundwater. They also theorized that some of the soluble reactive phosphorus was absorbed by the substrate. Concentrations of nutrients in the canals (dissolved inorganic nitrogen, 4.91 $\mu M$; soluble reactive phosphate, 0.43 $\mu M$) were elevated compared to control sites. Concentrations of nutrients in the canals were highest in the summer because of seasonally maximum tidal ranges and increased flushing during the summer wet season. Lapointe et al. (1990) used a groundwater flow meter to demonstrate that lateral rates of shallow groundwater flow increased by approximately three times during ebbing tides as compared to flooding tides. This observation was supported by Lapointe and Matzie (1997) who found the maximum concentration of dissolved inorganic nitrogen off the Avenue J Canal (Big Pine Key) when tidal ranges were the highest during the study period.

Shinn et al. (1994b) placed and sampled 24 wells beneath the Keys, nearshore areas, and outer reefs to determine if sewage effluent from Class V wells is reaching offshore reef areas via underground flow. Class V wells (drilled 90 ft and cased to 60 ft) are currently permitted by the Florida Department of Environmental Protection for disposal of wastewater. Sample wells were located in transects off Ocean Reef Club, Key Largo, and Saddlebunch Keys and were sampled quarterly for one year. Investigators found well water to be consistently hypersaline with a marked increase in ammonia in offshore groundwater. Other forms of nitrogen and phosphorus present in offshore groundwater were only slightly elevated above levels found in surface marine waters. Highest levels of nitrate, nitrite, and phosphorus were found in shallow onshore groundwaters.

Nearshore wells were observed to discharge water during falling tides and draw water into the wells during rising tides. This "tidal pumping" results in considerable water movement in and out of the upper few meters of limestone and is a likely mechanism for mixing and transferring nutrient-rich groundwater into overlying surface waters.

G. Shinn (pers. comm.) described Key Largo Limestone as having a consistency of Swiss cheese, and several other studies have confirmed the rapid connection of groundwaters with surface waters in the Key Largo Limestone matrix. Paul et al. (1995a) placed a man-made tracer virus in a septic tank and into a 13.7-m (45-ft) deep injection well in Key Largo and found the virus in the surface waters of an adjacent canal and the Atlantic Ocean in 11 and 23 hr, respectively. Rates of migration ranged from 0.57 to 24.2 m/hr (1.87 to 79.3 ft/hr). They concluded that current onsite disposal practices in the Florida Keys can lead to rapid nutrient enrichment and fecal contamination of subsurface and surface marine water in the Keys. Viral tracers were detected on falling tides, confirming the findings of tidal pumping by Shinn et al. (1994b), Lapointe et al. (1990), and Lapointe and Matzie (1997).

Paul et al. (1997) repeated the viral tracer experiment with 12.2-m (40-ft) deep injection wells on Key Largo and a permitted 27.4-m (90-ft) deep Class V injection well on Long Key. At both sites, viral tracers appeared in the groundwater within 8 hr after injection, and in marine surface waters within 10 hr in Key Largo and 53 hr in Long Key.

Chanton et al. (1998) used natural tracers to locate areas of groundwater discharge to surface waters surrounding the Florida Keys. They also used artificial tracers to quantify rates of flow of materials injected into groundwater to surface waters.

Chanton et al. completed two extensive surveys and have mapped areas of concentrations of natural tracers near the Keys. Groundwater seepage areas have been found on both the Florida Bay and Atlantic Ocean sides of the Keys. Two injection studies have been completed, one on Key Largo and one on Long Key. In both tests, the tracer was injected into groundwaters and was observed, greatly diluted (approximately one million times), within hours to days in nearby surface waters. At the Long Key site it was found in a canal located across U.S. 1 from the injection site. Wastewater injected into the groundwater at Long Key rapidly migrated toward the surface due to the fact that freshwater "floats" on the highly saline groundwater.

Kump (1998) sampled groundwater in wells drilled to various depths surrounding a wastewater injection well on Long Key. He confirmed the presence of a shallow, low-salinity lens floating on top of groundwaters. Distribution of nutrients away from the site of injection was variable, but phosphate, nitrate, and ammonia concentration appears to be highest nearest the injection well at a depth of 5 m; however, the elevated concentrations of these nutrients were observed in sampling wells located in different directions from the point of injection. The absence of phosphate in high-pH waters in shallow wells leads to the postulation that phosphate may be removed by adsorption onto the limestone substrate.

In October 1996, Kump injected phosphate at the same time that Chanton et al. injected a nonreactive tracer (sulfur hexafloride, $SF_6$) into a Class V injection well (60/90 ft) at Long Key. Within 4 hr tracers were elevated at the sampling well located between the injection well and the Atlantic Ocean. The peak of both tracers occurred after about 3 hr. After the peak, the ratio of the tracers fell because the concentration of $PO_4$ fell more rapidly than that of $SF_6$. Using data from one of the sampling wells, it was calculated that the tracer $SF_6$ appeared to be moving vertically at about 7 m/day. The pattern of early $SF_6$ peaks in some wells that are associated with phosphate peaks, and later SF6 increases with no increase in phosphate concentration at other wells, cannot be ascribed simply to dilution of phosphate by groundwater. The predicted phosphate concentrations based on the assumption of no preferential uptake and the observed tracer concentrations would be well above detection at many of the wells. These observations support the hypothesis that phosphate is being stripped from the groundwater. The rate and long-term capacity of substrates in stripping phosphate are topics that require additional research.

## FECAL COLIFORM BACTERIA AND DISEASE ORGANISMS

In addition to nutrient enrichment of subsurface and surface waters, on-site disposal systems and injection wells are known to be a source of microbial contamination of groundwater (Keswick, 1984). Because the groundwaters and surface waters are very closely linked in the Keys, it is not surprising that fecal coliform bacteria are common in canals and boat basins. As discussed above, fecal coliform violations were common in some studies (Florida Department of Environmental Regulation, 1987, 1990). To date, there has not been a systematic public health survey of canals and other confined waters of the Keys to determine their risk to human health. That is a topic that is currently undergoing study.

Paul et al. (1993) sampled the occurrence of viruses and bacteria in the vicinity of Key Largo. Water column viral counts were highest in Blackwater Sound and decreased to the shelf break;

lower salinity waters had higher numbers of viruses. Viral counts in sediments averaged nearly 100 times those found in the water column and did not correlate with salinity. Paul et al. concluded that viruses are abundant in the Key Largo environment, particularly on the Florida Bay side, and that processes governing their distribution in the water column are independent of those governing their distribution in sediments.

Shinn et al. (1994b) found fecal coliform and fecal *Streptococci* bacteria in several of their wells. At the Saddlebunch transect, they found that the inshore well and the wells farthest from the shore (>2 nm) tested positive for fecal coliform bacteria during several rounds of testing. The investigators speculated that the source of the bacteria in the well on shore may be from septic tank drainfields at a recreation vehicle park on Saddlebunch Key. The source of the bacteria in the offshore wells is unknown because the locations of the wells are remote from areas of large human populations. The authors speculated that contamination of the more offshore wells could be the result of rapid flow through the underlying Key Largo Limestone from a remote site, such as Marathon, where there is a large community built on Key Largo Limestone. The investigators theorized that if the bacteria are not some unknown, anoxic, nonfecal, nonhuman form indigenous to hypersaline groundwater, then their presence suggests a land source and considerable offshore groundwater movement.

In Key Largo, Shinn et al. (1994b) found fecal coliform and *Streptococci* bacteria consistently in the shallow well and once in a deep well on the island. The shallow well was within 50 ft of a septic tank drainfield. At Ocean Reef Club, the shallow onshore well also had fecal bacteria during all four sampling rounds. Fecal bacteria were also found in offshore wells, including a well located approximately 5 nm offshore.

Supporting evidence of nearshore contamination by fecal bacteria is provided by Paul et al. (1995b). They found two or all three fecal indicators for which they tested (fecal coliform, *Clostridium perfringens*, and *Enterococci*) in onshore shallow (1.8 to 3.7 m or 6 to 12 ft deep) monitoring wells at Key Largo. Deep wells (10.7 to 12.2 m or 35 to 40 ft deep) at the same sites contained few or no fecal bacteria. Fecal indicators were found in two of five nearshore wells that were 1.8 and 2.9 miles from shore. Wells farther offshore showed little signs of contaminations. All indicators were also found in surface waters of a canal in Key Largo and in offshore surface waters in March but not in August. These results suggest that fecal contamination has occurred in the shallow onshore aquifer, parts of the nearshore aquifer, and certain surface waters. Paul et al. (1995b) concluded that current sewage waste disposal practices may have contributed to the observed contamination.

Finally, Griffin et al. (1997) found fecal coliform *Escherichia coli* and *Clostridium* at most stations sampled in Boot Key Harbor June 8 to 13, 1997.

## EFFECTS ON BIOLOGICAL COMMUNITIES

Nutrient-rich, land-based sources of pollution from runoff and wastewater disposal practices in the Keys rapidly gets into surface waters. Data on nutrient enrichment in canals is compelling. Several investigators observed that canals were depauperate in marine life (Taylor and Saloman, 1968; Barada and Partington, 1972; USEPA, 1975), and fish kills in residential canals are common (Taylor and Saloman, 1968; Barada and Partington, 1972). Seagrasses, which are common in shallow waters around the Keys, are generally absent or reduced in density in stagnant canals because of canal depths and/or periodically high phytoplankton blooms, turbidity, and hydrogen sulfide gas (USEPA, 1975). Lapointe et al. (1994) observed a shift in community structure in an enriched canal and that the seagrass meadows adjacent to the mouth of the canal were eutrophic, as demonstrated by lush macroalgae growth, high epiphyte load on seagrass blades, and high phytoplankton (chlorophyll) concentration. A large area of seagrasses was stressed, and approximately 2.5 acres of seagrasses were replaced by benthic algae at the mouth of the Avenue J Canal (Lapointe and Matzie, 1997). Waters in Boot Key Harbor and adjacent canals had high nutrient and chlorophyll concentrations (Florida Department of Environmental Regulation, 1987, 1990). It would be beneficial, but costly, to have long-term water quality data from all canal systems and harbors in the Keys; however, there is no reason to believe that other deadend canal systems, enclosed marinas, and harbors are radically different from the ones that have been studied.

Natural gradients in community structure are related to depth, current flow, sediment types, and other environmental conditions. The marine life in many confined water bodies and some nearshore areas are dissimilar, structurally and functionally, to natural communities found in less disturbed, more oligotrophic waters. The causes of these community shifts are differences in physical conditions (e.g., circulation, temperature) and nutrient enrichment (eutrophication). Based on available information, it is reasonable to conclude that poorly flushed canals, other confined water bodies, and nearshore areas in the Keys have reached and exceeded their assimilative capacity for nutrient addition. If nutrient loading continues, impacted areas will become increasingly dysfunctional and the impacts will extend farther from shore.

Previous sections have summarized scientific information that demonstrated the effects of human-derived pollutants on marine waters. Perhaps even more compelling than the scientific data is the general acknowledgment by long-time residents and visitors that water quality has declined in the Florida Keys. Although long-term climatic cycles cannot be completely excluded, there is abundant, albeit anecdotal, evidence that deteriorating environmental conditions in the Keys are correlated with increased population and human activities. DeMaria (1996) interviewed 75 individuals who have spent many years on the waters surrounding the Florida Keys. These individuals were asked to identify changes in fisheries, seagrass, communities, the coral reef, algae blooms, and water quality. Each person interviewed was asked to comment on the most significant changes that they observed. The results included the following conclusions:

- Water quality has declined, particularly in canals, nearshore areas, Florida Bay, and the coral reefs.
- Algal blooms are larger, more frequent, and more persistent.
- Seagrass beds have fluctuated in extent and species composition throughout the area and have drastically declined in Florida Bay.
- Corals and coral reefs show signs of declining health; disease is more common and benthic algae have increased in abundance and spatial coverage.
- Populations of sponges, giant anemones, long-spine sea urchins, and queen conchs have declined in nearshore waters. Jellyfish have increased in abundance.
- Tropical fish, specifically butterfly fish, angel fish, and groupers, have declined.

With very few exceptions, DeMaria (1996) reported that long-term Keys residents observed changes for the worse.

More research is required before definitive statements can be made on the long-term health of the Florida Keys Reef Tract and the extent and effects of anthropogenic nutrients. Szmant and Forrester (1996) reported that reefs off Key Largo and Long Key are not receiving a nutrient subsidy via surface waters from land-based sources at the present time, although the potential exists due to observed higher nutrient concentrations near shore. It is their opinion that land-based nutrients are absorbed by algae and seagrasses within 0.5 km of shore.

Lapointe et al. (Lapointe and Clark, 1992); Lapointe and Matzie, 1996, 1997) have concluded that nutrient enrichment at offshore reefs is possible following heavy rains and/or high wind events. It is their opinion that sampling during storms is required to document rapid, episodic transport of nutrients. However, if land-based nutrient subsidy to the reef is common, Szmant and Forrester (1996) theorized that reef sediments should have elevated nutrient concentrations. Their findings demonstrated that nutrient concentrations in sediments decreased rapidly from the shore.

Upwelling of deep, relatively nutrient-rich oceanic water may be a source of nutrients to the outer reefs. Szmant and Forrester (1996) concluded that upwelling was probably responsible for elevated phosphorus observed in offshore waters in the Upper Keys. Upwelling events have also been reported at Looe Key during spring and summer and may be a source of nitrogen to at least the fore reef (Lapointe and Smith, 1987). However, coral reefs generally do not develop in areas influenced by persistent upwelling due to the cold temperatures and high nutrient content (Dubinsky and Stambler, 1996). The frequency, duration, geographic extent, and nutrient loading of upwelling events is an area that requires further study.

Worldwide, a marked acceleration of the deterioration of coral reefs has occurred. Wilkinson (1993, 1996) estimated that 30% of all coral reefs have reached the "no-return" critical stage, another 30% are seriously threatened, and less than 40% are stable. The main factors in the demise of coral reefs are human pressures, such as over-fishing, physical damage, nutrient enrichment, and sediment loading. Because measurable global-scale increase in oceanic productivity is not evident, as would be in the case for significant overall eutrophication, Dubinsky and Stambler (1996) concluded that human impacts on coral reefs are on a local and regional scale, rather than a global scale (excluding impacts of global warming and increases in ultraviolet light exposure).

Long-term, quantitative studies of coral reef community structure in South Florida have documented high coral loss rates. Porter and Meier (1992) monitored six coral reef locations between Miami and Key West in 1984 and 1991. They found that all six areas lost coral species and that these losses constituted between 13 and 29% of their species richness. Coral cover decreased at five of the six sites and net losses ranged between 7.3 and 43.9%. Porter and Meier (1992) concluded that loss rates of this magnitude cannot be sustained for protracted periods if the coral community is to persist in a configuration resembling historical coral reef community structure in the Florida Keys. Porter et al. (1994) suggested that regional patterns of decline are suggestive of large-scale flow of water masses (e.g., influence of Florida Bay or Gulf of Mexico waters).

A variety of diseases have caused coral decline and mortality (Antonius, 1981a,b; Peters, 1984; Santavy and Peters, 1997). These diseases have been reported worldwide from pristine as well as heavily polluted areas. Recent systematic monitoring in the Keys has revealed that the incidence of coral diseases may be increasing. Disease probably caused the decimation of the long-spine urchin populations throughout the Caribbean during 1983–1984. Determining the etiology and distribution patterns of diseases of corals and other marine organisms is a topic for future research and monitoring.

Mass mortalities of sea fans have been reported throughout the Caribbean for many years. The causative agent of sea fan mass mortalities has been determined to be a fungal pathogen (*Aspergillus*) that is typically a soil inhabitant. It is thought that the primary infection by the fungus is probably associated with sediment particles from land-based sources (runoff) (Smith et al., 1996) or African dust (Shinn, pers. comm.).

Bell (1992) critically reviewed case studies of eutrophication of coral reefs and noted that eutrophication typically causes phase shifts from slow-growing corals to faster growing macroalgae and phytoplankton. Macroalgal blooms have been correlated with nutrient enrichment of reefs in Jamaica (Lapointe, 1997; Lapointe et al., 1997), the southeast coast of Florida (Lapointe and Hanisak, 1997), Belize (Lapointe, et al., 1993), and the inner Great Barrier Reef Lagoon, Australia (Bell and Elmtri, 1993). Others have pointed out that over-fishing and reduction of algal grazers must be taken into account (Zieman and Szmant, pers. comm.; McClanahan et al., 1995).

The impacts of nutrient enrichment to coral reefs are not always clear cut or devastating to the coral community. Nutrient enrichment studies performed at One Tree Island, Southern Great Barrier Reef, demonstrated that daily additions of both nitrogen and phosphorus to a single patch reef for 8 months enhanced community primary production by approximately 25% and inhibited calcification of the system by approximately 50% (Kinsey, 1988). An extensive nutrient enrichment experiment (ENCORE, Enrichment of Nutrients on a Coral Reef Experiment; Larkum and Steven, 1994) was performed on the Australian Great Barrier Reef to quantify the response of the community to nutrient additions. Larkum and Koop (1997) found that fertilization had no effect on growth or primary production of epilithic algae; these results are contrary to the widely held opinion that enhanced levels of nutrients cause rapid growth of algae and problems for associated biota.

## EXAMPLES OF AREAS WITH SIMILAR PROBLEMS

This section is not meant to be a comprehensive analysis of coastal eutrophication, but rather to highlight several other areas that have experienced nutrient enrichment reminiscent of observations in the Florida Keys. It is hoped that we can learn from the actions taken in other locations to correct nutrient enrichment problems. In several instances, providing additional treatment of wastewater resulted in rapid improvement of degraded biological communities.

Coastal eutrophication is a national and worldwide problem (Valiela et al., 1992). It is most evident in enclosed and semi-enclosed seas and estuaries, and the main sources of nutrient enrichment are agriculture and urban runoff and domestic wastewater (Nixon, 1990; 1995). It is estimated that the input of nutrients to the coastal waters and oceans from human sources (via rivers) is currently equal to or greater than natural input (Windom, 1992). Some prominent examples of collapses of coastal ecosystems from anthropogenic nutrient loading include the loss of seagrasses and benthic fauna in the Chesapeake Bay (Officer et al., 1984), noxious algal blooms in the Adriatic Sea (Justic, 1987), anoxia problems in the Baltic Sea (Larsson et al. 1985) and off the Mississippi River delta (Gulf of Mexico) (Rabalais et al., 1996; Turner and Rabalais, 1994), and periodic toxic algal blooms in the North Sea (Underdal et al., 1989).

### AUSTRALIA

Australian coastal waters in the vicinity of the Great Barrier Reef are naturally nutrient poor. All coastal water bodies with long residence times (poor flushing) in the populated part of the Australian coast have experienced some measurable effects of enhanced eutrophication. Phytoplankton blooms and seagrass losses are the most prominent evidences of nutrient loading (Brodie, 1994). As is true in the Florida Keys, the farther offshore, the more difficult it becomes to link community changes with land-based sources of nutrients. However Bell (1991, 1992) and Brodie (1995) have proposed that the Great Barrier Reef is showing evidence of eutrophication, as evidenced by an increase between historic and current levels of phytoplankton.

The Australian government is sponsoring a nationally coordinated approach to monitoring and managing sources of nutrient enrichment (National Water Quality Management Strategy). The goals and objectives of the Australian Strategy are very similar to those in the Water Quality Protection Program for the Florida Keys. In Sydney, inadequate sewage treatment has resulted in significant degradation of nearshore waters. Improvements to the sewage treatment system are being partially financed through a household levy of $80/year (Brodie, 1994).

### ST. LUCIE COUNTY

Over the past 50 years, seagrass coverage in the Indian River Lagoon declined overall about 6%. However, large losses (60%) occurred in the Melbourne to Grant area (Woodward-Clyde Consultants, Inc., 1994). The Indian River National Estuary Program concluded that seagrass losses were predominantly due to nutrient enrichment from domestic wastewater and stormwater discharges (USEPA, 1996).

In 1993, St. Lucie County completed a Surface Water Improvement and Management Act project to identify areas where existing onsite sewage disposal systems were a threat to the water quality of the Indian River Lagoon. Ten high-priority areas were identified based upon their pollutant loads. Principal recommendations of the study included (Moses and Anderson, 1993):

1. Port St. Lucie should be considered a threat to the water quality of the Indian River Lagoon by way of the C-24 Canal and North Fork of the St. Lucie River. Expansion and improvement of sewage treatment facilities in this area should be pursued aggressively.

2. The entire county must be regarded as a single environmental area with respect to water and wastewater policies regarding effects of septic tanks and reducing potential for water degradation.
3. A full-time county position should be established that is responsible for identifying and procuring funding sources for program implementation.

In 1990, the State of Florida Legislature passed the Indian River Lagoon Act that required all domestic wastewater treatment facilities to cease discharges into the Lagoon by 1996. It is estimated that implementation of the Act resulted in a 60% reduction of nutrients entering the northern half of Indian River Lagoon. In 1997, seagrass beds at six fixed transects in the Vero Beach area have extended in length an average of approximately 260 feet past previous seagrass bed limits. In the Melbourne area, seagrass beds at seven transects have extended an average of 195 feet (Vernstein and Morris, Indian River National Estuary Program, pers. comm.).

## Tampa Bay

Between 1950 and 1982, seagrass coverage in Tampa Bay declined from approximately 40,000 acres to 21,600 acres. Associated with the seagrass loss were declines in commercial and recreationally important fishes. Three factors were believed responsible for the decline: dredging and filling of seagrass beds for residential, commercial, and port development; shading by algae, both phytoplankton and macroalgae, which bloomed in response to excessive nutrient inputs from sewage treatment plants and industrial discharges; and turbidity induced by dredging the main shipping canal. In 1987, the Florida Legislature passed the Grizzle-Figg Bill, which required that all discharges into Tampa Bay meet strict nutrient guidelines (advanced wastewater treatment). Also, in 1984, the Legislature established a bay study group which, in 1985, resulted in the formation of the Agency on Bay Management, which has become a vigilant guardian of Tampa Bay. From 1982 to 1992, seagrass coverage increased by about 4000 acres (18.5%). Most "new" grass has been shoalgrass (*Halodule wrightii*), an early colonizer that may eventually be replaced in many areas by turtle grass. Increase in seagrass coverage has been attributed to the substantial reduction in nitrogen loadings to the bay. Reduction in nitrogen has allowed more light to penetrate deeper into the water column, thus allowing seagrass to re-establish itself.

## OPTIONS FOR CORRECTING WASTEWATER PROBLEMS

### Onsite Disposal Systems

During the early development of the Keys, human wastes were disposed directly on the ground, into shallow holes, or in the water. Because the population of humans was very small, the ecosystem absorbed these additional nutrients with little or no change to ecosystem structure or function. During the period of rapid development in the Keys, cesspits were constructed under or immediately adjacent to residences and commercial establishments. Cesspits average approximately 4 to 5 ft deep and may be supported with timbers or stacked cement blocks. Cesspits are directly connected to groundwater, adjacent canals, or other surface waters through the porous limerock substrate. Because of the limited amount of land in the Keys, developments are generally crowded, and many early developments featured 50 × 50-ft lot sizes. That circumstance not only maximized the development of many areas, but also provided a concentrated source of nutrient enrichment of groundwater and surface waters. Cesspits provide no treatment of wastewater nutrients. They also do not provide any confinement or treatment of human fecal pathogens. The Florida Department of Health estimates that there are currently approximately 4000 cesspits in the Florida Keys (J. Teague, Florida Department of Health, pers. comm.).

In the mid-1960s, there was a gradual shift to the use of septic systems for onsite waste disposal. This shift was prompted by the newly formed State Board of Health, which recognized that use of cesspits was a public health concern. Septic systems consist of a concrete or fiberglass tank designed to hold waste material anaerobically. Some nutrients are removed through the production of biomass that settles in the tank. The accumulated organic sludge must be pumped periodically and disposed. Pumped sludge is currently disposed by transporting it to Dade County, where it is added to the wastewater entering sewage treatment plants. Liquid wastewater effluent exits from the outlet of the septic tank and is distributed to the surface substrate (drainfield). Ideally, the drainfield is composed of soils with cation exchange sites that trap and hold chemical nutrients. Unfortunately in the Florida Keys, the substrate is predominantly porous limestone and has few bonding sites for nutrients other than phosphorus, and the long-term capability of limestone to trap phosphorus is not known. Thus, in the Keys, wastewater from septic systems can rapidly seep into the groundwater with little nitrogen removal. Removal of nutrients by septic systems can vary greatly because of design, installation methods, and operation, but on the average septic systems remove approximately 4% of nitrogen and 15% of phosphorus (Table 31.8). Location of septic tanks near surface waters, as in closely spaced canal developments, represents a significant source of nutrient-rich wastewater to surface waters. Rain events and low tides can result in the rapid movement of septic tank effluent into surface waters. Approximately 25,000 parcels in the Keys have onsite disposal systems. Of those, 18,000 are permitted septic tank systems. Approximately 7900 lots have no record of sewage disposal method (J. Teague, Florida Department of Health, pers. comm.).

Since 1992, the State Department of Health has required that septic tank drainfields be underlined by 12 inches of clean sand. This requirement has little impact on nutrient removal in the effluent, but the sand may trap some pathogens by filtration through the sand bed.

**TABLE 31.8**
**Percent Nutrient Removal and Pollutants Remaining with Central Wastewater Treatment and Onsite Treatment Systems**

| | Central Treatment Systems | | | Onsite Treatment Systems | | |
|---|---|---|---|---|---|---|
| Pollutant | Secondary | Advanced Secondary | AWT | Septic Tank | Aerobic | Composting Toilet |
| **Percent Removal** | | | | | | |
| Carbon | 85 | 97 | 97 | 44 | 77 | 65 |
| Nitrogen | 12 | 13 | 93 | 4 | 17 | 82 |
| Phosphorus (no ban) | 10 | 13 | 93 | 15 | 9 | 40 |
| Phosphorus (with ban) | 20 | 26 | 86 | 30 | 18 | 81 |
| **Nutrients Remaining after Treatment (g/person/day)** | | | | | | |
| Carbon | 7.2 | 1.4 | 1.4 | 26.9 | 11.0 | 16.8 |
| Nitrogen | 7.0 | 7.0 | 0.6 | 7.7 | 6.6 | 1.4 |
| Phosphorus (no ban) | 3.6 | 3.5 | 0.3 | 3.4 | 3.6 | 2.4 |
| Phosphorus (with ban) | 1.6 | 1.5 | 0.3 | 1.4 | 1.6 | 0.4 |

*Source:* From Harkin, J., *Nutrients in Wastewater and Nutrient Removal*, U.S. Environmental Agency, Region 4, Atlanta, GA, 1996.

The elevations of the Keys are very close to sea level, so the groundwater is very close to the surface. The ground level of most dredge-and-fill subdivisions is only 3 to 4 ft above sea level. On most Keys, the groundwater is as salty as seawater. The groundwater responds to tidal action that connects it with Florida Bay and the Atlantic Ocean. The net movement of groundwater is toward the Atlantic Ocean (Shinn et al., 1994b).

Septic tanks are installed underground, and during installation they can float in areas where groundwater is high. Many of the early installed septic tanks had holes punched in their bottoms to sink them in the groundwater before they were covered over with fill material. Because of that practice, many septic systems function as cesspits, where wastewater is in direct contact with groundwater without settlement or treatment.

Chapter 381.0065(4)(k) of the Florida Statutes currently mandates that the Florida Department of Health (FDOH) permit only onsite systems capable of meeting advanced waste treatment (AWT) standards. AWT is defined in Section 403.086 of the Florida Statutes as 5 mg/l CBOD, 5 mg/l total suspended solids (turbidity), 3 mg/l nitrogen, and 1 mg/l phosphorus. In order to remove nutrients to those levels, additional processes must be incorporated into the treatment process. In the nitrification/denitrification process, ammonia is first converted to nitrate (aerobically) and then nitrate is converted to nitrogen gas (anaerobically) and released to the atmosphere. Phosphorus can be removed either biologically or chemically. In either case, excess phosphorus is removed from the effluent stream through settling and subsequent disposal of the solids.

Ayres and Associates, Inc., under contract to FDOH, is field testing five onsite systems at a test facility on Big Pine Key. This research has been funded through 1998. Pending the results of that research, the FDOH has been permitting onsite systems that meet current best available technology. FDOH has determined that aerobic onsite treatment systems that discharge to either a bore hole or a drainfield currently must meet BAT. A wide variety of designs of onsite aerobic treatment systems exists, but in general they are a small-scale version of a conventional secondary treatment plant. Some nutrients are removed through the growth of bacterial biomass and subsequent disposal of biosolids. Operation and maintenance is critical to the efficient performance of aerobic systems. Aerobic systems are much more efficient than septic tanks in removal of carbon (77%) but are only slightly better than septic systems in removing nitrogen (Table 31.8).

Currently, no AWT treatment facilities are located in the Florida Keys. By state statute, advanced wastewater treatment facilities are required in two locations in Florida, Tampa Bay and Indian River Lagoon. The statutes were enacted to address eutrophication of those waters due to excess nutrient loading and to reverse the pending collapse of the ecosystems.

## PACKAGE PLANTS

During the early development of the Keys, multifamily dwellings, motels, and resorts utilized cesspits and septic tank systems. Florida Administrative Code (F.A.C.) Rule 62-620.100 currently requires a valid permit from the Florida Department of Environmental Protection (FDEP) for construction and operation of domestic wastewater facilities with flows exceeding 10,000 gal/day (gpd) and for commercial establishments with wastewater flows greater than 5000 gpd.

At the present time, 250 FDEP-permitted wastewater treatment plants (WWTPs) are in operation in Monroe County. Approximately 14 additional FDEP wastewater permits have been issued for new WWTPs that have not yet been constructed. Most of these WWTPs consist of onsite facilities with permitted flows under 100,000 gpd and are commonly known as "package plants." However, it is important to note that discharges from package plants represents about 33% of the total wastewater flow from FDEP package plants. The remainder of the flow (67%) comes from discharges from a few large facilities with permitted capacities exceeding 100,000 gal/day, or 0.1 million gallons per day (mgd).

All FDEP-permitted WWTP systems are required to meet, at a minimum, secondary treatment and disinfection requirements of Chapter 62-600, F.A.C. That regulation requires supervision and monitoring of these facilities by a Florida licensed operator and submission of discharge monitoring reports, containing all required test results, for each month of operation. These facilities are also inspected by FDEP personnel to ensure compliance with permit requirements.

Secondary treatment provides up to 90% removal of the total suspended solids and organic (carbon) wastes producing oxygen demand (CBOD) in the wastewater. This process also removes organic nitrogen and phosphorus associated with the suspended solids, but does little to remove nutrients dissolved in the wastewater such as nitrates and phosphates. Chlorination is employed for disinfection of the effluent in order to protect public and environmental health. The wastewater sludge from the settled solids is periodically removed and transported to the mainland for disposal at FDEP-permitted treatment facilities. Alternatively, disinfected wastewater sludge can also be delivered to approved land application sites for disposal, as long as the wastewater sludge meets the treatment criteria specified in Chapter 62-640, F.A.C., and the Code of Federal Regulations Part 503.

Because of strict regulatory standards required for surface water discharges, the primary method of effluent disposal employed by the package plants is discharge to the groundwater by means of Class V wells. Currently, 750 FDEP-permitted Class V wells are in the Florida Keys. These disposal wells are required to be drilled to a depth of 90 ft and lined with cement (cased) to 60 ft. As of June 1997, Chapter 62-528, F.A.C., requires that all Class V wells designated to inject domestic wastewater in Monroe County, as part of their operation permit application, provide reasonable assurance that operation of the well will not cause or contribute to a violation of surface water standards as defined in Chapter 62-302, F.A.C.

At least nine WWTPs in Monroe County utilize wastewater re-use systems, such as subsurface or spray irrigation, as either their primary or secondary effluent disposal method. The use of drainfields and percolation ponds for groundwater effluent disposal is also allowed in accordance with Chapter 62-610, F.A.C., but the use of those systems is limited in Monroe County because they require large surface areas and because of the lack of soil and the high groundwater table in the Keys.

## CENTRAL SEWAGE SYSTEMS

Central sewage systems involve the collection of wastewater from multiple sources by means of a sewer system and pumping the wastewater to a sewage treatment facility for treatment and disposal. Construction of sewage collection systems for wastewater is difficult and expensive in the Keys because of the rock substrate. Central treatment of wastewater in a large-volume sewage treatment plant is very efficient because of the economy of scale and the presence of full-time operators. Two municipal central sewage collection and treatment systems are currently operating in the Keys: Key Colony Beach and Key West. In addition, several privately owned utilities in Monroe County have central collection systems, including Key West Resort Utilities (Stock Island), Key Haven Utilities, Key West Naval Air Station, and Key Largo Utility (Ocean Reef Club). All these facilities, with the exception of the City of Key West, use Class V wells and/or wastewater reuse systems for effluent disposal.

The City of Key Colony Beach operates a wastewater treatment plant with a current capacity of 0.22 mgd. This facility is over 28 years old and provides secondary treatment of wastewater. Wastewater is collected through a gravity sewer line system that includes 15 lift stations. The 20- to 30-year-old collection pipes are subject to infiltration of saline groundwaters, particularly during extreme high tides. Prior to 1994, effluent was discharged directly to the Atlantic Ocean. In late 1994, the discharge was rerouted to six Class V injection wells. The wastewater treatment plant services 1233 residential units and 96 business units. The facility has not utilized re-use of treated wastewater for irrigation because the cost of additional treatment facilities required and the need for increased operator attendance are much greater than the cost of potable water presently being used to irrigate greens of the nine-hole, par 3 golf course. Also, the high amounts of saline groundwater infiltration into the collection system would make the effluent too saline at times for irrigation use. Plans are being developed to replace the existing facility with a facility that can achieve AWT standards. Efforts to correct the infiltration problem are ongoing.

The City of Key West collects and treats wastewater at a central treatment plant with a permitted capacity of 7.2 mgd of secondary treated wastewater. Discharge is through a submerged ocean outfall located about 1000 m (328 ft) from the southern tip of Key West. The ocean outfall is the largest single source of nutrient pollution in the Keys. In 1997, the effluent included an average of 342 lb of nitrogen and 62 lb of phosphorus daily. However, because of the tremendous dilution at the location of the outfall, Ferry (undated) concluded that impacts from the ocean outfall are mainly limited to localized eutrophication and some sewage contamination of the benthos in the immediate vicinity of the outfall. The probability of transport of any significant amounts of pollutants or contaminants from the outfall to offshore bank reefs appears to be low.

Key West is currently under an enforcement action by FDEP for violations related to collection system failure and excessive infiltration. During 1996, a Consent Judgement was prepared by FDEP requiring the city to take corrective action to reduce the infiltration problem. The city has proposed a 5-year schedule and has initiated an aggressive sewer rehabilitation program. The level of treatment currently approaches advanced treatment standards; since 1995, the effluent has averaged 4.2 mg/l nitrogen and 1.1 mg/l phosphorus. Also, the city has decided to eliminate the ocean outfall and inject treated wastewater into a deep injection well drilled into the boulder zone (2500+ feet). The FDEP has issued and a deep well permit.

Key Haven Utility is a private system serving the subdivision of Key Haven on Raccoon Key, located just east of Stock Island. The plant is currently permitted for 0.20 mgd provided by two connected facilities. One unit was constructed in approximately 1970 and is currently in poor condition. The second unit was built in 1994. Plant upgrades have been undertaken to replace the original unit. The wastewater collection system consists of a gravity sewer with five lift stations. The collection system has a history of infiltration problems. None of the treated wastewater is re-used.

Key West Resort Utility provides service to approximately 90% of the Stock Island area south of U.S. Highway 1. This includes approximately 600 residences and several commercial establishments. Plant capacity is 0.499 mgd and the plant is in excellent condition. The collection system consists of a gravity sewer, force mains, and 13 lift stations. The primary disposal method is spray irrigation on the Key West Golf and Country Club golf course located on Stock Island north of U.S. 1. The effluent is treated to public access re-use standards as required in Part III, Chapter 62-610, F.A.C. Secondary disposal is to Class V injection wells during wet periods of the year. The plant does not currently serve the entire area of the utility district. Plans are currently underway for an expansion that will include the entire utility area. Also, the utility is interested in expanding its boundaries to include adjacent areas.

Key West Naval Air Station treatment plant has a capacity of 0.4 mgd and serves the Naval Air Station. The plant has a history of exceeding peak flow capacities and has significant infiltration problems. Improvements have been recently made to the plant and to the 11 lift stations. In addition, Class V wells have been installed to better manage effluent disposal.

The Key Largo Utility serves Ocean Reef and Anglers Club developments located at the extreme north end of Key Largo. The plant consists of two connected units, one older than the other, and has a capacity of 0.55 mgd. Both units are in good condition. The collection system consists of gravity sewers, force mains, and 37 lift stations. The collection system has infiltration problems that are currently being improved by replacing older clay and cast iron pipes with PVC pipes. Effluent is treated to secondary standards and disposed into Class V injection wells. The facility does not currently provide for re-use because infiltration problems result in an effluent with high chloride concentration. The three 18-hole golf courses in the development currently irrigate using a 1.7-mgd reverse osmosis plant ($1.75 per 1000 gallons). Complete rehabilitation of the collection system for re-use would be more costly than costs of water from the reverse osmosis plant.

## WASTEWATER FACILITIES PLAN FOR THE MARATHON AREA AND PHASED IMPLEMENTATION FOR LITTLE VENICE (VACA CUT TO 94TH STREET)

In February 1996, Monroe County completed a Marathon Area Facilities Plan. That plan originated in recognition of the need to develop a long-range wastewater management plan for Monroe County. Marathon was chosen as the first area in Monroe County for this planning because of the large number of high density developments with small lot sizes, a large number of identified cesspits, and documented degraded water in canals. The purpose of the plan is to define the most cost-effective, environmentally sound, and implementable program for the management of existing and future wastewater pollutants that currently act, or will act, to deteriorate the water quality in the Marathon area. The plan will be a part of a comprehensive Wastewater Master Plan for Monroe County. In general, three steps comprise implementation of a wastewater management system: planning, design, and construction. The Marathon Area Facilities Plan is the first step in the implementation of a wastewater management system for the Marathon area and includes:

- Evaluation of existing water quality
- Identification of existing point source pollution sources
- Documentation of existing background environmental conditions
- Preparation of an inventory of existing wastewater plants
- Estimation of future waste loads and flows
- Development and evaluation of collection, treatment, and disposal alternatives
- Identification of a potential site, or sites, for location of treatment facilities
- Selection of the most cost-effective, environmentally sound, and implementable wastewater management alternative
- Development of conceptual design and planning level cost estimates for the recommended plan
- Assessment of the recommended plan's environmental impact
- Discussion of the institutional framework and financial requirements needed to implement the plan

The Marathon Area Wastewater Facilities Plan concluded that a regional wastewater collection, treatment, and disposal system be implemented to serve the primary service area (Seven-Mile Bridge to Coco Plum, excluding Key Colony Beach). The recommended technology for the wastewater collection system is a vacuum system, comprised of vacuum collection mains, combination vacuum/conventional pumping stations, and force mains. Based on direction provided by the Monroe County Board of County Commissioners, the regional wastewater treatment plant will treat the wastewater to AWT standards to provide a high level of solids and nutrient removal. The recommended effluent management system is deep underground injection of highly treated effluent to the Boulder Zone (2500 ft). The Plan recommends that re-use of effluent be explored. The estimated capital and annual operation and maintenance costs for collection, treatment, and disposal are given in Table 31.9.

In February 1996, the Monroe County Board of County Commissioners approved the recommendations in the Facilities Plan provided that the connection fee per household does not exceed $1600 and monthly service fee does not exceed $35. Monroe County has applied for a loan from the State Revolving Fund for approximately $30 million for design and construction costs. In October 1997, Congress, with the assistance of the Governor's Office and EPA Region 4, appropriated $4.3 million of de-obligated Title II construction fund monies to be used in wastewater improvements in Monroe County. These funds will be used by amending the Marathon Area

**TABLE 31.9**
**Estimated Costs for Marathon Central Collection and Treatment**

| Item | Amount ($ Millions)[a] |
|---|---|
| Construction Costs | |
|     Collection/transmission system | 29.1 |
|     Wastewater treatment plant | 5.0 |
|     Effluent disposal system | 2.3 |
|     Solids management system | 1.7 |
|     Land acquisition | 3.5 |
|     Subtotal | 41.6 |
| Other Costs | |
|     Contingency (25%) | 10.4 |
|     Engineering, legal, administrative (15%) | 6.2 |
|     Financing (33% of costs @ 12%) | 2.3 |
|     Subtotal | 18.9 |
| Total Capital Costs | 60.5 |
|     Annual operation and maintenance | 1.4 |
|     Annual renewal and replacement | 0.1 |
|     Administrative costs | 0.4 |

[a] 1995 dollars.

Wastewater Facilities Plan to include a first phase for Little Venice (Vaca Cut to 94th Street on the ocean side of U.S. 1). An offer and award of these monies was made in October 1998, with design to begin in Fall 1999 after contract approval. Construction should begin in late 2001 and be completed by early 2003.

During the second session of the 105th Congress, Monroe County and others spent considerable time working with Congress to develop additional wastewater funding proposals for the Marathon Area Facilities Plan and for the projected needs for the remainder of Monroe County. Included in these discussions was the concept of developing a nontransportation toll on U.S. 1, located somewhere north of Key Largo. Efforts are ongoing and will continue in the 106th Congress and beyond.

## Monroe County Wastewater Master Plan

Monroe County initiated the development of a countywide Wastewater Master Plan in August 1997. The purpose of the Wastewater Master Plan is to identify environmentally acceptable and cost-effective wastewater treatment and disposal alternatives for geographic service areas within the Florida Keys. Different wastewater management practices, from onsite systems to community and/or regional collection and treatment systems, will be evaluated for each geographic service area and the costs and environmental benefits compared. The cost of development of the Wastewater Master Plan is approximately $2.2 million.

At the outset, Monroe County and the Water Quality Protection Program Steering Committee approved a Technical Advisory Committee (TAC) consisting of approximately 20 individuals with interest and/or special knowledge and expertise to oversee the development of technical documents produced in the project. The TAC will review work products and meet with the consultant approximately six times during the course of the project. When complete, the Wastewater Master Plan will be evaluated by the Water Quality Program Steering Committee and approved by the Board of County Commissioners (BOCC).

New legislation has been recently passed regarding the authority of the Florida Keys Aqueduct Authority (FKAA) to function as a wastewater utility. A Memorandum of Understanding has been finalized between Monroe County and the FKAA regarding the agencies' roles in wastewater treatment. The FKAA has become the utility authority for Monroe County. The wastewater Master Plan was completed in June 2000 and approved by the Monroe County Board of County Commissioners.

## CESSPIT IDENTIFICATION AND REPLACEMENT

In conformance with the Governor's Executive Order 96-108 and Polity 901.2 of the Monroe County Year 2010 Comprehensive Plan, Monroe County and the FDOH initiated a 5-year operating permit procedure for onsite disposal systems. The ordinance requires homeowners to have their onsite disposal system inspected within 30 days of notification. Notification dates are based upon the age of the structure; older structures are notified first. Inspection results must be submitted to the Florida Department of Health. Disposal systems found to be in compliance with current requirements will receive a 5-year operating permit. Disposal systems that are found to be in compliance with requirements in place when the structure was built, but do not meet current minimum standards, will receive a 2-year temporary operating permit; those systems must be replaced within 2 years. Structures found to have a cesspool or septic tank that does not meet the standards in place at time of construction must comply with current standards within 180 days of written notice.

Low-interest loans are available to assist homeowners in funding replacement of inadequate onsite disposal systems. Homeowners in the Marathon Service Area have been exempted from compliance with this ordinance because the method of central collection and treatment was determined to be the most cost effective and environmentally acceptable solution of wastewater disposal in that service area. In other geographic areas, homeowners with an approved system for which an operating permit has been obtained may continue to use the approved system so long as:

1. The system is properly maintained and remains in satisfactory operating condition.
2. The operating permit is properly renewed.
3. An approved sewage treatment plant has not been available for connection for longer than 365 days.
4. No alterations are made to the residence, commercial structure, or site that would change the sewage or wastewater characteristics, increase sewage flow, or impede the performance of the onsite disposal system.

## CANAL BEST MANAGEMENT PRACTICES

Homeowners can undertake many simple activities that will help improve water quality in canals adjacent to residences. Activities can be divided into two categories:

1. Reducing nutrient loading into canal water
2. Increasing circulation and flushing (where applicable)

### Reducing Nutrient Loading

Because canals generally exhibit poor circulation and flushing, they are very susceptible to eutrophication due to excess nutrients. The following activities are required to minimize nutrient loading into canals:

- Eliminate cesspits.
- Install adequate drainfields for septic systems that results in binding of nutrients.
- Pump out septic tanks on a regular basis to prevent organic loading from tanks full of sludge.
- Do not apply fertilizers on lawns or other vegetation adjacent to canals.

- Do not dispose of organic wastes into canals, including grass clippings, animal droppings, fish carcasses, etc.
- Slope lots adjacent to canals so that surface drainage is directed away from canals.
- Eliminate fast-growing exotic vegetation from canal banks (e.g., Australian pine, Brazilian pepper) and maintain native vegetation (e.g., buttonwood and mangroves) as a buffer.
- Use phosphate-free detergents.
- Do not discharge gray water onto soil or into canals.
- Eliminate live-aboard discharges into canals.

## Canal Circulation

Deep, deadend canal systems exhibit poor water quality due to the geometry of the canal system. Several physical alterations can be attempted that may improve canal water quality, including:

- Backfilling canals to a maximum of –6 ft MSL at the mouth of the canal and sloped to –4 ft MSL at its distal end
- Aerating canal waters to assist vertical circulation
- Dredging canals or otherwise treating canal bottoms to remove accumulation of organic, oxygen-demanding sediments
- Installing flushing channels/culverts in suitable areas if actions will not degrade receiving waters

The orientation of some canals make them susceptible to accumulation of wind-driven, floating organic matter, predominantly seagrass leaves. Physically preventing transport of floating organic matter into canals will improve quality of canal waters. Floating booms, air curtains, and other devises are used as weed gates at mouths of canals.

Several canal systems in the Keys were constructed but never connected to adjacent waters. Those canal systems are plugged with fill material at their mouths. Recently, there has been increased interest in removing the plugs from those canal systems to connect them with adjacent surface waters. Removal of plugs requires federal, state, and county permits. Permit agencies recognize that existing open canal systems represent a source of degraded water quality to receiving waters and that water quality within open canals may violate state water quality standards. Therefore, there is a great reluctance to consider requests to open additional canal systems. Before such a request can be considered, there must be overwhelming evidence that the canal currently does not violate water quality standards and that opening of the canal system will not degrade receiving waters. Generally, currently plugged canals systems will not meet those requirements.

## DISCHARGES FROM VESSELS

A large community in the Keys lives on boats. Many live-aboard vessels are permanently anchored in harbors and are not capable of movement. Transient vessels also anchor in harbors and other protected sites and are very numerous in winter months. The number of live-aboard vessels has increased dramatically in recent years. For example, the number of live-aboards in the Key West area increased from 235 in 1992 to 393 in 1995 (Monroe County Grand Jury Report). Over 300 anchored or moored vessels were observed in Boot Key Harbor (Marathon) in February 1995 (Kruczynski, pers. observ.). A Monroe County Grand Jury received testimony that up to 80% of live-aboard vessels do not use sewage dumping facilities.

The Clean Vessel Act (Florida Statute 327.53) prohibits the discharge of raw sewage from any vessel, houseboat, or floating structure into Florida waters. A houseboat is a vessel that is used primarily as a residence (21 days out of any 30-day period), and its use as a residence precludes its use as a means of transportation. Houseboats and floating structures must have permanently installed toilets attached to Type III Marine Sanitation Devices (MSDs) or connect their toilets directly to shoreside plumbing. A Type III MSD is one that stores sewage onboard in a holding tank for pumpout. Houseboats may also have other approved MSDs on board, but, if they do, the valve or other mechanism selecting between devices should be selected and locked to direct all sewage to the Type III device while in state waters. All vessels that have MSDs capable of flushing raw sewage directly overboard or into a holding tank should set and secure the valve directing all waste to the holding tank, so that it cannot be operated to pump overboard while in state waters. All waste from a Type III MSD or from portable toilets shall be disposed of in an approved sewage pumpout or waste reception facility.

While the Clean Vessel Act prohibits the dumping of raw sewage, treated wastewater from transient vessels may be discharged into state waters. Wastewater treatment (disinfection) by Type I and II MSDs does not remove nutrients from wastewater. Gray water does not have to be stored or treated from any vessel and may be discharged directly into water of the state.

There are few land-based pumpout facilities in the Keys and no mobile pumpout facilities. One land-based pumpout facility is located in Boot Key Harbor. Thus, many live-aboard vessels and most transient vessels discharge wastewater into surface waters. It is estimated that nutrients from vessel wastewater account for 2.8% of nitrogen and 3.0% of phosphorus loadings into nearshore waters of the Keys (Table 31.3). Although nutrient loadings from vessels may be relatively minor contributions to the total loading, loadings from vessels are a significant source to harbors and result in eutrophication of waters that typically exhibit poor circulation/flushing. Violations of fecal coliform standards are common in marinas and harbors (Florida Department of Environmental Regulation, 1987, 1990).

The USEPA, the State of Florida, Monroe County, and the City of Key West are pursuing the designation of marinas, harbors, and anchorages as "no-discharge zones" (NDZ). The NDZ designation will require all boats, live-aboards, and transients to use Type III holding tanks and have wastes pumped at approved facilities. Federal regulations require that adequate pumpout facilities be available before an area is designated as a NDZ. Plans are being developed for construction of land-based and mobile pumpout facilities and for strict enforcement of the prohibition against disposal of wastewater into surface waters of the state.

## Stormwater Treatment

Stormwater runoff can be successfully treated with the use of one or more best management practices (BMPs). Stormwater treatment BMPs in Florida typically are described in Florida Department of Environmental Regulation (undated) and usually consist of a retention or detention facility, such as a pond or large swale area. These facilities are designed to capture 80 to 95% of the runoff. Effectiveness of BMPs can vary, however. One national study of pond BMPs used in urban areas found that about half of the phosphorus was removed from the runoff and roughly one third of the nitrogen (Center for Watershed Protection, 1997). Therefore, even under the best of circumstances, treatment of the runoff downstream raises several questions:

1. What are the water quality criteria or success criteria to be met?
2. What are the sources of pollutants?
3. How much land is available to install BMPs?
4. What types of BMPs will be most effective?

Waters surrounding the Florida Keys are Outstanding Florida Waters (OFW), the most protected by state law. Where a discharge to an OFW is permitted, the South Florida Water Management District (SFWMD) requires that ambient water quality is not degraded. Specific water quality targets for some substances are listed in Appendix K of the OFW report to the Environmental Regulation

Commission (Florida Department of Environment Regulation, 1985). Currently, all stormwater systems permitted in the Keys by the SFWMD must retain from 1 to 2.5 inches of runoff on site. Most storm events are less than this, so discharge volumes are zero most of the time. Where an outfall discharges into the OFW, an additional 1 to 1.25 inches (50%) must be retained on site. These design criteria are presumed to achieve OFW water quality criteria, although a detailed analysis for the Keys has not been conducted.

Source control is an important issue. If pollutants can be prevented from entering the runoff stream at the source, it can greatly reduce the expense of treating runoff downstream. This can often be accomplished by implementing better housekeeping practices on individual properties and can sometimes save property owners money. A homeowner, for example, may use less fertilizer and get identical results. A business owner may find that captured wastes can be recycled and turned into an asset. Government and educational programs such as "Florida Yards and Neighbors" can assist property owners in identifying low-cost ways to reduce pollutant loads.

The very limited land available in the Florida Keys profoundly affects the types of BMPs that can be utilized. The typical land-intensive BMPs used elsewhere in Florida are not feasible in the Keys; however, some BMPs utilized in urban areas of South Florida can be implemented in the Keys. These BMPs take advantage of salt-intruded groundwater and high percolation rates of the soils. In some cases, pumps may be required because of low elevations. No single BMP is typically adequate to treat a runoff stream. Well-designed stormwater treatment systems include a series of BMPs to ensure that as much as possible of the pollutant load is removed.

These issues and the issue of where and how to best spend public funds to improve stormwater runoff in the Keys is best analyzed within the context of a master plan. A good stormwater master plan will include an objective evaluation and recommendations tied to specific outcomes. Some of these kinds of analyses have been conducted in specific locations, such as Key Colony Beach; however, a regional plan is needed. To this end, Monroe County and the SFWMD have established a partnership. Monroe County is developing the scope of work for a master plan. Once the project is "scoped," professional services can be retained to complete the plan. One of the areas of investigation in the plan will be the issue of hot spots of water quality degradation. Hotspot areas will be evaluated to determine what portion of the pollutant load could be related to stormwater runoff. The plan will recommend measures that will be effective for remediation in those areas. The plan will include a state-of-the-art load analysis for the Keys and will examine current design feasibility of various BMPs.

Implementing stormwater treatment measures in the Keys will be very expensive. The cost of stormwater improvements is estimated to be between $370 million and $680 million, depending on the percentage reduction in stormwater pollutant loadings to be achieved and areas selected for retrofitted treatment BMPs (USEPA, 1993).

## CARRYING CAPACITY

Ecosystems are able to assimilate and adjust to certain levels of stresses. When stresses reach threshold levels, changes to the ecosystem structure and function will occur. Some changes are acceptable or reversible once the stress is removed. Other changes are detrimental and permanent and can lead to the collapse of the existing ecosystem. Carrying capacity is an ecological concept that delineates acceptable limits of stresses to an ecosystem.

Carrying capacity analysis can define threshold limits of nutrients that will result in eutrophication of waters. As a result of a legal challenge of the Monroe County Comprehensive Plan, the State Hearing Officer in that case determined that the nearshore waters adjacent to the Florida Keys have exceeded the carrying capacity for assimilation of nutrients.

Determining the number of people a geographic area can support without irreversible or unacceptable damage to the ecosystem is a complex analysis. Carrying capacity has many components including socio-economic, aesthetic, public health and safety, as well as environmental.

Quantifying these elements requires defensible data and consensus on assumptions of thresholds, limiting factors, and acceptable limits.

The U.S. Army Corps of Engineers has completed a "Draft Scope of Work" for a carrying-capacity analysis of the Florida Keys. The Scope has been submitted to the Florida Department of Community Affairs and is being reviewed by experts in carrying capacity analysis. The results of this important study will be used by planners in setting acceptable limits of growth and use of this important and unique ecosystem.

## MONITORING

A long-term, comprehensive monitoring program is required by the Florida Keys National Marine Sanctuary and Protection Act. Monitoring is critical in maintaining and improving the ecological condition of the Sanctuary, as it will provide information on the status and trends of water quality and important biological parameters. Data generated by monitoring programs will provide managers the necessary information for identifying or confirming problem areas. In addition, monitoring is required to evaluate the effectiveness of corrective actions taken to reduce pollution sources. Water quality, coral reef and hard-bottom, and seagrass monitoring programs were designed in 1993 (USEPA, 1993) and finalized in 1995 (USEPA, 1995).

### WATER QUALITY

The Water Quality Monitoring Program uses a stratified random design based upon the EPA Environmental Monitoring and Assessment Program (EMAP) hexagonal grid (Overton et al., 1990). Strata were based upon variability of physical transport regimes, as described by Klein and Orlando (1994). Nearshore to offshore transects are randomly located within strata (Figure 31.1). Segment 1 includes the Tortugas and surrounding waters and is most influenced by the Tortugas gyre of the Loop Current (Lee et al., 1994). Segment 2 includes the Marquesas Keys and the Quicksands. Segment 4 is the shallow waters around the myriad of Keys in the "Back Country." Segment 6 is the Sluiceway that is heavily influenced by transport from Florida Bay and Gulf shelf waters. Segments 5, 7, and 9 include inshore, Hawk Channel, and Reef Tract waters on the Atlantic side of the Keys.

Approximately 150 stations have been sampled quarterly since March 1995. Data for 1997 are summarized in Figures 31.2 to 31.5. Several trends are apparent in the data. Silicate is an indicator of freshwater and was highest in the Sluiceway (6) and Back Country (Figure 31.2). Total phosphorus was highest in Back Country and Sluiceway and lowest in the and Upper Keys (9). Total inorganic nitrogen was highest in the Back Country and at stations adjacent to the Keys in the Upper(5) and Middle (7) Keys. Chlorophyll was highest in the Marquesas Quicksands (2), probably due to Gulf shelfwater input, and lowest in the Tortugas and Upper Keys. Turbidity was highest and most variable in the shallow waters of the Back Country and Sluiceway and lowest in the Tortugas.

Concentrations of total inorganic nitrogen, total phosphorus, silicate, and turbidity were highest inshore and declined toward the reef tract (Figures 31.3 and 31.4). The Lower and Middle Keys had much higher nearshore concentrations than the Upper Keys. Total inorganic nitrogen, total phosphorus, silicate, chlorophyll, and turbidity were highest in individual transects situated along passes between the Keys, indicating the prevalence of Sluiceway and shelf influence (Figure 31.5). Waters in Biscayne Bay passes had lower concentrations of nutrients compared to waters in passes between the Keys.

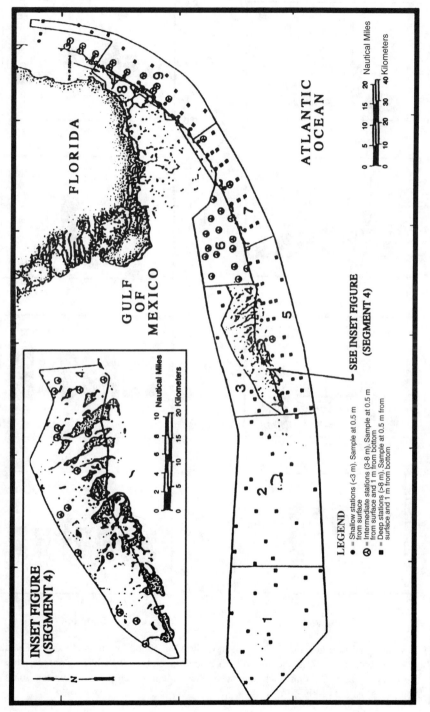

**FIGURE 31.1** Location of segments (strata) in the Florida Keys National Marine Sanctuary and location of water quality monitoring stations.

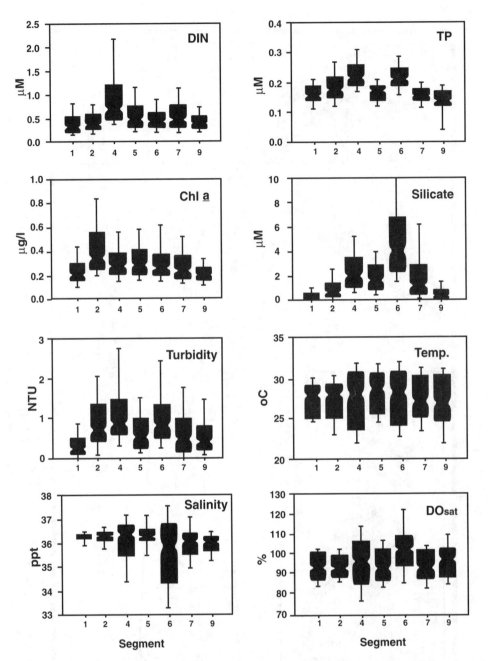

**FIGURE 31.2** Water quality values for 1997 by strata. Central horizontal line in the box is the median, the top and bottom of the box are the 25th and 75th quartiles, and the ends of the whiskers are the 5th and 95th percentiles. The notch is the 95% confidence interval of the median. When notches between boxes do not overlap, the medians are significantly different.

## Coral Reef and Hard Bottom

Prior to the establishment of the Water Quality Protection Program, little robust information on long-term changes in coral reef ecosystems of the Florida Keys existed. The Coral Reef and Hard Bottom Monitoring Program is designed to evaluate the status and trends of 40 permanently located

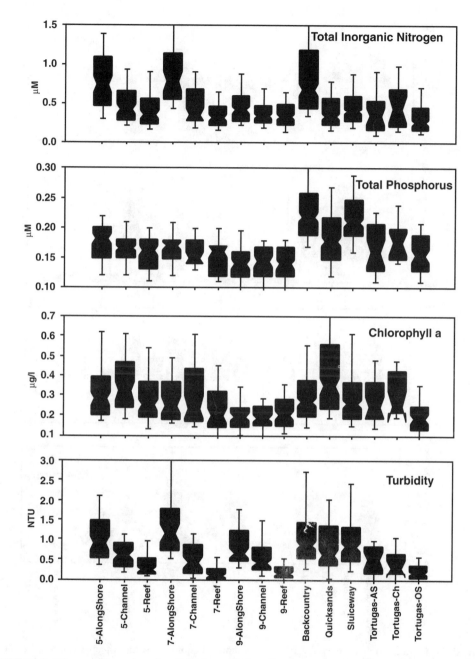

**FIGURE 31.3** Water quality values for 1997 for onshore–offshore transects in Upper (9), Middle (7), and Lower (5) Keys and other sites. See Figure 30.2 legend. When notches between boxes do not overlap, the medians are significantly different.

reef and hard-bottom sites. Stations have been observed annually using video techniques since 1996. A summary of the data on number of taxa by habitat type for 1996 and 1997 is shown in Figure 31.6. Although it is much too early to detect long-term trends and variability, mean species numbers declined for patch reef and offshore deep reef stations. In addition, coral diseases appear to have significantly increased, whether reported in terms of number of monitoring stations affected, number of coral species affected, or number of different diseases recorded.

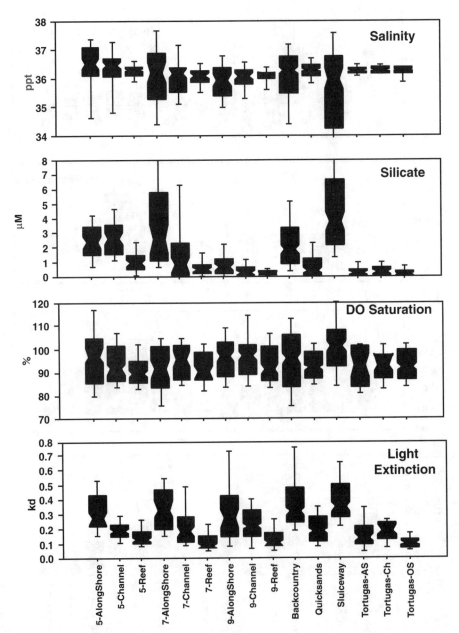

**FIGURE 31.4** Water quality values for 1997 for onshore–offshore transects in Upper (9), Middle (7), and Lower (5) Keys and other sites. See Figure 30.2 legend. When notches between boxes do not overlap, the medians are significantly different.

## SEAGRASSES

A comprehensive seagrass monitoring program in the Sanctuary has been in place since 1996. Distribution, productivity, and morphometrics of seagrasses are monitored quarterly throughout the Sanctuary. Sampling is performed at three levels. At level I sites, shoot morphometrics and productivity of turtle grass are measured quarterly. Level II sites are sampled annually to obtain shoot morphometrics. Level III sites are sampled annually to assess percent cover. Locations of sites sampled in 1996 and 1997 are shown in Figure 31.7. There are approximately 30 level I stations,

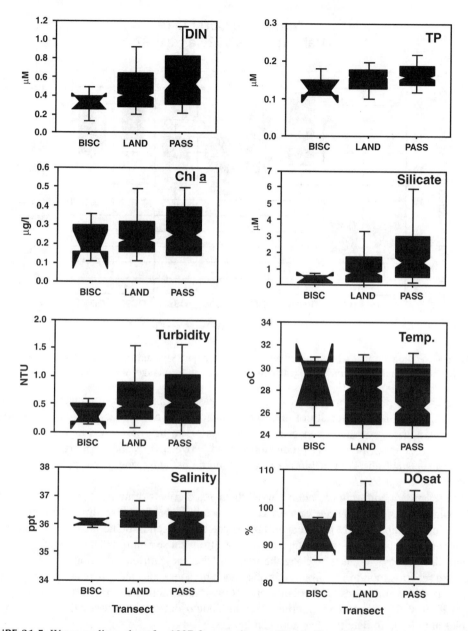

**FIGURE 31.5** Water quality values for 1997 from stations at Florida Keys tidal passes (pass), adjacent to Keys (land), and Biscayne Bay (Bisc). See Figure 31.2 legend. When notches between boxes do not overlap, the medians are significantly different.

87 level II stations, and 187 level III stations. Level I sites were selected to conform with water quality monitoring sites. Level II and III sites are randomly located within segments using the EMAP grid system. The mix of site types is designed to monitor trends through intense quarterly sampling of a few permanent locations (level I) and to annually characterize the broader seagrass population through less intensive, one-time sampling at more locations (level II and III).

Turtle grass and manatee grass are the most stable and widely distributed of the seagrasses within the Sanctuary (Figure 31.8). The overall average standing crop biomass 21.9 g/m$^2$ for turtle grass and 8.2 g/m$^2$ for manatee grass (aboveground, dry weight). Seasonal variations of

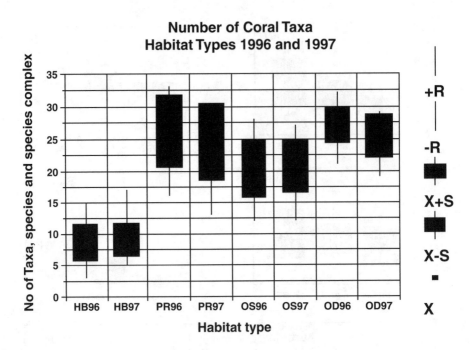

**FIGURE 31.6** Number of coral taxa observed by habitat type in 1996 and 1997. HB = hard-bottom stations; OS = offshore reef shallow stations; OD = offshore reef deep stations. Range, median, and ± one standard error.

standing crop and productivity are evident, with increases in third and fourth quarters of sampling (Figure 31.9). Short-shoot density of turtle grass ranged from 66 to 1025/m$^2$ for all sites. Above-sediment standing crop ranged between 5 and 93 g/m$^2$. Leaf mass exhibited high variation (21 to 415 mg/short shoot). Short shoot production ranged between 0.18 and 8.31 mg/short shoot/day. Higher values were observed in Florida Bay than on the Atlantic side. Areal productivity ranged between 0.07 and 3.37 g/m$^2$.

The seagrass monitoring has not observed the marked effects of nutrient enrichment described by Lapointe et al. (1994). The reason, at least in part, is probably due to differences in sampling methodologies employed. Lapointe et al. (1994) selected hypereutrophic areas associated with a known source of pollutants and sampled transects from those sources. They found gradients in nutrients and biological changes along the transects that they attributed to source pollutants. The long-term seagrass monitoring program is on a much broader scale and utilizes a random sampling pattern. If the observations of Lapointe et al. (1994) are more widespread than their selected sampling sites, the seagrass monitoring program should detect similar variations when enough samples are taken. To date this has not been the case.

## THE ECONOMICS OF CLEAN WATER AND NATURAL RESOURCES

Natural resources have market values and nonmarket values. Market values are the prices of commodities on the open market (e.g., an acre of land). Non-market values are less immediately tangible and include the values of being part of a balanced, self-sustaining ecosystem (e.g., habitat value). Effects of habitat loss and other nonmarket values may take years to become apparent, but these values have long-lasting socio-economic effects. A sustainable market economy depends on maintenance of nonmarket values over long time periods. For example, the tourist-based economy of the Florida Keys depends upon clean water and abundant natural resources. If nonmarket values of these resources decline, the market value will eventually decline.

**FIGURE 31.7** Location of seagrass sampling sites in 1996 and 1997. Large circles are level I sites sampled quarterly. Other sites are randomly located within strata and sampled annually.

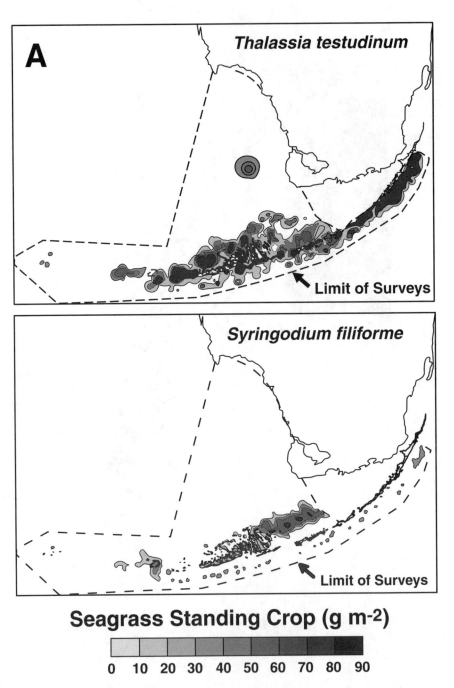

**FIGURE 31.8** Spatial variability of standing crop biomass of (A) *Thalassia testudinum* and (B) *Syringodium filiforme*.

Leeworthy and Bowker (1997) recently quantified the nonmarket value of natural resources in the Florida Keys. The study estimated values tourists receive from the natural resources that are over and above the costs for them to come to the Keys to use them. The study determined that the overall nonmarket user value for visitors to the Florida Keys is $654 per visitor per trip, or $1.2 billion annually. The study estimated that 76% of all activity days by visitors are spent in some sort of natural

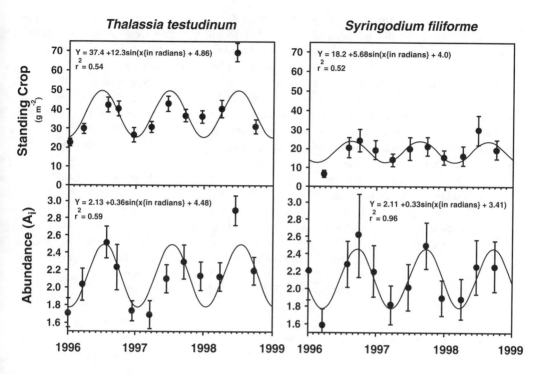

**FIGURE 31.9** Seasonal variation in standing crop and abundance of *Thalassia testudinum* and *Syringodium filiforme*.

resource-related activity. Thus, the amount of nonmarket value attributed to natural resources is $910 million annually (76% of $1.2 billion). When market values ($1.3 billion) are added to nonmarket user values ($1.2 billion), the total annual value of the Florida Keys to tourism is $2.5 billion.

In a sustainable economy, market values do not come at the expense of declines in nonmarket values. Nonmarket user values calculated on a sustainable basis are called asset values, which are long-term market values. The total nonmarket value of the Florida Keys to tourism was calculated to be $24.1 billion and the natural resource total market value was calculated to be $18.3 billion.

Nonmarket user values can be used in benefit–cost analysis of projects that impact natural resources. For example, the cost to improve wastewater and stormwater treatment in the Keys to improve water quality of surface waters may range from $500 million to $1 billion million depending on options selected (USEPA, 1996). Although that is a large cost, it is small compared to the estimated $2.5 billion in annual market and nonmarket values of tourism to the Keys. Even if just the annual natural resource nonmarket value of $910 million per year is used as a comparison, the investment to improve water quality still makes sound economic sense.

Cost of water quality improvements (assume $1 billion) are only 5.5% of the long-term asset value of the natural resource ($18.3 billion). Clearly, the costs of water quality protection and improvement measures are a relatively small proportion of the nonmarket economic user value of the resources they are designed to protect.

## CONCLUSIONS

The Florida Keys are a chain of tropical islands composed of several interdependent community types, including tropical hardwood forests, fringing mangrove wetlands, seagrass meadows, hard and soft bottoms, and coral reefs. The tropical setting and ecological diversity have made the Florida Keys a popular place to live and vacation.

The natural communities that make up the Florida Keys ecosystem exist in a dynamic equilibrium, which means that changes that result in a direct impact to one community type can have profound effects on adjacent communities. The continued existence of the Keys marine ecosystem is dependent upon maintenance of clear waters with relatively low nutrients.

Historically, development in the Keys relied on the use of cesspits and septic tanks, which provide little treatment of domestic wastewater in porous lime rock substrates. In addition, stormwater runs untreated into nearshore surface waters. Lack of nutrient removal from domestic wastewater and stormwater has resulted in the addition of nutrient-rich wastewaters into confined waters and adjacent nearshore areas. The cumulative effects of these discharges have led to water quality degradation of these inshore areas.

The following statements on water quality issues in the Florida Keys are supported by the literature and knowledge of scientists:

1. A rapid exchange of groundwater and surface waters in the Keys is driven by tidal pumping.
2. Cesspits are not appropriate for disposal of wastewater; they are illegal, provide very little treatment, and are a health hazard. Cesspit effluent can rapidly migrate to surface waters.
3. Properly functioning septic tank systems remove very little nutrients (4% N, 15% P) from wastewater, and, depending upon their location, effluent from septic tank drainfields can rapidly migrate to surface waters.
4. Sewage discharged from cesspits and septic tanks is a source of nutrients and human pathogens to ground and surface waters.
5. Contaminants in stormwater runoff contribute substantially to the degradation of nearshore water quality.
6. Water quality problems due to onsite sewage disposal practices and stormwater runoff have been documented in residential canals. Water quality parameters that are degraded include nutrient enrichment, fecal coliform contamination, and biochemical oxygen demand.
7. Long, deadend canal systems, deep canals of any length, and poorly flushed basins accumulate weed wrack and other particulate matter.
8. The water column of many canals over 6 ft deep is stratified, and bottom waters are usually in violation of Florida's Class III Surface Water Quality Standard for dissolved oxygen. Because they usually violate Class III Surface Water Quality Standards, canals were excluded from Outstanding Florida Waters designation.
9. Artificial aeration of canals does not eliminate the sources of excessive nutrients in canal waters but may result in better mixing which may facilitate nitrogen cycling.
10. Improving flushing of degraded canal systems may improve the water quality within the canal but will also result in adding additional nutrients to the adjacent waters.
11. Canal systems and basins with poor water quality are a potential source of nutrients and other contaminants to other nearshore waters.
12. Seagrass beds located near the mouths of some degraded canal systems exhibit signs of eutrophication, such as increased epiphyte load and growth of benthic algae.
13. Vessel-generated turbidity (resuspended sediments) is a growing concern in many areas with high boat traffic including canals and open waters.
14. Aerobic treatment units and package plants provide secondary treatment, removing 80 to 90% of the total suspended solids (TSS) and organic wastes that are responsible for biochemical oxygen demand. In poor soil conditions with high groundwater tables, where drainfields are rendered inefficient, secondary treatment systems are better than septic tanks at removing organically bound nutrients associated with the TSS. These systems, however, are not designed to remove dissolved nutrients.
15. Disposal of wastewater from package treatment plants or onsite disposal systems into Class V injection wells results in nutrient enrichment of the groundwater. However, it is

not known whether discharges into Class V wells results in substantial nutrient loading to surface waters. This question is currently under investigation.

16. In areas where groundwater is saline, injected wastewater is buoyant and rapidly rises to the surface.

17. Recent tracer studies have demonstrated rapid migration of Class V effluent to surface waters (hours to days). These studies demonstrated that tracers were greatly diluted before reaching surface waters and that some phosphorus was stripped from groundwater by the substrate. The long-term ability of the substrate to strip phosphorus is currently under investigation.

18. Sewage discharges from vessels degrade the water quality of marinas and other confined water anchorages.

19. Florida Bay discharge, oceanic and Gulf of Mexico upwelling and currents, rainwater, and other natural sources add nutrients to surface waters of the Keys.

20. Net water movement through the tidal passes between the Keys is toward the Atlantic Ocean. Once entering Hawk Channel, water direction and speed are controlled by prevailing winds and ocean currents.

21. Coral habitats are exhibiting declines in health; coral diseases are more common, and benthic algae have increased in abundance and spatial coverage.

22. No definitive studies on the geographic extent of the impact of anthropogenic nutrients have been conducted. Scientists agree that canal and other nearshore waters are affected by human-derived nutrients from sewage. Improved sewage treatment practices are needed to improve canal and other nearshore waters. Impacts farther from shore that may be due to anthropogenic nutrients may be reduced or eliminated by cleaning up nearshore waters.

23. Planning and implementation of improvements to wastewater treatment are underway. A cesspit identification and onsite disposal certification program has been initiated. A Marathon Area Feasibility Plan has been completed and a Monroe County Wastewater Master Plan has been completed. Funding is being sought for planning, design, and construction of wastewater and stormwater infrastructure.

24. A long-term monitoring program has been implemented to provide information on the status and trends of water quality, coral, and seagrass communities.

25. The costs of water quality improvements are a small fraction of the long-term asset value that natural resources, such as reefs, hard bottoms, and seagrasses, provide to the economy of the Florida Keys.

If sources of nutrient enrichment continue unabated, it is likely that the ecological balance of nearshore communities of the Keys will be changed. Changes in the structure and function of nearshore communities could result in stresses to other components of the Keys ecosystem. Because the economy of the Keys is directly linked to a healthy ecosystem, it is imperative that sources of excessive nutrients to this ecosystem be eliminated. In recognition of the warning signals of degraded water quality, the U.S. Environmental Protection Agency and the State of Florida, in conjunction with the National Oceanic and Atmospheric Administration, have, at the direction of Congress, prepared a Water Quality Protection Program for the Florida Keys National Marine Sanctuary, full implementation of which will help reverse the trend of environmental degradation and restore and maintain the Florida Keys marine ecosystem.

Several significant actions have occurred while this chapter was in press. Long-term monitoring has continued to show significant losses in coral diversity and coverage, particularly at deep reef stations. Between 1996 and 2000, there were significant losses of coral species richness at 67% of 160 stations sampled. Stony coral coverage decreased at 45% of the stations, and coral coverage decreased 38% Sanctuary-wide. Also, disease condition exhibited increased frequency of all coral infections from 1996 to 1997 and 1997 to 1998. There was significant decrease in black band disease in 1998 and 1999, but a significant increase in white-type diseases and other diseases from 1996 to 2000.

A significant increase in total phosphorus was observed in water samples for most of the FKNMS from 1995 to 2000. The trend was linear and ranged from 0.01 to 0.07 µ*M*/year. No long-term temporal trends in total phosphorus were observed in Florida Bay. It is thought that the increases in total phosphorus in the Sanctuary are driven by regional circulation patterns arising from the Yucatan and Loop Currents.

The Monroe County Sanitary Wastewater Master Plan and the Monroe County Stormwater Management Master Plan have been finalized and accepted by the Board of County Commissioners. Those plans provide quantitative analyses of sources of nutrients and recommendations for corrective actions. Monroe County continues to seek state and federal support to implement those plans. Monroe County published a "request for proposals" (RFP) for a design–build–operate solution to wastewater problems in Key Largo. A Technical Review Committee reviewed and ranked the proposals and recommended that a contract be negotiated with the submitter of the highest ranked response. A lawsuit was filed over the procedures used by the Technical Review Committee, and the results of that lawsuit have not yet been resolved.

The Village of Islamorada also published an RFP for a design–build–operate solution to wastewater problems. A Technical Review Committee ranked the proposals and recommended the highest ranked proposal to the village commission for funding. No final action has been taken to date on that recommendation.

## REFERENCES

Adams, C. 1992. *Economic Activities Associated with the Commercial Fishing Industry in Monroe County, Florida*, University of Florida Food and Resources Economics Dept., Institute of Food and Agricultural Science Staff Paper SP92-27, 20 pp.

Antonius, A. 1981a. The "band" diseases in coral reefs, *Proc. Fourth Int. Coral Reef Symp.*, 2:7–14.

Antonius, A. 1981b. Coral reef pathology: a review, *Proc. Fourth Int. Coral Reef Symp.*, 2:3–6

Barada, W. and W.M. Partington, Jr. 1972. *Report of Investigation of the Environmental Effects of Private Waterfront Canals*, Environmental Information Center, Florida Conservation Foundation, Inc., Winter Park, FL, 63 pp.

Bell, P.R.F. 1991. Status of eutrophication in the Great Barrier Reef lagoon, *Mar. Poll. Bull.*, 23:89–93.

Bell, P.R.F. 1992. Eutrophication and coral reefs: some examples in the Great Barrier Reef lagoon, *Water Res.*, 26:553–568.

Bell, P.R.F. and I. Elmetri. 1993. *Validation of 1928–1929 Phosphate and Microplankton Data for Low Isles as a Baseline for Eutrophication in the Great Barrier Reef Lagoon*, report to the Great Barrier Reef Marine Park Authority, Townsville, FL, 33 pp.

Brand, L. 1997. *Semi-Synoptic Sampling of Phytoplankton in the Florida Keys National Marine Sanctuary*, final report submitted to Water Quality Protection Program, 6 pp.

Brodie, J. 1994. *The Problems of Nutrients and Eutrophication in the Australian Marine Environment*, The State of the Marine Environment Report for Australia Dept. Environ., Sports and Territories Special Report, Canberra, 29 pp.

Brodie, J. 1995. *Nutrients and the Great Barrier Reef*, The State of the Marine Environment Report for Australia Dept. Environ., Sports and Territories, Special Report, Canberra, 17 pp.

Center for Watershed Protection. 1997. *Comparative Pollutant Removal Capability of Urban BMPs: A Re-Analysis*, Technical Note 95, Vol. 2, No. 4, 32 pp.

Chanton, J. 1998. *Use of Natural and Artificial Tracers To Detect Subsurface Flow of Contaminated Groundwater in the Florida Keys National Marine Sanctuary*, final report submitted to the Water Quality Protection Program, 52 pp.

Chesher, R.H. 1973. *Environmental Analysis: Canals and Quarries, Lower Florida Keys*, prepared for Charley Toppino and Sons, Inc., Rockland Key, FL, Marine Research Foundation, Inc., Key West, FL, 162 pp.

Cook, C. 1997. *Reef Corals and Their Symbiotic Algae as Indicators of Nutrient Exposure*, final report submitted to the Water Quality Protection Program, 37 pp.

DeMaria, K. 1996. *Changes in the Florida Keys Marine Ecosystem Based upon Interviews with Experienced Residents*, The Nature Conservancy, Key West, FL and the Center for Marine Conservation, Washington, D.C., 105 pp.

Dubinsky, Z. and N. Stambler. 1996. Marine pollution and coral reefs, *Global Change Biol.*, 2:511–526.

Dutka, B.J., A.S.Y. Chau, and J. Coburn. 1974. Relationship between bacterial indicators of water pollution and fecal sterols, *Water. Res.*, 8:1047–1055.

English, D.B.K., W. Kriesel, V.R. Leeworthy, and P. Wiley. 1996. *Economic Contribution of Recreating Visitors to the Florida Keys/Key West: Linking the Economy and Environment of Florida Keys/Florida Bay*, NOAA, The Nature Conservancy, Monroe County Tourist Development Council, 22 pp.

Ferry, R.E. Undated. *Key West Ocean Outfall Study: Synopsis of Results and Conclusions*, Draft Report, U.S. EPA Region 4, Atlanta, GA, 37 pp.

Florida Bay Watch Program. 1996. *Annual Report on Water Quality Data Collected by the Florida Bay Watch Volunteer Program*, July 1995–June 1996, The Nature Conservancy, Key West, FL, 16 pp.

Florida Bay Watch Program. 1997. *Annual Report on Water Quality Data Collected by the Florida Bay Watch Volunteer Program*, July 1996–June 1997, The Nature Conservancy, Key West, FL, 7 pp.

Florida Department of Environmental Regulation. Undated. *Stormwater Management: A Guide for Floridians*, Tallahassee, FL, 72 pp.

Florida Department of Environmental Regulation. 1985. *Proposed Designation of the Waters of the Florida Keys as Outstanding Florida Waters*, report to the Florida Environmental Regulatory Commission, Tallahassee, FL, 57 pp.

Florida Department of Environmental Regulation. 1987. *Florida Keys Monitoring Study: Water Quality Assessment of Five Selected Pollutant Sources in Marathon, Florida*, FDER, Marathon, FL, 187 pp.

Florida Department of Environmental Regulation. 1990. *Boot Key Harbor Study*, FDER, Marathon, FL, 21 pp.

Florida Department of Pollution Control. 1973. *Survey of Water Quality in Waterways and Canals of the Florida Keys with Recommendations*, Department of Pollution Control, Tallahassee, FL, 19 pp.

GESMAP. 1990. *The State of the Marine Environment*, UNEP Regional Seas Reports and Studies, UNEP, Nairobi, 115 pp.

Ginsburg, R.N. and E.A. Shinn. 1964. Distribution of the reef-building community in Florida and the Bahamas (abstr.), *Am. Assoc. Petrol. Geol. Bull.*, 48:527.

Ginsburg, R.N. and E.A. Shinn. 1994. Preferential distribution of reefs in the Florida reef tract: the past is the key to the present, in *Global Aspects of Coral Reefs, Health Hazards and History*, University of Miami, pp. H21–H26.

Glynn, P., A.M. Szmant, E.F. Corcoran, and S.V. Cofet-Shabica. 1989. Conditions of coral reef cnidarians from the northern Florida Reef tract: Pesticides, heavy metals, and histopathological examination, unpublished report, National Park Service, Contract No. CX5280-5-1447, U.S. Dept. of Interior, Washington, D.C.

Goodfellow, R.M., J. Cardoso, G. Eglinton, J.P. Dawson, and G.A. Best. 1977. A fecal sterol survey in the Clyde Estuary, *Mar. Poll. Bull.*, 8:272–276.

Griffin, D.W., J.B. Rose, and J.H. Paul. 1997. Summary of water quality testing in Marathon, Florida, June 8–13, 1997. (unpublished data).

Halley, R.B., H.L. Vacher, and E.A. Shinn. 1997. Geology and hydrogeology of the Florida Keys, in *Geology and Hydrogeology of Carbonate Islands: Developments in Sedimentology*, Vacher, H.L. and Quinn, T., Eds., Elsevier Science, New York, pp. 217–248.

Harkins, J. 1996. *Nutrients in Wastewater and Nutrient Removal*, Technical Support Team, Water Management Division, U.S. Environmental Protection Agency, Region 4, Atlanta, GA, 8 pp.

Hazen, T.C. 1988. Fecal coliforms as indicators in tropical waters: a review, *Toxic. Assess.*, 3:461–477.

Hoffmeister, J.E. and H.G. Multer. 1968. Geology and origin of the Florida Keys, *Geol. Soc. Am. Bull.*, 79:1487–1502.

Jaap, W.C. 1984. *The Ecology of the South Florida Coral Reefs: A Community Profile*, FWS OBS-82/08 and MMS 84-0038, U.S. Fish and Wildlife Service and Minerals Management Service, 139 pp.

Justic, D. 1987. Long-term eutrophication of the North Adriatic Sea, *Mar. Pollut. Bull.*, 18:281–284.

Keswick, B.H. 1984. Sources of groundwater pollution, in *Groundwater Pollution Microbiology*, Bitton, G. and Gerba, C.P., Eds., John Wiley & Sons, New York.

Kinsey, D.W. 1988. Coral reef system response to some natural and anthropogenic stresses, *Galaxea*, 7:113–128.

Klein, C.J. and S.P. Orlando, Jr. 1994. A spatial framework for water quality management in the Florida Keys National marine Sanctuary, *Bull. Mar. Sci.*, 54(3)1036–1044.

Kump, L. 1998. *Wastewater Nutrients in Groundwaters of the Florida Keys: Contrasting Behaviors of Phosphorus and Nitrogen*, final report submitted to the Water Quality Protection Program, 56 pp.

Lapointe, B.E. 1997. Nutrient thresholds for bottom up control of macroalgal blooms on coral reefs in Jamaica and southwest Florida, *Limnol. Oceanogr.*, 42:1119–1131.

Lapointe, B.E. and M.W. Clark. 1992. Nutrient inputs from the watershed and coastal eutrophication in the Florida Keys, *Estuaries*, 15:465–476.

Lapointe, B.E. and D. Hanisak. 1997. *Algal Blooms in Coastal Waters: Eutrophication on Coral Reefs of Southeast Florida*, final report, Florida Sea Grant Project R/C-E-34, 23 pp.

Lapointe, B.E. and W.R. Matzie. 1996. Effects of stormwater nutrient discharges on eutrophication processes in nearshore waters of the Florida Keys, *Estuaries*, 19:422–435.

Lapointe, B.E. and W.R. Matzie. 1997. *High-Frequency Monitoring of Wastewater Nutrient Discharges and Their Ecological Effects in the Florida Keys National Marine Sanctuary*, final report submitted to the Water Quality Protection Program, 25 pp.

Lapointe, B.E. and N.P. Smith. 1987. A preliminary investigation of upwelling as a source of nutrients to Looe Key National Marine Sanctuary, NOAA Technical Memorandum, NOS MEMD 9.

Lapointe, B.E., J.D. O'Connell, and G. Garrett. 1990. Nutrient couplings between on-site sewage disposal systems, groundwaters, and nearshore surface waters of the Florida Keys, *Biogeochemistry*, 10:2289–307.

Lapointe, B.E., M.M. Littler, and D.S. Littler. 1993. Modification of benthic community structure by natural eutrophication: the Belize Barrier Reef, *Proc. Seventh Int. Coral Reef Symp.*, 1:323–334.

Lapointe, B.E., D.A. Tomasko, and W.R. Matzie. 1994. Eutrophication and trophic state classification of seagrass communities in the Florida Keys, *Bull. Mar. Sci.*, 54:696–717.

Lapointe, B.E., M.M. Littler, and D.S. Littler. 1997. Macroalgal overgrowth of fringing coral reefs at Discovery Bay, Jamaica: bottom-up versus top-down control, *Proc. Eighth Int. Coral Reef Symp.*, 1:927–932.

Larkum, A.W.D. and A.D.L. Steven. 1994. ENCORE: The effect of nutrient enrichment on coral reefs. 1. Experimental design and research programme, *Mar. Poll. Bull.*, 29:112–120.

Larkum, A.W.D. and K. Koop. 1997. ENCORE: Algal productivity and possible paradigm shifts, *Proc. Eighth Int. Coral Reef Symp.*, 1:881–884.

Larson, U., R. Elmgren, and R. Wulff. 1985. Eutrophication and the Baltic Sea: causes and consequences, *Ambio*, 14:9–14.

Lee, T.N., M.E. Clarke, E. Williams, A.F. Szmant, and T. Berger. 1994. Evolution of the Tortugas gyre and its influence on recruitment in the Florida Keys, *Bull. Mar. Sci.*, 54:621–646.

Leeworthy, V.R. and J.M. Bowker. 1997. *Nonmarket Economic User Values of the Florida Keys/Key West: Linking the Economy and Environment of Florida Keys/Florida Bay*, NOAA, The Nature Conservancy, Monroe County Tourist Development Council, 41 pp.

Leeworthy, V.R. and P.C. Wiley. 1996. *Visitor Profiles, Florida Keys/Key West: Linking the Economy and Environment of Florida Keys/Florida Bay*, NOAA, The Nature Conservancy, Monroe County Tourist Development Council, 159 pp.

Leeworthy, V.R. and P.C. Wiley. 1997. *Visitor Profiles, Florida Keys/Key West — A Socioeconomic Analysis of the Recreation Activities of Monroe County Residents in the Florida Keys/Key West: Linking the Economy and Environment of Florida Keys/Florida Bay*, NOAA, The Nature Conservancy, Monroe County Tourist Development Council, 41 pp.

Lidz, B.H. and E.A. Shinn. 1991. Paleo shorelines, reefs and a rising sea: South Florida, *J. Coastal Res.*, 7:203–229.

Loh, P.C., R.S. Fujoika, and S. Lau. 1979. Recovery, survival and dissemination of human enteric viruses in ocean waters receiving sewage in Hawaii, *Water Air Soil Bull.*, 12:1997–2017.

Manker, J.P. 1975. Distribution and Concentration of Mercury, Lead, Cobalt, Zinc, and Chromium in Suspended Particles and Bottom Sediments, Upper Florida Keys, Florida Bay, and Biscayne Bay, Ph.D. dissertation, Rice University, Houston, TX.

McClanahan, T.R. 1992. Epibenthic gastropods in the Middle Florida Keys: the role of habitat and environmental stress on assemblage composition, *J. Exp. Mar. Biol. Ecol.*, 160: 169–190.

McClanahan, T.R., A.T. Kamukura, N.A. Muthiga, M. Gilagabher Yebio, and D. Obura. 1995. Effects of sea urchin reductions on algae, coral, and fish populations, *Conserv. Biol.*, 10:136–154.

Moses, J.H. and J.E. Anderson. 1993. *S.W.I.M. Program Septic Tank Study*, St. Lucie County, FL, 16 pp.

Nixon, S.W. 1990. Marine eutrophication: A growing international problem, *Ambio*, 19:101.

Nixon, S.W. 1995. Coastal marine eutrophication: a definition, social causes, and future concerns, *Ophelia*, 41:199–219.

Officer, C.B., R.B. Biggs, J.L. Taft, L.E. Cronin, M.H. Tyler, and W.R. Boynton. 1984. Chesapeake Bay anoxia: origin, development and significance, *Science*, 223:22–27.

Ogden, J.C., J.W. Porter, N.P. Smith, A.M. Szmant, W.C. Jaap, and D. Forcucci. 1994. A long-term interdisciplinary study of the Florida Keys seascape, *Bull. Mar. Sci.*, 53:1059–1071.

Overton, W.S., D. White, and D.L. Stevens, Jr. 1990. *Design Report for EMAP: Environmental Monitoring and Assessment Program*, Report EPA600/3-91/053 for U.S. Environmental Protection Agency, Corvallis, OR.

Paul, J.P., J.B. Rose, S.Jiang, C.A. Kellogg, and L. Dickson. 1993. Distribution of viral abundance in the reef environment of Key Largo, Florida, *Appl. Environ. Microbiol.*, 59:718–724.

Paul, J.P., J.B. Rose, J. Brown, E.A. Shinn, S. Miller and S.R. Farrah. 1995a. Viral tracer studies indicate contamination of marine waters by sewage disposal practices in Key Largo, Florida, *Appl. Environ. Microbiol.*, 61:2230–2234.

Paul, J.P., J.B. Rose, S. Jiang, C. Kellogg, and E.A. Shinn. 1995b. Occurrence of fecal indicator bacteria in surface waters and the subsurface aquifer in Key Largo, Florida, *Appl. Environ. Microbiol.*, 61:2235–2241.

Paul, J.P., J.B. Rose, S.C. Jiang, X. Zhou, P. Cochran, C. Kellogg, J. Kang, D. Griffin, S. Farrah, and J. Lukasik. 1997. Evidence for groundwater and surface marine water contamination by waste disposal wells in the Florida Keys, *Water Res.*, 31:1448–1454.

Peters, E.C. 1984. A survey of cellular reactions to environmental stress and disease in Caribbean scleractinian corals, *Hel. Meer.*, 37:113–137.

Porter, J.W. and O.W. Meir. 1992. Quantification of loss and change in Floridian reef coral populations, *Am. Zool.*, 32:625–640.

Porter, J.W., O.W. Meier, J.I. Tougas, and S.K. Lewis. 1994. Modification of the south Florida hydroscape and its effect on coral reef survival in the Florida Keys, *Bull. Ecol. Soc. Am. Abst. Suppl.*, 75:184.

Rabalais, N.N., R.E. Turner, D. Justic, Q. Dortch, W.J. Wiseman, Jr., and B. Sen Gupta. 1996. Nutrient changes in the Mississippi River and system responses on the adjacent continental shelf, *Estuaries*, 19:386–407.

Santavy, D.L. and E.C. Peters. 1997. Microbial pests: coral disease in the western Atlantic, *Proc. Eighth Int. Coral Reef Symp.*, 1:607–612.

Sargent, F.J., T.J. Leary, D.W. Crewz, and C.R. Kruer. 1995. *Scarring of Florida's Seagrasses: Assessment and Management Options*, Florida Department of Environmental Protection, Florida Marine Research Institute Technical Report TR-1, 37 pp.

Shinn, E.A. 1988. The geology of the Florida Keys, *Oceanus*, 31:47–53.

Shinn, E.A., J.H. Hudson, R.B. Halley, and B.H. Lidz. 1977. Topographical control and accumulation rate of some Holocene coral reefs, South Florida and Dry Tortugas, *Proc. Third Int. Coral Reef Symp.*, 2:1–7.

Shinn, E.A., B.H. Lidg, J.L. Kindinger, J.H. Hudson, and R.B. Halley. 1989. *Reefs of Florida and the Dry Tortugas: A Guide to the Modern Carbonate Environments of the Florida Keys and the Dry Tortugas*, Int. Geol. Cong., IDC Field Trip T176, Am. Geophysical Union, Washington, D.C., 54 pp.

Shinn, E.A., B.H. Lidz, and M.W. Harris. 1994a. Factors controlling distribution of Florida Keys reefs (abstr.), *Bull. Mar. Sci.*, 54:1084.

Shinn, E.A. R.S. Reese, and C.D. Reich. 1994b. *Fate and Pathways of Injection Well Effluent in the Florida Keys*, U.S. Geological Survey Open-File Report 94–276, 116 pp.

Skinner, R. and W.C. Jaap. 1986. Trace metals and pesticides in sediments and organisms in John Pennekamp Coral Reef State Park and Key Largo National marine Sanctuary, unpublished report to the Florida Department of Environmental Regulation, Office of Coastal Zone Management.

Smith, G. L.D. Ives, I.A. Nagelkerken, and K.B. Ritchie. 1996. Apergillosis associated with Caribbean sea fan mortalities, *Nature*, 382:487.

Smith, Milo, and Associates, and Hale and Kulligren, Inc. 1970. Environment and identity, a plan for development in the Florida Keys. Land use plan prepared for Monroe County, FL.

Strom, R.N., R.S. Braman, W.C. Jaap, P. Dolan, K.B. Donnelly, and D.F. Martin. 1992. Analysis of selected trace metals and pesticides offshore of the Florida Keys, *Florida Sci.*, 55:1–13.

Szmant, A.M. and A. Forrester. 1996. Water column and sediment nitrogen and phosphorus distribution patterns in the Florida Keys, *Coral Reefs*, 15:21–41.

Taylor, J.L. and C.H. Saloman. 1968. Some effects of hydraulic dredging and coastal development in Boca Ciega Bay, Florida, *Fishery Bull.*, 67:212–241.

Turner, R.E. and N.N. Rabalais. 1994. Coastal eutrophication near the Mississippi River delta, *Nature*, 368:619–621.

USEPA. 1975. *Finger-Fill Canal Studies: Florida and North Carolina*, EPA 904/9-76-017, U.S. Environmental Protection Agency, Washington, D.C., 232 pp.

USEPA. 1993. *Water Quality Protection Program for the Florida Keys National Marine Sanctuary: Phase II Report*, Contract No. 68-c2-0134, U.S. Environmental Protection Agency, Washington, D.C.

USEPA. 1995. *Water Quality Protection Program for the Florida Keys National Marine Sanctuary: Phase III Report, Implementation Plan for Water Quality Monitoring and Research Programs*, Contract No. 68-C2-0134, U.S. Environmental Protection Agency, Washington, D.C., 70 pp.

USEPA. 1996. *Indian River Lagoon Comprehensive Conservation and Management Plan*, Indian River National Estuary Program, U.S. Environmental Protection Agency, Washington, D.C.

Underal, B. O.M. Skulberg, E. Dahl, and T. Aune. 1989. Disastrous bloom of *Chrysochromulina polylepis* (Prymnesiophyceae) in Norwegian coastal waters 1988: mortality in marine biota, *Ambio*, 18:265–270.

Valiela, I., K. Foreman, M. LaMontagne, D. Hersh, J. Costa, P. Peckol, B. DeMeo-Anderson, C. D'Avanzo, M. Babione, Chi-Ho Sham, and K. Lajtha. 1992. Couplings of watersheds and coastal waters: sources and consequences of nutrient enrichment in Waquoit Bay, Massachusetts, *Estuaries*, 15:443–457.

Vaughn, T.W. 1918. The temperature of the Florida coral reef tract, *Papers from the Tortugas Laboratory of the Carnegie Institute of Washington*, 9:319–339.

Viele, J. 1996. *The Florida Keys: A History of the Pioneers*, Pineapple Press, Sarasota, FL, 80 pp.

Voss, G. 1988. *Coral Reefs of Florida*, Pineapple Press, Sarasota, FL, 80 pp.

Wilkinson, C.R. 1993. Coral reefs are facing widespread devastation: can we prevent this through sustainable practices, *Proc. Seventh Int. Coral Reef Symp.*, 1:11–21.

Wilkinson, C.R. 1996. Global change and coral reefs: impacts on reefs and human cultures, *Global Change Biol.*, 2:547–558.

Windom, H.L. 1992. Contamination of the marine environment from land-based sources, *Mar. Poll. Bull.*, 25:1–4.

Woodward-Clyde Consultants, Inc. 1994. *Historical Imagery and Seagrass Assessment of the Indian River Lagoon*, Indian River National Estuary Program, U.S. Environmental Protection Agency, Washington, D.C.

# 32 The Role of the Florida Keys National Marine Sanctuary in the South Florida Ecosystem Restoration Initiative

*Billy D. Causey*
Florida Keys National Marine Sanctuary

## CONTENTS

## BACKGROUND

The Florida Keys National Marine Sanctuary is administered by the National Oceanic and Atmospheric Administration in the U.S. Department of Commerce. The sanctuary is one of thirteen national marine sanctuaries that are managed as a system spread throughout the coastal U.S.

The Florida Keys extend approximately 404 km (220 miles) southwest from the southern tip of the Florida peninsula. Located adjacent to the Keys land mass are nationally significant marine environments, including seagrass meadows, mangrove islands, and extensive living coral reefs. These marine environments support rich biological communities possessing extensive conservation, recreational, commercial, ecological, historical, research, educational, and aesthetic values which give this area special national significance. The lure of the Florida Keys has attracted visitors for decades. The clear tropical waters, bountiful resources, and appealing natural environment were among the many fine qualities that attracted visitors to the Keys, in the past.

The National Marine Sanctuary Program has managed segments of the coral reef tract in the Florida Keys since 1975. The Key Largo National Marine Sanctuary was established in 1975 to protect 353 km$^2$ (103 nmi$^2$) of coral reef habitat stretching along the Reef Tract from just north of Carysfort Lighthouse to south of Molasses Reef, offshore of the Upper Keys. In 1981, the 18 km$^2$ (5.32 nmi$^2$) Looe Key National Marine Sanctuary was established to protect the very popular Looe Key Reef located off Big Pine Key in the Lower Keys. These two national marine sanctuaries were, and continue to be, managed very intensively. The installation of mooring buoys to protect the reefs from anchor damage, educational programs, research and monitoring programs, and various resource protection programs, including interpretive law enforcement, have been concentrated in these two marine protected areas. Because these two sanctuaries are located between 5 and 7 km offshore, the health of these coral reef resources has been affected by land-based sources of pollution and nutrients. Managing these two sites has been like trying to manage islands in the middle of the ecosystem. Obviously, the major threats come from outside the boundaries of the sanctuaries. In order to be successful at management, an ecosystem approach has to be implemented.

By the late 1980s it became evident that a broader, more holistic approach to protecting and conserving the health of the coral reef resources had to be implemented. Regardless of the intensity in managing small portions of the coral reef tract, sanctuary managers were witnessing declines in water quality and the health of corals from a wide range of sources. The more obvious causes of decline were from impacts due to point-source discharges, habitat degradation due to development and overuse, and changes in reef fish populations due to over-fishing. Clearly, less obvious sources of decline were affecting the health of the coral reefs and these had to be identified.

## SANCTUARY DESIGNATION

In 1989, mounting threats to the health and ecological future of the coral reef ecosystem in the Florida Keys prompted Congress to take action to protect this fragile natural resource. The threat of oil drilling in the mid- to late 1980s off the Florida Keys, combined with reports of deteriorating water quality throughout the region, occurred at the same time scientists were assessing the adverse affects of coral bleaching, the dieoff of the long-spined urchin, loss of living coral cover on reefs, a major seagrass dieoff, declines in reef fish populations, and the spread of coral diseases. These were topics of major scientific concern and the focus of several scientific workshops when three large ships ran aground on the coral reef tract within a brief 18-day period in the fall of 1989. Coincidental as it may seem, it was this final physical insult to the reef that prompted Congress to take action to protect the coral reef ecosystem of the Florida Keys. Although most remember the ship groundings as having triggered Congressional action, it was in fact the cumulative events of environmental degradation in conjunction with the physical impacts that prompted Congress to take action to protect the coral reef ecosystem of the Florida Keys. On November 16, 1990, President Bush signed into law the Florida Keys National Marine Sanctuary and Protection Act (FKNMS Act).

The Act designated 9600 km$^2$ (2800 nmi$^2$) of coastal waters off the Florida Keys as the Florida Keys National Marine Sanctuary and immediately addressed two major concerns of the residents of the Florida Keys. There was an immediate prohibition on any oil drilling, including mineral and hydrocarbon leasing, exploration, development, or production within the sanctuary. In addition, the legislation prohibited the operation of tank vessels (ships) greater than 50 m in length in an internationally recognized "Area To Be Avoided" (ATBA) within the boundary of the sanctuary.

Clearly, the greatest threat to the environment, the natural resources, and economy of the Keys has been the degradation of water quality over the past two decades, which has been a major concern for the residents of the Keys. Commercial and recreational users of the resources in the Keys, environmentalists, scientists, and resource managers are all in agreement that the water quality of the Keys is in sharp decline and the commercially and recreationally important resources are extremely threatened. Some of the reasons for the decline are believed to be the lack of freshwater entering Florida Bay; nutrients from domestic wastewater such as shallow-well injection, cesspits, septic tanks, etc.; stormwater runoff containing heavy metals, fertilizers, insecticides, etc.; marinas and live-aboards; poor flushing of canals and embayments; buildup of organic debris along the shoreline; sedimentation; lack of hurricanes; and environmental changes associated with global climate change and sea-level rise.

Congress recognized the critical role of water quality in maintaining the sanctuary resources when it directed the administrator of the U.S. Environmental Protection Agency, in conjunction with the governor of the State of Florida and in consultation with the Secretary of Commerce, to develop a comprehensive Water Quality Protection Program (WQPP) for the sanctuary.

The FKNMS Act called for the Secretary of Commerce, in consultation with appropriate federal, state, and local government authorities and with a Sanctuary Advisory Council, to develop a comprehensive management plan and implementing regulations to achieve protection and preservation of living and other resources of the Florida Keys marine environment.

Because approximately 65% of the FKNMS encompasses state waters and numerous state and federal areas of jurisdiction overlap or lie adjacent to the FKNMS boundary, it was imperative that the planning process for the sanctuary be an inter/intra-agency effort. Also, due to the high level and diversity of public utilization of the resources in the Florida Keys and the importance of tourism to the economy of the Keys, it was equally important that the public have a strong role in the development of the comprehensive management plan.

The FKNMS Act called for the public to be a part of the planning process and that a Sanctuary Advisory Council (SAC) be established to aid in the development of the comprehensive management plan. A 23-member Advisory Council was selected by the Governor of Florida and the Secretary of Commerce. The council consists of members of various user groups; local, state, and federal agencies; scientists; educators; environmental groups; and private citizens. Over the course of the planning process, numerous public workshops were held to get input from knowledgeable individuals on a wide range of topics that could be implemented in the management of the sanctuary. Development of the final management plan took six years of comprehensive planning and utilized an integrated approach with all the local, state, and federal agencies, as well as the public through the Sanctuary Advisory Council made up of a wide range of stakeholders.

The final management plan for the sanctuary contains ten action plans including: (1) channel and reef marking, (2) education and outreach, (3) enforcement, (4) mooring buoy, (5) regulatory, (6) research and monitoring, (7) submerged cultural resources, (8) volunteer, (9) water quality, and (10) marine zoning. The marine zoning plan represents a major departure from the traditional management actions in Sanctuaries. The Act mandated that the sanctuary program "consider temporal and geographical zoning, to ensure protection of sanctuary resources."

## PERSPECTIVE

Because declining water quality and ocean pollution were identified as the greatest threats to the continued health of the coral reef in the Florida Keys, Congress directed the U.S. Environmental Protection Agency to work with the State of Florida and the National Oceanic and Atmospheric Administration to develop a water quality protection program for the sanctuary. The planning effort was initiated parallel to development of sanctuary's management plan. Even though the geographic scope or spatial extent that managers were considering as important to addressing water quality problems was enormous, it was soon learned that it was not large enough.

At their first meeting in 1992, the Sanctuary Advisory Council pointed out that the problems affecting water quality in the Keys were not simply derived from the Keys themselves, but from upstream. Upstream was Florida Bay, South Florida, the west coast shelf of Florida, and tributaries that drain a vast portion of South Florida. It became quite clear that we had to look well beyond the boundaries of the sanctuary to address the source of water quality problems affecting the health of the coral reef. But how far should managers look for the source of impacts?

The answer to this question became clearer in 1993, when the U.S. Secretary of the Interior Bruce Babbitt convened a meeting of all the federal resource managers in South Florida. This action initiated the formation of the South Florida Ecosystem Restoration effort that is currently underway. Today, local, state, federal, and tribal interests are all members of the Task Force, whose primary objective is to "get the water right in South Florida."

Over the decades many mistakes have been made in the way we manage our freshwater and its runoff into our estuaries. Today, we are attempting to restore the quality, quantity, timing, and distribution of freshwater in the system so as to resemble historic patterns of flow through the built environment and ultimately to the ocean.

## THE SOUTH FLORIDA ECOSYSTEM RESTORATION STORY

This case study will focus on many of the lessons learned along the way in both the sanctuary planning effort for the Florida Keys National Marine Sanctuary and the South Florida Ecosystem Restoration project. A challenge in an ecosystem management approach is to get resource managers to create a vision that extends well beyond jurisdictional boundaries, both at national and international scales, and establish broader objectives in ecosystem management. Another challenge is to get scientists to re-think their classical definition of an "ecosystem" and apply the same broad vision of the ecosystem system as the managers. Important, too, is that managers and scientists alike recognize that human activities are an integral part of ecosystem management and their activities have to be included in an ecosystem management program.

### THE ECOSYSTEM

"There are no other Everglades in the world. They are, they always have been, one of the unique regions of the earth, remote, never wholly known. ...It is a river of grass." Marjory Stoneman Douglas wrote those words about the Everglades in 1947. Since then we have come to realize that the "River of Grass" is part of the much larger South Florida ecosystem.

This ecosystem covers an amazing diversity of landscapes, including the Upper Chain of Lakes, above Lake Okeechobee which are the headwaters for South Florida; the meandering Kissimmee River, which flows into Lake Okeechobee; the hardwood hammocks where both tropical and temperate species reside; the mangrove forests that line the coast and Florida Keys; all the estuaries that support numerous species of fish and wading birds; and all the way to and including the biologically rich coral reefs.

Before efforts were made to drain the South Florida wetlands, the landscape had three key qualities: First, it was extremely flat, with no more than a 20-foot drop in elevation over 100 miles from Lake Okeechobee to Florida Bay. Second, the landscape had varied flora, fauna, and habitats.

Finally, and most importantly, the landscape was a rainfall-driven system, characterized by dynamic water storage and sheet flow.

Because of its many natural assets, South Florida attracted people and money, which led to development, agriculture, tourism, and other growth industries. Today over 5 million people reside along South Florida's east coast. This number is expected to triple by 2050 if current trends continue. In the Florida Keys alone, the current census reports 80,000 permanent residents with a seasonal population of 130,000. In addition, 3 million visitors come to the Keys annually and spend 13.3 million visitor days.

The increase in population, combined with increasing development, agriculture, and other human activities, is putting all of the South Florida ecosystem in peril. From the headwaters through the Florida Keys, the natural system is being strained as never before. Urban and suburban areas also face equally severe problems, such as crime, underemployment, and water shortages. *This unique natural and human system is in trouble.*

## How Did We Get Here?

Funding efforts such as the Central and South Florida Project, which both opened the door for urban and agricultural growth and altered the timing and distribution of water through the South Florida ecosystem, contributed to the current situation. In addition we did the following:

1. Channelized the Kissimmee River
2. Polluted Lake Okeechobee with agricultural runoff
3. Damaged our coastal estuaries with excessive freshwater
4. Brought Florida Bay to the brink of collapse by altering freshwater and nutrient flows
5. Introduced harmful exotic plants
6. Intensified the effects of floods and droughts
7. Reduced the spatial extent of wetlands by 50%
8. Permitted development to sprawl farther and farther into the natural system

The collective consequences of these changes have affected all living beings in South Florida — plants, animals, and people. These changes also threaten the well-being of South Florida's multibillion dollar tourism, agricultural, trade, and fishing industries, which are the economic backbone of the region and the state.

Several observations stand out based on what is happening to the South Florida ecosystem. First, South Florida is a holistic, complex system that includes both the natural and the built environment. Second, the quality of life in South Florida is inextricably linked to the health of the natural system. And third, the health of the Everglades and the entire South Florida ecosystem depends on what actions all of us take or do not take.

The challenge we face today is to reconcile our human demands with the needs of the South Florida ecosystem. So, what is being done to address the problems we face? Over the past 50 years the state and federal governments have been taking actions to stem and reverse the downward trends. Lands and waters have been protected, laws and initiatives have been passed to manage growth and protect the natural environment, and partnerships have been established to restore the ecosystem.

Of particular note are three recent events that have helped create the foundation for the current restoration effort:

- In 1993, the Federal South Florida Ecosystem Restoration Task Force was established through an interagency agreement. This task force has focused primarily on the protection and restoration of natural systems. This group has worked to develop a consistent approach to addressing environmental concerns, to set priorities for federal restoration efforts, and to oversee and evaluate restoration efforts underway.

- In 1994, Governor Chiles established the Governor's Commission for a Sustainable South Florida. The commission has focused on making recommendations for achieving a healthy Everglades ecosystem. It also has formulated strategies to achieve a sustainable economy and quality communities.
- Finally, in 1996, Congress passed the Water Resources Development Act. Among its many provisions, four stand out for South Florida:
  - The Act formally established the South Florida Ecosystem Restoration Task Force and expanded its membership to include tribal, state, and local governments.
  - The Act accelerated the authorization for and funding of critical projects, such as the Central and South Florida Restudy Project.
  - The Act enabled federal and nonfederal partners to share costs (50-50) for South Florida restoration projects.
  - The act authorized the task force to address the full scope of restoration, including the interconnections of environment, economy, and society.

Two important premises have emerged from the Governor's Commission and the Task Force work. First, on its present course South Florida is not sustainable. Second, the important relationship between South Florida's environment, economy, and society cannot be ignored. A common vision is emerging from theses realizations and from the ongoing restoration efforts. It is a vision of a landscape whose health, integrity, and beauty is restored and is nurtured by its interrelationships with South Florida's human communities.

## WHAT ACTIONS DO WE TAKE?

Today we understand much more that we ever have about South Florida and its problems. But, in suggesting solutions, we should keep in mind that there is still much to learn about the South Florida ecosystem. Because there is still much to learn, the ecosystem restoration effort has adopted an *adaptive management* approach that stresses taking action where possible while also continuing to collect data, learn, and plan. More specifically, the restoration effort is stressing the need for:

- Systemwide management
- Integrated governance
- Broad-based partnerships
- Public outreach and communication
- Science-based decision making

Systemwide management means taking a holistic, systematic approach to address issues region-ally, not locally. It means placing an emphasis on obtaining results rather than on developing processes that may never be carried out. And, it means searching for long-term, holistic solutions to South Florida problems rather than finding easy, temporary "fixes" to our problems. *Integrated governance* is also critical to creating a shared vision for the restoration effort. Different levels of government need to work together to:

- Develop regulations that are based on common sense and sound science.
- Share funding and cut costs.
- Integrate budgets.
- Develop cooperative programs that enable action to be taken faster.
- Streamline red tape and other institutional barriers.

Broad-based partnerships are another key element of the restoration effort. Governments also need to work cooperatively with interested parties if we are to solve the problems facing South

Florida. Partnerships also are needed between federal, state, local, and tribal governments and other partners to:

- Advance a shared vision and commitment.
- Foster the mutual respect and trust needed for the restoration effort.

Public outreach and communication are essential to building support for the restoration effort. With the region's high degree of cultural diversity, communication is needed to:

- Connect people in meaningful ways with the effort.
- Foster a clear exchange of views, ideas, and information.
- Instill a broad sense of stewardship, ownership, and responsibility for the fate of South Florida.

Finally, sound restoration decisions must be based on sound science. The results of specific decisions and actions must be monitored to assess the effectiveness of the actions. Relevant scientific data need to be identified and collected. Predictive ecological and socio-economic models need to be developed to forecast and track trends. Science-based decisions also means coordinating research efforts and making them accountable. In other words, we need to make sure that we get the *best research, at the best price, and delivered on time*. Additionally, we need to encourage new, creative technology that integrates both human and natural needs.

## How Do We Attain This Vision?

Three overarching goals must be achieved by the South Florida ecosystem restoration effort. We need to get the water right, restore and enhance the natural system, and transform the built environment. Getting the water right means restoring more natural hydrologic functions while also providing adequate water supplies and flood mitigation. To do this we need to address:

- Quantity of water flowing through the ecosystem
- Quality of water and the timing and duration of water flows and levels
- Distribution of water through the system

More specifically, the restoration effort needs to:

- Re-establish both the sheetflow and groundwater flow that once were common throughout the system.
- Restore the natural variations in water flows and levels, without diminishing the region's water supply or flood control.
- Ensure that water supplies are clean enough for their intended use.

Other critical elements in getting the water right include:

- Reducing the amount of water lost to tide through stormwater drainage and agricultural runoff
- Replacing the lost water storage capacity of the system.

The second major goal of the restoration effort is restoring, protecting, and enhancing natural areas. Attention needs to be devoted to preservation efforts that allow for the recovery of threatened and endangered species. The physical and biological connections between natural areas must be re-established. Many more wetlands and other disappearing habitats need to be permanently set aside and protected.

The diversity and abundance of South Florida's native species must be re-established. The spread of exotic species, such as the melaleuca tree, has to be stopped and reversed. In addition, the productivity of coastal areas, estuaries, and fisheries should be revived; coral reefs must be protected, and commercial and recreation interests need to adopt practices that help sustain the natural system.

The third major goal of the restoration effort is to transform the built environment to sustain a prosperous economy, vibrant society, and a healthy natural environment. To achieve this goal, the restoration effort must address future development and the economy, including agriculture. Fostering sustainable development is key to achieving this goal. Unending urban sprawl must be stopped. Land-use decisions should be compatible with ongoing restoration efforts, and resources should be used efficiently for development. Government programs, incentives, and tax structures must be modified to support smart development.

A prosperous, diverse, and balanced economy also must be present to restore the ecosystem. Industries such as ecotourism should be supported and promoted. Also, necessary is work to ensure that the actions of resource-dependent industries are compatible with the restoration effort's goals. The support of business interests must be secured if these goals are to be achieved. Finally, a prosperous and sustainable agriculture should be supported by:

- Protecting disappearing farmlands
- Promoting research and best management practices that improve the sustainability of the agricultural industry
- Encouraging strong markets

## What Is Being Done to Achieve the South Florida Ecosystem Restoration Goals?

Many projects are underway. Some are nearing completion; others will take decades to complete. The following examples illustrate the nature and scale of these efforts.

### Performance Indicators and Models

The ecosystem restoration effort is encouraging the use of performance indicators and models to provide direction, feedback, and accountability for all of the projects going on or planned for South Florida. Performance indicators and models will help us keep track of changing hydrologic, ecological, water quality, and socio-economic conditions. They enable agencies to evaluate their performance, and they help the public identify the benefits and costs of the projects.

### Central and South Florida Review Study

The Central and South Florida (C&SF) Project Comprehensive Review Study is a massive under-taking aimed at assessing how well the C&SF Project is functioning. The restudy will determine what modifications should be made to the project to restore natural hydrologic conditions in natural areas, while still providing for the other water-related needs of the region. Together with other efforts, it is hoped that the restudy will improve water quality and help restore the historic abundance and diversity of native species.

### Water Preserve Areas

Another project underway is the creation of a series of water preserve areas along the eastern margin of the Everglades, spanning Miami–Dade, Broward, and Palm Beach Counties. The water preserve areas, consisting of an interconnected system of marshlands, reservoirs, and aquifer recharge areas, are intended to:

- Capture, store, and clean excess stormwater now lost to tide.
- Protect and conserve wetlands outside the Everglades.
- Provide a buffer between the expanding westward urban development and the Everglades.

## Everglades Construction Project

The Everglades Construction Project covers a number of actions that are being taken to:

- Improve the quality of runoff discharged from farms into the Everglades.
- Capture, store, and clean stormwater runoff now lost to tide.
- Re-establish sheetflow and increase the quantity of water delivered to the Everglades.
- Decrease excessive freshwater discharges into estuaries.

The project is focusing primarily on building man-made wetlands, improving the canal system, and encouraging the adoption of best management practices for agriculture.

Sixty-eight federally listed threatened and endangered species, as well as other species of special concern, occur in South Florida.

## Everglades Forever Act

As the result of the settlement of a law suit between the federal government and the State of Florida, polluters of the Everglades agreed to reduce phosphorus from agriculture runoff from 240 parts per billion (ppb) to 40 ppb.

## Multispecies Recovery Plan

To ensure the long-term survival of these species, the U.S. Fish and Wildlife Service is preparing a comprehensive multispecies recovery plan. This will be one of the first plans in the nation that meets the needs of multiple species on a regional basis. It will provide a blueprint that agencies can use in their work to restore the South Florida ecosystem.

## Eastward Ho!

One of the projects underway to transform the built environment is the Eastward Ho! initiative. The purpose of this initiative is to redirect growth back to the historical eastern corridor and away from natural and agricultural areas in South Florida. To redirect growth, federal, state, local, and private entities are looking at ways to enhance the appeal of older urban areas. It is hoped that Eastward Ho! will both raise the quality of life in urban centers and reduce the impacts urban areas have on South Florida's natural and agricultural areas.

## Florida Keys Carrying Capacity

The Florida Keys have experienced tremendous growth over the past several decades, which in turn has affected many of the natural resources of the Keys. In response to these impacts, the U.S. Army Corps of Engineers is directing a carrying capacity analysis for the Keys. Information from the study should enable planners to model different growth scenarios and determine when resource thresholds are being exceeded. This should improve the capability of agencies to plan for and manage future growth on the Keys. The model that comes out of the Keys study may have applicability elsewhere in South Florida and even internationally.

## Florida Keys National Marine Sanctuary Marine Zoning Plan

On July 1, 1997, the FKNMS implemented the National Marine Sanctuary Program's first comprehensive network of marine zoning. At the time, the use of marine zoning represented a major departure from the traditional management actions in National Marine Sanctuaries. Marine zoning is an effective and useful tool for managing marine protected areas, especially those as large as the FKNMS. Marine zoning allows managers the perfect opportunity to balance resource use with resource protection while providing a very common-sense approach to focusing protection in critical portions of sensitive habitats. The marine zoning plan for the FKNMS includes five different types of zones. Three of these zone types are "no take" areas that serve very specific functions directed at balancing visitor-use and resource protection; the sanctuary has 24 of these "no take" areas. The three types are Sanctuary Preservation Areas, Ecological Reserves, and Special Use Areas (Research Only). The other two zone types are Wildlife Management Areas and Existing Management Areas.

## South Dade Land Use/Water Management Planning Project

One area in South Florida that potentially could see much change in upcoming years is South Dade, located between Miami's suburbs and Biscayne National Park. The project entails three separate but linked components:

1. Agricultural and Rural Lands Retention Plan
2. South Biscayne Bay Watershed Management Plan
3. South Dade Wellfield Study

The results of the planning project will determine the future economic, social, and environmental sustainability for urban and rural Miami–Dade County.

## Environmental Impact Statement for Southwest Florida

Southwest Florida is an area experiencing rapid growth and development, as well as many impacts to the natural environment. The U.S. Army Corps of Engineers and Lee and Collier Counties have agreed in principle to prepare an environmental impact statement (EIS) that will take a holistic view of future development in the region. It will specifically be assessing the impacts of permits issued for development under Section 404 of the Clean Water Act. The EIS should enable the Corps to speed up the processing of development permits. It also should help ensure that the counties and the Corps take a consistent approach to new development, and it may generate new ideas for sustainable development.

## WHERE DO WE STAND?

The South Florida ecosystem today is facing many serious problems that directly or indirectly affect all of us. Building on the work that has been done over the past 50 years, we now have a blueprint to restore the ecosystem. We have a vision and goals for South Florida. Projects are underway and progress has been made in achieving these ends. But, we have a ways to go. All of us — businesses, governments, private citizens — need to continue to support and participate in the restoration effort. Working together, we can achieve our vision. South Florida's fate is in our hands.

## CAN WE AFFORD IT?

According to the results of the restudy, the cost of the restoration effort will be approximately $7.8 billion U.S. dollars. Although this cost seems high, the question has to be reversed. Can we afford to *not* do the restoration? Based on the following economic considerations, the decision is clearly, yes, we must make the investment:

- One of six U.S. jobs is related to the Oceans.
- The 1995 U.S. Fishing industry revenue was $20 billion.
- In 1995, coastal tourism accounted for $54 billion (beaches are the leading destination) in revenue.
- Recreational fishing brought in $30 billion.
- The Florida Keys National Marine Sanctuary annually has 3 million tourists who stay for 13.3 million visitor-days and spending $1.2 billion dollars.
- U.S. coastal and marine waters support 28.3 million jobs.
- U.S. coastal areas are the destination for 180 million people yearly.

## LESSONS LEARNED

The following is a list of lessons learned as they relate to ecosystem management planning for the Florida Keys National Marine Sanctuary. Their inclusion does not mean to imply they were not considered from the outset, but only to emphasize their importance to managers.

### ECOSYSTEM APPROACH

- Establish a comprehensive boundary for the ecosystem based on natural and physical processes and not political or jurisdictional boundaries (barriers). Strive to eliminate jurisdictional and administrative barriers to ecosystem management.
- Apply the principles of ecosystem-based management from the outset in the planning process. In other words, approach the planning process with an ecosystem perspective, focusing on watershed-based management. Include the appropriate spatial extent within the boundary of the ecosystem.
- Use a public process to establish ecosystem management objectives and restoration goals based on our best understanding of the concepts of sustainability. Establish an advisory group made up of stakeholders and local elected officials separate from an interagency core group to assist in the planning process.
- Utilize an adaptive management process and, in the absence of information, use the best science available upon which to base decisions.
- Support the planning process with analytical and technical expertise.

### INTEGRATED MANAGEMENT

- Establish an integrated planning process but do not let the rigor of the process dominate the activities, but rather treat the process as another adaptive management tool. Utilize to the fullest extent possible existing integrated coastal management programs.
- Bring all levels of government to the table for the planning process, from the local and regional level to the state, territorial, tribal, and national level. Consult international levels of government when feasible and necessary. Ensure that the integrated planning process moves vertically and horizontally through the structure of the agencies and that all levels of government can participate in the planning process.
- Require that participating representatives have adequate authority to make decisions in the planning process.
- Focus on ways to implement effective ocean governance within the confines of existing authorities, but be open to new legislation when necessary.

## SOCIOECONOMIC CONSIDERATIONS

- Recognize from the outset that humans are a part of the ecosystem and that our activities, or the effects of our activities, cannot be separated from any holistic approach to management.
- Although we continue to struggle with a true definition of sustainability, continue to apply the spirit of what we collectively think to be a sustainable approach on the most conservative side of management principles.
- Invest heavily in outreach efforts at all target audience levels with the recognition that the environment and economy are linked at the outset of the project. This is especially true of decision-maker and policy-maker audiences.
- It is absolutely essential to bring socioeconomic information into the planning process as a foundation for informed participation at an early phase. Treat this discipline with the level of importance that you would give the natural or physical sciences.
- Utilize the concept of marine zoning in the management planning process. This management tool is useful to eliminate or lessen visitor-use conflicts. Establish marines reserves or "no take areas" where marine life is fully protected in critical marine environments.
- Listen to and attempt to understand all points of view in an ecosystem management planning process.

## CONCLUSION

The list of lessons learned is more accurately the reflection of changing spatial perspectives. Clearly, the old paradigm of managing just within the boundaries of one's marine protected area does not and cannot succeed. It is critical that resource managers step back and take a broader perspective of the true spatial extent of the geographic and oceanographic boundaries that affect their areas. That's the easy step ... the next is to work with others in an integrated process that focuses on achieving sustainable goals.

## ACKNOWLEDGMENT

This chapter is amended from a paper presented by the author at the International Tropical Marine Ecosystems Management Symposium (ITMEMS) held in Townsville, Australia, November 23–26, 1998. The symposium was coordinated by the International Coral Reef Initiative (ICRI) and its Secretariat currently hosted in Australia at the Great Barrier Reef Marine Park Authority. Recognition for portions of this chapter must be given to the Working Group of the South Florida Ecosystem Restoration Task Force. The sections of this report that discuss the South Florida Ecosystem Restoration Project were in large part written through a facilitated process by members of the Working Group. The Working Group is made up of local, state, federal, and tribal representatives. The collaborative process to write this section makes this an even more powerful message.

# 33 The Role of a Nonprofit Organization, Reef Relief, in Protecting Coral Reefs

*DeeVon Quirolo*
Reef Relief

## CONTENTS

**FIGURE 33.1**  Coral reefs are the most biologically diverse marine ecosystems on Earth, rivaled only by the tropical rainforests on land. Coral reefs have existed for 400 million years and reached their current level of biodiversity 50 million years ago. They are the oldest, most complex ecosystems in the sea. (Photograph courtesy of Robert Gauthier.)

Reef Relief provides a case study highlighting the role of a local nongovernment organization (NGO) in the conservation of coral reefs in a developed country. Reef Relief is a nonprofit membership organization dedicated to preserving and protecting living coral reef ecosystems through local, regional, and international efforts. The lessons learned at the heavily used coral reefs of the Florida Keys provide a template that is being transferred throughout the Caribbean and the world. It demonstrates the success that a few motivated individuals with a small budget and a committed volunteer labor force can achieve.

Reef Relief began out of a love for coral reefs and a desire to take action to protect them. It was founded in 1986 in Key West, FL, by charterboat captain Craig Quirolo and a group of captains initially concerned about the damage their anchors caused to the coral reef. The group's first action was to install reef mooring buoys to eliminate anchor damage. The weight of an anchor being dropped on corals can crush the organisms and open the entire coralhead to bacteria and disease. Reef Relief led the first private effort to install reef mooring buoys in the Florida Keys and eliminated anchor damage at heavily visited sites.

Reef Relief's educational efforts began by teaching residents and visitors of the Keys to avoid harming the reef when boating, diving, or fishing. Physical impacts were reduced by educating divers and snorkelers to avoid contact with the reef from fins, hands, and equipment. Reef Relief organized marine debris cleanups that removed tons of trash from Keys reefs and shorelines. The group launched further educational programs to inform and enlist school students, community members, visitors, business owners, divers, fishermen, boaters, media, and policymakers to actively protect coral reefs. Over the past 15 years, this grassroots group has expanded its activities to include policy development, monitoring, restoration, and international efforts.

Other successful programs to which Reef Relief has made contributions include marine debris reduction, a ban on offshore oil drilling near the Florida Reef Tract, creation of the Florida Keys National Marine Sanctuary (FKNMS), passage of legislation that improves water quality, establishment of marine-life harvesting regulations, increased scientific monitoring of Florida's coral reefs, and restoration of reefs through treatment of diseases and creation of coral nurseries.

## COMMUNITY-BASED EFFORTS

Although the installation of mooring buoys is not a comprehensive coral reef management approach, it is a simple, effective strategy. Reef Relief enlists support for a program with broad conservation benefits that can be easily understood and supported by local residents and businesses, who recognize the links between their tourist-driven economy, the commercial fishing industry, and the continued health of their greatest natural resource — the coral reef ecosystem. The management program is easily adapted for implementation in developing countries with emerging tourism economies and limited funds for reef conservation. It illustrates how community-based efforts and citizen involvement are relevant and necessary in the U.S., where there is a long history of national legislation and government management of reefs and natural resources. Because they are composed of local residents, grassroots efforts are uniquely qualified to define issues and act on a local level. Those involved have a strong sense of place and are often the first to recognize that there is a problem, identify its source, and propose a viable solution. This clarity of mission enables Reef Relief to be effective.

## SITE PROFILE

### FLORIDA's CORAL REEF ECOSYSTEM

The coral reef of the Florida Keys is North America's only living coral barrier reef and the third longest coral barrier reef in the world. The reef extends for 128 miles, from North Key Largo to the Dry Tortugas, and provides habitat for fish, stony and soft corals, sponges, jellyfish, anemones, snails, crabs, lobsters, rays, moray eels, sea turtles, dolphins, sea birds, and other sealife. Coral reefs are the most biologically diverse marine ecosystems on Earth, rivaled only by the tropical rainforests on land. Florida's coral reefs are home to more than 150 species of tropical fish and 50 species of coral, representing over 80% of the coral species in the tropical western Atlantic. Coral reefs have existed for 400 million years and reached their current level of biodiversity 50 million years ago. They are the oldest, most complex ecosystems in the sea.

The coral reef ecosystem is a delicately balanced, interdependent marine environment comprised of coral reefs, mangroves, and seagrasses, each of which depends upon the health of the others to survive. Together with the hardwood hammocks, Florida's coral reefs are home to one third of Florida's threatened or endangered species, including sea turtles, crocodiles, manatees, conchs, Key deer, Schaus swallowtail butterflies, snook, wood storks, peregrine falcons, and roseate spoonbills.

The health of these reefs has declined dramatically in the past 15 years. A combination of factors are responsible, among them heavy use and physical damage from boats, anchors, divers, snorkelers, and fishing; loss of habitat from coastal development; declining water quality, especially an overabundance of nutrients; and global climate change and natural storm events. There has been a significant loss of coral species. From 1996 to 1999 stony coral cover decreased 38% on average in the 160 stations of the U.S. Environmental Protection Agency (EPA)-sponsored Florida Keys Coral Reef Monitoring Program, according to Dr. James Porter, lead investigator of the survey. The survey also detected an early exponential increase in diseased corals and the appearance of new diseases throughout the Keys. This large-scale, well-funded survey was the first to deliver irrefutable data documenting the decline of Keys coral reefs.

In order to heighten awareness of this decline, beginning in 1993 Reef Relief conducted a low-tech photo monitoring survey that led to the discovery of new coral diseases and documented significant coral loss due to disease, storm damage, bleaching, nuisance algal blooms, and boat groundings. Visibility has dropped from "gin-clear" waters in years past to an average of 30 feet. Sewage and stormwater pollution from the Keys, agricultural runoff from the Everglades, and poor water quality in Florida Bay are responsible for the harmful nutrients and pesticides found in the reefs downstream. Storms, accidental boat groundings, and heavy sedimentation add to the multiple stressors that severely compromise the coral reef.

### SOCIOECONOMIC SIGNIFICANCE

Although the Florida Keys coral reef is only one tenth the size of Australia's Great Barrier Reef, it is visited by ten times more people annually. The reef is essential to a growing tourism industry and commercial fisheries, and the local economy depends heavily these industries. The Keys generate a $2.5 billion annual economy by playing host to 4.5 million tourists who dive, fish, and boat in the Keys. In 1990, the hospitality industry (restaurants, catering, hotels) accounted for about one third of Monroe County's $1.6 billion in total sales and business. During the year 2000, 275,380 people flew into Key West Airport, with additional visitors driving down the overseas highway and still more arriving by boat.

To ensure the continued strength of the tourism industry, more than $9 million is raised annually by a county bed tax that is used for advertising to attract more visitors to the Keys. Marine-related industries accounted for almost 5% of total sales and business in 1990. In addition to many nearshore commercial charter, dive, and fishing operations, the Florida Keys are the southernmost link of the U.S. to the Caribbean basin, a major conduit for offshore freighter and cruiseship traffic. These warm Gulfstream waters have become a favorite cruiseship route, with stopovers in Key West. During the month of December 2000, for example, Key West hosted 70,000 cruiseship passengers, according to the Key West Chamber of Commerce. Plans are underway to expand cruiseship facilities due to the acquisition of the former Truman Annex Navy Base by the City of Key West.

### INSTITUTIONS

Jurisdiction over the coral reefs of the Florida Keys is comprised of a mosaic of local, state, and federal agencies. Monroe County is a string of islands under the authority of various local government jurisdictions as well as state and federal agency mandates. Key West, Marathon, Key Colony Beach, and the Village of Islamorada are the municipalities in the area. Each municipality has a mayor and a board of commissioners. Monroe County also has a Board of County Commissioners. The municipal county components form the local government.

The Florida Keys National Marine Sanctuary Program is the principle authority for marine environmental protection. The sanctuary is a partnership with the State of Florida, whose jurisdictional waters include much of the sanctuary. Other important agencies with some mandate over reefs, shoreline, and water quality are the Florida Department of Community Affairs, the Florida Department of Environmental Protection, the U.S. Interior Department, and the U.S. Environmental Protection Agency. Several national and local NGOs, in addition to Reef Relief, are involved in marine and environmental conservation in the area, which complicates any policymaking and tends to frustrate the local citizenry.

## THE REEF RELIEF ORGANIZATION

Reef Relief was incorporated as a not-for-profit corporation in the State of Florida in 1986 to preserve and protect the living coral reef of the Florida Keys. It is guided by a board of directors that was originally composed of local watersports operators, nature tour guides, and dive-boat skippers. An advisory board provided technical expertise when necessary, and members included representatives from the National Marine Sanctuary Program and a natural science educator. The board was expanded to bring in conservation leaders, retirees, and business representatives.

In 1998, Reef Relief combined forces with San Francisco-based Coral Forest and the mission statement was revised in recognition of its growing activities in the region and around the world.

At that time, Randy Hayes, founder of Rainforest Action Network, Wendy Weir, and her brother Bob Weir (of the Grateful Dead) were board members from Coral Forest that joined the Reef Relief board as part of the reorganization. A Scientific Advisory Board was formally created with long-term collaborative researchers Drs. James and Karen Porter, of the University of Georgia; Dr. Brian Lapointe, of Harbor Branch Oceanographic Institution, Dr. Drew Harvell, of Cornell University, Dr. Bill Alevizon, of the Wildlife Conservation Society; and Harold Hudson, of the Florida Keys National Marine Sanctuary. A Citizen Advisory Board was created as well, recognizing those local citizens and businesses that had been long-term supporters.

Reef Relief is staffed by eight paid personnel (three of whom work part-time in the Environmental Center/Gift Shop) and numerous volunteers. Although its headquarters are in Key West, its programs extend throughout the Keys, with an active director in Tallahassee, the state capital, who focuses on state issues that affect coral reefs. In addition, Reef Relief takes a stand to protect coral reefs by becoming involved in reef-related issues around the world.

Reef Relief depends upon memberships, contributions, in-kind donations of goods and services, foundation support, and volunteers. It also holds one major fundraising event each year in Key West: Cayo Carnival, a celebration featuring Caribbean music, various activities, food donated by over 30 local restaurants, and raffles for prizes donated by local businesses. Reef Awareness Week is held each July in Key West, as well, including a series of events related to reef protection and fundraising. Activities of past Reef Awareness Weeks included mooring buoy splicing parties, an environmental film festival on local cable television, a water quality luncheon presentation, a sand sculpture contest, a scientific panel discussion, and children's art projects. The board of directors holds their annual "advance" that week and the annual membership meeting features founder Craig Quirolo's annual State of the Reef Address.

Individuals and businesses provide in-kind contributions of goods and services, ranging from printing and artwork to boats and buoy installation equipment. The City of Key West donates office space and boat dockage. The group's budget has grown along with its activities, and fundraising is a constant challenge for the small staff. The executive director writes grants and coordinates fundraising efforts by the board of directors. The project director maintains the membership records

**FIGURE 33.2** Reef Relief organizes reef mooring buoy splicing parties to provide an opportunity for volunteers to get involved in reef protection. Pictured is Reef Relief founder Craig Quirolo teaching Key West resident Arnie Hermelin and his sons how to splice lines for the reef mooring buoy program. Pick-up lines and down-lines for the buoy assembly are spliced on land, then replaced when needed at the reef. (Photograph by DeeVon Quirolo. © Reef Relief.)

and helps expand memberships through local events and direct mail. A small gift shop and an online store of educational gift items help fund operation of the Reef Relief Environmental Center in Key West.

Reef Relief also helps start up similar coral reef conservation efforts around the world. Community-based programs have been launched in cooperation with local partners in Negril, Jamaica; Puerto Rico; Guanaja, Bay Islands, Honduras; Cuba; and the Abacos, Bahamas. The website at www.reefrelief.org spreads Reef Relief's message around the globe and includes educational materials, scientific papers, and information about coral reefs.

## PROJECT GOALS AND COMPONENTS

Reef Relief evolved out of a private effort to deploy mooring buoys to protect the reef from anchor damage. A total of 119 reef mooring buoys were installed at seven reefs near Key West. The message carried by Reef Relief staff and supporters is that coral reefs deserve the active protection of everyone. The community involvement initiated by the mooring buoy program has allowed Reef Relief to expand activities to address additional threats to the marine environment. Their success is a result of policy guidance, education, and direct action, based on direct knowledge of conditions at the reef that are interpreted by leading coral reef scientists.

**FIGURE 33.3**  Hudson, restoration biologist for the Florida Key National Marine Sanctuary, has devised a method of removing black-band diseases from infected coralheads. An aspirator is used to remove the infected leading edge of the disease, which is packed with modeling clay to prevent its reinfection. (Photograph compliments of Harold Hudson.)

## CORAL PHOTO MONITORING SURVEY

Reef Relief maintained the buoy system for 10 years, expanding it regularly as the number and size of vessels increased, before handing it over to the new sanctuary. During that time, visibility at the reef dropped each year and the water turned green. Corals began exhibiting disease symptoms and loss, while nuisance algal growth increased.

Yet, many residents and policymakers questioned Reef Relief's concern over the deterioration of the reef and whether it was actually declining. In 1993, Reef Relief was awarded the first Robert Rodale Environmental Achievement Award. With the funds from that award, Director of Marine Projects Craig Quirolo launched a low-impact coral photo monitoring survey. For the past 8 years, he has documented change in the coral reefs near Key West. Without installing pins or other markers, he returns to the same coral sites and captures various angles of the same coral using a video and slide format. Next, he digitizes the images in his computer lab, studies them in comparison to previous images of the same site, and shares his observations with coral reef scientists. Reef Relief provides the images to other researchers, media, students, and film-makers. The data from the survey are incorporated into Reef Relief educational materials.

## CORAL DISEASES

The survey has a led to the discovery of several new coral diseases and an ongoing collaboration with various scientists. Harold Hudson, who joined the sanctuary staff in 1989, taught Quirolo how to remove black-band disease from infected coralheads, and an effort began to restore corals. In May, 1995, Quirolo and Hudson cored out a sample of yellow-band disease, which Quirolo is credited with discovering. Reef Relief has produced and distributed many reports describing various coral diseases such as white diseases (white plague type 11, white-band, white-line), black-band disease, hyperplasia, yellow-band disease, *Aspergillus* fungus on sea fans and other soft corals, bleaching, and other anomalies. Drs. Drew Harvell and Kiho Kim worked with Reef Relief to tag and monitor sea fans infected with Aspergillus, a study that has implications for pharmaceuticals. A multiyear collaboration with Dr. Brian Lapointe of Harbor Branch Oceanographic Institution

**FIGURE 33.4** Yellow-band disease is characterized by a band of yellow diseased tissue that eats into the living coral polyps, leaving only exposed coral substrate behind. This coralhead is at Sand Key in the Florida Keys. (Photograph by Craig Quirolo. © Reef Relief.)

focused on the role of nutrients and the process of eutrophication that has occurred at Florida's coral reef ecosystem. Reef Relief provided field support for the research done by Dr. Larry Brand of the University of Miami Rosenstiel School of Marine Sciences that resulted in publication of a paper documenting the flow of green water from Florida Bay to the offshore coral reefs.

## CORAL NURSERY

Reef Relief's documentation and observations of storm damage to the reef led to the creation of a Coral Nursery at Western Sambo Reef in 1998. On February 2, 1998, a severe winter gale struck the Florida Keys, creating literally tons of rubble where branching and boulder corals had once been. A volunteer effort was mobilized to respond. In collaboration with Hudson, Quirolo led a team of volunteers who righted overturned boulder corals and elevated and secured fragments of elkhorn coral that were buried in rubble and would otherwise have died. The fragments were cemented onto concrete pads with hydraulic cement to form *rosettes* (as Hudson called them) that prevented the pieces from rolling about and becoming buried in subsequent storm activity. Quirolo monitored the coral nursery until recently, when the rosettes were moved by sanctuary staff to help restore a boat grounding site. This successful technique was repeated and a second nursery was created at No Name Cay in the Abacos, Bahamas, in 2000, using wire tires to secure large fragments of elkhorn coral. Elkhorn coral grows only in the Caribbean and is fast becoming endangered.

## KEY WEST MARINE PARK

At this writing, Reef Relief is launching an effort to create the Key West Marine Park. In cooperation with the City of Key West, Reef Relief will create a municipal underwater park with shoreline access to a swimming beach that excludes motorized watercraft along the oceanside of the island. City boundaries will extend out to 600 feet to establish a no-take zone to increase marine life in these nearshore waters. Buoys will mark the area, and access lanes will provide egress for existing watersports operators. Mooring buoys will be installed outside the boundaries for visiting boaters. The park will become part of the ongoing coral photo monitoring survey. Initial surveys of the area indicate a wide variety of living corals and other sealife. This focus will increase awareness of the need for stormwater treatment throughout Key West and will help Reef Relief build community support for its implementation so that sedimentation to nearshore waters is reduced. The shoreline is a juvenile nursery and breeding ground for numerous species that later migrate to the offshore coral reefs.

## INTERNATIONAL EFFORTS

Reef Relief is involved in international efforts to protect coral reefs as well, applying the lesson learned in the Florida Keys to grassroots efforts around the world. A multiyear collaboration began in 1990 with the Negril Coral Reef Preservation Society (NCRPS) led by founder Katy Thacker. Reef Relief held a reggae concert in Key West to raise funds for the installation of reef mooring buoys to protect Negril's coral reefs. Thereafter, the two groups worked together and helped local residents create the Reef Ranger Program, an educational patrol of the reef patterned after a similar Reef Relief program in the Keys. The Junior Ranger Program was begun to educate local elementary school students, and beach cleanups were organized. Years of effort resulted in the establishment of the Negril Marine Park with no-motor, no-take zones that protect swimmers from boaters and protect marinelife from harvesting. Reef Relief was instrumental in organizing workshops that brought in its Scientific Advisors to begin water quality and coral reef studies of Negril's coral reefs. The NCRPS manages the park and was able to attract foreign funding to help support these efforts.

In Puerto Rico, Reef Relief helped organize a workshop and provide strategic support to help launch an organization called Coralations, which is directed by Mary Ann Lucking, an activist for Puerto Rico's coral reefs.

In 1996, Reef Relief was contacted by the Tourism Association for Guanaja, Bay Islands, Honduras, to help the community organize a reef mooring buoy program. In cooperation with Sandra Bazley, a volunteer diver who organized the effort, 35 buoys were installed at the coral reefs that encircle the island. The project was funded by in-kind donations of goods and services, tourist taxes collected by volunteers at the tiny airport on the island, and contributions from local landowners. Educational programs were introduced to school students and dive operators and educational materials printed for distribution to visitors and residents. Reef Relief plans to return to Guanaja to restore the buoy program after several years of use and will provide the local environmental officer with a donated workboat to enable ongoing buoy maintenance. Despite severe hurricane damage to the island in recent years, the buoy program is intact.

Some of the scientists who have visited the Keys and met with Reef Relief to study the coral reefs include Dr. Rodolfo Claro and Dr. Juan Pablo Garcia of the Cuban Institute of Oceanography, who were hosted by Ken Lindeman of Coastal Research and Education, Inc. As a result of that meeting, Reef Relief directors sailed to Havana, met with these scientists, and entered into a 3-year agreement to survey the coral reefs of Cuba. Cuba's northwest coast features the Arrecifes Colorados, a coral reef ecosystem very similar to that of the Florida Keys. Many of these reefs had not been surveyed for years due to a lack of funds and/or foreign collaborators. Reef Relief's first- and second-year surveys were integrated into published reports that have been widely distributed. Reef Relief has also helped Cuban ecologists produce educational materials for school students in Havana. The embargo has made it difficult for this collaboration to continue. U.S. grant funds cannot be spent in Cuba and permits are difficult to obtain. Cuban authorities will not allow their scientists to accompany foreigners on their boats to perform field work. Yet, Dr. Claro's studies have documented that several species of fish that spawn in Cuban waters are carried by the Gulfstream into Florida's waters and are found in the Palm Beaches. This drives Reef Relief to try to return for more research.

For the past few years, Reef Relief has also been active in the Abacos, the fastest growing area of the Bahamas. With funding from the Green Turtle Cay Foundation, the Edith and Curtis Munson Foundation, and local individuals and businesses, Reef Relief trained local divers to install reef mooring buoys, visited the school to make educational presentations, and held workshops to share videos and information about the reef with local residents and visitors. In cooperation with Friends of the Environment of Hope Town and the Bahamas National Trust, Reef Relief refurbished existing mooring buoys at two marine protected areas, the Pelican Land and Sea Park and the Fowl Cay Preserve in the Abacos. Fundraising is underway for a return to install additional buoys, demark the protected areas, and produce educational materials.

## EDUCATION

Reef Relief's education efforts are intended to inform members, voters, activists, policymakers, educators, students, the 87,000 residents of the Florida Keys, and its more than 4 million annual visitors. Increasingly, its efforts are also focused on community-based initiatives around the Caribbean and Bahamas as well. The Reef Relief Coral Reef Awareness Campaign began at the Environmental Center when it opened in July 1990. This high-profile, multimedia outreach program was intended to inspire the public to abandon destructive habits and adopt an attitude of stewardship toward the coral reef ecosystem. Today, the program is called the Coral Reef Conservation Program and includes the Clean Water Campaign to improve water quality.

The Environmental Center features displays on the coral reef, mangroves, sea turtle nesting, marine debris, and reef mooring buoys and provides information on action that individuals can take to improve or protect the coral reef ecosystem. Reef Relief staff also give presentations throughout the state, participate in regional educational and trade shows, and host informational booths at local events, including Earth Day and Reef Awareness Week. They create television and radio announcements that air on six Keys stations, flyers, signs for charterboats, posters, handbooks, videos for television, and teacher kits. Reef Relief collaborated with FKNMS staff to produce a multilingual educational brochure on the Upper and Lower Keys. Recently, the Pump It, Don't Dump It campaign to support the Key West no-discharge zone has included English and Spanish language materials for all boaters. Reef Relief has built a website that features most of its educational materials, images of coral reefs, scientific studies, and an online gift shop. Recently, it has developed an online community that receives e-alerts on a regular basis.

A newsletter, *Reef Line*, is published quarterly, has a circulation of 8000, and is mailed to key conservation interests, policymakers, Reef Relief's media list, and members. Volunteers deliver copies to every dive shop and marina in the Florida Keys. *Reef Line* is mailed out in general information packets and is available at booths, events, online on the website, and at the Environmental Center.

Reef Relief provides the *Discover Coral Reefs* School Program to fourth grade students in the Keys, including a Teacher Kit of educational materials to prepare the class, a video presentation at the Environmental Center, an interpretive visit to the reef aboard a glassbottom boat, and a follow-up slide show and activity in the classroom.

Special events are planned and hosted to enhance public awareness of and to encourage participation in reef protection. For example, the Blueprint for Action Seminar was designed to address freighter groundings on the reef, and the Hospitality Symposium provided hoteliers with information on Earth-friendly business practices. A Carrying-Capacity Study Workshop introduced residents to an important planning study underway. Various skipper's meetings have been organized to advise local charterboat operators of new conditions or new information.

The Environmental Center activities are made possible by a staff that includes Executive Director DeeVon Quirolo, Director of Marine Projects Craig Quirolo, Educational Coordinator Joel Biddle, Project Director Michael Blades, and Environmental Center retail staff. They are supported by local volunteers, court-appointed community service volunteers, interns from the Monroe County summer school training program, and a competitive college internship that is offered to two qualified candidates per year.

## COALITION BUILDING

Reef Relief and representatives from a number of other environmental organizations formed the Coral Reef Coalition to "securing lasting preservation of the biologic diversity and productivity of Florida's coral reef ecosystem and wise use of its resources." By working together, the coalition enhanced its chances for success, and the sanctuary designation for the Florida Keys was approved in a record 1-year period. The Coral Reef Coalition comprised over a hundred individuals, environmental organizations, and industry associations. Members included Greenpeace, the Center for Marine Conservation, the Wilderness Society, Last Stand, Florida Audubon, Sierra Club, the Nature Conservancy, and Clean Water Action. The Coral Reef Coalition was influential at the national level in bringing pressure on the U.S. Congress to consider legislation protecting coral reefs and water quality. In the context of U.S. politics, such interest groups are important in that they show the substantial citizen concern necessary before national, state, or local action is possible.

## POLICY GUIDANCE

When Reef Relief pursues a campaign, it begins by identifying the issue, which must fall within the parameters of its mission and grassroots abilities. The staff does thorough research and stays abreast of new developments by communicating with scientists, users, policymakers, and citizens as well as reviewing their own data from field work. Staff members then develop an objective and identify tangible steps toward its achievement. The next step is to meet with, telephone, or write to policymakers at the local, county, state, and national levels regarding the issue and what can be done to address it. They then follow up regularly. Some issues require years of watchdog efforts, even after an initial success.

After this initial stage — the identification of a single issue and the gathering of information related to it — the staff prepares written materials and issues press releases highlighting action-oriented steps that can be taken to address the problem. "Action Alerts" are one-page reports sent to Reef Relief members, the Reef Relief E-Activists List, other conservation interests, and concerned citizens that asks them to write letters, make phone calls, write editorials, send e-mails, or attend public hearings to help effect a change. White papers and brochures that contain in-depth information on an issue are produced and shared with policymakers, NGOs, community groups, and businesses, enabling Reef Relief to build a coalition of supporters. Additional media coverage is scheduled, including editorials and advertisements in the local press, radio and television talk shows, magazine articles, and interviews. To make the public aware of the issue, the staff creates displays at the Environmental Center or at other public educational facilities. "Action Alerts" or petitions are provided at the displays so that interested citizens can become involved. Special meetings of grassroots activists are planned, public forums are hosted, and the issue is brought up at civic and business meetings to garner broad support at the local level.

Reef Relief maintains a dialogue with legislators, aides, and policymakers. Members have spoken at numerous public hearings, addressed committees of the House and Senate of the U.S. Congress, and reported to the governor of Florida. The organization has hosted several congressional visits and organized workshops for the National Marine Sanctuary Program. As noted before, through local citizen and government action, it influenced the National Marine Sanctuary designation for the Florida Keys.

Staff members attend and testify at public hearings and other forums, and they distribute materials throughout the geographic area affected by the issue. They prepare and submit written comments on issues pertaining to the marine environment in response to planning efforts at all levels of government. Reef Relief contacts policymakers in order to obtain their commitment to taking specific steps toward improvement. It contacts opponents to try to resolve their opposition to an initiative. Staff members continue to employ public pressure and provide scientific evidence to encourage the desired result. Reef Relief guides policy development and encourages public participation in policymaking by advising the public of hearings and encouraging people to get involved.

## LEGISLATION PROTECTING THE CORAL REEF ECOSYSTEM

### TEN-YEAR BAN ON OFFSHORE OIL DRILLING AND EXPLORATION

Reef Relief was an active participant in the Outer Continental Shelf (OCS), a national coalition that pushed for several congressional-mandated 1-year moratoriums on offshore oil drilling and exploration near the Florida Keys. Permanent protection remained the main objective, however. Opposition to offshore oil exploration and production is the most unifying issue to ignite public support in the Florida Keys. Numerous public hearings, petition drives, and other actions were taken, including Black Friday, a day when many Keys businesses shut down and draped their storefronts in black to demonstrate solidarity in opposing oil development. In 1992, President Bush announced a 10-year ban on leases held by nine oil companies for areas as close as 25 miles from Key West and the Dry Tortugas. The ban was extended by President Clinton for the Five Year Plan to be implemented in 2002. Newly elected President Bush was visited by his brother, Florida Governor Jeb Bush, to reinforce the overwhelming mandate for an indefinite extension of the ban. It may not become an issue again in the Florida Keys, but grassroots activists are gearing up for a new battle in other areas, including Alaska and the Eastern Gulf of Mexico.

## CANCELLATION OF NAVY EXPLOSIVES TESTING

The U.S. Navy has historically tested ordnance in coral reef areas and in the habitats of dolphins, sea turtles, and other marine life. The testing has now been canceled because of the public outcry catalyzed by both Reef Relief and a few outspoken dolphin activists. A similar controversy rages in Vieques, Puerto Rico, and recent testing in the Bahamas has been protested by dolphin activists who report harm to numerous dolphins.

## FLORIDA KEYS NATIONAL MARINE SANCTUARY LEGISLATION

Reef Relief was an early and continual supporter of the National Marine Sanctuary Program administered by the National Oceanic and Atmospheric Administration (NOAA). Founder Craig Quirolo addressed committees before both houses of Congress and was invited to the White House by President Bush when the legislation was signed into law in 1990. Reef Relief generated public input into the public discussions for development of a zoned management plan for the new sanctuary and hosted an environmental education workshop. It issued recommendations for a stronger plan when the first draft was issued after 6 years of public debate. Today the coral reefs are protected by regulations that prohibit anchoring, harvesting, and other damaging activities at heavily visited reefs and discharging pollutants into sanctuary waters.

For the first time in the U.S., the sanctuary included a U.S. EPA-led Water Quality Protection Program that supported scientific research of Florida's coral reef including studies of coral diseases. The EPA-sponsored scientists recently completed 5 years of studies and EPA is seeking to transfer important scientific surveys of the Keys coral reef to NOAA. These studies are important. Efforts currently underway to improve water quality in the Keys may result in future reports that will document that recent trends of dramatic degradation are being reversed. The Florida Keys coral reefs can return to healthy, vibrant coral reefs if we all do our part. Federal support through the Sanctuary Program can make a big difference if the funds are spent wisely.

## PHOSPHATE BAN FOR MONROE COUNTY

Reef Relief worked for over 3 years to obtain legislation limiting the levels of phosphate in cleaning products sold in the Florida Keys. After a challenge from the Soap and Detergent Manufacturers Association, the limits were eased slightly for commercial dishwashers. The legislation reduces nutrient loading to coral reefs that require clear, clean, nutrient-free waters to thrive.

## RESTRICTIONS ON HARVESTING MARINE LIFE

Reef Relief was a strong advocate for regulations governing the collection of marine life. It supported a ban on the use of quinaldine and other chemicals on the reef and an immediate halt to the harvesting of live fish and live rock. Regulations passed by the Florida Marine Fisheries Commission and ratified by the governor and cabinet in 1992 restrict the type and volume of organisms that tropical fish collectors may harvest. The gathering of live rock (coral substrate with living marine organisms attached) for collectors is being regulated for the first time. Reef Relief also supports stronger bag limits on the harvest of marine life so as not to deplete the breeder stock of various species. One success of the sanctuary management plan is the establishment of 18 Sanctuary Preservation Areas (SPAs) at the most frequently used coral reefs. The SPAs prohibit all spearfishing, shell collection, tropical fish collection, and other activities that result in the harvest of marine life within their boundaries. The areas are marked by large yellow buoys.

An ongoing debate has been generated by Reef Relief's efforts to end fish feeding by divers based on the premise that it creates health problems for the animals fed, it disrupts the natural ecological processes and biological communities, and it increases the risk of harm to humans from wildlife attacks by increasing the aggressiveness of sealife seeking a handout from divers. Shark attacks have risen sharply in recent years, with 23 such attacks reported in Florida in 2000. Shark rodeos, where diver operators chum the water to feed the animals for the amusement of onlookers, are a big industry. Reef Relief's Scientific Advisor and fisheries expert Dr. Bill Alevizon along with Board Member Paul Johnson have led the effort to create a coalition to end the practice in Florida that includes the World Wildlife Fund, Environmental Defense, Wildlife Conservation Society, Defenders of Wildlife, Caribbean Conservation Corps, the Humane Society of the United States, Recreational Fishing Alliance, *Florida Sportsman Magazine*, and Watchable Wildlife, Inc. The Florida Fish and Wildlife Commission approved rulemaking to ban such activities, but reversed their decision after an outcry from professional dive operators.

## FLOOD THE BAY

In the 1990s, a multi-agency effort to restore the River of Grass and save the Everglades resulted in a policy of releasing agricultural runoff containing mercury, nutrients, and pesticides through the Everglades and down man-made canals into Florida Bay. It precipitated a Reef Relief position best characterized by the slogan "we all live downstream." Dr. Brian Lapointe determined that the increased water flows into Florida Bay during the 1990s that were designed to reduce a perceived problem of high salinity were implicated in a massive dieoff of seagrasses in the bay. This "flood the bay" policy was instrumental in the creation of a hypoxic dead zone in Florida Bay characterized by peasoup-green water which had catastrophic downstream impacts on the coral reefs. Lapointe's viewpoint and Reef Relief's support of that viewpoint were unpopular in many political circles, especially those who supported the Everglades initiative. The negative impacts on Keys coral reefs have been demonstrated by Lapointe and Brand's research. Earlier, Dr. Ned Smith of Harbor Branch Oceanographic Institution had established that the net flow of water ran out of Florida Bay to the offshore coral reefs through the passes along the Florida Keys. The increased flows corresponded with the period of time that Quirolo documented the proliferation of coral diseases on the downstream reefs.

This is an example of common sense and factual data from onsite research flying in the face of policy decisions made to suit another agenda, one further upstream. Unfortunately, what happens in the Everglades is inextricably linked to the health of Florida's coral reefs. The Everglades restudy, incorporating "adaptive management," is looking for ways to clean up the water before releasing it into Florida Bay, and a recent drought has helped reduce the flow of this dirty water. Reef Relief's grassroots efforts were unable to overcome a strong, well-funded national environmental coalition that pursued a different agenda, and to date the coral reefs are not included in the massive federally funded cleanup plan for the Everglades.

## THE CLEAN WATER CAMPAIGN

The *Summary of Water Quality Concerns in the Florida Keys: Sources, Effects and Solutions*, by William Kruczynski (EPA Water Quality Protection Program, Florida Keys National Marine Sanctuary) noted that:

> Since maintenance of healthy natural communities of the Keys is dependent on low nutrient environments, localized sources of nutrients can have immediate negative impacts that can result in cascading effects throughout the ecosystem. ...Indeed, the main factors in the demise of coral reefs are human pressures, such as eutrophication and sediment loading.

Dubinsky and Stambler (1996) concluded that:

> Human impacts on coral reefs are on a local and regional scale, rather than global impacts (excluding global warming and increases in ultraviolet light). ...The survival of the existing Florida Keys marine ecosystem is dependent upon clear, low-nutrient waters.

In recent years, Reef Relief has worked to improve water quality in the Florida Keys through the Clean Water Campaign. Reef Relief supports local efforts to upgrade sewage infrastructure for the Florida Keys, encourages state and federal funding for such improvements, and helps guide state and federal policymaking in South Florida. Inadequately treated sewage is a major local source of nutrients to the coral reef ecosystem of the Florida Keys, which require clear, clean, nutrient-free waters to thrive. Illegal cesspits, shallow injection wells, and leaky septic systems deliver nitrates and phosphates into nearshore waters in a matter of hours. Most boaters used the ocean as a dumping ground for their sewage by discharging it overboard.

In 1999, Reef Relief led the effort to create a "no-discharge zone" for boater sewage in Key West. Reef Relief helped draft the legislation and obtain local passage of an ordinance that prohibits boaters from discharging sewage in Key West waters, in addition to supporting efforts to generate state grant funds that were used to install vessel pumpout facilities at city-owned marinas and encouraging private marinas to install such facilities. Reef Relief staff helped speed approval of the "no-discharge zone" designation through the Florida governor's office and EPA regional head-quarters. Thereafter, the group organized a local grassroots activists group that designed a public awareness campaign to spread word of the new regulation. A contest was held and the *Pump It, Don't Dump It* campaign was born. The slogan was incorporated into brochures printed in English and Spanish, in posters, in radio and television public service announcements and in ads in local boating publications.

Reef Relief worked with other local interests to spearhead an effort to upgrade sewage treatment in Key West. When beaches were closed due to high fecal coliform counts, the community joined in and approved a voter referendum by 83% to fund a 2-year accelerated program that will implement an advanced wastewater plan for Key West. The plan included upgrading the level of treatment to advanced wastewater treatment with nutrient removal, phasing out the ocean outfall, and replumbing city streets. Reef Relief is now working with the City of Key West to ensure that re-use is incorporated and is helping to build community support throughout the rest of the Florida Keys as the county and municipalities determine how to upgrade inadequate sewage treatment in these areas to meet state mandates. Reef Relief has helped garner state and federal funding for such improvements by mobilizing citizen support. A grassroots workshop for local volunteers is held regularly to design strategies to deal with water quality issues.

In cooperation with statewide citizen groups, Reef Relief is working to upgrade the standards for deep injection wells. One third of the state's wastewater is disposed of in deep wells that lack confinement and migrates through cracks and fissures upward into coastal waters. These efforts led to an EPA rule, published in 2000, that proposed that all deep injection wells meet advanced wastewater standards for nutrient removal. Post-election changes in the administration of the EPA have delayed final issuance of a rule, but Reef Relief will continue to be involved to ensure that all wastewater in South Florida is treated to high nutrient-removal standards leading to re-use. As freshwater supplies become scarce due to cyclical drought and overdevelopment, re-using highly treated wastewater will become a necessity.

## ACCOMPLISHMENTS AND LESSONS LEARNED

### ELIMINATION OF ANCHOR DAMAGE

In 1986, when Reef Relief issued "Public Announcement Number One to the Watersport Industry" to publicize how anchors were damaging the coral reefs, it enlisted the support of the county commission that approved use of Monroe County Boating Improvement Funds to pay for the first 60 mooring buoys at six reefs in the Key West area. In return for the funding, Reef Relief signed a non-paying contract with the county agreeing to maintain the buoys for 20 years. By 1992, Reef Relief had installed over 100 mooring buoys in the Keys. The mooring buoys originally used in the Key West area were designed by John Halas in collaboration with Hudson, who was with the U.S. Geologic Survey at the time. Halas had installed them at the Key Largo and Looe Key National

**FIGURE 33.5**   Anchors dropped on coral reefs can damage the living animals, but it is an easily resolved problem. Reef mooring buoys provide an easy way to secure a boat on a coral reef without damaging the fragile living organisms. (Photograph by Craig Quirolo. © Reef Relief.)

Marine Sanctuaries. Subsequent modifications by Reef Reliefs founder, Craig Quirolo, have increased buoy holding strength and allowed their use by larger vessels. Reef mooring buoys provide a visible sign of reef protection, leading to a greater appreciation of the fragile coral below the ocean surface.

Today, anchoring has been virtually eliminated at the many reefs where the buoys have been installed, and it is prohibited in the SPAs where buoys are available for use. Boaters who do anchor elsewhere generally seek a sandy area where the anchor will not grate on coral. Two other volunteer groups in the Florida Keys (Florida Keys Marine Sanctuaries, Inc., and the Coral Reef Foundation) installed buoys at other Keys reefs. All of the buoys are now maintained by the sanctuary as part of the Reef Mooring Buoy Action Plan.

The success of Reef Relief's buoy project demonstrates that the drive, vision, and leadership of a few committed individuals can have far-reaching and profoundly positive effects. A simple, cost-effective technology that requires little training can rapidly be adopted for reef protection. The effectiveness of the reef mooring buoys indicates that some reef problems can be addressed without extremely expensive mechanisms or restrictive policies.

The success of a strategy addressing a single issue, as revealed through tangible results, can spur initiatives beyond the original scope of the project. A variety of projects involving the elimination of additional threats to coral reefs (for example, water pollution, explosives testing, marine-life harvesting, offshore oil development) have evolved out of the success of the original single-issue focus.

## REDUCTION IN PHYSICAL IMPACTS TO CORAL

Most charter boat operators now deliver a "reef etiquette" speech to their guests before they enter the water to snorkel or dive. Reef Relief has provided phonetic translations of commonly used phrases in French, Spanish, German, Italian, and Japanese. Many charter boats provide "float coats" to their guests to eliminate harmful contact with the reef, another Reef Relief innovation. As a result of these precautions, most visitors are aware of the need to avoid standing on and touching coral formations and collection marine life. Trash is kept aboard the vessels and aluminum beverage cans are often recycled. By emphasizing the tourism industry's dependence on the coral reef ecosystem, it was possible to encourage active participation in reef conservation in practical ways.

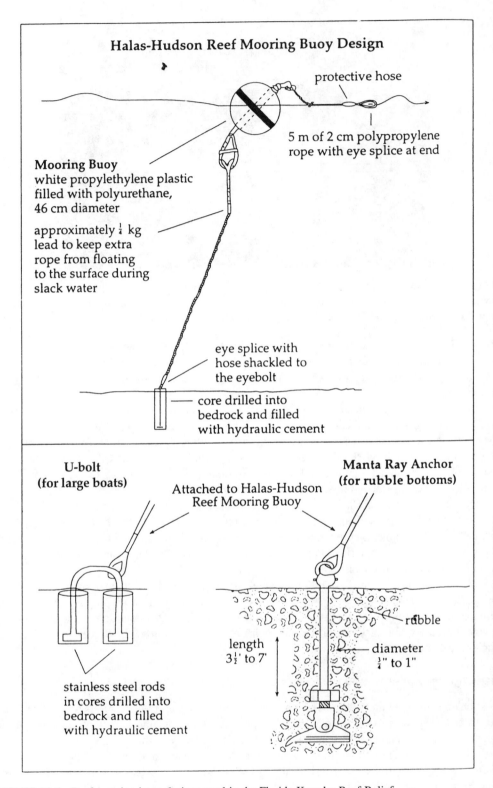

**FIGURE 33.6** Reef mooring buoy designs used in the Florida Keys by Reef Relief.

## INCREASED PARTICIPATION IN BEACH CLEANUPS

Reef Relief organized the first volunteer cleanup in the Florida Keys in 1987. During that cleanup, volunteers collected over 8.5 tons of debris from a beach that had not been cleaned in 18 years. Each year, tons of marine debris are collected by hundreds of volunteers in the Keys, and the number of cleanups grows annually. Reef Relief designed a campaign entitled *Don't Teach Your Trash To Swim*, promoted by placement of signs along the highway and at marinas and with radio and television public service announcements. Reef Relief helped found Clean Florida Keys, an affiliate of Keep America Beautiful that has adopted the issue as its sole focus in the Florida Keys.

Beach cleanups are simple, inexpensive ways to enhance public awareness of the threat and pervasiveness of marine debris. Beaches are left cleaner, and the participants feel as though they are contributing to environmental improvement. More important, participants become less tolerant of the deliberate inappropriate disposal of trash. Beach cleanups organized by numerous private groups show how a single event can spur other related activities and encourage additional participants (such as charter boat guests).

## PASSAGE OF PROTECTIVE LEGISLATION

Through the advocacy of Reef Relief and the hard work of its members to unite various groups, several legislative restrictions on practices destructive to coral reefs have been passed at national, state, and local levels. This success demonstrates that it is possible to influence national legislation once a substantial and powerful coalition is formed. When individuals and small groups operate

**FIGURE 33.7** Beach cleanups are an easy way to get involved in protecting coral reefs. Marine debris is a threat to all marinelife, especially from plastics items that never go away. (Photograph by Craig Quirolo. © Reef Relief.)

together, they can achieve far more than each one on its own. As the local community becomes sensitized to reef problems, an environment can be created in which much larger issues can be tackled: "Never doubt that a small group of thoughtful, committed citizens can change the world; indeed, it's the only thing that ever has" (Margaret Mead).

## MONITORING OF REEF CONDITIONS

Reef Relief's presence in the area and frequent documentation of reef conditions have been important because it has been a first alert for emerging reef conditions. For example, in April 1991, a report that the long-spined sea urchins were dying off prompted quick action to collect samples before they decomposed or were eaten. The samples were sent to Keys Marine Lab and Woods Hole Oceanographic Institution for analysis.

The grounding of the pleasure craft *Jacqueline L* was one of several instances when Reef Relief's quick action was instrumental in ensuring that the vessel was removed properly and as quickly as possible to reduce further damage. Reef Relief alerted the sanctuary staff that a major grounding had occurred and was then able to assist in the careful removal of the vessel from the site. They took possession of and obtained jurisdiction over the vessel in order to pursue prosecution and obtain funds, which were dedicated to restoring the damage caused to the coral reef. Such public/private cooperation should be encouraged.

Reef Relief founder Quirolo supports scientific research by taking scientists and others to the reef to inspect actual conditions and by participating in ongoing research projects designed to help understand and protect the ailing reef system.

His focus on the emergence of coral diseases encouraged the EPA to conduct scientific studies of coral diseases as part of the Water Quality Protection Program. Such a high-profile organization encourages responsible action among citizens, resource users, and government agencies. The accessibility of Reef Relief staff also enables the public and especially other researchers to draw easily upon its expertise and capabilities. This, too, should be encouraged by federal managers because it strengthens and enhances research efforts.

## FINAL LESSONS

The measurable progress of Reef Relief is the result of deep commitment on the part of a few individuals with an understanding of the coral reef ecosystem derived from first-hand knowledge. That knowledge is used to design and implement programs that increase public awareness of the importance and value of coral reefs, increase scientific understanding and knowledge of coral reefs, and strengthen grassroots, community-based efforts to protect coral reef ecosystems. Reef Relief designs, develops, and helps implement strategies for marine protected areas and encourages and supports eco-tourism as part of sustainable communities that protect coral reefs. Community participation and government commitment are integral parts of the overall conservation program facilitated by Reef Relief.

This approach has proven that people will abandon destructive behavior and gain a sense of stewardship for the coral reef ecosystem when provided with an opportunity to do so. As confidence in government ebbs, the role of nonprofit organizations will become more important to ensure that someone is paying attention and driving change for the better. Reef Relief sells a sticker that reads, "If you're not outraged, you're not paying attention." There is still a lot to be done to truly save the coral reefs of the Florida Keys and those around the world. As one challenge is resolved, another appears. The simplicity of installing reef mooring buoys has led to the complexity of questioning the restoration of South Florida's ecosystem.

Reef Relief's efforts have made a big difference in the Florida Keys and elsewhere. But have they succeeded? If the benchmark is the health of coral reefs, then the answer is "no," although they have, by any measure, prevented much damage to coral reefs and played an important role in their protection. This collaborative, community-based, nonprofit organization has succeeded in making a change for the better in our society — in this case, for the benefit of coral reefs, and that is the most difficult of all challenges.

## REFERENCES

Dubinsky, Z. and N. Stambler. 1996. Marine Pollution and coral reefs. *Global Change Biology*, 2:511–526.
U.S. Environmental Protection Agency. Water Quality Protection Program–Florida Keys National Marine Sanctuary. 1999. *The Summary of Water Quality Concerns in the Florida Keys: Sources, Effects and Solutions*. EPA 904-R-99-005, page 13.

# Section V

## International Analog: An Integrated Watershed Approach to Coral Reef Management in Negril, Jamaica

*"Spreading the word is just as important as doing the work"*
Craig Quirolo

by
Karen Porter

# 34 Patterns of Coral Reef Development in the Negril Marine Park: Necessity for a Whole-Watershed Management Plan

*Karen G. Porter, James W. Porter, and Delene W. Porter*
Institute of Ecology, University of Georgia

*Katy Thacker, Courtney Black, Webster Gabbidon, and Linval Getten*
Negril Coral Reef Preservation Society

*Craig Quirolo*
Reef Relief

*Douglas M. Marcinek*
Department of Fisheries and Aquatic Sciences, University of Florida

*Phillip Dustan*
Department of Biology, University of Charleston

## CONTENTS

0-8493-2026-7/02/$0.00+$1.50
© 2002 by CRC Press LLC

# INTRODUCTION

Historically, the coral reefs of Negril, Jamaica, were considered some of the best-developed reefs in Jamaica and the Caribbean (Goreau et al., 1997; T.F. Goreau, pers. comm.). Negril, with Jamaica's longest beach and "calm to near calm weather" prevailing 95% of the time (Jamaica Meteorological Service, 1973), has become a magnet for tourists with concomitant population growth and development. Once pristine, Seven Mile Beach is now ringed with high-rise tourist hotels, associated businesses, and residences. Coral reefs all around the island of Jamaica have undergone a series of shifts in their benthic flora and fauna following Hurricane Allen (1980), Hurricane Gilbert (1988), and the dieoff of the herbivorous spiny sea urchin, *Diadema antillarum* (Hughes, 1994). Recent studies have also identified a variety of anthropogenic stresses to Negril's coastal environments stemming from road construction and poor wastewater treatment (Goreau et al., 1997).

In 1992, a coalition of government and nongovernmental organizations (NGOs) proposed the establishment of the Negril Marine Park to encompass marine environments around Jamaica's West End, from Davis Cove on the North Coast to Little Bay on the South Coast. In addition, there has been an explicit understanding that protection of Negril's coastal habitats requires whole-watershed management due to substantial surface and subsurface drainage running into the nearshore waters (Negril Area Environmental Protection Trust, 1995).

Coral reef monitoring is an essential component of coral reef management. A well-designed monitoring program can demonstrate the status and trends of biotic resources within the park and reveal the causative agents of change within the protected area (Woodley et al., 1997; CARICOMP, 1997). Monitoring can also be used to test the effectiveness of zoning restrictions and enforcement efforts. Monitoring programs coupled with training programs for local citizens are an effective way to promote environmental awareness and sustained local conservation efforts (Wells, 1997). With funding from the European Economic Union administered by the Negril Coral Reef Preservation Society, we established a coral reef monitoring program for the proposed Negril Marine Park. Protocols established in the large-scale U.S. Environmental Protection Agency's Coral Reef Monitoring Program in the Florida Keys (J. Porter et al., this volume) were adapted to a scale and technical level appropriate for use by a small, nonprofit coral reef conservation organization in a developing country. The authors report the successful initial results of this monitoring program in detecting differences among reef sites and suggesting causative factors of stress that can be alleviated by whole-watershed management of water quality.

# MATERIALS AND METHODS

## SITE SELECTION AND STATION INSTALLATION

Five bays were chosen around Jamaica's West End as monitoring sites. Monitoring stations were installed at these sites using the following four criteria: (1) stations were located inside the boundaries of the proposed Negril Marine Park, (2) stations spanned the extent of the park,

(3) stations were located on reef crest habitats in approximately 2 to 3 m of water, and (4) stations were located roughly in the middle of the selected bay sites, which were the areas with the greatest coral reef development. Approximately 200 m of reef crest near the middle of each bay were surveyed by snorkel before the actual site was chosen. The presence of *Acropora palmata* in all five sites surveyed attests to the basic similarity of these sites. Sand channels and seagrass beds were excluded during the site selection process. These criteria were used to assure that the stations were located in comparable coral reef habitats and observed differences between reefs would reflect actual differences between locations and not differences between optimal habitats in one location and suboptimal habitats in another.

Station installation methods were adapted from protocols developed for the U.S. EPA's Coral Reef Monitoring Program. Survey markers were installed at either end of a 20-m transect line stretching from the reef crest toward the fore-reef. Roofer nails, 9 inches long, were drilled into the reef surface using an underwater pneumatic drill and later cemented into place for greater permanence. At Long Bay, the station was located within a short swim of the Marine Park mooring buoys (Quirolo, 1991; Halas, 1997). Station position was determined by a hand-held portable Magellan GPS (see Table 34.1 for the specific GPS location of existing stations).

Each underwater station was set up by placing 2-m-long T-bars on the survey pins at either end of the transect. The center of the T-bar was on the pin, making each arm of the T-bar 1 m long. Floating yellow polypropylene lines are stretched taught between the ends of the bars and directly down the middle, creating three parallel transect lines (Figure 34.1a). The belt transect created by these parallel lines encompasses a 40-m$^2$ area of reef surface.

## STATION SPECIES INVENTORIES

Coral species counts were made visually in each 40 m$^2$ by two trained census takers on SCUBA and were recorded on preprinted Mylar score sheets clipped to underwater slates (Figure 34.1b). Counters enumerated three parameters: (1) the number of scleractinian and milleporine coral species within the stations, (2) the total number of *Diadema antillarum* present inside the station boundaries, and (3) the presence or absence of coral disease for each species encountered. Three disease conditions were recognized: white diseases, black-band disease, and other diseases, plus bleaching. The independent counts of coral species, coral diseases, and *Diadema* density in the 40-m$^2$ areas at each station were then compared by the divers on site and any discrepancies were resolved. Final counts from each site represent the combination of information from the two divers' score sheets to produce cross-checked information on species number (Tables 34.2 and 34.3), urchin density (Table 34.4), and the incidence of coral disease (Table 34.5). Quality assurance and quality control are therefore maintained by initial training by coral experts and cross-checking of counts on site by paired divers.

## VIDEO TRANSECTING

Information on the absolute abundance of the biota is best collected using video surveys that can be quantified in the laboratory (Aronson and Swanson, 1997; Vogt et al., 1997). With our method, following the species inventories, a weighted chain is unrolled directly under the floating polypropylene lines (Figure 34.1a) to guide the video camera and then the yellow lines are removed. Facing the shore, the left line is designated transect 100, the middle line is transect 300, and the right line is transect 500 (see Table 34.1 for a description of the transect and file-labeling protocols). At the beginning of each video swim, the camera records basic station identification information using an underwater "clapper board" fitted with underwater, interchangeable Velcro numbers to ensure that each video segment is properly identified on the tape. The camera points straight down at all times, and the 40-cm wand on the front of the camera almost touches the guide chain as the camera traverses the bottom (Figure 34.1c). The swim is slow and steady, taking not less than 5 minutes

**TABLE 34.1**
**Site Locations in Negril, Jamaica, for Installed Coral Reef Monitoring Stations 2 to 6**

| Site Name | Station No. | Depth (m) | Latitude | Longitude | Image Files |
|---|---|---|---|---|---|
| Green Island (shallow) | 2 | 2 | 18°24.01′ N | 078°16.68′ W | |
| Transect 100 | | | | | JA2L7100.JPG |
| Transect 300 | | | | | JA2L7300.JPG |
| Transect 500 | | | | | JA2L7500.JPG |
| Orange Bay (shallow) | 3 | 2 | 18°22.60′ N | 078°18.71′ W | |
| Transect 100 | | | | | JA3L7100.JPG |
| Transect 300 | | | | | JA3L7300.JPG |
| Transect 500 | | | | | JA3L7500.JPG |
| Little Bloody Bay (shallow) | 4 | 2 | 18°21.35′ N | 078°20.68′ W | |
| Transect 100 | | | | | JA4L7100.JPG |
| Transect 300 | | | | | JA4L7300.JPG |
| Transect 500 | | | | | JA4L7500.JPG |
| Long Bay (shallow) | 5 | 2 | 18°19.07′ N | 078°21.08′ W | |
| Transect 100 | | | | | JA5L7100.JPG |
| Transect 300 | | | | | JA5L7300.JPG |
| Transect 500 | | | | | JA5L7500.JPG |
| Little Bay (shallow) | 6 | 2 | 18°13.05′ N | 078°15.94′ W | |
| Transect 100 | | | | | JA6L7100.JPG |
| Transect 300 | | | | | JA6L7300.JPG |
| Transect 500 | | | | | JA6L7500.JPG |

**File Naming Protocol for the J-PEG Files on the CD ROM:**

| | |
|---|---|
| Character position 1–2: | JA (Jamaica) |
| Character position 3: | 1 (Station number) (currently 2–6) |
| Character position 4: | L (Month of the year by letters) (A–L) |
| Character position 5: | 7 (Last digit of the year, 1997) |
| Character position 6–8: | 100 (Transect image numbers) (100–220; 300–420; 500–620) |
| | Image number 100 is the underwater site location "Clapper Board." |
| | Image numbers 101–220 are the images from the 20-m transect. |

Example: JA2L7100.JPG

and not more than 7 minutes. The entire procedure for videotaping each station includes the following videotape segments: (1) A left-to-right pan of the transect area from the offshore pin; (2) the transect 100 clapper board (Figure 34.1D); (3) a slow, steady swim up the transect 100 line; (4) the transect 300 clapper board; (5) a slow, steady swim up the transect 300 line; (6) the transect 500 clapper board; (7) a slow, steady swim up the transect 500 line; and (8) a left-to-right pan of the transect area from the nearshore pin.

## IMAGE AND STATISTICAL ANALYSES

Frames are grabbed from the 8-mm tape at uniform time intervals, and the 8-mm (analogue) images are transferred to CD-ROM (digital) files. Image files are named by the file-naming protocol outlined in Table 34.1. With the video camera held 40 cm above the reef, only about 60 images are required to cover the 20-m belt transect. For sampling purposes, however, twice as many frames are grabbed from the tape. This library of images is then viewed as a gallery of consecutive frames, and abutting frames with minimal overlap (defined as images with 15% or less area in common) are selected

**FIGURE 34.1** Each coral monitoring station is demarcated by two survey pins drilled into the reef surface 20 m apart. A T-bar is placed on top of these survey pins, and three floating yellow polypropylene lines are stretched taught between them with each line 1 m from the next. Swimming toward shore (that is, toward shallow water), the left line is transect 100, the middle line is transect 300, and the right line is transect 500; weighted chains run under each of the three transect lines (A). Identification of species within this region is conducted by two counters (B; see Tables 34.2 to 34.5 for results of these station species inventories). Video is taken along each of the three transect chains for enumeration of the percent cover of different corals and other substrate types (C). Each video segment is identified on the tape by an underwater clapper board (D). (Photographs by Craig Quirolo.)

**TABLE 34.2**
**Taxonomic List[a] of Hydrozoan and Scleractinian Hard Corals Recorded During the Underwater Species Inventories Inside the 40-m² Plots of the Negril Coral Monitoring Project**

| Coral | Computer Code Abbreviation |
|---|---|
| Phylum Cnidaria | |
|   Class Hydrozoa | |
|     Order Athecate | |
|       Family Milleporidae | |
|         *Millepora alcicornis* Linne, 1758 | *M. alc.* |
|         *Millepora complanata* Lamarck, 1816 | *M. com.* |
|   Class Anthozoa | |
|     Order Scleractinia | |
|       Suborder Astrocoeniina | |
|         Family Astrocoeniidae | |
|           *Stephanocoenia michelinii* Milne Edwards & Haime, 1848 | *S. mic.* |
|         Family Pocilloporidae | |
|           *Madracis mirabilis* sensu Wells, 1973 | *M. mir.* |
|         Family Acroporidae | |
|           *Acropora palmata* (Lamarck, 1816) | *A. pal.* |
|       Suborder Fungiina | |
|         Family Agaricidae | |
|           *Agaricia agaricites* (Linne, 1758) | *A. aga.* |
|           [Complex][b] | |
|           *Agaricia fragilis* (Dana, 1846) | *A. fra.* |
|           *Leptoseris cucullata* (Ellis & Solander, 1786) | *L. cuc.* |
|         Family Siderastreidae | |
|           *Siderastrea radians* (Pallas, 1766) | *S. rad.* |
|           *Siderastrea siderea* (Ellis & Solander, 1786) | *S. sid.* |
|         Family Poritidae | |
|           *Porites astreoides* Lamarck, 1816 | *P. ast.* |
|           *Porites porites* (Pallas, 1766) | *P. por.* |
|           [Complex][c] | |
|       Suborder Faviina | |
|         Family Faviidae | |
|           *Diploria clivosa* (Ellis & Solander, 1786) | *D. cli.* |
|           *Diploria labyrinthiformis* (Linne, 1758) | *D. lab.* |
|           *Diploria strigosa* (Dana, 1846) | *D. str.* |
|           *Favia fragum* (Esper, 1795) | *F. fra.* |
|           *Montastraea cavernosa* (Linne, 1767) | *M. cav.* |
|           *Montastraea annularis* (Ellis & Solander, 1786) | *M. ann.* |
|           [Complex][d] | |
|         Family Meandrinidae | |
|           *Dichocoenia stokesi* Milne Edwards & Haime, 1848 | *D. sto.* |
|         Family Mussidae | |
|           *Isophyllastraea rigida* (Dana, 1846) | *I. rig.* |
|           *Isophyllia sinuosa* (Ellis & Solander, 1786) | *I. sin.* |
|           *Mycetophyllia lamarckiana* (Milne Edwards & Haime, 1848) | *M. lam.* |

**TABLE 34.2 (CONTINUED)**
**Taxonomic List[a] of Hydrozoan and Scleractinian Hard Corals Recorded During the Underwater Species Inventories Inside the 40-m² Plots of the Negril Coral Monitoring Project**

| Coral | Computer Code Abbreviation |
|---|---|
| Suborder Caryophylliina | |
| Family Caryophylliidae | |
| *Eusimilia fastigiata* (Pallas, 1766) | *E. fas.* |

[a] Systematic arrangement follows Wells and Lang (1973) and Cairns et. al. (1991).

[b] *Agaricia agaricites* Complex includes: *agaricites* (Linne, 1785); *carinata* Wells, 1973; *danai* Milne Edwards & Haime, 1860; *purpurea* (Lesueur, 1821).

[c] *Porites porites* Complex includes: *porites* (Pallas, 1766); *clavaria* Lamarck, 1816; *furcata* Lamarck, 1816; *divaricata* Lesueur, 1821.

[d] *Montastraea annularis* Complex (Weil and Knowlton, 1994) includes: *annularis* (Ellis & Solander, 1786); *faveolata* (Ellis & Solander, 1786); *franksi* (Gregory, 1895).

for counting. It is important to skip overlapping frames and include adjacent frames in order to avoid double counts of the same area or, conversely, under-counts of other areas.

Random points are generated for each image. These random points are stored on the CD-ROM along with the image itself and are recalled every time the frame is displayed. The images are displayed using image analysis software by Image-Pro, Inc. A dedicated macro, Point-Count for Coral Reefs, is used to score the image. Point-Count for Coral Reefs displays ten random points on the image, and, by using hot-button keys, the identity of organisms under each point is entered. Corals are identified to species (Tables 34.2 and 34.3) and the remaining substrate types are identified as (1) sponge, (2) macroalgae, (3) zoanthids, (4) octocorals, or (5) substrate. The macroalgal category is defined as large fleshy or calcareous algae ≥4 cm in length (two chain links within the camera's view). This category is distinguished from substrate (which is usually also covered by algae) based primarily on the size of the attached plants.

The random point count is used to calculate the percent cover of the different substrate types and coral species diversity (Table 34.3). In all cases, means are calculated based on $N = 3$ transects for each station. Technically, these three parallel transects are subsamples and not true replicates. Because the images along each transect do not overlap, these parallel transects are not, however, pseudo-replicates.

## RESULTS

### STATION SPECIES INVENTORIES

A total of 23 species of hard corals were identified within the Negril Coral Reef Monitoring Stations (Table 34.2). Between 62 and 80% of these species were also identified during the point-count analyses (Table 34.3). Species missed by the point-count enumeration were either cryptic (e.g., *Agaricia fragilis* and *Leptoseris cucullata*), extremely small (e.g., *Siderastrea radians*), or extremely rare (e.g., *Isophyllastrea rigida, Isophyllia sinuosa, Dichocoenia stokesi,* or *Eusmilia fastigiata*). Thus, the point count analysis identifies common corals well and misses only rare or obscure species. The relatively uniform percentage of species identified by point-count, particularly at the less species-rich stations (3 to 6), suggests that point-count analyses might be used to predict total species number ($S$) with a fair degree of confidence, even in the absence of the underwater visual counts.

**TABLE 34.3**
**List of Coral Species Inside the Negril Coral Reef Monitoring Project Stations**

| | Station Name (Station No.) | | | | |
|---|---|---|---|---|---|
| | Green Island (2) | Orange Bay (3) | Little Bloody Bay (4) | Long Bay (5) | Little Bay (6) |
| *Millepora alcicornis* | (X) | (X) | | | (X) |
| *Millepora complanata* | (X) | (X) | (X) | (X) | (X) |
| *Stephanocoenia michelinii* | X | X | | X | X |
| *Madracis mirabilis* | | | | (X) | |
| *Acropora palmata* | (X) | (X) | (X) | (X) | (X) |
| *Agaricia agaricites* | (X) | (X) | (X) | (X) | (X) |
| *Agaricia fragilis* | X | X | X | | X |
| *Leptoseris cucullata* | X | | | | X |
| *Siderastrea radians* | X | X | X | X | (X) |
| *Siderastrea siderea* | (X) | (X) | (X) | (X) | (X) |
| *Porites astreoides* | (X) | (X) | (X) | (X) | (X) |
| *Porites porites* | (X) | (X) | (X) | (X) | (X) |
| *Diploria clivosa* | (X) | (X) | (X) | | (X) |
| *Diploria labyrinthiformis* | (X) | (X) | | | |
| *Diploria strigosa* | (X) | (X) | (X) | | (X) |
| *Favia fragum* | (X) | (X) | X | | |
| *Montastraea cavernosa* | X | (X) | (X) | | |
| *Montastraea annularis* | (X) | (X) | (X) | (X) | (X) |
| *Dichocoenia stokesi* | | X | | | |
| *Isophyllastraea rigida* | X | | | | |
| *Isophyllia sinuosa* | X | | | | |
| *Mycetophyllia lamarckiana* | (X) | | | | |
| *Eusimilia fastigiata* | X | | | | |
| Total *S* at each station by visual counts | 21 | 17 | 13 | 10 | 14 |
| Total *S* at each station by random points | 13 | 13 | 10 | 8 | 11 |
| *S* by video vs. visual count methods (%) | 62 | 76 | 77 | 80 | 79 |
| Mean *S* by video count | 10.0 ± 1.6 | 8.0 ± 1.6 | 6.7 ± 0.5 | 6.0 ± 0.8 | 7.7 ± 0.5 |
| Average *H'* | 1.81 ± 0.35 | 1.43 ± 0.19 | 1.60 ± 0.5 | 1.34 ± 0.17 | 1.62 ± 0.21 |
| Average *J* | 0.74 ± 0.18 | 0.70 ± 0.11 | 0. 84 ± 0.06 | 0.76 ± 0.14 | 0.79 ± 0.11 |

*Note:* Species are marked as X if encountered in the visual counts only, or as (X) if encountered in both the visual counts and the video counts. Average *S* for the video counts is calculated as the mean ± SD with *N* = 3 transects/station.

Species number correlates with the distance from Long Bay on the West End of Jamaica. *S* is lowest at the Long Bay site (*S* = 10) and increases heading east along either the north coast (to a maximum of 21 species at Green Island) or the south coast (to a maximum of 14 species at Little Bay) (Figure 34.2). These data demonstrate that coral reef development is least in Long Bay near the port city of Negril and greatest at increasing distances from the city.

Two of five stations (40%) had diseased coral (Table 34.5). Whereas black-band disease has been known in Jamaica, where it has caused extensive damage (Bruckner and Bruckner, 1997), this is the first report of the new coral disease, white pox, from Jamaica. All but one station (80%) had bleached coral. There is a tendency for bleaching and disease to be more common and to affect

**TABLE 34.4**
*Diadema antillarum* **Density, Macroalgal Abundance, and Live Coral Coverage Inside the 40-m² Negril Coral Reef Monitoring Project Stations During the December 1997 Survey**

| | Station Name (Station No.)/(Station Area) | | | | |
|---|---|---|---|---|---|
| | Green Island (2)/(44.5 m²) | Orange Bay (3)/(44.0 m²) | Little Bloody Bay (4)/(47.5 m²) | Long Bay (5)/(44.0 m²) | Little Bay (6)/(44.0 m²) |
| *Diadema* density (no./m²) | 0.00 | 0.70 | 0.38 | 1.39 | 0.04 |
| Macroalgal density (% cover) | 65.4 ± 6.2 | 10.7 ± 1.2 | 31.3 ± 5.0 | 63.3 ± 2.0 | 71.7 ± 4.5 |
| Live coral density (% cover) | 14.3 ± 0.9 | 12.6 ± 2.7 | 4.5 ± 2.0 | 5.7 ± 2.5 | 8.4 ± 3.2 |

**TABLE 34.5**
**The Number of Coral Species Exhibiting Disease During the December 1997 Survey of the 40-m² Negril Coral Reef Monitoring Project Stations**

| Station Name | Station No. | BL[a] | BB[b] | WH[c] | OD[d] |
|---|---|---|---|---|---|
| Green Island | 2 | 1 | 0 | 0 | 0 |
| Orange Bay | 3 | 0 | 0 | 0 | 0 |
| Little Bloody Bay | 4 | 2 | 0 | 1 | 0 |
| Long Bay | 5 | 1 | 1 | 0 | 1 |
| Little Bay | 6 | 1 | 0 | 0 | 0 |
| Total number of stations with each condition | | 4 | 1 | 1 | 1 |
| Total number of species with each condition | | 2 | 1 | 1 | 1 |

[a] BL = bleaching.
[b] BB = black-band disease.
[c] WH = White diseases, including white pox, white band, white plague type I, white plague type II.
[d] OD = Other diseases, including dark spot (purple blotch), yellow blotch (yellow band), red stripe, brain coral swelling disease.

more species the closer the station is to Long Bay (Table 34.5). Water temperatures during this December 1997 period were the same at all five localities, 28°C.

## IMAGE ANALYSES

The results of the point-count analyses are shown in Figures 34.2 to 34.5. The data show unequivocally that macroalgae are more common than corals on all of these reefs (Figures 34.4A–E, the relative abundance of different substrate types). Macroalgae are most abundant on reefs at Green Island (station 2), Long Bay (station 5), and Little Bay (station 6), whereas substrate is more abundant than macroalgae on the remaining two reefs (Orange Bay, station 3; and Little Bloody Bay, station 4). Only at Green Island and Orange Bay does the coral cover rise above 10% (14.3 ± 0.9% and 12.6 ± 2.7%, respectively; see Table 34.4). From a percent cover perspective, therefore, these "coral reefs" are actually "algal reefs."

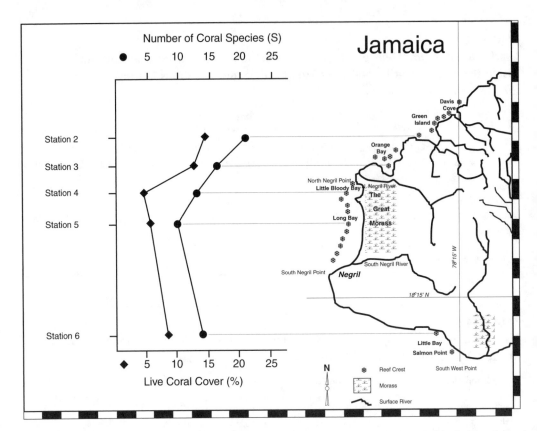

**FIGURE 34.2** Species richness and percent coral cover vs. latitude on Jamaica's West End in the Negril Coral Reef Marine Park (see Table 34.1 for GPS station location).

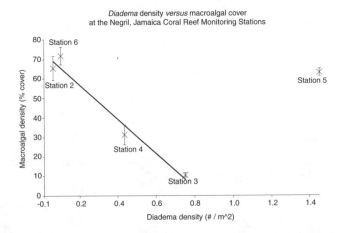

**FIGURE 34.3** *Diadema* density (no./m²) vs. macroalgal abundance (percent cover) at the Negril, Jamaica, Coral Reef Monitoring Stations (see Table 34.4). Exclusive of station 5 (Long Bay), the negative linear regression is highly significant ($p < 0.005$), with an $r^2 = 0.97$.

*Acropora palmata* is present in every station (Table 34.3). Initially, this might suggest that these are classic reef-crest/fore-reef localities. However, *A. palmata* is proportionately dominant only in Little Bay, where it constitutes 21% of the living coral, and nowhere is it as common as indicated in prior unpublished reports from the Negril area (T.F. Goreau and T.G. Goreau, pers. comm.).

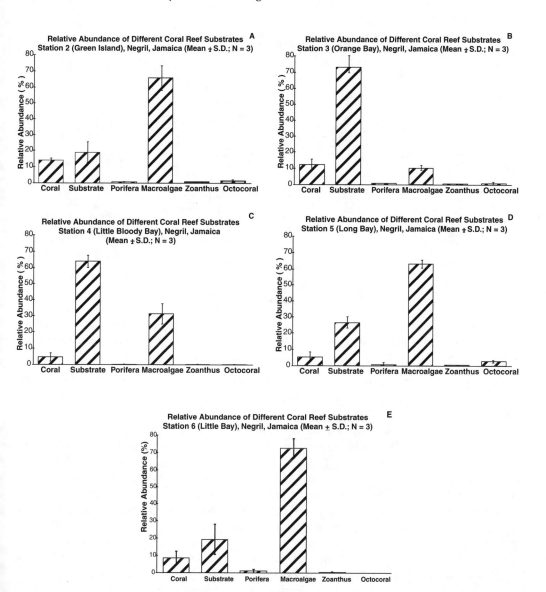

**FIGURE 34.4** Figure 4 (A) Relative abundance (mean ± SD; $N = 3$) of different coral reef substrates as determined by point-count analysis at station 2 (Green Island), Negril, Jamaica. Macroalgae are the most abundant cover type. (B) Relative abundance (mean ± SD; $N = 3$) of different coral reef substrates as determined by point-count analysis at station 3 (Orange Bay), Negril, Jamaica. Substrate is the most common cover type. (C) Relative abundance (mean ± SD; $N = 3$) of different coral reef substrates as determined by point-count analysis at station 4 (Little Bloody Bay), Negril, Jamaica. Substrate is the most common cover type. (D) Relative abundance (mean ± SD; $N = 3$) of different coral reef substrates as determined by point-count analysis at station 5 (Long Bay), Negril, Jamaica. Macroalgae are the most abundant cover type. (E) Relative abundance (mean ± SD; $N = 3$) of different coral reef substrates as determined by point-count analysis at station 6 (Little Bay), Negril, Jamaica. Macroalgae are the most abundant cover type.

In addition, *A. palmata* appears to have been more common at every site in the very recent past, as indicated by the fact that at all five sites the three-dimensional topography is dominated by large, dead stands of *A. palmata* trunks and branches. This also proves that the dieoff was not due to mechanical damage due to storms or boat groundings. Variance in *A. palmata* abundance between

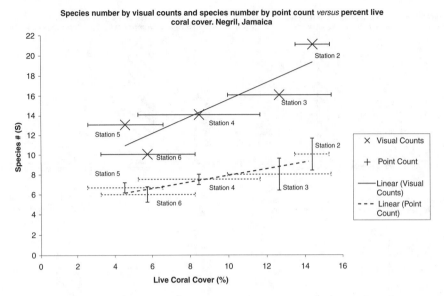

**FIGURE 34.5** Species number ($S$ by visual counts; see Table 34.3) and species number ($S$ by point-counts; mean $\pm$ SD; $N = 3$; see Table 34.3) vs. percent live coral cover (mean $\pm$ SD; $N = 3$; Table 34.4) both fit linear regression models well ($p < 0.05$) with $r^2 = 0.78$ and 0.82, respectively.

the three transects within a station is extremely high (greater than or equal to the mean at all five stations), demonstrating its patchy distribution in areas where once this species was almost completely dominant.

*Siderastrea siderea* and *Montastraea annularis* are relatively common at Green Island and Little Bay. *Madracis mirabilis* is found in abundance in the Long Bay station, but nowhere else. *Millepora complanata* dominates at Orange Bay, Little Bloody Bay, and Long Bay. These three bays are more frequently visited by tourists than the other sites. In general, therefore, there is a positive correlation between the presence of fire coral and the presence of tourists.

Percent live coral cover correlates strongly with coral species number (Figure 34.5). For instance, correlations between species number (by either visual counts or point-counts) and percent live coral cover fit linear regression models well ($p < 0.05$) with $r^2 = 0.78$ and 0.82, respectively (Figure 34.5). As a result, the striking relationship between species number and latitude demonstrated for Jamaica's West end also holds for percent live coral and latitude (Figure 34.2).

Four out of five stations demonstrate a simple linear relationship between macroalgal abundance and density of the herbivorous grazing sea urchin, *Diadema antillarum* (Table 34.4; Figure 34.3). Based on a tremendous amount of theoretical and empirical work on effects of this sea urchin on algal abundance, most of which was done in Jamaica (Morrison, 1988), one would expect this negative correlation to be universal. However, Long Bay is a completely unexpected exception to this generally inviolate rule (Figure 34.3). The general nature of the algal–sea urchin relationship, as well as the Long Bay exception, are shown in Figures 34.6A–C. These video images show high *Diadema* density and low algal cover at Orange Bay (predictable), low *Diadema* density and high algal cover at Little Bay (also predictable), and finally high *Diadema* density and high algal cover at Long Bay (unexpected). This macroalgal anomaly may also partly explain why macroalgal cover does not correlate well with percent live coral cover (Table 34.4).

Species diversity $H'$ patterns (Table 34.3) demonstrate that the highest average $H'$ is on the Green Island reef (1.81 $\pm$ 0.35) and the lowest value is at Long Bay (1.34 $\pm$ 0.17). The reduced species diversity value at Long Bay is due mostly to lower species number rather than lower species evenness (Table 34.3). As with species number and percent cover of live coral (Figure 34.2), species diversity rises as a function of distance from Long Bay (Table 34.3).

**FIGURE 34.6** Video images showing: (A) high *Diadema* density and low algal cover at Orange Bay, (B) low *Diadema* density and high algal cover at Little Bay, and (C) high *Diadema* density and high algal cover at Long Bay.

Sponges, zoanthids, and octocorals are less common than coral on every reef studied in the park. Octocorals are entirely absent from Little Bay (station 6), and zoanthids are almost absent from Green Island (station 2).

## DISCUSSION

### THE NEGRIL CORAL REEF PRESERVATION SOCIETY MONITORING PROJECT

We successfully identified and installed five permanent monitoring stations within Negril Coral Reef Marine Park. We also completed the initial monitoring of these stations and have completely analyzed all of the underwater counts and video images from them. All stations came from similar depths and similar habitats and all contained the major shallow-water, reef-building coral *Acropora palmata*.

### PATTERNS OF CORAL SPECIES NUMBER, CORAL SPECIES DIVERSITY, AND CORAL ABUNDANCE

Patterns of coral species number, coral species diversity, and coral abundance (Figure 34.2, Tables 34.3 and 34.4) and the incidence of coral disease (Table 34.5) are suggestive of a gradient of coral stress that is greatest in the Long Bay/Little Bloody Bay area. Tomascik and Sander (1987) also demonstrated a cline of coral species number in Barbados such that $S$ increased at increasing distance from point-source pollution. The presence at all sites of large, but recently dead, coral skeletons of *Acropora palmata* suggests that these reefs are currently undergoing coral loss and that the loss is not due to physical damage such as storms or boat groundings. The percent of living coral is so low (less than 6% in Long Bay and less than 5% in Little Bloody Bay) that we can safely conclude that coral loss on these reefs threatens ecosystem function.

### PATTERNS OF MACROALGAL AND DIADEMA ABUNDANCE

Macroalgal abundance is the product of (1) nutrient loading, which increases algal growth rates; and (2) grazer density, which decreases algal abundance by increasing algal removal rates. These two opposing forces have been referred to as top-down vs. bottom-up controls of reef community structure (Lapointe, 1994, 1997; Lapointe et al., 1997). The threshold nutrient concentrations required to sustain macroalgal blooms on Caribbean coral reefs are extremely low, approximately 1.0 μM DIN (dissolved inorganic nitrogen in the form of nitrate and ammonia) and 0.1 μM SRP (soluble reactive phosphorus, as orthophosphate and organophosphate) (Lapointe, 1997). These values are equivalent to 0.014 ppm N (or 0.040 ppm $NO_3$) and 0.003 ppm P (or 0.007 ppm $PO_4$), respectively. Levels higher than these promote algal growth (Bell and Elmetri, 1995; Lapointe, 1997) and depress coral growth (Hoegh-Guldberg et al., 1997; Mate, 1997). These threshold values are frequently exceeded in Long Bay and Little Bloody Bay, and led Wade (1994) to state, "Negril (nutrient) levels are higher than those recorded in formerly sewage-contaminated Carlisle Bay in Barbados (Turnbull and Lewis, 1981)."

Prevailing currents around Jamaica are from east to west (Goreau and Thacker, 1994), suggesting that the causes for water quality deterioration in Negril come from within the Negril watershed rather than from runoff from less inhabited sites to the east on either side of the island. Overland flows of water such as the North and South Negril Rivers are an obvious source of nutrient input to the nearshore area. These receive runoff from farms, cane fields, and livestock pastures. Wastewater treatment facilities at some of the hotels also discharge into the sandy beach substrate or immediately offshore. In a karstic region such as Jamaica's North End, groundwater contamination can also be a factor because surface waters are directly connected to groundwater (Figure 34.7). Cesspits, the common form of household waste disposal, make direct connections. Wade (1994) points out that nitrogen levels are frequently higher at depth near the reef's surface than at the

## Hydrological Transport of Nutrients
### Negril, West End, Jamaica

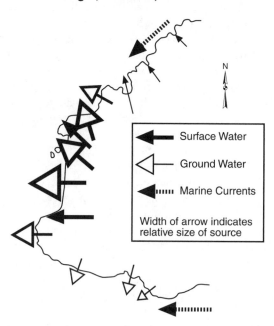

**FIGURE 34.7** Nutrients are transported hydrologically from both fresh and saltwater sources onto the fringing coral reefs of Negril, Jamaica. The multiple pathways of overland and underground flow onto Jamaican fringing reefs demonstrates the necessity for a whole-watershed approach to conserving these marine coastal resources. (Redrawn from Goreau and Thacker, 1994.)

water's surface. Data on the subsurface movement of nutrient-laden freshwater from the uplands into the marine coastal zone and out onto the reef at depth has been demonstrated both in Discovery Bay, Jamaica (D'Elia et al., 1981) and in Negril (Goreau and Thacker, 1994; Underground Water Authority, 1996). Enriched groundwater flowing through honeycomb limestone caves and karst formations creates the subsurface transport mechanisms by which shore-based pollution is transferred from beneath into the marine environment.

Phosphorous is also transported onto the reef by the same mechanism. By analyzing the C:P and N:P ratios of *Sargassum polyceratium* and *Turbinaria tricostata* plants in Long Bay, Lapointe (1994) concludes that "land-based human activities are a source of the P, contributing to eutrophication of Negril's coastal waters." Wade (1994) comments that, "The inescapable conclusion from these data is that the nutrient levels in Long Bay, both along the shore and over the reefs, are now above the levels considered desirable for maintaining ecological balance. ...The evidence is sufficiently strong and convincing for us to conclude that the recent enrichment of Long Bay leading to eutrophication with prolific growth of algae is due primarily to the leaching of nutrients from sewage-saturated soils along the Long Bay coastal strip and West End." These nutrient data provide a causative explanation for our high algal cover data (Table 34.4).

Mendes et al. (1997), studying an offshore pollution gradient emanating from Kingston Harbour, demonstrated that tissue from *Montastraea annularis* colonies closest to the source of pollution in Kingston Harbour is enriched in [15]N relative to colonies farther away from the source. We predict that this same stable isotopic gradient will exist for coral colonies at increasing distances from Long Bay.

The Mendes et al. (1997) study also shows a progressive increase in water transparency and decrease in the abundance of free-living algae along this Kingston Harbour gradient. Reduced water transparency increases coral respiration rates and reduces coral photosynthetic rates in colonies of *Montastraea annularis* from Negril (Dallmeyer et al., 1982; see also Te, 1997). Elevated nutrients also promote attached algal growth, and this can increase the abundance of fleshy algae which smother and kill adult and juvenile corals (Dubinsky and Stambler, 1996; Goreau et al., 1997).

Macroalgae are most common where coral is rarest, suggesting that elevated nutrient loads, emanating from the Negril watershed, are to blame for the excessive plant abundance present in Long Bay. This view is confirmed by the fact that, despite a healthy population of the grazing sea urchin *Diadema antillarum*, macroalgal abundance is still exceedingly high (72%) in Long Bay. This almost theoretically impossible juxtaposition of high plant density in the face of high grazer abundance is possible only if algal growth rate, accelerated by high nutrient inputs, greatly exceeds grazing rates, overwhelming the ability of the grazers to reduce algal numbers. At present, the authors are unaware of any published reports of high algal abundance in the presence of high grazer abundance and, therefore, high grazing rates. The simplest explanation for this anomaly is that nutrient loading in Long Bay accelerates algal growth to the point where even a healthy population of grazing sea urchins is unable to suppress macroalgal abundance. If true, Long Bay is one of the most nutrient-stressed tropical marine environments in the world.

Water quality in Long Bay is currently suboptimal for coral reef maintenance, let alone development. Whereas algae respond quickly to nutrient loading by increasing their growth rates and abundance, corals are more permanent features of the reef and take longer to respond and show a reduction in their abundance. The current coral cover may be a legacy of previous water quality conditions. If water quality has declined recently, the coral reef may still be dominated by older coral specimens that give the appearance of a reef, even if current conditions favor neither the maintenance of the existing reef nor the development of a new reef. These data suggest that coral populations within the Negril Coral Reef Marine Park may be in a state of transition and that they are being pushed to extinction by multiple stressors (Porter et al., 1999). Extension of this monitoring program beyond this initial study is needed to determine whether the area will be dominated by short-lived organisms like the marine algae or longer-lived organism such as corals. In aggregate, these data lead us to conclude that nutrient loading due to poor watershed management is killing Negril's coral reef.

## COMPARATIVE COMMENTS ON SPECIFIC REEF AREAS

Reefs transected by the five permanent monitoring stations installed during this project fall along a clinal gradation from reefs with the greatest biodiversity (Green Island, on the North Coast, and Little Bay, on the South Coast) to those with the least biodiversity (Long Bay, on the West End). On this cline, Orange Bay and Little Bloody Bay (both on the North Coast) have intermediate biodiversity values (Figure 34.2). This is not to suggest, however, that reefs at the high biodiversity end of this cline are not at risk. Goreau (1994) described coral reefs of Green Island by stating that, "Several well developed reef crest structures occur at the mouth of Green Island Bay. Those nearest the Bay [however] may suffer from high turbidity and from algal overgrowth. Some good reefs may exist around nearby headlands, away from the Bay's muddy plume."

Little Bay coral reefs are also at risk. Goreau (1994) commented that "Algal species diversity is high, but biomass is moderate [at Little Bay]." It seems likely that, given the high algal biomass recorded during our survey (Table 34.4), that the situation has become more eutrophic between Goreau's survey in 1992 and our survey in 1997. Because we have marked our survey locations precisely, future reef monitors will be able to track with certainty community changes through time.

For the intermediate reefs, Goreau (1994) noted that, "Water transparency [at Orange Bay] is very good, far better than at Negril. The garbage dump serving Negril and Lucea, [however], is a current or potential source of solid, liquid, and air pollution. Orange Bay has exceptional potential for diving and snorkeling if excellent water quality can be maintained." Little Bloody Bay "continues to show signs of excessive nutrient stress during 1992. This area may be subject to leakage of nutrients under the beach from the sewage plant outfall across the road in the adjacent Morass." At Long Bay, "[the] reefs have had so much of the coral killed that recovery, even in the absence of all further stress, would require settlement and growth of a whole new coral community, a process which could take centuries. These reefs could recover if stresses to them from excessive nutrients and turbidity were abated."

## TESTABLE HYPOTHESES AND RECOMMENDATIONS FOR FUTURE RESEARCH

To help guide future conservation research in the park, we offer the following testable hypotheses, which, if proven, would provide extremely valuable information for defining "best management practices" within the political and watershed boundaries of the park:

1. At the five stations, nutrient concentrations will be highest in Long Bay.
2. Nutrient concentrations will follow the species curve; that is, N:P at the Long Bay station > Little Bloody Bay > Orange Bay > Little Bay > Green Island.
3. Coral growth and photosynthesis will be diminished in Salmon Bay due to high water turbidity and low water clarity at this site.
4. If the nutrient addition problem is not solved, future patterns of coral loss will follow the species curve (Figure 34.2); that is, the coral loss rate at Long Bay > Little Bloody Bay > Orange Bay > Little Bay > Green Island.
5. If the nutrient addition problem is not solved, future patterns of species loss will also follow the species curve (Figure 34.2); that is, the species loss rate at Long Bay > Little Bloody Bay > Orange Bay > Little Bay > Green Island.
6. If the nutrient addition problem is not solved, the rarest corals will go locally extinct at all stations.
7. Caging experiments will show exceedingly high algal growth rates in the absence of grazing, especially in the Long Bay area.

## SPECIFIC RECOMMENDATIONS FOR THE NEGRIL MARINE PARK

1. Nutrient input to the Negril Marine Park must be substantially reduced. As Goreau (1997) points out, "Because coral reefs are directly or indirectly affected by so many human activities, their protection requires addressing all major stresses simultaneously, not picking and choosing a few for symbolic efforts, such as sewage treatment schemes which lack tertiary treatment."
2. Whole-watershed management is the best and only hope for the survival of Negril's coral reefs and for the protection of the Marine Park's living resources.
3. In order to protect the natural process of algal removal, regulations should be written and strictly enforced forbidding the collection or killing of sea urchins. Perhaps a public awareness campaign of "Don't shoot your lawnmower: Leave the little urchins alone!" should be started to let people know that the common practice of crushing urchins to attract brightly colored coral reef fish is highly deleterious to coral reef health.
4. The designated Marine Park authority should consider "re-seeding" other areas within the park boundaries with *Diadema*. Like wolves returning to the Yellowstone National Park, this would be reintroducing a species that was once in abundance throughout the park and that played a keystone role in the maintenance of the ecosystem's delicate ecological balance. Such a program should be accomplished by participation by the public, student groups, and eco-tourists. Every attempt possible should be made to have this conservation action covered extensively by the news media, including both print and broadcast journalists.

5. Our data demonstrate the absolute necessity of protecting from development both bays and reefs in the eastern-most sector of the Marine Park on Jamaica's North Coast. If these reefs are destroyed, the Park will no longer have the capacity to rejuvenate itself or to recover from disturbances, natural, or anthropogenic. If, on the other hand, these areas are conserved, then the Marine Park does have the capacity to self-seed and regrow. One could, of course, hope that restocking for the Negril Marine Park might occur from elsewhere in the Caribbean or from unprotected areas elsewhere in Jamaica, but that is a very risky strategy and not one that is on the list of approved conservation measures. The hopeful message here is that if these multiple assaults to the reef can be brought under control, reefs from only a short distance away, but still *inside the Marine Park*, can reseed the Long Bay area.

## TOP-DOWN VS. BOTTOM-UP CONTROLS OVER ALGAL GROWTH IN JAMAICA

Discovery Bay and Negril, Jamaica, are similar in that substrates once dominated by scleractinian corals have been taken over by macroalgae. Hughes (1994; Hughes et al., 1999) suggests that in Discovery Bay, algal growth is due to the dieoff of the major grazer, *Diadema* (i.e., top-down controls), whereas Lapointe (1997, 1999) emphasizes the addition of nutrients to Jamaica's North coast (i.e., bottom-up controls). Coral reefs on the west end of Jamaica provide evidence that both kinds of controls are operating in Negril. In Long Bay, it is clear that bottom-up controls predominate because the highest population of urchins (1.39 individuals per m²; Table 34.4) coexists with one of the highest densities of macroalgae (63.3%, Table 34.4). As predicted by Lapointe (1997) and Lapointe et al. (1990), this area of high nutrient loading and high herbivory also has a high density of calcareous algae such as *Halimeda* (Figure 34.4D). *Halimeda* and other crustose algae are somewhat more resistant to herbivory than fleshy green algae and bloom under these conditions (Morrison, 1988). The other four reefs surveyed demonstrate the predicted negative correlation between urchin and algal abundance (Figure 34.3), and may therefore be said to be under top-down controls; however, even these reefs are receiving substantial inputs of nutrients, as demonstrated by the massive macroalgal blooms on these reefs (Figure 34.4). Therefore, while top-down controls are occurring on these reefs, nutrient input and bottom-up controls are still setting macroalgal density at a high level. While algal populations on coral reefs in Jamaica may historically have been influenced primarily by "top-down" controls, this is most definitely not the case currently in Negril.

## CHRONIC STRESS VS. CATASTROPHIC STRESS

Goreau (1994) pointed out that, whereas in the past Jamaica's reefs recovered quickly from hurricanes, their recovery now is slow or nonexistent. Resilience to natural catastrophic stresses such as hurricanes requires optimal water quality conditions following the storm event. A reasonable speculation is that deteriorating water quality creates a chronic stress that retards or possibly, in the case of Negril, prevents recovery. In his review of coral population dynamics (Connell, 1997) pointed out that the best predictor of the course of reef recovery was whether the stress that initially caused the decline was chronic (such as pollution) or catastrophic (such as a hurricane). Chronically stressed reefs do not recover. Negril's coral reefs are now chronically stressed. Improving water quality will determine whether or not Long Bay's reef will recover.

## AN EXTENDED ANALOGY BETWEEN WESTERN JAMAICA AND SOUTH FLORIDA

Many similarities and close parallels exist between the situation described here for Jamaica and the South Florida hydroscape (Table 34.6). Coral reefs of both Negril and the Florida Keys are hydrologically linked to terrestrially based sources of pollution and nutrient enrichment. The porous nature of karstic geology means that both surface and subsurface flows can transport dissolved and

## TABLE 34.6
## An Extended Analogy Between the Coral Reefs of Negril, Jamaica, and Those of the Florida Keys

### Similarities

| Characteristic | Analogous Situation |
|---|---|
| Hydrological linkages to land | Pollution/nutrient enrichment from land |
| Transport mechanisms | Overland sheet flow/subterranean karst flow |
| Intensively fertilized agricultural areas | Sugar cane |
| Population pressures | Population growth/tourist development |
| Upslope swampland | Everglades/Great Morass |
| Oceanographic correlations | Salinity/nutrient concentration gradients from land |
| Lack of critical information | Frequency/duration of land–sea linkage unknown |
| | Nutrient budgets unquantified |

### Differences

| Negril | Florida Keys |
|---|---|
| Fringing reefs nearshore | Barrier reefs offshore |
| No barrier islands between land and reef | Barrier islands (Elliot Key/Key Largo) |
| Reefs dominated by macroalgae | Reefs dominated by turf algae |

suspended materials in freshwater into the marine environment. In both places, rapid population and tourist development is concentrated along the shoreline. In both cases, behind this intense coastal zone, development is intensive agriculture. In both cases, sugar cane is the dominant agricultural crop, involving massive expropriation of land from natural habitats and massive annual fertilization of the crop.

While no environment on earth approaches the awesome magnitude of the Everglades, Jamaica's Great Morass has many hydrological similarities to this unique Floridian environment, including the buildup of peat-like soils in the middle of the swamp and the removal, by filtration, of nutrients that enter the swamp along its upstream borders.

In both areas, there is a highly significant inverse correlation between salinity and the nutrient content of coastal marine waters (D'Elia et al., 1981; Goreau, 1994), demonstrating with absolute certainty the terrestrial origin and freshwater transport of dissolved nutrients onto the coral reef (Figure 34.7). In both cases, nutrient concentrations are often higher near the reef surface than at the ocean surface (D'Elia et al., 1981; Wade, 1994), demonstrating that seaward subsurface flows resulting in underwater "springs" are also occurring. While the frequency and duration of these land-to-ocean flows have not been quantified, the data definitely prove that this is occurring. As a conservation priority, these connections between ecosystems must be quantified so that upland management can address these land–sea couplings in both Jamaica and Florida. Material does move between the land and the reef. Studies on the mass balance and transport of nutrients are desperately needed in both places (Table 34.6).

Striking differences also exist between the Florida Keys and the Negril coral reefs (Table 34.6). The main difference is the distance between the reefs and the land: Negril's fringing reefs are close; Florida's barrier reefs are distant. Macroalgal growth and coral loss in Long Bay are obvious, but these indicators of reef stress are being hotly debated in Florida (Lapointe, 1997). While most knowledgeable observers in the Keys concede that less coral is there now than 20 years ago (Porter et al., this volume), especially in the shallow-water, reef-crest zone, many Floridian reef scientists and managers are demanding statistically valid results from extraordinarily detailed long-term datasets with broad spatial and temporal coverage before they will accept the statements that (1) the reefs are declining, and (2) that anthropogenic stress is playing any role whatsoever in the decline.

Floridian monitoring programs (Porter et al., this volume) have been designed with the knowledge that they will be challenged in a court of law and that they must be able to stand up under intense scientific scrutiny. With respect to the need for high-quality coral reef monitoring data, Jamaica and Florida are identical.

Negril's coral reefs should serve as a warning to reef managers in Florida, and worldwide, that there are nutrient thresholds that cannot be exceeded. If they are, then the swift and immediate consequences include, but are not limited to, the growth of massive amounts of attached fleshy algae and the loss of coral reefs and biotic resources that so frequently attract tourists (and tourist dollars) to these same areas.

Historically, the Great Morass acted as a natural biological filter. Now Negril is building a wastewater treatment plant (Lapointe and Thacker, this volume), but it is not clear whether this plant has been designed and constructed in a way so as to meet the wastewater treatment standards required to remove the root causes of coral reef decline on Jamaica's West End. Only a tertiary treatment plant built and operated in a way that eliminates both surface and groundwater connections to the coastal environment will work.

## Summary and Conclusions

A coral reef monitoring program was established in Jamaica within the boundaries of the Negril Marine Park. Five stations were located on shallow-water, reef-crest environments around Jamaica's western end (Green Island, Orange Bay, Little Bloody Bay, Long Bay, and Little Bay). Permanent markers, situated 20 m apart, were located on the five stations, and underwater visual counts of coral species richness, coral disease, and *Diadema* density were made within the 20-m belt transects following protocols established by the U.S. Environmental Protection Agency's Coral Reef Monitoring Program in Florida. Three parallel transect lines along the outside and down the middle of this rectangle were videotaped. Images from the 8-mm tape were grabbed and transferred to digital data files on CD-ROM. Digital images were enumerated using random point-counts, and the data were examined for patterns of coral species richness and percent cover of benthic groups such as hard corals, soft corals, sponges, and macroalgae.

Coral diversity and abundance increase significantly as one moves away from the Long Bay area at Negril. The relative abundance of the reef-building coral *Acropora palmata* also conforms to this trend, whereas the incidence of coral disease is highest at Long Bay. A highly significant inverse relationship exists between macroalgal density and the population density of the herbivorous sea urchin *Diadema antillarum*, except in Long Bay, where the highest density of macroalgae coincides with the highest density of sea urchins. We propose that algal growth, accelerated by nutrient loading in Long Bay, overwhelms the ability of grazing sea urchins to control algal abundance there. If true, Long Bay is one of the most nutrient-stressed tropical marine environments in the world. The percent live coral cover in Long Bay is so low (<6%) that we further conclude that coral loss on this reef threatens ecosystem function.

In aggregate, our data lead us to conclude that poor water quality is killing Negril's coral reefs and that action, rather than further debate (Risk, 1999), is warranted. The best and only hope for the survival of Negril's coral reefs is whole-watershed management in an attempt to reduce the flow of nutrients into the coastal zone. Our data also demonstrate the necessity of protecting bays and reefs in the eastern-most sector of the Marine Park on Jamaica's North Coast. These reefs are the richest in the park and hold the potential for reseeding the rest of the preserve if conservation efforts elsewhere in the park return the area to conditions favorable for coral growth and development.

## ACKNOWLEDGMENTS

We thank Dr. Jeremy D. Woodley and Judith M. Mendes for help in station setup and coral species counts on the Green Island Station; Everton Frame and Earl Murdock, Assistant Rangers, for logistical help with boats, SCUBA tanks, and transportation in Jamaica; and Buddha for technical assistance with the video camera.

# REFERENCES

Aronson, R.B. and D.W. Swanson. 1997. Video surveys of coral reefs: uni- and multivariate applications, *Proc. 8th Int. Coral Reef Symp.*, 2:1441–1446.

Bell, P. and I. Elmetri. 1995. Ecological indicators of large-scale eutrophication in the Great Barrier Reef Lagoon, *Ambio*, 24:208–215.

Bruckner, A.W. and R.J. Bruckner. 1997. The persistence of black-band disease in Jamaica: impact on community structure, *Proc. 8th Int. Coral Reef Symp.*, 1:601–606.

Cairns, S.D., D.R. Calder, A. Brinckmann-Voss, C.B. Castro, P.R. Pugh, C.E. Cutress, W.C. Jaap, D.G. Fautin, J.R. Larson, G.H. Harbison, M.N. Ari, and D.M. Opresko. 1991. Common and scientific names of aquatic invertebrates from the United States and Canada: Cnidaria and Ctenophora, *Am. Fish. Soc. Spec. Publ.*, 22:1–75.

CARICOMP. 1997. CARICOMP monitoring of coral reefs, *Proc. 8th Int. Coral Reef Symp.*, 1:651–656.

Connell, J.H. 1997. Disturbance and recovery of coral assemblages, *Proc. 8th Int. Coral Reef Symp.*, 1:9–22.

Dallmeyer, D.G., J.W. Porter, and G.J. Smith. 1982. Effects of particulate peat on the behavior and physiology of the Jamaican reef-building coral *Montastraea annularis*, *Mar. Biol.*, 68:229–233.

D'Elia, C.F., K.L. Webb, and J.W. Porter. 1981. Nitrate-rich groundwater inputs to Discovery Bay, Jamaica: a significant source of N to local reefs? *Bull. Mar. Sci.*, 31:903–910.

Dubinsky, Z. and N. Stambler. 1996. Marine pollution and coral reefs, *Global Change Biol.*, 2:511–526.

Goreau, T.J. 1994. Coral reef protection in western Jamaica. Extended Abstracts, Caribbean Water and Wastewater Association Conference, Kingston, Jamaica, October 3–7, 1994.

Goreau, T.J. and K. Thacker. 1994. Coral reefs, sewage, and water quality standards. Extended Abstracts, Caribbean Water and Wastewater Association Conference, Kingston, Jamaica, October 3–7, 1994.

Goreau, T.J., L. Daley, S. Ciappara, J. Brown, S. Bourke, and K. Thacker. 1997. Community-based whole-watershed and coastal zone management in Jamaica, *Proc. 8th Int. Coral Reef Symp.*, 2:2093–2096.

Halas, J.C. 1997. Advances in environmental mooring technology, *Proc. 8th Int. Coral Reef Symp.*, 2:1995–2000.

Hoegh-Guldberg, O., M. Takabayashi, and G. Moreno. 1997. The impact of long-term nutrient enrichment on coral calcification and growth, *Proc. 8th Int. Coral Reef Symp.*, 1:861– 866.

Hughes, T. 1994. Catastrophes, phase shifts, and large scale degradation of a Caribbean coral reef, *Science*, 265:1547–1551.

Hughes, T.P., A.M. Szmant, R. Steneck, R. Carpenter, and S. Miller. 1999. Algal blooms on coral reefs: what are the causes? *Limnol. Oceanogr.*, 44:1583–1586.

Jamaica Meteorological Service. 1973. *The Climate of Jamaica*.

Lapointe, B.E. 1994. Eutrophication thresholds for macroalgal overgrowth of coral reefs. Extended Abstracts, Caribbean Water and Wastewater Association Conference, Kingston, Jamaica, October 3–7, 1994.

Lapointe, B.E. 1997. Nutrient thresholds for bottom-up control of macroalgal blooms on coral reefs in Jamaica and southeastern Florida, *Limnol. Oceanogr.*, 42:1119–1131.

Lapointe, B. E. 1999. Simultaneous top-down and bottom-up forces control macroalgal blooms on coral reefs (reply to the comment by Hughes et al.), *Limnol. Oceanogr.*, 44:1586–1592.

Lapointe, B.E., J.D. O'Connell, and G.S. Garrett. 1990. Nutrient couplings between on-site sewage disposal systems, groundwaters, and nearshore waters of the Florida Keys, *Biogeochemistry*, 10:289–307.

Lapointe, B.E., M.M. Littler, D.S. Littler. 1997. Macroalgal overgrowth of fringing coral reefs at Discovery Bay, Jamaica: bottom-up versus top-down control, *Proc. 8th Int. Coral Reef Symp.*, 1:927–932.

Mate, J.L. 1997. Experimental responses of Panamanian reef corals to high temperature and nutrients, *Proc. 8th Int. Coral Reef Symp.*, 1:515–520.

Mendes, J.M., M.J. Risk, H.P. Schwarcz, and J.D. Woodley. 1997. Stable isotopes of nitrogen as measures of marine pollution: a preliminary assay of coral tissue from Jamaica, *Proc. 8th Int. Coral Reef Symp.*, 2:1869–1872.

Morrison, D. 1988. Comparing fish and urchin grazing in shallow and deeper coral reef algal communities, *Ecology*, 69:1367–1382.

Negril Area Environmental Protection Trust (NEPT), Natural Resources Conservation Authority (NRCA), and Negril Green Island Area Local Planning Authority. 1995. *The Negril and Green Island Area Environmental Protection Plan*, NEPT, Negril, Jamaica.

Porter, J.W., S.K. Lewis, and K.G. Porter. 1999. The effect of multiple stressors on the Florida Keys coral reef ecosystem: a landscape hypothesis and a physiological test, *Limnol. Oceanogr.*, 44:941–949.

Risk, M.J. 1999. Paradise lost: how marine science failed the world's coral reefs, *Mar. Freshwater Res.*, 50:381–387.

Quirolo, C. 1991. *Protecting Jamaica's Coral Reefs*, final report of the Negril Reef mooring buoy workshop and installation project, Negril Coral Reef Preservation Society, Negril, Jamaica.

Te, F.T. 1997. Turbidity and its effect on corals: a model using the extinction coefficient (K) of photosynthetic active radiance (PAR), *Proc. 8th Int. Coral Reef Symp.*, 2:1899–1904.

Tomascik, T. and F. Sander. 1987. Effects of eutrophication on reef-building corals. II. Structure of scleractinian coral communities on fringing reefs, Barbados, West Indies, *Mar. Biol.*, 94:53–75.

Turnbull, D. and J. Lewis. 1981. *Pollution Ecology of a Small Tropical Estuary in Barbados, West Indies*. 1. *Water Quality Characteristics*, Marine Science Center, Barbados.

Underground Water Authority. 1996. *Dumping of Garbage and Use of Untreated Water in Whitehall-Negril*, Underground Water Authority, Hope Gardens, Kingston, Jamaica.

Vogt, H., A.R.F. Montebon, and M.L.R. Alcala. 1997. Underwater video sampling: an effective method for coral reef surveys? *Proc. 8th Int. Coral Reef Symp.*, 2:1447–1452.

Wade, B. 1994. Nutrient levels in Negril wetland and coastal waters: the problem of nutrient enrichment and algal overgrowth in Long Bay. Extended Abstracts, Caribbean Water and Wastewater Association Conference. Kingston, Jamaica, October 3–7, 1994.

Wells, J.W. and J. Lang. 1973. Systematic list of Jamaican shallow-water Scleractinia, *Bull. Mar. Sci.*, 23:55–58.

Weil, E. and N. Knowlton. 1994. A multi-character analysis of the Caribbean coral *Montastraea annularis* (Ellis & Solander, 1786) and its two sibling species, *M. faveolata* (Ellis & Solander, 1786) and *M. franksi* (Gregory, 1895), *Bull. Mar. Sci.*, 55:151–175.

Wells, S. 1997. Capacity building for science and management in Belize: towards sustainable reef management, *Proc. 8th Int. Coral Reef Symp.*, 2:1991–1994.

Woodley, J.D., K. De Meyer, P. Bush, G. Ebanks-Petrie, J. Garzon-Ferreira, E. Klein, L.P.J.J. Pors, and C.M. Wilson. 1997. Status of reefs in the south central Caribbean, *Proc. 8th Int. Coral Reef Symp.*, 1:357–362.

# 35 Community-Based Water Quality and Coral Reef Monitoring in the Negril Marine Park, Jamaica: Land-Based Nutrient Inputs and Their Ecological Consequences

*Brian E. Lapointe*
Division of Marine Sciences, Harbor Branch Oceanographic Institution, Inc.

*Katy Thacker*
Negril Coral Reef Preservation Society

## CONTENTS

## THE NEGRIL CORAL REEF PRESERVATION SOCIETY: THE ROLE OF A NONGOVERNMENTAL ORGANIZATION IN MONITORING AND MANAGING CORAL REEF ECOSYSTEMS

Negril is located on the West End of the north coast of Jamaica — the third largest island in the Caribbean. Until the 1960's Negril was a small fishing and farming village of approximately 500 people and one of the most remote areas of Jamaica. Basic infrastructure was developed in the

8493-2026-7/02/$0.00+$1.50
2002 by CRC Press LLC

early 1960s which included a new road, expansion of electricity, a water-treatment plant (for drinking water), and drainage canals dug into the Great Morass, a wetland historically dominated by sawgrass, *Cladium jamaicense*. The Morass, located landward of the Long Bay beach (Figure 35.1), was drained and used for the production of sugarcane because of its nitrogen-rich peat (in a fashion similar to that which occurred in the Everglades in South Florida). With the rapid growth of tourist facilities in the 1960's and 1970s and construction of the Negril Beach Village in 1975, Negril became recognized as a "destination resort," attracting additional tourists. By the 1980s, tourism had become Jamaica's primary industry and Negril, with the highest growth rate of tourism in all of Jamaica, had become Jamaica's third largest resort area. In 1984, Negril's residential population reached 2440. By 1992, the town had over 60 accommodation facilities for visitors, ranging from all-inclusive hotels to small inns and villas located on Long Bay.[1]

The Negril Coral Reef Preservation Society (NCRPS) was formed in 1990 by a small group of volunteers concerned about the state of Negril's coral reefs. With the growth of tourism came an increase in the number of recreational boats in the area, especially those carrying visitors to the reefs located in Long Bay and along the West End. Boat operators were dropping anchors on the

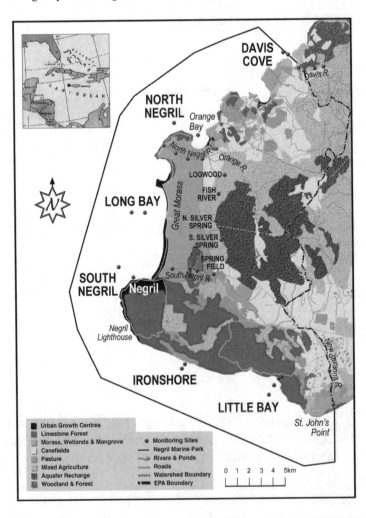

**FIGURE 35.1** Map showing the locations of 12 reef sampling stations (black dots) in the Negril Marine Park (NMP, delineated by boundary line) and in the Negril Environmental Protection Area (NEPA) on the NMP watershed. Various land-use cover classifications of the NEPA are shown.

reef, causing damage to the resources upon which their livelihoods depended. NCRPS's initial plan was to protect the coral reefs from further decline through the elimination of anchor damage and the creation of a national marine park, but it was later expanded to include efforts at whole-watershed and reef management.

The first major project of NCRPS was the installation of 35 state-of-the-art mooring buoys on Negril's reefs in 1991. The society received half of its funding and all of its technical support from Reef Relief, a U.S.-based nonprofit organization with experience in the installation and maintenance of reef mooring buoys and public awareness programs in Key West, Florida. The other half of the funding came from donations from the local hotel and business community. The buoys were the first of their kind in Jamaica and eliminated over 20,000 anchorages on reefs in Negril in the first year. A comprehensive awareness-building program followed, with brochures and posters printed for distribution in hotels and businesses. A workshop in 1991 entitled "Protecting Jamaica's Coral Reefs: the Negril Mooring Buoy Project" offered presentations on how to use the reef mooring buoys and other presentations by noted coral reef scientists and fisheries experts. It was during this workshop that NCRPS came to the realization that, although an important negative impact to the coral reefs had been eliminated by the installation of the reef mooring buoys, an even more important issue needed to be addressed: deteriorating water quality.

Coral reefs thrive in clear, clean waters with low nutrient levels. Negril's coastal waters were being negatively impacted by land-based pollution. Recommendations were made for the implementation of an ongoing coral reef and water quality monitoring program as a matter of urgency. Historically, sewage disposal and treatment in Negril consisted of pit latrines, and at best, septic tanks and tile fields, often discharging untreated and inadequately treated wastes directly into the groundwaters and coastal waters. Through extensive lobbying by the local Negril Chamber of Commerce (NCC), the European Union (EU) agreed to fund the construction of a central sewage system in Negril and hired a consulting firm to design a system appropriate for the Negril area.

In 1992, NCRPS's annual workshop in Negril, "Protecting Jamaica's Coral Reef Ecosystems," focused on water quality issues. Representatives from agencies and organizations who had data on water quality monitoring in Negril were invited to present the results of their studies. This included presentations from the National Water Commission (NWC), the Natural Resources Conservation Authority (NRCA), the Ministry of Health, and a private firm hired by the consulting engineers from DHV Burrow Crocker called Environmental Solutions, Ltd. (ESL). Participants learned that results of nutrient measurements in Long Bay reflected levels well above normal and this was seriously affecting the health of coral reefs, seagrass beds, and coastal waters. The high levels of nutrients were fertilizing nuisance seaweeds or macroalgae that were overgrowing living corals and threatening their survival. Representatives from DHV Burrow Crocker presented their design for the proposed Negril sewage collection and treatment project, and participants at the workshop expressed opposition to the proposed design. Objections were raised about the collection system as well as the treatment works, which involved only secondary sewage treatment that would not adequately remove nutrients. NCRPS drafted a resolution shortly after the workshop requesting changes in the collection system design and installation of a tertiary treatment system that would strip nutrients from the final effluent. These resolutions were directed to the EU, the Government of Jamaica (GOJ), and the NWC, which is the government agency responsible for the management of sewage treatment facilities in Jamaica. In response to the resolutions, some positive changes were made to the collection system design but the consulting engineers and the NWC maintained the position that tertiary sewage treatment would not be needed because the planned facultative ponds would remove nutrients, making tertiary treatment in the Negril scheme unnecessary. NCRPS was informed that an ongoing water quality monitoring program would be put in place, and, if results indicated a need for tertiary treatment, provisions could be made for expansion of the facultative pond system to include a final polishing step to remove nutrients.

At the 1992 workshop, the Minister of Tourism confirmed that the NCRPS proposal to develop and manage a marine park in Negril had been accepted by the GOJ. It was after this workshop that representatives from the EU delegation in Jamaica invited the NCRPS to submit a proposal to their environmental budget line for the establishment of the Negril Marine Park (NMP). This was done, and in 1994 the EU pledged funding to NCRPS for the establishment of a management structure for the NMP. This was the largest grant ever awarded to an nongovernmental organization (NGO) in the Caribbean region by the EU.

The NCRPS expanded its focus for coral reef protection with funding assistance from the Environmental Foundation of Jamaica (EFJ) and U.S. AID, launching a program of whole-watershed planning. On the advice of the NRCA, a local advisory committee was set up to oversee the NCRPS management plan for the NMP. In 1993, the Negril Area Environmental Protection Trust (NEPT) was formed as a local environmental management council from a committee made up of representatives from eleven community-based organizations. Led by NCRPS, NEPT expanded to become a broad-based umbrella organization with representatives from government, nongovernment, and community-based organizations.[2] In the process of the development of NEPT, more than 50 meetings were held in communities throughout the proposed watershed environmental protection area in an effort to build awareness and consult with community members regarding issues affecting the environment. On November 30, 1997, the Minister of Environment and Housing announced the creation of the Negril Environmental Protection Area (NEPA). This is the first protected area in Jamaica to include five area watersheds and a national marine park (NMP), which was officially established on March 4, 1998.

The EU-funded project to establish a management structure for the NMP officially began in September 1995 and involved a wide variety of programs, including training, education, fisheries management, mariculture, boat mooring/zoning, and public awareness targeted to schools, the general public, tourists, and government representatives. The initial management team for the NMP was hired from within the NEPA and consisted of a Project Manager, Administrative Assistant, Education Officer, Scientific Officer, and four rangers. Boats and equipment were purchased, and the NMP headquarters was set up in a 38-ft shipping container. After holding the 1992 workshop on water quality issues, the NCRPS knew that the health of the coral reefs depended on water quality monitoring, particularly levels of dissolved nutrients — nitrate, phosphate, and ammonium. A small laboratory was set up inside the headquarters for this purpose. Park rangers were trained in basic marine science and coral reef ecology, natural resource management, conflict management, computer skills, monitoring techniques, and interpretation of environmental data.

A baseline assessment of the coral reefs within the boundaries of the NMP was initiated using the expertise of noted marine biologists and coral reef experts. Local and overseas experts trained the park rangers in coral reef assessment and underwater videography. They were trained in U.S. Environmental Protection Agency (EPA) coral reef monitoring techniques, and five permanent stations were established within the boundaries of the NMP as part of the ongoing monitoring program. The rangers were also trained to quantitatively assess the biota on any reef and provide reports on the overall percentages of coral, macroalgae, turf algae, and other marine life. They are now able to identify approximately 60 species of marine macroalgae and correlate the presence of algal indicators with suspected land-based nutrient inputs into the coastal waters. The rangers also monitor beach erosion along Long Bay beach and sedimentation in turbid waters within the NMP boundaries. The data gathered from regular sampling are compiled and entered into a database, and information is shared with local organizations within the Negril area, Jamaica, and worldwide.

When the NCRPS was formed, it was recognized that fishermen would play a significant role in monitoring and managing the resources and a solid partnership with them would be needed to set up the NMP. NCRPS initiated dialogue with fishermen from five major fishing villages within the NEPA in an effort to establish this important relationship. More than 60 meetings with them were held to discuss coral reef and fisheries management issues and zoning of the park. Five fishermen were hired as part-time community liaison officers. Their job was to distribute educational materials and information to all fishermen, organize meetings and workshops, and act as volunteer rangers and game wardens. Training workshops included presentations on water quality issues and monitoring techniques, and the fishermen were trained to act as "watchdogs" to alert the NMP rangers about algae blooms, fish kills, and any other issues surrounding water quality, coral reefs, and fisheries. This partnership has proven to be very effective, as fishermen are on the sea day in and day out and provide valuable assistance in monitoring the natural marine resources.

A school training program was developed by NCRPS to increase awareness among the teachers and students within the area. A Junior Ranger training program was set up to train pupils between the ages of 10 and 17 from 12 schools within the NEPA. This involved the development of a training manual for students and teachers and a series of workshops and field trips. Study topics include coral reefs, seagrass beds and mangroves, wetlands and watersheds, fisheries management, forestry management, first aid, CPR, and lifesaving, among others. An important part of the training involves familiarizing the students and teachers with the natural resources within the NEPA, the environmental threats, the importance of monitoring, and what should be done to correct the problems. Once students complete all the requirements outlined in the training manual and attend workshops and field trips, they are eligible for certification as official NEPA Junior Rangers. Teachers take part in a similar program for certification as official NEPA Junior Ranger Trainers. After completing the basic program, students can pursue advanced certification in various areas, including environmental monitoring, which involves training in water quality monitoring techniques. Junior Rangers are provided the opportunity to work alongside the rangers in collecting water samples and performing monthly coral reef monitoring duties on the reefs. Over a thousand students and teachers have taken part in the Junior Ranger training program and approximately 500 have received official certification. It has proven to be a very effective method for increasing awareness within the communities, and the Junior Rangers have become important partners in the monitoring and management of the coral reef ecosystem.

Annual workshops on protecting Jamaica's coral reef ecosystem are ongoing with participation from government, NGOs, fishermen, local community members, and stakeholders. In February 1999, the NCRPS staged a second workshop focusing on coral reef and water quality issues. The central sewage collection and treatment works were completely set up at this point; however, only a small percentage of residential and commercial properties in the town of Negril were hooked up to the system. (Connection to the collection system is not mandatory under the law). Presentations were made on the results of extensive monitoring efforts under the EU funding. The results showed that coral reef health within the Negril Marine Park continues to decline. The reefs are still dominated by macroalgae and suffer from bleaching and coral diseases. Outstanding issues regarding the effective treatment of sewage collected in the Negril Scheme and the need for tertiary treatment remain. In addition, large residential areas within the NEPA have little or no sewage treatment at all. Environmental threats to the NEPA were summarized in the Negril Green Island Environmental Protection Plan:

Growth and development have brought ... considerable costs to the environment. Rural parts of the watershed lack safe and adequate water supplies or garbage removal, posing potential public health risks. Coastal water quality is deteriorating, coral reefs are dying, fish stocks are depleting, public beaches are scarce and prime natural areas have been lost. Prime reef areas have been destroyed by effluents from inadequately treated sewage and garbage dumps. The beach and West End cliffs are almost completely lined with hotels and guesthouses, which have largely blocked access to and even the view of the shore. Beaches have eroded and ground waters are polluted with sewage from "soak away" toilets (pits). Big trees are being cut down for charcoal kilns or sawmills on forested hill slopes which are vital groundwater recharge areas, increasing soil erosion.

The efforts of NCRPS to address these problems through a whole-watershed approach to coral reef protection is an example of the importance of grass roots efforts in complex environmental management. Valuable partnerships have been established among NGOs, community groups, governments, schools, and local citizens to address the many negative impacts threatening the health of the natural resources with the NEPA. Through training and education programs, the capacity of the local community has been strengthened to play a critical role in the monitoring of these national treasures. This has proven to be an invaluable tool for the management and protection of Negril's coral reef ecosystem, for our children's children.

## ASSESSING LAND-BASED NUTRIENT POLLUTION
## IN THE NEGRIL MARINE PARK

Pollution from land-based sources is now considered the single most important threat to the marine environment of the Caribbean and impediment to the sustainable use of its coastal resources.[3] Of the various types of land-based pollution impacting coastal waters, nutrient pollution is arguably the most rapidly growing problem in the Caribbean and other tropical regions of the world where even low levels of nutrient enrichment foster eutrophication and the degradation of seagrass and coral reef ecosystems.[4]

In Jamaica, a variety of land-based human activities represent potential sources of nutrient pollution to coastal waters. Deforestation, which began centuries ago in the West Indies region,[5] not only causes top soil loss and sedimentation, but also increases $NO_3^-$ concentrations of groundwaters and rivers[6] which ultimately discharge into downstream coastal waters. Although this non-point-source $NO_3^-$ enrichment potentially contributes to nitrogen-fueled eutrophication,[7] other, more concentrated anthropogenic nutrient sources are better known. Since the 1970s, sewage pollution from tourist hotels and resorts was recognized as a source of nutrients causing algal overgrowth of coral reefs in Jamaica;[8] similar problems associated with sewage-driven eutrophication and macroalgal blooms were documented during the 1960s in Kaneohe Bay, Hawaii.[9] UNEP[3] identified sewage as one of the most significant pollutants affecting Caribbean coastal waters, which not only accelerates eutrophication and algal overgrowth of coral reefs but also poses direct threats to human health.

More recently, fertilizer use for agriculture has become the most rapidly growing source of nutrient pollution to coastal waters globally[4] and is likely to be important in some areas of Jamaica. Over just the past two decades, the use of fertilizers has approximately doubled the amount of fixed nitrogen on the earth.[10] In Jamaica, fertilizer use increased from 15,962 to 27,000 metric tons between 1978 and 1995, a 60% increase over that 17-year period.[11] Although agriculture and tourism are both important to the economy of Jamaica, the increasing effects of agricultural nutrient pollution could become devastating to coastal resources, especially in Negril where they are inextricably linked to tourism and fisheries. To date, however, few studies in the Caribbean region have made direct links between agricultural activities, nutrient enrichment of coastal waters, and ecological impacts in coral reef communities.[11]

Historically, the calm, coastal waters around Negril had some of the best developed coral reefs in Jamaica and the Caribbean.[12] Following Hurricane Allen in 1980, coral communities at Discovery Bay on Jamaica's north coast began to shift away from corals and towards macroalgae.[13,14] This phase-shift towards blooms of macroalgae has since occurred on many reefs on Jamaica's north coast and began impacting the reefs in the Negril area in the early 1990s.[15] Although nutrient pollution had been known for decades to cause such phase-shifts on coral reefs,[16] some coral reef ecologists interpreted this phenomenon to result only from reduced grazing by reef herbivores.[14,17] According to this hypothesis, overfishing led to reduced herbivory of reef algae by fish, which, together with the massive dieoff of the long-spined sea urchin *Diadema antillarum*, led to "dramatic blooms" of macroalgae and the demise of hermatypic (reef-forming) corals. Scientists promoting this hypothesis categorically dismissed the role of nutrient enrichment,[17] despite considerable evidence that sewage and groundwater $NO_3^-$ enrichment were fueling the macroalgal blooms at

Discovery Bay and other locations in Jamaica.[18] A more balanced alternative hypothesis was that reduced herbivory and nutrient enrichment were both causative factors of the macroalgal blooms.[7,13,18] That view was supported by the Relative Dominance Model (RDM),[19] which predicted that coral reefs impacted by reduced rates of herbivory and high nutrient levels will tend towards dominance by macroalgae rather than corals, turf algae, or coralline algal crusts.

Because of the expansion of macroalgal blooms on coral reefs throughout Long Bay in 1991,[15] the NCRPS workshop on water quality issues specifically addressed the impact of nutrient enrichment from increased sewage inputs associated with the expansion of tourism. The EU was providing funding for the design of a central sewage collection and treatment system for Negril. Jelier and Roberts[20] described how the NWC engaged DHV Burrow-Crocker Consulting to design the sewage treatment works and also provided a scheme detailing the collection system and treatment ponds. Comments by a panel of scientists noted that the limited design features of the ponds would predictably result in escalated nutrient pollution of the South Negril River and downstream reefs in Long Bay and the West End.[15] Jelier and Roberts[20] emphasized, however, that other anthropogenic sources of nutrient pollution already existed in the Negril area, particularly fertilizers used in agriculture.

To characterize the status of nutrient pollution from land-based sources prior to the sewage diversion project, the NCRPS implemented an integrated water quality and coral reef monitoring program in October 1997. The NEPA provided a spatial framework for addressing nonpoint-source pollution originating from a variety of domestic, urban, agricultural and industrial land uses on Negril's watershed (Figure 35.1). The downstream coastal waters of the NMP encompassed an area up to ~5 km offshore and extended from Davis Cove on the northern park boundary to St. John's Point on the southern park boundary (Figure 35.1). A broad shelf with depths between 5 and 12 m below sea level extends offshore up to 2.0 km from the popular Long Bay beach (the longest beach in Jamaica) in the central area of the NMP, terminating offshore at a drowned cliff escarpment.[21]

The goal of the NCRPS monitoring program was to test several scientific and environmental management hypotheses concerning the potential impacts of various land uses on nutrient enrichment of groundwaters and rivers of the NEPA and coral reefs in the NMP. First, we hypothesized that agricultural and urban land uses on Negril's watershed would result in elevated dissolved inorganic nitrogen (DIN) and soluble reactive phosphorus (SRP) concentrations in groundwaters, rivers, and coastal waters of the NMP; as a corollary, we predicted that shallow reefs would have higher DIN and SRP concentrations compared to deeper reefs because of their closer proximity to groundwater discharges and land-based runoff. Second, because tourism and urban development have occurred primarily within the central part of the NEPA (Long Bay and the West End) compared to agricultural land-uses that are located mostly in the outlying areas of the NEPA (Figure 35.1), we hypothesized that stable nitrogen isotope values of reef macroalgae adjacent to the tourist resorts would reflect sewage inputs ($\delta^{15}N$, ~+5.0‰) compared to those downstream of agricultural land uses or on deeper reefs that would have lower values typical of fertilizers (~+1.85‰) or nitrogen-fixation (~+0.5‰) on relatively unimpacted deep reef sites.[22-24] Last, we hypothesized that higher DIN and/or SRP concentrations on shallow compared to deep reefs would result in lower coral cover due to multiple stress mechanisms associated with the eutrophication process on coral reefs.

## LINKING WATER QUALITY MONITORING IN THE NEGRIL ENVIRONMENTAL PROTECTION AREA WITH THE NEGRIL MARINE PARK

The NCRPS initiated monthly water sampling in October 1997 that continued through December 2000 at a network of stations in three subwatersheds of the NEPA and twelve downstream reef sites in the NMP (Figure 35.1). This spatial framework allowed assessment of several geographically distinct areas of the NEPA that link upland watersheds on the coastal ridge to reefs of the NMP.

The subwatersheds included the Davis River, the North Negril River, and the South Negril River. In all three subwatersheds, sampling stations were located in the uppermost reaches of the rivers, upstream and downstream of a variety of agricultural and urban land uses, and on six shallow (2 to 4 m) and six deep (9 to 18 m) reefs in the NMP (Figure 35.1). On the South Negril River, sampling sites were selected to assess the potential impact of nutrient pollution that was hypothesized to result from the discharge of partially treated effluent of the treatment ponds from the Negril sewage treatment system that became operational in summer of 1999. In addition to the river sampling, seven upland groundwater-fed springs and grottoes were sampled at Logwood, Fish River, North Silver Spring, South Silver Spring, Springfield, and "upstream" and "downstream" springs near the shoreline at Little Bay (Figure 35.1).

Following the monthly NEPA water sample collection and return to the NCRPS lab, the samples were analyzed for salinity (hand-held refractometer, ±1 ppt), filtered (0.45-μm GF/F filters), frozen, and subsequently analyzed for $NH_4^+$, $NO_3^-$ plus $NO_2^-$, and SRP on a Hach DR-2000 spectrophotometer calibrated using standard curves according to the manufacturer's methods. In contrast to the monthly NEPA sampling, seasonal sampling was performed at the reef stations in the NMP to provide two months of sampling in the winter "dry" season (January and February 1998; Figure 35.2) and two months in the summer "wet" season (July and August 1998; Figure 35.2). Because of the relatively low concentrations of dissolved inorganic nutrients at the NMP reef sites, these samples were analyzed for salinity, filtered (0.45-μm GF/F filters), frozen, and returned to the Harbor Branch Oceanographic Institution's (HBOI) Environmental Laboratory in Ft. Pierce, FL, where they were analyzed on a Bran and Luebbe TRAACS Analytical Console and an ALPKEM 500 Series autoanalyzer. The methods for collection, handling, and processing of the NMP samples for low level nutrient concentrations followed a quality-assurance/quality-control protocol developed by HBOI's Environmental Lab that prevented problems associated with sample contamination and excessive holding times. The GF/F filters used for filtering the NMP samples were frozen and subsequently analyzed for phytoplankton biomass (chlorophyll $a$) by Dr. Larry Brand at the Rosenstiel School of Atmospheric and Marine Science, University of Miami, FL.[24]

Salinity of the water samples ranged from freshwater (0 ppt) in the uppermost reaches of the various rivers and springs of the subwatersheds in the NEPA to almost full strength seawater (33 to 35 ppt) on the downstream reef sites in the NMP. The lowest salinity values of the twelve reef sites were measured at the shallow and deep stations in Long Bay, with the lowest salinity in near-bottom waters of the submarine cliffs at the deep station, which was also the station farthest from shore (~2 km; Figure 35.1). Salinity values were commonly lower in the near-bottom samples compared to surface water samples in the NMP, indicating considerable rates of submarine groundwater discharge. The relatively low salinity values throughout the NMP reef sites, compared to oceanic salinities of ~36 ppt, demonstrated a widespread hydrological coupling between the NEPA and reefs in the NMP.

Monitoring of the NEPA during 1998 showed that the freshwater runoff carries significant dissolved inorganic nitrogen and phosphorus into coastal waters of the NMP. DIN ($NH_4^+$ plus $NO_3^-$ plus $NO_2^-$), and SRP concentrations correlated significantly and negatively with salinity in the Davis Cove (Figure 35.3), North Negril (Figure 35.4), and South Negril (Figure 35.5) subwatersheds during this year-long sampling effort.[24] Both DIN and SRP concentrations decreased with increasing salinity, although SRP had more scatter in the data within the estuarine region (15 to 30 ppt), especially the North and South Negril Rivers, which drain agricultural and urban land uses (Figure 35.1). The highest DIN and SRP concentrations in the Davis Cove and North Negril subwatersheds were in the upper reaches adjacent to agricultural land uses. For example, the highest DIN and SRP concentrations in the Davis River did not occur at the uppermost station on this subwatershed, but rather at a lower station on the river adjacent to fertilized sugarcane fields (Figure 35.1). At this station, DIN concentrations averaged 20.0 ± 9.2 μM in the dry season and 14.8 ± 5.4 μM in the wet season with no significant seasonal difference ($n = 12$, $p > 0.05$). However, SRP concentrations at this station increased significantly ($n = 12$, $p < 0.05$) from 2.6 ± 1.2 μM in

RAINFALL IN THE NEGRIL MARINE PARK 1998

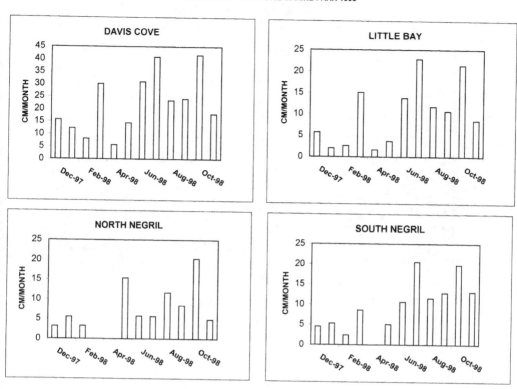

**FIGURE 35.2** Total monthly rainfall (cm/month) between December 1997 and November 1998 measured at Davis Cove, Little Bay, North Negril, and South Negril. Rainfall data provided by the National Meteorological Service.

the dry season to 5.3 ± 1.2 μM in the wet season. This seasonal increase in SRP was observed at many of the monitoring stations on the Davis, North Negril, and South Negril Rivers where mean SRP concentrations were higher in the summer wet season compared to the winter dry season.

The highest DIN concentrations of all three subwatersheds during 1998 occurred at the uppermost reaches of the South Negril River (Royal Palm Reserve) where high $NO_3^-$ concentrations occurred.[24] DIN concentrations (mostly $NO_3^-$) at this station averaged 56.0 ± 33.5 μM in the dry season and 46.9 ± 34.0 μM in the wet season (ANOVA, $n = 12$, $p > 0.05$). This station was located in the drainage basin of the springs subwatershed (region between Logwood and Springfield, Figure 35.1) where we also measured high $NO_3^-$ concentrations in groundwaters; the mean $NO_3^-$ concentration of five springs in the wet season was 33.8 ± 27.2 μM compared to 25.3 ± 19.2 μM in the dry season ($n = 46$, $p > 0.05$). In contrast to DIN, SRP concentrations of the groundwaters increased significantly ($n = 45$, $p < 0.01$) from a mean of 0.43 ± 0.28 μM in the dry season to 0.75 ± 0.44 μM in the wet season, respectively.

DIN and SRP concentrations at the downstream "bridge" stations of the three rivers varied as a function of land use on the different subwatersheds and season in 1998.[24] The highest DIN concentrations were at the South Negril River bridge station adjacent to urban land uses around Negril, which averaged 9.2 ± 4.2 μM in the dry season and 20.9 ± 13.7 μM in the wet season. At this location, SRP concentrations averaged 0.37 ± 0.19 μM in the dry season and increased to 0.72 ± 0.63 μM in the wet season. In comparison, DIN at the Davis River bridge, downstream of fertilized canefields, averaged 11.1 ± 5.4 μM in the dry season and 9.9 ± 2.0 μM in the wet season; however, this station had the highest SRP concentrations of all subwatersheds, averaging

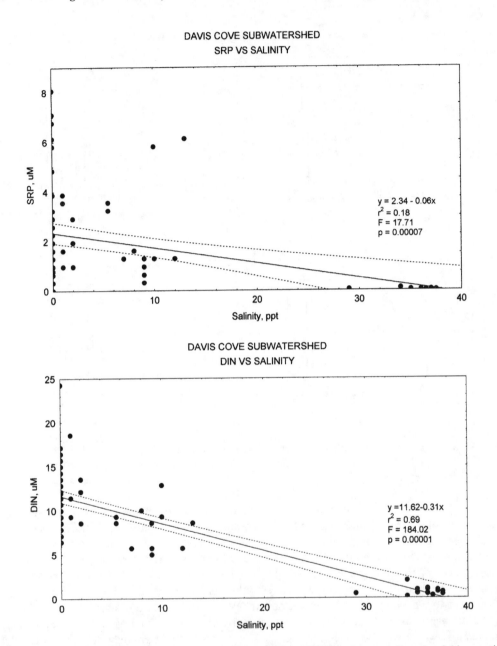

**FIGURE 35.3** Concentrations of dissolved inorganic nitrogen (DIN, bottom panel) and soluble reactive phosphorus (SRP, top panel) vs. salinity along mixing gradients extending from the upper reaches of the Davis Cove watershed to downstream fringing reefs in the Negril Marine Park.

1.6 ± 1.3 $\mu M$ in the dry season and 2.0 ± 0.9 $\mu M$ in the wet season. The lowest mean DIN (6.89 $\mu M$) and SRP (0.40 $\mu M$) concentrations occurred at the North Negril River bridge station, which is downgradient from the largest natural wetland area of all three subwatersheds.

Long-term sampling on the South Negril River (1997 to 2000) showed a significant trend of increasing nutrient concentrations that is attributable, in large part, to the recent discharge of partially treated sewage effluent from the new treatment ponds. At the upstream effluent station, annual mean $NH_4^+$ concentrations increased significantly (Mann Whitney, $p < 0.0001$) over the period of record, ranging from 6.6 $\mu M$ in 1998 to 12.4 $\mu M$ in 1999 to 37.5 $\mu M$ in 2000 (Figure 35.6).

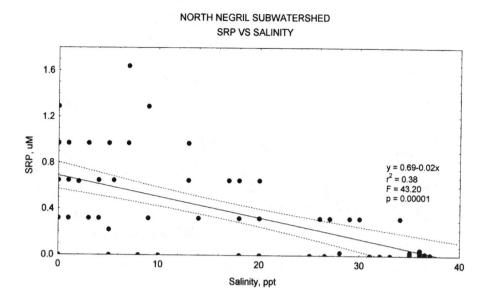

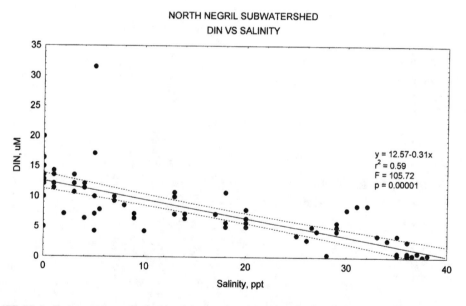

**FIGURE 35.4** Concentrations of dissolved inorganic nitrogen (DIN, bottom panel) and soluble reactive phosphorus (SRP, top panel) vs. salinity along mixing gradients extending from the upper reaches of the North Negril watershed to downstream fringing reefs in the Negril Marine Park.

For SRP, annual means also increased significantly (Mann Whitney, $p < 0.0001$) from 0.32 $\mu M$ in 1998 to 2.3 $\mu M$ in 1999 to 1.1 $\mu M$ in 2000 (Figure 35.7). The largest increase in nutrient concentrations on the South Negril River occurred at the downstream effluent station. Here, $NH_4^+$ increased significantly (Mann Whitney, $p < 0.0001$) from an annual mean of 6.2 $\mu M$ in 1998 to 27.5 $\mu M$ in 1999 to 58.8 $\mu M$ in 2000 — an almost tenfold increase (Figure 35.6). For SRP, annual mean concentrations at the downstream effluent station increased (Mann Whitney, $p < 0.0001$) from 0.35 $\mu M$ in 1998 to 5.10 $\mu M$ in 1999 to 7.18 $\mu M$ in 2000 — a 20-fold increase (Figure 35.7). At the downstream South Negril River Bridge, $NH_4^+$ increased significantly (Mann Whitney, $p = 0.01$) from an annual mean of 4.48 $\mu M$ in 1998 to 9.9 $\mu M$ in 1999 to 17.5 $\mu M$ in 2000 — a fourfold

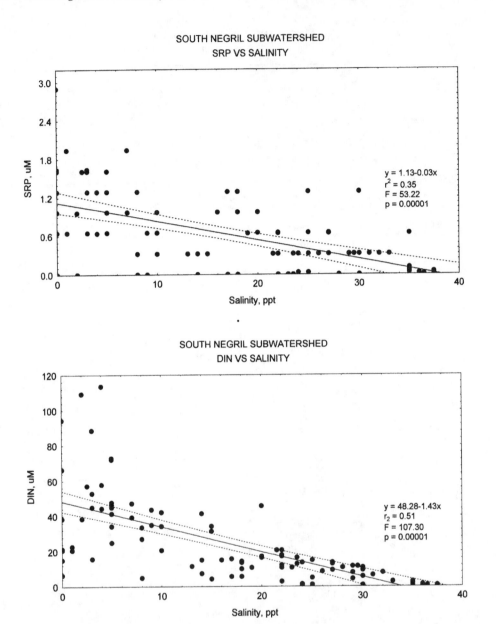

**FIGURE 35.5** Concentrations of dissolved inorganic nitrogen (DIN, bottom panel) and soluble reactive phosphorus (SRP, top panel) vs. salinity along mixing gradients extending from the upper reaches of the South Negril watershed to downstream fringing reefs in the Negril Marine Park.

increase (Figure 35.6). However, the annual mean SRP concentrations at this station increased significantly (Mann Whitney, $p = 0.04$) from 0.23 $\mu M$ in 1998 to 2.35 $\mu M$ in 1999 to 8.22 $\mu M$ in 2000 — a 36-fold increase (Figure 35.7). Parallel monitoring at four stations in the North Negril River (Figure 35.1) showed no significant increases in either $NH_4^+$ or SRP concentrations during the same time frame (NCRPS, unpublished data).

These data from the NEPA document the role of human activities — deforestation, sewage, and agriculture — in nutrient pollution of groundwaters and rivers that increases nutrient loads in downstream coastal waters of the NMP. The high $NO_3^-$ concentrations observed in the five springs and upstream waters of the South Negril River could result, in part, from historical deforestation

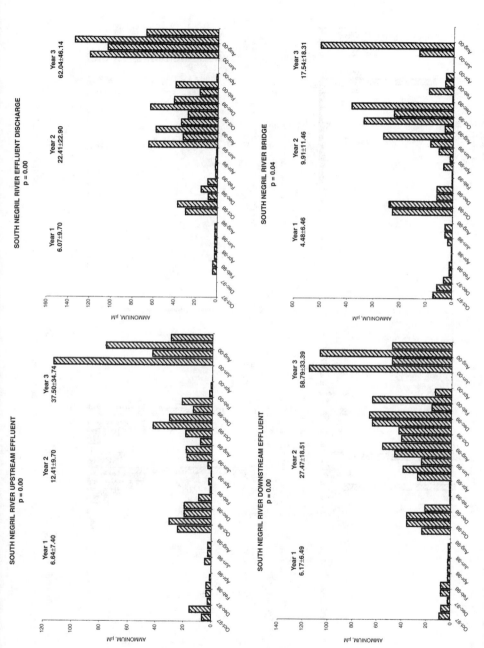

**FIGURE 35.6** Monthly mean ($n = 2$) concentrations of ammonium at four stations (upstream effluent, effluent discharge, downstream effluent, bridge) on the South Negril River over a 3-year period from October 1997 to September 2000.

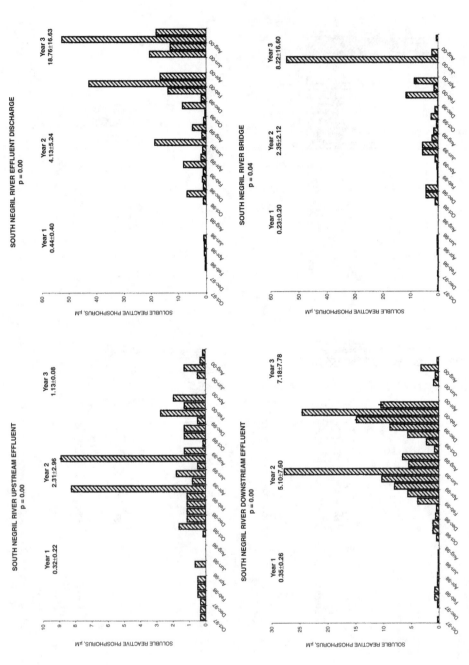

**FIGURE 35.7** Monthly mean ($n = 2$) concentrations of soluble reactive phosphorus at four stations (upstream effluent, effluent discharge, downstream effluent, bridge) on the South Negril River over a 3-year period from October 1997 to September 2000.

and use of fertilizers in agriculture.[4] Higher than expected $NO_3^-$ concentrations in groundwaters at Discovery Bay were attributed to natural rather than anthropogenic sources.[25] However, studies of fertilizer use associated with sugarcane and vegetable agriculture in Barbados have shown that its impact has been ubiquitous: No groundwater unaffected by agricultural fertilizers remains in the aquifer.[26] Very high $NO_3^-$ concentrations (up to ~750 $\mu M$) were reported, which is consistent with leaching of some 25 to 30% of the applied fertilizer nitrogen. Continued monitoring since that study has confirmed the impact of agriculture but suggests that the situation is stable with respect to nitrogen leaching. Considering the high concentrations of groundwater $NO_3^-$ and the high hydraulic conductivities and recharge rates in Barbados, discharges of polluted groundwater were concluded to be affecting the coastal zone.[26] Given the very steep hydraulic gradient of groundwaters in the NEPA and the reduced salinity values over broad areas of the NMP, the $NO_3^-$-rich groundwaters we observed in this study must also be impacting coastal waters of the NMP.

The dissolved inorganic nutrient data from the reef monitoring in the NMP during 1998 supported our hypothesis that land-based discharges of groundwaters and rivers result in considerable nutrient enrichment of reef communities. DIN concentrations were significantly (two-way ANOVA, $F = 4.89$, $p = 0.038$, $n = 49$) higher on shallow compared to deep reefs (annual mean of 1.15 vs. 0.78 $\mu M$, respectively), consistent with the hypothesized enrichment from land-based sources. However, DIN concentrations did not vary significantly ($F = 0.14$, $p = 0.71$, $n = 49$) between the dry and wet seasons, which points to an overriding importance of groundwater discharges in delivery of DIN to the NMP. On shallow reefs, DIN concentrations averaged $1.12 \pm 1.08$ $\mu M$ in winter compared to $1.19 \pm 0.89$ $\mu M$ in summer (Figure 35.8). On deep reefs, DIN concentrations averaged $0.74 \pm 0.49$ $\mu M$ in winter and $0.82 \pm 0.32$ $\mu M$ in summer (Figure 35.8). $NH_4^+$, rather than $NO_3^-$, was the dominant DIN species on both shallow and deep reefs, underscoring the importance of sewage and fertilizers as dissolved nitrogen sources to the NMP.[24]

In contrast to DIN, SRP concentrations varied as a function of wet vs. dry seasonality.[24] SRP averaged $0.017 \pm 0.010$ $\mu M$ on shallow reefs and $0.017 \pm 0.016$ $\mu M$ on deep reefs in the dry season (two-way ANOVA, $F = 0.476$, $p = 0.49$, $n = 49$) and increased significantly ($F = 17.89$, $p = 0.0004$, $n = 49$) to $0.047 \pm 0.029$ $\mu M$ on shallow reefs and $0.040 \pm 0.027$ $\mu M$ on deep reefs in the wet season, respectively (Figure 35.9). Because of the seasonal increases in SRP during summer but

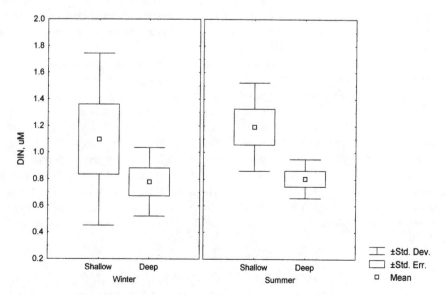

**FIGURE 35.8** Concentrations of dissolved inorganic nitrogen (ammonium plus nitrate plus nitrite, DIN) on six shallow and six deep reefs in the Negril Marine Park during winter (January, February) and summer (July, August) 1998. Values represent means ± SD and SE (n = 49).

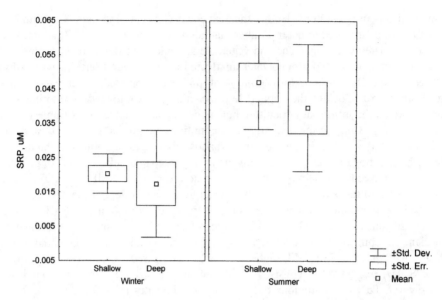

**FIGURE 35.9** Concentrations of soluble reactive phosphorus (SRP) on six shallow and six deep reefs in the Negril Marine Park during winter (January, February) and summer (July, August) 1998. Values represent means ± SD and SE ($n = 49$).

relatively similar DIN concentrations year-around, the DIN: SRP ratios decreased from 58.7 and 49.3 on shallow and deep reefs in the dry season to 38.5 and 33.4 during the wet season, respectively.

The seasonal increases in SRP concentrations on reefs in the NMP most likely resulted from rainfall infiltration/mobilization of fertilizer and sewage on the NEPA watershed. Our observations of increased SRP concentrations in groundwaters and rivers in the NEPA in parallel with the NMP during the summer wet season indicate a land-based linkage. In the Florida Keys, SRP concentrations in nearshore waters (canals) increased significantly from 0.15 ± 0.05 $\mu M$ in the winter dry season to 0.43 ± 0.38 $\mu M$ in the wet season due to increased discharge of groundwater-borne SRP derived from septic tank effluent.[27] In Guam, SRP concentrations increased on fringing reefs from 0.22 $\mu M$ to 0.59 $\mu M$ due to increased terrestrial runoff and groundwater seepage, which initiated a phytoplankton bloom in this nitrate-rich system.[28]

Phytoplankton biomass in the NMP, measured as chlorophyll *a*, averaged higher on the shallow inshore reefs compared to the more offshore deep reefs during 1998. However, there was no significant seasonal variation in chlorophyll a attributable to seasonal rainfall patterns. The chlorophyll *a* concentrations on the shallow reefs ranged from 0.12 µg/l at North Negril to 0.55 µg/l at Ironshore during the dry season and from 0.09 µg/l at Ironshore to 0.43 µg/l at Long Bay in the wet season. On deep reefs, chlorophyll *a* ranged from 0.16 µg/l at Long Bay to 0.35 µg/l at Davis Cove in the dry season and from 0.09 µg/l at Ironshore to 0.31 µg/l at North Negril in the wet season. These chlorophyll *a* values are well above values typical of Caribbean "blue water" (<0.1 µg/l) and indicate significant increases in phytoplankton biomass and associated light attenuation due to terrestrial nutrient runoff. However, the values are not exceptionally high, such as those reported for the hypereutrophic waters of Florida Bay (chlorophyll *a* values up to 20 µg/l; see Brand, this volume).

## INTEGRATING WATER QUALITY WITH CORAL REEF MONITORING IN THE NEGRIL MARINE PARK

To link nutrient enrichment from the Negril watershed to reef biota in the NMP, we integrated the water quality monitoring with a comprehensive coral reef monitoring program. The reef monitoring program included the same six shallow and six deep stations that were monitored for water quality during the dry and wet seasons. At monthly intervals between January and November 1998, the NCRPS reef rangers used SCUBA and underwater Hi-8 videotape to estimate percent cover of reef biota along 25-m belt transects. The repeated video transect imagery was analyzed using the randomized point-count method in the NCRPS lab to estimate annual mean biotic cover for the six shallow and six deep reef sites. The mean annual percent cover values for each station represent a total of 3300 point counts for the 11-month study.

Quantitative quadrat sampling was also used to estimate biomass and species composition of reef macroalgae at the shallow and deep stations. Each month, the NCRPS reef rangers collected fleshy (noncalcareous) and calcareous macroalgae from within a 25-m long × 2-m wide belt transect (50 m$^2$ area) at each station using a 0.06 m$^2$ PVC quadrat.[24] In addition, abundant fleshy macroalgae (*Sargassum polyceratium, Sargassum hystrix, Lobophora variegata, Cladophora fuliginosa, Chaetomorpha linum, Codium* spp.) were sampled (n = 5 per species) at the 12 stations during the winter dry season (February) and the summer wet season (August) for determination of tissue carbon:nitrogen:phosphorus ratios (C:N:P) on a Carlo-Erba Elemental Analyzer (Analytical Services Lab, University of Maryland; Solomons, MD) and $\delta^{15}N$ isotope values (Isotope Analytical Services; Los Alamos, New Mexico). During the monthly algal collections, one diver identified and counted sea urchins (echinoids) within each 50-m$^2$ study area.

Percent cover of macroalgae was significantly (ANOVA, $p = 0.017$, $n = 49$) lower on the shallow reefs (57.2 ± 11.2%) compared to the deep reefs (66.6 ± 12.5%) in the NMP (Figure 35.10). Mean annual cover of macroalgae ranged from a low of 47.6 ± 16.9% on the deep reef at Long Bay to a high of 82 ± 7.5% on the deep reef at Davis Cove. Similarly, percent cover of hard coral was significantly ($p < 0.00001$, $n = 49$) lower on shallow (5.04 ± 2.9%) compared to deep (10.2 ± 2.1%; Figure 35.10) reefs. Mean coral cover ranged from a low of 2.2 ± 2.0% on the shallow reef in Long Bay to a high of 12.3 ± 3.4% on the deep reef at North Negril.

In contrast to macroalgae and hard corals, cover of algal turf was significantly higher (ANOVA, $p = 0.0001$, $n = 49$) on shallow (18.0 ± 10.1%) compared to deep (8.8 ± 6.4%; Figure 35.10) reefs. Algal turf cover ranged from a low of 3.3 ± 4.6% on the deep reef at Davis Cove to a high of 32.7 ± 14.2% on the shallow reef at Long Bay.

Mean biomass of fleshy macroalgae was significantly higher (ANOVA $p < 0.00001$, $n = 66$) on shallow (203 ± 188 g dry wt m$^{-2}$) compared to deep reefs (97 ± 80 g dry wt m$^{-2}$) during 1998 (Figure 35.11). Mean fleshy biomass ranged from a low of 61.4 g dry wt m$^{-2}$ on the deep reef at Ironshore to a high of 304.1 g dry wt m$^{-2}$ on the shallow reef at Little Bay. In contrast, mean biomass of calcareous macroalgae was higher ($p < 0.00001$, $n = 66$) on deep (891 ± 1199 g dry wt m$^{-2}$) compared to shallow (127 ± 201 g dry wt m$^{-2}$; Figure 35.11) reefs. Mean biomass of calcareous macroalgae ranged from a low of 22.1 ± 38.9 g dry wt m$^{-2}$ on the shallow reef in Long Bay to a high of 1631 ± 1236 g dry wt m$^{-2}$ on the deep reef at Ironshore. The lowest biomass of calcareous macroalgae of the six deep reefs occurred at Long Bay, where it averaged 301 ± 250 g dry wt m$^{-2}$.

The fleshy Phaeophyte *Sargassum polyceratium* was particularly abundant on shallow reefs of the NMP during its growing season in the summer, fall, and early winter (Color Figure 1, parts C and D*). Other abundant fleshy species on shallow reefs included the Chlorophytes *Cladophora fuliginosa* (Color Figure 1, part E) and *Chaetomorpha linum* (Color Figure 1, part F), which formed massive blooms in Orange Bay (North Negril shallow reef) during the summer of 1998 when SRP concentrations increased to ~0.80 ± 0.02 µM ($n = 8$) at this location. The deep reefs (except for Long Bay) were dominated by calcifying Chlorophytes, which formed massive accumulations of carbonate "*Halimeda* hash" (Color Figure 1, part G) as a result of high biomass production of *Halimeda copiosa* (Color Figure 1, part H).

Sea urchin density was significantly (ANOVA, $p < 0.00001$, $n = 49$) higher on shallow (89.1 ± 65.9 · 50 m$^{-2}$) compared to deep (10.8 ± 26.1 · 50 m$^{-2}$) reefs of the NMP. The predominant echinoids

---

* Color figures follow page 648.

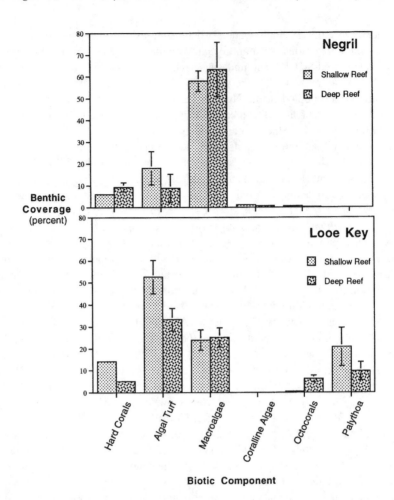

**FIGURE 35.10** Top panel: Mean biotic cover of turf algae, macroalgae, hard corals, coralline algae, and octocorals on shallow vs. deep reefs in the Negril Marine Park during 1998. Values represent means ± 1 SD ($n = 3300$). Bottom panel: Comparative mean biotic cover of turf algae, macroalgae, hard corals, coralline algae, and octocorals on shallow vs. deep reefs at the fore-reef of Looe Key National Marine Sanctuary during summer 2000. Data for coral cover provided by Dr. Jim Porter; other values represent means ± 1 SD ($n = 10$).

included the black, long-spined sea urchin *Diadema antillarum*, the reef urchin *Echinometra viridis*, and the West Indian sea egg *Tripneustes ventricosus*. Sea urchin density ranged from a low of 0.3 ± 0.8 · 50 m⁻² on deep reefs at Davis Cove to a high of 190 ± 154 · 50 m⁻² on the shallow reef in Long Bay. The highest urchin density of the six deep reefs was also at Long Bay, which averaged 66 ± 76 · 50 m⁻². Overall, mean sea urchin density was significantly (ANOVA, $p = 0.001$, $n = 79$) and negatively correlated with mean salinity of the reef sites.

Tissue levels of C, N, and P in predominant fleshy macroalgae on deep and shallow reefs showed significant spatial patterns during the study.[24] Overall, levels of percent C in fleshy macroalgae were similar between shallow and deep reefs (25.7 ± 3.8 vs. 25.4 ± 7.4% dry wt, respectively, in summer) and did not vary seasonally. Tissue percent N of macroalgae also did not vary seasonally but was significantly (two-way ANOVA, f = 15.148, $p = 0.0002$) higher on shallow compared to deep reefs. For example, percent N of macroalgae was ~63% higher on shallow compared to deep reefs in summer (1.66 ± 0.66 vs. 1.02 ± 0.21 dry wt, respectively) although levels were similar in winter (1.05 vs. 1.02, $p > 0.05$). These results support our hypothesis that land-based runoff and groundwater discharges of nitrogen are fueling eutrophication and macroalgal blooms over broad areas of the NMP, especially in the summer wet season.

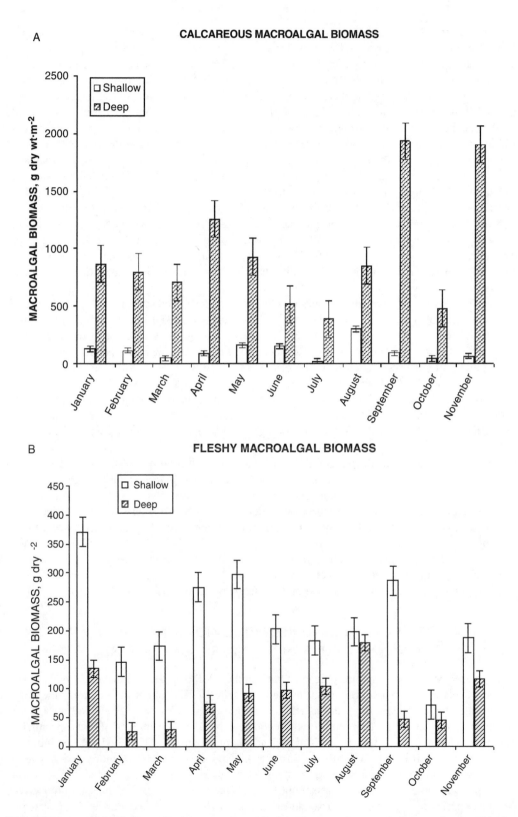

**FIGURE 35.11** Mean biomass of calcareous (A) and fleshy (B) macroalgae on shallow vs. deep reefs in the Negril Marine Park during 1998. Values represent means ± 1 SD (*n* = 110).

In contrast to N, percent P of macroalgae did not vary significantly with either season or depth (varied from $0.06 \pm 0.02$ to $0.04 \pm 0.01$ dry wt, respectively), although shallow reefs did have higher mean percent P values compared to deep reefs. These data indicate that the limiting tissue nutrient in macroalgae of the NMP (% P) may not be a good indicator of limiting nutrient availability in the water column. As described above, SRP concentrations increased significantly (almost threefold) in summer compared to winter (Figure 35.9). Accordingly, the increased water-column SRP concentrations during summer appear to initiate seasonal blooms of macroalgae (e.g., *Chaetomorpha linum*; see Color Figure 1, part F) by increasing growth and reproduction of opportunistic species without necessarily increasing the tissue storage of P in most macroalgae.

The changes with depth in tissue percent N and P of the macroalgae are reflected in increases in C:N and C:P molar ratios from shallow to deep reefs in summer (21.2 vs. 29.9 and 1364 vs. 1670, respectively) and winter (23.2 vs. 29.7 and 1467 vs. 1694, respectively). In contrast to the C:N and C:P ratios, the N:P ratio of macroalgae decreased from shallow to deep reefs in summer (65.4 vs. 54.0) and winter (64.2 vs. 59.4) and overall were greater than 35:1, indicating primary growth limitation by SRP, not DIN.[29,30] The primary SRP limitation of the macroalgal blooms in the NMP is in agreement with previous studies that concluded SRP limited macroalgal productivity at Discovery Bay, Jamaica.[7,18,25]

The $\delta^{15}N$ values of macroalgal tissue from the 12 reef sites varied spatially but not seasonally.[24] During winter, $\delta^{15}N$ ratios of macroalgae on shallow reefs averaged $+3.56 \pm 1.65$, a value significantly higher ($t = 2.65$, $p = 0.044$) than the average value of $+1.84 \pm 0.72$ for macroalgae on the deep reefs. The highest values (+4 to +6‰) occurred on the shallow South Negril and Ironshore reefs adjacent to or downstream of urban centers impacted by sewage pollution. In comparison, lower values (+2 to +3‰) were measured in macroalgae on shallow reefs in the rural, agricultural areas at Davis Cove, North Negril, and Little Bay. During summer, a similar spatial pattern was observed, and $\delta^{15}N$ ratios of macroalgae on shallow reefs averaged $+3.11 \pm 1.42$, a value significantly higher ($t = 2.68$, $p = 0.043$) than the average of $+1.81 \pm 0.71$ for deep reefs. The lowest $\delta^{15}N$ values during both seasons (~+0.5‰) were similar to those reported for nitrogen fixation on coral reefs[23] and occurred on the deep reef off North Negril — the deepest reef of the study. The highest value (>+6‰) occurred during the massive bloom of *Chaetomorpha linum* at Orange Bay (North Negril shallow reef station; see Color Figure 1, part F) in August 1998.

The high $\delta^{15}N$ values of macroalgae on shallow reefs at South Negril, Ironshore, and North Negril were similar to the values measured from human sewage in Negril (+5.12‰) and higher than pig waste sampled from a farm on the South Negril River (+3.34‰).[24] In contrast, the lower $\delta^{15}N$ values of macroalgae on shallow reefs downstream of agricultural areas at Davis Cove and North Negril, and on most deep reefs (except North Negril, deep) closely matched the values measured in commercial fertilizers used on the canefields of the NEPA (+1.85‰).

## UNDERSTANDING AND REDUCING NUTRIENT POLLUTION ON CORAL REEFS IN THE NEGRIL MARINE PARK

The replacement of corals by macroalgae on reefs in the NMP over the past decade is symptomatic of reefs experiencing increased nutrient availability.[4] The NCRPS monitoring data for the NEPA documented the contribution of various land-based nutrient sources, which included sewage, fertilizers, and top soils. DIN concentrations throughout the NMP now average >1 $\mu M$, a concentration adequate to saturate the growth demands of tropical reef macroalgae.[7,18] That the DIN concentrations ranged up to ~4 $\mu M$ in the NMP further illustrates the high degree of pulsed nutrient supply to the NMP during storm events and extreme low tides. Macroalgae often increase productivity following such episodic nutrient pulses.[7,18] For example, Schaffelke[31] recently showed that net productivity of three species of ephemeral macroalgae on inshore reefs of the Great Barrier Reef increased following short-term (24-hr) exposure to pulses of ammonium and/or phosphate. The high DIN:SRP ratios of the water column and N:P tissue ratios in macroalgae of the NMP clearly indicate that pulses of SRP would be most stimulatory to productivity and growth of macroalgae in coastal waters around Negril. In fact, the increased SRP concentrations in the discharge of the South Negril River over the past several years have correlated with a significant (ANOVA, F = 8.99, p < 0.01) 3.7-fold increase in biomass of macroalgae (both fleshy and calcareous) on the shallow reef at Yacht Club (shallow reef, South Negril) that is impacted episodically by the river's plume (Figure 35.12). Because fishing pressure did not alter herbivory or sea urchin density at this reef during this period, increased SRP availability resulting from sewage contamination of the South Negril River is the only factor that could explain this dramatic increase in macroalgal biomass. This finding underscores the need for policies to control phosphorus sources in sewage and fertilizers on the NEPA to moderate the ongoing macroalgal overgrowth of corals on reefs in the NMP.

The low cover of coral and high cover of microalgae in the NMP are similar to the bank reefs of the Florida Keys that have been impacted by increasing DIN concentrations over the past two decades (Lapointe et al., this volume). Reefs in the NMP have <10% coral cover and >60% macroalgae, compared to the fore-reef at Looe Key National Marine Sanctuary (LKNMS), that now has <12% coral cover and >30% macroalgal cover (Figure 35.10). The mean annual DIN concentrations on both reef systems is ~1.0 $\mu M$, with SRP concentrations at LKNMS being several-fold higher than those in the NMP. While many have argued that the development of macroalgal blooms and loss of coral on reefs in Jamaica and the Florida Keys resulted exclusively from overfishing and reduced grazing by herbivorous fish,[14,17] this hypothesis is not supported by this work or other reported scientific evidence. Overfishing of large herbivorous fish occurred in the 1960s in Jamaica, many years before the onset of macroalgal blooms and loss of coral cover.[7,14] Unlike Jamaica, herbivorous fish are not targeted by fishermen in the Florida Keys; LKNMS, like other bank reefs in the Keys, has extremely abundant populations of grazing fish.[32] Numerous experimental studies have shown that reduced grazing results in expansion of algal turf under oligotrophic conditions; in comparison, reduced grazing under eutrophic conditions, such as after the *Diadema* dieoff on nitrogen-rich Jamaican reefs, typically results in expansion of macroalgal blooms. The fact that macroalgal blooms and coral loss have co-occurred with nutrient enrichment and high grazing pressure at LKNMS over recent decades supports the view that bottom-up controls (nutrient availability) are of primary importance in forcing these biotic phase-shifts on coral reefs.[7,18] However, macroalgal biomass is generally lower on fore-reefs of the FKNMS compared to the NMP, most likely a result of the higher grazing rates by herbivorous fishes.

The hypothesis that increased density of sea urchins will lead to recovery of Jamaican coral reefs[14,17] is not supported by our reef monitoring data. Throughout the NMP, coral cover was significantly and inversely correlated with sea urchin density. In fact, the lowest coral cover coincided with the highest sea urchin density on shallow and deep reefs in Long Bay. Considering that the growth of herbivores in marine waters is often limited by dietary nitrogen,[33] it is probable that historical increases in nitrogen enrichment of Jamaican reefs has supported not only the development of macroalgal blooms but also abnormally large sea urchin populations as a cascading trophic response to enhanced food quality. This hypothesis is supported by our observations of higher sea urchin densities co-occurring with ephemeral macroalgae on shallow reefs that had significantly higher tissue nitrogen content compared to deep reefs. High nitrogen (protein) levels are reported to enhance the egg quality, larval survival, and somatic growth of sea urchins in controlled feeding studies.[34] In a review of algal nutrient limitation and the nutrition of aquatic herbivores, Sterner and Hessen[35] concluded that "herbivores with high nutrient demands appear frequently to be limited not by the food quantity or energy available to them but by the quantity of mineral elements in their food." These data and observations support recent interpretations that a "bottom-up template" is the most realistic approach for understanding trophic dynamics in ecosystems because plants have obvious primacy in food webs.[36] Both primary production and biochemical composition of prey items are recognized as fundamental controls on the production and nutrition of higher trophic levels.

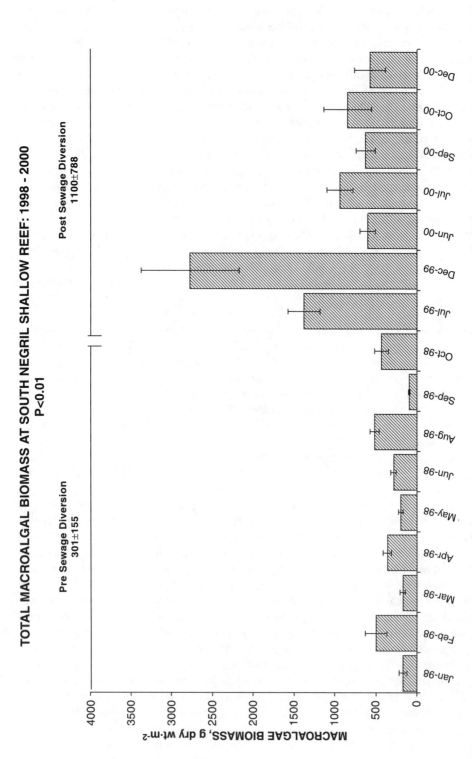

**FIGURE 35.12** Mean (± 1SD, n = 10) monthly biomass of total macroalgae (calcareous and fleshy) at the South Negril shallow reef (Yacht Club) from January 1998 through December 2000. Macroalgal biomass increased significantly (~ 3.7-fold, ANOVA, F = 8.99, p < 0.01) from the period of "pre-sewage diversion" (1998) to "post sewage diversion" (July 1999–December 2000).

That the two Long Bay reef sites had the highest densities of echinoids, the lowest cover of calcifiers (both corals and calcareous macroalgae), and the lowest salinities of all reefs sites sampled in the NMP indicates an advanced stage of eutrophication in this popular, sewage-impacted tourist area. Roobal and Gibb[21] described the geology of the dentate, drowned submarine sea cliffs and caves that occur several miles from shore in Long Bay at 10- to 20-m depths just east of the Long Bay deep station. Although no salinity measurements were made, they concluded that "today there are no freshwater springs occurring inside the cave." However, the low salinity levels (as low as 24 ppt) measured in near-bottom water at the deep reef site in Long Bay, ~0.5 km west of Roobal and Gibb's cave site, provided direct evidence that considerable groundwater discharges can occur along this geologic feature. These salinity measurements demonstrate that groundwater discharges of DIN and SRP can impact deep reef communities at considerable distances from shore. Hence, the high density of sea urchins and low cover of corals on both deep (Sand's Club) and shallow (Sandy Cay) reefs in Long Bay are ecological expressions of the groundwater-borne nutritional subsidy to this area. That dense sea urchin populations are indicative of local nitrogen enrichment, and the decline of coral reefs in the NMP, especially Long Bay, may be applicable to the Caribbean region in general. Recent worldwide surveys have shown that the Caribbean region has the highest densities of *D. antillarum* and the lowest coral cover of any coral reef region globally.[37]

Within the past decade the Long Bay beach has eroded by up to 20 m in some places (see Color Figure 35.1, part B). This recent beach erosion may be linked, in part, to the historical impacts of land-based nutrient enrichment of Long Bay. High densities of sea urchins fueled by nutrient enrichment accelerate the bioerosion of limestone reef frameworks,[38] and the recent demise of coral reefs at Sandy Cay in the middle of Long Bay would result in increased wave energy exposure of the leeward beach shoreline, particularly during storms. Studies of the composition of calcareous beach sand in Long Bay suggest that *Halimeda*, red calcareous algae, and foraminifera may all contribute significantly to the production of the beach sand (T. Robinson and S. Mitchell, University of the West Indies, Kingston, pers. comm.). Experimental studies have shown that elevated nitrogen and phosphorus concentrations directly reduce calcification in reef corals and overall coral reef growth.[38,39] In the Florida Keys, experimental nutrient enrichment with DIN and SRP enhanced the productivity of fleshy, ephemeral species of macroalgae and either had little stimulatory effect or inhibited productivity and calcification of calcareous macroalgae (*Halimeda, Penicillus* spp.).[41] In turtle grass (*Thalassia testudinum*) meadows of the Florida Keys, blade algal epiphyte communities shifted from fleshy, opportunistic species in nutrient-enriched, nearshore waters to calcareous species (*Fosliella* sp., *Melobesia* sp.) in more oligotrophic, offshore reef environments.[42] Considering the low cover of calcifying corals and macroalgae currently on reefs in Long Bay and the recent epiphytization of Long Bay turtle grass meadows by noncalcareous, fleshy algal species (K. Thacker, pers. observ.), the supply of biogenic calcium carbonate beach sand may have decreased in recent years in Long Bay. Such a reduced supply of beach sand, combined with increased wave energy exposure from the degraded coral reefs at Sandy Cay, could be contributing factors to the ongoing erosion of the Long Bay beach. This problem appears to be exacerbated by the massive accumulation of fleshy macroalgae on the beach following storms which formed seaweed mats up to 1.5 m high. Much of this seaweed is buried in the sand where it decomposes and leaches organic acids that would have dissolved the calcium carbonate beach.

The significant increases in $NH_4^+$ and SRP concentrations in the South Negril River since 1997 appear to be linked largely to the increased nutrient loads associated with the effluent from the sewage treatment ponds that have become operational in the past 2 years. This nutrient source can be easily controlled by providing tertiary sewage treatment (nutrient removal) at the treatment site, a level of treatment recognized long ago as necessary for protection of Jamaica's sensitive coral reefs.[8,15] As noted above, the phosphorus loads from sewage should be targeted for maximum treatment to achieve the most economically-viable protection of water quality and coastal resources. However, proper treatment of human sewage alone may not be adequate to restore water quality in the NMP to the levels necessary for coral reef restoration. Our data and observations indicate that deforestation and associated top-soil loss, agriculture, fertilizers, and animal feedlots represent other major contributors to nutrient enrichment of the NEPA and downstream waters of the NMP. Improved management and moderation of all these human activities and nutrient sources could lead to monetary savings (e.g., avoiding unnecessary or excessive use of fertilizers by farmers) and significant improvements in water quality.

As actions are taken to improve water quality on the NEPA, we stress that it will be necessary to continue the long-term monitoring of the biogeochemistry of the NEPA and downstream reefs in the NMP.[43] Haphazard collection of short-term data, so common in the field of coral reef ecology, rarely provides data that are useful to whole-watershed management — as needed in Negril. We hope that our study illustrates how informed, community-based, long-term research enhances the success of multidisciplinary teams tackling large and complex environmental problems.

## ACKNOWLEDGMENTS

The authors thank Hedonism II, Sandals, and Beaches for their generous accommodations and support that made this research possible. Linval Getten, Courtney Black, Everton Frame, Webster Gabbidon, and Everett Gill performed the bulk of the field sampling reported here. Dr. Peter Bell provided assistance with water sampling and nutrient analysis. We are especially grateful to Elsa Hemmings, Chantelle Black, Roberta Raigh Pryor, Peter Barile, Liliana Velasquez, and Carl Hanson for data management, statistical analysis, and graphics. Dr. Diane Littler, Craig Quirolo, and Dean Barrette provided photographic services. We also thank the National Water Commission for analytical assistance and the National Meteorological Service for the rainfall data. This research was supported by the Harbor Branch Oceanographic Institution, Inc., and a grant from the European Union to the Negril Coral Reef Preservation Society. This is Contribution No. 1357 from the Harbor Branch Oceanographic Institution, Inc.

## REFERENCES

1.  Olsen, B., Environmentally sustainable development and tourism:lessons from Negril, Jamaica, *Human Org.*, 56, 285, 1997.
2.  Jones Williams, M., Biodiversity, in *Development Project Case Study Series 9*, Jamaica-Negril coral reef preservation project, Department for International Development, 1999, p. 13.
3.  Regional overview of land-based sources of pollution in the wider Caribbean region, in *Caribbean Environment Program Technical Report 33*, UNEP Caribbean Environment Program, Kingston, 1994.
4.  National Research Council, *Clean Coastal Waters: Understanding and Reducing the Effects of Nutrient Pollution*, National Academy Press, Washington, D.C., 2000.
5.  Watts, D., *The West Indies: Patterns of Development, Culture, and Environmental Change Since 1492*, Cambridge University Press, Cambridge, U.K., 1987.
6.  Likens, G.E., Bormann, F.H., Pierce, R.S., and Reiners, W.A., Recovery of a deforested ecosystem, *Science*, 199, 492, 1978.
7.  Lapointe, B.E., Nutrient thresholds for bottom-up control of macroalgal blooms on coral reefs in Jamaica and southeast Florida, *Limnol. Oceanogr.*, 42(5, part 2), 1119, 1997.
8.  Barnes, E., Sewage pollution from tourist hotels in Jamaica, *Mar. Poll. Bull.*, 4(7), 102, 1973.
9.  Banner, A.H., Kaneohe Bay, Hawaii: urban pollution and a coral reef ecosystem, *Proc. 2nd Int. Coral Reef Symp.*, 2, 685, 1974.
10. Vitousek, P.M., Aber, J., Howarth, R.W., Likens, G.E., Matson, P.A., Schindler, D.W., Schlesinger, W.H., and Tilman, D.G., Human alteration of the global nitrogen cycle: sources and consequences, *Ecol. Appl.*, 7, 737, 1997.
11. Rawlins, B.G., Ferguson, A.J., Chilton, P.J., Arthurtone, R.S., Rees, J.G., and Baldock, J.W., Review of agricultural pollution in the Caribbean with particular emphasis on small island developing states, *Mar. Poll. Bull.*, 36, 658, 1998.
12. Goreau, T.J., Daley, L., Ciappara, S., Brown, J., Bourke, S., and Thacker, K., Community-based whole-watershed and coastal zone management in Jamaica, *Proc. 8th Int. Coral Reef Symp.*, 2, 2093, 1997.
13. Liddell, W.D. and Olhorst, S.L., Ten years of disturbance and change on a Jamaican fringing reef, *Proc. 7th Int. Coral Reef Symp.*, 1, 144, 1992.
14. Hughes, T.P., Catastrophes, phase-shifts, and large-scale degradation of a Caribbean coral reef, *Science*, 265, 1547, 1994.

15. Goreau, T.J., Negril: environmental threats and recommended actions, in *Water Quality Issues: Final Report*, Thacker, K., Ed., 1992, p. 9.

16. Johannes, R.E., Pollution and the degradation of coral reef communities, in *Tropical Marine Pollution*, Wood, E. and Johannes, R.E., Eds., Elsevier, New York, 1975, p. 13.

17. Hughes, T. et al., Algal blooms on coral reefs on coral reefs: what are the causes?, *Limnol. Oceanogr.*, 44, 6, 1583, 1999.

18. Lapointe, B.E., Simultaneous top-down and bottom-up forces control macroalgal blooms on coral reefs, *Limnol. Oceanogr.*, 44(6), 1586, 1999.

19. Littler, M.M. and Littler, D.S., Models of tropical reef biogenesis: the contribution of algae, *Prog. Phycol. Res.*, 3, 323, 1984.

20. Jelier, C.A. and Roberts, S., Negril sewerage and sewerage treatment, in *Water Quality Issues: Final Report*, Thacker, K., Ed., 1992, p. 131.

21. Roobal, M.J. and Gibb, J.S., Submarine sea cliffs and caves at Negril, W. Jamaica, *J. Geol. Soc. Jamaica*, 13, 14, 1973.

22. Heikoop, J.M. et al., Nitrogen-15 signals of anthropogenic nutrient loading in reef corals, *Mar. Poll. Bull.*, 40(7), 628, 2000.

23. France, R., Holmquist, J., Chandler, M., and Cattaneo, A., $\delta^{15}N$ evidence for nitrogen fixation associated with macroalgae from a seagrass-mangrove-coral reef ecosystem, *Mar. Ecol. Prog. Ser.*, 167, 297, 1998.

24. Lapointe, B.E. et al., Watershed linkages to coral reefs in the Negril Marine Park, Jamaica: anthropogenic nutrient inputs and their ecological consequences, in *Proc. Phase VIII Workshop: Protecting Jamaica's Coral Reef Ecosystem: Coral Reefs, Water Quality and Human Survival Issues*, Negril Coral Reef Preservation Society, Negril, Jamaica, 1999.

25. D'Elia, C.F., Webb, K.L., and Porter, J.W., Nitrate-rich groundwater inputs to Discovery Bay, Jamaica: a significant source of N to local reefs?, *Bull. Mar. Sci.*, 31, 903, 1981.

26. Chilton, P.J., Vlugman, A.A., and Foster, S.S.D., A groundwater pollution risk assessment for public water supplies in Barbados, in *Symposium on Tropical Hydrology and Caribbean Water Resources*, Puerto Rico, July 1990, American Water Resources Association, 1990, p. 279.

27. Lapointe, B.E., O'Connell, J.D., and Garrett, G.S., Nutrient couplings between on-site sewage disposal systems, groundwaters, and nearshore surface waters of the Florida Keys, *Biogeochemistry*, 10, 289, 1990.

28. Marsh, J.A., Terrestrial inputs of nitrogen and phosphorus to fringing reefs of Guam, *Proc. 3rd Int. Coral Reef Symp.*, 1, 331, 1977.

29. Atkinson, M.J. and S.V. Smith, C:N:P ratios of benthic marine plants, *Limnol. Oceanogr.*, 28, 568, 1983.

30. Lapointe, B.E., Littler, M.M., and Littler, D.S., Nutrient availability to marine macroalgae in siliciclastic versus carbonate-rich coastal waters, *Estuaries*, 15, 75, 1992.

31. Schaffelke, B., Short-term nutrient pulses as tools to assess responses of coral reef macroalgae to enhanced nutrient availability, *Mar. Ecol. Prog. Ser.*, 182, 305, 1999.

32. Littler, M.M, Littler, D.S., and Lapointe, B.E., *Baseline Studies of Herbivory and Eutrophication on Dominant Reef Communities of Looe Key National Marine Sanctuary*, National Oceanic and Atmospheric Administration, U.S. Department of Commerce, Washington, D.C., 1986.

33. Mattson, W.J., Herbivory in relation to plant nitrogen content, *Ann. Rev. Ecol. Syst.*, 11, 119, 1982.

34. de Jong-Westman, P.-Y.Q., March, B.E., and Carefoot, T.H., Artificial diets in sea urchin culture: effects of dietary protein level and other additives on egg quality, larval morphometrics, and larval survival in the green sea urchin, *Strongylocentrotus droebachiensis*, *Can. J. Zool.*, 73, 2080, 1995.

35. Sterner, R.W. and D.O. Hessen, Algal nutrient limitation and the nutrition of aquatic herbivores, *Ann. Rev. Ecol. Syst.*, 25, 1, 1994.

36. Hunter, M.D. and Price, P.W., Playing chutes and ladders: heterogeneity and the relative roles of bottom-up and top-down forces in natural communities, *Ecology*, 73, 3, 724, 1992.

37. Hodgson, G., A global assessment of human effects on coral reefs, *Mar. Poll. Bull.*, 38, 345, 1999.

38. Glynn, P.W., Bioerosion and coral reef growth: a dynamic balance, in *Life and Death of Coral Reefs*, Birkeland, C., Ed., Chapman & Hall, New York, 1997, p. 68.

39. Kinsey, D.W. and Davies, P., Effects of elevated nitrogen and phosphorus on coral reef growth, *Limnol. Oceanogr.*, 24, 935, 1979.

40. Marubini, F. and Davies, P.S., Nitrate increases zooxanthellae population density and reduces skeletogenesis in corals, *Mar. Biol.*, 127, 319, 1996.

41. Delgado, O. and Lapointe, B.E., Nutrient-limited productivity of calcareous versus fleshy macroalgae in a eutrophic, carbonate-rich tropical marine environment, *Coral Reefs*, 13, 151, 1994.

42. Lapointe, B.E., Tomasko, D.A., and Matzie, W.R., Eutrophication and trophic state classification of seagrass communities in the Florida Keys, *Bull. Mar. Sci.*, 54(3), 696, 1994.

43. Likens, G.E., Biogeochemistry, the watershed approach: some uses and limitations, *Mar. Freshwater Res.*, 52, 5, 2001.

# Index